The Earth's Crust and Upper Mantle

The Earth's Crust and Upper Mantle was prepared under the auspices of the International Upper Mantle Committee by the following editorial group:

V. V. BELOUSSOV, editor in chief
BRUCE C. HEEZEN
HISASHI KUNO
V. A. MAGNITSKY
TAKESI NAGATA
A. R. RITSEMA
GEORGE P. WOOLLARD

geophysical monograph 13

The Earth's Crust and Upper Mantle

*structure, dynamic processes,
and their relation to
deep-seated geological phenomena*

PEMBROKE J. HART
editor

American Geophysical Union
Washington, D. C.
1969

*Published with the aid of a grant
from the Charles F. Kettering Foundation*

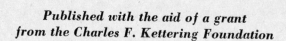

National Academy of Sciences—National Research Council Publication 1708

Worldwide Book Number 87590013

Copyright © 1969 by the American Geophysical Union
2100 Pennsylvania Avenue, N. W.
Washington, D. C. 20037

Library of Congress Catalog Card Number 75-600572

WILLIAM BYRD PRESS, RICHMOND, VIRGINIA

Foreword

AMONG the major results of the Upper Mantle Project, two are especially significant. First, it has been clearly demonstrated that the earth is not radially symmetric: important lateral inhomogeneities exist in the uppermost 700 kilometers. Second, it is now clear that large-scale motions of the upper parts of the earth are taking place now and have taken place in the past. These two ideas may well be coupled: the inhomogeneities may be the evidence for the driving mechanism for the motions; the motions, in turn, may provide the process whereby the inhomogeneities are produced.

These results, the details and full significance of which were unforeseen at the beginning of the Upper Mantle Project, led to a clearer understanding of the essential interrelation of geophysical and geological phenomena. Thus, they are in large part responsible for the decision of the International Union of Geodesy and Geophysics and the International Union of Geological Sciences jointly to propose a long-range program of studies of the solid earth—beginning in 1972 under the sponsorship of the International Council of Scientific Unions—with 'dynamics of the earth's interior' as a principal focal theme.

The Earth's Crust and Upper Mantle was prepared under the auspices of the International Upper Mantle Committee. It is designed to give a broad survey of modern developments in solid earth studies, with emphasis on the interrelation of geophysical, geochemical, and geological observations. Thus, the monograph serves not only as a review of achievements during the period of the Upper Mantle Project but also underscores the relations of these developments, with their uncertainties, to the future studies of the earth's interior.

Inspiration for the preparation of the monograph was provided by Professor V. V. Beloussov, whose vision and enthusiasm for international cooperation in geosciences led to the Upper Mantle Project, originally planned for 1962–1964 but twice extended, thus ending in 1970. Professor Beloussov also served as chief editor of the group appointed by the Upper Mantle Committee to prepare the monograph. Articles were written by the various authors upon invitation of this editorial group; responsibility for the statements in the articles rests, of course, with the authors. We hope that the diversity of views of the various authors will emphasize the complexity of the problems of the earth's interior.

On behalf of the Upper Mantle Committee, it is a pleasure to acknowledge the contributions to the UMC monograph, *The Earth's Crust and Upper Mantle,* by many persons throughout the world, especially the ninety-two authors, who devoted much time and energy to preparing the articles; Professor Beloussov and the other members of the editorial group, Professors Bruce C. Heezen, Hisashi Kuno, V. A. Magnitsky, Takesi Nagata, A. Reinier Ritsema, and George P. Woollard, who compiled the monograph; Dr. Pembroke J. Hart, who did the comprehensive editorial work; and Miss Mary Jane Miles, who was responsible for the effective copy editing. Finally, I wish to thank the officers and the Monograph Board of the American Geophysical Union for their cooperation and encouragement in publication of the monograph of the International Upper Mantle Committee.

LEON KNOPOFF

Preface

WE HAVE many reasons to believe that the history of the development of the earth's crust is fundamentally dependent on processes in the upper mantle to a depth not exceeding 1000 km. Because of this relation, the Upper Mantle Project was organized as an international program of geophysical, geochemical, and geological studies concerning the 'upper mantle and its influence on the development of the earth's crust.'

Many important results have emerged during the course of the Upper Mantle Project. Probably the most significant is the delineation in some detail of the 'fine structure'—the lateral and vertical variations of properties—of the upper mantle. These results have fully convinced us that there is a fundamental relation between this fine structure, the motions of materials in the upper mantle, and activities at or near the surface of the earth such as earthquakes (and the types of earthquake motions), volcanism, the building and deformation of mountains, and the differentiation of various rock types, including concentrations that are of economic interest.

We believe that recent results and work now under way will soon lead to a major step forward in our understanding of the interlocked processes of the crust and mantle. A great excitement pervades this field of study, yet there are diverse and contradictory opinions about the significance of the new developments.

The Upper Mantle Committee decided to prepare the present monograph to bring together in a series of articles a comprehensive presentation about the many aspects of investigations of the earth's crust and upper mantle, and their interrelations. The articles were written by specialists in the respective fields, and reflect their personal views of the nature of the problems, the significance of the results, and the directions for future research. The compilers did not try in any way to standardize the approaches or bring them under some uniform scheme; on the contrary, they believed that variations in methods of analysis and presentation by the different authors would help the reader to recognize the many contradictions in the study of the problems of the earth's crust and upper mantle.

It is the hope of the compilers that the monograph will serve not only as a reference and tutorial text, but that it will serve to implant in the reader the excitement and enthusiasm of the authors and will encourage him to make his personal contribution to the solution of some of the many unsolved problems of the earth's interior.

V. V. BELOUSSOV

Contents

Foreword ... v

Preface ... vii

1. Composition

Composition and Evolution of the Upper Mantle ... A. E. RINGWOOD ... 1

Density and Composition of the Upper Mantle: First Approximation as an Olivine Layer ... FRANCIS BIRCH ... 18

Chemical Composition of the Earth's Crust ... A. B. RONOV AND A. A. YAROSHEVSKY ... 37

Isotope Geochemistry of Crust-Mantle Processes ... STANLEY R. HART ... 58

2. Heat Flow

Thermal History of the Earth ... E. A. LUBIMOVA ... 63

Heat Flow in North America ... GENE SIMMONS AND ROBERT F. ROY ... 78

Heat Flow Map of Eurasia ... E. A. LUBIMOVA AND B. G. POLYAK ... 82

Heat Flow in Oceanic Regions ... R. P. VON HERZEN AND W. H. K. LEE ... 88

Terrestrial Heat Flow in Volcanic Areas ... KI-ITI HORAI AND SEIYA UYEDA ... 95

3. Seismology

Seismology and Upper Mantle Investigations ... A. R. RITSEMA ... 110

Seismicity of the Earth ... SETUMI MIYAMURA ... 115

Tectonic Activity in North America as Indicated by Earthquakes ... GEORGE P. WOOLLARD ... 125

Seismicity of Continental Asia and the Region of the Sea of Okhotsk, 1953–1965 ... E. F. SAVARENSKY AND N. V. GOLUBEVA ... 134

Seismicity of the European Area ... VÍT KÁRNÍK ... 139

Seismicity of Southeast Australia ... J. C. JAEGER AND LESLEY READ ... 145

Seismicity of the Mid-Oceanic Ridge System ... LYNN R. SYKES ... 148

Worldwide Earthquake Mechanism ... ANNE E. STEVENS ... 153

Mechanism of Earthquakes in and near Japan and Related Problems ... MASAJI ICHIKAWA ... 160

The Field of Elastic Stresses Associated with Earthquakes ... L. M. BALAKINA, L. A. MISHARINA, E. I. SHIROKOVA, AND A. V. VVEDENSKAYA ... 166

Zero Frequency Seismology ... FRANK PRESS ... 171

Prediction of Earthquakes ... TAKAHIRO HAGIWARA ... 174

Explosion Seismology: Introduction ... I. P. KOSMINSKAYA ... 177

Explosion Seismic Studies in Western Europe .... H. CLOSS 178

Seismic Crustal Studies in Southeastern Europe .... V. B. SOLLOGUB 189

Explosion Seismology in the USSR
    I. P. KOSMINSKAYA, N. A. BELYAEVSKY, AND I. S. VOLVOVSKY 195

Explosion Seismic Studies in North America
    JOHN H. HEALY AND DAVID H. WARREN 208

Seismic Model of the Atlantic Ocean .... JOHN EWING 220

Explosion Seismic Refraction Studies of the Crust and Upper Mantle in the Pacific and Indian Oceans .... GEORGE G. SHOR, JR., AND RUSSELL W. RAITT 225

Surface Waves and Crustal Structure .... JAMES N. BRUNE 230

Regional Variations of P-Wave Velocity in the Upper Mantle beneath North America .... EUGENE HERRIN 242

Upper Mantle Structure and Velocity Distribution in Eurasia .... JIRI VANEK 246

Anisotropy of the Upper Mantle .... RUSSELL W. RAITT 250

Seismic Surface-Wave Data on the Upper Mantle .... JAMES DORMAN 257

Higher-Mode Surface Waves .... Z. ALTERMAN 265

Attenuation of Seismic Waves in the Mantle .... LEON KNOPOFF 273

Seismic Models of the Upper Mantle .... JIRI VANEK 276

Earthquakes and Tectonics .... B. A. PETRUSHEVSKY 279

4. Gravity

Standardization of Gravity Measurements .... GEORGE P. WOOLLARD 283

Figure of the Earth and Mass Anomalies Defined by Satellite Orbital Perturbations
    M. A. KHAN 293

Gravity Anomalies as a Function of Elevation: Some Results in Western Europe
    S. CORON 304

The Relation between the Earth's Crust, Surface Relief, and Gravity Field in the USSR .... R. M. DEMENITSKAYA AND N. A. BELYAEVSKY 312

Regional Variations in Gravity .... GEORGE P. WOOLLARD 320

Gravity Field over the Atlantic Ocean .. MANIK TALWANI AND XAVIER LE PICHON 341

Gravity and Its Relation to Topography and Geology in the Pacific Ocean
    PETER DEHLINGER 352

Gravity Anomalies over Volcanic Regions .... ALEXANDER MALAHOFF 364

Recent Movements of the Earth's Crust and Isostatic Compensation
    E. V. ARTYUSHKOV AND YU. A. MESCHERIKOV 379

5. Magnetism

Reduction of Geomagnetic Data and Interpretation of Anomalies
    TAKESI NAGATA 391

The Relation of Magnetic Anomalies to Topography and Geologic Features in Europe .... A. HAHN AND A. ZITZMANN   399

Aeromagnetic Investigations of the Earth's Crust in the United States
ISIDORE ZIETZ   404

Relation of Magnetic Anomalies to Topography and Geology in the USSR
TATIANA SIMONENKO   415

Magnetic Intensity Field in the Pacific .... VICTOR VACQUIER   422

Geomagnetic Studies in the Atlantic Ocean .... J. R. HEIRTZLER   430

Magnetic Studies over Volcanoes .... ALEXANDER MALAHOFF   436

The Paleomagnetic Vector Field .... S. K. RUNCORN   447

Magnetic Anomalies and Crustal Structure .... NED A. OSTENSO   457

Conductivity Anomaly of the Upper Mantle .... TSUNEJI RIKITAKE   463

Magnetotelluric Studies of the Electrical Conductivity Structure of the Crust and Upper Mantle .... T. R. MADDEN AND C. M. SWIFT, JR.   469

6. *Magmatism and Metamorphism*

The Ultramafic Belts .... PETER J. WYLLIE   480

Batholiths and their Orogenic Setting .... AHTI SIMONEN   483

The Origin of Basalt Magmas .... D. H. GREEN AND A. E. RINGWOOD   489

Plateau Basalts .... HISASHI KUNO   495

Andesitic and Rhyolitic Volcanism of Orogenic Belts ... ALEXANDER R. MCBIRNEY   501

Mafic and Ultramafic Inclusions in Basaltic Rocks and the Nature of the Upper Mantle .... HISASHI KUNO   507

Petrology of the Precambrian Basement Complex .... K. R. MEHNERT   513

Age Relationships of the Precambrian Basement Rock Complex HISASHI KUNO   519

Metamorphism and Its Relation to Depth .... AKIHO MIYASHIRO   519

7. *Tectonics*

Crustal Movements and Tectonic Structure of Continents
V. E. KHAIN AND M. V. MURATOV   523

Continental Rifts .... V. V. BELOUSSOV   539

Wrench (Transcurrent) Fault Systems .... H. W. WELLMAN   544

Continental Margins .... CHARLES L. DRAKE   549

The Crust and Upper Mantle beneath the Sea
PETER R. VOGT, ERIC D. SCHNEIDER, AND G. LEONARD JOHNSON   556

8. *Experimental and Theoretical Geophysics*

The Electrical Conductivity of the Mantle .... D. C. TOZER   618

Heat Conductivity in the Mantle .... SYDNEY P. CLARK, JR.   622

Magnetic Properties of Rocks .... JOHN VERHOOGEN   627

Phase Equilibrium Studies Relevant to Upper Mantle Petrology .. M. J. O'HARA   634

Phase Transitions .... A. E. RINGWOOD AND D. H. GREEN   637

Equation of State at High Pressure .... V. N. ZHARKOV AND V. A. KALININ   650

The Mohorovicic Discontinuity .... D. P. MCKENZIE   660

Low-Velocity Layers in the Upper Mantle .. V. A. MAGNITSKY AND V. N. ZHARKOV   664

The Transitional Layer in the Mantle .... V. A. MAGNITSKY   676

9. Special Problems

Continental Drift and Convection .... LEON KNOPOFF   683

Problem of Convection in the Earth's Mantle .... E. N. LYUSTIKH   689

Convection in the Mantle .... S. K. RUNCORN   692

Interrelations between the Earth's Crust and Upper Mantle ..V. V. BELOUSSOV   698

Author Index   713

Subject Index   726

Contributors

ZIPORA ALTERMAN — Tel-Aviv University, Ramat-Aviv, Israel
265

E. V. ARTYUSHKOV — Institute of Physics of the Earth, USSR Academy of Sciences, Moscow, USSR
379

L. M. BALAKINA — Institute of Physics of the Earth, USSR Academy of Sciences, Moscow, USSR
166

V. V. BELOUSSOV — Institute of Physics of the Earth, USSR Academy of Sciences, Moscow, USSR
539, 698

N. A. BELYAEVSKY — All-Union Geophysical Research Institute, Ministry of Geology, Moscow, USSR
195, 312

FRANCIS BIRCH — Hoffman Laboratory of Experimental Geology, Harvard University, Cambridge, Massachusetts, USA
18

JAMES N. BRUNE — Seismological Laboratory, California Institute of Technology, Pasadena, California, USA
230

SYDNEY P. CLARK, JR. — Department of Geology, Yale University, New Haven, Connecticut, USA
178, 622

H. CLOSS — Bundesanstalt für Bodenforschung, Hannover, Federal Republic of Germany
178

S. CORON — Bureau Gravimétrique International, Faculté des Sciences, 9, Quai Saint-Bernard, Tour 14, Paris (5ème), France
304

PETER DEHLINGER — Institute Marine Sciences, University of Connecticut, Storrs, Connecticut, USA
352

R. M. DEMENITSKAYA — Scientific Research Institute of Arctic Geology, Leningrad, USSR
312

JAMES DORMAN — Lamont-Doherty Geological Observatory of Columbia University, Palisades, New York, USA
257

CHARLES L. DRAKE — Lamont-Doherty Geological Observatory of Columbia University, Palisades, New York, USA
549

JOHN EWING — Lamont-Doherty Geological Observatory of Columbia University, Palisades, New York, USA
220

N. V. GOLUBEVA — Institute of Physics of the Earth, USSR Academy of Sciences, Moscow, USSR
134

D. H. GREEN — Department of Geophysics and Geochemistry, Australian National University, Canberra, Australia
489, 637

TAKAHIRO HAGIWARA — Earthquake Research Institute, Faculty of Science, University of Tokyo, Tokyo, Japan
174

A. HAHN — Bundesanstalt für Bodenforschung, Hannover, Federal Republic of Germany
399

STANLEY R. HART — Department of Terrestrial Magnetism, Carnegie Institution of Washington, Washington, D. C., USA
58

JOHN H. HEALY — National Center for Earthquake Research, U. S. Geological Survey, Menlo Park, California, USA
208

J. R. HEIRTZLER — Woods Hole Oceanographic Institute, Woods Hole, Massachusetts, USA
430

EUGENE HERRIN — Department of Geology and Geophysics, Southern Methodist University, Dallas, Texas, USA
242

KI-ITI HORAI — Department of Geology and Geophysics, Massachusetts Institute of Technology, Cambridge, Massachusetts, USA
95

MASAJI ICHIKAWA 160	Seismological Section, Japan Meteorological Agency, Tokyo, Japan
J. C. JAEGER 145	Department of Geophysics and Geochemistry, Australian National University, Canberra, Australia
G. LEONARD JOHNSON 556	U. S. Naval Oceanographic Office, Washington, D. C., USA
V. A. KALININ 650	Institute of Physics of the Earth, USSR Academy of Sciences, Moscow, USSR
VÍT KÁRNÍK 139	Geophysical Institute, Czechoslovak Academy of Sciences, Prague, Czechoslovakia
V. E. KHAIN 523	Moscow State University, Moscow, USSR
M. A. KHAN 293	Hawaii Institute of Geophysics, University of Hawaii, Honolulu, Hawaii, USA
LEON KNOPOFF 273, 683	Institute of Geophysics and Planetary Physics, University of California, Los Angeles, California, USA
I. P. KOSMINSKAYA 177, 195	Institute of Physics of the Earth, USSR Academy of Sciences, Moscow, USSR
HISASHI KUNO 495, 507, 519	Geological Institute, Faculty of Science, University of Tokyo, Tokyo, Japan
W. H. K. LEE 88	National Center for Earthquake Research, U. S. Geological Survey, Menlo Park, California, USA
XAVIER LE PICHON 341	Centre National pour l'Exploitation des Océans, 39, Avenue Iéna, Paris (16ème), France
E. A. LUBIMOVA 63, 82	Institute of Physics of the Earth, USSR Academy of Sciences, Moscow, USSR
E. N. LYUSTIKH 689	Institute of Physics of the Earth, USSR Academy of Sciences, Moscow, USSR
ALEXANDER R. MCBIRNEY 501	Center for Volcanology, University of Oregon, Eugene, Oregon, USA
D. P. MCKENZIE 660	Department of Geodesy and Geophysics, University of Cambridge, Cambridge, England
T. R. MADDEN 469	Department of Geology and Geophysics, Massachusetts Institute of Technology, Cambridge, Massachusetts, USA
V. A. MAGNITSKY 664, 676	Moscow State University and Institute of Physics of the Earth, Moscow, USSR
ALEXANDER MALAHOFF 364, 436	Hawaii Institute of Geophysics, University of Hawaii, Honolulu, Hawaii, USA
K. R. MEHNERT 513	Mineralogisches Institut der Freien Universität, Berlin, Federal Republic of Germany
YU. A. MESCHERIKOV 379	Institute of Geography of the USSR Academy of Sciences, Moscow, USSR
L. A. MISHARINA 166	Institute of Physics of the Earth, USSR Academy of Sciences, Moscow, USSR
SETUMI MIYAMURA 115	Earthquake Research Institute, Faculty of Science, University of Tokyo, Tokyo, Japan
AKIHO MIYASHIRO 519	Geological Institute, Faculty of Science, University of Tokyo, Tokyo, Japan, and Lamont-Doherty Geological Observatory of Columbia University, Palisades, New York, USA
M. V. MURATOV 523	Geological Institute of the USSR Academy of Sciences, Moscow, USSR

TAKESI NAGATA *391*	Geophysical Institute, Faculty of Science, University of Tokyo, Tokyo, Japan
M. J. O'HARA *634*	Department of Geology, University of Edinburgh, Edinburgh, Scotland
NED A. OSTENSO *457*	Geology and Geophysics Program, Office of Naval Research, Washington, D. C., USA
B. A. PETRUSHEVSKY *279*	Institute of Physics of the Earth, USSR Academy of Sciences, Moscow, USSR
B G. POLYAK *82*	Institute of Physics of the Earth, USSR Academy of Sciences, Moscow, USSR
FRANK PRESS *171*	Department of Geology and Geophysics, Massachusetts Institute of Technology, Cambridge, Massachusetts, USA
RUSSELL W. RAITT *225, 250*	Scripps Institution of Oceanography, University of California, La Jolla, California, USA
LESLEY READ *145*	Department of Geophysics and Geochemistry, Australian National University, Canberra, Australia
TSUNEJI RIKITAKE *463*	Earthquake Research Institute, Faculty of Science, University of Tokyo, Tokyo, Japan
A. E. RINGWOOD *1, 489, 637*	Department of Geophysics and Geochemistry, Australian National University, Canberra, Australia
A. R. RITSEMA *110*	Royal Netherlands Meteorological Institute, de Bilt, Netherlands
A. B. RONOV *37*	Institute of Geochemistry and Analytical Chemistry of the USSR Academy of Sciences, Moscow, USSR
ROBERT F. ROY *78*	Department of Geology and Geophysics, University of Minnesota, Minneapolis, Minnesota, USA
S. K. RUNCORN *447, 692*	Department of Geophysics and Planetary Physics, School of Physics, The University, Newcastle upon Tyne, England
E. F. SAVARENSKY *134*	Institute of Physics of the Earth, USSR Academy of Sciences, Moscow, USSR
ERIC D. SCHNEIDER *556*	U. S. Naval Oceanographic Office, Washington, D. C., USA
E. I. SHIROKOVA *166*	Institute of Physics of the Earth, USSR Academy of Sciences, Moscow, USSR
GEORGE G. SHOR, JR. *225*	Scripps Institution of Oceanography, University of California, La Jolla, California
GENE SIMMONS *78*	Department of Geology and Geophysics, Massachusetts Institute of Technology, Cambridge, Massachusetts, USA
AHTI SIMONEN *483*	Geological Survey, Otaniemi, Finland
TATIANA SIMONENKO *415*	Institute of Terrestrial Magnetism, Ionosphere and Radiowave Propagation (Leningrad Branch) of the USSR Academy of Sciences, Leningrad, USSR
V. B. SOLLOGUB *189*	Institute of Geophysics of the Ukrainian Academy of Sciences, Kiev, USSR
ANNE E. STEVENS *153*	Division of Seismology, Observatories Branch, Department of Energy, Mines and Resources, Ottawa, Ontario, Canada
C. M. SWIFT, JR. *469*	Geophysics Division, Kennecott Copper Company, Salt Lake City, Utah, USA
LYNN R. SYKES *148*	Lamont-Doherty Geological Observatory of Columbia University, Palisades, New York, USA

MANIK TALWANI *341*	Lamont-Doherty Geological Observatory of Columbia University, Palisades, New York, USA
D. C. TOZER *618*	Department of Geophysics and Planetary Physics, School of Physics, The University, Newcastle upon Tyne, England
SEIYA UYEDA *95*	Geophysical Institute, Faculty of Science, University of Tokyo, Tokyo, Japan
VICTOR VACQUIER *422*	Scripps Institution of Oceanography, University of California, La Jolla, California, USA
JIRI VANEK *246, 276*	Geophysical Institute, Czechoslovak Academy of Sciences, Prague, Czechoslovakia
JOHN VERHOOGEN *627*	Department of Geology and Geophysics, University of California, Berkeley, California, USA
PETER R. VOGT *556*	U. S. Naval Oceanographic Office, Washington, D. C., USA
I. S. VOLVOVSKY *195*	All-Union Geophysical Research Institute, Ministry of Geology, Moscow, USSR
R. P. VON HERZEN *88*	Woods Hole Oceanographic Institute, Woods Hole, Massachusetts, USA
A. V. VVEDENSKAYA *166*	Institute of Physics of the Earth, USSR Academy of Sciences, Moscow, USSR
DAVID H. WARREN *208*	National Center for Earthquake Research, U. S. Geological Survey, Menlo Park, California, USA
H. W. WELLMAN *544*	Victoria University of Wellington, Wellington, New Zealand
GEORGE P. WOOLLARD *125, 283, 320*	Hawaii Institute of Geophysics, University of Hawaii, Honolulu, Hawaii, USA
PETER J. WYLLIE *480*	Department of the Geophysical Sciences, University of Chicago, Chicago, Illinois, USA
A. A. YAROSHEVSKY *37*	Institute of Geochemistry and Analytical Chemistry of the USSR Academy of Sciences, Moscow, USSR
V. N. ZHARKOV *650, 664*	Institute of Physics of the Earth, USSR Academy of Sciences, Moscow, USSR
ISIDORE ZIETZ *404*	U. S. Geological Survey, Silver Spring, Maryland, USA
A. ZITZMANN *399*	Bundesanstalt für Bodenforschung, Hannover, Federal Republic of Germany

1. Composition of the Crust and Upper Mantle

Composition and Evolution of the Upper Mantle

A. E. Ringwood

GEOPHYSICAL EVIDENCE

The seismic P-wave velocity in the mantle immediately beneath stable continental regions and deep oceanic basins is usually within the range 8.2 ± 0.2 km/sec. This property, together with certain broad petrological and chemical limitations, effectively restricts the mineralogical composition of the upper mantle to some combination of olivine, pyroxene, garnet, and perhaps, in restricted regions, amphibole. The two principal rock types carrying these minerals are peridotite (olivine-pyroxene) and eclogite (pyroxene-garnet). Both types may carry some amphibole. Complete mineralogical transitions between the two major rock types are uncommon and usually of local significance only. Possible petrological reasons for this dichotomy are mentioned below. These relationships have given rise to the alternative hypotheses that the upper mantle is of peridotitic or eclogitic composition.

The eclogitic hypothesis is often coupled with the hypothesis that the Mohorovicic discontinuity is caused by an isochemical phase change from a gabbroic lower crust into an eclogitic upper mantle. As is shown in another section of this Monograph (p. 639), this hypothesis is no longer tenable. Nevertheless, the possibility that the Moho is caused by a chemical change from an acidic-intermediate lower crust into an eclogitic upper mantle might still be considered. A critical test of this hypothesis could, in principle, be supplied by knowledge of the density of the upper mantle. The densities of peridotitic (ultramafic) rocks which might qualify as major upper mantle constituents range between 3.25 and 3.38 g/cm^3, with an average close to 3.32 g/cm^3. The densities of *fresh* eclogites, on the other hand, range between 3.4 and 3.65 g/cm^3, with an average of about 3.5 g/cm^3 [*Ringwood and Green,* 1966]. The types of eclogites which best qualify as upper mantle constituents (quartz-free varieties such as are found in diamond pipes) fall in the higher part of this range. Accordingly, it is desirable to consider evidence relating to the density of the upper mantle.

Limitations on the density of the upper mantle arise from the interpretation of gravity observations and from the theory of isostasy. Most interpretations of gravity data are made on the assumption of a mantle density close to 3.3 g/cm^3, although *Talwani et al.* [1959] preferred a density of 3.4 g/cm^3. *Worzel and Shurbet* [1955] reviewed available data on continental and oceanic crustal structures and densities and concluded that isostatic balancing of standard oceanic and continental sections required a mantle density of 3.27 g/cm^3 if the crustal density was 2.84 g/cm^3. Actually, it is the density contrast between crust and mantle that is obtained (subject to certain simplifying assumptions) from gravity observations. *Drake et al.* [1959] concluded that the average density difference between crust and mantle was about 0.43 g/cm^3. Limitations on the density of the crust are obtained from direct observations of the occurrence and densities of crustal rocks, combined with geologic inferences concerning their abundances. Another method of obtaining the mean crustal density is from the seismic velocity distribution in the crust, combined with knowledge of the relationship between seismic velocity and density for common rock types [*Birch,* 1961]. Arguments based upon the above methods have led to the widely accepted view that the mean density of the normal continental crust is between 2.8 and 2.9 g/cm^3. From these values, together with the density contrast between crust and mantle as deduced from gravity data, we can conclude that the most probable mean density of the upper mantle is between 3.3 and 3.4 g/cm^3. This is further supported by an independent method of density determination based upon inversion of surface wave data [*Dorman and Ewing,* 1962].

The density arguments as developed above thus point rather strongly toward a dominantly ultramafic rather than an eclogitic upper mantle. Perhaps some caution is needed before finally accepting this conclusion, since more complex models for isostatic balancing of continental and oceanic sectors could be devised. It is clearly of importance to refine the gravity arguments and to devise new and independent ways of estimating the mean upper-mantle density. Perhaps surface-wave dispersion studies may provide the answer. It is not always realized how powerful a discriminant is a simple property such as density in choosing between alternative ultramafic and eclogitic upper-mantle models. This is probably because most published densities of eclogites have been measured on altered samples, resulting in systematic underestimates, so that the published values overlap the range for ultramafics.

It should be emphasized that the above conclusion concerning the nature of the mantle is based largely upon geophysical observations in stable continental regions and deep oceanic basins; hence, the conclusion is strictly applicable only to these regions. It is possible that the mantle beneath tectonically unstable regions, e.g., continental margins and island arcs, is characterized by substantially different density and seismic velocity distributions and that eclogite may be more important in these regions. This possibility has been further discussed by *Ringwood and Green* [1966].

GEOLOGICAL EVIDENCE

An important source of information on the composition of the earth's mantle beneath stable continental regions is derived from the study of xenoliths in kimberlite pipes. Kimberlite pipes carrying diamonds are of frequent occurrence over 1,000,000 square miles of southern Africa, and are also known in India, the United States, and Siberia. In Africa, these pipes carry numerous xenoliths of crustal rocks that they are known to have intruded on their journey upward. They contain, also, large numbers of xenoliths of rocks that are not known to occur in the vicinity, particularly peridotites, pyroxenites, and eclogites. The presumption is that these xenoliths have been derived from the mantle and represent a random sample of mantle rock types that have been intersected by the pipes.

The occurrence of diamonds both in the pipes and in some of the inclusions implies that the pipes are derived from depths of at least 120 km. It is of great significance that in all cases but one where adequate sampling has been carried out, peridotitic inclusions are found to be much more common than eclogitic inclusions [*Wagner*, 1914, 1928; *Williams*, 1932; *Dawson*, 1962; *Nixon et al.*, 1963]. [The exception is the Roberts Victor pipe (I. D. MacGregor, personal communication).] If the sampling is representative, then the upper mantle must be of peridotitic composition, with eclogite a minor, but widely distributed constituent.

The kimberlite that forms the host rock to these inclusions is a rock of variable composition and degree of alteration. The dominant (50–75%) primary mineral is magnesian olivine. This suggests that the region in the mantle where kimberlite originates is one where magnesian olivine is also abundant.

A second source of information on the composition of the upper mantle derives from an interpretation of the occurrence and significance of alpine peridotites. *Benson* [1926], *Hess* [1939, 1955a, b], and *Thayer* [1960] have extensively discussed the relationships of this class of rocks. Alpine peridotites are characteristically intruded close to the axes of maximum deformation along mountain building belts and island arcs. Intrusion has frequently been controlled by major faults, which may extend for hundreds of miles and which almost certainly extend into the mantle. Ultramafic bodies may occur in the form of innumerable separated bodies along these fault zones, as in the Appalachians and the Great Serpentine Belt of New South Wales [*Turner and Verhoogen*, 1960]. Elsewhere, as in New Guinea, New Caledonia, Cuba, the Philippines, Newfoundland, and British Columbia, ultramafics occur in the form of large individual intrusions, covering hundreds of square miles, and closely connected with major tectonic features. An example of the latter class is the Nahalin Peridotite in British Columbia. This is an elongated body 100 miles long and about 5 miles wide, closely associated with a fault along its major axis (I. D. MacGregor, personal communication).

It is widely believed by geologists that many alpine peridotites of this type are derived directly from the upper mantle and are representative of the rocks occurring in that part of the earth. This hypothesis is supported by their tectonic setting, by the evidence of extreme solid-state deformation which they often display, and sometimes, by their mineralogy [*Green*, 1964]. It is also supported by their physical properties. They yield, when fresh, the required mantle seismic velocities and possess a density of 3.32 g/cm^2.

GEOCHEMICAL AND PETROLOGICAL EVIDENCE

The evidence discussed so far suggests a model according to which the subcontinental mantle consists dominantly of peridotite and dunite, with lesser amounts of eclogite probably occurring as widely scattered segregations [*Ringwood*, 1958]. Because of the similarity in physical properties between sub-continental and sub-oceanic mantles, the simplest step would be to assume that this model was also applicable to the sub-oceanic mantle.

This hypothesis, however, must be rejected because of its inability to explain the origin of basaltic magmas, which have been erupted abundantly on both oceanic and continental regions of the earth's surface throughout geologic time. The source region of basalt is believed on strong grounds to be the upper mantle. Clearly a fundamental property of much of the upper mantle is that it should be capable of yielding basaltic magmas when subjected to suitable (fractional) melting processes. Alpine peridotites, because of their low abundances of K, U, Th, Ba, Sr, La, and many other trace elements, do not possess the compositions required to yield basalts. Furthermore, only a very small proportion have sufficiently high abundances of Na, Ca, and Al to be capable of supplying the concentrations of these elements that are observed in basalts. The general abundance pattern of alpine peridotites is that of a highly fractionated refractory residuum of the type which might remain after a basaltic liquid had been removed.

A solution to the problem was suggested by two parallel developments. *Rubey* [1951, 1955] produced a series of powerful arguments showing that the earth's atmosphere, hydrosphere, and crust had been formed gradually, over geologic time, by degassing and fractional melting of the upper mantle. This implied that beneath continents there must exist a zone which has been deprived of its low-melting and volatile components. Might not the alpine peridotites be representative of this barren zone? If so, then beneath the residual, refractory ultramafic zone there should exist a more primitive material, which could yield crustal material on partial fusion, leaving behind refractory dunite and peridotite. This general model of continental evolution is of ancient lineage, but has received support from many sources during the past decade or so. Excellent accounts of geological aspects of the model have been given by *Wilson* [1954] and *Engel* [1963]. Further geophysical-geochemical consequences have been explored by *Ringwood* [1962a, b], *MacDonald* [1963], and *Clark and Ringwood* [1964]. Studies on the development of strontium 87 in the crust and mantle [*Gast*, 1960; *Hurley et al.*, 1962; *Hedge and Walthall*, 1963] have provided strong support for the general validity of the crustal evolution model while, at the same time, imposing some important constraints.

The notion that the crust is geochemically and tectonically coupled to a considerable thickness of the mantle immediately below finds its strongest support in Precambrian shields and stable continental platforms. It is now widely recognized that the heat flow over extensive regions of shields is substantially lower than the world average. *Ringwood* [1962a] attributed this to a net deficiency of U, Th, and K in the crust and mantle beneath shields, caused by the long history of fractional melting in the subjacent mantle, selective removal of U, Th, and K from the mantle, and concentration of these elements at high levels in the crust by magmatism, whence they were depleted by surface erosion. Detailed evidence supporting the operation of these processes has been described by *Heier and Adams* [1965], *Lambert and Heier* [1967], and by *Hyndman et al.* [1968]. Because of the deficiency in net radioactivity and the strong upward concentration of radioactivity in the crust, *Ringwood* [1962a, b] and *Clark and Ringwood* [1964] inferred that subcrustal temperatures would be substantially lower for 300-400 km beneath shields compared to off-shield areas, and that this would have a marked differential effect upon the seismic velocity distribution, particularly upon the occurrence and position of the low velocity zone. This inference was substantially verified by *Brune and Dorman* [1963] and *Toksöz et al.* [1967]. The former authors demonstrated that the S-wave velocities beneath the Canadian shield were substantially higher on the average than beneath off-shield areas down to depths of about 200 km, and that the low-velocity minimum was less pronounced. A similar situation occurs for P waves, which show early arrivals at shield stations [*Cleary and Hales*, 1966] and do not appear to display a velocity minimum. Yet another indication of the different seismic structure beneath shields is shown by the low attenuation of S waves (H. Doyle and J. Brune, personal communications).

Independent evidence that the mantle beneath shields is characterized by lower temperatures than off-shield areas comes from studies of the electrical conductivity of the upper mantle [*Everett and Hyndman*, 1968]. The temperature distribution beneath shields has an important effect upon their tectonic properties. It is probable that the temperatures are below the melting point at all depths. Many rheological properties, such as viscosity, strength, plasticity, and creep depend very sensitively upon the difference between the actual temperature and the melting temperature. The occurrence of deep-seated cooling has a major effect upon the rheological properties of the mantle in such regions when compared to the mantle in neighboring regions that have not been through this cycle. The mean 'viscosity' and strength of the upper 200 km of the mantle beneath shields may be higher, possibly by orders of magnitude, than beneath neighboring regions. Thus the rigidity and relative tectonic stability of shield regions, which is their outstanding geological characteristic, may have a simple ultimate explanation in terms of this model of crustal evolution.

The second development affecting the status of ultramafic models for the upper mantle has arisen from petrology and can be traced back to *Bowen* [1928], who maintained that the primary composition of the upper mantle was that of a felspathic peridotite and that basalt magmas were formed by fractional melting of this primary material leaving behind a 'barren' residuum. (According to previous discussion, we would now identify the barren residuum with some kinds of peridotites and dunites.)

Bowen's lead was not immediately pursued

by succeeding geologists, who were more interested in the fractional crystallization of basalt magma than in its ultimate origin. However, since the 1950's, petrologists have displayed more interest in the origin of basalt, and many papers appeared reviving Bowen's hypothesis that basalts were formed by fractional melting in the mantle [*Verhoogen*, 1954, 1956; *Powers*, 1955; *Kuno et al.*, 1957; *Wager*, 1958; *Wilshire and Binns*, 1961; *Yoder and Tilley*, 1962; *Kushiro and Kuno*, 1963; *O'Hara*, 1965]. These authors were somewhat vague about the nature of the parental peridotite, usually identifying it with types of alpine peridotites and selected ultramafic nodules from basalt, richer than usual in Na_2O, Al_2O_3, and CaO. Although a step in the right direction, this assumption is inadequate, since such rocks do not generally possess the correct trace element chemistry (particularly K, U, Th) to yield basalt magmas on fractional melting, even if the major element partitions are satisfactory.

Ringwood [1958, 1962a, b], *Green and Ringwood* [1963, 1967a], and *Clark and Ringwood* [1964] sought to follow more closely the logical consequences of the earlier hypotheses of *Bowen* [1928], *Rubey* [1951], and *Bullard* [1952]. It was found necessary to postulate a primitive, parental mantle material which was arbitrarily defined by the property that on fractional melting it would yield a typical basaltic magma and leave behind a residual refractory dunite-peridotite. The composition of this primitive material must lie between those of basalt and dunite-peridotite. Such a primitive material would be unlikely to contain less Al_2O_3 and CaO than are present (relatively to MgO and SiO_2) in chondrites and in the sun [*Ringwood*, 1966, Table 1]. This would indicate a limit of about 1 basalt to 3 dunite-peridotite. On the other hand, the ratio is hardly likely to be less than 1 basalt to 1 dunite-peridotite if basalt is to be produced by partial melting in the mantle [*Hess*, 1960; *Green and Ringwood*, 1967a].

Experimental investigations on the fractionation of basaltic magmas at high pressures and the formation of basalts by fractional melting of pyrolite by *Green and Ringwood* [1967a] have led to a broadly self-consistent interpretation of the relationships between the different major basalt types and their complementary peridotites and dunites (see section by Green and Ringwood in this Monograph, p. 489). It was shown that alkali basalts can be formed by a relatively small degree of fractional melting of pyrolite in the depth range 35–70 km. Alkali basalts formed under these conditions were found to be in equilibrium with a refractory assemblage of olivines, aluminous orthopyroxene, and aluminous clinopyroxene, which is identical with the mineralogy of peridotite nodules occurring as inclusions in alkali basalts all over the world. With a higher degree of fractional melting in this depth interval, olivine tholeiites are formed and the residual refractory crystals are olivine $\pm$ orthopyroxene (Al-poor). It was also shown that high alumina basalts and quartz tholeiites can be formed by fractional melting of pyrolite accompanied by segregation of magma from residual crystals at depth shallower than 35 km.

Green and Ringwood [1963] have studied primary ultramafic rocks in order to find how closely the hypothetical pyrolite composition is approached. Starting with a 'typical' alpine ultramafic containing virtually no Ca, Al, and Na and composed of olivine and orthopyroxene, one can find a complete continuum of compositions over to rocks containing about 4% Al_2O_3, 3% CaO, and 0.4% Na_2O. Such rocks are found as garnet peridotites in diamond pipes, as inclusions in alkali basalts, and as high-temperature peridotites [*Green*, 1964]. Rocks with this composition lie very close to a 3:1 peridotite:basalt composition, as do chondrites. Nevertheless, even in such rocks, the content of minor elements, particularly Na, K, Ti, P, Ba, Th, and U, is too low for them to be regarded as unfractionated pyrolite. Another interesting fact is that representative ultramafic rocks of the type under consideration possessing compositions corresponding to peridotite:basalt ratios smaller than 3:1 (based on major elements) are very rare. There seems to be some sort of discontinuity in natural occurrences at the 3:1 ratio. This is probably connected with the dichotomy previously mentioned, which appears to exist between eclogitic and peridotitic rocks. Recent experimental work and interpretation by *O'Hara* [1963, 1965], *Davis*, [1964], *Green and Ringwood* [1967a], and *Ito and Kennedy* [1967] on melting relationships in the system olivine-orthopyroxene-Ca-rich clinopyroxene-pyrope-rich garnet have contributed greatly toward an understanding of these observations. They are connected with the tendency of garnet-pyroxene (eclogite) and olivine-orthopyroxene (peridotite) to behave as high melting point residua or as refractory thermal barriers in this system at pressures above 30 kb, whereas intermediate compositions (equivalent to mixtures of peridotite and eclogite) melt at lower temperatures and may be removed from the system as picritic magmas.

It appears that nearly all ultramafic rocks exposed at the surface of the earth have been subjected to some degree of fractionation, and that rocks approaching the ideal pyrolite composition are extremely rare. (The only possible examples known to the author are some rock types described by *Wiseman* [1966] from St. Paul's Rocks.) The deficient elements (K, Ti, P, Ba, Sr, Rb, Th, U, and others) are unable to enter the major phases of the mantle because of their incompatible ionic charges and sizes, and they are accordingly fractionated rather readily [*Harris*, 1957]. It appears that they may be strongly concentrated in the liquid phase which forms at a comparatively low temperature and at a very early stage of fractional melting in the mantle when the amount of liquid formed is small. This would account for the high mobility of these elements under mantle conditions, which seems necessary to account for the formation of kimberlites and other highly alkali rock.

Thayer [1960] has recently examined some of the larger terrestrial ultramafic intrusions. He found in all of these that they do not consist of pure peridotite, but rather of mixed peridotite and gabbro. The two rocks clearly have a common origin. The most straightforward and in many ways the most satisfactory explanation of this association is to regard them as complementary differentiates of primary mantle material.

The logical development of the hypotheses of *Rubey* [1951] and *Bullard* [1952], as discussed in this section, leads to a chemically zoned upper mantle [*Ringwood*, 1958], as depicted in Figure 1. The depth of the refractory dunite-periodotite zone under continents would vary, the maximum thickness, perhaps about 200 km, being reached beneath Precambrian shields. Beneath deep oceanic basins, on the other hand, the primitive pyrolite may extend to the Moho. Alternatively, a thin layer of dunite-peridotite

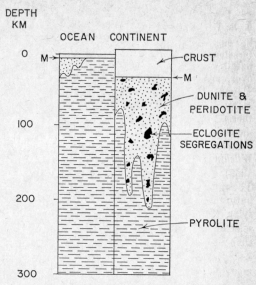

Fig. 1. Chemical model for the upper mantle.

a few tens of kilometers thick may be present.

A composition for the pyrolite model as derived by *Ringwood* [1966, p. 307] is given in Table 1. This is based upon a mixture of 3 parts alpine peridotite (20% orthopyroxene) to 1 part Hawaiian basalt. This composition is compared with a mantle composition derived from chondritic abundances (Table 1). It is seen that the 3:1 ratio gives quite a good match. Nevertheless, it is emphasized that this ratio is rather arbitrary, and that large variations are possible within the framework of the pyrolite model.

ZONING IN THE UPPER MANTLE

An interesting property of rocks approaching the pyrolite composition is their capacity to crystallize in four distinct mineral assemblages, as follows:

Olivine + amphibole   Ampholite
Olivine + Al-poor pyroxenes   Plagioclase pyrolite
+ plagioclase
Olivine + aluminous pyroxenes Pyroxene pyrolite
± spinel
Olivine + Al-poor pyroxenes   Garnet pyrolite
+ pyrope-rich garnet

The occurrence of the various pyrolite mineral assemblages in the mantle is largely determined by the intersection of geotherms with the stability fields of the assemblages. *Green and Ringwood* [1967c] have recently determined

TABLE 1. Comparison of Pyrolite Model and Mantle Compositions

Component	Model Pyrolite	Model Mantle*
SiO_2	45.16	43.25
MgO	37.49	38.10
FeO	8.04 ⎫	9.25
Fe_2O_3	0.46 ⎭	
Al_2O_3	3.54	3.90
CaO	3.08	3.22
Na_2O	0.57	1.78
K_2O	0.13	
Cr_2O_3	0.43	
NiO	0.20	
CoO	0.01	
TiO_2	0.71	
MnO	0.14	
P_2O_5	0.06	
	100.00	

* Derived from chondritic abundances [Ringwood, 1966].

several of the stability field boundaries experimentally and their results, together with two possible characteristic geotherms, are shown in Figure 2. Ito and Kennedy [1967] have carried out an analogous investigation on a natural garnet peridotite. The pyrolite mineral assemblages possess distinctly different physical properties. Calculated densities and P velocities of their assemblages, together with those of dunite and a typical fractionated alpine peridotite, are given in Table 2.

The seismic P and S velocity profiles in the upper mantle vary in a complex manner from region to region and are discussed elsewhere in this volume. This complex variation is due to an interplay of several factors.

Regional Variations in Temperature-Depth Distribution

In the uppermost mantle, temperature gradients are generally high enough to cause S-wave velocities to decrease initially with depth [Birch, 1952] (O. Anderson, unpublished measurements), unless this is prevented by mineralogical zoning (see below), and locally are high enough to cause P-wave velocities to decrease with depth. At deeper levels, below 100–300 km, temperature gradients are insufficient to offset the effect of pressure on seismic velocities, which therefore increase. Clearly regional variations of temperature profiles as suggested in Figure 2 will cause corresponding variations in velocity profiles.

Mineralogical Zoning

Because of the differing characteristic seismic velocities among pyrolite mineral assemblages (Table 2), seismic velocity distributions in the mantle will be affected by the depth at which geotherms intersect the pyrolite stability fields (Figure 2). These will cause intrinsic velocity changes, which will be superimposed upon those due simply to temperature-pressure effects on elastic properties. As *Green and Ringwood* [1967c] have shown, the effects of mineralogical zoning upon velocity distributions can be of a complex and subtle nature (Figure 3).

Chemical Zoning

The extensive fractional melting and upward segregation of low-melting components that is inferred to have occurred beneath highly evolved continental regions results in zoning caused by layers which differ in chemical composition (Figure 1). This may also contribute to seismic velocity differences.

Incipient Melting

In many oceanic regions, the uppermost 50 km of the upper mantle has an S-wave velocity of about 4.6 km/sec. Below this 'lid,' velocity decreases rather sharply, reaching a minimum of about 4.3 km/sec at about 100 km [Anderson, 1967b]. In some orogenically active regions, e.g.,

TABLE 2. Calculated Densities and P-Wave Velocities at Atmospheric Temperatures and Pressures for Pyrolite Mineral Assemblages and Peridotite*

Rock	Density, g/cm³	V_p, km/sec
Dunite	3.32	8.48
Peridotite †	3.31	8.32
Ampholite ‡	3.27	7.98
Plagioclase pyrolite	3.26	8.01
Pyroxene pyrolite	3.33	8.18
Garnet pyrolite	3.38	8.38

* Ringwood [1964].
† 20% orthopyroxene.
‡ 35% amphibole.

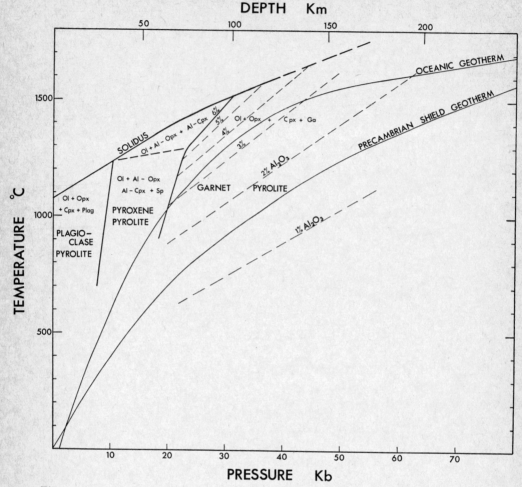

Fig. 2. Diagram illustrating the P, T fields of different mineral assemblages in pyrolite. The values 1% Al_2O_3, 2% Al_2O_3, etc., refer to the Al_2O_3 content of orthopyroxene in equilibrium with garnet in the garnet pyrolite field [from *Green and Ringwood*, 1967c].

the Western Mediterranean, the S velocity in the low velocity zone may be as low as 4.1 km/sec [*Berry and Knopoff*, 1967]. It is difficult to explain such low absolute velocities and particularly the magnitude of the negative velocity gradients in terms of a plausible model using the first three effects discussed above, particularly when the latest data on elastic properties of minerals (O. Anderson, personal communication) are used. (See also *Toksöz, et al.* [1967] for discussion of the problem.) In such regions several workers [*Anderson*, 1962, 1967b; *Oxburgh and Turcotte*, 1968a; *Aki*, 1968] have suggested that the low velocities imply the occurrence of a small degree of partial melting (incipient melting). *Press* [1959] and *Gutenberg* [1959] also came close to this explanation when they suggested that the low velocity zone was caused by a close approach to the melting point.

The conditions that permit large volumes of the upper mantle to remain in a quasi-stable state of incipient melting deserve some comment. It does not appear likely that this would be possible if the mantle were completely anhydrous. At pressures of 20–30 kb, the melting interval between solidus and liquidus of olivine tholeiite and picrite is smaller than 100°C [*Green and Ringwood*, 1967a]. It is probable that dry pyrolite would become 30% molten

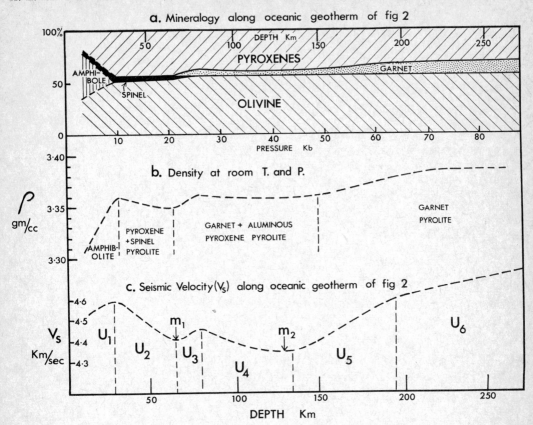

Fig. 3. The changes in mineralogy and density (atmospheric P and T), and the relative changes in seismic velocity V_s for pyrolite along the oceanic geotherm of Figure 2. Critical gradient for constant shear wave velocity with depth is assumed to be 4.5°C/km [from *Green and Ringwood*, 1967c].

within this interval (Figure 4). Temperature fluctuations as small as 20°C in the low-velocity zone would produce substantial (~5%) partial melting. Such a state would not remain stable for long; liquid would soon segregate into magma bodies, which would rise to the surface producing volcanism. Clearly, if the upper mantle were in this condition, widespread volcanism should occur on a much larger scale than is now observed. To account for a state of incipient melting throughout the low-velocity zone, an efficient stabilization mechanism must be assumed to be present.

It is possible that water is ultimately responsible for the stabilization. *Ringwood* [1966, p. 325] inferred from comparison with the degassing of nitrogen from the earth's interior that the mantle has probably retained at least 3 times as much water as has been degassed into the oceans; this would suggest an average content of 0.1% H_2O. Important evidence bearing upon the occurrence of water in the mantle has been obtained by *Bultitude and Green* [1968], who showed that the crystal-liquid equilibria that are responsible for the generation of olivine-nephelinite and olivine-melilite-nephelinite magmas in the mantle are largely controlled by the presence of water in their source regions.

The effect of a fixed amount of water (say 0.1%) in lowering the temperature of beginning of melting will be strongly influenced by the stabilities of hydrous phases, primarily amphibole and phlogopite. We can disregard the latter mineral, since in view of the probable abundance of potassium in the mantle, this mineral is capable of accounting for less than 0.02% of H_2O. The role of amphibole may be critical. *Green and*

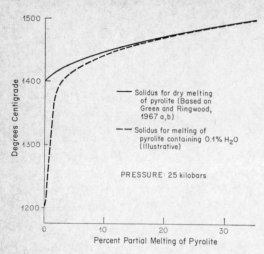

Fig. 4. Diagrammatic representation of degree of melting versus temperature in pyrolite under anhydrous conditions and in the presence of 0.1% water at 25 kb.

Ringwood [1967*a* and unpublished data] found that in olivine-rich rocks, amphibole was stable at temperatures not far below the dry solidus under conditions of $P_{H_2O} < P_{load}$ at load pressures up to 15 kb. Indeed, in the experiments cited, charges and furnace components had to be carefully dried before use to prevent formation of amphibole. Accordingly, in the mantle at depths up to about 60 km, the small amount of water present will be held mostly as amphibole, and the partial pressure of water will be much smaller than load pressures in most regions.

At pressures higher than 20 kb, and at 1100°C, *Green and Ringwood* [1967*b*] obtained evidence that amphibole was unstable in olivine-rich basaltic compositions under P_{H_2O} conditions similar to those that caused formation of amphibole between 10 and 15 kb. It was inferred that amphibole would decompose to an eclogitic mineral assemblage plus water under these conditions, and that the decomposition temperature would be expected to decrease with pressure (for constant P_{H_2O}). This general pattern of behavior has since been observed in many other systems varying from ultramafic [*Bultitude and Green*, 1968] to mafic-acidic [*Green and Ringwood*, 1966, 1968*a*, *b*]. This work suggests that even at relatively high water pressures, amphiboles are not generally stable above 1200°C (1000°C in some systems) at pressures greater than about 20 kb.

This has an important effect on melting behavior in the mantle below about 60 km, where most of the water present will therefore occur as a free phase. If there is only 0.1% H_2O present, it will occur mostly along grain boundaries and in small bubbles. Surface effects, combined with solution of other components, will lower the fugacity of water substantially, perhaps by a factor of 2 to 10 compared to the fugacity of a large volume of pure water under the load pressure. Nevertheless, water vapor at this fugacity will have a large effect upon the temperature of beginning of melting and may well lower it by ~200°C.

Melting occurs initially as suggested in Figure 4. After a very small degree (~1%) of partial melting (termed incipient melting to distinguish it from the 5–30% partial melting necessary to form magmas), most of the 0.1% water originally present occurs in solution in the melt. With further increase in temperature, the degree of melting will increase. However, this causes a lowering of the concentration of water in the melt, and hence a lowering of water pressure in the system. This has a directly opposed effect on the degree of melting, since it raises the crystallization temperature. Melting is controlled by a balance between these factors. Qualitatively it results in the situation shown in Figure 4, where a comparatively large increase in temperature causes only a small increase in the degree of incipient melting. Ultimately the 'wet' melting curve becomes asymptotic to the dry melting curve.

Wet melting, as is indicated in Figure 4, may well provide the stabilization required for an incipiently melted low velocity zone, since a comparatively large increase in temperature results in the formation of a relatively small amount of additional liquid until the dry solidus is approached, when the situation is reversed. Substantial temperature fluctuations in the low velocity zone (e.g., owing to differences in abundances of radioactive elements) will usually not lead to partial melting to the extent required for magma segregation and surface volcanism.

The foregoing discussion suggests that the velocity distributions in the low velocity zone are controlled by a number of factors which probably vary widely in their relative importance in different regions of the earth. The detailed unraveling of these patterns presents many difficulties and remains a task for the future. Nevertheless, a few broad generalizations can be made. It appears that the S low velocity zone beneath shields, which is relatively deep, can be interpreted in terms of thermal and chemical effects probably not involving incipient melting. (It can be seen that the shield temperature gradients in Figure 2 are *higher* at depths greater than 120 km than the sub-oceanic temperature gradients at similar depths). On the other hand, it will probably be necessary to invoke incipient melting as a major cause of the low velocity zone in oceanic and orogenically active regions. The tectonic and petrologic implications are far-reaching.

EVOLUTION OF THE CRUST AND UPPER MANTLE

The recent emergence of powerful evidence supporting the sea-floor spreading hypothesis [*Hess*, 1962; *Vine and Matthews*, 1963] is having a revolutionary impact upon previous models of crust-mantle evolution. Let us first consider a model that was receiving a wide measure of attention before the 'revolution.'

The discovery [*Revelle and Maxwell*, 1952; *Bullard*, 1952, 1954] that the mean heat flow from the deep oceanic crust was approximately equal to the mean continental heat flow was surprising to those who had previously advocated an oceanic upper mantle of alpine type peridotite, since it was known that the radioactivity of these rocks was far too low to yield the observed oceanic heat flow. Bullard proposed two alternative hypotheses in explanation: first, that the oceanic heat flow was due to convection; second, that, to a first approximation, the mean chemical composition of the mantle was the same over both continent and ocean when averaged down to depths of several hundred kilometers. The principal difference between continent and oceans was that continental regions had become more strongly differentiated, and the low melting components and radioactive elements were concentrated near the surface. The second hypothesis was probably the more popular, particularly as it fitted in rather well with *Rubey's* [1951, 1955] theories and with developing ideas relating to the origin of basalts and their relation to ultramafic rocks, as discussed above. Numerous models [*Ringwood*, 1958, 1962*a, b; MacDonald*, 1963; *Clark and Ringwood*, 1964] were developed along these lines. In my opinion these models continue to retain valuable attributes, particularly in explaining the geochemical, petrological, and geophysical relationships between evolved continental regions and the mantle immediately beneath them. Such notions must be incorporated in any finally successful theory.

The sea-floor spreading hypothesis strikes at one of the foundations of these earlier hypotheses, because it implies that a large proportion of the oceanic heat flow is brought up by convection. The source of part of this heat may be 'original heat' inherited from an early high temperature stage in the earth's history. Part may also come from deeply buried radioactivity, but there is no necessity that the ultimate source should permanently lie in the oceanic upper mantle. Accordingly, there is no longer any justification for assuming that the mean chemical compositions beneath continents and oceans when averaged down to considerable depths are identical. It would be more reasonable to construct models with much lower abundances of U, Th, and K in the oceanic upper mantle than previously appeared possible. Indeed, the discovery of the widespread occurrence of oceanic tholeiites with their distinctive trace element chemistry [*Engel et al.*, 1965] supported the assumption of much lower abundances of U, Th, and K than were previously acceptable in the suboceanic mantle.

We consider now the manner in which sea-floor spreading may perhaps be connected with crustal evolution. The following model is a development of that presented by *Ringwood and Green* [1966]. Needless to say, it is highly speculative in many aspects, as indeed are all current models of this class. A diagram of the model is presented in Figure 5. According to the model, gravitational instability develops in the mantle beneath mid-oceanic ridges. Pyrolite flows from the low-velocity zone (which is the immediate source region) toward the ridge axis and then rises upward. Fractional melting of pyrolite occurs during upward motion, leading to generation of basaltic magma, together with residual unmelted peridotite, as described by *Green and Ringwood* [1967*a*]. The axes of the ridges are characterized by high heat flow, and the subsurface temperatures are high enough to maintain the stability of the basaltic

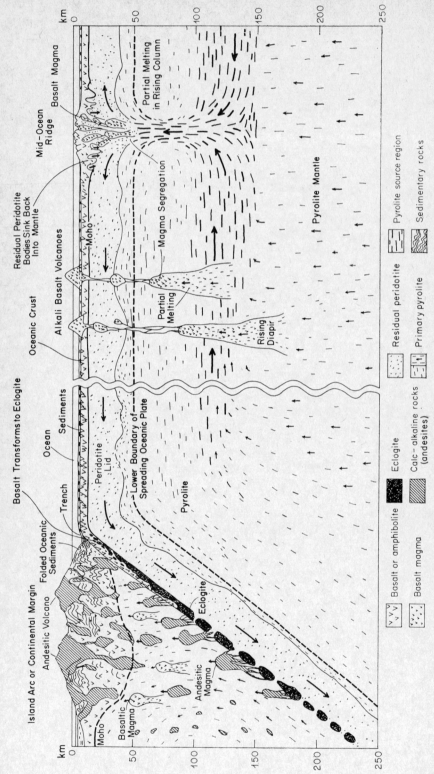

Fig. 5. Modified version of ocean-floor spreading hypothesis [*Hess*, 1962]. Not to scale.

mineral assemblage. The mid-oceanic ridges thus develop as expanding features composed of heterogeneous mixtures of gabbro, dolerite, peridotite, and pyrolite with surficial basalts. Differentiation accompanies the horizontal expansion, resulting in the formation of a rather uniform mafic crust, overlying a depleted ultramafic mantle [Oxburgh, 1965]. The upper layers of the mafic crust are basaltic, cool quickly, and may become partly oxidized by interaction with sea water. This altered layer may be the source of the linear magnetic anomalies. Deeper layers of the crust are protected from the sea water and hence remain relatively dry and unoxidized. Moreover, the deeper regions of the crust may cool through the Curie point on a time scale which is long compared to the period of magnetic field reversals, so that it does not become so strongly and uniformly magnetized as the upper crust.

The cooling oceanic crust, together with its coupled depleted ultramafic layer, then moves outward, sliding over the weak, incipiently melted low velocity zone. A thoughtful analysis of the thermal behavior of a closely related system has been provided by Oxburgh and Turcotte [1968a]. The crust and underlying plate of mantle cool as they move away from the ridge. The results of Ringwood and Green [1966], suggest that, at the same time, the dry basalt passes into the eclogite stability field. However, it does not transform initially because of kinetic difficulties.

Ultimately the crust reaches an oceanic trench. The reasons for the initial subsidence of the crust into the mantle to form a trench are not well understood. Ringwood and Green [1966] have suggested that this may happen initially near a continental margin, where sedimentation on the oceanic crust causes an increase in pressure and temperature and permits the lower dry mafic crust to transform to eclogite, whereas the upper wet regions of the crust transform to amphibolite. The high density of dry eclogite ($\sim$3.5 g/cm^3) compared to that of mantle peridotite ($\sim$3.3 g/cm^3) causes gravitational instability and results in crustal subsidence.

Once the process starts, it becomes self-sustaining, the driving force being supplied by the high density of the cool descending column. Also, once the process starts, it is possible that transformation of dry basalt to eclogite will not occur until depths of a few tens of kilometers and that it may be spread out over a rather large depth interval owing to the slowness of transformation at low temperatures. Ultimately, however, all the basalt in the descending column must transform to dense eclogite, thereby increasing the gravitational instability. Amphibolite will similarly transform to eclogite, but at somewhat greater depths [Green and Ringwood, 1967b].

Oxburgh and Turcotte [1968b] have suggested that viscous dissipation in the upper mafic boundary layer of the sinking slab may cause strong heating and may even lead to partial melting of this layer. Green and Ringwood [1966, 1968a,b] have shown experimentally that the partial melting of mafic rocks at high pressure provides an efficient means of generating calc-alkaline magmas. Two models were suggested. The first involved partial melting of quartz eclogite under relatively dry conditions at depths of 80–150 km. It was shown that the low melting fraction was andesitic and that the residuum would be a very dense refractory eclogite which would continue to sink. The second model involved the partial melting of amphibolite at lower pressures, yielding andesitic-dacitic magmas and leaving behind a silica-poor, amphibole-rich assemblage which might ultimately transform to eclogite. At pressures greater than about 20 kb, partial melting of amphibolite would also yield calc-alkaline magmas, leaving behind residual eclogite. Thus the two models are in fact closely related and represent particular cases of a generalized mechanism for obtaining calc-alkali rocks by partial melting (and fractional crystallization) of mafic rocks under high load pressures ($\sim$10 kb) and varying water pressures, which, however, are likely to be generally less than load pressures.

It appears that processes such as these may be capable of providing a mechanism for the growth of continents by the addition near continental margins of sialic calc-alkaline rocks derived ultimately from the oceanic mantle (see also Coats [1962]). Green and Ringwood [1968b] have also suggested that the upward vective instability in the wedge-shaped region between the surface and the sinking slab of oceanic crust, leading to fractional melting in this region and the generation of basaltic magmas which also enter the growing continental margin or island arc. Thus the region beneath the growing continental margin may ultimately become depleted in low-melting material, thereby reaching the chemical state of an alpine peridotite.

It is assumed that the descending plate continues to sink to depths of 700 km or greater. This residual, fractionated material enters the deep mantle and does not return in a second cycle. Rather, it displaces upward relatively unfractionated material into the low velocity zone, which acts as the immediate source region for the mid-oceanic ridges (Figure 4). Thus the process which is being advocated is a one-way, irreversible differentiation mechanism. This differs from some other current models which suggest remixing of sediments, crust, and mantle, followed by continental recycling of this material [e.g. *Armstrong*, 1968]. Such a process appears implausible, there being no obvious way in which blocks of fractionated eclogite, peridotite, and sediments can be intimately remixed in the solid state in the deep mantle. Furthermore, the trace element patterns of the material rising under ridges do not suggest a multistage history of prior, complex fractionation.

CONCLUDING REMARKS

It appears that the generation of crust from mantle has proceeded via two distinct types of differentiation processes characterized by vertical and lateral mass transport, respectively. Vertical differentiation involves fractional melting of the mantle and transport of the low melting components into the crust immediately above. The process leads to the formation of a residual refractory ultramafic zone in the mantle immediately beneath the crust, and has reached its maximum development in Precambrian shield regions. The lateral differentiation process is an intrinsic part of the major sea-floor spreading cycle. Basalts which are formed by fractional melting beneath oceanic ridges are transported to trenches in the neighborhood of continental margins and island arcs, where they are in turn fractionally melted as the oceanic crust sinks into the mantle. The product of this second stage of fractional melting is the calc-alkali igneous rock suite. It is likely that water incorporated in the oceanic crust plays an important role in this second stage of fractional melting by lowering the temperature of melting and also by influencing the nature of the crystal-liquid equilibria. The second process appears to be predominant in contributing to the formation of new crust in continental margins and island arcs at the present time. The ultimate source region of the calc-alkali rocks that finally enter continents by this mechanism is the oceanic upper mantle.

It is possible that the relative importance of these processes has varied over geologic time. At the present rates of sea-floor spreading, it can be shown that the entire continental crust could have been formed more or less uniformly over geologic time by this process alone. The uniformitarianism approach does not explain, however, the lack of observed continental crust rocks formed in the interval between 4.5 and 3.4 b.y. ago. This might suggest that sea-floor spreading did not commence until about a billion years after the earth was formed. *Armstrong* [1968] has pointed to another argument in apparent conflict with uniformitarianism. Many Precambrian shield regions appear to have evolved very rapidly between 3 and 2.5 b.y. ago, and have had a rather simple geologic history since then, apparently unaffected by major vertical movements and crustal deformation. Armstrong argues plausibly that the thickness of the crust (i.e., depth to the Moho) in these regions has not changed greatly during the last two billion years. Accordingly, the relative elevations of these continental regions with respect to ocean basins have also not changed, and the oceans must have been at about their present size early in the Precambrian. Since it is widely believed that oceans are derived from the interior by degassing processes connected with magmatism and differentiation [*Rubey*, 1955], these considerations suggest that between 3.5 and 2.5 b.y. ago continental evolution may have proceeded at a much greater rate than, on the average, has occurred since then. It is conceivable that vertical differentiation played a more important role in this early rapid phase of crust formation than lateral differentiation via sea-floor spreading.

Reasons that can be suggested for this non-uniformitarian behavior are necessarily highly speculative. It may well be significant that the formation of continental calc-alkali rocks appears to require a two-stage process. It is conceivable that during the first billion years of earth history extensive basaltic magmatism occurred over the earth leading to the formation of a primitive basaltic crust. Because of high thermal gradients, the primitive basaltic crust was prevented from transforming to eclogite and thus was unable to sink into the mantle, thereby preventing operation of the second stage of fractional melting required for the formation of calc-alkaline rocks. A second speculation concerns the inference that the luminosity of the sun may have increased by a factor of 1.6 over the past 4.5 b.y. [Schwarzschild, 1958]. Ringwood [1961] noted that, if this had occurred, the surface temperature over most of the earth before 3.5 b.y. ago may have been below 0°C, so that the normal geological cycle operating by erosional processes involving water would not have operated. Because of low salinity, the primitive oceans may have been frozen. This would largely prevent the hydration of basaltic rocks at the earth's surface. We have previously discussed the important role that water, carried into the mantle by hydrated rocks in the sinking column beneath trenches, may play in the evolution of calc-alkali rocks. It is thus conceivable that during the first billion years there was no way of causing water to re-enter the mantle in sufficient concentration to cause the formation of calc-alkali rocks.

Acknowledgments. I acknowledge the value of discussions with E. R. Oxburgh and D. L. Anderson, which have influenced my opinion on seafloor spreading and incipient melting in the mantle.

REFERENCES

Aki, K., Seismological evidence for the existence of soft, thin layers under Japan, *J. Geophys. Res., 73,* 585–594, 1968.

Anderson, D. L., The plastic layer of the earth's mantle, *Sci. Am., 205,* 2–9, July 1962.

Anderson, D. L., Latest information from seismic advances, in *The Earth's Mantle,* edited by T. Gaskell, chapter 12, pp. 355–420, Academic Press, London, 509 pp., 1967a.

Anderson, D. L., A review of upper mantle seismic data (abstract), *Trans. Am. Geophys. Union, 48,* 254, 1967b.

Armstrong, R. L., A model for the evolution of strontium and lead isotopes in a dynamic earth, *Rev. Geophys., 6,* 175–199, 1968.

Benson. W. H., The tectonic conditions accompanying the intrusion of basic and ultrabasic rocks, *Natl. Acad. Sci. Mem., 19*(1), 6, 1926.

Berry, M. J., and L. Knopoff, Structure of upper mantle beneath western Mediterranean basin, *J. Geophys. Res., 72,* 3613–3626, 1967.

Birch, F., Elasticity and constitution of the earth's interior, *J. Geophys. Res. 57,* 227–286, 1952.

Birch, F., The velocity of compressional waves in rocks to 10 kilobars, 2, *J. Geophys. Res., 66,* 2199–2224, 1961.

Bowen, N. L., *The Evolution of the Igneous Rocks,* Princeton University Press, Princeton, 1928.

Brune, J., and J. Dorman, Seismic waves and earth structure in the Canadian shield, *Bull. Seismol. Soc. Am., 53,* 167–210, 1963.

Bullard, E. C., Discussion of paper by Revelle and Maxwell, *Nature, 170,* 200, 1952.

Bullard, E. C., The flow of heat through the floor of the Atlantic Ocean, *Proc. Roy. Soc. London, A, 222,* 408–429, 1954.

Bultitude, R. J., and D. H. Green, Experimental study at high pressures on the origin of olivine rephelinite and olivine melilite rephelinite magmas, *Earth Planetary Sci. Letters, 3,* 325–337, 1968.

Clark, S. P., and A. E. Ringwood, Density distribution and constitution of the mantle, *Rev. Geophys., 2,* 35–88, 1964.

Cleary, J., and A. L. Hales, An analysis of the travel times of P waves to North American stations in the distance range 32° to 100°, *Bull. Seismol. Soc. Am., 56,* 467–489, 1966.

Coats, R. R., Magma type and crustal structure in the Aleutian Arc, in *The Crust of the Pacific Basin, Geophys. Monograph 6,* edited by G. A. Macdonald and H. Kuno, pp. 92–109, American Geophysical Union, Washington, D. C., 1962.

Davis, B. T. C., The system diopside-forsterite-pyrope at 40 kbar, *Carnegie Inst. Washington Year Bk., 63,* 165–171, 1964.

Dawson, J. B., Basutoland kimberlites, *Bull. Geol. Soc. Am., 73,* 545–560, 1962.

Dorman, J., and M. Ewing, Numerical inversion of seismic surface wave dispersion data and crust-mantle structure in the New York-Pennsylvania area, *J. Geophys. Res., 67,* 5227–5242, 1962.

Drake, C. L., M. Ewing, and G. H. Sutton, Continental margins and geosynclines, *Phys. Chem. Earth, 3,* 110–198, 1959.

Engel, A. E. J., Geologic evolution of North America, *Science, 140,* 143–152, 1963.

Engel, A. E. J., C. G. Engel, and R. G. Havens, Chemical characteristics of oceanic basalts and the upper mantle, *Geol. Soc. Am. Bull., 76,* 719–734, 1965.

Everett, J. E., and R. D. Hyndman, Geomagnetic variations and electrical conductivity structure in southwestern Australia, *Phys. Earth Planetary Interiors*, *1*, 24–34, 1968.

Faure, G., and P. M. Hurley, The isotopic composition of strontium in oceanic and continental basalts: Application to the origin of igneous rocks. *J. Petrol.*, *4*, 31–50, 1963.

Gast, P. W., Limitations on the composition of the upper mantle, *J. Geophys. Res.*, *65*, 1287–1297, 1960.

Green, D. H., The petrogenesis of the high-temperature peridotite intrusion in the Lizard area, Cornwall, *J. Petrol.*, *5*, 134–188, 1964.

Green, D. H., and A. E. Ringwood, Mineral assemblages in a model mantle composition, *J. Geophys. Res.*, *68*, 937–946, 1963.

Green, D. H., and A. E. Ringwood, Genesis of basaltic magmas, *Beitr. Mineral. Petrog.*, *15*, 103–190, 1967a.

Green, D. H., and A. E. Ringwood, An experimental investigation of the gabbro to eclogite transformation and its petrological applications, *Geochim. Cosmochim. Acta*, *31*, 767–833, 1967b.

Green, D. H., and A. E. Ringwood, The stability fields of aluminous pryoxene peridotite and garnet peridotite and their relevance in upper mantle structure, *Earth Planetary Sci. Letters*, *3*, 151–160, 1967c.

Green, T. H., and A. E. Ringwood, Origin of the calc-alkali igneous rock suite, *Earth Planetary Sci. Letters*, *1*, 307–316, 1966.

Green, T. H., and A. E. Ringwood, Crystallization of basalt and andesite under high pressure hydrous conditions, *Earth Planetary Sci. Letters*, *3*, 481–489, 1968a.

Green, T. H., and A. E. Ringwood, Genesis of the calc-alkali igneous rock suite, *Contrib. Mineral and Petrol.*, in press, 1968b.

Gutenberg, B., The asthenosphere low-velocity layer, *Ann. Geofis.*, *12*, 439–460, 1959.

Harris, P. G., Zone refining and the origin of potassic basalts, *Geochim. Cosmochim. Acta*, *12*, 195–208, 1957.

Hedge, C. E., Variations in radiogenic strontium found in volcanic rocks, *J. Geophys. Res.*, *24*, 6119–6126, 1966.

Hedge, C. E., and F. G. Walthall, Radiogenic strontium-87 as an index of geological processes, *Science*, *140*, 1214–1217, 1963.

Heier, K. S., and J. A. S. Adams, Concentration of radioactive elements in deep crustal material, *Geochim. Cosmochim. Acta*, *29*, 53–61, 1965.

Heier, K. S., and J. J. W. Rogers, Radiometric determination of thorium, uranium, and potassium in basalts and in two magmatic differentiation series, *Geochim. Cosmochim. Acta*, *27*, 137–154, 1963.

Hess, H. H., Island arcs, gravity anomalies, and serpentinite intrusions, *Proc. Intern. Congress, Moscow*, 1937, Rept. 17, vol. 2, pp. 263–283, 1939.

Hess, H. H., The oceanic crust, *J. Marine Res., Sears Found.*, *14*, 423–439, 1955a.

Hess, H. H., Serpentines, orogeny and epeirogeny, in *Crust of the Earth*, edited by A. Poldervaart, *Geol. Soc. Am. Spec. Paper*, *62*, 391–408, 1955b.

Hess, H. H., Stillwater igneous complex, *Geol. Soc. Am. Mem.*, *80*, 177–185, 1960.

Hess, H. H., History of ocean basins, in *Petrologic Studies: A Volume to Honor A. F. Buddington*, edited by A. E. Engel et al., pp., 599–620, Geological Society of America, New York, 1962.

Hurley, P. M., H. Hughes, G. Faure, H. W. Fairbairn, and W. H. Pinson, Radiogenic strontium-87 model of continent formation, *J. Geophys. Res.*, *67*, 5315–5336, 1962.

Hyndman, R. D., I. B. Lambert, K. S. Heier, J. C. Jaeger, and A. E. Ringwood, Heat flow and surface radioactivity measurements in the Precambrian shield of Western Australia, *Phys. Earth Planetary Interiors*, *1*, 129–135, 1968.

Ito, K., and G. C. Kennedy, Melting and phase relations in a natural peridotite to 40 kilobars, *Am. J. Sci.*, *265*, 519–538, 1967.

Kennedy, G. C., The origin of continents, mountain ranges and ocean basins, *Am. Sci.*, *47*, 491–504, 1959.

Kuno, H., K. Yamasaki, C. Iida, and K. Nagashima, Differentiation of Hawaiian magmas, *Japan J. Geol. Geography Trans.*, *28*, 179–218, 1957.

Kushiro, I., and H. Kuno, Origin of primary basalt magmas and classification of basaltic rocks, *J. Petrol.*, *4*, 75–89, 1963.

Lambert, I. B., and K. S. Heier, The vertical distribution of uranium, thorium, and potassium in the continental crust, *Geochim. Cosmochim. Acta*, *31*, 377–390, 1967.

Lee, W. H. K., and S. Uyeda, Review of heat flow data, in *Terrestrial Heat Flow, Geophys. Monograph 8*, edited by W. H. K. Lee, chapter 6, pp. 87–190, American Geophysical Union, Washington, D. C., 1965.

Lovering, J. F., The nature of the Mohorovicic discontinuity, *Trans. Am. Geophys. Union*, *39*, 947–955, 1958.

MacDonald, G. J. F., Calculations on the thermal history of the earth, *J. Geophys. Res.*, *64*, 1967–2000, 1959.

MacDonald, G. J. F., Surface heat flow in a differentiated earth, *J. Geophys. Res.*, *66*, 2489–2493, 1961.

MacDonald, G. J. F., The deep structure of continents, *Rev. Geophys.*, *1*, 587–665, 1963.

Nixon, P. H., O. von Knorring, and J. M. Rooke, Kimberlites and associated inclusions of Basutoland: a mineralogical and geochemical study, *Am. Mineral.*, *48*, 1090–1131, 1963.

O'Hara, M. J., Melting of garnet peridotite and eclogite at 30 kilobars, *Carnegie Inst. Wash. Yr. Book*, *62*, 71–77, 1963.

O'Hara, M. J., Primary magmas and the origin of basalts, *Scottish J. Geol.*, *1*, 19–40, 1965.

Oxburgh, E. R., Volcanism and mantle convection, *Phil. Trans. Roy Soc. London, A*, *258*, 162–166, 1965.

Oxburgh, E. R., and D. L. Turcotte, Mid-ocean ridges and geotherm distribution during mantle convection, *J. Geophys. Res.*, *73*, 2643–2661, 1968a.

Oxburgh, E. R., and D. L. Turcotte, The problem of heat flow and volcanism associated with zones of descending mantle convective flow, *Nature*, in press, 1968b.

Powers, H. A., Composition and origin of basaltic magma of the Hawaiian Islands, *Geochim. Cosmochim. Acta*, *7*, 77–107, 1955.

Press, F., Some implications on mantle and crustal structure from G waves and Love waves, *J. Geophys. Res.*, *64*, 565–568, 1959.

Revelle, R., and A. E. Maxwell, Heat flow through the floor of the eastern North Pacific Ocean, *Nature*, *170*, 199–200, 1952.

Ringwood, A. E., Constitution of the mantle, 3, *Geochim. Cosmochim. Acta*, *15*, 195–212, 1958.

Ringwood, A. E., Changes in solar luminosity and some possible terrestrial consequences, *Geochim. Cosmochim. Acta*, *21*, 295–296, 1961.

Ringwood, A. E., A model for the upper mantle, *J. Geophys. Res.*, *67*, 857–867, 1962a.

Ringwood, A. E., A model for the upper mantle, 2, *J. Geophys. Res.*, *67*, 4473–4477, 1962b.

Ringwood, A. E., Composition and origin of the earth, in *Advances in Earth Science*, edited by P. M. Hurley, pp. 287–356, MIT Press, Cambridge, 1966.

Ringwood, A. E., and D. H. Green, An experimental investigation of the gabbro-eclogite transformation and some geophysical consequences, *Tectonophysics*, *3*, 383–427, 1966.

Rubey, W. W., Geologic history of sea water, *Bull. Geol. Soc. Am.*, *62*, 1111–1147, 1951.

Rubey, W. W., Development of the hydrosphere and atmosphere with special reference to the probable composition of the early atmosphere, in *Crust of the Earth*, edited by A. Poldervaart, *Geol. Soc. Am. Spec. Paper*, *62*, 631–650, 1955.

Schwarzschild, M., *Structure and Evolution of the Stars*, p. 206, Princeton University Press, Princeton, 1958.

Talwani, M., G. H. Sutton, and J. L. Worzel, A crustal section across the Puerto Rico trench, *J. Geophys. Res.*, *64*, 1545–1555, 1959.

Thayer, T. P., Some critical differences between alpine-type and stratiform peridotite-gabbro complexes, *Intern. Geol. Congr.*, *21st, Copenhagen, 1960, Rept. Session, Norden*, Part 13, 247–259, 1960.

Tilton, G. R., and G. W. Reed, Radioactive heat production in eclogite and some ultramafic rocks, in *Earth Sciences and Meteoritics*, dedicated to F. C. Houtermans, chap. 2, pp. 31–43, North Holland Publishing Company Amsterdam, 1963.

Toksöz, M. N., M. A. Chinnery, and D. L. Anderson, Inhomogeneities in the earth's mantle, *Geophys. J.*, *13*, 31–59, 1967.

Turner, F. J., and J. Verhoogen, *Igneous and Metamorphic Petrology*, 2nd edition, p. 694, McGraw-Hill Book Company, New York, 1960.

Verhoogen, J., Petrological evidence on temperature distributions in the mantle of the earth, *Trans. Am. Geophys. Union*, *35*, 85–92, 1954.

Verhoogen, J., Temperatures within the earth, *Phys. Chem. Earth*, *1*, 1956.

Vine, F. J., and D. H. Matthews, Magnetic anomalies over ocean ridges, *Nature*, *199*, 947–949, 1963.

Wager, L. R., Beneath the earth's crust: Presidential address to Section C, British Association for the Advancement of Science, *Advan. Sci.*, *58*, 1–14, 1958.

Wagner, P. A., *The Diamond Fields of South Africa*, The Transvaal Leader, Johannesburg, 1914.

Wagner, P. A., The evidence of the kimberlite pipes on the constitution of the outer parts of the earth, *S. African J. Sci.*, *25*, 127–148, 1928.

Williams, A. F., *The Genesis of the Diamond*, 2 volumes, Ernest Benn, London, 1932.

Wilshire, H. G., and R. A. Binns, Basic and ultrabasic xenoliths from volcanic rocks of New South Wales, *J. Petrol.*, *2*, 185–208, 1961.

Wilson, J. T., The development and structure of the crust, in *The Earth as a Planet*, edited by G. P. Kuiper, pp. 138–214, University of Chicago Press, Chicago, 1954.

Wiseman, J. D. H., St. Paul rocks and the problem of upper mantle, *Geophys. J.*, *11*, 519–525, 1966.

Worzel, J. L., and G. L. Shurbet, Gravity interpretations from standard oceanic and continental sections, *Geol. Soc. Am. Spec. Paper*, *62*, 87–100, 1955.

Yoder, H. S., and C. E. Tilley, Origin of basalt magma, an experimental study of natural and synthetic rock systems, *J. Petrol.*, *3*, 362–532, 1962.

Density and Composition of the Upper Mantle: First Approximation as an Olivine Layer

Francis Birch

WE CONSIDER here certain aspects of the problem of determining density and composition in the upper mantle. The geophysical observations most useful for this purpose are measurements of the gravitational field, travel times and amplitudes of various seismic waves, and the periods of the earth's free oscillations. It is not possible, in general, to separate the treatment of the upper mantle and transition zone from that of the whole earth; important constraints, for example, the knowledge of the mean density and moment of inertia, or of the periods of the low-order harmonics, determine various integrals over the whole mass. On the other hand, there are no general solutions that do not depend, in varying degree, upon more or less arbitrary assumptions about special features, since in the absence of such assumptions, the distributions of density and velocity are indeterminate. Even in the zero-order approximation, with all quantities functions of radius alone, we have unlimited degrees of freedom and a finite number of observations. This situation is brought under control by assuming continuity within certain specified zones, by associating discontinuities of density with those of velocity, and by introducing physical or mathematical hypotheses about rates of radial change. With the accumulation of experimental data bearing on the behavior of known materials at pertinent temperatures and pressures, it becomes possible to examine the implications of some of these assumptions, and to discard those that are no longer physically acceptable.

The models of seismic structure, virtually unchanged for thirty years, are now undergoing substantial revision, particularly with respect to just those features, such as discontinuities and velocity gradients, that are most significant for a physical interpretation. Much of this material has been reviewed by *Anderson* [1966, 1967b] and *Caloi* [1967]. The rapid progress in experimental petrology has suggested important modifications of classical views; this has been considered in relation to mantle constitution by *Clark and Ringwood* [1964] and *Ringwood* [1966]. Regional studies have begun to show differences in upper mantle structure, not only between oceanic and continental mantle, but also within continental or oceanic mantle. With so much still in flux, it seems most profitable to consider only a few general problems of interpretation, which will have to be faced whatever solutions are found for the seismic structure. In view of the indeterminateness of the velocities in the 'low velocity' zone, it may turn out that a physical-petrological solution, consistent with the seismic evidence but not determined by it, will in the end prove most satisfactory. We devote most of this discussion to consideration of the upper 200 km of the mantle. Below this, at depths that vary according to different solutions, we encounter more or less rapid increases of velocity with depth, which imply transformations of crystal structure to denser, more tightly packed forms, now fairly well understood in the light of experimental mineralogy and of high-pressure shock compression of rocks. This transition region will not be discussed in this section; for recent work the reader may consult *The Earth's Mantle*, edited by *Gaskell* [1967].

The first question to be discussed is the effect of temperature and pressure on velocities and density in the uppermost 200 km of mantle, assuming that, to a first approximation, the elastic properties are those of olivine. The composition of the upper mantle is widely believed to resemble that of some variety of peridotite, a rock in which olivine is the major component. Accessory minerals such as orthopyroxenes have elastic properties approaching those of olivine; a few, garnets and spinels, may lead to higher velocities, feldspars and amphiboles to lower velocities. The calculations for an aggregate of olivine or dunite, of a typical composition, provide a starting point for the consideration of more complex peridotites.

Since the work of *Bowen* [1928], partial melting of peridotite has been the favored explanation of the origin of basaltic magma, and various authors have proposed that some fraction of melt may reside for prolonged periods as droplets or pockets distributed through the unmelted residue. The composition of the residue varies with the amount of melt, but approaches that of olivine. Thus a model for the elasticity of a partly melted peridotite can be constructed from the properties of olivine and of basaltic glass; it is not perfectly consistent with uniform chemistry, but again may serve for an initial survey of the possibilities of this model.

Finally, we consider briefly the problem of inferring density directly from velocities.

NOTATION

- ρ density
- T temperature
- P pressure
- S entropy
- α volume thermal expansion
- K_T isothermal bulk modulus
- K_S adiabatic bulk modulus
- G rigidity modulus
- V_P velocity of compression waves
- V_S velocity of shear waves
- V_ϕ 'hydrodynamical wave velocity'
- σ Poisson's ratio
- g acceleration of gravity
- ϕ $\equiv K_S/\rho$

P and S are used in two senses: as pressure and entropy in various derivatives, and as designations of the two velocities of body waves. Both usages are well established. A completely unambiguous notation seems to be unattainable except at the price of discarding familiar symbols.

ELASTIC PROPERTIES OF OLIVINE

A considerable mass of experimental data on the elastic properties of olivine can be found in recent papers, but no single authority gives all the required information, and different materials have served for the various experiments. *Verma* [1960] found the dynamic elastic constants for a single crystal of olivine (peridot) of gem quality. *Birch* [1960, 1961], *Simmons* [1964a], *Kanamori and Mizutani* [1965], and *Christensen* [1966b] studied dunites under pressures up to 10 kb; the least altered was the Twin Sisters dunite, which gives values in fair agreement with those calculated for an aggregate of Verma's olivine of virtually the same Mg/Fe ratio. *Schreiber and Anderson* [1967] found pressure coefficients, as well as other properties, for a synthetic sample of polycrystalline forsterite of 6% porosity. *Soga et al.* [1966] gave velocities as a function of temperature, based on measurements to 1000°K, with extrapolation for porous forsterite to 2500°K. *Birch* [1962] measured V_s up to 500°C for dunite held under a pressure of 9 kb during heating. *McQueen et al.* [1967] and *Trunin et al.* [1965] published shock-compression measurements on various dunites; these measurements are not useful in the range of pressure of the upper mantle, but are essential for understanding the transition zone and lower mantle.

Several methods are available for finding the elastic constants of quasi-isotropic homogeneous aggregates from single-crystal constants; *Hill* [1952] has shown that the two classical methods of Voigt and Reuss furnish upper and lower limits for K and G; (parameters calculated according to these schemes are marked (V) and (R) in Table 1.) The Reuss value for the compressibility of the aggregate is exactly that of the single crystal. The two limits on G differ by only 4%, and the velocities calculated from the Reuss averages will be taken for the following discussion, rounded to 8.40 and 4.90 km/sec, respectively. These are in good agreement with the values for the dunite, if we compare, not the zero pressure velocities, which are lowered by porosity, but the values at 10 kb, where this porosity is largely eliminated. The parameters of the synthetic porous forsterite differ in the expected direction; that is, the densities, velocities, and moduluses are all appreciably lower than for compact material.

Equations given by *Soga et al.* [1966] permit the calculation of velocities as a function of temperature for the synthetic forsterite; from these we find the temperature coefficients

TABLE 1. Elastic Properties of Olivine at 25°C

Parameter	Olivine (peridot)[a]	Dunite $P = 10$ kb (except density)	Forsterite (synthetic aggregate)	
			Porosity 6%[b]	Nonporous[c]
ρ, g/cm³	3.324	3.312[d]	3.021	3.213
		3.326[e]		
		3.33[f]		
		3.30[g]		
V_P, km/sec	8.403 (R)[h]	8.42[d]	7.586	
	8.560 (V)[h]	8.66[f]		
		8.52[g]		
V_S, km/sec	4.886 (R)	4.83[e]	4.359	
	4.979 (V)	4.74[f]		
		4.80[g]		
V_ϕ, km/sec	6.23	6.21[d,e]	5.67	
		6.69[f]		
		6.47[g]		
ϕ, (km/sec)²	38.8	39.8[d,e]		
		44.8[f]		
		41.9[g]		
K_S, mb	1.289	1.32[d,e]	0.974	1.275
		1.50[f]		
		1.38[g]		
G, mb	0.793 (R)	0.776[e]	0.574	0.776
	0.824 (V)	0.75[f]		
		0.760[g]		
σ	0.245 (R)	0.255[d,e]	0.254	0.247
	0.244 (V)	0.286[f]		
		0.267[g]		

[a] *Verma* [1960]; Burmese peridot, $2(Mg_{91.7}Fe_{8.3})O \cdot SiO_2$.
[b] *Schreiber and Anderson* [1967].
[c] *Soga and Anderson* [1967].
[d] *Birch* [1960]; Twin Sisters dunite (olivine, 92%; pyroxene, 7%).
[e] *Simmons* [1964a]; Twin Sisters dunite.
[f] *Christensen* [1966b]; Twin Sisters dunite.
[g] *Kanamori and Mizutani* [1965]; Horoman dunite (olivine, 88%; pyroxene, 11%).
[h] R, calculated according to method of Reuss; V, calculated according to method of Voigt.

and also velocities relative to room-temperature velocity. This is done below for the forsterite and for periclase and corundum, both synthetic aggregates of virtually zero porosity.

Examination of Tables 2, 3, and 4 suggests the following comments: Within the common interval of temperature, to 500°C, the variation of V_S is very nearly the same for the dunite and the synthetic forsterite, porosity having apparently little effect on the relative change; the variations of velocities in periclase and forsterite aggregates are much the same up to about 2000°K; all coefficients increase in absolute magnitude as the temperature increases; the coefficients for V_P and V_S in the forsterite are more nearly equal than for the other two materials. To what extent some of these results depend on the special assumptions and choices of data will not be discussed here. It will be supposed that the relative values of Table 2 for the forsterite aggregate can be applied to the velocities of Table 1 for a compact olivine aggregate; the resulting velocities are given in Table 5.

For the effect of pressure we have measurements of compression by *Adams* [1931] of a dunite to 12 kb, and by *Bridgman* [1948] of Red Sea peridot to 40 kb. Bridgman's measurements show a nearly linear change of volume with pressure, but a second-degree curve can be fitted which gives $K_T = 1.22$ mb, and a poorly determined $dK/dP \approx 2$. Adams obtained a mean compressibility of 0.79 mb⁻¹, or $K_t = 1.26$ mb, in good agreement with the dynamical value (Table 1). *Schreiber and Anderson* [1967] found the effect of pressure on

TABLE 2. Velocities in Synthetic Aggregates as Fractions of Velocity at 300°K*
($P = 0$)

T, °K	Corundum		Periclase		Forsterite, 6% porosity	
	V_P	V_S	V_P	V_S	V_P	V_S
300	1.000	1.000	1.000	1.000	1.000	1.000
800	0.9781	0.9704	0.9690	0.9598	0.9709	0.9657
1000	0.9688	0.9578	0.9556	0.9424	0.9580	0.9510
1500	0.9441	0.9247	0.9204	0.8961	0.9221	0.9119
2000	0.9173	0.8892	0.8821	0.8451	0.8813	0.8695

* Soga et al. [1966].

TABLE 3. Shear Velocities in Rocks as Fractions of Velocity at 20°C*
($P = 4$ to 9 kb)

T, °C	Granite	Diabase	Dunite	Eclogite
20	1.000	1.000	1.000	1.000
100	0.997	0.997	0.994	0.997
200	0.994	0.993	0.987	0.992
300	0.991	0.988	0.980	0.988
400	0.988	0.981	0.973	0.984
500	0.984	0.974	0.965	0.980

* Birch [1962].

TABLE 4. Temperature Coefficients of Velocity (parts per million per degree) for Synthetic Aggregates*
($P = 0$)

T, °K	Corundum		Periclase		Forsterite, 6% porosity	
	$-\partial \log V_P/\partial T$	$-\partial \log V_S/\partial T$	$-\partial \log V_P/\partial T$	$-\partial \log V_S/\partial T$	$-\partial \log V_P/\partial T$	$-\partial \log V_S/\partial T$
300	42	57	59	76	53	65
1000	49	67	71	94	70	79
1500	55	93	80	108	83	89
2000	61	107	90	125	98	102

* Soga et al. [1966].

velocities of the porous forsterite aggregate to approximately 2 kb; their coefficients are shown in Table 6. The measurements of velocity in rocks can also be used to find pressure coefficients; straight lines have been fitted to the data of Birch, Simmons, and Christensen for Twin Sisters dunite at pressures greater than several kilobars; that is, the initial curvature resulting from pore closure has been as far as possible eliminated. The results are included in Table 6. There are several discrepancies. The coefficients for dunite are all about the same, but are higher than those for the porous forsterite; the coefficients for V_P and V_S for the porous forsterite differ by what seems to be an abnormal amount.

TABLE 5. Velocities in Olivine Aggregate of Zero Porosity as a Function of Temperature
($P = 0$)

T, °K	V_P, km/sec	V_S, km/sec
300	8.400	4.900
800	8.156	4.732
1000	8.047	4.660
1500	7.746	4.468
2000	7.403	4.261
2500	(7.019)	(4.035)

TABLE 6. Pressure Coefficients of Velocity (per megabar) for Several Materials

Material	$3/2K_S$	$(\partial \log V_P/\partial P)_T$ Calc.[a]	Meas.	$(\partial \log V_S/\partial P)_T$ Calc.[b]	Meas.
Corundum	0.595[c]	0.59	0.48[c]	0.57	0.35[c]
Periclase	0.874[d]	0.83	0.79[d]	0.78	0.73[d]
Spinel	0.743[e]	0.75	0.49[e]	0.76	0.08[e]?
Garnet	0.847[f]	0.87	0.92[f]	0.91	0.46[f]?
Forsterite (porous)	1.54[g]	1.55	1.36[g]	1.56	0.56[g]?
Forsterite (extrapolated to zero porosity)	1.18[h]	1.17		1.17	
Olivine (s.c.)	1.16[i]	1.15		1.14	
Dunite			1.8[j]		1.7[k]
			1.9[l]		2.1[l]

[a] Calculated from equation 4.
[b] Calculated from equation 5.
[c] Schreiber and Anderson [1966b].
[d] Anderson and Schreiber [1965].
[e] Schreiber [1967].
[f] Soga [1967].
[g] Schreiber and Anderson [1967].
[h] Soga and Anderson [1967].
[i] Verma [1960].
[j] Birch [1961a].
[k] Simmons [1964a].
[l] Christensen [1966b].

Approximate values of these coefficients can be estimated in several ways. If we write $V_\phi = (K/\rho)^{1/2}$, we have

$$V_P = f_1 V_\phi \qquad V_S = f_2 V_\phi$$

$$f_1 = [(3 - 3\sigma)/(1 + \sigma)]^{1/2}$$

$$f_2 = [(3 - 6\sigma)/(2 + 2\sigma)]^{1/2}$$

where σ is Poisson's ratio. Then

$$\frac{1}{V_P}\frac{dV_P}{dP} = \frac{1}{V_\phi}\frac{dV_\phi}{dP} + \frac{1}{f_1}\frac{df_1}{d\sigma}\frac{d\sigma}{dP}$$

$$\frac{1}{V_S}\frac{dV_S}{dP} = \frac{1}{V_\phi}\frac{dV_\phi}{dP} + \frac{1}{f_2}\frac{df_2}{d\sigma}\frac{d\sigma}{dP} \qquad (1)$$

$$\frac{1}{V_\phi}\frac{dV_\phi}{dP} = \frac{1}{2K_S}\left[\frac{dK_S}{dP} - \frac{K_S}{K_T}\right]$$

$$\approx \frac{1}{2K_S}\left[\frac{dK_S}{dP} - 1\right]$$

If we take temperature derivatives instead of pressure derivatives, we have only to replace P by T in the first two equations; the third becomes

$$\frac{1}{V_\phi}\frac{dV_\phi}{dT} = \frac{\alpha}{2}\left[\frac{1}{\alpha K_S}\frac{dK_S}{dT} + 1\right] \qquad (2)$$

Now $df_1/d\sigma \approx -1.4$ and $df_2/d\sigma \approx -1.9$ in the range of interest for σ. With $\sigma = 0.25$, $f_1 = 1.342$, $f_2 = 0.775$, and for $dK_S/dP = 4$, we have

$$\frac{1}{V_P}\frac{dV_P}{dP} \approx \frac{3}{2K_S} - 1.04\frac{d\sigma}{dP}$$

$$\frac{1}{V_S}\frac{dV_S}{dP} \approx \frac{3}{2K_S} - 2.45\frac{d\sigma}{dP} \qquad (3)$$

Thus these two coefficients should be equal, except for the term in $d\sigma/dP$, and inversely proportional to K_S. For olivine, $3/2K_S = 1.16$ mb^{-1}; for the porous forsterite, 1.5 mb^{-1}. Schreiber and Anderson give the velocity coefficients shown in Table 6, with $d\sigma/dP = 0.74$ mb^{-1}. This appears to be abnormally high; at any rate, such a coefficient cannot be continued very far without yielding impossible values for σ. It seems probable that the pressure coefficients of velocity found for the dunites are somewhat too high because of residual porosity, and that porosity accounts for the large difference in coefficients of V_P and V_S found for the porous forsterite. The very compact corundum and periclase aggregates give nearly equal coefficients for these two velocities, with $d\sigma/dP$ of the order of 0.1 mb^{-1}. With $d\sigma/dP = 0.1$ mb^{-1}, we find for olivine

$$(\partial \log V_P/\partial P)_T = 1.1 \text{ mb}^{-1}$$

$$(\partial \log V_S/\partial P)_T = 0.9 \text{ mb}^{-1}$$

These are based on a 'normal' value of 4 for $(\partial K_S/\partial P)_T$. The value of 4.8 given by Schreiber and Anderson for the forsterite does not seem applicable here, as it depends upon the abnormally low rate of change of V_S. The rates found for the dunite give $dK/dP \approx 6$. The values below 400 kb found by McQueen, Marsh, and Fritz for dunite yield $\partial K/\partial P \approx 2.5$, in rough agreement with the value given by Bridgman's 40-kb data, but it is not certain that this applies to untransformed olivine. Thus the effect of pressure up to 100 kb is uncertain by perhaps 50%; a value of 1 mb^{-1}, for both velocities, is adopted for the following calculations.

Pressure coefficients have been given in several papers by Anderson and Schreiber for quasi-isotropic aggregates, based on the single-crystal constants; these are shown as relative values in Table 6. The coefficients for V_s for garnet and spinel appear to be too low. Even with the high precision of the measurements, the changes in 2 kb are exceedingly small, and bias in the method of finding the constants for aggregates from single-crystal constants may contribute to error. While we expect to find the coefficients slightly smaller for V_s than for V_P, the differences for these materials seem excessively large.

Another set of estimates of the pressure coefficients can be derived from the theory of finite strain [Birch, 1939]; for small compressions, the limiting values can be written:

$$\frac{1}{V_P} \frac{dV_P}{dP} = \frac{1}{6K} \frac{13\lambda + 14\mu}{\lambda + 2\mu}$$

$$= \frac{1}{6K} \frac{39K + 16G}{3K + 4G} \quad (4)$$

$$\frac{1}{V_S} \frac{dV_S}{dP} = \frac{1}{6K} \frac{3\lambda + 6\mu}{\mu}$$

$$= \frac{1}{6K} \frac{3K + 4G}{G} \quad (5)$$

Here λ and μ are Lamé's constants, and $\mu \equiv G$.

For $\lambda = \mu$, or $\sigma = 0.25$, the coefficients of V_P and V_s are both equal to $3/2K$; departures from this value increase as σ deviates from 1/4. Again we notice (Table 6) the abnormally low values for V_s for the spinel, garnet, and porous forsterite, and high values for the natural dunite. The agreement is relatively good for V_P, and it can be argued either that this confirms the measurements, or that it shows that the neglect of third-order terms in the elastic potential is justified.

From seismology we obtain a rough estimate of $d\sigma/dP$; in the lower mantle, σ increases from about 0.276 at roughly 1000 km to 0.298 at 2400 km, or 0.022 for about 0.7 mb. Thus we obtain $d\sigma/dP = 0.03$ mb^{-1}, but some of this may be the effect of rising temperature. Like many pressure coefficients, however, this can be expected to become smaller as pressure rises, and a value near 0.1 mb^{-1} may not be unreasonable at low pressures. Since $d\sigma/dT$ is usually positive, however, it might be anticipated that $d\sigma/dP$ would be negative, on the general expectation of opposite effects from pressure and temperature.

At high temperatures, the quantity $(\alpha K_S)^{-1} (dK_S/dT)$ is nearly independent of temperature, and for porous forsterite has the value -4.1 [Soga and Anderson, 1967]; then

$$\frac{1}{V_P} \frac{dV_P}{dT} = -3.1\alpha/2 - 1.1 d\sigma/dT \quad (6)$$

$$\frac{1}{V_S} \frac{dV_S}{dT} = -3.1\alpha/2 - 2.7 d\sigma/dT \quad (7)$$

With $d\sigma/dT = 10 \cdot 10^{-6}$ [Soga and Anderson, 1967] and thermal expansions as given by Skinner [1966] for olivine with Fa$_{10}$, we obtain values in substantial agreement with the coefficients of Table 4, derived from the equations of Soga, Schreiber, and Anderson.

The density of olivine as a function of temperature and pressure to 200 km can be calculated from the thermal expansion and bulk modulus at 1 atm, both as functions of temperature, plus an equation of state valid along isotherms to maximum compressions of about 7%. Skinner [1962] gives volume as a function of temperature based on X-ray measurements on forsterite to 1100°C; his formula is assumed to be valid to 2000°K. We then require the bulk modulus for the same range of temperature; this is computed from the velocities of Table 5, with a correction for the difference between K_S and K_T. The values so obtained are a little smaller than those found by applying the coefficient $(\partial K_S/\partial T)_P = -0.13$ kb/°C given by Soga and Anderson [1967] to the room temperature value of K_S, but the differences are not important for this discussion. For the isotherms, we use

$$P = (3/2) K_T (y^{7/3} - y^{5/3}) \quad (8)$$

where $y = \rho/\rho_0$, and ρ_0 is the density at temperature T and $P = 0$. The results are shown in Tables 7 and 8.

The conditions of constant density in the T-z plane are represented by nearly straight lines, with slopes of approximately 7 to 8°C/km. From these, the density of the isotropic olivine aggregate of composition approximately $Fo_{90}Fa_{10}$ can be found for any temperature distribution, with an error probably not exceeding 1%.

LOW VELOCITY ZONE:
THE HOMOGENEOUS LAYER

Velocity increases with depth throughout most of the mantle, but as early as 1926 Gutenberg proposed a reversal of this trend in the upper mantle as an explanation of anomalous amplitudes at distances of about 15°, and 'low velocity' layers have appeared in most recent velocity distributions. There are, however, as many variants as seismologists (rather more, if we count the numerical models recently generated by computers); if downward refraction takes place, a region exists in which no ray 'bottoms,' and the velocities cannot be determined by the usual methods. The dispersion of surface waves gives some indication of the velocity-density structure, but it does not appear that a unique solution can be reached. The problem is further confused by the existence of mathematical models having finite discontinuities of velocities and density. It is not always clear whether these are mere artifices of calculation intended as idealizations of more gradual changes, or whether first-order physical discontinuities are meant.

We first ask how much can be explained in terms of variations of temperature and pressure in a compositionally uniform layer. The pressure rises with depth z according to the hydrostatic relation, $dP = g\rho \, dz$, and we

TABLE 7. Density of Olivine as a Function of Temperature and Pressure ($\rho_{25} = 3.32$ g/cm³)

T, °K	K_T, kb	ρ_T/ρ_{25}		
		$P = 0$	$P = 31$ kb	$P = 65$ kb
298	1,270	1.000	1.024	1.048
1000	1,110	0.9748	1.001	1.028
1500	980	0.9524	0.981	1.010
2000	850	0.9276	0.960	0.992

TABLE 8. Temperatures (°C) for Constant Density of Olivine at Several Depths

ρ_T/ρ_{25}	0 km	100 km	200 km
1.02		160	950
1.01		490	1230
1.00	25	760	1500
0.99	350	1010	1780
0.98	610	1250	2060
0.97	840	1490	
0.96	1060	1730	

expect density and velocity to increase as pressure increases; the temperature increases with depth, and density and velocity decrease as temperature increases. The rate of increase of temperature with depth is probably greatest near the surface and must decrease as depth increases. At some depth, pressure and temperature effects exactly cancel. Several different critical temperature gradients must be distinguished.

First, the relation for constant density in a uniform medium that does not undergo any change of phase is, by definition of thermal expansion α and incompressibility K_T,

$$(\partial T/\partial P)_\rho \equiv -(\partial \rho/\partial P)_T/(\partial \rho/\partial T)_P$$
$$\equiv 1/(\alpha K_T) \quad (9)$$
$$(\partial T/\partial z) = (\rho g)/(\alpha K_T)$$

Similarly, the condition for constant velocity V (either V_P or V_S) is:

$$(\partial T/\partial z)_V = -\rho g (\partial V/\partial P)_T/(\partial V/\partial T)_P \quad (10)$$

This must be distinguished from the relation for critical refraction, derived from the condition for upward refraction at radius r [*Gutenberg*, 1959b, p. 25; *Bullen*, 1963, p. 112]: $dV/dr < V/r$. In the sphere, the condition $dV/dr = 0$ always means upward refraction; it can be identified with the condition at the minimum of a 'low velocity' layer. The gradients required for downward refraction, assuming continuity for all quantities [*Birch*, 1952, p. 260], must exceed $(dT/dz)_{crit}$ where

$$(dT/dz)_{crit} = -(dT/dr)_{crit} = -[1/r$$
$$+ \rho g (\partial \log V/\partial P)_T]/(\partial \log V/\partial T)_P$$
$$(11)$$

The term $1/r$ is sometimes neglected in discus-

TABLE 9. Critical Gradients (degrees per kilometer) for Several Materials

Gradient	Corundum		Periclase		Forsterite, 6% porosity		Olivine	
	300°K	1500°K	300°K	1500°K	300°K	1500°K	300°K	1500°K
$(\partial T/\partial z)_\rho$	9.7	6.6	6.8	5.4	14.4	10.2	11.2	(7)
$(\partial T/\partial z)_{V_P}$	5.5	4.2	4.8	3.5	7.8	5.0	6.3	4.0
$(\partial T/\partial z)_{V_P}$ (crit)	9.3	7.1	7.5	5.5	10.8	6.9	9.3	5.9
$(\partial T/\partial z)_{V_S}$	2.4	1.5	3.4	2.4	2.6	1.9	5.1	3.7
$(\partial T/\partial z)_{V_S}$ (crit)	5.3	3.2	5.5	3.9	5.1	3.7	7.6	5.5

sions of the critical gradient, but it is a substantial portion of the numerator.

Estimates of these gradients for several materials are given in Table 9; r is taken to be 6250 km, but the effect of pressure on the coefficients is neglected.

In general agreement with earlier calculations [*Birch*, 1952; *Valle*, 1956; *MacDonald and Ness*, 1961], we find the various critical gradients comparable in magnitude with geothermal gradients in the upper mantle. The critical gradients for a shadow zone are typically greater for V_P than for V_S, though the difference for olivine is small and possibly not significantly different from zero. The gradients for constant density may not differ significantly from those for critical refraction of V_P, when the uncertainties are considered.

TABLE 10. Temperatures Inferred from Velocities for a Uniform Olivine Layer

Depth, z	P, kb	V_P, km/sec		T, °K	dT/dz, °/km	V_S, km/sec		T, °K	dT/dz, °/km
		Obs.	Corr.			Obs.	Corr.		
Velocities after *Gutenberg* [1959a]									
40	11	8.08	7.99	1100		4.60	4.55	1280	
					23				18
50	15	7.97	7.85	1330		4.55	4.48	1460	
					21				14
60	18	7.87	7.72	1540		4.51	4.43	1580	
					11				16
70	21	7.82	7.65	1650		4.47	4.37	1740	
					8				10
80	25	7.80	7.60	1730		4.44	4.33	1840	
					2				6
100	31	7.82	7.57	1770		4.42	4.28	1960	
					0.2				4.6
150	48	7.95	7.56	1780		4.40	4.18	2190	
					−0.2				2.2
200	65	8.10	7.57	1770		4.43	4.13	2300	
Velocities after *Anderson* [1967b, model CIT11A]									
16.5	5	8.10	8.06	970		4.60	4.58	1200	
					2.0				2.5
36	10	8.12	8.04	1010		4.61	4.56	1250	
					17				12
51	15	8.01	7.89	1270		4.56	4.49	1430	
					25				18
71	22	7.76	7.58	1760		4.45	4.35	1790	
					15				17
91	28	7.60	7.37	2050		4.34	4.21	2120	
					−2				1.7
151	48	7.85	7.46	1930		4.34	4.12	2220	
					−21				−10
171	55	8.19	7.74	1510		4.50	4.25	2030	

With the data for the olivine aggregate assembled in the preceding section, we can examine the following questions: Given a distribution of velocity throughout the uppermost 200 km, what temperatures are implied if the material is wholly olivine? Given a temperature distribution, what velocities are implied? We take as examples *Gutenberg*'s [1959a] and *Anderson*'s [1967b, model CIT11A] velocities and *Ringwood*'s [1966] 'oceanic' temperatures.

The seismic velocities are given as functions of depth; assigning a nominal pressure to each depth, we first correct the velocities to zero pressure with the aid of the pressure coefficients discussed in the preceding section; the correction at 200 km or 65 kb is about 6% of the velocities. The corrected velocities are then used to find the corresponding temperatures (Table 10) from the curves based on Table 5. The two velocities V_P and V_S provide two temperature distributions that would coincide if the seismic values (and the supposed variations of velocity) were consistent with the hypothesis that olivine alone makes up this layer. Negative gradients indicate regions where the rise of velocity is greater than can be accounted for by isothermal compression.

With regard to the behavior of velocity just

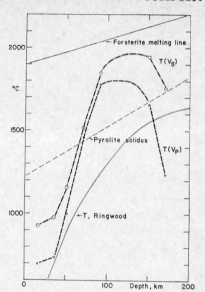

Fig. 2. Temperatures inferred from *Anderson*'s [1967b] model CIT11A velocities for an olivine layer. Other curves as in Figure 1.

below the Mohorovicic discontinuity, Gutenberg was of two minds. He first wrote 'On the basis of the new results there can be little doubt that the asthenosphere low velocity channel starts for both wave types immediately at the Mohorovicic discontinuity' [*Gutenberg*, 1959a, p. 351]. As *MacDonald and Ness* [1961, p. 1910] have remarked, this is to be expected with the usual form of temperature distribution and a uniform layer. With a little adjustment of velocities, it seems possible to reconcile *Gutenberg*'s [1959a] distribution with a plausible temperature curve and the assumption of uniformity of material to a depth of about 200 km; beyond this, a transition to material of intrinsically higher velocities is required.

But elsewhere, Gutenberg writes: 'Under ocean bottoms there may be just below the Mohorovicic discontinuity, more frequently than under continents, a relatively thin layer in which the velocity increases with depth, before the decrease of velocity with depth begins' [*Gutenberg* 1959b, p. 88]. This feature is shown in the model CIT11A distribution, where velocity rises with depth between 16 and 36 km; for a uniform layer of olivine, the gradient implied is only 2°C/km. If this detail of the velocity distribution is essential, then we have either an anomalous behavior of the temperature, or

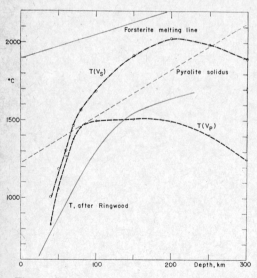

Fig. 1. Temperatures inferred from *Gutenberg*'s [1959a] velocities for an olivine layer. Also shown are the forsterite melting line [*Davis and England*, 1964], the 'oceanic' temperatures, and 'pyrolite' solidus [*Ringwood*, 1966].

a gradation of composition capable of offsetting the effect of a normal temperature gradient. *Ringwood* [1966, p. 368] suggests dehydration of an amphibole-bearing peridotite; deserpentinization might also serve.

As to the possibility of anomalous temperatures, *Lubimova* [1959] has given a temperature distribution which qualitatively resembles the curves derived from model CIT11A in the rising portions; this is based on an estimate of the variation of thermal conductivity with temperature, with allowance for a decrease of phonon conduction and an increase of radiative conduction as temperature increases. The combination produces a minimum of conductivity at about 100 km, with a calculated temperature of 1400°C; at this depth, the conductivity is taken to be 0.002 cal/cm sec °C. Lubimova's gradients are as follows:

Depth, km	dT/dz, deg/km
40	6
50	13
60	16
80	18
100	19
150	9
200	4

It seems unlikely, however, that the conductivity reaches so low a value at 1400°C. Recent measurements of infrared transmission at elevated temperatures [*Aronson et al.*, 1967] suggest a radiative contribution alone for peridot at 1240°C of 0.005 cal/cm sec °C; *Kanamori et al.* [1967] find the minimum conductivity of olivine (001 direction) to be about 0.009 cal/cm sec °C at 500°C. The maximum negative velocity gradient for the CIT11A solution occurs at 60 km, and the subsequent rise of velocities is not consistent with positive temperature gradients in a uniform layer. A less pronounced maximum of temperature gradient might exist, however, beneath the oceans where the heat-producing layer is presumably thickest.

The absolute values of temperature depend upon the olivine velocities, and would be lower for a peridotite having some fraction of more compressible minerals. It is to be remembered also that, beyond 1000°K, these velocities were obtained by a linear extrapolation of K and G; the higher temperatures found from V_S, as compared with those found from V_P, suggest that V_S may fall off with increasing temperature

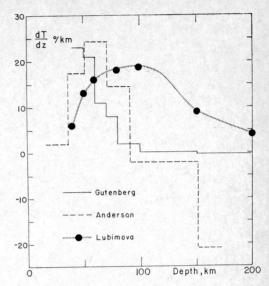

Fig. 3. Mean gradients of temperature implied by V_P curves of Figures 1 and 2 for an olivine layer; gradients after *Lubimova* [1959].

more rapidly than is shown in Table 5. However, the discrepancies are much greater for Gutenberg's distribution than for model CIT11A. No great change of velocity ratios is required to give consistent temperatures to a depth of 100 km.

The inferred temperatures are everywhere below the melting curve for forsterite [*Davis and England*, 1964], but rise above the 'pyrolite solidus' of *Ringwood* [1966, p. 362] at depths greater than 60 km. Temperatures not exceeding the pyrolite solidus can be obtained by taking the initial velocities as 8.2 and 4.7 km/sec, respectively; these are close to Ringwood's estimate for 'pyroxene pyrolite.'

LOW VELOCITY ZONE: PARTIAL FUSION

The existence of a molten fraction in the low velocity zone has been proposed by a number of writers, following the recognition that a completely molten substratum could not be reconciled with tidal deformation or with seismology. Partial fusion of peridotite is probably the most generally favored explanation for the origin of basaltic magma, and between the moment of partial fusion and the eventual expulsion of magma a stage can be visualized during which some part exists in the form of distributed films or globules of basaltic melt.

TABLE 11. Velocities in an Olivine Layer Based on Ringwood's 'Oceanic' Temperatures*

Depth, km	T, °C	V_P, km/sec	V_S, km/sec
50	880	8.08	4.67
100	1300	7.96	4.58
150	1550	7.93	4.56
200	1640	8.00	4.60

* *Ringwood* [1966, p. 362].

Several aspects of this hypothesis have been examined by *Shimozuru* [1963a, b]. *Beloussov* [1966] has given it a rather definite form, supposing that the volume fraction of liquid basalt may be of the order of 10%, that this lowers V_P in the low velocity zone by 0.3 km/sec, and the density by 0.1 g/cm³.

This hypothesis can be evaluated with the aid of formulas given by *Mackenzie* [1950], *Oldroyd* [1956], and *Hashin* [1959] for two-component materials, in which one component exists in small spherical cavities. Let c be the fraction by volume of the cavities filled with basaltic melt and having zero rigidity but the bulk modulus K_1 of basalt glass. The matrix has the rigidity G_2 and bulk modulus K_2 of olivine; possibly slightly different values should be used for 'peridotite,' but in any case the contrast of properties between crystalline peridotite and molten basalt will be great and the conclusions not significantly different. For the numerical estimates, we take $K_2 = 1.30$ mb, $K_1 = 0.65$ mb [*Birch and Bancroft*, 1942], $G_2 = 0.80$ mb; we suppose the density of the olivine to be 3.324 g/cm³, of the glass, 2.76 g/cm³. The Hashin relation for K of the composite material is then $K/K_2 = 1 - c(3 - 3\sigma)/(5 - 7\sigma)$, or approximately $1 - 0.7c$ for relevant values of σ. The Mackenzie-Hashin relation for G is very nearly $G/G_2 = 1 - 2c$. These are valid for small values of c, and we suppose that they hold for c up to 10%. The calculated values for K, G, and the velocities are shown in Table 12, all for ordinary temperature and pressure. We are interested in temperatures sufficient to melt basalt, at least 1200°C. A rise of temperature of 1200°C reduces the velocities of olivine by about 8%, and for a rough estimate of the velocities in the composite material at 1200°C we apply

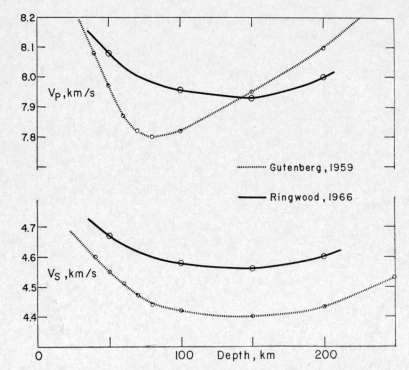

Fig. 4. Velocities in an olivine layer derived from *Ringwood*'s [1966] oceanic temperatures and *Gutenberg*'s [1959a] velocities.

TABLE 12. Properties of Olivine Plus Basaltic Glass

c*	K, mb	G, mb	ρ, g/cm³	V at 20°C, km/sec		V at 1200°C, km/sec	
				V_P	V_S	V_P	V_S
0	1.300	0.80	3.324	8.44	4.91	7.76	4.52
0.05	1.254	.72	3.296	8.20	4.67	7.54	4.30
0.10	1.209	.64	3.268	7.94	4.43	7.30	4.08
0.15	1.164	.56	3.239	7.68	4.16	7.07	3.83

* Fraction by volume of melt.

this correction, finding the velocities of the last two columns. The effect of pressure is nearly independent of c, with 30 kb raising V_P by about 0.3 km/sec, and V_S by half as much. The numbers of Table 12 can be used to form the ratio $\Delta V/\Delta\rho$ for members of this series. At 1200°C, this is 8.2 for V_P, 7.9 for V_S. These are unlike the values found for isothermal compression, where $\Delta V_P/\Delta\rho \approx 3$ and $\Delta V_S/\Delta\rho$ is about half as much. A reduction of density by 0.1 g/cm³ means a molten fraction of 15% and a reduction of both velocities by about 0.8 km/sec. If we require a reduction in velocities of 0.3 km/sec by this mechanism, then the molten fraction is about 6%, and the reduction of density is only 0.03 g/cm³, or about 1%. The ratio V_P/V_S increases with the fraction of melt. These features appear to be at least qualitatively consistent with the deductions by *Hales and Doyle* [1967] concerning the mantle below the western United States, for which they find $\Delta V/\Delta\rho$ between 9 and 13, and suggest 'an approach of one constituent to melting' in explanation. Actual partial melting appears to provide an interpretation.

With increasing uncertainty, we can estimate the properties of composites with different fractions of melt along the solidus, taking for this purpose the line given by *Ringwood* [1966, p. 362]. Pressure and temperature variations along this line have been taken into account with the application of coefficients for olivine. The corrected velocities are given in Table 13.

The main features of the model CIT11A low velocity layer can be accounted for by supposing that the solidus is reached at about 60 km, where the solidus temperature is 1400°C, and that the temperature then follows the solidus, rising at the rate of 3°C/km, to about 160 km, below which the rate of rise is less than 3°C/km. If between 60 and 160 km the amount of melt reaches 6% by volume, the velocity minima will come close to those given by Anderson. The mean gradient of temperature in the upper 60 km will then be about 23°C/km, however, and an explanation is required for the rise of velocities between 16.5 and 36 km; if a low gradient is supposed to exist in this layer, then still higher ones must be postulated above or below or in both regions. With the gradient of 23°C/km, the velocity V_P should decrease some 0.2 km/sec as depth increases from 16.5 to 36 km. Thus, if a rise of velocity is required for this interval, as in model CIT11A, the gradient must be low or a gradation of composition must offset the effect of temperature. Conceivably dehydration could play a part, as Ringwood suggested; the discrepancy between the calcu-

TABLE 13. Velocities in Olivine-Glass Composite along Solidus

Depth, km	T, °C	V at c* = 0, km/sec		V at c* = 0.05, km/sec		V at c* = 0.10, km/sec	
		V_P	V_S	V_P	V_S	V_P	V_S
0	1200	7.76	4.52	7.54	4.30	7.30	4.08
50	1350	7.81	4.55	7.59	4.33	7.35	4.11
100	1500	7.87	4.58	7.65	4.36	7.40	4.14
150	1650	7.92	4.62	7.70	4.39	7.45	4.17
200	1800	7.98	4.65	7.75	4.42	7.50	4.19

* Fraction by volume of basaltic melt.

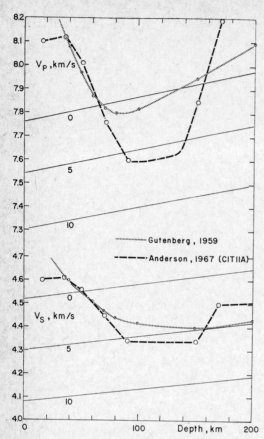

Fig. 5. Calculated velocities (solid lines) for olivine plus basaltic melt along the solidus, and velocities after *Gutenberg* [1959a] and *Anderson* [1967b]. The lines correspond to 0, 5%, and 10% of melt.

lated 8.06 km/sec for olivine at 36 km and 840°C and the CIT11A value of 8.12 km/sec is not in itself significant, and the calculated 8.27 km/sec for olivine at 16.5 km and 390°C can be reduced to a value below 8.06 by the substitution of a small amount of serpentine [*Birch*, 1961; *Christensen*, 1966b]. Evidently, if all features of this distribution are significant, no single simple hypothesis will suffice.

The explanation in terms of partial melting also encounters difficulties at the base of the low-velocity layer. If this is determined by the disappearance of melt and a return to the original olivine (or peridotite), the velocities should return to the line for $c = 0$, both rising about 0.3 km/sec. In CIT11A, however, V_P rises about 0.6 and V_S about 0.16 km/sec.

The implications of Gutenberg's velocities in terms of the 'temperature solution' in a homogeneous layer have been discussed above. The minimum for V_P is at 80 km, for V_S at 150 km; the shapes of the distributions do not seem to permit interpretation in terms of partial melting, the intervals in which melting must exist differing greatly for the two velocities.

Toksöz et al. [1967] show differences of the S velocities beneath oceans and continental shields reaching about 0.4 km/sec at depths of 100–120 km. If this is interpreted in terms of temperature alone, a lateral difference at 100 km of some 1000°C is required; this seems improbably large. An alternative explanation in terms of partial melting requires a difference of the molten fraction of about 10%, if the temperatures are at the solidus in both regions, somewhat less if cooler conditions exist beneath the continents. The localization of the largest difference between 80 and 140 km is roughly consistent with the interval of partial melting needed for the interpretation of the CIT11A low velocity zone.

The numbers used in these calculations are such that the estimates of the amount of melt are probably upper limits, since the solid matrix has been assumed to be olivine. Endless variations might be introduced by the substitution of other minerals, such as garnets, pyroxenes, and feldspars, and further work along these lines must be aided by petrological considerations, as in the work of *Clark and Ringwood* [1964] and *Ringwood* [1966].

RELATIONSHIP OF DENSITY TO VELOCITY

A number of proposals for deriving density directly from seismic velocity can be found in the literature. A curve compiled by *Nafe and Drake* [1957, 1965] has been much used in connection with seismic work at sea. *Woollard* [1959], *Talwani et al.* [1959], and *Steinhart and Meyer* [1961] have used empirical velocity-density curves for interpreting crustal structure. Additional data on velocities in rocks under compression in the laboratory [*Birch*, 1960, 1961; *Simmons*, 1964a, b; *Christensen* 1965; 1966a, b; *Kanamori and Mizutani*, 1965] can be used to establish more systematic relations among the variables of density, velocity, composition, pressure, and temperature.

We restrict the present discussion to silicate

rocks and minerals, considered as a subdivision of the oxides. In these materials, despite much diversity of composition and crystal structure, oxygen anions comprise most of the volume, with smaller cations occupying interstitial sites. Furthermore, in the common rocks, the mean atomic weight is virtually always between 20 and 22, exceeding this range only for relatively scarce iron-rich minerals. Thus density is an indicator of closeness of packing, which in turn determines such bulk properties as compressibility and elastic wave velocities.

In dealing with the elasticity of rocks, it is important to eliminate as far as possible extraneous effects of porosity before looking for intrinsic relationships. The experiments at high pressures are consequently of most interest. Measurements of V_P at 10 kb and room temperature [Birch, 1961a, p. 2218; Kanamori and Mizutani, 1965, p. 189], when plotted against density, show a clustering of points for the common rocks along a line that can be represented by $V_P = a + b\rho$, with b about 3(km/sec)/(g/cm^3). Points for minerals and rocks with mean atomic weight significantly greater than 22 fall off this line in a more or less systematic way, with lower velocity for a given density, or higher density for a given velocity. In an initial attempt to rationalize these measurements, a system of straight lines was fitted, each representing, in principle, materials of a fixed mean atomic weight m. Nearly all the points have $m = 20$ to 22; there are not enough points with higher values of m to determine lines, and they were drawn parallel to the well-determined line for $m \approx 21$.

These lines can be interpreted, first, as indicating roughly the effect of isothermal compression for individual materials; second, and more important, as suggesting the effect of changes of crystal structure involving closer packing, composition remaining the same; third, as showing the effect of substituting heavier elements, such as titanium or iron, in a given structure (olivine, ilmenite). The range of velocity and density represented by these measurements is especially suitable for discussing the upper mantle and transition zone. In the absence of more directly pertinent information, these lines can be used to find the changes of density associated with the phase changes of the transition zone; as isothermal compression curves, they are probably increasingly in error beyond a density of about 4 g/cm^3. There is no direct indication to be derived from these measurements as to the effect of temperature, though one may imagine that, acting in the opposite direction from an increase of pressure, an increase of temperature might decrease velocity and density in approximately the ratio represented by these lines. This is examined further below.

There are several suggestions for combining the separate effects of density and mean atomic weight to produce a single relation between velocity and a suitable parameter. The simplest method [Birch, 1961a] is to plot velocity versus the ratio ρ/m, which is inversely proportional to mean atomic volume; this produces a nearly linear relation, though with greater scatter than for V_P versus ρ at constant m. As a variant, Fairbairn's packing index $\bar{v}\rho/m$ can be used, with about the same degree of success; here $\bar{v}$ is the mean volume of the atoms, treated as spheres having the ionic radii as determined by Goldschmidt, for example (this is different from the mean atomic volume, which includes the 'empty space' between these spheres).

O. L. Anderson and Nafe [1965] plot bulk modulus K_0 versus mean atomic volume on logarithmic scales, finding a relation for oxides and silicates of the form $K_0 = \text{const} \, (\rho_0/m)^4$. This is equivalent to: $\phi_0 = K_0/\rho_0 = \text{const} \, (\rho_0^3/m^4)$, or to ϕ_0 proportional to ρ_0^3 for fixed m.

Similarly, D. L. Anderson [1967a] has combined m and ϕ in a relation that can be written $\rho = \text{const} \, (m^a \, \phi^b)$; the constants are determined from experimental data, and depend to some extent on the selection employed. Probably the most relevant solution [Anderson, 1967a] gives $a \approx 1$, $b \approx 1/3$. Thus, this relation becomes $\rho = \text{const} \, (m\phi^{1/3})$, or $\rho/m = \text{const} \, \phi^{1/3}$. For constant m, ϕ is proportional to ρ^3.

Also, the velocity $\phi^{1/2}$ is evidently nearly proportional to ρ/m. If $\phi \approx \rho^3$ is imagined to constitute the compression curve for a given material, it leads to $K \approx \rho^4$, and $\partial K/\partial P = 4$.

A more elaborate scaling law has been derived by Knopoff [1967]. This is designed to merge at very high pressures with the Thomas-Fermi solutions for elements, and a 'representative' atomic number Z is used in place of the natural mean atomic number or mean atomic weight. This is defined for compounds by

$$Z^{2/3} = \Sigma\, n_i Z_i^{5/3} / \Sigma\, n_i Z_i$$

where n_i is the percentage by number of the element with atomic number Z_i in the chemical formula. Knopoff then finds that the V_P, ρ data for rocks can be represented either by a linear relation of the form

$$V_P Z^{-2/3} = a + b(\rho/Z^2)$$

which for constant Z reduces to a linear relation between V_P and ρ, or by a power law, of the form

$$V_P Z^{-2/3} = \text{const}\, (\rho/Z^2)^n$$

with n found empirically to be 0.813 ± 0.067, or nearly 5/6. Again a close proportionality of V_P to ρ is indicated. This is of course a feature of the data on which all of these relations depend, and to the extent that they all represent the experimental data they must agree within the experimental range. The extrapolations to the higher densities of the lower mantle may differ to some extent, but the necessity for such extrapolations has been effectively eliminated by the work on shock compression of rocks. If density can be found independently, as from the solutions for free vibrations, then the velocity-density relation furnishes an important restriction on the heavy metal or iron content [*Birch*, 1964; *Anderson*, 1967a].

The experiments on rocks and other aggregates determine velocity as a function of pressure at constant temperature, or as a function of temperature at constant pressure, and in combination with the coefficients of compressibility or thermal expansion, give the ratio $dV/d\rho$ of velocity change to density change, pressure or temperature remaining constant. On the average, this ratio is roughly 3(km/sec)/

TABLE 14. $\Delta V_P/\Delta\rho$ [(km/sec)/(g/cm³)] for Constant Temperature and for Constant Pressure for Several Materials

Material	$(\Delta V_P/\Delta\rho)_T$	$(\Delta V_P/\Delta\rho)_P$
Corundum	3.3	5.7
Periclase	3.6	4.4
Spinel	2.7	5.8
Garnet	3.3	3.9
Forsterite (porous)	3.3	6.3
Olivine	(3.3)	(6.4)

TABLE 15. $dV_P/d\rho$ for Periclase as a Function of dT/dP

dT/dP, °C/kb	dT/dz, °C/km	$dV_P/d\rho$, (km/sec)/(g/cm³)
0	0	3.6
10	3.6	2.7
15	5.4	0.5
18	6.4	−13.8
18.8	6.7	−Inf.
20	7.2	17.4
30	10.7	5.8
50	17.9	4.9
Inf.	Inf.	4.4

(g/cm³) for V_P for changes of pressure at constant temperature, and somewhat larger for changes of temperature at constant pressure (Table 14). Considered as a ratio of finite differences, which may be formed for any pair of materials, however, $\Delta V/\Delta\rho$ can evidently have virtually any value, since there are pairs with equal density but different velocities, others with equal velocities and different densities, even while these points remain close to the average line.

For a layer of homogeneous material, when both temperature and pressure change, we have:

$$dV = (\partial V/\partial P)_T\, dP + (\partial V/\partial T)_P\, dT$$

$$d\rho = (\rho/K_T)\, dP - \alpha\rho\, dT$$

whence

$$\frac{dV}{d\rho} = \frac{K_T}{\rho}\, \frac{(\partial V/\partial P)_T + (\partial V/\partial T)_P (dT/dP)}{1 - \alpha K_T (dT/dP)}$$

In this form, it is obvious that a singularity exists for $(dT/dP) = 1/\alpha K_T$, the condition for constant density, unless this happens to coincide exactly with the condition for constant velocity. In general, this will not be true. As an example we consider the variations of $dV/d\rho$ as a function of dT/dP for MgO; we take, following O. L. Anderson and collaborators, $K_T = 1691$ kb, $\alpha = 31.4 \cdot 10^{-6}$ °C⁻¹, $\rho = 3.58$ g/cm³, $(\partial V_P/\partial P)_T = 7.71 \cdot 10^{-3}$ km/sec/kb, $(\partial V_P/\partial T)_P = -500 \cdot 10^{-6}$ km/sec/°C. With these numbers (which do not allow for variation of these coefficients with pressure or temperature), we find the values of Table 15; dT/dP has been multiplied by 0.358 kb/km to give dT/dz.

The values for dT/dP equal to zero and infinity are just the coefficients for constant temperature and constant pressure, respectively. In the application to the upper mantle, we begin at shallow depths with high temperature gradients: velocity and density both decrease as depth increases. Supposing the gradient to decrease gradually, the material remaining the same, a depth is reached at which compression balances expansion, the density gradient is zero, but the velocity is still decreasing; $dV_P/d\rho$ first becomes positively infinite, then negatively infinite when dT/dP becomes less than 18.8°C/kb. Between 18.8°C/kb and about 15.6°C/kb, density increases downward, while velocity continues to decrease. Finally, beyond this point, both density and velocity increase downward, the ratio gradually approaching the limiting value 3.6 (km/sec)/(g/cm³). In the special case $(dT/dP)_{V_P} = (dT/dP)_\rho$, we have $dV/d\rho = (K/\rho)(\partial V/\partial P)_T$, independent of (dT/dP). Evaluating this with the numbers tabulated above, we find the values of Table 14. Thus it appears that ratios of this order will be found except in the neighborhood of gradients leading to constant density. At very high pressures, as in the lower mantle, the ratio $\Delta V/\Delta \rho$ as found from solutions of the Adams–Williamson equation is about 2.4 (km/sec)/(g/cm³). A small part of this reduction is accounted for by the implied adiabatic rise of temperature, but it must be mainly the effect of smaller velocity coefficients. This example illustrates the hazards of adopting a one-to-one relationship of velocity to density for any region of high and variable temperature gradient. The error for density, as inferred from velocity, may not be serious, but we cannot give a unique significance to the ratio $\Delta V/\Delta \rho$. A large value for this may mean a change of material, or passage through a critical temperature gradient or partial melting. In the olivine series, the substitution of iron for magnesium increases density but decreases velocity, thus leading to negative values of $\Delta V/\Delta \rho$.

LOW VELOCITY ZONE: VARIATION OF IRON CONTENT

While retaining the assumption of an olivine composition, we now envisage the production of a low velocity zone by an increase, with depth, of the proportion of fayalite, followed at the base of the zone by a return to the original composition, which we take to be approximately $Fo_{90}Fa_{10}$. *Adams* [1931] found that fayalite, though denser than forsterite, was more compressible. Compression thus increases the density difference between these two olivines, and the velocities in fayalite are lower than in forsterite. This is supported by the measurements of velocity in dunites: the velocities in the hortonolite dunite from the Mooihoek mine in the Bushveld complex [*Birch*, 1960, V_P; *Simmons*, 1964a, V_S] of composition close to Fa_{50} are considerably lower than those of common dunites with about 10% of fayalite. The published numbers present a few problems, however, which we now examine.

The sample of fayalite measured by Adams was from blast-furnace slag, and the density, found from loss of weight in butyl ether, was 4.068 g/cm³, as compared with 4.392 g/cm³ for pure synthetic fayalite [*Yoder and Sahama*, 1957]. Adams mentioned 'some undetermined material of much lower refractive index,' but he believed this amounted to a few percent and made no correction to the compressibility. The difference of density between the sample and pure fayalite requires about 7.4% by volume of empty space (vesicles), or about 15% by volume of glass of density 2.2 g/cm³, or a still larger proportion of accessory material of higher density. An appreciable increase of compressibility might result. If the empty vesicles were sealed against the pressure medium, *Hashin*'s [1959] theory would give a correction of about 17%, and a compressibility of 0.76 mb⁻¹ for the solid fayalite, close to and slightly less than the compressibility of forsterite. A similar reduction of compressibility would follow a correction for a large proportion of relatively compressible glass. If, however, the low density resulted from voids open to the pressure fluid, the measured compressibility might be valid. There is evidently need for a redetermination upon a fayalite sample of the correct density.

It seems best to employ the direct determinations of velocity on hortonolite dunite; these and various derived quantities are given in Table 16. This is a coarse-grained rock with more than 90% hortonolite of composition about Fa_{50} and small but variable amounts of diallage, hornblende, and oxides. A chemical analysis of a sample from the Mooihoek mine is given by Wagner (1929; repeated by *Birch* [1961a]); the density of this sample was 3.83 g/cm³. For the six samples studied by Birch and

TABLE 16. Physical Properties of Olivine, as Dependent on Iron Content[a]

Property	Dunite, Twin Sisters	Hortonolite Dunite, Mooihoek Mine	Fayalite
Composition	Fa_{8-10}	Fa_{50}	Fa_{100}
Density, g/cm³	3.312^b	3.744^b	4.392^c
	3.326^d	3.760^d	$(4.068)^e$
V_P, km/sec (10 kb)	8.42^b	7.36^b	$(7.1, 15\ kb)^e$
V_S, km/sec (10 kb)	4.83^d	3.90^d	
ϕ, (km/sec)² (10 kb)	$39.8^{b,d}$	$33.9^{b,d}$	$(27)^e$
K_S, mb (10 kb)	$1.32^{b,d}$	$1.27^{b,d}$	$(1.1)^e$
β_S, mb⁻¹ (10 kb)	$0.78^{b,d}$	$0.81^{b,d}$	$(0.91)^e$
σ (10 kb)	$0.255^{b,d}$	$0.30^{b,d}$	
$\Delta V/V_0$, 20°C to 1000°C, %	3.32^f	3.2^f	2.64^f
$10^6\ \alpha$ at 800°C	39^f	35^f	31^f

[a] See also Table 1.
[b] Birch [1960].
[c] Yoder and Sahama [1957].
[d] Simmons [1964a].
[e] Adams [1931]; values in parentheses are for the sample of density 4.068 g/cm³.
[f] Skinner [1966].

Simmons, the range of density was from 3.707 to 3.828 g/cm³. There is a small and variable amount of alteration, probably leading to densities and velocities slightly low for unaltered hortonolite. The theoretical density of olivine of composition $Fo_{50}Fa_{50}$ is 3.80 g/cm³.

The thermal expansion of hortonolite is not appreciably different from that of forsterite [Skinner, 1962, 1966]. Thus, following the earlier discussion, we take the relative change of velocity with temperature to be the same for the hortonolite as for forsterite, and construct a table of velocities as a function of depth for hortonolite by multiplying the corresponding values for forsterite (Table 11) by the ratios of velocities at room temperature and 10 kb. The results are shown in Table 17 and Figure 6. The figure also shows the model CIT11A velocities, and an interpolated set of values for Fa_{30}.

TABLE 17. Velocities (kilometers per second) in an Olivine Layer, with Ringwood's 'Oceanic' Temperatures*†

Depth, km		50	100	150	200
T, °C		880	1300	1550	1640
Fa_{10}	V_P	8.08	7.96	7.93	8.00
	V_S	4.67	4.58	4.56	4.60
Fa_{30}	V_P	7.57	7.46	7.43	7.50
	V_S	4.22	4.14	4.12	4.16
Fa_{50}	V_P	7.06	6.96	6.93	6.99
	V_S	3.77	3.70	3.68	3.72

* Ringwood [1966, p. 362].
† See also Table 11.

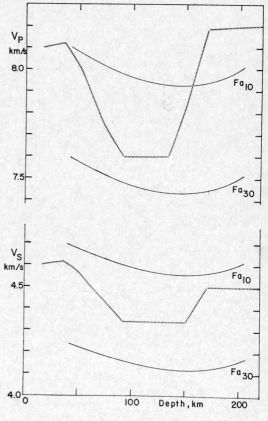

Fig. 6. Velocities for olivine aggregates of composition Fa_{10} and Fa_{30} as function of depth and Ringwood's oceanic temperatures (solid curves), and Anderson's CIT11A velocities (dotted curves).

The V_P distribution can be accounted for by an increase of fayalite content from 10% to about 24% at the lowest velocity, with a return to 10% at the base of the low velocity zone. For V_S, the interpolated composition at the minimum of velocity is about 21% fayalite. Thus with minor revisions of velocity or temperature, a consistent account of the velocity minimum can be given in terms of a layer of iron enrichment, to about 23 ± 2% of fayalite, between the depths of, roughly, 90 and 140 km.

With this interpretation, the low-velocity zone would be characterized by higher, rather than lower, density, as compared with the overlying and underlying layers. Remaining within the restriction to olivine composition, we find the density at the minimum of velocity to be about 3.48 g/cm³ (Fa_{23}), neglecting the small and nearly compensating corrections for pressure and temperature.

While the numbers used in these computations are strictly valid (within the many kinds of errors) only for the olivine series, the conclusions may still be qualitatively correct for the more complex peridotites, with accessory garnet, pyroxene, or other minerals, which probably constitute the upper mantle, since the dependence of density and velocity on iron content appears to be much the same for these as for the olivines [Birch, 1961a, b]. Like the interpretation of partial melting, the interpretation in terms of iron enrichment implies that the low velocity zones for P and S coincide with respect to depth.

REFERENCES

Adams, L. H., The compressibility of fayalite and the velocity of elastic waves in peridotite with different iron-magnesium ratios, *Gerlands Beitr. Geophys., 31*, 315–321, 1931.

Anderson, Don L., Recent evidence concerning the structure and composition of the earth's mantle, *Phys. Chem. Earth, 6*, 1–31, 1966.

Anderson, Don L., A seismic equation of state, *Geophys. J., 13*, 9–30, 1967a.

Anderson, Don L., Latest information from seismic observations, in *The Earth's Mantle*, edited by T. F. Gaskell, pp. 355–420, Academic Press, New York, 1967b.

Anderson, O. L., and J. E. Nafe, The bulk-modulus-volume relationship for oxide compounds and related geophysical problems, *J. Geophys. Res., 70*, 3951–3963, 1965.

Anderson, O. L., and E. Schreiber, The pressure derivatives of the sound velocities of polycrystalline magnesia, *J. Geophys. Res., 70*, 5241–5247, 1965.

Aronson, J. R., S. W. Eckroad, A. G. Emslie, R. K. McConnell, Jr., and P. C. von Thüna, Radiative thermal conductivity in planetary interiors, *Nature, 216*, 1096–97, 1967.

Beloussov, V. V., Modern concepts of the structure and development of the earth's crust and the upper mantle of continents, *Quart. J. Geol. Soc., London, 122*, 293–314, 1966.

Birch, Francis, The variation of seismic velocities within a simplified earth model in accordance with the theory of finite strain, *Bull. Seismol. Soc. Am., 29*, 463–479, 1939.

Birch, Francis, Elasticity and constitution of the earth's interior, *J. Geophys. Res., 57*, 227–286, 1952.

Birch, Francis, The velocity of compressional waves in rocks to 10 kilobars, 1, *J. Geophys. Res., 65*, 1083–1102, 1960.

Birch, Francis, The velocity of compressional waves in rocks to 10 kilobars, 2, *J. Geophys. Res., 66*, 2199–2224, 1961a.

Birch, Francis, Composition of the earth's mantle, *Geophys. J., 4*, 295–311, 1961b.

Birch, Francis, Investigations of the earth's crust, in *IUGG Monograph 22*, edited by M. Båth, p. 24, IUGG, Paris, 1962.

Birch, Francis, Density and composition of mantle and core, *J. Geophys. Res., 69*, 4377–88, 1964.

Birch, Francis, and Dennison Bancroft, The elasticity of glass at high temperatures and the vitreous basaltic substratum, *Am. J. Sci., 240*, 457–490, 1942.

Bowen, N. L., *The Evolution of the Igneous Rocks*, Princeton University Press, Princeton, 332 pp., 1928.

Bridgman, P. W., Rough compressions of 177 substances to 40,000 kg/cm², *Proc. Am. Acad. Arts Sci., 76*(3), 71–87, 1948.

Bullen, K. E., *An Introduction to the Theory of Seismology*, Cambridge University Press, 3rd ed., 381 pp., 1963.

Caloi, P., On the upper mantle, *Advan. Geophys., 12*, 80–211, 1967.

Christensen, N. I., Compressional wave velocities in metamorphic rocks at pressures to 10 kilobars, *J. Geophys. Res., 70*, 6147–6164, 1965.

Christensen, N. I., Shear wave velocities in metamorphic rocks at pressures to 10 kilobars, *J. Geophys. Res., 71*, 3549–3556, 1966a.

Christensen, N. I., Elasticity of ultrabasic rocks, *J. Geophys. Res., 71*, 5921–5931, 1966b.

Clark, Sydney P., Jr., and A. E. Ringwood, Density distribution and constitution of the mantle, *Rev. Geophys., 2*, 35–88, 1964.

Davis, B. T. C., and J. L. England, The melting of forsterite up to 50 kilobars, *J. Geophys. Res., 69*, 1113–1116, 1964.

Gaskell, T. F., editor, *The Earth's Mantle*, Academic Press, New York, 1967.

Gutenberg, B., Über Gruppengeschwindigkeit bei Erdbebenwellen, *Phys. Z., 27*, 111–114, 1926.

Gutenberg, B., Wave velocities below the Mohorovicic discontinuity, *Geophys. J., 2*, 348–352, 1959a.

Gutenberg, B., *Physics of the Earth's Interior*, Academic Press, New York, 240 pp., 1959b.

Hales, A. L., and H. A. Doyle, P and S travel time anomalies and their interpretation, *Geophys. J.*, *13*, 403–415, 1967.

Hashin, Z., *Proceedings of the IUTAM Symposium on Non-Homogeneity in Elasticity and Plasticity*, p. 463, Pergamon Press, London, 1959.

Hill, R., The elastic behavior of a crystalline aggregate, *Proc. Phys. Soc., London*, *65*, 349–354, 1952.

Kanamori, H., and H. Mizutani, Ultrasonic measurement of elastic constants of rocks under high pressures, *Bull. Earthquake Res. Inst.*, *43*, 173–194, 1965.

Kanamori, H., N. Fujii, and H. Mizutani, The measurement of thermal diffusivity of rock-forming minerals over the temperature range from 300°K to 1100°K, *J. Geophys. Res.*, *73*, 595–607, 1968.

Knopoff, L., Density-velocity relations for rocks, *Geophys. J.*, *13*, 1–8, 1967.

Lubimova, E. A., On the temperatures gradient in the upper layers of the earth and on the possibility of an explanation of the low-velocity layers, *Bull. Acad. Sci. USSR, Geophys. Series*, *1959*(12), 1300–1301, 1959.

MacDonald, G. J. F., and N. F. Ness, A study of the free oscillations of the earth, *J. Geophys. Res.*, *66*, 1865–1911, 1961.

MacKenzie, J. K. The elastic constants of a solid containing spherical holes, *Proc. Phys. Soc. London, B*, *63*, 2–11, 1950.

McQueen, R. G., S. P. Marsh, and J. N. Fritz, Hugoniot equation of state of twelve rocks, *J. Geophys. Res.*, *72*, 4999–5036, 1967.

Nafe, J. E., and C. L. Drake, Variation with depth in shallow and deep water marine sediments of porosity, density and the velocities of compressional and shear waves, *Geophysics*, *22*, 523–552, 1957.

Nafe, J. E., and C. L. Drake, in *Interpretation Theory in Applied Geophysics*, edited by F. S. Grant and G. F. West, pp. 199–200, McGraw-Hill Book Company, New York, 1965.

Oldroyd, J. G., The effect of small viscous inclusions on the mechanical properties of an elastic solid, *Deformation and Flow of Solids*, edited by R. Grammel, pp. 304–313, Springer, Berlin, 1956.

Ringwood, A. E., Mineralogy of the mantle, in *Advances in Earth Science*, edited by P. M. Hurley, pp. 357–399, in MIT Press, Cambridge, 1966.

Schreiber, Edward, Elastic moduli of single-crystal spinel at 25°C and to 2 kilobars, *J. Appl. Phys.*, *38*, 2508–2511, 1967.

Schreiber, Edward, and O. L. Anderson, Temperature dependence of the velocity derivatives of periclase, *J. Geophys. Res.*, *71*, 3007–3012, 1966a.

Schreiber, Edward, and O. L. Anderson, The pressure derivatives of the sound velocities of polycrystalline alumina, *J. Am. Ceram. Soc.*, *49*, 184–190, 1966b.

Schreiber, Edward, and O. L. Anderson, Pressure derivatives of the sound velocities of polycrystalline forsterite with 6% porosity, *J. Geophys. Res.*, *72*, 762–764, and correction, 3751, 1967.

Shimozuru, D., The low velocity zone and temperature distribution in the upper mantle, *J. Phys. Earth*, *11*, 19–24, 1963a.

Shimozuru, D., On the possibility of the existence of the molten portion in the upper mantle of the earth, *J. Phys. Earth*, *11*, 49–55, 1963b.

Simmons, Gene, Velocity of shear waves in rocks to 10 kilobars, 1, *J. Geophys. Res.*, *69*, 1123–1130, 1964a.

Simmons, Gene, Velocity of compressional waves in various minerals at pressures to 10 kilobars, *J. Geophys. Res.*, *69*, 1117–1121, 1964b.

Skinner, B. J., Thermal expansion of ten minerals, *U.S. Geol. Survey, Prof. Paper*, *450D*, 109–112, 1962.

Skinner, B. J., Thermal expansion, in *Handbook of Physical Constants, Geol. Soc. Am. Mem.*, *97*, 78–96, 1966.

Soga, N., Elastic constants of garnet under pressure and temperature, *J. Geophys. Res.*, *72*, 4227–4234, 1967.

Soga, N., and O. L. Anderson, High temperature elastic properties of polycrystalline MgO and Al_2O_3, *J. Am. Ceram. Soc.*, *49*, 355–359, 1966.

Soga, N., and O. L. Anderson, High temperature elasticity and expansivity of forsterite and steatite, *J. Am. Ceram. Soc.*, *50*, 239–242, 1967.

Soga, N., E. Schreiber, and O. L. Anderson, Estimation of bulk modulus and sound velocities of oxides at very high temperatures, *J. Geophys. Res.*, *71*, 5315–5320, 1966.

Steinhart, John S., and Robert P. Meyer, Explosion studies of continental structure, *Carnegie Inst. Wash. Publ. 622*, 409 pp., 1961.

Talwani, M., G. H. Sutton, and J. L. Worzel, A crustal section across the Puerto Rico trench, *J. Geophys. Res.*, *64*, 1545–1555, 1959.

Toksöz, M. N., M. A. Chinnery, and D. L. Anderson, Inhomogeneities in the earth's mantle, *Geophys. J.*, *13*, 31–59, 1967.

Trunin, R. F., V. I. Gonshakova, G. V. Simakov, and N. E. Galdin, A study of rocks under the high pressures and temperatures created by shock compression, *Bull. Acad. Sci. USSR, Phys. Solid Earth*, *9*, 579–586, 1965.

Valle, P. E., Sul gradiente di temperatura necessario per la formazione di 'low-velocity layers,' *Ann. Geofis.*, *9*, 371–377, 1956.

Verma, R. K., Elasticity of some high-density crystals, *J. Geophys. Res.*, *65*, 757–766, 1960.

Woollard, G. P., Crustal structure from gravity and seismic measurements, *J. Geophys. Res.*, *64*, 1521–1544, 1959.

Yoder, H. S., and Th. G. Sahama, Olivine X-ray determinative curve, *Am. Mineralogist*, *42*, 475–491, 1957.

Chemical Composition of the Earth's Crust

A. B. Ronov and A. A. Yaroshevsky

CLARKE [1924] and *Goldschmidt* [1933] calculated the volume of the earth's crust assuming a mean thickness of 16 km. *Poldervaart* [1955] made calculations based on the depth to Moho as the natural boundary between the crust and mantle; new data enable us to revise Poldervaart's results.

INITIAL DATA AND CALCULATION METHODS

The crust is divided into three fundamental types: continental, subcontinental, and oceanic. Within the continents, there are ancient platforms including the Precambrian shield and platforms with pre-Riphean folded basement, and orogenic (geosynclinal) folded zones of the Neogene, which are subdivided into the areas of Riphean-Paleozoic and Meso-Cenozoic folding (Figure 1, Table 1). Within the oceans, there are ancient and young depressions, medial oceanic ranges, and arched uplifts. The boundaries of the subdivisions are reasonably accurate on the continents, but are less precise for the oceans.

The volume in each crustal zone was calculated from *Demenitskaya*'s [1961] map of crustal thickness (see Figure 4, p. 318) and the tectonic map of the world [*Tugolesov et al.*, 1964]. Isopachous lines of the crust and outlines of the indicated tectonic zones were combined on a hemispheric equidimensional azimuthal projection (scale 1:50,000,000). Areas were measured by the weighting method within an accuracy of ±3% relative to the total area of the continents and oceans known from geodetic evidence. Crustal volumes of individual blocks and zones were calculated by the volumetric method [*Ronov*, 1949]. The volumetric measurements are somewhat less accurate than the areal, owing to the approximate method of compilation of Demenitskaya's map, which is based on the correlation between the relief, Bouguer effects, and crustal thickness; it does not take into consideration some possible local deviations of the crustal structure from the adopted average pattern. However, volumes calculated according to this map are presumably more accurate than those based on estimates of the mean thickness of the crust for the entire continental and oceanic crusts.

The values in Table 2 for volumes of sedimentary and volcanic rock in the sedimentary shell of the continents (stratisphere) were obtained by direct measurements of the volumes of the platform and geosynclinal Devonian, Carboniferous, Permian, Triassic, and Jurassic rock series according to world maps of lithologic associations [*Ronov and Khain*, 1954, 1955, 1956, 1961, 1962; *Ronov*, 1961], as well as by less accurate volumetric evaluation of the Riphean, lower Paleozoic, Cretaceous, and Tertiary sediments. The volumetric evaluation was based on average thicknesses of these sediments on different continents, taking into account the results of direct measurements of the rock volumes and their ratios in the rock series of corresponding ages on the Russian platform and the surrounding geosynclines of the Urals, Caucasus, Carpathians, and Balkans. These measurements were performed according to the paleotectonic maps from the *Atlas of Lithological-Paleogeographical Maps* [1961, 1962].

The composition of each rock type is derived from regular studies of many thousands of specimens and average samples of rocks from the Russian platform and the Caucasian geosyncline and from determinations for other regions [*Vinogradov and Ronov*, 1956; *Ronov et al.*, 1963, 1965, 1966; *Ronov and Migdisov*, 1960, 1965; *Ronov and Khlebnikova*, 1957]. The composition of platform volcanic rocks (Tables 2 and 3) is assumed as a mean value between the compositions of traps and plateau basalt developed in platform regions. Basalt and andesite are approximately equally widespread within geosynclines, acid effusive rocks being subordinate. Their abundance in sedimentary series was calculated on the basis of volumetric measurements of the geosynclinal volcanic formations

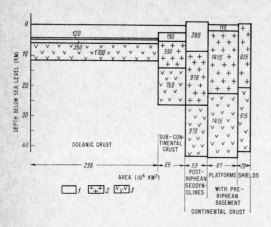

Fig. 1. Schematic subdivision of the earth's crust. Values indicate volume in 10^6 km³: (1) sedimentary layer; (2) granitic layer; (3) basaltic layer.

and of the adopted ratio basalt:andesite:rhyolite = 5:4:1 derived from *Daly*'s [1933] and *Soloviev*'s [1952] areal measurements. The average composition of the geosynclinal volcanics was inferred from this ratio (Table 2).

The total volume of the 'granitic' and 'basaltic' layers of the continents and oceans was calculated with satisfactory accuracy, but a precise separation of the total volume into its component volumes of the granitic and basaltic layers is hardly possible now, owing to different location of the boundary between these layers in orogenic areas of different ages, on platforms and in zones of tectonic activation [*Beloussov*, 1966; *Kosminskaya*, 1958]. We have tentatively divided the total volume of the crystalline part of the continental crust into two equal parts, on the basis of latest data [*Belyaevsky et al.*,

TABLE 1. Areas, Volumes, Thicknesses, and Masses of the Main Tectonic Zones of Continents and Oceans

Types of Crust	Large Structural Units of the Crust		Area, 10^6 km²	Volume, 10^6 km³	Average Thickness, km	Mass, 10^{24} g
Continental	Platforms with pre-Riphean basement	Shields	29.4	1,230	41.8	3.45
		Platforms	66.9	2.940	44.0	8.27
		Total platforms	96.3	4,170	43.3	11.72
	Geosynclines	Riphean-Paleozoic	24.4	1,040	42.7	2.87
		Mesocenozoic	28.3	1,290	45.6	3.60
		Total geosynclines	52.7	2,330	44.3	6.47
	Total		149.0	6.500	43.6	18.19
Subcontinental (transitional type)	Underwater regions of platforms with pre-Riphean basement		26.0	670	25.8	1.88
	Underwater regions of Riphean-Paleozoic geosynclines		14.9	370	24.8	1.02
	Underwater regions of Mezocenozoic geosynclines		24.0	500	20.8	1.40
	Total		64.9	1,540	23.7	4.30
Oceanic	Ancient parts of bottom of Pacific ocean (thalassocratons)		101.7	650	6.4	1.80
	Ancient depressions of Atlantic, Indian, and Arctic Oceans		82.1	460	5.6	1.26
	Volcanic medial ranges, arched uplifts, and block ranges of ocean bottom		82.6	670	8.1	1.90
	Regions of crust with reduced granitic layer		7.6	80	10.5	0.23
	Deep troughs of Cenozoic and Recent geosynclinal systems (including marginal oceanic trenches)		22.1	310	14.0	0.88
	Total		296.1	2,170	7.3	6.07
Total earth's crust			510.0	10,210	20.0	28.56

1967; *Demenitskaya*, 1967; *Pakiser and Robinson*, 1967].

An approximate estimate of the abundances of the most important types of intrusive and metamorphic rocks known from the granitic shell is based on results of measurements of their outcrop areas on the shields, i.e. in the upper part of the 'granitic' shell. We have taken as a basis the measurements performed by one of us (A.R.) for the Baltic and Ukrainian shields and the basement rocks of the Russian platform, according to L. A. Vardanyants' map in the *Atlas of Lithological-Paleogeographical Maps* [1961]. The results obtained are in adequately close agreement with the data of other authors (Table 4).

Data of *Nockolds* [1954] and *Soloviev* [1964, 1965] were used to characterize the chemical composition of intrusive rocks of the shields. The average composition of the metamorphic rocks and the abundance pattern of their chief types have not been determined, owing to their petrographical diversity. We have accepted a hypothesis that one half of the metamorphic rocks are composed of different types of metamorphosed igneous rocks (the proportion of acid to basic rocks taken to be the same as that on the surface of the shields), and the other half of metamorphosed geosynclinal sediments (including volcanics) with known proportions and compositions (Table 2). This assumption appears to be reasonable and in conformity with modern concepts that the old Precambrian rocks of the shields passed through the geosynclinal cycle of their development and were subjected to regional metamorphism and granitization at their final stages [*Beloussov*, 1962, 1966; *Pavlovsky*, 1962; *Ronov*, 1964; *Khain*, 1964].

A first quantitative estimate of the abundances of different types of metamorphic rocks (gneiss, schist, marble, and amphibolite) was obtained and a model of the petrographical structure and chemical composition of the 'granitic' shell was outlined on the basis of the concept of the geosynclinal origin of metamorphic rock series (Tables 3 and 5).

This model can be calculated by two independent methods: (1) assuming that the chemical composition of metamorphic rock series corresponds to that of nonmetamorphosed intrusive and sedimentary geosynclinal rocks and (2) taking as the initial data the actual compositions of intrusions and metamorphic rocks of the shield. Both methods result in identical compositions of the granitic shell (Table 4), which substantiates the adopted hypothesis of geosynclinal origin and composition of metamorphic pararocks of the shield. The significance of this conclusion can scarcely be exaggerated.

The most difficult problem concerning the chemical composition of the earth's crust relates to the deep structure and the composition of the basaltic shell of the continental crust. This problem is quite indeterminate; its solution may involve incompatible hypotheses consistent with the geophysical data. They are limited by two extreme alternatives: (1) the hypothesis of similarity of the chemical composition of the basaltic and granitic shells, and (2) the hypothesis of the basaltic (basic) composition of the basaltic shell. In our work we have proceeded from the assumption that the petrographical and chemical composition of the basaltic shell gradually changes with depth from a composition similar to that of the granitic shell in its upper layers, to a composition corresponding to continental basalt of a geosynclinal type in the lower horizons. At present it is impossible to show this tendency on a geochemical section of the earth's crust, and we have presented it only schematically, considering that the basic (basaltic) and acid (granitic) materials equally take part in the formation of the basaltic shell. No doubt the uncertainty of the shell composition is the principal cause of possible errors in the calculation of crustal composition.

Any attempt to construct a chemical model of the basaltic shell has to take into account: (1) the increase in the velocity of seismic waves, which indicates an increased density with depth from 2.5 in the sedimentary rocks to 2.7 in the granitic layer and to 2.9 in the basaltic [*Beloussov*, 1966; *Magnitsky*, 1965]; (2) the possibility that an increase in density is the result not only of the compaction of rocks but also of increase of their basicity. The latter assumption is supported by a significant increase in the amount of the basic rocks (amphibolite) in the deepest of visible zones of the crust in the ancient Lower Archean series (the so-called 'greenstone nuclei of the continents') [*Vinogradov and Tugarinov*, 1961; *Kolotukhina*, 1964; *Pavlovsky*, 1962; *Semikhatov*, 1964;

TABLE 2. Volume, Mass, and Average Chemical Composition of Sedimentary

Types of Crust	Large Structural Units of Crust and Layers	Volume, 10^6 km^3	Average thickness,* km	Mass, 10^{24} g	Types of Rocks and Sediments	Abundance, % of volume on continents and % of area in oceans	Basis of Values	References
Continental	Platforms	135	1.8	0.35	Sands	23.6	658 analyses of average samples of 7787 specimens	Ronov and Migdisov, unpublished data
			(1.2)		Clays	49.5	695 analyses of average samples of 10,746 specimens	
					Carbonates	21.0	371 analyses of average samples of 10,942 specimens	Ronov, unpublished data
					Evaporites: dolomites, 50%; sulfates, 25%; salts, 25%	2.0	52 analyses of average samples of 2928 specimens of dolomites and gypsum; anhydrite average of 11 analyses, halite average of 6 analyses	*Stewart*, 1963
					Volcanics	3.9	Average of 600 analyses	See Table 3
					Average composition of sedimentary series of platforms	100.0		
	Geosynclines	365	10.0	0.94	Sands	18.7	368 analyses of average samples of 7183 specimens	*Ronov et al.*, 1966
					Clays and shales	39.4	455 analyses of average samples of 11,151 specimens	Ronov et al., unpublished data
					Carbonates	16.3	267 analyses of average samples of 4678 specimens	Ronov et al., unpublished data
					Evaporites: dolomites, 50%; sulfates, 25%; salts, 25%	0.3	Same as for platforms	
					Volcanics	25.3	Average of 411 analyses	See Table 3

Content of Components, wt. %, and Their Masses in Stratisphere as a Whole, 10^{24} g

SiO$_2$	TiO$_2$	Al$_2$O$_3$	Fe$_2$O$_3$	FeO	MnO	MgO	CaO	Na$_2$O	K$_2$O	P$_2$O$_5$	C$_{org}$	CO$_2$	SO$_3$†	Cl†	H$_2$O†
75.75	0.49	6.90	2.55	1.31	0.06	1.43	3.40	0.58	1.83	0.16	0.30	2.75	0.29	0.09	2.11
55.09	0.86	16.30	4.17	1.87	0.05	2.46	4.75	0.75	3.01	0.11	0.99	3.92	0.43	0.12	5.16
9.80	0.18	2.54	0.85	0.54	0.06	6.83	38.93	0.24	0.76	0.07	0.33	35.05	2.36	0.08	1.40
3.02	0.05	0.76	0.27	0.13	0.01	6.16	28.25	13.06	0.28	0.01	0.09	17.55	18.15	14.78	0.64
49.22	1.53	15.74	3.33	8.02	0.18	6.11	10.00	2.51	0.73	0.18		0.01	0.03	0.005	1.81
49.21	0.65	10.88	2.97	1.65	0.06	3.37	12.28	0.81	2.11	0.12	0.63	10.31	1.13	0.40	3.42
62.93	0.52	12.12	2.30	3.20	0.11	2.28	5.69	1.92	1.69	0.11	0.25	4.22	0.12	0.09	2.47
55.76	0.71	17.56	3.61	3.35	0.08	2.52	4.08	1.27	2.76	0.15	0.78	2.80	0.11	0.12	4.37
13.30	0.14	2.70	0.43	0.94	0.09	2.92	42.40	0.56	0.49	0.10	0.35	34.44	0.03	0.08	1.05
3.02	0.05	0.76	0.27	0.13	0.01	6.16	28.25	13.06	0.28	0.01	0.09	17.55	18.15	14.78	0.64
55.62	1.00	16.12	4.17	4.52	0.24	4.22	6.91	3.33	2.02	0.34		0.01	0.03	0.007	1.46

TABLE 2. (Continued)

Types of Crust	Large Structural Units of Crust and Layers	Volume, 10^6 km^3	Average thickness,* km	Mass, 10^{24} g	Types of Rocks and Sediments	Abundance, % of volume on continents and % of area in oceans	Basis of Values	References
					Average composition of sedimentary series of geosynclines	100.0		
	Total	500	4.2 (3.4)	1.29				
Sub-continental	Shelf and Continental slope	190	2.9	0.48				
Oceanic	Sediments of seismic layer I	120 70‡	0.4	0.19	Terrigenous	7.3	Average of 3 analyses	El Wakeel and Riley, 1961; Landergren, 1964
					Calcareous	41.5	Average of 8 analyses	
					Siliceous	17.0	Average of 2 analyses	
					Red deep-water clays	31.2	Average of 464 analyses	
					Volcanic	3.0	Average of 2 analyses	
					Average composition of pelagic sediments	100.0		
	Sediments of seismic layer II	175	0.6	0.44				
	Total	295 (245)‡	1.0	0.63				
	Total for sedimentary shell, including volcanics	985 935‡	2.0 (1.8)‡	2.40				
	Total for sedimentary shell, excluding volcanics	800‡		2.00				

* Thicknesses including areas of shields are shown in parentheses.
† The sum of analyses exceeds 100% by O-Cl$_2$.
‡ Recalculated for consolidated sediment with a density of 2.5.
§ In calculating average values, the content of O$_2$ and Cl in marine sediments was taken to be the same as in continental ones.

TABLE 2. (*Continued*)

Content of Components, wt %, and Their Masses in Stratisphere as a Whole, 10^{24} g

SiO$_2$	TiO$_2$	Al$_2$O$_3$	Fe$_2$O$_3$	FeO	MnO	MgO	CaO	Na$_2$O	K$_2$O	P$_2$O$_5$	C$_{org}$	CO$_2$	SO$_3$	Cl†	H$_2$O†
50.00	0.65	13.69	2.98	3.21	0.13	2.99	11.42	1.83	2.00	0.18	0.42	7.56	0.12	0.12	2.73
49.82	0.65	12.97	2.97	2.81	0.11	3.09	11.64	1.57	2.03	0.17	0.47	8.26	0.38	0.19	2.91

Composition considered to be similar to that of sedimentary rocks of continents

SiO$_2$	TiO$_2$	Al$_2$O$_3$	Fe$_2$O$_3$	FeO	MnO	MgO	CaO	Na$_2$O	K$_2$O	P$_2$O$_5$	C$_{org}$	CO$_2$	SO$_3$	Cl†	H$_2$O†
52.92	0.81	15.73	5.04	1.92	0.33	3.82	4.68	1.59	3.09	0.08	0.30	3.35			6.34
18.84	0.29	6.09	2.55	0.46	0.19	1.98	35.61	0.70	1.03	0.11	0.28	28.90			2.97
62.80	0.66	13.21	4.92	1.15	0.14	3.02	1.47	1.52	2.70	0.16	0.26	1.08			6.91
54.22	0.81	16.10	6.99	0.85	0.63	3.42	3.33	1.28	2.85	0.21	0.22	2.45			6.64
44.30	2.48	12.41	5.83	6.40	0.41	8.85	9.79	1.88	0.78	0.20	0.32	3.36			2.99
40.63	0.62	11.31	4.62	0.97	0.34	2.95	16.70	1.13	2.03	0.15	0.26	13.27			5.02

Composition considered to be similar to that of sediments of seismic layer I

Composition considered to be similar to that of sediments of seismic layer I

SiO$_2$	TiO$_2$	Al$_2$O$_3$	Fe$_2$O$_3$	FeO	MnO	MgO	CaO	Na$_2$O	K$_2$O	P$_2$O$_5$	C$_{org}$	CO$_2$	SO$_3$	Cl†	H$_2$O†
46.29	0.64	12.32	3.58	2.11	0.20	3.05	13.51	1.40	2.02	0.17	0.39	10.12	0.38§	0.19§	3.69
1.111	0.015	0.295	0.086	0.051	0.005	0.073	0.324	0.034	0.048	0.004	0.010	0.242	0.010	0.005	0.088
46.20	0.58	10.50	3.32	1.95	0.16	2.87	14.00	1.17	2.07	0.13	0.49	12.10	0.50§	0.24§	3.85
0.924	0.012	0.210	0.066	0.038	0.003	0.056	0.280	0.023	0.041	0.003	0.010	0.242	0.010	0.005	0.077

TABLE 3. Distribution of Volumes and Masses of Magmatic and Metamorphic Rocks of Continental

Types of Crust	Shells	Volume, 10^6 g	Mass, 10^{24} g	Types of Rocks	Abundance, % of total volume of shell	Mass, 10^{24} g	Number of Analyses Used and References
Continental and Subcontinental	Sedimentary	690	1.77	Traps and plateau-basalts of platforms	1.1	0.02	Average of 600 analyses [Daly, 1936; Nockolds, 1954; Walker and Poldervaart, 1949; Nesterenko et al., 1964]
				Basalts of geosynclines	9.2	0.18	Average of 198 analyses [Daly, 1933]
				Andesites of geosynclines	7.4	0.15	Average of 87 analyses [Daly, 1933]
				Rhyolites of geosynclines	1.8	0.03	Average of 126 analyses [Daly, 1933]
	Granitic	3,590	9.81	Granites	18.1†	1.75	Average of 72 analyses [Nockolds, 1954] and of 170 analyses [Soloviev, 1964]
				Granodiorites, diorites	19.9†	1.93	Average of 137 analyses [Nockolds, 1954] and of 151 analyses [Soloviev, 1964]
				Syenites, nepheline syenites	0.3†	0.03	Mixture 4:1 average of syenite (24 analyses) and average of nepheline, syenite (80 analyses) [Nockolds, 1954]
				Gabbro	3.7†	0.38	Average of 160 analyses [Nockolds, 1964] and of 146 analyses [Soloviev, 1964]
				Dunites, periodotites	0.1†	0.01	Average of 23 analyses [Nockolds, 1954]
				Gneisses	37.6	3.66	Average of 365 analyses [Poldervaart, 1955]
				Crystalline schists	9.0	0.88	Average of 214 analyses [Poldervaart, 1955]
				Marbles	1.5§	0.15	Average of 4 analyses, [Struve, 1940]
				Amphibolites	9.8	1.02	Average of 200 analyses [Poldervaart, 1955]
				Average composition of magmatic rocks of granitic shell	42.1	4.10	
				Average composition of metamorphic rocks of granitic shell	57.9	5.71	
	Basaltic	3,760	10.91	Acid magmatic and metamorphic ortho- and para-rocks	50.0	5.26	
				Basic magmatic and metamorphic ortho- and para-rocks	50.0	5.65	
Oceanic	Basaltic plus Volcanic Rocks of Seismic Layer II	1,875	5.44	Oceanic tholeiitic basalts	99.0	5.39	Average of 10 analyses [Engel et al., 1965]
				Alkaline differentiates of oceanic basalts	1.0	0.05	Average of 10 analyses [Engel et al., 1965]

* The content of carbon, sulfur, and chlorine was taken from Vinogradov [1962].
† The estimate of relative abundance of magmatic rocks is based on the results of measurement and the map compiled by L. A. Vardanyants [Atlas of Lithological-Paleogeographical Maps, 1961] (acid, 38.3%; base 3.8%), as well as the data obtained by Fleischer and Chao [1960] (Daly's figures) on the relations of granites, granodiorites, and syenites in the acid group of rocks and basic and ultrabasic rocks in the basic group.

Subcontinental, and Oceanic Types of Crust and Chemical Compositions of Rocks

							Content of Components, wt. %							
SiO₂	TiO₂	Al₂O₃	Fe₂O₃	FeO	MnO	MgO	CaO	Na₂O	K₂O	P₂O₅	C*	CO₂	S*	Cl*	H₂O⁺
9.22	1.53	15.74	3.33	8.02	0.18	6.71	10.00	2.51	0.73	0.18	0.01		0.03	0.005	1.81
49.04	1.36	15.69	5.38	6.37	0.31	6.17	8.94	3.11	1.52	0.45	0.01		0.03	0.005	1.62
59.57	0.77	17.30	3.33	3.13	0.18	2.75	5.79	3.58	2.04	0.26	0.01		0.03	0.005	1.26
72.74	0.33	13.47	1.45	0.88	0.08	0.38	1.20	3.38	4.45	0.08	0.03		0.04	0.02	1.47
72.33	0.31	14.00	0.95	1.50	0.05	0.53	1.36	3.14	5.07	0.15	0.03		0.04	0.02	0.61
65.66	0.55	15.72	1.54	2.63	0.10	1.73	3.71	4.10	3.05	0.23	0.03		0.04	0.02	0.98
57.73	0.66	18.70	2.51	2.02	0.15	1.77	3.80	4.85	6.97	0.20	0.01				0.61
48.73	1.17	16.80	2.93	7.09	0.21	7.38	11.13	2.35	0.70	0.24	0.01		0.03	0.005	1.24
43.54	0.81	3.99	2.51	9.84	0.21	34.02	3.46	0.56	0.25	0.05					0.76
69.6	0.5	14.7	1.7	2.2	0.1	1.3	2.1	3.2	3.7	0.2					0.7‡
62.0	0.9	18.5	2.7	4.8	0.1	2.8	1.5	1.9	3.8	0.2					0.8‡
12.38	—	1.18	0.55	—	—	17.89	26.74	0.03	0.05	0.11		40.71			0.36
49.6	1.6	15.5	3.5	7.7	0.2	6.9	9.4	2.9	1.1	0.3					1.3‡
66.91	0.50	15.07	1.41	2.54	0.09	1.78	3.34	3.52	3.72	0.19	0.03		0.04	0.02	0.84
63.72	0.74	15.12	2.13	3.48	0.11	2.91	3.88	2.87	3.21	0.22		1.05			0.56
49.60	1.50	17.12	2.00	6.85	0.17	7.23	11.78	2.74	0.16	0.16					0.69
47.78	2.89	18.16	4.20	5.84	0.16	4.83	8.72	4.02	1.67	0.93					0.80

‡ Water content is an average of 30 analyses of USSR gneisses [*Morkovkina*, 1964], 28 analyses of USSR schists [*Struve*, 1940], and 8 analyses of USSR amphibolites [*Morkovkina*, 1964].
§ The percentage of marbles was reduced to ⅛ the amount of carbonate rocks in geosynclines, because of their less widespread occurrence in Early Archean rocks of the Precambrian.

TABLE 4. Average Chemical Composition (wt %) and Abundances of the Main Rock Types (vol %) of Shields and Basement

Oxides	Baltic Shield in Finland [Sederholm, 1925]	Canadian Shield [Grout, 1938]	Canadian Shield [Shaw et al., 1967]	Magmatic Rocks of Baltic and Ukrainian Shields [Soloviev, 1952]	Baltic and Ukrainian Shields and Basement of Russian Platform (this work) (1)*	(2)*
SiO_2	68.4	63.9	66.7	68.6	65.6	66.0
TiO_2	0.4	0.8	0.5	0.6	0.6	0.6
Al_2O_3	14.8	17.0	15.0	14.7	15.6	15.3
Fe_2O_3	1.3	2.4	1.4	1.5	2.1	1.9
FeO	3.2	3.0	2.8	2.6	2.9	3.1
MnO			0.1	0.1	0.1	0.1
MgO	1.7	1.8	2.3	1.6	2.3	2.4
CaO	3.4	4.1	4.2	2.9	4.1	3.7
Na_2O	3.1	3.7	3.6	3.7	3.1	3.2
K_2O	3.6	3.1	3.2	3.6	3.4	3.5
P_2O_5	0.1		0.2	0.1	0.2	0.2
Total	100.0	100.0	100.0	100.0	100.0	100.0
Granites and granodiorites	53	70		84		38.3
Alkaline rocks				3		
Gabbro and basalts	8	10		13		3.8
Migmatites	22					
Metamorphic rocks	17	17				57.9
Total	100			100		100.0

* Our data calculated by two alternatives: (1) assuming that the chemical composition of metamorphic rocks corresponds to the composition of nonmetamorphosed intrusive and sedimentary geosynclinal rocks, and (2) using actual chemical compositions of intrusive and metamorphic rocks of shields as initial ones.

Tugarinov and Voitkevich, 1966; Tugarinov, 1956; Frolova, 1950, 1962; Gastil, 1960; Michot, 1955, 1956, 1957; Wilson, 1949, 1952], as well as by the data on the geochemical balance of the elements and isotopes in the earth's crust [Vinogradov, 1956, 1962b; Faure and Hurley, 1963; Taylor, 1964], which indicate that a large part of the basic (basaltic) material is required as a source of these elements and isotopes in sediments and sea water.

However, it is quite probable that not only basic but also strongly consolidated, more acid metamorphic rocks participate in the formation of the basaltic layer [Beloussov, 1966; Rezanov, 1960, 1962]. We have assumed that these metamorphic rocks make up about half of the shell's total volume, while the second half is composed of the basic rocks (gabbro, amphibolites, eclogites, etc.; Tables 3 and 5). With such proportion of the rocks in the basaltic shell, the average composition of the crust of the continental block approaches that of the mixture of granite and basalt at a ratio of 1:1, and the average composition of the whole earth's crust at a ratio of 2:3. From this point of view we admit the possibility of errors in determining the average composition of the continental crust. It can be estimated separately for each component considering two limiting alternatives: (1) the composition of the basaltic shell is identical with that of the granitic one, and (2) the composition of the basaltic shell is similar to that of the continental geosynclinal basalts. The results of the calculations performed show that the limiting possible divergences from the average composition of the continental crust adopted by us are generally not great (Table 6).

Three layers are distinguished in the oceanic crust. The first seismic layer, corresponding to nonconsolidated sediments, is characterized by low seismic velocities (1.5–1.8 km/sec). Its thickness varies from 0.3 to 0.8 km and reaches greater values in the oceanic troughs and depressions. We have accepted 0.4 km as its average thickness [Zverev et al., 1961]. The lithological composition of bottom sediments has

TABLE 5. Volumes, Masses, and Average Chemical Compositions of the Crust

Types of Crust	Shell	Volume, 10^6 km^3	Average Thickness, km	Mass, 10^{24} g	SiO$_2$	TiO$_2$	Al$_2$O$_3$	Fe$_2$O$_3$	FeO	MnO	MgO	CaO	Na$_2$O	K$_2$O	P$_2$O$_5$	C	CO$_2$	S	Cl	H$_2$O$^+$
Continental	Sedimentary	500	3.4	1.29	49.95	0.65	13.01	2.98	2.82	0.11	3.10	11.67	1.57	2.04	0.17	0.47	8.28	0.15	0.19	2.92
					0.645	0.008	0.168	0.038	0.036	0.001	0.040	0.151	0.020	0.026	0.002	0.006	0.107	0.002	0.002	0.038
	Granitic	3,000	20.1	8.20	63.94	0.57	15.18	2.00	2.86	0.10	2.21	3.98	3.06	3.29	0.20	0.17	0.84	0.04	0.05	1.53
					5.243	0.047	1.245	0.164	0.234	0.008	0.181	0.326	0.251	0.270	0.016	0.014	0.069	0.003	0.004	0.125
	Basaltic	3,000	20.1	8.70	58.23	0.90	15.49	2.86	4.78	0.19	3.85	6.05	3.10	2.58	0.30	0.11	0.51	0.03	0.03	1.00
					5.066	0.078	1.348	0.249	0.416	0.016	0.335	0.526	0.270	0.224	0.026	0.009	0.044	0.003	0.003	0.087
	Total continental	6,500	43.6	18.19	60.22	0.73	15.18	2.48	3.77	0.14	3.06	5.51	2.97	2.86	0.24	0.16	1.21	0.04	0.05	1.38
					10.954	0.133	2.761	0.451	0.686	0.025	0.556	1.003	0.541	0.520	0.044	0.029	0.220	0.008	0.009	0.250
Sub-continental	Sedimentary	190	2.9	0.48	49.95	0.65	13.01	2.98	2.82	0.11	3.10	11.67	1.57	2.04	0.17	0.47	8.28	0.15	0.19	2.92
					0.240	0.003	0.062	0.014	0.013	0.001	0.015	0.056	0.007	0.010	0.001	0.002	0.040	0.001	0.001	0.014
	Granitic	590	9.1	1.61	63.94	0.57	15.18	2.00	2.86	0.10	2.21	3.98	3.06	3.29	0.20	0.17	0.84	0.04	0.05	1.53
					1.029	0.009	0.244	0.032	0.046	0.002	0.036	0.064	0.049	0.053	0.003	0.003	0.013	0.001	0.001	0.025
	Basaltic	760	11.7	2.21	58.23	0.90	15.49	2.86	4.78	0.19	3.85	6.05	3.10	2.58	0.30	0.11	0.51	0.03	0.03	1.00
					1.287	0.020	0.342	0.063	0.106	0.004	0.085	0.134	0.068	0.057	0.007	0.002	0.011	0.001	0.001	0.022
	Total subcontinental	1,540	23.7	4.30	59.45	0.74	15.08	2.53	3.85	0.16	3.17	5.91	2.89	2.79	0.26	0.17	1.49	0.06	0.06	1.42
					2.556	0.032	0.648	0.109	0.165	0.007	0.136	0.254	0.124	0.120	0.011	0.007	0.064	0.003	0.003	0.061
Oceanic	Sedimentary (layer I)	120	0.4	0.19	40.63	0.62	11.31	4.62	0.97	0.34	2.95	16.70	1.13	2.03	0.15	0.26	13.27	—	—	5.02
					0.077	0.001	0.021	0.009	0.002	0.001	0.006	0.032	0.002	0.004	0.0003	0.001	0.025	—	—	0.009
	Volcanic sedimentary (layer II)	350	1.2	0.96	45.50	1.09	14.46	3.20	4.15	0.25	5.27	14.03	2.00	1.02	0.16	0.12	6.08	—	—	2.67
					0.437	0.010	0.139	0.031	0.040	0.002	0.051	0.135	0.019	0.010	0.001	0.001	0.058	—	—	0.026
	Basaltic	1,700	5.7	4.92	49.58	1.51	17.13	2.02	6.84	0.17	7.21	11.75	2.75	0.18	0.17	0.01	—	0.03	0.03	0.69
					2.439	0.074	0.843	0.099	0.336	0.008	0.355	0.578	0.135	0.009	0.008	—	—	0.001	0.001	0.034
	Total oceanic	2,170	7.3	6.07	48.65	1.40	16.52	2.29	6.23	0.18	6.79	12.28	2.57	0.37	0.15	0.03	1.37	0.02	0.02	1.14
					2.953	0.085	1.003	0.139	0.378	0.011	0.412	0.745	0.156	0.023	0.009	0.002	0.083	0.001	0.001	0.069
Total Crust		10,210	20.0	28.56	57.64	0.88	15.45	2.43	4.30	0.15	3.87	7.01	2.87	2.32	0.23	0.13	1.29	0.04	0.05	1.33
					16.463	0.250	4.412	0.699	1.229	0.043	1.104	2.002	0.821	0.663	0.064	0.038	0.367	0.012	0.013	0.380

* The mean composition of sedimentary shells differs from that in Table 2 by recalculation of S for SO$_3$. For each shell, the upper line is weight percent; the lower line is mass (10^{24} g).

TABLE 6. Average Chemical Composition of Continental Crust (wt %)

Components	Calculation Alternatives			
	Assuming That Composition of Basaltic Shell is Similar to That of Granitic One	Assuming That Composition of Basaltic Shell is Similar to That of Continental Basalts	Adopted in Present Paper	Limit Deviations
SiO_2	65.2	57.4	61.9	±4.0
TiO_2	0.6	1.0	0.8	±0.2
Al_2O_3	15.6	15.7	15.6	±0.1
Fe_2O_3	2.1	3.8	2.6	±0.8
FeO	2.8	4.7	3.9	±1.0
MnO	0.1	0.2	0.1	±0.05
MgO	2.3	4.3	3.1	±1.0
CaO	4.7	7.1	5.7	±1.2
Na_2O	3.1	3.1	3.1	±0.1
K_2O	3.3	2.4	2.9	±0.4
P_2O_5	0.2	0.3	0.3	±0.05
Total	100.0	100.0	100.0	

been illustrated by numerous samples and cores to a depth of as much as 30 meters. Areal distribution of the main genetic groups of sediments within the oceans has been relatively well investigated. An estimate of their distribution pattern has been made from the outcrop areas of each group and type of sediments on the map compiled by *Bezrukov et al.* [1964]. The chemical characteristics of each group of sediments are based on the analyses of *El Wakeel and Riley* [1961], which we have classified in accordance with the Map of Sedimentation, whereas the average composition of the oceanic sediments has been calculated with due account of the areal distribution of each group (Table 2). The values obtained are generally rather similar to the values of *Goldberg and Arrhenius* [1958] for the Pacific Ocean area and the values of *Landergren* [1964] for the Pacific, Atlantic, and Indian Oceans.

The second seismic layer is characterized by velocities of 2.1–5.5 km/sec. Its thickness varies within 1–2 km [*Engel and Engel*, 1968; *Zverev et al.*, 1961], and its mean value is considered to be 1.2 km. The petrographical and chemical composition of the rocks that comprise the layer is unknown. We assumed that the second layer is half composed of consolidated sedimentary rocks, with composition similar to that of the sediments in the first layer (Table 2), and half made up of tholeiitic basalts whose composition is considered by analogy to be close to the average composition of oceanic tholeiites (Table 3).

The third, proper basaltic layer has high seismic velocities (6.5–7.0 km/sec). The volume of this layer was calculated according to the difference between the total volume of the oceanic crust block measured on the map (Table 1) and the individual volumes of the first and second seismic layers (Tables 2 and 5). The rock composition of the proper basaltic layer was well studied during recent years [*Engel and Engel*, 1968; *Engel et al.*, 1965], the quantitative relations between the tholeiitic and alkali basalts being taken into consideration in the calculations (Tables 3 and 5).

The total volume of the subcontinental-type crust was obtained by direct measurements (Table 1). The volume of the sedimentary layer was calculated approximately on the assumption that its thickness gradually decreases from the continents towards the oceans. The continental average thicknesses were established with a satisfactory accuracy for the platform (1.8 km) and geosynclinal (10 km) areas. The average thickness of the sedimentary layer in the oceans was considered to be equal, on the average, to 1.6 km. The volumes of the sedimentary rocks and their average thickness were calculated on the basis of these values, as well as the areal measurements of the platforms and geosynclines of different ages within the shelf and continental slope (Table 2). The relations between the volumes of the granitic and basaltic layers were obtained on the assumption that the granitic layer completely pinches out toward the ocean (Table 5). The compositions of the sedimentary, granitic, and basaltic shells are considered to be analogous to those of the corresponding shells on the continents.

The masses of the earth's crust as a whole and of its different structural units (Table 1) were calculated with reference to the total volume of the crust and the adopted relationship between the volumes of the sedimentary, granitic, and basaltic layers in different structural zones of the crust, as well as the data on the average densities of the layers (2.5, 2.7, and 2.9, respectively, and 1.6 for nonconsolidated sediments of the first seismic layer; the average porosity is assumed to be 35% [Zverev et al., 1961]).

All the data obtained are summarized in Table 5, which shows the volumes, average thicknesses, masses, and average chemical compositions of the different layers of the continental, subcontinental, and oceanic crust. The information on its average petrographic composition is generalized in Table 7; qualitative mineralogical composition, based on the above information, appears in Table 8.

CHEMICAL CONSTITUTION OF THE SHELLS

Sedimentary Shell

The total volume of the earth's sedimentary shell is equal to 940,000,000 km³ (taking into account the consolidation of recent sediments) and 800,000,000 km³ without volcanic rocks, i.e., about 10% the volume of the crust and 0.1% the volume of the whole earth. The average thickness of the sedimentary shell is 1.8 km; if the area of the shields not covered by sediments is excluded, 2 km. On the whole, our values are close to those obtained by Kuenen [1950] (volume 1,000,000,000 km³ and average thickness 2 km) and *Poldervaart* [1955] (volume 710,000,000 km³ and average thickness 1.4 km). Our measurements have shown that the bulk of the sediments is accumulated on the continents (500,000,000 km³) and continental margin of the oceans (190,000,000 km³), whereas only 250,000,000 km³ is in the oceans. This distribution of the sediments according to *Kuenen's* [1950] scheme is 200,000,000 km³ for continents and 800,000,000 km³ for oceans with shelves; *Poldervaart's* [1955] values are 180,000,000 km³ for continents, 160,000,000 km³ for the pelagic part of the ocean, and 370,000,000 km³ for the suboceanic area. Because of the measuring method, we think that our determinations of the volumes of sediments within the continental block are more precise.

On the continents, about 75% of the volume of all sedimentary rocks is found in geosynclinal areas and only 25% in the platforms, their average thicknesses being 10 km and 1.8 km, respectively. These values take into account a reduction, over time, of the sedimentation regions in the geosynclines and corresponding accretion on the platforms (Figure 2).

Clay and shale are the most widespread sedimentary rocks on the continents (42%). Arenaceous, volcanic, and carbonate rocks are approximately equally abundant (20, 19, and 18%, respectively). All other rock types, mainly evaporites, comprise about 1%. The average composition of the sedimentary shell with or without volcanic rocks was calculated on the basis of the data on the volume distribution pattern of individual rock types (Table 2).

A fundamental peculiarity of sedimentary rocks is the clearly pronounced difference between their composition (Tables 2 and 5) and

TABLE 7. Abundance and Mass of Main Petrographic Types of Rocks in the Crust

Rocks	% of Total Volume of Crust	Mass, 10^{24} g
Sands	1.7	0.43
Clays and shales	4.2	1.07
Carbonates (including saltbearing deposits)	2.0	0.51
Granites	10.4	2.95
Granodiorites, diorites	11.2	3.11
Syenites, nepheline syenites	0.4	0.11
Basalts, gabbros, amphibolites, eclogites	42.5	12.70
Dunites, peridotites	0.2	0.06
Gneisses	21.4	5.96
Crystalline schists	5.1	1.41
Marbles	0.9	0.25
Total	100.0	28.56

TABLE 8. Abundance of Main Minerals in the Crust

Minerals	Modal Composition [Clarke, 1924]	Normative Composition [Fersman, 1934]	Normative Composition [Mason, 1958]	Modal Composition of Igneous Rocks [Wedepohl, 1967]	Modal Composition (this work)	Normative Composition (this work)
Quart	12.0	12.0	10.2	18	12	14
Potassic feldspar	59.5	55.0	18.35	22	12	14
Plagioclase			47.78	42	39	49 } 63
Micas	3.8	3.0		4	5	
Amphiboles	16.8	15.0		5	5	
Pyroxenes			15.09	4	11	20
Olivines				1.5	3	
Clay minerals (+ chlorites)		1.5			4.6	
Calcite (+aragonite)		1.5			1.5	
Dolomite		0.1			0.5	
Magnetite (+ titanomagnetite)	1.5	3.3	6.39	3	1.5	3
Others (garnets, kyanite, andalusite, sillimanite, apatite, etc.)	6.4	8.6	2.19	0.5	4.9	
Total	100.0	100.0	100.00	100.0	100.0	100

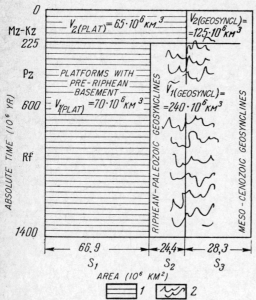

Fig. 2. Average thicknesses of (1) platform sediments and (2) geosynclinal sediments. Calculated according to the formulas:

$$h_{plat} = \frac{V_{1(plat)}}{S_1} + \frac{V_{2(plat)}}{S_1 + S_2}$$

$$h_{geosyncl} = \frac{V_{1(geosyncl)}}{S_2 + S_3} + \frac{V_{2(geosyncl)}}{S_3}$$

the average composition of rocks of the granitic shell (Tables 3 and 5), which was the chief source of sediment matter (at least for the last 1.5 billion years). This difference is reflected in a strongly heightened content of water, carbon dioxide, and organic carbon as well as of sulphur, chlorine, fluorine, boron, and other 'excess volatiles' [Rubey, 1951, 1955] in the stratisphere and the hydrosphere. This is an indication of direct release of these excess volatiles from the mantle during the process of its degassing [Vinogradov, 1959b, 1964, 1967; Goldschmidt, 1933, 1954; Rubey, 1951, 1955].

The other important peculiarity of the composition of sedimentary rocks is their high calcium content, which is a most enigmatic feature of the geochemistry of the outer shells. Very typical is the displacement of the ratio of potassium to sodium in favor of potassium which is not compensated by sodium excess in the ocean. This leads to some deficiency of sodium in the stratisphere and hydrosphere, taken together, relative to the granitic shell. This peculiarity is also a problem that requires special investigation. The oxidizing conditions on the earth's surface have determined the higher ferric to ferrous (Fe_2O_3/FeO) ratio in sedimentary rocks, as well as the heightened content of sulfate sulfur in them.

These peculiarities are most clearly manifested in platform sediments, since these are products of deep weathering and strongly developed surface differentiation. In contrast, geosynclinal sediments have undergone less extensive alteration (especially sands), and their composition approaches the composition of the parent rocks [Ronov et al., 1965, 1966].

Granitic Shell

The granitic shell is completely concentrated on the continents, and its volume and mass are approximately 3,600,000,000 km³ and 9.8×10^{24} g, respectively (Tables 3 and 5). Acidic granitoids and metamorphic rocks are the main rock types of the granitic shell; the basic and ultrabasic rocks make up less than 15% of the shell's volume. These relations determine its generally acidic chemical composition: a typical high content of silica, and concentrations of alkalis (K > Na) and of the majority of rare elements (uranium, thorium, rare earths, zirconium, niobium and others).

The average composition of the granitic shell differs considerably from that of the Neogene sedimentary rocks. This indicates that the material of the crustal granitic shell could not be a single source of the sediments; basic lava entering the geosynclines during volcanic eruptions from deeper zones of the crust and the upper mantle must have played a significant role. On the other hand, the admission of a considerable participation of old sediments in the formation of the granitic shell leads to the conclusion that a substantial amount of these sediments were changed by surface and hypogenic processes. Since the most ancient sediments appear to have been the products of weathering of more basic rocks than granites, the assumption is inevitable that the excess of bases and iron had to be moved, by weathering processes, into the oceans and pelagic sediments [Ronov, 1964]. At the same time, the deficiency of silica and alkalis in them should have been made up as a result of a fundamental reworking of sediments by processes of regional metamorphism developed at a certain stage in the open system and accompanied by the supply of solutions of silica and alkalis; according to modern concepts, this appears to be the essence of the granitization process.

Basaltic Shell

The basaltic shell consists of two parts, the continental and the oceanic, differing in structure and apparently in composition. In our opinion (see Tables 3 and 5) the basaltic shell of the continental crust is formed by deeply metamorphosed rocks of a basic and acid composition with a significant portion of magmatic rocks. The amount of the basic rocks reaches 50% of the total volume, while their composition is considered to be similar to that of geosynclinal basalts.

The continental crust has considerable thickness and a diversity in composition; the more homogeneous oceanic crust is 90% composed of original oceanic tholeiitic basalts. These basalts are characterized by a low content of potassium, rubidium, strontium, barium, phosphorus, uranium, thorium, and zirconium, and high ratios of K/Rb and Na/K [[*Engel and Engel*, 1968; *Engel et al.*, 1965], which strongly distinguish them from continental rocks and can be associated with the absence of granitization here and less intensive processes of differentiation and fusion from the mantle.

The oceanic crust is essentially characterized by the occurrence of ultrabasic rocks in the zones of deep faults (rift valleys of medial ranges) [*Vinogradov*, 1967; *Dmitriyev*, 1966; *Udintsev and Chernyshova*, 1966; *Engel and Engel*, 1968], these rocks being considered outcrops of mantle rocks. This implies a possible participation of ultrabasic rocks in the formation of the lower horizons of the oceanic crust [*Engel and Engel*, 1968]. By analogy they may also be located at the base of the basaltic layer of the continental block [*Beloussov*, 1966], but at present it is practically impossible to estimate quantitatively their contribution to the composition of the crust. Perhaps we greatly underestimate their abundance (0.2%, Table 7).

The Crust as a Whole

About 64% of the crustal volume is continental, or 79% when the subcontinental crust is included; 21% is oceanic crust. The ratio of the crustal volumes of the Riphean platforms and of the post-Riphean geosynclines is 1.8:1, whereas the crustal average thickness of the latter exceeds that of the platform by 1 km (Table 1). However, this poorly pronounced tendency is more clearly manifested by seismic refraction profiles of the platforms and geosynclines. These profiles show that the maximum thickness of platforms, as a rule, does not exceed 45 km, while that of the geosynclines reaches 75 km [*Demenitskaya*, 1961; *Kosminskaya*, 1958]. From Table 1 another tendency can also be traced, namely, the average thicknesses of the crust grow from ancient to young geosynclinal zones. This regularity is closely related to the surface relief: the highest areas belong to young folded zones (Alpine and Mesozoic), while lower regions are associated with more ancient ones (Hercynian, Caledonian, and Baikalian). A similar tendency to an increase in the thickness from ancient troughs toward those of the Cenozoic and Recent geosynclinal systems (Table 1) is also observed in the oceanic crust. The average thickness of the entire crust (including the continents and oceans) amounts to about 20 km.

The chemical composition of the crust (Table 5) as a whole approaches that of intermediate rocks, though it is impossible to find its close analog among them. In rough approximation the crust's composition can be described as a mixture of the two prevailing types of rocks: granite and basalt (geosynclinal basalt + oceanic tholeiite) at a ratio of 2:3. The average chemical composition of the crust changes with depth from the sedimentary shell to the basaltic one (Table 5), with a continuous increase in the content of iron, magnesium, alumina and a decrease in the amount of combined water (H_2O^+), whereas the content of alkalis (potassium) and silica first increases from the sedimentary shell toward the granitic, and then decreases toward the basaltic shell of the continents and oceans. These changes in the distribution of individual components through the crustal section are manifested in a regular decrease from shell to shell of the ratios K_2O/Na_2O, CaO/MgO, and F_2O_3/FeO (Figure 3) and a progressive growth of the alumina-silica modulus (Al_2O_3/SiO_2) which reflects a general tendency of the increase in basicity with depth.

The most widespread minerals in the crust are feldspars (over 50%), then quartz, pyroxenes, amphiboles, and micas (Table 8).

The comparison of the results of our calculations of the crust's mass (28.56×10^{24} g) with those obtained by *Poldervaart* [1955] (23.67×10^{24} g) reveals some differences caused by different initial data and calculation methods. Some differences are also noted in the determination of the average chemical composition of the crust; our estimation takes the intermediate place between those of *Poldervaart* [1955] and *Vinogradov* [1962b] (Table 9). The existence of different estimates is quite natural; it reflects the uncertainty of our knowledge about the composition and structure of the deep horizons of the crust. The spread in estimates is confined to the range of possible alternatives to the composition of the basaltic shell (see Table 6).

The problem of the amount of water contained in the crust and hydrosphere as a whole is complicated and unsolved, but very important for geochemistry. Our value (0.38×10^{24} g, Table 5) represents only chemically combined water usually designated in analyses as H_2O^+. Taking into consideration hygroscopic (H_2O-), interstitial, and stratal water, whose mass at present it is impossible to determine with satisfactory accuracy, the total amount of water in the crust probably should at least be doubled (about 0.8×10^{24} g). This makes up approximately half the water in the oceans equal to 1.4×10^{24} g [*Poldervaart*, 1955]. Thus, the total amount of water in the crust and oceans appears to be approximately 2.2×10^{24} g. *Poldervaart* [1955] estimated the total amount of water in the crust and hydrosphere to be in

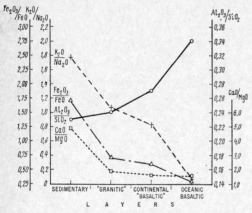

Fig. 3. Changes in the ratios of rock-forming elements versus depth in the crust.

TABLE 9. Average Chemical Composition of the Crust (wt %)

Oxides	Clarke [1924]	Goldschmidt [1954]	Vinogradov [1962b]	Taylor [1964]	Poldervaart [1955] Continental Crust	Poldervaart [1955] Entire Lithosphere	Pakiser and Robinson [1967] Continental Crust	This Work Continental Crust	This Work Entire Lithosphere
SiO_2	60.3	60.5	63.4	60.4	59.4	55.2	57.8	61.9	59.3
TiO_2	1.0	0.7	0.7	1.0	1.2	1.6	1.2	0.8	0.9
Al_2O_3	15.6	15.7	15.3	15.7	15.5	15.3	15.2	15.6	15.9
Fe_2O_3	3.2	3.3	2.5		2.3	2.8	2.3	2.6	2.5
FeO	3.8	3.5	3.7	7.2*	5.0	5.8	5.5	3.9	4.5
MnO	0.1	0.1	0.1	0.1	0.1	0.2	0.2	0.1	0.1
MgO	3.5	3.6	3.1	3.9	4.2	5.2	5.6	3.1	4.0
CaO	5.2	5.2	4.6	5.8	6.7	8.8	7.5	5.7	7.2
Na_2O	3.8	3.9	3.4	3.2	3.1	2.9	3.0	3.1	3.0
K_2O	3.2	3.2	3.0	2.5	2.3	1.9	2.0	2.9	2.4
P_2O_5	0.3	0.3	0.2	0.2	0.2	0.3	0.3	0.3	0.2
	100.0	100.0	100.0	100.0	100.0	100.0	100.0	100.0	100.0

* Fe_2O_3 and FeO given as FeO.

the range 1.8–2.7×10^{24} g; *Vinogradov* [1967] estimated 1.8×10^{24} g.

SOME CONSEQUENCES

The difference between the average chemical composition of the crust (and its separate layers) and the presumed primary composition of the mantle (silicate fraction of chondrite) leads directly to the ideas about differentiation and rising of abyssal material to the surface, resulting in the formation of the earth's outer shells including the crust [*Vinogradov*, 1959a, b, 1961, 1962a, 1964, 1967; *Beloussov*, 1966; *Urey*, 1952; *Poldervaart*, 1955; *Rubey*, 1951, 1955]. As for its composition, the crust differs from the mantle and primary (meteoritic) material by a higher concentration of silica, alkali metals, and the majority of rare and radioactive elements (uranium, thorium, rare earths, zirconium, niobium, beryllium, lithium, rubidium and others) and by a lower content of magnesium and elements of the iron group (iron, cobalt, nickel, and chromium).

Comparison of the masses of the components removed to the surface with their content in initial (meteoritic) and residual (dunitic) matter makes it possible to determine the depth of the mantle affected by differentiation [*Vinogradov*, 1959b; *Vinogradov and Yaroshevsky*, 1967]. Calculation yielded the values given in Table 10. The total mass of the material removed from the mantle into the continental (22.49×10^{24} g) and oceanic (6.07×10^{24} g) crust reflects the intensity of the processes under these structures; the differences in the degree of differentiation are presumably directly connected with its dissimilar energy sources under the continents and oceans. The values show that the mantle has been affected to different depths by the removal of different elements. To explain these differences is difficult, but it is possible to interpret them as a result of nonuniform mobility of alkalis and and other elements during zone melting of the mantle material [*Vinogradov and Yaroshevsky*, 1967].

At present we have no data to consider quantitatively the evolution of the composition and mass of the material supplied from depth to the crust. Since the bulk of the juvenile substance enters the continental crust from the mantle at the geosynclinal stage of evolution, the decrease of the total area of the geosynclines during the last 1.5 billion years undoubtedly indicates a decrease in the intensity of this process during this period.

TABLE 10. Depths Affected by Differentiation

Component	Continental and Subcontinental Crust, km	Oceanic Crust, km
Si	60	15
Al	140	30
Mg	60	10
Ca	50	40
Na	180	30
K	1,300	30

Along with predominant centrifugal tendencies in the removal of the material from the mantle, there are some traces of a reversely directed centripetal process in the movement of matter from the surface layers into the depth of the earth. This conclusion is suggested by the paradoxical fact that there is no balance between the weathering of subsurface parent rocks and sedimentation. The idea of geochemical balance suggests that the ratios and average chemical composition of the sedimentary rocks are determined by the chemical composition of the weathered material [*Mead*, 1907; *Clarke*, 1924; *Goldschmidt*, 1933; *Barth*, 1962, 1965]. The measured ratios of sedimentary rocks (clay:sands:carbonates = 53:25:22) and their average chemical composition (see Table 2) reveal, however, the absence of such balance if compared with the average compositions of the igneous rocks and granitic and basaltic shells. There is a significant surplus of calcium in sediments and some deficiency of sodium (if sodium in the ocean is taken into account). *Kuenen* [1941] and *Ronov* [1949] suggested that a part of the sedimentary mass (poor in calcium) escaped the cycle and accumulated on the bottom of the world ocean. However, the thinness of ocean sediments excludes this interpretation; our calculations indicate that the thickness of the sediments completely free of calcium on the bottom of the ocean (with its present area) should be 5 km. One must assume that a considerable amount of sediments poor in calcium was assimilated by the oceanic crust and perhaps by the underlying mantle. This assumption is supported by the fact of large tectonic subsidences, which took place at the end of the Paleozoic and in Mesozoic time in the region of the Indian and southern Atlantic oceans, and assimilation by the upper mantle of a considerable part of Gondwanaland that once existed there. In this connection it is important to note that Paleozoic and Mesozoic sedimentary series of Gondwanaland, where they have been preserved (South Africa, South America, West Australia, and India), are to a considerable extent poorer in carbonate sediments (13.6%) than the sedimentary series of Laurasia (24.8%) that includes the northern block of the continents (North America and Eurasia) [*Ronov*, 1961].

The main characteristic feature of the crust as a whole is the homogeneity of its material. A large diversity of compositions of the products of the intracrustal differentiation, which are of exceptional importance for practical human activity, does not exceed 0.5% of the total mass of the crust. They include the extreme products of magmatic differentiation (nepheline syenites, pegmatites, chromitic and titanomagnetitic ores, etc.), products of metasomatic and hydrothermal minerogenesis, in which most of the elements are concentrated. Quantitatively, all these varied and complex manifestations of differentiation are controlled in the crust by other, more powerful processes of homogenization of its material, i.e. metamorphism, granitization, and palingenesis.

We stress again our conclusion on the similarity between the average chemical compositions of metamorphic rocks and geosynclinal sediments (see Table 4 and *Ronov and Yaroshevsky* [1967]). In all probability this indicates that the processes of metamorphism in the crust take place in a closed system without significant supply of new material. At the same time a direct comparison of the composition of geosynclinal sedimentary rocks with that of the granitic shell as a whole points to a significant supply of silica and alkalis (especially potassium) during the granitization of sediments. These two processes—metamorphism of geosynclinal rocks and their subsequent granitization, associated in space and time—are, from the geochemical point of view, the principal hypogenic processes influencing the crust. The structure and composition of the earth's crust result from the supply of material and energy from the mantle and the fundamental reworking of the crustal material by surface processes.

Note added in proof. Since this article was prepared, new data have led to revised estimates of the thickness of seismic layer II [*Engel and Engel*, 1968] and of the abundance and chemical composition of evaporites (A. B. Ronov, unpublished data). Recalculation gives more precise estimates:

sedimentary shell
  sodium, 1.51%
  chlorine, 0.21%
  SO_3, 0.46%

sediments (excluding volcanics)
volume, 900×10^6 km^3
mass, 2.25×10^{24} g

These new values lead to insignificant changes in the calculated composition and mass of the components of the crust as a whole (see Table 11).

TABLE 11. Crustal Composition (revised calculations added in proof)

Component	Mass, 10^{24} g	Weight, %
SiO_2	16.414	57.60
TiO_2	0.240	0.84
Al_2O_3	4.360	15.30
Fe_2O_3	0.722	2.53
FeO	1.218	4.27
MnO	0.045	0.16
MgO	1.105	3.88
CaO	1.992	6.99
Na_2O	0.822	2.88
K_2O	0.667	2.34
P_2O_5	0.062	0.22
C	0.040	0.14
CO_2	0.399	1.40
S	0.012	0.04
Cl	0.014	0.05
H_2O^+	0.389	1.37
Total crust	28.50	

REFERENCES

Atlas of Lithological-Paleogeographical Maps of the Russian Platform and Its Geosynclinal Surroundings, vol. 1, Gosgeoltekhizdat, Moscow, 47 pp., 1961.

Atlas of Lithological-Paleogeographical Maps of the Russian Platform and Its Geosynclinal Surroundings, vol. 2, Gosgeoltekhizdat, Moscow, 48 pp., 1962.

Barth, T. F. W., Ideas on the interrelation between sedimentary and igneous rocks, Geokhimiya, 1962(4), 296-299, 1962.

Barth, T. F. W., Relationship of sodium in igneous and sedimentary rocks, in Problemy Geokhimii. Devoted to the 70th Anniversary of Academician A. P. Vinogradov, pp. 424-428, Izd. Nauka. Moscow, 1965.

Beloussov, V. V., Principal Problems in Geotectonics, Gosgeoltekhizdat. Moscow, 604 pp., 1962.

Beloussov, V. V., Earth's Crust and Upper Mantle of Continents. Izd. Nauka, Moscow, 121 pp., 1966.

Belyaevsky, N. A., A. A. Borisov, and I. S. Volvovsky, The depth structure of the USSR territory, Sov. Geol., 1967(11), 56-84, 1967.

Bezrukov, P. L., A. P. Lisitsyn, V. P. Petelin, and N. S. Skornyakova, Map of sedimentation in the world ocean, in Physical-Geographical Atlas of the World, pp. 16-17, Akad. Nauk SSSR, Moscow, 1964.

Clarke, F. W., The Data of Geochemistry, U. S. Geol. Surv. Bull. 770, 841 pp., 1924.

Daly, R. A., Igneous Rocks and the Depths of the Earth, McGraw-Hill Book Company, New York, 598 pp., 1933.

Demenitskaya, R. M., Principal Features in Structure of the Earth's Crust according to Geophysical Data, Gostoptekhizdat, Moscow, 222 pp., 1961.

Demenitskaya, R. M., Crust and Mantle of the Earth, Izd. Nedra, Moscow, 279 pp., 1967.

Dmitriyev, L. V., Chemical composition of intrusive rocks of rift valleys of the Indian Ocean, 2nd International Oceanographic Congress, Abstracts of Papers, p. 98, Nauka, Moscow, 1966.

El Wakeel, S. K., and J. P. Riley, Chemical and mineralogical studies of deep-sea sediments, Geochim. Cosmochim. Acta, 25, 110-146, 1961.

Engel, A. E. J., and C. G. Engel, Rocks of the ocean floor, in Fundamental Problems of Oceanology, 2nd International Oceanographic Congress, Plenary Lectures, pp. 183-217, Izd. Nauka, Moscow, 1968.

Engel, A. E. J., C. G. Engel, and R. G. Havens, Chemical characteristics of oceanic basalts and the upper mantle, Bull. Geol. Soc. Am., 76(7), 719-734, 1965.

Faure, G., and P. M. Hurley, The isotopic composition of strontium in oceanic and continental basalts: Application to the origin of igneous rocks, J. Petrol., 4(1), 31-50, 1963.

Fersman, A. E., Geokhimiya, vol. 1, Izbrannye Trudy, 1934.

Fersman, A. E., Geokhimiya, vol. 3, Izd. Akad. Nauk SSSR, Moscow, 1955.

Fleischer, M., and E. T. C. Chao, Some problems in the estimation of abundances of elements in the earth's crust, Intern. Geol. Congr., 21st, Report of Session, Norden, Probleme 1, 106-131, 1960.

Frolova, N. V., On the most ancient sedimentary rocks of the earth, Priroda, 1950(9), 15-21, 1950.

Frolova, N. V., Problems of stratigraphy, regional metamorphism, and granitization of Archean in South Yakut and East Siberia, Tr. Vost.-Sibirsk. Geol. Inst., Ser. Geol., 5, 13-49, 1962.

Gastil, G., The distribution of mineral dates in time and space, Am. J. Sci., 258(1), 1-35, 1960.

Goldberg, E. D., and G. O. S. Arrhenius, Chemistry of Pacific pelagic sediments, Geochim. Cosmochim. Acta, 13, 153-212, 1958.

Goldschmidt, V. M., Grundlagen der quantitativen geochemie, Fortschr. Mineral. Krist. Petrog., 17, 112-156, 1933.

Goldschmidt, V. M., Geochemistry, Clarendon Press, Oxford, 730 pp., 1954.

Grout, F. F., Petrographic and chemical data on the Canadian shield, J. Geol., 46, 486-504, 1938.

Khain, V. E., General Geotectonics, Izd. Nedra, Moscow, 479 pp., 1964.

Kolotukhina, S. E., Principal features of tectonic

development of Africa in the Precambrian, *Izv. Akad. Nauk SSSR, Ser. Geol.*, *1964*(4), 20–37, 1964.

Kosminskaya, I. P., Structure of the earth's crust according to seismic data, *Bull. Mosk. Obshchestva Ispytatelei Prirody, Otd. Geol.*, *33*(4), 25–38, 1958.

Kuenen, P. H., Geochemical calculations concerning the total mass of sediments in the earth, *Am. J. Sci.*, *239*, 161–190, 1941.

Kuenen, P. H., *Marine Geology*, John Wiley & Sons, New York, 568 pp., 1950.

Landergren, S., On the geochemistry of deep-sea sediments, *Rept. Swedish Deep-Sea Expedition*, *10*(4), 57, 1964.

Magnitsky, V. A., *Internal Structure and Physics of the Earth*, Izd. Nedra, Moscow, 379 pp. 1965.

Mason, Brian, *Principles of Geochemistry*, John Wiley & Sons, New York, 310 pp., 1958.

Mead, W. J., Redistribution of elements in the formation of sedimentary rocks, *J. Geol.*, *15*, 238–256, 1907.

Michot, P., L'anatexie leuconoritique, *Bull. Acad. Roy. Belg., Cl. Sci.* [5]*41*(3), 374–385, 1955.

Michot, P., La geologie des zones profondes de l'écorce terrestre, *Ann. Soc. Geol. Belg.*, *80*, 19–59, Oct. 1956.

Michot, P., Phénomènes géologique dans la catazone profonde, *Geol. Rundschau*, *46*(1), 147–173, 1957.

Morkovkina, V. F., *Chemical Analyses of Igneous Rocks and Rock-Forming Minerals*, Izd. Nauka, Moscow, 249 pp., 1964.

Nesterenko, G. V., N. S. Avilova, and N. P. Smirnova, Rare elements in traps of the Siberian platform, *Geokhimiya*, *1964*(10), 1015–1021, 1964.

Nockolds, S. R., Average chemical composition of some igneous rocks, *Bull. Geol. Soc. Am.*, *65*(10), 1007–1032, 1954.

Pakiser, L. C., and R. Robinson, Composition of the continental crust as estimated from seismic observations, in *The Earth beneath the Continents, Geophys. Monograph 10*, edited by J. S. Steinhart and T. J. Smith, pp. 620–626, American Geophysical Union, Washington, D. C., 1967.

Parker, R. L., Composition of the earth's crust, in *Data of Geochemistry, U. S. Geol. Survey Prof. Paper 440-D*, 17 pp., 1967.

Pavlovsky, E. V., On the specific style of tectonic development of the earth's crust in early Precambrian, *Tr. Vost.-Sibirsk. Geol. Inst., Ser. Geol.*, *5*, 77–108, 1962.,

Poldervaart, A., Chemistry of the earth's crust, *Geol. Soc. Am., Spec. Paper 62*, 119–144, 1955.

Rezanov, I. A., On the question of geological interpretation of data on deep seismic sounding, *Sov. Geol.*, *1960*(6), 65–77, 1960.

Rezanov, I. A., On the structure of the earth's crust of platform areas, *Byul. Mosk. Obshchestva Ispytatelei Prirody, Otd. Geol.*, *37*(1), 25–42, 1962.

Ronov, A. B., History of sedimentation and epeirogenic movements of the European part of the USSR (according to data of the volumetric method), *Tr. Geofiz. Inst. Akad. Nauk SSSR*, *1949*(3), 130, 1949.

Ronov, A. B., Some general laws in the development of eperirogenic movements of the continents, in *Collected Works, Problemy Tektoniki*, pp. 118–164, Moscow, 1961.

Ronov, A. B., General tendencies in evolution of composition of the earth's crust, ocean, and atmosphere, *Geokhimiya*, *1964*(8), 715–743, 1964.

Ronov, A. B., Yu. P. Girin, G. A. Kazakov, and M. N. Ilyukhin, Comparative geochemistry of geosynclinal and platform sedimentary strata, *Geokhimiya*, *1965*(8), 961–976, 1965.

Ronov, A. B., Yu. P. Girin, G. A. Kazakov, and M. N. Ilyukhin, Sedimentary differentiation in platform and geosynclinal basins, *Geokhimiya*, *1966*(7), 763–776, 1966.

Ronov, A. B., and V. E. Khain, Devonian lithological formations of the world, *Sov. Geol.*, *1954*(41), 46–76, 1954.

Ronov, A. B., and V. E. Khain, Carboniferous lithological formations of the world, *Sov. Geol.*, *1955*(48), 1955.

Ronov, A. B., and V. E. Khain, Permian lithological formations of the world, *Sov. Geol.*, *1956*(54), 20–36, 1956.

Ronov, A. B., and V. E. Khain, Triassic lithological formations of the world, *Sov. Geol.*, *1961*(1), 27–48, 1961.

Ronov, A. B., and V. E. Khain, Jurassic lithological formations of the world, *Sov. Geol.*, *1962*(1), 9–34, 1962.

Ronov, A. B., and Z. V. Khlebnikova, Chemical composition of main genetic clay types, *Geokhimiya*, *1957*(6), 527–552, 1967.

Ronov, A. B., and A. A. Migdisov, On the relationship between normal clark and ore concentrations of aluminum in the depositional cycle, *Reports of Soviet Geologists, 21st Session of the Intern. Geol. Congress*, Problem 1, pp. 157–177, Gosgeoltekhizdat, Moscow, 1965.

Ronov, A. B., and A. A. Migdisov, Principal features of geochemistry of hydrolysate elements in weathering and sedimentation, *Geokhimiya*, *1965*(2), 131–158, 1965.

Ronov, A. B., M. S. Mikhailovskaya, and I. I. Solodkova, Evolution of the chemical and mineralogical composition of arenaceous rocks, in *Khimiya Zemnoi Kory (Proceedings of the Geochemical Conference Devoted to the 100th Anniversary of V. I. Vernadsky)*, vol. 1, 201–252, Izd. Akad. Nauk SSSR, 1963.

Ronov, A. B., and A. A. Yaroshevsky, Chemical composition of the earth's crust, *Geokhimiya*, *1967*(11), 1285–1309, 1967.

Rubey, W. W., Geologic history of sea water: An attempt to state the problem, *Bull. Geol. Soc. Am.*, *62*, 1111–1147, 1951.

Rubey, W. W., Development of the hydrosphere and atmosphere with special reference to probable composition of the early atmosphere, *Geol. Soc. Am. Spec. Paper 62*, 631–650, 1955.

Sederholm, J. J., The average composition of the earth's crust in Finland, *Comm. Geol. Finlande Bull.*, *12*, 70, 1925.

Semikhatov, M. A., On the problem of the Proterozoic, *Izv. Akad. Nauk SSSR, Ser. Geol.*, *1964*(2), 66–84, 1964.

Shaw, D. M., G. A. Reilly, J. R. Muysson, G. E. Pattenden, and F. E. Campbell, An estimate of the chemical composition of the Canadian Precambrian shield, *Can. J. Earth. Sci.*, *4*, 829–853, 1967.

Soloviev, S. P., *Distribution of Magmatic Rocks in the USSR*, Geogeltekhizdat, 215 pp., 1952.

Soloviev, S. P., Some problems on chemism of igneous rock, *Zap. Vses. Mineralog. Obshchestva*, *93*(6), 613–640, 1964.

Soloviev, S. P., Main chemical peculiarities of basic magmatic rocks of the USSR, *Zap. Vses. Mineralog. Obshchestva*, *94*(6), 625–641, 1965.

Stewart, F. H., Marine evaporites, in *Data of Geochemistry, U. S. Geol. Surv. Prof. Paper 440-Y*, 1963.

Struve, E. A., *Collected Analyses of Igneous and Metamorphic Rocks of the USSR*, Izd. Akad. Nauk SSSR, Moscow, 540 pp., 1940.

Taylor, S. R., Abundance of chemical elements in the continental crust: A new table, *Geochim. Cosmochim. Acta*, *28*(8), 1273–1285, 1964.

Tugarinov, A. I., Epochs of mineral formation in Precambrian, *Izv. Akad. Nauk SSSR, Ser. Geol.*, *1956*(9), 3–26, 1956.

Tugarinov, A. I., and G. V. Voitkevich, *The Precambrian Geochronology of the Continents*, Izd. Nedra, Moscow, 385 pp., 1966.

Tugolesov, D. A., and G. B. Udintsev, compilers; M. V. Muratov and A. L. Yanshin, editors; Tectonic regions of continents (map), in *Physical-Geographical Atlas of the World*, pp. 10–11, Akad. Nauk SSSR, Moscow, 1964.

Udintsev, G. B., and V. I. Chernyshova, Upper mantle rocks from the rift zone of the Indian Ocean, *2nd International Oceanographic Congress, Abstracts of Papers*, pp. 375–376, Nauka, Moscow, 1966.

Urey, H. C., *The Planets, Their Origin and Development*, Yale University Press, New Haven, 245 pp., 1952.

Vinogradov, A. P., Distribution patterns of chemical elements in the earth's crust, *Geokhimiya*, *1956*(1), 1–43, 1956.

Vinogradov, A. P., Meteorites and the earth's crust, *Izv. Akad. Nauk SSSR, Ser. Geol.*, *1959*(10), 5–27, 1959a.

Vinogradov, A. P., *Chemical Evolution of the Earth: 1st Lecture Dedicated to V. I. Vernadsky*, Izd. Akad. Nauk SSSR, Moscow, 43 pp., 1959b.

Vinogradov, A. P., The origin of the material of the earth's crust, 1, *Geokhimiya*, *1961*(1), 1–32, 1961.

Vinogradov, A. P., Origin of the earth's shells, *Izv. Akad. Nauk SSSR, Ser. Geol.*, *1962*(11), 3–17, 1962a.

Vinogradov, A. P., Average content of chemical elements in main types of igneous rocks of the earth's crust, *Geokhimiya*, *1962*(7), 555–571, 1962b.

Vinogradov, A. P., Gaseous regime of the earth, in *Khimiya Zemnoi Kory, Proceedings of the Geochemical Conference Devoted to the 100th Anniversary of V. I. Vernadsky*, vol. 2, pp. 5–21, Izd. Nauka, Moscow, 1964.

Vinogradov, A. P., The formation of the ocean, *Izv. Akad. Nauk SSSR, Ser. Geol.*, *1967*(4), 3–9, 1967.

Vinogradov, A. P., and A. B. Ronov, Composition of sedimentary rocks of the Russian platform in relation to the history of its tectonic movements, *Geokhimiya*, *1956*(6), 533–559, 1956.

Vinogradov, A. P., and A. I. Tugarinov, Geochronology of Precambrian, *Geokhimiya*, *1961*(9), 723–731, 1961.

Vinogradov, A. P., and A. A. Yaroshevsky, Further investigation of the differentiation mechanism of the earth's mantle: The problem of heat-mass transfer in connection with zone melting in the mantle, IUGG General Assembly, 14th, Abstracts of Papers, IIa, A-2, 1967.

Walker, F., and A. Poldervaart, Karroo dolerites of the Union of South Africa, *Bull. Geol. Soc. Am.*, *60*, 591–706, 1949.

Wedepohl, K. H., *Geochemie*, Walter de Gruyter and Company, Berlin, 220 pp., 1967.

Wilson, J. T., The origin of continents and Precambrian history, *Trans. Roy. Soc. Can.*, [3] *43*(3), 1949.

Wilson, J. T., Some consideration regarding geochronology with special reference to Precambrian time, *Trans. Am. Geophys. Union*, *33*(2), 1952.

Zverev, S. M., V. M. Kovylin, and G. B. Udintsev, Thickness of bottom sediments in the ocean, in *Sovremennye osadki morei i okeanov*, pp. 292–316, Izd. Akad. Nauk SSSR, 1961.

Isotope Geochemistry of Crust-Mantle Processes

Stanley R. Hart

THE general field of isotope geochemistry encompasses a wide spectrum of research and viewpoints; from these I will try to highlight only a few of the currently interesting problems and controversies bearing on crust-mantle interactions. Fortunately, there are a number of fine review papers covering many aspects of this field [*Armstrong*, 1968; *Doe*, 1967; *Gast*, 1967; *Kanasewich*, 1968; *Wasserburg*, 1966]. *Doe and Marvin* [1967] list 365 references for the period 1963–1966 relating to isotopic research in North America alone.

Isotopic research can be classified in two groups:

Geochronologic studies: Use of radioactive age determinations for evaluation of time relationships on a regional basis, as in orogenic belts; development of an absolute time scale, usually related to tectonic markers; study of the time evolution of the continents and ocean basins, and the related questions of continental growth, continental drift, and global relationships of tectonic features.

Isotope tracer studies: Use of radiogenic daughter product (strontium and lead) abundances as 'fingerprints' to follow the derivation and differentiation of igneous rocks and to trace the evolution of mantle and crust through geologic time.

The analytical methods depend upon several radioactive decay schemes (see Table 1). Measurements can be made on whole rocks as well as individual minerals. A brief review of the applicability of measurements to whole rocks and various minerals, as well as the corresponding uncertainties, is given by *Hart* [1966]; a more detailed discussion appears in the review by *Wetherill and Tilton* [1968].

Geochronologic Approach

While development and understanding of techniques are still proceeding, it is possible now to determine ages for most types of rock with useful precision. Many of the techniques are laborious; as for example, separating and analyzing zircons from basic rocks. Future improvements in technique may be aimed more at efficiency than accuracy. One important exception is the problem of dating the abyssal basalts of the ocean basins. The areal distribution of ages of the oceanic crust is crucial to the concept of sea-floor spreading. Because of the young age of these rocks, the Rb-Sr and U-Pb dating techniques are of little use. K-Ar dating of individual minerals from abyssal basalts is hampered by the difficult mineral separation problem and the low potassium content of these minerals. Whole-rock K-Ar dating is complicated by the alteration of the fine-grained material and the possible contamination by sea water alkalies. Demonstrably fresh material, such as unaltered volcanic glass, will frequently have initial argon trapped or quenched-in, thus producing anomalously old ages [*Dalrymple and Moore*, 1968]. Deep-water drilling in the oceans will probably provide samples of oceanic crust before there are reliable techniques available for dating them.

At present the delineation of age patterns on the continents is well advanced and for some continents is fairly complete and comprehensive. These are prime data necessary for resolution of the question of continental growth or accretion. The age data from North America, which define belts of progressively decreasing age in most directions toward the continental margins, have provided strong support for the concept of continental accretion. Now, however, with the abundant evidence substantiating continental mobility, one has to consider continental age patterns in context with those on other continents. The remarkable contribution of *Hurley et al.* [1967] shows that tectonic belts of two ages are perfectly aligned when the continents of South America and Africa are reconstructed into a single continent.

TABLE 1. Commonly Used Radioactive Decay Schemes

Parent	Daughter	Half-life, years	Probable Uncertainty, %	Isotope Abundance, at. %
K^{40}	Ca^{40} (β decay)	1.47×10^9	5	0.0119
	Ar^{40} (K capture)	1.19×10^{10}	3	
Rb^{87}	Sr^{87}	$4.7, 5.0 \times 10^{10}$	5	27.8
Th^{232}	Pb^{208}	1.39×10^{10}	2	100
U^{235}	Pb^{207}	7.13×10^8	1–2%	0.720
U^{238}	Pb^{206}	4.51×10^9	<1	99.27

With complete geochronologic data available for all the continents, a simple and unambiguous reconstruction may be possible. We should consider the possibility of older periods of continental mobility, such as in the Precambrian, which will considerably complicate the task of unraveling patterns of continental growth or regeneration.

Delineation of continental age patterns is but a first step toward an understanding of continental evolution (and mantle evolution as well, for the mantle is not an infinite reservoir when supply of crustal elements in large quantities is considered). *Hurley et al.* [1962] show that, for North America, the area of surface crust of a given age is approximately constant throughout geologic time, and the simplest interpretation of this is one of continuous continental growth. However, since most radioactive clocks are started by a melting-crystallization event, it is also possible that old crust has simply been regenerated by a fusion cycle to look like new crust. Age determinations by themselves may not be able to distinguish these alternatives. Here the use of radiogenic daughter product abundances as natural tracers is a very powerful tool.

Isotope Tracer Approach

A typical sialic shield area will have a Rb/Sr ratio of about 0.25. In 2.7 b.y., this rubidium will produce strontium 87 in an abundance such that the ratio of total Sr^{87} to a common nonradiogenic strontium isotope (such as Sr^{86}) will be $Sr^{87}/Sr^{86} \sim 0.725$. This Sr^{87}/Sr^{86} ratio of 0.725 will not be altered during any process in which old crust is regenerated into 'new' crust. Even if the lower crust, which may have a lower Rb/Sr and Sr^{87}/Sr^{86} ratio, is included in this process, the overall crustal system will still be tagged with $Sr^{87}/Sr^{86} > 0.71$. Since the igneous rocks of young orogenic belts usually have Sr^{87}/Sr^{86} ratios less than 0.71, they cannot be formed by the simple closed-system regeneration of older crust. For a closed-system model, these arguments, as presented originally by *Hurley et al.* [1962], appear as compelling evidence for continuous continental growth.

The case against continuous continental growth, based on lead isotope arguments, has been presented by *Patterson* [1964] and *Patterson and Tatsumoto* [1964]. They conclude that continent formation was confined largely to the period 2.5–3.5 b.y. ago, and that the presence of crust of younger age is explained by periodic regenerations of this initial crust. The radiogenic strontium in this regenerated crust is presumed to equilibrate with a large deeply buried reservoir of primitive (nonradiogenic) strontium; the model is thus an open-system model, at least with respect to strontium.

The Patterson-Tatsumoto model is based on a comparison of model lead ages and Rb-Sr ages of feldspar composites from various beach and river sediments. For their 2.5-b.y. feldspar composite, the ages were in agreement, but successively younger composites showed larger and larger discordances between the two ages, the lead isotope ages always being younger. This 'K-feldspar lead isotope aberration' was interpreted in terms of a failure of the usual homogeneous closed-system lead model, but was shown to be eliminated by use of a multistage lead evolution model in which a crustal system of higher U/Pb ratio was separated from the mantle system at an early time (2.5–3.5 b.y.).

This concept of an early formation of the continental crust is not supported, however, by recent lead isotope work on oceanic and continental igneous rock.

First, feldspars have been shown to incorporate radiogenic lead from their surroundings during metamorphic episodes, whereas the Rb-Sr ages are less affected [*Doe et al.*, 1965]. This would produce a K-feldspar lead isotope aberration. Secondly, feldspars of 2.7-b.y. age show a reversed lead isotope aberration, the model ages being considerably older than the Rb-Sr ages [*Tilton and Steiger*, 1965]. This cannot be explained by the early crustal generation model. Finally, there is evidence now to suggest that the crust and oceanic mantle do not have grossly dissimilar U/Pb ratios or lead isotope ratios. The present weathering crust, as deduced from pelagic sediment leads [*Chow and Patterson*, 1962], has lead isotope ratios that are completely embraced by the leads from oceanic volcanic rocks [*Gast et al.*, 1964; *Tatsumoto*, 1966]. In other words, new crust derived from an oceanic-type mantle could not be distinguished from crust formed by regeneration of average surface sialic crust (though there are areas such as the Superior province shield for which the average lead is more radiogenic than any of the oceanic leads). The oceanic volcanic leads appear to require a multistage history for their mantle source region, the latest separation occurring 1–2 b.y. ago [*Tatsumoto*, 1966; *Gast*, 1967]; this is further indication of the similarity between continental and oceanic leads.

Recently an evolutionary model attempting to reconcile the strontium and lead isotope data has been proposed by *Armstrong* [1968]. The part of this model dealing with continental evolution is essentially similar to the Patterson-Tatsumoto lead model: the crust and mantle were almost completely differentiated before 2.5 b.y., and the crust and mantle periodically mixed, allowing equilibration of radiogenic crustal strontium with the large mantle reservoir of nonradiogenic strontium. So far no realistic mechanism has been advanced that could produce such large-scale equilibration of strontium.

Perhaps the resolution of some of these controversies lies in more complete knowledge of the way in which strontium and lead isotope ratios vary throughout geologic time. A great many data exist for lead from ore deposits of different ages, but almost all the lead data for major rock units (e.g., volcanic and plutonic rocks) are on relatively young material. A good start was made by *Hedge and Walthall* [1963] in tracing strontium isotope evolution trends, but the body of data is still insufficient to define these with any certainty. A schematic representation of possible evolutionary paths for a given region, say the upper mantle, is given in Figure 1. The common starting point (taken as 'primordial' or 'initial' strontium of meteorites) is at $Sr^{87}/Sr^{86} \sim 0.699$. The end point is taken as $Sr^{87}/Sr^{86} \sim 0.704$, the value of average oceanic basalts.

Model I. The mantle starts with the Rb/Sr ratio of chondrites; an initial rapid rise of Sr^{87}/Sr^{86} is followed shortly (point b) by ~ 1 b.y. of differentiation with upward loss of rubidium from the mantle. The resultant low Rb/Sr ratio of mantle produces only a very slow increase in Sr^{87}/Sr^{86} for the rest of the earth's history. This is essentially an early crustal differentiation model.

Model II. There is a very early major differentiation of earth (point a) and establishment of an upper mantle with a Rb/Sr ratio ~ 0.04. Continental evolution at a constant rate starts at about 3.5 b.y. (point c) and continuously depletes rubidium from the upper mantle.

Model III. There is very early major differentiation of the earth (point a); the Rb/Sr ratio of the upper mantle is established at about 1/10th the value of chondrites. The upper mantle is not further differentiated. This is an infinite reservoir model, as extraction of crustal material produces no change in the mantle Rb/Sr ratio. In terms of the material arriving in the crust, it is as if new untapped mantle is being sampled for each crustal addition.

Model IV. The earth starts with a low Rb/Sr ratio, like that in achondrites. The Rb/Sr ratio of the upper mantle increases as rubidium is added from below by differentiation. Crustal evolution takes place simultaneously.

Obviously these models are not all equally plausible when other restraints such as thermal history and major element chemistry are considered; other more complicated models can be easily constructed. These four basic models

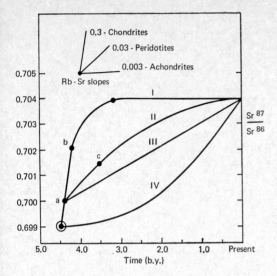

Fig. 1. Strontium evolution diagram showing possible trends of various mantle evolution models.

serve to illustrate the possible use of strontium isotope ratios in studying mantle-crust evolution. Though the total change in the Sr isotope ratio is little more than ½%, present analytical techniques are adequate. The real difficulty is the need to analyze many samples in order to average out the geographic and geologic variations between samples. For example, young oceanic basalts show a range in Sr^{87}/Sr^{86} ratio from 0.702 to 0.705, clear indication that the mantle source regions of these volcanics are heterogeneous areally or with depth. It is of course necessary to identify in the older crust those rocks that can be presumed to be derived from similar mantle source regions, though at different times. It seems probable that the large layered basic intrusives are one such group of rocks. Another group might be the volcanic rocks of the orogenic belts, which are similar to present island-arc and continental margin volcanics.

The sparse data at hand appear incompatible with the two extreme models, I and IV. The distinction between the other two models will be more difficult. A similar approach can be made using whole-rock lead istotope-U/Pb ratios, though we do not have the decade of experience that lies behind the whole-rock strontium studies.

REFERENCES

Armstrong, R. L., A model for the evolution of strontium and lead isotopes in a dynamic earth, *Rev. Geophys.*, *6*, 175–199, 1968.

Chow, T. J., and C. C. Patterson, The occurrence and significance of lead isotopes in pelagic sediments, *Geochim. Cosmochim. Acta*, *26*, 263–308, 1962.

Dalrymple, G. B., and J. G. Moore, Argon 40: Excess in submarine pillow basalts from Kilauea volcano, Hawaii, *Science*, *161*, 1132–1135, 1968.

Doe, B. R., The bearing of lead isotopes on the source of granitic magma, *J. Petrol.*, *8*, 51–83, 1967.

Doe, B. R., and R. F. Marvin, Radioactive and radiogenic isotope research in North America, *Trans. Am. Geophys. Union*, *48*, 672–686, 1967.

Doe, B. R., G. R. Tilton, and C. A. Hopson, Lead isotopes in feldspars from selected granitic rocks associated with regional metamorphism, *J. Geophys. Res.*, *70*, 1947–1968, 1965.

Gast, P. W., Isotope geochemistry of volcanic rocks, in *Basalts*, pp. 325–358, Interscience Publishers, New York, 1967.

Gast, P. W., G. R. Tilton, and C. Hedge, Isotopic composition of lead and strontium from Ascension and Gough Islands, *Science*, *145*, 1181–1185, 1964.

Hart, S. R., Current status of radioactive age determination methods, *Trans. Am. Geophys. Union*, *47*, 280–286, 1966.

Hedge, C. E., and F. G. Walthall, Radiogenic strontium 87 as an index of geologic processes, *Science*, *140*, 1214–1217, 1963.

Hurley, P. M., F. F. M. de Almeida, G. C. Melcher, U. G. Cordani, J. R. Rand, K. Kawashita, P. Vandoros, W. H. Pinson, and H. W. Fairbairn, Test of continental drift by comparison of radiometric ages, *Science*, *157*, 495–500, 1967.

Hurley, P. M., H. Hughes, G. Faure, H. W. Fairbairn, and W. H. Pinson, Radiogenic strontium 87 model of continent formation, *J. Geophys. Res.*, *67*, 5315–5334, 1962.

Kanasewich, E. R., The interpretation of lead isotopes and their geological significance, in *Radiometric Dating for Geologists*, pp. 147–223, Interscience Publishers, New York, 1968.

Patterson, C., Characteristics of lead isotope evolution on a continental scale in the earth, in *Isotopic and Cosmic Chemistry*, pp. 244–268, North-Holland Publishing Company, Amsterdam, 1964.

Patterson, C., and M. Tatsumoto, The significance of lead isotopes in detrital feldspar with respect to chemical differentiation within the earth's mantle, *Geochim. Cosmochim. Acta*, *28*, 1–22, 1964.

Tatsumoto, M., Genetic relations of oceanic basalts as indicated by lead isotopes, *Science, 153,* 1088–1093, 1966.

Tilton, G. R., and R. H. Steiger, Lead isotopes and the age of the earth, *Science, 150,* 1805–1808, 1965.

Wasserburg, G. J., Geochronology and isotopic data bearing on development of the continental crust, in *Advances in Earth Science,* pp. 431–459, MIT Press, Boston, 1966.

Wetherill, G. W., and G. R. Tilton, Geochronology, in *Researches in Geochemistry,* vol. 2, edited by P. H. Abelson, pp. 1–28, John Wiley & Sons, New York, 1968.

2. *Heat Flow*

Thermal History of the Earth

E. A. Lubimova

THE classical method of estimating temperature distribution in the deep interior of the earth is based on solving the equation of heat conduction with (1) assumed distribution of heat sources versus depth and (2) assumed initial and boundary conditions. All modern models of thermal balance of the earth are based on the radioactive decay of the long-lived isotopes of uranium, thorium, and potassium. Other possible sources, such as release of energy resulting from gravitational reorganization during the separation of the iron core, or the heat caused by tidal friction, have been recognized in calculations of the thermal history of planetary bodies. An extensive discussion of the thermal history of the earth is given by *Birch* [1965] and *Lubimova* [1967a].

The thermal state of the earth was discussed by *Kelvin* [1897], who made an attempt to estimate the age of the earth using thermal calculations. *Strutt* [1906], *Holmes* [1915], *Adams* [1924], *Tykhonov* [1937], and *Jeffreys* [1959] regarded radioactive heat as important for heat flow. *Leibenson* [1939a] calculated folding properties from the hypothesis of contraction; *Slichter* [1941] considered the influence of thickness of the radioactive layer on the value of terrestrial heat flow. *Urry* [1949] showed that heat flow decreased due to radioactive decay in time. *Jacobs* [1956] suggested a plausible mechanism for the formation of the solid internal core of the earth in the process of its secular cooling. *Jacobs and Allan* [1954] calculated the thermal history of the earth from its initial hot state; their calculations have shown that a small percentage of radioactive elements occurring at great depth, where cooling can be neglected, is quite sufficient to bring the mantle into a state of new melting. *Verhoogen* [1960] doubted that part of the mantle could be molten at present because of the high elasticity.

Distribution of initial temperature is given either in the form of the melting point curve or in the form of a curve obtained from considering the formation of the earth by accretion of particles.

In the latter case, initial temperature is assumed to be lower than the melting point, and the main differences between the existing thermal models results from the fact that neither the chemical composition of the mantle and the core nor the heat transfer processes are adequately known. Several thermal models of the earth with many fluctuating parameters, such as generation of heat, thermal conductivity factors, and initial temperature, have been subjected to thorough analysis [*Clark*, 1961; *MacDonald*, 1959]. The development of this problem suggests that both the curves of internal temperature distribution and the value of the surface heat flow change greatly with variations in the distribution of radioactivity.

The effect of upward differentiation during the history of the earth was taken into account by *MacDonald* [1959] and *Lubimova* [1956] for models of discontinuous migration, and by *Lee* [1967], *Mayeva* [1967], and *Fricker et al.* [1967] for continuous migration. Calculations show that development of heat flow for different geological periods must be different and depends on the velocity of the concentration of sources into the crust and on the degree of stabilization of the crust. This leads to the conclusion that the thermal evolution of the earth was not uniform for the sections of crust of different age and for the layers of the upper mantle underlying them.

Beloussov [1966] is of the opinion that the processes taking place in the earth's upper mantle should leave traces in the form of peculiarities in the structure of the tectonosphere. These peculiarities may be observable as variations of heat flow.

Indeed, experimental data on heat flow, particularly those concerning ancient crystalline shields on the one hand, and the regions of Alpine folding on the other, seem to indicate that regional background values of heat flow depend on the age of tectogenesis. If tectogenesis depends on inflow of sialic material from the mantle, we can expect a decrease in the heat flow with geologic time in the zones with an almost stabilized crust (of the Pre-Cambrian type).

There is not yet a satisfactory theory of the thermal history of the earth based on convective transfer of heat. Convection covering the whole of the mantle (its viscosity being $>10^{24}$ poise) seems to be doubtful [*Knopoff*, 1964; *Magnitsky*, 1965; *McConnell*, 1965].

DATA ON HEAT GENERATION

For the chondrite model, the abundance of radioactive isotopes in the earth or its mantle is assumed to be, on the average, equal to the occurrence in chondritic meteorites [*Birch*, 1964, 1965]. This model has been popular for thermal calculations. Though various authors usually suggest different combinations of radioactive elements [*Lubimova*, 1956a, 1958; *MacDonald*, 1959; *Levin and Mayeva*, 1960; *Fujii and Uyeda*, 1966; *Lee*, 1967; *Reynolds et al.*, 1966], the total content of radioactive elements is approximately that associated with chondrites.

The following arguments are given to support the chondrite model: the occurrence of chemical elements is about the same in chondrites and in the solar system, and the compositions of various chondrites are almost identical. Thus an ordinary medium chondrite can be assumed to be a sample of undifferentiated earth matter. In fact, the internal generation of heat according to the chondrite model is almost equal to the heat flow observed on the surface [*Birch*, 1965].

The argument against the model [*Gast*, 1960; *Wasserburg et al.*, 1964] is the fact that the K/U ratio for chondrites is close to 8×10^4, in spite of great difference in concentration of each element, while the value K/U = 10^4 is more typical of terrestrial rocks. *Heier and Rogers* [1963] found that the arithmetical mean ratio K/U for 755 samples of granites is equal to 0.8×10^4, the maximum value being 10^4; for basalt the ratio is 1.7×10^4.

In the rocks available for investigation, potassium and uranium are usually found together, though their correlation is different. Thus, assuming that uranium, in contrast to potassium, would mostly migrate from the mantle into the crust, which can be expressed by the equation $K/U = 10^4$, one would expect the ratio K/U to be still higher for the rocks of the mantle after such a differentiation than for chondrites. The only solution to this difficult problem is found by assuming a concentration of potassium in the lower mantle, whereas about 80% of uranium must be concentrated within the upper 400 km of the mantle, and 85% of potassium deeper than 400 km. The mechanism of such a differentiation is not clear. Because of the short period of its half-life, potassium is given a special place in the theory of the thermal history of the earth. There is a great difference in the relative contribution of K^{40} isotopes to heat generation in the chondrite model and in earth materials with the ratio $K/U = 8 \times 10^4$: in chondrites, 59% of the heat generation at the present time; in rocks with the ratio $K/U = 10^4$, only 16%.

If the amount of potassium is somewhat decreased and that of uranium increased so that their ratio K/U is close to 10^4, with concentrations $U = 2.25 \times 10^{-8}$ g/g, $Th = 8.3 \times 10^{-8}$ g/g, $K = 2.25 \times 10^{-4}$ g/g ($Th/U = 3.7$), a 'Wasserburg mixture' [*Wasserburg et al.*, 1964] will be obtained, according to which heat generation in the mantle corresponds at present to that resulting from chondrite, as obtained by recent measurements. *Lubimova's* [1958] C concentration decreased by half (and designated $C_{1/2}$) is approximately equal to the Wasserburg mixture: $C_{1/2}$ is $U = 2.5 \times 10^{-8}$ g/g, $Th = 10.5 \times 10^{-8}$ g/g ($K/U = 1.37 \times 10^4$, $Th/U = 4$).

In this case, heat generation for 4.5×10^9 years ago must be equal to only 63% of the heat generation for chondrites. This lower heat generation is attributed to the shorter half-life of K^{40} (1.3×10^9 years) in comparison with longer half-lives of U^{238} (4.5×10^9 years) and Th^{232} (13.7×10^9 years). According to the chondrite model, during the first 2×10^9 years of the earth's existence, almost the entire heat effect was due to the very intensive generation of heat from potassium and U^{235}. In the model where a smaller amount of potassium was dominated by uranium and thorium, heat liberation must have been more uniform in the history of the earth. The relationship between the generation value 4.5×10^9 years ago and that of the present time in chondrites is equal to 8.2, and for terrestrial matter according to *Wasserburg et al.* [1964] the value is as low as 4.5 (Figure 1). Thus, the difference in the ratio K/U in chondrites and terrestrial rocks suggests that either the chondrite model is unacceptable for describing the thermal history of the earth, or else potassium has separated from the upper layers of the earth; in the latter case potassium (as well as rubidium and chemically similar cesium) must be buried deeper in the earth's mantle than uranium (and chemically similar strontium and barium).

The phase transformations of the minerals containing potassium have recently been investigated by *Markov et al.* [1968]. According to these authors, potassium sodium feldspar ($Na_2K_8AlSi_3O_8$) disintegrates into the minerals jadeite, hydrosanidine, and coesite under pressure of 80 kb at a temperature of 1500°C.

Alkali elements react by redistribution in two phases having different density. The less dense newly formed mineral must rise into upper layers, often accompanied by potassium. This explains why potassium is concentrated in the upper layers; such an explanation, however, increases the difficulties presented by the chondrite hypothesis. The mechanism of potassium differentiation into the lower mantle being not very clear, it seems reasonable to give up the chondrite model in favor of an Orgueil meteorite, or some other model. Indeed, the K/U ratio is much closer to 1×10^4 in achondrites (heavily calcined meteoric stone) than in chondrites, but the uranium and thorium

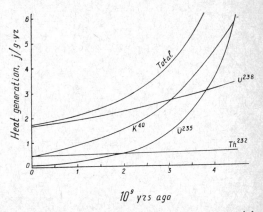

Fig. 1. Heat production by important terrestrial isotopes in geological time [after *Lee*, 1967].

concentrations in achondrites are much higher than in chondrites.

One more model based on actual isotope ratios of Th/U and K/U in the earth's crust was suggested by *Wasserburg et al.* [1964]. The uranium concentration in the mantle accepted for this model, estimated by *Birch* [1965], is true for the entire earth. Assuming that radioactive elements have been withdrawn from the central core into the mantle and the crust, one can obtain the value:

$$U = 3.3 \times 10^{-8} \text{ g/g}$$

The abundance of uranium, thorium, and potassium in chondrites and Orgueil meteorite and terrestrial rocks is shown in Table 2.

Lovering and Morgan [1964] found that the carbonaceous chondrite meteorites (Orgueil) satisfy the terrestrial correlation K/U and provide for the amount of heat flow 1.2×10^{-6} cal/cm^2 sec better than does any other material.

Birch [1965] compared the generation of heat that might be supplied by the matter assumed to be present in the mantle with the thermal flow ascribed to the entire earth or to its mantle only. This comparison indicates that the upper mantle cannot contain pure peridotites or enstatite chondrite only. Heat generation in peridotites is too low to account for the heat flow in the oceanic areas if the suboceanic mantle is made up exclusively of pure peridotites.

Ultrabasic rocks reveal a wide range of radioactivity [*Tilton and Reed*, 1963; *Lovering and Morgan*, 1964]. From the known potassium concentrations and given the K/U ratio, *Heier* [1963] estimated the average content of these elements in different rocks. These estimates indicated that the content of uranium, thorium, and potassium in the mantle must be much higher than in peridotites. Regarding Wasserburg's mixture as an initial abundance of uranium, thorium, and potassium, *Birch* [1965] calculated that a typical continental crust must contain 70% of the radioactivity beneath the continental area; the remaining 30% must be in the mantle; thus the following concentrations in the entire mantle are estimated: $U = 0.05 \times 10^{-6}$ g/g; $K = 0.05\%$. *Birch* [1965] suggests the probability of even greater concentrations of uranium, thorium, and potassium in the upper part (400 km) of the mantle: $U = 0.1-0.2 \times 10^{-6}$ g/g; $K = 0.1-0.2\%$.

Ringwood [1962] proposed the name 'pyrolite' for the mantle rock to emphasize that this is a very special kind of peridotite that contains the small but essential amounts of alkalis, alumina, and lime required to yield basalt on fractional fusion (see Table 1, p. 7). *Clark and Ringwood* [1964] emphasize that one feature of the pyrolite model is the removal of the basaltic fraction of the pyrolite from the mantle through magmatic processes leaving residual dunite or peridotite. Beneath the oceans this zone of residual material is thin or absent. Beneath normal continents its thickness may be of the order of 100 km, and beneath Precambrian shields it is perhaps twice as thick.

Clark and Ringwood [1964] propose a concentration of 0.13 ppm of uranium for pyrolite immediately beneath the M discontinuity in the oceanic model. The concentration decreases linearly to 0.03 ppm at 400 km, below which it is small. A variation of potassium with depth is assumed similar to that of uranium, with a mean potassium content of 0.13%. Accordingly, the pyrolite immediately beneath the M discontinuity is taken to contain 0.22% potassium; it decreases linearly to 0.05% at 400 km.

Masuda [1965] assumed that the earth encountered a complete melting at its primeval age; the mantle grew upwards from the bottom through the fixation of crystals segregated from the melt and the partition coefficients for minor elements between the melt, and only the forming crystals were approximately constant for the process of the mantle growth. The distribution of radioactive heat sources was estimated as a function of the depth within the earth.

If the abundance in the earth's primitive material is given, it is possible to evaluate the partition coefficients by comparison of it with

TABLE 1. Radioactive Heat Production as a Function of the Depth of the Partition Coefficient for Potassium 0.18 [after *Masuda*, 1965]

Depth, km	Heat Production, ergs/g year
2900	0.171
1690	0.240
500	0.660
103	2.55
37	6.59
(37)	49.6

TABLE 2. Isotope Abundance and Heat Generation in Various Models of the Mantle*

Pattern	Isotope Abundance, 10^{-9} g/g				Heat Generation, 10^{-1} erg/g year				
	U^{238}	U^{235}	Th^{232}	K^{40}	U^{238}	U^{235}	Th^{235}	K^{40}	Total
Orgueil									
Present	46.6	0.34	120	71	13.9	0.6	10.0	6.3	30.8
Initial†	93.2	26.70	150	843	27.7	48.0	12.6	74.0	162.3
Wasserburg									
Present	32.8	0.24	122	39	9.8	0.4	10.2	3.5	23.9
Initial	65.6	18.79	153	464	19.5	33.8	12.8	40.7	106.8
Ordinary chondrites									
Present	18.9	0.14	58	146	5.6	0.3	4.9	12.8	23.6
Initial	37.8	10.81	73	1726	11.2	19.5	6.1	151.7	188.5

* After *Lee* [1967].
† Initial time, 4.5×10^9 years ago.

average abundance in basalts. The radioactive heat production as a function of the depth is given in Table 1. According to the calculation, the total heat source for mantle plus crust is 58–66% of that calculated for the perfect chondritic earth model, and 18.5–22% of total heat source is reserved below 600 km in the mantle.

Uranium content in the mantle after differentiation according to *Levin and Mayeva* [1960] is 2×10^{-8} g/g.

A crucial problem in the theory of the thermal history of the earth, which has recently attracted the attention and efforts of many investigators, is how did the uniform distribution of sources according to the Orgueil or Wasserburg or chondrite schemes change into the observable distribution, as shown in Table 2. The author has calculated integral heat production (H) and integral heat flow (Q) according to the models of various authors. The values H and Q are found by the formulas:

$$H = \int_0^t d\tau \int_V \sum_i H_i \exp(-\lambda_i \tau) \, dV$$

$$Q = 4\pi R^2 \int_0^t q(t) \, dt$$

where V is the volume of the earth, t is the age of the earth, λ_i are constants of radioactive decay, H_i is the heat generated by a radioactive element i. It was found that the integral heat flow Q due to molecular processes is the part of the model expressed by the formula $Q = 1-8 \times 10^{37}$ ergs, whereas H, according to different authors, is within the range $H = 6-20 \times 10^{37}$ ergs. Thus, for the earth, the inequality $H \geq Q$ holds with the accuracy of $\pm 2 \times 10^{37}$ ergs.

The majority of models for the thermal history of the earth lead to the conclusion that some heat accumulates, and that the interior of the earth is heated in the course of geological time. This conclusion is substantiated by *Lee* [1967], who found that the total heat generation for the various models Orgueil (12 × 10^{37} ergs), Wasserburg (9 × 10^{37} ergs), and chondrite (14 × 10^{37} ergs) exceeds the total heat flow (4.5 × 10^{37} ergs) by a factor of 2 or 3. Comparing the values of H and Q with the total heat capacity of the globe Λ as a whole at the melting point, we supplemented the inequality $H \geq Q$. The available melting point curves for the mantle and the core were used (see

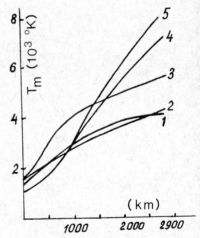

Fig. 2. Melting-point curves: (1) diopside [*Boyd and England*, 1963]; (2) *Zharkov*'s [1959b] calculations; (3) *Uffen*'s [1952] calculations; (4) *Clark*'s [1963] melting curve for a gradient $\gamma_m^0 = 15°C/kbar$; (5) the same, for $\gamma_m^0 = 25°C/kbar$ in the zone of olivine-spinel transition.

Figure 2). The formula for Λ can be written as

$$\Lambda = 4\pi \int_0^R c\rho T_m(r) r^2 \, dr \geq 30 \times 10^{37} \text{ ergs}$$

Therefore $Q \leq H < \Lambda$.

This inequality indicates that the generation of heat in the earth exceeds the loss of heat, but it is not great enough to have melted the earth completely. Complete melting could have been possible had the earth been older than 5.5×10^9 years.

INITIAL HEATING OF THE EARTH

The first studies of the temperature of the interior of the earth were based on the hypothesis that the earth was initially in a molten state. In spite of numerous attempts to study the process of solidification of the earth from an initial molten state, the problem remains practically unsolved.

Well known investigations were carried out by *Kelvin* [1897], *Jeffreys* [1959], *Adams* [1924], *Slichter* [1941], *Urey* [1952], *Jacobs* [1956], and others. *Leibenson* [1939b] devoted attention to the problems of the consolidation of the earth from the initially molten state and the dynamic and temperature conditions that caused folding on the surface of the earth during cooling. When considering this problem under conditions of emission on the surface according to Stefan's law, a method was used similar to that of the Karman integral condition employed in aerodynamics. An integral condition of consolidating (from top and from bottom) has been found for a nonuniform earth, as in the Wiechert model. Leibenson suggests that according to the contraction theory, only low and dense foldings are possible. He disregards the influence of internal generation of radiogenic heat.

Distribution of temperature suggested by *Holmes* [1915], *Adams* [1924], *Jeffreys* [1959], and *Jacobs and Allan* [1954] rested on the assumption that the initial temperature T_0 is equal to the melting point. *Verhoogen* [1960] maintains that this distribution introduces some difficulties into the problem of melting. Such distribution is based on the assumption of different decreases in the content of radioactive elements with depth, and neglects slight emission of radiogenic heat likely to occur at great depths where cooling can be disregarded. This heat would inevitably cause a new melting of the mantle after some time and by now the temperature of the whole of the mantle would have been higher than the melting point; *Jacobs and Allan* [1954] showed that the temperature of the lower layer of the mantle would now exceed the melting point by 300°. It is very unlikely, however, that such a state is possible in view of the high elasticity of the mantle. The theory of planet accumulation [*Shmidt*, 1958] from solid protoplanet matter enables us to explain many fundamental facts of the planetary system such as the rotation of planets and their satellites, the distribution of the moments of inertia, the formation of the asteroid belt, and the abnormal inclination of the axis of Uranus. Formation of the earth would be accompanied by heating due to impacts, radioactive heat, and energy converted into heat by compression of the internal region of the growing earth.

The estimates of the influence of the compression effect on initial temperature are based on the distribution of density, the acceleration of gravity g, the thermal capacity C, and the coefficient of thermal expansion a in the hypothetical model of an undifferentiated primary earth [*Lubimova*, 1958]. The result of the estimates is governed by the accepted hypothesis of the composition of the inner part of the core.

According to the hypothesis of the iron core, the temperature rises with an increase in density up to the present state (which is to be considered an advanced stage of the thermal evolution of the earth) with differentiation and gradual formation of the iron core.

If we adopt the theory of the silicate core, we assume that the formation of the core took place during the period of accretion; hence the effect of the core compression should be considered directly in estimating the initial temperature. Therefore, in estimating initial temperature, this hypothesis is mainly assumed to refer to the composition and origin of the earth's core. Using thermodynamical parameters C, a, ρ, and g for a non-iron silicate core in accordance with *Zharkov* [1959a], coupled with the following approximation for the Grüneisen parameter $\gamma(\rho)$:

Mantle $\quad \gamma_1 = (a_1/\rho) - b_1$

$$a_1 = 6.72 \quad b_1 = 0.13$$

Core (silicate) $\gamma_2 = b_2 - a_2\rho$

$$b_2 = 2.18 \quad a_2 = 0.146$$

we can write the following equation for estimating the compression effect at the initial temperature of the earth with the silicate core:

$$\ln \frac{T_0(r)}{T_1^{\circ}(r) + T_2^{\circ}(r)} = \int_{3.38}^{5.68} \frac{\gamma_1 d\rho}{\rho} + \int_{9.69}^{12.17} \frac{\gamma_2 d\rho}{\rho}$$

where T° is the temperature due to impacts on the surface of the growing earth, and

$$T_2^{\circ}(r) = a\Delta t$$

is the temperature of radioactive heating during the period of mass accumulation ($a = 300^{\circ}/10^8$ years).

When calculating $T_0(r)$ for the earth with an iron core, we consider first the undifferentiated earth before the development of the iron core; when estimating $T_0(r)$ for the primary earth, we can adopt density distribution in the form of Rosh's law used by Birch [1964]. In the second case, by the time of the earth's core formation, the value of $T_0(r)$ will be lower than 1000°C throughout, i.e., lower than the melting point.

According to calculations made by *Safronov* [1965] for the silicate core, the maximum temperature of the upper layers of the mantle might have been as high as 2500°C (see Figure 3). Separation of large bodies might have been responsible for the initial thermal heterogeneities in the mantle, which led to the formation of the continents.

GRAVITATIONAL ENERGY AND THE PROBLEM OF AN IRON CORE

Energy liberation in the gravitational reorganization of the earth is regarded by *Lyustikh* [1948], *Urey* [1952], and *Birch* [1965] as an energy source comparable to radioactivity; it is attested by the formation of a dense iron core. Difference in potential energy between differentiated and undifferentiated earth is estimated as 1.5–2.0×10^{37} ergs. *Birch* [1965] regards the formation of the earth's iron core as the main event in the thermal life of the planet. *Iriyama and Shimazu* [1967] suggest that the released potential energy serves as a heat source for the subsequent thermal state of the moon.

Shock-wave experiments on adiabatic compression of iron have exceeded the range of the

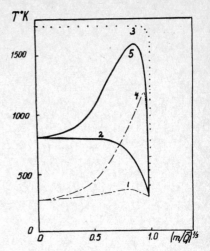

Fig. 3. Distribution of initial temperature versus accumulating mass: $\bar{Q}$ is the present mass of the earth; m is the accumulated mass. Curves 1, 2, 4, and 5 according to the hypothesis of silicate core: (1 and 4) calculations for small protoplanet particles; (2) calculation for large protoplanet bodies, with the changes in their velocities during impact taken into account; (5) maximum for the initial temperature during accretion [after *Lubimova*, 1958]. Curve 3 is according to hypothesis of iron core formaton [*Reynolds et al.*, 1966]. Curves 1, 2, and 4 after *Safronov* [1959, 1965].

pressure within the inner part of the earth's core [*Altschuler and Kormer*, 1961]. The equation of the state of iron based on these experiments is in fair agreement with geophysics [*Birch*, 1964]. The hypothesis of a silicate core has been neither substantiated nor rejected experimentally [*Magnitsky*, 1965].

The hypothesis of an iron core requires a disproportionate amount of iron in a comparative analysis of the earth's density and that of other planets [*Levin et al.*, 1956]. Besides, it cannot provide an acceptable explanation for the mechanism of core formation.

The mechanism suggested by *Elsasser* [1963], which explains the formation of the iron core by an accumulation of vast drops of iron in the unstable layers and by the forcing of these drops through the mantle, seems difficult in view of recent data on the great viscosity (10^{24} or even 10^{26} poise) of the lower layers of the mantle [*McConnell*, 1965].

The arguments about the origin of the earth's core make the estimation of the energy source connected with differentiation rather uncertain. It is not clear where the main part of this energy was liberated. *Urey* [1952] maintains that it is the region of the core where gravitational energy must have been liberated, but the process of forcing the iron drops through a very viscous mantle must surely have been accompanied by considerable friction and dissipation of heat. In this case distribution of energy of gravitational differentiation is likely to be comparatively uniform all over the earth and not concentrated in the core only. The rate of energy liberation is of the order of 10^{-14} cal/cm³ sec, which can practically be compared with radioactive generation of heat.

The distribution of the energy of gravitational differentiation would not be precisely uniform: variations in the energy density over the radius may be proportional to the acceleration of gravity, indicated by the following expression:

$$E_r = \begin{array}{l} \text{const} = 6 \times 10^{10} \text{ ergs for } r_0 < r < R \\ ar/R \quad\quad\quad\quad\quad\quad\quad \text{for } r < r_0 \end{array}$$

where $a = 11 \times 10^{10}$ ergs/g year, and r_0 is the radius of the earth's core. *Beck* [1961] has estimated the density of energy distribution during the gravitational reorganization; he found it was equal to the difference between the gravitational potential energy $E_2 = (GPE)$ for the layer R_1R_2 of the contemporaneous earth and the accretion energy E_1 in the layer having the same mass. His formula can be written as follows for the layer with boundaries R_1 and R_2:

$$E_1 = \frac{4}{5} G\pi\rho_0 \left[\frac{R_2^5 - R_1^5}{R_2^3 - R_1^3} \right]$$

where G is the gravitational constant. The curve in Figure 4 indicates that the lower layers of the mantle, not the core, are more likely to have been melted in the process of gravitational differentiation.

Iriyama and Shimazu [1967] proposed for the moon that, when the liquid iron phase formed, it sank toward the center, releasing gravitational potential energy:

$$\int_0^R 4\pi r^2 \rho \bar{g} h \, dr$$

where $\rho = 7.65$ g/cm³ is the density of iron, $\bar{g} = 100$ cm/sec², $R = 690$ km (radius of the hypothetical lunar core), and $h = 1210$ km is the mean distance of fall of the iron phase.

Let us consider the other possible heat sources.

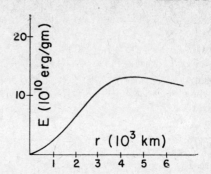

Fig. 4. Energy density: the release of gravitational energy as a function of radius [after *Beck*, 1961].

A dissipation of energy caused by the interaction of the moon and earth would occur in the interior of the earth if the solid earth is considered to be incompletely elastic. This problem was considered by Munk and MacDonald, and later by *MacDonald* [1964], *Zotov* [1960], and *Ruskol* [1963]. The mean value of dissipation is expected to be equivalent to the value of the heat flow: 4×10^{-8} cal/cm² sec (the dissipation constant being $I/Q = 0.05-0.01$). This is, on the average, 30 times less than the heat flow observed. According to *Ruskol* [1963], the emission of energy may not have occurred uniformly in space and time. It is possible that a period in which the effect of tidal friction was more intensive came in the initial period of the earth's existence (provided that the moon had been formed). This period is supposed to have been comparatively short, of the order of 2×10^8 years. The effect of the other sources is minor [*Lubimova*, 1967a].

CURRENT THEORIES ON THE THERMAL HISTORY OF THE EARTH

The content of radioactive elements in various layers, the thermal conductivity coefficient, and the initial and boundary conditions are the main parameters for any theory on the thermal history of the earth. The major goal is to construct a distribution of temperature in the earth that would be in agreement with geophysical data. For example, the mantle of the earth is essentially solid, while the upper part of the core is liquid; the calculated heat flow should be equal to the observed value (1.2–1.5 HFU), and the age of the earth is not more than 4.5 $\times 10^9$ years, while that of the crust is approximately 2–3 $\times 10^9$ years. At present a great number of curves for temperature distribution in models of the earth have been calculated from the heat conductivity equation, variations in the initial data being taken into consideration.

Models of an undifferentiated earth, which assume that the content of radioactive elements is characteristic of chondrites, of Orgueil-type carbonaceous meteorites, or of the Wasserburg mixture (or terrestrial model) have been considered, as well as some intermediate-content models suggested by *Lubimova* [1958], *MacDonald* [1959], *Lee* [1967], and *Mayeva* [1967].

Uniform models reveal very distinctly a trend of secular heating in the interior of the earth at the present time. The process of cooling will reach the center 5×10^9 years in the future [*Lubimova*, 1958]. The calculations for a uniform model give the times and depths at which melting begins. In the most recent theories for a uniform terrestrial model, the time at which melting occurs is 3–4×10^9 years ago and the depth is in the range of 250–600 km (Figure 5). The uniform model probably describes the thermal history adequately until a significant amount of melting has occurred. These calculations can serve as a starting point for a study of the problem of differentiation. If the radioactive materials become concentrated toward the surface, the heat flow from the surface will increase, and the increased loss of heat from the planet will permit the molten regions within the mantle to resolidify.

Thus, there are two main events in the thermal evolution of the earth according to the uniform model: (a) a separation of the core by gravitational reconstruction of the earth; and (b) a fractionation of radioactive elements in the crust as a result of vertical geochemical differentiation.

Another theory for the early stage in the earth's history is considered: it is assumed that the earth underwent extensive melting in its early stage and that the formation of the core and mantle with a protocrust was completed 4.5×10^9 years ago. These thermal history calculations have been made by *Fujii and Uyeda* [1966] on a once-molten earth model. They determined the conditions needed for a once-molten earth to cool. The value of thermal conductivity, including the radiative components necessary for such an earth to cool to the present state, has been obtained in terms of the opacity value as $\epsilon = 1$–3 cm^{-1}, instead of the former $\epsilon = 100$ cm^{-1}. From recent detailed measurements, however, the value of opacity for forsterite was estimated to be several cm^{-1} or less.

Hypotheses for the composition of the core are fundamental to any theory of the thermal history of the earth. The thermal history is much more easily estimated for a silicate core than for an iron core, since the mechanism of an iron core formation is not clear. Such estimates for a silicate core were first made by *Lubimova* [1958] and by *Mayeva* [1967], who included an analysis of melting parameter variations (Figure 6a).

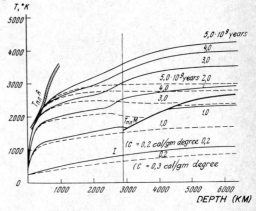

Fig. 6a. Temperature distribution for a silicate core (according to *Mayeva* [1967]). Chondrite composition is accepted with: $U = 2 \times 10^{-8}$ g/g; $c_m = 0.3$ cal/g °C; $\lambda_m^0 = 0.012$ cal/cm sec °C; $\lambda_c = 0.1$ cal/cm sec °C. Solid curves are for $C_c = 0.2$ cal/g °C; dashed curves are for $C_c = 0.3$ cal/g °C; $T_0(r)$ is variant 2 in Figure 3.

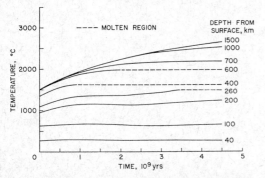

Fig. 5. The variation of temperature with time for a uniform terrestrial model using the modified Clark melting curve [after *Reynolds et al.*, 1966].

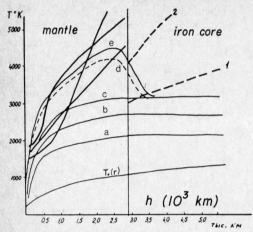

Fig. 6b. Temperature distribution for the chondrite model of the mantle and iron core.

Positive heat flow from the silicate core into the lower mantle is expected: (1) if the heat capacity of the matter of the core C_c is less than the heat capacity of the mantle C_m (for instance $C_c = 0.2$, $C_m = 0.3$, as shown by *Mayeva* [1967]): (2) for concentration of gravitational differentiation energy in the core only. For the latter, however, the temperature of the whole core would be much higher than the melting point of iron, thus making it difficult to explain the solid inner core.

The inner reorganization of the planet caused by the formation of an iron core is expected to have been accompanied by the absorption of energy of gravitational differentiation. If, for simplicity, the density of this energy is assumed to be constant ($E =$ const), the solution of the equation of thermal conductivity for temperature can be written as:

$$T_E(r, t^*) = \frac{E t^*}{c\rho} - \frac{ER}{c\rho r}$$
$$\cdot \int_0^{t^*} \left\{ 1 - \mathrm{erf}\left[\frac{R-r}{2\sqrt{\kappa(t^*-\tau)}}\right] \right\} d\tau$$
$$= \frac{E t^*}{c\rho} \left\{ \left(\frac{R}{r} - 1\right) \right.$$
$$+ \frac{R}{r}\left[\mathrm{erf}\left(\frac{R-r}{2\sqrt{\kappa t^*}}\right)\left(1 + 2\frac{R-r}{2\sqrt{\kappa t^*}}\right)\right.$$
$$- 2\left(\frac{R-r}{2\sqrt{\kappa t^*}}\right)^2 + \frac{2}{\sqrt{\pi}}\frac{R-r}{2\sqrt{\kappa t^*}}$$
$$\left.\left. \cdot \exp\left(-\frac{(R-r)^2}{4\kappa t^*}\right)\right]\right\}$$

where t^* is a period of activity for the process of gravitational differentiation in the earth. This term must be added to the temperature governed by the radioactive decay during the period in which the process of iron core separation is likely to have taken place. If this temperature is disregarded and the core region is considered to be poor in sources, the curve of the temperature distribution will have a hump in the mantle, as shown in Figure 6b, and heat flow from the mantle into the core is then expected. Humped curves of this kind for the upper layers of the mantle result from considering the greatly impoverished lower layers of the mantle according to *MacDonald*'s [1959] and *Lee*'s [1967] calculations, as shown in

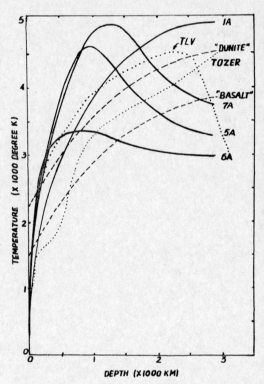

Fig. 7. Present temperature distribution in the mantle from nonfractionated models of different radioactivity distributions [after *Lee*, 1967]: (1A) uniform Orgueil; (5A) concentrated Orgueil, radioactivity in outer 1000 km; (6A) superconcentrated Orgueil, radioactivity in outer 400 km; (7A) oceanic tholeiite, radioactivity concentrated in outer 2/3 of mantle; (TLV) calculations of Tykhonov, Lubimova and Vlasov for fractionated model of the earth with an iron core. Melting curves of basalt and dunite are also shown. Present temperature is estimated by *Tozer* [1959].

Figure 7; they maintained that heat sources are to be found only in the upper layers and not deeper than 1000 km, and they did not analyze the problems of temperature distribution in the core and core composition. Temperature distribution is given only for the mantle. Plotting the curves for the mantle neglecting the problems of the core seems rather inconsistent, because the core evolution has a great effect on temperature distribution in the mantle.

Assuming a core of iron without any heat sources, the numerical estimates of the thermal history of the earth will inevitably lead to the humped curves of the type given in Figure 6b or 7, provided that the lower layers of the mantle do not contain sources either. These humps are likely to be lowered with added gravitational differentiation energy, but this problem remains to be investigated.

Of greater interest is an attempt to calculate 'fractionation' of radioactive elements from a melting region of the mantle into the crust and, in this way, to simulate the process of the earth's crust formation. This stage is quite necessary to proceed to the modern stratified distribution. The starting point for this study is the data on the age of the crust, which is assumed to be 2 or 3 billion years. The processes of the formation of the earth's crust should have a great effect on the history of heat flow. The processes taking place in the earth's interior should leave traces in the form of peculiarities in the structure of the tectonosphere [*Beloussov*, 1960, 1962, 1966]. These peculiarities may be observable in the form of heat flow variation.

THE CORRELATION OF HEAT FLOW WITH OTHER GEOPHENOMENA

At present, more than 2600 observations of heat flow are available. According to *Lee and Uyeda*'s [1965] analysis, the arithmetic mean heat flow of the earth is equal to $1.5 \pm 10\%$ HFU (HFU = 10^{-6} cal/cm² sec). Usually this arithmetic mean is accepted as normal for terrestial heat flow. *Girdler* [1966] suggested that the geometric mean may be more representative than the arithmetic mean because nearly all histograms of terrestrial heat flow observations show positive skew distributions. The geometric mean for all the continental areas is 1.36, and for oceans, 1.27. The geometric mean for the whole world is 1.29 [*Girdler*, 1966].

It was also suggested that weighted values of heat flow should be taken into account, and that heat flow data be classified according to major geological features such as the age of foldings [*Polyak and Smirnov*, 1966]. This resulted in a mean value of heat flow for the whole world of 1.18. According to all these considerations, one can accept a global 'normal value' of the heat flow within the range 1.2–1.5 HFU.

The high values observed in the mid-oceanic crests are undoubtedly related to recent volcanism, but the low values are difficult to explain. Analysis of nearby and repeated measurements suggests that in general, regional heat flow variations >0.2 HFU are significant. The average heat flow over the continents does not differ significantly from that over the oceans. This equality of heat flow suggests that radioactivity is approximately the same beneath land and oceans, but that there are some differences between the upper mantle under the continents and under the oceans.

Global analysis of heat flow observations shows that on the whole the linear correlation of heat flow with topography is insignificant (correlation coefficient -0.1 by *Lee and Uyeda* [1965]). Lee and Uyeda concluded that the correlation between heat flow and gravity is not yet well established because of uncertainties in: (*a*) our knowledge of both the heat flow and gravity fields, and (*b*) their interelationships. The gravity field reflects the present mass distribution, whereas the surface heat flow field may lag millions of years in indicating the temperature distribution of the corresponding depth in the earth's interior.

A more definite correlation seems to be the relationship between heat flow and the ages of basement rocks. *Lee and Uyeda* [1965] noted that heat flow values are well correlated with major geological features. In fact, heat flow values appear very uniform in all shield areas, the average being 0.92 ± 0.17. They gave 1.54 ± 0.38 for post Precambrian nonorogenic areas and 1.92 ± 0.49 for Mesozoic-Cenozoic orogenic areas. These values are shown by asterisks in the plot of heat flow versus age in Figure 8. Other results shown in Figure 8 indicate that heat flow values decrease as age increases. One can see the closeness of the curves of Lubimova and Verma et al., on the one hand,

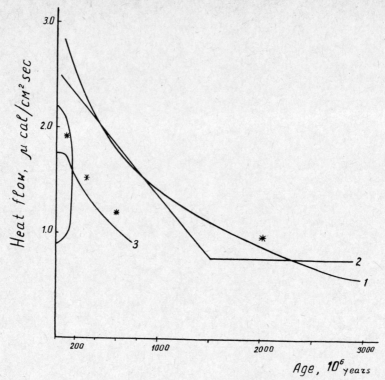

Fig. 8. The relationship of heat flow to age of the main tectonic structures: *asterisks*, Lee and Uyeda [1965]; (1) *Verma et al.* [1967]; (2) *Lubimova* [1967b]; (3) *Makarenko et al.* [1967], who give high heat flow for active tectonic regions, low heat flow for other regions.

with those of Lee and Uyeda and Makarenko et al. on the other hand: the first group depends on nearby and repeated measurements of heat flow and individual geochemical dating for the same sites; the second group used the statistical averaging and indirect evidence on the ages of areas. *Lubimova* [1967b] suggested a simple linear approximation for the relationship of heat flow q to age τ^*:

$$q(\tau^*) = q_0 + b(\tau_0 - \tau^*)$$

where q_0 is uniform heat flow for Precambrian shields, τ_0 is age of the late Precambrian epoch, and b is a constant for $\tau^* \leq \tau_0$ and is zero for $\tau^* > \tau_0$. The equation relation corresponds to curve 2 in Figure 8.

The role of radioactivity in the formation of surface heat flow differs in two hypotheses. In the first, distribution of radioactive sources of heat is stable up to a certain period in which the layers of the crust and mantle are considered to have been stable. This period began either with formation of the entire earth 4–4.5 billion years ago [*Jacobs and Allan*, 1954] or at the time when the most ancient sections of the earth's crust are assumed to have formed, i.e., from two to three billion years ago [*Urry*, 1949; *Lubimova*, 1958]. The latter case would imply a pattern of slow fractionation of radioactive elements from the interior of the earth to its surface.

In the first hypothesis, the function of heat liberation in the crust rapidly acquires great significance for a comparatively short period, during which the earth's formation must have occurred; then its role diminishes. The type of generation curves in the latter case is fundamentally different: heat loss continuously becomes more and more intensive.

Estimations suggest that the history of surface heat flow and, consequently, temperature changes in the upper layers of the earth with time depend on the degree of 'stabilization' of the crust. If a stable crust is defined as sections formed long ago with no further inflow of sialic matter or radioactive elements from below, calculations can be based on the assumption that stratification of the earth occurred 'instantaneously' from a geological point of view, and took place a few billion years ago. The surface heat flow of these sections is expected to have decreased in the course of geological time. These variants were subjected to a thorough investigation by *Urry* [1949], *Lubimova* [1956b], *Jacobs and Allan* [1954], and *MacDonald* [1963]. Some of them are shown in Figure 9 (curves 3, 4, 5).

According to the other hypothesis, with the 'unstable' crust, a gradual fractionation of sources and their inflow into the crust are believed to be occurring. Depending on the scale of fractionation, its decrease can be either compensated and a constant-in-time value obtained, or else the result will be a continuous increase in heat flow with time (curves 1 and 2 in Figure 9).

It is reasonable to attribute curves 1 and 2 in Figure 9 to the regions of modern orogenesis, where inflow of mantle material is expected to be most intensive.

Comparison of these approximate calculations suggests that heat flow changes with time in a much different way for regions of tectogenesis of dissimilar age, provided that the deepest levels of the upper mantle are regarded as responsible for the formation of tectonic zones and tectogenesis of dissimilar age is characterized by various intensities in the inflow of sialic material from the mantle. As regions of ancient shields and platforms can be considered relatively stable areas of the earth's mantle, curves of the type 3, 4, and 5 given in Figure 9 can be interpreted as the history of heat flow for these regions. In this case, curves 1 and 2 in Figure 9 explain the development of heat flow in young tectonic zones. Our assumption concerning the dependence of heat flow on the age of foldings has been corroborated by experimental data, as well as by data of statistical analyses [*Lee and Uyeda*, 1965; *Polyak and Smirnov*, 1966].

CONCLUSION

The chondrite model of distribution of radioactive elements involves serious difficulties because of the difference in the values of the K/U ratio for isotopes in rocks and meteorites. The

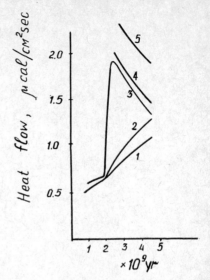

Fig. 9. Changes in heat flow with time for different assumptions of rates of fractionation for radioactive elements from the mantle into the continental crust: (1 and 2) for gradual fractionation [*Mayeva*, 1967]; (3) for sudden fractionation 2.5×10^9 years ago [*Lubimova*, 1958]; (4 and 5) the heat flow for *MacDonald*'s [1963] model of the continental crust.

Wasserburg and Orgueil meteorite models seem to be more acceptable.

All the analyzed models show an excess of total heat production in the interior of the earth versus total heat flow for the entire age of the earth.

According to the majority of the thermal history calculations, temperature at the mantle-core boundary is close to $T = 4000° \pm 1000°K$. However, the range of the assumed melting-point curves permits a greater range of estimated temperatures at the base of the mantle—as great as $T = 6000° \pm 2000°K$.

Temperature distribution in the earth's mantle cannot be investigated thoroughly when divorced from the problem of the structure and composition of the earth's core, because the iron core hypothesis implies that the effect of gravitation differentiation energy in the mantle is significant in comparison with radiogenic energy.

The history of heat flow for certain sections of the earth's crust is closely connected to the age of tectogenesis. In the ancient stabilized sections, heat flow decreases with time; in the regions of Cenozoic folding, heat flow is expected to increase with time because of the migration of radioactive sources of heat into the crust.

Acknowledgments. It is a pleasure to thank Dr. W. H. K. Lee for the use of unpublished material; Dr. A. Beck for his information concerning the calculations of the distribution of the gravitational energy differentiation; and Professor V. A. Magnitsky for his valuable comments on the manuscript.

REFERENCES

Adams, L. H., Temperatures at moderate depths within the earth, *J. Wash. Acad. Sci.*, *14*(20), 459–472, 1924.

Altschuler, L. V., and S. B. Kormer, On the internal structure of the earth, *Izv. Akad. Nauk SSSR, Ser. Geofiz., 1961*, 33–37, 1961.

Beck, A. E., Energy requirements of an expanding earth, *J. Geophys. Res.*, *66*(5), 1485–1490, 1961.

Beloussov, V. V., Development of the earth and tectogenesis, *Sov. Geol. 1960*(7), 1960.

Beloussov, V. V., *Basic Problems in Geotectonics*, McGraw-Hill Book Company, New York, 816 pp., 1962.

Beloussov, V. V., *Earth's Crust and Upper Mantle of Continents*, Nauka, Moscow, 124 pp., 1966.

Birch, F., Density and composition of mantle and core, *J. Geophys. Res.*, *69*(20), 4377–4388, 1964.

Birch, F., Speculations on the earth's thermal history, *Bull. Geol. Soc. Am.*, *76*(2), 133–154, 1965.

Boyd, F. R., and J. L. England, Effect of pressure on the melting points of diopside and albite in the range up to 50 kilobars, *J. Geophys. Res.*, *68*, 311–323, 1963.

Clark, S. P., Heat flow from a differentiated earth, *J. Geophys. Res.*, *66*(4), 1231–1234, 1961.

Clark, S. P., Variation of density in the earth and melting curve in the mantle, in *The Earth Sciences*, p. 5–42, Chicago University Press, Chicago, 1963.

Clark, S. P., and A. E. Ringwood, Density distribution and constitution of the mantle, *Rev. Geophys.*, *2*, 35–88, 1964.

Elsasser, W. M., Early history of the earth, in *Earth Science and Meteoritics*, edited by J. Geiss, pp. 1–30, North-Holland Publishing Company, Amsterdam, 1963.

Fricker, P. E., R. T. Reynolds, and A. L. Summers, On the thermal history of the moon, *J. Geophys. Res.*, *72*(10), 2649–2663, 1967.

Fujii, N., and S. Uyeda, Conditions for a once-molten earth to cool, *J. Phys. Earth*, *14*(1), 15–24, 1966.

Gast, P. W., Limitations on the composition of the upper mantle, *J. Geophys. Res.*, *65*, 1287–1297, 1960.

Girdler, R. W., Statistical analyses of terrestrial heat flow observations, *Trans. Am. Geophys. Union*, *47*(1), 182, 1966.

Heier, K. S., Uranium, thorium and potassium in eclogitic rocks, *Geochim. Cosmochim. Acta*, *27*(8), 849–860, 1963.

Heier, K. S., and J. J. W. Rogers, Radiometric determination of thorium, uranium, and potassium in basalts and in two magmatic differentiation series, *Geochim. Cosmochim. Acta*, *27*, 137–154, 1963.

Holmes, A., Radioactivity and the earth's thermal history, 1 and 2, *Geol. Mag.*, *2*, 60–71, 102–112, Dec. 6, 1915.

Iriyama, J., On the thermal history of the earth, *J. Seismol. Soc. Japan*, *19*, 11–22, 1966.

Iriyama, J., and Y. Shimazu, A note on the thermal history of the moon, *Icarus*, *6*, 453–457, 1967.

Jacobs, J. A., The earth's interior, in *Handbuch Phys.*, *47*, 364–406, 1956.

Jacobs, J. A., and D. W. Allan, Temperatures and heat flow within the earth, *Trans. Roy. Soc. Can.*, *48*, 33–39, 1954.

Jeffreys, H., *The Earth, Its Origin, History, and Constitution*, 4th edition, Cambridge University Press, London, 420 pp., 1959.

Kelvin, L., Age of the earth as an abode fitted for life, 3, *Smithsonian Rept.*, 337–357, 1897.

Knopoff, L., The convection current hypothesis, *Rev. Geophys.*, *2*, 89–122, 1964.

Lee, W. H. K., Thermal history of the earth, Ph.D. thesis in Planetary and Space Physics, University of California at Los Angeles, 1967.

Lee, W. H. K., and S. Uyeda. Review of Heat Flow Data, in *Terrestrial Heat Flow, Geophys. Monograph 8*, edited by W. H. K. Lee, pp. 87–191, American Geophysical Union, Washington, D.C., 1965.

Leibenson, L. S., On the dynamic and thermal conditions of the origin of folding of the earth's crust during its cooling, *Izv. Akad. Nauk SSSR, Ser. Geograph. Geofiz. 1939*(6), 597–624, 1939a.

Leibenson, L. S., On the solidification of the earth from the initial molten state, *Izv. Akad. Nauk SSSR, Ser. Geograph. Geofiz., 1939*(6), 625–660, 1939b.

Levin, B. Yu., and S. V. Mayeva, Thermal history of the earth, *Izv. Akad. Nauk SSSR, Ser. Geofiz., 1960*(2), 243–252, 1960.

Levin, B. Yu., S. V. Kozlovskaya, and A. G. Starkova, The composition of the earth, *Meteorite Sci.*, *14*, 38–53, 1956.

Lovering, J., and J. W. Morgan, Uranium and thorium abundances in stony meteorites, 1, The chondritic meteorites, *J. Geophys. Res.*, *69*, 1979–1988, 1964.

Lubimova, E. A., Influence of redistribution of radioactive sources on the thermal history of the earth, *Izv. Akad. Nauk SSSR, Ser. Geofiz., 1956*(10), 1145–1160, 1956a.

Lubimova, E. A., Thermal history of the earth and its geophysical effect, *Dokl. Akad. Nauk SSSR*, *107*(11), 55–58, 1956b.

Lubimova, E. A., Thermal history of the earth

with consideration of the variable thermal conductivity of the mantle, *Geophys. J.*, *1*(2), 115–134, 1958.

Lubimova, E. A., On the temperature gradient in the upper layers of the earth and the possibility of an explanation of the low-velocity layers, *Izv. Akad. Nauk. SSSR, Ser. Geofiz.*, *1959*, 1861–1863, 1959.

Lubimova, E. A., On processes of heat transfer in the earth's mantle and on conditions of magmatism origin and role of volcanic activity, *J. Phys. Earth*, *8*(2), 11–16, 17–21, 1960.

Lubimova, E. A., *Problems of Terrestrial Heat Flow*, pp. 3–31, Izd. Nauka, Moscow, 1965.

Lubimova, E. A., Theory of thermal state of the earth's mantle, in *The Earth's Mantle*, edited by T. F. Gaskell, pp. 232–323, Academic Press, New York, 1967a.

Lubimova, E. A., Terrestrial heat flow for the USSR and its connection with other geophenomena, 14th IUGG General Assembly, Abstracts, vol. 2a, p. B-3, 1967b.

Lyustikh, E. N., *Trans. USSR Acad. Sci.*, *69*(8), 1417, 1948.

MacDonald, G. J. F., Calculations on the thermal history of the earth, *J. Geophys. Res.*, *64*(11), 1967–2000, 1959.

MacDonald, G. J. F., Stress history of the moon, *Planetary Space Sci.*, *2*(4), 249–255, 1960.

MacDonald, G. J. F., The deep structure of continents, *Rev. Geophys.*, *1*, 587–665, 1963.

MacDonald, G. J. F., Tidal friction, *Rev. Geophys.*, *2*, 467–541, 1964.

MacDonald, G. J. F., Geophysical deductions from observations of heat flow, in *Terrestrial Heat Flow, Geophys. Monograph 8*, edited by W. H. K. Lee, pp. 191–211, American Geophysical Union, Washington, D.C., 1965.

Magnitsky, V. A., *Interior Structure and Physics of the Earth*, Izd. Nedra, Moscow, 379 pp., 1965.

Makarenko, F. A., B. G. Polyak, and Ya. B. Smirnov, Thermal regime of the upper parts of the lithosphere, 14th IUGG General Assembly, Abstracts, vol. 2a, p. B-9, 1967.

Markov, V. K., Yu. N. Ryabinin, I. S. Delitsin, and V. P. Petrov, The possible causes of migration of potassium from the earth's interior and its bearing on geothermal studies, *Izv. Akad. Nauk SSSR, Fiz. Zemli*, *1968*(2), 3–7, 1968.

Masuda, A., Geothermal and petrogenetic implications of the analysis of the distributional relationship between thorium and uranium, *Tectonophysics*, *2*, 69–81, 1965.

Mayeva, S. V., On the thermal history of the earth, *Izv. Akad. Nauk, Ser. Fiz. Zemli*, *1967*(3), 3–17, 1967.

McConnell, R. K., Jr., Isostatic adjustment in a layered earth, *J. Geophys. Res.*, *70*, 5171–5188, 1965.

Polyak, B. G., and Ya. B. Smirnov, Heat flow on continents, *Dokl. Akad. Nauk SSSR*, *168*(1), 170–172, 1966.

Reynolds, R. T., P. E. Fricker, and A. L. Summers, Effects of melting upon thermal models of the earth, *J. Geophys. Res.*, *71*(2), 573–582, 1966.

Ringwood, A. E., The chemical composition and origin of the earth, in *Advance in Earth Science*, edited by P. M. Hurley, pp. 284–356, MIT Press, Cambridge, 1962.

Ruskol, E. L., Tidal evolution of the system earth-moon, *Izv. Akad. Nauk. SSSR, Ser. Geofiz.*, *1963*(2), 216–222, 1963.

Safronov, V. S., On the primeval temperature of the earth, *Izv. Acad. Nauk SSSR, Ser. Geophys.*, *1*, 139–143, 1959.

Safronov, V. S., Original inhomogeneities in the earth's mantle, *Izv. Akad. Nauk SSSR, Fiz. Zemli*, *1965*(7), 1–8, 1965.

Shmidt, O. Yu, A theory of the earth's origin, in *Four Lectures on the Origin of the Earth*, Foreign Language Publishing House, Moscow, 138 pp., 1958.

Slichter, L. B., Cooling of the earth, *Bull. Geol. Soc. Am.*, *52*(4), 561–600, 1941.

Strutt, R. J., On the distribution of radium in the earth's crust and on the earth's internal heat, *Proc. Roy. Soc. London*, *77*, 472–485, 1906.

Tilton, G. R., and G. W. Reed, Radioactive heat production in eclogite and some ultramafic rocks, in *Earth Science and Meteoritics*, edited by I. Geiss and E. D. Goldberg, pp. 31–43, North-Holland Publishing Company, Amsterdam, 1963.

Tozer, D. C., The electrical properties of the earth's interior, in *Phys. Chem. Earth*, *3*, 414–436, 1959.

Tykhonov, A. N., On the influence of radioactive decay on the temperature of the earth's crust, *Izv. Akad. Nauk SSSR, Ser. Geofiz. Geog.*, *1937*(3), 1937.

Uffen, R. J., A method of estimating the melting-point gradient in the earth's mantle, *Trans. Am. Geophys. Union*, *33*(6), 893–896, 1952.

Urey, H. C., *The Planets, Their Origin and Development*, Yale University Press, New Haven, Connecticut, 245 pp., 1952.

Urry, W. D., *Trans. Am. Geophys. Union*, *30*(2), 171–180, 1949.

Verhoogen, J., Temperatures within the earth, *Am. Sci.*, *48*(2), 134–159, 1960.

Verma, R. K., R. U. M. Rao, M. L. Gupta, V. M. Hamza, and G. V. Rao, Terrestrial heat flow measurements in various parts of India, 14th IUGG General Assembly, Abstracts, vol. 2a, p. B-5, 1967.

Wasserburg, G. J., G. J. F. MacDonald, F. Hoyle, and W. A. Fowler, Relative contributions of uranium, thorium, and potassium to heat production in the earth, *Science*, *143*, 465–467, 1964.

Zharkov, V. N., Thermodynamics of the earth's mantle, *Izv. Akad. Nauk SSSR, Ser. Geofiz.*, *1959*(9), 1414–1419, 1959a.

Zharkov, V. N., Fusion temperature of the earth's mantle and the fusion temperature of iron under high pressures, *Izv. Akad. Nauk SSSR, Ser. Geofiz.*, *1959*(3), 465–470, 1959b.

Zotov, P. P., *Nature USSR*, *1960*(3), 122–124, 1960.

Heat Flow in North America[1]

Gene Simmons and Robert F. Roy

THE NUMBER of measurements of heat flow on land has increased dramatically during the Upper Mantle Project. In 1962 there were only 11 reliable determinations in the United States. In 1965, Lee and Uyeda [*Lee*, 1965] catalogued 76 values. At the present time, 350 measurements are published or are in various stages of completion.

In this chapter we present the regional distribution of heat flow in North America. We use the available data as control points and extrapolate on the basis of regional geology and geophysics. Measurement techniques are not discussed; for these, one is referred to *Lee* [1965]. The details (often complex) of individual measurements also are not treated here, but reference is made to the original publications. The present discussion is limited to broad features of continental scale, and no interpretation is made of the many interesting smaller areas.

Regional variations of heat flow at the surface of the earth may be caused by one or more of several factors. Without doubt, most of the heat flowing to the surface of the earth today is due to the energy released by the decay of the radioactive elements uranium, thorium, and potassium. Regional variations in the amount of radioactivity will thus produce regional variations in heat flow (we discuss below an example of this mechanism in the data for New England). A second major cause of regional variation is the presence of hot material injected in various parts of the upper mantle at some time in the geologic past. Other factors, such as refraction due to contrasts of thermal conductivity or local mass transfer, are in our opinion likely to be insignificant on the broad scale with which we are here concerned.

Use of the heat flow data to study the upper mantle requires the separation of the contributions from the crust and mantle. As *Birch* [1954] pointed out, the individual units of heat arriving at the surface of the earth are not tagged as to their origin. Nor are they tagged as to the depth of origin. Some large lateral variations of heat flow can be attributed to lateral variations in crustal radioactivity. Because some changes of heat flow occur over short horizontal distances, it is necessary that the associated sources lie within the crust. Other large variations of regional heat flow occur over longer horizontal distances, and, because they are associated with geophysical anomalies whose source is very likely in the upper mantle, they are very likely due to regional differences in the mantle. Some of the high values observed in western North America are good examples. The values of heat flow determined at the surface really provide boundary conditions that must be satisfied by geophysical models of the crust and mantle.

Because the collection of heat flow data is expensive, the characterization of the heat flow in large areas must be done by interpolating relatively few measurements. Such interpolation is best done on the basis of regional geology and geophysics. Knowledge of the variation of such seismically determined properties as P delays and the velocity of P_n is useful for guessing the geographical extent of heat flow anomalies. The lateral variation of electrical conductivity in the upper mantle is particularly significant for the purpose of outlining regions of anomalously high temperatures because of the high sensitivity of electrical conductivity to changes in temperature. Crustal studies by seismic refraction, now available for many areas, provide other data useful for extrapolation.

The general pattern that is beginning to emerge from heat flow studies in North America is that of two provinces (Figure 1). The eastern heat flow province is characterized by normal heat flow values ($\sim$1.0–1.2 HFU[2]) with

---

[1] Contribution 1556, Division of Geological Sciences, California Institute of Technology.
[2] The heat flow unit (HFU) is 10^{-6} cal sec^{-1} cm^{-2}.

occasional high values that are clearly associated with an abnormal concentration of radioactivity in near-surface rocks. The western heat flow province is characterized by generally high values (~2 HFU). Some of these high values may be due to variations in crustal radioactivity but most, because they are associated with low seismic velocities and high electrical conductivities in the upper mantle, probably arise from anomalously high temperatures in the mantle.

EASTERN PROVINCE

Heat flow in the eastern province is normally about 1.1 HFU, the total range being 0.8 to about 2.3 HFU. *Roy and Decker* [1965], on the basis of closely spaced stations, showed that the variations in New England are due mainly to variation in the radioactive content of near-surface rocks. The high values are associated with highly radioactive granites and the low values with other igneous rocks that are lower in radioactivity. The anomalies (from a background of 1.1 HFU) are readily interpreted as due to the local radioactivity of the crustal rocks. We would expect similar relations to hold throughout the Appalachian region.

Measurements on the exposed Canadian shield reveal a rather uniform, but somewhat lower than normal, heat flow field. The average value is about 0.8 HFU and the departures from this value are small. If the contribution to heat flow from the mantle is the same as that for the rest of the eastern province, then the value of 0.8 HFU implies that the contribution from crustal radioactivity is somewhat lower than in those areas characterized by a heat flow of 1.1 HFU.

It appears to us that radioactivity is concentrated in the upper levels of crustal rocks. Continued erosion in the Canadian shield has removed enough of the radioactive upper layer to lower the average regional value by about 0.3 HFU and enough of the small bodies (if ever present) of abnormally radioactive granite (a major source of large variation in heat flow in New England) to produce a more uniform heat flow field.

The part of the United States between the Appalachian and Cordilleran systems is characterized by rather uniform heat flow of about 1.2 HFU. Variations are again small. The surface rocks are sedimentary (very low in radioactivity), and rather few boreholes reach the Precambrian basement. Rocks like those exposed in the Canadian shield are believed to extend beneath the area, and the heat flow might be expected to be the same as that in the Canadian shield rather than the observed value (which is based on roughly thirty-five heat flow stations). Several explanations are possible. It may be that erosion of the Precambrian rocks has been less and therefore the contributions to heat flow from the radioactivity of the outer crustal rocks is more nearly normal or there may have been an initial difference in the average radioactivity of the crust or mantle between the two areas.

WESTERN PROVINCE

Heat flow in the western province is higher (about 2 HFU) and is more variable geographically than in the eastern province. In addition to actual measurements of heat flow, other phenomena indicating high heat flow in the western province (and absent in the eastern province) are the presence of widespread Tertiary volcanism and the existence of many hot springs throughout the region. The boundary between the two provinces coincides with the decrease toward the west of electrical conductivity of the upper mantle observed by *Schmucker* [1964] in New Mexico, *Reitzel* [1967] in Colorado, and Reitzel and Simmons (unpublished) in Montana. The western boundary in the south is the Peninsula Ranges and Sierra Nevada of California, but it is poorly defined further north because of insufficient data. It is possible that large regions of normal heat flow are included within the boundaries of the western province. Although two normal values (~1.2 HFU) have been observed in the Colorado Plateau, insufficient data exist to determine the geographical extent of the normal area.

There are several indications that most of the heat flow in excess of 1.2 HFU in the western province originates in the mantle. Already cited is the fact that large electrical conductivity contrasts (very likely related to thermal differences) associated with the region are observed in the mantle. *Herrin* [1966] reported low velocities of P_n in the western United States that correlate well with the high regional heat

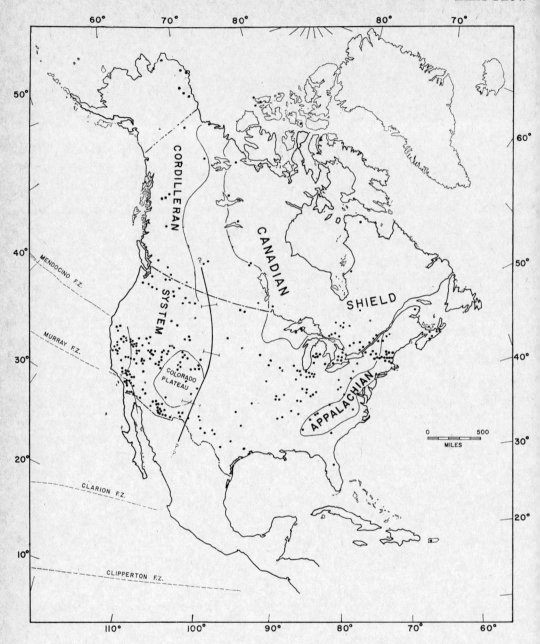

Fig. 1. Location of heat flow stations in North America. Final values are available (as of August 1967) for fewer than 15% of the locations. The line just east of the Colorado Plateau separates the eastern heat flow province from the western heat flow province.

flow. The crustal studies by *Pakiser* [1963] and others confirm that the velocity of the upper mantle is low (~7.8 km/sec) over much of the western province. *Zietz et al.* [1966] interpreted the spectrum of static magnetic anomalies observed along several east-west profiles near latitude 37°N to imply that the Curie point was reached at a shallower depth beneath the western province than beneath the eastern province. From these associations of high heat flow at the surface of the earth with anomalous properties of the upper mantle, we infer that the origin of the excess heat flow lies in the upper mantle and not in the crust.

The part of the excess heat flow that originates in the mantle may be attributed to two causes, lateral variation of the radioactive content of the upper mantle and the previous injection of hot material in the mantle near the crustal boundary. In large parts of this province widespread volcanism and deformation began in the early Tertiary after a long period of relative quiescence. For this reason we place less emphasis on the hypothesis of radioactive heat sources in the mantle, which would not account for the sudden onset of tectonic activity in the early Tertiary, and believe the emplacement of hot material to be the chief cause of high heat flow in the western province. It may be that the hot material is also enriched in radioactive elements. The present high heat flow of the Basin and Range province could be explained, for example, by an injection perhaps 10 million years ago, whereas the heat flux at the surface in the Colorado Plateau is still low because the injection there occurred only within the last 2 million years. We do not visualize a large scale single-shot injection under the whole area as the major cause, but rather a series of plutonic events that have occurred at different times. While it is possible mathematically to attribute all the variations of heat flow to the lateral distribution of radioactivity in crustal rocks, the amounts of radioactivity required are geologically unreasonable. Furthermore, the temperatures attained in the mantle in such models are insufficient to account for the other geophysical observations.

In the present set of data, two regions of low heat flow appear along a linear trend near the coast and parallel to it, the Sierra Nevada and Southern California batholiths. Data are not yet sufficient to indicate whether there is a zone of low, or normal, heat flow that is continuous along the entire western margin of the continent.

CONCLUSION

Although large areas in North America still remain for which there are no heat flow measurements—for example, we know of no values in Mexico—the regional pattern is becoming clear. There is an eastern heat flow province in which a few anomalies clearly associated with crustal variations of radioactivity are superimposed on a rather uniform value of about 1.2 HFU. The western province is characterized by generally high values of heat flow greater than, or equal to, 2 HFU, about 1 HFU being attributable to phenomena in the upper mantle.

Acknowledgment. We acknowledge with thanks the help given by F. Birch, D. D. Blackwell, E. R. Decker, A. M. Jessop, A. M. Lachenbruch, and J. H. Sass in allowing us to see unpublished data. We appreciate the critical comments of A. J. Erickson, A. W. England, D. W. Strangway, T. L. Henyey, and D. L. Anderson. This work was supported by National Science Foundation Grant GA-715 at the California Institute of Technology.

REFERENCES

Birch, Francis, The present state of geothermal investigations, *Geophysics, 19*(4), 645–659, 1954.

Herrin, Eugene T., Travel-time anomalies and structure of the upper mantle (abstract), *Trans. Am. Geophys. Union, 47*(1), 44, March 1966.

Lee, William H. K., editor, *Terrestrial Heat Flow, Geophys. Monograph 8,* American Geophysical Union, Washington, D. C., 1965.

Pakiser, L. C., Structure of the crust and upper mantle in the western United States, *J. Geophys. Res., 68*(20), 5747–5756, October 15, 1963.

Reitzel, John, Magnetic deep sounding near the Rocky Mountain front: Preliminary results (abstract), *Trans. Am. Geophys. Union, 48*(1), 210, March 1967.

Roy, R. F. and E. R. Decker, Heat flow in the White Mountains, New England (abstract), *Trans. Am. Geophys. Union, 46*(1), 174, March 1965.

Schmucker, Ulrich, Anomalies of geomagnetic variations in the southwestern United States, *J. Geomag. Geoelec., 15,* 193–221, 1964.

Zietz, Isidore, Elizabeth R. King, Wilburt Geddes, and Edward G. Lidiak, Crustal study of a continental strip from the Atlantic Ocean to the Rocky Mountains, *Bull. Geol. Soc. Am., 77*(12), 1427–1447, 1966.

Heat Flow Map of Eurasia

E. A. Lubimova and B. G. Polyak

HEAT flow measurements were started in 1939 in Europe and in 1947 in Asia. The review by *Lee and Uyeda* [1965] greatly facilitated our preparation of a Eurasian heat flow map. Moreover, heat flow measurements are increasing at such a rate that we were able to add much new information, especially for the USSR, Czechoslovakia, and the German Democratic Republic. In addition to a description of the amount, quality, geographical distribution, averaging, probable errors, and allowances, we shall try to classify the material, point out the most prominent anomalies, and offer a tentative geological-geophysical interpretation.

SCOPE AND DISTRIBUTION OF MEASUREMENTS

To date, heat flow has been estimated at more than 400 points of the Eurasian continent, including Great Britain and Iceland (excluding Japan, see p. 101). There are about 300 independent heat flow values (counting as one those measurements within 10–20 km of one another). They are extremely unevenly distributed, the highest concentration being in central Europe, southern European USSR, and Japan (Figures 1, 2, and 3). In Europe there are some 250 independent measurements; in Asia, apart from Japan, there are 34 independent values in the USSR, 4 in India, and 1 in Iran.

QUALITY OF THE MATERIALS

The available materials are far from being uniform in quality. Most are highly reliable, as they were obtained by measuring the equilibrium geothermal gradient in long-abandoned boreholes and the thermal conductivity of rock samples extracted from the same holes. In some measurements made in England (data by Mullins and Hinsley) and in Czechoslovakia (6 points), there is no correspondence between the observed distribution of temperature and the natural thermal field. In this connection we should like to mention the unfortunate omission in many publications of the length of waiting period before temperature measurements were taken in a borehole. In some cases, the vertical thermal gradient was estimated on the basis of observations in mines (German D. R., F. R. Germany) and tunnels (Austria and Switzerland.) Some data were based on the values of thermal conductivity of rock extracted from adjacent openings, and finally, a few data are the so-called 'estimated values' calculated from thermal conductivities of similar rock types (Norway, Sweden, Poland, and part of the Volga-Ural region of the USSR).

PERTURBATIONS AND CORRECTIONS OF THE TEMPERATURE GRADIENT AND HEAT FLOW

The greater part of the observation points are located in areas of tectonic stability characterized by almost unbroken relief, but some measurements were taken in mountainous areas that have experienced intensive orogeny in recent geological times (the Alps, Tien-Shan, and others). In these cases the observed results were corrected by the authors with regard to the effects of the relief, crustal uplift, and erosion, in accordance with geologically based assumptions concerning the rate of formation of modern relief. Such allowances may be necessary for some other areas.

Past climatic variations are another important factor. From a comparison of climatic changes with the geologically recent past, some authors [*Jaeger*, 1965] introduce certain corrections into the observed values of heat flow. Since the time of glaciation is usually uncertain, these corrections are disputable; hence, the values on the map do not include these corrections.

Ground water circulation can distort heat-flow measurements, either increasing or lowering them, depending on the source. In fact, *Kappelmeyer* [1957] discussed the use of surface measurements in studying underground water movements. However, in areas of poorly permeable rocks (e.g., crystalline shields) measurements made below the uppermost 50 meters are probably representative. Since most of the

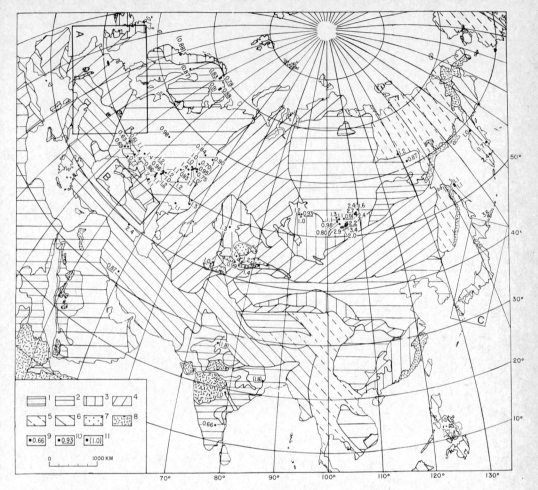

Fig. 1. Heat flow in Eurasia: (1) Precambrian shields; (2) platforms with Precambrian crystalline and folded foundation; (3) Caledonian folded regions; (4) Hercynian folded regions; (5) Mesozoic folded regions; (6) Cenozoic folded regions; (7) Meso-Cenozoic superimposed basins; (8) areas of Meso-Cenozoic effusive rock. Heat flow (in HFU = 10^{-6} cal/cm² sec): (9) individual measurement; (10) mean for a group of adjacent measurements; (11) tentative value. Insets show location of detailed maps: (A) Figure 2; (B) Figure 3; (C) Figure 2 of article by Horai and Uyeda in this Monograph, p. 101. Base map for Figures 1–3 from *Yanshin*, [1966].

measurements were made at greater depths, one should not consider the (disregarded) hydrogeologic effect as a probable source of major errors.

Finally there are local peculiarities of the geological structure whose influence is insignificant in the great majority of cases; in areas of saline domes it may become quite appreciable because of a sharp (severalfold) difference in the thermal conductivity of the salt and sedimentary rock enveloping the dome. The unusually large values of the flow in the areas of Roneberg, Benthe, Hanza, and Bergwerk Riedel (Germany) have most probably been caused by redistribution of heat flow due to salt plugs piercing the sedimentary rocks. The same is probably true of the relatively high values of heat flow in some areas of the north German lowland.

THE HEAT FLOW FIELD AND ITS ANOMALIES

From Figure 1, we can see that heat flow results in Eurasia are well correlated with major geological features. We can distinguish some areas of approximately similar values. These

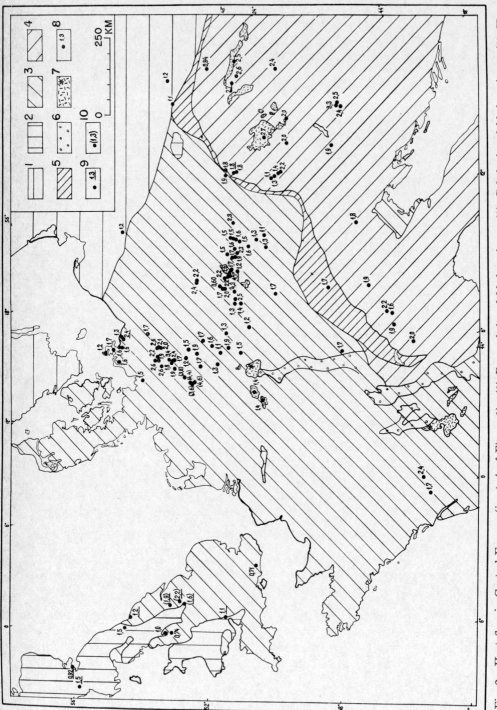

Fig. 2. Heat flow in Central Europe (inset A of Figure 1): (1) Precambrian folded regions; (2) Caledonian folded regions; (3) Hercynian folded regions; (4) Cenozoic folded regions; (5) foredeeps of Cenozoic folded regions; (6) Cenozoic superimposed basins; (7) areas of Cenozoic effusive rock. Heat flow in HFU: (8) individual; (9) mean for groups of adjacent stations; (10) tentative.

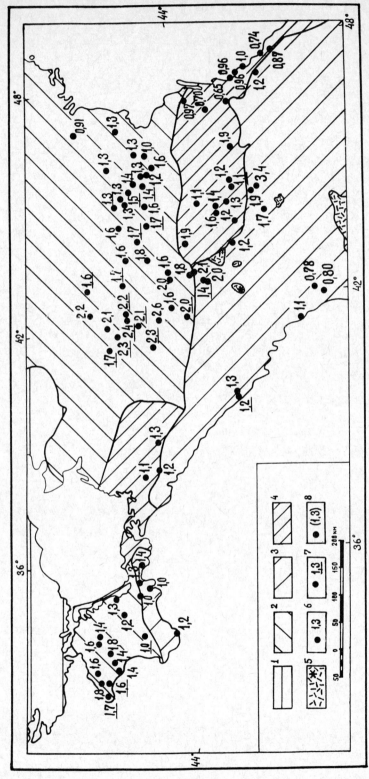

Fig. 3. Heat flow in the Caucasus and Crimea (insert B of Figure 1): (1) Precambrian folded regions; (2) Hercynian folded regions; (3) Cenozoic folded regions; (4) foredeeps of Cenozoic folded regions; (5) areas of Cenozoic effusive rock. Heat flow in HFU: (6) individual; (7) mean for groups of adjacent stations; (8) tentative.

include primarily the low heat flow zones associated with Precambrian shields: Kola peninsula (Baltic shield), 0.86 HFU; Ukrainian shield, 0.70; Indian shield, 0.92. These areas are characterized by a very small range of heat-flow values, although this regularity does not apply to the areas of Khetri copper mines and Mosabani (India).

Platforms of Precambrian folded floors, such as the east European and Siberian, are also characterized by very uniform fields of normal heat flow (1.0–1.1).

Quite prominent are the higher heat flow fields in the areas of Alpine folded mountain structures (for the Alps the values reach as high as 2.2; the Carpathians, 2.7; and the Caucasus, 2.0) and of the regions that were rejuvenated during the Cenozoic phase of tectogenesis (Tien-Shan, 1.8). Similar values have not yet been found for the mountainous part of the Crimea. Close to this group is the zone on the Baikal rift (values up to 3.1) and the areas of recent volcanic activity (Kamchatka, 2.4; and near Iturup island, 3.6).

Anomalies evidently related to underground water movement have been found in central England (Nottinghamshire, 2.2) and in northern Crimea (Novoselovskaya, 1.8; Tarchankutskaya, 1.7), and also probably in the Hungarian basin (up to 3.3), as well as in some other regions. Especially marked are the anomalies related to recent hydrothermal activity in the zones of abyssal fractures and Cenozoic volcanism in Czechoslovakia (Banska Stjavnica, 2.7; Teplice, 4.4), in Armenia (Jermuk, 2.4–5.0), and Italy (Larderello, 6–14).

The effects of underground water movement on heat flow are well marked in the variations of the geothermal gradient in geothermal areas, for instance, in the eastern Cis-Caucasus (Figure 3), where the area of the Fore ranges with their numerous underground water discharge zones is clearly marked.

The higher values of heat flow in the zone of Zhigulevsky dislocations (east European platform), the Rhine graben (central Europe), and, probably, in some other areas must have been due to the effects of Alpine orogenesis. Such areas are, in fact, geothermal anomalies and reflect the regional heterogeneity in the thermal field.

Various structural effects, in addition to those of the relief, have a strong bearing on the heat flow field in the areas adjoining Lake Baikal. Estimates of topography corrections based on electrical models have been made by *Lubimova and Shelagin* [1966]. Yet, an analysis of the various corrections for this zone cannot affect the conclusion that a considerable positive anomaly of heat flow exists here (up to 3.2). The results of magnetotelluric investigations for higher electrical conductivity under Lake Baikal provide an independent proof of considerable local heating in the mantle. Consequently, this anomaly is considered to originate from considerable depths and not to be related to the effects of the surface factors. A specific feature of this anomaly is the clear boundary separating it from the relatively uniform field of somewhat lower values in the adjacent part of the Siberian Platform, that of Irkutsk plateau ($\sim$1.0 HFU).

We have not yet succeeded in finding a clear correlation between the distribution of heat flow and the thickness of the crust. Still, some evidence of an indistinct block structure of the crust can be obtained from analysis of the geophysical data of southern European USSR. Here the alternation of densities in the upper mantle indicated by gravity data seems to conform to the alternation of more or less intense heat flows.

A correlation between heat flow and vertical crustal movements was discussed by *Magnitsky* [1965], who showed that more intense heat flows in the upheaving parts of the crust might result from the transfer of subcrustal masses and changes in the volume of the mantle and crust. More marked divergencies occur in the values of flow observed in areas differing in the sign of longer-term vertical motion (e.g., in adjoining areas of piedmont and intermontane depressions, and in mountain structures of the Alpine phase) in the recent phase of their geological history (starting with Neogene). For instance, in the Caucasus the difference in the flow is as high as 100% (from 1.1 to 2.0).

ELEMENTS OF DATA CLASSIFICATION

The amount and distribution of the available materials enables one, apart from the correlation with the hydrodynamic factor, which is usually of limited value, to correlate the materials with the major features of geological history and the structure of various parts of the continent of Eurasia. The theory of the earth's thermal his-

TABLE 1. Classification of Heat Flow Values for Major Geological Features in Eurasia.

Tectonic Zone	Number of Measurements	Arithmetic Mean, HFU	Range of Values, HFU	Comments
Precambrian folded regions	49	0.96	0.61–1.8	
Shields	11	0.86	0.61–1.4	Two values on Deccan excluded
Platforms	38	0.99	0.70–1.4	
Paleozoic folded regions	113	1.6	0.6 –2.6	
Caledonian folded regions	11	1.1	0.8 –1.5	
Hercynian folded regions	102	1.6	0.6 –2.6	
Cis-Caucasus	32	1.6	0.91–2.6	Including Stavropol area; mean higher, owing to saline domes, Cenozoic tectonic activity, and volcanism
Crimea	13	1.4	1.0 –1.8	
Western Europe	57	1.7	0.60–2.6	
Cenozoic folded regions	82	1.7	0.65–3.3	
Foredeeps	33	1.2	0.65–1.9	
Foredeeps, not including Cis-Alpine foredeep	30	1.12		
Folded mountain structures	28	1.8	1.2 –2.7	
Cenozoic volcanic areas	21	2.1	1.4 –3.3	Data on geothermal areas excluded

tory shows that the ancient relatively stabilized areas of the crust are characterized by values of heat flow decreasing with time, whereas the younger ones are characterized by rising values. The theory concerning correspondence between heat flow and the age of tectogenesis is supported by statistics [Lee and Uyeda, 1965; Polyak and Smirnov, 1966].

Data on the heat flow in Eurasia were classified by variations in the age of tectogenesis. The values for the arithmetical mean heat flow (see Table 1) in the first approximation support this classification.

If we regard all materials in areas of Paleozoic folding summarily, heat flow approaches 1.6. Yet, one cannot help feeling that the estimate is unreasonably high owing to perturbations introduced into the normal heat field of these structures by the effects of saline-dome tectonics (northern Germany), more active later tectonics (southern Germany, northwestern Czechoslovakia), and Cenozoic magmatism. In the Scythian platform (Cis-Caucasus) one can suspect some effect of Cenozoic magmatism on the prominently high heat flow (up to 2.4–2.6) in the Stavropol region, which is near the still active volcano Elbrus, and the magmatic and hydrothermal manifestations of the region Mineralniye Vody in the Caucasus. A group of data has been selected to describe the structures of the Caledonic phase of folding (England and Norway). They are limited and not always satisfactory, but one can see even now that the values of heat flow in these areas are significantly lower than the Paleozoic average (Table 1).

In structures formed and rejuvenated during the Alpine phase of folding, heat flow seems very similar (1.7), yet three groups of data are clearly distinguished. One of these, characterizing foredeeps (the Cis-Carpathians, Indolo-Kuban, Tersko-Caspian, and the Rion-Kura and Balkhash intermontane depressions) displays a relatively low heat flow (1.2). Another, which belongs to the folded mountain structures of the Alps, Carpathians, and Caucasus, shows an average flow close to 1.8. The third, which includes the results measured in the areas of Cenozoic volcanism in the Central Caucasus, East Kamchatka, the Hungarian depression, the southern parts of the Rheno-Herz and Saxon-Thüringen zones, is prominent with its maximum values of heat flow (2.2 on the average). Thus, we see an unquestionable differentiation of heat flow within this area.

Acknowledgments. The authors are much indebted for valuable contributions of materials by E. Bullard (England), T. Boldiszar (Hungary), J. Goguel (France), D. Malmqvist (Sweden), H. Militzer and W. Oelsner (GDR), W. Chermak (Czechoslovakia), R. Verma (India), F. Makarenko, J. Smirnov, A. Gamalova, F. Lebedev, R. Kutas, W. Gordienko, G. Sucharev, S. Vlasova, and J. Taranucha (USSR).

REFERENCES

Jaeger, J. C., Application of the theory of heat conduction to geothermal measurements, in *Terrestrial Heat Flow, Geophys. Monograph 8,* edited by W. H. K. Lee, pp. 7–23, American Geophysical Union, Washington, D.C., 1965.

Kappelmeyer, O., The use of near-surface temperature measurements for discovering anomalies due to causes at depths, *Geophys. Prospecting, 5,* 239–258, 1957.

Lee, W. H. K., and S. Uyeda, Review of heat flow data, in *Terrestrial Heat Flow, Geophys. Monograph 8,* edited by W. H. K. Lee, pp. 87–190, American Geophysical Union, Washington, D.C., 1965.

Lubimova, E. A., and V. A. Shelagin, Heat flow through the bottom of Lake Baikal, *Dokl. Akad. Nauk SSSR, 171*(6), 1321–1324, 1966.

Magnitsky, V. A., *Internal Structure and Physics of the Earth,* Izd. Nedra, Moscow, 379 pp., 1965.

Polyak, B. G., and Ya. B. Smirnov, Heat flow on continents, *Dokl. Akad. Nauk SSSR, 168*(1), 170–172, 1966.

Yanshin, A. L., editor, *Tectonic Map of Eurasia,* Moscow, 1966.

Heat Flow in Oceanic Regions[1]

R. P. Von Herzen and W. H. K. Lee

THE study of terrestrial heat flow was pursued by committees of the British Association in 1868–1883 and 1935–1939. The second period of study resulted in the publication of the first reliable heat-flow measurements on land by *Bullard* [1939] and *Benfield* [1939] and established that the heat conducted through the continental crust is orders of magnitude greater than the heat loss from volcanic eruptions or energy dissipated in earthquake waves. It also initiated Bullard's attempts to measure heat flow through the ocean floor as early as 1939, although facilities for doing so were not available until a decade later [*Bullard,* 1965]. The result of the first successful oceanic heat-flow measurements by *Revelle and Maxwell* [1952], subsequently confirmed, constitutes one of the most important discoveries in this field of geophysics: the approximate equality of heat flux through continental and oceanic crusts, despite their marked structural differences.

The rapid expansion of oceanic heat-flow measurements in more recent years has resulted from the development of reliable instrumentation and the increased availability of suitable vessels for oceanic research. Instrument development was facilitated by two favorable factors: (*a*) large thermal inertia of the deep ocean waters; and (*b*) relatively soft sediments over most of the ocean floor that are penetrable by a temperature probe. Oceanic heat-flow measurements are relatively easily made and inexpensive in comparison to those on continents; as a result, the present ratio of oceanic to continental measurements is about 10:1.

MEASUREMENT TECHNIQUES

Heat flow through the ocean floor is determined by measuring the temperature gradient and thermal conductivity in the ocean sediment. Temperature gradients are measured by one of two methods, both of which are the same in principle but differ somewhat in physical design: the cylindrical probe technique developed by Bullard, and the attachment of small temperature sensors to a coring barrel by the method of Ewing. Thermal conductivity is usually measured by the transient method using a fine needle on samples of sediments brought to the surface by coring devices either at or near the site of the temperature gradient measurement. These measurement techniques have been described in detail by *Langseth*

---

[1] Contribution from Woods Hole Oceanographic Institution No. 1927 and Institute of Geophysics and Planetary Physics, University of California at Los Angeles, Publication No. 592.

[1965]; only the most recent advances will be briefly reviewed here.

Although the thermal conductivity of ocean sediments does not vary greatly from location to location (the mean at 586 locations is 1.98 mcal/cm sec °C with a standard deviation of 0.22 [*Lee and Uyeda*, 1965, p. 114]), in situ conductivity measurements were made only recently by a refined Bullard-type probe introduced by *Corry et al.* [1968]. After penetration, thermal conductivity of the surrounding sediments is measured at the same time as the temperature gradient by heating a thin probe. This slider, which is mounted outside the main probe, also measures the depth of penetration and takes a small core of the sediment. Preliminary results indicate that the in situ values of thermal conductivity may be somewhat higher than those measured on core specimens. The validity of the in situ method, however, is still under investigation [*Vacquier et al.*, 1967].

RESULTS OF MEASUREMENTS

The growth of the number of measurements of heat flux through ocean floors has been extremely rapid. The first measurements were made only slightly more than 15 years ago; the latest compilation taken from *Lee and Uyeda* [1969] includes about 2600 individual observations. Most of these have been made during the past five years; new values are being reported at a rate of about 500 per year.

Following the procedure of Lee and Uyeda, heat-flow data that were published or in press are catalogued; nearby stations (less than 10 km apart) are usually grouped together as an average value. However, only selected data from the catalog are analyzed, and data were rejected if one or more of the following applied: (*a*) original authors considered the data unreliable; (*b*) insufficient information as to the quality of data or station location; (*c*) partial penetration of temperature gradient probe into the sediment such that the measured temperature gradient might be uncertain; (*d*) oceanic measurements made in water less than 500 meters deep; or (*e*) continental stations located in geothermal areas. It is difficult to apply the criterion of geothermal areas to oceanic measurements. This may explain the substantially larger standard deviation for oceanic values than for continental ones.

A convenient way to summarize the presently available heat-flow data is by means of maps containing $5° \times 5°$ averages, as are shown in Figures 1. These averages are based on 2225 selected catalogued data (out of a total of 2526), plus 359 selected uncatalogued data. At sea, the areal coverage has been advanced relatively rapidly, although substantial gaps still remain. Between latitudes of 45°N and 45°S at sea, there are about 915 areas $5° \times 5°$, of which 561 have at least one heat-flow measurement, representing a coverage of somewhat more than 60% between these latitudes. In these latitudes on continents, and at high latitudes for both continents and oceans, the percentage of areal coverage is much lower.

Statistical analyses of 2584 selected heat-flow data do not differ substantially from previous analyses by *Lee and Uyeda* [1965], as Table 1 shows. Because of the uneven geographical distribution of heat-flow measurements, the results must be interpreted with caution. In order to compensate for this possible bias, weighted statistics according to surface area were introduced by *Lee and Uyeda* [1965, p. 136]. Since grid elements formed by equally spaced intervals of latitude and longitude have unequal surface area, geographical grid averages do not give an unbiased sample. To avoid this difficulty, averages for grid elements of equal area (9×10^4 square nautical miles, or $5° \times 5°$ at the equator) were computed and statistically analyzed. Again, the results do not differ greatly from the previous analysis of Lee and Uyeda, as Table 2 shows.

Comparison of Tables 1 and 2 reveals that the arithmetic mean and standard deviation for grid averages are less than that for the original data. The likely interpretations are that geographic density of measurements is greater in regions of high heat flow, and that some local variability is removed by grid averaging. The continental grid averages do not extend beyond 3 μcal/cm² sec (Figure 2), because values from geothermal areas on land have been excluded. Detailed comparison between oceanic and continental histograms is premature because of the limited coverage on land. However, the *mode* of both oceanic and continental heat-flow grid averages is 1.3 μcal/cm² sec; it was 1.1 μcal/cm² sec in the analysis by *Lee and Uyeda* [1965]. The arithmetic mean is 1.46 μcal/cm² sec for

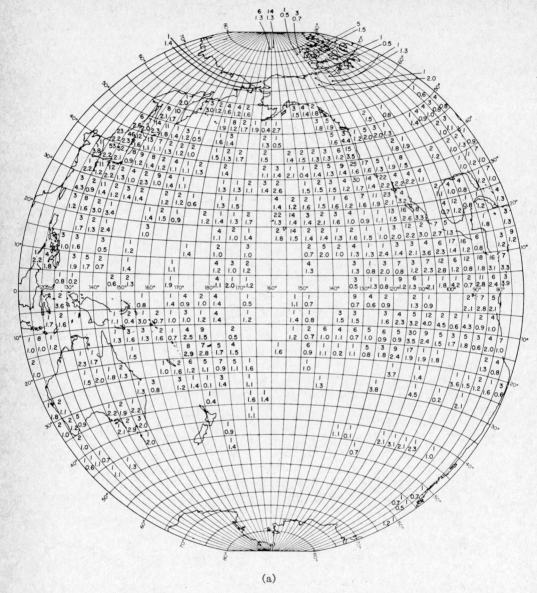

(a)

Fig. 1. Selected heat-flow data in 5° × 5° grid. In each grid element, the upper value is the number of slected data, the lower value is the arithmetic mean in $\mu cal/cm^2$ sec.

oceanic heat-flow grid averages, 1.45 $\mu cal/cm^2$ sec for continental. Clearly the difference is insignificant, which further confirms the equality of heat flux through continental and oceanic crusts.

Tables 1 and 2 also give heat-flow statistics for various oceans, whose boundaries are defined by *Sverdrup et al.* [1942]. In Table 1, we note that for all oceanic regions except marginal seas, the standard deviation increases with increasing arithmetic means, and the standard deviations are much larger than can be explained by measurement errors. This result suggests that areas of high heat-flow also have larger local variations. The low standard deviation in marginal seas is due to the relatively high but fairly uniform heat-flow values from the Sea of Japan. Among the major oceans, the highest arithmetic

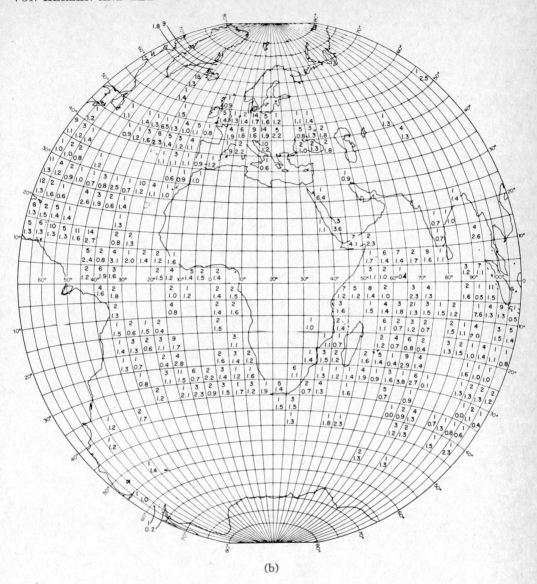

(b)

mean is for the Pacific Ocean and the lowest for the Atlantic Ocean. This may be due in large part to the continued concentration of measurements in the East Pacific in association with the East Pacific rise; otherwise, the differences between oceans may not be significant. The ratio of number of measurements from ocean ridges to those from basins varies from about 1:1 in the Atlantic and Indian Oceans to about 2:1 in the Pacific Ocean. When statistics for grid averages are compared (Table 2), heat-flow differences between various oceans become smaller. This again demonstrates the need for caution in interpreting data from uneven geographic distributions.

RELIABILITY AND VARIABILITY OF DATA

Because of the large variability in the heat-flow values from oceans over relatively short distances at some localities, the reliability of such measurements has been discussed in several recent papers. There seems to be some confusion in distinguishing between the *accuracy* of heat-flow values and the *representativeness* of any given value. *Von Herzen and Langseth* [1965] have noted that accuracy may vary with loca-

TABLE 1. Statistics of Selected Heat-Flow Data*

Area	Lee and Uyeda [1965, p. 141]			This Work (July 1968)		
	N	$\bar{q}$	S.D.	N	$\bar{q}$	S.D.
Atlantic Ocean	206	1.29	1.00	406	1.43	1.07
Indian Ocean	210	1.47	0.89	331	1.44	1.09
Pacific Ocean	497	1.79	1.31	1232	1.71	1.24
Arctic Ocean				29	1.23	0.33
Mediterranean seas	†			71	1.33	0.89
Marginal seas	†			260	2.13	0.63
All oceans and seas	913	1.60	1.18	2329	1.65	1.14
Continents	131	1.43	0.56	255	1.49	0.54

*N number of data.
$\bar{q}$ arithmetic mean in $\mu cal/cm^2$ sec.
S.D. standard deviation from the mean in $\mu cal/cm^2$ sec.
† Data were few and were included under oceans.

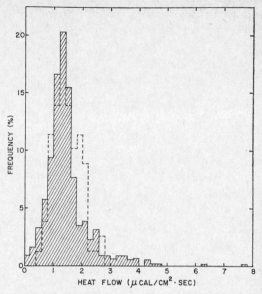

Fig. 2. Histograms of heat-flow grid averages (9×10^4 square nautical miles per grid element). The solid-line histogram is for oceanic data and the dashed-line histogram is for continental data.

tion, and depends on the magnitude of the temperature gradient and the variability in type of sediments penetrated. An unfavorable combination of these factors can occasionally produce errors of 20%, although for most stations, errors should be 10% or less. The accuracy can be determined from repeated or nearby measurements where the regional heat flow is constant or varies only slowly with distance.

Lee and Uyeda [1965, pp. 89–91] obtained a general estimate of heat-flow measurement

TABLE 2. Statistics of Heat-Flow Grid Averages*
(9×10^4 square nautical miles per grid element)

Area	Lee and Uyeda [1965, p. 141]			This Work (July 1968)		
	N	$\bar{q}$	S.D.	N	$\bar{q}$	S.D.
Atlantic Ocean	65†	1.21	0.64	127	1.32	0.56
Indian Ocean	94†	1.35	0.67	110	1.36	0.82
Pacific Ocean	181†	1.53	0.87	305	1.50	0.78
All oceans and seas	340	1.42	0.78	577	1.46	0.78
Continents	51	1.41	0.52	79	1.45	0.47

*N number of data.
$\bar{q}$ arithmetic mean in $\mu cal/cm^2$ sec.
S.D. standard deviation from the mean in $\mu cal/cm^2$ sec.
† Because there were few data from Mediterranean and marginal seas, they were included under oceans.

errors by analyzing repeated and closely spaced observations less than 10 km apart (the uncertainty in location of most measurements at sea is about 5 km). They noted that because of measurement errors the mean heat-flow value for a region determined from n observations can be in error by $\pm 0.66/(n)^{1/2}$ $\mu cal/cm^2$ sec at the 95% confidence level.

In addition to measurement errors, one must consider several other factors (e.g., bias in station location and environmental effects) in determining the reliability of the value of regional heat flux from any group of heat-flow measurements. In many oceanic regions, especially on and near oceanic ridges, the local variability may be substantial [Von Herzen and Uyeda, 1963; Langseth et al., 1966]. In such regions there is a relatively high probability that any one measurement may not be representative of the region. It is thought that much of this variability is due to local environmental effects: a variable thickness of sediment cover overlying basement rock, rapid sedimentation or local sediment slumping, recent magmatic intrusions, or a combination of such factors [Von Herzen and Uyeda, 1963]. In addition, small localized fluctuations of temperature at the sea floor may contribute to the variability, especially for short probes [*Lubimova et al.*, 1965]. From time-series analysis of ocean water temperatures, *Lee and Cox* [1966] found that semidiurnal temperature fluctuations of ocean waters may also introduce some errors in oceanic heat-flow measurements, especially in rough topography. Lateral variations of thermal conductivity in crust and mantle may also cause variations between small regions [*Lachenbruch and Marshall*, 1966]. However, no definitive field studies have been made into any of these possible sources of variability, and these will likely be emphasized in future work.

Birch [1966] attempts to estimate the reliability of oceanic measurements by comparing them with techniques used in measurements on continents. By comparison of the relative length of hole utilized in continental measurements with the length of oceanic temperature probes, it is concluded that *one* continental measurement may be comparable to *ten* oceanic determinations. However, such a comparison seems difficult to support in general, since the considerably greater precision in measurement of temperature difference and greater uniformity of thermal conductivity tend to enhance the oceanic method relative to continental determinations. Furthermore, the bias in selection of continental heat-flow stations (usually in existing oil wells or mines) is far more unfavorable in ascertaining regional heat flow than the more random selection of oceanic stations.

DISCUSSION

The most important discovery in terrestrial heat flow is that the average heat flow over the continents does not differ significantly from that over the oceans. The classic interpretation of this result given by *Bullard* [1952] is now commonly accepted: since the outflow of heat from the earth's interior is mainly radiogenic in origin, the equality of heat flow suggests that radioactivity is approximately the same beneath land and sea. This further implies that the upper mantle under the continents differs from that under the oceans. Continents and oceans therefore are different not only near the surface, but extending several hundred kilometers deep [*MacDonald*, 1963].

A correlation between heat flow and major geological features has been found by *Lee and Uyeda* [1965]. The average and standard deviation of oceanic heat-flow values are 0.99 ± 0.61 μcal/cm^2 sec from trenches, 1.28 ± 0.53 μcal/cm^2 sec from basins, and 1.82 ± 1.56 μcal/cm^2 sec from ridges. Although such analysis has not been repeated with the currently available data, we expect no substantial changes in the results, as is evidenced by comparisons in Tables 1 and 2 between the two data analyses.

The most notable variation of oceanic heat flow is the high values associated with oceanic ridges. Most of the highest values measured are near the crests of these ridges; nevertheless, many normal and even low values are also found on ridge crests. The detailed distribution of heat flow across ridge crests differs considerably with location. The Mid-Atlantic ridge seems to exhibit high heat flow mostly within a band about 200 km wide centered at its crest [*Vacquier and Von Herzen*, 1964; *Langseth et al.*, 1966]. Parts of the East Pacific rise are associated with a considerably wider band of high heat flow [*Von Herzen and Uyeda*, 1963; *Langseth et al.*, 1965]. Plots of heat flow versus distance from ridge crest can be found in *Lee and Uyeda* [1965, Figures 15, 20, and 29].

The apparent distribution of heat flow over the ocean floors and its relationship to processes in the earth's mantle have changed as more data are accumulated. After only a few tens of measurements were available, it appeared that the heat-flow measurements directly supported the concept of an oceanwide convective pattern [*Bullard et al.*, 1956; *Von Herzen*, 1959]. As the data and coverage increased, some difficulties with this simple model became apparent: (1) the relatively broad band of high heat flow on the East Pacific rise is quite different from the narrow and sometimes nonexistent anomalies of the Mid-Atlantic and Mid-Indian Ocean ridges [*Von Herzen and Langseth*, 1965; *Langseth et al.*, 1966]; furthermore, a long but narrow and sinuous heat-flow pattern coincident with ridge crests was not particularly congruous with a simple convective pattern; (2) profiles across oceanic ridges indicate an extremely variable pattern near the crests, even on the relatively smooth topography of the East Pacific rise [*Von Herzen and Uyeda*, 1963]; (3) not all high heat-flow regions are associated with oceanic ridges, and at the boundaries of some anomalous heat-flow regions the transition to a region of different heat flow occurs over relatively short distances [*Von Herzen*, 1967].

For the most part, these complexities have led to more complex hypotheses regarding the distribution of convection currents in space and time. The relatively narrow and variable band of high heat flow on most ridges may indicate a relatively shallow convection, perhaps with associated intrusions of magma near the surface [*Von Herzen and Uyeda*, 1963; *Von Herzen and Vacquier*, 1966]. Another interpretation is that the convection pattern and upwelling of material beneath some ridges has been rejuvenated only recently in the geologic past, after a longer period of dormancy [*Langseth et al.*, 1966].

Alternatively, processes other than convection may account for the heat-flow pattern. For example, *Lee* [1968] emphasized the effects of selective fusion on the earth's thermal history and its bearing on the evolution of various geological provinces. Since pronounced fractionation of radioactive elements usually occurs in selective fusion, the complex heat-flow pattern may be due in part to variations in chemical differentiation of the mantle and crust.

The rapid accumulation of oceanic heat-flow data is leading to a more uniform geographic distribution. Detailed investigations of the environmental effects on measurements are needed to clarify the variability of the heat-flow pattern. The solutions to the broad problems of the earth's interior to which heat flow is intimately related (history, composition, dynamics, etc.) depend upon the interpretation of heat-flow data in conjunction with many other geophysical and geological studies.

Acknowledgments. We thank the following colleagues for sending us their heat-flow data in advance of publication: Drs. G. Bodvarsson, R. E. Burns, P. Grim, M. G. Langseth, J. Sclater, S. Uyeda, V. Vacquier, T. Watanabe, and M. Yasui. Support from the U.S. Office of Naval Research, contract CO-241(RVH), and the National Aeronautics and Space Administration, grant NsG-216 (WHKL), during the analysis of data and preparation of this paper is acknowledged.

REFERENCES

Benfield, A. E., Terrestrial heat flow in Great Britain, *Proc. Roy Soc. London, A, 173,* 428–450, 1939.

Birch, F., Earth heat flow measurements in the last decade, in *Advances in Earth Science,* edited by P. M. Hurley, pp. 403–430, MIT Press, Cambridge, 1966.

Bullard, E. C., Heat flow in South Africa, *Proc. Roy. Soc. London, A, 173,* 474–502, 1939.

Bullard, E. C., Heat flow through the floor of the eastern North Pacific Ocean, *Nature, 170,* 202–203, 1952.

Bullard, E. C., Historical introduction to terrestrial heat flow, in *Terrestrial Heat Flow, Geophys. Monograph 8,* edited by W. H. K. Lee, pp. 1–6, American Geophysical Union, Washington, D.C., 1965.

Bullard, E. C., A. E. Maxwell, and R. Revelle, Heat flow through the deep sea floor, *Advan. Geophys., 3,* 153–181, 1956.

Corry, C. E., C. Dubois, and V. Vacquier, Instrument for measuring terrestrial heat flow through the ocean floor, *J. Marine Res., 26,* 165–177, 1968.

Lachenbruch, A. H., and B. V. Marshall, Heat flow through the arctic ocean floor: the Canada basin–Alpha rise boundary, *J. Geophys. Res., 71,* 1223–1248, 1966.

Langseth, M. G., Techniques of measuring heat flow through the ocean floor, in *Terrestrial Heat Flow, Geophys. Monograph 8,* edited by W. H. K. Lee, pp. 58–77, American Geophysical Union, Washington, D.C., 1965.

Langseth, M. G., P. Grim, and M. Ewing, Heat flow measurements in the East Pacific Ocean, *J. Geophys. Res., 70,* 367–380, 1965.

Langseth, M. G., X. Le Pichon, and M. Ewing, Crustal structure of the mid-ocean ridges, 5, Heat flow through the Atlantic Ocean floor and convection currents, *J. Geophys. Res., 71,* 5321–5355, 1966.

Lee, W. H. K., Effects of selective fusion on the thermal history of the earth's mantle, *Earth Planetary Sci. Letters, 4,* 270–276, 1968.

Lee, W. H. K., and C. S. Cox, Time variation of ocean temperatures and its relation to internal waves and oceanic heat flow measurements, *J. Geophys. Res., 71,* 2101–2111, 1966.

Lee, W. H. K., and S. Uyeda, Review of heat flow data, in *Terrestrial Heat Flow, Geophys. Monograph 8,* edited by W. H. K. Lee, pp. 87–190, American Geophysical Union, Washington, D.C., 1965.

Lee, W. H. K., and S. Uyeda, manuscript in preparation, 1969.

Lubimova, E. A., R. P. Von Herzen, and G. B. Udintsev, On heat transfer through the ocean floor, in *Terrestrial Heat Flow, Geophys. Monoraph 8,* edited by W. H. K. Lee, pp. 78–86, American Geophysical Union, Washington, D.C., 1965.

MacDonald, G. J. F., The deep structure of continents, *Rev. Geophys., 1,* 587–665, 1963.

Revelle, R., and A. E. Maxwell, Heat flow through the floor of the eastern north Pacific Ocean, *Nature, 170,* 199–200, 1952.

Sverdrup, H. U., M. W. Johnson, and R. H. Fleming, *The Oceans*, Prentice-Hall, New York, 1087 pp., 1942.

Vacquier, V., S. Uyeda, M. Yasui, J. G. Sclater, T. Watanabe, and C. E. Corry, Heat flow measurements in the northwestern Pacific, *Bull. Earthquake Res. Inst., Tokyo Univ.*, 44, 375–393, 1967.

Vacquier, V., and R. P. Von Herzen, Evidence for connection between heat flow and the Mid-Atlantic ridge magnetic anomaly, *J. Geophys. Res.*, 69, 1093–1101, 1964.

Von Herzen, R. P., Heat flow values from the southeastern Pacific, *Nature*, 183, 882–883, 1959.

Von Herzen, R. P., and M. G. Langseth, Present status of oceanic heat flow measurements, in *Phys. Chem. Earth*, 6, 367–407, 1965.

Von Herzen, R. P., and S. Uyeda, Heat flow through the eastern Pacific ocean floor, *J. Geophys. Res.*, 68, 4219–4250, 1963.

Von Herzen, R. P., and V. Vacquier, Heat flow and magnetic profiles on the Mid-Indian Ocean ridge, *Phil. Trans. Roy. Soc. London, A*, 259, 262–270, 1966.

Von Herzen, R. P., Surface heat flow and some implications for the mantle, in *The Earth's Mantle*, edited by T. F. Gaskell, pp. 197–230, Academic Press, New York, 1967.

Terrestrial Heat Flow in Volcanic Areas

Ki-iti Horai and Seiya Uyeda

HEAT discharge at the earth's surface can be expressed generally by

$$Q_T = Q_C + Q_M \quad (1)$$

where Q_T is the total heat flux, and Q_C and Q_M are heat fluxes due to ordinary conduction and mass transportation, respectively. At the earth's surface, Q_M is usually negligibly small. However, in some areas, such as active volcanoes, hot springs, and geothermal areas, Q_M is important. We shall review the data on Q_C in volcanic and geothermal areas and examine the relationship between Q_C and Q_M in these areas. We show below that volcanic and geothermal areas are the regions where Q_C is generally high. Therefore, we suggest that practically all the anomalously high heat flows are related to magmatic activities even though they are not manifested on the surface.

TERRESTRIAL HEAT FLOW Q_C IN VOLCANIC AREAS

Measurements of Q_C on land and at sea enable us to characterize the thermal features of the major tectonic units of the earth: the average heat flow values are almost equal on the continent and the ocean (Table 1).

The areas where Q_C is regionally higher than 2 HFU (1 HFU = 10^{-6} cal sec^{-1} cm^{-2}, defined by *Simmons* [1966]) are indicated in Figure 1a and 1c. These figures show that high heat flow regions have a fairly distinctive distribution: (1) land areas on the circum-Pacific orogenic belt and the region of the Alpine orogeny; (2) mid-oceanic ridges and their possible extensions such as the Gulf of California and the Red Sea; (3) basins such as the Hungarian basin, Sea of Japan, and the Sea of Okhotsk; and (4) other areas such as the east Australian highland.

Figures 1b and 1d show the geographical distribution of volcanoes and Cenozoic volcanics. About two-thirds of the modern volcanoes are on the circum-Pacific belt. Though we lack heat flow data on the large part of the belt, the data from western North America, the Japanese islands, and Kamchatka suffice to suggest that the area of more than 2 HFU is closely related to the volcanoes. This relation holds well on the land areas, except in Hungary [*Boldizsar*, 1964].

An example of the more detailed relations in a volcanic area is given by the data from the Japanese area (Figure 2). From a macroscopic point of view, the Japanese area as a whole is in the circum-Pacific Mesozoic-Cenozoic orogenic belt, but, on closer examination, it is ap-

TABLE 1. Statistics of Heat Flow Values for Major Geological Features on Land and at Sea*

Geological Feature	Number of Values	Mean, HFU	Standard Deviation	Standard Error	Modes
On Land					
Precambrian shields	26	0.92	0.17	0.03	0.9
Post-Precambrian nonorogenic areas	23	1.54	0.38	0.08	1.3
Post-Precambrian orogenic areas†	68	1.48	0.56	0.07	1.1
Paleozoic orogenic areas	21	1.23	0.40	0.09	1.1
Mesozoic-Cenozoic orogenic areas	19	1.92	0.49	0.11	1.9, 2.1
Island arc areas	28	1.36	0.54	0.10	1.1
Cenozoic volcanic areas (excluding geothermal areas)	11	2.16	0.46	0.14	2.1
All land values	131	1.43	0.56	0.07	1.1
At Sea					
Ocean basins	273	1.28	0.53	0.03	1.1
Ocean ridges	338	1.82	1.56	0.09	1.1
Ocean trenches	21	0.99	0.61	0.13	1.1
Other areas	281	1.71	1.05	0.06	1.1
All sea values	913	1.60	1.18	0.04	1.1

* After *Lee and Uyeda* [1965].
† Excluding Cenozoic volcanic areas.

parent that the volcanoes, hot springs ($T > 30°C$), and geothermal areas are located exclusively in the inner zone of the island arc and that this zone is almost exactly coincident with the zone of high heat flow [*Uyeda and Horai*, 1964]. The average of the eight heat flow values in the volcanic zone of Japan is 2.16 HFU. This value is significantly higher than either the world average (1.5 HFU) or the average of the entire Japanese area (1.54 HFU). It must be remembered that all these Q_c values are taken far enough from active volcanoes and geothermal areas to be unaffected by mass transport. Q_c is known to be much higher locally in the vicinity of active volcanoes or in geothermal areas. For example, $Q_c = 11$ HFU at Kusatsu-Shirane volcano (Figure 3), and $Q_c = 15$ HFU at the Matsukawa geothermal area.

We can thus write the total heat flux Q_T as

$$Q_T = Q_{CO} + Q_{CR} + Q_{CL} + Q_M \quad (2)$$

where Q_{CO} is the average heat flow, and Q_{CR} and Q_{CL} are the 'regional' and 'local' excess heat flux.

In the volcanic zone in Japan, $Q_{CR} = 0.62$ HFU $= 2.16$ (average heat flow in volcanic zone) minus 1.54 (average heat flow of all the Japanese islands). In other high heat flow areas in Figure 1, Q_{CR} is also roughly 0.5–1.0 HFU. In the Kusatsu-Shirane volcano $Q_{CL} = 9$ HFU, and in the Matsukawa geothermal area $Q_{CL} = 13$ HFU. *Boldizsar* [1963] reported that Q_{CL} in the Larderello area is 6–14 HFU, whereas *Dawson and Fisher* [1964] find Q_{CL} as high as 40 HFU in the Wairakei area. Thus, Q_{CL} shows large local variations, and also may vary greatly from one locality to another. The local variation of Q_{CL} in the vicinity of an active volcano or geothermal area, as well as the total amount of heat transferred as Q_{CL} at a given locality, are important problems. Although extensive heat flow measurements have been made on active volcanoes [*Facca*, 1964; *Polyak*, 1966] or geothermal areas [*Benseman*, 1959; *Banwell*, 1963; *Dawson*, 1964], much more work seems needed for the clarification of the above problems. In local hot spots, where $Q_M > Q_C$ [*Robertson and Dawson*, 1964], Q_C measurement has often been neglected because Q_{CL} is unimportant from the engineering point of view. Application of the oceanic heat flow techniques to volcanic islands has so far yielded no reliable data: 1.5 HFU at about 30 km from the erupting volcano Miyake-jima [*Yasui et al.*, 1963] and 0.4–1.8 HFU on the flank of Stromboli about 12 km from the island [*Birch and Halunen*, 1966] were obtained from probes that partially penetrated into hard bottom.

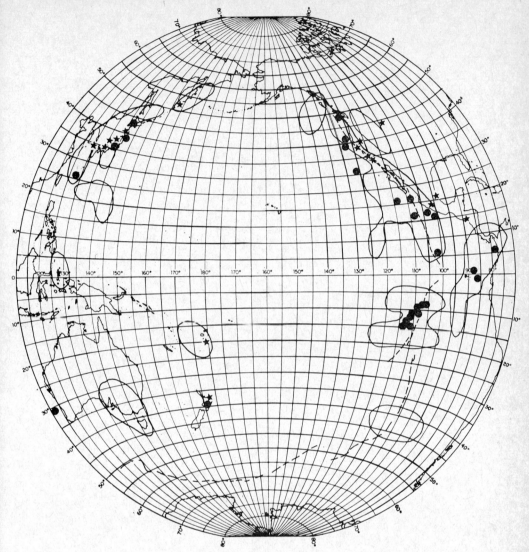

Fig. 1a. High heat flow regions: mid-oceanic ridges are shown by dashed lines; regions with $Q > 2$ HFU are enclosed by solid lines; areas with $Q > 5$ HFU are shown by heavy dots; geothermal areas are shown by stars. Data from *Lee and Uyeda* [1965] and *Von Herzen* [1967].

A few examples of Q_C measurement in the crater of an active volcano have been reported. On the floor of the crater of volcano Mutnovka, an average heat flow Q_{CL} of 1000 HFU was measured by *Polyak* [1965]. Estimated heat flow on the surface of cooling crust of the lava lake (Figure 4) in Alae crater, Hawaii, is 6000 HFU (temperature data from *Peck et al.* [1964], combined with the estimated thermal conductivity 2×10^{-3} cal/cm sec °C).

These results show that Q_{CL} can be 10^3 times larger than Q_{CO}.

In oceanic areas, high values of heat flow are concentrated almost exclusively on mid-oceanic ridges and rises. In Figure 5, heat flow profiles across the East Pacific rise, mid-Indian Ocean ridge, and Mid-Atlantic ridge are shown. The typical profile of heat flow across the oceanic ridge could be interpreted as being composed of a moderate increase in Q_{CR} (2–3

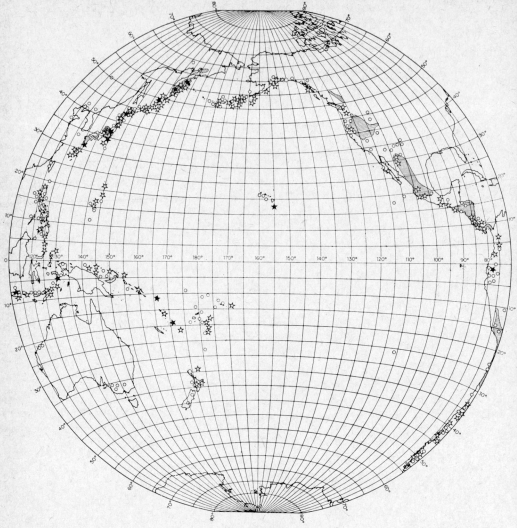

Fig. 1b. Distribution of volcanoes: closed stars, active volcanoes; open stars, recent volcanoes; open circles, extinct volcanoes; shaded areas, Cenozoic volcanic rocks. Redrawn from *Vening Meinesz* [1964].

HFU) associated with the ridge and, superposed on it, peaks of high heat flow (Q_{CL} = 6-8 HFU) localized near the crest of the ridge. The moderately high Q_{CR} associated with the mid-oceanic ridges extends a considerable distance away from the crest (Figure 6). These broad features may be caused by deep structures beneath the oceans. For example, it was noticed that heat flow is generally higher in the eastern Pacific than in the western Pacific [*Uyeda and Vacquier*, 1968; *Vacquier et al.*, 1967]. This large-scale anomaly may be related to a large mantle convection cell with the lateral dimension of the entire Pacific, but the extremely high heat flow (Q_{CL}) on the crest of the rise must have a localized origin.

The nature of submarine hot springs and submarine geothermal areas is little known. If the thermal conductivity of the substratum is assumed to be 5×10^{-3} cal/cm sec °C, a value of 5 HFU implies that the geothermal gradient is as high as 100°C/km (unless oceanic uppermost mantle is extremely radioactive); this

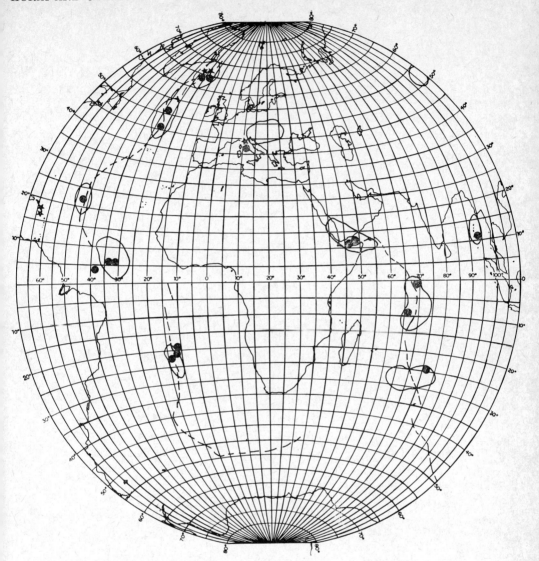

Fig. 1c. High heat flow regions; see Figure 1a for legend.

gradient would certainly cause some melting in the uppermost layer of the mantle. All the localities on land where Q_C has been reported in excess of 5 HFU (Figure 1) are in geothermal areas. Thus, a locality in the ocean with heat flow in excess of 5 HFU should also be regarded as a geothermal area. *Von Herzen and Uyeda* [1963] showed that the two narrow belts of extremely high heat flow observed over the crest of the East Pacific rise could be caused by belts of magmatic bodies at a depth of about 10 km beneath the sea floor.

Local high heat flow Q_{CL} on the crest of the Mid-Atlantic ridge and the Mid-Indian Ocean ridge can be interpreted in the same way. Studies of the bottom core samples gave much evidence of recent volcanism over the crest of the East Pacific rise [*Peterson and Griffin,* 1964; *Bostrom and Peterson,* 1966]. The ocean floor spreading hypothesis (see articles in this Monograph by Heirtzler, p. 430, and Vogt et al., p. 556) is generally favored by the thermal point of view: if the magma extrudes continually at a rate of 4 cm/year to form the

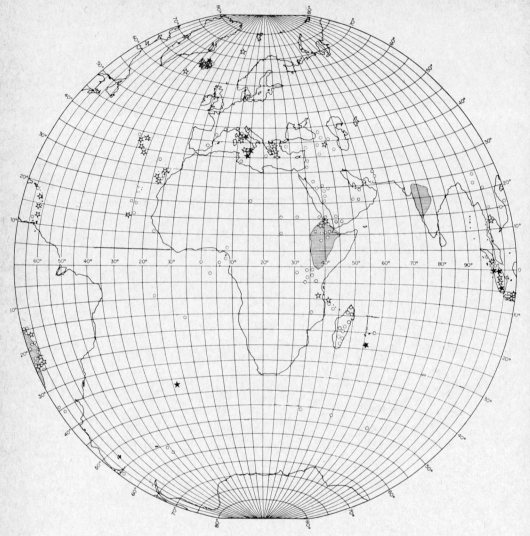

Fig. 1d. Distribution of volcanoes; see Figure 1b for legend.

basalt layer 2 km thick on the crest of the East Pacific rise (40 km wide) (Figure 6), heat carried by the magma will be observed as $Q_{CL} \sim 10$ HFU in the crestal zone. This Q_{CL} (calculated by equation 3 in the next section) is close to the observed values.

Evidence of the magmatic activity of the oceanic ridge is found in Iceland, on the mid-Atlantic ridge (Figure 7). *Bodvarsson* [1955] measured Q_{CR} of 4 to 5 HFU (average 4.8) in areas distant from active geothermal areas (Table 2). In the vicinity of the geothermal area, the local anomaly of heat flow Q_{CL} would naturally be much higher. In Iceland, volcanism has been taking place in the zone that overlies the Mid-Atlantic ridge. The rate of heat release due to mass transportation (Q_M) averaged over the total area of Iceland 103,000 km²) amounts to 1.8 HFU (heat due to effusive volcanic materials in the post-glacial period) plus 0.8 HFU (heat due to thermal water), a total of 2.6 HFU [*Bodvarsson*, 1955]. Q_T in Iceland would then be more than 7 HFU.

The Gulf of California, the Gulf of Aden, and the Red Sea represent another type of oceanic high heat flow area (Figure 1). If these areas are the extensions of mid-oceanic ridges

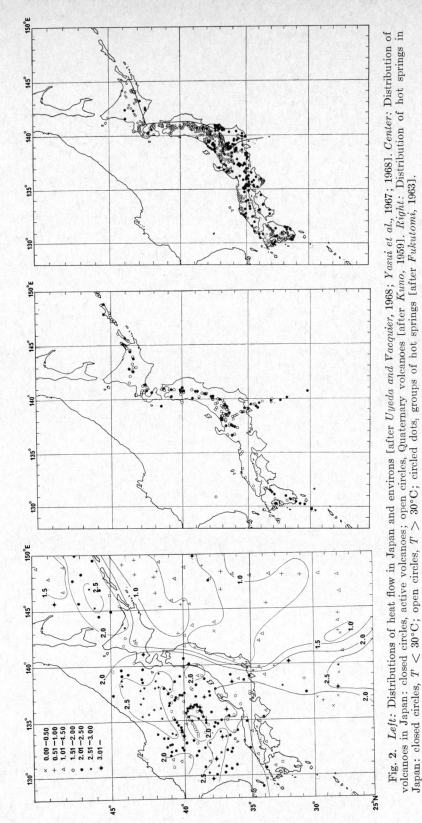

Fig. 2. *Left*: Distributions of heat flow in Japan and environs [after *Uyeda and Vacquier*, 1968; *Yasui et al.*, 1967; 1968]. *Center*: Distribution of volcanoes in Japan: closed circles, active volcanoes; open circles, Quaternary volcanoes [after *Kuno*, 1959]. *Right*: Distribution of hot springs in Japan: closed circles, $T < 30°C$; open circles, $T > 30°C$; circled dots, groups of hot springs [after *Fukutomi*, 1963].

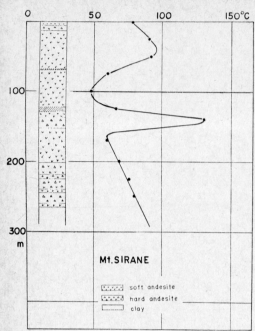

Fig. 3. Temperature-depth curve in a bore hole on the volcano Kusatsu-Shirane. The temperature is disturbed at shallow depth by the hydrothermal activity. Q_c was estimated from the gradient below 170 meters [after *Uyeda et al.*, 1958].

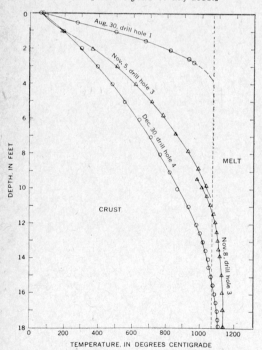

Fig. 4. Temperature profile for Alae crater, Hawaii. Measurement was made immediately after drilling [after *Peck et al.*, 1964].

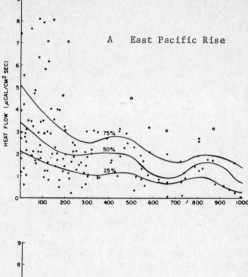

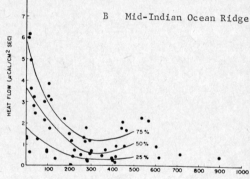

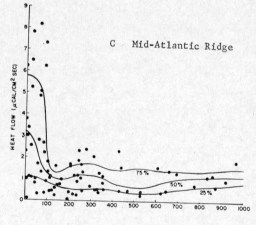

Fig. 5. Heat flow profiles across mid-oceanic ridges. Percentile line separates x % of data points under the line and $(100 - x)$ % above the line. Abscissas are distances from the crest, in kilometers [after *Lee and Uyeda*, 1965].

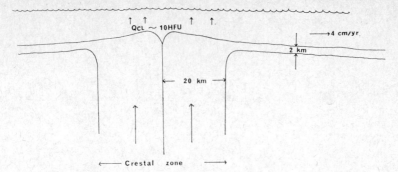

Fig. 6. Schematic representation of the mid-oceanic ridge crest according to the ocean floor spreading hypothesis. Anomalous local high heat flow $Q_{CL} = 10$ HFU can be expected on the crest.

into continental areas [*Girdler*, 1962], then the high heat flow should be explained in the same manner as for the mid-oceanic ridges. The East African rift and the rift in Siberia (passing through Lake Baikal) may be extensions of the oceanic ridges deep into continents. Active volcanoes are found in the former rift, and high heat flow values are found in both rifts [*Von Herzen and Vacquier*, 1967; *Lubimova*, 1966]. Numerous extinct volcanoes are known in the Baikal rift, though not indicated in Figure 1.

Still another type of oceanic high heat flow is found in the marginal seas like the Sea of Japan and the Sea of Okhotsk. The average heat flow of the deep basins in these seas is 2.46 HFU [*Yasui et al.*, 1968] and 2.23 HFU [*Yasui et al.*, 1967], respectively. Since these basins possess a thin oceanic crust [*Murauchi et al.*, 1968; *Kosminskaya et al.*, 1963], crustal radioactivity is a most unlikely cause for the high values. Because active volcanoes and high heat flows are confined to the continental side of island arcs (Figure 2), the high heat flow in the marginal seas may indicate subsurface magmatic activity [*Uyeda and Sugimura*, 1968].

NONCONDUCTIVE HEAT TRANSFER Q_M

In the immediate vicinity of an active volcano or geothermal area, Q_{CL} is found to be more than, say, five times greater than the normal heat flow value Q_{CO}. Because thermal conductivity of most rocks is of the same order of magnitude, the high Q_C is caused essentially by larger geothermal gradients. These high gradients, however, cannot persist to great depths without causing melting. Clearly, such high gradients near the surface are brought about by the rise of isotherms due to nonconductive processes in the deeper interior.

In geothermal areas, energy transfer is dominated by the behavior of the water-vapor system; hydrothermal processes are undoubtedly the principal cause of raising the local isotherms. As a result, Q_C is much enhanced, but the enhancement is much more important in Q_M in these areas. Hydrothermal heat trans-

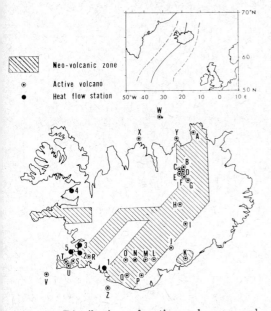

Fig. 7. Distribution of active volcanoes and heat flow stations in Iceland. For heat flow values, see Table 2. Inset shows the location of Iceland in the Atlantic Ocean; solid line indicates the crest, and the dashed lines the approximate extent of the Mid-Atlantic ridge [*Bodvarsson*, 1955].

TABLE 2. Temperature Measurements in Wells in Iceland*

Locality	Depth, meters	Thermal gradient, °C/meter	Heat flow,† HFU	Geology and Distance from Nearest Thermal Activity
Thykkvibaer	91	0.093	3.7	50 meters dense sand and 41 meters basalt; distance 25 km.
Korpulfsotaoir	260	0.095	3.8	Basalt; distance 5 km.
Arnarholt	210	0.16	6.4	Basalt, probably near a major fault; distance 9 km.
Tindar	106	0.115	4.6	Basalt, on Skarosstrond fault; distance 17 km.
Reykjavik‡	55–105	0.11§	4.4	Basalt.

* After *Bodvarsson* [1955].
† Obtained from the temperature data, assuming the thermal conductivity of the strata concerned to be 0.004 cal/cm sec °C.
‡ Four wells in the neighborhood of Reykjavik.
§ Average of four values in the locality.

fer systems have been studied in geothermal areas from the standpoint of geothermal energy utilization. *Elder* [1965] and *McNitt* [1965] give extensive reviews of these problems. Table 3 is based on a compilation by *White* [1965] of the values of total heat discharge in various geothermal fields. As can be seen, the total energy output from a geothermal area is generally 10^6–10^8 cal/sec, depending on both the areal extent and the specific intensity Q_M.

The near-surface discharge in a hydrothermal system is maintained by convection of hot water beneath the earth's surface. In the Wairakei hydrothermal system in New Zealand, the total energy stored as heat content of rocks and waters is estimated to be about 2×10^{18} cal [*Elder*, 1965]. This could be exhausted within 250 years at the present discharge rate of 2.5×10^8 cal/sec, but there is geologic evidence that the thermal system in this region has existed for more than 10^6 years [*Healy*, 1962] and that in the last 10^3 years no gross changes in temperature have occurred. The body of hot water is supposed to be maintained largely by the influx of ground water. The time necessary for recharging the reservoir by the contact of surface water with deep-seated heat sources is about 10^5 years. Thus, to maintain the Wairakei hydrothermal system, the supply of energy from beneath should be continuous for a long period, although the rate of supply need not be constant. The convective system of hot water in a permeable medium would not likely extend more than a few kilometers beneath the surface. At greater depths, some other process of heat supply is needed.

As *Elder* [1965] noted, geological and seismic evidence shows that the penetrative convection of mantle material may be the most likely mechanism. The most typical process of this kind is magmatic intrusion. Therefore, it seems certain that the ultimate cause of geothermal anomalies is magmatic activity.

Volcanoes are the direct surface expressions of magmatic activity; geothermal phenomena are the surface expressions of magma that cools at shallow depths (Figure 8). If magma ascends slowly from depth and surrenders heat to the surroundings, it acts as a moving heat source. Unless the original quantity of magma is large, it would tend to solidify. The generally high Q_{CR} in volcanic areas as well as in the areas such as the Sea of Japan and the Sea of Okhotsk may be attributed to such subsurface volcanic activities (Figure 8c). The high regional heat flow Q_{CR} in the Hungarian basin and the East Australian highlands may be due to the same underground activities. An alternative explanation for high heat flow is the anomalously high radioactive heat generation in the crust or the upper mantle; examples are the high heat flow (~2.0 HFU) in Rum Jungle, Australia [*Howard and Sass*, 1964], and in the White Mountains, New England [*Birch*, 1966; *Simmons*, 1967]. Thermal conductivity contrasts in the earth may influence the heat flow anomaly, but they will not produce widespread anomalies.

When magma extrudes onto the surface, a great amount of energy is removed from the interior of the earth in a very short interval of time. Sugimura (personal communication,

TABLE 3. Natural Heat Flow of Some Hot Spring Areas of the World*

Area	Approximate Size, km²	Total Heat Flow, 10⁶ cal/sec	Q_M, 100 HFU
British West Indies			
Qualibou, St. Lucia	~0.1	8.6	~86
St. Vincent	~1	18	~18
Dominica	~1	17	~17
Montserrat	~0.1	1.6	~16
El Salvador, total		200	
Northern belt, total		50	
Southern belt, total		>150	
Ahuachapán group	80	80	1
El Playón de Ahuachapán	~0.25	0.46	~1.3
Agua Shuca	~0.25	0.32	~1.8
Fiji Islands			
Savusavu	~1	2	~2
Iceland			
Steam fields, total flow		630	
		1,000	
Henghill, total	50	55– 80	1.1– 1.6
		25–125	
Henghill, southern part only		28	
Torfajökull	100	500	5
Reykjanes	1	5– 25	5 –25
Trölladyngja	5	5– 25	1 – 5
Krysuvik	10	5– 25	0.5– 2.5
Kerlingafjöll	5	25–125	5 –25
Vonarskard	?	5–125	
Grimsvötn	12	125–750	13 –63
Kverkfjöll	10	25–125	2.5–12.5
Askja	25	5– 25	0.2– 1
Námafjall	2.5	25–125	10 –50
Krafla	0.5	5– 25	10 –50
Theistareykir	2.5	5– 25	2 –10
Low temperature areas; about 250 areas		100	
Six lines of thermal springs, each		5– 25	
Reykjavík	~5	1.7	~0.3
Reykir	~5	11	~2.2
Deilartunga line, total		25–125	
Deilartunga spring		24	
Italy			
Larderello	(~50)	(5)	(0.1)
Ischia and Flegreian Fields	~10	?	
Mont' Amiata	(~3)	(?)	
Vulcano	~1	?	
Japan			
Otaki, Kyūshū		?	
Atami, Shizuoka-ken	5	16	3.2
		22	
Ito, Shizuoka-ken		44	
Obama, Nagasaki-ken	1.5	57	38
Beppu, Ōita-ken	~10	19	~1.9
Kawayu, Hokkaido	0.7	8	11.4
Yunokawa, Hokkaido	~1	4.0	~4
Yachigashira, Hokkaido	?	.5	
Shikabe, Hokkaido	~0.5	1.2	~2.4
Toyako, Hokkaido	~3	2.2	~0.7
Noboribetsu, Hokkaido			
Hot Lake area, total	~0.2	14	
Jigokudani Valley (variable)	~0.3	~6–11.2	~70
Matsukawa, N. Honshu		?	~20–37

TABLE 3. (continued)

Area	Approximate Size, km²	Total Heat Flow, 10⁶ cal/sec	Q_M, 100 HFU
Onikobe, N. Honshu	~80	?	
Narugo, N. Honshu		?	
Mexico			
Pathé, Hidalgo	~2	?	
Ixtlan, Michoacan		?	
New Zealand			
Wairakei, 1951, 1952	7	133	19
1954	7	82	11.7
1956?	7	143	20.4
1958, 1959	7	163	24.0
1958	7	101	14.4
Waiotapu	~15	272	~18.1
Orakei Korako	~5	130	~26
Tikitere	5	40	8
Tokopia		30	
Waikiti		20	
Ngatamariki	~1	12.6	~12.6
Rotokaua	~5	52	~10.4
Ohaki	~1	12.8	~12.8
Taupo Spa	~3	36	~12.0
Kawerau (Onepu) 1959?	?	25	
1962	?	18	
Rotorua	?		
Union of South Africa			
Seven scalding springs	?	1.7	
United States			
California			
The Geysers	~1	0.4	~0.4
Sulphur Bank	~2	0.2	~0.1
Wilbur Springs area	~5	0.4	~0.8
Casa Diablo-Hot Creek	>25(?)	70	~2.8
Alkali Lakes area			
(Salton Sea)	(~50)		
Nevada			
Steamboat Springs	5	7	1.4
Bradys Springs	~2	?	
Beowawe	~3	?	
Wyoming			
Yellowstone Park	9,000		
Total, discharging water	~70	207	~1.5
Total, calculated	~70	500	~7.1
Norris Geyser Basin	~3	8	~2.7
Upper Geyser Basin	~10	90	~9.0
Mammoth-Hot River	~8	34	~4.3
USSR			
Pauzhetsk, Kamchatka	~1	18	~18

* Compiled by *White* [1965] from other sources.

1967) noted that the amount of heat E removed by M grams of effusives is given by

$$E = (1.4 \pm 0.3) \times 10^{10} M \text{ erg} \quad (3)$$

For the 1777–1792 eruption of Mihara volcano, Oshima, M was estimated to be 0.65×10^{15} grams [*Nakamura*, 1964], thus $E = 0.8 \times 10^{25}$ ergs. However, heat release Q_M due to volcanic eruption occurs intermittently through the history of an active volcano. For Mihara volcano, the value of Q_M averaged with time over 1500 years, during which the volcano has experienced more than ten major eruptions, becomes 5×10^7 cal/sec [*Nakamura*, 1964].

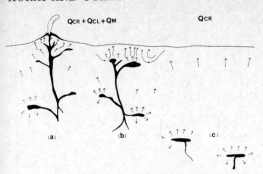

Fig. 8. (a) Surface volcanic activity; (b) geothermal activity; (c) regional high heat flow.

The similarity of this rate to those of hydrothermal areas as listed in Table 3 suggests a causative relation between volcanism and geothermal activity.

ENERGETICS OF VOLCANISM

Energy discharge accompanying volcanic eruptions has been estimated by several scientists [*Verhoogen*, 1946; *Yokoyama*, 1956, 1957a, b; *Sugimura et al.*, 1963; *Polyak*, 1966]. Clearly, the major fraction of energy is transported in the form of the thermal energy of the ejecta. Transport by volcanic earthquakes and tremors or by kinetic energy of the ejecta is negligible.

The contribution of volcanism to heat flow can be estimated from the size of the crust. The total volume of the earth's crust (7×10^9 km^3) divided by the age of the oldest rock (3.5×10^9 years) gives the average rate of crustal growth as 2 km^3/year. The present rate, on the other hand, of the production of volcanic rocks is estimated to be approximately 1 km^3/year [*Kuenen*, 1950]. In view of the uncertainties of these estimates and also the possibility that a significant fraction of magma may solidify in the crust as intrusives, these values are consistent with the view that the crust has been formed by volcanic activity throughout time. According to equation 3, the rate of 2 km^3/year is equivalent to a thermal discharge of about 2×10^{18} cal/year. This estimate is about an order of magnitude greater than previous estimates [*Verhoogen*, 1946], but is still only about 1% of the conducted heat flow of 2.4×10^{20} cal/year [*Lee and Uyeda*, 1965]. These values have generally been taken as indicating the small importance of volcanism as an energy carrier on a global scale. If we look at the Cenozoic volcanic zones only, the rate of energy discharge should be greater. *Bodvarsson* [1955] estimated that in Iceland for the postglacial period Q_M (volcanic effusive) $= 1.8$ HFU; *Yokoyama* [1957b] estimated that in the Izu volcanic archipelago south of Tokyo for the last 10^6 years $Q_M = 0.06$ HFU; *Sugimura et al.* [1963] estimated that in the volcanic zones of Japan for the late Cenozoic activity $Q_M = 0.04$ HFU for the Quarternary and 0.3 HFU for the last 50 years. On the other hand, *Polyak* [1966] estimated that in the Kamchatka volcanic zone the average $Q_M = 2.5$ HFU. These estimates, based on the volume estimates of volcanic ejecta, cannot be greatly in error. The large differences among the Japanese, Icelandic, and Kamchatka estimates might be related to the difference in the basic mechanism of magmatic activity in these areas.

An interesting possibility that volcanic processes may transmit much larger quantities of thermal energy has been suggested by Rikitake and Yokoyama [*Rikitake and Yokoyama*, 1955; *Yokoyama*, 1956, 1957a, b]. On the basis of the observation on changes of the local geomagnetic field associated with the activity of the volcano Mihara, Rikitake and Yokoyama deduced that material beneath the volcano must have lost its magnetization because of an increase of temperature during the active phase and that the material then cooled off after the activity and recovered the magnetization. The estimated amount of thermal energy discharged during this process was about 2 orders of magnitude greater than the thermal energy of ejecta. If this relationship is general, one must multiply the volcanic energy hitherto estimated [*Hédervári*, 1963] by 100. This would mean that the regional energy discharge of 0.04 HFU of the Japanese Cenozoic volcanism (quoted above) should be changed to 4 HFU. *Yokoyama* [1957b] in fact stated that the energy discharge for the Izu volcanic archipelago should be 9 HFU on this basis. Of course, these values are averaged over the volcanic zone, but such an averaging may not be appropriate, since this type of energy accompanies eruptions; it is essentially intermittent and is spatially limited to actual volcanoes. It cannot be detected by regional Q measurements away from active volcanoes. Therefore, if the ratio of the areas of the actual volcanoes to the volcanic zone is 100, the time-averaged heat discharge at volcanoes will be 400 to 900 HFU. The Q_T from volcanoes in action should be higher than these values by orders of magnitude, depending on the ratio of the periods of activity and inactivity. As was noted previously, Q_{CL} alone was found to be 10^3-10^4 HFU in volcanoes in action. Q_T would be much higher than Q_{CL} because of large Q_M. In fact, *Polyak* [1965] reported that Q_M transported by gases and vapors in the volcano Mutnovka amounted to 2×10^5 HFU in an interparoxysmal stage.

Delsemme [1960] and *Bonnet* [1960] found that the thermal radiation from the lava lake of volcano Nyiragongo is about 1×10^6 HFU. Energy transfer by radiation Q_R has not been considered in the present review, but when the temperature concerned is high, as in the case of active lava, Q_R may also be important.

The main objection to the analysis of Rikitake and Yokoyama is that there is no carrier for that very large quantity of thermal energy. Since the amount of energy is 100 times more than the ejecta can carry, the subsurface carrier must also be at least 100 times more than the ejecta in amount. Validity of the Rikitake-Yokoyama hypothesis must be carefully examined on various volcanoes. Basaltic volcaones, such as the Hawaiian volcanoes, being highly magnetic, will be suitable for such a study. This study is important because the multiplication of volcanic energy by a factor of 100 will bring the present rate of volcanic heat discharge to equivalence with the world's conducted heat flow; this will affect the total heat budget of the earth seriously. For example, the chondrite coincidence (see section by Lubimova in this Monograph, p. 64) would not hold any longer.

Acknowledgment. The authors thank Dr. Arata Sugimura, University of Tokyo, and Dr. Gene Simmons, Massachusetts Institute of Technology, who critically read the manuscript. Financial support was provided in part to one of us (Horai) by the Sakkokai Foundation, Japan. The other of the authors (Uyeda) was supported partly by a grant from the Earthquake Research Institute, University of Tokyo, and by the National Science Foundation Contract Ga 1077.

REFERENCES

Banwell, C. J., Thermal energy from the earth's crust, 1, Natural hydrothermal systems, *New Zealand J. Geol. Geophys.*, *6*, 52–69, 1963.

Benseman, R. F., Estimating the total heat output of natural thermal regions, *J. Geophys. Res.*, *64*, 1057–1062, 1959.

Birch, F., Earth heat flow measurements in the last decade, in *Advances in Earth Science*, edited by P. M. Hurley, pp. 403–429, MIT Press, Cambridge, 1966.

Birch, F. S., and A. J. Halunen, Jr., Heat-flow measurements in the Atlantic Ocean, Indian Ocean, Mediterranean Sea, and Red Sea, *J. Geophys. Res.*, *71*, 583–586, 1966.

Bodvarsson, G., Terrestrial heat balance in Iceland, *Timarit Verkfraeoingafelags Islands*, *39*, 1–8, 1955.

Boldizsar, T., Terrestrial heat flow in the natural stream field at Larderello, *Geofis. Pura Appl.*, *56*, 115–122, 1963.

Boldizsar, T., Terrestrial heat flow in the Carpathians, *J. Geophys. Res.*, *69*, 5269–5276, 1964.

Bonnet, G., Le rayonnement thermique du lac de lave du volcan, Nyiragongo, *Bull. Seances Acad. Roy. Sci. Outre-Mer.*, [6] *4*, 709–714, 1960.

Bostrom, K., and M. N. A. Peterson, Precipitates from hydrothermal exhalations on the East Pacific rise, *Econ. Geol.*, *61*, 1258–1265, 1966

Dawson, G. B., The nature and assessment of heat flow from hydrothermal areas, *New Zealand J. Geol. Geophys.*, *7*, 155–171, 1964.

Dawson, G. B., and R. G. Fisher, Diurnal and seasonal ground temperature variations at Warakei, *New Zealand J. Geol. Geophys.*, *7*, 144–154, 1964.

Delsemme, A., Première contribution à l'étude du débit d'énergie du volcan Nyiragongo, *Bull. Seances Acad. Roy. Sci. Outre-Mer*, [6] *4*, 699–707, 1960.

Elder, J. W., Physical processes in geothermal areas, in *Terrestrial Heat Flow, Geophys. Monograph 8*, edited by W. H. K. Lee, pp. 211–239, American Geophysical Union, Washington, D.C., 1965.

Facca, G., Geothermal studies on Etna (in Italian), *Riv. Mineraria Siciliana*, *15*, 66–75, 1964.

Fukutomi, T., *Physics of Hot Springs* (in Japanese), Iwanami, Tokyo, 1936.

Girdler, R., Initiation of continental drift, *Nature*, *194*, 521–524, 1962.

Healy, J., Structure and volcanism in the Taupo volcanic zone, New Zealand, in *Crust of the Pacific Basin, Geophys. Monograph 6*, edited by G. A. Macdonald and H. Kuno, pp. 151–157, American Geophysical Union, Washington, D. C., 1962.

Hédervári, P., On the energy and magnitude of volcanic eruptions, *Bull. Volcanol.*, *25*, 373–385, 1963.

Howard, L. E., and J. H. Sass, Terrestrial heat

flow in Australia, *J. Geophys. Res., 69*, 1617–1625, 1964.

Kosminskaya, I. P., S. M. Zverev, P. S. Weizman, Yu.V. Tulina, and R. M. Krakshina, Basic features of the crustal structure of the Okhotsk Sea and the Kuril-Kamchatka zone of the Pacific Ocean, according to deep seismic sounding data, *Izv. Akad. Nauk SSSR, Ser. Geofiz., 1963* (1), 20–41, 1963.

Kuenen, P. H., *Marine Geology*, John Wiley & Sons, New York, 568 pp., 1950.

Kuno, H., Origin of Cenozoic petrographic provinces of Japan and surrounding areas, *Bull. Volcanol.,* [2]*20*, 37–76, 1959.

Lee, W. H. K., and S. Uyeda, Review of heat flow data, in *Terrestrial Heat Flow, Geophys. Monograph 8*, edited by W. H. K. Lee, pp. 87–190, American Geophysical Union, Washington, D. C., 1965.

Lubimova, E. A., On construction of temperature profiles for the earth's crust and reduction of seismic ambiguity, *J. Phys. Earth, 14,* 27–36, 1966.

McNitt, J. R., Review of geothermal resources, in *Terrestrial Heat Flow, Geophys. Monograph 8*, edited by W. H. K. Lee, pp. 240–266, American Geophysical Union, Washington, D. C., 1965.

Murauchi, S., N. Den, S. Asano, K. Ichikawa, T. Asanuma, H. Hotta, K. Hagiwara, T. Sato, and M. Yasui, The crustal structure of the Japan Sea derived from the deep sea seismic observation, *J. Phys. Earth,* to be published, 1968.

Nakamura, K., Volcano-stratigraphic study of Oshima Volcano, Izu, *Bull. Earthquake Res. Inst., Tokyo Univ., 42,* 649–728, 1964.

Peck, D. L., J. G. Moore, and G. Kojima, Temperatures in the crust and melt of Alae lava lake, Hawaii, after the August 1963 eruption of Kilauea volcano—A preliminary report, *U. S. Geol. Survey, Prof. Paper 501-D,* D1–D7, 1964.

Peterson, M. N. A., and J. J. Griffin, Volcanism and clay minerals in the southeastern Pacific, *J. Marine Res., 22,* 13–21, 1964.

Polyak, B. G., Thermal power of active volcano Mutnovka in an interparoxysmal stage (in Russian), *Dokl. Akad. Nauk SSSR, 162*(3), 643–646, 1965.

Polyak, B. G., *Geothermal Characteristics of the Region of Contemporary Volcanism* (in Russian), Izd. Nauka, Moscow, 179 pp., 1966.

Rikitake, T., and I. Yokoyama Volcanic activity and changes in geomagnetism, *J. Geophys. Res., 60,* 165–172, 1955.

Robertson, E. I., and G. B. Dawson, Geothermal heat flow through the soil at Wairakei, *New Zealand J. Geol. Geophys., 7,* 134–143, 1964.

Simmons, G., Heat flow in the earth, *J. Geol. Educ., 14,* 105–110, 1966.

Simmons, G., The interpretation of heat flow anomalies, 1, Contrast in heat production, *Rev. Geophys., 5,* 43–52, 1967.

Sugimura, A., T. Matsuda, K. Chinzei, and K. Nakamura, Quantitative distribution of late Cenozoic volcanic materials in Japan, *Bull. Volcanol., 26,* 125–140, 1963.

Uyeda, S., and K. Horai, Terrestrial heat flow in Japan, *J. Geophys. Res., 69,* 2121–2141, 1964.

Uyeda, S., and A. Sugimura, Island arcs (in Japanese), *Kagaku* (Science), *38,* 91–97, 138–145, 269–277, 322–331, 382–390, 443–447, 499–504, 1968.

Uyeda, S., and V. Vacquier, Geothermal and geomagnetic data in and around the Island Arc of Japan, in *The Crust and Upper Mantle of the Pacific Area, Geophys. Monograph 12*, edited by L. Knopoff et al., pp. 349–366, American Geophysical Union, Washington, D. C., 1968.

Uyeda, S., T. Yukutake, and I. Tanaoka, Studies of the thermal state of the earth, 1, Preliminary report of terrestrial heat flow in Japan, *Bull. Earthquake Res. Inst., Tokyo Univ., 36,* 251–273, 1958.

Vacquier, V., J. Sclater, and C. Corry, Heat flow in the eastern Pacific, *Bull. Earthquake Res. Inst., Tokyo Univ., 45*(2), 375–393, 1967.

Vening Meinesz, F. A., *The Earth's Crust and Mantle,* Elsevier Publishing Company, Amsterdam, 124 pp., 1964.

Verhoogen, J., Volcanic heat, *Am. J. Sci., 244,* 745–771, 1946.

Von Herzen, R. P., Surface heat flow and some implications for the mantle, in *The Earth's Mantle,* edited by T. F. Gaskell, Academic Press, New York, 1967.

Von Herzen, R. P., and S. Uyeda, Heat flow through the eastern Pacific Ocean floor, *J. Geophys. Res., 68,* 4219–4250, 1963.

Von Herzen, R. P., and V. Vacquier, Terrestrial heat flow in Lake Malawi, Africa, *J. Geophys. Res., 72,* 4221–4226, 1967.

White, D. E., Geothermal energy, *U. S. Geol. Survey Circ. 519,* 1965.

Yasui, M., K. Horai, S. Uyeda, and H. Akamatsu, Heat flow measurements in the western Pacific during the JEDS-5 and other cruises in 1962 abroad M/S Ryofu Maru, *Oceanog. Mag., Japan Meteorol. Agency, 14,* 147–156, 1963.

Yasui, M., T. Kishii, and K. Sudo, Territorial heat flow in the Okhotsk Sea, 1, *Oceanogr. Mag., 19,* 87–94, 1967.

Yasui, M., T. Kishii, S. Uyeda, and T. Watanabe, Heat flow in the Japan Sea, in *The Crust and Upper Mantle of the Pacific Area, Geophys. Monograph 12,* edited by L. Knopoff et al., pp. 3–16, American Geophysical Union, Washington, D. C., 1968.

Yokoyama, I., Energetics in active volcanoes, 1, *Bull. Earthquake Res. Inst., Tokyo Univ., 34,* 190–195, 1956.

Yokoyama, I., Energetics in active volcanoes, 2, *Bull. Earthquake Res. Inst., Tokyo Univ., 35,* 75–98, 1957a.

Yokoyama, I., Energetics in active volcanoes, 3, *Bull. Earthquake Res. Inst., Tokyo Univ., 35,* 99–108, 1957b.

3. Seismology

Seismology and Upper Mantle Investigations

A. R. Ritsema

THE contents of the chapter on seismology reflect the present rapid developments in seismic investigations of the crust and upper mantle. Some of the authors stress theoretical points, others practical implications and results. Inevitably, more attention is paid to the more intensely investigated areas of the world.

Until recently, models of the earth were based on the assumption of lateral homogeneity of the mantle. It is now clear that there is significant lateral inhomogeneity not only in the crust, but also in the mantle. With improved data acquisition and automatic processing, there is no doubt that many important facts about the structure and the processes of the upper mantle will be revealed in the near future.

INSTRUMENT DEVELOPMENT

Seismological Network

The period since the beginning of the Upper Mantle Project (1962) coincides with the rapid development of the world seismological network, with standardization of instruments and with the widening of their frequency range.

The U. S. Coast and Geodetic Survey installed more than one hundred standard stations over the world. Each station consists of a three-component short-period set (0.5–1 sec) with a magnification up to 500,000, a three-component long-period set (15–90 sec) with a magnification of a few thousand, and a crystal clock with time accuracy better than 0.05 sec. The sensitivities are dependent only on the microseismic noise amplitudes, which on the sites concerned vary from 1 to 50 mμ in the short period band, and from 1 to 2 μ in the long period band. A similar network of more than twenty stations has been installed in Canada.

The USSR developed several local networks with high-sensitivity instruments for detailed investigation of separate seismic regions and installed teleseismic stations in remote territories, making the world seismological network much more even.

Array Stations

With the latest instrument developments there is no problem in obtaining magnifications of the order of one or ten million in the frequency range that is suitable for the detection of faint earthquake signals. Movements of the order of 1 A can be detected, but normally these are drowned in the natural earth noise of greater amplitude.

The array station, consisting of a number of spatially distributed identical sensors, was developed to increase the detectability of such faint signals relative to the ambient and signal-generated noise. By a filtering in the space domain, the noise is suppressed significantly. Array systems of the United Kingdom type, about twenty short-period vertical seismometers in a cross ($\sim$20 km), are being used in Scotland, Canada, India, and Australia.

Arrays with several dozen identical seismometers are being used at six places in the United States. At the Large Aperture Seismic Array in Montana, which has 525 sensors in an array about 200 km in diameter, events of $m = 3\frac{1}{2}$ can be detected up to distances of 90°. With suitable instruments, array techniques can be used also for surface-wave studies. At several other places in the world, new array stations are under construction (Japan, Scandinavia, Brazil, Africa).

Ocean Bottom Seismographs

The distribution of continents and oceans on the globe, and consequently that of the existing stations, is far from optimal for seismic observations. The development of ocean-bottom seismometers was a natural consequence.

Although the Lamont station (OBS) 100 miles off the coast of California, at an ocean depth of more than 4 km, with recording facilities on land, has been recording on a routine basis since 1965, it seems doubtful that routine data from deep oceans will become available on an extended scale in the near future. The maintenance costs are high; moreover, the most recent findings show that the noise level at the ocean bottom is larger than at quiet places on the continents.

Self-contained instruments in the short-period band with a dynamic range up to 72 db and with built-in tape recording have been successfully tested for periods of 30 days. These less elaborate instruments can be used for the recording of aftershocks of a severe earthquake, or for other special purposes.

Improvement of Data Acquisition

The traditional way of recording is on photographic paper or film with a dynamic range of only 40 db. The modern requirements of processing (filtering in space-and-time domain, Fourier analysis, mode discrimination, etc.) are extremely time consuming. For this reason, tape registration is used more and more widely, the output being in a convenient form for a direct entry in universal and special electronic computers (analog, digital, and hybrid). Moreover, this type of recording permits an increase of the dynamic range of the individual sensor to 120 db.

DEVELOPMENTS IN THEORY AND INTERPRETATION: THE USE OF HIGH-SPEED COMPUTERS

Data Processing

Computer processing of seismic records permits real-time acquisition of a complete spectrum of the earthquake waves for frequencies between 10 and 4×10^{-3} cps, and many sophisticated kinds of analysis. Other important practices are the automatic determination with a higher internal accuracy of focal locations, origin times, phase identifications, and focal mechanism.

Moreover, the computer opened new possibilities in upper mantle studies, such as the accumulation of data on regional and local anomalies in travel-time and amplitude, and the use of secondary phases. It has become possible to compute travel times and amplitudes of body waves, the dispersion of surface waves, the periods of free oscillations, and the steady state and impulse response of the medium and theoretical seismograms for a given model of the earth and the earthquake source.

Inverse Problems

Several years ago the inverse problem was formulated: to find the set of the earth models which satisfy a given set of seismic observations (allowing for observational errors). These investigations have been emphasized in connection with the Upper Mantle Project in the form of a special working group led by V. I. Keilis-Borok.

The first results showed a great ambiguity; for instance, several dozens of upper mantle velocity distributions fitted with the observed data for Europe. On the other hand, some quite definite properties of the velocity-depth distribution could be obtained by a joint interpretation of travel-time, amplitude, and dispersion data, such as the presence or absence of a low-velocity layer. Ultimately, the observations of other geophysical fields, as gravity, magnetics, and thermal data, should be included in the calculations to obtain the actual picture of the distribution in depth of all kinds of physical parameters that we want to know.

SOME RESULTS

Structure of the Upper Mantle

Although the mantle is still considered to be laterally homogeneous below 700–800 km, the critical experiments have not yet been done. Our experience with lateral inhomogeneities in the crust and upper mantle suggests that we should expect to find lateral inhomogeneities at greater depths also.

At short distances, important variations in P amplitudes and waveform occur. Travel-time anomalies (up to 2 sec in P and S waves) clearly show a pattern that divides the crust and mantle into regions and structural blocks of different properties.

In shield areas like Scandinavia and northeast Canada, the P wave arrives early; the same seems to be true in island arcs. The greatest delays are found in mid-oceanic ridges and young mountain ranges of the Basin and Range type. The transition between two types of upper mantle indicated by these anomalies is often abrupt, sometimes less than 100 km wide. P_n velocities around 8.2 km/sec seem to be indicative of stable regions, the lower values around 7.8–8.0 of island arcs and young mountain ranges; the still lower values of mid-oceanic ridges are indicative of mobile zones and an unstable mantle. The Alps, with a P_n velocity of 8.15 km/sec, seem to be underlain by a stable mantle.

On the other hand, important local anomalies in the upper mantle have been found by seismic refraction work in apparently uncomplicated regions such as Lake Superior.

From the study of the dispersion characteristics of surface waves, it follows that the upper mantle layer, in which significant differences in a horizontal sense occur, is several hundred kilometers thick. Under most of the continents and oceans, a low-velocity channel for S waves has been found with a velocity decrease that may amount to 6%. The P-wave channel, when present, is narrower (vertically) and higher in the mantle. Under oceans, the S channel is placed higher in the mantle (100–200 km) than under continents (100–350 km). Under most young mountain ranges and mid-oceanic ridges, the low velocity zone has not been traced. The surface-wave dispersion data for the region of the Alps are an exception, pointing to the existence of an S-wave channel connected with a velocity decrease of -0.4 km/sec at depths of 80 to 220 km.

It must be kept in mind that the conclusions on the precise form of the waveguide are not unique, as investigations of the inverse problem have revealed.

Seismicity as an Indication of the Dynamics of Crust and Mantle

The distribution of shallow, intermediate, and deep earthquakes in seismic zones is related to the dynamics of the upper mantle. The new methods in epicenter determination are yielding much more reliable seismicity maps.

In the mid-ocean ridge zones, earthquakes occur only in the crust and uppermost parts of the mantle. Regions with major transcurrent faulting are characterized by the absence of intermediate and deep earthquakes. Examples of the latter are the San Andreas, Alpine (New Zealand), and North Anatolian faults. Where deep shocks do occur in the neighborhood of transcurrent faults, these seem to be genetically connected to neighboring zones of another type, such as the Masterfault of the Philippines (transcurrent) and the deep shocks of the North Celebes-Sangi Islands; the transcurrent Taiwan fault zone and the intermediate shocks of the Ryukyu arc. The converse, however, that regions with only shallow shocks are regions with major transcurrent faulting, is not true.

Intermediate earthquakes seem to be closely linked with shallow earthquake activity. In most regions with intermediate activity there is a gradual transition from one to the other. The zone with intermediate shocks seems to be narrower than that of shallow activity, and there is a tapering toward deeper levels. Sometimes, however, the number and magnitude of intermediate shocks do surpass greatly those of shallow shocks in the same region, as, for example, in the Hindu Kush and Romanian earthquake nests.

Deep earthquakes are very restricted geographically. In all zones with deep shocks, a hiatus exists in the frequency-depth and in the strain-release-depth distribution where activity is roughly less than 1/10 of that above or below. The depth of this level of reduced activity is not the same everywhere, ranging from 300–550 km in South America, 250–550 km in the Sunda arc, and 300–450 km in the North Celebes-Philippines arc and the Tonga-Kermadec arc, to 250–350 km in the New Hebrides, 200–350 km in the Solomon Islands, and 200–300 km in zones around Japan. In the Tonga-Kermadec seismic zone, the earthquake foci to a large extent concentrate on a single plane surface that dips into the upper mantle and extends to depths of 600 km and more. In most zones, however, deep earthquake centers are not continuous along the zone, the foci being concentrated in a limited number of clusters of relatively small dimensions (for example, South America and the Sunda arc).

There is some indication that zones with intermediate and deep earthquakes mark the transitions between upper mantle regions with different properties. Conversely, not all transition zones are marked by intermediate and deep earthquakes.

The strain-release accumulated in the course of time shows that for a single seismic arc, earthquakes are to a certain extent interdependent. This is also shown in the aftershocks of severe earthquakes. Sometimes a single arc or zone that from a geological viewpoint forms a single entity seems to include several compartments that act more or less independently of each other. A severe earthquake in one part will not produce aftershocks in another. There are indications that the boundaries of the compartments do not change place over several tens of years.

On the other hand, important differences in density of earthquake centers do occur over small distances within separate zones. The regular strain release curves for the zone as a whole imply that the movements are not restricted to the parts where earthquakes actually occur, but that the parts with small numbers of earthquakes are also involved in this general movement. This could be in the form of continuous creep.

It seems extremely unlikely that shocks in restricted volumes should not be expressions of a single process that is continuous in time. Thus the relatively irregular strain release curves for very localized earthquake nests such as Hindu Kush, Romania, and the South Tyrrhenian Sea suggest that not all the movement finds its expression in the form of earthquake displacements.

The study of micro-earthquakes is of great importance for reliable delineation of seismic zones. Methods for the observation of micro-shocks in the frequency band 20–30 cps have extended the range of observed earthquake magnitudes down to -2 or -3.

Indications of migration of earthquake foci in the course of time have been reported for the Sunda arc, the North Anatolian fault zone, and for the Solomon and New Hebrides arcs. Because of the shortness of the time series, it seems doubtful that these indications form a sufficient basis for earthquake prediction.

Type of Movements in Earthquake Foci

The classical earthquake mechanism studies result in the determination of the position of a pair of mutually perpendicular nodal planes of longitudinal waves in the earthquake center, and in the position of the main stress components causing the earthquake. A likely interpretation is that one of these nodal planes represents the actual fault plane. The direction of motion can be derived from the distribution of the compression and dilatation data around the focus.

Pressure Areas. In general the position of the nodal planes in crust earthquakes is such (both being vertical) that transcurrent faults prevail and principal compressive stresses are roughly perpendicular to the active tectonic zone. Often the variation in azimuth of a curved tectonic zone does not find its counterpart in a change of direction of the principal compressive stress, the last one being more uniform.

In the upper mantle, the position of the nodal planes is generally such (one about horizontal, the other about vertical and parallel to the zone) that dip-slip fault movements prevail. The sense of the motion seems to reverse sign from intermediate to deep shocks.

These results can be given some unity by assuming a horizontal flow of mantle material from the ocean side toward the continent at a depth of roughly 200 or 300 km. Normally flow will take place in the form of continuous creeplike relative movement subparallel to the flow direction. Obstruction of this continuous flow where the gradient of the flow is strongest, from 300 km downward or from 200 km upward, gives rise to sudden earthquake movements in the general direction indicated by the earthquake fault plane solutions. At higher levels in mantle and crust, the dragging effect of the flow exerts a horizontal pressure parallel to the flow direction, giving rise to reverse faulting and transcurrent faulting in subcrust and crust. The gap in the distribution of earthquake foci in depth follows naturally from the location of the low-velocity channel where the rigidity of the material seems to be least.

According to this model, the basic cause of earthquakes in the Japan region is a transfer of mantle material from the ocean block in the east to the continental block in the west. In South America, earthquakes seem to be caused by a flow of mantle material at the appropriate depth from west-southwest to east-northeast; in the Tonga-Kermadec most likely from east-northeast to west-southwest, although a considerable variation in direction does occur from one earthquake to the next (complications may arise from the bending of the seismic zone at its northern end over an angle of more than 90°); and in the Sunda and Celebes-Philippines arcs, roughly in a northwest-southeast direction.

In regions with intermediate but with no deep earthquakes, the gradient in flow is probably much greater upward than downward. The earthquakes in the Aleutian arc seem to be caused by a southeast-northwest flow of mantle material; in the Ryukyu arc by a flow south-southwest to north-northeast; in Assam and Hindu Kush by a flow south-southwest to north-northeast; Mexico by a flow southwest to northeast; and in Romania and the South Tyrrhenian Sea by a flow west-northwest to east-southeast.

In regions with only shallow shocks, the crust itself moves in concurrence with the flow, giving rise to very strong disturbances of the transcurrent fault type that often affect the complete crust from bottom to surface.

Tension Areas. In the mid-ocean ridges, the earthquake movements are of the dip-slip kind with horizontal tension for epicenters in the rift, and of the transcurrent fault type for epicenters on the fault lines traversing the rift. In the last case one of the nodal planes has the same orientation as the cross faults. Both types of motions are expressions of horizontal tension in a direction about perpendicular to the local rift and parallel to the strike of the cross faults. This points to a sideways flowing off of the originally adjacent ridge sectors, and consequently to an upwelling of mantle material under the rift.

The model of movements described here for pressure and tension areas of the earth is not the only possible one. It must be considered as a working hypothesis that needs verification by data from other geo-disciplines.

REFERENCES

Anderson, Don L., Recent evidence concerning the structure and composition of the earth's mantle, *Phys. Chem. Earth, 6,* 1–131, 1966.

Caloi, P., On the upper mantle, *Advan. Geophys., 12,* 79–211, 1967.

Discussion on Recent Advances in the Technique of Seismic Recording and Analysis, organized by Sir Edward Bullard, F.R.S., and Sir William Penny, F.R.S., London, 28–29 January 1965, *Proc. Roy. Soc., London, A, 290*(1422), 1966.

Environmental Science Services Administration Symposium on Earthquake Prediction, Rockville, Maryland, 7–9 February 1966, Environmental Science Services Administration, Washington, D.C., 1966.

International Union of Geodesy and Geophysics, 14th General Assembly, Switzerland, 1967, Abstracts of Papers, vol. 2 (IASPEI) and vol. 8 (UMC), 1967.

Mantles of the Earth and Terrestrial Planets, Proceedings of NATO Advanced Study Institute, Interscience Publishers, New York, 584 pp., 1967.

Nuclear Test Detection, Special issue of *Proc. IEEE*, 53(12), 1813–2172, 1965.

Nuttli, O., Seismological evidence pertaining to the structure of the earth's upper mantle, *VESIAC Report*, Institute of Science and Technology, University of Michigan, Ann Arbor, 1963.

Solid-Earth Geophysics, Survey and Outlook, by the Panel on Solid-Earth Problems, *Natl. Acad. Sci—Natl. Res. Council Publ. 1231*, 198 pp., 1964.

Symposium on Seismic Models, *Studia Geophys. Geodaet. Praha*, 10(3), 240–400, 1966.

Upper Mantle Committee 1st Symposium on Geophysical Theory and Computers, Moscow-Leningrad, May 1964, Proceedings, *Rev. Geophys.*, 3, 1–210, 1965.

Upper Mantle Committee 2nd Symposium on Geophysical Theory and Computers, Rehovoth, 13–23 June 1965, Proceedings, *Geophys. J.*, 11 (1-2), 1–266, 1966.

Upper Mantle Committee 3rd Symposium on Geophysical Theory and Computers, Cambridge, 27 June to 5 July 1966, Proceedings, *Geophys. J.*, 13(1-3), 1–375, 1967.

Upper Mantle Committee 4th Symposium on Geophysical Theory and Computers, Trieste, 18 to 22 September 1967, Proceedings, *Nuovo Cimento, Suppl.*, 6(1), 165 pp., 1968.

Upper Mantle Committee 5th Symposium on Geophysical Theory and Computers, Tokyo-Kyoto, August 1968, Proceedings, *J. Phys. Earth*, 16(special issue), 1968.

Upper Mantle Committee Symposium on Non-Elastic Processes in the Mantle, Newcastle, 21–25 February 1966, Proceedings, *Geophys. J.*, 14(1-4), 1–450, 1967.

Seismicity of the Earth

Setumi Miyamura

SEISMICITY was in earlier days treated as the geographical distribution of earthquake intensity. Very often local conditions confused the determination of epicenters; instrumental seismology led to precise determination of epicenters indepedent of local intensity effects. Discovery of deep-focus earthquakes added a third element. Finally the concept of instrumental magnitude of *Richter* [1935] was brought into the study of seismicity by *Gutenberg and Richter* [1941, 1945].

Gutenberg and Richter [1954] compiled an exhaustive earthquake catalog, which has been repeatedly used in the intervening years. However, their data are not necessarily complete; *Duda* [1965] has provided a more comprehensive list of the large earthquakes. Development of a world seismological network and of international data processing services in the last several years has enabled us to decrease the magnitude limit for the study of world seismicity to $M = 5$ or less, a limit that was attained only in very restricted periods and regions for the statistics given by Gutenberg and Richter.

For this worldwide review of seismicity, two catalogs were used: (1) *Duda's* [1965] catalog of large earthquakes, $M \geq 7$, for the period 1897–1965; and (2) the Preliminary Determination of Epicenters (PDE) of the U. S. Coast and Geodetic Survey for the period January 1963 to June 1966.

Duda's catalog supplies the most reliable data now available for large earthquakes and seems homogeneous over the whole world for the magnitudes and periods considered. The PDE includes the magnitude m calculated by the maximum ground amplitude of P-wave group and its period at different stations from 1963 to date. It seems to supply homogeneous data over the whole world for magnitude $m \geq 5$, except antarctic regions. The PDE also contains many epicenters of earthquakes with magnitudes $m < 5$ (even down to $m = 4$ for some regions).

LARGE EARTHQUAKES

Accuracies of the epicenter coordinates, focal depths, and magnitudes of earthquakes in Duda's catalog are presumably better than ± 200 km, ± 50 km, and ± 0.5 magnitude unit, respectively; these accuracies are taken into consideration in the following statistical treatments. Duda's data are presented in Table 1 and Figure 1 according to magnitudes and depths. Three relative minima of depth frequency, 30–60, 270–300, and 420–480 km, allow us to classify the earthquakes into shallow

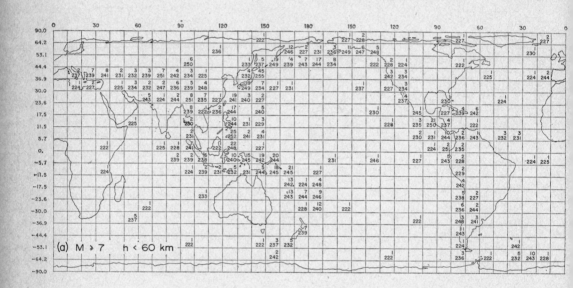

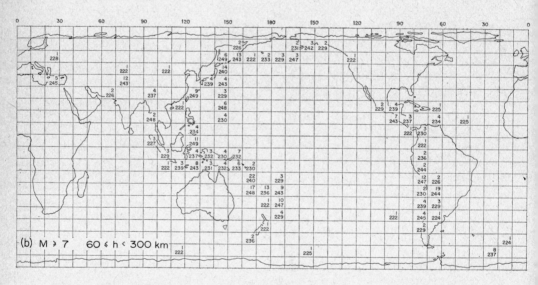

Fig. 1. Large earthquakes ($M \geq 7$): number of earthquakes (upper number) and energy index K (lower number) for equal area divisions ($\sim 7.1 \times 10^5$ km^2) on Lambert's conical projection for the period 1897–1964, based on *Duda's* [1965] catalog. Equal areas are bounded

(<60 km), intermediate (<300 km), transient (<450 km), and deep (≥450 km) groups. The classification differs slightly from the traditional definitions used by *Gutenberg and Richter* [1954] and *Båth and Duda* [1963]. Because of uncertainty in determination of depth, the minimum at 30–60 km depth may not be real, but the division into shallow and intermediate at 60 km is adopted in order to cause no drastic change in the definitions of shallow and intermediate earthquakes. The minimum at 420–480 km seems to be very important as Duda emphasized. However, he ignored the minimum at 270–300 km, which corresponds to the traditional boundary of intermediate and deep shocks. The present author believes that

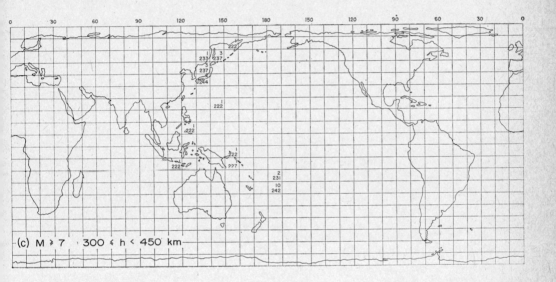

(c) M ≥ 7, 300 ≤ h < 450 km

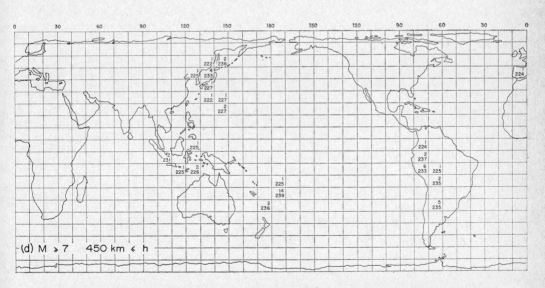

(d) M ≥ 7, 450 km ≤ h

by 10° of longitude and by latitudes 0, 5.74, 11.54, 17.46, 23.58, 30.00, 36.88, 44.43, 53.13, 64.16, and 90°. Epicenters just on the boundary are included in the southern or eastern neighboring division.

TABLE 1. Worldwide Seismicity for Large Earthquakes, $M \geq 7$, 1897–1964.*
Magnitude, frequency, and energy relations for 30-km intervals, at 33 km, and for the four depth groups.

Depth, km	Frequency						Energy index K
	$M = 7.0$	$M = 7.5$	$M = 8.0$	$M = 8.5$	Undefined	Total	
0–29	425	191	79	19	0	714	26.5
30–59	107	7	0	0	0	114	24.7
60–89	98	19	10	5	0	132	25.7
90–119	53	7	6	2	0	68	25.4
120–149	37	9	0	0	0	46	24.4
150–179	25	7	0	2	0	34	25.2
180–209	24	6	4	0	0	34	24.9
210–239	14	4	3	0	0	21	24.5
240–269	13	2	0	0	0	15	23.8
270–299	1	0	0	0	0	1	22.3
300–329	5	0	0	0	0	5	23.2
330–359	7	3	1	0	0	11	24.6
360–389	7	0	0	0	0	7	23.3
390–419	3	0	2	0	0	5	24.2
420–449	1	1	0	0	0	2	23.1
450–479	1	1	0	0	0	2	23.2
480–509	4	0	0	0	0	4	23.1
510–539	2	0	0	0	0	2	22.8
540–569	2	2	0	0	0	4	23.9
570–599	7	0	0	0	0	7	23.5
600–629	14	8	0	0	0	22	24.3
630–659	10	2	0	0	0	12	24.0
660–689	1	0	0	0	0	1	22.3
>690	0	0	0	0	0	0	0.
Total	861	269	105	28	0	1263	26.6
33	12	0	0	0	0	12	23.5
0–59	532	198	79	19	0	828	26.5
60–299	265	54	23	9	0	351	26.0
300–449	23	4	3	0	0	30	24.8
>450	41	13	0	0	0	54	24.6
Total	861	269	105	28	0	1263	26.6

* After *Duda* [1965].

this boundary cannot be disregarded; thus he adopts four depth ranges in order to divide the traditional deep focus earthquakes (≥ 300 km) into transient and deep groups.

The energy-depth relation is similar to the frequency-depth relation, as can be seen by the energy index K in Table 1 ($K = 10 \log E$, where $E = \Sigma E_i$, and E_i is the energy of the individual earthquakes; and $\log E_i = 11.8 + 1.5 M_i$ according to *Gutenberg*'s [1956] relation).

Although Duda noted that his catalog might be incomplete for deeper shocks, it seems suggestive indeed that none of the largest earthquakes ($M \geq 8$) is found in the deep-focus group and only one (January 21, 1906, 138°E, 34°N, $h = 340$ km, $M = 8.4$) is found in the transient-depth group.

The magnitude-frequency relation $\log N = a + b (8 - M)$ seems to hold. There are no appreciable differences among the coefficients b calculated for the respective depth groups, except for $M = 7$–8.5 of shallow shocks (see Figure 2).

MEDIUM AND SMALL EARTHQUAKES

Accuracies of focal coordinates in the PDE data for $m \geq 5$ are undoubtedly better than in Duda's catalog. The PDE data for January 1963 to June 1966 are summarized in Table 2. The minimum seismicity at depth 300–450 km is clear in Table 2 for the smaller earthquakes;

TABLE 2. Worldwide Seismicity, January 1963 to June 1966.*
Magnitude, frequency, and energy relations for 30-km intervals, at 33 km, and for the four depth groups.

Depth, km	Frequency												Energy Index K
	$M=1.0$	$M=3.0$	$M=3.5$	$M=4.0$	$M=4.5$	$M=5.0$	$M=5.5$	$M=6.0$	$M=6.5$	$M=7.0$	Undefined	Total	
0–29	1	7	76	518	691	456	172	56	2	0	490	2469	22.6
30–59	0	23	597	2837	3172	1559	547	119	21	1	1255	10131	23.4
60–89	0	2	51	311	407	247	109	28	1	2	178	1336	23.4
90–119	0	3	48	229	256	155	72	18	3	0	128	912	22.6
120–149	0	0	42	181	143	123	43	11	3	0	109	655	22.3
150–179	0	1	49	133	128	69	23	4	3	0	71	481	22.1
180–209	0	1	25	68	69	38	13	8	0	0	34	256	21.5
210–239	0	1	16	52	58	27	19	2	1	0	45	221	21.7
240–269	0	1	21	46	30	9	4	1	1	0	15	128	21.5
270–299	0	1	14	27	20	8	3	0	0	0	10	83	19.8
300–329	0	0	6	21	16	10	1	1	0	0	4	59	20.7
330–359	0	0	5	17	22	7	0	0	0	0	4	55	19.2
360–389	0	0	6	21	19	6	4	1	0	0	9	66	20.3
390–419	0	1	4	25	13	7	3	0	0	0	4	57	19.9
420–449	0	0	7	21	12	6	3	0	0	0	5	54	20.0
450–479	0	0	8	25	19	10	6	0	0	0	3	71	20.1
480–509	0	0	14	21	27	10	1	3	0	0	12	88	20.8
510–539	0	1	14	22	38	23	4	0	1	0	19	122	21.4
540–569	0	0	23	43	39	27	19	4	0	0	24	179	21.2
570–599	0	1	10	28	23	20	12	1	0	0	13	108	20.9
600–629	0	3	14	40	22	21	7	4	0	0	14	125	21.3
630–659	0	1	4	12	9	10	2	2	0	0	6	46	21.2
660–689	0	0	0	1	5	1	0	0	0	0	1	8	18.6
>690	0	0	0	0	0	2	0	0	0	0	0	2	18.5
Total	1	47	1054	4699	5238	2851	1067	263	36	3	2453	17712	23.8
33	0	21	513	2016	2063	874	277	57	8	0	909	6738	23.0
0–59	1	30	673	3355	3863	2015	719	175	23	1	1745	12600	23.4
60–299	0	10	266	1047	1111	676	286	72	12	2	590	4072	23.5
300–449	0	1	28	105	82	36	11	2	0	0	26	291	20.9
>450	1	6	87	192	182	124	51	14	1	0	92	749	22.0
Total	1	47	1054	4699	5238	2851	1067	263	36	3	2453	17712	23.8

* Compiled from PDE of USCGS. After *Mizoue* [1967].

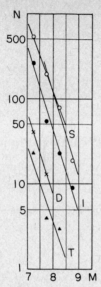

Fig. 2. Magnitude-frequency relation of large earthquakes ($M \geq 7$) of the whole world for the period 1897–1964 based on the catalog of *Duda* [1965]: S, shallow, $b = 0.83 \pm 0.01$ ($M = 7$–8.5) or $b = 1.02 \pm 0.09$ ($M = 7.5$–8.8); I, intermediate, $b = 0.96 \pm 0.08$; T, transient, $b = 0.89 \pm 0.26$; D, deep, $b = -1.00 \pm 0.00$.

therefore the same depth classification as for the large earthquakes is appropriate.

The magnitude determined by USCGS is not indisputable, but the accuracy of ±0.5 may be tenable. The number of earthquakes with undefined magnitude is 13.5% of all reported earthquakes; it reaches 20% for some depths, including very shallow ones. The increase of the number of earthquakes with decreasing magnitude is seen down to $m = 4.5$ for depths shallower than 120 km and to $m = 4.0$ for levels deeper than 120 km, except a few narrow depth ranges. However, the linear relation between log N and m is seemingly untenable, as can be seen in Fig. 3, and if we assume the linear relation, it is difficult to define the lower magnitude limit for all four depth groups. This may be partly due to the too small estimate of m by USCGS definition in the larger magnitude range and partly because of the insufficient detection of earthquakes in the smaller magnitude range. It may also be attributable to the different limit of detection and the different slope of the linear relation in the various regions of the world, which will be discussed in the next section.

Figure 4 shows the worldwide distribution of epicenter density and energy index* of medium earthquakes for the four depth ranges. Because medium earthquake data are almost homogeneous over the world, Figure 4 beautifully indicates the main seismic zones of the world: (1) the mid-oceanic seismic zones, ranging along the Mid-Arctic, Mid-Atlantic to Mid-Indian, and southeast Pacific Ocean ridges via circum-

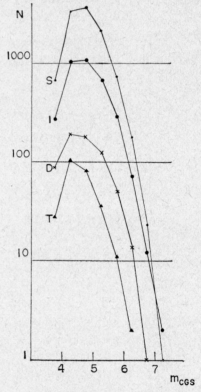

Fig. 3. Magnitude-frequency relation of earthquakes of the whole world for the period January 1963 through June 1966 after the monthly PDE summaries of USCGS. S, shallow; I, intermediate; T, transient; D, deep [after *Mizoue*, 1967].

---

* To calculate the energy from USCGS magnitude, we tentatively applied the relation between M and unified magnitude m given by *Gutenberg* [1957], i.e. $M = 1.59m - 3.97$. Unified magnitude and USCGS magnitude are not equal. While the latter is to be written M_{PZ}, the former is defined as the weighted mean of M_{PZ}, M_{PPZ}, M_{SH}, etc. This relation is also necessary to compare the magnitude-frequency relations obtained by M and m (or M_{PZ}), as is discussed in the next section.

antarctic oceanic rises; (2) the circum-Pacific seismic zone including the Carribean and Southern Antilles loops; (3) Alps-Himalayan seismic zone; and (4) the continental seismic zones (Tien-Shan-Baikal zone, Mongol-China region and East Africa rift zone). Data for small earthquakes are quite insufficient in many regions; therefore the corresponding maps [Mizoue, 1967] show the present detectability of earthquakes at the level $m = 4$ rather than the seismicity of small shocks.

REGIONAL SEISMICITY

In order to investigate the regional seismicity by PDE data, the world is divided into 65 regions bounded by meridians and parallels on the basis of epicenter distribution and seismotectonic viewpoints (Figure 5). Seismicity tables (similar to Table 1 and 2) have been published for all regions and some regional groups by Mizoue [1967].

Oceanic stable regions (1-3), oceanic seismic zones (4-16, except 7), continental seismic zones (46-47), and continental stable regions (48-65) include only shallow earthquakes. A few exceptional intermediate earthquakes in these regions are presumably due to the error in epicenter or depth determination. Although both oceanic and continental earthquakes are geometrically shallow, they are geophysically different, because the latter are crustal and the former are subcrustal, considering the thickness of the crust in the respective regions.

Deep and transient earthquakes are found only in particular regions of the circum-Pacific and Alps-Himalayan seismic zones (18, 19, 20, 24, 25, 27, 28, 29, 36, 37, 41, 44, 45). Common relative minimums of seismicity at the transient depth are observed in the different regions with slight shift of the range. Remarkable exceptions are the Kamchatka-Kuril (18) and Japan-Bonin (19) regions, where a relative maximum is found at the transient depth.

Intermediate earthquakes are found widely in the circum-Pacific and Alps-Himalayan seismic zones. Exceptions are western Canada (31), western United States (32), Caroline Islands (24), and Palmer Peninsula (40).

Magnitude-frequency relations for different depth groups and different regions are calculated, assuming a linear relation between $\log N$ and m; however, this relation is often of doubtful validity.

Older tectonic regions, such as continental seismic zones (46, 47) and some regions in the circum-Pacific zone (30 + 7, 38) and continental stable zones (61), indicate smaller b values (0.5-0.7) relative to younger tectonic regions, such as oceanic seismic zones (4-16; $b = 0.9$-1.5), circum-Pacific regions (17-38; $b = 0.8$-1.5), and Alps-Himalayan regions (41-45; $b = 0.9$-1.1). Variation of b values in the regions of the latter groups is certainly observed, but does not have the clear seismotectonic implications that were found in statistics for larger earthquakes by surface wave magnitude [Miyamura, 1962]. Further study in magnitude determination and seismotectonic classification of the regions would be necessary.

CONCLUSIONS

The map of the large shallow earthquakes (Figure 1a) depicts clearly the circum-Pacific and Alps-Himalayan seismic zones and central Asian earthquake area, which are shown on the map of the medium earthquakes (Figure 4). Only two large shocks are in the east African rift zone. The mid-oceanic seismic zones, which are quite beautifully shown by medium earthquakes, include only a small number of large earthquakes.

Isolated epicenters of large earthquakes off the west coast of Australia, Yakutsk, Baffin Bay, St. Lawrence River, and off Newfoundland occur in areas where very few small and almost no medium earthquakes are found by USCGS. Seven earthquakes in the north and central Pacific Ocean are very surprising. No medium or small earthquakes are found near them in USCGS data, although several medium and small shocks are located near the Hawaiian Islands and one small shock east of Wake Island.

On the other hand, while Duda's catalog does not supply any data for north China around Peking, USCGS data give many epicenters that resulted from the activity after the large earthquakes on March 22, 1966 ($M = 7$). This is supposedly an indication of a long recurrence time of seismic activity and small value of b in the stable regions.

The distributions of intermediate and deep

earthquakes do not show important differences, except the famous large deep shock in Spain in 1957. Medium and large earthquakes in the transient group are remarkably restricted to the western periphery of the Pacific Ocean; small shocks in this depth group are also found in South America and around Italy.

FUTURE PROBLEMS

It is still necessary to complete the epicenter determination for the isolated old earthquakes in the Pacific basin and compare them with the recent determination of small shocks. As *Båth* [1964] pointed out, it is of paramount importance to intensify the network in the oceanic regions. The lack of detectability in the antarctic region is clear in the maps given here.

A continuation of the study of past earthquakes is very necessary, because not all available data have been included in Duda's catalog. (For example, the shock on September 9, 1920, 18h 56m, 15°S, 171.5°E, was almost equally re-

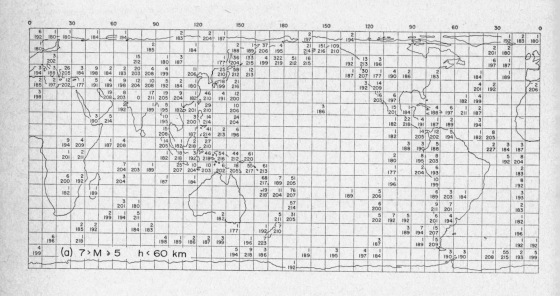

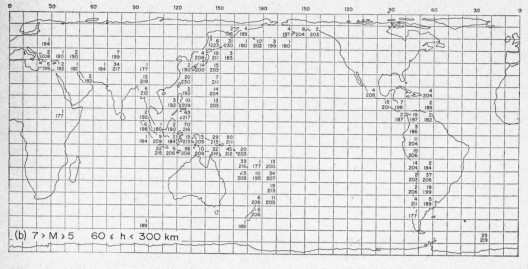

Fig. 4. Medium earthquakes ($7 > M \geq 5$): number and K index for the period January 1963 through

ported in ISS as the shock on February 2, 1920, 11h 22m, 7°S, 150°E, $M = 7.7$, but the former is not listed by Duda.) It is also desirable to compile a comprehensive catalog down to $M = 6$ at least from 1950 onward. The importance of the standardization and practical improvement of magnitude or magnitudes is indisputable. For the comparison of past and present magnitude data, the relation of M and M_{PZ} should also be established more precisely.

Increase of the accuracy of depth determination of shallow earthquakes is very necessary in order to separate the crustal and subcrustal seismicity.

Seismicity study to date has been based only on focal position and magnitude; focal mechanism should be the next important element. Mass processing of focal mechanism by routine service should be the next step to develop the seismicity study after (or parallel with) the perfection of hypocenter and magnitude determination.

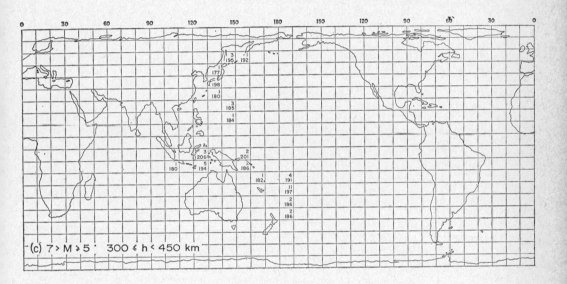

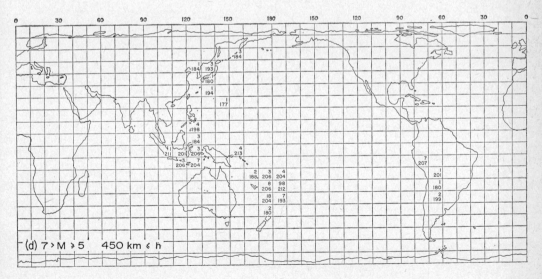

June 1966, based on PDE of USCGS. See Figure 1 for complete explanation [after *Mizoue*, 1967].

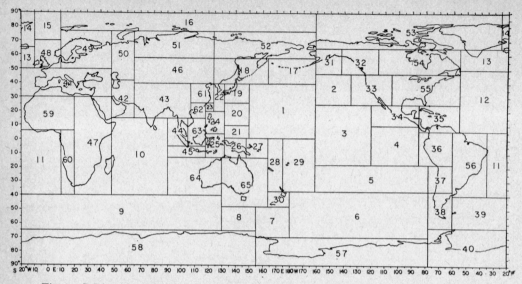

Fig. 5. Regions bounded by meridians and parallels for the study of world seismicity: oceanic stable regions are 1–3; oceanic seismic regions are 4–16, except 7; circum-Pacific seismic regions are 17–40 and 7; Alps-Himalayan seismic regions are 41–45; continental seismic regions are 46–47; continental stable regions are 48–65. [after *Mizoue*, 1967].

Finally, the seismotectonic point of view should not be neglected in seismicity studies.

Acknowledgement. The most comprehensive study of seismicity ever published appears in *Gutenberg and Richter* [1954]. The present review is based in large part on studies of *Miyamura* [1962, 1968] and *Mizoue* [1967], to which the reader is referred. The author is indebted to Dr. M. Mizoue for the computer processing in this review.

REFERENCES

Båth, M., Futur développment des réseaux de stations séismoloques, *Scientia*, pp. 1–8, September–October 1964.

Båth, M., and S. J. Duda, Strain release in relation to focal depth, *Geofis. Pura Appl., 56*, 93–100, 1963.

Duda, S. J., Secular seismic energy release in the circum-Pacific belt, *Tectonophysics, 2*, 409–452, 1965.

Gutenberg, B., The energy of earthquakes, *Quart. J. Geol. Soc. London, 112*, 1–14, 1956.

Gutenberg, B., and C. F. Richter, Seismicity of the earth, *Geol. Soc. Am., Spec. Paper 34*, 131 pp., 1941.

Gutenberg, B., and C. F. Richter, Seismicity of the earth, 2, *Bull. Geol. Soc. Am., 56*, 103–668, 1945.

Gutenberg, B., and C. F. Richter, *Seismicity of the Earth and Associated Phenomena*, 2nd edition, Princeton University Press, Princeton, 273 pp., 1954.

Gutenberg, B., and C. F. Richter, Magnitude and energy of earthquakes, *Ann. Geofis., 9*, 1–15, 1956.

Miyamura, S., Magnitude-frequency relation of earthquakes and its bearings on geotectonics, *Proc. Japan Acad., 38*, 27–30, 1962.

Miyamura, S., Seismicity of island arcs and other arc tectonic regions of the circum-Pacific zone, in *Crust and Upper Mantle of the Pacific Area*, *Geophys. Monograph 12*, edited by L. Knopoff et al., pp. 60–69, American Geophysical Union, Washington, 1968.

Mizoue, M., Variation of earthquake energy release with depth, 1, *Bull. Earthquake Res. Inst., Tokyo Univ., 45*, 679–709, 1967.

Richter, C. F., An instrumental earthquake magnitude scale, *Bull. Seismol. Soc. Am., 25*, 1–32, 1935.

Tectonic Activity in North America as Indicated by Earthquakes[1]

George P. Woollard

EARTHQUAKES are evidence of mechanical adjustment to stress. More than 90% of all earthquakes occur at relatively shallow depths and, on the continents, mostly within the crustal layer defined by the Mohorovicic seismic discontinuity; it is clear that the crust behaves in part as a brittle shell encasing the earth. Any change in overall planetary volume or shape would presumably result in a global pattern of crustal strain and fracturing, i.e., the loci of earthquakes. Similarly, any local set of distortional forces should result in a superimposed local stress pattern, possible local crustal fractures, and attendant earthquakes.

The present distribution of earthquakes does define a global pattern, but it is a peculiar one; most of the activity is confined to the perimeter of the Pacific Ocean basin in association with island arcs and trenches, and continental border trenches. The pattern of focal depths suggests shear planes dipping away from the ocean. There are also localized centers of marked seismic activity not associated with the Pacific area, such as those found in the Hindu Kush, Greece, and Turkey, which constitute part of a separate belt of seismicity extending from the Strait of Gibraltar to Burma. Belts of seismicity are associated with each of the major oceanic rises that are believed to be centers of crustal spreading.

Geologic evidence indicates cycles of tectonic orogeny that suggest periodic changes in volume, as well as continental growth by accretion (through mountain formation in former islands arcs and trenches). There is also evidence of localized crustal expansion and tensional forces that resulted in long-term uplifts, often accompanied by crustal rifting with associated vulcanism. Similarly, there is evidence of crustal subsidence and compressional forces resulting in overthrusting. In some areas the geologic record indicates that uplift and subsidence occurred simultaneously in adjacent areas. Crustal shear is evidenced by nonseismic oceanic ridges of volcanic origin and fracture zones extending continuously over hundreds of miles on the ocean floor, as well as similar features on the continents. In addition, paleomagnetic data suggest migration of the continents relative to each other and crustal spreading away from major fracture zones in the ocean basins.

At one time these areas of crustal fracturing must have been loci of seismicity; the present stress pattern indicated by seismicity may well shift with time to other areas. Present seismicity, therefore, is related to the pattern of stress relief rather than the mechanism responsible for the overall pattern of geological fractures. Geology and seismicity, however, can not be divorced, but should be considered jointly.

The Geologic Structural Pattern

North America can be represented as having a central core of ancient crystalline rocks, in part covered by a veneer of sedimentary rocks, flanked by north-south trending mountain ranges. The mountains on the west flank (the Rocky Mountain system) are of Mesozoic age; those on the east flank (the Appalachian Mountains in the United States and the Laurentian, Caledonian, and Labrador Mountains in Canada) are of Paleozoic and late Precambrian age. The same basic structural pattern is found in Mexico, except that the central area consists of an elevated plateau flanked by mountain ranges. This plateau merges into the area immediately west of the Rocky Mountains—the Colorado Plateau and the Basin and Range block faulted structure. West of the Basin and Range area is another orogenic zone represented by the Sierra Nevada (a major batholith) and the Cascade Mountains of Cenozoic volcanic ori-

---

[1] Hawaii Institute of Geophysics Contribution 245.

gin. The west coast area, another region of recent tectonic disturbance, is separated from the Sierra Nevada by a major basin, the Central Valley of California.

Superimposed on this basic structural pattern is a series of uplifts and basins. In the eastern region between the Appalachian Mountains and the Atlantic coast, there is a series of Triassic basins of block fault origin. In the Appalachian province and the central shield area, as well as in the Colorado Plateau, there are major basins and uplifts that developed from epeirogenetic subsidence and uplift of the crust. Major normal faults hundreds of miles long, with marked vertical and horizontal displacement, are known from Mexico through Alaska in all of the various geologic provinces. There are similarly marked overthrust faults, especially along the western side of the Appalachians and the eastern side of the Rockies, as well as in association with the east-west trending Ouachita Mountains of Paleozoic age in Oklahoma and Arkansas. The Appalachians plunge to the south and disappear beneath the coastal plain of the Gulf of Mexico, which is believed to be a continental land mass that foundered in Jurassic time. The Rocky Mountains change from a fold belt in Canada to a block faulted complex of major horsts and grabens in the northern United States. The Sierra Nevada plunges to the north and is replaced by the volcanoes of the Cascade range.

Although the general structural grain of the continent is north to south, there are several major east-west striking features. The ancient Penokean Mountains in Wisconsin have this strike, as do the Lake Superior graben, the Snake River graben in Idaho, the Uinta Mountains in Utah and Colorado, the Ouachita Mountains, and the Valley of Mexico graben. There are also major east-west trending faults along which there has been apparent transcurrent movement of more than one hundred miles, such as the Saltillo-Monterrey fault in northern Mexico, and the faults responsible for the embayments in the northern Appalachians.

With such a complex tectonic pattern and evidence of recurrent base leveling and uplift of both the Appalachian and the Rocky Mountains, one would expect a widespread pattern of earthquakes throughout the continent in response to regional stresses. Actually seismic activity is rather restricted and confined for the most part to the United States (see Figure 1).

The Pattern of Seismicity

The pattern of seismicity shown in Figure 1 may reflect the distribution of seismological observatories and population, especially for minor earthquakes. However, the writer feels that the map is an essentially true representation of actual conditions, because there has been an adequate seismograph station net since the International Geophysical Year (1957–1958) to record any earthquake of magnitude 4 or higher almost anywhere in North America. During this period, remote moderate magnitude earthquakes have been regularly reported throughout the area, but none from the blank areas shown in Figure 1; however, the greater density of earthquakes shown in the United States may in part reflect the longer history of recording and the distribution of observatories.

The west coast of North America is generally regarded as part of the circum-Pacific belt of seismicity; however, this belt is not continuous and has no consistent geological association.

The seismic belt associated with the Aleutian island arc and trench ends in central Alaska. A second segment extends from the southern Alaska coast (about 140°W) to the Washington-Oregon boundary. A third segment made up of two independent branches, starts at about the Oregon-California boundary: one branch follows the coast and is associated with the San Andreas fault system of California; the other turns inland, follows the east flank of the Sierra Nevada batholith, and then merges with the coastal branch in northern Baja California, where the two terminate. Although there is some seismicity in the Gulf of California, intense activity is not resumed until about latitude 20°N, where the Middle America trench begins. This trench ends near the Costa Rica–Panama boundary, where seismic activity appears to end also.

There are thus several different geologic associations with the North American segment of the circum-Pacific seismic belt. In the Aleutian area it is related to an island arc and trench, which is believed to represent the surface expression of a major crustal shear zone (indicated by a zone of earthquake foci) dipping away from the ocean, i.e., toward the concave

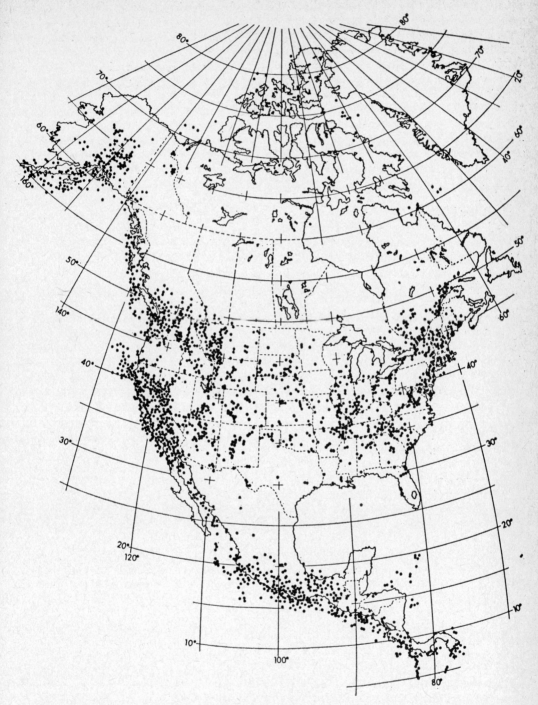

Fig. 1. Map of earthquake epicenters in North America, showing all earthquakes reported through 1964 by the U. S. Coast and Geodetic Survey, Dominion Observatory of Canada, and the Instituto de Geofisica of the University of Mexico.

side of the arc [*Benioff*, 1955]. As *Tobin and Sykes* [1966] pointed out, the intermediate-depth earthquakes (60–150 km) in Alaska are restricted to a narrow belt and cease at about 63°N; the shallow earthquakes have a more diffuse pattern and continue to about 67°N. The Canadian segment is associated with continental border horsts and graben structure. The California coastal branch follows the San Andreas fault system, along which it has been estimated that there have been more than 400 miles of translational movement (seaward side moving north) [*Allen et al.*, 1965]; the interior branch follows the fault system associated with the east flank of the Sierra Nevada. In the Central American area, the seismicity is associated with a continental boundary trench and an apparent shear zone dipping beneath the continent; this zone is defined by a progressive increase in depth of foci from 30 to 150 km under the Plateau of Mexico.

The segment off the coast of southern Alaska and Canada is believed by *Menard* [1964] to be an extension of the crest of the East Pacific rise that loses its identity as an oceanographic bathymetric and seismic feature in the Gulf of California and is thought to disappear beneath the continent and reappear in the vicinity of Vancouver Island. The seismicity associated with the east flank of the Sierra Nevada and in western Nevada may be related to uplift associated with the subcontinental continuation of the forces responsible for the East Pacific rise. The Sierra Nevada block is tilted down to the west and is bounded by faulting on the east. However, as is evident from Figures 1 and 2, there is only a tenuous connection from Nevada to Vancouver Island, unless one postulates that the causative forces continue northeast into Idaho or that the connection is masked in Oregon by the Columbia basalts. The load of these basaltic lavas, averaging more than 1000 meters in thickness, is conceivably sufficient to inhibit movement along the fractures beneath them. In Antarctica, the superimposed load of ice, about 3500 meters in thickness, appears to inhibit seismicity despite the fact that the Trans-Antarctic Mountains appear to be bounded by one of the world's major faults in length and displacement [*Woollard*, 1967]. The lack of seismicity in eastern and central Canada might be related to the burden of the Pleistocene ice cap, which disappeared only in the last 10,000 years; gravity data suggest that the crust is still under compressive stress even though ice is no longer present. It appears clear that this area, characterized by negative free air and isostatic gravity anomalies, is not in hydrostatic equilibrium with the mantle [*Innes and Weston*, 1966; *Woollard*, 1966). Seismicity would be expected south of the flexure line between the area that was depressed by the ice and the area to the south that is under tension from the buckling upward caused by the loading to the north. Such differential distortion of the crust is indicated by drowned valleys (e.g., Chesapeake Bay), and the series of exposed Tertiary and Pleistocene marine terraces south of Cape Fear, such as those found in the coastal plain of Georgia. That this displacement is now being reversed is indicated by raised beaches in the Hudson Bay region, by progressive uplift of the north shore relative to the south shore of the Great Lakes, and by a relative change in mean sea level from Maine to Florida. There are also well defined zones of extraordinary seismicity that are apparently not related to this differential stress pattern caused by the emplacement and removal of the Pleistocene ice cap. Figure 2 shows a well defined seismic belt that follows the Appalachian Mountains from Alabama through Maine into New Brunswick, Canada, that is not identified with any particular tectonic segment of the Appalachian system. The general trend cuts diagonally across from the younger Appalachian fold and overthrust belt in Alabama of late Paleozoic age to the older Appalachian province in New England of early Paleozoic age. In the south there appears to be a parallel segment associated with known faulting in the Piedmont area of Georgia. In the North there is a bifurcation; one segment is associated with the Taconic fault on the western boundary of Vermont, the other follows the New Hampshire-Vermont state boundary, with an aseismic area between the two. A third independent seismic belt appears to extend from central Maine to the tip of Long Island.

The marked local area of seismicity extending from the Atlantic coast in South Carolina to the Appalachians has no known geological association. The Charleston, South Carolina, earthquake of 1886 was one of the major earth-

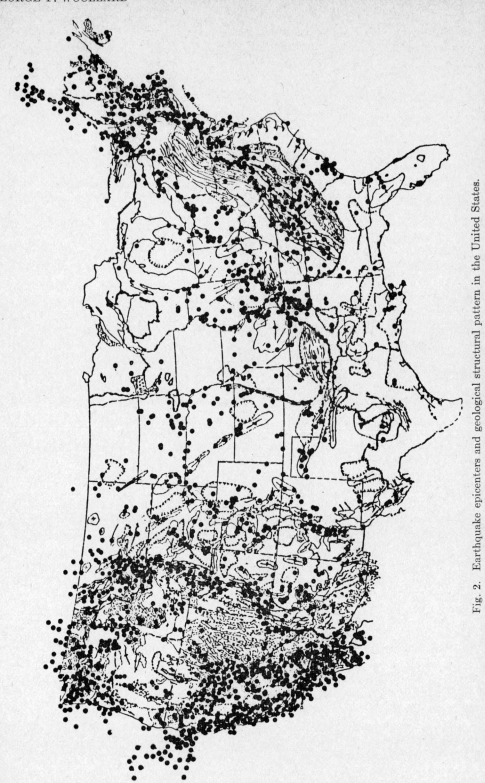

Fig. 2. Earthquake epicenters and geological structural pattern in the United States.

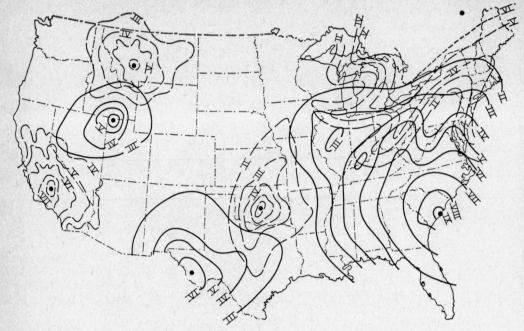

Fig. 3. Isoseismal contours in the United States for seven major earthquakes: Charleston, 1886; Quebec, 1935; Oklahoma, 1952; Texas-Mexican border, 1931; Helena, Montana, 1935; Kosomo, Utah, 1934; Kern County, California, 1952.

quakes on record and was felt over an area of 2,000,000 square miles. Although *Heck* [1963] reports that the focal depth was about 40 km, the isoseismal pattern (Figure 3) suggests a depth of nearer 60 km.

An essentially continuous zone of seismicity, parallel to the Appalachian seismic belt, extends from northern Arkansas to eastern Quebec. This belt follows a major fault system marked by the St. Lawrence River Valley, which has been likened by *Kumarapeli and Saull* [1966] to the Rift Valley of Africa. Several major earthquakes have occurred in this belt. An earthquake in the St. Lawrence River Valley in 1663 was felt over an area of at least 750,000 square miles. Other earthquakes of similarly large magnitude have occurred in the same general region since 1870, at about ten-year intervals. At the southern end of this seismic belt, the New Madrid, Missouri, earthquake of 1811 was felt over an area of 2,000,000 square miles; major earthquakes have occurred in the same general region at intervals ranging from five to ten years. Except in western New England where the seismic belts of the Appalachian, Missouri, and St. Lawrence Rivers are interconnected by the seismic zone associated with the Taconic fault and Green Mountain anticline in western Vermont, the two seismic belts are separated by an area that is mostly aseismic. This area of seismic quiesence corresponds to the Cumberland Plateau in the South; the region of Paleozoic coal basins in West Virginia and Pennsylvania; and the Adirondack massif in New York, which appears to be an independent unit, bounded by intermittent seismicity.

The eastern interior belt of seismicity has an apparent relationship not only to the St. Lawrence River, but also the Mississippi and Ohio Rivers, which form the western and southern boundaries of the state of Illinois, and correspond geologically to the boundaries of the Illinois basin, which subsided some 3500 meters during Paleozoic time. The fact that the earthquake pattern also conforms to the northern boundary of the Illinois basin, where it is in juxtaposition to the Wisconsin uplift, as well as to the western boundary, where it is in juxtaposition to the Missouri uplift, suggests that differential movement of this crustal block is still in progress.

In the mid-continent region there is an east-west pattern of earthquakes associated with the boundary between South Dakota and Nebraska; it merges into an area of scattered seismicity over the eastern halves of these two states and the adjoining area of Kansas. As with the eastern mid-continent seismic zone, there is no specific geological association throughout, but rather partial association with several tectonic features. Farther south, in the region of the Texas-Oklahoma boundary, there is an apparent east-west trending zone of seismicity that can be identified with the late-Paleozoic Amarillo-Wichita uplift, which is bounded by major faults that still appear to be active.

A narrow zone of seismicity in central New Mexico extends northward, but becomes diffuse north of the New Mexico-Colorado state boundary, apparently following the eastern boundary of the Rockies. Parallel to this zone, another, more strongly defined belt of seismicity follows the western boundary of the Rocky Mountain–Colorado Plateau block, where it adjoins the Basin and Range province; this belt cuts across the Rockies in Wyoming and follows their eastern boundary in Montana to the Canadian border, where it appears to swing westward to the Pacific coast. There is an apparent bifurcation in southern Idaho at the Utah-Idaho boundary, with earthquakes following the western flanks of the Idaho batholith as well as the Rocky Mountain front: the two branches swing back toward the Pacific coast to join again in the vicinity of Vancouver Island.

This brief summary indicates a definite association of seismicity with several major geological features, but no consistent pattern of association of seismicity with specific geologic features or tectonic elements. In fact, it appears that much of the major fracture pattern at depth defined by seismicity is not related except by coincidence to the surficial tectonic pattern. Although seismicity is associated with both the Appalachian and Rocky Mountain provinces, in each province the belt of epicenters cuts across the geologic strike. Even in California, where seismicity is associated with a major fault system, the epicentral pattern is not confined to the fault pattern, but follows in part an independent cross-cutting pattern

[*Niazi*, 1964; *Allen et al.*, 1965; *Ryall et al.*, 1966].

Energy Transmissibility

Although isoseismal patterns for individual major earthquakes showed that the energy was not radiated uniformly, it was the Gnome subsurface atomic explosion that called attention to the marked inequalities of energy transmission through the mantle in different regions. *Herrin and Taggart* [1962] showed that energy attenuation and travel times in various azimuths from the Gnome explosion were related: regions of greater travel time (lower velocity) correspond to higher attenuation, and vice versa. These relations are discussed in the article by Herrin in this monograph; Figure 1, p. 243, shows the general pattern of mantle velocity under the United States. The surprising feature of this map is that the Appalachian and Rocky Mountain systems both overlie areas with gradients rather than closures in the velocity contours.

In Figure 3, the Charleston earthquake of 1886 shows a number of interesting features. There is a well-defined extension of the isoseismal contours to the northwest, and marked flexures in the intensity contours to the northeast that indicate poor energy transmissibility along the strike of the Appalachians, but higher transmisibility on both flanks of the mountains. The higher transmissibility zone on the western flank corresponds to *Herrin and Taggart*'s [1962] map showing a high mantle velocity zone crossing Ohio; there is also striking agreement between high mantle velocity and the pronounced bulge in intensity contours across the states of Missouri and Iowa. (Neither of these features appears in the more recent version of this map, Figure 1, p. 243).

These correlations are supported by the even more pronounced flexures of the isoseismal contours for the Quebec earthquake of 1935. There is thus good evidence in the eastern half of the country that intensity is significantly affected by mantle velocity.

In other areas, the isoseismal pattern also tends to be elongated in the direction defined by maximum mantle velocity. In some areas, this corresponds to the structural grain defined by the geology; in others, it cuts across the geologic trends. Nor does energy transmissibility correspond systematically to seismicity. The isointensity pattern for the Oklahoma earthquake of 1952, for example, is oriented at right angles to both the geologic strike and the seismic belt with which it appears to be associated. In addition, well-defined extensions in the isointensity pattern lie in the aseismic areas of central Texas to the south and Iowa to the north; however, both areas have high mantle velocity values. Similarly, the Texas-Mexican border earthquake of 1931 shows a well-defined extension in the isointensity pattern into the aseismic area of high mantle velocity in central Texas. This overlap of isointensity contour pullouts for energy arriving from different azimuths in aseismic areas (as well as seismic areas) where there is a high mantle velocity is convincing evidence of the importance of mantle velocity in defining the intensity of a given earthquake with distance.

Depth of Foci

The depths of earthquake foci show two patterns: one apparently relates to shear zones dipping away from the Pacific Ocean (in connection with the Aleutian arc and trench and the Middle America trench off Central America), and one, over the rest of the continent, apparently relates to the depth of the Mohorovicic discontinuity. In Table 1 the dominant

TABLE 1. Depth of Earthquake Foci

Geographical Location	Geological Association	Depth of Foci,* km
Aleutians	Island arc and trench	*20*, **30**, *40*, *50*, *60*, up to 150
Alaska Peninsula and Central Alaska	Mountain and intermontane platform	20, **30**, 40, 50, 60, 100
Southern Alaska and West Coast of Canada	Coastal range and horst and graben structure	*20*, **30**, 40, up to 150
Vancouver Island, Washington	Puget trough graben	30 and *40*
Southern Oregon, central and southern California	San Andreas fault and Sierra Nevada	*17*, *31*, 45, 60, 100
Southern California, Nevada	Winnemucca fault, flank Sierra Nevada, and Basin and Range	*17*, **31**, 43
Arizona, Utah, Idaho	Colorado Plateau boundary, Wasatch front, and west flank of Idaho batholith	30, **38**, 52, 61
Wyoming, Montana, British Columbia	Rocky Mt. front and east and north flanks of batholith complex	20, **30**, *45*
New Mexico, Colorado, Wyoming	Rocky Mt. front, middle Rocky horsts and graben	20, *30*, **45**, 70
Oklahoma, northern Texas	Amarillo-Wichita uplift	*30*, 50
Kansas, Nebraska, South Dakota, Minnesota	Central shield	**30**, 46
Southern Wisconsin, Missouri, Illinois	Northern and western flank of Illinois basin	*20*, **36**, 58
Southern Missouri, western Tennessee, Indiana, Ohio, St. Lawrence River	In part southern and eastern flank of Illinois basin, St. Lawrence fault	20, **36**, *43*, 65
Alabama, eastern Tennessee, North Carolina, Virginia, New England	Appalachian Mt. system	25, **43**, 89
South Carolina	Coastal plain	57
Louisiana	Gulf Coastal plain	*18*
Western Coast of Mexico	Offshore trench and coastal mountains	20, **30**, 40, *50*, *60*, **100**, **150**
Central Mexico	Plateau	*30*, 50, 60, **100**, **150**
Guatamála, Panama	Offshore trench and volcanic mts.	20, **30**, *40*, 50, *60*, 70, *100*, *150*

* Boldface, most frequent; italic, frequent; roman, occasional.

focal depths are given for the various areas of seismicity.

It is not known why earthquakes should occur at or near the depth of the Mohorovicic discontinuity unless it represents a zone of polymorphic phase transformation. Earthquakes at greater depth suggest that the rate of stress application is higher in these regions because the tendency to yield by fracture rather than by plastic flow is related to the rate at which stress is applied. The analysis of explosion seismic refraction studies of crustal structure in the United States by Woollard [1968] indicates that the M discontinuity is geochemically controlled. This study shows that there is a definite relationship between crustal thickness as defined by the Mohorovicic discontinuity and mantle velocity values, and that changes in crustal thickness can in general be related to the degree of development of a basal crustal layer having velocity values of 6.8 to 7.4 km/sec. Where the mantle velocity is 7.8 to 8.0 km/sec, this high-velocity basal layer is usually absent, and the crust is abnormally thin for the surface elevation. Where the mantle velocity is 8.2 to 8.4 km/sec, the basal layer is well developed and the crust is abnormally thick for the surface elevation. That these variations in crustal structure and composition have attendant variations in mean crustal density resulting in epeirogenic uplift and subsidence is indicated by gravity anomaly values and by the surface geology. Where the crust is thick and the high-velocity basal layer is well developed, free air and isostatic gravity anomalies are generally positive. Where the basal layer is missing and the crust thin for the surface elevation, the free air and isostatic gravity anomalies are negative. The areas having a high-density thick crust are usually areas of geologic subsidence (basins); those of subnormal crustal density where the crust is thin are usually areas of geologic uplift. In the block-faulted areas, however, the relations of gravity to structure and crustal thickness are the reverse. Nevertheless, in both situations, the boundaries of the mobile crustal elements constitute zones of flexure and possible shear, i.e., potential zones of seismicity. The seismicity outlining the Illinois basin, the Rocky Mountain–Colorado Plateau block, and the Adirondack massif are probably of this origin.

REFERENCES

Allen, C. R., P. St. Amand, C. F. Richter, and J. M. Nordquist, Relationship of seismicity and geologic structure in the southern California region, Bull. Seismol. Soc. Am., 55, 753–797, 1965.

Benioff, H., Seismic evidence for crustal and tectonic activity, in Crust of the Earth, Geol. Soc. Am. Spec. Paper 62, 61–73, 1955.

Heck, N. H., Earthquake history of the United States, U. S. Coast Geodet. Surv. Spec. Publ. 609, 84 pp., 1938.

Herrin, E., and J. Taggart, Regional variations in P_n velocity and their effect on the location of epicenters, Bull. Seismol. Soc. Am., 52(5), 1037–1046, 1962.

Innes, M. J. S., and A. A. Weston, Crustal uplifts of the Canadian shield and its relation to the gravity field, Ann. Acad. Sci. Fennicae, A III, 90, 169–176, 1966.

Kumarapeli, P. S., and V. A. Saull, The St. Lawrence Valley system: A North American equivalent of the East African rift valley system, Can. J. Earth Sci., 3, 639–6557, 1966.

Menard, H. W., Marine Geology of the Pacific, McGraw-Hill Book Company, New York, 271 pp., 1964.

Niazi, M., Seismicity of northern California and western Nevada, Bull. Seismol. Soc. Am., 54, 845–850, 1964.

Ryall, Alan, D. B. Slemmons, and L. D. Gedney, Seismicity, tectonism, and surface faulting in the western United States during historic time, Bull. Seismol. Soc. Am., 56(5), 1105–1135, 1966.

Tobin, D. G., and L. R. Sykes, Relation of hypocenter of earthquakes to the geology of Alaska, J. Geophys. Res., 76(6), 1659–1667, 1966.

Woollard, G. P., Regional isostatic relations in the United States, in The Earth beneath the Continents, Geophys. Monograph 10, edited by J. Steinhart and T. Smith, pp. 557–594, American Geophysical Union, Washington, D.C., 1966.

Woollard, G. P., Continental structure of Antarctica, in Antarctic, Ann. Intern. Geophys. Yr., 44, 122–143, 1967.

Woollard, G. P., The inter-relationship of the crust, upper mantle, and isostatic gravity anomalies in the United States, in The Crust and Upper Mantle of the Pacific Area, Geophys. Monograph 12, edited by Leon Knopoff et al., pp. 312–341, American Geophysical Union, Washington, D.C., 1968.

Seismicity of Continental Asia and the Region of the Sea of Okhotsk, 1953-1965

E. F. Savarensky and N. V. Golubeva

THIS review is a continuation of the summary by *Gutenberg and Richter* [1954] of the seismicity from 1904 to 1952 in the Asian part of the Asian-Mediterranean seismic belt and the northwestern part of the Pacific seismic belt. They compiled maps of earthquake epicenters from instrumental and historical macroseismic data. In our study for the period 1953–1965, only instrumental observations were used; the results are plotted in Figures 1–3, and the larger earthquakes ($M > 7$) are catalogued in Table 1. We found that the distribution of earthquake epicenters from 1953–1965 in continental Asia is essentially the same as the distribution described for the earlier period by Gutenberg and Richter.

Of the 673 earthquakes recorded for this period in continental Asia and the adjacent regions: 37 belong to classes a and b (see Figure 1 for explanation of earthquake classes); 92 have an intermediate depth of focus (70–300 km); 7 with epicenters in the Sea of Okhotsk or along the Asian coast have foci at depths greater than 300 km.

The magnitudes of the earthquakes are the averages of magnitudes from many stations reported in the *Bulletins Mensuels*. The time of occurrence, coordinates of epicenter, and the depth of focus have been taken from several sources (listed at end of article).

In the northwestern part of the Pacific seismic belt, the epicenters of earthquakes are in deep-ocean basins and mountain systems that lie along the island arcs. In the central part of the Sea of Okhotsk are epicenters of deep-focus earthquakes, with $H = 300$–650 km. On the west side of the Kuril ridge there is a band of foci with depth 100–160 km. To the east of the Kuril Islands and Kamchatka, there is a belt of earthquakes with focal depth ranging from 30 to 60 km [*Savarensky et al.*, 1961]. Sakhalin appears less subject to earthquakes; from 1953 to 1965 only one earthquake of class c and two weak ones of class d were recorded on the island. More than half of all the strong earthquakes included in the catalog occurred in the Kuril-Kamchatka region.

The Asian-Mediterranean seismic belt begins in the east as a wide band, with the epicenters of earthquakes distributed mainly along mountain ridges. Earthquakes of the magnitude classes a, b, and c have been observed here. Their foci are mainly in the earth's crust; a few have foci in the range 70–300 km. For this period, only one deep earthquake was recorded (near the Pacific coast). The band of epicenters grows thinner as it approaches the seismic region of the Pamirs and the Hindu-Kush, which are characterized by frequent earthquakes of classes b and c (depths 60–300 km).

The wide seismoactive area of eastern Asia is bordered on the south by a curved chain of mountain ridges, convex to the south. In the Burma arc, earthquakes of class c predominate, with depth of focus 60–200 km; the epicenters lie along the arc.

West of the Hindu-Kush, the seismoactive belt widens again, but then becomes thinner in the Caucasus, Asia Minor, and farther to the west in Europe. From 1953 to 1965, four shocks were recorded in Iran with $M = 7$–7.2. A strong shock occurred in Asia Minor with $M = 7.2$.

Of the 37 strong earthquakes in the catalog, two were catastrophic; the Muiskoje earthquake, $M = 7.8$, on June 27, 1957, in Pribaikalye and the Gobi-Altai earthquake, $M = 7.8$–8, on December 4, 1957, in Mongolia. The Gobi-Altai earthquake was one of the strongest in this region.

In spite of the extension of the network of

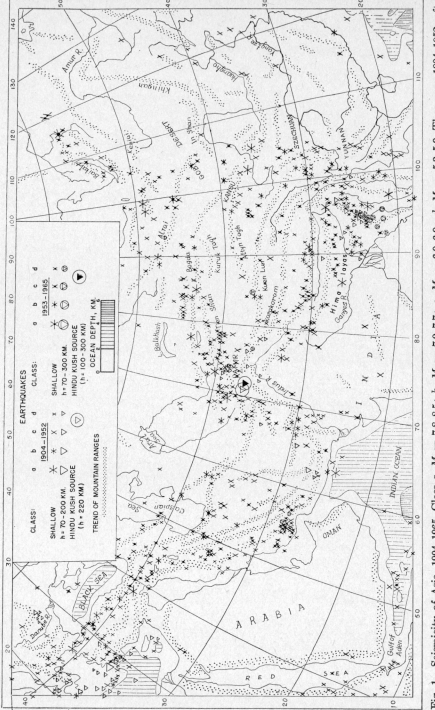

Fig. 1. Seismicity of Asia, 1904–1965. Class a, $M = 7.8$–8.5; b, $M = 7.0$–7.7; c, $M = 6.0$–6.9; d, $M = 5.3$–5.9. The years 1904–1952, after *Gutenberg and Richter* [1954]; 1953–1965, present authors; see end of article for sources of data.

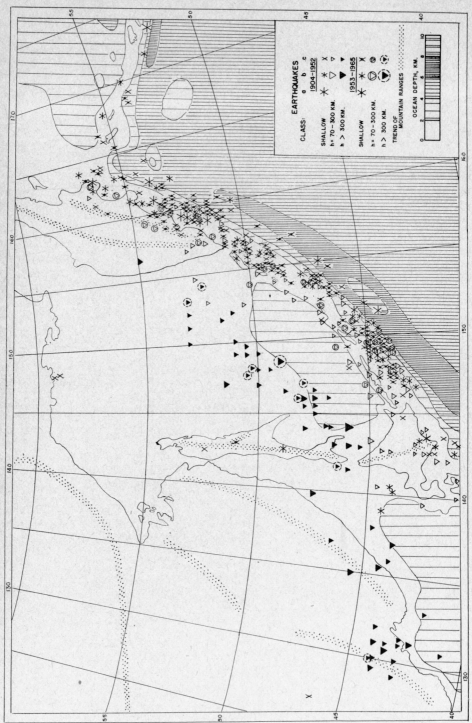

Fig. 2. Seismicity of the Kuril Islands, Kamchatka region, 1904–1965 ($M \geq 6$). For sources and explanation of classes, see Figure 1.

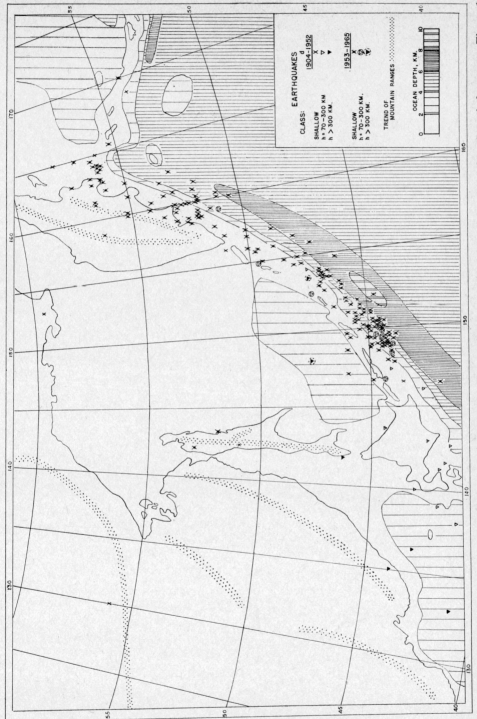

Fig. 3. Seismicity of the Kuril Islands, Kamchatka region, 1904–1965 ($M = 5.3$–5.9). For sources and explanation of classes, see Figure 1.

TABLE 1. Catalog of Class a and b Earthquakes ($M = 7.0$ to 8.5) in Continental Asia and Adjacent Regions, 1953–1965

Date	Time (GMT) h	m	s	Coordinates Lat., °N	Long., °E	M	H	Region
1953								
Jan. 5	07	48	17	54	170	7.2	Crust	Kamchatka
Jan. 5	10	06	25	49	156	7	Crust	Kuril Islands
Jan. 12	17	23	39	49.5	156	7	60	Kuril Islands
Sept. 4	07	23	05	50	156.5	7	60	Kuril Islands
Sept. 23	02	14	36	50.5	156	7	60	Kuril Islands
Nov. 10	23	40	20	50.5	157	7	60	Kamchatka
Dec. 25	01	51	26	52	159.5	7	Crust	Kamchatka
1954								
Feb. 11	00	30	16	39.5	101	7.5	Crust	China
March 21	23	42	05	24.5	95	7.2	150	Burma
July 31	00	59	57	39	104	7	Crust	China
1955								
March 18	00	06	42	54.5	161	7.2	Crust	Kamchatka
April 14	01	28	58	30	101.5	7.4	Crust	China
April 15	03	40	52	40	74.5	7.1	Crust	Middle Asia
April 15	04	13	23	40	75	7	Crust	Middle Asia
April 17	18	35	27	52	159.5	7	60	Kamchatka
Sept. 23	15	06	19	27	101.5	7.1	Crust	China
Nov. 23	06	29	29	50.5	157	7.1	60	Kamchatka
1956								
June 9	23	13	51	35.5	67.5	7.4	Crust	Afganistan
Oct. 11	02	24	33	46	150.5	7.3	60–100	Kuril Islands
1957								
May 26	06	33	30	40.7	31.2	7.2	Crust	Turkey
June 27	00	09	28	56.5	116	7.8	Crust	Baikal
July 2	00	42	23	36	53	7.2	Crust	Iran
Nov. 17	05	57	48	49	148.5	7.2	320–350	Okhotsk
Dec. 4	03	37	45	45.5	99.5	7.8–8	Crust	Mongolia
Dec. 13	01	44	59	34.5	48	7	Crust	Iran
1958								
March 28	12	06	24	37	71	7	200	Middle Asia
Nov. 6	22	58	06	44.5	148.5	8	60–100	Kuril Islands
Nov. 12	20	23	26	44.5	148.5	7	33–60	Kuril Islands
1959								
May 4	07	15	42	52.5	159.5	8	60	Kamchatka
1960								
Dec. 3	04	24	18	42.9	104.4	7.2	60	Mongolia
1961								
Feb. 12	21	53	43	43.7	147.6	7	45	Kuril Islands
June 11	05	10	26	27.9	54.6	7	37	Iran
1962								
Sept. 1	19	20	39	35.6	49.9	7.2	21–27	Iran
1963								
Oct. 13	05	17	57	44.8	149.5	8	20–60	Kuril Islands
Oct. 20	00	53	07	44.7	150.7	7	25	Kuril Islands
1964								
July 24	08	12	41	47.4	153.5	7	Crust	Kuril Islands
1965								
June 11	03	33	47	44.8	148.8	7.2	61	Kuril Islands

seismic stations during the past few years in continental Asia, the distribution is still inadequate to show the total picture of the seismicity of continental Asia, especially regarding smaller shocks.

SOURCES OF DATA FOR FIGURES 1–3

For the period up to April 1964: *Bulletins Mensuels*, Bureau Central International de Séismologie, Strasbourg; *International Seismological Summary*, Oxford; and *Seismological Bulletin*, U.S. Coast and Geodetic Survey, Washington. For the period April 1964 to January 1965: Bulletins of the Network of USSR Seismic Stations; and *Operative Seismological Bulletin* of the Institute of the Physics of the Earth, Moscow.

REFERENCES

For extensive bibliographies on the seismicity of Asia, see *Savarensky et al.* [1961; 1962].

Golubeva, N. V., Seismic zones of the earth and their energy characteristics for the period 1950–1960, *Izv. Akad. Nauk SSSR, Fiz. Zemli, 1965*(5), 1965.

Gutenberg, B., and C. F. Richter, *Seismicity of the Earth*, 2nd edition, Princeton University Press, Princeton, New Jersey, 310 pp., 1954.

Savarensky, E. F., I. E. Gubin, and D. A. Kharin, editors, *Earthquakes in the USSR*, Izd. Akad. Nauk SSSR, Moscow, 409 pp., 1961.

Savarensky, E. F., S. L. Soloviev, and D. A. Kharin, editors, *Atlas of Earthquakes in the USSR, 1911–1957*, Izd. Akad. Nauk SSSR, Moscow, 337 pp., 1962.

Seismicity of the European Area

Vít Kárník

THE European area is defined here to include the Mediterranean seismic belt of Tertiary folding and the provinces of Variscan and Caledonian folding, which have some minor seismic activity and stable shields with very low seismicity (Figure 1).

A systematic investigation of the seismicity of the European area was initiated by the European Seismological Commission (IASPEI, IUGG) in 1951 [*Båth*, 1960; *Kárník*, 1961, 1963, 1965a, 1960]. Cooperation among European seismologists enabled the Commission to accumulate basic information on earthquakes with $I_0 \geq$ VI (or equivalent M) from 1901 to 1955.

Homogeneous data are necessary for any kind of comparative earthquake study. The main problem in unifying data of different origin was the classification of about 5500 earthquakes included in the 1901–1955 catalog. For this purpose, the quantity M, magnitude, was employed on the basis of the procedure suggested by *Kárník* [1965a, 1968, 1969].

To obtain a good picture of the geographical distribution of seismic activity, different types of seismic maps were constructed on a scale of 1:5,000,000. They are: four epicenter maps ($M = 4.1$–4.6, incomplete; $M = 4.7$–5.1; $M = 5.2$–6.2; and $M = 6.3$–8.3, Figure 2); an energy release map (specific seismicity map); a maximum intensity map, and a seismic zoning map. The principles for the preparation of all these maps are discussed by *Kárník* [1965a, 1966, 1968]. For statistical investigation, the whole European area was divided into 39 seismic zones (Figure 1) for each of which earthquake recurrence curves have been constructed [*Kárník*, 1965b].

All the seismic maps constructed so far confirm a close relation between the highest seismic activity and zones of Tertiary folding or Quaternary differential movement (Figure 2) [*Kárník*, 1968, 1969]. Most epicenters and isolines of macroseismic intensity follow Tertiary mountain ridges or large fault systems. It is, however, impossible to relate all faults to epicenters and vice versa. There are also epicenters that cannot be simply associated with faults or ridges. They might correspond to boundary lines separating two regions with different movements;

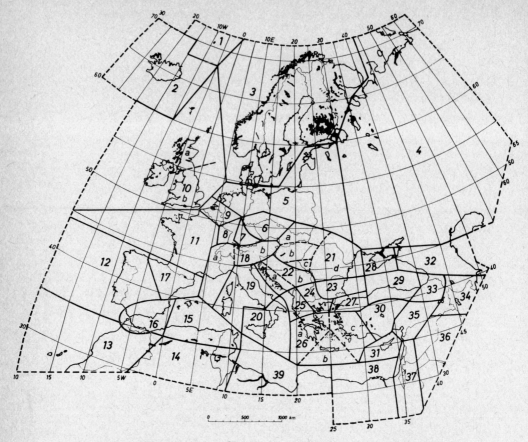

Fig. 1. European area, showing the thirty-nine seismic regions.

such information is, however, rarely given in tectonic maps. When relating seismic events to tectonic lines, we must take into account that the average accuracy of epicenter determination is about ±20–30 km for continental shocks and about ±50–100 km for oceanic ones. Only occasionally can the depth of the focus within the crust be calculated with a relative error less than 50%. The situation is more favorable for shocks deeper than 80 km.

The best example of a close relationship between seismicity and faulting is the North Anatolian fault system, the largest strike-slip fault in the area under discussion. Some strong earthquakes located along the fault were accompanied by horizontal movements [*Richter*, 1958]. It appears likely that at both ends the fault splits into several branches. In the east the epicenters indicate the continuation toward the Akhalkalaki region (Georgia) and Armenia and Iran. In the west one branch may continue to Bulgaria; the other probably crosses southern Khalkidiki and central Greece to the Ionian Islands [*Galanopoulos and Papazachos*, 1966]. Epicenters also delineate a belt joining eastern Anatolia with the Gulf of Iskenderun, continuing to Cyprus and possibly on to the Jordan valley.

There are no quantitative data available about horizontal or vertical movements in earthquake zones. Repeated leveling or triangulation has been carried out recently, but only in a few regions in northern, western, and central Europe. Sudden subsidence was observed during a few very strong shocks, e.g. in Bulgaria (April 14 and 18, 1928, $M = 7$) and in Greece (April 27, 1894).

Seismic maps reveal other seismogenetic lines that are not as clearly evidenced by geologic data. One distinct line begins at the western edge of the Carpathians, follows the valleys of the rivers Váh, Leitha, and Mur, and crosses

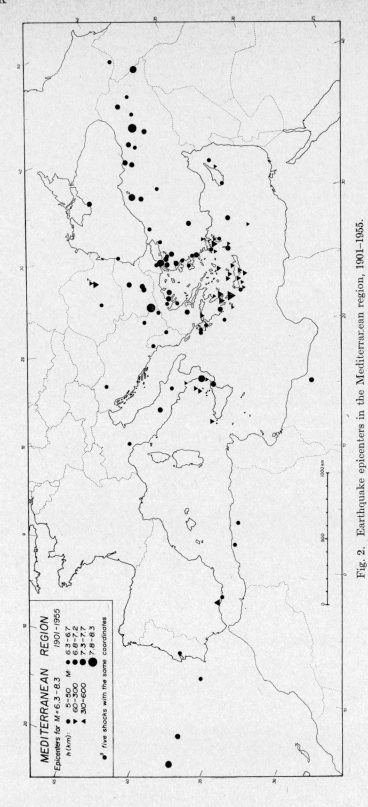

Fig. 2. Earthquake epicenters in the Mediterranean region, 1901–1955.

the Carnic Alps in northeastern Italy; its southwestern end is marked by the highest activity. Another belt follows the eastern edge of the Adriatic Sea, continuing past Greece and Crete toward Rhodes. There is an indication of a line joining the Bay of Lisbon with the Mid-Atlantic ridge. Other remarkable zones, although not as extended, are located south of the Crimea, southeast of Shabla (Bulgaria), in northwestern Libya, etc. The area in Figure 1 covers part of the North Atlantic ridge in the region of Iceland and Jan Mayen. The epicenters follow the ridge closely, including the part that intersects Iceland.

Seismic activity is not limited to the provinces of alpine folding. Minor shocks occur in the Rhine valley ($M_{max} = 5.75$), in the periphery of the Bohemian massif ($M_{max} = 5$), along the coast of Brittany, in Wales and Scotland ($M_{max} = 5.75$), in the North Sea, and along the western coast of Scandinavia as far as Oslo Fjord ($M_{max} = 6$).

The trend and intensity of seismic activity can be demonstrated by using the values of energy or strain release and the magnitude-frequency graphs (recurrence curves). Seismic energy release is dominated by the largest shocks; the total sum is two orders higher in the provinces of the alpine folding than in the older tectonic units [Kárník, 1965b]. The most striking phenomenon is that more seismic energy was released in the upper mantle than in the crust, a situation that does not agree with the worldwide pattern [Gutenberg and Richter, 1954]. There is, however, some discrepancy between the two formulas used for the calculation of energy ($\log E = 11.8 + 1.5M$; $\log E = 5.8 + 2.4m$), because M is based on surface-wave amplitude, m on body waves.

The most concentrated zone of intermediate shocks is related to the sharp bend of the eastern Carpathians [Radu, 1965]. The crustal activity is very small in this zone compared to activity at depths of 100-200 km. The strongest shock occurred on November 10, 1940 ($m = 7.3$, $h = 130$ km). It is remarkable that strong intermediate shocks in Rumania originate in the same time sequence as intermediate shocks in the Tyrrhenian Sea [Kárník, 1965b]. At present, there is no simple explanation for this surprising coincidence. In the Tyrrhenian Sea, some foci approach or even exceed the depth of 300 km. The foci form a conical surface with the axis inclined toward the northwest; the base intersects northern Sicily and Calabria, where surficial foci were located [Peterschmitt, 1956]. The deepest shock in the Mediterranean area occurred on March 29, 1954 ($m = 7$, $h = 650$ km) in southern Spain, i.e. in a region with no previous deep activity (see section by Miyamura, p. 122). This shock and one in the Tyrrhenian Sea (February 2, 1955; $m = 5.6$, $h = 450$ km) are the only deep shocks known outside the Pacific area. The distribution of cumulative sums of strain released at different depths (Figure 3) has only one important maximum for the Vrancea region in Rumania ($h = 130$ km), two maximums for the Aegean region ($h = n$ and 100 km), and four for the Tyrrhenian Sea ($h = n$, 100, 200, and 300 km).

The graphs of the cumulative strain released by shallow- and intermediate-depth shocks in the Mediterranean area (Figure 4) reveal some interesting phenomena. The trends of strain release in the crust and upper mantle are the

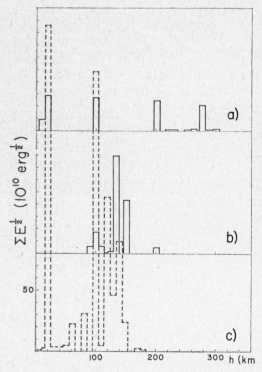

Fig. 3. Cumulative strain release versus depth: (a) Tyrrhenian Sea; (b) Vrancea region, Rumania; (c) Greece, Crete, and western Turkey.

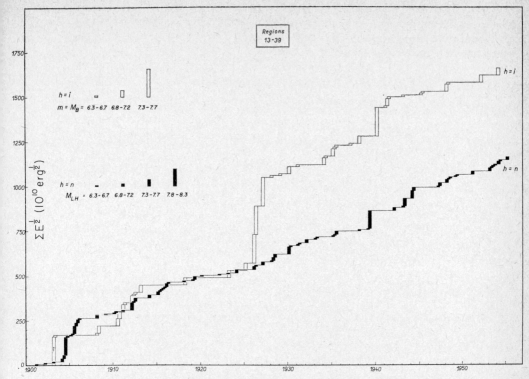

Fig. 4. Strain released in the Mediterranean seismic belt in the period 1901–1955 by shallow and intermediate earthquakes (regions 13–39 in Figure 1).

$h = n$, normal depth of focus, i.e. $h = 5$–50 km.
$h = i$, intermediate depth of focus, i.e. $h = 60$–300 km.
$m = M_B$, magnitude determined using body waves.
M_{LH}, magnitude determined using surface waves (LH).

same and relatively monotonous. There are only two periods of increased activity of shallow shocks: 1904–1906 and 1939–1944. The curve for intermediate shocks follows a similar shape but jumps suddenly in 1926–1927 to a higher level. This shift is caused by three strong shocks in the Aegean region.

The magnitude-frequency curves, usually expressed by the formula $\log N = a - bM$, are important for the study of the recurrence of shocks. Laboratory experiments of *Mogi* [1962, 1963] and *Vinogradov* [1962] show that the parameter b of the curve decreases with increasing homogeneity of material, stress rate, and bulk modulus. The deviations from the linear logarithmic magnitude-frequency law can also be interpreted in terms of the structure of the building material [*Mogi*, 1963]. Usually $b = 0.9$, and some authors argue that the parameter is stable [*Riznichenko*, 1959]. Others, however, explain its regional variation by the stage of tectonic development in an area [*Miyamura*, 1962] (see section of this Monograph, p. 121). Ordinary as well as cumulative recurrence curves plotted for European seismic zones confirm the existence of differing values of b. They also demonstrate that the lowest slopes (lowest b) correspond to the intermediate shocks ($b = 0.3$–0.6, m converted to M) and to regions 24, 27, 29, and 33, which are characterized by the highest magnitudes ($M = 7$–8). Intermediate values, $b = 0.7$–0.8, coincide with the zones of medium activity (zones 2, 21, 23, 26, 34) and foci in the crust, with $M_{max} = 7.25$. The highest slopes ($b = 0.9$–1.0) relate to regions with prevailing shallow or surficial foci ($h = 1$–4 km); zones 15, 16, 18, 19, 20, 22, 25, and 32, with $M_{max} = 6.25$–6.75.

We can derive an approximate relation between b and the average focal depth, or between

b and M_{max}. The result corresponds to increasing homogeneity and strength with depth, and agrees with laboratory experiments.

One interesting phenomenon is the migration of activity observed in the North Anatolian fault system and in southern and central Italy. In the former, the cycle started with a shock in 1930 in eastern Anatolia ($M = 7.3$, 38°N, 44.5°E); this was followed by shocks shifting successively westward in the following sequence: 1939, $M = 8$ (39.7°N, 39.7°E); 1942, $M = 7.0$ (40.7°N, 36.6°E); 1943, $M = 7.3$ (41.4°N, 33.8°E); 1944, $M = 7.3$ (41.5°N, 33.5°E); 1953, $M = 7.2$ (40.1°N, 27.3°E). These shocks dominated the energy release in the period 1930–1953; weaker shocks did not follow this trend but occurred more randomly. The velocity of migration of activity (progressive faulting) was irregular; it increased from 50 km/year to 145 km/year, and then decreased to 60 km/year.

A similar phenomenon occurred in southern Italy, where the main shock, in 1905 ($M = 7.3$, Messina), initiated a series of aftershocks of decreasing magnitude, with epicenters that moved along Calabria toward central Italy during 1908–1920. Earlier, in 1783–1785, the epicenters of five destructive shocks in Calabria were similarly distributed successively eastward.

This migration of activity must be carefully studied. Identification of cycles might help in a more realistic prediction of recurrence of destructive shocks.

Investigation of the seismicity of the European area continues. The present program deals with problems of earthquake statistics and earthquake mechanisms. The recent installations of sensitive stations in the Balkan area is encouraging, for it is highly desirable to concentrate serious efforts on a detailed study of the most active zone, i.e., the North Anatolian fault.

REFERENCES

Båth, M., *Seismicity of Europe, A progress report*, IUGG Monograph 1, International Union of Geodesy and Geophysics, Paris, 1960.

Galanopoulos, G. A., and B. C. Papazachos, Progress Report 1964–1965, Seismol. Inst. Natl. Obs. Athens, Seismol., Lab., Univ. of Athens, p. 11, 1966.

Gutenberg, B., and C. F. Richter, *Seismicity of the Earth*, Princeton University Press, Princeton, 1954.

Kárník, V., *Seismicity of Europe, 2*, IUGG Monograph 9, International Union of Geodesy and Geophysics, Paris, 1960.

Kárník, V., *Seismicity of Europe, 3*, IUGG Monograph 23, International Union of Geodesy and Geophysics, Paris, 1963.

Kárník, V., *Seismicity of Europe, 4*, IUGG Monograph 29, International Union of Geodesy and Geophysics, Paris, 1965a.

Kárník, V., Magnitude, frequency and energy of earthquakes in the European area, *Travaux Inst. Géophys. Acad. Tchecosl. Sci., 1965*(222), 247–273, 1965b.

Kárník, V., *Seismicity of Europe, 5*, IUGG Monograph, International Union of Geodesy and Geophysics, Paris, in press, 1968a.

Kárník, V., *Seismicity of the European Area, 1*, Academia, Praha, 1968b.

Kárník, V., *Seismicity of the European Area, 2*, in press, Academia, Praha, 1969.

Miyamura, S., Magnitude-frequency relation and its bearings on geotectonics, *Proc. Japan Acad., 38*(1), 27–30, 1962.

Mogi, K., Study of elastic shocks caused by the fracture of heterogeneous materials and its relation to earthquake phenomena, *Bull. Earthquake Res. Inst., Tokyo Univ., 40*, 125–173, 851–853, 1962.

Mogi, K., Study of elastic shocks caused by the fracture of heterogeneous materials and its relation to earthquake phenomena, *Bull. Earthquake Res. Inst., Tokyo Univ., 41*, 615–668, 1963.

Peterschmitt, E. Quelques données nouvelles sur les séismes profonds de la Mer Tyrrhénienne, *Ann. Geofis., 9*(3), 1956.

Radu, C., Regimul seismic al regiunii Vrancea, *Studii Cercetari Geol. Geograf., Ser. Geofiz., 3*(2), 231–279, 1965.

Richter, C. F., *Elementary Seismology*, W. H. Freeman and Company, San Francisco, 1958.

Riznichenko, J. V., On quantitative determination and mapping of seismic activity, *Ann. Geofis., 12*, 227–237, 1959.

Vinogradov, S. D., Experimental study of the distribution of the number of fractures in respect to the energy liberated by the destruction of rocks, *Izv. Akad. Nauk SSSR, Ser. Geofiz., 1962*(2), 171–180, 1962.

Seismicity of Southeast Australia

J. C. Jaeger and Lesley Read

THE last decade has witnessed a steadily increasing interest in the study of earthquakes of low magnitude. *Hagiwara* [1964] defined microearthquakes as having magnitudes in the range $1 < M < 3$ and ultra-microearthquakes as having magnitude $M < 1$. Low-magnitude earthquakes can be investigated most efficiently in regions in which the density of seismographic stations is high; it is thus not surprising that a large proportion of the work in this field has been carried out in Japan and the western United States (see the summary by *Oliver et al.* [1966]). Since these areas are of high seismic activity, attention has been especially directed towards the relationship between microearthquakes and tectonic activity. Microearthquake studies in more stable regions (such as that by *Isacks and Oliver* [1964] in New Jersey) have usually been for special purposes and short periods of time.

In a very stable region like Australia, 80 to 90% of the earthquakes now being located are of magnitude <3, and thus most of the work being done concerns microearthquakes. This review is primarily a report on microearthquakes in southeastern Australia, where the density of stations is highest.

Up to 1951, no more than six widely separated stations were operating in Australia; only 80 seismic events, with magnitudes between 3 and 7, had been located, primarily in the south and eastern part of the continent [*Burke-Gaffney*, 1951].

In the last decade, instrumentation at the main stations has been improved, and several local networks have been installed. In southeastern Australia, the Snowy Mountains Authority and the Sydney Metropolitan Water Board, with Dean S. Carder as consultant, each decided in 1958 to establish a network of four stations around their installations, partly because of historic reports of seismic activity in these regions, and partly to study shocks that might take place during the filling of large dams, as happened at Hoover dam [*Carder*, 1945]. All eight stations were equipped with Benioff seismometers and were operated in cooperation with the Australian National University and its station at Canberra. The eight stations, together with the stations at Sydney, Canberra, and Melbourne, are shown by large triangles in Figure 1.

After a few years of observations, it became apparent that the greatest concentration of earthquakes was in the Dalton-Gunning area north of Canberra; there was also reason to expect significant activity in Victoria, east of Melbourne. Accordingly, three additional stations were installed in these two areas (small triangles in Figure 1).

When possible, epicenters and focal depths are determined by *Flinn's* [1960] computer program. To date, about 700 tremors have been recorded in southeastern Australia. Figure 1 shows events of magnitude 1 or greater for the period 1958–1966. Unfortunately, the disposition of the stations is such that fault-plane solutions can be obtained for only a few of the larger events.

Detailed studies have been made in two regions: Dalton-Gunning and the Snowy Mountains. In his study of the Dalton-Gunning regions up to 1961, *Cleary* [1967] concluded that the activity can be explained by compressive forces in a northwest direction that cause a wedge in the northwest sector to override the surrounding country. Many of the events occur at the boundaries of granite masses. Considering the short time during which observations have been made, the variation in activity is quite marked. There is a strong suggestion that activity oscillates between the two sides of the wedge, and that when there is activity within the region, it is diminished in the surrounding area. It is apparently for this reason that few events have been recorded on the Lake George fault (just south of the Dalton-Gunning region), which might have been expected to be active.

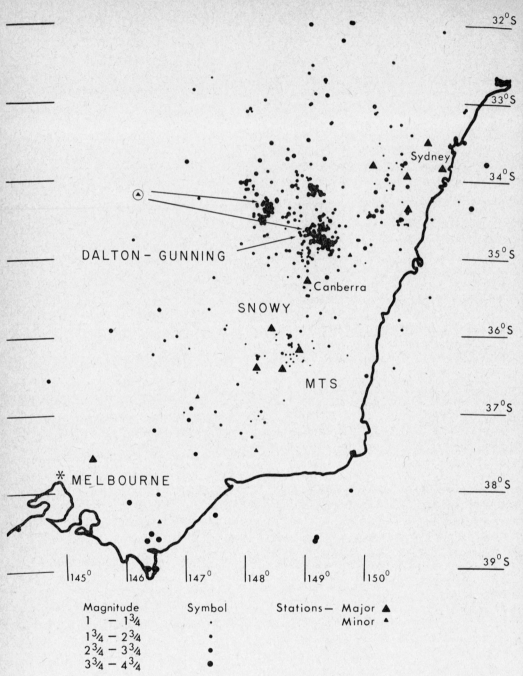

Fig. 1. Epicenters of tremors in southeastern Australia in the years 1958–1966.

In the Snowy Mountains region, activity was reported around 1900 and again more recently in the period 1958–1962, when 44 tremors were located, the largest being of magnitude 5 and depth 17 ± 2 km [*Cleary et al.*, 1964]. The fault-plane solution for this shock was consistent with the stress pattern inferred for the Gunning area. It was probably associated with the Crackenback fault, a high-angle reverse fault reaching the surface in the Crackenback Valley in the Snowy Mountains. Although the usual aftershock sequence did not occur, a series of tremors having a typical aftershock strain-release pattern took place in the same area, beginning 100 days after the event. A second tremor of magnitude 4 was probably associated with the Murrumbidgee fault. Since this period, there has been a considerable decrease in activity in the region.

In addition, activity appeared around the edge of the Sydney basin. A relatively large shock of magnitude 5.5 and at a depth of 19 ± 4 km occurred near Robertson (about 50 miles south of Sydney) on the edge of the coastal escarpment, and was followed by a well-defined series of aftershocks; these events have been studied in some detail by Cleary and Doyle [1962]. There has been relatively little other activity in this region.

Few events have been recorded by the Victorian stations, although previous reports had indicated that the region was active. In fact, throughout southeastern Australia there have been remarkable changes in activity during the short period involved. This, and the finding that all foci are crustal, are the most noteworthy features of the observations. In addition, the decrease in activity throughout southeastern Australia in the last few years suggests that the tectonic activity of the area is controlled by a single stress field.

There are several other studies of local seismicity in Australia. The University of Adelaide operates a network of three stations in its neighborhood. Since 1959, Mundaring Geophysical Observatory in Western Australia, operated by the Bureau of Mineral Resources, has located 46 tremors of magnitude greater than 3.5 and 150 of magnitudes between 1.6 and 3.5; most of these shocks lie in an area about 500 km long and 50 km wide on the southwestern part of the shield associated with changes in geology and gravity anomalies [Everingham, 1966]. The University of Tasmania, in collaboration with the Hydro-Electric Commission of Tasmania, has operated a four-station telemetered network since 1957; some 40 epicenters have been located, mostly in the western part of the island, in the years 1961–1967 [Green et al., 1967].

A useful amount of accurate information is now being accumulated; in a few years results of real tectonic significance can be expected.

REFERENCES

Burke-Gaffney, T. N., The seismicity of Australia, J. Proc. Roy. Soc. N. S. Wales, 75, 47, 1951.

Carder, D. S., Seismic investigations in the Boulder Dam area, 1940–1944, and the influence of reservoir loading on local earthquake activity, Bull. Seismol. Soc. Am., 35, 175, 1945.

Cleary, J. R., The seismicity of the Gunning and surrounding areas, 1958–1961, J. Geol. Soc. Australia, 14, 23, 1957.

Cleary, J. R. and H. Doyle, Application of a seismograph network and electronic computer in near earthquake studies, Bull. Seismol. Soc. Am., 52, 673–682, 1962.

Cleary, J. R., H. A. Doyle, and D. G. Moye, Seismic activity in the Snowy Mountains region and its relationship to geological structures, J. Geol. Soc. Australia, 11, 89, 1964.

Everingham, I. B., Seismicity of Western Australia, Bur. Min. Resources Australia, Record No. 1966/127, 1966.

Flinn, E. A., Local earthquake location with an electronic computer, Bull. Seismol. Soc. Am., 50, 467–470, 1960.

Green, R., C. Dampney, W. D. Parkinson, V. P. St John, B. D. Johnson, P. A. Watt, and G. The, Upper Mantle Project: Second Australian Progress Report (1965–67), pp. 32–35, Australian Academy of Science, Canberra, 1967.

Hagiwara, T., Proceedings of the U.S.–Japan Conference on Research Related to Earthquake Prediction Problems, pp. 10–12, Japan Society for the Promotion of Science, Tokyo, 1964.

Isacks, B., and J. Oliver, Seismic waves with frequencies from 1 to 100 cycle per second recorded in a deep mine in northern New Jersey, Bull. Seismol. Soc. Am., 54, 1941–1979, 1964.

Jaeger, J. C., and W. R. Browne, Earth tremors in Australia and their geological importance, Australian J. Sci., 20, 225, 1958.

Oliver, J., A. Ryall, J. N. Brune, and D. B. Slemmons, Microearthquake activity recorded by portable seismographs of high sensitivity, Bull. Seismol., Soc. Am., 56, 899, 1966.

Seismicity of the Mid-Oceanic Ridge System[1]

Lynn R. Sykes

A NARROW belt of seismic activity is associated with the crest of the mid-oceanic ridge system and its extensions into East Africa and western North America (Figure 1). Although seismic belts had previously been identified with certain parts of this ridge system, it is only within the last 12 years that the continuity of this tectonic feature was recognized on a worldwide scale [*Ewing and Heezen*, 1956].

Several lines of geophysical and geological evidence indicate a system of extensional tectonics along the oceanic ridge system and a system of compressional tectonics in island arcs. These include investigations of the mechanism and distribution of earthquakes [*Honda*, 1962; *Sykes*, 1967], evidence from magnetic anomalies for sea-floor spreading [*Vine*, 1966], and analyses of the types of faulting that are most often associated with these tectonic systems. Island arcs and similar arcuate structures of the circum-Pacific and Alpide belts account for a very large percentage of the world's shallow earthqukes (depths 0–70 km) and for nearly all intermediate- and deep-focus earthquakes. In contrast, only about 5% of the world's shallow earthquakes are associated with the mid-oceanic ridge system or its continental extensions; earthquakes of intermediate and deep focus have not been detected on the oceanic ridge system. The smaller numbers of earthquakes along the ridge system and the fact that much of this system remained relatively unexplored until 10 or 20 years ago probably explain why it was only recently recognized as a global system of extensional tectonics.

The increase in the number, distribution, and sensitivity of the world network of seismograph stations and the use of computers in the processing of the data have led to a marked increase in the number of earthquake epicenters recorded on the mid-oceanic ridge system and on its continental extensions. In many regions, the number of events reported for the last five or ten years is greater than the number reported during the preceding fifty years. The epicenters of many earthquakes can now be located on the ridge system with a precision of 20 km or better. Previously, uncertainties in the location of epicenters may have exceeded 100 km. Hence, many features on the ridge system can now be recognized that were not previously resolvable because of uncertainties in the epicentral locations.

GEOGRAPHIC DISTRIBUTION OF SEISMIC ZONES ON THE RIDGE SYSTEM

The well-developed oceanic ridge system can be traced through the Atlantic, Indian, Pacific, and Arctic Oceans (Figure 1). Although it maintains a median position throughout much of the Atlantic and Indian Oceans, the ridge is decidedly nonmedian in large areas of the Pacific and Arctic Oceans. In the Arctic Ocean, however, a belt of earthquakes (and presumably the ridge crest) occupies a median position between the Lomonosov ridge and the continental shelf of Eurasia [*Sykes*, 1965].

A large number of linear zones of rough topography (or fracture zones) intersect the crest of the ridge [*Menard*, 1965; *Heezen and Tharp*, 1965]. Undoubtedly, other fracture zones will be detected in other parts of the ridge system as exploration continues. The ridge system is not continuous in detail, but seems to be composed of a series of linear segments interrupted by fracture zones. Nearly all earthquakes on the ridge system occur either along the ridge crest or along those parts of fracture zones that are between crests of the ridge [*Sykes*, 1967].

A well-developed rift zone has been detected along many crossings of the ridge crest in the Atlantic and Indian Oceans [*Heezen and Ewing*, 1963]. Much of the seismic activity and all postglacial volcanic activity in Iceland

---

[1] Contribution 1232 from the Lamont-Doherty Geological Observatory of Columbia University.

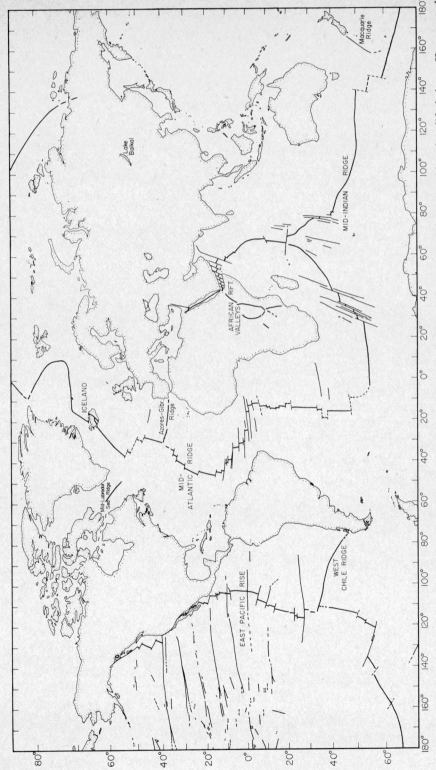

Fig. 1. Crest of the oceanic ridge system (heavy lines) and fracture zones (light lines) that intersect the ridge (modified from *Heezen and Tharp* [1965], *Menard* [1965, 1967], and other sources). Nearly all earthquakes on the ridge system are located either along the crest or along those parts of fracture zones that are between crests of the ridge.

are confined to a prominent central graben that appears to be an extension of the central rift zone of the Mid-Atlantic ridge [*Heezen and Ewing*, 1963; *Bodvarsson and Walker*, 1964; *Sykes*, 1967]. With the exception of the Gorda ridge [*Menard*, 1964] off the coast of Oregon and northern California, the median rift is typically absent on the East Pacific rise. Ridge system earthquakes that do not appear to be on fracture zones are characterized by a predominance of normal faulting [*Sykes*, 1967]. Thus far, all the solutions of this type are confined to those parts of the ridge system that possess a prominent median rift. Earthquakes on fracture zones, however, are characterized by a predominance of strike-slip motion on a steeply dipping plane. The strike of this plane is nearly coincident with the strike of the fracture zone [*Sykes*, 1967]. Most of the earthquakes on the East Pacific rise are confined to the parts of fracture zones that are between crests of the ridge. Events on the crest of the East Pacific rise itself seem to be quite rare [*Menard*, 1966].

Earthquakes on the oceanic parts of the ridge system are typically confined to single, narrow, linear zones less than a few tens of kilometers wide. In contrast, earthquakes on the continental extensions of the ridge system are normally scattered over a region at least 100 km wide [*Sykes*, 1965]. Geological structures in the continental parts of the ridge system are often multi-branched and are usually much more complex than the structures in the oceanic parts. Whether these differences are related to a difference in the basic tectonic process or to a difference in the response of the crust and upper mantle in the two regions is not known.

One feature common to both the oceanic and continental parts of the ridge is the absence of earthquakes of intermediate and deep focus. In island arcs, however, the number of earthquakes N per unit depth within the upper 200 km of the earth seems to be characterized by the relation

$$\log_{10} N = \text{constant} - 0.008Z$$

where Z is the depth in kilometers [*Sykes*, 1966]. This distribution can be contrasted with that found in California. Nearly all earthquakes in California appear to be confined to the upper 20 km of the earth. Seismic activity near the western coast of North America between central Alaska and central Mexico is less than that in many other parts of the circum-Pacific zone. The absence of intermediate- and deep-focus earthquakes [*Gutenberg and Richter*, 1954; *Tobin and Sykes*, 1968] and the presence of fracture zones and segments of ridge crest in this area strongly suggest that the seismicity and tectonics of this part of western North America are related to the East Pacific rise [*Girdler*, 1964; *Menard*, 1965]. The continuity of seismic belts and morphologic features indicates that the oceanic ridge system extends into the continents in eastern Africa and western North America.

EVIDENCE FOR TRANSFORM FAULTS AND SEA-FLOOR SPREADING

Figure 2 illustrates the distribution of ridges, fracture zones [*Heezen and Tharp*, 1965], and earthquakes [*Sykes*, 1967] along part of the Mid-Atlantic ridge. Between latitude 15°N and 5°S the crest of the ridge is displaced to the east a total of nearly 3500 km. The apparent displacements are such that the ridge crest maintains its median character between Africa and South America. *Heezen and Tharp* [1965] concluded that the ridge crest in this area had been offset by sinistral transcurrent faults. They noted that the fracture zones appeared to be parallel to flow lines for the continental drift of Africa relative to South America.

Nevertheless, in this and other parts of the ridge system, nearly all the seismic activity on fracture zones is confined to the region between the ridge crests. This is particularly well illustrated for the Chain fracture zone near 1°S, 15°W (Figure 2). Only a very few earthquakes are found on other parts of fracture zones. The seismicity of the ocean basins appears to be even less. The distribution of seismic activity along fracture zones is a strong argument against the hypothesis of a simple offset of the ridge crest by transcurrent faulting.

Wilson [1965] has recently proposed a separate class of horizontal shear faults, the transform fault. Transform faults that join two segments of oceanic ridge are compared with their transcurrent counterparts in Figure 3. In the transcurrent models it is tacitly assumed that the faulted medium is continuous and con-

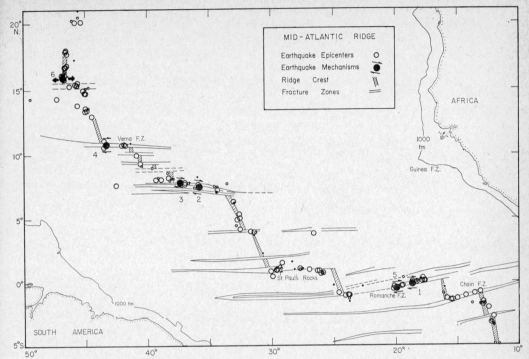

Fig. 2. Relocated epicenters of earthquakes (1955–1965) and mechanism solutions for six earthquakes along part of the Mid-Atlantic ridge. Ridges crests and fracture zones are from *Heezen and Tharp* [1965]. Events 1–5 were characterized by a predominance of right-lateral strike-slip motion on steeply dipping planes that strike approximately east. Event 6 was characterized by a predominance of normal faulting; the axis of maximum tension for this event (heavy arrows) is nearly horizontal and nearly perpendicular to the strike of the ridge. Large circles denote more precise epicentral determinations; smaller circles, poorer determinations; large solid circles denote earthquakes for which the mechanism is illustrated.

served. In the transform hypothesis the ridges expand to produce new crust. Thus, a transform fault may terminate abruptly at both ends even though great displacements may have occurred either on the central parts of the fracture zone or at some time in the past on parts of the fracture zone that are no longer between the two ridge crests. Note also the difference in sense of motion along the fault between offsets of the ridge: the relative motion for transform faults is opposite the motion required to offset the ridge by transcurrent faulting (Figure 3).

Sykes [1967] demonstrated that earthquakes on fracture zones are characterized by a predominance of strike-slip motion; the strike of one of the nodal planes for P waves is nearly coincident with the strike of the fracture zone. The sense of strike-slip motion is in agreement with that predicted for transform faults; i.e., it is opposite that expected for a simple offset

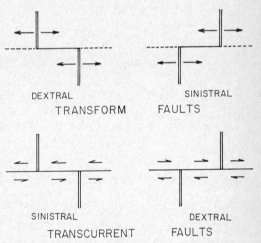

Fig. 3. Sense of relative displacements for transform faults and for transcurrent faults (modified from *Wilson* [1965]). Double line represents crest of mid-oceanic ridge; single line denotes fracture zone. The terms dextral and sinistral describe the sense of motion on the fracture zone.

of the ridge by transcurrent faulting. Earthquakes that do not appear to be on fracture zones are characterized by a predominance of normal faulting. Thus, both the distribution and the mechanisms of earthquakes support the concept of the transform fault.

Evidence that magnetic anomalies on the ocean floor can be identified with past reversals in the earth's magnetic field [*Vine*, 1966] has added strong support to the hypothesis of seafloor spreading as postulated by *Hess* [1962] and *Dietz* [1961]. The distribution and mechanisms of earthquakes on the oceanic ridge system are in agreement with the magnetic and other evidence for sea-floor spreading.

IMPORTANT PROBLEMS

The concept of the transform fault bears a relatively simple relation to continental drift. If the rate of spreading on two sides of a transform fault has remained the same throughout a sequence of spreading, then the pattern of ridges and fracture zones should correspond to the shape of the two blocks that have drifted apart [*Wilson*, 1965]. Fracture zones (large parts of which will be aseismic at the present time) should connect points that were once together. The orientations of fracture zones, the shapes of continental margins, and the mechanisms of earthquakes suggest that a process of this type has occurred in the Gulf of Aden, in the Gulf of California, and in the central and south Atlantic Ocean.

Unfortunately, neither the seismological evidence nor the magnetic evidence for sea-floor spreading gives information directly pertinent to the tectonics of the ridge system in the third dimension. At the present time the only statement that can be made about the depths of focus for most of the earthquakes on the ridge system is that they are within the upper 50 or 100 km of the earth. If the focal depths for the entire ridge system are similar to those in California, however, then the seismicity is of very shallow origin. More accurate determinations of these depths are urgently needed.

Computations of the epicentral locations for earthquakes on the oceanic ridge system are limited by uncertainties in the seismic travel times. When a standard set of travel-time tables is used, the standard errors for latitude and longitude are usually less than 20 km for most well-recorded earthquakes. If the actual travel times are biased and are a complicated function of position, azimuth, and distance, then the standard errors may not be indicative of the accuracy (as opposed to the precision) of the computations. Accurately determined travel times would provide information about any bias that may exist in the locations; in addition, they could be used to compute more accurate locations for earthquakes on the ridge. With the precision now available, it is usually impossible to determine whether an earthquake occurred along the sides or in the center of the median rift.

Seismically active belts are also associated with the Macquarie, Azores-Gibraltar, West Chile, and Mid-Labrador Sea ridges (Figure 1). Although these ridges appear to join the mid-oceanic ridge system, their tectonic significance is not understood. Likewise, the geographic extent and tectonic significance of the Galápagos rise [*Menard*, 1965] is not well known. Analyses of magnetic anomalies that might be associated with these ridges and investigations of the mechanisms of earthquakes should provide more definitive information about these features. The possible continuation of the ridge system into northeastern Asia, the relationship of the Baikal rift to this system, and the northern continuation (or termination) of the East Pacific rise off western Canada are problems that need elucidation.

CONCLUSIONS

An active belt of shallow-focus earthquakes is associated with the mid-oceanic ridge system. Although this belt accounts for only about 5% of the world's earthquakes, the distribution and mechanism of these events provide considerable information about tectonics on both a regional and a global scale. Seismological and other geophysical data suggest that the mid-oceanic ridges and their continental extensions are characterized by transform faults, sea-floor spreading, and other aspects of extensional tectonics. As a result of better instrumentation, the number of earthquakes on the ridge system that can be located and studied has increased markedly during the last few years. Thus, many features can be investigated that were not previously resolvable either because of uncertainties in the epicentral locations or because of the small number of events that had been detected from a given region.

Acknowledgment. This study was partially supported under Department of Commerce ESSA grants E-22-96-67G and E-22-100-68G.

REFERENCES

Bodvarsson, G., and G. P. L. Walker, Crustal drift in Iceland, *Geophys. J.*, 8, 285–300, 1964.
Dietz, R. S., Continent and ocean basin evolution by spreading of the sea floor, *Nature*, 190, 854–857, 1961.
Ewing, M., and B. C. Heezen, Some problems of Antarctic submarine geology, in *Antarctica in the International Geophysical Year, Geophys. Monograph 1*, edited by A. Crary et al., pp. 75–81, American Geophysical Union, Washington, D. C., 1956.
Girdler, R. W., How genuine is the circum-Pacific belt, *Geophys. J.*, 8, 537–540, 1964.
Gutenberg, B., and C. F. Richter, *Seismicity of the Earth*, 2nd edition, Princeton University Press, Princeton, New Jersey, 1954.
Heezen, B. C., and M. Ewing, The mid-oceanic ridge, in *The Sea*, vol. 3, edited by M. N. Hill, pp. 388–410, Interscience Publishers, New York, 1963.
Heezen, B. C., and M. Tharp, Tectonic fabric of the Atlantic and Indian Oceans and continental drift, *Phil. Trans. Roy. Soc. London, A, 258*, 90–106, 1965.
Hess, H. H., History of ocean basins, in *Petrologic Studies: A Volume to Honor A. F. Buddington*, edited by A. E. J. Engel et al., pp. 599–620, Geological Society of America, New York, 1962.
Honda, H., Earthquake mechanism and seismic waves, *J. Phys. Earth (Tokyo)*, 10, 1–97, 1962.
Menard, H. W., *Marine Geology of the Pacific*, McGraw-Hill Book Company, New York, 1964.
Menard, H. W., Sea floor relief and mantle convection, *Phys. Chem. Earth*, 6, 315–364, 1965.
Menard, H. W., Fracture zones and offsets of the East Pacific rise, *J. Geophys. Res.*, 71, 682–685, 1966.
Menard, H. W., Extension of northeastern-Pacific fracture zones, *Science*, 155, 72–74, 1967.
Sykes, L. R., The seismicity of the Arctic, *Bull. Seismol. Soc. Am.*, 55, 501–518, 1965.
Sykes, L. R., The seismicity and deep structure of island arcs, *J. Geophys. Res.*, 71, 2981–3006, 1966.
Sykes, L. R., Mechanism of earthquakes and nature of faulting on the mid-oceanic ridges, *J. Geophys. Res.*, 72, 2131–2153, 1967.
Tobin, D. G., and L. R. Sykes, Relationship of hypocenters of earthquakes to the geology of Alaska, *J. Geophys. Res.*, 71, 1659–1667, 1966.
Tobin, D. G., and L. R. Sykes, Seismicity and tectonics of the Northeast Pacific Ocean, *J. Geophys. Res.*, 73, 3821–3845, 1968.
Vine, F. J., Spreading of the ocean floor: new evidence, *Science*, 154, 1405–1415, 1966.
Wilson, J. Tuzo, A new class of faults and their bearing on continental drift, *Nature*, 207, 343–347, 1965.

Worldwide Earthquake Mechanism

Anne E. Stevens

THIS ARTICLE reviews observational data on the mechanism of energy release at the focus of an earthquake. Particular attention is paid to a recent catalog of P nodal solutions and to evidence furnished by S and surface waves for type 1 and type 2 focal mechanisms. References are given to some theoretical focal models and to more detailed papers on earthquake mechanism. No attempt is made to give a comprehensive review of all recent papers pertaining to focal mechanism.

The mechanism of energy release at the focus of an earthquake can be studied by considering radiation patterns in longitudinal (P), shear (S), and surface waves over the surface of the focal sphere. Initial displacements in the P wave exhibit a quadrantal pattern of zones in which motion is alternately toward the epicenter (dilatational) and away from the epicenter (compressional). The theoretical amplitude of the initial displacement increases toward the center of the zones and decreases to a minimum at the boundaries of the zones, across which the displacement changes sign. The zones are separated by a pair of orthogonal planes intersecting in the focus, called 'P nodal planes.' Figures

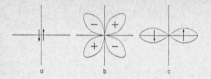

Fig. 1. Type 1 force system. Polar diagrams of theoretical displacement patterns from (a) a single couple for (b) P waves and (c) S waves [after *Hodgson and Stevens*, 1964].

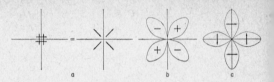

Fig. 2. Type 2 force system. Polar diagrams of theoretical displacement patterns from (a, left) a double couple or (a, right) a double dipole for (b) P waves and (c) S waves [after *Hodgson and Stevens*, 1964].

1 and 2 show two simple force systems, a single couple and a double dipole, respectively, acting at a point focus, that produce the quadrantal pattern. For the type 1 force system, one nodal plane corresponds to the fault on which the single couple produces motion in an earthquake. For the type 2 system, the nodal planes lie at 45° to the orthogonal double dipole axes, which represent axes of maximum compressive and tensile stress in the focal region. Figure 2 also shows the double couple force system, which is mathematically equivalent to the double dipole.

Several years ago it appeared that the problem of the mechanism of energy release would soon be resolved by a study of shear and surface waves that could remove the ambiguity in interpretation of P nodal planes. Unfortunately, the problem still remains. First, shear and surface wave results are yet too few to define a typical mechanism in each seismic region. Second, other models of the focal mechanism have been developed that predict the same radiation patterns and that are at least as physically reasonable as the single couple and double dipole models. Therefore, at present a fundamental ambiguity exists in the physical interpretation of both body- and surface-wave radiation patterns.

MATHEMATICAL MODELS

The reader is referred to review papers of *Honda* [1962] and *Stauder* [1962a] for a detailed treatment of various mathematical models of focal mechanism. The type 1 focal model has been regarded as a simple model of faulting, although the mathematical analysis assumes that the single couple acts in a continuous medium without a discontinuity. Type 2 represents three orthogonal principal stresses in the focal region. (The intermediate stress axis is perpendicular to the plane of the compressive and tensile stress axes.) There is a tendency to expect that two of the stresses are horizontal and the other vertical. This is undoubtedly reasonable for shallow focal depths, but not necessarily true at greater depths.

Subsequently, a more realistic fault model, in which the presence of the fault is included in the mathematical analysis, was derived with dislocation theory by several seismologists, including *Vvedenskaya* [1959] and *Knopoff and Gilbert* [1960]. The resulting equations are identical with the type 2 equations for the double dipole. A sudden change in volume and density in the focal region can also be represented by the same equations [*Randall*, 1966]. Therefore, a type 2 displacement pattern can be produced by any of three different focal mechanisms. The stress axes of greatest pressure and tension may represent the double dipole axes, but could equally well represent stresses accompanying faulting or initiating a phase change. In this article, the term 'type 2' means the mathematical description of the radiation patterns, not a specific physical model.

P NODAL PLANE SOLUTIONS

A review of P nodal plane studies to the end of 1963 has been given by *Hodgson and Stevens* [1964]. More recent P studies are given by *Wickens and Hodgson* [1967], who describe computer-redetermined nodal plane solutions, and by *Sykes* [1967], who describes mid-oceanic ridge earthquakes.

Wickens-Hodgson Catalog

Wickens and Hodgson have compiled a catalog of 618 earthquakes from 1922 to 1962, for which they have calculated nodal plane solutions by the same computer program. Solutions for most of these earthquakes had been determined previously by visual methods of analysis by many investigators, including the Ottawa group. The prime purpose in recalculating these solutions by a standard method was to discover why, in the past, different solutions had often been obtained for earthquakes with similar foci and for the same earthquakes studied by different investigators.

The writer found that agreement between visual and computer solutions in the catalog was within 10° for almost 50% and within 20° for about 70% of the earthquakes and did not depend consistently on such factors as seismic region, focal depth, magnitude, year of occurrence, and original investigator [Stevens and Hodgson, 1968]. The significant factor was the distribution of data in azimuth with respect to the epicenter. It appears to be essential to have at least two reliable and consistent data points in every 90° segment of azimuth. Only 30% of the solutions met this azimuth requirement.

In addition, well-defined solutions with shallow-dipping planes require many measurements at short distances. Those with steeply dipping planes require data from PKP phases. However, the azimuth requirement is more important than the distance requirement. All previous discrepancies between visual solutions were due simply to inadequate distributions of data. No more than 10% of the 618 solutions were closely defined by the data. Each of these solutions satisfied the above azimuth requirement and, in addition, would not have been altered by more than a few degrees by the arbitrary addition or deletion of several data points. It is significant that whenever the data were numerous and well-distributed in azimuth and distance, the nodal planes had to be orthogonal. In theory, adequate data can now be obtained for earthquakes in most seismic regions, inasmuch as the Worldwide Standard Seismograph Network (WWSSN) of the United States and the standard networks of Russia and Canada provide the necessary coverage, when supplemented by data from other reliable stations. Data from only one or two of these networks rarely are sufficient for well-defined solutions.

Previous analyses of P nodal solutions have shown that the pressure axis is perpendicular to the trend of the continental seismic belts, but that the tension axis tends to be perpendicular to the mid-oceanic belts. Discovery of this relation did not require extremely well-defined P nodal solutions. The trend of any tectonic feature is not constant over great distances, and variations of ten or twenty degrees in the azimuth of a stress axis did not conceal the tendency to perpendicularity.

Many years ago Honda [1932] suggested that in Japan the dip of the pressure axes was near zero for shallow earthquakes and increased to about 45° for deep earthquakes. Some solutions in the catalog supported this idea. However, any systematic dependence of the dip angle of the stress axes on focal depth cannot be determined until many very well-defined P nodal solutions are available.

Mid-Oceanic Ridges

Sykes [1967] studied earthquakes of the mid-oceanic seismic belts and discovered that earthquakes occurring in fracture zones exhibit properties different from those occurring on the ridges. Three earthquakes that occurred on the crest of a mid-oceanic ridge, not associated with a known fracture zone, showed normal faulting, the tension axis being horizontal and perpendicular to the local strike of the ridge. One of the two nodal planes in each of ten earthquakes that occurred in fracture zones of the Mid-Atlantic ridge and East Pacific rise dipped steeply parallel to the strike of the fracture zone. Motion on these planes was strike-slip, but in the opposite sense to that expected if the fracture zone represented a transcurrent displacement of the ridges.

Wilson [1965] has shown that the offset of ridges in a left-lateral sense can be explained either by sinistral transcurrent faulting or by dextral transform faulting (and conversely for ridges offset in a right-lateral sense). The assumed mechanism for transform faulting is that mantle convection currents rise beneath the ridge and move outward, creating horizontal tensile stresses across the ridge. Where the ridge is offset by a fracture zone, the tensional forces are parallel to the zone and act in opposite senses in the fracture zone between the ridges and in the same sense beyond the ridges. The relative displacement beyond the ridges is zero. The relative displacement between the ridges is termed transform faulting and is always in the opposite sense to the transcurrent faulting that would be required to offset the ridges. Sykes found eight solutions indicating dextral transform faulting and two indicating sinistral transform faulting. Therefore, the P nodal solutions of Sykes have shown that present tectonic movements indicated by earthquakes along some parts of the mid-oceanic seismic belts can be explained by a transfer of mass upward in the mantle beneath the ridges.

FOCAL MECHANISM

S Waves

Types 1 and 2 focal mechanisms can be distinguished from each other by their S radiation patterns, as illustrated in Figures 1 and 2, but the physical interpretation of these types is not unique. A literature review of the use of transverse waves in mechanism studies has been published by *Stefánsson* [1966]. Most focal mechanism solutions have been determined with S polarization angles by Stauder and his students.

The polarization angle S, defined by arc tan (SH/SV), is measured in the plane of S between the direction of S and a vertical plane through the great circle path between epicenter and station. When the focal sphere is mapped in an azimuthal projection on a plane, the polarization angles lie along curves that intersect in one point for the type 1 mechanism and in two points for the type 2 mechanism. Polarization angles can be measured only for epicentral distances from about 20° to 80°, which corresponds to a maximum range of about 40° in emergent angle at the focus. Thus S measurements are available for less than one-quarter of the surface of the focal sphere, in contrast to more than half the sphere for P data. It is often difficult to distinguish type 1 from type 2 by a visual inspection of the S data, when the point or points where the polarization curves intersect lie outside the range of the data. This difficulty is partly overcome by replacing the subjective visual analysis by an objective computer analysis [*Stevens*, 1964, 1967].

Neither method produces meaningful results, however, when good quality data are not available in all four quadrants about the epicenter. An inadequate azimuth distribution of S data tends to favor a type 1 mechanism, as such data readily lie on curves intersecting in one point. Many of the type 1 mechanism determinations reported in the literature are probably invalid for this reason. Data in many azimuths are required to show whether the polarization curves intersect in a second point as well.

Table 1 lists 33 earthquakes in whose mechanism solutions at least one S data point was available in every 90° segment of azimuth. Despite this azimuth requirement, the focal type of three earthquakes is uncertain, since analysis of the same S data by different investigators and/or methods yielded different solutions. In addition, Stauder noted that a type 1 mechanism could fit the P and S data of 17 Alaskan earthquakes of 1964 and 1965, although the measures of error were notably smaller for a type 2 [*Stauder and Bollinger*, 1966a].

The list in Table 1 represents about 25% of the earthquakes for which mechanism solutions using S polarization angles have been published. In most of the remainder, both P and S data were used to determine P nodal planes and either S or both P and S data had an inadequate azimuth distribution. The addition of P data was of limited help in determining focal mechanism if the azimuth distribution of S was not adequate. Too few focal regions are represented in Table 1 to draw any general conclusions about mechanism type.

Surface Waves

Table 2 describes the four earthquakes for which surface waves have been used to choose between types 1 and 2 and one for which P waves have been so used. The Banda Sea earthquake of March 21, 1964, had a type 2 mechanism determined by a method which equalized body-wave spectra to the source for a number of stations well distributed in azimuth about the epicenter. The focal depth of this earthquake is the largest of those for which a mechanism type has been determined.

The focal mechanism of about 100 earthquakes has been examined with surface waves, chiefly by Aki, by Brune, and by Ben-Menahem and their colleagues in papers dating from 1960. For many of these earthquakes observational data were not compared with both types 1 and 2. *Aki* [1964] used P nodal solutions for some surface wave studies, most of which the writer feels were not well enough defined. This undoubtedly explains why Aki found little agreement with type 1 or 2.

Two basically different approaches were used for the mechanism determinations given in

TABLE 1. Earthquake Focal Mechanism from S Waves

Date and Time, GMT	Location	Depth, km	Richter Mag.	Focal Type	References
Nov. 6, 1951, 16h 40m 06s	47°N, 154°E, Kurils	Normal	$7-7\frac{1}{4}$	2	Stauder, 1962b; Udias and Stauder, 1964
Mar. 5, 1953, 21h 01m 23s	51°N, 158°E, Kamchatka	60	$6\frac{3}{4}$	2	Stauder, 1962b; Udias and Stauder, 1964
Sept. 4, 1953, 07h 23m 05s	50°N, 156.5°E, Kamchatka	60	$6\frac{3}{4}-7$	2	Stauder, 1962b; Udias and Stauder, 1964
Sept. 23, 1953, 02h 14m 36s	50.5°N, 156°E, Kamchatka	60	7	2	Stauder, 1962b; Udias and Stauder, 1964
Apr. 7, 1958, 15h 30m 38s	66.5°N, 157°W, Alaska	Shallow	7	1*	Ritsema, 1962; Stevens, 1964
July 10, 1958, 06h 51m 51s	58.6°N, 137.1°W, Alaska	15	8	1†	Stauder, 1960
Aug. 15, 1958, 19h 55m 39s	53°N, 160.5°E, Kamchatka	60	$6\frac{3}{4}$	2	Stauder, 1962b; Udias and Stauder, 1964
Nov. 12, 1958, 20h 23m 26s	44.5°N, 148.5°E, Kurils	Normal	$6\frac{3}{4}-7$	2	Stauder, 1962b; Udias and Stauder, 1964
Apr. 26, 1959, 20h 40m 38s	25°N, 122.5°E, Taiwan	150	$7\frac{1}{2}-7\frac{3}{4}$	1, 2	Ritsema, 1962; Stevens, 1964
Oct. 13, 1960, 14h 52m 34.7s	55.0°N, 161.2°E, Kamchatka	35	$6\frac{3}{4}$	1, 2	Stauder, 1962b; Udias and Stauder, 1964
Mar. 16, 1963, 08h 44m 51.1s	46.6°N, 154.8°E, Kurils	46	7	2	Stauder and Bollinger, 1966b
May 22, 1963, 13h 56m 47.5s	48.7°N, 154.8°E, Kurils	54	$6\frac{1}{2}$	2	Stauder and Bollinger, 1966b
June 28, 1963, 21h 55m 36.8s	46.7°N, 153.3°E, Kurils	12	$6\frac{3}{4}$	2	Stauder and Bollinger, 1966b
Oct. 12, 1963, 11h 26m 57.9s	44.8°N, 149.0°E, Kurils	40	$6\frac{3}{4}-7$	2	Stauder and Bollinger, 1966b
Nov. 15, 1963, 21h 06m 34s	44.3°N, 149.0°E, Kurils	50	$6\frac{1}{4}-6\frac{1}{2}$	2	Stauder and Bollinger, 1966b
Nov. 17, 1963, 00h 48m 02.6s	7.6°N, 37.4°W, N. Atlantic	33	$6\frac{1}{4}-6\frac{1}{2}$	2	Stauder and Bollinger, 1966b
Feb. 6, 1964, 13h 07m 25.2s	55.7°N, 155.8°W, Kodiak	33	$6\frac{3}{4}-7$	2	Stauder and Bollinger, 1966a
Mar. 28, 1964, 20h 29m 08.6s	59.8°N, 148.7°W, Prince Wm.	40	6.6	2	Stauder and Bollinger, 1966a
Mar. 29, 1964, 06h 04m 44.5s	56.1°N, 154.3°W, Kodiak	30	5.8	2	Stauder and Bollinger, 1966a
Mar. 30, 1964, 02h 18m 06.3s	56.6°N, 152.9°W, Kodiak	25	6.6	2	Stauder and Bollinger, 1966a
Mar. 30, 1964, 07h 09m 34.0s	59.9°N, 145.7°W, Prince Wm.	15	6.2	2	Stauder and Bollinger, 1966a
Mar. 30, 1964, 16h 09m 28.4s	56.6°N, 152.1°W, Kodiak	25	5.5	2	Stauder and Bollinger, 1966a
Apr. 4, 1964, 17h 46m 08.6s	56.3°N, 154.4°W, Kodiak	25	$6\frac{1}{2}$	2	Stauder and Bollinger, 1966a
Apr. 5, 1964, 01h 22m 13.3s	56.2°N, 153.5°W, Kodiak	25	6	2	Stauder and Bollinger, 1966a
Apr. 16, 1964, 19h 26m 57.4s	56.4°N, 152.9°W, Kodiak	30	$6\frac{1}{2}-6\frac{3}{4}$	2	Stauder and Bollinger, 1966a
May 12, 1964, 18h 16m 41.9s	56.6°N, 152.4°W, Kodiak	10	$5\frac{1}{2}-5\frac{3}{4}$	2	Stauder and Bollinger, 1966a
May 17, 1964, 00h 50m 17.9s	59.4°N, 142.7°W, Prince Wm.	35	$5\frac{3}{4}$	2	Stauder and Bollinger, 1966a
Aug. 2, 1964, 08h 36m 16.9s	56.2°N, 149.9°W, Kodiak	31	6	2	Stauder and Bollinger, 1966a
Aug. 24, 1964, 21h 56m 54.2s	58.4°N, 150.3°W, Kodiak	22	$4\frac{3}{4}-5$	2	Stauder and Bollinger, 1966a

TABLE 1 (continued)

Date and Time, GMT	Location	Depth, km	Richter Mag.	Focal Type	References
Sept. 27, 1964, 15h 50m 54.7s	56.6°N, 152.0°W, Kodiak	27	$5\frac{1}{4}$	2	Stauder and Bollinger, 1966a
June 23, 1965, 11h 09m 15.7s	56.5°N, 152.8°W, Kodiak	33	$6\frac{1}{4}$–$6\frac{1}{2}$	2	Stauder and Bollinger, 1966a
Sept. 4, 1965, 14h 32m 46.7s	58.2°N, 152.7°W, Kodiak	10	$6\frac{3}{4}$–7	2	Stauder and Bollinger, 1966a
Dec. 22, 1965, 19h 41m 23.1s	58.4°N, 153.1°W, Kodiak	51	$6\frac{3}{4}$–7	2	Stauder and Bollinger, 1966a

* No focal type was determined by second author.
† See Table 2 and text for surface-wave analyses of this earthquake.

Table 2. The method of Ben-Menahem and Toksöz required records from a single station, whereas the other required records from a number of stations well distributed in azimuth about the epicenter. Surface waves recorded at a single station perhaps cannot reliably determine mechanism; data from as many stations as possible should be used to avoid undue reliance on the interpretation of surface waves along a single path.

The mechanisms of only two earthquakes have been examined both with S polarization angles and surface waves. For the earthquake of July 6, 1962 (Table 2), both S-wave [Stevens, 1966] and surface-wave analyses agreed on a type 2 mechanism. (This earthquake is not included in Table 1, because there is one azimuth segment of 92° without S data.) For the earthquake of July 10, 1958 (Tables 1 and 2), both S and P data are sufficient to choose type 1 unambiguously and to define P nodal planes closely. Mechanism determination by surface waves is not conclusive. The Rayleigh-wave radiation pattern observed by Brune could be produced by either type 1 or 2, according to the theoretical results of Haskell [1963]. Brune presented some qualitative data on Love-wave amplitudes at two stations that favored type 1 rather than 2, but these data are not sufficient to choose type 1 unambiguously. The quantitative Love-wave analysis of Ben-Menahem and Toksöz, which was based on one station, showed a type 2 mechanism. However, some aspects of their work have recently been criticized by Savage [1967]; their mechanism determination may require modification.

TABLE 2. Earthquake Focal Mechanism

Date and Time, GMT	Location	Depth, km	Richter Mag.	Focal Type	References
From Surface Waves					
Nov. 4, 1952, 16h 58m 22s	52.6°N, 160.3°E, Kamchatka	Normal	$8\frac{1}{4}$	2	Ben-Menahem and Toksöz, 1963a
July 10, 1958, 06h 15m 51s	58.6°N, 137.1°W, Alaska	15	8	1, 2	Brune, 1961, 1962; Ben-Menahem and Toksöz, 1963b
July 6, 1962, 23h 05m 32.2s	36.6°N, 70.4°E, Hindu Kush	203	$6\frac{3}{4}$	2	Chander and Brune 1965
Sept. 1, 1962, 19h 20m 39s	35.6°N, 50.0°E, Iran	21	7–$7\frac{1}{2}$	2	Wu and Ben-Menahem, 1965
From P Waves					
Mar. 21, 1964, 03h 42m 19.6s	6.4°S, 127.9°E, Banda Sea	367	7–$7\frac{1}{4}$	2	Teng and Ben-Menahem, 1965

CONCLUSIONS

Tables 1 and 2 list those earthquakes whose mechanism solutions with shear and surface waves appear to the author to be based on reasonably reliable data. They represent less than 20% of the published mechanism solutions, which illustrates the difficulty in securing adequate data and an unambiguous choice between types 1 and 2. Most of the solutions in the tables are classified as type 2, but one solution is almost certainly type 1, and more than a dozen may have type 1 mechanisms. Only three earthquakes lie outside the circum-Pacific belt, and only three have focal depths greater than 60 km. Therefore, the nature of the focal mechanism is still quite uncertain, despite progress made in the last few years in methods of data analysis. Observational data appear to conform to either a type 1 or a type 2 focal model whenever sufficient data are available. No other models appear necessary. However, various physical interpretations of the two models exist.

The most important task in earthquake mechanism studies at present is the application of seismic data to an accurate determination of the mechanism type in each seismic region. Type 2 should not be considered to apply to all tectonic earthquakes until the data show this conclusively. A further refinement of theoretical focal models or even the choice of the proper physical meaning of type 2 requires some guidance from observations.

An accurate interpretation of S- and surface-wave data frequently requires a good P nodal solution. Therefore, every effort should be made to ensure that published P nodal solutions are based on sufficient data. First-motion data should be read from copies of seismograms from not only the Canadian and WWSSN networks, but also from the standard network of the USSR, as stations of the latter frequently span an important range of azimuths and distances. They are even more essential in S-wave studies in which S data are limited in distance range, and a clear distinction between the two focal models requires an adequate azimuth distribution. Surface-wave data from many stations rather than from only a single station should be used in order to ensure reliable mechanism determination.

REFERENCES

Aki, K., Study of Love and Rayleigh waves from earthquakes with fault plane solutions or with known faulting, 2, Application of the phase difference method, *Bull. Seismol. Soc. Am.*, *54*, 529–558, 1964.

Ben-Menahem, A., and M. N. Toksöz, Source mechanism from spectrums of long-period surface waves, 2, The Kamchatka earthquake of November 4, 1952, *J. Geophys. Res.*, *68*, 5207–5222, 1963a.

Ben-Menahem, A., and M. N. Toksöz, Source-mechanism from spectra of long-period seismic surface waves, 3, The Alaska earthquake of July 10, 1958, *Bull. Seismol. Soc. Am.*, *53*, 905–919, 1963b.

Brune, J. N., Radiation pattern of Rayleigh waves from the Southeast Alaska earthquake of July 10, 1958, *Publ. Dom. Obs., Ottawa*, *24*, 373–383, 1961.

Brune, J. N., Correction of initial phase measurements for the Southeast Alaska earthquake of July 10, 1958, and for certain nuclear explosions, *J. Geophys. Res.*, *67*, 3643–3644, 1962.

Chander, R., and J. N. Brune, Radiation pattern of mantle Rayleigh waves and the source mechanism of the Hindu Kush earthquake of July 6, 1962, *Bull. Seismol. Soc. Am.*, *55*, 805–819, 1965.

Haskell, N. A., Radiation pattern of Rayleigh waves from a fault of arbitrary dip and direction of motion in a homogeneous medium, *Bull Seismol. Soc. Am.*, *53*, 619–642, 1963.

Hodgson, J. H., and A. E. Stevens, Seismicity and earthquake mechanism, in *Research in Geophysics*, vol. 2, pp. 27–59, edited by H. Odishaw, MIT Press, Cambridge, 1964.

Honda, H., On the types of seismograms and the mechanism of deep earthquakes, *Geophys. Mag.*, *5*, 301–326, 1932.

Honda, H., Earthquake mechanism and seismic waves, *Geophys. Notes, Suppl.*, *15*, 1–97, 1962.

Knopoff, L., and F. Gilbert, First motions from seismic sources, *Bull. Seismol. Soc. Am.*, *50*, 117–134, 1960.

Randall, M. J., Seismic radiation from a sudden phase transition, *J. Geophys. Res.*, *71*, 5297–5302, 1966.

Ritsema, A. R., P and S amplitudes of two earthquakes of the single force couple type, *Bull. Seismol. Soc. Am.*, *52*, 723–746, 1962.

Savage, J. C., Spectra of S waves radiated from bilateral fracture, *Bull. Seismol. Soc. Am.*, *57*, 39–54, 1967.

Stauder, W., S. J., The Alaska earthquake of July 10, 1958: Seismic studies, *Bull. Seismol. Soc. Am.*, *50*, 293–322, 1960.

Stauder, W., S. J., The focal mechanism of earthquakes, *Advan. Geophys.*, *9*, 1–76, 1962a.

Stauder, W., S. J., S-wave studies of earthquakes

of the North Pacific, 1, Kamchatka, *Bull. Seismol. Soc. Am., 52*, 527–550, 1962b.

Stauder, W., and G. A. Bollinger, The focal mechanism of the Alaska earthquake of March 28, 1964, and of its aftershock sequence, *J. Geophys. Res., 71*, 5283–5296, 1966a.

Stauder, W., and G. A. Bollinger, The S-wave project for focal mechanism studies, earthquakes of 1963, *Bull. Seismol. Soc. Am., 56*, 1363–1371, 1966b.

Stefánsson, R., The use of transverse waves in focal mechanism studies. *Tectonophysics, 3*, 35–60, 1966.

Stevens, A. E., Earthquake mechanism determination by S-wave data, *Bull. Seismol. Soc. Am., 54*, 457–474, 1964.

Stevens, A. E., S-wave focal mechanism studies of the Hindu Kush earthquake of July 6, 1962, *Can. J. Earth Sci., 3*, 367–387, 1966.

Stevens, A. E., S-wave earthquake mechanism equations, *Bull. Seismol. Soc. Am., 57*, 99–112, 1967.

Stevens, A. E., and J. H. Hodgson, A Study of P nodal solutions (1922–1962) in the Wickens-Hodgson Catalogue, *Bull. Seismol. Soc. Am., 58*(3), 1071–1082, 1968.

Sykes, L. R., Mechanism of earthquakes and nature of faulting on the mid-oceanic ridges, *J. Geophys. Res., 72*, 2131–2153, 1967.

Teng, T., and A. Ben-Menahem, Mechanism of deep earthquakes from spectrums of isolated body-wave signals, 1, The Banda Sea earthquake of March 21, 1964, *J. Geophys. Res., 70*, 5157–5170, 1965.

Udias, A., S. J., and W. Stauder, S. J., Application of numerical method for S-wave focal mechanism determinations to earthquakes of Kamchatka-Kurile Islands region, *Bull. Seismol. Soc. Am., 54*, 2049–2065, 1964.

Vvedenskaya, A. V., The displacement field associated with discontinuities in an elastic medium, *Bull. Acad. Sci. USSR, Geophys. Ser.*, pp. 357–362, 1959.

Wickens, A. J., and J. H. Hodgson, Computer re-evaluation of earthquake mechanism solutions, 1922–1962, *Publ. Dom. Obs., Ottawa, 33*, 1–560, 1967.

Wilson, J. T., A new class of faults and their bearing on continental drift, *Nature, 207*, 343–347, 1965.

Wu, F. T., and A. Ben-Menahem, Surface wave radiation pattern and source mechanism of the September 1, 1962, Iran earthquake, *J. Geophys. Res., 70*, 3943–3949, 1965.

Mechanism of Earthquakes in and near Japan and Related Problems

Masaji Ichikawa

MANY Japanese seismologists have made statistical studies of the focal mechanism of Japanese earthquakes based on the abundant, reliable data accumulated during the past decade and the recent results of investigations in the geological and geodetic fields. This paper reviews the general tendency of stresses that produce large earthquakes in Japan and its vicinity and describes the results of recent studies on the stress system that produces shallow and very shallow earthquakes in central and southwestern Japan.

From data obtained by near and distant stations, more than 150 nodal plane solutions have been determined for large earthquakes occurring in and near Japan. The results of statistical studies based on these solutions will be briefly described [*Ichikawa*, 1966; *Honda and Ichikawa*, 1968].

The pressure directions of stresses that produce earthquakes are, in general, distributed regularly. More specifically, the pressure directions of earthquakes occurring in the upper mantle are nearly normal to deep and intermediate earthquake belts (Figure 1), and those for events taking place in the uppermost part of the mantle and in the earth's crust are nearly parallel to the trend of the island arcs in central and southwestern Japan and perpendicular to the island arc of northern Japan.

It is notable that the inclinations of pressure axes vary systematically with the focal depth:

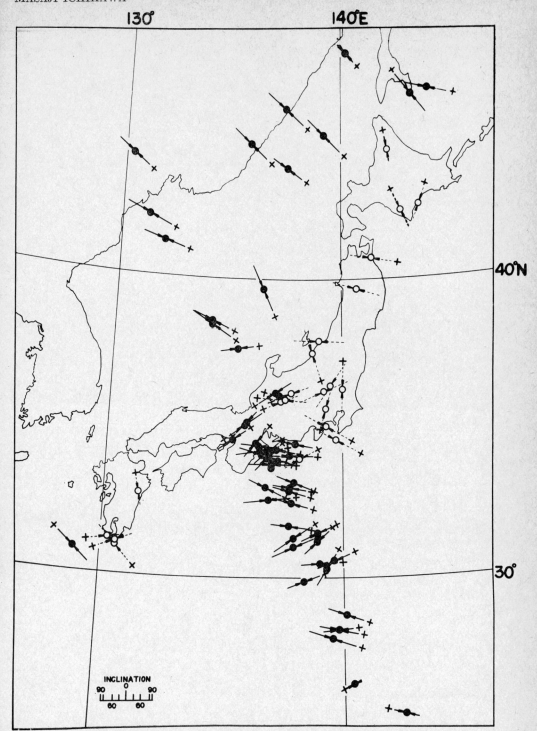

Fig. 1. Pressure directions of deep (solid arrows) and intermediate (broken arrows) earthquakes in the period 1927–1962. The length of the arrows is proportional to $\sin i$ (i is inclination of pressure axis), and $+$ indicates the upside of the dipping pressure axis. Taken from unpublished data of H. Honda et al. (1927–1949) and M. Ichikawa (1950–1962).

while pressure axes for earthquakes occurring in the crust lie horizontally, those for events in the uppermost part of the mantle dip gently downward from the inland side of Honshu toward the Pacific side and are nearly normal to the tendency of spatial distribution of earthquake foci. Furthermore, the pressure axes of deep earthquakes are approximately parallel to the tendency of spatial distribution of earthquake foci.

The tension directions of stresses that produce earthquakes in the upper mantle do not exhibit the same regularity in geographical distribution as the pressure directions. However, the tension directions appear generally to lie on a plane peculiar to each seismic zone.

These results are in good agreement with Aki's [1966] conclusion, which he obtained statistically from the data of small earthquakes that occurred in a short interval of time.

While the earthquakes in northern Japan occur mostly in the uppermost part of the mantle, many earthquakes in central and southwestern Japan take place in the crust. Shallow earthquakes also occur in a few confined areas in central and southwestern Japan: southeast of Kwanto in central Japan, the Kii peninsula, and the area between Shikoku and Kyushu in western Japan.

The radiation pattern of P waves for most of the earthquakes in southern Japan is separated by two straight lines perpendicular to

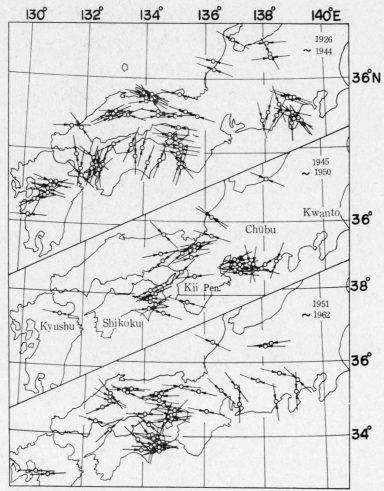

Fig. 2. Pressure directions of very shallow earthquakes occurring in the period 1926–1962 [*Ichikawa*, 1965].

each other. This suggests that the directions of tension as well as pressure are parallel to the earth's surface, and their directions are perpendicular to each other. The earthquakes in northern Japan, however, generally exhibit a more complicated radiation pattern of P waves on the earth's surface, since they are generated by a pressure-tension system whose axes are not always parallel to the earth's surface.

It is worthwhile to describe in detail the state of pressures for very shallow earthquakes in connection with tectonic and geodetic evidences. The nodal plane solutions of 247 earthquakes that occurred in central and southwestern Japan in the period 1926–1962 were analyzed and were compared with the results of investigations in geological and geodetic fields.

The geographical distribution of the pressure directions of very shallow earthquakes is rather irregular in central and southwestern Japan, although the directions are usually parallel to each other in each district (Figure 2). (In the Chubu district, the pressure directions are somewhat irregular.) The pressure directions of earthquakes in the crust and upper mantle in the region between Shikoku and Kyushu are nearly normal to those of neighboring areas. This phenomenon can be explained satisfactorily by the geological and tectonic structures.

For the period 1926–1944, the pressure directions in southwestern Japan, excluding the area between Shikoku and Kyushu and the eastern part of Shikoku, are essentially parallel to the Honshu arc. The pressure directions in the central and eastern Chubu district appear to be irregular at first sight, but they are also nearly parallel to the Honshu arc or to the Izu-Mariana arc, both of which run through Chubu district. The peculiar pressure directions in the area between Shikoku and Kyushu are probably related to the Ryukyu arc (Figure 3).

The pressure directions in the eastern part of Shikoku continue to be variable, but this can be explained by the influence of the stress that produced the large earthquake of Nankaido in 1946. In fact, the pressure directions for earthquakes occurring in the area after that earthquake changed by nearly 90°. The prevailing

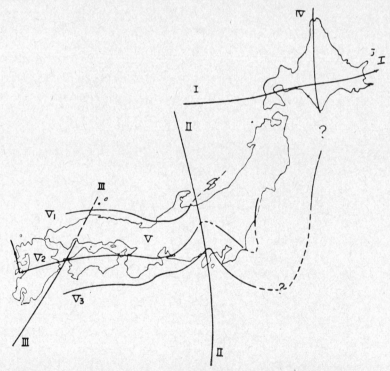

Fig. 3. Island arcs that meet at the Japanese Islands [after *Miyamura*, 1962]. I, Kuril arc; II, Izu–Bonin arc; III, Ryukyu arc; IV, Hidaka–Sakhalin arc; V, Honshu arc (1, axis of Akiyoshi orogeny; 2, axis of Sakawa orogeny; 3, axis of latest orogeny).

direction is approximately parallel to the Honshu arc and to the trend of pressure directions in neighboring areas.

Furthermore, pressure directions exhibit good harmony with geodetic evidence. The pressure directions in most areas of central and southwestern Japan generally agree in their trend with directions of minimum principal axes (contraction directions) of land deformation, which were calculated by *Kasahara and Sugimura* [1964] from retriangulation data.

In the three shallow earthquake areas described above, the geographical distribution of pressure directions of many local earthquakes is almost consistent with the results of statistical studies on the nodal plane solutions of large earthquakes; however, the inclination and direction of the pressure axes for shallow earthquakes are more irregular than those for very shallow or deep earthquakes.

It is particularly interesting that the pressure axes in the area between Shikoku and Kyushu lie in the north-south direction. The minimum principal axes of the land deformation and the trend of the island arc passing through the area also lie in the north-south direction. This direction is normal to the prevailing pressure directions in the neighboring zones. Thus, it appears that the movements in the crust and in the uppermost part of the mantle in this

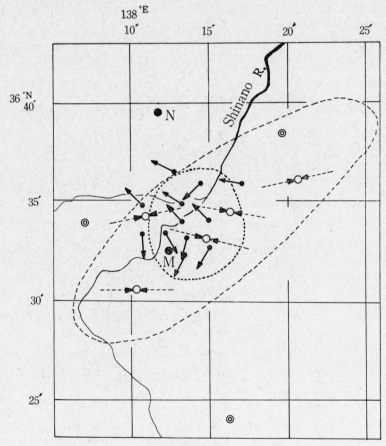

Fig. 4. Distribution of mean pressure directions and directions of ground movement accompanying Matsushiro earthquake swarm. Broken arrows show mean direction of pressure, and solid arrows show direction of ground movement obtained geodetically. Small dotted circle encloses area in which earthquakes occurred August 1965 to August 1966. The area in which earthquakes occurred September 1966 to December 1966 is enclosed by a broken line. N, Nagano weather station; M, Matsushiro seismological station. Small circles within circles are temporary stations.

area are closely related to the same force system.

Since August 1965 a great number of earthquakes have been taking place at depths of a few kilometers in an area including Matsushiro Seismological Observatory in the northern part of Chubu district, central Japan. The seismology, geology, and geodesy in this area are temporarily being observed in order to discover the cause of the earthquake swarm.

Data obtained by the permanent and temporary networks clearly show that the regions of condensation and rarefaction of P waves for these earthquakes are separated by two perpendicular straight lines, and that the polarization of the initial motion of S waves is consistent with that given by Honda's type II force system (double couple).

The mechanisms are determined for more than 200 earthquakes whose magnitudes are 3 or larger. The solutions show that the pressure directions lie east-west, which is consistent with the predominant directions of earthquakes that have occurred in this region in the past.

The pressure directions agree very well with the minimum principal axis of land deformation in the epicentral area, determined on the basis of geodetic data. Analysis of the triangulation data also reveals the occurrence of a strike-slip fault accompanying the earthquake swarm. The fault strike (approximately northwest-southeast) coincides with one of the nodal lines determined seismologically.

Still unexplained is the distribution of epicenters. From August 1965 to August 1966, the earthquakes occurred in a circular area that had a diameter of several kilometers. After September 1966, the area of the epicenters expanded in a northeast-southwest direction, which is not the strike direction of the fault revealed geodetically (Figure 4).

The geographical distribution of pressure directions exhibits a systematic pattern, irrespective of the time lapse and earthquake magnitude. This suggests the existence of a persistent stress field in the crust and the upper mantle, which may be produced by a compressive force.

Good harmony exists between the prevailing pressure directions and the geodetic and geotectonic evidence. A peculiar, abrupt change of pressure system at the time of the large earthquake of Nankaido in 1946 ($M = 8.1$) and the occurrence of the earthquake swarm in a narrow confined area in Chubu district may be explained by these forces.

Acknowledgment. The author wishes to thank Professor H. Honda, Tokyo University, who read the manuscript and offered many helpful suggestions.

REFERENCES

Aki, K., Earthquake generating stress in Japan for the years 1961 to 1963 obtained by smoothing the first motion radiation patterns, *Bull. Earthquake Res. Inst., Tokyo Univ., 44,* 447-471, 1966.

Honda, H., and M. Ichikawa, Mechanism of earthquakes in and near Japan, 1927-1966, in preparation, 1968.

Ichikawa, M., The mechanism of earthquakes occurring in central and southwestern Japan, and some related problems, *Papers Meteorol. Geophys., Tokyo, 16,* 104-155, 1965.

Ichikawa, M., Mechanism of earthquakes in and near Japan, 1950-1962, *Papers Meteorol. Geophys., Tokyo, 16,* 201-229, 1966.

Kasahara, K., and A. Sugimura, Horizontal secular deformation of land deduced from retriangulation data, land deformation in central Japan, *Bull. Earthquake Res. Inst., Tokyo Univ., 42,* 479-490, 1964.

Miyamura, S., Seismicity and geotectonics (in Japanese), *J. Seismol. Soc. Japan, 15,* 23-52, 1962.

The Field of Elastic Stresses Associated with Earthquakes

L. M. Balakina, L. A. Misharina, E. I. Shirokova, and A. V. Vvedenskaya

INSTRUMENTAL observations show that the mechanism of most earthquake foci, irrespective of their depth, is a shear due to the sliding of one fault face relative to another. The development and propagation of the slip surface is due to shear stresses along this surface. The shear stresses at each point of the slip surface are equivalent to two main stresses, tension and compression, acting on two mutually perpendicular planes at 45° angles to the slip surface. The third main stress, the intermediate one, is equal to zero.

In an earthquake, the release of shear stresses (which we call stresses acting in foci) takes place at points of the slip area. The directions of the main axes of these stresses are related to the position of the nodal surfaces of the displacements fields of *P, SV,* and *SH* waves generated in the process of propagation of the slip area [*Vvedenskaya*, 1956, 1960]. The position of the nodal surfaces of the wave field does not depend on the rate and duration of propagation of the slip region as a source [*Vvedenskaya*, 1965, 1967]. Hence the observations of the displacements in seismic waves may serve as the basis for determining the directions of the three principal stresses in earthquake foci.

The spatial distribution of the principal stress axes in earthquake foci of various seismic zones of the earth are discussed by *Balakina et al.* [1967]. In each seismic region, definite, characteristic regularities have been found in the orientation of these axes. Analysis of the established regularities permits the determination of the character of the stressed state and some conclusions concerning fields of elastic stresses associated with earthquakes.

REGULARITIES IN THE ORIENTATION OF STRESSES IN EARTHQUAKE FOCI

Regularities in the orientation of stresses acting in earthquake foci are discussed for several seismic belts (see Figure 1).

The Mediterranean-Asiatic Belt

This belt is divided into several regions according to the type of stress orientation.

In the eastern Mediterranean [*Shirokova*, 1963], the compression in the investigated earthquake foci (which are not systematically distributed) is not characterized by a distinct predominant orientation relative to the horizontal plane. The tension has a clear horizontal orientation. Regularities in the orientation of the stress axes relative to surface structures have not been established.

In the earthquake foci near the bend of the Carpathian arc [*Vvedenskaya and Ruprekhtova*, 1961], the compressional axis is nearly parallel to the horizontal plane and normal to the arc; the tension and the intermediate stresses are in a plane whose intersection with the earth's surface forms a tangent to the arc and coincides with the strike of axes of the observed folds in this region.

In the central part of the belt (the Caucasus and Asia to 90°E longitude) [*Shirokova*, 1959, 1961, 1962], the compression is approximately horizontal and is normal to the trend of the surface structures. The predominant orientation of tension is nearly vertical. The direction of the intermediate stress is close to horizontal and coincides with the strike of structures. Thus the eastern Mediterranean and the central part of the belt differ sharply as to the orientation of stresses. Thus tension appears to be horizontal in the Mediterranean zone, while compression is horizontal in the adjacent region to the east, with no area of mixed orientation between. The boundary separating the two regions is near 40°E longitude.

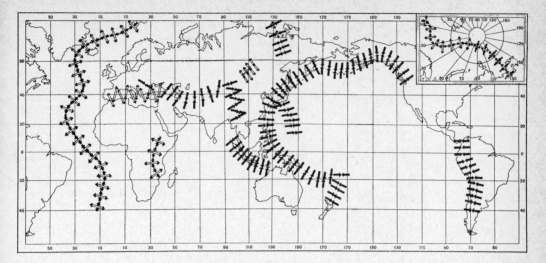

Fig. 1. The orientation of the principal stress axes in the elastic stress field of the earth: the greatest relative compression acting in the horizontal plane transversely to tectonic structures is indicated by converging black arrows; the greatest relative tension is indicated by diverging white arrows. White and black circles similarly indicate that another of the principal stresses, which gives the maximum difference with the greatest relative compression or tension stress acting transversely to the tectonic structure, has vertical direction. Small white or black arrows indicate the second principal stress directed horizontally and parallel to the tectonic structure. Uniaxial compressions or tensions are denoted by an additional symbol adjacent to large converging or diverging arrows.

In the eastern Asiatic part of the belt (east of 90°E longitude) [*Shirokova*, 1967], the orientation of compression is the same as in the central part of the belt. The tension and the intermediate stress do not have any distinct predominant orientation.

In the Sunda Islands [*Shirokova*, 1967], the compression is approximately horizontal and directed across the strike of the island arc, i.e. the orientation of these stresses is the same as within the central part of the belt (the Caucasus and Asia). The tension is characterized by a slight predominance of the horizontal orientation. The intermediate stress is not marked by any predominant orientation.

Earthquakes with a focal depth of about 150 km (Carpathians) and 200 km (the Hindu Kush) reveal the same regularities in stress orientation as earthquakes whose foci are located in the crust.

The Mongolian-Baikal Zone

Within the Mongolian-Baikal seismic zone [*Vvedenskaya and Balakina*, 1960; *Vvedenskaya*, 1961; *Misharina*, 1964b; *Misharina and Pshennikov*, 1965], two distinct regions can be established according to the stress peculiarities in earthquake foci. One of them, including the Baikal depression and extending northeast along the system of ranges and depressions to the Stanovoi ridge, is characterized by a nearly horizontal tension oriented across the strike of structures, and a nearly vertical compression. In the second region, to the west of Baikal, the direction of tension is close to vertical, while the compression is practically horizontal and normal to the trend of structures. In both the regions the direction of the intermediate stress is close to horizontal. To the south and southwest of Baikal a third region can be distinguished where the directions of the compression and tension form angles not greater than 40° with the horizontal plane, and the intermediate stress is characterized by various orientations. Because of the lack of observational data, the boundaries of this region are not well established. It appears to include the seismic zones of central and western Mongolia and to be bordered by the Gobi-Altai mountain system to the south. The earthquake foci of the Gobi-Altai reveal nearly horizontal compression normal to the structures and almost vertical tension; the orientation of the intermediate stress is close to horizontal.

The Arctic-Atlantic Belt

The earthquake foci of the Arctic-Atlantic seismic belt [*Misharina*, 1964a; *Lazareva and Misharina*, 1965] are almost all characterized by the nearly horizontal orientation of compression and tension and nearly vertical orientation of the intermediate stress.

In the Atlantic Ocean the tension is generally oriented across the trend of structures, while the direction of the compression coincides with the axis of the sea ridges, as traced by earthquake epicenters. The same orientation of the stress axes appears to be typical for the earthquake foci of eastern Africa.

In the Arctic part of the Arctic-Atlantic belt, the orientation of stresses varies. In the region of the Norwegian and Greenland seas, the nearly horizontal tension is oriented normally to the strike of the sea ridge (an extension of the Atlantic Ocean ridge), while the direction of the nearly horizontal compression coincides with the strike of the sea structures. In the eastern part of the arctic seismic belt (in the region of the Lomonosov ridge, the Laptev Sea, and the delta of the Lena River), the predominant directions of compression and tension remain almost horizontal, but the belt trend is pursued only by the tension orientation, while the compression is generally normal to the line of epicenters. Tentative results show that in the region of the Verkhoyansk and Chersky ranges the compression is horizontal and approximately normal to the trend of surface structures. The orientation of the tension and intermediate stress in this region is not definite.

The Pacific Belt

For the majority of the investigated foci in the marginal Pacific zone (excluding the California coast and probably the coast of Central America) [*Balakina*, 1959, 1960, 1962], the direction of compression is close to horizontal and normal to the trend of the earthquake zones and to the main surface structures. These relations between the orientation of compression and the trend of structures are observed in the foci of both surface and deep earthquakes (600–700 km). The orientation of tension and intermediate stress varies.

CHARACTERISTICS OF THE STRESSED STATE IN THE PRINCIPAL SEISMIC REGIONS OF THE EARTH

The regularities found in the orientation of the principal stresses acting in earthquake foci permit us to outline the character of the stressed state in the seismic regions investigated.

The stressed state at each point of the deformed medium is determined by the position of the three main axes and by the values of the three principal stresses. Our method was to find the position of the main axes of the stressed state in a seismic zone as a whole and then to determine the condition connecting its principal values, assuming that the location of the planes of the maximum tangent stresses in this zone is determined by the established direction of the discharged shear stresses in the entire system of earthquake foci.

First, let us consider this condition for those seismic regions in which the position of the principal axes of released stresses for the entire system of foci remains constant (e.g., central Asia and Pribaikalye). The spatial directions of the compression, intermediate stress, and tension axes are given the indices i, j, and k, respectively. The following relations among the main stresses σ_i, σ_j, σ_k in the foci

$$|\sigma_i - \sigma_k| > |\sigma_k - \sigma_j| \quad |\sigma_i - \sigma_k| > |\sigma_j - \sigma_i|$$
$$|\sigma_i - \sigma_k| = 2\tau_{max}$$

show that the tangential stresses attain maximum absolute values τ_{max} on planes making 45° angles with the axes with the subscripts i and k. Proceeding from these conditions we establish two relations among the values of the main stresses: $\sigma_i < \sigma_j < \sigma_k$ and $\sigma_i > \sigma_j > \sigma_k$. Of these, only the first is valid.

Now let us determine the relations among the main values of the total stressed state in those seismic regions where successive earthquakes show that only the position of one main axis of stresses in the foci remains constant, while the directions of the other two main stresses change from focus to focus (e.g., the Mediterranean Sea, the Carpathian Mountains, the Pacific belt). We assume that the compressional axis is constant. The axis of the total stressed state coinciding with compression is given by the subscript i, while the subscripts j and k refer to the other two stress axes. The following ratios must be written for the main values σ_i, σ_j, σ_k of the total stressed state:

$$|\sigma_i - \sigma_k| > |\sigma_k - \sigma_j| \quad |\sigma_i - \sigma_k| \approx |\sigma_j - \sigma_i|$$

$$|\sigma_i - \sigma_k| = 2\tau_{\max}$$

They show that the tangential stresses τ_{ik} and τ_{ij}

$$\tau_{ik} = \frac{\sigma_i - \sigma_k}{2} \quad \tau_{ij} = \frac{\sigma_j - \sigma_i}{2}$$

under the condition of the total stressed state have the greatest absolute values and are equal. In this case the main values of the total stressed state are related by $\sigma_i < \sigma_j \approx \sigma_k$, and the two orthogonal principal axes corresponding to the approximately equal two main stresses can be arbitrarily oriented in the plane normal to σ_i.

If the direction of the tension axis appears to be constant in a seismic region, the main values of the total stressed state will be related by $\sigma_i \approx \sigma_j < \sigma_k$, where the subscript k refers to the principal axis of the stressed state coinciding with the direction of the axis of the released tension in foci.

Each of the dependences $\sigma_i < \sigma_j < \sigma_k$, $\sigma_i < \sigma_j \approx \sigma_k$, and $\sigma_i \approx \sigma_j < \sigma_k$ is characterized by a variety of stressed states. Hence, the total stressed state cannot be unambiguously determined from the known orientation of the main axes of stresses acting in foci.

Here we introduce one more set of subscripts for the main axes of the stressed state according to the orientation of the axes relative to the earth's surface and the trend of structures. Subscript 1 will refer to the principal axis that lies in the horizontal plane normal to the trend of structures; 2, to the axis located in the same plane, and parallel to the structures; and 3, to the vertical axis. If we apply this set of subscripts to the ratios established between the values of the main stresses and give the same sign to all three stresses, then the values of the ellipsoid axes of stresses for different seismic regions of the earth will be related as shown in Table 1.

TABLE 1. The Stress State in Seismically Active Zones of the Earth

Seismic Region	Main Stresses	Relation of Stresses to Geological Structures	Type of Stress State on Surface of the Earth and in Deep Sections Relative to σ_3
Pacific zone, Sunda Islands, Southeast Asia, the region of the Verkhoyansk Chersky ranges; Carpathians	$\sigma_k \approx \sigma_j > \sigma_i$	$\sigma_3 \approx \sigma_2 > \sigma_1$	Uniaxial horizontal compression, perpendicular to the tectonic structures
Eastern Arctic	$\sigma_k > \sigma_j > \sigma_i$	$\sigma_2 > \sigma_3 > \sigma_1$	Horizontal tension and compression parallel and normal to the tectonic structures, respectively
Central part of the Mediterranean-Asian zone (from 40° to 90°E long.)	$\sigma_k > \sigma_j > \sigma_i$	$\sigma_3 > \sigma_2 > \sigma_1$	Nonuniform horizontal compression, predominantly normal to the tectonic structures
Mediterranean	$\sigma_k > \sigma_j \approx \sigma_i$	$\sigma_k > \sigma_j \approx \sigma_i = \sigma_3$	Uniaxial horizontal tension; orientation relative to tectonic structures not clear
Western Arctic, Atlantic ridge, East Africa	$\sigma_k > \sigma_j > \sigma_i$	$\sigma_1 > \sigma_3 > \sigma_2$	Horizontal compression and tension parallel and normal to tectonic structures, respectively
Pribaikal zone	$\sigma_k > \sigma_j > \sigma_i$	$\sigma_1 > \sigma_2 > \sigma_3$	Nonuniform horizontal expansion, mainly normal to tectonic structures

An interesting system of possible states is the one for which the value σ_3, i.e. the stress along the vertical axis, becomes zero. This additional condition on one of the main values of stress is possible near the earth's surface. Its introduction into the ratio of the main values of the stressed state in a seismic region permits us to determine the signs of the values σ_1 and σ_2 and thus to establish the type of the stressed state near the earth's surface in that region. In general, the positive main stress corresponds to tension; the negative one to compression.

The value σ_3 in the deep parts of seismic regions does not equal zero but remains unknown. However, the relation among the main values of stresses of a seismic region as a whole can give us the values of σ_1 and σ_2 algebraically greater than, smaller than, or equal to σ_3. Let us suppose that the main stress having a value algebraically greater than σ_3 is a relative tension, while the main stress with the value less than σ_3 is a relative compression. Then the type of the stressed state near the earth's surface can be generalized and extrapolated to the deep parts of the seismic region.

The results of the determination of the type of stressed states for different seismic regions are presented in the last column of Table 1.

THE ELASTIC STRESS FIELD OF THE EARTH

Table 1 shows that the type of stressed state changes from one seismic region to another. In each seismic region of the earth either the greatest relative compression or the greatest relative tension is parallel to the horizontal plane. In most regions, a certain relation is observed between the trend of surface structures and the direction of the principal axes of the stressed state: either the axis of the greatest relative compression or the axis of the greatest relative tension is in the horizontal plane and is normal to the trend of structures.

The connections between the seismic zones and the regions of modern tectonic processes, as well as the relations between the direction of the main axes of stresses acting in foci and the trend of tectonic structures, imply that both the modern development of tectonic structures and the local discharge of stresses in earthquake foci are due to the field of elastic stresses of the earth shown in Figure 1. Both processes are maintained by the continuous generation of stresses in this field caused by processes in the interior of the earth.

REFERENCES

Balakina, L. M., On the distribution of stresses acting in earthquake foci in the northwestern Pacific, *Izv. Akad. Nauk SSSR, Ser. Geofiz., 1959*(11), 1599–1604, 1959.

Balakina, L. M., Some results of the study of earthquake foci of May 4 and June 18, 1959, according to the instrumental data, *Bull. Sov. Seismol., 1960*(11), 25–31, 1960.

Balakina, L. M., General regularities in the directions of the main stresses acting in the earthquake foci of the Pacific seismic belt, *Izv. Akad. Nauk SSSR, Ser. Geofiz., 1962*(11), 1471–1483, 1962.

Balakina, L. M., A. V. Vvedenskaya, L. A. Misharina, and E. I. Shirokova, The stressed state in earthquake foci and the elastic stress field of the earth, *Izv. Akad. Nauk SSSR, Fiz. Zemli, 1967*(6), 3–15, 1967.

Lazareva, A. P., and L. A. Misharina, On the stresses in the earthquake foci of the Arctic seismic belt, *Izv. Akad. Nauk SSSR, Fiz. Zemli, 1965*(2), 5–10, 1965.

Misharina, L. A., On the stresses in the earthquake foci of the Atlantic Ocean, *Izv. Akad. Nauk SSSR, Ser. Geofiz., 1964*(10), 1527–1534, 1964a.

Misharina, L. A., On the problem of stresses in the earthquake foci of Prybaikalye and Mongolia, *Tr. Inst. Zemnoy Kory SO Akad. Nauk SSSR, 1964*(18); *The Problems of the Seismicity of Siberia*, pp. 50–69, Novosibirsk, 1964b.

Misharina, L. A., and K. V. Pshennikov, The process of the discharge of stresses in the earth's crust from the data of the Prybaikal and Mongolian earthquakes, in *Collected Papers: Geological Results of Applied Geophysics, 22nd Intern. Geol. Congr.*, Nedra, Moscow, pp. 75–84, 1965.

Rustanovich, D., and E. I. Shirokova, Some results of the study of the Ashkhabad earthquake of 1948, *Izv. Akad. Nauk SSSR, Ser. Geophys., 1964*(12), 1782–1788, 1964.

Shirokova, E. I., Determination of stresses acting in the foci of the Hindu-Kush earthquakes, *Izv. Akad. Nauk SSSR, Ser. Geofiz., 1959*(12), 1739–1744, 1959.

Shirokova, E. I., On the stresses acting in the foci of central Asia earthquakes, *Izv. Akad. Nauk SSSR, Ser. Geofiz., 1961*(6), 876–881, 1961.

Shirokova, E. I., On the stresses acting in the foci of the earthquakes of the Caucasus and the adjacent region, *Izv. Akad. Nauk SSSR, Ser. Geofiz., 1962*(10), 1297–1306, 1962.

Shirokova, E. I., On the stresses acting in the foci of the earthquakes of the northwestern part of the Mediterranean-Asiatic seismic belt, *Bull. Sov. Seismol., 1963*(15), 72–80, 1963.

Shirokova, E. I., General regularities in the

orientation of the main stresses in the earthquake foci of the Mediterranean-Asiatic seismic belt, *Izv. Akad. Nauk SSSR, Fiz. Zemli, 1967* (1), 22–36, 1967.

Vvedenskaya, A. V., The determination of displacement fields in earthquakes by means of the dislocation theory, *Izv. Akad. Nauk SSSR, Ser. Geofiz., 1956*(3), 277–284, 1956.

Vvedenskaya, A. V., The determination of stresses acting in earthquake foci from the observations of seismic stations, *Izv. Akad. Nauk SSSR, Ser. Geofiz., 1960*(4), 513–519, 1960.

Vvedenskaya, A. V., The peculiarities of the stressed state in the foci of the Prybaikal earthquakes, *Izv. Akad. Nauk SSSR, Ser. Geofiz., 1961*(5), 668–669, 1961.

Vvedenskaya, A. V., The determination of the displacements in body waves depending on the rate and the duration of the spread of dislocation, *Izv. Akad. Nauk SSSR, Fiz. Zemli, 1965* (1), 3–11, 1965.

Vvedenskaya, A. V., On the possibility of determining stresses in earthquake foci, *Izv. Akad. Nauk SSSR, Fiz. Zemli, 1967*(7), 14–19, 1967.

Vvedenskaya, A. V., and L. M. Balakina, The technique and results of the determination of stresses acting in earthquake foci of Prybaikalye and Mongolia, *Bull. Sov. Seismol., 1960*(10), 73–84, 1960.

Vvedenskaya, A. V., and L. A. Ruprekhtova, The peculiarities of the stressed state in the earthquake foci near the bend of the Carpathian arc, *Izv. Akad. Nauk SSSR, Ser. Geofiz., 1961*(7), 953–965, 1961.

Zero Frequency Seismology

Frank Press

THE mechanical response of the earth to transient sources such as earthquakes and explosions can be described in terms of the free oscillations, the lowest mode being S_2^0, with a period of 53 minutes. 'Zero frequency seismology' is a somewhat inaccurate, recently coined name that usually refers to the observation and interpretation of long term (compared to 53 minutes) displacements, strains, and tilts. Perhaps its most important application has been in the study of the residual displacement, strain, and tilt fields associated with earthquakes. Another important aspect is the monitoring of changes of these fields as manifestations of tectonic strain accumulation. Both approaches are important in the study of orogeny, volcanism, and earthquake mechanism, and they may be essential in earthquake prediction research. The response of the earth to tidal forces and to variable oceanic and atmospheric loading has been included as part of zero frequency seismology by some investigators.

Some representative 'signal' levels are shown in Table 1 for various sources, for near-field (<10 km) and far-field (1000–5000 km) observations. Some seismologists believe that for tectonic strain accumulation 10^{-7}/yr characterizes aseismic belts and 10^{-5}/yr characterizes highly seismic belts. These values are obtained from surveying and from strain seismographs. The strain change of 10^{-4} for major earthquakes is derived from surveying in the near-source region. This corresponds to a stress jump near the source of 10–100 bars, if we assume reasonable values for rigidity. The far-field observations of 10^{-9} come from strain meter data. The examples of tilts and displacements accompanying the filling and depletion of shallow magma chambers was taken from *Eaton and Murata*'s [1960] study of the Kilauea eruptions. The premonitory strain change associated with the Niigata earthquake of 1964 was reproduced on many instruments.

Table 1 is an incomplete list. Eventually one would hope to include noise levels due to variable atmospheric and ocean loading, winds, thermoelastic effects of solar energy, inherent instrumental noise, etc. Other signal sources, such as strains and displacements accompanying continental drift and mantle convection, are subjects for future exploration.

TABLE 1. Some Signals Pertinent to Zero Frequency Seismology

Source	Strain and Tilt per Year		Displacement	
	Near Field	Far Field	Near Field	Far Field
Tectonic accumulation	10^{-7}–10^{-5}	?	0.1–10 cm/yr	?
Premonitory event (Niigata, 1964)	10^{-5}	?	?	?
Stress jump ($M \sim 8$)	10^{-4}	10^{-9}	100–1000 cm	0.1–1.0 cm
Tidal yielding	10^{-8}	?	30 cm	—
Volcanism (Kilauea, 1959)	10^{-4}	?	200 cm	?

The requirements of zero frequency seismology tax the skills of the most sophisticated instrument designers. The residual fields must be detected in the presence of the violent motions associated with the radiation fields. The effects of temperature, atmospheric pressure, rainfall, sea level changes, and heterogeneous crustal blocks are difficult to evaluate and remove. Long term stability is required to sense changes having characteristic times ranging from minutes and hours to years.

Table 2 lists the capabilities of several techniques for measuring the signals of zero frequency seismology. First-order surveying can measure distance to 1 part in 10^5. Thus, about 10 years of observations are needed to detect regional movements associated with the San Andreas fault. The newer electronic surveying tools provide an order of magnitude improvement when carefully used. Indeed, laser surveying devices may ultimately yield a capability of 1 part in 10^7 when procedures are developed for correcting for atmospheric changes.

Although strain seismographs and tiltmeters have a short term sensitivity several orders of magnitude better than the values quoted, instrumental drift and noise from diverse sources limit the long term stability.

For long term displacements (>1 hour), the gravimeter responds not as an accelerometer, but as a displacement meter measuring changes in distance from the center of the earth from the changing gravity field.

Displacement seismographs, using exotic sensors such as gyroscopes and vibrating string accelerometers (doubly integrated), are currently being designed. The sensitivity quoted in Table 2 is expected but has not yet been achieved.

Despite the fact that in situ stress measurements can reveal absolute stress levels, the technique has been employed primarily in engineering applications. Strain gages are inserted in boreholes and are subsequently overcored. The resulting changes in the strain indicators are then related to the pre-existing absolute stress level.

Instrumented water wells measure volumetric strain changes. Although difficult to calibrate, and of uncertain long term stability, they seem to have the same sensitivity as strain gages.

The radiation field of seismic waves has been properly emphasized by seismologists in recent decades. Long term stresses, strains, tilts, and displacements associated with the accumulation and release of stress is a promising field, one that awaits the further development of a new class of instruments. Strain, stress, and tilt sensing devices operating in deep boreholes (remote from sources of noise and close to the sources of signal) are required. The sensitivities quoted in Table 2 have been achieved, yet there

TABLE 2. Signal Threshold for Instruments of Zero Frequency Seismology

Instrument	Strain and Tilt	Displacement
Surveying equipment	10^{-5} to 10^{-6}	10 cm/100 km
Strain seismograph	10^{-7} to 10^{-8}	—
Tiltmeter	10^{-6} to 10^{-7}	—
Gravimeter	—	1 mm
Displacement seismograph	—	<1 mm
In situ strain gages	(10% of stress)	—
Tide gages	—	1 to 10 cm
Instrumented water wells	10^{-7} to 10^{-8}	—

are few documented cases to indicate that tectonic changes over a period of years have actually been monitored. Electronic surveying devices measuring (on-line) regional strain over many kilometers are technically feasible but have yet to be operated successfully. Arrays of sensors monitoring slow changes will be required to separate regional changes from the effects of temperature, atmospheric and ocean loading, winds, etc. Although the contrast in absolute stress between stable areas and tectonic belts can probably be examined with existing methods, this important study has not yet been made.

The following is only a partial list of problems that remain to be solved:

1. Dynamic response of solid earth to transient loading (atmospheric, oceanic, etc.).
2. Strain accumulation and release in seismic belts, with seismicity as a parameter.
3. Earthquake mechanism and nature of faulting as revealed by near and distant measurements of strain, tilt, and displacement fields.
4. Premonitory strains, tilts, and displacements associated with earthquakes and volcanic eruptions.
5. Large scale strains and tilts related to mantle convection, phase changes, continental drift, orogeny, sea floor spreading, and changes in the earth's radius.

BIBLIOGRAPHY

Aki, K., Scaling law of seismic spectrum, *J. Geophys. Res.*, 72(4), 1217–1231, 1967.

Archambeau, C., Elastodynamic source theory, Ph.D. thesis, California Institute of Technology, June 1964.

Benioff, H., Movements on major transcurrent faults, in *Continental Drift*, edited by S. K. Runcorn, pp. 103-134, Academic Press, New York, 1962.

Benioff, H., Source wave forms of three earthquakes, *Bull. Seismol. Soc. Am.*, 53, 893–903, 1963.

Berckhemer, H., and G. Schneider, Near earthquakes recorded with long-period seismographs, *Bull. Geol. Soc. Am.*, 54, 973–987, 1964.

Blayney, J. L., and R. Gilman, A semi-portable strain meter with continuous interferometric calibration, *Bull. Seismol. Soc. Am.*, 55, 955–970, 1965.

Bonchkovsky, V. F., Deformation of the earth's surface accompanying certain disastrous earthquakes, *Bull. Acad. Sci. USSR, Geophys. Ser.*, English Transl., 2, 190–193, 1962.

Byerly, P., and J. DeNoyer, Energy in earthquakes as computed from geodetic observations, *Contributions in Geophysics*, vol. 1, pp. 17–35, Pergamon Press, London, 1958.

Chinnery, M. A., The deformation of the ground around surface faults, *Bull. Seismol. Soc. Am.*, 51, 355–372, 1961.

Chinnery, M. A., The stress changes that accompany strike-slip faulting, *Bull. Seismol. Soc. Am.*, 51, 921–932, 1963.

Eaton, J. P., and K. J. Murata, How volcanoes grow, *Science*, 132, 925–938, 1960.

Judd, W. R., editor, *State of Stress in the Earth's Crust*, Elsevier Publishing Company, New York, 1964.

Kasahara, K., The nature of seismic origins as inferred from seismological and geodetic observations, *Bull. Earthquake Res. Inst., Tokyo Univ.*, 35, 473–532, 1957.

Knopoff, L., Energy release in earthquakes, *Geophysics*, 1, 44–52, 1958.

Maruyama, T., Statical elastic dislocations in an infinite and semi-infinite medium, *Bull. Earthquake Res. Inst., Tokyo Univ.*, 42, 289–368, 1964.

Nishimura, E., On some destructive earthquakes observed with a tiltmeter at a great distance, *Bull. Disaster Prevent. Res. Inst., Kyoto Univ.*, 6, 1–16, 1953.

Nishimura, E., *Geophysical Papers Dedicated to Professor K. Sassa*, Geophysical Institute, Kyoto University, Kyoto, 1963.

Plafker, G., Tectonic uplift, subsidence, and faulting associated with Alaska's Good Friday earthquake, *Science*, 148, 1675–1687, 1965.

Press, F., Displacements, strains, and tilts at teleseismic distances, *J. Geophys. Res.*, 70, 2395–2412, 1965.

Press, F., and C. Archambeau, Release of tectonic strain by underground nuclear explosions, *J. Geophys. Res.*, 67, 337–343, 1962.

Press, F., and W. F. Brace, Earthquake prediction, *Science*, 152, 1575–1584, 1966.

Slippage on the Hayward Fault: Six related papers, *Bull. Seismol. Soc. Am.*, 56, 257–323, 1966.

Stauder, W., and G. A. Bollinger, The focal mechanism of the Alaskan earthquake of March 28, 1964, and its aftershock sequence, *J. Geophys. Res.*, 71, 5283–5296, 1966.

Steketee, J. A., On Volterra's dislocations in a semi-infinite medium, *Can. J. Phys.*, 36, 192–205, 1958.

Tsuboi, C., K. Wadati, and T. Hagiwara, *Report by the Earthquake Prediction Research Group in Japan*, Earthquake Research Institute, Tokyo University, Tokyo, 1962.

Vvedenskaya, A. V., The determination of displacement fields by means of dislocation theory, *Bull. Acad. Sci. USSR, Geophys. Ser.*, English Transl., 3, 277–284, 1956.

Witkind, J. J., W. B. Meyers, J. B. Hadley, W. Hamilton, and G. D. Fraser, The earthquake at Hebgen Lake, Montana, on August 18, 1959: Geological features, *Bull. Seismol. Soc. Am.*, 52, 163–180, 1962.

Prediction of Earthquakes

Takahiro Hagiwara

MAN has suffered frequently from great earthquakes throughout history; many lives and much property have been lost each time. Prediction of earthquakes is ardently desired by people who live in an active seismic region where destructive earthquakes can be expected to occur in the future as they have in the past. In spite of the long endeavor by scientists, however, no reliable method for predicting earthquakes has yet been discovered.

For prediction to be of practical value, the time, location, and magnitude must be specified rather accurately. However, earthquake phenomena are essentially different from other natural phenomena that are continuous or reversible. Earthquakes occur abruptly and sporadically, which makes their prediction very difficult.

Pessimistic views have long prevailed among scientists who thought that earthquake prediction could be only speculative until a perfect knowledge of earthquake phenomena was obtained, and that attention should therefore be concentrated only on basic research of the phenomena. In recent years, however, the situation has changed considerably; many earth scientists are beginning to believe that some measures for predicting earthquakes can eventually be achieved. This relatively optimistic view is based on recent exciting progress in three areas; first, improved instruments are successfully measuring microearthquakes, small strain, tilt, and small anomalies of geomagnetism and gravity. Second, laboratory studies of fracture and creep of rocks under pressure and temperature are revealing fracture mechanisms that suggest the possibility of earthquake prediction. Third, geodetic and geophysical evidence, such as unusual land deformation or an unusual series of microearthquakes before a major earthquake, though not yet decisive, have actually been reported as a result of painstaking observations over a long period of time in active seismic regions.

In 1962, a group of Japanese scientists published a report, *Prediction of Earthquakes—Progress to Date and Plans for Further Development* [1962], in which they described what kinds of measurements should be made by what method in order to predict earthquakes most effectively. This report attracted the attention of scientists all over the world, especially in those countries where casualties have been caused by earthquakes. In 1965, as a result of the great Alaskan earthquake of 1964, a formal committee in the United States presented a report entitled *Earthquake Prediction—A Proposal for a Ten Year Program of Research* [1965]. In the same year, the intergovernmental meeting on seismology and earthquake engineering, under the auspices of Unesco, requested Unesco to support further advanced studies aimed at earthquake prediction. In 1964 and 1966, conferences held by U. S. and Japanese scientists provoked valuable and enthusiastic discussions concerning earthquake prediction problems. Interest in earthquake prediction has become world wide, with the USSR and the European countries also eager to develop their own research in this area.

PRESENT KNOWLEDGE AND FUTURE RESEARCH

A number of scientists believe that some crustal deformation can be detected before a major earthquake. In Japan, some remarkable land deformations have been observed by people near the epicenter several hours or tens of minutes before an earthquake. In these cases the epicenter was near the coast; people standing on the shore witnessed the sea water noticeably retreating. This implies that, before the earthquake, the land slowly upheaved. Phenomena of this type though very few, affirm that at least some earthquakes are preceded by specific crustal deformation. We therefore hope that methods for predicting earthquakes by precise instrumental measurement of crustal deformations will be developed in seismic regions.

In other instances an anomalous crustal deformation has been detected by precise leveling before a large earthquake. In these cases revision surveys had been made along a route that passed quite accidentally through the epicentral area just before the earthquake. The most explicit example was the land deformation before the Niigata earthquake in the Sea of Japan ($M = 7.3$) in 1964. Precise levelings had been made along the coast near the epicenter in 1898, 1930, 1955, 1958, and 1961. The surveys demonstrated that the land near the epicenter had been upheaving slowly at a constant speed since the first survey, had accelerated its upheaving by several times five or six years before the earthquake, and had largely subsided during the earthquake. These facts suggest that earthquakes of at least the magnitude of the Niigata earthquake might be predicted by repeated precise leveling in regions subject to destructive earthquakes. Geodetic surveys of other kinds, especially with newly developed instruments such as the geodimeter, would also be useful for this purpose. The Japanese research group has emphasized that such works are of utmost importance, and has made them the basis of their proposed project.

In addition to geodetic surveys, some kind of routine instrumental observation that would give continuous information on the crustal deformation with respect to time is needed. Ground tiltmeters and strainmeters are available for this purpose. There are two types of tiltmeters: the old pendulum type and the more recent water-tube type. Each has advantages, but only the latter can measure slow secular variation, as the pendulum type is apt to be disturbed by the drift of neutral position. Strainmeters that use a quartz standard are also stable, in principle, for recording secular variations.

Although there have been many reports, especially in Japan, that anomalous changes in tilt or strain have been observed before a large earthquake, some of them lack sufficient reliability, and more careful study will be needed before it can be concluded that such changes are really forerunners of earthquakes. Observations of this kind are frequently affected by weather conditions and, especially with pendulum tiltmeters, the neutral line on the records is usually unstable. One of the most reliable observations of remarkable changes in ground tilt before a large earthquake is related to the Tottori earthquake ($M = 7.3$) in 1943. A pendulum tiltmeter in a deep mine 60 km from the epicenter registered a gradual but large change in tilt of the ground inclining to the opposite side of the epicenter, starting six hours before the earthquake.

When a station is established for continuous observation by tiltmeters and strainmeters, it would be desirable to set up a number of bench marks in the vicinity and to repeat precise levelings and geodimeter measurements so as to check results of the observations.

The activity of small earthquakes is expected to be, to some extent, related to the occurrence of large ones. Therefore, accumulation of data on the activity of small earthquakes would be of great importance for research related to earthquake prediction. There have been a few reports that people have felt anomalously frequent earthquakes before a destructive earthquake. It appears, however, that large earthquakes have usually taken place without any such warning signs. In recent years, the marked advance in observation techniques using highly sensitive seismographs has enabled us for the first time to detect small earthquakes called microearthquakes. It is reasonable to anticipate that there would be an anomalous pattern in the activity of microearthquakes before a large earthquake; therefore, systematic observations of microearthquakes in seismic regions have been recommended.

The Matsushiro earthquake swarm provided some useful information for the study of earthquake prediction. This swarm started in the central part of Japan in August 1965, and gradually extended its active area. Observations of microearthquakes were carried out in and around this area in conjunction with the ordinary seismographic network. Mircoearthquakes began a few months before the larger earthquakes, which included the destructive ones.

The elastic properties of the crust near the hypocenter, where the rocks are placed under a highly stressed state, are presumed to undergo some modification. Although there have been reports of changes in seismic wave velocity before great earthquakes, all such reports have been made from observations of natural earthquakes and, consequently, are not of the highest degree of precision. To obtain more conclusive data, investigations using the techniques of explosion seismology are required. Very recently, U. S. scientists conducted a preliminary seismic field investigation aimed at developing a method for predicting earthquakes in the San Andreas fault zone in California. Their results demonstrated that the first arrival time of seismic waves caused by repetitive explosions in the same hole could be determined with a precision of more than 1 msec at distances up to 40 km.

Detailed laboratory experiments on the fracturing of rocks have revealed certain important phenomena that precede the outbreak of the main fracture. When the rocks that compose most of the earth's crust are deformed by an external force, their deformation rate is generally accelerated as fracture approaches. This increase in the deformation rate occurs both for uniform external force and for a force that increases with time. Such experimental data suggest the value of continuous observation of the earth's crust for predicting earthquakes. In the same laboratory experiment, a cluster of small elastic shocks (microshocks) was generated. Numerous small cracks created in the rock sample before the main fracture occurred were observed with a sensitive vibrograph attached to the sample. The frequency of occurrence of such microshocks depends on the brittleness and structural nonuniformity of the rocks. The main fracture develops abruptly unaccompanied by marked elastic shocks when the rock is very uniform. In contrast, frequent elastic shocks can be measured when the rock is not of uniform structure, as is the case in the crust. These experiments suggest the possibility of predicting a major earthquake by observing small foreshocks with the use of highly sensitive seismographs.

Close examination of geomagnetic data leads us to conclude that a change in the earth's magnetic field of about 10 γ may accompany an earthquake of moderate magnitude. This conclusion was reached by observations with recently developed proton precession magnetometers and optical pumping magnetometers, which are practically free from drift. Compression experiments on basaltic rocks also indicate changes in the magnetic susceptibility as well as the remanent magnetization of the order of 10^{-4}/bar; thus it is not unreasonable to expect a geomagnetic change of the order cited in association with a stress change of 100 bars in the earth's crust.

Although there have been a number of reports that suggest local anomalous changes in the geomagnetic field before an earthquake, whether or not such changes actually occur is still to be verified. As arrays of modern magnetometers for detecting seismomagnetic effects have been set up in a number of countries in recent years, it seems highly probable that any geomagnetic change before an earthquake will be observed with high degree of accuracy.

BIBLIOGRAPHY

Earthquake prediction—A proposal for a ten-year program of research (report of the ad hoc panel on earthquake prediction, Office of Science and Technology, Washington, D. C.), 39 pp., September 1965.

ESSA Symposium on Earthquake Prediction (February 1966), U. S. Government Printing Office, Washington, D. C., 167 pp., 1966.

Hagiwara, T., and T. Rikitake, Japanese program on earthquake prediction, *Science, 157*(3790), 761–768, 1967.

Oliver, Jack, Earthquake prediction: United States—Japan Cooperative Science Program, *Science, 164*, 92–93, 1969.

Prediction of earthquakes—Progress to date and plans for further development (English translation of a report by the Earthquake Prediction Research Group in Japan), Earthquake Research Institute, Tokyo University, 1962.

Press, Frank, and W. F. Brace, Earthquake prediction, *Science, 152*(3729), 1575–1584, June 1966.

Proceedings of the U. S.-Japan Conference on Research Related to Earthquake Prediction Problems, 105 pp., Lamont Geological Observatory, Palisades, New York, March 1964.

Proceedings of the Second U. S.-Japan Conference on Research Related to Earthquake Prediction Problems, 106 pp., Earthquake Research Institute, Tokyo, Japan, June 1966.

Proposal for a ten-year national earthquake hazards program (report of ad hoc interagency working group for earthquake research, Federal Council for Science and Technology, Washington, D. C.), 81 pp., December 1968.

Rikitake, T., A five-year plan for earthquake prediction research in Japan, *Tectonophysics, 3*(1), 1–15, 1966.

Symposium on Earthquake Prediction (XIV IUGG General Assembly), collected papers, *Tectonophysics, 6*(1), 7–87, July 1968.

Explosion Seismology: Introduction

I. P. Kosminskaya

A SPECIAL method of seismic exploration, very similar to seismic prospecting, has been developed for the study of the earth's crust. This technique is called 'explosion seismology' (in contrast to earthquake seismology). In the USSR, it is called 'deep seismic sounding' (DSS). The DSS method is based on registration of series of comparatively small (hundreds to several thousand kilograms) explosions with various combinations of the explosion and receiver patterns. Large industrial explosions may also be utilized.

The possibility of exact formulation of the aims of the experiment with precise knowledge of the time and location of the seismic source is the major advantage of explosion seismology. However, the development of the observational techniques and the improvement of the accuracy in the determination of epicenters of earthquakes and depths of foci makes possible detailed seismological observations in seismoactive zones.

Explosion seismic profiles (seismic velocities and depths of seismic boundaries) of the earth's crust are made on land and at sea. Explosion seismology is based primarily on registration of longitudinal waves within the range of 5–15 Hz and at distances from 0 to 500 km. Explosion seismic profiles reach depths of 50–70 km on land and 15–20 km at sea, thus including the crustal structure and the M discontinuity.

The detail of seismic profiles depends considerably on the registration patterns. In turn, the choice of pattern depends not only on the purpose of the research, but on other important factors: conditions of the region, possibilities for detonating explosions, background of microseisms, instrumental and other facilities. The variations of all these circumstances in different countries lead to different modifications of techniques of explosion seismology; although the physical principles are presumably universal, the results differ considerably in detail.

Observations to study the general features of crustal structure with minor detailing are being conducted in the USA, Canada, Western Europe, and Japan, and to a more limited extent in other areas and on seas and oceans. In the USSR, where the basis of the special DSS method has been elaborated, explosion seismology is conducted by different detailing systems: for the study of large territories, in order to find the deep foundation of tectonic zoning, reconnaissance observations with minor details are used; detailed observations are applied for the study of the structure of small blocks (less than 200 km) and the fault zones between them and for special studies of actual correspondence (or discrepancy) between the structures, revealed by geological methods, and deep structures in the crust. These researches are often conducted in conjunction with practical tasks: seismic prospecting for oil, gas, and minerals.

The effect of the methods of the experiment on the character of the resultant crustal profile is highly significant; failure to take this into account can lead to basic misunderstandings in interpretation of seismic profiles. This difficulty can best be resolved by the seismologists themselves, who are specialists in this field. In this connection, an urgent need arises for international cooperation.

In the following articles, several authors present results of the study of the earth's crust in different countries. Unfortunately, we cannot yet give the generalized concepts in this field with due allowance for differences in methods and in approaches to the interpretation of data. This is a task for the future.

Explosion Seismic Studies in Western Europe

H. Closs

SEISMIC explosion studies of the crust and upper mantle have been made in three principal ways in western Europe: (1) refraction and reflection surveys by petroleum companies (reflections down to the upper mantle); (2) explosions by mining companies in quarries; and (3) explosions for scientific purposes in the sea, in alpine lakes, and in boreholes.

Many varieties of instrumentation have been used, ranging from instruments without amplifier (geophones and galvanometers) to modern magnetic-tape instruments, and including truck-mounted and portable equipment. On the basis of earlier experience, thirteen geophysical institutes in the Federal Republic of Germany decided to adopt uniform instrumentation: a portable magnetic-tape recorder with three seismic channels.

Although methods of interpretation of data have gradually changed, a uniform method of interpretation has not yet been achieved. Within the European Seismological Commission, a working group is studying this problem. Four methods are found in the literature:

1. Interpretation based on first arrivals.
2. Interpretation based on travel-time diagrams with straight lines.
3. Interpretation based on seismogram compilation and employment of reduced travel-time curves to make curvature more discernible. By means of an iterative procedure, a velocity-depth function is derived from the travel-time diagram on the basis of the Wiechert-Herglotz method [*Giese*, 1966].
4. Preparation of the travel-time curve as in method 3; interpretation by multi-layer computer programs, with trial and error calculation. Figure 1, for example, shows one of 12 calculated models with the best fit of calculated and recorded travel-time curves (simplified) [*Behnke*, 1967].

In view of the heterogeneity in the methods of interpretation, all velocities of layers, depths of discontinuities, and even the existence of the discontinuities reported in this review must be regarded as a preliminary communication. If a uniform procedure for evaluation of data is eventually found, an enormous effort will be necessary to evaluate existing data. A compilation for the Alps was presented at the IUGG General Assembly in Zurich by *Choudhury et al.* [1967].

LOW VELOCITY CHANNEL

A low velocity channel (for P waves) below the Moho has been traced in Western Europe by explosion seismic techniques on two profiles extending northward from Nice more than 200 km (St. Mueller, personal communication) and northwestward through France. Low-velocity channels for longitudinal waves in the crust have been postulated by several authors. *Mueller and Landisman* [1966] cite the termination of the branch of the wave penetrating into the granitic layer (Figure 2, P_g), the beginning of a travel-time branch with large amplitudes at a short distance from the shot point (Figure 2, P_e), reflections from within the granitic layer, and the fact that the velocities known from earthquakes for P_g are often lower than those from explosion seismology. They are of the opinion that this low-velocity channel is a worldwide phenomenon in the continental crust, mainly occurring at a depth of 8–12 km.

Giese [1966] and *Giese et al.* [1967] think that almost all results of explosion seismic work in western Europe can be characterized by the scheme shown in Figure 3 (a schematic diagram with a low-velocity channel is also shown in Figure 3). *Meissner* [1967] has determined velocities from a common reflection-point line along a 145-km profile; he finds a low-velocity channel between 12 and 18 km (Figure 12).

MODEL SEISMOLOGY

Figure 4 shows a model that has been used for studying amplitude ratios in the northern alpine forelands [*Waniek and Schenk*, 1966],

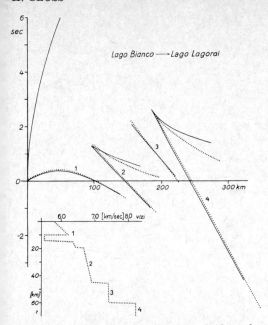

Fig. 1. $v(z)$ diagram for the Central Alps; observed (solid line) and calculated (dotted line) travel times [after *Behnke*, 1967].

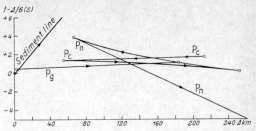

Fig. 2. Diagram for reduced travel times; P_g, for arrivals from crystalline basement; P_c, velocity increase below the sialic low-velocity channel; P_n, Mohorovicic discontinuity [after *Mueller and Landisman*, 1966].

based on variations of the system water-glycerine-gelatin. Model experiments are continuing; see section in this Monograph by Vanek, p. 276.

REFRACTION METHODS

Most of the western European surveys have relatively large separations (2–5 km or more) between recording points along the profiles. The density of observation points in eastern Europe is much higher.

REFLECTION METHODS

In the Alps, reflection recording has been carried out near some refraction shot points; reflections have been observed down to the Moho [*Fuchs and Kappelmeyer*, 1962; *Closs and Labrouste*, 1963]. Vertical reflections were recorded in the Rhenish massif from dynamite charges of 60–100 kg [*Dürbaum et al.*, 1967]. The zone of the Conrad discontinuity (depth 17–20 km) is characterized by a sequence of three reflections between 5.5 and 6.8 sec (Figure 5). In the same area, this result can also be derived from the reflection statistics of about 80 seismograms (Figure 6). In this figure, there is a pronounced maximum at 8 sec, the Moho (depth about 30 km). The quality of the individual reflections indicates that the Moho in

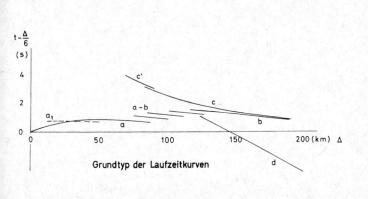
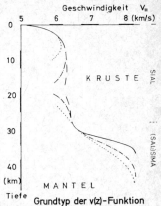

Fig. 3. *Left*: Scheme of travel-time diagrams for the crust and upper mantle in western Germany and the Alps. *Right*: Scheme of velocity-depth diagrams in western Germany and the Alps (dashed lines indicate range of variability) [after *Giese et al.*, 1967].

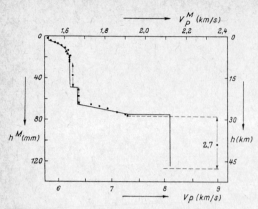

Fig. 4. Crustal model for amplitude control. h^M and v_p^M are thicknesses and velocities of the model substance (water-glycerine-gelatin); h and v_p are the corresponding crustal parameters. The line represents the $v(z)$ function that had to be tested; the points represent the velocity control of the model substance. v_p^M for the upper mantle was 2.7 km/sec, instead of 2.1 [after *Waniek and Schenk*, 1966].

this region is a relatively sharp discontinuity. (Field procedures were such that multiple reflections, or reflections of transverse waves, can be excluded.)

Nearly vertical reflections from deep discontinuities have been observed by *Reich* [1958], *Båth and Tryggvason* [1962], and *Geneslay et al.* [1956].

The work of Dürbaum et al. provides important evidence that a statistical interpretation of normal reflection records can give valuable information about crustal structure. Additional evidence was supplied by the common-reflection-point method in the northern foreland of the Alps [*German Research Group*, 1966; *Meissner*, 1966].

The statistical interpretation of many thousands of reflection seismic records (10–15 sec) of petroleum companies [*Dohr*, 1957; *Dohr et al.*, 1967; *Liebscher*, 1964] has proved to be an important tool for crustal research because

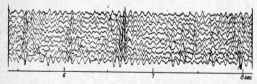

Fig. 5. Vertical reflections from Conrad discontinuity (5.5–6.7 km/sec) and Moho (~8 km/sec) in the Rhenish massif [after *Dürbaum et al.*, 1967].

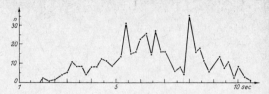

Fig. 6. Histogram of vertical reflections in the Rhenish massif; n, number of records [after *Dürbaum et al.*, 1967].

the resolving power is greater (frequencies 10–30 cps) than that of deep refraction seismic sounding. Results of these reflection seismic surveys have not been adequately taken into account in the compilation of velocity-depth curves.

SEISMIC AND GRAVITY METHODS

Efforts were made relatively early to make seismic interpretations in combination with gravity data [*Closs and Labrouste*, 1963]. It appeared that there were only very limited possibilities of using gravity data to select a most probable model from among several seismically equivalent ones. There is no doubt, however, that such a combination of independent geophysical observations can be helpful [*Closs and Plaumann*, 1966; *Hinz et al.*, 1967].

PROPAGATION OF ENERGY

The relations between amplitude and charge size (for explosions in a lake) have been studied for distances up to 34 km and charge weights from 2 to 760 kg [*Mueller et al.*, 1962]. Studies of frequency spectra have shown that most of the energy is contained in the frequencies 0.5–20 cps and that the predominant frequencies decrease from 6 cps at 75 km to 3 cps at 300 km in a special case [*Closs and Labrouste*, 1963].

AREAS UNDER INVESTIGATION

The Alpine Orogene

Figure 7 shows the refraction seismic observation lines (about 9000 km) and the shot points in the alpine region. The main part of this survey was done in collaboration by institutes of France, Germany, Italy, and Switzerland. Some contributions came from the United Kingdom. A preliminary interpretation of the results is given in *Closs and Labrouste* [1963]. On the basis of new surveys and improved in-

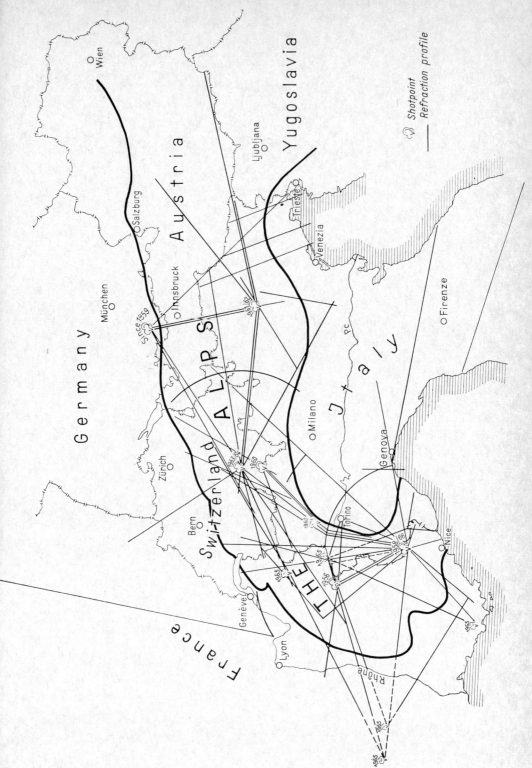

Fig. 7. Refraction shot points and observation lines in the alpine region.

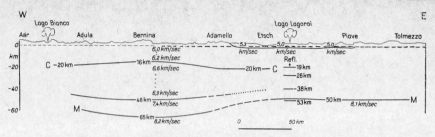

Fig. 8. Cross section through the Central Alps showing vertical reflection horizons below Lago Lagorai [after *Behnke*, 1967].

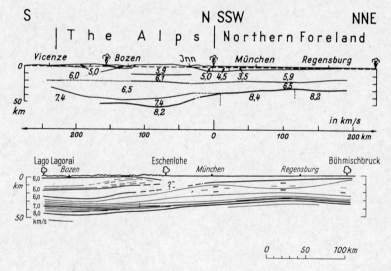

Fig. 9. Cross sections from the Bohemian massif (right) to the southern foreland of the Alps (Po Valley); 3 shot points. *Top,* after *Prodehl* [1965]; *bottom,* after *Giese* [1966].

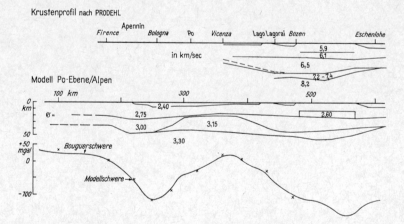

Fig. 10. Gravimetric calculation and extrapolation for cross section in Figure 9 (top) into the Po Valley [after *Closs and Plaumann*, 1966].

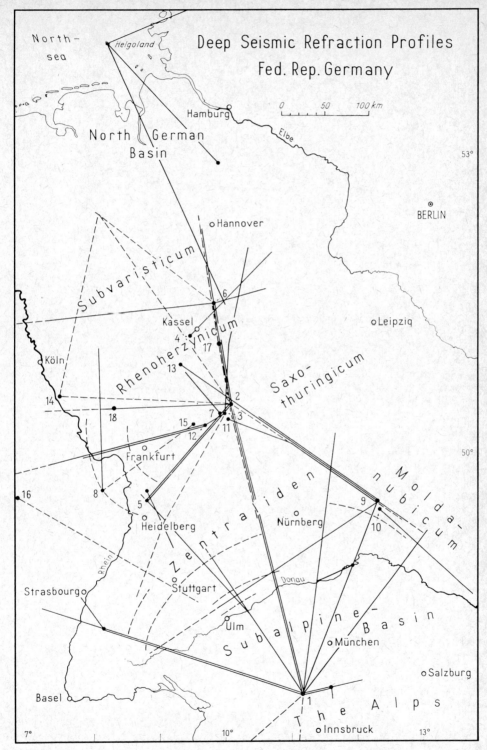

Fig. 11. Shot points and observation lines in western Germany (dashed lines are profiles planned or not yet fully covered).

terpretation techniques, it is probable that new models for this area will soon be presented. Figure 8 shows an example of results from the Central Alps. Figure 9 shows two different interpretations of the same data of one profile crossing the eastern Alps and the forelands in a north-south direction. This profile has been used for a gravimetric extrapolation into the southern foreland of the Alps, the Po Valley (Figure 10). It seems that there is a thickening of the intermediate layer, especially the lower part, along the southern inner margin of the Alps. The width of the presumably basic body at depth corresponds to the width of the Alpine body.

Central Europe

Figure 11 shows observation lines and shot points (areas of investigation by reflection statistics are not indicated). Figure 12 gives an idea of the uncertainties of interpretation: curve a represents the result of a 'classical' interpretation of earlier years; b is a very modern interpretation; and c shows the result of a long-range reflection study by using a common depth point of the Moho. Figure 13 shows the histogram of 130 records of normal vertical reflection seismograms from an area of 100 km² in the same region. Concepts in regard to the granitic layer differ and are contradictory in important points. The depth of the Conrad discontinuity differs by about 4 km in the three interpretations in Figure 12; approximately the same is true of the Moho. An earlier presentation of the discontinuities in western Germany [*German Research Group*, 1964] is currently being revised.

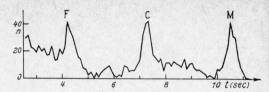

Fig. 13. Histogram of vertical reflection records: n, number of records; F, Förtsch discontinuity; C, Conrad discontinuity; M, Moho; t, two-way vertical reflection time [after *Liebscher*, 1964].

Northern Europe

Figure 14 shows the refraction lines in northern Europe with some data regarding velocities and depths of the main discontinuities. Figure 15 shows the results of a profile in northern Denmark: an extremely thin granitic layer, an exceptionally shallow depth to the Conrad discontinuity. The thickness of the intermediate layer is unusual for northern Europe; however, *Aric* [1968] published a cross section of the northern end of this profile (with reflections down to the Moho) showing normal crustal conditions.

A shallow depth to the Moho is reported in the northern part of the Norwegian coast in the zone of high positive gravity anomalies [*Sørnes and Bleie*, 1966].

Southern and Western Europe

Finetti et al. [1966] report preliminary values from the Adriatic (offshore, near Bari) for the Conrad and Moho: 7.10 km/sec, 23-25 km depth, and 8.25 km/sec, 38-41 km depth, respectively. In addition to the work in the Alps and adjacent regions, much work has been done in the Central massif of France; preliminary results have been published by *Perrier* [1963, 1965]. From shots in the Irish Sea and the North Sea, *Agger and Carpenter* [1964] deduced a model of the earth's crust without an intermediate layer. Profiles near the British coast and in the North Sea (Dogger bank) gave different types of crust: P_n velocity, 7.99-8.3 km/sec; P_b, 6.5 km/sec; P_g, 6.15 km/sec [*Colette et al.*, 1965]. Possibly the intermediate layer is relatively thin, but most of these data seem not to be definitive.

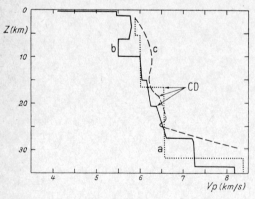

Fig. 12. $v(z)$ diagrams for southern Germany. Curve a after *Förtsch* [1952]; curve b after *Mueller and Landisman* [1966]; curve c after *Meissner* [1967].

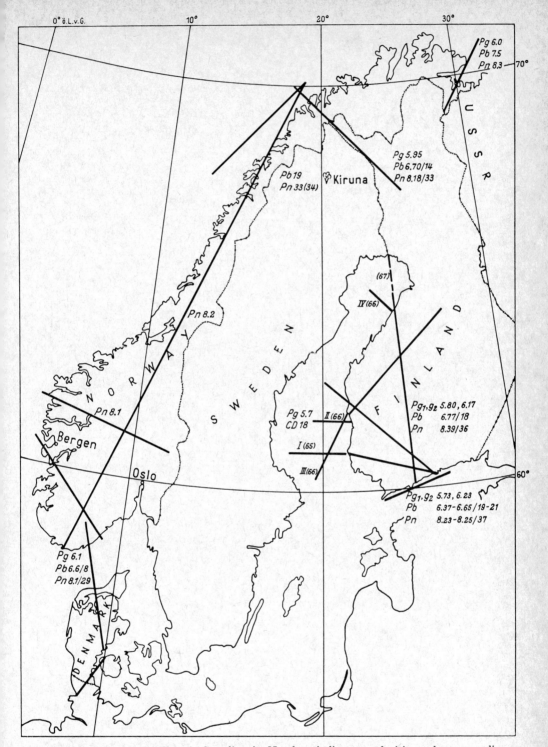

Fig. 14. Observation lines in Scandinavia. Numbers indicate p velocities and corresponding depths of discontinuities thus: P_g 6.1, p velocity of granitic layer is 6.1 km/sec; P_b 6.6/18, p velocity of intermediate layer is 6.6 km/sec, depth of Conrad discontinuity is 18 km; P_n 8.1/29, p velocity below Moho is 8.1 km/sec, depth of Moho is 29 km. Shotpoint near Kiruna, reflection observation [after *Båth and Tryggvason*, 1962; *Hirschleber et al.*, 1966; *Sørnes and Bleie*, 1966; *Porkka*, 1966; *Penttilä*, 1965; *Vesanen et al.*, 1960; *Sellevoll*, 1964.]

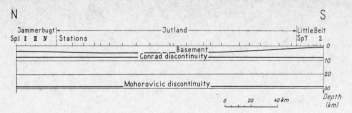

Fig. 15. Cross section of northern Denmark [after *Hirschleber et al.*, 1966].

CRUST AND MANTLE IN WESTERN EUROPE

Granitic layer

If only one value is given for the velocity of the granitic layer, or for the intermediate layer, it must be regarded as a first approach. By a special refraction survey in a granite-gneiss area, *Giese* [1963] showed that from the surface to a depth of 1.72 km there is an increase in velocity from 5000 to 5820 m/sec (Figure 16).

The velocity in the granitic layer varies from 5.0 to 6.3 km/sec; the higher velocities usually occur near the intermediate layer, i.e., at depths of 15–20 km. There are some exceptions [e.g., *Meissner*, 1967]. Several authors assume low velocity channels in the granitic layer.

The smallest reported thickness of the granitic layer occurs in Denmark (<4 km); the greatest thickness is shown by the 'Paris' model of the western Alps (>50 km) [*Closs and Labrouste*, 1963]. Other interpretations in the Alps show a normal thickness of about 20 km.

Intermediate Layer

It is not yet certain whether it is justified to speak of an intermediate layer everywhere in Europe. In some areas, a layer between the granitic layer and the Moho was not found by seismic observations [*Hinz et al.*, 1967]. Most geophysicists take for granted an intermediate layer that can often be subdivided into (usually two) parts. Other geophysicists consider that a transition zone of several kilometers thickness (up to 5 km and more) is probable or proved. Velocities of 7–7.4 km/sec in the lower part of the intermediate layer are widely reported. In some areas, the intermediate layer seems to begin with high velocities (Ivrea high, Italy; northern Norway). The range in which a velocity of 6.3 km/sec or somewhat higher is found with a positive velocity gradient can be defined as the top of the intermediate layer. It is customary in western Europe to designate this zone as the Conrad discontinuity. In and outside the Alps, the Conrad discontinuity lies between 15 and 20 km. Exceptions are known, especially the gravity highs at the southern inner margin of the Alps (Ivrea, Lago di Garda), and northern Jutland.

The inner structure of the intermediate layer seems to vary remarkably; the greatest thickness of the intermediate layer has been found in the Central Alps of Switzerland and western Austria. Until now, there has been no success

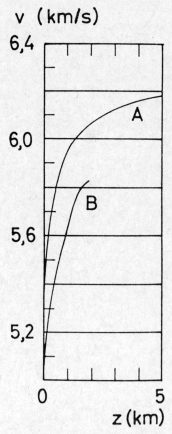

Fig. 16. $v(z)$ diagram: A, granite [after *Birch*, 1958]; B, field measurement [after *Giese*, 1963].

in finding common characteristics of the intermediate layer in the Alps and the older orogenetic belts in Europe [Closs, 1965].

The Upper Mantle

In continental western Europe, the sub-Moho velocity is generally greater than 8.0 km/sec (exception, see Perrier [1963, 1965]). Often it reaches values near 8.2 km/sec, independent of the depth of the Moho. Higher velocities of the upper mantle are reported from Finland. The depth of the Moho ranges from >60 km in the Alps to <20 km in the Central Massif. In many places, the depth is about 30 km; it may be smaller in parts of Scandinavia and England than in other areas of western Europe.

Acknowledgments. I wish to thank the presidents of the two groups for explosion seismology of the European Seismological Commission, Professor Jensen, Copenhagen (Group Northern Europe) and Professor Morelli, Trieste (Group Southern and Western Europe) for their kind assistance, and many authors who made available unpublished data. Professor Willmore, Edinburgh, supported this compilation by a bibliography of British papers for the period 1936-1966. In addition to the sources cited, some information for this report was obtained from the Progress Reports for 1967 for the Upper Mantle Project of the Federal Republic of Germany, Finland, France, Italy, Netherlands, Norway, and the United Kingdom.

REFERENCES

Agger, H. E., and E. W. Carpenter, A crustal study in the vicinity of the Eskdalemuir Array Station, *Geophys. J.*, *9*, 69-83, 1964.

Aric, K., Reflexionsseismische Messungen im Skagerrak, *Z. Geophys.*, *34*(2), 223-226, 1968.

Båth, M., and E. Tryggvason, Deep seismic reflections experiments at Kiruna, *Geofis. Pura Appl.*, *51*, 79-90, 1962.

Behnke, Cl., *Seismic Velocities in the Central Alps's Crust*, pp. 97-101, European Seismological Commission, 1966, Copenhagen, 1967.

Birch, Francis, Interpretation of the seismic structure of the crust in the light of experimental studies of wave velocities in rocks, in *Contributions in Geophysics in Honor of Beno Gutenberg*, pp. 158-170, Pergamon Press, New York, 1958.

Choudhury, M., P. Giese, and G. de Visintini, Crustal structure of the Alps—Some general features from explosion seismology, paper presented at the 14th General Assembly, IUGG (IASPEI), Zurich, 1967.

Closs, H., Results of explosion studies in the Alps and the German Federal Republic, in *Upper Mantle Symposium, New Delhi, 1964*, edited by C. H. Smith and T. Sorgenfrei, pp. 94-102, Intern. Union Geol. Sci., Copenhagen, 1965.

Closs, H., and Y. Labrouste, Recherches séismologiques dans les Alpes occidentales qu moyen de grandes explosions en 1956, 1958, et 1960, *Seismologie*, [12] *2*, 241, 1963.

Closs, H., and S. Plaumann, Ein Erdkrustenmodell Ostalpen—Poebene, unpublished report, Geol. Surv., Fed. Rep. Germany, 1966.

Colette, B. J., R. A. Lagaay, and A. R. Ritsema, Depth of the Mohorovicic discontinuity under the North Sea basin, *Nature*, *205*, 688-689, 1965.

Dohr, G., Zur reflexionsseismischen Erfassung sehr tiefer Unstetigkeitsflächen, *Erdoel und Kohle*, *10*, 278-281, 1957.

Dohr, G., B. Hadjebi, and Kl. Hehn, *Beobachtungen von Tiefenreflexionen in Norddeutschland*, pp. 87-96, European Seismological Commission, 1966, Copenhagen, 1967.

Dürbaum, H., J. Fritsch, and H. Nickel, *Deep Seismic Sounding in a Part of Rhenish Massif*, pp. 265-273, European Seismological Commission, 1966, Copenhagen, 1967.

Finetti, I., S. Bellemo, and G. de Visintini, Preliminary investigations on the earth's crust in the South Adriatic Sea, *Boll. Geol. Teor. Appl.*, *29*, 1966.

Förtsch, O., Analyse der seismischen Registrierungen der Grosssprengung bei Haslach im Schwarzwald am 28.4.1948, *Geol. Jahrb.*, *66*, 65-80, 1952.

Fuchs, K., and O. Kappelmeyer, Report on reflection measurements in the Dolomites, Sept. 1961, *Boll. Geof. Teor. Appl.*, *4*, 133-141, 1962.

Fuchs, K., and M. Landisman, Detailed crustal investigation along a north-south section through the central part of Western Germany, in *The Earth beneath the Continents, Geophys. Monograph 10*, edited by J. Steinhart and T. J. Smith, pp. 433-453, American Geophysical Union, Washington, D. C., 1966.

Geneslay, R., Y. Labrouste, and J. P. Rothe, Reflexions à grande profondeur dans les grosses explosions (Champagne, Oct. 1952), *Bur. Centr. Seismol. Intern.*, *A*, *19*, 331-334, 1956.

German Research Group for Explosion Seismology, Crustal structure in Western Germany, *Z. Geophys.*, *30*, 209-232, 1964.

German Research Group for Explosion Seismology, Seismic wide angle measurements in the Bavarian Molasse Basin, *Geophys. Prospecting*, *14*, 1-6, 1966.

Giese, P., Die Geschwindigkeitsverteilung im obersten Bereich des Kristallins, abgeleitet aus Refraktionsbeobachtungen auf dem Profil Böhmischbruck-Eschenlohe, *Z. Geophys.*, *29*, 197-214, 1963.

Giese, P., Versuch einer Gliederung der Erdkruste im nördlichen Alpenvorland, in den Ostalpen und in Teilen der Westalpen mit Hilfe charakteristischer Refraktions-Laufzeitkurven sowie eine geologische Deutung, *Habilitationsschrift Math.-Naturwiss. Fak. Freien Universität Berlin*, 143 pp., 1966.

Giese, P., C. Prodehl, and Cl. Behnke, Ergebnisse refraktions-seismischer Messungen 1965 zwischen dem französischen Zentralmassiv und den Westalpen, *Z. Geophys., 33,* 215–267, 1967.

Hänel, R., Die Erdkrustenstruktur im rechtsrheinischen Gebirge und in der hessischen Senke nach neueren seismischen Untersuchungen, unpublished report, Geol. Surv., Fed. Rep. Germany, 1965.

Hinz, K., S. Plaumann, and A. Stein, *Seismic and Gravimetric Exploration of the Ringkjöbing-Fyn-High*, pp. 285–292, European Seismological Commission, Copenhagen, 1967.

Hirschleber, H., J. Hjelme, and M. Sellevoll, A refraction profile through the northern Jutland, *Geodaet. Inst., 41,* paper 1, Copenhagen, 1966.

Jensen, H., Report to the European Seismological Commission from the Sub-Commission on Explosion Work in Northern Europe, in press, 1968.

Landisman, M., and St. Müller, Seismic studies of the earth's crust in continents, 2, Analysis of wave propagation in continents and adjacent shelf areas, *Geophys. J., 10,* 539–548, 1966.

Liebscher, H. J., Deutungsversuche für die Struktur der tieferen Erdkruste nach reflexionsseismischen und gravimetrischen Messungen im deutschen Alpenvorland (Teil I u. II), *Z. Geophys., 30,* 51–96, 115–126, 1964.

Meissner, R., An interpretation of the wide angle measurements in the Bavarian Molasse basin, *Geophys. Prospecting, 14,* 7–16, 1966.

Meissner, R., Zum Aufbau der Erdkruste, Ergebnisse der Weitwinkelmessungen im bayerischen Molassebecken, *Gerlands Beitr. Geophys., 76,* 211–254, 295–314, 1967.

Morelli, C., Activity Report 1964–1966, Subcommission for Explosions in Southern and Western Europe, in press, 1968.

Mueller, St., and M. Landisman, Seismic studies of the earth's crust in continents, 1, Evidence for a low-velocity zone in the upper part of the lithosphere, *Geophys. J., 10,* 525–538, 1966.

Mueller, St., A. Stein, and R. Vees, Seismic scaling laws for explosions on a lake bottom, *Z. Geophys., 28,* 258–280, 1962.

O'Brien, P. N. S., Seismic observations 20 km from explosions in a lake, *Boll. Geof. Teor. Appl., 26,* 144–164, 1965.

Penttilä, E., On seismological investigations of crustal structures in Finland, *Geoteknillisiä Julkaisuja, 68,* 28–35, 1965.

Perrier, G., Ondes séismiques enregistrées dans les monts du Forez, *Compt. Rend., 257,* 1321–1322, 1963.

Perrier, G., Variations d'épaisseur de la croûte terrestre dans le centre et le Nord-Est de la France, *Compt. Rend., 261,* 493–496, 1965.

Porkka, M. T., *Progress Report of 1965, Inst. Seismol. Univ. Hels. Publ. 80,* 1966.

Prodehl, C., Struktur der tieferen Erdkruste in Südbayern und längs eines Querprofiles durch die Ostalpen, abgeleitet aus refraktionsseismischen Messungen bis 1964, *Boll. Geof. Teor. Appl., 25,* 35–88, 1965.

Reich, H., Über seismische Beobachtungen der PRAKLA von Reflexionen aus grossen Tiefen in Blaubeuren, *Geol. Jahrb., 68,* 225–240, 1953.

Reich, H., Seismische und geologische Ergebnisse der 2-to-Sprengung im Tiefbohrloch Tölz I am 11.XII.1954, *Geol. Jahrb., 75,* 1–46, 1958.

Sellevoll, M. A., A publication summary on crustal and upper mantle structure in Fennoscandia, Iceland and the Norwegian Sea, Vesiac Rept. 4410-75-X, pp. 111–124, Ann Arbor, 1964.

Sørnes, A., and J. Bleie, A brief review on the state of the crust and upper mantle studies in Norway for the period 1955–1966, unpublished report, 1966.

Vesanen, E., A. Metzger, M. Nurmia, and M. T. Porkka, Explosion seismic determination of P_g and S_g velocities in Finland, *Geofiz. Közlemenyek, 9,* 69–71, 1960.

Waniek, L., and V. Schenk, Modellseismischer Beitrag zur Deutung des Krustenaufbaus in der bayerischen Molasse, *Z. Geophys., 32,* 482–487, 1966.

Seismic Crustal Studies in Southeastern Europe

V. B. Sollogub

SINCE 1958, systematic crustal studies have been carried out in southeastern Europe, including a significant portion of a plan of investigations along eight profiles crossing the main geostructural regions (Figure 1) [*Subbotin et al.*, 1965] outlined in 1963 by the Geophysical Commission of the Carpatho-Balkan Geological Association.

The investigations in general used the technique of continuous profiling to obtain fully correlated travel-time curve systems for the waves from all main interfaces within the crust [*Sollogub et al.*, 1966a, b].

In the studies of the upper crustal layer the distance between shots was 20 to 25 km; sometimes it was reduced to 10–15 km; in the deeper crustal studies, the separation was 40–50 km. The lengths of profiles varied correspondingly, from 100–110 km for the first case to 160–180 km for the second case, on both sides of the shot point. Figure 2 shows a diagram of travel-time curves along profile I of Figure 1 and types of the waves recorded.

In the USSR, Czechoslovakia, Romania, Bulgaria, and other countries, the recording equipment consisted of geophones and low-frequency 60-channel seismic stations whose amplifiers provided the possibility of recording waves with frequencies of 6 to 15 cps. In Yugoslavia 24-channel seismic equipment of American manufacture (7000 B) was also used. Explosives were generally detonated in bore-

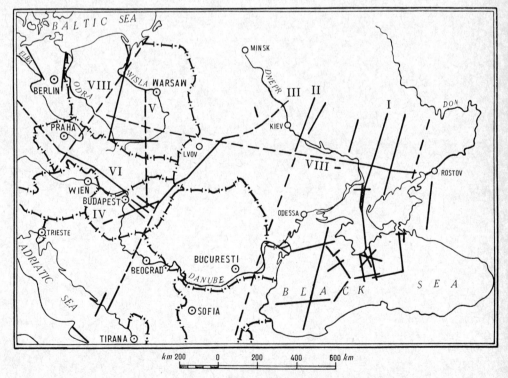

Fig. 1. Seismic profiles in southeastern Europe: solid lines, completed profiles; dashed lines, planned profiles.

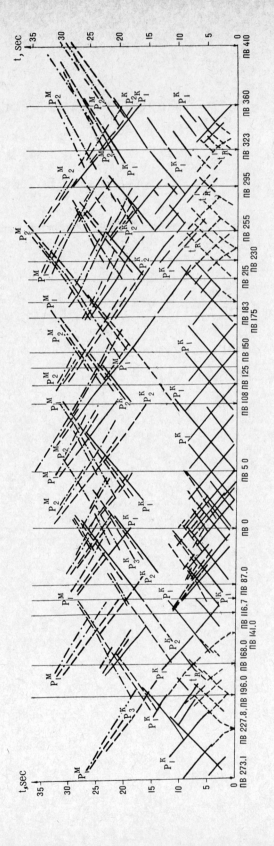

Fig. 2. Travel-time curves for profile I of Figure 1 (the Black Sea to the Voronezh massif): (1) waves in sediments; (2) waves from the basement surface and interfaces within the granitic layer; (3) waves from the Conrad surface; (4) waves from the Moho; shot points (ПВ).

holes with an average depth of 25 to 30 meters; the charge size at great distances was 400 to 700 kg, sometimes 1 to 2 tons. As a rule, 50 kg of TNT was used for one borehole; for larger explosions, a grouping of shots was used.

In the course of field studies numerous waves were recorded which can be classified into the following groups: (1) waves associated with the horizons in the sedimentary complex; (2) waves characterizing the basement surface and the interfaces within the granitic layer; (3) waves related to the Conrad surface and the interfaces within the basaltic layer; (4) waves characteristic of the Mohorovicic discontinuity or the transition zone from the crust to upper mantle. The above waves can be of different types: head, refracted, subcritical, and beyond-critical reflections, diffracted and converted. In addition, deep fractures are identified in the records by the following indications: (1) breaks in travel-time curves, with a time shift of the waves from the footwalls of a fault; (2) anomalous attenuation of the waves in fracture zones; (3) general change in the wave pattern; (4) appearance of diffracted waves; and (5) appearance of local sporadic waves, suddenly emerging and vanishing, having high or even negative apparent velocities [*Sollogub et al.*, 1966a, b].

The longest and most characteristic of the eight crustal cross sections are observed along profiles I and III. A cross section along profile I from the Black Sea to the Voronezh massif (Figure 3) displays considerable differences for the individual geological structures, crustal thickness, and structure, thickness, and relationships of the crustal layers. The difference in structure of the sub-oceanic crust of the Black Sea depression and the continental crust of the Ukraine is especially pronounced. The crust is much thinner under the Black Sea (20-30 km) than in the continental area (35-50 km).

The granitic layer is found throughout the continental zone. In many areas, nearly horizontal seismic interfaces are revealed at shallow depths (2-5 km) within the granitic layer; this contrasts with the predominantly vertical bedding of the near-surface part of the basement. The basaltic layer surface under the Dnieper-Donetz depression generally follows the basement topography. Under the Ukrainian shield, the Conrad surface occurs at average depths of 17-20 km; in one area its depth is 5-8 km. Under the Crimean Mountains an anticlinal bend (anti-root) of the basaltic layer surface is observed [*Sollogub and Chekunov*, 1966]. The Moho under the Black Sea occurs at depths of 22-30 km. Under the Crimean Mountains its depth increases to 50-53 km, which is indicative of the presence of mountain roots. Under the Ukrainian shield, crustal thickness varies from 40 to 50 km [*Semenenko et al.*, 1964]. Under the lowest part of the Dnieper-Donetz depression, the Moho forms an anticlinal bend. In some areas two Moho discontinuities are traced, which seems to indicate the presence of a transition zone up to 5 km in thickness between the crust and the mantle. Many deep fractures are distinguished along the profile, their planes generally dipping to the north.

Profile III (Figure 4) crosses the southwestern part of the Russian platform, the Transcarpathian depression, the Carpathians (USSR), the Pannonian middle massif (Hungary), and the Dinarides (Yugoslavia). Figure 3 shows a cross section along part of this profile. Along the profile the basement surface rises in the zone of the Ukrainian shield and the Voronezh massif; in the depressions it descends to 5-12 km. The surface of the basaltic layer occurs at an average depth of 20 km; only in one area of the Ukrainian shield (region of the Korosten Pluton) does it occur at rather shallow depths, 3 to 6.0 km (Figure 4). The Moho in the Dnieper-Donetz depression occurs at a depth of 35 km (forming an anticlinal bend); under the Ukrainian shield and in the southwestern margin of the Russian platform, 40 to 47 km; under the highland Carpathians, 55 to 57 km; under the Pannonian middle massif, 25 km; and in the Dinarides (Yugoslavia), 45 km [*Dragašević and Andrić*, 1968].

In the zone where the Carpathians and Dinarides join the Pannonian middle massif, crustal thickness varies widely, from 21 km in the Pannonian middle massif [*Vojtczak-Gadomska et al.*, 1965; *Hrdlička*, 1967] to 45-57 km in the Carpathians and Dinarides, where the roots of these mountain structures obviously exist. Such sharp variations in the crustal structure along the profile seem to have been caused by a vertical displacement of the Pannonian block, which is separated from the Carpathians and the Dinarides by deep fractures. The vertical

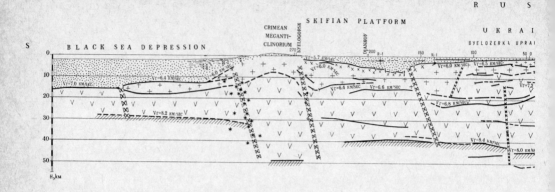

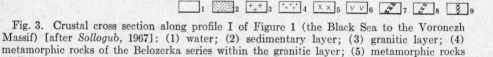

Fig. 3. Crustal cross section along profile I of Figure 1 (the Black Sea to the Voronezh Massif) [after *Sollogub*, 1967]: (1) water; (2) sedimentary layer; (3) granitic layer; (4) metamorphic rocks of the Belozerka series within the granitic layer; (5) metamorphic rocks of Paleozoic-Triassic age within the granitic layer; (6) basaltic layer; (7) deep fractures

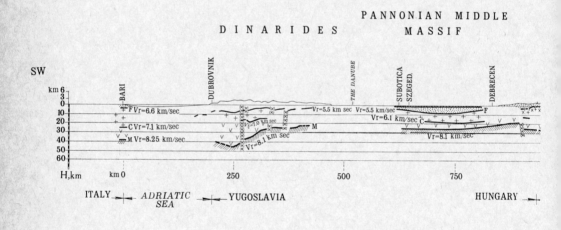

Fig. 4. Crustal cross section along profile III (the Adriatic Sea to the Voronezh massif) [after *Sollogub et al.*, 1967a, b]: (1) water; (2) sedimentary layer; (3) granitic layer; (4) metamorphic rocks of Paleozoic age; (5) basement surface; (6) basaltic layer; (7) basic rocks

displacement of the Pannonian block, in turn, caused horizontal displacements in the uppermost crustal portion toward the northeast in the Carpathians, and to the southwest in the Dinarides.

In addition to the two profiles discussed, much work has been done in Hungary, Czechoslovakia, East Germany, and Poland. Quite good reflections from the Moho have been recorded throughout Hungary [*Mituch and Pozhgay*, 1965], and reliable crustal cross sections can consequently be constructed for that territory. In some areas the Hungarian geophysicists used nonlongitudinal profiling which is an original technique for recording beyond-critical reflections from the Moho, allowing small dis-

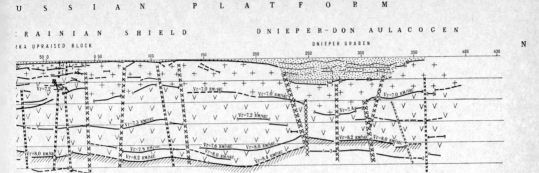

separating the main regions; (8) deep fractures separating individual blocks; (9) same as 8, but less reliable; (10) fractures within individual blocks; (11) dislocations along the basement surface and within the sedimentary complex; (12) seismic interfaces; (13) Mohorovicic discontinuity; (14) Conrad discontinuity; (15) boundary velocities (velocity below the boundary); (16) boreholes ($R-1$); (17) earthquake foci; (18) diffraction points.

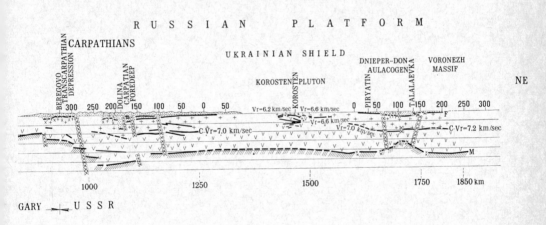

(gabbro) on the Ukrainian shield; (8) Mohorovicic discontinuity; (9) Conrad discontinuity; (10) boundary velocities; (11) deep fractures separating main regions; (12) deep fractures separating individual blocks; (13) dislocations along the basement surface and within the sedimentary series.

locations along this surface to be found. Crustal thickness in the Pannonian middle massif varies only slightly, from 20 to 25 km.

Seismic studies in Czechoslovakia were carried out along profile VI crossing the Bohemian massif and the western margin of the Carpathians. In the area of the massif the crust is 35 km thick.

In the southwestern part of East Germany the crust reaches 35–40 km in thickness. Experimental studies have also been performed in Bulgaria (Misian platform) [*Dachev and Petkov*, 1967] and in Romania.

Summary

The earth's crust in the Carpatho-Balkan region is a layered structure in which, besides the main Mohorovicic and Conrad discontinuities, intermediate seismic interfaces exist in many areas. The main seismic interfaces are, as a rule, unconformable. The existence of mountain roots under the Crimean structures, the Carpathians, and the Dinarides has been proved, and there is evidence for the existence of 'anti-roots' along the Conrad discontinuity.

In the depressions (Dnieper-Donetz, Black Sea, Indolo-Kuban), the main interfaces within the crust are also unconformable; the Moho has a reversed topography compared to the upper seismic interfaces. The unconformable occurrence of the Moho relative to the overlying interfaces, and the sharp decrease or increase in thickness of the earth's crust and its individual layers appear to have been caused primarily by vertical movements of crustal areas; these movements have also caused horizontal displacements of the upper crust, as manifested in the Carpathians and Dinarides in the form of thrusts and overthrust sheets.

In connection with individual blocks of the shields and massifs, steeply dipping rock masses to depths of 1 to 2 km are associated with shallow-dipping interfaces. Structures of this type can be developed only by horizontal movements of the upper crustal layers.

Crustal thickness ranges from 20 to 60 km. These thickness values are also characteristic of genetically different geological provinces. For this reason, crustal thickness cannot be considered a characteristic feature of any one geological region (platform, shield, geosyncline) as was hitherto believed.

The thickness of individual layers also varies widely: the sedimentary cover, 0 to 17 km; the granitic layer, 0 to 18 km; and the basaltic layer, 5 to 40 km. Analysis of the sedimentary cover and granitic layer thicknesses suggests that they are genetically related to each other. Variations in crustal thickness are primarily accounted for by the thickness of the basaltic layer.

The transition from the earth's crust to mantle has a range of characteristics, from a sharp seismic discontinuity to a smooth transition zone, sometimes many kilometers thick.

The earth's crust is broken by numerous deep fractures that originate in the upper mantle and transect the entire crust. These fractures are responsible for the block structure of the earth's crust and control the formation of structural units.

REFERENCES

Betlej, K., B. Gadomska, L. Gorczyński, A. Guterch, A. Mikolajczak, and I. Uchman, Deep seismic soundings on the profile Strachowice-Radzyn Podlaski, in *Selected Problems of Upper Mantle Investigations in Poland*, pp. 41–47, Lodz, 1967.

Dachev, Ch. I., and I. N. Petkov, On registration of seismic depth waves on the territory of the People's Republic of Bulgaria and experience of their interpretation, in *Subbotin et al.* [1967], pp. 68–72, 1967.

Demidenko, Ju. B., M. G. Manyuta, V. A. Lysenko, and L. M. Spikhina, Results of deep seismic studies within Eastern Ukraine, *Geophys. Commun. Inst. Geophys., Acad. Sci. Ukr. SSR, Kiev*, 7(5), 107–115, 1963.

Dragašević, T., Some characteristics of deep seismic soundings in Dinarides, in *Subbotin et al.* [1967], pp. 53–63, 1967.

Dragašević, T., and B. Andrić, Deep seismic sounding of the earth's crust in the area of the Dinarides and the Adriatic sea, *Geophys. Prospecting*, 16(1), 54–77, 1968.

Galfi, G., and L. Stegena, Deep reflections and the structure of the earth's crust in the Hungarian plain, *Geofiz. Közlemen*, 8(4), 45–51, 1960.

Hrdlička, A., Deep seismic crustal studies on the territory of Czechoslovakia, *Studia Geophys. Geodaet.*, 11(3), 348–351, 1967a.

Hrdlička, A., Deep seismic sounding investigations on the territory of Czechoslovakia and their preliminary results, in *Subbotin et al.* [1967], pp. 63–66, 1967b.

Knothe, Ch., H. Henschel, and B. Lange, Short review of state of deep seismic sounding in the German Democratic Republic, in *Subbotin et al.* [1967], pp. 66–68, 1967.

Mituch, E., and K. Pozhgay, Über die Entwicklung und Ergebnisse der seismischen Tiefensondierungen in Ungarn, in *Sonderband über Seismische Tiefensondierungen, Brno*, pp. 51–61, 1965.

Mituch, E., and K. Pozhgay, On results of crustal investigations on Hungarian sections of international profiles, in *Subbotin et al.* [1967], pp. 39–48, 1967.

Prosen, D., Deep seismic studies in the Carpatho-Balkan region and adjacent areas, *Trans. Carpatho-Balkan Geol. Assoc., 7th Congr., Sofia*, part 6, pp. 107–118, 1965.

Prosen, D., and T. Dragašević, Seismische Tiefensondierungen zur Erforschung der Erdkruste in der SFR Jugoslawien, in *Sonderband über Seismische Tiefensondierungen, Brno*, pp. 62–72, 1965.

Prosen, D., and T. Dragašević, On results of deep seismic soundings in Yugoslavia, in *Subbotin et al.* [1967], pp. 33–39, 1967.

Semenenko, N. P., S. I. Subbotin, V. B. Sollogub, M. N. Ivantishin, A. V. Chekunov, and V. D. Dadieva, Structure of deep crustal zones in the Ukrainian crystalline shield, *Sov. Geol., 1964* (11), 48–60, 1964.

Sollogub, V. B., Results of deep seismic soundings on the territory of the Ukraine, in *Subbottin et al.* [1967], pp. 8–19, 1967.

Sollogub, V. B., and A. V. Chekunov, On crustal structure in the area of Mountainous Crimea, *Dokl. Akad. Nauk Ukr. SSR, 1966*(9), 1194–1197, 1966.

Sollogub, V. B., A. V. Chekunov, L. T. Kalyuzhnaya, and L. A. Khilinskiy, On deep structure of the Korosten Pluton from seismic data, *Dokl. Akad. Nauk SSSR, 152*(5), 1215–1217, 1963a.

Sollogub, V. B., A. V. Chekunov, L. T. Kalyuzhnaya, L. A. Khilinskiy, and G. E. Kharechko, Inner structure of the crystalline basement in the southwestern part of the Korosten Pluton from seismic data, *Geophys. Commun. Acad. Sci., Ukr. SSR, 7*(5), 115–122, 1963b.

Sollogub, V. B., A. V. Chekunov, L. T. Kalyuzhnaya, and L. A. Khilinskiy, Structure of the upper crystalline crust in the area of the Ovruch synclinorium from seismic data, *Geophys. Commun. Inst. Geophys., Acad. Sci. Ukr. SSR, 12*(1), 18–26, 1965.

Sollogub, V. B., N. I. Pavlenkova, A. V. Chekunov, and L. A. Khilinskiy, Deep crustal structure along meridional line the Black Sea, the Voronezh massif, *Geophys. Commun. Inst. Geophys. Acad. Sci. Ukr. SSR, 1966*(15), 46–59, 1966a.

Sollogub, V. B., A. V. Chekunov, L. T. Kalyuzhnaya, L. A. Khilinskiy, I. A. Garkalenko, and P. G. Trifonov, Deep crustal structure of the Byelozerka iron ore area from seismic studies, *Geophys. Commun. Inst. Geophys. Acad. Sci. Ukr. SSR, 1966*(18), 3–18, 1966b.

Sollogub, V. B., A. V. Chekunov, L. P. Livanova, M. V. Chirvinskaya, and N. Z. Turchanenko, Deep crustal structure of East Carpathians and adjacent territories of the Ukraine according to DSS data, in *Subbotin et al.* [1967], pp. 49–55, 1967a.

Sollogub, V. B., E. Mituch, A. V. Chekunov, and K. Pozhgay, On experimental Soviet-Hungarian deep seismic studies in the area of Debrecen-Beregovo, *Dok. Akad. Nauk Ukr. SSR*, 1967(2), 120–123, 1967b.

Subbotin, S. I., V. B. Sollogub, and A. V. Chekunov, Deep crustal structure of main geostructural units on the territory of the Ukraine, *Dok. Akad. Nauk Ukr. SSSR, 153*(2), 440–443, 1963.

Subbotin, S. I., V. B. Sollogub, V. I. Slavin, and A. V. Chekunov, On studies of the deep crustal zones in the Carpatho-Balkan region, *Materials of the 6th Congress of the Carpatho-Balkan Geol. Assoc.*, pp. 86–97, Naukova Dumka, Kiev, 1965.

Subbotin, S. I., V. B. Sollogub, and T. S. Lebedev, editors, *Results of Researches of International Geophysical Projects: Upper Mantle Series, No. 5, Geophysical Researches of Crustal Structure of Southeastern Europe*, 162 pp., Nauka, Moscow, 1967.

Vojtczak-Gadomska, B., A. Guterch, and I. Uchman, Ergebnisse der seismischen Tiefensondierungen in der Poland, *Sonderband über Seismische Tiefensondierungen, Brno*, pp. 83–90, 1965.

Explosion Seismology in the USSR

I. P. Kosminskaya, N. A. Belyaevsky, and I. S. Volvovsky

THE study of the earth's crust in the USSR is being carried out to investigate the deep structures of large tectonic zones and to establish the relation between composition and structure of the crust and geophysical fields. The methods are closely related to those of prospecting and exploration for oil, gas, and other minerals.

This paper reviews the results obtained with the deep seismic sounding (DSS) method designed by G. A. Gamburtsev in 1948–1955; it is based on the correlation of seismic waves generated by small (hundred kilogram) explosions. The depth range of the DSS method is related to the recording range, which in practice is limited by the background of regional microseisms. The optimal frequency range for land investigations that use explosions of 1–3 tons

TABLE 1. Average Thinness of Layers and Predominant Seismic Velocities for Tectonic Zones of the USSR (see Figure 9)*

Section in Figure 9	Tectonic Zones	Thickness of the Layers of the Crust, km						Velocity, km/sec						
		Layer 1	Layer 2	Layer 3	Layer 4	Total Crust	Total Crust (average)	Layer 1	Layer 2	Layer 3	Layer 4	Average in Consolidated Crust	Upper Mantle (below M)
1	Baltic shield			13.0	25.0	38.0	6.3–6.5			5.4–6.3	6.6–7.0	6.4	8.0
2	Ukrainian shield		0.5	16.0	35.5	52.0	6.3–6.4		3.0	6.0	6.6–7.0	6.5	8.1–8.2
3	Russian platform, Volga Ural elevation		3.0	18.0	17.0	38.0	6.0–6.2		4.2	5.8–6.4	6.8	6.3	8.0–8.2
4	Precaspian depression		16.0	11.0	14.0	41.0	6.0		4.2–4.6	6.4	7.2	6.7	8.0–8.4
5	Dniepr-Donetsk aulocogen†		7.0	17.0	12.0	36.0	5.9–6.1		4.5	6.0–6.2	6.6–7.1	6.4	8.2
6	Turanian platform		4.0	21.0	16.0	41.0	6.0–6.3		3.5–4.0	5.8–6.3	6.6–7.2	6.4	8.2
7	Zone neotectonics (Pamirs)			37.0	28.0	65.0	6.0–6.2			5.5	6.3–6.4	5.9	8.1
8	Fergana downwarp		12.0	22.0	18.0	52.0	6.2		4.1–4.2	6.4	7.4	6.8	8.2
9	Zone of alps folding (Caucasus)			20.0	35.0	55.0	5.6–5.8			5.6–6.0	6.9–7.4	6.5	8.3–8.4
10	Kurinsk depression		4.0	12.0	29.0	45.0	5.7–6.0		3.3–3.6	5.6	7.0–7.2	6.2	8.2–8.3
11	Mesozoides			22.0	14.0	36.0	6.1			6.0–6.4	6.7	6.4	8.1
12	South Caspian depr.	1.0	20.0		23.0	44.0	5.3–5.5	1.5	3.4–3.8		6.3–6.7	6.5	8.0
13	Black Sea depr.	2.0	10.0		13.0	25.0	4.4–5.0	1.5	3.0–3.5		6.4–6.8	6.6	8.0
14	South Okhotsk depr.	3.0	4.0		6.5	13.5	4.6–5.5	1.5	2.4–3.0		6.5–6.7	6.6	7.9–8.0
15	Japan depr.	4.0	0.5		8.5	13.0	4.6–5.1	1.5	1.5–2.0		6.5	6.5	8.2
16	Kuril Islands	4.0	7.0		25.0	36.0	5.6	115	2.8		6.6	6.6	8.0
17	Pacific Ocean	5.0	0.5		7.5	13.0	5.6–6.0	1.5	2.1–2.4		6.6	6.6	8.1–8.5

* Average layer velocity in consolidated crust of continents is 6.5 km/sec; in oceans, 6.6 km/sec; and in inland seas, 6.5 km/sec. There are some contradictions between the values in the table and the values in Figure 9 because the columns in the figure relate to the fixed place of the DSS cross section, whereas the table gives the dominant values.

† Aulocogen(e): 'a deep, narrow, very extensive sedimentary depression within a platform, bounded by faults,' N. S. Shatsky, as quoted by N. A. Belyaevsky, A. A. Borisov, I. S. Volvovsky, and Yu. K. Schukin, Transcontinental crustal sections of the USSR and adjacent areas, *Can. J. Earth Sci.*, 5, 1067–1078, 1968.

appears to be 5–15 cps; the depth of investigation is about 40–50 km. Under especially favorable conditions of small background of microseisms (up to 1–10 Å) in regions like Kara-Kum and Tien Shan, it is sometimes possible to increase the distance of recording to 400–600 km and the depth of investigation to 70–120 km, and consequently to obtain useful information on the structures of the upper part of the mantle. Sea investigations using the DSS method were started in the USSR in 1956 and were carried out mainly in the near-shore zones, Kuril Island zone, and in the inland seas and the Indian Ocean. In sea investigations using explosions of 100 kg, the dominant frequencies are 3–5 cps; recording range is 100–150 km and depth of investigation is up to 20–35 km.

Instruments

Seismic waves generated by explosions are recorded on the land by multichannel seismic detector stations of the prospecting type with a frequency range of 5–30 cps. The resonant frequency of seismographs is 3–5 cps. In recent years seismic detector units with intermediate magnetic recorders have been used. For sea studies, lower frequency units were used. Sonic waves have been recorded within the frequency range 50–200 cps.

Principal Observational Method

Seismic profiles sometimes several hundreds of kilometers long have been recorded across different tectonic zones (Figure 1). In longitudinal profiling, the shot points are located along the profile. Nonlongitudinal profiling (shot points not along the profile) is used to establish and trace deep anomaly zones or to locate various structures [*Radzhabov and Agranovsky*, 1960]. In sparsely populated regions, spacing between detectors is 100–200 meters. These investigations usually include studies of the sedimentary formations and the boundary of the basement complex [*Godin*, 1958].

To obtain reversed and overtaking travel-time curves corresponding to main types of crustal and mantle waves, the distance between the explosion points must be 50–100 km, and the length of travel-time curves 200–300 km [*Zverev et al.*, 1962]. In regions of complex relief the observations are made at discrete intervals along the profile (called piece-continuous profiling).

Point profiling (mobile shot points) is used mainly in the sea investigations. The optimal distance between two adjacent shot points is 3–5 km, between recording stations, is 30–70 km [*Aksenovich et al.*, 1962].

Point sounding is the observation of one explosion at one point only at a distance con-

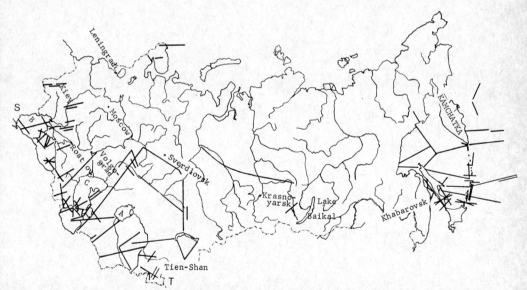

Fig. 1. DSS profiles in the Soviet Union, 1967. Total length of profiles is 53,000 km; this total is made up of 19,000 km of continuous profiling, 9000 km of piece-continuous profiling, 5000 km of point sounding on land, and 20,000 km of point profiling on the sea. *A*, Aral Sea; *B*, Black Sea; *C*, Caspian Sea; *ST*, profile from the Black Sea to Tien Shan (Figure 8).

sidered optimal for observing a single or several predominant deep waves. Data from such observations along a profile or within an area make possible the construction of a specific time-field diagram instead of the usual travel-time curves. This time-field diagram shows the change of relief of the main deep boundaries. The seismic data from different localities are correlated on the basis of specific features in the records, along with other geophysical data. Point sounding is used in the almost inaccessible taiga regions of Siberia and the Far East [*Puzyrev et al.*, 1965].

Principal Methods of Interpretation

DSS data are interpreted according to the following scheme:

1. Analyses of the records and correlation of single waves, from which travel-time curves can be constructed.

2. Determination of the parameters of velocities in the medium, and the compilation of a cross section, assuming a uniformly layered medium with continuous boundaries.

3. Comparison of the kinematic and dynamic characteristics of waves in order to construct a model corresponding to the observed wave field. It is necessary to identify and discuss all contradictions between suggested and observed features of the wave field, as well as to obtain more precise information on the model of the medium, taking into consideration the irregular character of velocities in the crust and upper mantle.

4. Geological and geophysical interpretation of the results.

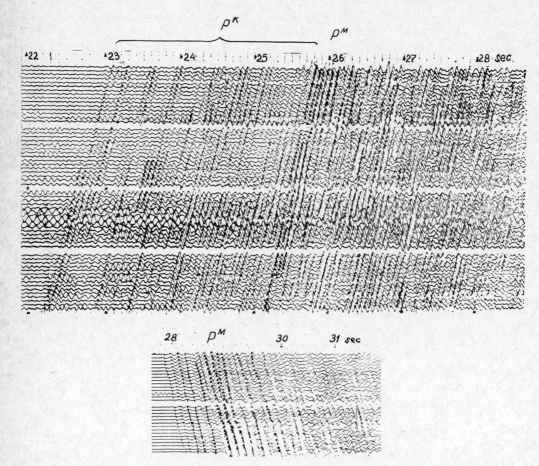

Fig. 2. Seismograms observed (*top*) in Fergana intermontane basin, distance $R = 130$ km, and (*bottom*) in West Kyzylkum, $R = 160$ km. P^k waves belong to the boundaries of the crust, M in mantle.

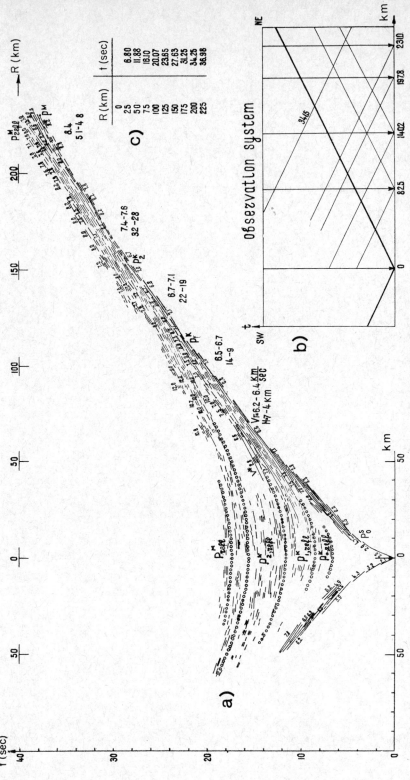

Fig. 3. (a) Phase travel-time curves of deep waves on profile lines DSS Kopetdag-Central Kyzylkum [Beloussov et al., 1962]; (b) system of observation; (c) time table for the first waves on the hodograph. $P_0{}^s$, 'sedimentary' waves; $P_i{}^k$, refracted waves in the earth's crust; $P_{i,\,refl}{}^k$, reflected waves; PM and $P_{refl}{}^M$, refracted and reflected waves on the M boundary; V^*, apparent velocities in kilometers per second; circles, theoretical travel-times of reflected waves; V_r, boundary velocities; and H, depth to the boundary with velocity V_r.

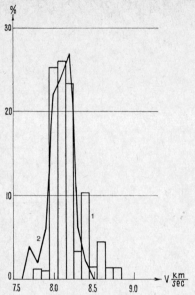

Fig. 4. Histogram of boundary velocities on the M discontinuity for the continental crust: (1) DSS data for the USSR; (2) for continents [*McConnell and McTaggart-Cowan*, 1963].

Special attention should be paid to initial studies of the seismic records and to primary analysis of the wave field.

The prevalent feature of the wave field appears to be regular waves (Figure 2). For the continental type of crust with frequencies of 5–15 cps, many waves of uninterrupted phase correlation (on the average, less than 10 km) could be traced on the seismograms.

Figure 3, the combined travel-time curves for these waves, shows regularities similar to those of families of reflected and refracted waves in uniform or partly nonuniform multilayered media. The velocity of the first arrivals increases with distance. The apparent velocity of successive later arrivals increases to distances up to 150–200 km; at greater distances, the apparent velocity of successive later arrivals decreases. On typical profiles recorded in the USSR, phase correlations cannot be made across the entire profile; they must be made in interrupted segments. Therefore the segments must be combined to provide a schematic, continuous travel-time curve.

It is assumed that each group is characterized by specific features of a definite type of wave; hence there is a relation between this characteristic and its representation by a family of uninterrupted curves, which serve as a basis for compiling cross sections that assume continuous boundaries.

Thus, these cross sections are schematic and

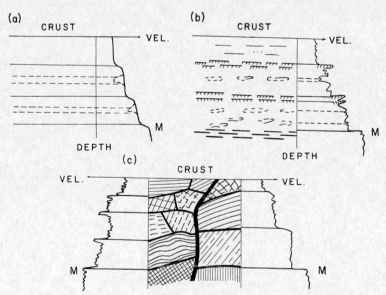

Figure 5. Crustal models: (*a*) continuous boundaries; (*b*) piecewise-continuous boundaries corresponding to observed piecewise-continuous travel-time curves; (*c*) assumed block-layered crust, corresponding to the geophysical (Kosminskaya and Riznichenko, 1964) and geological (Beloussov, 1965) points of view [after *Kosminskaya and Zverev*, 1968].

reflect only approximate and average characteristics of the layering of the crust. Schematizing of the multilayered medium even in the case of uninterrupted boundaries could lead to systematic errors caused by the interference of waves connected with adjacent boundaries. If there are insufficient observations to determine the uppermost structure in detail, the depth to deeper boundaries will be underestimated.

Types of Waves

The seismic records show mainly longitudinal waves. For most regions of the USSR, certain refracted waves are stable and are easily traced. These waves correspond to the upper boundary of the consolidated crust (the boundary of metamorphic or crystalline basement rocks, where the velocity appears to be 6.2 ± 0.4 km/sec); to intermediate boundaries in the earth's crust, where the velocity appears to be 6.8 ± 0.3; and to the boundary of the M discontinuity, where the velocity appears to be $8.0-8.2 \pm 0.3$ km/sec. Less stable are the groups of refracted waves that correspond to a lower boundary in the earth's crust, where the velocity is 7.5 ± 0.3 km/sec, and to boundaries in the upper part of the mantle, where the velocity appears to be 8.4, 8.6–8.8, and 9.0 km/sec. Figure 4 shows a histogram of boundary velocities on the M discontinuity.

The deep reflections from those boundaries can be traced with certainty beginning at a distance of 60–80 km (critical reflection from the explosion outward to 150–200 km [*Beloussov et*

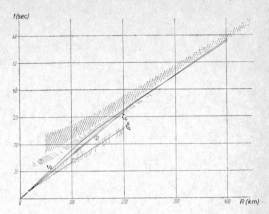

Fig. 7. Summary of travel-time curves of reflected and refracted M waves for different tectonic zones: (1) Pacific Ocean; (2) marginal seas (Sea of Okhotsk, Bering Sea); (3) platforms and shields; (4) folded mountain areas; t_D, travel-time curve by Jeffreys and Bullen; t_c, mean statistical sections of continental ($t_c{}^k$) and oceanic ($t_c{}^o$) crust.

al., 1962]. For incidence angle close to normal, many unstable reflections are observed that are difficult to correlate with the reflections at greater distance.

Only in a few regions, such as the Sub Ural foredeep, the Baltic shield, and the Kyzyl-Kum area, can a well-defined, compact group of wavelets that correspond to the deep reflections be observed; these are stable at large intervals and at a great distance (up to 100–200 km) from the explosion along the profile.

Cross Sections

The deep seismic boundaries are outlined by methods similar to those applied in seismic prospecting and according to systems of travel-time curves of refracted and reflected waves. For reflected waves, the most widely used principle is *Puzyrev's* [1959] method of circles; *Riznichenko's* [1946] field-time construction method and others [*Gamburzev*, 1952, 1960] are used for refracted waves.

In the USSR, the symbols used to designate the waves on the hodographs and legends for the cross sections are as follows: the crust waves, longitudinally refracted, $P_0{}^k$, $P_1{}^k$, $\cdots$; longitudinally reflected, $P_{0,\,refl}{}^k$, $P_{1,\,refl}{}^k$, $\cdots$. Mantle waves, longitudinally refracted, $P_0{}^M$, $P_1{}^M$; longitudinally reflected, $P_{0,\,refl}{}^M$; $P_{1,\,refl}{}^M$. For the corresponding boundaries, the following indexes

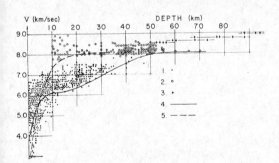

Fig. 6. P-wave velocities in the crust and mantle: (1) in the crust; (2) at the Moho; (3) in the upper mantle; (4) mean curve for continents; (5) mean curve for oceans (1–3 compiled in 1968 by I. S. Volvovsky and N. I. Razinkova from DSS data for the USSR; 4 and 5 after *McConnell and McTaggart-Cowan* [1963]).

are applied: $d_{0,1,2}{}^k$, $\cdots$; $d_{0,1,\,refl}{}^k$, $\cdots$, and $d_{0,1,2}{}^M$ or $d_{0,1,2,\,refl}{}^M$.

In some cases for geological purposes of classification, the seismic layers of the crust are divided into the following groups: W, water; S, sediments; K_1, upper part of consolidated crust with velocity from 5.5 to 6.5 km/sec; K_2, lower part of consolidated crust with velocity more than 6.5 km/sec; and M, mantle with velocity greater than 8.0 km/sec.

Model of Crust

Geometric interpretation of the kinematic characteristics of the wave field should assume an ordinary multilayer medium with monotonic or nonmonotonic change of velocity with depth and with uninterrupted boundaries between layers. Supplementary analyses of the dynamic features of waves makes possible an evaluation of how well this model reflects the real features of the medium.

Figure 5 shows a schematic diagram of the two principal features observed by the DSS method: continuous and piecewise-continuous boundaries; and block structure in the crust. Thin low-velocity layers are indicated. Such local decreases in velocity are probable because of irregularities in the rock formations; at present, however, these layers are only indirectly indicated (by discrepancies in mean layer and boundary velocities as observed in several regions of the USSR).

Conclusions

The structure of the earth's crust was investigated by seismic methods in different tectonic zones (Figure 1): old shields, Baltic and Ukraine [*Litvinenko*, 1963; *Sollogub et al.*, 1966]; old platforms, Russian and Siberian [*Godin*, 1958; *Puzyrev et al.*, 1965]; young platforms, Tura and West Siberian [*Beloussov et al.*, 1962]; the foredeeps, Cis-Uralian [*Khalevin et al.*, 1966], Cis-Caucasus (Terso-Sunjensk) [*Yurov*, 1963], and Cis-Kopetdag [*Beloussov et al.*, 1962; *Ryaboi*, 1966]; sub-montane basins in zones of Paleozoic folding in Fergana [*Zverev et al.*, 1962; *Alekseev et al.*, 1963, 1964] and Kuznetsk; regions of Baikalids distribution, on the Yenisei range; areas of Caledonides and Hercynides, i.e., in the Tien Shan mountains [*Zverev et al.*, 1962] and Kazakhstan and the Urals [*Khalevin et al.*, 1966]; the area of Mesozoic and Alpine folded formations, i.e. the Carpathians, the Caucasus [*Yurov*, 1963], the Pamirs ranges, and mountains on the northeast of the USSR, i.e., in the regions of the Kolyma and Magadan Rivers [*Zverev*, 1962]; the inland seas, Black and Caspian [*Aksenovich et al.*, 1962]; the marginal seas, Barents, Sea of Japan, Sea of Okhotsk, and the Bering Sea, and the water areas adjacent to the Kuril and Commander Islands [*Galperin and Kosminskaya*, 1964]. Separate investigations were carried out in the Indian and Pacific Oceans. The comparison with the mean curve (Figure 4) for continents shows that the velocities in the USSR are higher than average on the continents: this may be due to differences in applied methods and also to differences in crustal structure in different regions.

Figure 7 summarizes the travel-time curves of refracted and reflected M waves and illustrates the principal differences between the wave fields observed under the continents and oceans. This summary can be used to correct regional seismological travel-time curves.

Statistical data characterizing the M discontinuity in the USSR coincide with the middle continental value for the rest of the world, once again emphasizing the 'middle' stability of this boundary.

Figure 8 shows the generalized seismic cross section for the area from the Black Sea to the Tien Shan range. This cross section is based on all available DSS data for the southern part of the USSR; it shows the conformity between tectonic zones of different types and the relief of the main seismic boundaries. These relations are shown in Figure 9.

The depth of the M discontinuity varies within a wide range for different zones: the roots of the young growing Caucasus mountains are shallow in comparison with deep active folded systems of the Pamirs and the Tien Shan.

Figure 10 shows the dependence and relation of the hypsometry of the surface of the solid earth to the depth of the M discontinuity. It appears evident that the higher the mountains, the deeper their roots, as well as the deeper the sea, the thinner the crust. However, this linear relation cannot be applied to plain areas of con-

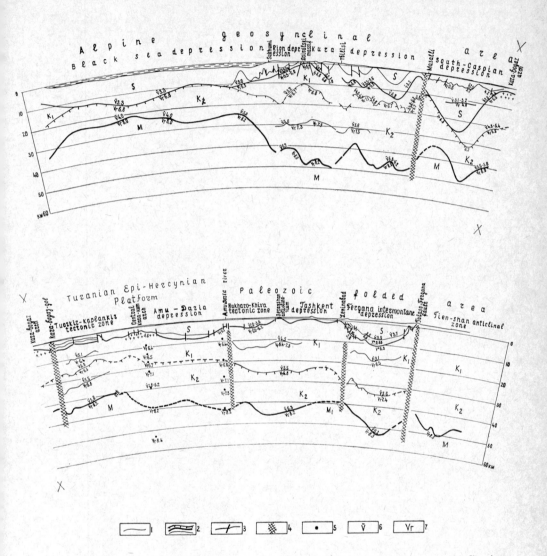

Fig. 8. Section of the earth's crust from the Black Sea through the Caucasus and Caspian Sea to central Kara-Kun-Tien Shan: (1) seismic boundaries; (2) the surface of consolidated crust and M boundary; (3) fractures according to boreholes and seismic prospecting data; (4) zones of anomaly in the seismic records; (5) the depths according to DSS profiles intersecting a given profile; (6) average velocity up to the seismic boundaries; and (7) boundary velocities. For S, M, K_1, and K_2 see Figure 9.

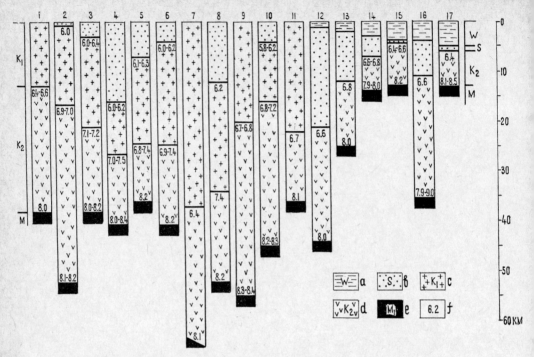

Fig. 9. Models of the earth's crust for different tectonic zones of the USSR on the basis of DSS data: (a) water; (b) unconsolidated crust (sediments); (c) consolidated crust, upper layers 'granite'; (d) consolidated crust, lower layers; (e) upper mantle; (f) boundary velocity in kilometers per second. See Table 1.

tinent; i.e., on platforms it is impossible to predict the existence of deep structures on the basis of the type of relief or according to the broad geophysical anomalies, for instance, the Bouguer anomaly.

Figure 11 shows one of the most recent relief maps of the M discontinuity for the USSR and the adjacent territory. It was compiled on the basis of DSS data available at the beginning of 1967 and takes into account regional interrelations and other geological-geophysical data. The accuracy of the map varies according to the area: it is more accurate for west and southwest and less so for the north, especially the northeast.

The accuracy of the isolines for the whole USSR is not as great as the accuracy in the regions of the seismic profiles. That is why this approximate type of map can be regarded as a basis for future detailed seismic and other geophysical investigations. The map emphasizes the main regional geological structures of the USSR.

The study of the seismic sections of the earth's crust and comparison with geological data of the main structures and their development, and with geophysical fields leads to these conclusions:

1. The crust of continents and oceans is characterized by layered block structure.

2. Definite differences exist among sections of the earth's crust: continental and oceanic platforms, folded structures activated recently, land and marginal seas, recent geosynclines, etc. The most strongly pronounced example of such differences appears to be the depth of occurrence of the M discontinuity, as well as the relation of sedimentary cover formations to the thickness of the consolidated crust. Occurrence of layers with velocities equal to the velocities in 'acid' and 'basic' rocks within the consolidated crust makes it possible to divide the earth's crust into types of geological structures.

3. Statistical relations between the depth of the M discontinuity and elevations of the surface of the solid earth and Bouguer anomaly, as well as other parameters, not only indicate that a general (global) dependence exist, but also suggest definite interrelations between the main geological structures (with the exception of con-

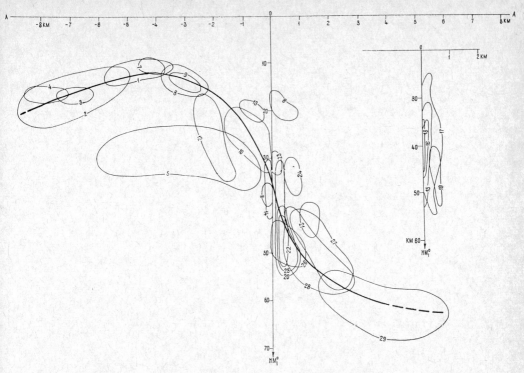

Fig. 10. Diagram of correlation and dependence of the earth's surface and the depth of the M discontinuity; range of relief in meters is given below: (1) The Far Eastern margin of the Pacific Ocean, 0 to −5000; (2) the Kuril trench, to −10,542; (3) the Aleutian trench, to −7822; (4) the Japan trench, to −8412; (5) outer slope of the Kuril island arc, 0 to −1000; (6) Kuril island arc, 0 to 700. Marginal Seas; (7) the Bering Sea, to −3782; (8) the Sea of Japan, 0 to −3669; (9) South Okhotian basin, −2000 to −3921; (10) central and southern parts of the Sea of Okhotsk, 0 to −2000; (11) the Barents Sea, 0 to −326. Inland seas: (12) the Black Sea, 0 to −2185; (13) south Caspian basin, to −985; (14) central and northern parts of the Caspian Sea, 0 to −800. Plains and platforms: (15) East European (Russian) plain, 0 to 300; (16) plains of Crimea and the Pre-Caucasus region, 0 to 200; (17) Turanian lowland, −26 to 200; (18) West Siberian platform, 200 to 600; (19) Central Siberian platform, 200 to 1200. Low mountains: (20) Kazakhstan Mountains, 200 to 1200; (21) Crimea Mountains, 0 to 1500; (22) the Urals, 200 to 1600; (23) Kopet-Dagh, 200 to 1700. Middle mountains: (24) mountains of the northeast of the USSR, 200 to 2200; (25) Carpathian Mountains, 200 to 2200; (26) mountains of Transcaucasus, 200 to 2700. High mountains; (27) Great Caucasus, 200 to 5633; (28) Tien-Shan, 500 to 5494; (29) Pamirs, 500 to 7495. See Figure 3, p. 317.

tinental and oceanic platforms of any age where correlation is interrupted).

4. The roots of mountains located on the margin of the Pacific Ocean are shallower than the roots of mountains located in the inland areas of Asia, in accordance with the general decrease of thickness of the earth's crust from the inland regions of Asia (65–70 km deep) toward its marginal areas. The regions of island arcs and the shields of the easterrn European platform (Ukraine and Voronezh massifs) are confined to the zone of sublatitudinal dislocations in the Mediterranean belt (Tethys), which includes the main part of the Tura platform. The Carpathians, the Kopetdag Mountains, and probably some other folded areas confined to the Mediterranean orogenic belt, of maximum thickness of the earth's crust, are displaced from the axial line toward the foredeeps. All the rock formations of the earth's crust in this region are involved in a process of wave-like dislocations and, consequently, are folded into large synclines, the most pronounced being in the Fergana intermontane basin. But, at the same time, in all the depressions in the Dniepr-Donetsk aulacogen, the depth of the M discontinuity de-

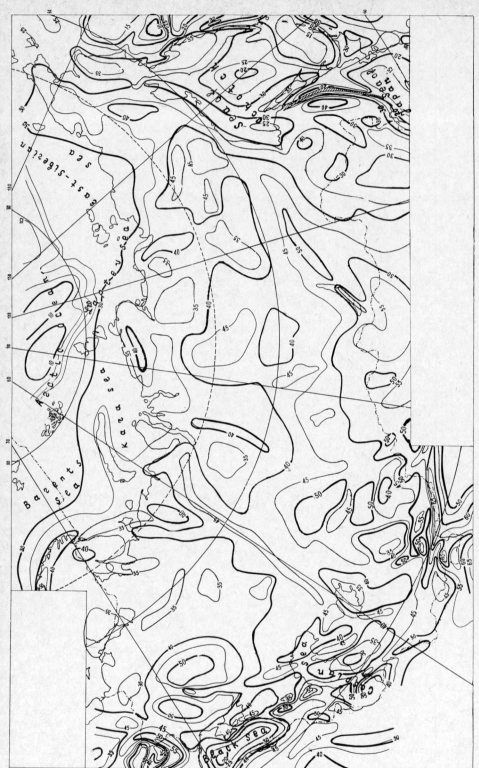

Fig. 11. Relief map of the M boundary for the USSR and adjacent territory. Numbers are depth below sea level in kilometers (see Figure 4, p. 318).

creases according to the shield depth of M, indicating the different relations of the crust to the upper mantle.

5. The values of velocities are persistent on the M boundary, varying only within the range $8.0–8.2 \pm 0.3$ km/sec, independent of depth. At the same time, the seismic velocities in the crust vary with depth and composition. Comparison with experimental data of laboratory studies of rock velocities helps to establish that the lower horizons of the earth's crust are characterized by a more basic composition than the upper ones.

6. The upper mantle is characterized by lateral rock alteration, illustrated by correlation of the DSS section with Bouguer anomalies. The calculation of residual gravity anomalies suggests that the material of the mantle under high mountains is less dense than under plains, while in sedimentary basins the material is more densely consolidated.

7. All investigations by seismic methods in Central Asia, the Urals, and eastern Siberia indicate that subhorizontal lamination layering of crust and mantle occur to depths of tens of kilometers below the M discontinuity.

REFERENCES

Aksenovich, G. I., L. E. Aronov, A. A. Gagelganz, E. I. Galperin, M. A. Zaionchkovsky, I. P. Kosminskaya, and R. M. Krakshina, *Deep Seismic Sounding in the Central Part of the Caspian Sea*, Izd. Akad. Nauk USSR, Moscow, 1962.

Alekseev, A. S., Some inverse problems of the theory of wave distribution, 1 and 2, *Izv. Akad. Nauk SSSR, Ser. Geofiz.*, 1962(11), 1514–1522, 1523–1531, 1962.

Alekseev, A. S., I. S. Volvovsky, et al., On the problem of the physical nature of some waves recorded by DDS methods, 1, Characteristics of experimental data, *Izv. Akad. Nauk SSSR, Ser. Geofiz.*, 1963(11), 1620–1630, 1963.

Alekseev, A. S., I. S. Volvovsky, et al., On the problem of the physical nature of some waves recorded by DDS methods, 2, Theoretical analysis of models of the earth's crust for Central Asia, *Izv. Akad. Nauk SSSR, Ser. Geofiz.*, 1964(1), 3–19, 1964.

Alekseev, A. S., I. S. Volvovsky, et al., On the problem of the physical nature of some waves recorded by DDS methods, 3, Comparison of theoretical calculations and experimental results, *Izv. Akad. Nauk SSSR, Ser. Geofiz.*, 1964(2), 184–195, 1964.

Alekseev, A. S., I. S. Volvovsky, et al., *Methodic Recommendations Concerning the Investigation by Method of Deep Seismic Sounding*, Institute of Physics of the Earth, Moscow, 47 pp., 1966.

Beloussov, V. G., B. S. Volvovsky, I. S. Volvovsky, and V. Z. Ryaboi, Experimental data on recording of deep reflected waves, *Izv. Akad. Nauk SSSR, Ser. Geofiz.*, 1962(8), 1034–1044, 1962.

Beloussov, V. V., *The Crust and Upper Mantle of the Continents*, Izd. Nauka, Moscow, 123 pp., 1966.

Galperin, E. I., and I. P. Kosminskaya, editors, *The Structure of the Earth's Crust in the Transition Zone from the Asiatic Continent to the Pacific Ocean*, Izd. Nauka, Moscow, 308 pp., 1964.

Gamburzev, G. A., *Selected Works*, Izd. Akad. Nauk SSSR, Moscow, 461 pp., 1960.

Gamburzev, G. A., Yu. V. Riznichenko, I. S. Berzon, A. M. Epinatieva, I. P. Kosminskaya, E. V. Karus, *The Correlation Refraction Method*, Akad. Sci. USSR, Moscow, 239 pp., 1952.

Godin, U. N., Complex geophysical investigations concerning deep structures of the earth's crust in example of the Caspian Sea, *20th Intern. Geophys. Congress*, vol. 1, *Reports of Soviet Geologists*, Gostoptekhizdat, Moscow, 1958.

Gurvich, I. I., *Seismic Prospecting*, Gostoptekhizdat, Moscow, 1960.

Khalevin, N. I., V. S. Druzhinin, V. M. Rybalka, E. A. Nezolenova, and L. N. Chudakova, Results of deep seismic sounding of the earth's crust on the Middle Urals, *Izv. Akad. Nauk SSSR, Phys. Earth*, 1966(4), 36–44, 1966.

Kosminskaya, I. P., and Yu. V. Riznichenko, Seismic studies of the earth's crust in Eurasia, in *Research in Geophysics*, vol. 2, edited by H. Odishaw, pp. 81–122, MIT Press, Cambridge, 1964.

Kosminskaya, I. P., Modern problems of deep seismic sounding, in *Problems of the Physics of the Earth, Geophys. Collection*, No. 15, pp. 34–45, Izd. Naukova dumka, Kiev, 1965.

Kosminskaya, I. P., and S. M. Zverev, Abilities of explosion seismology in oceanic and continental crust and mantle studies, *Can. J. Earth Sci.*, 5, 1091–1100, 1968.

Litvinenko, I. V., Seismic method investigation of the deep structure of the Baltic Shield, *Proc. Leningrad Mining Inst.*, 16(2), 1963.

McConnell, R. K., Jr., and G. H. McTaggart-Cowan, *Crustal Seismic Refraction Profiles—A Compilation*, University of Toronto, Toronto, 1963.

Puzyrev, N. N., *Interpretation of Seismic Prospecting Obtained Data by Method of Reflected Waves*, Gostoptekhizdat, Moscow, 1959.

Puzyrev, N. N., S. V. Krilov, et al., Point seismic soundings, in *Methods of Seismic Prospecting*, edited by N. N. Puzyrev, pp. 5–70, Nauka, Moscow, 1965.

Radzhabov, M. M., and L. E. Agranovsky, Determination of depth and relief of an interface by means of single transverse travel-time curves of refracted waves, *Izv. Akad. Nauk SSSR, Ser. Geofiz.*, *1960*(6), 854–862, 1960.

Riznichenko, Yu. V., Geometrical lows characteristic of layered seismic medium, *Trans. Inst. Theoret. Geophys.*, *2*, sec. 1, 114 pp., 1946.

Ryaboi, V. Z., Kinematic and dynamic characteristics of deep waves connected with boundaries in the earth's crust and upper mantle, *Izv. Akad Nauk SSSR, Phys. Earth*, *1966*(3), 74–82, 1966.

Sollogub, V. B., A. V. Chekunov, L. T. Kalyuzhnaya, L. A. Khilinsky, I. A. Garkalenko, and P. G. Trifonov, Deep structures of the earth's crust in the Byelozersk iron-bearing region according to seismic data, *Geophys. Collection*, *1966*(18), 3–18, Kiev, 1966.

Yegorkin, A. V., Techniques for determining velocity parameters of the cross section of the earth's crust from travel-time curves of reflected waves, *Izv. Akad. Nauk SSSR, Phys. Earth*, *1966*, 108–114, 1966.

Yurov, Yu. G., Structure of the earth's crust in the Caucasus and isostasy, *Sov. Geol.*, *1963*(9), 113–118, 1963.

Zverev, S. M., G. G. Mikhota, I. V. Pomerantseva, and M. V. Margotieva, editors, *Deep Seismic Sounding of the Earth's Crust in the USSR*, Gostoptekhizdat, Moscow, 495 pp., 1962.

Explosion Seismic Studies in North America

John H. Healy and David H. Warren

SEVERAL comprehensive reviews of seismic crustal studies have been published in recent years: a worldwide review and extensive bibliography [*James and Steinhart*, 1966] and supplement to the preceding [*Steinhart*, 1967]; compilation of numerical results [*McConnell et al.*, 1966]; and studies in the Transcontinental Geophysical Survey [*Pakiser and Zietz*, 1965]. The present article is devoted primarily to representative studies in North America.

NORMAL CRUSTAL STRUCTURE

The Vela Uniform Program and the Upper Mantle Project have provided a strong impetus to explosion seismology in North America. Many new profiles have become available, and it is now possible to present crustal structure in some detail, particularly for the area between latitudes 35°N and 39°N, where the U. S. Upper Mantle Committee has recommended a concentration of effort to provide a Transcontinental Geophysical Survey (TGS). There are about fifty profiles in North America along which upper mantle arrivals have been recorded adequately to determine the thickness of the crust; about twenty of these lie within the TGS. Fence diagrams of crustal structure prepared for the TGS (Figures 1–4) illustrate some of the conclusions that can be drawn about the structure of the crust and upper mantle.

Between approximately 20 and 130 km from the source, the first refracted arrival propagates near the top of the crust. The speed of this phase, P_g, is nearly constant throughout the continent, in most places, 5.9 to 6.2 km/sec; the average of all determinations is slightly higher than 6.0 km/sec. A statistical study (Borcherdt and Healy, unpublished data) shows that the standard deviation of the velocity determination for P_g is about 0.1 km/sec; therefore, most of the scatter in the reported velocities is about as large as the precision of the measurement. In view of the wide range in velocities appropriate for crystalline rocks observed at the earth's surface, the small variation in the velocity of P_g is remarkable.

The velocity of the phase refracted along the top of the mantle varies between 7.8 and 8.3 km/sec. This phase, P_n, reveals a marked variation across the United States, velocities in the west averaging about 7.8 km/sec and those in the east about 8.1 km/sec. The standard deviation of these determinations is about 0.1 to 0.15 km/sec (Borcherdt and Healy, unpublished data). *Pakiser and Steinhart* [1964] have used

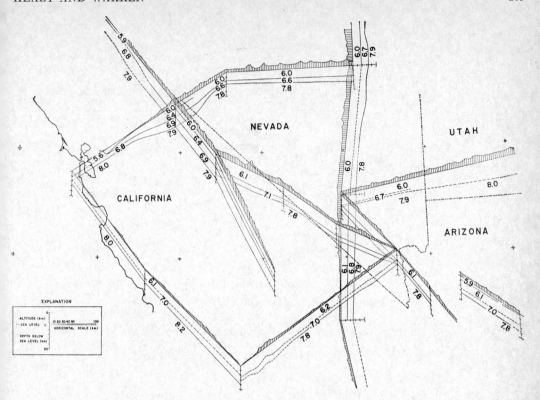

Fig. 1. Fence diagrams of crustal structure in the western portion of the Transcontinental Geophysical Survey. Numbers refer to the average velocity (kilometers per second) of compressional waves in the crustal and upper mantle layers; thicknesses show no vertical exaggeration. Altitude above sea level is shaded and is exaggerated 5:1.

these differences to suggest the existence of two superprovinces in North America, separated by a line running along the front of the Rocky Mountains. The western superprovince is characterized by high heat flow, volcanism, and Cenozoic tectonic activity. The eastern superprovince is characterized by low-to-average heat flow and tectonic stability.

Between the upper crust and the top of the mantle, the velocity structure is more variable. Phases refracted through deeper layers either do not appear as first arrivals or are observed as first arrivals over relatively short distances. Intermediate layers are often deduced from the study of secondary arrivals, which are more difficult to follow. Therefore, much of the variation reported for the properties of the lower crust might be accounted for by difficulties in interpretation, but some similarities and differences can be established on a regional basis. Almost all the profiles in the TGS show a lower crustal layer or zone, with a P-wave velocity between 6.6 and 7.1 km/sec. Some of the profiles indicate a layer at the base of the crust with a P-wave velocity between 7.1 and 7.4 km/sec. The lower layers in the crust are usually thicker in the eastern United States than in the western United States; the layer at the base of the crust with velocity 7.1 to 7.4 km/sec is detected more frequently in the east than in the west.

The gross velocity structure of the crust can be deduced from the refraction-profile method, but the detailed velocity structure is more obscure. The transition between velocity layers is qualitatively deduced from the character of first and later phases on the seismograms. In particular, the phases that have been reflected near the critical angle (computed for an appropriately layered model) have amplitude variations much larger than those of the refracted phases. These amplitude variations suggest that the

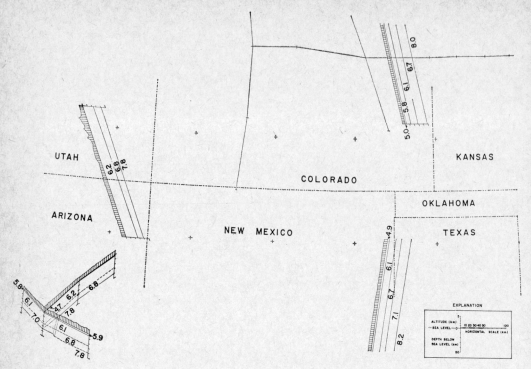

Fig. 2. Fence diagrams of crustal structure in the central-western portion of the Transcontinental Geophysical Survey. See legend of Figure 1 for explanation.

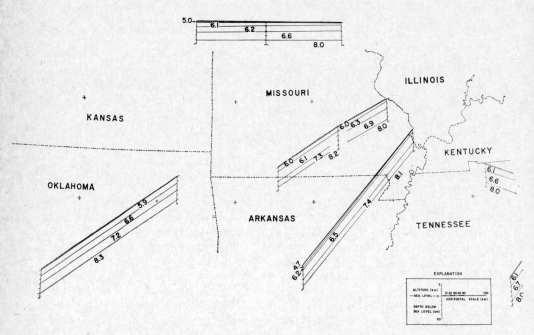

Fig. 3. Fence diagrams of crustal structure in the central-eastern portion of the Transcontinental Geophysical Survey. See legend of Figure 1 for explanation.

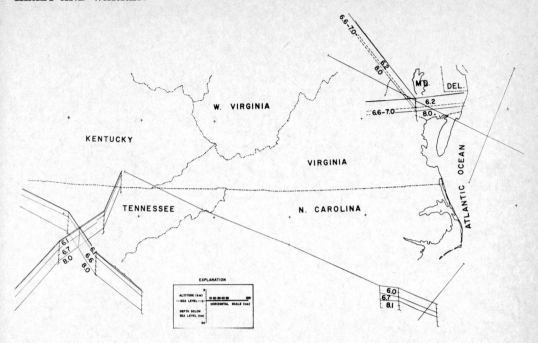

Fig. 4. Fence diagrams of crustal structure in the eastern portion of the Transcontinental Geophysical Survey. See legend of Figure 1 for explanation.

boundaries between zones of different velocities are much sharper in some areas than in others. For instance, profiles in the Basin and Range province usually show large-amplitude reflected phases, and profiles from the Great Plains province usually show relatively weak reflected phases. Factors other than the sharpness of discontinuities can control the amplitude of reflected phases, but the effect is so pronounced in some areas that qualitative conclusions can be made.

The amplitude variations are well illustrated by two record sections (Figures 5 and 6) for data recorded on two north-trending profiles, one in central Colorado in the Southern Rocky Mountains [*Jackson and Pakiser*, 1965], the other in eastern Colorado in the Great Plains [*Jackson et al.*, 1963]. The Rocky Mountain profile (Figure 5) plotted on a reduced travel time section shows two large-amplitude secondary events between 120 and 200 km from the shot point. These events arrive at times appropriate for reflections from the top of the upper mantle (P_mP) and from the top of a lower crustal layer (P_lP) with a velocity of 6.8 km/sec. The Great Plains profile (Figure 6) also shows events that arrive at times appropriate for reflected events, but these are of about the same amplitude as other events and thus are less prominent than the reflections on the Rocky Mountain profile. These two profiles, close to the superprovince boundary along the Rocky Mountain front, typify the differences seen on many (but by no means all) surveys in the eastern and western United States.

Future work on secondary arrivals may permit the calculation of the fine structure of the crust. At present it can be concluded that there are marked regional differences in the internal structure of the crust, and that the crust in tectonically active areas tends to contain sharper discontinuities.

Similar descriptions of crustal structure were given in some of the earliest papers on the subject, and the vast amount of new information gives support to some of the conclusions of pioneer workers. The importance of the recent work lies in the new capability to describe the crust over large areas in sufficient detail to recognize those properties that are character-

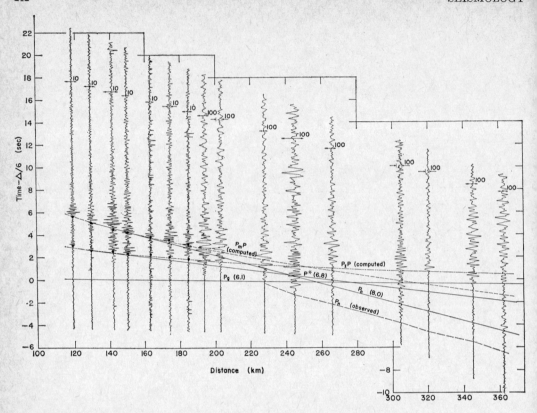

Fig. 5. Time-distance layout of seismograms from the southern Rocky Mountains in Colorado. A reduction time, equal to distance divided by 6.0 km/sec, has been subtracted from each recording. The first arrivals for P_g are weak and are generally not seen. The phase P_n is easily recognized, but is seemingly complicated by changes in crustal thickness. Large-amplitude secondary arrivals are interpreted to be reflections from the top of the intermediate layer (P_IP) and from the top of the upper mantle (P_mP). Refractions from within the intermediate layer (P^*) never appear as first arrivals. Relative calibration amplitudes are shown by arrows near the top; the change from 10 to 100 marks a change of explosion size. Reprinted from *Jackson and Pakiser* [1965].

istic of an 'average' crust, to look for anomalous regions, and to study small variations between geologic provinces.

ANOMALOUS CRUSTAL STRUCTURE

Three seismic profiles reveal anomalous crustal structure in widely separated regions of North America. They are at Vancouver Island [*White and Savage*, 1965], in the Snake River Plain [*Hill and Pakiser*, 1966], and at Lake Superior [*Smith et al.*, 1966]. Crustal columns for these regions (Figure 7) show the velocity in the upper crust to be quite different from the usual velocity of 6.0 km/sec. Beneath the upper crust a thick layer of about 6.8-km/sec velocity extends to the M discontinuity at depths of 40 to 55 km. Basalt flows are exposed at the surface in all three regions. The Snake River Plain underwent extensive Cenozoic volcanism; Eocene basalts are exposed on the southern part of Vancouver Island. Much of the surficial geology of the Lake Superior region is obscured under the lake, but the Duluth gabbro complex and extensive volcanism of Precambrian (Keweenawan) age suggest that the rocks under the lake may be of basaltic composition.

Some speculations can be made about these anomalous areas. The top of this type of crust has some similarity to oceanic crust in that there is no significantly thick layer with velocity appropriate for a granitic upper crustal layer. Because of the basaltic composition that

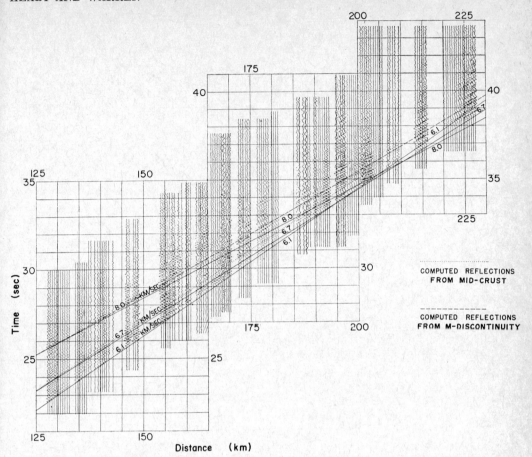

Fig. 6. Time-distance layout of seismograms from the Great Plains in eastern Colorado. The first arrivals for P_g give a well-determined upper crustal velocity of 6.1 km/sec. The first arrivals for P_n are weak but recognizable beginning 210 km from the source. Refractions from the layer of 6.7-km/sec velocity possibly appear as first arrivals over a very short distance range near 200 km. Certain secondary phases can be interpreted as reflections from the top of the 6.7-km/sec layer and the 8.0-km/sec layer, but they do not differ greatly from other unexplained phases. Reprinted from *Jackson et al.* [1963].

occurs at the surface, it seems logical to associate this composition with a thick lower crustal layer. Because of the thinness or absence of a granitic part of the crust, these areas may be regions where basalt has intruded into a rift in the granitic crust.

Another type of anomalous crustal structure is revealed by attempts to fit the earth's gravitational field with a density model based on seismic crustal studies. If the density of the mantle is constant, and if the densities of rocks in the crust are related to seismic velocities by a velocity-density law, the variations in crustal structure should produce variations in the gravity field. However, there are several locations where marked variations in crustal structure do not produce commensurate variations in the gravity field. This phenomenon has been observed in Mississippi near the Gulf Coast [*Warren et al.*, 1966], in Colorado at the boundary between the Rocky Mountains and the Great Plains, in Nevada at the boundary between the Basin and Range province and the Snake River Plain [*Hill and Pakiser*, 1966], and in Arizona at the boundary between the Colorado Plateaus and the Basin and Range province. A comparison of the crustal model [*Smith et al.*, 1966] and the gravity field [*Weber and Goodacre*,

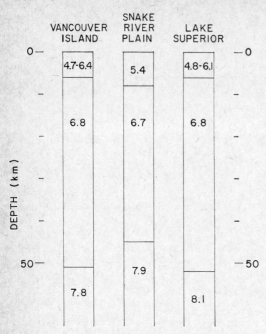

Fig. 7. Average crustal columns from Vancouver Island [*White and Savage*, 1965], the Snake River Plain [*Hill and Pakiser*, 1966], and Lake Superior [*Smith et al.*, 1966]. Numbers are average P-wave velocities (kilometers per second). In these regions the upper crust tends to be thin and to exhibit low velocity. Mafic material appears to compose most of the crust.

1966] at Lake Superior suggests that a similar problem exists in the Lake Superior region.

Arizona furnishes a particularly good example of this phenomenon because the seismic coverage is dense enough to delineate the crustal structure in some detail, and because two profiles made at approximately right angles give three-dimensional control. One profile crosses from the Basin and Range province to the Colorado Plateaus province (Figure 8). The measured Bouguer anomaly, free-air anomaly, and surface altitude are compared with the Bouguer anomaly calculated from crustal structure (Fig. 9). The measured and computed Bouguer anomalies differ by 150 mgal along the refraction profile. The change of measured Bouguer anomaly is not sufficient to account for the magnitude of the change in crustal structure indicated by the seismic measurements. The differences between the measured and calculated Bouguer anomalies can be explained by density variations in the upper mantle that would be difficult to detect by seismic measurements alone. One possible model to explain the measured gravity can be constructed by assuming that under the upper mantle layer, which has a velocity of 7.85 km/sec, there is a layer of higher velocity and higher density. The difference between the measured and computed Bouguer gravity (Figure 9) was smoothed, and the depth variations necessary to match the smoothed residual anomaly were computed (Figure 10).

The Arizona profile illustrates that reasonable variations in velocity and density in the upper mantle, together with variations in crustal structure, can explain the regional gravity field measured at the surface.

This type of anomalous structure occurs in zones of rapidly changing crustal structure, including some of the transition zones between long-recognized geologic provinces. It appears that major changes in the crust are accompanied by changes in the structure of the upper mantle that tend to bring the local regions of anomalous crustal structure closer to isostatic equilibrium.

LAKE SUPERIOR EXPERIMENTS

The Carnegie Institution of Washington coordinated a large cooperative international seismic experiment centered in Lake Superior in 1963 [*Steinhart*, 1964]. One of several surprises from this experiment was the transmission of seismic energy from 1-ton shots to distances of more than 2000 km. This unusual efficiency apparently results from a combination of factors, including the water depth of about 600 feet and a relatively hard bottom. The possibility of using shot points in Lake Superior to record very long seismic refraction profiles was recognized immediately. The first long-offset work was done by the Geological Survey in 1964 [*Roller and Jackson*, 1966]; the results led to another experiment, Project Early Rise. Thirty-eight 5-ton shots were detonated in the lake; groups from North American universities and government agencies participated in a recording program along profiles radiating in many directions from Lake Superior (Figure 11). The data from this experiment are currently being analyzed. Preliminary results show a refracted arrival with an apparent velocity of 8.5 km from a layer at a

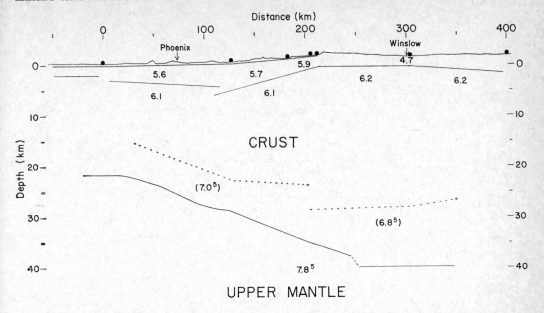

Fig. 8. Profile of crustal structure in central Arizona, extending from the Basin and Range province in the southwest (left) to the Colorado Plateaus province in the northeast (right). Solid circles show locations at which explosions were fired. Numbers represent the velocity (kilometers per second) in the crust and upper mantle layers. A lower crustal layer is present but is poorly defined; dashed lines show its upper permissible limits. The crustal thickness changes by almost a factor of 2: from 22 km in the Basin and Range to 42 km in the Colorado Plateaus.

depth of about 100 km. Variations in the distance at which this refracting arrival emerges as a first arrival indicate a variable depth for the refracting horizon. On the profile between Lake Superior and Colorado (Figure 12), the phase emerges as a first arrival at about 900 km. This refracting layer has been reported from an analysis of the data of the 1964 experiment [Roller and Jackson, 1966] and from an analysis of seismograms of earthquakes [Carder et al., 1966.]

DISCUSSION

In the last five years explosion seismology has led to substantial improvements in our understanding of the earth's crust and upper mantle. Enough data are available to make practical a discussion of variations in crust and upper mantle structure on the continents. Some of the earlier differences of opinion about the structure of the crust resulted from disagreements among investigators who were studying different regions of the earth's crust and who had assumed that the properties of the earth's crust were universal. Variations are illustrated by the extensive measurements of crustal structure that have been completed within the TGS (Figures 1 to 4). This work has brought us to a point where it is now possible to study regions of anomalous crustal structure and to relate such regions to processes acting within the crust and upper mantle.

Perhaps most important is the regional variation of the velocity of the phase P_n in the western and the eastern United States. This regional variation in the upper mantle must have great tectonic significance. A comparison of crustal thickness, surface altitude, and measured gravity shows that isostatic compensation must involve the mantle as well as the crust; it is therefore impossible to use observed gravity to calculate crustal thickness, as has been attempted many times in the past. The apparent correlation of tectonic activity and heat flow with lower velocity in the upper mantle also suggests that we are observing a process within the upper mantle.

The discovery of regions of the crust that

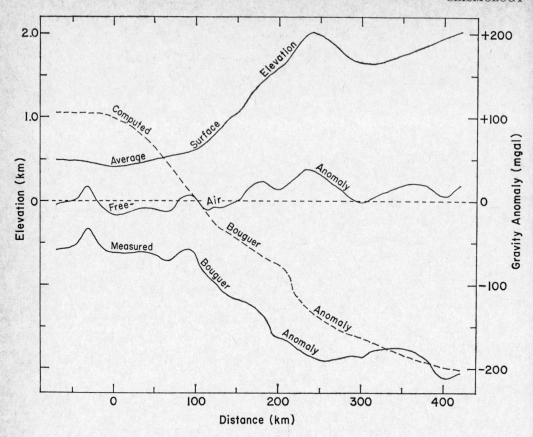

Fig. 9. Measured and computed gravity anomalies in central Arizona. The starting point of the comparison is the measured Bouguer anomaly. The smoothed, average surface altitude was used to arrive at a regional free-air anomaly. The free-air anomaly is near zero; the region is in approximate isostatic equilibrium. A velocity-density relation was assumed to compute the Bouguer anomaly from the crustal model of Figure 8. The computed anomaly was arbitrarily set equal to the measured anomaly at the right-hand (Colorado Plateaus) end of the profile; the deviation between the measured and the computed anomalies suggests that the model in Figure 8 has a mass deficiency under the Colorado Plateaus province relative to the Basin and Range province to the southwest.

are made up almost entirely of mafic material may also be very significant. Together with the discovery that regions of recent tectonic activity may have sharper discontinuities within the crust, these observations suggest that the mobility of the lower crust may play a large role in tectonics.

The lower crust may also become immobile, as is illustrated at Lake Superior where a major density anomaly has apparently existed since Precambrian time. A striking contrast is shown between the crustal structure at Lake Superior, which has been frozen for almost a billion years, and the crustal structure in the tectonic regions of the West, where the intermediate layer has been extremely mobile, as is illustrated at Vancouver Island and in the Snake River Plain. Apparently the mechanical properties of the crust vary greatly across North America; these differences may be related to variations in the temperature of the crust. If the crust exceeds some critical temperature it may become mobile, whereas if the temperature of the crust is below some critical temperature, density anomalies can be maintained for long periods of time.

Seismic observations have shown that models

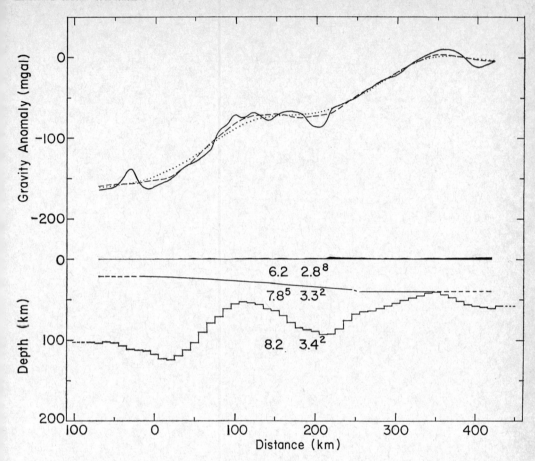

Fig. 10. Possible variation in the upper mantle for central Arizona. The upper curves are Bouguer gravity anomalies, as follows: the solid line is the difference between the measured and computed Bouguer anomalies of Figure 9, the dashed line is a smoothed version of the solid line, and the dotted line is the anomaly that would be removed by the model in the lower part of the figure. An upper mantle interface was assumed and arbitrarily set at a 60-km depth at the right-hand (Colorado Plateaus) end of the profile. In the model the numbers to the left represent velocity (kilometers per second), and the numbers to the right represent density (grams per cubic centimeter).

of isostatic equilibrium involving the crust alone cannot be applied to all regions. This raises the question: Where is isostatic equilibrium brought about? If the crust is mobile, isostatic equilibrium may be accomplished within the crust itself, but for regions where the crust is not mobile, isostatic equilibrium must be accomplished at some depth within the mantle. The dissemination of the earth's mantle under viscous flow could account for the degree of isostatic equilibrium observed at the surface, but this mechanism does not explain the uplift and depression of large areas of the earth's crust that have occurred during geologic time. Some active mechanism is required to account for regional vertical changes. One such mechanism could be a pressure-temperature sensitive phase at a depth of about 100 km within the mantle. Thus, the capability of recording long-range refraction profiles on the continents and the recent discovery of a layer within the mantle at a depth of about 100 km seem most significant. The nature of this layer and its relation to the low-velocity zone in the upper mantle may be one of the major factors in the tectonic mechanism.

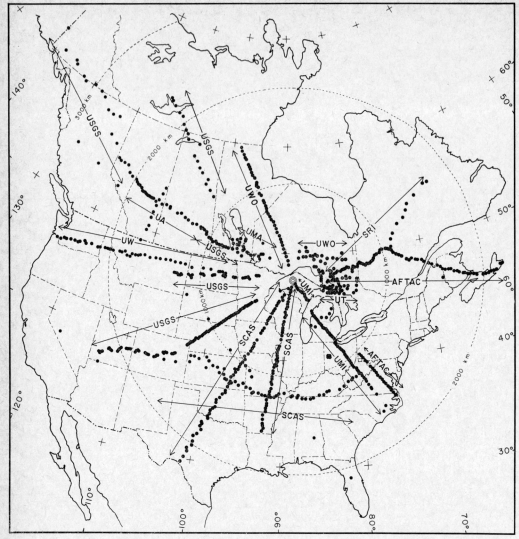

Fig. 11. Location of recording stations for Project Early Rise. Shots were fired from the same location in Lake Superior. Groups that recorded are indicated by the following code: AFTAC, Air Force Technical Applications Center; UA, University of Alberta; USGS, U. S. Geological Survey; UMA, University of Manitoba; UMI, University of Michigan; SCAS, Southwest Center for Advanced Studies; SRI, Stanford Research Institute; UT, University of Toronto; UWO, University of Western Ontario; UW, University of Wisconsin.

SUMMARY

Extensive new seismic measurements have supported many traditional geologic concepts and have made it possible to study regional variations of crustal and upper mantle structure in North America. Most areas contain a silicic upper crust in which the P-wave velocity ranges between 5.9 and 6.2 km/sec. A more mafic lower crust is present everywhere, although its properties appear to be much more variable. The P-wave velocity of this lower layer is 6.6 to 7.1 km/sec. In some regions a second deep crustal layer or zone with a P-wave velocity of 7.1 to 7.4 km/sec seems to lie just above the M discontinuity. The P-wave velocity in the uppermost mantle ranges from 7.8 to 8.3 km/

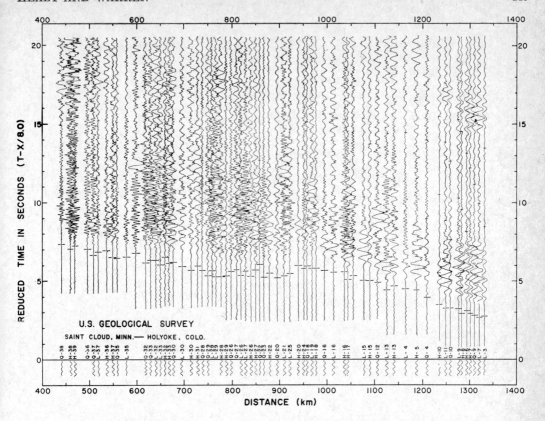

Fig. 12. Seismograms from shots fired in Lake Superior for Project Early Rise, recorded toward Colorado. Each trace was selected from an array of seismometers. The entire arrays were used to detect the first arrivals shown by dashes. The apparent velocity increases abruptly just beyond 900 km.

sec; higher upper mantle velocities predominate in eastern North America. Anomalous crustal and upper mantle structures have been discovered. In some regions the crust is composed almost entirely of mafic material. In some of the major transitions between geologic provinces and other areas of rapidly changing crustal structure, the measured gravity field cannot be explained without postulating local density variations in the upper mantle. The extension of explosion techniques to very long range refraction profiles has opened the possibility of exploring the detailed structure of the upper mantle; these experiments have confirmed the existence of a transition zone in the upper mantle at a depth of about 100 km.

Explosion seismology research during the past five years has been largely directed toward the descriptive aspects of the earth's crust and upper mantle. Emphasis is now shifting to attempts to learn about the processes acting within the crust and upper mantle through study of the details of crustal and upper mantle structure. The large quantity of data that has recently accumulated, particularly from the Lake Superior experiments, is now being analyzed by several research groups in the United States and Canada. Those involved in these efforts cannot help being excited over the possibility of major new understanding of the tectonics of North America. Research is now proceeding rapidly enough to ensure that any review, such as this one, will be obsolete within a few years.

REFERENCES

Carder, D. S., D. W. Gordon, and J. N. Jordan, Analysis of surface focus travel times, *Bull. Seismol. Soc. Am.*, 56(4), 815–840, 1966.

Hill, D. P., and L. C. Pakiser, Crustal structure between the Nevada test site and Boise, Idaho, from seismic-refraction measurements, in *The Earth beneath the Continents, Geophys. Monograph 10,* edited by J. S. Steinhart and T. J. Smith, pp. 391–419, American Geophysical Union, Washington, D. C., 1966.

Jackson, W. H., and L. C. Pakiser, Seismic study of crustal structure in the southern Rocky Mountains, in *Geological Survey Research, U. S. Geol. Surv. Prof. Paper 525-D,* pp. 85–92, 1965.

Jackson, W. H., S. W. Stewart, and L. C. Pakiser, Crustal structure in eastern Colorado from seismic-refraction measurements, *J. Geophys. Res., 68*(20), 5767–5776, 1963.

James, David E., and John S. Steinhart, Structure beneath the continents: A critical review of explosion studies 1960–65, in *The Earth Beneath the Continents, Geophys. Monograph 10,* edited by J. S. Steinhart and T. J. Smith, pp. 293–333, American Geophysical Union, Washington, D. C., 1966.

McConnell, R. K., Jr., R. N. Gupta, and J. T. Wilson, Compilation of deep crustal structure seismic refraction profiles, *Rev. Geophys., 4*(1), 41–100, 1966.

Pakiser, L. C., and J. S. Steinhart, Explosion seismology in the Western Hemisphere, in *Research in Geophysics, Vol. 2, Solid Earth and Interface Phenomena,* edited by H. Odishaw, pp. 123–147, chapter 5, MIT Press, Cambridge, 1964.

Pakiser, L. C., and Isidore Zietz, Transcontinental crustal and upper-mantle structure, *Rev. Geophys., 3*(4), 505–520, 1965.

Roller, J. C., and W. H. Jackson, Seismic wave propagation in the upper mantle: Lake Superior, Wisconsin, to Central Arizona, *J. Geophys. Res., 71*(24), 5933–5941, 1966.

Smith, T. J., J. S. Steinhart, and L. T. Aldrich, Lake Superior crustal structure, *J. Geophys. Res., 71*(4), 1141–1172, 1966.

Steinhart, J. S., Lake Superior seismic experiment —Shots and travel times, *J. Geophys. Res., 69* (24), 5335–5352, 1964.

Steinhart, J. S., Explosion studies on the continents, *Trans. Am. Geophys. Union, 48*(2), 412–415, 1967.

Warren, D. H., J. H. Healy, and W. H. Jackson, Crustal seismic measurements in southern Mississippi, *J. Geophys. Res., 71*(14), 3437–3458, 1966.

Weber, J. R., and A. K. Goodacre, A reconnaissance underwater gravity survey of Lake Superior, in *The Earth beneath the Continents, Geophys. Monograph 10,* edited by J. S. Steinhart and T. J. Smith, pp. 56–65, American Geophysical Union, Washington, D. C., 1966.

White, W. R. H., and J. C. Savage, A seismic refraction and gravity study of the earth's crust in British Columbia, *Bull. Seismol. Soc. Am., 55* (2), 463–486, 1965.

Seismic Model of the Atlantic Ocean[1]

John Ewing

A CONCERTED effort to study the crustal structure of the Atlantic Ocean by explosion seismology began shortly after the end of World War II; by 1960 well over 100 refraction profiles had been recorded in various parts of the deep basins. A comparable number of measurements were made on the continental shelves and in the mediterranean seas. Most of these measurements have been reported in the scientific literature, and compilations of the data have been published in various forms [*Hill*, 1957; *McConnell et al.,* 1966; *Lee and Taylor,* 1966].

The first deep water refraction profile in the Atlantic [*Ewing et al.,* 1950] produced a highly significant result: it demonstrated that the oceanic crust is only 1/4 to 1/3 as thick as the average continental crust. We have confirmed this with many subsequent observations and have determined that the average crustal section in ocean basins is approximately that pictured in the central part of Figure 1. Although the section shown is a generalized one for the North American basin, there are no strong indications that the other Atlantic basins differ significantly.

Though indicating almost monotonous thick-

---

[1] Contribution 1236 from the Lamont-Doherty Geological Observatory of Columbia University.

nesses of the two deeper crustal layers, the refraction profiles demonstrated a marked variation in the thickness of sediments from one part of the basin to another. Owing to the averaging effect of the refraction method, however, it was not possible to measure local variations in sediment thickness or to observe details of its stratification until the techniques and equipment for continuous reflection profiling were developed. This occurred early in the present decade, and the operation of seismic 'profilers' has been so rewarding that substantially more effort has since gone into reflection than into refraction studies. Present profilers, operated essentially as recording echo sounders, are capable of measuring with remarkable detail the total thickness and stratification of the low-velocity sediment layers; the measurements can be made at normal cruising speeds of the survey ships (10 to 12 knots). The amount of data in hand is now measured in hundreds of thousands of kilometers of traverse in the Atlantic alone.

THE MANTLE

The mantle under the oceans is recognized, as it is under continents, by a compressional wave velocity of about 8 km/sec. Velocities appreciably above and below this average value have been recorded, but if we delete measurements near such features as oceanic ridges, large seamounts, or islands, and continental margins, where mantle velocities tend to be low, a velocity histogram shows that 80% of the values fall between 7.7 and 8.3 km/sec. There is no obvious relationship between upper mantle velocity and depth or between mantle velocity and thickness of overlying crustal material. However, there does appear to be an inverse relationship between mantle depth and water depth. In the Atlantic, no specific experiments have been conducted to test for anisotropy, but where such tests have been made north of Hawaii in the Pacific, preliminary analysis indicates a measurable variation in velocity with variation in azimuth of shot line (see section by Raitt, p. 250, this monograph).

The average depth of the M discontinuity in the Atlantic is 12 km, as Figure 1 shows. One of the most significant deviations from this average is found on the flanks of the Mid-Atlantic ridge, where mantle velocities have been recorded at depths of the order of 9–10 km. Despite the recognized difficulties of obtaining accurate measurements in the rough topography of the ridge province, this shallow mantle is strongly enough indicated to be considered a significant feature of the ridge structure [*Le Pichon et al.*, 1965]. Also characteristic of the Mid-Atlantic ridge is the apparent disappearance of the M discontinuity in the crestal zone. Both explosion and earthquake seismology indicate a thick layer of intermediate velocity in the place of the normal oceanic crust and upper mantle [*Ewing and Ewing*, 1959; *Tryggvason*, 1962].

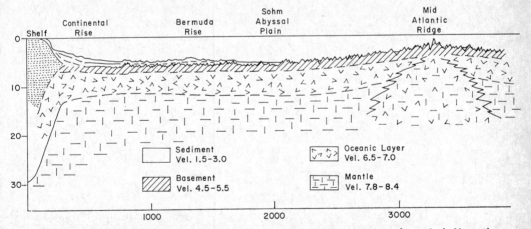

Fig. 1. Generalized structure section from the North American continental shelf to the Mid-Atlantic ridge. Horizontal and vertical scales in kilometers; velocities in kilometers per second.

THE OCEANIC LAYER (LAYER 3)

The main crustal layer under the oceans is so uniform in compressional wave velocity and thickness that it can justifiably be given a name. Besides oceanic layer, it is often called the third layer, or layer 3. Owing to its shallow depth and appreciable thickness, refracted waves from the oceanic layer usually appear as first arrivals over 10 to 15 km of the profile length, and in properly shot profiles its velocity can be accurately measured. A histogram of velocities in Atlantic basins shows 80% of the values falling between 6.5 and 7.1 km/sec. Some substantially lower velocities have been reported, but many of these are from profiles with insufficient shot points, in which arrivals from overlying, lower velocity layers were probably included with oceanic layer arrivals in the travel-time interpretations.

The average thickness of the oceanic layer is about 5 km. Except on the Mid-Atlantic ridge, there is a tendency for areas of thick oceanic layer to correspond to topographic highs. This is particularly noticeable in the Caribbean Sea. The oceanic layer thins on the upper flank of the Mid-Atlantic ridge and appears to merge with the mantle in the crestal zone where the intermediate velocity is found. The anomalous structure associated with the ridge crest is generally assumed to be some consequence of the volcanic outpourings and intrusions, possibly related to upwelling convection currents. According to the ocean floor spreading hypothesis of *Hess* [1962], the crestal zone is the site of the material upwelling from deep in the mantle, which then spreads laterally away from the crest and develops the stratification into mantle and crustal layers. The ridge structure is approximately compensated isostatically [*Talwani et al.*, 1965], and so the intermediate crustal material must be of appreciable thickness if we assume that its density, like its velocity, is intermediate between that of the mantle and that of the oceanic layer.

THE BASEMENT (LAYER 2)

Above the oceanic layer is a layer in which the compressional wave velocity is between 4.5 and 5.5 km/sec. Although more difficult to measure, this layer is apparently as basic a component of the oceanic crust as is the oceanic layer. Its widespread occurrence was first demonstrated in the Pacific [*Raitt*, 1956; *Gaskell et al.*, 1958], where the combination of water depth and sediment thickness is more favorable for its detection. In the Atlantic, most of the early refraction profiles were shot in deep areas with thick sediment accumulation, and the basement layer was not observed in the refraction data because of masking by the reflected arrivals from the sea floor and by refracted arrivals from the oceanic layer. Its presence is strongly indicated, however, by comparison of reflection with refraction data and by evidence from transformed shear waves [*Ewing and Ewing*, 1959]. In the shallower areas of the Atlantic it is observed by first arrival refracted waves, as it is in much of the Pacific Ocean. The average thickness of this layer, throughout the world's oceans, is close to 1.5 km.

Although we know only the general form of the topography of the mantle and of the oceanic layer, we can map the topography of the basement with considerable detail by reflection techniques. Our inability to map the deeper interfaces in a similar manner is probably due to a combination of smaller impedance contrasts and greater attenuation. It is also possible that the transitions between basement, oceanic layer, and mantle are more gradual than that between sediments and basement.

Reflection seismology has shown that the top of the basement is characteristically rough, both on a regional and on a local scale. The regional roughness is typically expressed by undulations in the basement topography with wavelengths of the order of 5–50 km (excluding major physiographic features). The small scale roughness is evidenced by noncoherent echo patterns for wavelengths of the order of 15–150 meters. Reflections from the basement are usually quite strong, except in some areas near continents where the older sediments may be so compacted and lithified that their acoustic impedance is near that of the basement.

The composition of the basement layer has been the subject of much speculation. It is most commonly judged to be composed of basaltic volcanic rock on the basis of samples cored and dredged from seamounts and certain areas of the sea floor that are devoid of sedimentary cover. There is still reason to give serious consideration to the thesis developed by *Hamil-*

ton [1959] that this layer is composed of old, compacted sediments that may be extensively intruded by volcanic rock of comparable seismic velocity and acoustic impedance. Accepting this hypothesis, we would have to assume that the dredging always samples the intruding, and not the intruded, rock.

THE SEDIMENTS (LAYER 1)

Because of the concentration of effort in seismic profiling, we have added more in recent years to our knowledge of the sediments than to that of the deeper layers. Not only does the seismic profiler give us the means to measure rapidly and in considerable detail the distribution and 'seismic stratigraphy' of the sediments, but also it serves as a most valuable guide to sampling. Detailed surveys of the sea floor have indicated faults or erosional areas where layers of sediments that are normally deeply buried are exposed for sampling. Thus the seismic profiler and the sediment corer used jointly are particularly effective tools in the study of marine geology. In addition to recording thickness and stratification of the sediments, the profiler also provides useful information about the processes by which sediments are transported and deposited. For example, there is strong evidence that deep sea currents not only may control deposition of sediments but also may erode those already deposited. They may also be active agents in the building of major topographic features such as the Blake-Bahama outer ridge [*Ewing et al.*, 1966a].

The major sediment accumulations are found at the foot of the continental slopes, consisting essentially of very large wedge-like bodies of well stratified, acoustically opaque material often several kilometers thick. There is considerable difficulty in this area in distinguishing between the oceanic basement, the continental basement, and the deeper sediment layers. The shallower sediments appear to have been deposited in coalescing outwash fans or aprons, perhaps partially shaped by along shore deep currents as suggested by *Heezen et al.* [1966]. Unconformities are common features in these deposits, apparently caused in some places by erosion, in others by slumping, and in still others by sediment ponding.

Seaward from the continental slope and rise, we find a variety of sediment types and forms. Figure 2 is a reproduction of the records from a traverse in an area about 500 km north of Bermuda showing the characteristic appearance of the basement, homogeneous sediments, and stratified sediments. The layers constituting a large percentage of the total volume of the basin deposits are relatively homogeneous. They appear in profiler records (e.g., Figure 2) as layers with few internal reflectors and are remarkably 'transparent' to the seismic waves. From comparison of seismic with core evidence, these layers are judged to be composed mainly of lutite. It is clear that the homogenous sediments shown in Figure 2 were deposited under dynamic conditions; otherwise, they would be more uniform in thickness. The bottom homogeneous layer filled in the large depressions in the basement and formed a moderately level sea floor, apparently having been transported from their source, distributed, and deposited largely under the influence of gravity. The variations in thickness of the upper homogeneous layer appear to have resulted from deposition controlled by deep currents or from erosion.

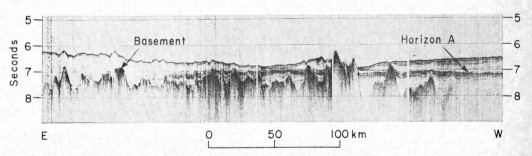

Fig. 2. Seismic profiler section in the North American basin north of Bermuda. Vertical scale is reflection time; each second corresponds approximately to 0.75 km in water and 1.0 km in sediment.

In other areas, generally farther removed from continents, sediments similar to these in acoustic properties have essentially blanketed the basement topography, indicating the absence of deep currents during the deposition and also indicating that such areas were not accessible to turbidity or density currents flowing out from the continents. Some parts of the Mid-Atlantic ridge are uniformly blanketed by sediments of this type, but more commonly most of the sediments are ponded in the depressions, and the intervening peaks or ridges are essentially bare.

The opaque zone between the two principal homogeneous layers, horizon A in Figure 2, is a widespread sub-bottom reflector in the Atlantic basins. Its acoustic properties closely resemble those of the turbidites that form the modern abyssal plains, and this has led to speculation that it is the surface of a fossil abyssal plain of very large areal extent. Recent sampling in an area where this reflector crops out on the sea floor has confirmed that it is a turbidite layer of upper Cretaceous age [*Ewing et al.*, 1966b].

In summary, the sediments of the Atlantic Ocean fit the following generalized description. The main factor governing the thickness of accumulation is distance from a well-drained continental margin. Most of the deep basin sediments are probably terrigenous and most of them, particularly the coarser fraction, are deposited relatively near the continental margin. The finer sediments are transported in nepheloid layers [*Ewing and Thorndike*, 1965], and their final place of deposition is influenced both by gravity flow and by deep sea currents. Hence, the thickness of such deposits will depend strongly on topographic setting and on oceanographic conditions responsible for deep water transport. In addition to the lutites, some coarse sediment may be transported far from its source by large-scale turbidity currents. Transport by turbidity currents may have either a smoothing or an erosive effect on the topography, probably erosive near the source and smoothing at greater distances. Transport in nepheloid layers probably results in smoothing of existing topography in areas where deep currents are weak or absent, but otherwise significant depositional features may be built. The amount of organic material deposited obviously depends on productivity and to some extent on the amount dissolved by the sea water, generally considered to be a function of depth. Because sedimentation in the Atlantic basins is dominated by accumulation of terrigenous, rather than biotic, material, it is difficult to detect a correlation between sediment thickness and biological productivity, but there does appear to be some such correlation on the Mid-Atlantic ridge, sediments being generally thicker in equatorial and high latitudes than in midlatitudes. Particularly in areas of low biological productivity, the ridge flanks have only a modest amount of sediment covering, generally less than 200 meters. A strip 100–200 km wide along the crest, even in the regions of high productivity, is essentially barren of sediments, indicating removal by currents, burial by recent lava flows, spreading of the crust away from the ridge axis, or some other process not yet recognized.

Acknowledgments. The data summarized here are mainly those collected by the scientific staffs of the Lamont Geological Observatory, Woods Hole Oceanographic Institution, and the Department of Geodesy and Geophysics of Cambridge University. The Lamont program has been supported by the U. S. Navy, Office of Naval Research and Bureau of Ships, and by the National Science Foundation under contracts Nonr 266 (48), Nobsr 85077, NSF-Y/914/77, and NSF GA-550.

REFERENCES

Ewing, J., and M. Ewing, Seismic refraction measurements in the Atlantic Ocean basins, in the Mediterranean Sea, on the Mid-Atlantic ridge, and in the Norwegian Sea, *Bull. Geol. Soc. Am.*, 70, 291–318, 1959.

Ewing, J., M. Ewing, and R. Leyden, Seismic-profiler survey of the Blake Plateau, *Bull. Am. Assoc. Petrol. Geologists*, 50, 1948–1971, 1966a.

Ewing, J., J. L. Worzel, M. Ewing, and C. Windisch, Ages of horizon A and the oldest Atlantic sediments, *Science*, 154, 1125–1132, 1966b.

Ewing, M., and E. M. Thorndike, Suspended matter in deep ocean water, *Science*, 147, 1291–1294, 1965.

Ewing, M., J. L. Worzel, J. B. Hersey, F. Press, and G. R. Hamilton, Seismic refraction measurements in the Atlantic Ocean basin, 1, *Bull. Seismol. Soc. Am.*, 40, 233–242, 1950.

Gaskell, T. F., M. N. Hill, and J. C. Swallow, Seismic measurements made by HMS *Challenger* in the Atlantic, Pacific and Indian oceans, and in the Mediterranean Sea, 1950–1953, *Phil. Trans. Roy. Soc. London, A*, 251, 23–83, 1958.

Hamilton, E. L., Thickness and consolidation of deep sea sediments, *Bull. Geol. Soc. Am., 70,* 1399–1424, 1959.

Heezen, B. C., C. D. Hollister, and W. F. Ruddiman, Shaping of the continental rise by deep geostrophic contour currents, *Science, 152,* 502–508, 1966.

Hess, H. H., in *Petrologic Studies,* edited by A. E. J. Engel et al., Geological Society of America, New York, 1962.

Hill, M. N., Recent geophysical exploration of the ocean floor, *Phys. Chem. Earth, 2,* 129–63, 1957.

Lee, W. H. K., and P. T. Taylor, Global analysis of seismic refraction measurements, *Geophys. J., 11,* 389–413, 1966.

Le Pichon, X., R. E. Houtz, C. L. Drake, and J. E. Nafe, Crustal structure of the mid-ocean ridges, *J. Geophys. Res., 70,* 319–339, 1965.

McConnell, R. K., R. N. Gupta, and J. T. Wilson, Compilation of deep crustal seismic refraction profiles, *Rev. Geophys., 4,* 41–100, 1966.

Raitt, R. W., Seismic-refraction studies of the Pacific Ocean basin, 1, Crustal thickness of the central equatorial Pacific, *Bull. Geol. Soc. Am., 67,* 1623–1640, 1956.

Talwani, M., X. Le Pichon, and M. Ewing, Crustal structure of the mid-ocean ridges, *J. Geophys. Res., 70,* 341–352, 1965.

Tryggvason, E., Crustal structure of the Iceland region from dispersion of surface waves, *Bull. Seismol. Soc. Am., 52,* 359–388, 1962.

Explosion Seismic Refraction Studies of the Crust and Upper Mantle in the Pacific and Indian Oceans

George G. Shor, Jr., and Russell W. Raitt

SUMMARIES of seismic refraction data on the crust and mantle under the oceans were published by *Raitt* [1963] and *Ewing* [1963]; since that time, considerable additional data have been obtained in the Pacific Ocean, and the body of data in the Indian Ocean has been increased manyfold. In order to derive averages for the 'normal' ocean basins, Raitt excluded all stations in water depths less than 3.0 km, as well as all stations in anomalous areas such as trenches and oceanic rises. Enough differences in structure or velocity can be found in many areas to justify excluding them from the 'normal,' but not enough data have yet been obtained to permit detailed mapping of the oceans by structural provinces. As a result, the present review makes no attempt to define the 'normal' stations, and data for all structural provinces are included where the water depth exceeds 2 km.

Most of the refraction studies performed in the Pacific and Indian Oceans have been made by the Scripps Institution of Oceanography. Though an increasing number of data have been obtained by the Lamont Observatory, Japanese groups, the University of Hawaii, and groups in the USSR, they have not been included here, since they are not all accessible to the authors, and would probably not change the statistics greatly. The Scripps data, both published and unpublished, are shown in the accompanying figures. The locations of the 229 stations analyzed are shown in Figure 1.

VELOCITIES

Figure 2 shows the velocities of all layers recorded on all Scripps stations in water depths greater than 2000 meters. The values shown are the computed 'true velocities' for reversed or split profiles, the observed velocities for single-ended profiles. Determinations by refraction methods of velocity in the sedimentary layer are few. The sedimentary layer is generally so thin that, when observations are made at the sea surface, the waves through the sedimentary layer appear later on the record than arrivals from the deeper layers. In addition, there is good evidence for variation of velocity

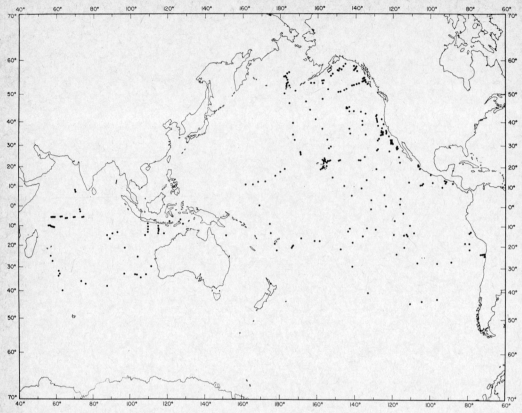

Fig. 1. Location of Scripps refraction stations in water depths greater than 2000 meters.

with depth within the sediments, so that the arrivals, if recorded, would form a short cuspate travel-time plot, observable over a short distance. The principal exceptions occur in areas that have a considerable thickness of land-derived sediments, such as the Gulf of Alaska and the Bering Sea. In such areas, refracted arrivals from within the sedimentary section are detected and can be used to derive a velocity. Unfortunately, such values cannot be considered typical of the sedimentary section in the oceans as a whole. Recent wide-angle reflection studies have shown that the velocity in deep-sea sediments increases from about 1.5 km/sec at the sea floor at a rate near 1.0/sec. An average velocity for the normal deep-sea sediments would be near 1.65 km/sec.

The next layer below the sediments, called the 'second layer' to avoid petrological implications, has a wide scatter of velocity values. These are partly due to the inaccuracy of determination, since this layer is observed as a first arrival over only a short distance if at all. In twenty-two of the stations, no second layer was observed; if it was present, it would cause a misinterpretation of the thickness of the sediment above. Second-layer velocities cover the range from 3.4 to about 6.0 km/sec; in one case (off the north coast of the Hawaiian Islands) there are definite indications of two layers within this section, with velocities near 4 and $5\frac{1}{2}$ km/sec. This situation may occur in other areas and be undetected. It is probable that this layer includes both older sediments and rock of igneous origin.

The 'oceanic layer' has the least scatter in velocity. Values are closely grouped around 6.8 km/sec. The scatter is little greater than the presumed accuracy of determination, suggesting that this layer is of uniform composition and physical properties throughout the ocean basins.

The mantle, with median velocity 8.1 km/sec, has a slightly greater scatter of velocity than

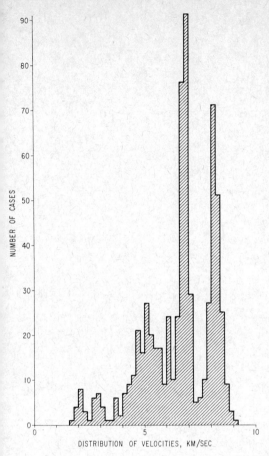

Fig. 2. Distribution of velocities at all stations. True velocities are used from reversed and split profiles, observed velocities from single-ended profiles.

the oceanic layer, although the accuracy of determination of mantle velocity is approximately the same as that for the crust. An appreciable part of the greater variability of mantle velocity may be caused by anisotropy of velocity in the mantle. Recent field measurements of the amount of the anisotropy are discussed in another section of this Monograph by Raitt (p. 250).

LAYER THICKNESSES

The sediments are generally thin; almost all Pacific and Indian Ocean values are less than 1 km, and the mode is at 300 meters (Figure 3). A few areas show practically no sediments: north of Hawaii the thickness becomes vanishingly small on the Hawaiian arch (an area probably covered by lava flows from the Hawaiian ridge). Most of the thicker sediments are found in the Gulf of Alaska and the Bering Sea, where the abyssal plains are underlain by thick sediments derived from the coast of Alaska. A few stations with thick sediments have been found in the northern part of the Indian Ocean. There are undoubtedly many other areas of thick sedimentary cover along the edges of the ocean basins; in the central parts of the basin only the thickening in the carbonate zones along the equator is found.

The median for the second layer is 1.4 km, but the thickness is extremely variable (Figure 4). Some of this variability probably comes from the inaccuracies of determination of the layer velocity; much of it is probably real.

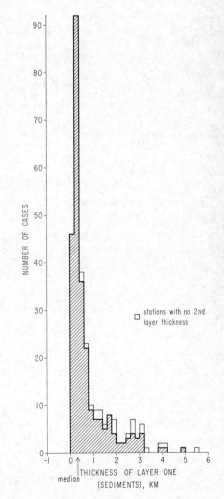

Fig. 3. Thickness of sediments, oceanic stations.

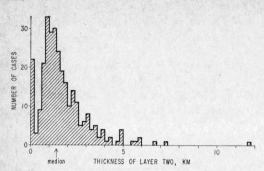

Fig. 4. Thickness of second layer, oceanic stations.

There is evidence from reflection profiling work that material assigned to layer two in refraction work in some cases has visible internal stratification and represents older sedimentary horizons. Layer two, as observed, sometimes undoubtedly represents relatively recent lava flows (as at the Guadalupe experimental Mohole site, and probably on the Hawaiian arch); in such cases probably older sediments are concealed beneath the flows. The nature of layer two continues to be one of the intriguing problems of oceanic crustal structure; its composition may vary from place to place. In a few places two layers have been found between the sediments and the oceanic crustal layer.

The oceanic layer has a median thickness of 4.7 km (Figure 5). Most of the determinations fall between 2.4 and 7.6 km thickness, although a significant number of stations have a greater layer thickness. Thick oceanic layers have been found beneath trenches, at the foot of the slope of insular ridges (such as the Aleutian and Hawaiian ridges), and near the edges of some enclosed basins (such as the Coral Sea). A thin oceanic layer occurs on the crest of the East Pacific rise, the outer ridge seaward of some of the trenches, and on the Hawaiian arch. Island ridges seem to have a welt of thickened oceanic layer beneath the ridge itself and extending out beneath the ocean floor on either side.

The median depth to the Mohorovicic discontinuity (Figure 6) is 11.1 km; the average would be somewhat deeper, owing to the influence of the trench stations. The error in determination of any mantle depth is at least 0.5 km, and in some stations may be more. In the plot of depth to Mohorovicic discontinuity against sea depth, for all stations in water 2000 meters deep or more (Figure 7), a slight correlation of water depth to mantle depth is suggested, but in the wrong direction: the deeper the water, the deeper the mantle. A crust-mantle density contrast of 0.5 g/cm^3 explains adequately the differences in structure between continents and oceans, but it does not seem to account for the variations in ocean depths. There must be lateral variation in the density of the upper mantle under the ocean basins.

DIFFERENCES BETWEEN PACIFIC AND INDIAN OCEANS

No significant differences appear between the structures and velocities of the Pacific and Indian oceans on the average, other than those that can be ascribed to varying percentages of geomorphic features such as abyssal plains, rift zones, abyssal hills, trench areas, and so on. The average velocities in the two oceans are:

	Pacific and Indian	Indian Only
Layer two	5.16 ± .68	5.19 ± .64
Oceanic layer	6.77 ± .22	6.73 ± .30
Mantle	8.15 ± .30	8.14 ± .18

It can be seen that the values for the Indian Ocean lie well within the variance of the data as a whole, so that they can be considered as samples drawn from the same population.

ANOMALOUS AREAS

The most prominent anomalous area in the Pacific is the East Pacific rise, a portion of

Fig. 5. Thickness of oceanic layer, oceanic stations.

Fig. 6. Depth below sea level to the Mohorovicic discontinuity.

the world rift system. Velocities near 7.5 km/sec are found along its crest in the southeast Pacific, and similar values have been found in the Gulf of California (not included in the illustrations) and along the Juan de Fuca ridge off the coast of Oregon. The distinctive features of the rise seem to be decreased mantle velocity, shallow mantle, severe attenuation of seismic waves through the mantle, and high heat flow in narrow bands along the crest.

The Bering Sea resembles a portion of the Pacific basin that has been cut off, partially filled with sediment, and depressed under the load. Near the foot of the continental slope in the eastern part of the deep basin and along the Aleutian ridge, the oceanic layer thickens greatly, providing a gradual transition to the structure beneath the continent and the island arc. Beneath the western part of the Coral Sea a similar thickening of the oceanic layer is found. The Sea of Okhotsk is reported to have some areas of shallow mantle within the Kuril arc. Within the complex area of the Philippines and the Indonesian archipelago there is ap-

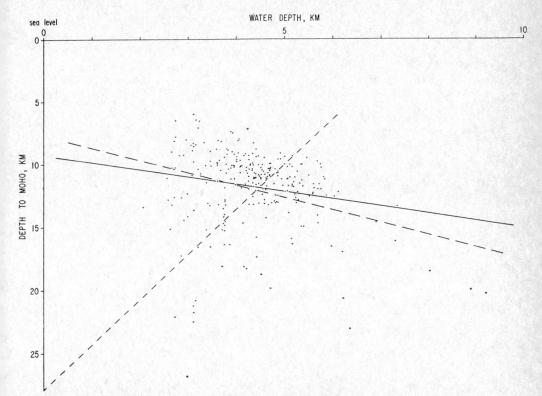

Fig. 7. Correlation of depth to the Mohorovicic discontinuity with water depth, oceanic stations. Solid line: statistical best fit. Long dashes: depth for constant crustal thickness (independent of water depth). Short dashes: expected depth for isostatic compensation (Airy method; density contrast at crust-mantle interface is 0.5 g/cm^3).

parently considerable variation of structure, not yet surveyed in sufficient detail for generalization.

The trench areas have uniformly deeper mantle than the ocean basin adjacent; the outer ridge seaward of many trenches generally has shallower mantle than normal.

SUMMARY

The averages presented by *Raitt* [1963] for the world oceans as a whole continue to be valid; differences in averages depend primarily on the definition of normal area for analysis. For the data given herein, the median values (not the mean) are:

Sediments	0.3 km; velocity variable between 1.5 and 3.4 km/sec.
Layer two	1.4 km; velocity between 3.4 and 6.0 km/sec, median near 5.1 km/sec.
Oceanic layer	4.7 km; median velocity 6.8 km/sec.
Mantle	median velocity 8.1 km/sec.

The only consistent regional variations are: sediment is thicker in areas that can receive turbidite flows from the continents; mantle velocities and mantle depth are frequently less than normal along the world rift system; mantle depth becomes significantly greater than normal beneath trenches and some areas near the foot of the continental slope where no trench is visible. Mantle depth does not vary with water depth in a manner consistent with isostasy; there is, rather, a tendency for the thickness from the seafloor to the Mohorovicic discontinuity to remain constant despite changes in water depth. This strongly suggests that the variations in water depth within the deep basins are controlled by variations in the density of the upper mantle rather than by the thickness of the crust.

Acknowledgment. This work was supported by the office of Naval Research, the U. S. Navy Bureau of Ships, and the National Science Foundation. Contribution from the Scripps Institution of Oceanography.

REFERENCES

Ewing, J. I., The mantle rocks, in *The Sea*, vol. 3, edited by M. N. Hill, pp. 103–109, Interscience Publishers, New York, 1963.

Raitt, R. W., The crustal rocks, in *The Sea*, vol. 3, edited by M. N. Hill, pp. 85–102, Interscience Publishers, New York, 1963.

Surface Waves and Crustal Structure[1]

James N. Brune

THE lower boundary of the earth's crust was discovered by A. Mohorovičić using the seismic refraction technique. He found that the seismic velocities at a certain depth in the earth underwent an abrupt increase. This discontinuity is now called the Mohorovicic discontinuity, or Moho. By 'crustal structure' we mean the distribution of elastic constants, seismic wave velocities, and densities in the outer parts of the earth down to and immediately below the Moho.

The practical use of surface waves in the study of crustal structure has only come in the last twenty years. Early theoretical studies of surface waves by Rayleigh, Lamb, and Love explained the existence and character of longitudinally/vertically polarized surface waves (Rayleigh waves) and transversely polarized surface waves (Love waves) observed on seismograms. Layering was found to cause dispersion, i.e., velocity is dependent on frequency. Thus seis-

---

[1] Division of Geological Sciences, California Institute of Technology, contribution 1452.

mologists were challenged to explain the dispersion observed on seismograms by layering in the earth.

By 1950 the study of surface waves had entered a quantitative stage and could be used in conjunction with refraction to give a detailed picture of crustal structure. Refined techniques for measuring group and phase velocity were developed in both the time domain and the frequency domain. Observations were extended to higher modes and leaking modes. The advent of high-speed computers made it possible to compute precise theoretical dispersion curves for complicated structures using matrix multiplication. This allowed the development of specialized techniques of dispersion fitting to solve for crustal structure (inversion). As a consequence of these developments, more than one hundred papers published in the last fifteen years have used surface wave dispersion in quantitative studies of crust and mantle structure.

CRUSTAL TYPES

The important crustal types that have emerged from surface wave studies are compared here in terms of dispersion curves, crust and mantle structure, and other geophysical and tectonic characteristics, e.g., heat flow and Bouguer gravity anomaly. The earliest and simplest classification of crustal types was a division into oceanic and continental types, and this division remains fundamentally correct today. The fraction of the earth's surface that cannot be clearly classified as either oceanic or continental is very small.

The crustal parameters that are most important in making a more refined classification of crustal types than simply oceanic or continental are: crustal thickness, upper mantle (P_n) velocity, tectonic characteristics, sediment thickness, and water depth. Consideration of these parameters leads to a division of the earth's crust into the following types: (A) shield, (B) mid-continent, (C) basin-range, (D) Alpine, (E) great island arc, (F) deep ocean basins, and (G) mid-ocean ridge. Two additional crustal types, which have not been adequately studied by surface wave techniques, are continental plateau and deep ocean trench. The classification is illustrated in Table 1.

Since there is often considerable variation in crustal structure within a given crustal type, and because of inaccuracies in measuring dis-

TABLE 1. Crustal Types

Crustal Type	Crustal Thickness, km	P_n Velocity, km/sec	Heat Flow, $\mu cal/cm^2$ sec	Bouguer Anomaly, mgal	Tectonic Characteristic	Geologic Features
A. Shield	35	8.3	0.7–0.9	−10 to −30	Very stable	Little or no sediment, exposed batholithic rocks of Precambrian age
B. Mid-continent	38	8.2	0.8–1.2	−10 to −40	Stable	Moderate thicknesses of post-Precambrian sediments
C. Basin-range	30	7.8	1.7–2.5	−200 to −250	Very unstable	Recent normal faulting, volcanism, and intrusion; high mean elevation
D. Alpine	55	8.0	Variable 0.7–2.0	−200 to −300	Very unstable	Rapid recent uplift, relatively recent intrusion; high mean elevation
E. Great island arc	30	7.4–7.8	Variable 0.7–4.0	−50 to +100	Very unstable	High volcanism, intense folding and faulting
F. Deep ocean basin	11	8.1–8.2	1.3	+250 to +350	Very stable	Very thin sediments overlying basalts, linear magnetic anomalies, no thick Paleozoic sediments
G. Shallow water mid-ocean ridge	10	7.4–7.6	High and variable 1.0–8.0	+200 to +250	Unstable	Active basaltic volcanism little or no sediment
Continental plateau						
Deep-ocean trench						

persion, the curves presented below are only approximate; they should not serve as rigid definitions of crustal types, but only as typical characteristics.

Crustal Type A: Shield

Shields are tectonically stable parts of the continents, with few volcanoes or earthquakes. Erosion has exposed deep metamorphic and plutonic rocks, and there are no thick sedimentary deposits younger than Precambrian. Heat flow values are low, about 0.7 μcal/cm^2 sec.

The typical shield structure has a crustal thickness of about 35 km, and the P-wave velocities increase with depth from 6.1 to about 6.8 km/sec at a depth of 30 km. The upper mantle velocities are high, about 8.3 km/sec for P waves and 4.7–4.8 km/sec for S waves.

The distinctive features of the dispersion curves for shields (Figure 1) are the relatively high group and phase velocities near periods of 60 seconds, a consequence of the high upper mantle velocities, and the relatively high group and phase velocities near periods of 10 seconds, a consequence of lack of low velocity sediments. The position of the higher mode curves reflects primarily the crustal thickness (shear-wave travel time in the crust).

Crustal Type B: Mid-Continent

The mid-continent crustal type is similar to the shield type, but shows somewhat greater instability; broad downwarps with sediment accumulation have occurred since Precambrian time, especially near the continental margins. A small amount of seismic and volcanic activity occurs. Heat flow values are of the order of 1 μcal/cm^2 sec. The typical mid-continent structure has a crustal thickness of about 38 km and upper mantle velocities of about 8.2 km/sec for P waves and 4.6 km/sec for S waves.

The group and phase velocities at a period of about 60 seconds are slightly lower than for shields, primarily a result of lower mantle velocities (see Figure 2). Velocities near periods of 10 seconds are variable, depending on the amount of sediment accumulation.

Crustal Type C: Basin-Range

The basin-range crustal type is characterized by recent normal faulting that has broken the crust into a series of basins and ranges. This has been accompanied by numerous volcanic extrusions and many earthquakes. The heat flow is very high, of the order of 2 μcal/cm^2 sec. The average elevation is high, about 1.3 km.

Recognition of the basin-range as a distinctive crustal type has only come in the last

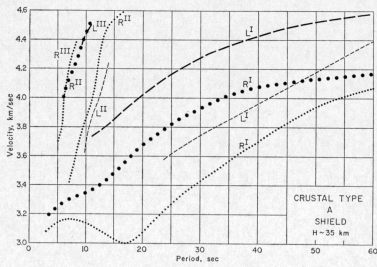

Fig. 1. Dispersion curves for crustal type A, shield. Small dots and dashes represent Rayleigh-wave and Love-wave group velocities. Large dots and dashes represent Rayleigh-wave and Love-wave phase velocities. R^I signifies the first Rayleigh mode, L^{II} signifies the second Love mode, etc.

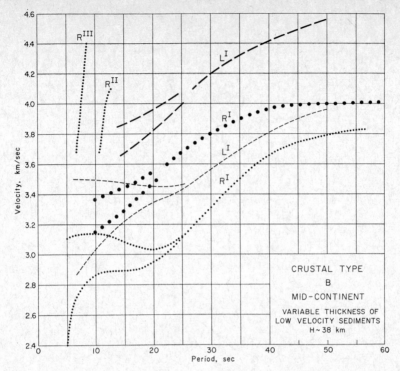

Fig. 2. Dispersion curves for crustal type B, mid-continent. See Figure 1 for legend.

few years. The crustal surface-wave velocities are considerably lower than for the shield and mid-continent, and originally this was interpreted as being due to a simple thickening of the crust. However, detailed refraction work in the Basin and Range province indicated a crustal thickness of only about 25–30 km and a low P_n velocity of about 7.8 km/sec. No sharp boundary with velocity 8.0 km/sec has been found. Thus a further complexity was introduced in the interpretation of crustal structure from surface waves: even in gross comparisons it was necessary to allow for both variations in crustal thickness and variations in upper mantle velocities.

Fundamental mode Love- and Rayleigh-wave group velocities at periods near 40 seconds are considerably lower than for shields, a result of the low upper mantle velocities (see Figure 3). The velocities at periods near 10 seconds are also low because of large sediment thicknesses.

There remains some uncertainty over the interpretation of the basin-range crustal structure, since Love-wave phase velocities have not been accurately measured, and since it is not known for certain whether the layer with P-wave velocity of 7.8 km/sec corresponds (in a strict geophysical sense) to the Mohorovicic discontinuity found in adjacent regions.

Crustal Type D: Alpine

This crustal type is characterized by high mountains created by rapid uplift. The formation of the mountains was preceded by crustal downwarping, accumulation of large thicknesses of sediments, and formation of batholithic rocks. There may be one or more stages of uplift between periods of quiescence. Finally, the mountains are eroded to gentle hills. In this study the crust is considered to be Alpine, provided that the mountain ranges are still relatively high with many peaks greater than 3 km in height. The Bouguer gravity anomalies for Alpine regions are low, reflecting isostatic compensation. Heat flow values appear to be relatively high (1.7 to 2 μcal/cm² sec) in certain younger mountain ranges, e.g., the Alps and Rocky Mountains, and low (0.7 to 1.0 μcal/cm² sec) in older mountains, e.g., the Sierra Nevada of California.

The thickness of the crust in Alpine regions

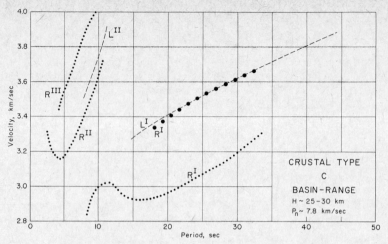

Fig. 3. Dispersion curves for crustal type C, basin-range.

is of the order of 45 to 55 km; P-wave velocities range from about 6.0 km/sec near the surface to about 7.0 km/sec at a depth of 40 km. The P_n velocity is about 8.0 km/sec. The upper mantle in Alpine regions has a low velocity channel with shear-wave velocities as low as about 4.3 to 4.4 km/sec at depths of about 150 km.

The fundamental mode Rayleigh-wave group velocities at periods of 40 seconds are lower for the Alpine crustal type than for any other crustal type, a result of both the thick crust and the pronounced low velocity channel in the upper mantle (see Figure 4).

Crustal Type E: Great Island Arc

This type of crustal structure is represented by large tectonically active islands such as

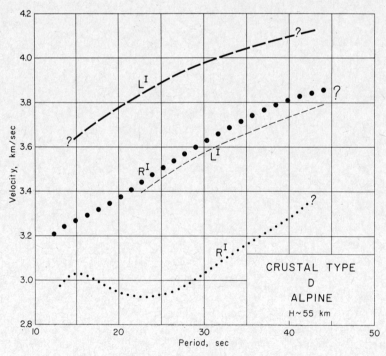

Fig. 4. Dispersion curves for crustal type D, Alpine.

Japan and New Zealand. Many large earthquakes occur and volcanic eruptions are common. Heat flows are locally very high or low, and isostatic anomalies are often high.

The crustal thickness for the great island arcs is about the same as that for normal continents, but rapid lateral variations occur. The upper mantle velocities are low, of the order of 7.6 to 7.8 km/sec for P waves and 4.3 to 4.5 km/sec for S waves.

The dispersions for island arcs are uncertain because the crustal structure is so variable and because the narrow dimensions make it difficult to eliminate the effects of surrounding regions. Within a given island arc structure, it is difficult to find paths long enough to permit accurate measurements of group velocity. For phase velocity measurements, it is often necessary to use waves that originate outside the island arc; thus uncertainties due to mode conversion at the boundaries are introduced.

The fundamental Rayleigh-wave group and phase velocities near periods of 60 seconds are much lower than for shields, a result of low upper mantle velocities (see Figure 5). However, the observed fundamental Love-wave group and phase velocities do not agree with this result and suggest relatively high velocities in the lower crust and upper mantle. This discrepancy has been interpreted as indicating strong anisotropy in the crust in Japan.

Crustal Type F: Deep Ocean Basin

Deep ocean basins are topographically stable parts of the earth's crust overlain by water 4–5 km deep. Heat flow values average about 1.3 μcal/cm^2 sec. Rocks are primarily basaltic, and there is no evidence that ordinary orogenic disturbances have ever occurred. The recent discovery of remarkable linear magnetic anomalies paralleling oceanic ridges has been interpreted to represent ocean floor spreading in

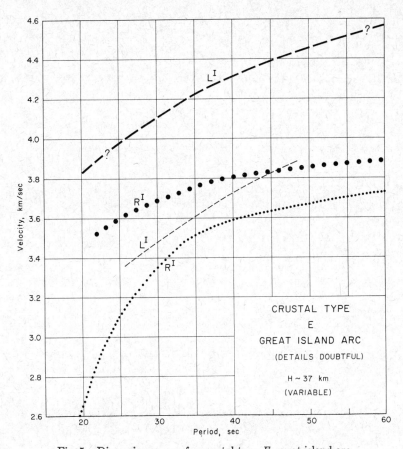

Fig. 5. Dispersion curves for crustal type E, great island arc.

the reversing magnetic field of the earth, implying that the present ocean floor is a relatively young feature. The most plausible explanation yet presented for the magnetic anomalies and the ocean floor spreading is the existence of convection (in a general sense) in the mantle, the upward convecting part of the mantle being under oceanic ridges.

The approximate oceanic crustal structure consists of 5 km of water overlying 0–1 km of low rigidity sediment (shear-wave velocity 0.5–1.0 km/sec) and 5 km of rock with P-wave velocity 6.41 km/sec. The Moho at about 11 km depth is represented by an abrupt increase in velocity to a P-wave velocity of about 8.1 km/sec and an S-wave velocity of 4.7 km/sec. The upper mantle has a low velocity channel with shear wave velocities of about 4.4–4.5 km/sec at a depth of about 100 km.

The fundamental Rayleigh-wave group velocities for periods between 15 and 25 seconds are primarily controlled by the thickness of water and low rigidity sediments. Because of the thin crust, the Rayleigh-wave group velocities reach high values (4.1 km/sec) at periods of only 35 seconds. The Love waves are not influenced by the water layer, and the group velocities rise steeply at a period of about 8–9 seconds and flatten out to a velocity of 4.4–4.5 km/sec at a period of about 15 seconds. However, the period range of the steep part of the Love-wave curve is critically dependent on the thickness of low rigidity sediments and can be as short as 5–7 seconds when little or no low rigidity sediment is present, and as long as 11–13 seconds when a large thickness ($\sim$1.5 km) of low rigidity sediment (shear velocity 0.2–1.0 km/sec) is present, e.g., in the Argentine basin. The part of the fundamental mode Rayleigh-wave group velocity curve between 6 and 11 seconds has only been observed from nuclear explosion sources. At high frequencies this curve approaches the velocity of the T phase, as Figure 6 indicates. It should be noted that the observed velocities for both Love and Rayleigh waves between 30 and 60 seconds

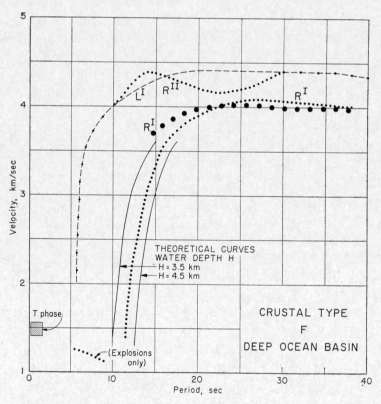

Fig. 6. Dispersion curves for crustal type F, deep ocean basin.

will almost always appear lower than shown, since some continental margin, ridge, or rise, or other crustal type with lower velocities will be included in the path.

Crustal Type G: Shallow Water Mid-Ocean Ridge

This crustal type is represented by the mid-Atlantic ridge near Iceland, the Easter Island rise, the Hawaiian Islands, and other large mid-ocean volcanic islands. The surface rocks are basaltic lava flows from recently active volcanoes. Heat flows are often very high. Small earthquakes and volcanic eruptions are common. Very little sediment is present. As was mentioned in the preceding section, recent interpretation of magnetic anomalies has led to the theory that certain mid-ocean ridges form the origin for ocean floor spreading.

The typical structure of this type has 0–3 km of water overlying an undetermined thickness of basaltic lava flows. The crustal thickness is uncertain, but the underlying mantle is anomalously low in both P-wave and S-wave velocities; about 7.4–7.7 km/sec and 4.2–4.4 km/sec, respectively.

Typical fundamental mode Love- and Rayleigh-wave dispersion curves for this type of crustal structure are given in Figure 7. The Love-wave group velocity is similar to that for the mid-ocean basin, since the water layer has no effect. However, because of the low mantle velocities the curve approaches velocities of only 4.2–4.3 km/sec at longer periods (~30 sec). For the shallowest water depths, the steep portion of the Rayleigh-wave group velocity curve occurs in the period range 5–9 seconds and occurs at progressively longer periods as the water depth is increased.

Crustal Types Not Investigated: Transitional and Composite Crustal Types

Most areas of the earth's surface can be classified into one of the seven crustal types described above. There are certain other crustal types that might be identified tectonically, but for which there are no surface-wave data available, e.g., deep ocean trenches and high continental plateau regions. The remaining parts of the earth's surface can be considered either transitional between two of the crustal types or composites of crustal types with such small

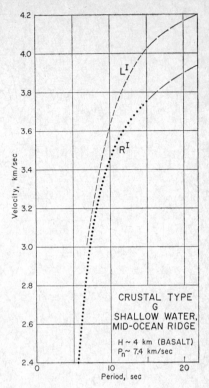

Fig. 7. Dispersion curves for crustal type G, shallow water mid-ocean ridge.

dimensions as to be experimentally difficult to study by surface-wave techniques.

COMPARISON WITH SANTO'S CURVES

Santo [1963] has recently divided the observed fundamental mode group velocity dispersion for Rayleigh and Love waves into a continuous distribution of curves between extremes, as indicated in Figures 8 and 9. The individual curves do not necessarily apply to large or distinctive tectonic areas, and most are for transitional or composite paths. However, certain members of each set correspond approximately to distinctive tectonic types as described in this paper: 0, 1, and I correspond approximately to crustal type F, deep ocean basin; 7 and VIII correspond approximately to crustal type E, great island arc.

SHEAR-VELOCITY STRUCTURES

The most recent interpretations of shear velocity structure for the various crustal types are compared in Figure 10. The structure for the

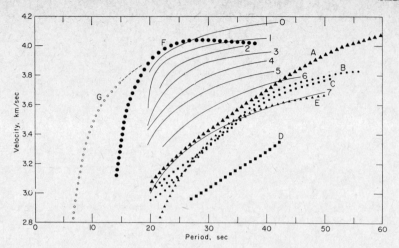

Fig. 8. Comparison of fundamental mode Rayleigh-wave group velocity curves for various crustal types with each other and with *Santo*'s [1963] curves.

Canadian shield, CANSD, is that obtained by *Brune and Dorman* [1963]. The structure for the mid-continent of the United States, as obtained by *McEvilly* [1964], is not shown in Figure 10, but it is similar to that for the Canadian shield with a slightly thicker crust and slightly lower upper mantle velocity. However, a better fit to the data was obtained by allowing 8% anisotropy in the uppermost mantle (SH ve-

locity greater than SV velocity), and this reduced the upper mantle SV velocity somewhat. The basin-range structure shown in Figure 10 is that obtained by *Smith* [1962], with upper mantle velocities about as low as for mid-ocean ridges. The Alpine structure, SPLAN-50, was obtained by *Knopoff et al* [1966]; however, the particular area studied did not have quite as thick a crust as the structure corresponding

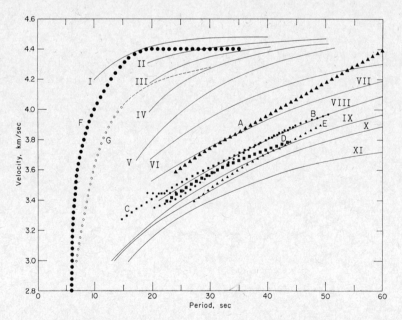

Fig. 9. Comparison of fundamental mode Love-wave group velocity curves for various crustal types with each other and with *Santo*'s [1963] curves.

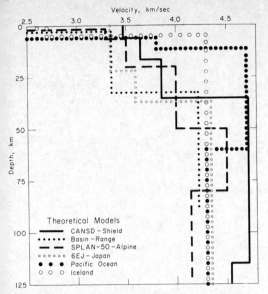

Fig. 10. Comparison of crust and upper mantle shear velocity distributions for various crustal models.

to the curves in Figure 4, and no Love-wave data were used in their study. The structure for Japan, 6EJ, was obtained by *Aki* [1961] using Rayleigh-wave phase velocities, but is inconsistent with the Love-wave phase velocity data unless about 7% anisotropy is introduced into the upper mantle [*Kaminuma*, 1966]. The Pacific Ocean structure is that obtained by *Saito and Takeuchi* [1966] and has somewhat higher upper mantle velocities than were obtained in earlier studies, evidently because most of the paths considered in earlier studies were contaminated by areas with low mantle velocities such as ridges, volcanic island areas, and continental margins. The Iceland crustal structure is that obtained by *Tryggvason* [1962], and the low mantle velocities in this structure agree with results of *Talwani et al.* [1965].

Aside from the data on typical crustal types given in Table 1, it is of interest to compare the mean crust and upper mantle shear velocities for the theoretical models in Figure 10. Table 2 gives the mean crustal shear velocity, the mean upper mantle shear velocity above 125 km depth, and the mean shear velocity above 125 km depth including both the crust and mantle for these models. Both the shields and ocean (water layer excluded) have relatively high average shear velocities, whereas the basin-range and island arc types have low average shear velocities. The Alpine structure has a relatively high mean shear velocity in the crust, but a low mean shear velocity in the mantle.

CRUSTAL TYPES AND CRUSTAL EVOLUTION

Most of the earth's surface can be divided unambiguously into either continental types (A, B, C, D, E) or oceanic types (F, G). The oceanic and continental structures are essen-

TABLE 2. Parameters for Theoretical Models Shown in Figure 10

Model and Reference		Crustal Type	Mean Crustal Shear Velocity, km/sec	Mean Upper Mantle Shear Velocity to 125-km Depth, km/sec	Mean Crust and Upper Mantle Shear Velocity to 125-km Depth, km/sec
CANSD, *Brune and Dorman* [1963]	A.	shield	3.72	4.70	4.42
Smith [1962]	C.	basin-range	3.27	4.40	4.11
SPLAN-50, *Knopoff et al.* [1966]	D.	Alpine	3.75	4.28	4.07
6EJ, *Aki* [1961]	E.	great island arc	3.44	4.34	4.07
Pacific Ocean, *Saito and Takeuchi* [1966]	F.	deep ocean basin	3.80	4.48	4.45
Iceland, *Tryggvason* [1962]	G.	shallow-water mid-ocean ridge	3.2	4.30	4.26

tially distinct, as indicated by the comparison of fundamental mode group velocities in Figures 8 and 9.

The crustal structures derived for the various crustal types, along with the parameters in Table 1, suggest a classification based on two crustal types and two mantle types. Thus we can say there are two basic crustal types, continental and oceanic, and two basic mantle types, stable and unstable. The two crustal types are distinguished by the thickness of crustal rocks, i.e. greater than 25 km for continental and less than about 15 km for oceanic; the two mantle types are distinguished by stability and upper mantle velocities, the unstable mantle being characterized by low P_n velocity or a pronounced low velocity channel and the stable mantle being characterized by high P_n velocity and a less pronounced low velocity channel.

In this classification, areas of the earth's surface would be identified as: (1) continental crust overlying a stable mantle, i.e., shield and mid-continent types, A and B; (2) continental crust overlying an unstable mantle, i.e., basin-range, Alpine, and great island arc types, C, D, and E; (3) oceanic crust overlying a stable mantle, i.e., deep ocean basin type, F; and (4) oceanic crust overlying an unstable mantle, i.e., mid-ocean ridge type, G, and deep ocean trenches.

The fact that most of the earth's surface can be clearly classified as oceanic or continental suggests that either: (1) the two major crustal types were formed as distinct structures at some early stage in the earth's history and have remained distinct in recent geologic time or have changed only very slowly; (2) if one type is actively evolving into the other, then the transition time for a given area is very short on a geologic time scale, so that a large area is not transitional at a given geologic time; or (3) continental blocks remain as distinct structures but drift relative to one another or relative to the ocean basis. The third possibility is consistent with recent speculations about continental drift and ocean floor spreading, with new oceanic crust created near the mid-ocean ridges as two bordering continents drift apart.

The recently discovered magnetic anomaly patterns in the oceans promise to revolutionize our thinking on tectonics and evolution of the crust, and many of its implications are now in the process of being understood. Thus it is not possible to reach final conclusions on the nature of crustal evolution.

IMPORTANT PROBLEMS FOR FUTURE RESEARCH

Completion of Dispersion Curves

Love-wave phase velocities have not been reliably measured for deep ocean basin, basin-range, and Alpine crustal types. Higher-mode phase velocity curves are missing for most crustal types.

Higher Modes, Body Waves, and Channel Waves

The superposition of very high normal modes in a layered crustal structure represents the various crustal refractions and multiple reflections occurring with phase velocities less than the critical phase velocity for refraction into the underlying medium. These refractions and multiple reflections may give pulse-like arrivals on a seismogram. A reliable interpretation of higher mode crustal dispersion may give detailed information about the presence of gradients or low velocity layers in the crust. However, this will probably require refined analysis techniques, for example, phase velocity filtering. From geologic considerations and consideration of velocity as a function of temperature and pressure, we can conclude that it is quite reasonable to have local low velocity layers in the crust.

Anisotropy

The importance of crust and upper mantle anisotropy (SH velocity different from SV velocity at a given depth) is still uncertain, even with the large amount of refraction and surface wave data available, because exhaustive studies have not been carried out in a given area to eliminate other possible causes of anomalous dispersion. In some regions (for example, Japan and the central United States), observed Love-wave phase velocities are anomalously high compared to Rayleigh-wave phase velocities, suggesting the presence of anisotropy. On the other hand, for shield areas the observed Love- and Rayleigh-wave phase velocities do not appear to require anisotropy. Phase velocity data adequate to test for the presence of anisotropy are lacking in most other regions. Anisotropy of the type that causes S_v- and P-wave velocities to be dependent in direction may also be present in some areas.

PL Waves, Leaking Modes

Multiple reflecting crustal waves at phase velocities greater than the critical value for loss of energy into the underlying medium may exhibit dispersion characteristics similar to those of normal modes. The observed dispersion and energy decay can be used to interpret crustal structure. Although few quantitative observational studies using leaking modes to determine crustal structure have been made, the leaking modes promise to give additional constraints in interpreting crustal structure, especially in interpreting the compressional-wave velocity structure.

Acknowledgments. The author's primary contribution to the paper has been that of collection, synthesis, and speculation. Most of the data analysis was performed by other investigators. This paper is not meant to be a historical review, and therefore no attempt has been made to reference the contributions of each investigator. For the benefit of those who wish to pursue the subject further, a list of articles that are review articles or that have extensive reference lists is presented in the bibliography.

I am indebted to Miss Gladys Engen for assistance in the collecting, plotting, and analysis involved in the study.

BIBLIOGRAPHY

For Review

Arkhangel'skaya, V. M., Dispersion of surface waves and crustal structure, *Izv. Akad. Nauk SSSR, Geophys. Ser., 1960*, 1360–1391, 1960.

Ewing, M., J. Brune, and J. Kuo, Surface-wave studies of the Pacific crust and mantle, in *Crust of the Pacific Basin, Geophys. Monograph 6*, pp. 30–40, American Geophysical Union, Washington, D.C., 1962.

Ewing, M., W. Jardetsky, and F. Press, *Elastic Waves in Layered Media*, McGraw-Hill Book Company, New York, 1957.

Kovach, R., Seismic surface waves: Some observations and recent developments, *Phys. Chem. Earth, 6,* 251–314, 1966.

Oliver, J., A summary of observed seismic surface wave dispersion, *Bull. Seismol. Soc. Am., 52,* 81–86, 1962.

Press, F., and M. Ewing, Earthquake surface waves and crustal structure, in *Crust of the earth, Geol. Soc. Am. Spec. Paper 62,* 51–60, 1955.

For Reference

Aki, Keiiti, Crustal structure in Japan from the phase velocity of Rayleigh waves, 1, Use of the network of seismological stations operated by the Japan Meteorological Agency, *Bull. Earthquake Res. Inst., Tokyo Univ., 39,* 255–283, 1961.

Alexander, Shelton S., Surface wave progagation in the western United States, Ph.D. Thesis, California Institute of Technology, 1963.

Brune, James, and James Dorman, Seismic waves and earth structure in the Canadian shield, *Bull. Seismol. Soc. Am., 53,* 167–210, 1963.

Cisternas, Armando, Crustal structure of the Andes from Rayleigh wave dispersion, *Bull. Seismol. Soc. Am., 51,* 381–388, 1961.

Ewing, Maurice, and Frank Press, Detemination of crustal structure from phase velocity of Rayleigh waves, 3, The United States, *Bull. Geol. Soc. Am. 70,* 229–244, 1959.

Gabriel, V. G., and J. T. Kuo, High Rayleigh wave phase velocities for the New Delhi, India–Lahore, Pakistan, profile, *Bull. Seismol. Soc. Am., 56,* 1137–1145, 1966.

Kaminuma, Katsutada, The crust and upper mantle structure in Japan, 3, An anisotropic model of the structure in Japan, *Bull. Earthquake Res. Inst., Tokyo Univ., 44,* 511–518, 1966.

Knopoff, L., S. Mueller, and W. L. Pilant, Structure of the crust and upper mantle in the Alps from the phase velocity of Rayleigh waves, *Bull. Seismol. Soc. Am., 56,* 1009–1044, 1966.

McEvilly, T. V., Central U.S. crust-upper mantle structure from Love and Rayleigh wave phase velocity inversion, *Bull. Seismol. Soc. Am., 54,* 1997–2015, 1964.

Oliver, Jack, and James Dorman, On the nature of oceanic seismic surface waves with predominant periods of 6 to 8 seconds, *Bull. Seismol. Soc. Am., 51,* 437–455, 1961.

Pomeroy, Paul, and Jack Oliver, Seismic waves from high-altitude nuclear explosions, *J. Geophys. Res., 65,* 3445–3457, 1960.

Saito, M., and H. Takeuchi, Surface waves across the Pacific, *Bull. Seismol. Soc. Am., 56,* 1067–1091, 1966.

Santo, T. A., Division of the south-western Pacific area into several regions in each of which Rayleigh waves have the same dispersion characteristics, *Bull. Earthquake Res. Inst., Tokyo Univ., 39,* 603–630, 1961.

Santo, T. A., Division of the Pacific area into seven regions in each of which Rayleigh waves have the same group velocities, *Bull. Earthquake Res. Inst., Tokyo Univ., 41,* 719–741, 1963.

Santo, T. A., and M. Båth, Crustal structure of the Pacific Ocean area from dispersion of Rayleigh waves, *Bull. Seismol. Soc. Am., 53,* 151–165, 1963.

Smith, Stewart W., A reinterpretation of phase

velocity data based on the Gnome travel time curves, *Bull. Seismol. Soc. Am.*, *52*, 1031–1035, 1962.

Sykes, Lynn R., and Jack Oliver, The propagation of short-period seismic surface waves across oceanic areas, 1, Theoretical study, *Bull. Seismol. Soc. Am.*, *54*, 1349–1372, 1964.

Sykes, Lynn R., and Jack Oliver, The propagation of short-period seismic surface waves across oceanic areas, 2, Analysis of seismograms, *Bull. Seismol. Soc. Am.*, *54*, 1373–1415, 1964.

Talwani, M., X. Le Pichon, and M. Ewing, Crustal structure of the mid-ocean ridges, 2, Computed model from gravity and seismic refraction data, *J. Geophys. Res.*, *70*, 341–352, 1965.

Tryggvason, Eysteinn, Crustal structure of the Iceland region from dispersion of surface waves, *Bull. Seismol. Soc. Am.*, *52*, 359–388, 1962.

Regional Variations of P-Wave Velocity in the Upper Mantle beneath North America

Eugene Herrin

STUDIES during the last ten years have clearly shown that the physical properties of the upper mantle beneath North America vary significantly from east to west and that the details of this variation correlate with the major tectonic provinces of the continent. The property most readily measured is the P-wave velocity of the rocks that compose the uppermost mantle. We define the phase P_n as the first arrival of seismic energy in the range from a few degrees to a distance where the travel-time function begins to show measurable curvature; thus the P_n phase can be used to estimate P-wave velocity in the uppermost mantle.

Before 1960 seismologists generally agreed that P_n velocity was 8.1 ± 0.1 km/sec in the United States [*Steinhart and Meyer*, 1961, p. 13]. Almost no measurements using explosion sources contradicted this conclusion. P_n velocities as low as 7.5 km/sec were reported from Japan, and values as high as 8.3 km/sec were measured in Scandinavia, but these determinations were from earthquakes and were considered to be less reliable.

A few experiments seem to argue against a uniformity of upper mantle velocity under the North American continent. *Tatle and Tuve* [1955] showed that crustal thickness under the Basin and Range province of Nevada was much thinner than required to allow isostatic compensation by the Airy-Heiskanen mechanism. It follows that compensation in the region must result, at least in part, from a Pratt-Hayford mechanism; thus the density and, by inference, the seismic velocity of the upper mantle should be lower there than beneath the plains areas of the central United States. The implications of the work of Tatel and Tuve were largely ignored because there was not yet enough evidence to compel seismologists to abandon the hypothesis of lateral homogeneity in the upper mantle.

RESULTS SINCE 1960

Berg et al. [1960] reported anomalously low P_n velocities in Utah from quarry blast studies. They found a thin crust, as had Tatel and Tuve, and a P_n velocity of only 7.6 km/sec. *Ryall* [1962] studied the Hebgen Lake, Montana, earthquake in 1959 and showed that travel time from the source region varied with azimuth.

The Gnome underground nuclear explosion in New Mexico was recorded at more than 100 stations in North America. Analysis of the data [*Herrin and Taggart*, 1962; *Romney et al.*, 1962] showed that significant variations in upper mantle velocities existed beneath the North American continent. This discovery was fol-

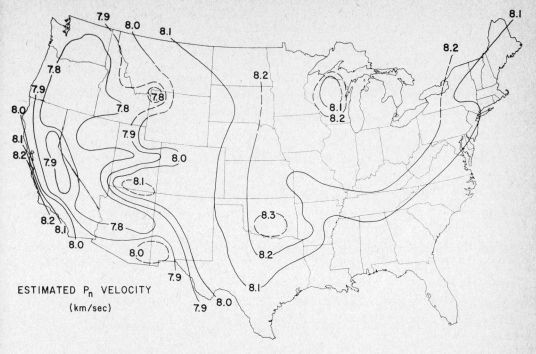

Fig. 1. Estimated P_n velocity in the United States, based on data from deep seismic soundings, underground nuclear explosions, and earthquakes (compiled by E. Herrin and J. Taggart).

lowed by an extensive program of deep seismic soundings across the United States and Canada, the results of which have recently been summarized by *James and Steinhart* [1966]. They included on page 297 a previously unpublished map by Herrin and Taggart which synthesizes knowledge of P_n velocities in the United States. This map, reproduced here as Figure 1, shows a strong regional difference in velocity in the uppermost mantle, with velocities being significantly lower in the mountainous western states than in the north-central and northeastern states. The contours shown on Figure 1 should accurately portray regional velocity differences; the details of this map are, of course, subject to revision as more data become available. The 8.3 contour in Oklahoma may result from structural complexity of the M discontinuity or from the fact that no correction was made for curvature in computing the velocities; thus the true P_n velocity there may be only 8.2 km/sec. Details of contours in Utah, where strong horizontal gradients are shown, may be questionable. Investigations are now under way to provide more information concerning the uppermost mantle beneath central and southern Utah.

P_n velocities are less well known in Canada, but the regional pattern there is similar to that in the United States. Velocities beneath the interior regions are about 8.2 km/sec, and beneath the mountainous western region they are generally lower than 7.9 km/sec.

SEISMIC DELAY TIMES

The most important effect of lateral variations in upper mantle velocity concerns P travel times to distances less than 20°. For travel paths longer than 30°, the rays reach depths greater than 750 km, i.e., below the region within which significant lateral variations of velocity are known to occur. For such paths, the delays presumably depend only on the velocity structure traversed in the upper part of the ray path.

Figure 2 shows relative seismic delays based on earthquake studies by E. Herrin and J. Taggart. The delays have been corrected for

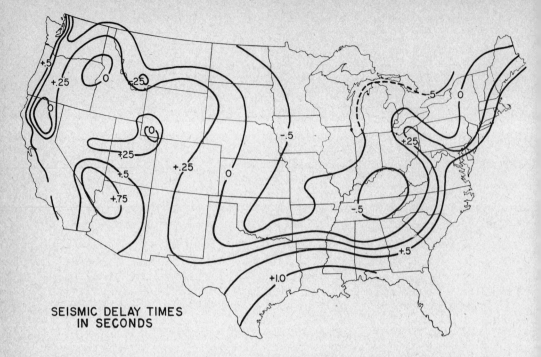

Fig. 2. Relative seismic delay time in the United States. This map is based on data recorded at 80 stations from 300 earthquakes in the period 1961–1964. Contours represent mean seismic delay with allowance made for azimuthal dependence (compiled by E. Herrin and J. Taggart).

azimuthal effects of the form described by *Cleary and Hales* [1966], *Herrin* [1966], *Otsuka* [1966], and *Bolt and Nuttli* [1966]. Regional patterns in Figure 2 are generally similar to the patterns for P_n velocity in Figure 1. Areas having low velocities in the upper mantle, such as the Basin and Range province of Nevada and Arizona, also show a positive delay time (relatively late arrivals), whereas P arrivals from distant events tend to be early in the midcontinent region where P_n velocities are known to be high. Thus the two maps, based on independent analyses of different types of data, show general agreement.

BODY WAVE AMPLITUDES

Nuclear explosions are very useful for studying variations in P-wave amplitudes, because there is a much more uniform radiation of energy from explosions than from earthquakes. As might be expected, areas that show strong regional variations in travel times show very strong variations in amplitudes [*Herrin and Taggart*, 1962; *Romney et al.*, 1962]. Regions having fast P_n travel times generally show higher amplitudes. For some explosions, such as Salmon [*Springer*, 1966], it has been possible to predict the amplitudes with fair accuracy. Over most of the United States, however, large, generally unpredictable variations in amplitude have been observed. With the accumulation of data from a large number of explosions, it has been possible to contour amplitudes for several regions [*Stepp et al.*, 1963; *Jordan et al.* 1966].

CONCLUSIONS

During the last decade, seismologists have established the fact that the velocity structure of the upper mantle beneath North America shows significant lateral variations. Effects of this inhomogeneity are seen in regional variations of P_n velocities and of seismic delay times for P waves from distant sources. The variations are strongly correlated with one another and generally correlate with broad-scale surface geology and topography.

Low velocity in the upper mantle implies abnormally low density; thus we should expect to find low density mantle material beneath the mountainous regions of western North America. Crustal studies in this region [*Pakiser and Steinhart*, 1964] show that the crust is only 30 to 40 km thick, rather than the 50 to 60 km required for compensation by the Airy-Heiskanen scheme. *Woollard* [1966] concludes that a Pratt-Hayford type of compensation involving relatively low density material in the upper mantle is required. Therefore, we conclude that the rocks which comprise the upper mantle beneath the mountainous regions of the western United States have relatively low density as well as seismic velocity. This conclusion probably applies to the western part of the North American continent.

Recent measurements by a number of workers have shown that the surface heat flow in Nevada and Utah is approximately twice that found in the northeastern United States. This greater heat flow implies that temperatures in the uppermost mantle are several hundred degrees higher in the west than in the eastern states. Increased temperature would cause relatively lower velocities and densities in the upper mantle, although it is not yet certain that the magnitude of the temperature difference is sufficient to explain the observed velocity contrasts. Perhaps there are also lateral variations in composition within the upper mantle under the North American continent.

Geophysicists studying upper mantle properties anywhere in the world must be concerned with lateral inhomogeneity and, in particular, with differences from one broad geologic region to another. It is now possible to search for lateral variations in the lower mantle [*Chinnery*, 1967]. *Alexander and Phinney* [1966] have suggested that variations in structure occur at the core-mantle boundary; thus perhaps the entire mantle must be examined for large scale lateral inhomogeneity.

A comprehensive set of articles on crustal structure appears in AGU Monograph 10, including the review and extensive bibliography in *James and Steinhart* [1966]. A more complete list of references pertaining to P-wave travel times in North America is given by *Herrin* [1967].

REFERENCES

Alexander, Shelton S., and R. A. Phinney, A study of the core-mantle boundary using P waves diffracted by the earth's core, *J. Geophys. Res.*, 71(24), 5943–5958, 1966.

Berg., J. W., K. L. Cook, H. D. Narens, and W. M. Dolan, Seismic investigation of crustal structure in the eastern part of the Basin and Range province, *Bull. Seismol. Soc. Am.*, 50(4), 511–535, 1960.

Bolt, Bruce, and Otto Nuttli, P wave residuals as a function of azimuth, 1, Observations, *J. Geophys. Res.*, 71(24), 5977–5985, 1966.

Chinnery, Michael A., Evidence for lateral variations in the lower mantle (abstract), *Trans. Am. Geophys. Union*, 48(1), 194, 1967.

Cleary, J., and A. L. Hales, Azimuthal variation of U. S. station residuals, *Nature*, 210, 619–620, 1966.

Herrin, Eugene, Travel-time anomalies and structure of the upper mantle (abstract), *Trans. Am. Geophys. Union*, 47(1), 44, 1966.

Herrin, Eugene, Travel times and amplitudes of body waves, *Trans. Am. Geophys. Union*, 48, 403–407, 1967.

Herrin, Eugene, and J. Taggart, Regional variations in P_n velocity and their effect on the location of epicenters, *Bull. Seismol. Soc. Am.*, 52(5), 1037–1046, 1962.

James, David E., and John S. Steinhart, Structure beneath continents: A critical review of explosion studies, 1960–1965, in *The Earth beneath the Continents, Geophys. Monograph 10*, pp. 293–333, American Geophysical Union, Washington, D. C., 1966.

Jordan, J. N., W. V. Mickey, Wayne Helterbran, and D. M. Clark, Travel-times and amplitudes from the Salmon explosion, *J. Geophys. Res.*, 71, 3469–3482, 1966.

Otsuka, M., Azimuth and slowness anomalies of seismic waves measured on the central California seismographic array, 1, Observations, *Bull. Seismol. Soc. Am.*, 56, 223–239, 1966.

Pakiser, L. C., and J. S. Steinhart, Explosion seismology in the western hemisphere, in *Research in Geophysics*, edited by Hugh Odishaw, vol. 2, chapter 5, MIT Press, Cambridge, 1964.

Romney, Carl, B. G. Brooks, R. H. Mansfield, D. S. Carder, J. N. Jordan, and D. W. Gordon, Travel times and amplitudes of principal body phases recorded from Gnome, *Bull. Seismol. Soc. Am.*, 52(5), 1057–1074, 1962.

Ryall, Alan, The Hebgen Lake, Montana, earthquake of August 18, 1959; P waves, *Bull. Seismol. Soc. Am.*, 52(2), 235–271, 1962.

Springer, D. L., Calculation of first-zone P-wave amplitudes for Salmon event and for decoupled sources, *J. Geophys. Res.*, 71, 3459–3467, 1966.

Steinhart, John S., and Robert P. Meyer, *Explosion Studies of Continental Structure*, Carnegie Inst. Wash. Publ. 622, 409 pp., 1961.

Stepp, J. C., W. J. Spence, S. T. Harding, R. W. Sherburne, and S. T. Algermissen, A study of P_n velocities, amplitudes and travel-time residuals for a portion of the western United States, *Earthquake Notes, 34*(3-4), 1963.

Tatel, Howard E., and Merle A. Tuve, Seismic exploration of a continental crust, in *The Crust of the Earth*, edited by Arie Poldervaart, *Geol. Soc. Am. Special Paper 62*, 35-50, 1955.

Woollard, G. P., Regional isostatic relations in the United States, in *The Earth beneath the Continents, Geophys. Monograph 10*, pp. 557-594, Americal Geophysical Union, Washington, D. C., 1966.

Upper Mantle Structure and Velocity Distribution in Eurasia

Jiří Vaněk

NEW seismological evidence on the upper mantle structure and velocity distribution $v(h)$ in Eurasia comes mainly from two sources: velocity-depth determinations by means of travel-time curves for deep earthquakes and amplitude curves of seismic body waves.

STUDY OF DEEP EARTHQUAKES

One of the most important investigations was carried out by *Lukk and Nersesov* [1965] for Central Asia; they obtained vertical and horizontal travel-time curves by observing deep Pamirs-Hindu Kush earthquakes ($h = 70$-270 km) along a 3500-km profile of special stations extending from Central Asia across Eastern Kazakhstan, Altai, and Sayan to the Lena River. The mean distance between individual stations was between 70 and 100 km. These observations were interpreted by well developed methods of seismic prospecting. The resulting scheme of the upper mantle structure (Figure 1) is characterized by a low-velocity channel at depths between 110 and 150 km observed for both P and S waves, by a second channel at 240-400 km for S waves only, and by a region of high velocity gradient between 700 and 780 km. The velocity-depth distribution $v(h)$ can be expressed as follows, where h is depth in kilometers, v_P and v_S are velocity of P and S waves in kilometers per second, and the arrow means velocity change at the indicated depth: $(h, v_P, v_S) \equiv (0, 6.0, 3.5)$, $(45, 6.0 \to 8.0, 3.5 \to 4.6)$, $(85, 8.0 \to 8.6, 4.6 \to 4.75)$, $(110, 8.6 \to 8.2, 4.75 \to 4.6)$, $(150, 8.2 \to 8.6, 4.6 \to 4.8)$, $(200, 8.6 \to 8.8, 4.8 \to 5.0)$, $(240, \cdots, 5.05 \to 4.8)$, $(300, \cdots, 4.6)$, $(400, 9.0 \to 9.05, 5.0 \to 5.05)$, $(700, 10.2 \to 10.4, 5.2 \to 5.4)$, $(780, 11.4, 6.55)$, $(900, 11.4, 6.55)$.

Yanovskaya [1963] showed that the present accuracy of travel-time curves does not permit a unique solution for velocity-depth distribution $v(h)$. However, a family of admissible $v(h)$ values can be selected by solving the inverse problem on computers, especially when a set of travel-time curves for different focal depths can be used [*Matveeva and Alekseev*, 1963]. This method was applied for the interpretation of horizontal travel-time curves derived from observations of deep earthquakes in the region covering central and northern Japan, the central and southern Kurils, the southern part of the Okhotsk Sea, and the adjacent seismoactive region of the Pacific Ocean [*Fedotov et al.*, 1964; *Tarakanov*, 1965]. Two different methods of optimization were used to determine the range of probable velocity distribution $v(h)$ in this region (Figure 2). The existence of a low-velocity channel at depths to 100 km is not excluded except in the region of the island arc with active volcanism in the southern Kurils, where the type of both the crust and the mantle changes; the velocity distribution $(h, v_P) = (20, 7.7)$, $(82, 7.7 \to 7.8)$, $(90, 7.8 \to 7.9)$, $(124, 7.9 \to 8.1)$ in the latter region was derived

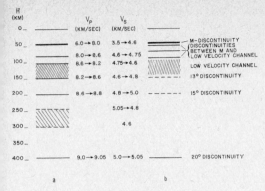

Fig. 1. Scheme of the upper mantle structure in Euro-Asia: (*a*) for central Asia from deep earthquakes [*Lukk and Nersesov*, 1965]; (*b*) for southeastern Europe from amplitude-distance curves of P waves [*Vaněk*, 1968].

by the interpretation of vertical travel-time curves [*Fedotov and Kuzin*, 1963].

An indication of the possible existence of a low-velocity channel at depths of 100–150 km was also found for the region of the eastern Carpathians on the basis of horizontal travel-time curves for deep earthquakes from the Vrancea region [*Iosif*, 1965].

Future study of deep earthquakes should combine travel times with dynamic parameters of seismic body waves, thus narrowing substantially the set of possible velocity-depth distributions [*Yanovskaya and Asbel*, 1964].

AMPLITUDE CURVES OF SEISMIC BODY WAVES

The study of the amplitude curves of seismic body waves is closely related to the investigation of earthquake magnitude. Magnitude influences the seismological definition of the amplitude curve [*Vaněk and Stelzner*, 1962]. If earthquakes of different magnitude are considered, normalized amplitudes

$$A_i^* = \log (A_i/T_i) - M$$

must be used, where A_j is the maximum ground amplitude in microns, T_j is the corresponding period in seconds of the individual wave group j, and M is the earthquake magnitude and has the function of a normalizing quantity. Two kinds of amplitude curves can be constructed: the amplitude-distance curves A^* (Δ) and the amplitude-depth curves A^* (h). The curves A^* (h) have been applied for determining the attenuation of seismic waves in the mantle and are not discussed in this section.

The first amplitude-distance curves derived by *Gutenberg* [1948], *De Bremaecker* [1955], *Ruprechtová* [1958], and *Romney* [1959] succeeded in expressing only the general behavior of amplitudes of body waves. A step forward was made with the discovery by *Vaněk and Stelzner* [1960] of the oscillatory character of the fine structure of amplitude curves. This made it possible to begin serious attempts to apply the amplitude curves of body waves for the detailed study of the upper mantle structure.

Before examples of amplitude-distance curves are discussed, the method of constructing the amplitude curves of seismic body waves should be explained briefly. During the investigations it appeared very suitable to use observations of a network of near seismic stations for constructing the amplitude-distance curves. The stations should be equipped with instruments of comparable frequency characteristics, and the amplitude observations of the stations in question should be homogeneous. The last point appears to be very important, because amplitude observations at two different stations are generally not comparable. The neglecting of this fact usually leads to a large scatter of

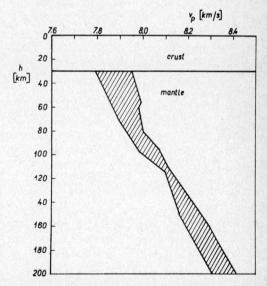

Fig. 2. Range of the probable velocity distribution in the region of the Kuril Islands [*Fedotov et al.*, 1964].

observations, which makes the derivation of the fine structure of the amplitude curve practically impossible. The simplest method of homogenization is to determine appropriate station corrections relative to one station of the network. The proximity of the stations also minimizes the influence of the earthquake mechanism on the observed amplitudes.

If, for the region investigated, a calibrating function for determining the magnitude M exists, it is not difficult to collect a set of values (A^*, Δ), on the basis of which the amplitude-distance curve $A^*(\Delta)$ can be constructed by the method of successive approximation. The process of approximation can be controlled by the corresponding mean quadratic errors. Usually three or four steps are sufficient.

If no magnitude calibrating functions exist for the region in question, another method must be used for deriving the first approximation of the amplitude-distance curve. For a network of near stations we have simultaneous observations of amplitudes and periods in a small distance interval $d\Delta$ at several stations for every earthquake. After applying the appropriate station corrections, we can determine for every earthquake the gradient of the amplitude curve $\alpha = da/d\Delta$, where $a = \log(A/T)$, and coordinate its value to the mean Δ in the interval $d\Delta$. Thus the gradient curve $\alpha(\Delta)$ can be constructed, and by integrating this we obtain the amplitude curve $A(\Delta)$. Then the method of successive approximation can be applied. The method of gradient curves is entirely independent of the magnitude M.

As an example, the amplitude-distance curve $A^*(\Delta)$ of the vertical component of P waves derived from observations of shallow earthquakes for the northern part of Asia Minor is shown in Figure 3 with the corresponding observations.

The shape of the amplitude-distance curves of seismic body waves is influenced by the velocity-depth distribution $v(h)$, as well as by the distribution of attenuation of seismic waves in different parts of the upper mantle. It seems, however, that the main peculiarities of the amplitude-distance curves, as, for example, their oscillatory character, are influenced rather by the velocity distribution.

The interpretation of amplitude-distance

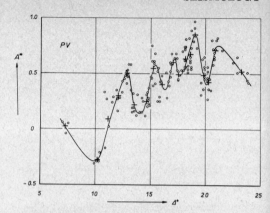

Fig. 3. Amplitude curve $A^*(\Delta)$ of the vertical component of P waves for Asia Minor [Vaněk, 1966b]; observations of selected Caucasian stations are denoted by circles, and representative points of natural intervals by crosses.

curves in connection with the velocity distribution $v(h)$ is a difficult problem. Theoretical investigations, which are mostly limited to the application of the ray theory [Keilis-Borok, 1966], do not at present allow for the solution of some basic problems of the propagation of elastic waves in inhomogeneous media. For this reason, measurements were performed on both two-dimensional [Riznichenko and Shamina, 1964; Shamina, 1966] and three-dimensional seismic models [Vaněk, 1966a] of the upper mantle with controlled velocity-depth distribution. On the basis of these investigations, the scheme of the possible upper mantle structure in southeastern Europe can be presented.

The amplitude-distance curve $A^*(\Delta)$ of P waves for southeastern Europe (Figure 4) is characterized by large oscillations for distances smaller than 10°. It follows from theoretical considerations and from the study of amplitude

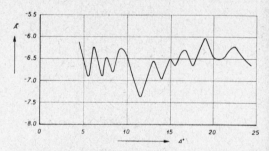

Fig. 4. Amplitude curve $A^*(\Delta)$ of P waves for southeastern Europe [Vaněk and Radu, 1964; revised Vaněk and Stelzner, 1962].

curves on seismic models that this part of the amplitude-distance curve can be interpreted as a mixture of amplitude curves of waves reflected at several discontinuities beneath the Mohorovicic discontinuity (reflected waves in the broad meaning of the word [Červený, 1961]). For the interpretation of the amplitude curve at distances greater than 10°, several different types of velocity distribution were tested. It was found that only the variant combining the low-velocity channel at the depth of about 100 km (a relatively weak negative gradient is admissible) and a discontinuity at the depth of about 400 km (discontinuity of the second order is sufficient) could explain the shadow zone around 11°, the increase of amplitudes between 12° and 19°, and the two maximums at 19° and 21-22° [Vaněk, 1966a]. In this connection, it is well to mention the famous discussion between *Gutenberg* [1948] and *Jeffreys* [1936, 1937, 1952] on the nature of the so-called '20° discontinuity.' It appears that their variants of the velocity distribution in the upper mantle are not mutually inconsistent. On the contrary, a combination of them seems to be necessary for explaining the features of the amplitude-distance curve observed for P waves. Two further oscillations with clear maximums at 13° and 15° can be observed in Figure 4. If these two maximums are interpreted as maximums of amplitude curves for reflected waves (by analogy with the oscillations at distances smaller than 10°), two further discontinuities may occur at depths of about 150 and 200 km. The interpretation of the maximum near 17° is not yet clear.

Comparison of the scheme of the possible upper mantle structure in southeastern Europe with the results obtained for central Asia by the study of deep earthquakes (Figure 1) leads to two suggestions, namely, that the actual structure of the upper mantle is much more complicated than that given by the Bullen model of the earth, and that the main lateral variations of the upper mantle structure in continental areas can be expected in the region between the Mohorovičić discontinuity and the upper boundary of the low-velocity channel, including the depth of this boundary. The latter suggestion is supported also by the amplitude curves for the northern part of Asia Minor [*Vaněk*, 1966b], which are in good accordance with those for southeastern Europe except for the position of the minimum in the shadow zone (see Figure 3). Its shift to smaller distances indicates that the top of the low-velocity channel occurs at smaller depths in Asia Minor (compare with the results of deep seismic sounding at the Black Sea [*Neprochnov*, 1959] and with the results of *Shebalin* [1959]).

The possibilities of the method of amplitude-distance curves are certainly not exhausted. The study of amplitude curves in different frequency ranges can contribute to a more detailed knowledge of the nature of discontinuities and the magnitude of velocity gradients. The study of amplitude curves in different regions could be important for investigating the character of lateral inhomogeneities in the structure of the upper mantle.

REFERENCES

Červený, V., The amplitude curves of reflected harmonic waves around the critical point, *Studia Geophys. Geodaet.*, *5*, 319-351, 1961.

De Bremaecker, J. C., Use of amplitudes, 1, P_n from 3° to 23°, *Bull. Seismol. Soc. Am.*, *45*, 219-244, 1955.

Fedotov, S. A., and I. P. Kuzin, Velocity section of the upper mantle in the region of the south Kuril Islands, *Izv. Akad. Nauk SSSR, Ser. Geofiz.*, *1963*(5), 670-686, 1963.

Fedotov, S. A., N. N. Matveeva, R. Z. Tarakanov, and T. B. Yanovskaya, On the velocities of longitudinal waves in the upper mantle in the regions of the Japanese and Kuril Islands, *Izv. Akad. Nauk SSSR, Ser. Geofiz.*, *1964*(8), 1185-1191, 1964.

Gutenberg, B., On the layer of relatively low wave velocity at a depth of about 80 kilometers, *Bull. Seismol. Soc. Am.*, *38*, 121-148, 1948.

Iosif, T., On the question of the existence of a layer of low velocity in the upper parts of the earth's mantle in Rumania, *Rev. Roumaine Geol. Geophys. Geograph., Ser. Geophys.*, *9*, 13-27, 1965.

Jeffreys, H., The structure of the earth down to the 20° discontinuity, *Monthly Notices Roy. Astron. Soc., Geophys. Suppl.*, *3*, 401-422, 1936.

Jeffreys, H., The structure of the earth down to the 20° discontinuity, *Monthly Notices Roy. Astron. Soc., Geophys. Suppl.*, *4*, 13-39, 1937.

Jeffreys, H., The times of P up to 30°, *Monthly Notices Roy. Astron. Soc., Geophys. Suppl.*, *6*, 348-364, 1952.

Keilis-Borok, V. I., editor, *Computer Seismology*, vol. 2, *Machine Interpretation of Seismic Waves*, Izd. Nauka, Moscow, 1966.

Lukk, A. A., and I. L. Nersesov, Structure of the upper parts of the mantle from observations of earthquakes with intermediate focal depths, *Dokl. Akad. Nauk SSSR, 162*, 559–562, 1965.

Matveeva, N. N., and A. S. Alekseev, Machine test of variants of velocity structure of the upper part of the mantle by the system of travel times of deep-focus earthquakes, *Questions of the Dynamic Theory of Seismic Waves, 8*, 130–143, Leningrad, 1963.

Neprochnov, J. P., Deep structure of the earth's crust beneath the Black Sea southwest of the Crimea, according to seismic data, *Dokl. Akad. Nauk, SSSR, 125*, 1119–1122, 1959.

Riznichenko, Yu. V., and O. G. Shamina, Model study of the upper mantle shadow zone, *Tectonophysics, 1*, 443–448, 1964.

Romney, C., Amplitudes of seismic body waves from underground explosions, *J. Geophys. Res., 64*, 1489–1498, 1959.

Ruprechtová, L., Dependence of amplitudes of seismic body waves on the distance, *Studia Geophys. Geodaet., 2*, 397–399, 1958.

Shamina, O. G., Experimental investigation of necessary and sufficient characteristics of a waveguide, *Studia Geophys. Geodaet., 10*, 341–350, 1966.

Shebalin, N. V., Correlation between magnitude and intensity of earthquakes; asthenosphere, *Publ. BCIS, Trav. Sci., A, 20*, 31–37, 1959.

Tarakanov, R. Z., Travel-time curves of P and S-P waves and a velocity cross section of the earth's upper mantle according to data obtained from observations of earthquakes in the Kurils and Japan, *Izv. Akad. Nauk SSSR, Fiz. Zemli, 1965*(7), 90–101, 1965.

Vaněk, J., Amplitude curves of longitudinal waves for several three-dimensional models of the upper mantle, *Studia Geophys. Geodaet., 10*, 350–359, 1966a.

Vaněk, J., Amplitude curves of seismic body waves for the region of Asia Minor, *Trav. Inst. Geophys. Acad. Tchecosl. Sci., 1966*(246), *Geofysikalni sbornik*, 1966b.

Vaněk, J., Amplitude curves of seismic body waves and the structure of the upper mantle in Europe, *Tectonophysics, 5*, 235–243, 1968.

Vaněk, J., and C. Radu, Amplitude curves of seismic body waves at distances smaller than 12°, *Studia Geophys. Geodaet., 8*, 319–325, 1964.

Vaněk, J., and J. Stelzner, Oscillatory character of amplitude curves of seismic body waves, *Nature, 187*, 491–492, 1960.

Vaněk, J., and J. Stelzner, Amplitudenkurven der seismischen Raumwellen, *Beitr. Geophys., 71*, 105–119, 1962.

Yanovskaya, T. B., Calculating velocity cross sections from travel-time curves as an inverse mathematical problem, *Izv. Akad Nauk SSSR, Ser. Geofiz., 1963*(8), 1171–1177, 1963.

Yanovskaya, T. B., and I. J. Asbel, The determination of velocities in the upper mantle from the observations on P waves, *Geophys. J., 8*, 313–318, 1964.

Anisotropy of the Upper Mantle

Russell W. Raitt

ROCK-FORMING minerals tend to be anisotropic in crystal structure. The anisotropy in elastic wave velocity associated with the crystallographic structure can be very large. For example, in olivine the difference between maximum and minimum velocities is greater than 20% [*Hess*, 1964; *Verma*, 1960]. Hence, the presence of a statistical tendency for orientation of anisotropic crystals in a horizontal plane would show an effect on the velocities observed in seismic refraction experiments.

Even if the crystal lineations were not randomly distributed, inequality of horizontal stress can produce velocity anisotropy. Observations of this effect are not numerous. Measurements of *Tocher* [1957] on Barre granite show a velocity difference of the order of 9% for a stress difference of the order of 1 kb. From these observations, Tocher concluded that measurements of seismic wave anisotropy could be used as an indication of strain energy accumulation in seismically active areas, such as the San Andreas fault.

The practical consequences of this possible horizontal anisotropy of seismic wave velocity on the studies of crustal structure by seismic

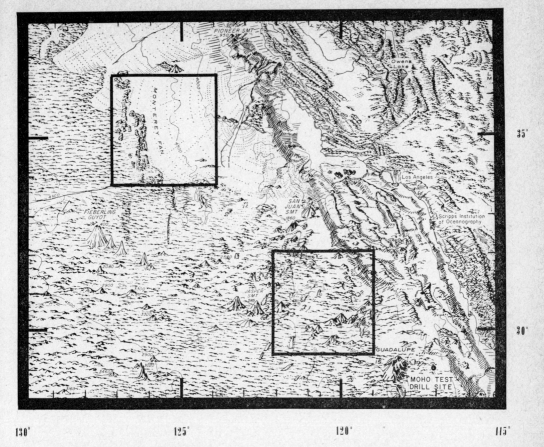

Fig. 1. Areas of anisotropy observations. The Flora observations were made within the upper left square, and the Quartet observations were made in the lower right square.

refraction methods have been largely ignored by the investigators in the field. In the comprehensive summary of worldwide seismic refraction results prepared by *Steinhart and Meyer* [1961], there was no mention of the existence of anisotropy. Refraction observations were rarely, if ever, made with the objective of determining this effect. Whatever velocity variations were observed were attributed to geographical variation of the composition or physical state.

However, *Hess* [1964] pointed out that Shor and Raitt's observations in the northeastern Pacific south of 45°N show correlation of velocity with profile direction. It appeared that nearly all the variation could be explained by anisotropy of velocity, with a maximum value of about 8.6 km/sec in the eastward direction and a minimum of 8.0 km/sec northward. These observations are widely scattered throughout the region between California and Hawaii, and it can be argued that the variation of velocity is caused by a variation of position. In other words, while anisotropy is consistent with the observations, it is not proved by them. In order to prove the existence of anisotropy, the observations showing it should represent different directions of propagation through the same material.

ANISOTROPY STUDIES

Backus [1965] has shown that the most general function describing the anisotropy of horizontal compressional velocity c as a function of angle ϕ is given by

$$c = c_0 + A \sin 2\phi + B \cos 2\phi + C \sin 4\phi + D \cos 4\phi \quad (1)$$

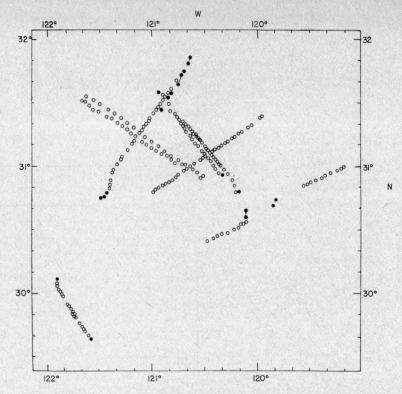

Fig. 2. Shot and receiver positions in the Quartet area. Open circles, shot positions; solid circles, receiver positions.

where ϕ is the angle measured counterclockwise from the direction east.

In principle, then, five measurements of velocity at five different angles would serve to determine the five constants of equation 1. In practice, however, the error of measurement, which is considerable, requires many more than five measurements for satisfactory description. Furthermore, these measurements should be distributed reasonably uniformly over all angles if possible.

Anisotropy studies have been made on two operations of the Scripps Institution of Oceanography, Quartet and Flora. The areas for these studies, shown in Figure 1, were chosen primarily because they are the areas nearest San Diego that are comparatively free from seamounts, and because each area had been the site of previous seismic refraction studies. Extensive anisotropy studies have also been made in the Hawaiian area. They are now being analyzed.

Shot and receiver positions are shown in Figures 2 and 3. Receiver positions were too numerous to be shown as individual points and are indicated by drift tracks for separate profiles. Only shots giving mantle arrivals are shown. Many short-range shots recorded for shallower arrivals are not shown.

The detailed travel time and position data of the observations depicted in Figures 2 and 3 are given by *Raitt et al.* [1967].

ANALYSIS OF DATA

Conventional methods of interpretation are not satisfactory for a study of this type of data. Solutions that assume constant velocity layers dipping uniformly apply to simple structure studies with short lines of shots and receivers. They are not easily generalized to the present situation, in which velocity is anisotropic and shots and receivers are widely distributed.

A method described by *Gardner* [1939], which was designed for areal seismic refraction

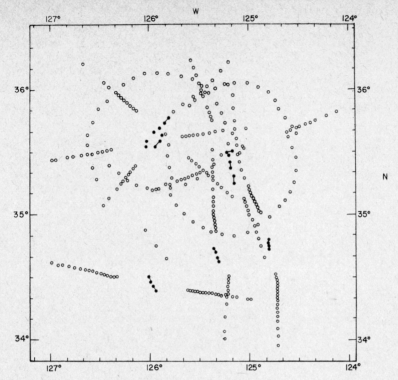

Fig. 3. Shot and receiver positions in the Flora area. Open circles, shot positions; solid circles, receiver positions.

studies, is readily adapted to anisotropy observations, which necessarily are distributed over a large area. Similar methods have been discussed by *Willmore and Bancroft* [1960] and *Scheidegger and Willmore* [1957]. An extensive application of the method was made recently by *Berry and West* [1966] and by *Smith et al.* [1966] for the data of the Lake Superior experiment of 1963.

In the Gardner-Willmore method, the mantle-refracted travel time T between surface points A and B separated by a distance D is given, approximately, by

$$T \cong D/c + \tau_A + \tau_B \qquad (2)$$

where c is the mantle velocity, and the quantities τ_A and τ_B are called delay times by Gardner, because they represent time delays in traversing the crust at points A and B. If the velocity function of depth within the crust is known at A and B, the depths to the upper mantle are determined from the delay times.

Solution of an areal study by this method involves the recording of travel times at many receiving positions from many shots properly distributed with respect to the receiving positions. Each measured travel time gives an equation of the form of equation 2. As each shot is received at several receiving positions, there are many more equations than there are values of delay time τ. These equations are solved by least squares for the values of τ and velocity c. If the velocity c depends on angle ϕ, equation 2 becomes nonlinear because of the velocity variation in the denominator of the first term of the right-hand side. However, as the anisotropy is small, equation 2 can be linearized for purposes of least squares solution by a simple modification:

$$T = D/c_0 + (2F - D)(A' \sin 2\phi \\ + B' \cos 2\phi + C' \sin 4\phi + D' \cos 4\phi) \\ + \tau_A + \tau_B \qquad (3)$$

where $A' = A/c_0^2$, $B' = B/c_0^2$, etc., and F is the average horizontal distance between the shot or receiver position and the point when the refracted wave enters or leaves the mantle.

There is no difficulty in reoccupying shot or receiver positions as many times as is necessary on land, but in seismic refraction work at sea, because of navigational problems and the drift of receiving ships, it is virtually impossible to do so. Hence each shot and receiving position are different, and, strictly speaking, the number of unknowns always exceeds the number of equations.

For the solution of the Quartet and Flora data, this problem was avoided by assuming that the delay times are described by a function of position that is fully specified by a limited number of parameters. Limiting the number of parameters to a quantity that is small compared to the number of positions is equivalent to assuming that the slopes and derivatives of the slopes of the refracting interface are small, an assumption that is already inherent in the delay time approach.

This method also has the advantage of reducing the size of the matrix to be inverted in the least squares solution. In the conventional method, the size of the matrix is determined by the number of shot and receiving positions. The solution of *Berry and West* [1966] was limited to a total of 117 shot and receiver positions by their matrix inversion program. In the Quartet and Flora data, satisfactory solutions were obtained using only 24 parameters, even though 289 shots were recorded in the Quartet operation and 303 shots in the Flora operation.

The functions used to describe the delay time surface can have a variety of forms, provided only that they are amenable to solution by least squares. The 19-parameter surface used to represent the Quartet and Flora data is formed by a uniform slope plus a double Fourier series with 16 terms. The three terms required to describe the slope plus the 16 terms of the Fourier series plus five terms of the velocity function represented 24 parameters altogether.

The equation of this surface is

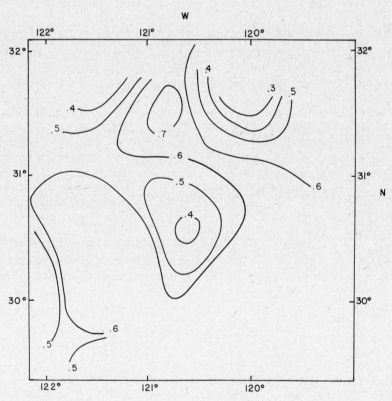

Fig. 4. Delay time surface for Quartet area. Contours represent delay times in seconds.

$$(x, y) = \tau_0 + ax + by$$
$$+ \sum_{n=1}^{2} \sum_{m=1}^{2} A_{nm} \sin(nkx) \sin(mky)$$
$$+ B_{nm} \sin(nkx) \cos(mky)$$
$$+ C_{nm} \cos(nkx) \sin(mky)$$
$$+ D_{nm} \cos(nkx) \cos(mky) \quad (4)$$

The delay surfaces representing these solutions are shown in Figures 4 and 5, which depict delay times in seconds for the areas of the Quartet and Flora studies. Figures 2 and 3 show that the shot and receiver locations are distributed very unequally. The significance of the contours in areas not containing shots or receivers is very uncertain. The gentle functions of these solutions represent considerable smoothing of the delay information, and they also give interpolation into vacant areas. This interpolation may be questionable; as a prediction of unobserved structure, it is probably very poor.

ANISOTROPY

The form of the delay time surface is worth a special study, but, in anisotropy studies aimed primarily at determining velocity variations, the topography is primarily a perturbation of the velocity determinations, whose effects must be eliminated from the velocity measurements. The results of the determinations of the five velocity coefficients for the two areas are listed in Table 1.

From this table, the magnitude of A and B, the coefficients of the sin 2ϕ and cos 2ϕ terms in the Backus anisotropy function, are much larger than the estimated standard errors. Furthermore, they agree approximately for the Quartet and Flora areas. The C and D terms, however, are not large compared to the standard error, and they are different in the two areas. The magnitudes of D in the Quartet area and of C in the Flora area appear to be statistically significant if the errors are distributed at random. However, the possibility of systematic

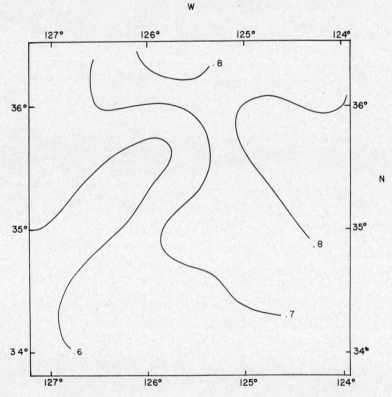

Fig. 5. Delay time surface for Flora area. Contours represent delay times in seconds.

TABLE 1. Velocity Coefficients

Coefficient	Magnitude, km/sec	Standard Error, km/sec
	For Quartet Data	
c_0	7.980	0.025
A	0.079	0.014
B	0.109	0.011
C	−0.007	0.011
D	−0.044	0.014
	For Flora Data	
c_0	8.139	0.015
A	0.054	0.013
B	0.159	0.009
C	0.025	0.009
D	−0.005	0.009

error makes the 4ϕ terms much less reliable than the 2ϕ terms.

If only the 2ϕ terms are considered significant, the description of the anisotropy is simplified, for only the A and B coefficients are involved. Defined by constants A and B, the directions of the anisotropy are nearly the same for the two areas. The angle ϕ of the direction of maximum velocity is 18° in the Quartet area and 9° in the Flora area. These correspond to azimuth angles of 72° and 81°, respectively. They agree well with the azimuth of maximum velocity calculated by *Backus* [1965], considering all of the stations west of California and Oregon. The magnitude of the effect is significantly less for Quartet and Flora, however. The velocity difference between maximum and minimum (as determined by the A and B coefficients alone) is 0.27 km/sec for the Quartet data, 0.34 km/sec for the Flora data, and 0.54 km/sec over-all.

This is reasonable, for even if the anisotropy were uniform throughout the Pacific west of California and Oregon, the apparent anisotropy could be biased by a geographical effect associated with a systematic dependence of azimuth on position. Hence, although the Quartet and Flora results support the presence of anisotropy in these areas, they do not necessarily prove its existence through the over-all area. From present results the over-all effect is about twice the Quartet and Flora values. One might reasonably conclude that about half the observed over-all velocity variation is geographical and about half could be attributed to anisotropy.

Acknowledgment. This work was supported by the Office of Naval Research under contract with the University of California. Contribution from the Scripps Institution of Oceanography.

REFERENCES

Backus, G. E., Possible forms of seismic anisotropy of the uppermost mantle under oceans, *J. Geophys. Res., 70,* 3429–3439, 1965.

Berry, M. J., and G. F. West, An interpretation of the first arrival data of the Lake Superior experiment by the time-term method, *Bull. Seismol. Soc. Am., 56,* 141–171, 1966.

Gardner, L. W., An areal plan of mapping subsurface structure by refraction shooting, *Geophysics, 4,* 247–259, 1939.

Hess, H., Seismic anisotropy of the uppermost mantle under the oceans, *Nature, 203,* 629–631, 1964.

Raitt, R. W., G. G. Shor, Jr., T. J. G. Francis, and G. B. Morris, Anisotropy of the Pacific Upper Mantle, Scripps Inst. Oceanog. Rept., 67–8, 1967.

Scheidegger, A. E., and P. L. Willmore, The use of a least squares method for the interpretation of data from seismic surveys, *Geophysics, 22,* 9–21, 1957.

Smith, T. Jefferson, John S. Steinhart, and L. T. Aldrich, Lake Superior crustal structure, *J. Geophys. Res., 71,* 1141–1172, 1966.

Steinhart, J. S., and R. P. Meyer, Explosion studies of continental structure, *Carnegie Inst. Wash. Publ. 622,* 1961.

Tocher, Don, Anisotropy in rocks under simple compression, *Trans. Am. Geophys. Union, 38,* 89–94, 1957.

Verma, R. K., Elasticity of some high-density crystals, *J. Geophys. Res., 65,* 757–766, 1960.

Whittaker, E. T., and G. Robinson, *The Calculus of Observations,* Blackie and Son, London, 1937.

Willmore, P. L., and A. M. Bancroft. The time-term approach to refraction seismology, *Geophys. J., 3,* 419–432, 1960.

Seismic Surface-Wave Data on the Upper Mantle[1]

James Dorman

WITHIN the past decade, surface-wave studies have become an important supplement to the body-wave methods applied by earlier workers. It was recognized about 25 years ago that various seismic body-wave velocity solutions were in good agreement for regions C and D, but differed significantly from site to site for region B (upper mantle, M discontinuity to about 400 km). The discrepancy was an early clue to the important geographic variations that undoubtedly exist in the upper mantle, and to the problems that have been the subject of the International Upper Mantle Project. Another important aspect of upper mantle investigations is the question of the existence of a low-velocity channel, which may cause a seismic shadow zone. Since body-wave methods are indirect in this case (see *Nuttli* [1963] for an excellent summary of this problem), a supplementary method for exploring this region is needed.

Surface waves assume considerable importance because they sample the elastic properties of the upper mantle, including a low-velocity channel, uniformly along the entire path between epicenter and recording station. The precise weighting of the sampling process with respect to depth in the earth is proportional to the amplitude of wave motion at depth. This generally is negligible below a depth equal to about one horizontal wavelength and is most important only within a half wavelength of the earth's surface. The horizontal sampling property of surface waves makes them valuable for reconnaissance, and provides more uniform geographic coverage at shallow depth than is possible with earthquake body waves. Methods of using surface waves for the study of the earth's upper layers depend mainly on the phenomenon of dispersion, whereby an individual Fourier component propagates with a phase velocity that is nearly the weighted average of shear velocity in the depth range sampled by its particular wavelength. Thus, since shear velocity varies strongly with depth, the velocity of surface waves, in turn, varies strongly with wavelength or period. Data discussed in this paper were recorded on long-period inertial seismographs of the Press-Ewing type and on strain seismographs of the Benioff type. Dispersion data are read from the seismograms by a simple measurement of frequency, or phase, versus arrival time (the peak-and-trough method) or by Fourier analysis of digitized seismograms. The close theoretical relationship between the dispersion characteristic (phase velocity versus period) and shear velocity structure (shear velocity versus depth) is then available for estimating the latter. *Nuttli* [1963], *Oliver and Dorman* [1963], *Bolt* [1964], *Anderson* [1965], and *Kovach* [1965] have reviewed the literature describing the general properties of Love and Rayleigh waves, and the methods for gathering and interpreting data.

DATA

No further summary of this literature will be attempted here. Rather, it appears useful to compile selected phase-velocity dispersion data, shown in Figures 1, 2, and 3, for a range of wavelengths which permits a qualitative comparison of important upper mantle effects for many regions of the world. The period range 20 to 300 sec covered in the figures corresponds roughly to wavelengths of 70 to 1500 km. These waves are all long enough to have appreciable amplitudes in region B, but the longest of them are short enough to vanish rapidly in region C. The figures are a compilation of data from published tables and graphs. Graphical data were transferred to these figures by use of an x-y digitizing table upon which copies of the published graphs were fastened. The locations of points in table coordinates were converted to the scale of the figures by computer, and the

---

[1] Contribution 1071 from Lamont-Doherty Geological Observatory of Columbia University.

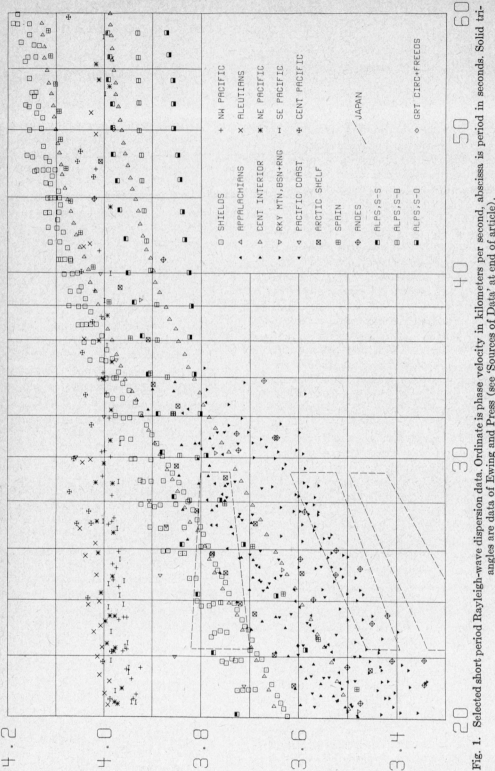

Fig. 1. Selected short period Rayleigh-wave dispersion data. Ordinate is phase velocity in kilometers per second, abscissa is period in seconds. Solid triangles are data of Ewing and Press (see 'Sources of Data' at end of article).

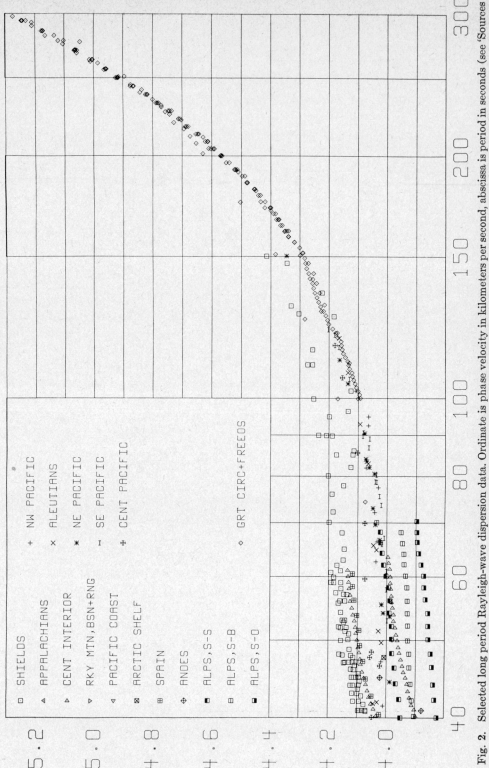

Fig. 2. Selected long period Rayleigh-wave dispersion data. Ordinate is phase velocity in kilometers per second, abscissa is period in seconds (see 'Sources of Data' at end of article).

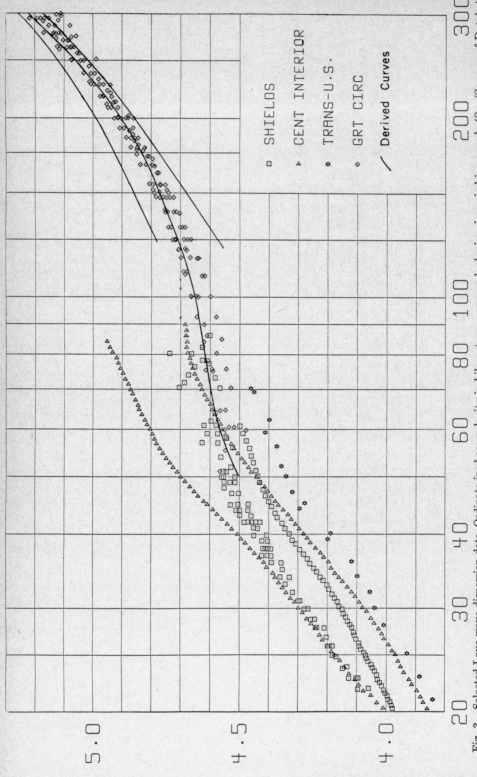

Fig. 3. Selected Love-wave dispersion data. Ordinate is phase velocity in kilometers per second, abscissa is period in seconds (See 'Sources of Data' at end of article). Derived curves by *Toksöz and Anderson* [1966] are calculated from great circle data, which are similar, though not identical, to the data shown. The derived curves represent, from top to bottom, shield, oceanic, and mountain-tectonic pure paths.

figures were plotted by an on-line incremental plotter. Plotting errors are judged to be small, probably less than the width of the plotting symbol in most cases; they arise mainly from the small scale and rarity of grid lines on some of the published figures. Fortunately, a minimum of only three identifiable grid intersections is required to relate the coordinates of points on a graph to the table coordinates. Thus, the figures permit reliable comparison of data from numerous sources.

This presentation of data is intended to facilitate a discussion of differences in gross geology or physical conditions of the crust and upper mantle between specific regions or tectonic provinces. A summary of dispersion data by *Oliver* [1962] covers a broader spectrum than that of upper mantle surface waves, and gives group-velocity as well as phase-velocity data. Oliver's figures show average observed dispersion curves and indicate the range of regional variations. The present figures make these variations more specific by identifying data collected from particular regions. Group-velocity data have been omitted here because greater scatter of the points obscures the clear relationships between various geologic provinces which emerge from the phase-velocity data. The greater scatter of group-velocity data is probably due mainly to the more complex relationship between layered structure and group-velocity dispersion. Unfortunately, world coverage for phase-velocity data is less complete than for group-velocity data. The most obvious gaps are lack of oceanic data for any pure paths except in deep basins of the Pacific and scarcity of data for Asia, Africa, and South America. Nevertheless, various tectonic types are fairly well represented. The shorter period data were selected to represent pure paths, i.e., paths lying only in one tectonic province, as much as possible.

PRINCIPAL DISPERSION EFFECTS

The major features of surface-wave dispersion are well known and can be seen clearly in the figures. Phase velocities throughout the range considered increase with period because of the general increase of rigidity with depth in the crust and upper mantle. However, flattening in the middle range of the curves, which occurs between 40 and 100 sec for Rayleigh waves and between 60 and 150 sec for Love waves, is due to the relatively slow increase or possible slow decrease of shear velocity with increasing depth in region B. The steep rise of both curves at longer periods is due to the strong downward increase of shear velocity in region C. Phase velocities of oceanic and continental Rayleigh waves are similar for periods greater than about 50 sec, reflecting the similarity of average continental and oceanic upper mantle structure. However, in Figure 1 the nearly constant phase velocity of oceanic Rayleigh waves, as contrasted with the strongly dispersed continental Rayleigh waves, is due to the much thinner oceanic crust. The continental Rayleigh-wave phase velocities thus approach the Rayleigh velocity of the upper mantle material as wavelength increases in this period range, while the wavelength of even the shortest oceanic Rayleigh waves shown is considerably greater than the crustal thickness, thus giving these waves a velocity nearly equal to the Rayleigh velocity of the upper mantle material.

Some other major effects visible in these figures may not be generally recognized. The first of these is the great spread of continental Rayleigh-wave velocity values as compared with the remarkably narrow band of oceanic velocities. This effect is accentuated in the figures by the selection of data, but is undoubtedly real. The oceanic data are mainly from undisturbed ocean basins, though paths that cross the Easter Island rise are included in the southeast Pacific category, while data from many tectonic provinces are included for the continents. However, if we consider only the old, undisturbed continental areas, i.e., shields, Appalachians, central interior of North America, and arctic shelf, we have stable regions that differ in average elevation of the continental surface by only a few hundred meters, which is comparable to the differences in average depth of the various ocean basins. Among the dispersion data for these continental regions we find a spread of at least 0.2 km/sec, or more than twice the spread of values for the oceanic basins.

In addition to the greater spread of continental dispersion values, the persistence of this spread out to very long periods also emerges clearly. The total spread of Rayleigh-wave values is greatest between 20 and 30 sec, about 0.5 km/sec, because of greater abundance of data in that range; few data for Japan, the area of the Rocky Mountains and Basin and Range province, and Peru are available for periods greater than 30 sec. The sum of available Rayleigh-wave dispersion data between 30 and 70 sec forms a band that is uniformly about 0.3 km/sec wide. The lower boundary of this band depends only on the path from Stuttgart to Oropa, which passes partly through the high Alps. Therefore, it seems virtually certain that the lower boundary would be lowered and the band widened by further data. Even without placing great weight on the surprisingly high Love-wave velocities observed by McEvilly, a similar uniform spread nevertheless exists throughout the Love-wave spectrum from 20 to 80 sec, though this is based on scantier data.

The observation concerning the small spread of oceanic dispersion values suggests remarkable lateral uniformity of the crust and upper mantle in deep basins of the Pacific. If these portions of the crust-upper mantle plate were formed along mid-ocean ridges in late geologic time as required by the concept of sea floor spreading, then the plate formed by this process is very uniform in space and time. The surface wave evidence does not preclude this, but an alternative hypothesis serves equally well here: specifically, that these regions owe their uniformity to their being remnants of the primitive outer layers of the earth. The latter would agree with the well known conclusion from heat flow data that the upper mantle beneath ocean basins is petrologically undifferentiated. In contrast with the ocean basins, the older portions of continents mentioned above clearly differ more among themselves in terms of dispersion. This may reflect greater heterogeneity or geographic variability resulting from differentiation which progressed to various stages of completion in different areas.

The persistent spread of continental dispersion values out to long periods means that the strong contrasts in surface conditions between old, stable areas and young, disturbed areas of the continents are reflected at great depth by equally important contrasts in physical properties. The strong differences probably persist to depths of at least 200 km, judging by the wavelengths involved. In any case, it is clear that the important seismic contrasts between disturbed and undisturbed tectonic provinces are not confined mainly to the crust, nor are they mainly differences in crustal thickness; rather they are differences distributed through a much thicker zone.

COMPARISONS OF CONTINENTAL TECTONIC PROVINCES

The high, middle, and low Alpine dispersion curves represent, respectively, a path north of the Alps crossing the Schwarzwald and the relatively low elevation of the Rhine graben, a more southerly path over the same crustal elements, and a path crossing the western Alps to Oropa near Turin.

By contrast, the three groups of Japanese data do not arrange themselves in Figure 1 in the inverse order of topographic elevation of the regions represented. The Chubu district has the highest elevations in Japan as well as the lowest phase velocity observed anywhere, but the southwest Kanto district, with intermediate phase velocities, has the lowest topographic elevations of the three areas represented. Also, the apparent cross-cutting trend of the east Tohoku data is a striking feature. The Alpine data exhibit a similar tendency with respect to data from regions of stable crust. This phenomenon, as noted by Aki, indicates that a single model of velocity structure with a depth scale that varies from region to region cannot satisfactorily represent crust-mantle structure for all continental areas.

The highest phase velocities observed for continents are those of the Canadian shield. The Finnish Rayleigh-wave data (the uniform line of 'shield' data points in Figure 1, beginning at 3.62, 20) rise to approach the Canadian shield data as period increases. This indicates that conditions in the upper mantle are similar in the two regions.

Love- and Rayleigh-wave data can be compared for three areas: Finland, Canada, and the central interior of North America. The Love-wave curves are not parallel to one another, and the same remark applies, though to a lesser degree, to the Rayleigh-wave curves. Also, the difference between the Love and Rayleigh velocities is two or more times greater for the central interior than for Finland and increases with period. The Canadian shield is intermediate in this respect. These phenomena suggest some interesting, though yet unknown, regional variations in crust-mantle layering and possible effects of widespread elastic anistropy in the upper mantle. The latter topic has been investigated theoretically by Anderson [1961].

GREAT CIRCLE PATHS

At periods greater than 100 sec, nearly all available data represent long, mixed paths or, in the case of free oscillations, an averaging on a worldwide scale. A range of velocities of about 0.1 km/sec, which does not appear to decrease appreciably with increasing period, is present in the Love-wave data. The precision of these data is sufficient to establish that all great circle paths are not identical in average phase velocity. Toksöz and Anderson [1966] have applied a method for determining pure, long-period phase velocity curves for each of three tectonic categories, oceanic, mountain-tectonic, and shield, using great circle Love-wave data; they find the three derived dispersion curves shown in Figure 3. The mountain-tectonic curve is lowest, the shield curve highest, and the ocean curve intermediate.

INTERPRETATION OF DATA IN TERMS OF MANTLE SHEAR VELOCITY MODELS

The three derived curves of Toksöz and Anderson correspond to the situation evident between periods of 40 and 70 sec in Figure 2, where the Stuttgart-Oropa data fall about 0.3 km/sec below the shield data, and the oceanic data have intermediate velocities. Layered shear-velocity models shown in Figure 4 fit the indicated Rayleigh wave data throughout the spectrum of available points, as is shown in Figures 1 and 2. The interpretations of dispersion data in terms of layered shear velocity models give a reliable comparison of shear velocity structure in the three regions down to depths of about 200 km. The shield model has the greatest mechanical stability, while the more pronounced low-velocity channel of the oceanic model, or especially the Alpine model, favors differential vertical mobility for the corresponding regions. The Japanese phase velocity data, though limited to very short periods, are the lowest measured and thus indicate lower mantle shear velocities for that region than in the ALPS model. Many other regional contrasts can be inferred by comparison of the other data in Figures 1, 2, and 3. Basically, all must be temperature and/or chemical-mineralogical differences related to the differences in rigidity and density of the materials that affect surface wave velocity directly. Below 200 to 300 km, it is clear that shear velocity increases downward in each model. However, owing to the wavelength limitation of the Alpine data, the differences in methods of deriving the three models, and the smaller shear velocity differences indicated, the relative positions of the three curves may not be reliable at greater depths.

Thus, based on present incomplete data, the principal regional differences seem to occur in the upper few hundred kilometers. Moreover,

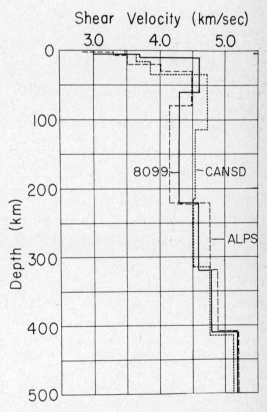

Fig. 4. Three shear velocity models for the upper mantle derived from Rayleigh-wave dispersion. The ALPS model [Seidl et al., 1966] fits the Stuttgart-Oropa data shown in Figures 1 and 2. The 8099 model fits the Pacific ocean data [Kuo et al., 1962]. The CANSD model [Brune and Dorman, 1963] fits the Canadian shield data.

the regional differences in upper mantle shear velocity may be as great as any corresponding differences in the crust. The M discontinuity is not an upper or lower bound to these regional contrasts, but is only incidental to them. Thus the entire zone from the surface to a few hundred kilometers depth can be called the active geologic layer.

SOURCES OF DATA FOR FIGURES 1–3

Abbreviations in the paragraphs below are as follows: BSSA, Bulletin of the Seismological Society of America; BGSA, Bulletin of the Geological Society of America; JGR, Journal of Geophysical Research; BERI, Bulletin of the Earthquake Research Institute; ZG, Zeitschrift für Geophysik. Following the journal abbreviation are two numbers: year of publication and page number. Four repeated references are:

EP, Ewing and Press, BGSA–59.
BNO, Brune, Nafe, and Oliver, JGR–60.
SMK, Seidl, Mueller, and Knopoff, ZG–66–479, Figure 5.
KBM, Kuo, Brune, and Major, BSSA–62–342 et seq.

The following list of sources corresponds to the sequence of symbols in the legends of Figures 1–3:

Shields. Canada, Brune and Dorman, BSSA-63-175, 179; Finland (single lines of regularly spaced Love and Rayleigh points), Noponen, BSSA-66-1101.

Appalachians. EP-242-Figure 9.

Cent. Interior. Central U. S., EP-241-Figure 8 except Tuc., Bou., Lub.; McEvilly, BSSA-64-2006 (represented by lines of open triangles defining the upper and lower limits of many scattered points).

Rocky Mountains and Basin and Range. EP-240-Figure 6 except Pas., Reno., Bo.C., 241-Figure 7, Figure 8 Tuc., Bou., Lub. only; Nevada to Laramie, BNO-293.

Pacific Coast. EP-240-Figure 5, Figure 6 Pas., Reno, Bo.C. only; San Francisco Bay region, Evernden, BSSA-54-178.

Trans-U. S. BNO-298.

Arctic Shelf. BNO-295 Novaya Zemlya to Uppsala, 297 Novaya Zemlya to Resolute.

Spain. Payo, BSSA-65-736.

Andes. Peru, Cisternas, BSSA-61-385.

Alps, S-S. Stuttgart to Strasbourg, SMK.

Alps, S-B. Stuttgart to Besancon, SMK.

Alps, S-O. Stuttgart to Oropa, SMK.

Japan. East Tohoku district (upper polygon), Aki, BERI-61-274-Figure 8; Southwest Kanto district (middle polygon), Aki; Chubu district (lower polygon), Aki.

NW Pacific. Japan, Kurils, Kamchatka to Suva, KBM-Figures 10, 11, 12.

Aleutians. Fox and Andreanof Islands to Suva, KBM-Figures 13, 14.

NE Pacific. Coast to California and Mexico to Suva, KBM-Figures 15, 16, 17.

SE Pacific. Peru, Easter Island, Chile, and Drake Passage to Suva, KBM-Figures 18, 19, 20, 21.

Cent. Pacific. Honolulu to Mt. Tsukuba, KBM-Figure 26; Marshall Islands to Honolulu, Pomeroy, BSSA-63-130.

Great Circle and Free Oscillations. Rayleigh waves: great circles from Assam to Pasadena and Alaska to Uppsala, etc., Brune et al., JGR-61-2902-Figure 5; Chile-Isabella great circles, Press et al., JGR-61-3482-Figure 18; Mongolia-Pasadena great circles, Ben-Menahem and Toksöz, JGR-62-1948-Table 2; Assam-Pasadena great circle (independent analysis of the same event reported by Brune et al.), Toksöz and Ben-Menahem, BSSA-63-746-Table 3; free oscillations $_0S_{25}$ through $_0S_{42}$, Alsop, BSSA-64-768-Table 4. Love waves: Satô's data for great circles from Kamchatka and New Guinea (1938) to Pasadena, great circles from Chile to Nana, Isabella and Ogdensburg, path from Rio de Janiero to Mt. Tsukuba (New Guinea earthquake of 1958), Brune et al., JGR-61-2905-Figure 7; Peru-Uppsala great circles, Båth and Lopez Arroyo, JGR-62-1938-Table 1; great circles from Mongolia, Assam, Kamchatka, Alaska, and New Guinea to Pasadena and from Alaska to Wilkes (New Guinea data from the same seismograms analysed by Satô, and Kamchatka data from phases not used by Satô), Toksöz and Ben-Menahem, BSSA-63-758-Figure 17 (all data); Gaulon's data, Anderson and Toksöz, JGR-63-3498-Figure 13.

Acknowledgements. This work was sponsored by the Advanced Research Projects Agency, Project Vela Uniform, under Contract Air Force 19 (628) 4082. The author is indebted to Drs. J. Oliver, P. W. Pomeroy, L. R. Sykes, and M. Talwani for reading the manuscript critically.

REFERENCES

Anderson, Don L. Elastic wave propagation in layered anisotropic media, *J. Geophys. Res., 66,* 2953–2963, 1961.

Anderson, Don L., Recent evidence concerning the structure and composition of the earth's mantle, *Phys. Chem. Earth, 6,* 1–131, 1965.

Bolt, Bruce A., Recent information of the earth's interior from studies of mantle waves and eigenvibrations, *Phys. Chem. Earth, 5,* 55–119, 1964.

Brune, James, and James Dorman, Seismic waves and earth structure in the Canadian shield, *Bull. Seismol. Soc. Am., 53,* 167–210, 1963.

Dorman, James, Maurice Ewing, and Jack Oliver, Study of shear-velocity distribution in the upper mantle by mantle Rayleigh waves, *Bull. Seismol. Soc. Am., 50,* 87–115, 1960.

Kovach, Robert L., Seismic surface waves: some observations and recent developments, *Phys. Chem. Earth, 6,* 251–314, 1965.

Kuo, John, James Brune, and Maurice Major, Rayleigh wave dispersion in the Pacific Ocean for the period range 20 to 40 seconds, *Bull. Seismol. Soc. Am.*, *52*, 333–357, 1962.

Nuttli, Otto, Seismological evidence pertaining to the structure of the earth's upper mantle, in *Rev. Geophys.*, *1*, 351–400, 1963.

Oliver, Jack, A summary of observed seismic surface wave dispersion, *Bull. Seismol. Soc. Am.*, *52*, 81–86, 1962.

Oliver, Jack, and James Dorman, Exploration of sub-oceanic structure by the use of seismic surface waves, in *The Sea*, vol. 3, edited by M. N. Hill, pp. 110–133, Interscience Publishers, London, 1963.

Seidl, D., St. Müller, and L. Knopoff, Dispersion von Rayleigh-wellen in Südwestdeutschland und in den Alpen, *Geophys.*, *32*, 472–481, 1966.

Toksöz, M. Nafi, and Don L. Anderson, Phase velocities of long-period surface waves and structure of the upper mantle, 1, Great-circle Love and Rayleigh wave data, *J. Geophys. Res.*, *71*, 1649–1658, 1966.

Higher-Mode Surface Waves

Z. Alterman

THEORY predicts an infinity of normal modes for waves of the Rayleigh type, both for spherical models of the earth and for plane layered models of the continental crust-mantle system.

Recent papers indicate increasing interest in the use of higher modes of surface waves as a tool for obtaining information about the low velocity zone in the upper mantle, about fault plane orientation, the focal depth, and the nature of the source. These papers deal with the observation of higher modes, approximations to spherical geometry, and with occurrence of higher modes in complete theoretical seismograms for a source in a sphere.

The importance of the second shear mode was pointed out by *Oliver et al.* [1959]. Their comments also apply to higher modes. Some examples are: (1) higher mode waves make up a significant portion of some seismograms, particularly when seismographs favoring the intermediate period range are used; (2) when the dispersion data for these waves are coupled with data for the lower modes, improved models of crustal structure can be deduced; (3) relative excitation of the various modes may ultimately prove useful in deducing the nature of the source.

OBSERVED HIGHER MODES

The fundamental Rayleigh mode and the second Rayleigh mode (first shear, or M_{21}) were identified by *Oliver and Ewing* [1957, 1958] on records of several North American stations from the arctic earthquake of June 3, 1956, and from several later events. Waves corresponding to the next higher mode were identified by *Oliver et al.* [1959] for two paths, one from Belgian Congo to Pietermaritzburg, South Africa, the other from Oklahoma to Palisades, New York. The identification of this mode is based on surface particle motion and an agreement between observed dispersion of these waves and theoretical calculations of this dispersion for assumed crust-mantle models. *Ewing et al.* [1959] found higher modes for oceanic, continental, and mixed paths, mainly the continental second shear mode, in an analysis of transients with an electronic sound spectrograph.

A search for higher modes through several year's records of Swedish stations was performed by *Crampin* [1964]. The records from all Swedish stations for the interval June 1961 to July 1963 were examined for the presence of higher modes. Crampin found that, of the 207 earthquakes recorded with appropriate amplitude on long period seismograms, 99 (43%) showed evidence of higher modes. They were mainly from earthquakes whose epicenters were in Eurasia. In these observations the second Rayleigh mode was the commonest higher mode. However, some records showed higher Love modes with no corresponding second mode. Other records showed third modes with no corresponding second mode. Records from deep-focus earthquakes showed higher modes when the fundamental mode surface waves were nearly absent. The Rayleigh modes have much larger amplitude on the vertical component than on the horizontal components, and consequently at times appear only on the vertical. The theoretical seismograms of *Alterman and Abramovici* [1966] and the studies of *Anderson and Toksöz* [1963] and *Kovach and Anderson* [1964] agree with these findings.

The higher modes of seismic surface waves do not appear as a pure wave train on observed seismograms. They are usually carried as a rider on the larger-period fundamental modes, and each higher mode may be confused with other higher modes and with microseisms whose periods are usually only slightly smaller. *Crampin and Båth* [1965] examined the possibility of mode separation by using a digital computer to filter seismograms in order to get a pure higher-mode wave train. As a result of filtering, the authors found third- and fourth-order modes that were hidden by the second modes.

Further improved techniques used by Landisman (private communication) reveal several of the higher modes. A review of observation of channel waves, such as the seismic phases S_a, G, L_i, L_{g_2}, and π_g, is given by *Kovach* [1965]. The interpretation of the channel waves by higher-mode group velocity dispersion curves was first suggested by *Oliver and Ewing* [1958] and received detailed verification by *Kovach and Anderson* [1964]. See also *Båth and Crampin* [1965].

PLANE LAYERED MODELS AND INFLUENCE OF SPHERICITY

In obtaining surface waves from a normal mode analysis, different ranges of periods can be analyzed on appropriate assumptions [*Alterman et al.*, 1961]:

1. The propagation of waves of periods T of 600 sec and higher is governed by the free modes of oscillation of the earth of spherical harmonic order n ranging from 2 to 10. The amplitude of these modes extends from the surface of the earth down into the core. Their motion is governed not only by the elastic restoring forces, but also by forces arising from the perturbed gravitational field. These modes must therefore be analyzed by the complete normal mode method by solving the system of six differential equations given by *Alterman et al.* [1959, equations 28, 33, 35–39].

2. The amplitude of higher spheroidal modes in the range $17 < n < 25$ becomes negligibly small throughout the core, so that it is sufficient to carry out the integrations only from the bottom of the mantle to the surface.

3. Increasing the mode number to $25 < n < 200$, the analysis can be considerably simplified by taking account of the approach to flat earth conditions, without, however, altogether neglecting the still appreciable effect of the sphericity of the earth. This is accomplished by the use of the earth-flattening approximation.

4. Finally, the standard flat earth approximation with neglect of gravity is valid for computation of dispersion curves for periods $T < 50$ sec for the fundamental spheroidal mode, or for even smaller periods in the case of higher modes. This method has been used extensively by *Stoneley* [1953], *Press and Takeuchi* [1960], *Dorman et al.* [1960], *Harkrider* [1964], and *Mooney and Bolt* [1966] and in further papers mentioned in Mooney and Bolt's bibliography. *Sykes and Oliver* [1964] used this method for the analysis of short-period seismic surface waves and their higher modes, which come across oceanic areas.

While the emphasis in this investigation on ranges of applicability of various approximations was on first-mode surface waves, the ranges of approximation were verified again by *Kovach and Anderson* [1964]. They calculated higher mode Love and Rayleigh waves in the period range 1 to about 50 sec for several continental and oceanic models in order to demonstrate the behavior and sampling ability of the higher mode waves. Comparison of dispersion results for equivalent spherical and flat models showed that sphericity should be considered for the higher Rayleigh modes if accurate deductions are to be made about the crust-mantle system. For example, the effect on the M_{21} mode is surprisingly quite pronounced for wave periods down to about 8 sec, and the flat and spherical results have not merged even for periods as short as 4 sec. However, the effect on the group velocity curve of the M_{21} mode is found to be negligible for periods less than 30 sec.

Kovach and Anderson [1964] found that the shape of the higher-mode Love- and Rayleigh-wave group velocity curves is very sensitive to details of the layered wave guide. For example, the plateau structure determined for the higher modes in the Jeffreys-Bullen A model becomes oscillatory with broad maximums and sharp minimums for a Gutenberg-type velocity distribution. Mooney and Bolt [1966] carried out an extensive investigation of the dispersive characteristics of a single elastic layer overlying an elastic half-space for the fundamental and the first and second higher modes of Rayleigh waves. Phase velocity, group velocity, and the ratio of horizontal to vertical surface displacement are computed for several layered models. The most important parameter of the structure for Rayleigh wave dispersion is found to be the shear velocity ratio. Variations in the Poisson's ratio in the surface layer and the density contrast may produce substantial effects. Poisson's ratio in the half-space is of least significance. Crampin [1967] studied the possibility of coupled Rayleigh-Love second modes. He observed that second-mode seismic wave trains along many paths in Eurasia have a phase relationship between the vertical motion and the transverse horizontal motion that cannot exist for elastic waves in isotropic layered media. These coupled, generalized, surface waves are assumed to be caused by an anisotropic layer immediately beneath the crust.

HIGHER-MODE SURFACE WAVES IN SEISMOGRAMS COMPUTED FOR A SOURCE IN A SPHERE

The following questions arise: (1) Which modes among the infinite sequence of possible higher modes are expected to occur for a given source? (2) What is the dependence of their amplitude on the mechanism, depth of source, and earth model? (3) Is it possible to deduce the nature of the source and the structure of the earth from the higher modes?

Let us consider first the simple example of a homogeneous elastic sphere. The exact theoretical response of a sphere to a P pulse was obtained by Alterman and Abramovici [1965, 1966]. Figure 1 shows the group velocity dispersion curves for the first ten modes. The first mode, $j = 1$, is separate and shows in the limit of period $T = 0$ a velocity U equal to the velocity of Rayleigh waves c_R in a homogeneous half-space. The limit for $j = 2$ is at the shear wave velocity c. Dispersion curves for $j > 3$ each have a minimum and a maximum near $U/c = 1.3$. The first curve in Figure 2 shows the radial displacements w of the surface of a sphere of radius a, at an epicentral distance of 135° from a P-wave source located at $a/32$ depth in the sphere. The second curve is for a similar source at $a/8$ depth. They approximate

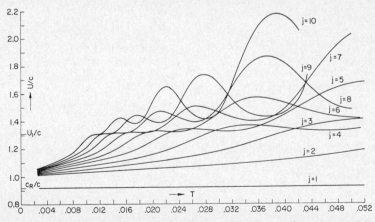

Fig. 1. The group velocity U in the elastic solid sphere as a function of the period T. c is the shear velocity, c_R is the Rayleigh-wave velocity in a half-space, and U_1 is the group velocity of the lowest Airy phases.

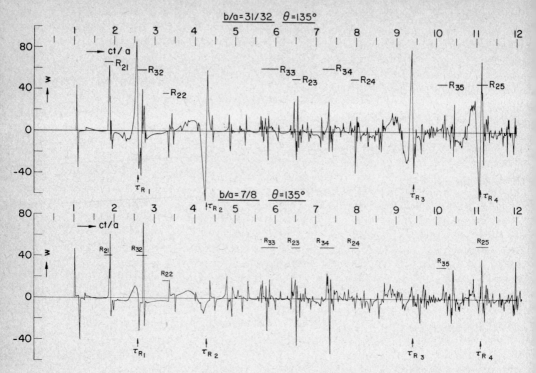

Fig. 2. Radial displacement w at distance $\theta = 135°$ from a P-wave point source at $a/32$ and $a/8$ depth in a sphere of radius a. Arrival of Rayleigh waves is marked τ_{R_i}, higher-mode surface waves are indicated by horizontal lines.

sources at 200 and 800 km depth in the earth. Figure 3 shows the angular component of displacement at the same distance. The time τ is measured in units of radius per shear-wave velocity, $\tau = ct/a$. A surface wave of group velocity $1.3c$ is found at a time $\tau = 1.8$ and denoted R_{21} on its first arrival in a clockwise direction. Coming in an anticlockwise direction, it is denoted R_{22}. After a complete circuit in either direction, it is denoted R_{23} or R_{24}, respectively.

The dispersion curves in Figure 1 have a next set of extremum values near $U/c = 1.4$. The seismograms show a surface wave R_{3i}, for $i = 2, 3, 4$, of velocity $U/c \sim 1.4$. One single surface wave is here associated with extremum values of several dispersion curves.

The dispersion curves indicate which higher-mode surface waves are expected to occur in a given model. However, it depends on the nature of the source which surface waves will actually occur on a seismogram.

The influence of source properties on the higher modes of surface waves is shown in the seismograms computed by *Alterman and Abramovici* [1967] in a comparison of three types of sources located at several depths in the homogeneous sphere: (1) the SV-P source consists of sudden application of a force at a point in the sphere; the force, being in a radial direction, causes propagation of both an initial P pulse and an initial SV pulse; (2) a spherically symmetric point source of compressional waves, the P-wave source; and (3) an SH-torque source.

We shall discuss here the differences in the higher modes only. The R_{2i} surface wave has a larger angular than radial component, both for the SV-P and the P sources. The R_{2i} are the largest pulses in the angular component from a deep P source where the fundamental Rayleigh R_{1i} is very small (see Figures 2 and 3). R_{2i} decreases slightly with decreasing depth of source. For a source at 200 km the modes R_{1i} and R_{2i} have comparable amplitude.

For the SV-P source, the angular R_{2i} are

relatively smaller than for the P source, they do not increase with decreasing depth of source, and there are additional guided waves, denoted GV_r, which have about the same angular amplitude and are even larger in the radial component.

The GV_r waves are groups of n-times reflected S_n pulses. The connection between these guided waves and reflected pulses is similar to the connection between reflected SH waves and the Love waves G_r from the SH-torque source. The GV_r waves have both radial and angular component. However, the radial component is two to three times larger than the angular component.

GV_r is the largest wave from the deep SV-P source. Its radial component is two to three times larger than the main other pulses from the deep SV-P source. It is the largest wave after R_{14} in the radial direction from the medium and shallow SV-P sources. The angular component of GV_r is about equal to the angular component of R_{24}. The GV_r wave is absent from the results for the P source. GV_r is clearer in the results from the source at medium depth than from the shallow source. The GV_r waves are comparable to results obtained by Oliver and Ewing [1958] for the M_2 mode, which is controlled primarily by SV waves. Kovach and Anderson [1964] found waves of similar properties in the vertical component seismograms. Sykes and Oliver [1964] found a family of shear modes controlled primarily by multiply reflected SV waves in a sedimentary layer across oceanic areas, in the same manner as Love modes are controlled by SH waves.

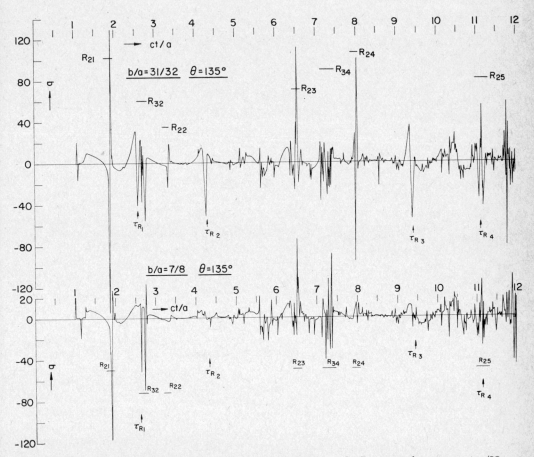

Fig. 3. Angular displacement q at distance $\theta = 135°$ from a P-wave point source at $a/32$ and $a/8$ depth in a sphere of radius a. Arrival of Rayleigh waves is marked τ_{R_i}, higher-mode surface waves are indicated by horizontal lines.

The main properties of the R_{2i} and GV_r waves are:

1. The angular component of R_{2i} is larger than the radial component for $10° < \theta < 170°$. In this range of distances the ratio of the components of R_{2i} from the SV-P source is in most cases 3:1. For the P source it is from 2:1 to 4:1.

2. The R_{2i} waves are clearest for the sources at medium depth. The R_{2i} from the SV-P source are similar to the higher modes of surface waves shown by Oliver and Ewing [1957] and by Båth and Crampin [1965].

3. As the GV_r waves have larger radial than angular component, the computed seismograms show the R_{2i} from the SV-P sources clearly in the angular component. The amplitude of the angular R_{2i} is comparable to the amplitude of the angular GV_r. The radial component of the GV_r is from three to six times larger than the radial R_{2i}.

4. The P source does not emit GV_r waves. The R_{2i} are the largest pulses in the angular component from the deep and middle sources. Their amplitude increases with decreasing depth of source.

The R_{3i} waves are found mainly in the radial component for both the P source and the SV-P source. From the source at medium depth, the radial R_{3i} are larger than the radial R_{2i} and smaller than the radial GV_r, the ratio of amplitude being about 1:4. From the deep source and from the shallow source, the radial R_{3i} are comparable to the radial R_{2i}. The ratio of amplitudes of R_{3i} to R_{2i} from the P source is similar.

Figure 1 shows that modes of different j contribute to a given wave R_{ki} at different frequencies. However, the properties of R_{ki} are independent of the mode number j and are unchanged through all the range of admissible frequencies. Usami et al. [1965] obtained theoretical seismograms for a surface source in an earth model based on the velocities of Gutenberg and Bullen's A' density distribution. The theoretical seismograms represent the superposition of spheroidal oscillations through the tenth radial mode, for all orders from the gravest, with periods approaching one hour, to those with periods slightly larger than one minute. Among the various interesting results obtained, the authors discussed the occurrence of a higher-mode surface wave which is found to correspond primarily to the group velocity minimum and adjacent maximum of the third radial mode for periods between 300 and 500 sec. Taking a summation of contributions from only the higher modes of the elastic model considered, a high velocity, dispersive higher-mode signal is found and named R_h. Usami et al. find that the colatitudinal component of R_h is mainly related to the third mode, whereas the radial component is the effect of a mixture of contributions, primarily from radial modes 2 and 3. The wave form of the colatitudinal component exhibits more regularity than does the radial component, and the beginning of this wave can clearly be identified on their curves. The travel time of the onset of this wave coincides well with that to be expected for a wave having maximum group velocity associated with the radial mode 3 for periods near 500 sec, which is approximately 7.0 km/sec. Usami et al. conclude, judging from the character of the R_h wave described above, that it might be possible that this wave could be found on the horizontal component of an ordinary teleseismic recording.

It stands to reason that the colatitudinal component of R_h is in fact the R_{2i} wave in the homogeneous model and that the radial component of R_h, which is found to be composed of modes 2 and 3, really includes the effect of GV_r and R_{2i} mentioned above.

HIGHER MODES REPRESENTED BY BODY WAVES

The exact seismograms for a point source emitting P waves in a sphere show that the higher modes are composed of reflected and diffracted P and SV pulses. The connection is similar to the surface waves composed of n-times-reflected P waves due to an explosion in a homogeneous fluid sphere and to the surface waves, including reflected SH waves from a torque source in an elastic homogeneous sphere [Alterman and Kornfeld, 1965]. In the latter cases, when all reflected pulses have only one velocity of propagation, the connection between the surface wave and the reflected pulses can be described as follows: n-times-reflected rays from a surface source reach an observer at an angular distance θ after a given number of circuits m around the sphere of radius a. With increasing n, the length of the ray path more closely approximates the arc

$d = (2\pi m + \theta)a$ or $d = [2\pi m + (2\pi - \theta)]a$. The travel time of the surface wave, determined by d/c, equals the limit for $n \to \infty$ of the travel times of the reflected rays. For an SV-P source, the mixed pulses $P_\mu S_\nu$, which have been reflected μ times as P and ν times as S, do not converge to an arc on the spherical surface. The shortest distance in which a ray starting as P can be reflected ν times as S (taking elastic constants $\lambda = \mu$) is

$$2\nu\, a \text{ arc cos } (1/\sqrt{3}) \quad (1)$$

We can now consider an increasing number of reflections as P. They will take place on the remaining part of the path from the source, given by

$$d - 2\nu a \text{ arc cos } (1/\sqrt{3})$$

and for $\mu \to \infty$ the travel time of the reflected pulse is

$$\tau_\nu^* = (1/\sqrt{3})\,[d/a - 2\nu \text{ arc cos }(1/\sqrt{3})]$$
$$+ 2\nu \sin\,[\cos^{-1}(1/\sqrt{3})] \quad (2)$$

In terms of surface waves, the group-velocity connected with $P_\mu S_\nu$ is

$$U/c = \frac{d/a}{\tau_\nu^*} \quad (3)$$

For example, the surface waves R_{2i} having a group velocity $U = 1.3c$ are here identified as the reflected rays that arrive at the observer after a time corresponding to this value of U/c. Specifically, R_{21} appears only at distances larger than 110° and consists of $P_\mu S$ pulses up to 215°. As R_{21} in the range of distances θa, 180° $\leq \theta \leq$ 215°, is equivalent to R_{22} in the range 145° $\leq \theta \leq$ 180°, R_{22} consists of $P_\mu S_2$ pulses. R_{23} consists of $P_\mu S_2$ for distances 0° to 10° and of $P_\mu S_3$ for distances 45° to 180° coming in clockwise direction after a complete circuit. R_{24} consists of $P_\mu\, S_4$ at all distances smaller than 170°, where it changes to $P_\mu\, S_3$.

The third mode of surface waves is found again to consist of mixed reflected pulses. R_{31} appears only at distances $\geq$225° and is the same as R_{32} in the range of 0° $\leq \theta \leq$ 135°. R_{31} and R_{32} consist of $P_\mu S$ pulses. R_{33} consists of $P_\mu S$ at small angles and of $P_\mu S_2$ for all 45° $< \theta <$ 180°.

Higher-mode Love waves in a layered sphere, and their representation by certain reflected and diffracted body waves, are discussed by *Alterman and Kornfeld* [1966].

Brune [1964, 1966] showed the connection between dispersion curves and reflected body waves, interpreting the dispersion curves as interference conditions for multiply reflected body waves, SH waves in the case of torsional oscillations, and P and SV waves in the case of spheroidal oscillations.

PARTITION OF ENERGY AMONG SURFACE-WAVE MODES

Crampin [1965] measured the energy of second Rayleigh-mode wave trains in relation to known variables of the earthquake: the geographic position and path, the depth of focus, and the magnitude.

Harkrider and Anderson [1966] computed the partition of energy among various surface-wave modes for horizontal and vertical point sources at various depths in oceanic and in shield layered earth models. They demonstrate the increasing importance of the higher modes in the total energy budget at short periods and for channel depth sources, and the influence of source orientation and depth on the shapes of the spectrums. They find that, in general, increasing the depth decreases the amount of energy in the fundamental mode. However, a focus at 50 km, which is the top of the low-velocity zone, can generate Rayleigh waves of slightly more energy for certain frequencies than can a surface source. The higher modes all have critical depths at which they are excited most efficiently and also periods at which they are not excited at all. For a given source orientation, these zeros are diagnostic of source depth. The zeros shift with frequency for increasing source depth. The successively deeper sources have zeros at successively longer periods. A source at 250 km, which is in the low velocity channel, is a very efficient generator of higher mode waves. The vertical and horizontal forces show differences in the shift in the location of the energy zeros.

CONCLUSION

Recently an increasing amount of investigation is being done on observed higher modes. Methods of analysis are being improved. Higher modes in both plane layered models and spherical models of the earth are being studied.

Theoretical seismograms from sources in a sphere are computed and show the relative excitation of higher modes for various sources and earth models. Combined with data from the lower modes and from travel times, the study of higher modes is expected to give a more accurate determination of earthquake sources and of the structure of the earth.

Acknowledgments. This article was written while the author was at the Courant Institute of Mathematical Sciences, New York University, on leave from the Weizmann Institute of Science, Israel, under a Ford Foundation fellowship.

REFERENCES

Alterman, Z., and F. Abramovici, Propagation of a P-pulse in a solid sphere, *Bull. Seismol. Soc. Am.*, 55, 821–861, 1965.

Alterman, Z., and F. Abramovici, Effect of depth of a point source on the motion of the surface of an elastic solid sphere, *Geophys. J., Roy. Astron. Soc.*, 11, 189–224, 1966.

Alterman, Z., and F. Abramovici, The motion of a sphere caused by an impulsive force and by an explosive point source, *Geophys. J., Roy. Astron. Soc.*, 13, 117–148, 1967.

Alterman, Z., H. Jarosch, and C. L. Pekeris, Propagation of Rayleigh waves in the earth, *Geophys. J., Roy. Astron. Soc.*, 4, 219–241, 1961.

Alterman, Z., H. Jarosch, and C. L. Pekeris, Oscillations of the earth, *Proc. Roy. Soc. London, A*, 252, 80–95, 1959.

Alterman, Z., and P. Kornfeld, Propagation of an SH-torque pulse in a sphere, *Rev. Geophys.*, 3, 55–82, 1965.

Alterman, Z., and P. Kornfeld, Normal modes and rays in the propagation of a seismic pulse from a point source in a layered sphere, *Israel J. Technol.*, 4, 198–213, 1966.

Anderson, D. L., and M. N. Toksöz, Surface waves on a spherical earth, 1, Upper mantle structure from Love waves, *J. Geophys. Res.*, 68, 3483–3500, 1963.

Brune, J. N., Travel times, body waves and normal modes of the earth, *Bull. Seismol. Soc. Am.*, 54, 2099–2128, 1964.

Brune, J. N., P- and S-wave travel times and spheroidal normal modes of a homogeneous sphere, *J. Geophys. Res.*, 71, 2959–2965, 1966.

Båth, M., and S. Crampin, Higher modes of seismic surface waves: Relations to channel waves, *Geophys. J.*, 9, 309–321, 1965.

Crampin, S., Higher modes of seismic surface waves: Preliminary observations, *Geophys. J.*, 9, 37–57, 1964.

Crampin, S. Higher modes of seismic surface waves: Second Rayleigh mode energy, *J. Geophys. Res.*, 70, 5135–5143, 1965.

Crampin, S., Coupled Rayleigh-Love second modes, *Geophys. J., Roy. Astron. Soc.*, 12, 229–238, 1967.

Crampin, S., and M. Båth, Higher modes of seismic surface waves: Mode separation, *Geophys. J.*, 10, 81–92, 1965.

Dorman, J., M. Ewing, and J. Oliver. Study of shear velocity distribution by mantle Rayleigh waves, *Bull. Seismol. Soc. Am.*, 50, 87–115, 1960.

Ewing, M., S. Mueller, M. Landisman, and Y. Satô, Transient analysis of earthquake and explosion arrivals, *Geofis. Pura Appl.*, 44, 83–118, 1959.

Harkrider, D. G., Surface waves in multilayered elastic media, 1, Rayleigh and Love waves from buried sources in a multilayered elastic halfspace, *Bull. Seismol. Soc. Am.*, 54, 627–679, 1964.

Harkrider, D. G., and Don L. Anderson, Surface wave energy from point sources in plane layered earth models, *J. Geophys. Res.*, 71, 2967–2980, 1966.

Kovach, R. L., Seismic surface waves: Some observations and recent developments, *Phys. Chem. Earth*, 6, 251–314, 1965.

Kovach, R. L., and Don L. Anderson, Higher mode surface waves and their bearing on the structure of the earth's mantle, *Bull. Seismol. Soc. Am.*, 54, 161–182, 1964.

Mooney, H. M., and B. A. Bolt, Dispersive characteristics of the first three Rayleigh modes for a single surface layer, *Bull. Seismol. Soc. Am.*, 56, 43–67, 1966.

Oliver J., J. Dorman, and G. Sutton, The second shear mode of continental Rayleigh waves, *Bull. Seismol. Soc. Am.*, 49, 379–389, 1959.

Oliver J., and M. Ewing, Higher modes of continental Rayleigh waves, *Bull. Seismol. Soc. Am.*, 47, 187–204, 1957.

Oliver, J., and M. Ewing, Normal modes of continental surface waves, *Bull. Seismol. Soc. Am.*, 48, 33–49, 1958.

Press, F., and H. Takeuchi, Note on the variational and homogeneous layer approximations for the computation of Rayleigh-wave dispersion, *Bull. Seismol. Soc. Am.*, 50, 81–85, 1960.

Stoneley, R., The transmission of Rayleigh waves across Eurasia, *Bull. Seismol. Soc. Am.*, 43, 127, 1953.

Sykes, L. R., and J. Oliver, The propagation of short-period seismic surface waves across oceanic areas, I, theoretical study, *Bull. Seismol. Soc. Am.*, 54, 1349–1372, 1964.

Usami, T., Y. Satô, and M. Landisman, Theoretical seismograms of spheroidal type on the surface of a heterogeneous spherical earth, *Bull. Earth Res. Inst., Tokyo Univ.*, 43, 641–660, 1965.

Attenuation of Seismic Waves in the Mantle

Leon Knopoff

MOST information on the nature of the earth's interior comes from an analysis of nondissipative fields; for example, densities and elastic-wave velocities are derived from fields in which energy is conserved. Additional information on the nature of the earth's interior can be found by studying the physical properties associated with processes that dissipate energy, including attenuation of seismic waves, the distribution of thermal and electrical conductivities, and the distribution of the parameters which govern rheological properties in the earth's interior; of these, the parameters related to seismic attenuation can be most directly measured.

There are two fundamental types of experiments by which attenuation factors can be measured: standing wave and propagating wave (a detailed discussion and list of references is given by *Knopoff* [1964]). In the standing-wave experiments (in which the free modes of oscillation of the earth are observed), one wishes to determine the rate of decay of the standing-wave pattern for the different frequencies of the free oscillations. The amplitude of a given spectral line should decay exponentially with time according to the expression $A \exp(-\gamma t)$, where γ is the attenuation factor. In propagating-wave experiments, we can consider the attenuation of seismic body and surface waves. If there in no effect of attenuation due to geometrical spreading, scattering, diffraction, etc., amplitudes should diminish exponentially with distance according to the expression $B \exp(-\alpha x)$, where α is the distance attenuation factor.

In the expressions above, A is the amplitude of the monochromatic wave motion at the origin of time, and B is the amplitude at the origin of coordinates in the two experiments. In the basic experiment of either type, one compares amplitudes either of the standing or propagating waves at two different times or at two different distances. This comparison removes the unknown spectral composition of the source. The ratio of the amplitudes at the two coordinates x_1 and x_2 is $\exp(-\alpha[x_2 - x_1])$ and is independent of the amplitude at the source. Thus, measurements at two coordinates or two times can be used to determine the attenuation factors α or γ without regard to the properties of the source. This procedure is applicable, in principle, as long as no loss in amplitude has taken place in the interval between the two points of observation or the two times of observation.

In the propagating-wave experiments, measurement of intrinsic attenuation is difficult if significant amounts of energy have been removed from the beam, for example, by scattering or reflection at discrete boundaries, as when surface waves cross a continental margin or when body waves are imperfectly transmitted or reflected at the Moho. One method of removing the effect of scattering by inhomogeneities is to use wavelengths that are large compared with the dimensions of the inhomogeneity.

Another problem in making body-wave or surface-wave observations is that the frequency response of the instruments at the two stations is not the same, owing both to instrumental characteristics and the effects of local geology. Often the effect of attenuation is much smaller than variations in amplitudes due to difference in the characteristics of nominally identical instruments at the stations. Moreover, it is generally difficult in body-wave experiments to ensure that two seismographs are located along the same ray path. All these difficulties can, in principle, be overcome by multiple observations of one ray at one station, i.e., by observing a seismic ray which arrives at the station at least twice, but without losing signal strength by imperfect transmission (other than attenuation) along the way: for example, S waves that reverberate along radii of the earth (i.e., to a station directly above the focus) and surface waves from great earthquakes which execute complete circuits of the earth.

Let the core-mantle boundary and the outer surface of the earth both be presumed to be perfect reflectors for S waves at normal incidence; i.e., the core is assumed to be a nearly ideal fluid at the frequencies of interest and the effect of the crust is neglected. Then the attenuation factor for body S waves in the mantle can be calculated from the ratios of the amplitudes of successive multiple reflections for rays that started from the source either upward or downward. *Anderson and Kovach* [1964] have used a deep focus earthquake for this purpose. The difference in path length for successive alternate arrivals is twice the thickness of the mantle. The attenuation factor for the mantle has been found to be approximately inversely proportional to the period of the seismic waves. The attenuation factor can therefore be expressed as $\alpha = \pi/cQT = \pi/Q\lambda$, where Q is a dimensionless specific attenuation factor, T is the period, c is a mean velocity and λ is a mean wavelength. The value of Q for the mantle over the period range 11-25 sec is of the order of 500. If one further postulates that S waves leaving the focus in opposite directions are equal in amplitude, a postulate consistent with models of focal mechanism in which the focus does not move with finite velocity or is not located in an inhomogeneous medium, one can compare the attenuation for the mantle above the source with that below. For the deep focus earthquake, the value of Q for the upper mantle was about 160; for the lower mantle, about 1450. Hence, the upper mantle appears to attenuate S waves considerably more than does the lower mantle.

Interpretation of surface-wave attenuation data is somewhat more complicated than interpretation of body-wave data. Surface waves penetrate to depths that are dependent on the wavelength. Hence, surface waves of short wavelength contain information about attenuation in the upper part of the earth; surface waves of longer wavelength contain information concerning the attenuation both at greater depths and in the upper part of the earth as well. To perform the interpretation, one needs a theory describing the way in which the attenuation factor in a homogeneous sample of material deep in the interior of the earth might vary with the frequency of the excitation. Results of experiments upon a large number of solid materials in the laboratory—metals and nonmetals, granular rocks, glasses and single crystals—and upon relatively homogeneous formations at shallow depths in the field indicate that the parameter Q is essentially independent of frequency or wavelength over a wide range of frequencies. The fact that the ratio of the specific attenuation factor to frequency is independent of frequency for a large number of solids at low frequencies led *Knopoff and MacDonald* [1958, 1960] to postulate that the basic physical mechanism underlying this type of loss is nonlinear in character. Independent of this consideration, the postulate that Q, or the ratio of the attenuation factor to the frequency, is independent of frequency can be used as a starting point for an interpretation of the surface-wave data. It can be shown that the surface-wave data interpreted on the basis of this postulate lead to attenuation factors for S waves that are consistent with the values obtained by Anderson and Kovach.

If it is recognized that the attenuation factor is the imaginary part of the complex wave number, and if it is postulated that the attenuations are small, the attenuation factor can be expressed as small, imaginary perturbations upon a real wave number. Since the real wave number depends strongly on the phase velocity at a given frequency, it follows that the perturbations can be derived by a Taylor series expansion about the lossless condition. Accordingly, in the interpretation, it is necessary to introduce a set of partial derivatives of the phase velocity with respect to each of the parameters in the interpretation. The broad outline of the theory is given by *Knopoff* [1964]. The attenuation as a function of depth is obtained by inverting a set of simultaneous equations where the matrix coupling the observed values of attenuation as a function of frequency and the values of Q as a function of depth are the partial derivatives of the phase velocity with respect to the elastic coefficients. The resolution in this interpretation is not great at present. Broadly, however, the lower mantle has a high Q, the upper mantle has a lower Q, in complete agreement with the body-wave results. The upper mantle probably can be split into several regions, the lower of which has a lower Q than the regions above it. *Knopoff* [1964] provides a model that fits the data reasonably, as follows: from zero to 325 km depth, $Q = 120$; from 325 to 650 km, $Q = 75$; from 650 to 2900 km, $Q = \infty$. The data are not sufficiently precise to warrant specifying these values with any greater precision. The difference between $Q = \infty$ and $Q = 1450$ (the value obtained from body-wave data) is not significant at present precision, since one is in fact comparing reciprocals of these quantities.

The attenuation factors in the free oscillations have been measured for those earthquakes that were strong enough to excite the earth into observable resonant free modes of oscillation. There are two basic methods of measuring the attenuation factors. In one method, a spectral analysis of the motion is made and the width at half power $\Delta\omega$ of the spectral line at frequency ω yields Q by the formula $Q = \omega/\Delta\omega$. In the other method, the record is subdivided into several parts and spectral analysis is made of each of the parts. The decrement in the energy content at the frequency of the line can be directly computed to obtain the attenuation factor.

Q values have been reported in the literature for spheroidal modes $_0S_0$, $_0S_2$, $_0S_3$, $_0S_9$, and $_0S_{12}$ and for torsional modes $_0T_2$, $_0T_3$, and $_0T_5$. A summary of the Q values is given by *Knopoff* [1964]. *Gilbert and Backus* [1965] indicate that, because of the splitting of the line structure due to the rotation of the earth, the measurement of Q for spheroidal modes of order 4 and higher is difficult because each of the split lines is broadened due to the intrinsic Q, with consequent overlap upon each other. Hence, the measurement of the apparent breadth of the spectral line does not lead to a real estimate of the Q in that line because of inability to resolve the fine structure. Similar arguments can be applied to the difficulty of measuring attenuations by the second method. In this case, the standing wave pattern in fact rotates relative to the observatory because of the rotation of the earth. The amplitudes may fluctuate at a given point of observation for reasons other than attenuation, especially in the vicinity of a nodal surface. This will lead to an incorrectly determined value of Q. Indeed, after the nodal surface has passed, the amplitude will appear to increase.

Q measured in standing-wave experiments is not the same as in propagating-wave experiments, especially if the medium is dispersive as in the case of the earth. In these cases, π/Q_T measured in standing-wave experiments is the attenuation factor per unit period, and π/Q_x obtained from propagating-wave experiments is the attenuation factor per unit wavelength. The two types of Q values are related by the ratio of phase velocity c and group velocity U:

$$cQx = UQ_T$$

Comparison of the Q's obtained from the two types of experiments must take this relation into account. Group velocities in the free modes have not been obtained; therefore, it is preferable to reduce the Q values obtained in propagating-wave experiments to the same basis as those obtained in standing-wave experiments.

It is possible to indicate some of the areas where additional work needs to be done. Thus far, Q values have been obtained for the earth as though it were radially symmetric. No evidence on variations in Q due to lateral heterogeneity have been obtained. Considerably greater resolution as to the distribution of Q with depth should be obtained to provide a better picture of the variation of properties of material with depth. To date, most of the interpretation has been on Q values in shear-wave experiments; calculations of intrinsic attenuation factors from Rayleigh waves and the spheroidal free modes of oscillation of the earth should be improved. Q values should be reliably obtained for the crust. Q values should be determined from body P-wave experiments. What is the influence of partial melting or total melting (such as in the core) on attenuation factors? In the laboratory, the attenuation factor in fluids varies as the square of the frequency, and in solids as the first power of the frequency; attenuation in liquids is very likely due to viscosity. Is Q independent of frequency at high pressures and temperatures in solids?

REFERENCES

Anderson, D. L., and R. L. Kovach, Attenuation in the mantle and rigidity of the core from multiply reflected core phases, *Proc. Natl. Acad. Sci. U.S., 51,* 168–172, 1964.

Gilbert, F., and G. Backus, The rotational splitting of the free oscillations of the earth, 2, *Rev. Geophys.*, *3*, 1–9, 1965.
Knopoff, L., Q, *Rev. Geophys.*, *2*, 625–660, 1964.
Knopoff, L., and G. J. F. MacDonald, Attenuation of small amplitude stress waves in solids, *Rev. Mod. Phys.*, *30*, 1178–1192, 1958.
Knopoff, L., and G. J. F. MacDonald, Models for acoustic loss in solids, *J. Geophys. Res.*, *65*, 2191–2197, 1960.

Seismic Models of the Upper Mantle

Jiří Vaněk

SEISMIC modeling, a relatively new experimental method, has grown in importance for the solution of those seismological problems that are too complex for theoretical treatment. Although this experimental technique has been applied in several branches of seismology and seismic prospecting [*Riznichenko*, 1966; *Berckhemer and Waniek*, 1966], we deal here mainly with the models of the earth's upper mantle. Of the different methods of seismic modeling, the pulse methods using elastic waves of ultrasonic frequencies appeared to be most effective for simulating the propagation of seismic waves in the earth.

One of the main difficulties in fabricating seismic models of the upper mantle is the technology of producing continuously variable velocity in the model. This problem was attacked first by means of two-dimensional models. Several methods were developed [*Ivakin*, 1966]: bimorph models (a set of sheets from different materials with variable thickness connected by gluing, melting, galvanization etc.); perforated models (sheets with a network of openings with different diameters and variable hole density); and thermic models (plates of special materials, the velocity distribution in which is controlled by temperature). Recently, three-dimensional models with continuously variable velocity have been realized by a large number of thin layers with negligible velocity differentiation between the individual layers. As modeling media, multicomponent gels [*Vaněk et al.*, 1964, 1966] or epoxy resin filled with quartz sand [*Shamina*, 1965] are used.

It must be emphasized that the present possibilities of seismic modeling and the theory are limited. It is very difficult (almost impossible) to find a modeling medium that is able to simulate all the properties of the prototype. If, for example, the velocities are in correct modeling scales, the attenuation characteristics may not satisfy the laws of physical modeling. Therefore, only qualitative solution of partial problems is usually possible on seismic models.

At present, the main significance of seismic modeling is the possibility of investigating dynamic parameters of seismic waves in those problems for which theory cannot give an appropriate solution. Two cases are of fundamental importance for the upper mantle: low-velocity channels, and discontinuities in media with continuously variable velocity (gradient media). Both problems were investigated by the seismic modeling technique.

Low-Velocity Channels

Because a low-velocity layer may act as a waveguide, several types of waveguides with sharp and weak boundaries were investigated on seismic models; both seismic body waves of P type and guided waves of P_a type were considered. Most experiments have been performed on two-dimensional seismic models.

In a series of papers, Riznichenko and Shamina studied the amplitude curves of P waves using two-dimensional bimorph models with various velocity-depth distribution. They found that in a medium with negative velocity gradient, the amplitudes decreased with distance

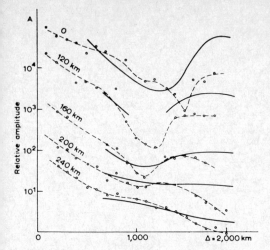

Fig. 1. Amplitude-distance curves observed on a two-dimensional model (dashed line) compared with magnitude calibrating curves (solid line) of *Gutenberg and Richter* [1956] for different depths of the source.

more intensively than in a medium with constant velocity or with positive velocity gradient [*Riznichenko and Shamina*, 1963]. On models with a waveguide, a clear minimum was observed at the amplitude curve, if the source was located above or inside the waveguide. The position of this minimum corresponded to distances at which a shadow zone should occur and was shifted to shorter epicentral distances with increasing depth of the source [*Riznichenko and Shamina*, 1964]. The observations on the models were in general agreement with the magnitude calibrating curves given by *Gutenberg and Richter* [1956] for different depth levels (Figure 1).

A relatively extensive model research has been carried out in connection with the nature of P_a and S_a waves, which were interpreted as waves guided by the low-velocity channel in the upper mantle. Using a two-dimensional model composed of a paraffin-polyethylene layer between two sheets of organic glass, *Chorosheva* [1962] demonstrated the possibility of recording the channel waves P_a and S_a on the earth's surface. To clarify the mechanism of generation of these waves, several two-dimensional perforated models were investigated by *Kapcan and Kislovskaja* [1966]. A clear oscillatory character of the amplitude curves of P_a waves was observed both for the models and for five deep earthquakes with magnitudes greater than 6.5. The experimental data confirmed that Caloi's hypothesis on the nature of P_a and S_a waves was more probable than the 'whispering gallery' hypothesis of Press and Ewing.

There is no doubt that the results obtained by the two-dimensional modeling technique are very interesting. However, it is questionable, in principle, whether the representation of wave propagation in three-dimensional space by that in a plate is a sufficient approximation of actual conditions, especially if dynamic parameters of waves are considered [*Petrashen*, 1964].

Therefore, *Shamina* [1966] investigated necessary and sufficient characteristics of a waveguide using three-dimensional models of epoxy resin filled with variable amounts of quartz sand. After confirming several results obtained previously on two-dimensional models, she paid special attention to the study of amplitude-depth curves $A(h)$ of P waves in the shadow zone. It was observed that these curves had a minimum at the depth corresponding roughly to the minimum of the velocity-depth distribution. This phenomenon was compared with the

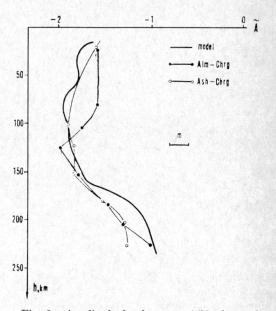

Fig. 2. Amplitude-depth curve $A(h)$ observed for deep Hindu Kush earthquakes and for a three-dimensional model with a waveguide; $\tilde{A} = \log(A/T)_{\Delta=1000 \text{ km}} - \log(A/T)_{\Delta=100 \text{ km}}$, where T is period. Alm, Alma Ata; Ash, Ashkhabad; Chrg, Khorog; m, mean quadratic error of seismological observations. For details see *Shamina* [1966].

amplitude-depth curves derived for deep Hindu Kush earthquakes (Figure 2). This comparison indicates a possible low-velocity channel with the velocity minimum at 140–180 km in Central Asia.

Discontinuities in Gradient Media

Few papers deal with modeling of the important problem of discontinuities in gradient media. Recent experiments on three-dimensional models of the earth's crust containing layers with continuously variable velocity [*Holub et al.*, 1966; *Waniek and Schenk*, 1966] pointed up the necessity of using the amplitudes of seismic waves for the interpretation of deep seismic sounding.

A series of three-dimensional gel models of the upper mantle, in which both the low-velocity channel and discontinuities of the second order were included, was investigated for explaining the course of the amplitude-distance curves of P waves derived from seismological observations for shallow earthquakes [*Vaněk*, 1966]. As an example, the mean amplitude curve of P waves observed on three models with identical velocity-depth distribution is shown in Figure 3.

The velocity distribution characterized by a weak low-velocity channel and by a discontinuity of the second order is schematically shown as an inset (upper left) in Figure 3, the scale factors governing the model being $s_{vP} = 0.185$, $s_T = 1 \times 10^{-6}$, $s_\lambda = 1.85 \times 10^{-7}$, where T is period, and λ is wavelength. On the basis of this investigation, a possible scheme of the upper mantle structure in southeastern Europe could be suggested (see section by Vaněk in this Monograph, p. 246).

Although the final quantitative interpretation of seismological observations cannot be completely solved by methods of seismic modeling, further research in this direction is urgently needed. This point was also emphasized at the first Symposium on Seismic Models held in 1965, where the development and outlook of this method in the scope of seismological research was summarized [*Symposium*, 1966].

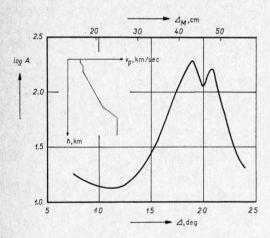

Fig. 3. Amplitude-distance curve of P waves observed on three-dimensional models corresponding to the upper mantle section $(h, v_P) \equiv (0, 8.1)$, (30, 8.1) (60, 8.25), (110, 8.24), (430, 9.17), (530, 9.85), (690, 9.85); depth h in kilometers, velocity v_P in kilometers per second. A, maximum amplitude in P-wave group; Δ_M, epicentral distance on the model in centimeters; Δ, actual epicentral distance in degrees, corrected for the influence of the actual earth's crust and curvature. Source in the 'crust.'

REFERENCES

Berckhemer, H., and L. Waniek, *Bibliography on Seismic Modeling*, European Seismological Commission, Upper Mantle Sub-Commission, Prague, 1966.

Chorosheva, V. V., Issledovanie volnovoda na tverdoj dvumernoj modeli s rezkimi granicami, *Izv. Akad. Nauk. SSSR, Ser. Geofiz., 1962*(8), 1025–1033, 1962.

Gutenberg, B., and C. F. Richter, Magnitude and energy of earthquakes, *Ann. Geofis., 9*, 1–15, 1956.

Holub, K., V. Kárník, and V. Tobyáš, A simple three-dimensional model of the earth's crust, *Studia Geophys. Geodaet., 10*, 370–375, 1966.

Ivakin, B. N., Methods of seismic modeling, *Studia Geophys. Geodaet., 10*, 253–259, 1966.

Kapcan, A. D., and V. V. Kislovskaja, Investigation of a waveguide with weak boundaries in two-dimensional perforated models, *Studia Geophys. Geodaet., 10*, 360–370, 1966.

Petrashen, G. I., O modelirovanii processov rasprostranenija sejsmicheskich voln, *Vopr. Dinam. Teorii Rasprostranenija Sejsmicheskich Voln, 7*, 7–35, 1964.

Riznichenko, J. V., Seismic modeling, development and outlook, *Studia Geophys. Geodaet., 10*, 243–253, 1966.

Riznichenko, J. V., and O. G. Shamina, Modelirovanie prodolnych voln v verchnej mantii Zemli, *Izv. Akad. Nauk SSSR, Ser. Geofiz., 1963*(2), 223–247, 1963.

Riznichenko, J. V., and O. G. Shamina, Model study of the upper mantle shadow zone, *Tectonophysics, 1*, 443–448, 1964.

Shamina, O. G., Metodika trechmernogo modelirovani aja volnovodnoj mantii na tverdych sredach, *Izv. Akad. Nauk SSSR, Fiz. Zemli*, *1965*(7), 102–105, 1965.

Shamina, O. G., Experimental investigation of necessary and sufficient characteristics of a waveguide, *Studia Geophys. Geodaet.*, *10*, 341–350, 1966.

Symposium on seismic models, *Studia Geophys. Geodaet.*, *10*, 239–400, 1966.

Vaněk, J., Amplitude curves of longitudinal waves for several three-dimensional models of the upper mantle, *Studia Geophys. Geodaet.*, *10*, 350–359, 1966.

Vaněk, J., L. Waniek, Z. Pros, and K. Klíma, Three-dimensional seismic models with continuously variable velocity, *Nature 202*, 995–996, 1964.

Vaněk, J., L. Waniek, Z. Pros, and K. Klíma Modellseismische Untersuchungen an dreidimensionalen Modellen des oberen Erdmantels, *Trav. Inst. Géophys. Acad. Tchécosl. Sci.*, *1966*(247), Geofysikální sborník, 247–289, 1966.

Waniek, L. and V. Schenk, Modellseismischer Beitrag zur Deutung des Krustenaufbaues in der bayerischen Molasse, *Z. Geophys.*, *32*, 482–487, 1966.

Earthquakes and Tectonics

B. A. Petrushevsky

THE vast majority of earthquakes are generated within the earth's crust by tectonic movements in the crust or in the uppermost part of the mantle. Earthquakes due to processes at greater depths, down to 700–800 km, are less frequent and occur mainly in the zone of the Pacific mobile ring.

Our knowledge of geological conditions responsible for earthquakes is not adequate. It is universally recognized today that earthquakes are the result of fracturing occurring at various depths of the crust and the upper mantle. Geological and physical investigations indicate that many zones of faults associated with earthquakes do not extend up to the earth's surface; therefore they cannot be directly associated with, or compared to, surface faults. Epicentral areas of some severe earthquakes provide no evidence of large faults related to the earthquakes (for example, the Gansun earthquake in China in 1920).

Although the immediate cause of earthquakes is fracturing, the tectonic movements causing seismogenetic fracturing at great depth occur under different geological conditions. The zones of high seismicity are related to regions of widely varying structural characteristics, including border zones of continents and oceans; recent geosynclines; zones of young folding of earlier geosynclines; and platforms, including ancient ones involved in intensive recent tectonic movements.

Therefore, study of geological history can contribute to our understanding of the existing seismo-geological relationship. We inquire whether it is possible to establish any correlations between the data on shifts along faults resulting in earthquakes and the data on the particular geological conditions. We distinguish three groups of earthquakes.

In the first group are earthquakes with very shallow foci (order of from 3 to 5 km); sometimes it is possible to establish definite relationship between them and the movements of surface structures. However, earthquakes generated by movements at greater depths have been recorded in almost all the seismic areas; thus it is very difficult to determine whether the shallow earthquakes are caused by the shifts in the surface structure or by shifts in the upper part of a deeper structure. Because of these peculiar features, the first group should be regarded as a particular case.

The earthquakes of the second group, which is considered to be the main one, are the result of shifts occurring at various depths, sometimes as deep as the base of the crust or the upper parts of the upper mantle. In this case we usually cannot speak of direct conformity between the movements at great depths that result in an earthquake and certain surface structures, no matter how large they are. This relationship is more reliable when several structures, their complexes, and even whole zones are considered.

Earthquakes with foci in the upper mantle that are caused by the extremely abyssal movements are classified as the third group. As a rule these bear no traces of relationship with any one surface structural component, except with the largest such as the zones of the deep-sea troughs in the Pacific Ocean.

Earthquakes of the second group are the most destructive ones. For this reason they have been studied much better than the deep shocks, and so their relationship with geological events is better understood. Geological investigations on the earth's surface are not adequate for establishing these relationships; we therefore resort to the analyses to reconstruct the history of the deeper part of the earth. We know that both earthquakes and large structural complexes of the earth's crust reflect processes at depth. Thus estimation of seismicity should be based both on the data of seismo-statistics and the information on the history of development and tectonic characteristics of great structures of the region under consideration.

Large structural complexes are directly related to abyssal tectonic processes. Earthquakes are secondary results, i.e., the result of recent processes, whereas structural complexes are related to processes of immeasurably greater duration.

Thus, according to this concept, large structural complexes are indicators of the geological processes operating at great depths, whereas earthquakes are considered to be the result of the initial, and, in some greater degree, of the latest fracturing of these structures. This provides a basis for genetic analysis of seismicity. The studies must therefore embrace a vast territory during a significantly long geologic interval.

Many geologists maintain that recent movements are the only geological factor to be taken into consideration when estimating the degree of seismicity. The latest period of geologic history (Neogene-Quaternary) is undoubtedly worth the greatest attention. Movements that occurred during that time and the resulting structure should be given the most detailed investigation, as the earthquakes are the best indication of recent tectonic activity and recent movements of the earth's crust. However, we should not confine ourselves to analyzing only the recent epoch; knowledge of past ages allows us to find valid regularities governing the recent structural reconstruction that is responsible for many earthquakes.

Historical and structural analyses used to find the relation between seismicity and geological conditions allow us a wide range of extrapolation. Structural complexes of a uniform structure and development are presumed to be characterized by more or less equal seismicity, even though severe earthquakes have been recorded on some sections within their boundaries.

Broad investigations by Soviet scientists during the last 20 years indicate that generalizations on the dependence of seismic and geological phenomena can be made. A number of particular geological factors are associated with qualitative determination of seismic activity whose degree may vary. To cite an example, let us consider the zones of abruptly jointed large structural complexes differing in their structure and movements. It is along these zones of high rate of movement gradients that shifts causing earthquakes often occur. Other examples of this kind are the zones characterized by a long development of large faults and by the movements which continue for several geological epochs or even periods, or the instances of unconformable superposition of the newly formed structures on those existing previously, accompanied by a severe crushing at various depths.

These relations are not the same in regions differing in their geological structure. The available data subjected to the analyses from this viewpoint prove that we can also project more general geological criteria of seismicity. We distinguish the following groups: (1) young folded regions, (2) young platforms, (3) platforms of any age involved during recent time in intensive tectonic movements, and (4) rift zones of the continents.

1. Local structural reconstructions are the most characteristic features of the development of folded regions (the alpine included). Deep depressions have been inverted into uplifts, the place was repeatedly subject to an intensive folding (the direction of fold structure chang-

ing from epoch to epoch), and repeated fracturing resulted in the formation of lasting faults. These tectonic features result in an extremely high seismicity in the alpine folded regions. Numerous earthquakes, some of them of intensity up to 8–9, occur here. The earthquake foci are usually in the earth's crust, more often in the upper half of the crust. Examples are the alpides of the Mediterranean (the Alps, the Apennines, the Dinarides, and others) and of the Caucasus, Turkey, Iran, and Beluchistan.

From the theoretical point of view, the regions classified by some geologists as geosynclines should be included in this group. They are the regions lying in the western and southwestern parts of the Pacific mobile ring, i.e. those in Japan, the Philippines, and the Indonesian Islands. These regions, however, are in the immediate vicinity of the deep-sea troughs. Recent movements in the trough zones are notable for their great intensity and extend as deep as 700–800 km. They result in a great number of earthquakes (some of which are intermediate and deep) determining seismicity of the given regions. It proves extremely difficult to distinguish under these conditions the earthquakes due to the development of the geosyncline process, especially because much of the area is under water.

2. In contrast to the developing folded regions, platforms do not undergo any structural reconstruction; platform structures develop in one direction during a very long geologic time. Of great importance for young platforms with a folded base of Paleozoic and Mesozoic age is their inherited development, the structural features of the platform mantle adapting themselves to the tectonic plan of the base. The inherited features are most evident in the zones of recent folding of the geosyncline stage. In the sedimentary mantle of the platform of certain sections, they manifest themselves by the greatest structural differentiation and highest tectonic mobility. It is here that seismicity is higher, but, general mobility being rather slow, earthquakes are few and weak; their intensity is seldom as high as 7, never higher than 8. To this group belong the Hercynides of west Europe, the Urals, and some regions of central Asia, and the Mesozoides of the northeastern parts of the Soviet Union.

3. Seismo-geological relations for the platforms involved in recent intensive tectonic movements are very complicated. The sections of the latest folding of the geosyncline stage are characterized by numerous earthquakes, some of which are very severe, i.e. activity of the recent movements is added here to the higher-than-average mobility due to the inherited development. Exemplifying this is the Tien Shan.

Platform sections, which have undergone an early stabilization during the geosyncline stage (for example, the platforms with Paleozoic base were subject to stabilization during the Caledonian epoch of tectogenesis) are of a higher than average rigidity; during the platform stage they remained almost immobile. Increase in seismicity depends mainly on the character of the recent movements. Differentiated and contrast movements result in the formation of a number of small blocks whose relative movements are highly varied. These conditions favor an abrupt rise in seismicity. Though the total number of earthquakes occurring here is moderate, severe shocks (intensity 9 to 10 and more) are relatively frequent. Examples are the 1911 Vernensk earthquake and the 1905 Mongolian earthquakes. Judging by the vast areas in which the earthquakes are felt, foci of many earthquakes are in the upper part of the earth's crust. Included in this group are northern and central Tien Shan, Nan Shan, and other mountain formations of central Asia, Mongolia, and many peneplain regions of the Chinese platform which is undergoing reconstruction.

4. The characteristic feature of rift zones is their great length and small width. Recent movements have a significant contrast. In conformity with this, geological criteria of seismicity of continental rifts are very similar to those for the platforms undergoing reconstruction, but the earthquakes are largely confined in narrow regions close to the rift zones. Examples are the Baikal and East African rifts.

Our present knowledge is insufficient for judging of geological criteria of seismicity for the regions lying beneath the ocean bottom. The same can practically be said about the zones where earthquakes occur at great depths, particularly, about the deep-sea troughs. This is so in spite of the fact that earthquakes in troughs are the most numerous and most intensive of all the earthquakes on the globe. The geophysical data available suggest that focal zones of earthquakes under troughs are not continuous. The corresponding systems of fracturing are presumed to be interrupted too.

However, we need much more knowledge to be able to produce reliable criteria of seismicity. Some scientists tried to establish some relationship between deep earthquakes of the northwestern part of the Pacific, the Hindu-Kush region, the Vrancea regions in the Carpathians, and various peculiarities of superficial and near-surface geological structures. These attempts are very interesting but the results obtained cannot provide us with even most general criteria as far as the above characteristics of structural components are concerned.

Progress in the study of the relationship between seismicity and tectonics is important not only for the theory of tectonics but also for practical purposes. Great structural complexes whose development influences the degree of seismicity can be fairly definitely found on the surface. This provides some grounds for an objective mapping of the relations between seismic events and geological data. These maps provide a geological basis for maps showing the division into seismic zones and contribute to their reliability, for example, the seismo-tectonic map of Europe [*Beloussov et al.*, 1966, 1968].

In conclusion, I want to draw your attention to an important gap in the seismo-geological research, namely to the lack of correspondence between the nature of events causing earthquakes and the methods employed to study them. Both the tectonic processes governing formation of structures (among these are faults which are considered to generate earthquakes) and earthquake foci involve a significant volume not only of the earth's crust, but also of the mantle. The geological structures under investigation lie on or near the surface. The results of seismic events are projected on the same surface irrespective of the depth of their occurrence. It is important to project our investigations to the depths of the earthquakes and tectonic processes. To do this, we must begin estimating earthquake energy, considering layer after layer; this will allow us to obtain data on intervals lying at great depths and characterized by the highest intensity. Geologists will have to explain these phenomena by the peculiar development of large structural components of the given region and by the correlation of these components. The statistical picture evolved should clarify our concept of the nature and periods of the processes operating in the deep parts of the earth's crust.

Some recent references are cited; the reader can find additional bibliographies in these references.

REFERENCES

Beloussov, V. V., and M. V. Gzovsky, Tectonic conditions and mechanism of earthquake origin, *Trans. Inst. Fiz. Zemli, Akad. Nauk SSSR, 1954*(152), 25–35, 1954.

Beloussov, V. V., A. A. Sorsky, and V. I. Bune, *Seismo-Tectonic Map of Europe* (1:5,000,000), 1966, and *Explanatory Notes* (in Russian and English), 39 pp., Nauka, Moscow, 1968.

Florensov, N. A., and V. P. Solonenko, editors, *The Gobi-Altai Earthquake*, Izd. Akad. Nauk SSSR, Moscow, 391 pp., 1963.

Goryatchev, A. B., *Main Regularities of Tectonic Evolution of the Kuril-Kamchatka Zone*, Nauka, Moscow, 235 pp., 1966.

Gubin, I. E., *Regularities of Seismic Occurrrences in the Tajik SSR (Geology and Seismicity)*, Izd. Akad. Nauk SSSR, Moscow, 464 pp., 1960.

Kirillova, I. V., E. N. Lyustikh, V. A. Rastvorova, A. Sorsky, and V. E. Khain, *Analysis of Geotectonic Development and Seismicity of the Caucasus*, USSR Academy of Sciences, Moscow, 340 pp., 1960.

Medvedev, S. V., editor, Seismic Regionalization of the USSR, Nauka, Moscow, 476 pp., 1968.

Petrushevsky, B. A., Geological conditions responsible for earthquakes, *Sov. Geol., 1960*(2), 74–82, 1960.

Petrushevsky, B. A., N. I. Nikolayev, and R. G. Garetsky, editors, *Activated Zones of the Earth's Crust, Latest Tectonic Movements, and Seismicity (Proceedings of the 2nd All-Union Conference on Tectonics, Dushanbe)*, Nauka, Moscow, 256 pp., 1964.

Rezanov, I. A., *Tectonics and Seismicity of Turkmen-Khorosan Mountains*, USSR Academy of Sciences, Moscow, 246 pp., 1959.

Solonenko, V. P., *Seismic Regions of East Siberia*, Siberian Department of the USSR Academy of Sciences, Irkutsk, 31 pp., 1963.

4. Gravity

Standardization of Gravity Measurements[1]

George P. Woollard

THE successful application of gravity to studies of the crust and upper mantle, as well as to studies of the geoid, depends upon: (a) a realistic theoretical representation of the gravity field to be expected at any point on the earth's surface, and (b) observed values of gravity on an absolute standard that are compatible with those defined by the theoretical model. These requirements led to the adoption in 1930 of the Potsdam absolute value of gravity as a world gravity datum and of the international gravity formula $g_\varphi = 978.049$ $(1 + 0.0052884 \sin^2 \varphi - 0.0000059 \sin^2 2\varphi)$ for defining the theoretical sea level value of gravity at any latitude φ. However, this formula applies to an earth whose equatorial radius is 6,378,388 meters and whose polar flattening is 1/297. Recent satellite orbital studies and new astro-geodetic determinations suggest that these parameters are incorrect by significant amounts. It is also now evident that the Potsdam absolute gravity datum is in error; that there are differences in gravity calibration standards in use and a significant discrepancy in the mean density of the crust used in the Bouguer anomaly reduction. Our purpose in this paper is to examine the magnitude of the changes in gravity anomaly values that would result from revisions that seem desirable in the light of current knowledge, and to consider other factors having a bearing on the interpretation of gravity data.

---

[1] Contribution 246 from the Hawaii Institute of Geophysics.

Revision of the Metric Parameters of the Earth

Although most geodesists agree that the metric parameters of the earth should be revised, they have reached no agreement about the actual changes or whether the reference figure of the earth is best described as a triaxial ellipsoid or an ellipsoid of revolution. It is probable, however, that the latter form and the corresponding gravity formula will be retained. In 1967, the International Association of Geodesy (IAG) adopted two resolutions designed to provide a modern reference system for both geometric and dynamic geodesy [*Moritz*, 1968]. Since the international ellipsoid and the international gravity formulas were retained as the basic reference system, the adopted values are called 'reference ellipsoid, 1967,' and 'gravity formula, 1967.' The official values are yet to be published by the IAG, and the following must be considered preliminary:
Reference ellipsoid, 1967

$$a = 6,378,160 \text{ meters}$$
$$f = 1/298.25$$

Gravity formula, 1967

$$g_\phi = 978.0318(1 + 0.005\ 302\ 4 \sin^2 \varphi$$
$$- 0.000\ 005\ 8 \sin^2 2\varphi) \text{ gal}$$

The new value for gravity at the equator (978.0318 gal) is a derived quantity based on definitions of a, f, GM (gravitation constant times mass of the earth), and the angular rotational velocity of the earth. It also incorporates a correction (approximately −13 mgal) to the absolute value of gravity at Potsdam. The need for correction of the Potsdam gravity datum will be discussed in the next section. The changes in theoretical gravity values resulting from adoption of the 'geodetic reference system, 1967,' show a difference of about −17 mgal at the equator and about −4 mgal at the poles. Since approximately −13 mgal can be attributed to the change of the absolute standard at Potsdam, the change due to the other geometric and dynamic parameters would be of the order of −4 mgal at the equator and +9 mgal at the poles.

These changes in the reference system would also slightly affect the vertical gradient of gravity, which is now assumed to be −0.3086 mgal/meter at the earth's surface. For the reference system, 1967, the normal vertical gradient is about −0.3083 mgal/meter (a change of 0.3 mgal/1000 meters).

Revision of the Potsdam Gravity System

The Potsdam gravity system is defined by the value of gravity at the Pendelsaal of the Geodetic Institute in Potsdam (East Germany) as determined by F. Kühnen and Ph. Furtwängler from reversible pendulum measurements during the period 1898–1904. The parameters of the Potsdam system were defined as follows:

$$g = 981.27400 \text{ gal (exactly)}$$
$$\varphi = 52°22.86'\text{N (approximately)}$$
$$\gamma = 13°04.06'\text{E (approximately)}$$
$$h = 86.24 \text{ meters (exactly)}$$

Modern examination of the reduction procedures employed by Kühnen and Furtwängler led to the belief that the established value of Potsdam was in error by some 10–13 mgal. Comparison of absolute values observed at other sites since 1900 through accurate relative pendulum and gravimeter ties confirmed that the absolute value of gravity at Potsdam was in error; consequently, the whole gravimetric system contains a scale error. For relative gravity measurements and for those computations in which only gravity anomalies will be used, it is sufficient to know the intensity of gravity in a relative system. For the computation of satellite orbits, atmospheric pressures, temperature scales, and other scientific units that are functions of mass, however, it is necessary to know the actual intensity of gravity in a system that is as nearly correct as possible.

Although it is not possible to establish an exact correction to the Potsdam system, a correction of −13 ± 2 mgal appears reasonable.

The development of portable absolute gravity measuring systems, such as that developed by J. E. Faller at Wesleyan University and that under development by J. C. Rose at the University of Hawaii, will make it possible to define absolute gravity values at widely separated points on the earth with the same apparatus; the establishment of a new absolute gravity system can therefore be expected within the next three years.

Revision of the Density Value for the Crust

At present the mean density of the crust commonly used in the Bouguer anomaly reduction is 2.67 g/cm³. This value was adopted when it was thought that granite was representative of crustal material. However, it is now known that even the so-called 'granitic' surface layer of the crust (the basement crystalline rock complex lying immediately beneath the sedimentary rock section) has a mean density of about 2.74 g/cm³ [Woollard, 1962]. The actual value that should be used can be approximated from what we have learned through seismic studies of the crust and upper mantle. One of the significant findings of such studies is that there is a regular relation between surface elevation and the depth of the crust-mantle interface defined seismically. It appears that the seismic crust is supported hydrostatically by the mantle in accordance with the Airy concept of isostasy. If this is the mechanism for isostatic compensation, then the mass of mantle material displaced equals that of the overlying crust ($\Delta H_m \sigma_m = H_c \bar{\sigma}_c$), and the surface elevation and free board will be dependent only on the thickness of the crustal column (H_c) and the density contrast between the crust and mantle ($\Delta \sigma = \sigma_m - \bar{\sigma}_c$). As the mean density of the crust (σ_c) and the density of the mantle (σ_m) are important, whereas the density distribution within the crust is not, the critical factor becomes the thickness of the actual crust, which probably extends well below the M discontinuity and possibly to the seismic low-velocity zone at about 150 km. Unfortunately, there is very little information as yet about seismic layering below the M discontinuity and even less information about the density significance of seismic velocities in excess of 8.2 km/sec. However, on an interim basis we can adopt the relations defined seismically for the composition of the crust and upper mantle and that between surface elevation and the depth of the M discontinuity to derive a first approximation of the density of the overlying crust. We can also use this model to evaluate Bouguer anomalies in terms of the mass distribution associated with the crust and upper mantle, and the more nearly actual values can be approximated for $\bar{\sigma}_c$ and $\bar{\sigma}_m$ under normal conditions, the more meaningful the interpretation of the gravity anomalies will become.

If the data from explosion seismological studies of the crust are combined with laboratory studies of the physical properties of rocks under confining pressures corresponding to those to be expected down to the depth of the mantle, the mean density of the crust appears to be in the range 2.87 to 2.92 g/cm³, and the density contrast between the crust and mantle appears to be about 0.45 g/cm³. An alternate solution is obtained if values of mantle depth below sea level as defined by seismic measurements are plotted as a function of surface elevation, and sea data are incorporated by using synthetic elevations derived by condensing the water column present to equivalent rock material having a density corresponding to the granitic layer on the continents with a density of 2.74 g/cm³. The resulting plot [Woollard, 1962] defines a linear relation that can be written as $H_m = -(33.2 + 7.5h)$, where h is the surface elevation. The slope factor (7.5) requires a density difference between the crust and mantle of ≈ 0.395 g/cm³ for equilibrium conditions. Depending on what value of density is assumed for the mantle, which is probably in the range 3.32 g/cm³ (olivine peridotite) to 3.40 g/cm³ (garnet peridotite), the derived mean crustal density ranges from 2.93 to 3.00 g/cm³. Either approach defines a crust having a significantly higher density than 2.67 g/cm³, and a value of $\sigma_c = 2.93$ g/cm³ appears reasonable, with $\Delta \sigma \approx 0.39$ g/cm³.

Since a difference of 0.1 g/cm³ in mean crustal density will result in about 4.2 mgal in the included mass correction for each 1000-meter change in elevation, this change would significantly affect Bouguer anomaly values.

From the standpoint of isostatic anomalies, use of a more correct value for the crustal density is not significant, as the isostatic model need only give agreement between the compensating mass at depth and the surficial mass above sea level for changes in surface elevation. However, from the standpoint of using gravity anomalies for quantitative studies of crustal and upper mantle structure and the correlation of gravity data with seismic data, the use of a more correct density is important.

Corrections for the Gravitational Effect of Major Undulations of the Geoid

Corrections for major geoidal undulations are

not now considered in studies of the crust and upper mantle, but they should be considered if the undulations have a deep-seated source, since the associated gravity contribution is of the order of 10 to 20 mgal. The basic question is whether the mass anomalies giving rise to undulations of the geoid covering areas of 30° to 40° in size are at depths of 1000 km or more. This is not known to be the actual case. As most of our information on geoidal heights is derived from satellite data, it can be argued that the satellite-defined geoidal undulations and associated anomalous gravity field (as sensed at satellite altitude) could arise from the integrated gravity effect of discrete regionally disposed mass anomalies located in the crust and upper mantle or from regional changes in upper mantle composition. At this time, the only point that is clear is that the geoidal anomalies defined by satellite data (Figure 1) bear no relationship to the surface mass distribution defined by continents and ocean basins. However, as *Strange* [1966] pointed out, the pattern of geoidal 'highs,' as well as a harmonic representation of available surface gravity data, suggests that there is a correlation with areas of recent tectonic activity. If this apparent genetic relationship between geoidal highs, anomalous gravity, and tectonic activity is real, the depth of the mass anomalies could well originate in the crust and upper mantle. The resolution of this problem may take several years, as the tectonic associations are so diverse as to lend credence to a deep-seated source. For example, the geoidal high associated with the North Atlantic region centers over the Mid-Atlantic ridge, which is believed to define a locus of crustal spreading and addition to the crust of new material derived from the mantle. The geoidal high associated with the southwest Pacific region, on the other hand, centers over the New Guinea-Solomon Islands region which appears to be a locus of crustal convergence. If deep-seated mantle convection is a factor, it is of opposite sign in the two areas. The only established common denominators are volcanic activity, shallow- to intermediate-depth earthquake activity, high heat flow, and positive free-air anomaly values. The relations suggest that stress, tensional or compressive, in combination with temperature abnormality may result in a mass anomaly, perhaps a mineralogic phase transformation in the upper mantle, whereby anomalous mass is created through a change in atomic weight and density; however, this is only a conjecture. The significant point is that there is a valid argument for not assuming that the satellite-defined geoidal undulations are related solely to deep-seated mass anomalies; therefore, removing the gravitational effect associated with geoidal undulations in studying the crust and upper mantle could remove a significant portion of pertinent data.

Another important factor concerning satellite data is that the derived data are spherical harmonic representations, and that an 8th-degree representation corresponds approximately to 22° × 22° areas at the earth's surface. Even a 15th-degree harmonic representation would embrace 12° × 12° areas. There are only a few areas of this size that are characterized by one-sign free-air gravity anomalies at the earth's surface. These are the North Atlantic region, the high plains of Montana, Wyoming, North Dakota and adjacent Canada, eastern Canada, peninsular India, and western Europe. It is significant that some of these areas are portrayed in the satellite data, others are not. For example, *Strange* [1966], in investigating the degree of agreement between satellite and surface gravity data, showed (Figure 2) that a harmonic representation of surface gravity data, while agreeing in general pattern with that defined from satellite data, disagreed sufficiently in detail to raise some doubts as to the degree that satellite data can be directly used to define corrections for geoidal undulations. Although part of the discrepancies in pattern shown in Figure 2 can be attributed to a difference in degree (6 versus 8) for the two harmonic representations, it is significant that neither indicates the area of pronounced isostatic anomalies and anomalous crustal structure found in the high plains region east of the Rocky Mountains, in the northern United States, and southern Canada (see Figure 3). It would appear that at least a 15th-degree representation will be required before any detailed correlations with the actual surface field can be established.

For these reasons, it does not appear warranted to ascribe mass inequalities at great depth to areas of anomalous gravity defined by satellites.

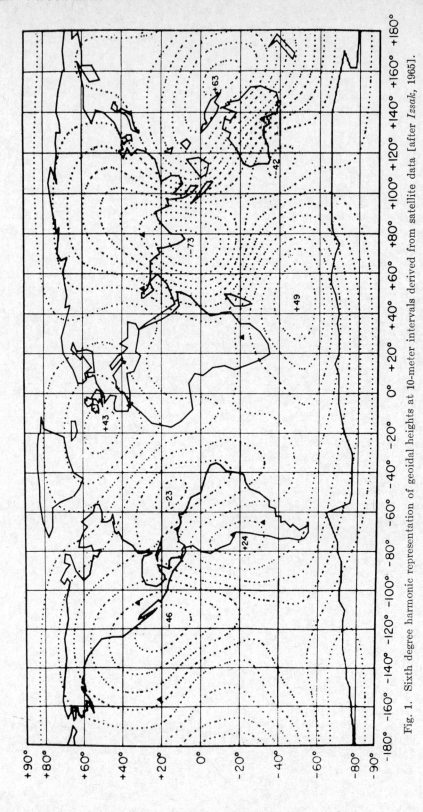

Fig. 1. Sixth degree harmonic representation of geoidal heights at 10-meter intervals derived from satellite data [after Izsak, 1965].

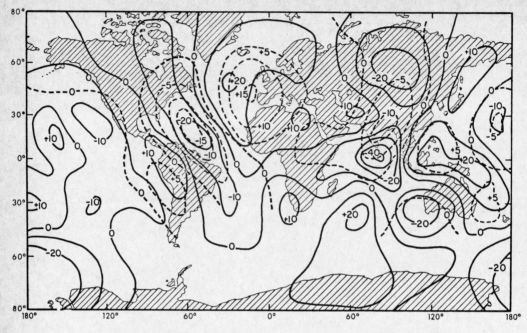

—— 8th Degree Representation Satellite Derived Gravity
--- 6th Degree Representation Surface Gravity Data

Fig. 2. Comparison of harmonic representation of surface gravity data to 6th degree and satellite-derived anomalous gravity field for the earth to 8th degree [after *Strange*, 1966].

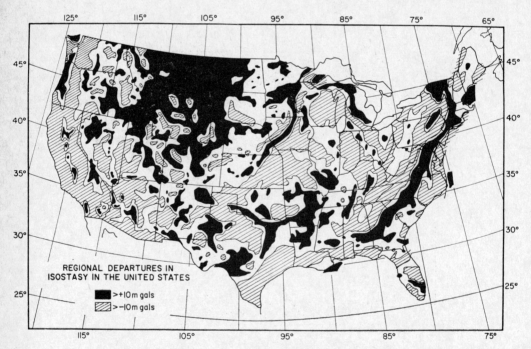

Fig. 3. Simplified isostatic anomaly map of the United States.

Revision of Intuitive Thinking in Interpreting Gravity Anomalies

In general, it is assumed that regional changes in Bouguer gravity anomaly values are related to changes in the thickness of the earth's crust, and that where there are anomalous relations for the surface elevation there is an anomalous value of crustal thickness. If the anomalous gravity values are positive, it is assumed the crust is too thin, and if the anonomalous gravity values are negative, it is assumed that the crust is too thick.

Local anomalous gravity values, on the other hand, are attributed to the near-surface mass distribution. Negative values are identified with sedimentary basins, granite outcrop areas, and grabens; positive values are identified with uplifts, horsts, and mafic rock areas. This interpretation of local anomalies is logical, for, with a short wavelength feature (even if it is perfectly compensated), the sign of the anomaly will always show a dependence on the near surface and surficial mass distribution. Free air anomalies therefore show a direct correlation with surface relief. However, the interpretation of regional anomalous gravity values in terms of anomalous crustal thickness is predicated on the constancy of the mean density of the crust and of the mantle. If these densities do not remain constant, the anomalous value of crustal thickness can be the reverse of the relation usually assumed.

From available data, it appears that approximately 65% of the seismically defined crustal thicknesses have a 'normal' relation to surface elevation; this suggests that there is a fairly uniform density contrast between the crust and underlying mantle. For such areas one can use either surface elevation or Bouguer anomaly values to predict the depth of the M discontinuity with a reliability better than ±3 km 50% of the time and ±6 km 90% of the time. However, for the remaining 35% of examples, predictions of crustal thickness from surface elevation and Bouguer anomalies are not reliable within 10 km. The seismic evidence in these areas suggests that the anomalous values of crustal thickness for the most part are not caused by a lack of isostatic equilibrium, but rather by significant changes in the density contrast between the crust and mantle. *Woollard* [1968] showed that there is a direct relationship between gravity, seismic velocity of the upper mantle, mean velocity and thickness of the overlying crust, and, in some areas, surface subsidence and uplift as evidenced geologically by basins and uplifts of the basement crystalline rock complex. This tectonic association might seem anomalous in that the areas of subsidence are areas of excess crustal thickness, high crustal and mantle velocity, and are characterized by an excess of gravity (positive free air and isostatic anomalies up to +35 mgal). The areas of uplift are areas of crustal thinning and subnormal crustal and mantle velocity, and are characterized by a deficiency of gravity (negative free-air and isostatic anomalies up to 30 mgal).

The apparently anomalous gravity relations are not due to a lack of isostatic compensation, but to earth curvature and the proximity effect inherent in gravitational attraction where there are horizontal discontinuities in mass distribution. This is well demonstrated by the fact that 95% of the mass effect of the crustal column above sea level is usually realized within 20 km of an observation point, whereas only 75% of the compensation effect is achieved at a distance of 167 km (the distance at which earth curvature becomes effective).

Although gravity relations contrary to the above are noted for all basins and uplifts that have originated through block faulting, the fact remains that many basins and uplifts have resulted, apparently, from the transformation of crustal and mantle material, with consequent changes in the mass distribution and density contrast between the crust and mantle resulting in subsidence or uplift. Although the area involved in either block faulting or known uplifts and basins of the type described are small (up to 3° × 4°), it is conceivable that broad areas of the crust could undergo deformation through a similar, if not the same, mechanism. A change with time in upper mantle environment (temperature, pressure, or stress) also might well reverse the process. Certainly there are examples of apparently significant changes in crustal parameters for which there is no inherited evidence of the causative mechanism or of any mass anomalies that might have existed during the period of active epeirogenic deformation. For example, recent drilling (1968) in the

Sigsbee Deep (Gulf of Mexico; water depth, 3500 meters) indicates that it must have foundered since Jurassic time. Seismic refraction data [Ewing et al., 1955] indicate that the velocity structure of the crust beneath the sedimentary cover is identical (to ±0.2 km/sec) to that found on the adjacent continent [Cram, 1961] and is not oceanic in character. The thickness of the crust (14 km) is normal for the depth of water (3.5 km) [Woollard, 1962] and the free-air gravity anomaly (+60 to −70 mgal) [Dehlinger and Jones, 1965] indicates that, on the average, isostatic equilibrium now prevails.

Geological evidence similarly indicates that all plateaus have risen, and whereas some are characterized by anomalous gravity, others are not. The plateau of Mexico is characterized by a mass anomaly having a −20 to −50 mgal effect; repeated gravity measurements over a 15-year period between Acapulco and Mexico City indicate that the plateau is rising at a rate of several centimeters a year (Woollard and Monges, unpublished data). The Colorado Plateau, on the other hand, appears on the whole to be in isostatic equilibrium [Woollard, 1966]. In this respect it is similar to the Gulf of Mexico, but it is not clear whether uplift of the Colorado Plateau and rejuvenation of the Rocky Mountains in Miocene time represent an opposite phase of the mechanism which caused subsidence in the Gulf of Mexico. Our best clue on such problems appears to be crustal and upper mantle relations in areas of anomalous gravity.

Revision in Gravity Standards

The number of modern absolute gravity measurements is relatively small (fourteen since 1900), and the range in gravity covered is limited to about 2200 mgal, or only about one-third the total range in surface gravity. Furthermore, there is only one modern measurement in the southern hemisphere. Most of our knowledge of the gravity field of the earth is derived from relative gravity measurements made either with pendulums or spring-type gravimeters. Before 1953 all relative pendulum measurements were made as loops usually with a closure only on the starting base (A-B-C-D-E-A). Starting in 1953, Woollard and Rose [1963] began a series of global pendulum observations using a ladder sequence (A-B-C-D-C-B-A) in order to obtain closures on each site. The resulting closures showed that pendulum measurements were subject to the same type of aberrations that affect gravimeter observations; namely, creep (like drift, except that the change in period is related to use rather than time), tares (jumps in period resulting in a permanent offset in values), and environmental effects (transient, nonpermanent changes in period that are related to either changes in ambient temperature, even though operating at constant temperature, or microseismic activity). Although the closure values defined the nature of the aberration and the sites affected, they did not define the leg at which the aberration occurred. Companion gravimeter observations were therefore adopted as a means of identifying which leg of a series of observations was in error. The gross closure (y) on the starting site could then be distributed so that $y = mX + b$, where m is the creep rate, X is the number of sites affected by creep, and b is the sum of the tares with regard to sign. Environmental effects are not considered because only a single station is affected; they represent transient effects equivalent to tares on adjacent stations of opposite sign but of the same magnitude.

It is clear that with closures on every site the general equation $y = mX + b$ applies to all observations between the first and second occupations of that site, so that there is continuous monitoring for aberrations, as well as a mechanism for introducing realistic corrections throughout a survey.

Attention is called to this procedure because the normal method has been to average pendulum interval values between sites on the assumption that all differences are related to either creep or observational uncertainty. As long as the creep rate is constant this is permissible, but averaging when there are tares and environmental effects can only bias the resulting values. The threshold of average observational reliability, however, must also be considered, but for most modern sets of pendulum apparatus, this is of the order of 0.15 to 0.2 mgal.

The above arguments that much of the existing pendulum gravity data were in significant error [Woollard and Rose, 1963] encouraged Study Group 5 of the International Association of Geodesy to propose a series of multiple pendulum and gravimeter observations along three main north-south traverses in different parts of the world. This program was also prompted by the fact that there were several different gravity standards in use for calibrating the response of high range gravimeters being used for global studies with differences up to 3.9 mgal/1000 mgal, and the fact that there was no general unanimity as to the error in either national gravity base values or the Potsdam datum.

The gravity standardization ranges agreed upon by Study Group 5 were: (*a*) Hammerfest, Norway, via Rome, Italy, and Nairobi, Kenya, to Capetown, South Africa; (*b*) Point Barrow, Alaska, via Denver, Colorado, and Mexico City, Mexico, to La Paz, Bolivia; and (*c*) Sapporo, Japan, via Manila and Singapore to Melbourne, Australia. The participating groups included Cambridge University, using the Cambridge compound Invar pendulums; Milan Polytechnic Institute, using the Italian multiple molybdenum pendulums; La Plata University, using a set of compound Askania pendulums; the Geographical Survey Institute (Tokyo), using the Japanese multiple quartz pendulum apparatus; and the University

Fig. 4. Pendulum gravity connections under the international gravity standardization program.

of Hawaii, using the Gulf compound quartz pendulums and two LaCoste-Romberg geodetic gravimeters. Auxiliary gravimeter measurements were also carried out by the Institute of Experimental Geophysics (Trieste) and other groups over limited areas, but principally by the U. S. Air Force 1381st Geodetic Survey Squadron using four LaCoste-Romberg gravimeters.

Because of accidents and other problems, only the Cambridge and Gulf pendulums occupied all sites, and only the Gulf pendulums were used for a complete set of interline connections as well as a multiple connection to Potsdam. These data are now being worked up to give a single set of improved values for all sites occupied, and, once accepted, will provide a single international gravity standard that can be used for both adjusting the national gravity bases to a common standard and for the calibration of gravimeters. Figure 4 shows the Gulf pendulum net.

Although this program showed that the Bad Harzburg alternate base value for Potsdam was +0.9 mgal in error, it did not resolve completely the error in the Potsdam datum, because there is an apparent systematic difference in gravity intervals between absolute sites and the intervals defined by relative pendulum gravity measurements. This problem, however, can be expected to be resolved soon with the aid of portable sets of absolute apparatus for measuring absolute gravity. One set of such apparatus is already operational, and at least two other sets should be available for measurements in 1969.

Conclusions

We stand on the threshold of a new era in the study of gravity, in which there will be:

1. A revised formula for defining the theoretical value of gravity at sea level.
2. A modification of the vertical gradient of gravity.
3. The establishment of a more precise value of absolute gravity.
4. The establishment of a world standard for all gravity measurements.
5. The definition of a more realistic value of crustal density.
6. The definition of a more realistic model for evaluating departures in mass distribution associated with the crust and upper mantle.
7. The recognition that apparent departures in isostasy can be caused by changes in the composition of the crust and/or upper mantle, as well as by actual departures in isostatic equilibrium.
8. The recognition that crustal parameters are not stable, but change with time in response to changes in the environment (temperature, pressure, stress) associated with the upper mantle, as well as in response to the effect of external factors (erosion, deposition, ice loading) on the crust, and tectonic displacement of both the crust and upper mantle through faulting and crustal spreading.

Finally, we are at a point when significant additions to gravitational knowledge are to be gained through both airborne gravimetry and more precise tracking of satellites and the determination of higher-order coefficients in analyzing the gravity significance of satellite orbital perturbations.

REFERENCES

Cram, I. H., Jr., A crustal structure refraction survey in south Texas, *Geophysics, 26*, 560–573, 1961.

Dehlinger, P., and B. R. Jones, Free air gravity anomaly of the Gulf of Mexico and its tectonic implications, 1963 edition, *Geophysics, 30*(1), 102–110, 1965.

Ewing, M., J. L. Worzel, B. Ericson, and B. C. Heezen, Geophysical and geological investigations of Gulf of Mexico, *Geophysics, 20*, 1–18, 1955.

Izsak, I. G., A new determination of nonzoned harmonics by satellites, in *The Use of Artificial Satellites for Geodesy, Proceedings of the Second International Symposium on the Use of Artificial Satellites for Geodesy, Athens, Greece*, vol. 2, edited by G. Veis, pp. 223–229, 1965.

Moritz, H., The geodetic reference system, 1967, *Allgem. Vermessungs-Nachrich.*, 2–7, 1968.

Strange, W. E., Comparisons with surface gravity data, in *Geodetic Parameters for a 1966 Smithsonian Institution Standard Earth*, edited by C. A. Lundquist and G. Veis, vol. 3, *Smithsonian Astrophys. Obs. Spec. Report 200*, 15–20, 1966.

Woollard, G. P., The relation of gravity anomalies to surface elevation, crustal structure and geology: *Univ. Wisc. Geophys. Polar Res. Center Rept. 62–9*, 350 pp., 1962.

Woollard, G. P., and J. C. Rose, *International Gravity Measurements*, Soc. Exploration Geophys. Spec. Publ., Tulsa, Oklahoma, 518 pp., 1963.

Woollard, G. P., and J. C. Rose, *International Gravity Measurements*, Soc. Exploration Geophys. Spec. Publ., Tulsa, Oklahoma, 518 pp., 1963.

Woollard, G. P., Regional isostatic relations in the United States, in *The Earth Beneath the Continents, Geophys. Monograph 10,* edited by J. Steinhart and T. Smith, pp. 557–594, American Geophysical Union, Washington, D.C., 1966.

Woollard, G. P., The inter-relationship of the crust, the upper mantle, and isostatic gravity anomalies in the United States, in *The Crust and Upper Mantle of the Pacific Area, Geophys. Monograph 12,* edited by L. Knopoff et al., pp. 312–341, American Geophysical Union, Washington, D.C., 1968.

Figure of the Earth and Mass Anomalies Defined by Satellite Orbital Perturbations[1]

M. A. Khan

FIGURE OF THE EARTH

The idea that the earth is spherical was first presented in the sixth century B.C. by Pythagoras, a Greek mathematician. The first scientific estimate of its size was made sometime in the third century B.C. by Eratosthenes. In 1617, Snell found that the length of a degree of meridional arc was 69 miles. Norwood, in 1636, and Picard, in 1671, gave relatively more accurate estimates of the length of a degree of meridian. The notion of a spherical earth persisted until Richer found in 1671 that the frequency of a pendulum beat varied with latitude. Newton explained this by postulating that, under the effects of self-gravitation and rotation, the earth would assume the form of an oblate spheroid. However, Cassini found from the measurements of an arc extending both northward and southward from Paris that the length of a degree computed for the northern part of the arc was smaller than that obtained from the southern part. Since the geodetic measurements refer to the normal to the ellipsoidal surface and *not* the geocentric radius vector of the ellipsoid, Cassini's results pointed to a prolate ellipsoid. This started a controversy between the English and the French scientists. In 1735–1736, the French Academy of Sciences, in an attempt to settle the controversey, sent geodetic expeditions to Peru and Lapland to measure the length of a meridian degree close to the equator and the Arctic Circle, respectively. The results showed that the length of a degree of meridional arc increased progressively from equator to pole. This proved conclusively that the earth was an oblate spheroid. Later, *Clairaut* [1743] published his classical work on the figure of the earth which provided, in principle, the essential mathematical basis for the real and equilibrium figures of the earth.

The methods of obtaining information on the figure of the earth can be divided into 'geometric' and 'dynamic' methods.

Geometric Methods

These methods can be divided into two subclasses: ground-oriented and space-oriented. The ground-oriented geometric methods consist of geodetic triangulation, trilateration, and traverse methods. Ground triangulation starts with the determination, with all possible accuracy, of the lengths and azimuths of suitably selected base lines, from which a continuous net of triangles is extended over the area of interest. Trilateration is like triangulation, but it involves measurement of distances instead of angles. For an accurate determination of the figure of the earth, the triangulation nets should be extended over very large areas of continental or even greater dimensions, but the nature of

---

[1] Contribution 248 from the Hawaii Institute of Geophysics.

error propagation, the difficulty of establishing vertical control by measuring heights as a function of elevation angles, and the necessity of a precise knowledge of the geoid heights in order to make a global fit to the observations diminish the accuracy of results. In addition, apart from being expensive and time-consuming, these methods are frequently difficult or impossible because the measurements must be carried across national boundaries, topographical obstacles, and oceans.

These difficulties can be overcome by using space-oriented geometric methods, usually involving artificial earth satellites. Simultaneous range or angular measurements to the satellite from several tracking stations separated by intercontinental distances will give their relative locations. A geodetic tie can then be established between any two of them by the method of long arc computation. In this case the satellite is used as an unknown survey point. If the orbit of the satellite is accurately defined, it can be used as a moving survey point whose location is known as a function of time. Angular or range measurements then define the locations of the different tracking stations separately in the same geodetic datum, without simultaneous observations by other tracking stations. These data give information on the figure of the earth directly.

Dynamic Methods

In this class of methods the information on the figure of the earth is obtained from the surface gravity data, the constant of precession, the satellite-determined spherical harmonic coefficients of geopotential and the earth's rate of rotation. The information yielded by these methods provides a useful check on the results of geometric methods and defines the geoid, an equipotential surface, a knowledge of which is important in the computation of triangulation chains in fixing the orientation of the reference ellipsoid. In addition, the methods give information on the equilibrium figure of the earth and hence an opportunity to investigate whether the earth is in hydrostatic equilibrium. Such information is important from the geophysical point of view, because the equilibrium figure provides the best reference for computing the anomalous gravity field which would be indicative of the hydrostatic stresses existing in the crust and the upper mantle.

For determination of the real flattening of the earth by these methods, a knowledge of the geopotential coefficient J_0 is essential. ($J = 3J_2/2$, where J_2 is the coefficient of the second harmonic in the spherical harmonic expansion of geopotential. Throughout this paper, we denote hydrostatic J as J_h and J for the real earth as J_0. Wherever it is not necessary or possible to make a distinction between J_h and J_0, we have used the symbol J). Before the launching of satellites, accurate determinations of J_0 were not available. Consequently, various attempts were made to find a method that would give a reliable estimate of J_0. The approach adopted by most investigators was to assume that the actual earth was in hydrostatic equilibrium, i.e., $J_0 = J_h$. This made it possible to estimate J_0 on the basis of *Clairaut*'s [1743] hydrostatic theory, which had been simplified by *Radau* [1885] by means of an important substitution. Before we proceed further, we give a brief account of Clairaut's theory.

Theory of Clairaut's Method

By considering the outer potential of a body, symmetrical with respect both to an equatorial plane and to an axis perpendicular to this plane, and by stipulating that the surface of the body be an equipotential surface, we can obtain [*de Sitter*, 1924; *Jones*, 1954; *Jeffreys*, 1952; *Khan*, 1968c]

$$J = f - \frac{1}{2} m - \frac{1}{2} f^2 + \frac{1}{7} mf \quad (1)$$

which gives

$$f = J + \frac{1}{2} m + \frac{5}{14} mJ + \frac{1}{2} J^2 + \frac{3}{56} m^2 \quad (2)$$

Equation 1 or 2 gives f_0 if J_0 is known (m is treated as known) and vice versa.

The condition of hydrostatic equilibrum can be represented analytically by the differential equation

$$dp = \rho \, d\Psi \quad (3)$$

where Ψ is the sum of gravitational and rotational potential, p the pressure at any point within the body, and ρ its density. This implies that p must be a function of Ψ and ρ either a constant or a function of Ψ also. Since the equipotential surfaces are given by $\Psi = $ constant, p and ρ will also assume a constant value on them, as is obvious from equation 3. Thus, the condition of hydrostatic equilibrium for the earth's interior is established by stipulating that the equipotential surfaces also be the surfaces of equal density. By considering the potential at an interior point of a body that is symmetrical with respect to its equatorial plane and an axis perpendicular to this plane, and by imposing the condition of hydrostatic equilibrium, we obtain two boundary conditions that together give the shape of the outer surface of the earth if it is in hydrostatic equilibrium. These boundary conditions [Khan, 1968a, c] are:

$$q = 1 - \frac{2}{3} f_h + \frac{2}{3} J_h + \frac{4}{9} f_h^2$$

$$- \frac{2}{5}\left(1 - \frac{2}{3} f_h + \frac{4}{9} f_h^2\right) \frac{\sqrt{1 + \eta_s}}{1 + \lambda_s} \quad (4)$$

$$\eta_s f_h' = 3 f_h' - \frac{6}{7} f_h^2 + \frac{4}{7} m f_h$$

$$- J_h\left(5 + \frac{10}{21} f_h + \frac{20}{21} m\right) \quad (5)$$

where

$$q = \frac{J_h}{H} = \frac{3}{2} \frac{C}{M a_e^2}$$

$$H = \frac{C - A}{C} \qquad J_h = \frac{3}{2} \frac{C - A}{M a_e^2}$$

C is the earth's moment of inertia about the polar diameter, A is the earth's moment of inertia about the equatorial diameter, M is the mass of the earth, a_e is the earth's equatorial radius, $m = \omega^2 r_m^3 / GM$, ω is the earth's angular velocity, r_m is the mean radius of the earth at $\sin^2 \phi = 1/3$ where ϕ is latitude, G is the constant of gravitation, η_s is the surface value of a parameter η which depends upon the earth's internal density distribution, λ_s is the surface value of a parameter λ which is indicative of the departure from unity of a certain function $F(\eta)$ introduced by Radau in the hydrostatic theory, and

$$f_h' = f_h - \frac{5}{42} f_h^2.$$

A simultaneous solution of equations 4 and 5 gives an explicit expression [Khan, 1968a, c] for f_h, i.e.

$$f_h = \frac{1}{M}\left[N + \frac{N \delta_1}{M} + \frac{25 A J_h^2}{M^2}\right] \quad (6)$$

where

$$M = 3 - \eta_0 \qquad N = 5 J_h + \delta_2$$

$$\eta_0 = \frac{25}{4} F^2 q'^2 - 1$$

$$F = 1 + \lambda_s \qquad q' = 1 - q \quad (7)$$

$$\delta_1 = \frac{25}{3} F^2 q' J_h - \frac{4}{7} m + \frac{10}{21} J_h$$

$$\delta_2 = \frac{20}{21} m J_h$$

$$A = \frac{17}{14} - \frac{5}{42} \eta_0 - \frac{25}{3} F^2 q q'$$

Equation 6 determines f_h if J_h, m, and λ_s are known.

Actual Figure of the Earth

Equation 1 or 2 gives the flattening of the real earth f_o if J_o is known. Corresponding to the satellite-determined $J_o = 1623.969 \times 10^{-6}$ [Kozai, 1964], $f_o = 1/298.25 \pm 0.05$. This value of flattening was recommended by the International Union of Geodesy and Geophysics in 1967 in preference to the value 1/297.0 (flattening of the international reference ellipsoid). Satellite studies have also shown that, for the real earth, the third degree zonal harmonic coefficient J_3 lies between -2×10^{-6} and -3×10^{-6}. There is a nonzero third degree zonal harmonic in the earth's potential because the earth is slightly more flattened at the south pole than at the north pole (the so-called 'pear shape'). Satellite determinations of the geopotential also show the second degree and second order sectorial harmonic coefficient J_{22} to be nonzero. This harmonic coefficient is related to the ellipticity of the equator. However, the investigations carried out to date have shown the equatorial ellipticity to be negligible. Table 1 lists some of the important results on the actual

TABLE 1. Figure of the Earth

Name	Year	Equatorial Radius, meters	f_0^{-1}	Where used as Datum	Remarks
Everest	1830	6,377,276	1/300	India	Arc measurement data
Airy	1830	6,376,542	1/299	Great Britain	Arc measurement data
Bessel	1841	6,377,397	1/299	Japan	Arc measurement data
Clarke	1866	6,378,206	1/295	North America	Arc measurement data
Clarke	1880	6,378,249	1/293	France, Africa	Arc measurement data
International*	1924	6,378,388	1/297	Europe	Arc measurement data
Krasovskii†	1940	6,378,245	1/298.3	Russia	Arc measurement data
Jeffreys	1948	6,378,099	1/297.7		Arc measurement and gravimetric data
Heiskanen and Uotila	1957		1/297.4		Gravimetric data
Smithsonian Standard Earth	1966	6,378,165	1/298.25		Satellite data
Satellite Ellipsoid‡	1967	6,378,160	1/298.25		Satellite data

* *Hayford* [1910].

† Krasovskii's ellipsoid is, in fact, a triaxial ellipsoid with equatorial ellipticity of 1/30,000 ± 2300, $\lambda = +15° \pm 2°$, where λ is the longitude of the largest meridian.

‡ Adopted by the assembly of the International Astronomical Union at Hamburg in 1964 and recommended by the assembly of International Union of Geophysics and Geodesy at Lucerne in 1967, in preference to the international reference ellipsoid.

figure of the earth obtained from geometric and dynamic methods from time to time.

Equilibrium Figure of the Earth

Before the satellite era, the geopotential coefficient J_0 was not known accurately; hence, Clairaut's method was used to find the best approximation to the shape of the earth, presuming that the actual earth was in hydrostatic equilibrium. *de Sitter* [1924] slightly modified *Darwin's* [1899] formulas of Clairaut's theory carried to the second order, expressing them in terms of mean radius of the earth instead of its equatorial radius. His final equations can be obtained by eliminating J_h (de Sitter did not differentiate between J_0 and J_h) appearing on the right-hand side of equations 4 and 5 with the help of equation 1 and are given as:

$$q = 1 - \frac{1}{3} m + \frac{1}{9} f_h^2 + \frac{2}{21} m f_h$$

$$- \frac{2}{5}\left(1 - \frac{2}{3} f_h + \frac{4}{9} f_h^2\right) \frac{\sqrt{1 + \eta_s}}{1 + \lambda_s} \quad (8)$$

$$\eta_s f_h' = \frac{5}{2} m - 2 f_h' + \frac{10}{21} m^2 + \frac{4}{7} f_h^2 - \frac{6}{7} m f_h \quad (9)$$

The method of *de Sitter* [1924, 1938], followed by *Bullard* [1948], consisted of estimating q on the basis of hydrostatic theory, from a knowledge of the internal density distribution, computing J from the relation $J = qH$ (H is known), and using this J to find the flattening from equation 1. A complete discussion of the method is given by *de Sitter* [1924], *Bullard* [1948], *Jones* [1954], and *Khan* [1968a, c]. The results obtained from this method by various investigators are given in Table 2.

With the satellite-determination of J_0, it became possible to compute f_0 and f_h independently of each other. A majority of the post-satellite investigators [*Henriksen*, 1960; *O'Keefe*, 1960; *Khan*, 1967] used de Sitter's equations 8 and 9 to determine hydrostatic flattening. They

TABLE 2. Equilibrium Figure of the Earth

Reference		f_h^{-1}	$f_h^{-1} - f^{-1}$
Pre-Satellite Method			
de Sitter,	1924	296.2 ± 0.136	
de Sitter,	1938	296.753 ± 0.086	
Bullard,	1948	297.228 ± 0.050	
Jeffreys,	1952	297.229 ± 0.071	
Jeffreys,	1963	296 75 ± 0.05	
Post-Satellite Method			
(f_h = hydrostatic flattening; $f = 1/298.25 \pm 0.05$)			
By de Sitter's equations			
Henriksen,	1960	300.0	+1.75
O'Keefe,	1960*	299.8	+1.55
Jeffreys,	1963	299.67 ± 0.05	+1.42
Khan,	1967	299.86 ± 0.05	+1.61

* Henriksen's calculations.

put $\lambda_s = 0$, and obtained η_s from (8), substituted it in equation 9, and solved the resulting quadratic for f_h. *Jeffreys* [1963] gave an excellent method that made use of the numerical integration procedure to evaluate the corrections for the second-order terms, avoiding completely the use of de Sitter's equations. This method gave a clear picture of the structure of the problem, but the analytical development was given only to the first order. The results obtained by various investigators are listed in Table 2. These results show that the hydrostatic flattening of the earth is smaller than its real flattening. However, there was some confusion as to the correct structure of the solution based on de Sitter's development of the hydrostatic theory, and it was suggested that in the solution based on de Sitter's development f_h is obtained from a solution of that part of the 'hydrostatic equations' which does not make use of any relation derived from the external potential theory as, for example, equations 1 and 2. However, *Khan* [1968a] pointed out that, in the course of deriving equations 8 and 9, de Sitter had already used equation 1 to eliminate J_h from the right-hand side of his equations; because de Sitter had done this at an early stage in the development of his hydrostatic equations, it probably did not stand out very clearly and may have led to the confusion. Khan modified *de Sitter's* [1924] equations, reinstating J_h, and gave the modified hydrostatic equations in the form of equations 4 and 5 or, preferably, in the form of the single hydrostatic equation 6. Since the J_h appears explicitly on the right-hand side of these modified equations, and since this J_h must be treated as an unknown to obtain a geophysically meaningful solution, the system of modified hydrostatic equations 4 and 5 is 'incomplete' and one must look for an additional boundary condition to solve them. This was done by *Khan* [1968b] in a paper that outlined the correct structure of the solution of the problem of hydrostatic equilibrium. This solution is briefly described below.

In the derivation of equations 1 and 2, no assumption is made as to the conditions existing in the earth's interior; hence, the equations are valid for both hydrostatic and nonhydrostatic equilibrium conditions. Thus, equations 1 and 2 should relate J_0 to f_0 for a nonhydrostatic case and J_h to f_h for a hydrostatic case. Consequently, equation 1 or 2 will assume the form (m is treated as known)

$$G(J_0, f_0) = 0 \quad (10)$$

for a nonhydrostatic case and

$$G(J_h, f_h) = 0 \quad (11)$$

for a hydrostatic case.

The hydrostatic equation 6 can be written in the form

$$F(m, H, J_h, f_h) = 0 \quad (12)$$

By considering the condition that equations 11 and 12 must match at the outer boundary of the earth for the hydrostatic equilibrium to exist, we obtain

$$F(m, H, J_h, f_h) = G(J_h, f_h) \quad (13)$$

This equation gives the general solution to the problem of hydrostatic equilibrium. The particular numerical solutions of equation 13 depend upon what parameters are held fixed in the solution of the hydrostatic equation. If the polar moment of inertia of the hydrostatic model is taken to be the same as that observed for the real earth (as computed from observed J and H), $f_h = 1/299.75 \pm 0.05$ ($\lambda_s = 0$). This value of f_h is about the same as that given by *O'Keefe* [1960], *Henriksen* [1960], *Jeffreys* [1963], and *Khan* [1967], but the structure of the solution differs slightly [*Khan*, 1968b, c] from that of the above investigators. If the dynamic flattening H of the hydrostatic earth is constrained to be the same [*Khan*, 1968b, c] as that observed for the real earth, the solution of equation 13 gives $f_h^{-1} = 297.29 \pm 0.05$ ($\lambda_s = 0$). This value is the same as that obtained by *Bullard* [1948], though the method used here is simpler. This is the hydrostatic model of presatellite times. A third solution of equation 13 can be obtained [*Khan*, 1968b, c] by treating the satellite-determined J as the initial datum. The flattening of the hydrostatic figure then turns out to be equal to that of the real earth. However, in the last two cases the polar moment of inertia of the hydrostatic models comes out to be greater than that observed for the real earth. This introduces serious dynamic complications. The results obtained from this solution are summarized in Table 3. *Ledersteger* [1967] is of the

TABLE 3. Equilibrium and Actual Figures of the Earth*

Hydrostatic Model	$J_h \times 10^6$	$H \times 10^6$	C/Ma^2	A/Ma^2	f_h^{-1}	$f_h^{-1} - f^{-1}$
Possible Hydrostatic Figures Obtained from the General Solution†						
(ω, M, a, H)	1635.225	3273.64	0.33300851	0.33191836	297.29 ± 0.05	−0.96
(ω, M, a, C)	1607.49	3240.43	0.33071598	0.32964432	299.75 ± 0.05	+1.50
(ω, M, a, J)	1623.969	3260.50	0.33204876	0.33096611	298.29 ± 0.05	0
Data for the Real Earth						
	1623.969‡	3273.64§	0.33071598	0.32963333	298.25 ± 0.05	

* After *Khan* [1968b].
† Based on $m = 0.00344980$ [*Khan*, 1967; *Jeffreys*, 1963].
‡ *Kozai* [1964].
§ *Khan* [1967].

opinion that the equilibrium figure of the earth has a greater rotational velocity (sidereal day 23h 44m 25.4s) and a greater surface flattening of about 1/295.8. In the beginning when the earth was a fluid, it was able to adjust its flattening to the reduced centrifugal force. With increasing rigidity, it lagged somewhat, so that there was a slight departure from the equilibrium state. Only the water cover was able to follow the lessening of the rotational velocity and, consequently, assumed the present flattening of 1/298.3 while the solid surface has a flattening of 1/295.8.

The hydrostatic model of geophysical significance is the one that has hydrostatic flattening of $1/299.75 \pm 0.05$, i.e. the one obtained by constraining the moment of inertia of the hydrostatic model to be the same as for the real earth.

MASS ANOMALIES DEFINED BY SATELLITE ORBITAL PERTURBATIONS

If the earth were perfectly spherical, its potential V could be expressed as

$$V = GM/r$$

The motion of a satellite around such an earth would be Keplerian, i.e. in closed orbits, the satellite would move in an ellipse, with the earth occupying one of the foci of this ellipse and the center of mass of the earth lying in the same plane as the elliptical orbit. However, potential of the actual earth differs from that due to spherical mass distribution by a small quantity R, which is called the disturbing potential. The potential V of the actual earth is therefore

$$V = (GM/r) + R$$

Because of this disturbing potential, the satellite is subjected to small disturbing forces. Since R is small in comparison to the potential of spherical attraction GM/r, the satellite orbit can still be regarded as an ellipse, but now the orbital elements of this ellipse undergo a slow change. Such a change in any of the orbital elements is commonly known as the 'perturbation' in that element.

The disturbing potential R can be expressed in terms of spherical harmonics as

$$R = -\frac{GM}{a_e} \sum_{n=2}^{\infty} \sum_{m=0}^{n} \left(\frac{a_e}{r}\right)^{n+1}$$
$$\cdot (J_{nm} \cos m\lambda + K_{nm} \sin m\lambda) P_{nm}(\sin \phi)$$

where P_{nm} are the associated Legendre functions. For $m = 0$, they are generally written as $P_n (\sin \phi)$ and are called Legendre polynomials. The functions $\cos m\lambda \, P_{nm} (\sin \phi)$ and $\sin m\lambda \, P_{nm} (\sin \phi)$ are the surface spherical harmonics. For $m = 0$, they are called zonal harmonics. For $m \neq 0$, they are called tesseral harmonics. J_{nm} and K_{nm} are the constant coefficients of the various spherical harmonics occurring in the expansions of R. The effect of the disturbing potential R on the various orbital elements of a satellite orbit is given by the following equations:

$$\frac{da}{dt} = \frac{2}{na} \frac{\delta R}{\delta M}$$

$$\frac{de}{dt} = \frac{1}{na^2 e} \left[(1 - e^2) \frac{\delta R}{\delta M} - (1 - e^2)^{1/2} \frac{\delta R}{\delta \omega} \right]$$

$$\frac{dM}{dt} = n - \frac{(1-e^2)}{na^2 e}\frac{\delta R}{\delta e} - \frac{2}{na}\frac{\delta R}{\delta a}$$

$$\frac{d\Omega}{dt} = \frac{1}{na^2(1-e^2)^{1/2}\sin i}\frac{\delta R}{\delta i}$$

$$\frac{d\omega}{dt} = \frac{(1-e^2)^{1/2}}{na^2 e}\frac{\delta R}{\delta e} - \frac{\cot i}{na^2(1-e^2)^{1/2}}\frac{\delta R}{\delta i}$$

$$\frac{di}{dt} = \frac{1}{na^2(1-e^2)^{1/2}}\left[\cot i \frac{\delta R}{\delta \omega} - \csc i \frac{\delta R}{\delta \Omega}\right]$$

where Ω is right ascension of the ascending node, a is semimajor axis of the satellite orbit, n is mean motion defined by $GM = n^2 a^3$, e is eccentricity, ω is argument of perigee, i is inclination, and M is mean anomaly defined by $M = n(t - \tau)$, where τ is the time of perigee passage.

The effect of zonal harmonics on the satellite orbit is much greater than that of the tesseral harmonics, which define longitude-dependent variation. The zonal harmonics cause secular and long-period changes in the right ascension of the ascending node Ω and argument of perigee ω, and only long-period changes in the eccentricity e and inclination i of the orbit. Further, long-period changes caused by the odd zonal harmonics are more pronounced in e and i, while the secular changes caused by the even zonal harmonics are more defined in Ω and ω. This facilitates the determination of zonal harmonics. The tesseral harmonics cause short-period changes in the orbital elements and are more difficult to determine because of the necessity of a greater and more even distribution of the observations in space and time in order to detect the short period variations caused by them in the orbital elements. However, by observing these changes, it is possible, in principle at least, to determine the tesseral harmonics. Because of the difficulties involved in the determination of these coefficients, many attempts have been made to compare the spherical harmonic coefficients obtained from various satellite solutions with each other and with coefficients obtained from surface gravity data in order to check their reliability. It is generally agreed now that the low degree harmonics have been determined with reasonable accuracy. Satellite-determined values of the higher degree zonal and tesseral harmonics, however, should be accepted with caution.

These harmonic coefficients give information on the mass anomalies, which are usually expressed in terms of either gravity anomalies or geoidal undulations. Both the geoidal undulations and the gravity anomalies are computed with respect to a reference. For geophysical studies involving small-scale features, this reference can be set up arbitrarily, but for investigations on a global scale, the reference should be uniform and is usually provided by the international reference ellipsoid (see Table 1). In 1967, the International Union of Geodesy and Geophysics recommended the use of the 'satellite ellipsoid,' whose parameters are based on improved information provided by artificial earth satellites. However, for a realistic appraisal of the large-scale mass anomalies associated with the crust and the upper mantle, and for consequent geophysical studies based thereon, the best reference is provided by the equilibrium figure of the earth, the departures from which are truly indicative of the hydrostatic stresses existing in the earth, especially in the crust and the upper mantle. The departure of the real earth from this equilibrium figure also gives a measure of the minimum strength which the earth should have in order to support these hydrostatic stresses. These stresses are not significant in localized studies but become important on the global scale. Since the satellite gravity results reflect the long wavelength component of the gravity field more accurately, it becomes all the more important that the satellite-derived anomalous gravity be computed with reference to the equilibrium figure, so that the satellite gravity results can be fully exploited. If some other reference is adopted instead, there is a likelihood that an unreal, long-wavelength component may be introduced in the gravity anomaly, the amplitude of such a component depending on the variation of the selected reference from the equilibrium figure.

The zonal harmonic coefficient J_{20} and J_{40} (usually written as J_2 and J_4) for the equilibrium figure can be calculated, if f_h is known, from equation 1 and the equation

$$J_{40} = -\frac{4}{35} f_h (7 f_h - 5m)$$

These two coefficients define the equilibrium figure to the order of accuracy required here.

Geoidal Undulations

The geoidal undulation N at a point (r, ϕ, λ) is obtained from

$$N = -\frac{GM}{a_e g_0} \sum_{n=2}^{\infty} \sum_{m=0}^{n} \left(\frac{a_e}{r}\right)^{n+1}$$
$$\cdot (\Delta J_{nm} \cos m\lambda + \Delta K_{nm} \sin m\lambda) P_{nm}(\sin \phi)$$

where

$$\Delta J_{nm} = \text{observed } J_{nm} - \text{reference } J_{nm}$$
$$\Delta K_{nm} = \text{observed } K_{nm} - \text{reference } K_{nm}$$

and g_0 is the theoretical gravity on the reference ellipsoid at (r, ϕ, λ). In the present case, the only nonzero reference J_{nm}'s are J_{20} and J_{40}. Hence, except for these two coefficients, all other ΔJ_{nm} and ΔK_{nm} are equal to the observed J_{nm} and K_{nm}.

Gravity Anomalies

The gravity anomaly Δg at point (r, ϕ, λ) is obtained from

$$\Delta g = -\frac{GM}{a_e^2} \sum_{n=2}^{\infty} \sum_{m=0}^{n} (n-1) \left(\frac{a_e}{r}\right)^{n+2}$$
$$\cdot (\Delta J_{nm} \cos m\lambda + \Delta K_{nm} \sin m\lambda) P_{nm}(\sin \phi)$$

where ΔJ_{nm} and ΔK_{nm} are defined as above.

DISCUSSION OF RESULTS

Minimum Strength of the Earth

The stress differences arising because of the departure of the earth from hydrostatic equilibrium are given by *Jeffreys* [1943, 1963]. Their main component is due to the P_2 inequality, i.e., the departure of the second coefficient in the spherical harmonic expansion of the actual geopotential from its hydrostatic counterpart. On the supposition that the stresses due to the P_2 inequality are supported by strength (1) down to the core and (2) to a depth of 0.1 of the earth's radius, the strength S needed for the equilibrium figure with $f_h^{-1} = 299.75$ is:

$$S = 4.7 \times 10^7 \text{ dynes/cm}^2 \text{ on supposition 1}$$
$$S = 8.7 \times 10^7 \text{ dynes/cm}^2 \text{ on supposition 2}$$

Gravity Field

Figure 1 shows the geoidal undulations, and Figure 2 shows the gravity anomalies referred to an ellipsoid with the flattening 1/299.75. These are computed from *Kozai*'s [1964] zonal harmonic coefficients and *Gaposhkin*'s [1966] tesseral harmonic coefficients. The satellite-derived gravity results give generalized global representation of the earth's gravity field. The spherical harmonic representation of the gravity field is given only to the eighth degree. Thus

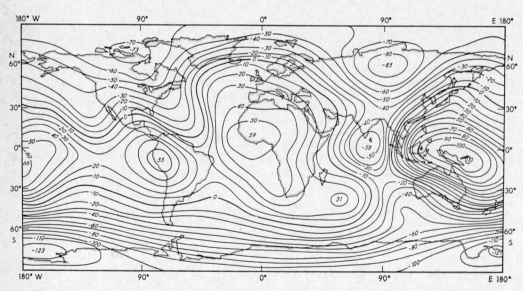

Fig. 1. Geoidal undulations in meters referred to the equilibrium figure with a flattening of $f = 1/299.75$.

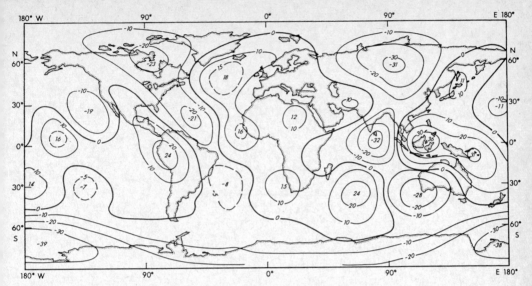

Fig. 2. Gravity anomalies in milligals referred to the equilibrium figure with a flattening of $f = 1/299.75$.

the gravity anomalies of Figure 2 should be comparable to roughly $22.5° \times 22.5°$ mean anomalies derived from surface gravity data. On this scale they represent adequately only the long wavelength component of anomalous gravity. The short wavelength component, which is of great interest to the geophysicist, is poorly represented in these results [Khan and Woollard, 1968]. This can be seen more clearly by considering the variance of the satellite-determined gravity field and comparing it with the variance of the free-air anomalies obtained from surface gravity measurements. If Δg denotes the gravity anomaly at any point in the field, the variance of the field is defined as

$$\text{var}(\Delta g) = \frac{1}{4\pi} \int_s \Delta g^2 \, ds$$

Var (Δg) for the satellite representation of the anomalous gravity field is about 145 mgal² [var (Δg) of the same field is 136 mgal² when referred to the international reference ellipsoid and 119 mgal² when referred to the ellipsoid with flattening of 1/298.25], while that for the anomalous gravity field (free-air anomalies) defined by the surface gravity measurements is about 1201 mgal² [Kaula, 1959]. The satellite representations of the gravity field to the degree of expansion considered here, therefore, can only outline regional areas of anomalous gravity such as the positive gravity anomaly bordering on the south-southwest coast of Iceland, the two negative gravity anomalies flanking the southern part of North America, the negative gravity anomaly in the Indian Ocean near Ceylon, and the most pronounced positive gravity anomaly over the Solomon Islands and the New Guinea areas.

The significance of the anomalous areas of gravity, however, is not clear at this stage. Because of the long wavelengths portrayed, the anomalous mass could be deep seated or represent the integrated effect of a number of shallow mass anomalies located in the upper mantle or crust. In either case there would also be a contribution from surface topography. The topographic effect appears to be only of secondary importance, and the correlation of the global variations of topography with those of the gravity field appears to be low (as Jeffreys [1952] pointed out). However, a better insight into this sort of relationship is obtained by making a power spectrum analysis [Kaula, 1967] of the satellite representation of the gravity field and the global variations of topography. Such an analysis is performed in terms of the degree variances of the functions in question. If $\bar{J}_{nm}$ and $\bar{K}_{nm}$ are the fully normalized constant coefficients in the spherical harmonic expansion of a function over a spherical surface, the degree variance D_n^2 is defined as

$$D_n^2 = \sum_m (J_{nm}^2 + \bar{K}_{nm}^2)$$

The degree correlation ρ_n of any two functions, say the anomalous gravity Δg and topography T, is then given by

$$\rho_n(\Delta g, T) = \frac{D_n^2(\Delta g, T)}{[D_n^2(\Delta g) \; D_n^2(T)]^{1/2}}$$

where $D_n^2(\Delta g)$ is the degree variance of Δg, $D_n^2(T)$ is the degree variance of topography T, and $D_n^2(\Delta g, T)$ is the cross degree variance of Δg and T. A plot of the degree correlation as a function of n, the degree of the appropriate spherical harmonic, is given in Figure 3. The correlation is somewhat erratic for $n \leq 5$. For $6 \leq n \leq 10$, however, the correlation is more systematic. For $n > 10$, the correlation between gravity and topography again becomes insignificant.

Figure 4 shows the power spectra of three versions of the anomalous gravity field. Because the vertical scales of the three plots in Figure 4 are different, the vertical separation of any two points on the different plots is not a linear measure of the difference in the power of the fields in question. The most important feature of Figure 4 is that the spectral curve B lies between the spectral curves A and C. This can be interpreted to mean that while isostasy does prevail at all wavelengths [*Kaula*, 1967], a considerable part of the anomalous gravity vector originates somewhere in the upper mantle.

For some of the areas it is difficult to integrate the available geophysical information into a meaningful framework. Some of the areas of anomalous gravity such as the positive anomaly area over the Mid-Atlantic ridge are known to be characterized by anomalous geophysical relations: a subnormal mantle velocity, pronounced magnetic anomalies, and high heat flow along the crest of the ridge but subnormal heat flow along the flanks. However, the anomalous gravity field conforms closely with the regional topographic relief; the Bouguer anomalies indicate that the relief is compensated. It is difficult to reconcile all these facts without postulating that the subnormal mantle velocity values are indicative of higher than normal density values or that there is deeper, as yet undiscovered, layering in the upper mantle. *Worzel* [1965] has shown three possible theoretical mass distributions, all in the upper 30 km of the crust, to explain the observed gravity relations over the Mid-Atlantic ridge. *Cook* [1962] has postulated that the apparent subnormal mantle velocities are due to a mixture of crustal and mantle materials as a result of convection with attendant high heat flow. While eminently reasonable for the Mid-Atlantic ridge, these suggestions do not explain the relations in the Indian Ocean area where the satellite data define a broad negative anomaly area that appears to be related to a stable ocean basin region lying between a narrow volcanic ridge and a rise (of Mid-Atlantic ridge type) which has many of the geophysical associations noted for the Mid-Atlantic ridge.

It is this lack of consistency between gravity and other geophysical relations on a regional scale that raises doubts as to interpretations that have been placed on the data and points up the need for more extensive geophysical studies in areas of anomalous gravity. In particular, the negative gravity anomaly in the Indian Ocean just south of Ceylon and the positive gravity anomaly over the Solomon Islands seem to be

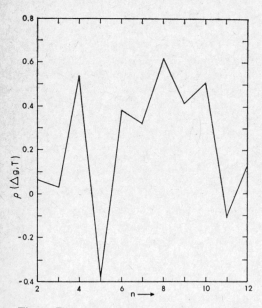

Fig. 3. Degree correlation function $\rho_n (\Delta g, T)$ of gravity Δg and topography T [data from *Kaula*, 1967].

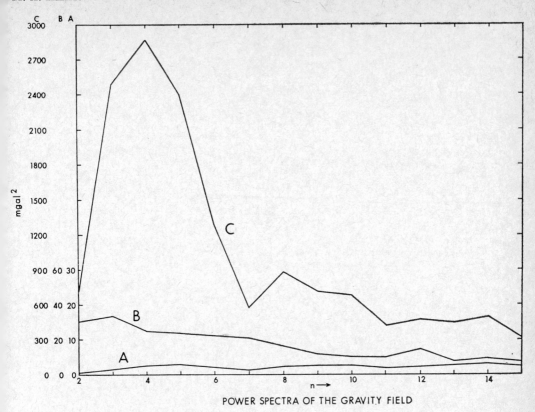

Fig. 4. Power spectra of the gravity field: (A) topographic features compensated at a depth ~30 km; (B) satellite determination of the anomalous gravity field (including departure of the real earth from hydrostatic equilibrium due to the P_2 inequality); (C) topography supported as a surficial load by a rigid crust (i.e., no compensation). Note the three different vertical scales for A, B, and C [data from Kaula, 1967].

the two most interesting features for detailed geophysical study because of their tectonic nature and surface geology association.

REFERENCES

Bullard, E. C., The figure of the earth, *Monthly Notices Roy. Astron. Soc., Geophys. Suppl., 5*, 186–192, 1948.

Clairaut, A. C., *Theorie de la Figure de la Terre*, Paris, 1743.

Cook, K. L., The problem of the mantle-crust mix: later inhomogeneity in the uppermost part of the earth's mantle, *Advan. Geophysic., 9*, 295–360, 1962.

Darwin, G., The theory of the figure of the earth carried to the second order of small quatities, *Monthly Notices Roy. Astron. Soc., 60*, 82–124, 1899.

de Sitter, W., On the flattening and the constitution of the earth, *Bull. Astron. Inst. Neth., 2*, 55, 97–108, 1924.

de Sitter, W., On the system of astronomical constants, *Bull. Astron. Inst. Neth., 8*, 307, 213–229, 1938.

Gaposhkin, E. M., A dynamical solution for the tesseral harmonics of the potential and for station coordinates (abstract), *Trans. Am. Geophys. Union, 47*, 47, 1966.

Hayford, J. F., *The Figure of the Earth and Isostasy*, U. S. Coast and Geodetic Survey Washington, D. C., 1909.

Hayford, J. F., *Supplementary Investigations in 1909*, U. S. Coast and Geodetic Survey, Washington, D. C., 1910.

Helmert, F. R., and S. B. Preuso, *Akad. Wiss., 1911*, 10–19, 1911.

Henriksen, S. W., The hydrostatic flattening of the earth, *Ann. Intern. Geophys. Yr., 12*, 197–198, 1960.

Jeffreys, H., The stress differences in the earth's shell, *Monthly Notices Roy. Astron. Soc., Geophys. Suppl., 5*, 71–89, 1943.

Jeffreys, H., *The Earth*, Cambridge University Press, Cambridge, 1952.

Jeffreys, H., On the hydrostatic theory of the figure of the earth, *Geophys. J., 8,* 196–202, 1963.
Jones, H. S., Dimensions and rotation, in *The Earth as a Planet,* edited by G. P. Kuiper, chapter 1, University of Chicago Press, Chicago, 1954.
Kaula, W. M., Statistical and harmonic analysis of gravity, *J. Geophys. Res., 64,* 2401–2421, 1959.
Kaula, W. M., Geophysical implications of satellite determinations of the earth's gravitational field, *Space Sci. Rev., 7,* 769–794, 1967.
Khan, M. A., Some parameters of a hydrostatic earth (abstract), *Trans. Am. Geophys. Union, 48*(1), 56, 1967.
Khan, M. A., A re-evaluation of the theory for the hydrostatic figure of the earth, *J. Geophys. Res., 73*(16), 5335–5342, 1968a.
Khan, M. A., General solution of the problem of hydrostatic equilibrium of the earth, in press, 1968b.
Khan, M. A., On the equilibrium figure of the earth, Hawaii Inst. Geophys. Sci. Rept. HIG-68-10, 1968c.
Khan, M. A., and G. P. Woollard, A review of perturbation theory as applied to the determination of geopotential, Hawaii Ins. Geophys. Sci. Rept. HIG-68-1, 1968.
Kozai, Y., New determination of zonal harmonic coefficient of the earth's gravitational potential, *Smithsonian Astrophys. Obs. Spec. Rept. 165,* 1964.
Ledersteger, K., Critical notes on mass functions derived from orbital perturbations of artificial satellites, in *The Use of Artificial Satellites for Geodesy,* vol. 2, edited by G. Veis, National Technical University, Athens, 1967.
O'Keefe, J. A., Determination of the earth's gravitational field, in *Space Research* vol. 1, edited by H. Kallmann, pp. 448–457, North-Holland Publishing Company, Amsterdam, 1960.
Poincare, H., *Bull. Astron., 27,* 321–356, 1910.
Radau, R., *Compt. Rend., 100,* 972–974, 1885.
Tisserand, F., *Mecanique Celest,* 4 volumes, Paris, 1889–1896.
Woollard, George P., A. S. Furumoto, G. H. Sutton, J. C. Rose, A. Malahoff, and L. W. Kroenke, Cruise report on 1966 seismic refraction expedition to the Solomon Sea, Hawaii Inst. Geophys. Rept. HIG-67-3, 1967.
Worzel, J. L., *Pendulum Gravity Measurements at Sea, 1939–1959,* John Wiley & Sons, New York, 1965.

Gravity Anomalies as a Function of Elevation: Some Results in Western Europe

S. Coron

FOR a long time it has been known that there is a relationship between gravity anomalies and surface topography: free-air anomalies are directly related to the elevation of the observation points (H_s), and Bouguer anomalies are principally related to regional elevations (isostasy). On the other hand, the isostatic anomalies bear more resemblance to some of the residual anomalies connected more or less with the local geology.

Several authors have already made studies of these correlations (in particular, *Heiskanen* [1950] and *Woollard* [1962]); here we will confine ourselves to the Bouguer anomalies of some characteristic regions of western Europe.

Bouguer anomalies (if possible, with terrain corrections) can be plotted as a function of station elevation (H_s). Figures 1, 2, 3, and 4 show such plots for four regions of contrasting relief.

The utilization of H_s in mountainous regions is not very desirable, as the observations are generally made in valleys, often even at a lower elevation than the mean elevation surrounding the observation point. Bouguer anomalies are, as a first approximation, independent of local conditions (valleys, peaks) [*Coron,* 1954, 1967].

In this representation, the dispersion of points around a mean line always remains very great,

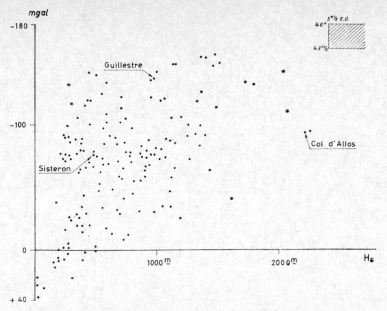

Fig. 1. Bouguer anomaly versus station elevation, western Alps.

even if it is possible to compute a proportionality factor between anomalies and elevations H_s.

Furthermore, for the islands (Corsica, Sardinia), coastal stations (at low elevation) group themselves poorly around a mean value, because of the different influence of the adjacent oceanic regions (continental shelf or oceanic depths).

Bouguer anomalies for each individual point can also be plotted as a function of the mean elevation around each observation point. The integration surface is variable according to the authors; some have chosen the mean elevation of the surface corresponding to the Hayford zones, for example, to zones *A-J* (radius of 12.4 km) or to zones *A-K* (radius of 18.8 km) [*Holopainen*, 1947; *Tanni*, 1942].

One could also adopt the mean elevation of

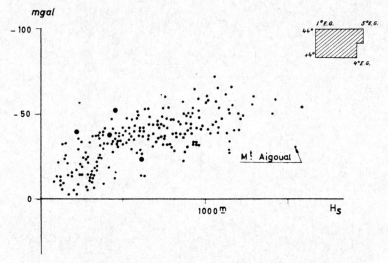

Fig. 2. Bouguer anomaly versus station elevation, Central massif.

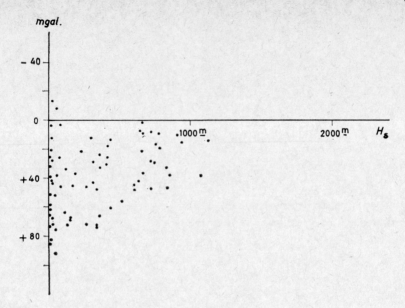

Fig. 3. Bouguer anomaly versus station elevation, Corsica.

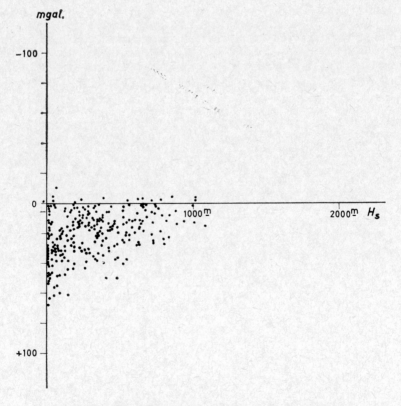

Fig. 4. Bouguer anomaly versus station elevation, Sardinia.

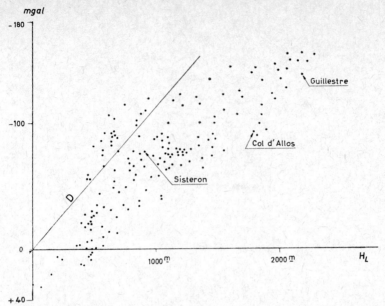

Fig. 5. Bouguer anomaly versus mean elevation of zone L (19 to 29 km from station), western Alps.

a single Hayford zone if it gives an exact enough idea of a regional elevation; this method is the most rapid when isostatic anomalies have already been computed. The elevation of zone L (19 to 29 km from the station), which was adopted in Figures 5, 6, and 7, refers to the regions already presented in Figures 1–4.

The comparison of Figures 1 with 5 and 2 with 7 shows that the dispersion of points is clearly less in the series for which H_L has been used in place of H_S.

For a general study of the deep terrestrial zones, it seems preferable to consider the mean values for somewhat larger areas.

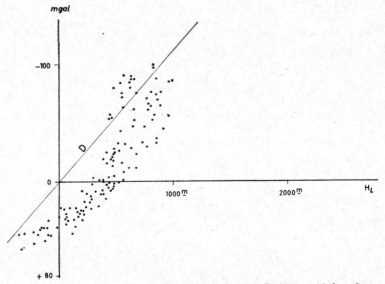

Fig. 6. Bouguer anomaly versus mean elevation of zone L (19 to 29 km from station), western Alps near the Mediterranean Sea.

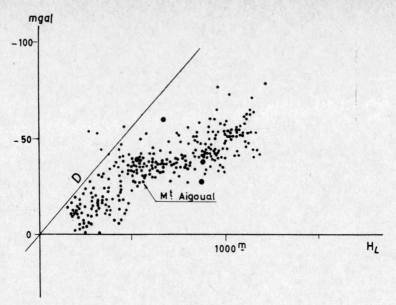

Fig. 7. Bouguer anomaly versus mean elevation of zone L (19 to 29 km from station), Central massif.

For the figures pertaining to Norway and Sweden, we chose quadrangles whose dimensions are $5' \times 10'$ (Figures 8 and 9) and $30' \times 30'$ (Figures 10 and 11). The general trend of the points is essentially the same for the two sizes of subdivision Figure 12 (western French Alps) was made with $20' \times 20'$ quadrangles.

Finally, we consider the mean values ($\bar{B}$) of the anomalies for the groups of stations of the same elevation interval $\bar{H}_S$. In Figure 13 (western Alps), groups of stations were chosen for 100-meter intervals of elevation. For each group (up to 1900 meters), the average was based on approximately 30 values.

The proportionality factor between anomalies and elevations is very small, of the order of

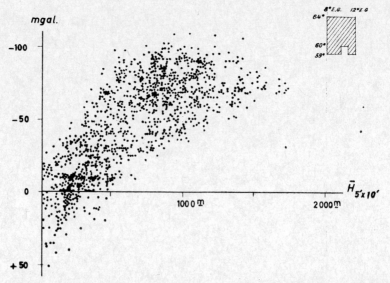

Fig. 8. Bouguer anomaly versus mean elevation of $5'$ by $10'$ quadrangle, southern Norway.

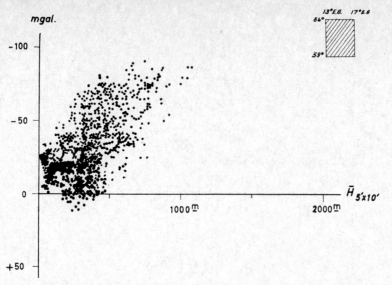

Fig. 9. Bouguer anomaly versus mean elevation of 5' by 10' quadrangle, southern Sweden.

3 mgal/100 meters; it remains essentially the same if one considers the regional elevation (radius of about 5 km) instead of $\bar{H}_S$ [Coron, 1967].

CONCLUSIONS

Anomaly data for western Europe are summarized in Tables 1 and 2.

1. As a first approximation, the proportionality factor c between Bouguer anomalies and elevations can be considered constant; this is actually so (1) for zones of limited extent and (2) for the variations of mean values related to large areas (Europe $1° \times 1°$) or to groups of stations of the same elevation.

2. In practice, this factor c varies with altitude: it decreases for high elevations. Thus, it proceeds from 1.2 mgal/10 meters ($0 < H_L < 1000$ meters) to 0.6 ($1000 < H_L < 2000$ meters) in the westerrn Alps, and from 0.6

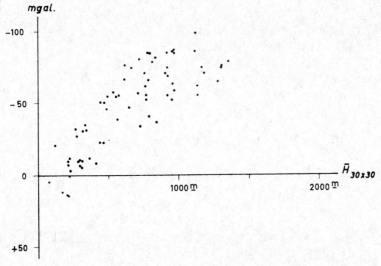

Fig. 10. Bouguer anomaly versus mean elevation of 30' by 30' quadrangle, southern Norway.

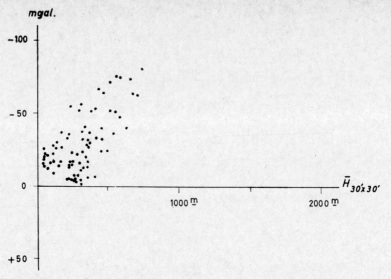

Fig. 11. Bouguer anomaly versus mean elevation of 30′ by 30′ quadrangle, southern Sweden.

($0 < H < 500$ meters) to 0.35 ($500 < H < 1000$ meters) in Norway.

3. It also varies according to the geographic conditions of the neighboring regions. In Figure 7, the stations of the central massif distribute themselves around two arcs in the inverse direction following the positions of the stations with reference to the plains or adjacent plateau.

4. The value of the factor c rarely equals the ideal value 1.1 (density 2.67), which corresponds to the isostatic compensation of a plate layer. This theoretical variation is indicated by

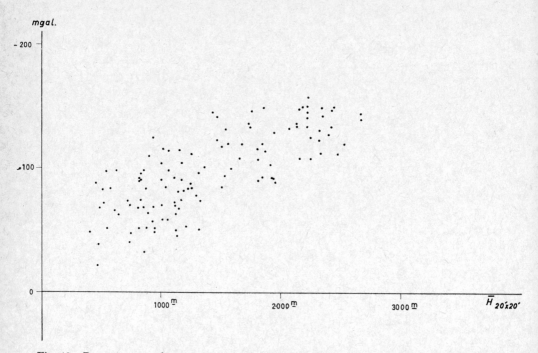

Fig. 12. Bouguer anomaly versus mean elevation of 20′ by 20′ quadrangle, western Alps.

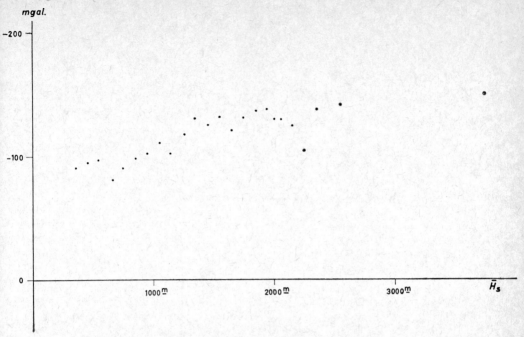

Fig. 13. Mean values of Bouguer anomalies for stations at same mean elevation intervals (100-meter intervals), western Alps.

TABLE 1. Bouguer Anomalies in Milligals as a Function of Station Elevation H_S or Mean Elevation*

Region	$H = 0$	$H = 500$ meters	$H = 1000$ meters	$H = 1500$ meters	$H = 2000$ meters	$H = 2500$ meters
Western Alps						
H_L	+40	−20	−80	−110	−140	−160
$H_{20'\times20'}$	(0)	−50	−80	−105	−125	−140
$\bar{H}_S$	(−75)	−90	−105	−120	−130	−140
Central massif						
H_L	+12	−30	−50	−(60)		
Southern Norway						
$H_{5'\times10'}$	+20	−40	−75	−85		
$H_{30'\times30'}$	+15	−35	−75	−(90)		
Southern Sweden						
$H_{5'\times10'}$	+10	−55	−85			
$H_{30'\times30'}$	+10	−55				
Corsica						
H_S	+50	+35	+20			
Sardinia						
H_S	+30	+20	+5			
Western Europe						
$H_{1°\times1°}$	+10	−30	−75	−115		
Eastern Alps†						
H_{A-J}	−14		−91			
Western Carpathian‡						
H_{A-K}	+8		−45			

* Based on data in the figures and other published data. H_L, mean elevation of zone 19 to 29 km; $H_{20'\times20'}$, mean elevation of quadrangle $20' \times 20'$; $\bar{H}_S$, mean elevation in intervals.
† Holopainen, 1947.
‡ Tanni, 1942.

TABLE 2. Anomaly Values in Some Characteristic Geologic Regions

Regions	Mean H, meters	Bouguer Anomalies			AIRY Anomalies (30 km)		
		Min.	Max.	Mean	Min.	Max.	Mean
Brittany	150	−26	+35	+1	−27	+34	+3
Central massif	550	−71	0	−31	0	+49	+17
Carpathians*		−62	0		−1	+46	
Ireland	(160)	−20	+35	+15			
Corsica	(550)	0	+90	+35	−20	+60	+26
Sardinia	450	0	+70	+35			
Netherlands	20	−24	+16	+3			
Paris basin	130	−41	0		−29	+20	
Po Valley	300	−150	0		−110	+20	

*Valek, 1954.

the line D in Figures 5, 6, and 7. The theoretical line D and the mean line traced through the points are almost parallel in Figure 6, for, in this example the coastal stations influenced by the Mediterranean were added to the cisalpine stations. In the other examples, the value of the factor is between 0.3 and 1.2.

5. There is a great diversity of results depending upon the mode of representation and the distribution of stations. It is preferable to always use the same mode of representation (for instance, the area in which the mean height is evaluated).

REFERENCES

Coron, S., Contribution à l'étude du champ de la pesanteur en France, *Sci. Terre*, 2(4), 1954.

Coron, S., Quelques relations entre anomalies de la pesanteur et altitudes dans les régions montagneuses, in *Aus der Geodätischer Lehre und Forschung, Festschrift zum 70. Geburtstag von Prof. W. Grossmann*, pp. 20–27, Konrad Wittwer, Stuttgart, 1967.

Heiskanen, W., On the isostatic structure of the earth's crust. *Publ. Isos. Inst. Helsinki*, 24, 60 pp., 1950.

Holopainen, P. E., On the gravity field and the isostatic structure of the earth's crust in the east Alps, *Publ. Isos. Inst. Helsinki*, 16, 90 pp., 1947.

Tanni, L., On the isostatic structure of the earth's crust in the Carpathian countries and the related phenomena, *Publ. Isos. Inst. Helsinki*, 11, 100 pp., 1942.

Valek, R., Gravimetric observations in the central part of the Slovak Carpathians and their interpretation, *Geofiz. Sb.*, 14, 44 pp., 1954.

Woollard, G. P., The relation of gravity anomalies to surface elevation, crustal structure and geology, Res. Rept. Ser. 62-9, University of Wisconsin, 264 pp., 1962.

The Relation between the Earth's Crust, Surface Relief, and Gravity Field in the USSR

R. M. Demenitskaya and N. A. Belyaevsky

GRAVITY data alone cannot yield a unique solution for density distributions in the earth's interior. Seismic crustal studies have assisted in the evolution of empirical formulas for the relation between crustal thickness, gravity anomalies, and topography (Table 1).

Calculations based on gravity measurements show that isostatic balance is approximately maintained over extensive areas of the earth's crust. On the basis of analysis of Bouguer and free-air anomaly distribution, *Lyustikh* [1957] found that the approximate factor of compensation is 91% for continents, 99% for oceans.

The distribution of Bouguer anomalies over the world correlates with topography; free-air anomalies are in general near zero. Thus, the crust seems to be essentially in isostatic equilibrium, and visual analysis of the gravitational field and relief features give little indication about the structure of the crust and upper mantle. Only for special regions (such as transitional zones from oceans to continents, deep inland seas, oceanic troughs and submarine ridges) are the free-air and Bouguer anomalies sufficiently clear to give evidence of sharp changes in mass distribution in the crust and upper mantle, as well as disturbances of isostatic balance.

Woollard and Strange [1962] found a relationship between the Bouguer anomalies and the crustal thickness very close to that obtained earlier by *Demenitskaya* [1958] (Figure 1). Recently completed studies by Belyaevsky (on the basis of more abundant data, about 1000 measurements) confirm the reliability of the curves in Figure 1. Figure 2 shows the distribution of the Bouguer anomalies associated with several structural features. These results indicate that gravity field and surface relief are closely connected with crustal thicknesses. However, the equations describing relationships reflect only average conditions.

The authors propose the name 'normal crust' for the crust described by the equations; this normal crust serves as a useful basis in connection with theoretical calculations of the relation between surface elevation, crustal thickness, Bouguer anomalies, and the mass distribution associated with the crust and upper mantle. Thus, the maps of M discontinuity compiled on the basis of the derived relationship $\Delta g(H_M)$ should be called 'maps of the thicknesses of normal crust.'

Crustal thicknesses determined by means of these general relationships usually differ to some degree from those determined by seismic investigations. These differences may be attributed to: (a) possible errors in seismic determinations (in excess of ± 2 km); (b) significant changes in crustal structure associated with major geological structures; and (c) inhomogeneity in the mantle. For example, relatively high positive gravity values of the Uralian folded system correspond to increased (not diminished) crustal thicknesses, while for the Tien-Shan folded system the great crustal thicknesses correspond to negative gravity values. Maps of '*M* anomalies' need to be constructed on the basis of the difference between thickness of the normal crust and that defined by seismic investigations.

Relationships between Bouguer anomalies and depths of the M discontinuity for the USSR, and adjacent inland and marginal seas, where many seismic investigations of the crust have been carried out, are shown in Figure 3. The outlines of groups of points peculiar to various tectonic structures define a general relationship between Bouguer anomalies and depth to M thus: the floor of the Pacific edge with the lowest values of crustal thickness and the highest gravity values; basins of the far eastern marginal seas (Japan and Okhotsk); basins of the Black and the Caspian Seas; intermontane areas (Chyuisk, Yeisk, Issyk, Fergana, and others) in zones of young activity of Paleozoic folded systems; intermontane areas characterized by thick crust; folded mountains such as the Tien-Shan and Pamirs that have the greatest known crustal thickness in the world. Some isostatic balance can be assumed only for extreme members of this progression; the intermediate ones probably deviate from isostatic equilibrium, except for the platforms, which are in rather high isostatic compensation.

In Figure 3, the curves representing platforms have been plotted separately. These curves are elongated along the depth axis, i.e., they do not show correlation between Bouguer anomaly and depth to Moho. The main deviations are associated with local thickening of the crust within the Saronatian shield and margins of the Turanian plate where it borders with the zones of young (Neogene-Quaternary) activation. The vast areas of platforms have rather constant crustal thicknesses and small variation of gravity values. The average value of the Bouguer anomaly is close to −10 mgal; this value and the crustal thicknesses of the platform correspond closely to those of the general relation in Figure 3. The data on folded mountains of Mesozoic and Cenozoic ages (the Caucasus, Car-

TABLE 1. Formulas for Determining Thickness of the Earth's Crust or Depth of M Discontinuity from Δg and Δh

	Formula	Symbols	Region	Reference
(1)	$H = 35(1 - \tanh 0.0037\ \Delta g)$	H thickness of the earth's crust in kilometers Δg Bouguer anomaly in megagals Δh surface elevation in kilometers (positive in continents, negative in oceans)	For the whole earth	Demenitskaya, 1958
(2)	$H = 33 \tanh(0.38\ \Delta h - 0.18) + 38$		For continental plains	
(3)	$H \approx 35 - 0.126\ \Delta g$			
(4)	$H = 30 - 0.1\ \Delta g$		For continental plains	Andreev, 1958
(5)	$H = 32 - 0.08\ \Delta g$		For the whole earth	Woollard, 1959
(6)	$H_M = (32 - 4.08h) \pm 3$ km	H_M depth of Moho in kilometers h depth of water	For shelves and sea borders	
(7)	$H_M = [21 - 2.2(9 - h)] \pm 3$ km		For trenches	
(8)	$H_M = 40.5 - \left(32.5 \tanh \dfrac{\Delta g + 75}{275}\right)$	H_M depth of Moho in kilometers Δg Bouguer anomaly	For the whole earth	Woollard and Strange, 1962
(9)	$\Delta H(a) = 39.76 \times 10^{-2} \sum_i \Delta g_i \phi_{a-i}$ where $\phi_{a-i} = \dfrac{c}{\pi(a^2 + c^2)}\ [\pm e^{c\pi} - 1] \quad c = \dfrac{H}{L}$	$\Delta H(a)$ variation of crustal thickness at a fixed point a in kilometers Δg_i Bouguer anomaly at a point i H crustal thickness taken as original L grid spacing	For any crustal structures	Tsuboi, 1938-1962
(10)	$H_0 = 35.5 - 32.5 \sin [0.212 v_z(\sigma)]$	$v_z(\sigma)$ zonal gravity anomalies σ_0 density of the granitic layer σ_1 density of the basaltic layer σ_2 density of the peridotite layer ΔH thickness of the basaltic layer $v_z(x, y, 0)$ attraction of anomalous mass k gravitational constant c constant, depending on the choice of normal formula of gravity	For Siberia and Far East	Karataev, 1960
(11)	$H = H_0 + \dfrac{2\sigma_1 - \sigma_2 - \sigma_0}{\sigma_2 - \sigma_1}\ \Delta H$ $ - \dfrac{v_z(x, y, 0)}{2\pi k(\sigma_2 - \sigma_1)} + c$			

(12)	$H = 33 - 0.055\ \Delta g$	H depth of M discontinuity in kilometers Δg Bouguer anomaly h elevation from sea level	For plains	Worzel and Shurbet, 1955
(13)	$H = 33 + 7.21 h$			
(14)	$H = -0.12\ \Delta g + (h_1 + h_2) + 3\text{ km}$	H crustal thickness in kilometers H_M depth of M discontinuity Δg Bouguer anomaly h_1 thickness of the granitic layer in kilometers h_2 thickness of the basaltic layer in kilometers d factor depending on the density and elevation of mountains	For different regions of middle Asia	Antonenko, 1961
(15)	$H_M = 16 - 0.33\ \Delta g$ $\Big\}$ Balkhash, Temir-Tau			
(16)	$H_M = 36 - 0.14\ \Delta g$			
(17)	$H = {}_M d - 0.104\ \Delta g$ Temir-Tau, Petropavlovsk			
(18)	$H_M(r) = 23.7 - 0.079\ \Delta g_p(r)$ $\qquad\qquad - 0.358 H_\phi(r) + 0.744 H_\delta$	$H_M(r)$ depth of M discontinuity in kilometers H_δ depth of Conrad discontinuity in kilometers H_ϕ depth to surface of consolidated crust Δg_p regional gravity anomaly	For Siberia and the Far East	Fotiadi and Karataev, 1963
(19)	$H_M(r) = 33.4 - 0.088\ \Delta g_p(r)$ $\qquad\qquad + 0.651 H_\phi(r) + 0.003 H_\delta$			
(20)	$H_\delta \approx 18.6 - 0.031\ \Delta g$			
(21)	$H_{pz} = (0.116\ \Delta g + 16.6)$	H_{pz} depth of paleozoic H_M depth of M discontinuity Δg Bouguer anomaly	For Fergana	Volvovsky et al., 1964
	$H_M = 2.44\ \Delta g + 4.54$			
(22)	$C = 0.8 h + 1.20$	C crustal thickness in kilometers h depth of water in kilometers	For the Mid-Atlantic ridge	Le Pichon et al., 1965

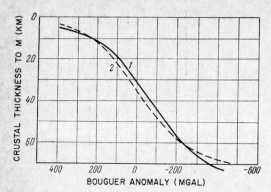

Fig. 1. Crustal thickness versus Bouguer anomalies [after *Demenitskaya*, 1967]: (1) *Woollard and Strange* [1962]; (2) Demenitskaya.

pathians, Kopet-Dag, etc.), as well as their intermontane areas (Kura, Rion, etc.) fall below the line of the general relationship.

The relation of major tectonic structures is not random. In the progression from intermontane areas to mountains of low and medium heights to high mountains, deep seismic sounding has indicated considerable thickening of the basaltic layer and thinning of the granitic layer. The structure of the Kuril island arc and its outer (oceanic) slope appears to have the same relationship. In Figure 3, the granitic layer gradually thickens to the right and thins to the left along the line of the general relationship. Thus the diagram of the relationship $\Delta g(H_M)$ gives some possibilities for prediction of crustal structure, in particular, the relative thicknesses of the granitic and basaltic layers.

As a result of the similarity of the formulas obtained by the authors for the USSR and by Woollard for the entire earth (see Table 1), and taking into account the large areal extent of the USSR, the authors believe that the formula for the general relation between Bouguer anomalies and depths to Moho can be used to predict depths to Moho in areas in which there are few or no seismic data. This projection was made by *Demenitskaya* [1958, 1967] for all the continents and ocean basins. The new map of the Moho under the USSR [*Belyaevsky et al.*, 1967] based on all available seismic and gravity data indicates that maximum crustal thicknesses occur in the southern regions associated with recent (Neogene-Quaternary) reactivation of the folded Paleozoic mountains, as well as the Pamirs. From this central Asiatic maximum, crustal thicknesses gradually decrease toward the north-northwest and northeast. Within platforms, crustal thicknesses are in the range of 35–38 km, but decrease to 20–30 km in the region of the arctic shelf. The considerable variation of crustal thickness from almost 50 km under the mountain regions and their foreland areas to less than 20 km under the Black Sea basin is characteristic for the zone of alpine folding in the south of the USSR. Considerable variations of thicknesses are also found in the transition from continent to the Pacific

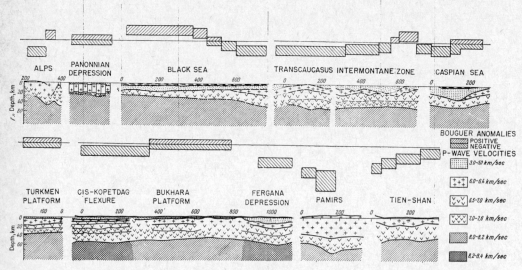

Fig. 2. Comparison of crustal thickness and structure with the Bouguer anomalies [after *Kosminskaya and Scheinmann*, 1965].

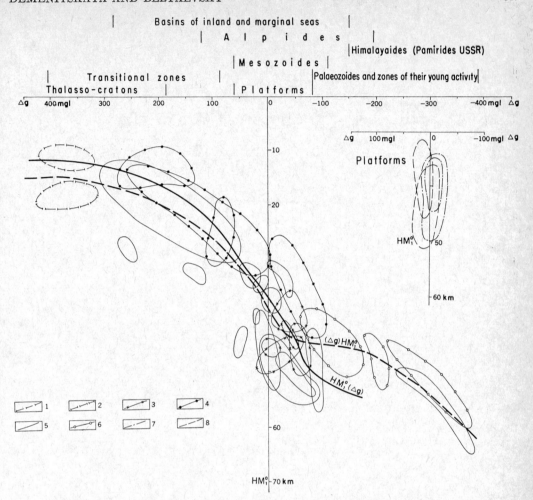

Fig. 3. Graph of distribution of point of cloud outlines, expressing relationships of gravity anomalies Δg and depth of the Mohorovicic discontinuity according to data of seismic deep sounding $HM_1°$ (composed by N. A. Belyaevsky, 1967): (1) the Pacific Ocean; (2) marginal troughs; (3) basins of inland and marginal seas; (4) piedmont and intermontane flexures; (5) folded mountain systems in the zone of Mesozoic and Alpine foldings; (6) folded mountains in the zone of young activization of ancient foldings; (7) young platforms; (8) ancient platforms. (See Fig. 10, p. 205.)

Ocean, where the modern geosynclines are situated; the greatest thickness probably does not exceed 35–37 km in this region.

The general relationship of $\Delta g(H_M)$ can also be used for preliminary investigations of the upper mantle. One can estimate the approximate temperature and pressure for the mantle surface on the basis of the derived values of crustal thickness. These data can be related to variations in the velocity at the M discontinuity and possible heterogeneities in mantle material. Variations in the density of mantle material can be estimated from gravity data. Additional characteristics of the upper mantle can be deduced from magnetotelluric measurements (depth of high conducting layer), seismological data about layering, petrological and geochemical studies of xenoliths of 'mantle' material, geochemical study of olivine nodules from volcanic basalts, etc. Figure 4 shows a physical map of the upper mantle based on such a combination of information. On this map, the mantle relief is contoured (ΔH) relative to the depth of 70 km by subtracting the crustal thickness

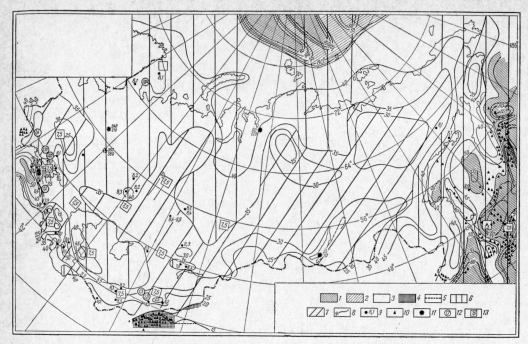

Fig. 4. Physical map of the upper mantle of the USSR [*Demenitskaya*, 1967]: (1) ~150°C, 1 kbar, crystalline; (2) ~300–600°C, 8–10 kbar, crystalline; (3) intermediate, ~700–900°C, 10–15 kbar; (4) ~1000+°C, ~20 kbar, possibly not crystalline, density may decrease with depth; (5) regions of possible chemical change; (6) regions where conditions allow phase changes; (7) regions of possible coexistence of phase and chemical changes; (8) isolines of relief, in kilometers, relative to surface 70 km below the geoid; (9) P-wave velocity; (10) epicenter of earthquakes with foci beneath the surface of the mantle; (11) depths of boundaries determined by magnetotelluric methods, in kilometers below the surface of the mantle and kilometers below the geoid; (12) heat flow in heat flow units; (13) regions where the 7.5 layer (7.4–7.8 km/sec) has been located. (See Figure 11, p. 206.)

(equations 1 and 2 in Table 1) from 70 km, thus:

$$\Delta H(km) = 70 - H \text{ (crustal thickness)}$$
$$= 32 - 33 \tanh (0.38\Delta h - 0.18)$$
$$= 35 [1 + \tanh (0.0037\Delta g)]$$

The most conspicuous changes in mantle relief occur in the north under the Arctic Ocean, in the east under the transitional zone from continent to the Pacific Ocean, and in the south under folded mountains of middle Asia, the Caucasus, and Caspian and Black Seas. Average undulations on the upper mantle surface have gradients to 5–7°. The pressure varies from less than 5 kbar to 20 kbar on the mantle surface. It does not exceed 8–10 kbar for the major part of platform areas. The mantle surface has temperatures of 700–900°C under the continental platforms [*Demenitskaya*, 1967), but, according to calculations by *Magnitsky* [1965], these temperatures should be rather lower, about 500–700°C. Under the oceans in regions of the shallowest depth of the mantle surface, the temperatures are of the order of 150–200°C. Conversely, in the region of greatest depth to the mantle (the Pamirs and adjoining regions), the temperatures probably exceed 1000°C. Seismic studies indicate a thinner crust in western Europe than is implied by the general relation of Bouguer anomalies versus depth derived for the USSR. However, seismic evidence for western Europe indicates that the basaltic layer is generally thicker than the granitic layer (as in the USSR).

Under the platform areas of the USSR, the basaltic layer usually constitutes about 60% of the total crust; the proportion of the basaltic layer increases under zones of Alpine folding and apparently reaches its maximum under eugeosynclines (perhaps reaching 80% under the Urals). There are some regions (Khantamansysky, Sredneokhotsky, Pamirs, and Tien-Shan) in which the granitic layer prevails over the basaltic one. Although it appears that there are differences in metallogenetic characteristics of mountain ranges of different proportions of granitic and basaltic crust, it is not known whether these differences relate to crustal structure or the underlying mantle.

The seismic layer with velocities 7.3-7.7 km/sec, which is sometimes described as a 'mantle-crust' mix, was found under mid-oceanic ridges and some other rift zones (California). In these areas, this material was presumed to represent anomalous mantle characteristic of activated areas of these types. However, seismic investigations in the USSR indicate that material with this velocity range is found in such regions as the Caucasus and Trans-Caucasus, Kazakhstan, Gissar ridge in middle Asia, the Urals, and Kola Peninsula, and others that have different tectonic structures and as a rule are not associated with rift zones. In many areas this layer is in the lower part of the crust, but in the Kola Peninsula its depth was found to be only 7-10 km from the surface. It is probable that the geological significance of this layer is different for oceans and for continents.

REFERENCES

Belyaevsky, N. A., A. A. Borisov, and I. S. Volvovsky, Deep structure under the USSR, *Sov. Geol.*, *1967*(11), 56-84, 1967.

Belyaevsky, N. A., A. A. Borisov, I. S. Volvovsky, and Yu. K. Schukin, Transcontinental crustal sections of the USSR and adjacent areas, *Can. J. Earth Sci.*, *5*, 1067-1078, 1968.

Borisov, A. A., and V. V. Fedynsky, Geophysical characteristics of geosynclinal regions of Central Asia, in *Activated Zones of the Earth's Crust, Latest Tectonic Movements, and Seismicity (Proceedings of the Second All-Union Conference on Tectonics, Dushanbe)*, edited by B. A. Petrushevsky et al., Nauka, Moscow, 256 pp., 1964.

Demenitskaya, R. M., Planetary structures and their reflection in Bouguer anomalies, *Sov. Geol.*, *1958*(8), 1958.

Demenitskaya, R. M., *Crust and Mantle of the Earth*, Nedra, Moscow, 280 pp., 1967.

Fedynsky, V. V., and Yu. V. Riznichenko, Study of the earth's crust, *Vestn. Akad. Nauk SSSR*, *1962*(6), 1962.

Gamburtsev, G. A., *Collected Works*, Izd. Akad. Nauk SSSR, Moscow, 1960.

Kosminskaya, I. P., and Yu. M. Sheinmann, Some laws of the structure and development of the earth's crust, *Bull. Moscow Soc. Naturalists*, *40*(3), 5-16, 1965.

Le Pichon, Xavier, R. E. Houtz, Charles L. Drake, and John E. Nafe, Crustal structure of the mid-ocean ridges, 1, Seismic refraction measurements, *J. Geophys. Res.*, *70*, 319-339, 1965.

Lyustikh, E. N., On convection in the mantle according to Pekeris' calculations, *Izv. Akad. Nauk SSSR, Ser. Geofiz.*, *1957*(5), 604-615, 1957.

Magnitsky, V. A., *Internal Structure and Physics of the Earth*, Nedra, Moscow, 379 pp., 1965.

Sazhina, N. B., Thickness of the earth's crust and its relation to gravity anomalies, *Sov. Geol.*, *1962*(8), 151-157, 1962.

Shurbet, G. L., and J. L. Worzel, Gravity observations at sea in USS *Conger*, cruise 3, *Trans. Am. Geophys. Union*, *38*(1), 1-7, 1957.

Subbotin, S. I., Gravitational anomalies of the Ukraine and their interpretation, *Geol. Zh. Akad. Nauk SSSR*, *10*(3), 1950.

Tsuboi, C., Crustal structure in northern and middle California from gravity pendulum data, *Bull. Geol. Soc. Am.*, *67*(12), 1641-1646, 1955.

Woollard, G. P., Crustal structure from gravity and seismic measurements, *J. Geophys. Res.*, *64*(10), 1521-1544, 1959.

Woollard, G. P., and W. E. Strange, Gravity anomalies and crust of the earth in the Pacific basin, in *The Crust of the Pacific Basin*, *Geophys. Monograph 6*, edited by G. A. Macdonald and Hisashi Kuno, pp. 60-80, American Geophysical Union, Washington, D. C., 1962.

Worzel, J. L. Deep structure of coastal margins and mid-oceanic ridges, submarine geology and geophysics, *Proc. Symp. Colston Res. Soc.*, *17*, 1965.

Regional Variations in Gravity

George P. Woollard

SYMBOLS

σ_c average density of crust
H crustal thickness
σ_m density of mantle substratum
H_s thickness of standard sea-level column
h surface elevation
ΔR root increment relative to base of standard column
θ angle from vertical at observation site
D slant distance from point of observation
Δg gravity effect of topographic mass above sea level
Δh increment in elevation
ΔM_c crustal mass increment for change in h
σ_s density of standard crust
σ_0 density of actual crust; $\sigma_s + \Delta \sigma_c$
$\Delta \sigma_c$ density difference between actual and standard crust
ΔH increment in crustal thickness (equation 1)
M_0 mass of actual crustal column (equation 2)
Δg_r gravitational increment due to increment of root ΔR (equation 11)
Δg_1 corresponds to Bouguer correction based on actual crustal density (equation 13)
Δg_2 correction term for departure of crustal density σ_0 from σ_s (equation 13)
γ gravitational constant
R_s thickness of mantle displaced by standard column H_s
R_0 thickness of mantle displaced by actual column H_0
H_0 actual crustal column
F_s freeboard of standard column; $H_s - R_s$
F_0 freeboard of actual column; $H_0 - R_0$
Δg_B Bouguer correction for mass above sea level

Under ideal conditions of isostatic equilibrium, all crustal columns exert equal pressure at some depth, and the column extending above sea level with elevation h has a mass equal to that of the compensating mass at depth. Under the Pratt concept of isostasy, the fundamental isostatic relation is expressed by $H_s \sigma_s = (H_s + h)\bar{\sigma}_c$, where H_s is the thickness of a standard sea level column with density σ_s, and h is the surface elevation of any other column $(h + H_s)$ having a mean density of $\bar{\sigma}_c$. Under these conditions, $\bar{\sigma}_c = \sigma_s[H_s/(H_s + h)]$.

Under the Airy concept of isostasy, whereby a crust of constant density σ_c but variable thickness H is hydrostatically supported by a denser substratum of density σ_m, $H_s \sigma_c + \Delta R \sigma_m = \sigma_c(H_s + h)$, where H_s is the thickness of a standard sea level crustal column, and h is the surface elevation of any other column having a root increment (ΔR) relative to the base of the standard column. Under these conditions, $h\sigma_c = \Delta R(\sigma_m - \sigma_c)$ and $\Delta R = h\sigma_c/(\sigma_m - \sigma_c)$.

Seismic studies suggest that isostasy is achieved through hydrostatic buoyancy in accordance with the Airy concept of isostasy, but approximately 35% of the time, there are significant variations in the values for σ_c and σ_m, at least relative to the M discontinuity. Also, as is brought out elsewhere in this volume (p. 285), it appears that the normal value for σ_c above the M discontinuity is approximately 2.93 g/cm³ and σ_m is approximately 3.32 g/cm³, rather than as the values now used in the Airy isostatic reduction, $\sigma_c = 2.67$ g/cm³ and $\sigma_m = 3.27$ g/cm³. It also appears from seismic studies that to a first approximation $H_s = 33 \pm 2$ km [*Woollard*, 1962]. The usual bias noted in gravity anomalies, when there are departures in σ_c and σ_m, is related to the change in σ_c, the density of the crust [*Woollard*, 1966, 1968]. This dependence on the value of σ_c is a consequence of earth curvature and the proximity effect whereby the gravity contribution from mass elements at different depths and lateral distances varies as $\sin \theta/D^2$, where θ is defined in terms of the depth to each mass, and D is the slant distance from the point of observation. It is for this reason that 95% of the gravity

---

[1] Contribution 250 from the Hawaii Institute of Geophysics.

effect of the topographic mass above sea level ($\Delta g = 2\pi\gamma h\sigma_o$) is usually realized within a distance of 20 km, while only about 13% of its compensation is realized at this range. Only about 75% of the compensation effect is realized at 166.7 km, the range at which earth curvature must be considered (Figure 1). This is reasonable because the compensation mass lies at a depth greater than H_s ($\approx$33 km). It explains why free-air anomalies show a direct dependence on local topographic relief and why the Bouguer reduction removes the short-wavelength irregularities due to topography, because in effect they are not compensated gravitationally. For large topographic masses, such as mountain ranges, there is also a correlation (although much less pronounced than that for local topography) between free-air anomalies and surface relief. This, however, is not always noted, as most gravity traverses across mountainous regions follow valley routes, and the negative effect associated with the valleys compensates for the positive effect of the mountain mass as a whole. In discussing the relation of free-air anomalies to topography, *Garland* [1965] has shown that an effective test of isostasy in areas of topographic relief is to use the ratio between a spherical harmonic expansion for the free-air anomaly and that for the topography. For a compensated topographic mass 1 km above sea level with a crustal column extending down to 50 km below sea level and normal crustal density, Garland shows that, if the harmonic degree of the topography (n) is 2, the free-air anomaly to be expected is 1.4 mgal. If $n = 10$, the anomaly to be expected is 11.8 mgal, and, if $n = 20$, the anomaly is of the order of 25 mgal. If there are har-

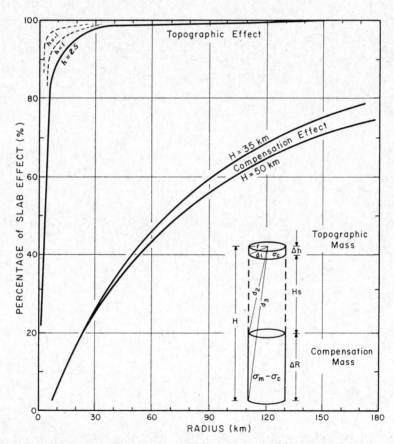

Fig. 1. Comparison of gravitational attraction of topographic mass and its compensation for different diameter crustal blocks expressed as percent of the attraction of a slab ($\Delta g = 2\pi\gamma h\sigma$).

monics in the expansion of the free-air gravity values that are not related to corresponding harmonics in the topography, they must be caused by uncompensated masses.

This test for isostatic compensation is valid for the conditions specified; however, there are many areas of low surface relief where free-air and isostatic anomalies indicate departures from isostasy and there are reasons to believe that there is no departure from isostasy. Seismic crustal studies in these areas show there is abnormality in crustal thickness that can be related to anomalous crustal composition. The magnitude of the gravity effect associated with such areas is appreciable (25 to 50 mgal), and equals that found in association with horsts and grabens that represent areas that are unquestionably out of isostatic equilibrium. There is a significant difference: the signs of the anomalies are reversed for similar changes in crustal thickness.

The writer has re-examined the applicable theory and tried to find some criteria in the gravity data, particularly in the relation of Bouguer anomalies to changes in surface elevation, that might provide a clue as to which type of control exists where there are regional changes in gravity. The solution is complicated because: (1) there are no consistent geological associations with areas of anomalous gravity; (2) the number of seismic measurements defining the thickness and composition of the crust and upper mantle are relatively few; and (3) the relation between density and seismic velocity is not unique.

THEORETICAL CONSIDERATIONS

Crust in Isostatic Equilibrium with the Mantle

For a crust and mantle having 'normal' density values as envisioned in the Airy concept of isostasy, $\Delta R(\sigma_m - \sigma_c) = \Delta h \sigma_c$. However, if the actual density of the crust is not that of the standard crust (σ_s) but rather $\sigma_o = \sigma_s + \Delta \sigma_c$, then ΔM_c, which is the crustal mass increment for any change in elevation (Δh), and which is normally expressed for equilibrium conditions as

$$\Delta M_c = \Delta H \sigma_c \qquad (1)$$

becomes

$$\Delta M_c = M_0 - M_s \qquad (2)$$
$$= (H_s + \Delta H)(\sigma_s + \Delta \sigma_c) - H_s \sigma_s$$

or

$$\Delta M_c = \Delta \sigma_c H_s + \sigma_0 \Delta H \qquad (3)$$

Similarly, the relation for Δh normally expressed as

$$\Delta h = \Delta H - \Delta R = \Delta H - \Delta M_c / \sigma_m \qquad (4)$$

for an anomalous crust becomes

$$\Delta h = \Delta H - (1/\sigma_m)(\Delta \sigma_c H_s + \sigma_0 \Delta H) \qquad (5)$$

or

$$\Delta h = \Delta H \frac{(\sigma_m - \sigma_0)}{\sigma_m} - \frac{\Delta \sigma_c}{\sigma_m} H_s \qquad (6)$$

The relation for ΔR normally given as

$$\Delta R = \frac{\Delta M_c}{\sigma_m} = \frac{\Delta H \sigma_c}{\sigma_m} = \frac{(\Delta R + \Delta h)}{\sigma_m} \sigma_c \qquad (7)$$

for an abnormal crust becomes

$$\Delta R = \frac{\sigma_0}{\sigma_m} \Delta H + \frac{\Delta \sigma_c}{\sigma_m} H_s \qquad (8)$$

and

$$\Delta R(\sigma_m - \sigma_0) = \Delta h \sigma_0 + H_s \Delta \sigma_c \qquad (9)$$

or

$$\Delta R = \frac{\Delta h \sigma_0}{\sigma_m - \sigma_0} + \frac{H_s \Delta \sigma_c}{(\sigma_m - \sigma_0)} \qquad (10)$$

Since the gravitational root effect for a normal crust is

$$\Delta g_r = 2\pi\gamma \, \Delta R(\sigma_m - \sigma_c) \qquad (11)$$

we find that

$$\Delta R = \frac{\Delta g_r}{2\pi\gamma(\sigma_m - \sigma_c)} \qquad (12)$$

Similarly, for a normal crust, $\Delta g_r = 2\pi\gamma\Delta h\sigma_c$ (the Bouguer correction), since $\Delta R \, (\sigma_m - \sigma_c) = \Delta h \sigma_c$.

It follows from (10) that when there is an abnormal crust the gravitational counterpart for ΔR will be

$$\Delta R = \frac{2\pi\gamma \, \Delta h \bar{\sigma}_0}{2\pi\gamma(\sigma_m - \bar{\sigma}_0)} + \frac{2\pi\gamma H_s \, \Delta\sigma_c}{2\pi\gamma(\sigma_m - \bar{\sigma}_0)} \quad (13)$$

$$= \frac{\Delta g_1 + \Delta g_2}{2\pi\gamma(\sigma_m - \bar{\sigma}_0)}$$

Thus $\Delta g_r = \Delta g_1 + \Delta g_2$, where Δg_1 corresponds to the normal Bouguer correction ($2\pi\gamma\Delta h\sigma_0$), but is based on the actual mean density of the crust (σ_0), and Δg_2 is a correction term for the departure ($\Delta\sigma_c$) of the actual density of the crust (σ_0) from that of the standard crust (σ_s).

To illustrate the magnitude of the gravitational effect when there is a change in crustal composition but no departure in isostasy, we assume a plateau with $H = 45.7$ km having an elevation of 2320 meters and a mean crustal density ($\bar{\sigma}_0$) = 2.87 g/cm³. The parameters for the standard sea level block are $H_s = 33$ km, $\sigma_s = 2.93$ g/cm³, and the density of the mantle is 3.32 g/cm³.

The mantle displaced by the standard column H_s with $h = 0$ is

$$R_s = \frac{H_s \sigma_c}{\sigma_m} = \frac{33 \times 2.93}{3.32} = 29.12 \text{ km}$$

The freeboard $F_s = H_s - R_s = 33 - 29.12 = 3.88$ km.

The mantle displaced by the column with $H_0 = 45.7$ km is

$$R_0 = \frac{H_0 \bar{\sigma}_0}{\sigma_m} = \frac{45.7 \times 2.87}{3.32} = 39.50 \text{ km}$$

The freeboard $F_0 = H_0 - R_0 = 45.7 - 39.5 = 6.2$ km.

The surface elevation for equilibrium conditions (Δh) is therefore $F_0 - F_s = 6.2 - 3.88 = 2.32$ km.

The root increment ΔR relative to the standard column is $R_0 - R_s = 39.50 - 29.12 = 10.38$ km.

If (10) had been used to solve for ΔR,

$$\Delta R = \frac{\Delta h \bar{\sigma}_0}{(\sigma_m - \bar{\sigma}_0)} + \frac{H_s \, \Delta\sigma_c}{(\sigma_m - \bar{\sigma}_0)}$$

where

$$\Delta\sigma_c = (\bar{\sigma}_0 - \sigma_s)$$

$$= (2.87 - 2.93) = -0.06 \text{ g/cm}^3$$

$$\Delta R = \frac{(2.32 \times 2.87) + (33 \times -0.06)}{3.32 - 2.87}$$

$$= \frac{6.68 - 1.98}{0.45} = 10.38 \text{ km}$$

The gravitational effect of the root increment (ΔR) from equation 11 is

$$\Delta g_r = 2\pi\gamma \times \Delta R \times (\sigma_m - \bar{\sigma}_0)$$

$$= 41.85 \times 10.38(3.32 - 2.87)$$

$$= 196 \text{ mgal}$$

However, as can be seen from (13), this value actually represents the sum of two components, $\Delta g_1 + \Delta g_2$:

$$\Delta g_1 = 2\pi\gamma \, \Delta h \bar{\sigma}_0$$

$$= 41.85 \times 2.32 \times 2.87 = 279 \text{ mgal}$$

$$\Delta g_2 = 2\pi\gamma H_s \, \Delta\sigma_c$$

$$= 41.85 \times 33 \times -0.06 = -83 \text{ mgal}$$

The root effect is

$$\Delta g_r = \Delta g_1 + \Delta g_2 = 279 - 83 = 196 \text{ mgal}$$

For the case under consideration, therefore, the actual gravitational root effect is 83 mgal less than that defined by the Bouguer correction Δg_1 for the crustal column above sea level.

If there had been no change in crustal density from that for the standard column ($\bar{\sigma}_0 = \sigma_s$), under isostatic conditions, the Bouguer correction would have defined the gravitational effect of the root increment and ΔR would have equaled $\Delta h R_s / F_s$. For the parameters defined for the standard column, the ratio $R_s/F_s = 29.12/3.88 = \approx 7.5$. For a surface elevation of 2.32 km, therefore, ΔR would have been 17.4 km rather than 10.38 km; or, conversely, if $\Delta R = 10.38$ km, the surface elevation should have been 1384 meters rather than 2320 meters. The fact that the values of Δh and ΔR are not compatible with those to be expected with a crust having a standard density (σ_s) is therefore not necessarily evidence of a departure from isostatic equilibrium, but only that $\bar{\sigma}_0$ differs from σ_s.

The only clues that can be used for recognizing the actual condition, if there are no seismic data, are the sign of the free-air and isostatic anomalies and geologic evidence for uplift or subsidence [*Woollard*, 1966]. If the regional free-air and isostatic gravity anomaly values are negative and there is evidence of uplift, the crust probably has a subnormal density and thickness for the surface elevation; if the regional anomalies are positive and there is evidence of subsidence, it is probable that the crust has an abnormal density and thickness.

Crust Not in Isostatic Equilibrium with the Mantle

Just as there are anomalous changes in gravity associated with anomalous changes in crustal composition and structure under equilibrium conditions, there are anomalous changes due to a lack of isostatic equilibrium. Examples of areas of this type are eastern Canada; Cook Inlet, Alaska; the Po Valley, Italy; the Rift Valleys of Africa; the Bighorn Mountains, Wyoming; the Harz Mountains, Germany; Puget Sound trough, Washington; and the Fergana basin, Asia.

The Fergana basin will be used for illustration, since there are both gravitational data and seismic data for the area (see Table 1). Using the same parameters as before for a standard crust, we find

$H_s = 33$ km

$\bar{\sigma}_s = 2.93$ g/cm^3 $\sigma_m = 3.32$ g/cm^3

$R_s = 29.12$ km $F_s = 3.88$ km

$R_s/F_s = 7.5$

The normal value for H when $h = 425$ meters would be $H_s + 7.5(h) + h = 33 + 3.19 + .425 = 36.6$ km.

The excess crustal thickness over normal conditions is therefore $52.2 - 36.6 = 15.6$ km. The anomalous gravitational effect assuming a slab of semi-infinite extent is

TABLE 1. Seismic Data for the Fergana Basin (elevation, 425 meters; isostatic anomaly, −195 mgal)

Layer	ΔH, km	v_p, km/sec	σ, g/cm*
1	7.2	2.0	2.2
2	3.0	5.0	2.60
3	8.0	6.2	2.74
4	16.0	6.75	3.0
5	18.0	7.45	3.18 $\bar{\sigma}_c = 2.89$ g/cm^3
	$H = 52.2$	$V_m = 8.3$	3.32 (assumed)

*Based on *Woollard*, 1968.

$$\Delta g_r = 2\pi\gamma[\Delta R \times (\sigma_m - \bar{\sigma}_0)]$$
$$= 41.85 \times 15.6(3.32 - 2.89)$$
$$= 281 \text{ mgal}$$

As can be seen from Figure 1, under normal conditions about 72% of this slab effect would be expected within a radius of 166.7 km from the center of the basin if there were no horizontal changes in mass distribution. However, the Fergana basin is physically restricted to dimensions of about 100×280 km, and in area corresponds to a crustal block having a radius of ≈ 90 km. We can approximate roughly its effective radius by equating the observed isostatic anomaly (-195 mgal) to the anomalous crustal increment effect (281 mgal); $K = 195/281 = 0.69$. As is shown in Figure 1, this corresponds to a radius of about 130 km. This procedure, of course, assumes that the full isostatic anomaly is generated locally. The actual departure from isostatic equilibrium can be shown using the equations presented for equilibrium conditions.

Using equation 10,

$$\Delta R = \frac{\Delta h \sigma_0}{(\sigma_m - \sigma_0)} + \frac{H_s \Delta \sigma_c}{(\sigma_m - \sigma_0)}$$

$$\Delta \sigma_c = (\sigma_0 - \sigma_s)$$
$$= 2.89 - 2.93 = -0.04 \text{ g/cm}^3$$

$$\Delta R = \frac{(0.425 \times 2.89) + (33 \times -0.04)}{3.32 - 2.89}$$

$$= \frac{1.228 - 1.32}{0.43} = 0.21 \text{ km antiroot}$$

Theoretically, therefore, for the surface elevation of 0.425 km and the seismically defined mean crustal density of 2.89 g/cm^3 under equilibrium conditions, the crustal thickness should have been $H_s + h + \Delta R = 33 + 0.425 - 0.21 = 33.2$ km rather than 52.2 km.

If we use (13) to define the theoretical attraction of the crustal root increment for equilibrium conditions,

$$\Delta g_1 = 2\pi\gamma \Delta h \sigma_c$$
$$= 41.85 \times 0.425 \times 2.89 = 50.6 \text{ mgal}$$

$$\Delta g_2 = 2\pi\gamma H_s \Delta \sigma_c$$
$$= 41.85 \times 33 \times -0.04 = -55.2 \text{ mgal}$$

$\Delta g_r = \Delta g_1 + \Delta g_2 = -4.6$ mgal

If we used the actual value of $\Delta R = H - (H_s + h) = 52.2 - (33 + 0.425) = 18.77$ km, the actual root increment effect is

$\Delta g_r = [2\pi\gamma \Delta R(\sigma_m - \bar{\sigma}_c)]$

$= 41.85 \times 18.77 \times 0.43 = 337$ mgal

If we use the anomalous value of $\Delta R = 18.98$ km,

$\Delta g_r = 41.85 \times 18.98 \times 0.43 = 340$ mgal

$\Delta g_2 = \Delta g_r - \Delta g_1$

$= (337 - 50.6) = 286$ mgal

This agrees closely with the value of 281 mgal estimated on the basis of the anomalous value of crustal thickness (15.6 km) if normal isostatic relations are assumed with no change in crustal density. The difference of 5 mgal corresponds to the 4.6 mgal derived for the crustal root increment under equilibrium conditions using the actual density of the crust. If we assume the same K factor (0.69) derived for the percentage of the root increment not realized within the physical boundaries of the basin under normal isostatic conditions and apply this to the anomalous root increment value (340 mgal), the actual departure in isostasy is $\Delta g_1 - 0.69\Delta g_a = 50.6 - 0.69(340) = -184$ mgal.

This is 11 mgal less than the isostatic anomaly of -195 mgal as derived by regular procedures (full world coverage and the normal relations between surface elevation and changes in crustal parameters). As no account was taken in the analysis of either the surface mass distribution or compensation beyond an effective radius of 130 km, this difference (11 mgal) has no critical significance. The results do serve to illustrate the origin of one of the world's largest gravity anomalies, and, as shown, three effects are represented: (1) anomalous crustal composition; (2) nonequilibrium due to tectonic displacement; and (3) the effects inherent with physically constrained (short wavelength) mass anomalies where the surficial effect (Δg_1) is fully realized and that of the crustal root increment which normally provides compensation is only partly realized.

Most areas involving actual departures from isostasy are of limited area as the examples considered. Major exceptions are areas that have been subject to glaciation in the recent past, as eastern Canada; some plateaus that are apparently undergoing current uplift, as the plateau of Mexico; and mountain areas where crustal adjustment lags erosional degradation, as the Appalachian Mountains.

Although it is not possible to explain satisfactorily all areas of anomalous gravity, it is instructive to examine where they occur, and the balance of this paper is devoted to such a review for four continental areas.

REPRESENTATION OF REGIONAL DEPARTURES IN GRAVITY RELATIONS

Isostatic anomalies are usually used for defining areas of abnormal gravity relations, but unfortunately the number of data available are relatively few. Free-air anomalies can also be used in all but areas of marked local topographic relief if allowance is made for the regional dependence of the anomaly value on surface elevation. This dependence of the free-air anomaly value on elevation, however, varies with the areal dimensions of the superimposed topographic blocks that make up the surface relief pattern of a continent. In general, three linear sets of relations are defined representing: (a) the coastal areas (sea level to 250 meters); (b) the continental platform (250 to 1800 meters); and (c) the mountain regions (elevations >1800 meters). That the mean relations do not differ significantly with the incremental size of the areas considered where regional equilibrium conditions prevail can be shown for the relations obtained in the United States using mean elevation and anomaly values for $1° \times 1°$, $2° \times 2°$, and $3° \times 3°$ squares. As might be expected, the spread in values for any elevation is greatest for $1° \times 1°$ size areas (± 30 to 35 mgal), but not significantly different for $2° \times 2°$ and $3° \times 3°$ areas ($\approx \pm 20$ mgal). The relations are given in Table 2. The high degree of agreement for the $1° \times 1°$, $2° \times 2°$, and $3° \times 3°$ relations suggests that the United States as a whole is in isostatic equilibrium, and that the relations defined are due solely to topography and not biased to any major extent by abnormalities in crustal parameters. To test this conclusion, we can use the Bouguer anomalies in combination with the results from seismic crustal studies.

TABLE 2. Relation of Free-Air Anomaly to Elevation in the United States

Height Interval, meters	Equation	h, meters	FA, mgal
	1° × 1° areas		
Sea level to 200	$FA = -0.103h + 18$	100	+ 8
		200	− 3
200 to 1800	$0.009h - 3$	1000	+ 6
1800+	$0.047h - 74$	2000	+20
	2° × 2° areas		
Sea level to 250	$-0.069h + 10$	100	+ 3
		200	− 4
250 to 1800	$0.008h - 3$	1000	+ 5
1800+	$0.038h - 58$	2000	+18
	3° × 3° areas		
Sea level to 225	$-0.066h + 12$	100	+ 5
		200	− 1
225 to 1900	$0.01h - 2$	1000	+ 8
1900+	$0.03h - 40$	2000	+20

Bouguer anomalies by their very nature have a marked dependence on elevation, and in order to use them for analyzing regional anomalous changes in gravity, it is necessary to define first what should be the normal relation. If we use the empirical expression based on seismic crustal studies for the depth of the M discontinuity with changes in surface elevation in areas where there appears to be isostatic equilibrium ($M = -33 + 7.5h$), we can derive a theoretical model. As was shown earlier, when $R/F = 7.5$, the required density contrast between the crust and mantle is 0.39 g/cm³, and, as under equilibrium conditions, the root gravitational effect [$\Delta g_r = 2\pi\gamma\Delta R(\sigma_m - \sigma_c)$] equals the Bouguer correction ($\Delta g_B = 2\pi\gamma h\sigma_c$), it is necessary only to specify σ_c such that $\Delta g_B = \Delta g_r$ when $\Delta R = 7.5h$. The required value of σ_c is 2.924 g/cm³. This agrees closely with the value of 2.93 g/cm³ deduced from other considerations (see p. 285), and, as will be seen, gives reasonable agreement with observed Bouguer anomalies if allowance is made for the observed free-air anomaly values, since $BA = FA - \Delta g_B$.

The relation of observed Bouguer anomaly values to elevation in the United States for areas of 1° × 1°, 2° × 2°, and 3° × 3° with $\sigma = 2.67$ g/cm³ are given in Table 3. The difference in observed Bouguer anomaly values over a range of 1900 meters is, on the average, no more than ±2 mgal for the 2° × 2° and 3° × 3° areas. It thus appears that both size areas are compensated. The somewhat larger differences in values for the 1° × 1° areas for elevations below 1000 meters suggest that mass distributions associated with this size area are not completely compensated, or else there is bias from anomalous changes in crustal composition and associated thickness.

If we adopt the 2° × 2° and 3° × 3° values as being compensated and correct the observed Bouguer anomaly to a density of 2.92 g/cm, so that the value $(FA - BA)$ can be compared with the value of Δg_r when $\Delta \sigma = 0.39$ g/cm³, the results are as shown in Table 4 for the 3° × 3° values and Table 5 for the 2° × 2° values.

Up to 2000 meters, there is little to choose between the two representations of the observed data, and the departures from the theoretical value defined by the isostatic model adopted over this range of elevation nowhere exceeds 5 mgal, and for both representations of the data averages 3 mgal. Above 2000 meters only the 2° × 2° representation indicates equilibrium conditions. The failure of the 3° × 3° values to indicate equilibrium conditions above 2000 meters can be related to the fact that in the United States there are no areas of 3° × 3° size having an average elevation of 2500 to 3000 meters, and, as a result, the slope of the empirical curve above 2000 meters is not well defined.

If the model defined by a crust of density 2.92 g/cm³ in isostatic equilibrium with a mantle of density 3.31 g/cm³ and with $\Delta R = 7.5h$ is used as a standard for evaluating Bouguer anomalies in conjunction with the free-air anomalies, the standard can be written as $BA = $

TABLE 3. Relation of Bouguer Anomaly to Elevation in the United States ($\sigma = 2.67$ g/cm³)

Height Interval, meters	Equation	h, meters	BA, mgal
	1° × 1° areas (see Figure 2 for control)		
Sea level to 200	$BA = -0.163h + 13$	100	− 3
		200	− 14
200 to 1700	$-0.109h - 1$	1000	−110
1700+	$-0.062h - 80$	2000	−204
	2° × 2° areas (see Figure 3 for control)		
Sea level to 350	$-0.169h + 12$	100	− 5
		200	− 22
350 to 1800	$-0.0995h - 10$	1000	−110
1800+	$-0.078h - 46$	2000	−202
	3° × 3° areas (see Figure 4 for control)		
Sea level to 250	$-0.150h + 8$	100	− 7
		200	− 22
250 to 1750	$-0.104h - 4$	1000	−108
1750+	$-0.062h - 76$	2000	−200

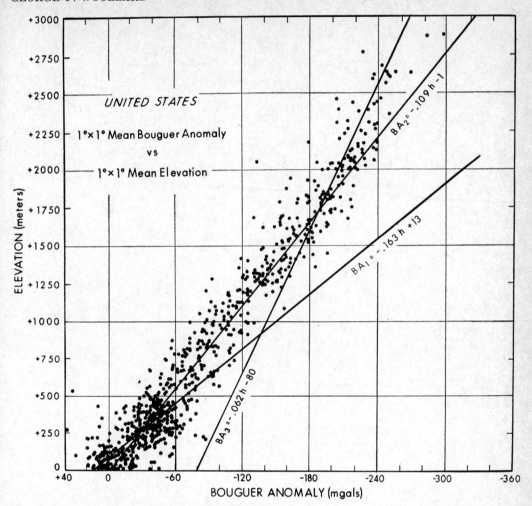

Fig. 2. Relation of $1° \times 1°$ mean Bouguer anomaly values to mean elevation h in the United States. BA_1, sea level to 200 meters; BA_2, 200 to 1700 meters; BA^3, 1700+ meters.

$-0.1222h + FA$. However, most Bouguer anomalies are computed on the basis of $\sigma_c = 2.67$ g/cm³ rather than 2.92 g/cm³. The standard can be written as $BA = -0.1118h + FA$ if $\sigma = 2.67$ g/cm³. To illustrate the validity of this transposition, consider the values for 1500 meters and 3000 meters in Table 5, which both show essentially zero difference between Δg_r, Δg_B, and $BA - FA$ for $2° \times 2°$ values. When $h = 1500$ meters, Δg_B with $\sigma_c = 2.67$ g/cm³ would be 168 mgal. $BA = -159$ mgal; $FA = +9$ mgal; $(FA - BA) = +168$ mgal. When $h = 3000$ m, $\Delta g_B = 335$ mgal; $BA = -280$ mgal; $FA = +56$ mgal; $(FA - BA) = +336$ mgal. The agreement is thus the same as when σ is 2.92 g/cm³.

The critical value for evaluating abnormalities in gravity using the Bouguer anomalies is therefore not the value of σ_c used in the Bouguer reduction but the value of the free-air anomaly. It is only when the Bouguer anomaly is used for evaluating abnormalities in crustal parameters that the value used for σ_c becomes significant.

As there is no general theoretical model relating the free-air anomaly to elevation, since it depends on the topographic structure of each continental mass, the nearest approach to a normal set of values is to use data from areas having apparently isostatic equilibrium as defined by isostatic anomalies. The sample of data used was world wide in distribution and restricted to sites for which there was apparent isostatic equilibrium to within ±10 mgal. The resulting set of relations between elevation and free-air anomalies are as follows:

Sea level to 200 meters $\quad FA = -0.102h + 13$
200 to 1800 meters $\quad\quad\quad\quad\quad 0.0075h - 6$
1800 to 3000 meters $\quad\quad\quad\quad\quad 0.047h - 77$

The relations portrayed are very close to those defined by free-air anomaly values for the United States and confirm the conclusion reached earlier that the United States as a whole is in isostatic equilibrium. The anomaly comparisons are as shown in Table 6.

It appears the 2° × 2° free-air and Bouguer anomaly values for the United States can be used as a standard for evaluating regional variations in gravity defined elsewhere.

OBSERVED FREE-AIR AND BOUGUER ANOMALIES IN OTHER AREAS

Equations were derived to describe the relation of free-air anomaly and Bouguer anomalies ($\sigma = 2.67$ g/cm^3) to elevation on a 1° × 1° and 3° × 3° basis in different parts of the world for which there are adequate data. Both free-air and Bouguer anomaly departures from the standard values described were examined, since the relations are not exact and each set of relations represented a best apparent fit to rather widely dispersed data. The 3° × 3° values are included, since crustal blocks of this size should be compensated, whereas 1° × 1° blocks may be only partially compensated. The

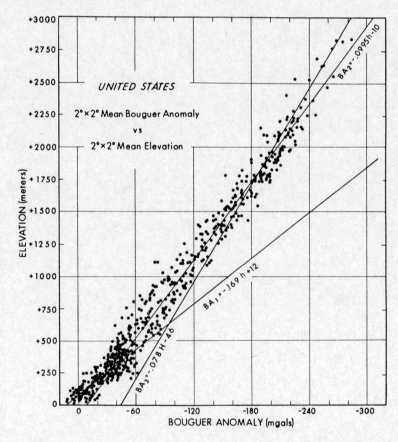

Fig. 3. Relation of 2° × 2° mean Bouguer anomaly values to mean elevation h in the United States. BA_1, sea level to 200 meters; BA_2, 200 to 1700 meters; BA_3, 1700+ meters.

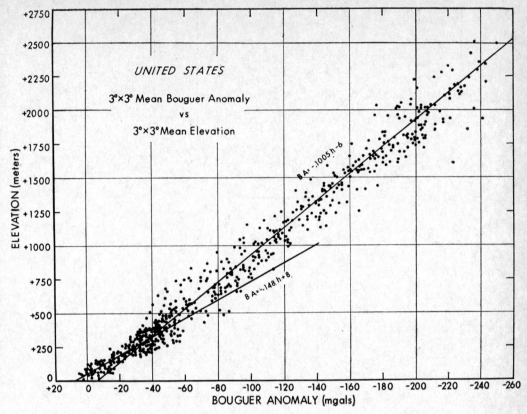

Fig. 4. Relation of $3° \times 3°$ mean Bouguer anomaly values to mean elevation h in the United States. BA_1, sea level to 200 meters; BA_2, 200 to 1700 meters; BA_3, 1700+ meters.

error in fitting equations to the data is least with the Bouguer anomalies and is not believed to exceed +5 mgal.

Alaska

The relations for Alaska are shown in Table 7. Both the free-air and Bouguer anomaly values have a positive bias that pertains to both $1° \times 1°$ values and $3° \times 3°$ values at all elevations. The $1° \times 1°$ Bouguer anomaly departure agrees with the $3° \times 3°$ Bouguer anomaly departure and also the $1° \times 1°$ free-air anomaly departure up to 1000 meters in defining a general systematic deviation that can be expressed as $\theta = 0.037h + 12$ mgal.

There is a positive bias in the Alaska values because of the restricted size of the Alaska Peninsula. In addition, the available seismic data on the crust show that the Chugach Mountains along the south coast of Alaska have a crust of abnormal thickness (+6 km) for the surface elevation of about +1200 meters if the expression $H = 33 + 7.5h + h$ is used to define the normal value of crustal thickness. As the observed Bouguer anomaly (−70 mgal) corresponds closely to that defined by the expression $BA = 0.059h − 4$ (−67 mgal) for Alaska as a whole, it would appear that most of Alaska is characterized by a crust of abnormal density and thickness. The marked abnormal positive gradient of 0.037 mgal/meter defined by the departure of the observed anomaly values from the standard thus appears to be related in large measure to the dominant effect of a high-density crust over that of its root increment.

Canada

There is no single set of relations that describe the relation of free-air and Bouguer anomalies to elevation in Canada. Three independent sets are defined for the free-air anomalies and four sets for the Bouguer anomalies. The relations

TABLE 4. Comparison of Theoretical and Observed Bouguer Anomalies in the United States Based on 3° × 3° Relations

Elev., meters	ΔR $(R/H = 7.5)$, km	Δg_r $(\Delta \sigma = 0.39)$, mgal	Δg_B $(\sigma = 2.92)$, mgal	Obs. BA $(\sigma = 2.67)$, mgal	Corr. for $\sigma = 2.92$, mgal	$BA(\sigma = 2.92)$, mgal	Obs. FA, mgal	$FA - BA$ $(\sigma = 2.92)$, mgal	$[FA - BA(\sigma = 2.92)] - \Delta g_r$, mgal
0	0	0	0	+8	0	+8	+12	+4	+4
200	1.5	24	24	−22	−2	−24	−1	+23	−1
500	3.57	61	61	−55	−5	−60	+3	+63	+2
1000	7.5	122	122	−108	−10	−118	+8	+125	+4
1500	11.25	184	184	−160	−16	−176	+13	+189	+5
2000	15.0	244	244	−200	−21	−221	+20	+241	−3
2500	18.75	306	306	−231	−26	−257	+35	+292	−14
3000	22.5	366	366	−262	−31	−293	+50	+343	−23

TABLE 5. Comparison of Theoretical and Observed Bouguer Anomalies in the United States Based on 2° × 2° Values

Elev., meters	ΔR $(R/H = 7.5)$, km	Δg_r $(\Delta \sigma = 0.39)$, mgal	Δg_B $(\sigma = 2.92)$, mgal	Obs. BA $(\sigma = 2.67)$, mgal	Corr. for $\sigma = 2.92$, mgal	$BA(\sigma = 2.92)$, mgal	Obs. FA, mgal	$FA - BA$ $(\sigma = 2.92)$, mgal	$[FA - BA(\sigma = 2.92)] - \Delta g_r$, mgal
0	0	0	0	+12	0	+12	+10	−2	+2
200	1.5	24	24	−22	−2	−24	−4	+20	−4
500	3.75	61	61	−60	−5	−65	+1	+66	+5
1000	7.5	122	122	−110	−10	−120	+5	+125	+3
1500	11.25	184	184	−159	−16	−175	+9	+184	0.0
2000	15.0	244	244	−202	−21	−223	+18	+241	−3
2500	18.75	306	306	−241	−26	−267	+37	+304	−2
3000	22.5	366	366	−280	−31	−311	+56	+367	+1

TABLE 6. Comparison of Free-Air Anomaly Values under Isostatic Conditions and Regional Values in the United States

h, meters	FA Isostatic Anom. ±10	FA 1° × 1°	FA 2° × 2°	FA 3° × 3°	FA − Isostatic Anom. 1° × 1°	FA − Isostatic Anom. 2° × 2°	FA − Isostatic Anom. 3° × 3°
0	+13	+18	+10	+12	+5	−3	−1
100	+3	+8	+3	+5	+5	0	+2
200	−7	−3	−4	−1	+4	+3	+6
500	−2	+2	+1	+2	+4	+3	+4
1000	+2	+6	+5	+8	+4	+3	+5
1500	+6	+10	+9	+13	+4	+3	+7
2000	+17	+20	+18	+20	+3	+1	+3
2500	+40	+44	+37	+35	+4	−3	−5
Numerical average					4.2	2.4	3.5
Average with regard to sign					+4.2	+1.0	+2.8

on a geographic basis are given in Table 8. Eastern Canada shows a negative offset in values of about −30 mgal which decreases with increase in elevation. The bias can be attributed to the lag in crustal adjustment after the removal of the Pleistocene ice cap. Central Canada appears on average to be in equilibrium, and western Canada shows a slight positive bias that increases somewhat with elevation. Seismic studies suggest this is related to abnormal crustal density east of the Rocky Mountain front, and possibly too thin a crust under the mountains proper. The relations in northwest Canada are very similar to those noted in Alaska. The available seismic data, which are all in the coastal region of southern Alaska, suggest an abnormal value of crustal thickness (+3 km) for the St. Elias Mountains, but a near normal value for the Coast Range. The principal point brought out by the relations in Canada is that there are marked local and regional changes in crustal parameters as evidenced by the marked difference in 1° × 1° and 3° × 3° relations found here as contrasted with the United States

TABLE 7. Relation of Anomaly Values to Elevation in Alaska

Height Interval, meters	Equations	h, meters	Anomaly, mgal	Departure from Standard Isos. Val.	Departure from U.S. 2° × 2° Ave.
	1° × 1° Free-Air Anomaly Values				
Sea level to 100	$FA = -0.224h + 38$	100	+16	+13	+13
100 to 900	$0.024h + 20$	500	+32	+34	+30
900+	$0.104h - 52$	1000	+52	+50	+47
		1500	+104	+98	+95
	1° × 1° Bouguer Anomaly Values				
Sea level to 350	$BA = -0.12h + 18$	100	+6		+11
350+	$0.059h - 4$	500	−34		+26
		1000	−63		+47
		1500	−93		+66
	3° × 3° Free-Air Anomaly Values				
Sea level to 1500	$FA = 0.06h - 7$	100	−1	−2	+4
		500	+23	+21	+22
		1000	+53	+51	+48
		1500	+83	+76	+74
	3° × 3° Bouguer Anomaly Values				
Sea level to 350	$BA = -0.131h + 21$	100	+8		+13
350+	$0.058h - 3$	500	−32		+28
		1000	−61		+49
		1500	−90		+69

TABLE 8. Relation of Anomaly Values to Elevation in Canada

Height Interval, meters	Equation	h, meters	Anomaly, mgal	Difference from Standard Isos. Val.	Difference from U.S. 2° × 2° Ave.
1° × 1° Free-Air Anomaly Values					
Eastern Canada and Central Canada					
Sea level to 400	$FA = 0.095h - 36$	100	−26	−29	−29
		200	−17	−10	−13
Western and Central Canada					
Sea level to 300	$0.095h + 4$	100	+14	+11	+11
300+ (Western Canada)	$0.008h - 8$	500	−4	−2	−5
		1000	0	−2	−5
		1500	+4	−2	−5
Northwest Canada					
750+	$0.0085h + 30$	1000	+39	+37	+34
		1500	+45	+37	+34
		2000	+47	+30	+29
1° × 1° Bouguer Anomaly Values					
Eastern Canada					
Sea level to 500	$BA = -0.06h - 32$	100	−38		−35
		200	−42		−20
		500	−62		−2
Central Canada					
Sea level to 500	$-0.176h + 12$	100	−5		0.0
		500	−76		−16
Western Canada					
750 to 1100	$-124h + 8$	1000	−116		−6
1100+	$-0.061h - 58$	1500	−150		+9
		2000	−180		+22
Northwest Canada					
750	$0.0335h - 40$	1000	−74		+36
		1500	−90		+69
3° × 3° Free Air-Anomaly Values					
Eastern Canada*					
Sea level to 350	$FA = 0.074h - 33$	100	−26	−29	−29
		100	−18	−11	−14
Central Canada*					
350 to 750	$0.074h - 33$	500	+4	+6	+3
Northwest Canada*					
750 to 1500	$0.074h - 33$	1000	+41	+39	+36
		1500	+78	+72	+69
Western Canada					
1000+	$0.39h - 40$	1000	−1	−3	−6
		1500	+14	+8	+5
		2000	138	+21	+20
3° × 3° Bouguer Anomaly Values					
Eastern Canada					
Sea level to 350	$BA = -0.051h - 34$	100	−39		−34
		200	−44		−22
Central Canada					
100 to 500	$-0.064h - 18$	200	−31		−9
		500	−50		+10
Western Canada					
500 to 1200	$-0.125h + 20$	500	−43		+17
1200+	$-0.08h - 32$	1000	−105		+5
		1500	−153		+6
		2000	−192		+10
Northwest Canada					
750 to 1500	$-0.081h + 15$	1000	−66		+44
		1500	−105		+54

* FA equations are the same for eastern, central, and northwest Canada, but elevation ranges differ.

Mexico

Only one basic set of relations is defined for Mexico by the mean 1° × 1° and 3° × 3° free-air anomaly values, although it should be noted that the scatter in values is appreciable and, at least on a 1° × 1° basis, two approximately parallel patterns separated by about 40 mgal can be described. The relations given in Table 9 are for average values.

All the values show a slight positive bias. However, this cannot be interpreted as evidence of excess gravity here, since the standard was set on the basis of continental relations, whereas Mexico is sub-continental in size (9° in width). In Central America south of Mexico, this bias increases as the width of the land mass decreases. The only seismic measurement of crustal thickness in the plateau of Mexico (Durango) suggests the crust has a subnormal thickness of $\approx$1.7 km for the surface elevation of 2200 meters. The isostatic anomaly of -25 mgal suggests the crust has a subnormal density. This appears to apply to the entire plateau.

Central America

Only 1° × 1° values can be considered for Central America, as the width of the land mass is too restricted to obtain any meaningful 3° × 3° values (see Table 10). Also, only one relation is defined for each type of anomaly.

The positive bias in values increases slightly with elevation and averages about 50 mgal. If we consider the above departures, along with those in Mexico, as being due strictly to the width of the topographic block, the following equation describes the departure to be expected as a function of the width of the block expressed in equivalent degrees of latitude: $\Delta g = (11 - x)6$, where x is the width of the continental block. This suggests that the standard adopted applies for all continental blocks exceeding 11° in width. The isostatic anomaly in the plateau of Mexico averages about -25 mgal and that in Central America about -15 mgal; for the standard to be meaningful, corrections

TABLE 9. Relation of Anomaly Values to Elevation in Mexico

Height Interval, meters	Equation	h, meters	Anomaly, mgal	Difference from Standard Isos. Val.	Difference from U.S. 2° × 2° Ave.
	1° × 1° Free-Air Anomaly Values				
Sea level to 1300	$FA = 0.014h - 2$	200	+1	+8	+5
		500	+5	+7	+4
		1000	+14	+10	+9
130+	$0.0225h - 14$	1500	+20	+14	+11
		2000	+33	+16	+15
		2550	+42	+2	+5
	1° × 1° Bouguer Anomaly Values				
Sea level to 1100	$BA = -0.111h + 1$	200	-21		+1
		500	-54		+6
		1000	-110		0.0
1100+	$-0.079h - 35$	1500	-153		+6
		2000	-193		+9
		2500	-232		+9
	3° × 3° Free-Air Anomaly Values				
Sea level to 2000	$FA = 0.0125h - 13$	200	-9	-2	-5
		500	-2	0.0	-3
		1000	+9	+7	+2
		1500	+19	+13	+10
		2000	+30	+13	+12
	3° × 3° Bouguer Anomaly Values				
Sea level to 2000	$BA = -0.092h - 10$	200	-28		-6
		500	-36		+24
		1000	-102		+8
		1500	-148		+11
		2000	-194		+8

TABLE 10. Relation of Anomaly Values to Elevation in Central America

Equation	h, meters	Anomaly, mgal	Difference from Standard Isos. Val.	Difference from U.S. 2° × 2° Ave.
	1° × 1° Free-Air Anomaly Values			
$FA = 0.02h + 32$	200	+36	+43	+40
	500	+42	+44	+41
	1000	+52	+50	+47
	1500	+62	+56	+53
	1° × 1° Bouguer Anomaly Values			
$BA = -0.096h + 38$	200	+20		+42
	500	−10		+50
	1000	−58		+52
	1500	−105		+54

should be included for the isostatic anomaly. A revised expression would be $\Delta g = (16 - x)5$. For Mexico, with $x = 9°$, the correction to be applied to the difference between the observed values and the standard would be −35 mgal. In Central America, with $x = 3°$, the correction would be −65 mgal.

As the width of Alaska is about 11°, a similar correction to the markedly positive values there would reduce the large observed differences from the standard by 25 mgal.

South America

Although the gravity coverage in South America is limited, it is possible to examine the relations for the Andes and Argentina. Because of variable geologic structure associated with the Andes Mountain system and apparent changes in crustal composition in the lowlands of Argentina, the spread in 1° × 1° values is quite large. The relations given in Table 11 represent average conditions as shown by most of the data. The other 1° × 1° Bouguer anomaly relations are $BA = 0.07h - 26$ (Northern Argentina and the altiplano), and $BA = -0.077h + 55$ (northern Andes).

Although about 10 mgal of the observed positive bias in values in Argentina can be attributed to the restricted width of the continental block (14°), the balance appears to be related to an abnormal crustal density. In the northern Andes the marked positive bias of 70 mgal, as well as the available isostatic anomaly values in the area which are as high as +90 mgal in places, suggest the Cordillera Central is a horst.

Africa

The relations in Africa vary considerably with location. In general, these locations are South Africa, equatorial east Africa, and central and northern Africa (see Tables 12–14).

Except for South Africa, where the width of the continental block (14°) suggests that the values may incorporate about a 10-mgal positive bias, there is no reason to suspect that the values shown are not truly representative. Certainly the marked difference between north and central Africa and east Africa appears to be real. The available isostatic anomaly values indicate that, except for the Atlas Mountains, north and central Africa are characterized by positive isostatic anomalies and east Africa by negative isostatic anomalies. The data for South Africa, which define relationships falling between the other two regions, are mixed. The available seismic data (all in South Africa) suggest the crust has a subnormal thickness of about 5 km for the surface elevation east of a line having a strike of about north-northeast passing through Johannesburg. West of this line there is a normal thickness of the crust ($\approx$42 km), for the surface elevation of about 1400 meters. As the isostatic anomalies are, on the average, positive in the region of subnormal crustal thickness (which lies west and north of the Drakensberg Mountains), it would appear this area is out of equilibrium.

India

Most of India is a shield area characterized by negative isostatic anomalies. In the northern part south of the Ganges plain at approximately 22°N, however, there is a discontinuous zone of positive isostatic anomaly areas that can be identified in part with geological features that appear to be horsts. There is also a continuous zone of positive isostatic anomalies associated with the Himalaya Mountains. As there are no seismic crustal measurements of which the writer is aware, the actual crustal significance of the isostatic anomaly pattern is not known. Because of marked changes in geology, the restricted range in elevation encountered over most of the peninsula, and the limited amount of data for the Himalaya region, it is difficult to establish reliable relationships for the change in free-air anomaly with elevation. Two well defined groups of values, however, are indicated. The Bouguer anomalies, because of their larger magnitude and the dominance of the crustal root effect over that of surficial geology, permit more meaningful relations to be established.

Although the 3° × 3° free-air anomaly values agree well with the 1° × 1° values in southern India, the agreement between the two sets of relations elsewhere in India is poor (see Table 15). This can be explained by the fact that most of the anomaly changes defined by the 1° × 1° values outside southern India either have a short wavelength or that the half wavelength (peak to trough distance) approximates 3° and is consequently eliminated in the 3° × 3° averages. An examination of the isostatic anomaly pattern for India (Figure 5) substantiates this explanation.

As the width of peninsular India south of 16°N (including the continental shelf) is about

TABLE 11. Relation of Anomalies to Elevation in Argentina and the Andes

Height Interval, meters	Equation	h, meters	Anomaly, mgal	Difference from Standard Isos. Val.	Difference from U.S. 2° × 2° Ave.
	1° × 1° Free-Air Anomaly Values				
150 to 500	$FA = 0.050h + 6$	200	+16	+23	+20
		500	+31	+33	+30
500 to 2500	$0.054h - 2$	1000	+52	+54	+51
		1500	+79	+74	+71
		2000	+106	+89	+88
		2500	133	+90	+93
	1° × 1° Bouguer Anomaly Values				
Sea level to 3000	$BA = -0.072h + 6$	200	−8		+14
		500	−30		+30
		1000	−66		+44
		1500	−102		+57
		2000	−138		+64
		2500	−174		+67
	3° × 3° Free-Air Anomaly Values				
Sea level to 250	$FA = -0.084h + 20$	100	+12	+9	+9
		200	+3	+10	+7
250+	$0.058h - 17$	500	+12	+14	+11
		1000	+41	+39	+36
		1500	+70	+64	+61
		2000	+99	+81	+80
	3° × 3° Bouguer Anomaly Values				
Sea level to 200	$BA = -0.196h + 20$	100	0.0		+5
		200	−20		+2
200+	$-0.077h - 2$	500	−40		+20
		1000	−78		+32
		1500	−117		+42
		2000	−156		+46
Cordillera Central of Andes Mountain in Colombia					
1000 to 3000	$BA = -0.089h + 46$	1000	−43		+67
		1500	−89		+70
		2000	−132		+70
		2500	−177		+64

TABLE 12. Relation of Anomaly Values to Elevation in Central and North Africa

Height Interval, meters	Equation	h, meters	Anomaly, mgal	Difference from Standard Isos. Val.	Difference from U.S. 2° × 2° Ave.
	1° × 1° Free-Air Anomaly Values				
Sea level to 350	$FA = -0.088h + 34$	100	+25	+22	+22
		200	+16	+23	+20
350+	$0.062h - 20$	500	+11	+13	+10
		1000	+42	+40	+37
		1500	+73	+66	+64
		2000	+104	+87	+86
	1° × 1° Bouguer Anomaly Values				
Sea level to 350	$BA = -0.172h + 20$	100	+3		+8
350+	$-0.05h - 20$	500	−45		+15
		1000	−70		+40
		1500	−95		+64
		2000	−120		+82
	3° × 3° Free-Air Anomaly Values				
Sea level to 400	$FA = -0.085h + 29$	100	+21	+18	+18
400+	$0.086h - 45$	500	−2	0	+3
		1000	+41	+39	+36
		1500	+84	+78	+75
	3° × 3° Bouguer Anomaly Values				
Sea level to 200	$BA = -0.184h + 24$	100	+6		+11
200+	$-0.072h + 4$	500	−32		+28
		1000	−68		+42
		1500	−104		+55

7°, the observed values should have a positive bias of about 45 mgal if we use the expression $\Delta g = (16 - x)5$, defined on the basis of relations in Mexico and Central America. On this basis the anomalous value of −4 mgal for the 1° × 1° Bouguer anomalies at an elevation of 500 feet should actually be nearer −50 mgal. This would agree well with the observed average isostatic anomaly of about −55 mgal.

The explanation of the anomaly pattern in India is not simple. Satellite data indicate a large negative mass effect centered just south of

TABLE 13. Relation of Anomaly Values to Elevation for South Africa

Height Interval, meters	Equation	h, meters	Anomaly, mgal	Difference from Standard Isos. Val.	Difference from U.S. 2° × 2° Ave.
	1° × 1° Free-Air Anomaly Values				
Sea level to 300	$FA = -0.137h + 23$	100	+9	+6	+6
300+	$0.057h - 32$	500	−4	−2	−5
		1000	+25	+23	+20
		1500	+54	+48	+45
	1° × 1° Bouguer Anomaly Values				
Sea level to 350	$BA = -0.232h + 20$	100	−3		+2
350+	$-0.061h - 37$	500	−68		−8
		1000	−98		+12
		1500	−129		+30
	3° × 3° Free-Air Anomaly Values				
700+	$FA = 0.0505h - 36$	1000	+15	+13	+10
		1500	+40	+34	+31
	3° × 3° Bouguer Anomaly Values				
700+	$BA = -0.073 - 20$	1000	−93		+17
		1500	−130		+29

TABLE 14. Relation of Anomaly Values to Elevation for Equatorial East Africa

Height Interval, meters	Equation	h, meters	Anomaly, mgal	Difference from Standard Isos. Val.	Difference from U.S. 2° × 2° Ave.
	1° × 1° Free-Air Anomaly Values				
Sea level to 300	$FA = -0.144h + 1$	100	-13	-15	-15
300+	$0.046h - 53$	500	-30	-28	-31
		1000	-7	-9	-12
		1500	$+16$	$+10$	$+5$
	3° × 3° Free-Air Anomaly Values				
150+	$FA = 0.05h - 56$	500	-31	-29	-32
		1000	-6	-8	-11
		1500	$+19$	$+13$	$+10$
	1° × 1° and 3° × 3° Bouguer Anomaly Values				
350+	$BA = -0.063h - 57$	500	-89		-29
		1000	-120		-10
		1500	-152		$+7$

Ceylon that may well have a deep source. The isostatic anomaly values range from −80 mgal on Ceylon and the southern tip of India up to zero at about 21°N latitude, where positive anomalies are found in association with the Satpura and Vindhya Mountains. These mountains may represent horsts. Certainly the Arvalli Mountains south of Jodhpur, where there is a local positive anomaly closure of about 40 mgal, appear to be a horst. However, this effect is superimposed on a pronounced positive anomaly belt that extends from the Gulf of Cambay up to the vicinity of Lahore in northern Pakistan. As the entire area is characterized by low elevations (200 to 350 meters), it probably is one of abnormal crustal density and thickness. The pronounced minimum associated with the Ganges valley, which in the vicinity of Benares reaches an isostatic value of −100 mgal, is undoubtedly in part related to a thick section of low density alluvium, but it is also probable that the main effect is due to tectonic displacement of the crust having surface expression as a graben.

Other Areas

Although data from other areas, particularly Europe and Australia, could be presented, our purpose is not a world review. We wish only to examine the types of regional changes that are encountered; to recognize them as normal or anomalous using Bouguer anomalies, in particular, since most of the world's gravity data have been reduced in this form; and to examine factors that can lead to anomalous gravity values. As there are significant variations in anomaly relations within each continent, it was necessary to compare relations for several continents. The degree of variability encountered is brought out in Figure 6, which shows the mean curves for the 3° × 3° Bouguer anomaly values. If the 1° × 1° values had been used, the variability would have been even greater.

CONCLUSIONS

The relations considered, though restricted to continental areas, demonstrate the following:

1. Significant changes (25 to 50 mgal) in isostatic anomalies are to be expected where there are changes in crustal composition with isostatic equilibrium, as well as where there are actual departures from isostasy.

2. The gravity effect of the surface and near-surface mass distribution dominates the compensation mass distribution at depth such that under isostatic conditions 90% of the topographic effect is realized within a radius (r) of 5 to 6 km from an observation site and the equivalent degree of compensation is not realized until $r = 3°$. As a result, there is a strong correlation between gravity values and topographic relief, changes in geology, and also areas of change in crustal composition and anomalous crustal thickness, particularly when the areas involved are of restricted width.

3. Free-air anomalies, while extremely sensitive to short wavelength variations in the surface and near surface mass distribution on 1° ×

TABLE 15. Relation of Anomalies to Elevation in India

Height Interval, meters	Equation	h, meters	Anomaly, mgal	Difference from Standard Isos. Val.	Difference from U.S. 2° × 2° Ave.
1° × 1° Free Air Anomaly Values					
Southern Peninsula					
Sea level to 750	$FA = 0.120h - 81$	200	-58	-51	-54
		500	-21	-19	-22
Northwest India, Satpura, Vindhya Mountain region					
Sea level to 750	$0.128h - 48$	200	-23	$+16$	$+19$
		500	$+16$	$+18$	$+15$
1° × 1° Bouguer Anomaly Values					
Southern Peninsula					
Sea level to 600	$BA = -0.162h + 16$	200	-16		$+6$
		500	-64		-4
Central India					
300+	$-0.118h - 78$	500	-137		-64
Himalaya Mountains					
500 to 2000	$-0.112h + 12$	500	-44		$+16$
		1000	-100		$+10$
		1500	-156		$+3$
		2000	-212		-10
Northwest India, Satpura, Vindhya Mountain region					
500 to 1600	$-0.06h - 12$	500	-42		$+18$
		1000	-72		$+38$
		1500	-101		$+58$
Ganges Valley					
Sea level to 300	$-0.2684h - 32$	200	-85		-63
300+	$-0.118h - 78$	500	-137		-77
		1000	195		-85
3° × 3° Free-Air Anomaly Values					
South of 22°N latitude					
200 to 700	$FA = 0.101h - 70$	200	50	-43	-46
		500	-19	-17	-20
North of 22°N latitude					
200 to 600	$0.138h - 54$	200	-26	-19	-27
		500	$+16$	$+14$	$+15$
3° × 3° Bouguer Anomaly Values					
India as a whole					
200 to 750	$BA = -0.218h + 42$	200	-1		$+23$
		500	-67		-7

1° and 2° × 2° average bases, show only a relatively slight direct dependence on elevation. The actual dependence on elevation is governed more by the width (wavelength) associated with the topographic blocks making up the over-all relief pattern of the continent. In general, the dependence on elevation can be described by three linear functions corresponding to the coastal area (sea level to ≈250 meters), the continental platform lying between 250 and 1700 meters, and mountain areas rising above 1700 meters. Whereas the sign of the function describing the relations of free-air anomalies in the coastal areas is negative, that for the other two areas is positive. Average relations for areas where there is isostatic equilibrium define the following equations:

Sea level to 200 meters $\quad FA = -0.102h + 3$
200 to 1700 meters $\quad FA = 0.0075h - 6$
1700 + meters $\quad FA = 0.0465h - 77$

Minor variations can be expected because of variations in boundary conditions (surface slope) between the three elevation zones, as well as variations in over-all relief pattern associated with different continents. Major changes result

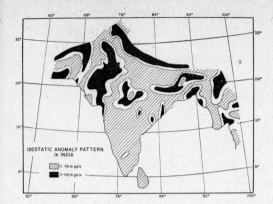

Fig. 5. Simplified isostatic anomaly map of India and adjacent areas [after *Gulatee*, 1956].

from variations in crustal and upper mantle parameters, and in some areas deep mass anomalies. In such cases the gradient can change significantly, and the sea level intercept values vary by as much as ±50 mgal. Major variations are noted on a continent-to-continent basis, as well as over distinct regions within a given continent.

4. Because the free-air anomaly is incorporated in all other anomalies, ($BA = FA - \Delta g_B$ and the isostatic anomaly $IA = BA + \Delta g_r$), any regional departure in free-air anomaly values will be reflected in the other anomalies.

5. From the standpoint of evaluating normal isostasy (no abnormalities in crustal or mantle density), the value of density (σ_c) used in the Bouguer reduction is not critical, and the value of $\Delta \sigma = (\sigma_m - \sigma_c)$ used in the isostatic reduction need only conform so that the mass of the topography above sea level ($\Delta h \sigma_c$) equals that of the compensating mass ($\Delta R \Delta \sigma$). How-

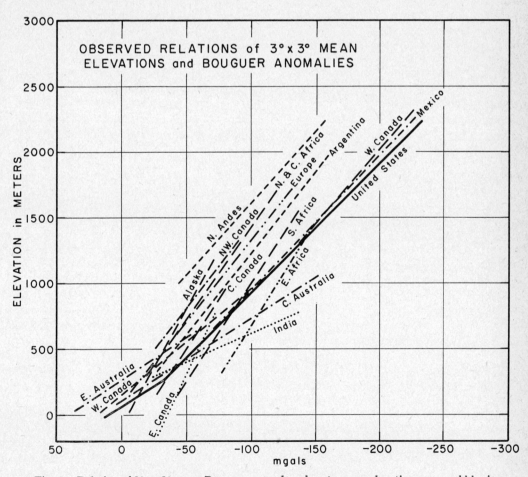

Fig. 6. Relation of 3° × 3° mean Bouguer anomaly values to mean elevation on a world basis.

ever, if agreement is to be obtained with seismic data concerning variations in crustal thickness with surface elevation, the following approximate parameters are indicated:

Sea level standard crust $H_s = 33$ km.
$\sigma_o = 2.92$ g/cm^3.
$\Delta\sigma = 0.39$ g/cm^3.

6. Because there is a marked dependence of Bouguer anomalies on elevation and a large change in values (300+ mgal for 3000 meters change in elevation), which significantly exceeds that attributable to unknown geological effects (± 10 to 30 mgal), these anomalies usually describe best the relation of gravity anomaly values to elevation. The change in free-air anomaly values over the same range of elevation usually does not exceed 60 mgal, and as the geological effect is the same, the uncertainty in establishing the relation of free-air anomalies to elevation is much greater, particularly where the range in elevation is limited. It is best in such areas to derive the free-air anomaly relation from the Bouguer anomaly relation using the expression $FA = BA + \Delta g_B$.

7. Although the effect of the geologic contribution can be suppressed and even eliminated by averaging data over an area, the degree to which the geologic effect is suppressed is a function of the wavelength of the disturbing mass distribution and that of the size area chosen in averaging. Most anomaly profiles have a sinusoidal pattern alternating between positive and negative values about some mean. If the peak-to-peak wavelength is used to define the dimensions of the anomaly pattern, the effect of averaging in suppressing the total amplitude (positive to negative peak values) can be written as $\theta = 120x - 20$, where x is the percent of the wavelength used in averaging, and θ is the percent suppression. For a 4° wavelength anomaly caused by changes in crustal parameters, 1° × 1° averaging reduces the magnitude of the total anomaly by 10%, 2° × 2° averaging reduces the total amplitude by 40%, and 4° × 4° averaging would eliminate the sinusoidal anomaly pattern and give the mean value. These factors are significant in considering the relation of free-air and Bouguer anomalies to elevation, since the crust has sufficient strength to sustain mass inequalities of limited area without complete compensation. For example, it is questionable whether a 1° × 1° area is totally compensated locally, whereas it is fairly certain that a 3° × 3° area is. Any set of 'normal' relations between elevation and free-air and Bouguer anomalies in lieu of isostatic anomalies (which assume local compensation) should be based on the minimum size area that does appear to be compensated.

The 1° × 1°, 2° × 2°, and 3° × 3° areas in the United States all give results that are in good agreement. However, the 2° × 2° values agree best (2 to 3 mgal) with the relations of free-air and Bouguer anomaly to elevation at sites where the isostatic anomalies are within ± 10 mgal of zero on a worldwide basis.

8. Although the relations for 2° × 2° Bouguer anomalies to elevation in the United States provide a normal isostatic standard for evaluating the relation of Bouguer anomaly to elevation in other areas, this only applies where the size of the continental block exceeds 16° in width, as defined in equivalent degrees of latitude. For smaller continental blocks, there is a positive bias related to bias in the free-air anomaly values. The degree of bias can be approximated from the expression $\Delta g = (16 - x)5$, where x is the width of the land mass in equivalent degrees of latitude. The effect is that of datum offset that is incorporated in the constant c in the general expression $BA = \theta h + c$.

9. Where there are changes in crustal density and thickness with no lack of equilibrium, the bias in the anomaly values conforms to that of the departure in crustal density. The gravity anomaly gradient with change in elevation exceeds that for normal conditions where the sign of the departure is positive and is less than normal where the departure is negative.

10. Where there is an actual lack of equilibrium due to tectonic displacement of the crust, the bias in the anomaly values conforms to that of the tectonic displacement. The gravity anomaly gradient with change in elevation where there is a graben exceeds the normal value. Where there is a horst, the gradient is subnormal.

Wherever there is anomalous gravity, the departure in sign and gradient is a function of the width (wavelength) of the mass disturbance. The shorter the wavelength, the more pronounced the anomalous relations.

In addition to the two cases considered (anomalous crustal composition and tectonic displacement), there are some areas where there are apparent combinations of the two whereby a graben is invaded by high density diabase that may or may not have surface expression as basalt flows. These areas are always characterized by positive gravity values. Examples are the Red Sea, the Snake River downwarp in Idaho, and the Lake Superior syncline which has gravity expression from Minnesota into Kansas in the United States.

11. The relations established and their interpretation are believed to be correct in principle. However, it must be recognized that outside of the United States and Mexico the samples of data available were restricted in areal extent and in general were based on only those areas where there were forty or more observations in a degree square (except where reliable anomaly contours could be extended across an area of limited control on the basis of surrounding data). This was necessary if the effect of local changes in geology were to be minimized. As more data become available, the relations can be expected to be modified. Europe was not considered in this paper, even though it is one of the more anomalous continents, as it is being discussed elsewhere in this volume by Coron (p. 304).

Acknowledgments. The writer acknowledges the cooperation of the many oil companies, government agencies, and individuals who made data available for this study. The work was supported by the National Science Foundation, the Gravity Division of the Army Map Service, the National Science Foundation, and the Aeronautical Chart and Information Center of the U. S. Air Force. In particular, thanks are due the latter for mean $1° \times 1°$ elevation values in all the areas studied. Although published data are not cited, the writer obviously depended heavily on this material, and is also grateful to the many authors who supplied unpublished data.

REFERENCES

Garland, G. D., *The Earth's Shape and Gravity*, Pergamon Press, London, 175 pp., 1965.

Gulatee, B. L., Gravity data in India, *Surv. of India, Tech. Paper 10*, 195 pp., 1956.

Woollard, G. P., The relation of gravity anomalies to surface elevation, crustal structure and geology, *Univ. Wisc. Geophys. Polar Res. Center, Rept. 62–9*, 292 pp., 1962.

Woollard, G. P., Regional isostatic relations in the United States, in *The Earth beneath the Continents, Geophys. Monograph 10*, edited by J. Steinhart and T. Smith, pp. 557–594, American Geophysical Union, Washington, D.C., 1966.

Woollard, G. P., The inter-relationship of the crust, the upper mantle, and isostatic gravity anomalies in the United States, in *The Crust and Upper Mantle of the Pacific Area, Geophys. Monograph 12*, edited by L. Knopoff et al., pp. 312–341, American Geophysical Union, Washington, D.C., 1968.

Gravity Field over the Atlantic Ocean[1]

Manik Talwani and Xavier Le Pichon

IN RECENT years considerable attention has been paid to the broad variations in the earth's gravity field as determined from satellite observations and to the geophysical implications of these variations (for example, see *Kaula* [1967]). Apparently the satellites are, at present, able to determine the earth's gravitational field in terms of spherical harmonics up to degree and order about twelve.

In this paper the regional gravity field over the Atlantic Ocean is obtained by averaging surface ship and submarine pendulum gravity data over 5° squares. These data represent much higher spherical harmonic information on the gravity field than is obtainable at present from analyses of satellite orbits. Correlation of the gravity field with topography suggests that iso-

[1] Contribution 1252 from the Lamont-Doherty Geological Observatory of Columbia University.

static compensation prevails for the wavelengths considered here and that the compensation must take place in part at depths much greater than 30 km. In addition, nonisostatic lateral inhomogeneities also exist in the upper mantle.

The regional gravity data are averaged over 20° squares for comparison with gravity data based on satellite observations.

SURFACE-SHIP GRAVITY MEASUREMENTS

Previous studies of regional surface gravity at sea have been based on submarine pendulum observations (see Figure 1). The number of pendulum observations in the South Atlantic Ocean is very small. In the past decade the gravity coverage of the Atlantic Ocean has been greatly increased by continuous measurements made with surface-ship gravimeters.

Measurements were made aboard Lamont Geological Observatory ships RV *Vema* and RV *Robert D. Conrad* in the North and South Atlantic during the period 1961–1967. In addition we have used the measurements made aboard H. Neth. MS *Snellius* in the North Atlantic as part of the Navado survey [Strang van Hees, 1967] and some measurements made by the British Admiralty aboard HMS *Hecla* in the South Atlantic. The tracks of the surface ships and the locations of the discrete submarine pendulum measurements are shown in Figure 1.

All the surface ship gravity measurements were made with the Graf Askania Gss2 gravity meters mounted on gyrostabilized platforms. The errors caused principally by the effect of horizontal accelerations have, for the Lamont measurements, been discussed by *Talwani* [1966]. No corrections for these errors were applied. Because the values used in the present study are preliminary and are subject to correction, a detailed estimate of the accuracy of measurements has not yet been made. We believe that the errors in the free-air anomalies averaged over 5° squares are less than 10 mgal.

Recent developments of surface ship gravimeters, stable platforms, and navigation systems (see for instance *Talwani et al.* [1966] and *LaCoste et al.* [1967]) have markedly improved the accuracy of surface ship gravity measurements (to $\sim$1 or 2 mgal). With extensive utilization of newer systems, considerably more accurate gravity maps over the oceanic areas are now possible.

AVERAGED SURFACE GRAVITY VALUES

Figure 2 shows the measured free-air anomalies averaged over 5° squares in the North and South Atlantic Oceans. These are based on surface ship gravimeter measurements along the tracks and the submarine pendulum values at the stations shown in Figure 1. In calculating the averages, the pendulum values were arbitrarily given a weight of 10 and the gravimeter values were given unit weight. Thus one pendulum measurement is given as much weight as surface ship gravimeter measurements made along 40 km of track.

In the North Atlantic a prominent feature of the surface gravity map is the minimum in the western basin having values more negative than −60 mgal in the vicinity of the Puerto Rico trench. Gravity values over the Caribbean Islands have not been used in calculating the 5° averages. The large positive gravity values on these islands will make the minimum less pronounced; however, the large area bounded by the −40-mgal and −20-mgal contours indicates that the marginal areas of the western basin still remain strongly negative even if the island values are taken into account. In contrast, the gravity anomalies in the eastern basin, though negative (lying in the range 0 to −20 mgal), do not attain values as large as in the western basin. The Mid-Atlantic ridge north of 20°N is characterized by positive values, which are greater than +40 mgal in some places over the crest. South of 20°N, gravity values over the ridge are much closer to zero. Note also in Figure 1 that the ridge is poorly developed south of 20°N, being intersected by fracture zones [*Heezen and Tharp*, 1961].

Because of fewer measurements, the gravity field is not as well defined in the South Atlantic as it is in the North Atlantic. As in the North Atlantic, there is a gravity minimum along the western margin (Figure 2). The marginal minimum is continuous over the whole western Atlantic Ocean, extending from Newfoundland in the North Atlantic to the Falkland Plateau near 50°S in the South Atlantic. The −60-mgal contour near 15°S, 25°W is based on a single track and may possibly be in error. Owing to limited control, the exact configuration of the

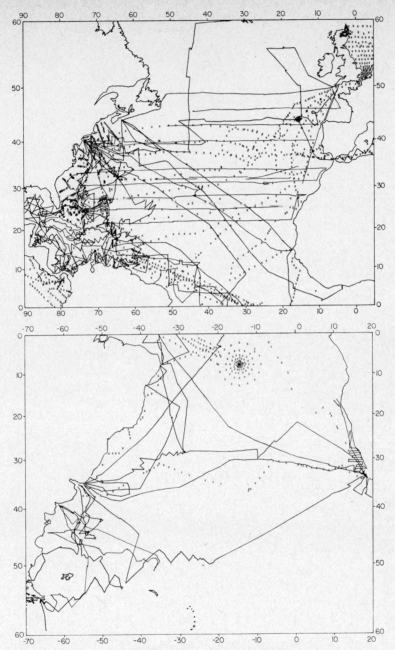

Fig. 1. Atlantic gravity index maps. Pendulum measurement locations (crosses) from *Vening Meinesz* [1948] and *Worzel* [1965]. Ships' tracks along which continuous gravity measurements were obtained are shown by solid lines. Figure 1 serves as a control chart for Figure 2. *Top:* North Atlantic. *Bottom:* South Atlantic.

−20-mgal contour may not be as shown in this figure. The gravity anomalies are not significantly positive over the Mid-Atlantic ridge in the South Atlantic, and the 0-mgal contour lies east of the ridge crest (north of 30°S). The eastern basin has slightly positive values; over the Walvis ridge area the anomalies approach +40 mgal. Thus the gravity field over the

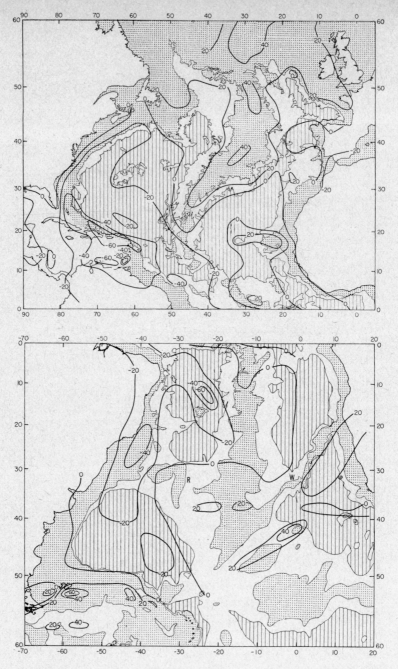

Fig. 2. Atlantic 5° average surface free-air anomaly maps. A total of 280 5° averages define the contours at 20-mgal interval. The accuracy of the 5° average values is believed to be better than 10 mgal. Generalized bathymetry is taken from the U. S. Navy Hydrographic Office World Chart. Areas with depths less than 2000 fathoms are stippled; areas with depths larger than 2500 fathoms are indicated by vertical lines. *Top:* North Atlantic. Note that the western basin has a greater extent than the eastern basin. Note also the progressive shallowing north of 50°N. *Bottom:* South Atlantic. Note that the Rio Grande rise (*R*) and Walvis ridge (*W*) near 30°S divide, respectively, the western and eastern basin into two parts.

South Atlantic appears to differ from that over the North Atlantic in that the ridge in the south has no large positive values over it, whereas the area to the east is essentially positive. However, there are very few measurements between 10°S and 30°S and south of 40°S on the Mid-Atlantic ridge and in the eastern basin; hence, more measurements are required before the conclusion stated above can be considered definitely proved.

In many profiles, the mid-oceanic ridge system is characterized by anomalies that tend to be positive with respect to the ocean basins on either side; the limited data in the South Atlantic do not invalidate this general conclusion.

Also in the North Atlantic, somewhat less than half the area has anomalies between -20 and $+20$ mgal, whereas in the South Atlantic a much larger portion of the area has anomalies in this range. Areas with anomalies more negative than -20 mgal or more positive than $+20$ mgal are not only small in total extent but generally lie in narrow patches.

SIGNIFICANCE OF THE REGIONAL GRAVITY FIELD

Munk and MacDonald [1960] first pointed out that the nonhydrostatic, low-order, zonal harmonics of the earth's gravitational field are not due to the combined effects of topography and its isostatic compensation. Similarly, one is led to ask whether the regional gravity field (Figure 2), which involves harmonics of much higher degree, can be explained by the effect of topography and its compensation.

In the North Atlantic, in particular, there appears to be a strong correspondence between topography and regional gravity. The 20-mgal contour over the Mid-Atlantic ridge, the -20-mgal and -40-mgal contours in the western basin, and the 0-mgal contour in the eastern basin are remarkably parallel to the topographic contours. In the South Atlantic the correspondence between topography and compensation is less apparent (but the gravity field is not determined very accurately in this region).

If isostatic equilibrium were attained locally and at a shallow depth, we would expect that free-air anomalies averaged over 5° squares would be close to zero and would not correlate with topography. That, obviously, is not the case. The question arises whether a different isostatic hypothesis would explain the free-air anomalies. Clearly we should investigate hypotheses that involve regionality as well as greater depths of compensation.

Kivioja [1963] has used averaged values over 5° squares of topography over the entire world to compute the combined effects of topography and compensation. He employed the Airy-Heiskanen hypothesis ($T = 30$), which is equivalent to a regional topographic isostatic correction, since the average elevation of 5° squares was used (see Figure 3). His values for the Atlantic Ocean have been contoured at 20-mgal intervals and were shown in Figure 3. A comparison of Figures 2 and 3 shows that the regional anomalies in the North Atlantic Ocean are generally reduced, though not eliminated, by applying the isostatic correction (that is, by substracting the values in Figure 3 from those in Figure 2). In the South Atlantic, applying the isostatic correction will reduce the anomalies in the western basin but makes them larger over the eastern basin. The absolute values over the ridge will be increased north of 30°S but reduced south of 30°S.

The average anomaly (without respect to sign) is 22 mgal in the North Atlantic and 16 mgal in the South Atlantic and becomes 15 and 16 mgal, respectively, after the isostatic correction is made. This implies that, at least in the North Atlantic, part of the 5° average surface gravity anomalies arises from isostatic effects. However, the mean anomaly of 15 mgal (without respect to sign) after this crude isostatic correction is applied is not mainly due to random 'noise,' as areas of several millions km^2 have anomalies in excess of 10 mgal consistently of one sign. Thus, apparently, a considerable part of these anomalies cannot be attributed to topography and its isostatic compensation (with a 5° regionality and $T = 30$).

We need next to examine whether a different isostatic hypothesis will explain the free-air anomalies. We are at present in the process of making detailed calculations with the assumption of various isostatic hypotheses. Some general conclusions for the North Atlantic can be obtained from an examination of Figures 2 and 3. If the depths of compensation were greater, the combined effect of topography and its compensation would tend to yield more negative values in areas where the values are negative and more positive where they are positive in

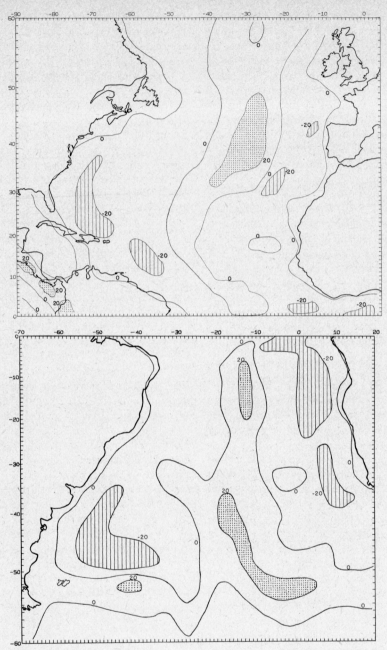

Fig. 3. Effect of the attraction of topography and its compensation in milligals in the Atlantic [*Kivioja*, 1963]. Compare with Figs. 1, 2, 4, and 5. *Top:* North Atlantic. *Bottom:* South Atlantic.

Figure 3. Therefore, an assumption of deeper compensation will yield a better approximation of the anomalies in Figure 2. However, the east-west asymmetry in the regional free-air maps of the North and South Atlantic (Figure 2), especially as expressed in the minimum along the western margin, would not be removed by any obvious isostatic correction. Tentatively, therefore, we can conclude that the isostatic mechanism operates to large depths in the upper

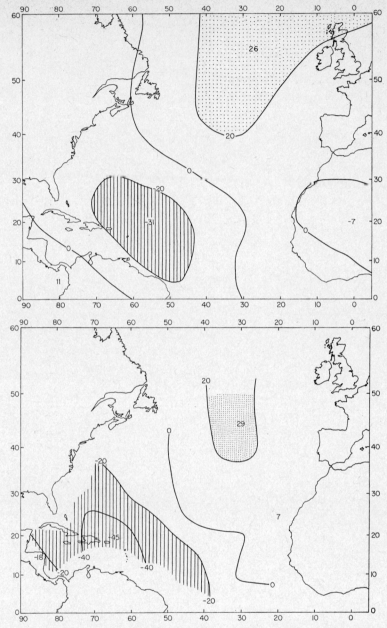

Fig. 4. *Top:* North Atlantic free-air anomaly map from satellite data (G8 solution, see text). *Bottom:* Surface ship gravity data average over 20° squares. Contours are at 20-mgal intervals. Some spot values are also given.

mantle, and that there are additional lateral inhomogeneities in the upper mantle which are nonisostatic in character.

Other geophysical data also point to inhomogeneities in the mantle. These are: variations in P_n and S_n velocity, variation in depth and extent of the low velocity layer in the upper mantle, and the near equality of heat flow in continental and oceanic areas despite the fact that the continental crust possesses more radiogenic rocks than the oceanic crust. In addition, density inhomogeneities would be associated

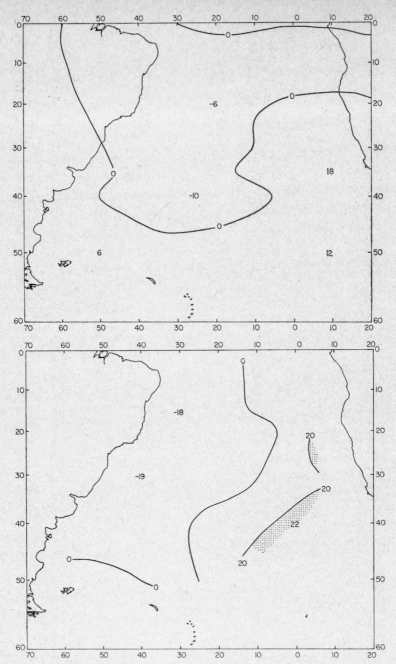

Fig. 5. *Top:* South Atlantic free-air anomaly map from satellite data (GS solution, see text). *Bottom:* Surface ship gravity data averaged over 20° squares. Contours are at 20-mgal intervals. Some spot values are also given.

with convective motions probably present in the mantle. Several attempts have been made in the past to correlate the gravity field described by low degree harmonics with various geophysical parameters. When data such as those described here are acquired for all the oceans, higher harmonics will become well defined, leading to more meaningful correlations.

In previous studies [*Worzel and Shurbet*, 1955; *Drake et al.*, 1959; *Talwani et al.*, 1959; *Woollard and Strange*, 1962], in which the crustal structure (or total crustal thickness) was obtained from gravity anomalies, it was generally assumed that no density variations exist below about 50 km. Acceptance of lateral changes in mantle density will lead to changes in these crustal sections. In general, the changes will be small, because in most of these studies crustal columns obtained from seismic refraction have been balanced using the Nafe-Drake density-velocity relationship and because large gradients in the Bouguer anomalies precluded very deep density inhomogeneities. These criteria still have to be considered. Of course, uncertainties exist in density-velocity relationships, and, even if Bouguer gradients are steep, part of the contribution to Bouguer anomalies can come from deeper inhomogeneities. In the type of crustal studies mentioned above, changes in Bouguer anomaly of 200–300 mgal are involved. Regional gravity anomalies discussed in this paper can amount to 20–30 mgal. Thus it would appear that changes in mantle density might account for perhaps 10% of the Bouguer anomaly, which in the studies above is attributed solely to changes in depth of M or other shallow density inhomogeneities. It should also be pointed out that an anomaly of about 20 mgal implies very small changes in mantle density: a layer 100 km thick with differential density of 0.005 g/cm³ would give rise to anomalies of this magnitude.

FREE-AIR ANOMALY FROM SATELLITE DATA

Kaula [1966] has discussed the various solutions for the spherical harmonic coefficients of the geopotential obtained from analyses of

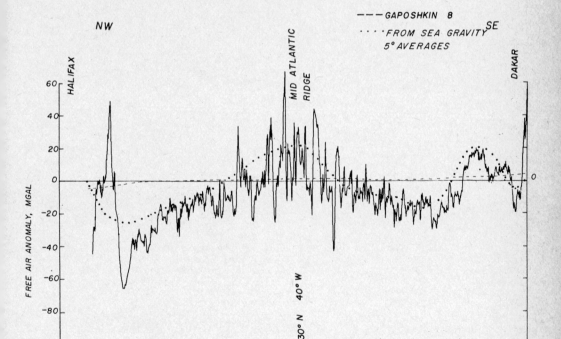

Fig. 6. Free-air anomaly profile across the North Atlantic projected along a small circle passing through 30°N, 40°W with a northwest-southeast azimuth. The profile comes from *Vema* cruise 17, track from Dakar to Halifax (Figure 1). Note the maximum associated with the Mid-Atlantic ridge, the small eastern gravity minimum, and the larger western gravity minimum. For comparison the gravity field as defined by Figures 2 (sea gravity) and 4 (satellite) are shown by dotted and dashed lines, respectively.

satellite orbits. Figures 4 and 5 show free-air anomalies obtained from satellite-derived spherical harmonic coefficients. We have used zonal harmonic coefficients (up to degree seven) given by Kaula (which are averages for four sets of solutions) and tesseral harmonic coefficients up to degree twelve obtained by Gaposhkin (G8 solution in *Kaula* [1966]. In order to obtain free-air anomalies on the international ellipsoid, we have subtracted the values of the second and fourth zonal harmonic coefficients associated with the international gravity formula from the corresponding harmonics obtained from satellite data. In the computations we assumed a constant value for $\gamma(a/r)^n = 980$ cm/sec^2 (using the notation of *Kaula* [1963]).

To compare the surface ship data with the satellite data, we have averaged the surface ship data over 20° squares. Roughly, then, the surface ship data will represent terms in a spherical harmonic expansion up to degree and order 9, and thus be comparable to the satellite-derived data. Figures 4 and 5 show the satellite-derived gravity data, as well as surface ship and sub-

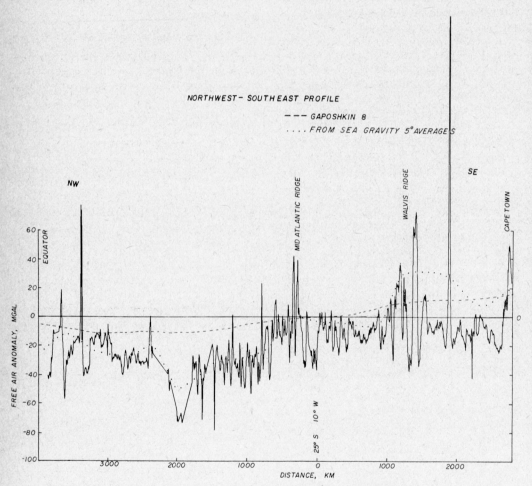

Fig. 7. Free-air anomaly profile over the South Atlantic projected along a small circle passing through 25°S, 10°W with a northwest-southeast azimuth. The profile comes from *Conrad* cruise 8, track from San Juan, Puerto Rico, to Cape Town (see Figure 1). Instrument problems are responsible for the gap in data near 2000 km northwest and render the existence of this −70 mgal minimum doubtful. This minimum is responsible for the −60 mgal contour of Figure 2. The very large maximum near 1800 km southeast corresponds to the *Vema* seamount. For comparison, the gravity field as defined by Figures 2 (sea gravity) and 4 (satellite) are shown by dotted and dashed lines, respectively.

marine pendulum gravity data, averaged over 20° squares. In the North Atlantic, especially, the agreement is very good. The more negative values from the surface ship and submarine pendulum data in the western North Atlantic are in part due to the fact that positive values on the Caribbean Islands have not been used in the surface ship averages. In the South Atlantic the agreement is perhaps less remarkable. Nevertheless, the negative values in the western basin and the positive values over the ridge and to the east appear in both the satellite and surface ship maps.

However, if Figure 2, in which surface ship data are averaged over 5° squares, is compared with Figures 4 and 5, it is clear that there is much more detail in the surface ship gravity maps. The correspondence with topography is less obvious in Figures 4 and 5 than in Figure 2. It is clear that surface ship gravity data would be very useful in studies of the higher harmonics of the earth's gravity field and its relation to the upper mantle.

The greater detail given by surface ship gravity data is illustrated also by northwest-southeast profiles in the Atlantic Ocean (Figures 6 and 7). We compare here profiles of actual shipboard gravity as well as profiles constructed from the 5° average map and from the satellite-derived gravity map. The profile in Figure 6 is at an azimuth which is unfavorable for the description of the satellite derived gravity data; a northeast-southwest profile would make the agreement much better.

Acknowledgments. In this study we use surface ship gravity data collected by Lamont ships RV *Vema* and RV *Robert D. Conrad* since 1961. We acknowledge the help of Captain H. Kohler and many other colleagues at Lamont who helped in the acquisition of the data. Dr. J. L. Worzel initiated this program and Drs. D. Hayes and R. Wall helped in many phases of data acquisition and analysis. E. Skinner and J. Hastings reduced the data. Financial support for the work came principally from the Office of Naval Research and the National Science Foundation. We are also grateful to the Hydrographer, Royal Navy, for sending us data from the South Atlantic obtained aboard HMS *Hecla*. We thank Drs. Dorman, Hayes, Nafe, and Worzel for critical review of the manuscript.

REFERENCES

Drake, C. L., M. Ewing, and G. H. Sutton, Continental margins and geosynclines: The east coast of North America north of Cape Hatteras, *Phys. Chem. Earth*, *3*, 110–198, 1959.

Heezen, B. C., and M. Tharp, *Physiographic Diagram of the South Atlantic Ocean*, Geological Society of America, New York, 1961.

Kaula, W. M., Determination of the earth's gravitational field, *Rev. Geophys.*, *1*, 507–551, 1963.

Kaula, W. M., Tests and combination of satellite determination of the gravity field with gravimetry, *J. Geophys. Res.*, *71*, 5303–5314, 1966.

Kaula, W. M., Geophysical implications of satellite determinations of the earth's gravitational field, *Space Sci. Rev.*, *7*, 769–794, 1967.

Kivioja, L. A., The effect of topography and its isotatic compensation of free-air anomalies, *Inst. Geod. Photogram. Cartog. Rept.* *28*, 134 pp., Ohio State University, Columbus, 1963.

LaCoste, L., N. Clarkson, and G. Hamilton, LaCoste and Romberg stabilized platform shipboard gravity meter, *Geophysics*, *32*, 99–109, 1967.

Munk, W. H., and G. J. F. MacDonald, Continentality and the gravitational field of the earth, *J. Geophys. Res.*, *65*, 2169–2172, 1960.

Strang van Hees, G. L., Gravity measurements on the Atlantic *Navado III*, paper presented at the General Assembly, IUGG, 14th Assembly, Lucerne, Switzerland, 1967.

Talwani, M., G. H. Sutton, and J. L. Worzel, A crustal section across the Puerto Rico Trench, *J. Geophys. Res.*, *64*, 1545–1555, 1959.

Talwani, M., Some recent developments in gravity measurements abroad surface ships, in *Gravity Anomalies: Unsurveyed Areas*, Geophys. Monograph 9, pp. 31–47, American Geophysical Union, Washington, D. C., 1966.

Talwani, M., J. Dorman, J. L. Worzel, and G. M. Bryan, Navigation at sea by satellite; *J. Geophys. Res.*, *71*, 5891–5902, 1966.

Vening Meinesz, F. A., *Gravity Expeditions at Sea, 1923–1938*, vol. 4, Netherlands Geodetic Commission, Delft, 1948.

Woollard, G. P., and W. E. Strange, Gravity anomalies and the crust of the earth in the Pacific Basin, *The Crust of the Pacific Basin*, Geophys. Monograph 6, pp. 60–80, American Geophysical Union, Washington, D. C., 1962.

Worzel, J. L., *Pendulum Gravity Measurements at Sea, 1936–1959*, John Wiley & Sons, New York, 1965.

Worzel, J. L., and G. L. Shurbet, Gravity anomalies at continental margins, *Proc. Natl. Acad. Sci.*, *41*, 458–469, 1955.

Gravity and Its Relation to Topography and Geology in the Pacific Ocean

Peter Dehlinger

FREE-AIR gravity anomalies have been measured in limited areas of the Pacific Ocean with gravity meters on surface ships and with pendulums in submarines. While the total area in the Pacific that has been surveyed gravimetrically is relatively small, the measurements and analyses indicate that both topography and regional geologic structures affect the free-air anomalies in a rather unpredictable manner.

The analyses in this discussion are based on measurements that the author and his associates have made with a LaCoste and Romberg gimbal-suspended surface-ship gravity meter. The root-mean-square uncertainty of these measurements is ± 5.5 mgal when navigation was based on Loran A and is somewhat lower when based on Loran C. Some of the included anomalies were measured by the U.S. Coast and Geodetic Survey, with estimated uncertainties between 5 and 10 mgal.

The areas described are: (1) off the coast of northern California, particularly the eastern Mendocino escarpment; (2) off the coast of Oregon, including the continental margin, the adjacent Cascadia abyssal plain, and the Ridge and Trough province west of the Cascadia plain; and (3) the Hawaiian archipelago and a continuation northward in the Pacific Ocean.

FREE-AIR ANOMALIES

The free-air anomalies indicate, as would be expected, that the broader areas which do not include large topographic features are nearly in isostatic equilibrium. This is illustrated by comparing the areas of smooth topography off northern California and Oregon in Figure 1 with the corresponding smooth gravity fields of near-zero free-air anomalies in Figures 2 and 3. A similar observation is made for the region north of the Hawaiian Islands.

In Figure 2, near isostatic equilibrium is indicated in the areas to the northwest and to the south of the eastern part of the Mendocino escarpment (the escarpment is represented by the east-west lineations in Figures 1 and 2 west of Cape Mendocino, California). In Figure 3, near equilibrium is indicated west of Oregon in the Cascadia abyssal plain (central and north-central parts of the map). The Hawaiian Islands in Figure 4 are seen to be out of equilibrium, although there is approximate equilibrium in the northern part of the figure. Figures 5 and 6 show that equilibrium is essentially achieved north of Hawaii along a north-south profile near 161°W longitude. Off the coasts of Washington and British Columbia, it has similarly been observed that near-zero free-air anomalies occur over most deep-water areas where topography is relatively flat (R. W. Couch, personal communication).

Much of the coasts of Oregon (Figure 3) and northern California (Figure 2), where anomalies are small, are also nearly in equilibrium. Even across the eastern part of the Mendocino escarpment, *Dehlinger et al.* [1967] concluded that the escarpment is produced by two juxtaposed different density mantle materials, overlain by crustal layers, in which the sections on both sides of the escarpment are essentially in isostatic equilibrium. West of the Cascadia plain (Figure 1), a buried mountain range extends northwesterly that is characterized by local free-air anomalies in Figure 3. It is the Ridge and Trough province of *Menard* [1964], which includes the Juan de Fuca Ridge of *Wilson* [1965]. Anomalies over the range vary with topography, but have an average value near zero. This suggests that the range as a whole is nearly in isostatic equilibrium. Such equilibrium is notably absent near the islands and large seamounts along the Hawaiian archipelago, where free-air anomalies are very large (Figure 4), demonstrating the presence of large mass excesses along the archipelago. Large positive anomalies across other portions of the

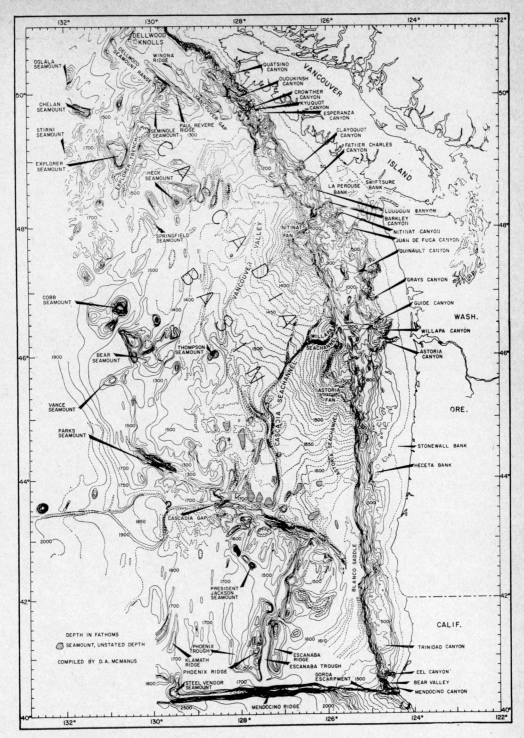

Fig. 1. Bathymetric map of the region west of the coast of northern California, Oregon, Washington, and Vancouver Island [*McManus*, 1964].

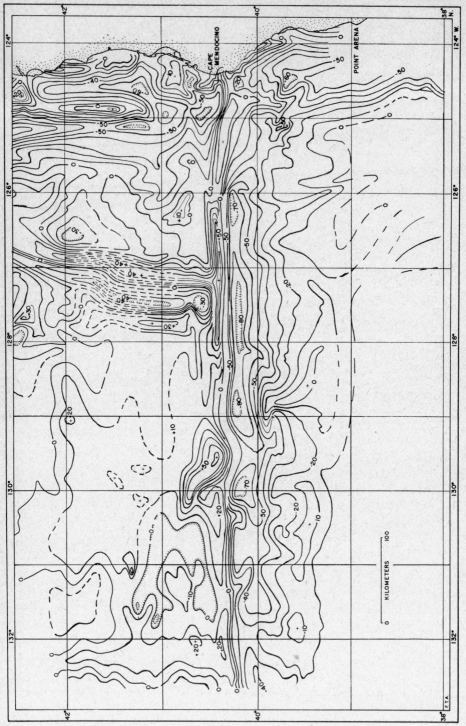

Fig. 2. Free-air anomaly map of the Mendocino area west of northern California [*Dehlinger et al.*, 1967]; contour interval, 10 mgal.

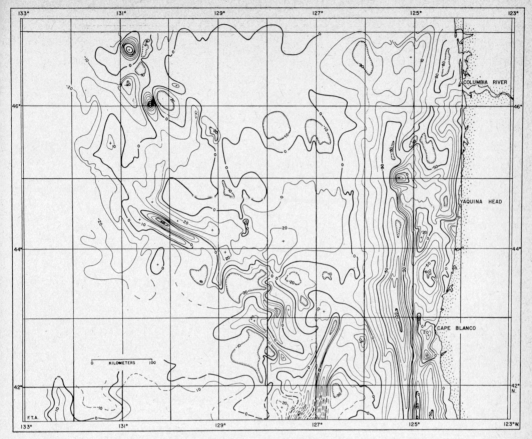

Fig. 3. Free-air anomaly map west of the Oregon coast; contour interval, 10 mgal.

Hawaiian Islands and the archipelago have been reported previously [*Vening-Meinesz*, 1941; *Woollard*, 1951, 1966; *Worzel*, 1965; *Strange et al.*, 1965].

Free-air anomalies at sea often conform approximately with local variations in bottom topography, although there are significant exceptions, especially in tectonically and volcanically active areas. The anomalies appear to be produced in part, at least, by geologic variations. Usually the geologic effects are smaller than the topographic effects because of the large density contrast between water and rock (approximately $2.8/1.0 = 1.8$ g/cm^3), which is 6 or more times greater than density contrasts of rocks (e.g., $2.84/2.67 = 0.27$ g/cm^3). The associated geologic variations may be large, however, and may extend to the lower crust and upper mantle.

There is, in general, a normal correlation between free-air anomaly and bottom topography; Figures 5 and 6 illustrate the relation in the North Pacific. Most of the topographic highs, usually seamounts, are characterized by positive free-air gravity anomalies; however, some of the gravity anomalies cannnot be correlated with topographic features.

Combined topographic and geologic effects on gravity anomalies are illustrated by crustal sections across the eastern part of the Mendocino escarpment in Figures 7 and 8. These are hypothetical two-dimensional crustal sections [*Dehlinger et al.*, 1967] that are consistent with bottom topography, seismic refraction data, and free-air anomalies. A large part of the anomaly across the escarpment is caused by bottom topography, as is observed by comparing the Bouguer [*Dehlinger et al.*, 1967, p. 1240] and free-air anomaly maps (Figure 2) of the area. The Bouguer anomaly variation across

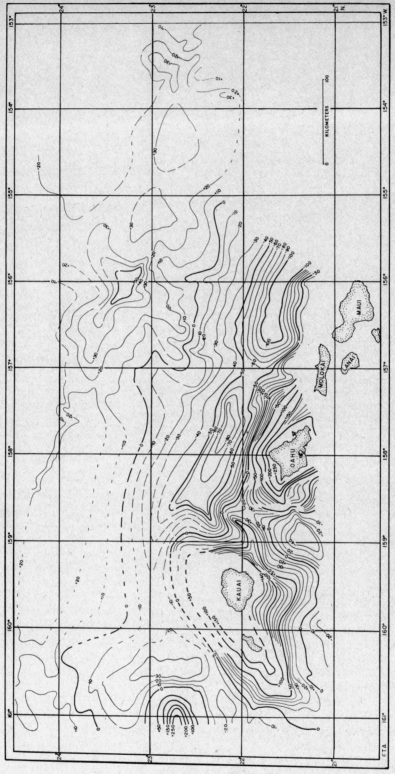

Fig. 4. Free-air anomaly map of the Hawaiian area; contour interval, 10 mgal.

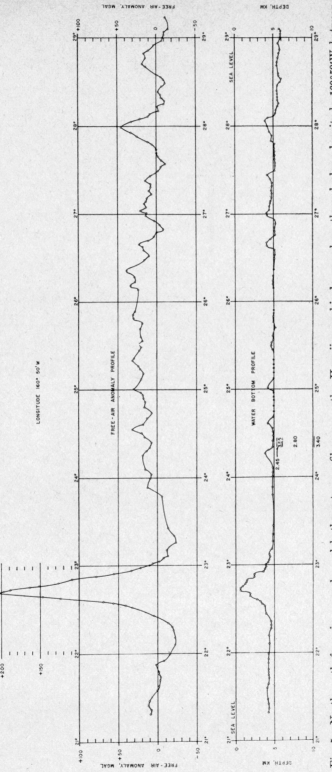

Fig. 5. North-south free-air anomaly and bathymetry profiles across the Hawaiian archipelago and northward along longitude 160°50'W between latitudes 21° and 29°N, line 10 of *Surveyor* 1964 measurements.

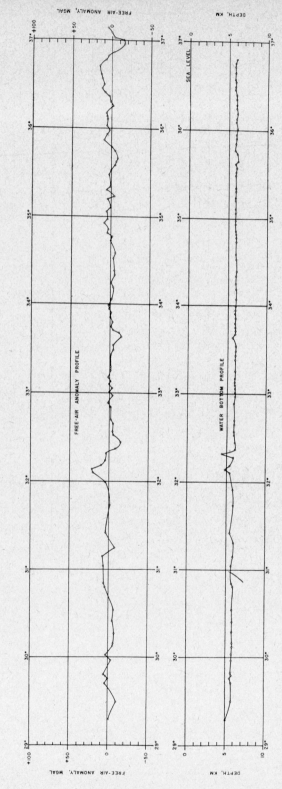

Fig. 6. North-south free-air anomaly and bathymetry profiles in the North Pacific Ocean along longitude 160°35'W between latitudes 29° and 37°N, line 6 of *Surveyor* 1964 measurements.

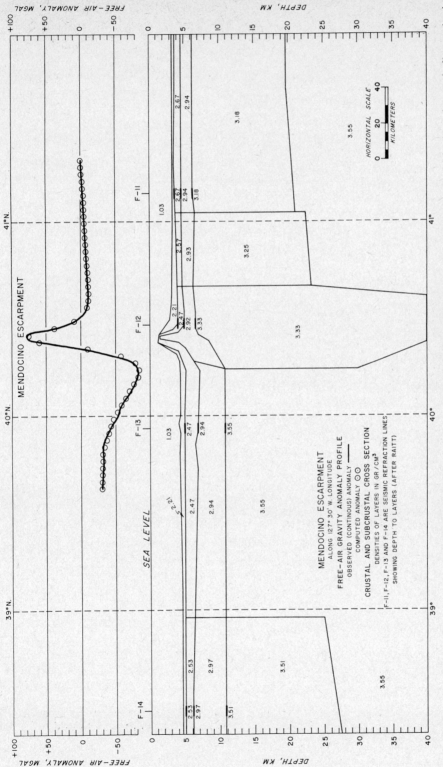

Fig. 7. North-south cross section over the Mendocino escarpment, along 127°30'W meridian, that is consistent with free-air anomalies, seismic refraction data [Raitt, 1963], and bathymetry [Dehlinger et al., 1967].

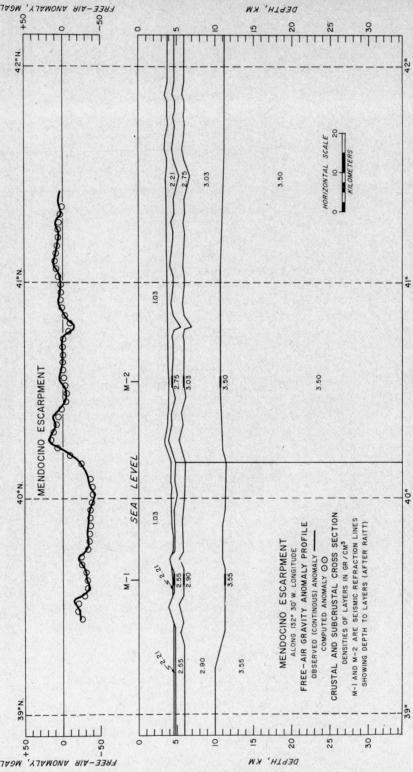

Fig. 8. North-south cross section over the Mendocino escarpment, along 132°30'W meridian, that is consistent with free-air anomalies, seismic refraction data [*Raitt*, 1963], and bathymetry [*Dehlinger et al*, 1967].

the escarpment is about 25 mgal, while the corresponding maximum free-air variation is 160 mgal. The two maps show that the Bouguer anomalies, hence the geologic variations, account for about 1/5 as much of the free-air anomalies over the escarpment as do the topographic variations. Figures 7 and 8 illustrate how large the geologic changes across the escarpment appear to be. (Alternative interpretations of these cross sections, based on minor changes in geologic variations, have been discussed by *Dehlinger et al.* [1967a].)

In Figure 1, the mountain range (Ridge and Trough province) that extends northwesterly along the western part of the map is irregular in height, averaging about 0.5 to 1 km above the adjacent abyssal plains, and contains large seamounts and one significant deep. (Cobb Seamount, which rises within 40 meters of sea level near 46°N latitude, 130°25'W longitude, has a free-air anomaly of +175 mgal, as is shown in Figure 3. The Blanco fault, trending southeasterly toward Cape Blanco on the Oregon coast, is characterized by an oceanic deep and a −40-mgal anomaly near 44°25'N latitude 130°10'W longitude.)

Most of the free-air anomalies over the mountain range do not exceed 40 mgal and correlate approximately with local topography. Despite the higher elevation of the range, the average free-air anomaly is nearly zero. This average anomaly is interpreted to mean that a low-density compensating material exists beneath the range. A series of five in-line seismic refraction lines [*Shor et al.*, 1968] across the range have indicated that the crust is unusually thin near the center of the range (thickness of 7 km, including a water depth of 2.7 km), and that the shallow mantle has a low seismic velocity. The low velocity is interpreted as a low-density upper mantle layer under the range. Preliminary Bouguer anomalies (unpublished) over the range indicate the presence of a minimum anomaly along the center of the range.

The Cascadia abyssal plain off Oregon (Figure 1) is relatively flat because of sediment deposition. The free-air anomalies (Figure 3) are near zero and the gravity field is smooth, indicating that geologic structures beneath the plain are uniform. This is substantiated by several seismic refraction lines shot in the plain [*Shor et al.*, 1968].

Along the continental margin west of Oregon (Figure 1), the free-air anomalies (Figure 3) are typically negative on the seaward side of the continental slope and positive on the shelf side. This is caused by the combined effect of the continental margin and the transition from continental to oceanic crusts. Calculations are currently being made across the transition zone, tying the gravity anomalies to seismic refraction data in order to determine the variations in geologic structures across the continental margin [*Dehlinger et al.*, 1968].

Along the shelf off Oregon (Figures 1 and 3) between the coast and continental slope, large negative free-air anomalies extend northward from Cape Blanco, a coastal headland that consists of dense volcanic rock. These linear negative anomalies occur over relatively smooth topography and are caused by a thick section of low-density sedimentary rock that is located between the coast and the continental slope.

In contrast to relations observed off Oregon and northern California, the free-air anomalies in the Hawaiian area depend more upon regional topographic and geologic variations. Figure 4 shows large positive free-air anomalies near the Hawaiian Islands. These anomalies continue as very large positive Bouguer anomalies on the islands [*Woollard*, 1951; and personal communication]. The positive anomalies are not due to topography alone; they indicate that the islands represent large mass excesses.

A simplified two-dimensional north-south hypothetical crustal section, shown in Figure 9, has been constructed across the archipelago at 161°20'W, west of the island of Kauai. This section is consistent with bottom topography and free-air anomalies. The layering is assumed to be the same as that obtained from refraction shooting [*Shor and Pollard*, 1964] to the east of the profile. Geologic variations in the figure are assumed to be in the form of changes in layer thicknesses, rather than due to changes in densities of layers. On this assumption, the crustal section shows a shallow depth to mantle under the archipelago, where the free-air anomalies are very large (+175 mgal). The resultant mantle depth of 18 km in the figure is quite similar to mantle depths estimated by *Vening-*

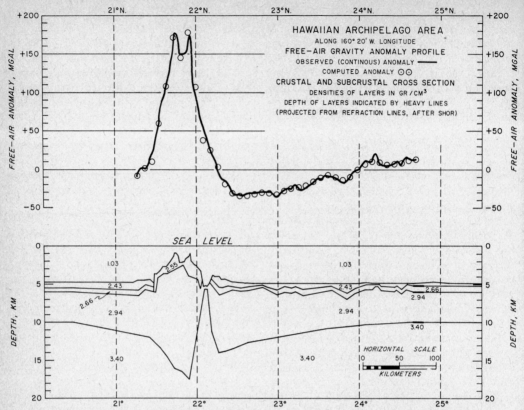

Fig. 9. North-south cross section across the Hawaiian archipelago, along 160°20'W meridian, that is consistent with gravity anomalies and bathymetry, and tied to the seismic refraction data of *Shor and Pollard* [1964].

Meinesz [1941], *Shor* [1960], *Worzel* [1965], and *Strange et al.* [1965] at other parts of the archipelago. For isostatic equilibrium to exist, mantle depths would need to be approximately 30 km with the assumed density contrasts.

Figure 9 also shows a narrow oceanic deep on the north side of the archipelago (22°10'N latitude) which is not associated with a local negative free-air anomaly. For this trough to exist without a negative anomaly, it must be compensated by heavy underlying rocks. The compensating material would be high-density rock within the crust or mantle, or a mantle that extends upwards, possibly up to the ocean bottom itself. The figure illustrates the latter condition.

A broad negative free-air anomaly extends along the north side of the archipelago (Figure 4). North of the islands of Oahu and Maui this anomaly is associated with a moat of deep water. Northwest of the island of Kauai (Fig-

ure 4), the negative anomaly (22°40'N latitude) is not associated with a topographic low. There it is due to geologic features alone, probably to a relatively deep mantle under the moat as shown in the figure. It is quite likely that the negative anomaly north of Oahu and Maui is produced by both a relatively deep mantle and the effect of the moat.

North of the moat lies the northwesterly trending Hawaiian arch, characterized by positive free-air anomalies (between +20 and +50 mgal in Figure 4). Seismic refraction results [*Shor and Pollard*, 1964] have indicated that the mantle along the Hawaiian arch is relatively shallow, and that the velocities in the lower crust (6.7–7.0 km/sec) and upper mantle (8.0 to 8.1 km/sec) have rather typical values. This suggests that the crustal and mantle rocks have correspondingly typical densities along the arch. The positive free-air anomalies appear to be produced by a shallow mantle with a rather normal composition. Indeed, a mantle 2 km shallower than normal would produce a positive anomaly of 10 to 30 mgal, depending on density contrasts and the shape of the arch. On this basis, it appears that the shallowest depth to mantle on the arch might occur where the gravity anomaly is largest, i.e., at +50 mgal, near 23°20'N latitude, 156°W longitude.

CONCLUSIONS

Both bottom topography and geologic variations affect the free-air anomalies in the Pacific, at least in the parts that are included in this study. The anomalies analyzed indicate that they are affected by geologic variations, some of which may be quite complex and may even extend throughout the crust and into the upper mantle. The free-air anomalies are frequently correlative with local topography, but there are many exceptions. The anomalies are generally not correlative with regional topography, the relationships varying with area and being dependent on the extent to which a region is in isostatic equilibrium. In the Ridge and Trough province west of Oregon and northern California, an underwater mountain range produces local but not regional free-air anomalies, because of the presence of a relatively low-density upper mantle layer under the range. In the Hawaiian area, mountains forming the archipelago produce large regional anomalies because the region is not in isostatic equilibrium. Across the eastern part of the Mendocino escarpment, where two juxtaposed mantle and crustal sections appear to be nearly in isostatic equilibrium, rather large anomalies are due to large local topographic features and to geologic variations that extend well into the mantle.

It is thus concluded that in the Pacific Ocean, local topography is a poor to unsatisfactory guide for estimating free-air anomalies, and regional topography is a still poorer guide for estimating these anomalies, even where the extent of isostatic equilibrium has been determined.

Analyses of gravity anomalies provide a powerful tool for estimating thicknesses of any existing anomalous-density upper mantle layers in areas where crustal thicknesses and upper mantle seismic velocities have been established from crustal refraction measurements.

Acknowledgments. The author wishes to thank Dr. Hyman Orlin of the U. S. Coast and Geodetic Survey for providing values of free-air anomalies northwest of Oahu and northeast of Kauai that were used in contouring a part of Figure 4. He wishes to acknowledge his associates, Messrs. R. W. Couch, M. Gemperle, W. A. Rinehart, Robey Banks, and John N. Gallagher, for their participation in the Oregon State University sea gravity program.

REFERENCE

Dehlinger, P., R. W. Couch, and M. Gemperle, Gravity and structure of the eastern part of the Mendocino escarpment, *J. Geophys. Res., 72*, 1233–1247, 1967.

Dehlinger, P., R. W. Couch, and M. Gemperle, Continental and oceanic structure from the Oregon coast westward across the Juan de Fuca ridge, *Can. J. Earth Sci., 5*, 1079–1090, 1968.

McManus, Dean A., Major bathymetric features near the coast of Oregon, Washington, and Vancouver Island, *Northwest Sci., 38*, 65–82, 1964.

Menard, H. W., *Marine Geology of the Pacific*, McGraw-Hill Book Company, New York, 271 pp., 1964.

Raitt, R. W., Seismic refraction studies of the Mendocino Fracture zone, *Abstracts of Papers, Intern. Assoc. Phys. Oceanog., 13th General Assembly*, Berkeley, vol. 6, p. 71, 1963.

Shor, G. G., Jr., Crustal structure of the Hawaiian ridge near Gardner Pinnacles, *Bull. Seismol. Soc. Am., 50*, 563–573, 1960.

Shor, G. G., Jr., P. Dehlinger, H. K. Kirk, and W. S. French, Seismic refraction studies off Oregon and northern California, *J. Geophys. Res., 73*(6), 2175–2194, 1968.

Shor, G. G., Jr., and D. D. Pollard, Mohole site selection studies north of Maui, *J. Geophys. Res., 69*, 1627–1637, 1964.

Strange, W. E., G. P. Woollard, and J. C. Rose, An analysis of the gravity field over the Hawaiian Islands in terms of crustal structure, *Pacific Sci., 19*, 381–389, 1965.

Vening-Meinesz, F. A., Gravity over the Hawaiian archipelago and over the Madeira area: Conclusions about the earth's crust, *Koninkl. Ned. Akad. Wetenschap. Proc., 44*, 1, 1941.

Wilson, J. T., Transform faults, oceanic ridges, and magnetic anomalies southwest of Vancouver Island, *Science, 150*, 482–485, 1965.

Woollard, G. P., A gravity reconnaissance of the island of Oahu, *Trans. Am. Geophys. Union, 32*, 558–568, 1951.

Woollard, G. P., Crust and mantle relations in the Hawaiian area, in *Continental Margins and Island Arcs, Can. Geol. Surv. Paper 66-15*, 294–310, 1966.

Worzel, J. L., *Pendulum Gravity Measurements at Sea, 1936–1959*, John Wiley & Sons, New York, 1965.

Gravity Anomalies over Volcanic Regions[1]

Alexander Malahoff

CENTERS of volcanism are typified by mass anomalies that can be related to the lithology of the rocks comprising the volcanic pile and the internal structure and mass distribution within and beneath the pile. Because most volcanoes have significant relief, terrain corrections are important. Complete Bouguer gravity anomalies are commonly used, because they eliminate the gravity effect of short wavelength changes in topographic relief that are incorporated in free-air gravity anomalies. If the density of the volcanic rocks is not known, the method of density profiling described by *Nettleton* [1940] can be used to define the probable density. This density value can then be applied to models of the entire region under study, theoretical gravity anomalies can be computed for different structural models, and the results can be compared with the observed anomalies. By successive approximations, the model is altered until a reasonable fit is obtained between the observed and calculated anomalies. The observed anomalies are usually residual Bouguer gravity anomalies determined by subtracting an empirically derived regional Bouguer gravity value from the actual Bouguer gravity anomaly value.

Volcanic regions and their gravity characteristics can be classified as follows:

1. Continental Regions such as the Basin and Range area of the United States, the plateau of Mexico, and the rift valleys of Africa, are made up of cinder cones and tuffaceous cones representing the volcanic or embryonic stage of volcanism. These have little if any gravity expression. As they are formed by explosive activity, the surficial material has a low density, and any gravity effect is negative in sign.

Composite andesitic volcanoes made up of ash, tuff, and rhyolite and andesite flows usually give rise to a negative gravity field that may be accentuated in the crater area if there has been a caldera formed by explosive action and collapse and foundering of the former cone into the magma chamber within the volcano. These commonly occur around the Pacific Ocean rim and mark the so-called Andesite line. They also are representative of the high volcanic mountains such as Mt. Etna in Sicily and rift valley volcanoes.

2. Plateau basalt areas and some composite volcanoes and volcanic centers have slightly negative or near zero residual Bouguer gravity anomalies associated with them. Although dense plugs like the Kimberly eclogites are found in association with some continental volcanoes, the gravity effect is usually small because of the limited diameter of the plug.

3. Mid-oceanic ridges and islands that are composed of tholeiitic basalt are all characterized by markedly positive Bouguer anomalies. This in large measure is due to the density contrast with the surrounding water. However, there is a secondary gravity effect in association with the caldera areas that in places exceeds +110 mgal. Local gravity highs of +10 mgal or higher are also found over dike-intruded rifts.

Interpretation is critical in considering gravity data for volcanoes, since no solution is unique. Because a volcanic pile is built up by accretion on an existing land surface, it is customary to regard the base area as being either underformed or at most regionally bowed down under the superimposed load of volcanic rock. Any change in crustal structure in association with volcanism would therefore normally be either that from crustal flexure or possibly formation of an anti-root due to melting and extrusion of former basal crustal material. Neither of these simple situations is found in Hawaii, the only area where it has been possible to obtain a crustal measurement beneath a major volcano. Crustal measurements at three widely separated

---

[1] Contribution 249 from the Hawaii Institute of Geophysics.

locations (Hawaii, Oahu, and Gardner Pinnacles) all show that there has been a marked thickening of the underlying crust (≈ 10 km). The seismic layering suggests that only about 2 km of this thickening can be attributed to subsidence; the rest appears to be a transformation of mantle material into crustal material [*Woollard*, 1965]. Whether there is a similar situation on the continents is not known. The only analogy is the Sierra Nevada batholith, which does have a well defined root.

Gravity Anomalies over Marine Volcanic Features

The Hawaiian ridge represents one of the most thoroughly studied island ridges. *Woollard* [1951] showed that each of the principal caldera areas of the Island of Oahu had an associated local Bouguer gravity high of about +110 mgal and that these anomalies could not be related to topography but had to be related to subsurface sources, presumably volcanic feeder pipes. Becaues the large gravity anomaly of +110 mgal associated with the caldera required a density contrast of about 1.0 g/cm³ and as the parent magma probably had a composition similar to that of peridotite (density about 3.3 g/cm³), the basaltic lavas could not have a density in excess of about 2.3 g/cm³. This low bulk density for Hawaiian basalts has since been verified by density measurements of both surface and borehole material [*Manghnani and Woollard*, 1965; *Kinoshita et al.*, 1963]. Three hundred samples of dike rocks collected from the exposed parts of the volcanic plugs of the Hawaiian Islands have densities between 3.0 and 3.2 g/cm³.

The normal coastal Bouguer gravity anomaly value, due to the topography of the Hawaiian ridge rising above the ocean floor, is about +200 mgal; superimposed upon this are the local anomaly effects associated with the centers of volcanism (calderas and rifts). The observed Bouguer anomaly relations along the Hawaiian ridge are summarized in Table 1 and shown in Figure 1.

Along the Hawaiian ridge, free-air anomalies at sea level range from +190 to +314 mgal,

TABLE 1. Bouguer Anomalies along the Hawaiian Ridge

Locality	Investigator	Anomalies, mgal	
		Coastal	Volcanoes
Hawaii	*Kinoshita*, 1965	+210 to +260	
Hualalai			+267
Kohala			+320
Mauna Kea			+332
Mauna Loa			+331
Kilauea			+316
Maui	*Kinoshita and Okamura*, 1965	+190 to +220	
West Maui			+252
Haleakala			+282
Lanai	*Krivoy and Lane*, 1965	+200 to +220	+257
Kahoolawe	*Furumoto*, 1965	+210 to +240	+245 to +252
Molokai	*Moore and Krivoy*, 1965	+235 to +240	
West Molokai			+273
Oahu	*Woollard*, 1951; *Strange et al.*, 1965	+195 to +203	
Waianae			+312
Koolau			+313
Kauai	*Krivoy et al.*, 1965	+240 to +270	+343
Niihau	*Krivoy*, 1965	+260 to +280	+292*
Nihoa	*Kroenke and Woollard*, 1965	+253 to +286	
French Frigate Shoals	*Wellman*, 1967	+285 to +305	
Laysan	*Kroenke and Woollard*, 1965	+270 to +290	
Lisianski	*Kroenke and Woollard*, 1965	+308 to +314	
Pearl and Hermes Reef	*Kroenke and Woollard*, 1965	+277 to +285	
Midway	*Kroenke and Woollard*, 1965	+284 to +307	+309†

* Rift.
† Closure in central lagoon.

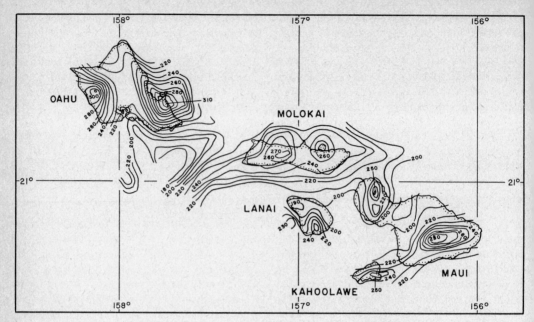

Fig. 1. Bouguer gravity anomalies in milligals of the southern Hawaiian ridge (based on a density of 2.35 g/cm³) [after *Strange et al.*, 1965].

the range noted on Oahu. On Oahu it is possible to differentiate the contribution of the volcanic pile (+190 to +200 mgal) from that of the plug plus the subsurface mass associated with the caldera (+313 mgal). It is reasonable to assume, therefore, that the subsurface mass distributions at French Frigate Shoals, Lisianski Atoll, and Midway Atoll, where anomaly values exceed +300 mgal, have the same origin as those found in association with the volcanic centers on Oahu, where supporting seismic evidence defines the existence of a dense volcanic plug at a depth of 2 km and the thickness of the crust underlying the island is ≈21 km.

As *Malahoff and Woollard* [1968] have shown, most of the volcanic masses of the Hawaiian ridge are essentially in hydrostatic equilibrium with each other. Changes in surface elevation and crustal thickness northwest of Oahu correspond closely to changes that would be expected hydrostatically on the basis of Archimedes principle. For example, if the mean surface elevation of Oahu is taken to be 0.4 km and the seismic depth of 21 km below sea level to the mantle is used with the density differential of 0.40 g/cm³ between crust and mantle, the section of mantle rock displaced in accordance with Archimedes principle [$R = H \times \sigma_c/\sigma_m$] is 17.9 km. The freeboard ($F = H - R$) is 2.5 km. (See p. 320 for explanation of symbols). If the same density values are used for Gardner Pinnacles, with surface elevation of −0.2 km at a crustal thickness of 16.9 km (obtained from seismic refraction data), the section of mantle displaced is 14.8 km and the freeboard value is 2.0 km. As the flotation level will be the same, the difference in freeboard values should correspond to the difference in surface elevation if both areas are in equilibrium. This difference is 0.5 km, which agrees closely with the actual value of 0.6 km.

If crustal migration has been a factor in forming the Hawaiian ridge, then the thickness of the crust involved beneath the ridge apparently has not been influenced by such migration; this implies translation of crustal layers in excess of 20 km in thickness. This is also substantiated by the similarity in gravity anomaly values for the caldera areas (+309 mgal), which agree closely with those on Oahu (+312 mgal). *Strange et al.* [1965] have shown that the anomaly on Oahu can be reconciled with the seismic data.

From a gravity study of the Cook Island group, which consists of eroded basaltic volcanoes, *Robertson* [1967a] concluded that most of his data were consistent with the calculated gravitational effects of uncompensated island platform (no thickening or deformation of the original crust) of average density 2.32 g/cm³ containing volcanic plugs of density 2.88 g/cm³. One island, in the northern Cook group, Rakahanga Atoll, exhibited anomalies consistent with an uncompensated island platform of uniform density (2.51 g/cm³) and no core. In computing the Bouguer anomalies, Robertson used the concept of 'modified' Bouguer anomalies, which were computed from the observed values of gravity by the normal methods. However, in order to aid in the interpretation of the gravity data, the sea water was not replaced by rock in the calculations. The terrain and Bouguer corrections were based on a density of 2.67 g/cm³ for coral areas. The necessity of having to apply a volcanic plug model within the island mass model is illustrated by Figure 2. In all his calculations, Robertson extended the volcanic plugs only to the level of the sea floor. Table 2 summarizes his observations.

In another paper, *Robertson* [1967b] points out that the gravity observations over the Cook island group are incompatible with the theory of local isostatic compensation.

The observation that oceanic volcanic masses do not obey the theory of local isostatic compensation is further substantiated by comparing Bouguer gravity anomalies and isostatic anomalies (Figure 3).

Surface-ship gravity measurements over isolated seamounts and seamount chains have not resolved plug effects, probably because of the depth of the source beneath the seamount and the integration of the plug effect with that of the topography. Seamounts with a general dome shape and with summits in water depths of a few kilometers have free-air gravity anomalies

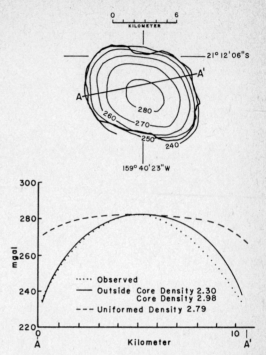

Fig. 2. Observed and calculated Bouguer gravity anomalies in milligals over Raratonga Island [after *Robertson*, 1967b].

ranging from a few milligals to a few tens of milligals. In their study of Jasper seamount, a conical seamount about 600 meters high in 4400 meters of water, 500 kilometers off the coast of San Diego, California, *Harrison and Brisbin* [1959] noted that an average density of 2.3 g/cm³ for the seamount gave the best fit to the observed free-air anomaly (Figure 4). This low apparent density is the same as that derived by *Woollard* [1951] for the Hawaiian ridge, and presumably the underlying crustal structure in the two areas is similar.

The development of reliable, continuous-read-

TABLE 2. Gravity Anomalies and Interpretations of Anomalies for the Cook Island Group*

Island	Peripheral Anomaly, mgal	Central Anomaly, mgal	Island Platform Density, g/cm³	Island Plug Density, g/cm³
Mangaia	220–230	260	2.21	2.79
Mauke	210–215	245	2.37	2.87
Mitiaro	220–230	255	2.40	2.90
Raratonga	230–240	280	2.30	2.98
Manihiki	185–200	225	2.34	2.84
Rakahnanga	182.5	187.5	2.52	None

* After *Robertson*, 1967a.

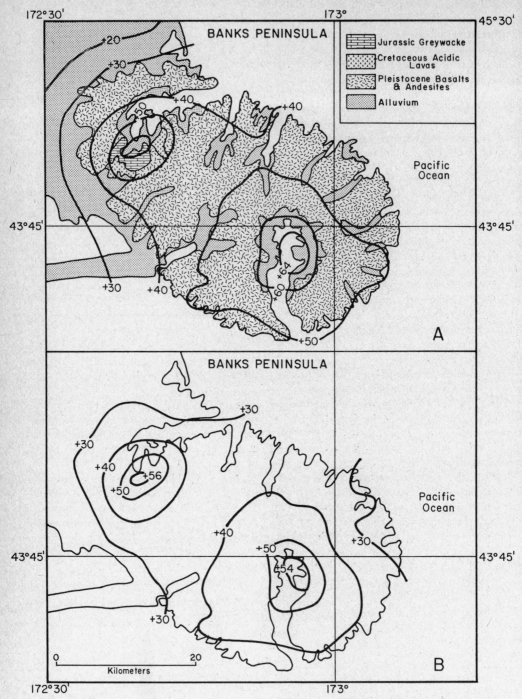

Fig. 3. Banks Peninsula, South Island, New Zealand: (A) geology and Bouguer gravity anomalies in milligals; (B) isostatic anomalies (based on a density of 2.67 g/cm³ and normal crustal thickness $T = 30$ km in the Airy-Heiskanen System) [after *Oborn and Suggate*, 1959; *Reilly*, 1966].

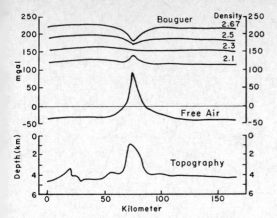

Fig. 4. Topographic, free-air, and Bouguer gravity profiles across seamount Jasper (anomaly values in milligals) [after *Harrison and Brisbin*, 1959].

ing gravimeters for surface ships has resulted in extensive marine gravity coverage over some of the mid-oceanic rises and ridges. The crests of rises such as the Mid-Atlantic ridge are areas of active volcanism. According to the theory of sea-floor spreading [*Vine*, 1966; see also pp. 578 ff. in this monograph], new crust is being generated by volcanic action and intrusion at the crests of the ridges, pushing the older parts of the surrounding oceanic crust laterally away from the crest of the ridge.

The relation of gravity anomalies to crustal structure over the Mid-Atlantic ridge has been discussed by *Talwani et al.* [1965], as well as by *Worzel and Harrison* [1963]. Over the Mid-Atlantic ridge, the free-air anomaly values are dominantly positive. The values tend to become negative as the ocean basins are approached. The Bouguer gravity anomalies (Figure 5) were computed by Talwani et al. using a differential density (basement rock density — water density) of 1.57 g/cm^3. The Bouguer anomalies in effect correct the free-air anomalies for the gravity deficit caused by the water layer and thus reflect more directly the effect of the sub-bottom structure. The short wavelength anomalies (up to 50 mgal) that are prominent in the free-air anomaly and are located largely over seamounts disappear almost completely in the Bouguer anomaly curve. This indicates that the small-scale anomalies are caused by the bathymetric features alone, not by internal or deep-seated structures. The minimum in the Bouguer anomaly over the crest of the Mid-Atlantic ridge suggests isostatic compensation of the ridge, but, as Worzel and Harrison have shown, this can be either through a crustal root, or an anti-root. The seismic data are indeterminate. There is an absence of a prominent Bouguer gravity anomaly over the central rift valley; thus the gravitational effect of any subsurface masses associated with the rift valley is very small at the sea surface.

The gravity anomaly pattern over the East Pacific rise (Figure 6) in its broad characteristics is similar to that over the Mid-Atlantic ridge [*Talwani et al.*, 1965]. The free-air anomalies average zero with small positive values over the crest of the East Pacific rise, but tend to become negative on either side. The Bouguer anomaly has a minimum over the crest of the rise and maxima on either side. The free-air anomalies and the pattern of Bouguer anomalies therefore also suggest isostatic compensation of the East Pacific rise. The crustal structure, though, shows upbowing of the crust, which would suggest that the compensation is achieved through low density upper mantle material. Again, no distinctive anomalies are present over the volcanically active crest of the rise.

In recent years many papers have been written relating heat flow and volcanism over the East Pacific rise. *Von Herzen and Uyeda* [1963] found that there were two narrow belts of high heat flow along the crest of the East Pacific rise. There appears to be no obvious correlation of the heat-flow anomalies with either the gravity or the magnetic anomalies.

Gravity Anomalies over Continental and Island Arc Regions of Volcanism

The most intensive and thorough studies of the volcanoes of the circum-Pacific volcanic belt have been carried out in Japan (Figure 7).

In a gravity survey of the Kuttyaro caldera lake of Hokkaido, *Yokoyama and Tajima* [1959] observed a conspicuously low Bouguer gravity anomaly over the central parts of the caldera (Figure 8). This Bouguer anomaly amounted to —46 mgal and extended over 20 km. This caldera is typical of the so-called Krakatoa type, having been formed after the

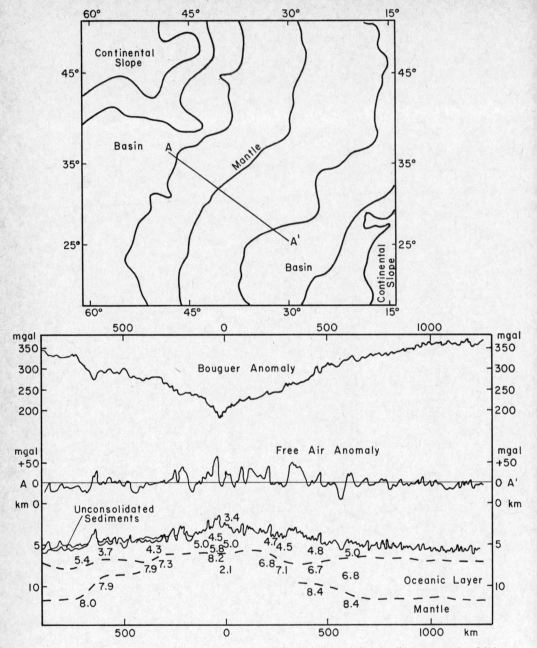

Fig. 5. Bouguer (assumed density, 2.60 g/cm³) and free-air anomalies across the Mid-Atlantic ridge. *Top:* location of profile. *Bottom:* crustal structure from seismic evidence and gravity profiles. Velocities of crustal layers in kilometers per second [after *Talwani et al.,* 1965].

ejection of a large volume of pumice and breccia. Interpretation of the Bouguer gravity anomaly suggests that coarse acidic breccia of negative density contrast of 0.3 to 0.5 g/cm³ has accumulated within the caldera to a depth of 3-4 km beneath the caldera floor. This interpretation is further substantiated by the fact that the gravity anomaly is concentric with the caldera and of the same wavelength. The total mass defect calculated from the Bouguer anomaly matches the calculated mass of the ejecta consisting of pumice and ignimbrite flows.

The Aira caldera in the southern part of Kyushu is one of the largest calderas in the world, measuring 20 kilometers in diameter. An active volcano, Sakurajima (1118 meters), stands on the southern edge of the caldera. No obvious Bouguer gravity anomalies are associated with Sakurajima; apparently it does not have a dense plug. Pumice and ignimbrite flows cover the area surrounding the caldera and are assumed to have been ejected during the Pliocene. The −45-mgal local Bouguer gravity anomaly is concentric with the caldera and can be interpreted in terms of accumulated breccia within the caldera reaching down to a depth of 3–4 km. A negative density contrast of 0.3

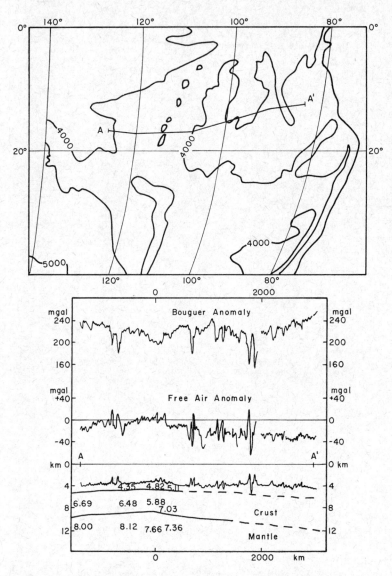

Fig. 6. Bouguer (assumed density, 2.60 g/cm³) and free-air anomalies across the East Pacific rise. *Top:* location of gravity profile. *Bottom:* crustal structure from seismic evidence and gravity profiles. Velocities of crustal layers in kilometers per second [after *Talwani et al.*, 1965].

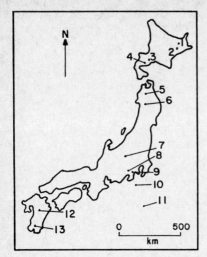

Fig. 7. The principal calderas of Japan: (1) Kuttyaro, (2) Akan, (3) Sikotu, (4) Toya, (5) Towada, (6) Tazawa, (7) Asama, (8) Fuji, (9) Hakone, (10) Oosima, (11) Hatizyo, (12) Aso, (13) Aira.

to 0.5 g/cm³ of the breccia with the basement rock has been calculated. The calculated mass deficiency as derived from the Bouguer gravity anomaly is 4.0×10^{10} metric tons and matches closely the mass of the ejected material.

Low residual Bouguer gravity anomalies (using a basement rock density of 2.67) concentric with the caldera and equivalent to a mass deficiency nearly equal to the mass of ejecta were also reported for the Oosima, Toya, Hakone, Aso, and Sikotu calderas of Japan [*Yokoyama*, 1963b]. *Yokoyama and Tajima* [1960] reported no obvious residual anomaly over Fuji (3776 meters). They concluded that the volcano, which is composed of mixed andesitic-basaltic rocks and is 30 km in diameter at the base, is not isostatically compensated, is supported by the rigidity of the earth's crust, and has no obvious dense plug or pipe effects. No Bouguer gravity anomalies were observed over Asama (2542 meters) or Fuji volcanoes [*Yokoyama*, 1961].

Yokoyama [1958, 1963a] was able to quantitatively compute the mass deficiencies over the principal calderas of Japan by using the theorem of Gauss [the mass deficiency $\Delta M = (1/2\pi G) \int_{-\infty}^{+\infty} \int_{-\infty}^{+\infty} \Delta g\ (x,\ y)\ dx\ dy$, where Δg is the residual Bouguer gravity anomaly associated with the mass deficiency]. Usually

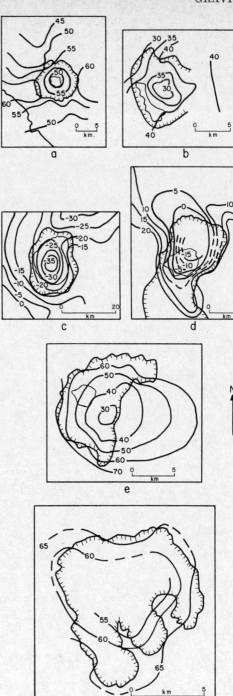

Fig. 8. Bouguer gravity anomalies in milligals over the principal calderas of Japan: (a) Toya, (b) Hakone, (c) Aso, (d) Aira, (e) Kuttyaro, (f) Towada.

the mass deficiency equalled the mass of the ejecta. The results of the Japanese studies can be summarized as follows:
 1. The low gravity anomalies (concentric with the caldera) observed over the Japanese calderas are due to accumulations of breccia to a depth of 3-4 km beneath the floor of the caldera.
 2. Faults on the rims of the calderas generally slope inward.
 3. Violent ejection of breccia and tuff and caldera subsidence account for the structures of the calderas.

The gravity data and interpretations of the observed anomalies over the Japanese volcanic centers are summarized in Table 3.

Medi and Morelli [1952] made a detailed gravity study of the island of Sicily. Although the island is dominated by a large negative Bouguer gravity anomaly of −90 mgal (a 2.67 density reduction is assumed), no distinctive anomalies are associated with the volcanic center of Mt. Etna (3100 meters), although it does lie along the eastern extension of the gravity low (Figure 9). The negative anomaly observed over the central part of Sicily may reflect a volcano or tectonic depression; Medi and Morelli suggest that it is due to a tectonic depression filled with low-density laminated clays.

Mt. Vesuvius is along the northern edge of a Bouguer gravity low that is concentric with Naples Bay, reaching a minimum of about −45 mgal at the central edge of Naples Bay. It has no distinct gravity anomaly, since it is a composite volcano; the vent is filled with ash and breccia, as well as denser feeder dikes.

Mt. Egmont (North Island, New Zealand) is an isolated andesitic cone, nearly 2500 meters high and 44 km wide at its base; it does not have an appreciable residual Bouguer anomaly (as deduced from the 1:4,000,000 scale gravity map of New Zealand [*Reilly*, 1965]).

Mount Rainier, (a 4350-meter dormant volcano) is on the western edge of a −70 mgal regional low, representing an acidic batholith underlying the Cascade Mountain Range in the United States. *Danes* [1964] calculated the batholith to be 30 km wide and extending for a depth of 14 km. Locally Mt. Rainier is associated with a residual Bouguer gravity low of about 30 mgal. This low is interpreted by Danes as being due to a granodioritic core beneath Mt. Rainier.

In many cases the individual volcanoes and their associated gravity anomalies are within a broader volcanic region characterized by many centers of volcanism and often by general graben-like subsidence of the crust. One such well studied area is the Central volcanic region of the North Island of New Zealand. The Central volcanic region, 150 km long and 40 km wide, is the center of the North Island. Quaternary basaltic, andesitic, and rhyolitic volcanism is present in this region. However, ignimbrite, rhyolite, and pumice deposits cover the greatest

TABLE 3. Gravity and Related Data for Japanese Volcanic Centers

Caldera	Diameter	Age	Principal Ejecta Material	Maximum Residual Bouguer Anomaly, mgal	Volume of Ejecta, km³	Mass Deficiency, 10¹⁰ tons
Oosima	4	Pleistocene	Basalt	+15		
Kuttyaro	20–26	Pleistocene	Ignimbrite	−45	100	7.8
Akan	13–24	Pleistocene	Ignimbrite	−25		
Aso	17–25	Pliocene	Ignimbrite	−20	200	4.0
Aira	23–24	Pleistocene	Ignimbrite	−45	230	10
Towada	10		Ignimbrite	−13		
Sikotu	30		Ignimbrite	−20	120	2.8
Hakone	11		Ignimbrite	−10	8	0.66
Toya	12		Ignimbrite	−14	20	0.66
Volcano	Height		Rock Type			
Asama	2542	Holocene	Andesite	None		
Fuji	3776	Holocene	Basalt and andesite	None		

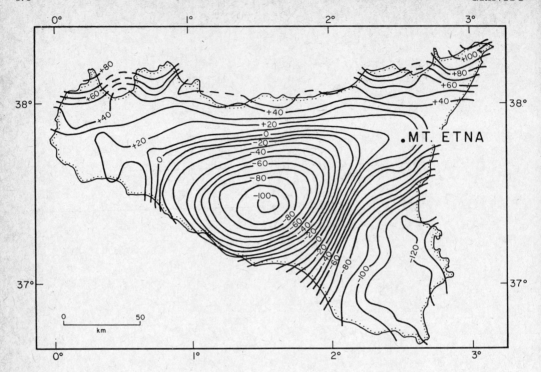

Fig. 9. Terrain-corrected Bouguer gravity anomalies in milligals over Sicily (assumed density, 2.67 g/cm³) [after *Medi and Morelli*, 1952].

area. The central part of this volcanic region is characterized by a regional Bouguer gravity low (Figure 10) and by a graben-like crustal structure. Within the volcanic region, upthrown rocks are characterized by residual Bouguer anomalies above −35 mgal and the downthrown depressions by anomalies below −35 mgal. *Modriniak and Studt* [1959] interpreted the negative residual Bouguer gravity anomalies in this region in terms of mixed volcanics of density 2.12 overlying greywacke basement rock of density 2.67 (Figure 10). Generally, the saturated mean densities used were as follows: pumice, tuff, breccia, 1.77 g/cm³; ignimbrite, 2.21 g/cm³; rhyolite, 2.18 g/cm³; dacite, 2.37 g/cm³; andesite, 2.47 g/cm³; basalt, 2.79 g/cm³. The relationship between low Bouguer anomalies and accumulations of low-density acidic or intermediate volcanic material with a horst and graben-like tectonic environment is obvious in the Central volcanic region of New Zealand. Another area of continental volcanism showing similar relationships is the Mono basin of California [*Pakiser et al.*, 1960]. Here a −60-mgal residual Bouguer gravity low is associated with a Cenozoic graben-like depression 21 by 15 km. The anomaly contours are concentric with the basin and can be interpreted in terms of a 3.5-km-thick accumulation of ejecta, tuff, and sediments of average density 2.3 g/cm³.

A residual Bouguer gravity anomaly −58 mgal in amplitude occurs in the hydrothermally active Yellowstone National Park in Wyoming (Figure 11). The source of the anomaly appears to be a breccia- and rhyolite-filled subsidence basin with a basaltic basement rock [*Malahoff and Moberly*, 1968]. The basin is roughly 68 km in width, and the associated anomaly is 150 km in wavelength.

The calderas of the volcanic regions of the world contrast with the volcanically active African rift valleys. The rift valleys are structurally similar to the mid-oceanic ridges; in terms of the hypothesis of sea-floor spreading, they are the centers of an expanding crust. *Sutton and Grow* [1969] found that the regional Bouguer gravity anomalies over the rift valleys surrounding Lakes Albert, Edward, Kiva, and

Tanganyika range between −100 and −140 mgal. The residual Bouguer gravity anomalies over the rift valleys are sharply defined, ranging between −70 and −80 mgal. Sutton and Grow interpret these anomalies in terms of a sediment- or breccia-filled graben about 30 km wide bounded by high-angle normal faults 60° or steeper in dip (Figure 12) penetrating to the mantle. The sharpness of the negative anomaly over the rift valleys suggests that the source of the anomaly is within the upper 5 km of the graben floor and is due to the mass deficiency created by the density contrast between the sediments (average density 2.17) and the basement rock (assumed average density 2.67 g/cm³). Some of the normal faults marginal to the rift valleys penetrate to the mantle and act as conduits for magma; they are marked on the surface by volcanic eruptives. Sutton and Grow's interpretation of the origin of the Bouguer gravity anomalies now seems more realistic than the interpretation of *Bullard* [1936], who analyzed the gravity data in terms of isostatic anomalies and therefore assumed that the negative isostatic anomalies had been generated by 'keystone'-like downthrusting of crustal material into the mantle due to lateral compression. The additional gravity data of

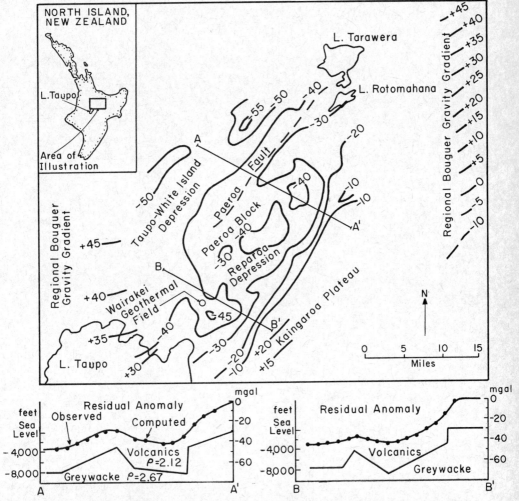

Fig. 10. Central volcanic region, North Island, New Zealand. *Top:* Bouguer gravity anomalies in milligals (assumed density, 2.0 g/cm³). *Bottom:* Structural interpretations of the gravity anomalies [after *Modriniak and Studt*, 1959].

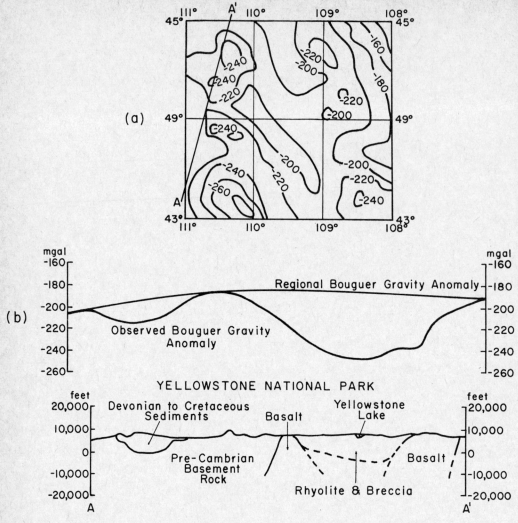

Fig. 11. Terrain-corrected Bouguer gravity anomalies (assumed density, 2.67 g/cm³) over the Yellowstone National Park, Wyoming, and the suggested geological profile [after *Malahoff and Moberly*, 1968].

Sutton and Grow indicate that rift valleys are tensional features, that the Mohorovicic discontinuity beneath the entire region is probably uniform at relatively uniform depths (36–39 km), and that the underlying mantle has normal density of 3.4 g/cm³.

Techniques Used in Analysis of Anomalies

1. As was mentioned previously, Bouguer gravity anomalies located over volcanoes and volcanic regions can be interpreted in terms of mass deficiency by using the Gaussian theorem.

2. Accurate rock densities are essential to logical interpretations of the gravity anomalies in terms of geological structure. Direct sampling is desirable; however, approximate densities can be obtained from seismic velocities by using the Nafe-Drake curve [*Talwani, Sutton, and Worzel*, 1959] or various modifications of the Nafe-Drake curve, utilizing laboratory density-seismic velocity relationships as illustrated by *Rose et al.* [1968].

3. Theoretical Bouguer gravity anomalies can be computed from geological sections con-

structed as density models by the use of rapid calculation techniques utilizing computers [*Talwani, Worzel, and Landisman*, 1959]. The theoretical and the terrain-corrected observed Bouguer anomalies can then be compared, and the density model can be adjusted until there is a reasonable fit between the two anomaly curves.

4. The use of free-air and isostatic anomalies in the study of the source of gravity anomalies over the volcanic regions is marginal in value. Volcanic features are usually of rugged terrain; therefore the free-air gravity anomalies usually reflect terrain rather than geological structure. The volcanic regions and their associated vol-

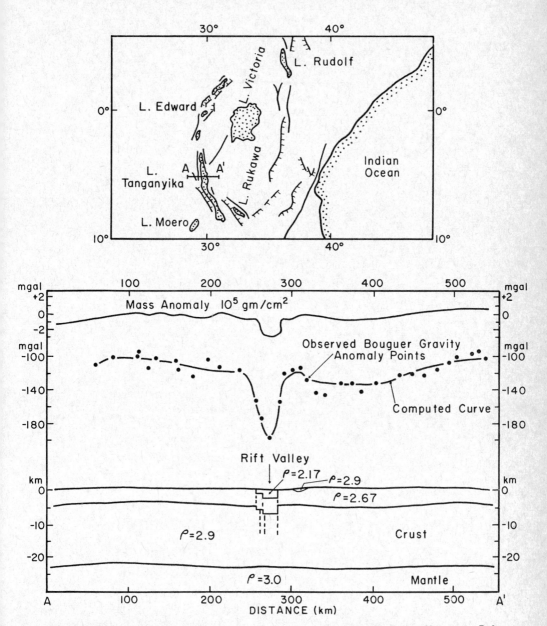

Fig. 12. African rift valleys. Observed and computed gravity anomaly profiles across Lake Tanganyika. Gravity values in milligals (assumed density, 2.67 g/cm³) [after *Sutton and Grow*, 1969].

canic features are usually too small to be reflected in any isostatic anomalies. Over submarine volcanic features, however, it is best to use the free-air anomaly observed at sea level for the study of the internal geological structure of these features.

Conclusions

1. The terrain-corrected Bouguer anomaly can be usefully employed in the study of the internal geological structure of volcanic features.

2. Basaltic volcanic islands and oceanic island ridges are characterized by Bouguer gravity anomalies of more than +200 mgal and by central residual anomalies of more than +30 mgal. The central residual anomalies can be interpreted in terms of dense dike-intruded volcanic plugs with densities about 0.4 g/cm³ higher than the average density of the volcanic cone or mass.

3. Usually no distinct Bouguer gravity anomalies are observed above continental acidic or intermediate volcanoes.

4. Continental volcanic depressions are characterized by local concentric negative Bouguer anomalies −10 mgal in amplitude or greater. The source of the anomalies lies in the low densities of the ejecta and breccia that fills these features and their vents to depths of 5 km.

5. Gravity anomalies observed over the Mid-Atlantic ridge can best be interpreted in terms of density variations in the lower crust and upper mantle. There are no obvious Bouguer anomalies associated with the central, volcanically active rift.

6. Bouguer gravity anomalies observed over the African rift valleys show local lows as large as −70 mgal. These anomalies can best be interpreted in terms of low density sediments 3–4 km thick occupying the graben floor, which was formed by the downward movement of the crust along high angle normal faults extending probably to the mantle. A local positive Bouguer anomaly of about +10 mgal over the normal faults suggests the presence of dense dike-like bodies within the fault zones; the data are too sparse to verify this suggestion.

REFERENCES

Bullard, E. C., Gravity measurements in East Africa, *Phil. Trans. Roy. Soc. London, A, 235,* 445–531, 1936.

Danes, Z. F., Gravity survey of Mount Rainer, Washington, *Trans. Am. Geophys. Union, 45*(4), 640, 1964.

Furumoto, A. S., A gravity survey of the Island of Kahoolawe, Hawaii, *Pacific Sci., 19*(3), 349, 1965.

Harrison, J. C., and W. C. Brisbin, Gravity anomalies off the west coast of North America: Seamount Jasper, *Bull. Geol. Soc. Am., 70,* 929–934, 1959.

Kinoshita, W. T., A gravity survey of the Island of Hawaii, *Pacific Sci., 19*(3), 339–340, 1965.

Kinoshita, W. T., H. L. Krivoy, D. R. Mabey, and R. R. MacDonald, Gravity survey of the Island of Hawaii, *U. S. Geol. Surv. Prof. Paper 475-C,* 114–116, 1963.

Kinoshita, W. T., and R. T. Okamura, A gravity survey of the Island of Maui, Hawaii, *Pacific Sci., 19*(3), 341–342, 1965.

Krivoy, H. L., A gravity survey of the Island of Niihau, Hawaii, *Pacific Sci., 19*(3), 359–360, 1965.

Krivoy, H. L., M. Baker, and E. E. Moe, A reconnaissance gravity survey of the Island of Kauai, Hawaii, *Pacific Sci., 19*(3), 354–358, 1965.

Krivoy, H. L., and M. P. Lane, A preliminary gravity survey of the Island of Lanai, Hawaii, *Pacific Sci., 19*(3), 346–348, 1965.

Kroenke, L. W., and G. P. Woollard, Gravity investigations on the Leeward Islands of the Hawaiian ridge and Johnston Island, *Pacific Sci., 19*(3), 361–366, 1965.

Malahoff, A., and R. Moberly, Effects of structure on the gravity field of Wyoming, *Geophysics, 33*(5), 781–804, 1968.

Malahoff, A., and G. P. Woollard, Magnetic anomalies and the geological structure of the Hawaiian archipelago and the Molokai and Murray fracture zones, in *The Sea,* vol. 4, John Wiley & Sons, New York, in press, 1968.

Manghnani, M. H., and G. P. Woollard, Ultrasonic velocities and related elastic properties of Hawaiian basaltic rocks, *Pacific Sci., 19*(3), 291–295, 1965.

Medi, E., and C. Morelli, Rilievo gravimetrico della Sicilia, *Ann. Geofis., 5*(2), 209–245, 1952.

Modriniak, N., and F. E. Studt, Geological structure and volcanism of the Taupo-Tarawera district, *New Zealand J. Geol. Geophys., 2*(4), 654–684, 1959.

Moore, J. G., and H. L. Krivoy, A reconnaissance gravity survey of the Island of Molokai, Hawaii, *Pacific Sci., 19*(3), 343–345, 1965.

Nettleton, L. L., *Geophysical Prospecting for Oil,* McGraw-Hill Book Company, New York, 444 pp., 1940.

Oborn, L. E., and R. P. Suggate, Sheet 21, Christchurch—Geological map of New Zealand, 1:250,000 series, New Zealand Dept. Sci. Ind. Res., Wellington, 1959.

Pakiser, L. C., F. Press, and M. F. Kane, Geophysical investigation of Mono basin, California, *Bull. Geol. Soc. Am., 71,* 415–448, 1960.

Reilly, W. I., Gravity map of New Zealand, 1:4,000,000 Bouguer anomalies, New Zealand Dept. Sci. Ind. Res., Wellington, 1965.
Reilly, W. I., Sheet 21, Christchurch—Gravity map of New Zealand, 1:250,000 series, New Zealand Dept. Sci. Ind. Res., Wellington, 1966.
Robertson, E. I., Gravity survey in the Cook Islands, New Zealand J. Geol. Geophys., 10(6), 1484–1498, 1967a.
Robertson, E. I., Gravity effects of volcanic islands, New Zealand J. Geol. Geophys., 10(6), 1466–1483, 1967b.
Rose, J. C., G. P. Woollard, and A. Malahoff, Marine Gravity and magnetic studies of the Solomon Islands, in The Crust and Upper Mantle of the Pacific Area, Geophys. Monograph 12, edited by L. Knopoff et al., pp. 379–410, American Geophysical Union, Washington, D. C., 1968.
Strange, W. E., L. F. Machesky, and G. P. Woollard, A gravity survey of the Island of Oahu, Hawaii, Pacific Sci., 19(3), 350–353, 1965.
Strange, W. E., G. P. Woollard, and J. C. Rose, An analysis of the gravity field over the Hawaiian Islands in terms of crustal structure, Pacific Sci., 19(3), 381–389, 1968.
Sutton, G. H., and J. A. Grow, A gravity study of the Western Rift Valley of Africa, to be published, 1969.
Talwani, M., X. Le Pichon, and M. Ewing, Crustal structure of the mid-ocean ridges: Computed model from gravity and seismic refraction data, J. Geophys. Res., 70(2), 341–352, 1965.
Talwani, M., G. H. Sutton, and J. L. Worzel, A crustal section across the Puerto Rico Trench, J. Geophys. Res., 64(10), 1545–1555, 1959.
Talwani, M., J. L. Worzel, and M. Landisman, Rapid gravity computations for two-dimensional bodies with application to the Mendocino submarine fracture zone, J. Geophys. Res., 64(1), 49–59, 1959.
Vine, F. J., Spreading of the ocean floor: New Evidence, Science, 154, 1405–1415, 1966.
Von Herzen, R. P., and S. Uyeda, Heat flow through the Eastern Pacific Ocean Floor, J. Geophys. Res., 68(14), 4219–4249, 1963.
Wellman, P., The aeromagnetic anomalies and the bathymetry of the central part of the Hawaiian ridge, New Zealand J. Geol. Geophys., 10(6), 1407–1423, 1967.
Woollard, G. P., A gravity reconnaissance of the Island of Oahu, Trans. Am. Geophys. Union, 32, 358–368, 1951.
Woollard, G. P., Crust and mantle relations in the Hawaiian area, in Continental Margins and Island Arcs, Can. Geol. Surv. Paper 66-15, 294–310, 1965.
Worzel, J. L., and J. C. Harrison, Gravity at sea, in The Sea, vol. 3, pp. 134–174, John Wiley & Sons, New York, 1963.
Yokoyama, I. Gravity survey on Kuttyaro Caldera Lake, J. Phys. Earth, 6(2), 75–79, 1958.
Yokoyama, I., Gravity survey on the Aria Caldera, Kyushu, Japan, Nature, 191(4792), 966–967, 1961.
Yokoyama, I., Gravity anomaly on the Aso caldera, in Geophysical Papers Dedicated to Prof. K. Sassa, pp. 687–692, Geophysical Institute, Kyoto University, Kyoto, 1963a.
Yokoyama, I., Structure of caldera and gravity anomaly, Bull. Volcanol., 26, 67–72, 1963b.
Yokoyama, I., and H. Tajima, Gravity survey on the Kuttyara caldera by means of a Worden gravimeter, Nature, 183(4663), 739–740, 1959.
Yokoyama, I., and H. Tajima, A gravity survey on Volcano Huzi, Japan, by means of a Worden gravimeter, Geofis. Pura Appl., 45(1), 1–12, 1960.

Recent Movements of the Earth's Crust and Isostatic Compensation

E. V. Artyushkov and Yu. A. Mescherikov

RECENT movements of the earth's crust include slow (secular) tectonic uplifts and subsidences, displacements and tilting that can be detected by oceanographic (tide-gage), geodetic, geophysical, geomorphological, and astronomical methods. The term *recent movements* usually refers to movements that can be detected in a period not longer than 50–70 years, though in some cases 250–300 years (the Netherlands settling, the Fennoscandia uplift).

Vertical Tectonic Movements

Vertical movements are probably the best studied type; geodetic and tide gage observa-

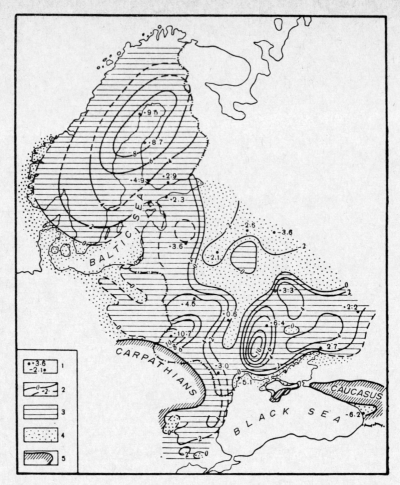

Fig. 1. Schematic map of recent movements of the earth's crust in eastern Europe and Fennoscandia: (1) rates of modern uplift (+) and subsidence (−) of the earth's crust in millimeters per year; (2) isolines showing the rates of recent movements of the earth's crust (contour interval, 2 mm/year); (3) areas of crustal uplift; (4) areas of crustal subsidence; (5) outlines of young (Alpine) mountain structures (compiled by Mescherikov).

tions of such movements in the western part of the European territory of the USSR [*Gerasimov and Filipov*, 1958], Finland [*Kääriäinen*, 1953], Poland [*Niewiarowski and Wyrzykowski*, 1962], and Bulgaria [*Hristov and Galabov*, 1962] have provided information for describing secular movements of the earth's crust over a vast territory from the north of the Scandinavian Peninsula to the Black Sea basin. Figure 1 shows that the rate of recent movements of the earth's crust in this area is on the average 2–4 mm/year. The maximum rate of uplift (∼10 mm/year) occurs in the central regions of Fennoscandia (the Baltic shield), in the region of Krivoy Rog (the Ukrainian shield), and in the near-Carpathian region.

In Fennoscandia and eastern Europe, uplift is associated mainly with positive elements of geological structure (arches), subsidence mainly with basins. There are, however, areas where recent movements do not coincide with geological structure but are of the opposite character. Of the areas involved in recent movement on the Russian platform, it is estimated that two thirds correspond in sign to the geological structure, one third are opposite (Table 1). Analysis of these movements in the Russian platform suggests block structure, with maximum differ-

TABLE 1. Correlation of Areas of Tectonic Basins and Arches with Areas of Modern Uplifts and Subsidences for the Western Part of the Russian Platform

Geostructures	Modern Movements	Area, %	Area Totals, %
Arches (+)	Uplifts (+)	44	69
Basins (−)	Subsidences (−)	25	
Arches (+)	Subsidences (−)	12	31
Basins (−)	Uplifts (+)	19	
		100	100

ential movement occurring at the boundaries between blocks.

The distribution of modern uplifts and subsidences of the earth's surface shows that the development of geostructures is still in progress. Glacio-isostatic processes are not solely responsible for the distribution of modern uplifts: rates of uplift in some extraglacial regions are at least as great as those in glacial regions. Geological and geomorphological data indicate that a glacio-isostatic factor is of great importance only during the first 5–7 thousand years after the melting of the glacier.

Small [1963] indicates that, in North America, the velocity of movement in the eastern (platform) part is ±3–5 mm/year (the Gulf Coast basin is subsiding while the Appalachian Mountains, the Ozark, and the Canadian shield are in the process of uplifting). In the western part, which is orogenic in character, the rate of movement is as great as 10–15 mm/year (the Rocky Mountains, Great basin, Coast Ranges).

Figure 2, a detailed map of recent uplifts and subsidences in the orogenic zone of the Japanese Islands, suggests that rates of vertical move-

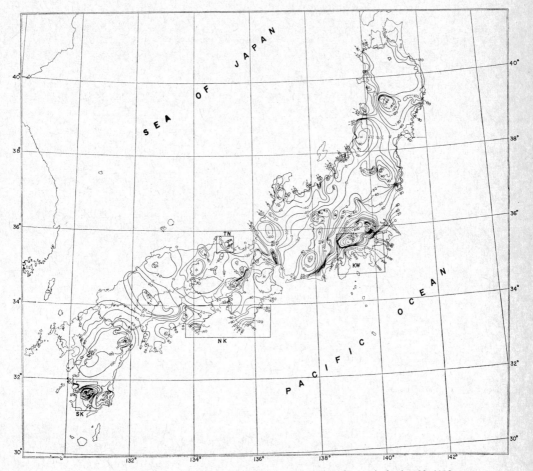

Fig. 2. Vertical movements of the earth's crust in Japan for the period of 1900–1928: contour interval, 20 mm [after *Miyabe et al.*, 1966].

ments in Japan, disregarding the areas subject to strong earthquakes where deformation of the earth's surface results from seismic processes, do not exceed 7–8 mm/year, the average value being ±3 mm/year, which approximates the rates of movement of the Russian platform. The movements in the Japanese Islands vary considerably, with correspondingly high differential gradients. High gradients of recent movements (10–100 or more times those of platform areas) seem to be a characteristic feature of orogenic zones. Thus prevailing gradients of the movements within the Russian platform are 10–30″. 10^{-3}/year, while values for orogenic areas of the Carpathians, Tien Shan, the Caucasus, and the Japanese Islands are often as high as 100–200″ · 10^{-3}/year. Areas where gradients of recent crustal movements are the highest are often associated with earthquake activity.

A harmonic analysis of the curves of recent movements of the earth's crust plotted in several profiles within the Russian platform and the Japanese Islands (with expansion in Fourier integral) showed that waves about 670, 290, 125, 90, and 45 km long can be distinguished on almost all the profiles [*Magnitsky*, 1965]. The occurrence of similar values of wavelengths for movements in platforms and orogenic regions suggests that recent movements of the earth's crust in the two regions with quite different tectonics have a common cause. However, the higher gradients in orogenic regions indicate a relative predominance of shorter wavelengths.

Intensity and even the direction of crustal movements undoubtedly change with time. Instrumental observations have been available for only the last two centuries. Indirect (archeological, geomorphological) data suggest that the period of secular oscillation is at least 500 to 700 years, or even longer.

The sharp changes in the sign and intensity of crustal movements detected in the mobile, orogenic regions are connected with earthquakes. The most detailed information on these changes was obtained in the Niigata earthquake region (Japan), where systematic releveling has been carried out since 1898 (Figure 3). These data, along with information, scanty as yet, from other earthquake regions, make it possible to distinguish three types of crustal movements in seismic regions: α, slow secular movements occurring in the long periods between the outbursts of seismic activity; β, accelerated move-

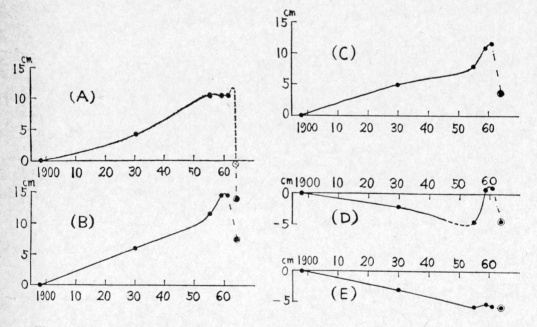

Fig. 3. Changes in the height of five typical points (*A, B, C, D, E*) in the Niigata earthquake region from 1900 to the end of 1964 [after *Tsubokawa et al.*, 1965].

ments that are forerunners of earthquakes and reflect deformation of the earth's crust during the preliminary stage of an incipient earthquake; and γ, quick movements connected with the earthquake and lasting till the seismic activity attenuates. The effect of β deformations that operate for a long period of time (10–15 years and longer) before the strong earthquakes is an important factor in connection with the problem of earthquake forecasting.

Horizontal Movements

Recent horizontal movements of the earth's crust have been studied less than vertical ones. Attempts to record the drift of continents during the recent epoch have yielded no conclusive results as yet; this problem remains to be solved on a global scale. Horizontal deformations of the earth's crust for certain areas, however, have been investigated rather well. Thus, observations over a period of many years in the region of the San Andreas fault in California show that blocks lying on opposite sides of the fault tend to move in opposite directions. The mean velocity of horizontal movements is 1 cm/year [*Whitten*, 1956]. A map of vectors of horizontal movements has been designed for the Japanese Islands, the velocity of the movements being several centimeters per year.

Horizontal displacements are, as a rule, one order greater than vertical movements. At the same time it is often possible to trace a connection between the two types of movement; this allows us to distinguish between areas of uplift and extension and areas of subsidence and compression. Vertical and horizontal movements can be regarded as components of three-dimensional deformations of the earth's crust [*Mescherikov*, 1963].

RECENT GLACIAL ISOSTATIC MOVEMENTS

The most recent epoch of continental glaciation in North America and Europe lasted about 40,000 years. The crust was depressed under the load of the ice sheet (several kilometers thick), and then began to recover its position after the ice sheet melted. The post-glacial uplifts of Fennoscandia, Canada, Alaska, and the islands of the Barents Sea have developed very quickly, reaching by now several hundred meters [*Flint*, 1957].

The Fennoscandia uplift has been investigated by *Sauramo* [1958] and others. The uplift has the form of a dome gradually rising from the outlying regions of Fennoscandia to the center, where its height is about 300 meters.

The outer boundary of the earth's crustal uplifts is parallel to the boundary of the latest glaciation, and its isobases (i.e., lines connecting points of equal uplift of shoreline) are concentric to the region where, judging by other characteristic properties, ice must have been the thickest. The rate of uplift (soon after the melting of the glacier) is estimated to be about 10 cm/year, whereas now it is only 1 cm/year in the center of the area. From an extrapolation of the curve of the uplift as a function of time, an additional uplift of 150–200 meters is expected [*Niskanen*, 1939].

There is a negative isostatic anomaly of several tens of milligals over the central part of the area [*Heiskanen*, 1936]. Nearer the border of the region of uplift, the isostatic anomalies are positive. Therefore, there is a qualitative correlation between modern movements and the gravity field. Post-glacial uplift of Canada shows a similar correlation [*Flint*, 1957].

Uplifts of Fennoscandia and Canada have for a long time been considered by most authors to be the result of isostatic rebound on the basis of the above distribution of isobases, the attenuation of their movements with time, and the tendency of gravity to decrease toward the center of the uplifts [*Gutenberg*, 1941]. Development of Holocene uplifts immediately after deglaciation in Alaska and on the islands in the Barents Sea [*Grossvald*, 1963] suggest that they are of isostatic nature also.

The isostatic nature of the uplifts of Fennoscandia and Canada is supported by other evidence [*Artyushkov*, 1966b]. In the first section of the paper, characteristic properties of platform tectonic movements were reviewed. Both Canada and Fennoscandia are platform areas, but the character of their uplifts differs from usual platform movements:

1. The rate of uplift at the initial stage ($\gtrsim 10$ cm/year) is at least an order of magnitude greater than typical rates of platform movements ($\lesssim 1$ cm/year).
2. The uplifts developed several hundred meters in about 10^4 years. Crustal displacements of this kind over large areas usually result from

platform movements lasting hundreds of thousands of years or longer.

3. The dimensions of the rising areas, particularly in Canada, considerably exceed typical dimensions of the platform areas characterized by intensive movements of one sign.

The total amplitude of post-glacial movements (movement to date, including the uplift under the glacier, plus expected future uplift) is in full agreement with the amplitude of isostatic settling of the crust (600–900 meters) under the load of the ice sheet.

CORRELATION BETWEEN RECENT MOVEMENTS AND GEOPHYSICAL FIELDS

Seismic Data

Many data have been accumulated by now on the correlation of recent movements with various geophysical fields. One can establish a rather definite dependence between seismic activity, on the one hand, and intensity and degree of contrast of recent movements, on the other [*Gzovsky et al.*, 1958].

Distribution of movements in the crust and the mantle causing recent movements of the crust are indicated by the depth of earthquake foci. The limited information on rheologic properties of various layers prevents us from reaching precise conclusions.

Comparatively new data on distribution of stresses in the earthquake foci for some areas of the earth (see p. 166) reveal, among other things, that regions of island arcs are zones of compression (which is on the average very close to horizontal), whereas central regions of middle ocean rises and continental rift valleys are under the influence of tensile stresses (which are close to horizontal). These results are of particular interest for interpreting recent movements.

Gravity Data

Artemyev [1966] found a relation between isostatic anomalies and various types of recent movements. The Alpine geosyncline zone and the regions of island arcs are characterized by extended belts of isostatic anomalies along the basic structural elements. The belts of negative anomalies (up to 200 mgal) are observed along deep-sea trenches and foredeeps. The axes of these anomalies are usually shifted to the inner parts of geosyncline zones. The island arcs and uplifts adjoining the foredeeps are characterized by positive anomalies of the order of 50–100 mgal. Positive isostatic anomalies are found in the area of the inner massifs of the Alpine geosyncline. These regions seem to have recently undergone considerable subsidence [*Kuenen*, 1950; *Beloussov*, 1962].

The distributions of isostatic anomalies in the region of island arcs and in the Alpine geosyncline zones are similar, the main difference being the magnitude of the anomalies. Thus the pattern of the gravity field suggests the similarity of the tectonic movements in the regions of island arcs and geosynclines.

Isostatic anomalies in platform areas are usually small (±40 mgal). We can thus distinguish between vast areas of tectonic activation (for example, central Asia, central regions of North America, eastern Africa) and platforms in the usual sense (for example, the Alpine platform of Europe, the northern and eastern parts of North America). Activated platforms have recently experienced considerable uplift, accompanied by intensive crushing and minor block faulting. They are characterized, on the average, by weak negative anomalies of the order of several tens of milligals or by anomalies close to zero, as is the case of the central part of North America. There is also some regularity of distribution of isostatic anomalies on ordinary platforms, which have not been subjected to activation. Uplifts are often accompanied by positive anomalies of the order of several tens of milligals, subsidences by negative anomalies of the same order but somewhat smaller. Finally, the areas of uplifts in Canada and Fennoscandia, where the isostatic equilibrium has recently been disturbed, are on the average characterized by negative anomalies.

Investigations of the isostatic state of the earth suggest that recent uplifts and subsidences are usually connected with anomalies that are, respectively, positive and negative, the exception being such structures as activated platforms, inner massifs, the regions of recent glaciation, and some others. Isostatic anomalies usually become more intensive with the increase in the amplitude of neotectonic movements.

The intensity of isostatic anomalies seldom exceeds 50–100 mgal, so that on the average they do not disturb the equilibrium of the earth's crust.

Vertical displacements of the crust can result from compression or expansion of the material in the mantle and crust, without causing any considerable horizontal displacement. Inflow and outflow of material can also be responsible for this movement; these two processes may develop together.

Let us assume that the earth's crust has been subjected to vertical displacement of ΔZ over a vast area (uplifts are given a positive sign). The change in the thickness of sediments (density is ρ_1) during this period equals Δh (for erosion Δh is negative).

Neither compression nor expansion changes the isostatic anomaly. If $\alpha \cdot \Delta Z$ is the part of the displacement attributed to the influx of the matter whose density is ρ_2, gravity will change by the value of $2\pi f \rho_2 \alpha \cdot \Delta Z$, where f is the gravity constant. With erosion and accumulation of deposits, it will result in additional Faye (free-air) anomaly equal to

$$\delta(\Delta g) = 2\pi f(\rho_1 h + \rho_2 \alpha \cdot \Delta z) \quad (1)$$

Let us consider the Fennoscandia uplift from this point of view. The mean free-air anomaly for the Fennoscandia region is now close to zero. When it began to melt, the glacier (approximately 2.5 km thick) must have been well compensated isostatically. Therefore the free-air anomaly was also close to zero at that time. Hence, $\delta(\Delta g) = 0$.

Introducing $h_1 = 2.5$ km, $\rho_1 = 0.9$, $\Delta Z = 0.5$ km (an approximate value of the complete uplift), and $\rho_2 = 3.4$ into (1), we obtain $\alpha \approx 1$. This indicates that the Fennoscandia uplift is connected with the flow of mantle material. A number of other recent movements can be subjected to similar analysis.

VISCOSITY OF THE UPPER MANTLE AND ISOSTATIC ADJUSTMENT OF THE CRUST

Most recent movements of the crust appear to be connected with the movement in the mantle. Determination of the type of these movements by various geophysical data requires information on distribution of viscosity at least in the upper layers of the mantle. Deviations of the crust from the equilibrium level and rate of its movements have been accurately known for glacial-isostatic uplifts. Thus information on viscosity in the upper mantle can also be obtained [Artyushkov, 1966a, b].

To define the character of the viscosity change with depth, let us consider the damping of disturbances on the free upper surface of a horizontal layer of the thickness H in an incompressible fluid, having high viscosity η and bounded below by a rigid layer. From hydrodynamic equations the characteristic time T for the damping of the disturbance is connected with the characteristic dimension L in the following way:

$$T \sim \frac{\eta}{\rho g} \cdot \frac{1}{L} \quad \text{if} \quad L \lesssim H \quad (2)$$

$$T \sim \frac{\eta}{\rho g} \cdot \frac{1}{H} \cdot \frac{L^2}{H^2} \quad \text{if} \quad L \gg H \quad (3)$$

where ρ is the density of fluid, and g is the acceleration of gravity.

Suppose that on the fluid surface there is a disturbance of dimension L with strong nonuniformities of dimension l ($l \ll L$). Because of the linearity of the hydrodynamic equations at high viscosities, the behavior of small nonuniformities can be considered independently of the general disturbance. As can be seen from equations 1 and 2, when $L \lesssim H$, small disturbances damp more slowly than the general disturbance. On the other hand, when $L \gg H$ (but $l \gtrsim H$), small nonuniformities will damp much more quickly than the general disturbance. Similarly, for $L \lesssim H$ the velocity of movement of the fluid surface under the influence of external loading increases with the area of load application; when $L \gg H$ (but $l \gtrsim H$), the inverse relation occurs.

Consider the evolution of the Fennoscandia uplift from the above point of view. Soon after the glacier disappeared the depression profile was rather irregular. However, small nonuniformities rapidly smoothed out and by the period of the Ancylus Lake, the profile became smooth. In particular, the deep subsidence at the center of the region had smoothed within a period not exceeding 700 years, at a mean velocity of 13 cm/year [Sauramo, 1958], while an appreciable decrease in the amplitude of the general depression required of the order of 10^4 years.

The analysis of late-glacial movements in Fennoscandia shows the similar rapid reaction of the earth's crust to a relatively slight change in glacial loading. At the periphery of the region under consideration soon after the glacier disappeared, the crust rapidly uplifted, the movement becoming slower as the glacier retreated. For instance, at the end of the first Yoldia Sea period in the northwest sea coast of Norway, the glacier retreated for a distance of the order of 100 km. During the same period of time (not exceeding 2000 years) the released territory uplifted by 100–150 meters, whereas its uplift during the subsequent 11,000 years was only 50–100 meters.

The next characteristic feature of the uplift is its intermittence in time as well as great independence of movement in adjacent regions. For example, in the period of the Echineis Sea, the velocity of the Lauhavuori hill uplift in Finland reached 40–50 cm/year at some periods (mean, 13–15 cm/year) [*Sauramo*, 1939]. The present movement is also more complicated than the average during the last thousand years [*Kääriäinen*, 1953].

The above facts indicate the existence in the upper mantle of an asthenosphere, i.e., a zone of low viscosity. A more detailed analysis shows that above the asthenosphere there is a considerably more viscous layer whose thickness does not exceed 100 km. This layer, including the crust and substratum, will be called the lithosphere.

The nonuniformity of the lithosphere does not allow the exact definition of the boundaries of the asthenosphere. The thickness of the asthenosphere is probably not more than 200 km. It is quite likely that this layer coincides with the low-velocity layer that is 80–200 km deep.

Detailed analysis of late-glacial Fennoscandia uplift suggests that the isostatic adjustment is due to the movement of mantle material through the asthenosphere. Large-scale isostatic movements can be described with sufficient accuracy by the following equation:

$$\frac{\partial \zeta}{\partial t} = C\left(\frac{\partial^2}{\partial x^2} + \frac{\partial^2}{\partial y^2}\right)\left(\zeta + \frac{\sigma}{\rho g}\right) \quad (4)$$

where

$$C = \frac{\rho g H^3}{12\bar{\eta}} \quad (5)$$

Here t is time, x and y are rectangular coordinates on the earth's surface, ζ is the surface deflection from the isostatic equilibrium level, σ is surface load on the crust, ρ is asthenosphere density, H is its thickness, and $\bar{\eta}$ is mean viscosity. For Fennoscandia, $C \sim 5.10^3$ cm^2/sec; if the thickness of the athenosphere is 100 km, $\bar{\eta} \sim 10^{20}$ poises. Data on other post-glacial uplifts testify to the fact that viscosity of the asthenosphere in platform regions is at least of the same order of magnitude. It should be noted that the above estimations refer to the diffusion viscosity associated with tectonic deformations but not to the intergranular viscosity responsible for the damping of seismic waves [*Zharkov*, 1960].

There is a sharp increase in mantle viscosity beneath the asthenosphere. Its value can be estimated as $\eta \gtrsim 10^{23}$ poises provided that tangential tension is lower than 100 kg/cm^2.

The existence of an asthenosphere was also concluded by *Takeuchi and Hasegawa* [1965] from investigations of gravity anomalies and the periods of uplift of Fennoscandia and Lake Bonneville. Their conclusion is based on the assumption that the second harmonic of the earth's gravitational potential is connected with the retardation in the adjustment of the crust to the decreasing rate of the rotation of the earth. As this adjustment is due to the flow of the mantle substance, it is possible to formulate some definite conclusions about its viscosity. Gravity anomalies, however, can also result from nonuniformity of the mantle underlying the asthenosphere [*McKenzie*, 1966]. We shall show here that it is the latter case that is more typical.

Investigations of isostatic uplift of the ancient Lake Bonneville in the Rocky Mountains in the United States result in the estimated value of $\eta \lesssim 10^{21}$ poises for the mantle viscosity for this region [*Crittenden*, 1963]. A more precise estimate cannot be made because the period of the uplift is known only approximately. The accuracy is further limited because Lake Bonneville was comparatively small and was in a region of intensive tectonic movements.

The lithosphere has little effect on isostatic movements of large areas. It is in fact a spherical shell whose thickness δ is small compared to the earth's radius R. Therefore, an external normal load of the order of σ per surface unit for the area of the characteristic dimension of $L \gg \delta$ results in stresses of the order of $(R/2\delta)\sigma \gtrsim 30\sigma$ in the lithosphere. Hence, when $L \gg \delta$, because of the finite strength of the lithosphere and creep, it is not able to sustain the load $\sigma \gtrsim 10$ kg/cm^2 for a long time without subsidence (isostatic compensation).

Therefore, in areas with dimensions of several hundreds of kilometers or more, the lithosphere, and consequently the earth's crust, must be essentially in isostatic equilibrium.

Analysis of post-glacial movements of Fennoscandia as well as of the uplift of the ancient Lake Bonneville suggest that the lithosphere effect on isostatic movements is limited to a distance of the order of 10^2 km or less. Isostatic movements over such distances occur mainly because of the displacement of individual blocks, which are apparently bounded by ancient faults.

SOME POSSIBLE MECHANISMS OF RECENT MOVEMENTS

Numerous mechanisms have been suggested to explain tectonic movements. We shall consider here only some of the mechanisms, making a brief reference to others. More detailed analysis can be found in works by *Beloussov* [1962, 1966], as well as in a paper by *Chadwick* [1962].

One of the causes suggested for explaining recent movements of the earth's crust is compression and expansion of the lithosphere [*Magnitsky*, 1965]. Thermal expansion or compression of lithospheric rocks can be responsible for movements of the order of 0.5 km for a period of the order of 10^8 years. Phase transitions with temperature variations result in the displacement of the surface of the crust up to 1–1.5 km in a period of the order of 10^6 years. Isostatic factors can increase the range of these movements by several times.

Inflow into the crust of some light material which migrated from the mantle and separation of the densified lower layers of the crust have also been suggested as causes. These processes lead, respectively, to increase and decrease in the crustal thickness, and are accompanied, respectively, by isostatic rising and settling. Formation of uplifts on activated platforms, as well as mid-ocean rises, seem to be the result of the inflow into the crust of large masses of comparatively light material that migrated from the mantle [*Artemyev and Artyushkov*, 1967].

At present, many authors connect the origin of main tectonics with large-scale convection current in the earth's mantle [*Heiskanen and Vening Meinesz*, 1958; *Runcorn*, 1962]. These currents are assumed to bring to the earth's surface (to be more exact, to the crust-mantle boundary) heat liberated from the core and the lower mantle. Tension of the crust in the zone of upward currents causes the formation of rift valleys, whereas downward currents result in the formation of island arcs and geosynclines in the compression zones. It has been suggested by many authors that convection currents in the mantle carry the crust along and bring about drift of continents.

The linear character of main tectonic formations of the earth, such as the Pacific Ocean belt and mid-ocean rises, is well explained by the thermal convection hypothesis. The conclusions drawn from this hypothesis are in qualitative agreement with the data on the distribution of stress in the earth (see p. 166). However, further detailed consideration of the problem reveals a number of contradictions.

A quite different mechanism of large-scale movements, 'gravity convection,' should exist if the earth's core is silicate [*Artyushkov*, 1968]. This mechanism is associated with the density differentiation of the earth substance. The main differentiation takes place at the core-mantle interface and is caused by metallization and liquifaction of silicates in the lower mantle. Heavy material subsides in the core and forms the inner core. Light material rises, thus making the lower mantle move, and forms the upper mantle.

Penetration of large masses of light, heated material into the upper layers of the earth may explain many tectonic processes, volcanism, and great inhomogeneity of the upper mantle in tectonically active regions. Gravity convection is a much more powerful mechanism than heat convection; if it exists it should practically exclude the latter phenomenon.

To conclude this section, let us analyze the tectonic movements and isostatic anomalies occurring in the platform regions where anomalies frequently reach a few tens of milligals, extending for hundreds of kilometers [*Artyushkov*, 1966a]. The crust in such regions is expected to be in a state very close to isostatic equilibrium. Therefore, the isostatic anomalies observed there should not be associated with nonuniformities in the crustal density. It is doubtful that such nonuniformities exist in the asthenosphere; because of its low viscosity, hydrostatic equilibrium could be restored in a period of the order of 10^3–10^5 years. Thus the main source of extensive isostatic anomalies occurring in the platform regions should be the mantle beneath the asthenosphere [*Magnitsky*, 1960]: in this region, the viscosity is greater than, or of the order of, 10^{23} poises, and density nonuniformities can have existed for of the order of 10^7 years.

The rate of the recent tectonic movements occurring on the platforms normally varies in the range from a few millimeters to one centimeter per year. If the direction (sign) of recent movements did not change, displacements of the crust in 10^5–10^6 years would be measured in kilometers; actually the displacements amount to only a few hundred meters. This can be explained only by a frequent change of the sign of the motion, even though one sign may predominate. Morever, platform uplifts are usually related to positive isostatic anomalies, while subsidences are associated with the negative anomalies.

All the above considerations suggest the following mechanism of platform movements, which agrees with the analyzed features. Let it be assumed that a block of a few hundred kilometers in width is rising in the mantle beneath the asthenosphere. The asthenospheric matter will move upward and spread sideways, causing the lithosphere to rise slightly. If the mantle beneath the asthenosphere is more dense, gravity over the block, in the region where the lithosphere rises, increases (we disregard now the processes pertaining to the lithosphere reorganization, including the crust; they cannot cause any appreciable changes in the isostatic anomalies).

Let us further assume that the block beneath the asthenosphere rises at a variable rate, for example, because of nonuniform rheological properties of the mantle (intermittence is typical of most tectonic movements). How much the lithosphere goes up does not depend on the displacement of the block but is governed by how quickly the asthenosphere over it spreads, since it is the viscous tension in the latter that supports the lithosphere. The spreading rate of the asthenosphere, in turn, depends on the rate with which the block goes up. Therefore, a more rapid upheaval of the block causes an increased displacement of the crust from its equilibrium; slowing in the upheaval of the block leads to a decrease in the displacement. Thus, an intermittent process of upheaval of the block will cause an alternate uplift and subsidence of the crust that is much more rapid than the average motion over a long period.

To cause crustal upheaval whose amplitude is of the order of a hundred meters and which is not compensated isostatically, the viscous tension in the asthenosphere must be tens of kilograms per square centimeter. To maintain such tensions for $\bar{\eta} \sim 10^{20}$ and L of the order of a few hundreds kilometers, the asthenospheric flows should be of the velocity of $\sim$1–10 cm/year. To achieve this, the amplitude of the total displacement of the block beneath the asthenosphere would be tens of kilometers. This example indicates that immense movements in the mantle can correspond to comparatively small movements of the crust.

QUATERNARY GLACIAL ISOSTATIC MOVEMENTS

An earlier section dealt with the recent glacial isostatic movements. Having knowledge of the manner in which the crust deforms under the surface load, one can maintain that similar movements are likely to have also occurred in connection with older Quaternary glaciations. During the Quaternary period, large ice sheets repeatedly appeared in North America and in the north of Eurasia. These glaciations lasted for rather short periods. The time of their advances and retreats can be compared with the characteristic time of isostatic adjustment of extensive regions. Therefore, the glaciations can be assumed to be accompanied by rapid movements that restored isostasy and by a considerable variation in the level of adjacent territories. Let us consider some consequences of the glacial isostatic movements to the periglacial regions [*Artyushkov*, 1967].

The thickness of the glacier rapidly increases from the edge, reaching 1 km or so 50–100 km from the edge [*Vyalov*, 1960]. Since the isostasy is of a regional nature, the weight of the glacier causes crustal subsidence beyond the glacier, the width of the zone of subsidence extending over at least a few tens of kilometers [*Gladun et al.*, 1963]. The regions farther away, however, become uplifted owing to the movement of asthenospheric substance squeezed from beneath regions of the glacier. Such changes in the relief occurring on the plains near the glacier result in formation of periglacial basins.

The retreating glacier leaves behind an extensive and deep subglacial depression. As can be seen from the Fennoscandia example, at a distance of several hundred kilometers from the former location of the glacier edge, the regions are released which are shifted in relation to the equilibrium state by several hundreds of meters. This value is sufficient even for formation of marine transgression.

Thus, simultaneously with glaciations in flat regions in the Quaternary period, there existed large basins where most of the Quaternary sediments of periglacial areas were formed. The material contained in the glacier was either deposited directly or was carried away by icebergs and streams flowing down from the glacier. In the deep-water regions of the basins ungraded deposits of the moraine type were formed. Far from the glacier in shallow water fine-grained deposits were laid down which were the materials for formation of covering loam and loess. As the glacier retreated, deep regions of the basin changed into shallow marginal regions. This causes the transition from ungraded deposits to coarsely grained deposits, and further to fine-grained deposits. This transition is typical of the Quaternary deposits. Those formed in relatively small basins should be of rather variegated character.

Let us determine the possible location of large Quaternary basins in the USSR. The European glacier, when at its maximum stage, seems to have reached the hills of northern Germany and to have joined the glaciation of the Polar Urals [Markov et al., 1965]. One can assume the formation of a vast fresh-water basin between the glacier and the near-glacier uplift. Its southern boundary is apparently thought to coincide with the boundary of the maximum extent of glaciation.

In the north of the Russian plain as well as in the north of the West Siberian lowland and on the North Siberian lowland, Quaternary marine transgressions occurred at about the same time as the glaciations [Strelkov, 1965]. These transgressions lasted for a very short period of time—one can suppose for not longer than a hundred thousand years or even a few tens of thousand years. Within a period not longer than 10^5 years, an area exceeding a million square kilometers subsided and then was lifted up again a few hundred meters. This extremely high rate of movement testifies to its isostatic character, since the time required for nonisostatic movements of comparable size occurring in the platform regions measures millions of years. Moreover, the marine transgressions that took place here resulted from the crustal warping due to the weight of the ice sheets.

Acknowledgment. The authors thank V. A. Magnitsky for valuable comments in connection with this review.

REFERENCES

Artemyev, M. E., *Isostatic Gravity Anomalies and Some Problems of Their Geological Interpretation,* Nauka, Moscow, 138 pp., 1966.

Artemyev, M. E., and E. V. Artyushkov, Isostasy and tectonics, *Geotectonics, 1967* (5), 41–57, 1967.

Artyushkov, E. V., On the character of viscosity changes in the upper mantle with depth, *Izv. Akad. Nauk SSSR. Ser. Fiz. Zemli, 1966* (8), 8–21, 1966a.

Artyushkov, E. V., On the isostatic equilibrium of the earth's crust, *Ann. Acad. Sci. Fennicae, A III, 90,* 455–466, 1966b.

Artyushkov, E. V., On the isostatic adjustment of the earth's crust, *Izv. Akad. Nauk SSSR, Ser. Fiz. Zemli, 1967* (1), 3–16, 1967.

Artyushkov, E. V., Gravity convection in the earth's interior, *Izv. Akad. Nauk SSSR, Ser. Fiz. Zemli, 1968* (9), 3–17, 1968.

Beloussov, V. V., *Main Problems of Geotectonics,* Moscow, 608 pp., 1962.

Beloussov, V. V., *The Earth's Crust and the Upper Mantle of the Continents,* Nauka, Moscow, 123 pp., 1966.

Beloussov, V. V., Some problems of evolutions of the earth's crust and the upper mantle of the oceans, *Geotectonics, 1967* (1), 3–14, 1967.

Chadwick, P., Mountain building hypotheses, in *Continental Drift,* edited by S. K. Runcorn, pp. 195–234, Academic Press, New York, 1962.

Crittenden, M., New data on the isostatic deformation of Lake Bonneville, *Geol. Surv. Profess. Papers, 454-E,* 31 pp., 1963.

Ewing, M., X. Le Pichon, and J. Ewing, Crustal structure of the mid-ocean ridges, 4. Sediment distribution in the South Atlantic Ocean and the Cenozoic history of the Mid-Atlantic ridge, *J. Geophys. Res., 71,* 1611–1636, 1966.

Flint, R. F., *Glacial and Pleistocene Geology,* John Wiley & Sons, New York, 553 pp., 1957.

Florensov, N. A., Neotectonics of the near-Baikal area associated with its seismic characteristics, *Bull. Seismol. Board, 10,* 11–20, Publishing House for USSR Academy of Sciences, Moscow, 1960.

Gerasimov, I. P., and Yu. V. Filipov, editors, Recent vertical movements of the earth's crust over the eastern half of the European part of the USSR, *Trans. Central Res. Inst. Geodesy, Aerial Photog., Cartography*, 123, Moscow, 298 pp., 1958.

Gladun, V. A., P. A. Stroev, S. A. Ushakov, and A. I. Frolov, Geophysical investigations on the earth's crust structure in the zone of transition from the Antarctic continent to the Indian Ocean between the 55° and 100° eastern longitude. *Dokl. Akad. Nauk SSSR*, 153(2), 427–428, 1963.

Grossvald, M. G., Ancient coastal lines of Franz Josef Land and late-Quaternary history of its glacial cover, *IGY Glaciological Research*, 9, pp. 119–144, Publishing House for USSR Academy of Sciences, Moscow, 1963.

Gutenberg, B., Changes in sea level, postglacial uplift and mobility of the earth's interior, *Bull. Geol. Soc. Am.*, 52, 721–772, 1941.

Gzovsky, M. V., V. N. Krestnikov, I. L. Nersesov, and G. I. Reisner, Comparison between tectonics and seismicity of the Garmsky region of the Tajik Soviet Socialist Republic, *Izv. Akad. Nauk SSSR, Ser. Geofis.*, 1958(8), 959–976; (12), 1425–1442, 1958.

Heezen, B. C., The deep-sea floors, in *Continental Drift*, edited by S. K. Runcorn, Academic Press, New York, pp. 235–288, 1962.

Heiskanen, W. A., Das problem der isostasie. *Handbuch Geophys.*, 1(4), 878–951, 1936.

Heiskanen, W. A., and F. A. Vening Meinesz, *The Earth and its Gravity Field*, McGraw-Hill Book Company, New York, 470 pp., 1958.

Hristov, K., and I. Galabov, Mitteilungen über eine vorlaufige Untersuchung über die vertikalen Bewegungen der Erdkruste in Bulgarien, *1st Intern. Symp. uber rezente Erdkrusten bewegungen*, pp. 314–316, Berlin, 1962.

Kääriäinen, E., On the recent uplift of the earth's crust in Finland, *Fennia*, 77(2), 106 pp., 1953.

Kuenen, P. H. *Marine Geology*, John Wiley & Sons, New York, 568 pp., 1950.

Lyustikh, E. N., Gravimetric investigations on oscillatory movements of the earth's crust and some results obtained thereby, *Izv. Akad. Nauk SSSR. Ser. Geol.*, 1948(6), 1948.

Lyustikh, E. N., Neomobilism and convection in the earth's mantle, 1, Pros and cons of mobilism and convection, *Byul. Mosk. Obshchestva Ispytatelei Prirody, Otd. Geol.*, 40(1), 5–26, 1965a.

Lyustikh, E. N., Neomobilism and convection in the earth's mantle, 2, Agreement between the convection and continental drift hypothesis, *Byul. Mosk. Obshchestva Ispytatelei Prirody, Otd. Geol.*, 40(2), 5–21, 1965b.

Magnitsky, V. A., On the interpretation of the main irregularities of the earth's gravity field, *Trans. Inst. Phys. Earth, USSR Acad. Sci.*, 11(178), 79–85, 1960.

Magnitsky, V. A., *Internal Structure and Physics of the Earth*, Nauka, Moscow, 379 pp., 1965.

Markov, K. K., G. I. Lasukov, and V. A. Nikolaev, *Quaternary Period*, 3 vols., Moscow University Press, Moscow, 1965.

McKenzie, D. P., The viscosity of the lower mantle, *J. Geophys. Res.*, 71(16), 3995–4010, 1966.

Mescherikov, Yu. A., Secular movements of the earth's crust, Some results and further aims of investigations, in *Recent Movements of the Earth's Crust*, vol. 1, pp. 7–24, Publishing House for USSR Academy of Sciences, Moscow, 1963.

Miyabe, N., S. Miyamura, and M. Mizoue, A map of secular vertical movements of the earth's crust in Japan, *Ann. Acad. Sci. Fennicae, A III*, 90, 287–289, 1966.

Niewiarowski, E., and T. Wyrzykowski, Die Bestimmung gegen-wartiger vertikaler Erdkrusten bewegungen im polnischen Raum durch Vergleich von Ergebnissen wiederholter Prazisions nivellements, *1st Intern. Symp. uber rezente Erdkrustenbewegungen*, pp. 119–125, Berlin, 1962.

Niskanen, E. On the upheaval of land in Fennoscandia, *Ann. Acad. Sci. Fennicae, A III*, 53(10), 30 pp., 1939.

Runcorn, S. K., Convection currents in the earth's mantle, *Nature*, 195, 1248–1249, 1962.

Sauramo, M., The mode of the land upheaval in Fennoscandia during late Quaternary time, *Fennia*, 66(2), 26 pp., 1939.

Sauramo, M., Die Geschichte der Ostsee, *Ann. Acad. Sci. Fennicae, A III*, 51, 9–522, 1958.

Small, J. B., Interim report on vertical crustal movement in the United States, Washington, 1963, paper presented at the 13th General Assembly, International Union of Geodesy and Geophysics, Berkeley, California, 1963.

Strelkov, S. A., *The North of Siberia*, Nauka, Moscow, 336 pp., 1965.

Takeuchi, H., and Y. Hasegawa, Viscosity distribution within the earth, *Geophys. J.*, 9, 503–508, 1964.

Tsubokawa, J., J. Ogawa, and T. Hayashi, Crustal movements before and after the Niigata earthquake, *J. Geodetic Soc. Japan*, 10(3-4), 165–171, 1964.

Vening Meinesz, F. A., The determination of the earth's plasticity from the postglacial uplift of Scandinavia, isostatic adjustment, *Koninkl. Ned. Akad. Wetenschap. Proc.*, 40(8), 654–662, 1937.

Vyalov, S. S., On the theory of the viscous plastic flow of ice sheets, *Soviet Antarctic Expedition No. 10 (Second Continental Expedition, 1956–1958)*, pp. 324–336, Morskoi Transport, Leningrad, 1960.

Whitten, C. A., Crustal movement in California and Nevada, *Trans. Am. Geophys. Union*, 37(4), 393–398, 1956.

Zharkov, V. N., Viscosity of the earth's interior, *Trans. Inst. Phys. Earth, USSR Acad. Sci.*, 11(178), 30–60, 1960.

5. Magnetism

Reduction of Geomagnetic Data and Interpretation of Anomalies

Takesi Nagata

THE major parts of the geomagnetic field originate within the earth's core, where electric currents are generated by heat energy and the earth's rotation. However, in the uppermost part of the earth, where temperature is below the Curie point, magnetic minerals are secondly magnetized by the major magnetic field. Natural rocks that contain a large amount of magnetic minerals are more magnetized than those having fewer such minerals. In general, the magnetization of rocks decreases with an increase of their acidity. Thus, the real distribution of the geomagnetic field over the earth's surface is extremely complicated because of the magnetic fields caused by the secondary magnetization of individual geological masses.

There is a wide range in the secondary magnetization of natural rocks, from 10^{-6} to 10 cgs emu/cm^3. This large range allows us to use data on the geomagnetic field distribution to identify the nature and configuration of geological structures. Some key points in analyzing geomagnetic data for the use of geological identification are briefly summarized here.

MAGNETIZATION OF THE EARTH'S CRUST AND THE UPPER MANTLE

Analysis of the geomagnetic field based on potential theory has shown that more than 98% of the field originates within the earth. The geomagnetic potential W can therefore be approximately expressed on a spherical coordinate system (r, θ, λ) as

$$W(r, \theta, \lambda) = R_e \sum_{n=1}^{\infty} \sum_{m=0}^{n} \left(\frac{R_e}{r}\right)^{n+1} C_n^m$$
$$\cdot \cos(m\lambda + \varphi_n^m) P_n^m(\cos\theta) \quad (1)$$

where r is geocentric distance, R_e is the earth's mean radius, θ is colatitude, and λ is longitude.

The coefficient $(R_e/r)^{n+1} C_n^m$ in (1) becomes C_n^m on the earth's surface, where $r = R_e$. The actual values of $(R_e/r)^{n+1} C_n^m$ on the earth's surface are approximately expressed [*Nagata*, 1965a] by

$$n \leq 8 \quad (R_e/R_e)^{n+1} C_n^m = A k^{-(n+1)}$$
$$k \simeq 2.7 \quad (2)$$
$$n \geq 10 \quad (R_e/R_e)^{n+1} C_n^m = B k'^{-(n+1)}$$
$$k' < 1.1$$

Using practically continuous data of a magnetic profile along an approximate great circle, *Alldredge et al.* [1963] have examined the actual values of k' for various values of n up to $n = 2000$. The result shows that $k' \simeq 1$ for $n = 10$ to 2000.

If we assume that the geomagnetic field is produced either by magnetization or by electric current on a spherical surface of $r = r_c < R_e$, the coefficients $(R_e/r_c)^{n+1} C_n^m$ on the surface for various values of n may have roughly the same order of magnitude, namely,

$$(R_e/r_c)^{n+1} C_n^m \simeq C \quad (3)$$

Then, putting (3) into (2), we get

$$C(R_e/r_c)^{-(n+1)} = A k^{-(n+1)} \quad (4)$$

which indicates that $R_e = 2.7 r_c$, and, therefore, $r_c = 2300$ km for $n \leq 8$ and $r_c \simeq R_e$ for $n \geq 10$. These results may show that the origin of lower harmonic components of the geomagnetic field of $n = 8$ is within the earth's core. On the other hand, the origin of components of shorter wavelength ($n \geq 10$) is just beneath the earth's surface, that is, either in the earth's crust or in the upper mantle.

Since $0 \leq m \leq n$ for each harmonic, the representative wavelengths of the harmonic components from $n = 10$ to $n = 2000$ range from 4000 to 20 km. Hence, possible correspondence between geomagnetic spatial variations of the shorter wavelength and constitution of the uppermost part of the earth can be different for different wavelengths. Consequently, there is no rigorous definition of 'geomagnetic anomaly.' In practical research dealing with an area of l in linear dimension, one can consider that the geomagnetic field distribution, which can be represented by a quadratic formula, represents the general field distribution in the area concerned, whereas shorter wavelength components (e.g., less than $l/3$ in wavelength), are 'local anomalies' within the specific area. Then, potential W can be expressed, in general, as

$$W = W_o + \Delta W \quad (5)$$

where W_o and ΔW denote, respectively, potential for the general field and that for the local anomaly. The local anomaly potential, ΔW, will be the main object for geomagnetic research of the subterranean structure in a specified area.

MAGNETIC ANOMALY PATTERNS

In most magnetic surveys on land, one or some combinations of the north (X), east (Y), and downward (Z) components of the geomagnetic field, and sometimes its horizontal intensity (H), declination (D), and inclination (I), are used to represent geomagnetic anomalies. In most cases of recent extensive magnetic surveys by airborne or shipborne magnetometers, only the total force F is measured.[1]

Interpretation of the anomaly component ($\Delta \mathbf{F}$) of the total force ($\mathbf{F}$) is a little complicated, because the direction cosines (l, m, n) of a vector $\Delta \mathbf{F}$ are proportional to a vector ($\Delta X, \Delta Y, \Delta Z$), and are not unique even within a small area. Sometimes an arithmetical difference $\Delta f = |\mathbf{F}| - |\mathbf{F}_0|$ is essentially different from $|\Delta \mathbf{F}| = |\mathbf{F} - \mathbf{F}_0|$. An approximate method of interpretation in such a case is as follows:

Provided that $|\Delta \mathbf{F}/\mathbf{F}_0| \ll 1$,

$$\Delta f = \frac{X_0 \Delta X + Y_0 \Delta Y + Z_0 \Delta Z}{F_0} \quad (6)$$

the right-hand side of which is equal to the anomaly-field component along the direction of F_0, namely,

---

[1] The total force F is often called T, and the anomaly of total force is called $\triangle T$ ($\triangle T_a$ in the Russian literature).

$$\Delta f = \left(l_0 \frac{\partial}{\partial x} + m_0 \frac{\partial}{\partial y} + n_0 \frac{\partial}{\partial z} \right) \Delta W \qquad (7)$$

where (l_0, m_0, n_0) denote the direction cosines of $\mathbf{F}_0$.

Figure 1 shows an example of the Δf pattern, in which a remarkable structure of long lineation of geomagnetic anomalies can be observed. Such patterns of long lineation of anomalies have been found over many areas in the Atlantic, Pacific, and Indian Oceans [Heirtzler, 1965]. Figure 2 shows an example of a geomagnetic anomaly in the neighborhood of a volcano.

MAGNETIC FIELDS PRODUCED BY MAGNETIZED BODIES

A geologic body having magnetization J causes a magnetic field $\Delta \mathbf{F}$ around the body. As an elemental unit of magnetization of a small volume (dv) is of a dipole character, expressed by $\mathbf{J}\,dv$, a direct illustrative representation of the magnetic field produced by a magnetized body of even a simple shape is still fairly complicated. Putting

$$\Delta \mathbf{F} = \operatorname{grad} \Delta W \qquad (8)$$

potential ΔW caused by a magnetized body (v) is expressed as

$$\Delta W = -\int_v \frac{(\mathbf{J} \cdot \mathbf{P})}{r^3} \, dv \qquad (9)$$

where $\mathbf{P}$ denotes a radial vector from an 'auspunkt' (ξ, η, ζ) at $dv = d\xi d\eta d\zeta$ to an 'aufpunkt' (x, y, z), namely,

$$r^2 = (x - \xi)^2 + (y - \eta)^2 + (z - \zeta)^2$$

In most geologic bodies, $\mathbf{J}$ can be considered homogeneous at every point within v. Expressing, then,

$$\mathbf{J} = (\alpha J, \beta J, \gamma J) \qquad \alpha^2 + \beta^2 + \gamma^2 = 1$$

$$\frac{\partial}{\partial S} = \alpha \frac{\partial}{\partial x} + \beta \frac{\partial}{\partial y} + \gamma \frac{\partial}{\partial z} \qquad (10)$$

ΔW is rewritten as

$$\Delta W(x, y, z) = \frac{\partial}{\partial S} U(x, y, z) \qquad (11)$$

where

$$U = \int_v \frac{J}{r} \, dv \qquad (12)$$

Obviously, U represents the potential caused by a distribution of a single pole of J in intensity within the same body v, which can be more intuitively evaluated than ΔW itself. Equation 11 is called Poisson's relation.

Combining (8), (10), and (11), one can get mathematical expressions of magnetic field components in terms of U in such a form, for example, as

$$\Delta X = \alpha \frac{\partial^2 U}{\partial x^2} + \beta \frac{\partial^2 U}{\partial x \, \partial y} + \gamma \frac{\partial^2 U}{\partial x \, \partial z}$$

$$\Delta Y = \alpha \frac{\partial^2 U}{\partial x \, \partial y} + \beta \frac{\partial^2 U}{\partial y^2} + \gamma \frac{\partial^2 U}{\partial y \, \partial z} \qquad (13)$$

$$\Delta Z = \alpha \frac{\partial^2 U}{\partial x \, \partial z} + \beta \frac{\partial^2 U}{\partial y \, \partial z} + \gamma \frac{\partial^2 U}{\partial z^2}$$

where

$$\frac{\partial^2 U}{\partial x^2} + \frac{\partial^2 U}{\partial y^2} + \frac{\partial^2 U}{\partial z^2} = 0.$$

In practical evaluation of the force component, $\partial^2 U / \partial x^2$, etc., in (13) are computed by the numerical integration of

$$\frac{\partial^2 U}{\partial x^2} = \int_v J \frac{\partial^2}{\partial x^2} \left(\frac{1}{r} \right) d\xi \, d\eta \, d\zeta \quad \text{etc.} \qquad (14)$$

Various convenient numerical computations have been proposed [Haalck, 1930; Jakosky, 1950; Henderson and Zietz, 1957; Affleck, 1958]. Applying these formulas on reasonably assumed geological bodies, the distribution of $\Delta \mathbf{F}$ can be obtained. The trial and error approach to a satisfactory agreement between the assumed model structure and the observed distribution of $\Delta \mathbf{F}$ or Δf is usually adopted. However, the problem of determining the distribution of magnetization from a surface distribution of $\Delta \mathbf{F}$ cannot, in principle, have a unique solution.

MAGNETIZATION OF ROCKS

Magnetization $\mathbf{J}$ of rocks is generally composed of two parts, induced magnetization $\kappa \mathbf{F}_0$ and natural remanent magnetization $\mathbf{J}_n$:

$$\mathbf{J} = \kappa \mathbf{F}_0 + \mathbf{J}_n \qquad (15)$$

where κ denotes the magnetic susceptibility of rocks. The κ of a rock depends on its contents of magnetic minerals and their chemical composition and size. In most rocks, the magnetic minerals are related to magnetite-rich titaniferous magne-

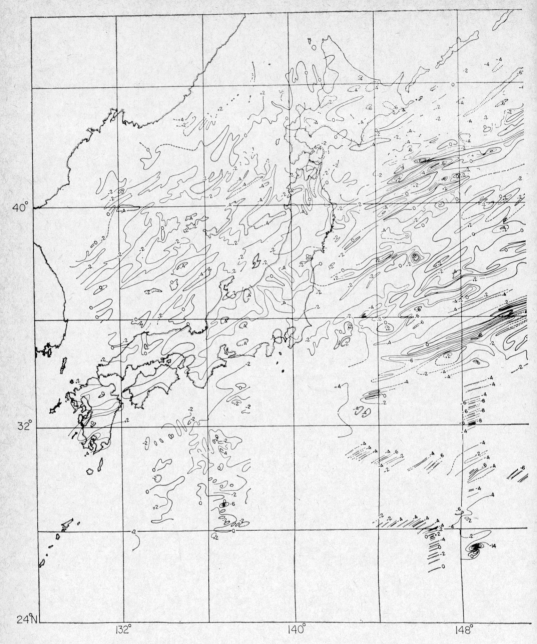

Fig. 1. Anomalies in the geomagnetic total force in the northwestern Pacific Ocean. Units, 100 γ; contour interval, 200 γ [after *Uyeda and Vacquier*, 1968].

tite whose chemical composition is given as $0 \leq x \leq 0.3$ (in mole %) in $x \, Fe_2TiO_4 \cdot (1-x)Fe_3O_4$. The magnetic susceptibility (κ) of rocks is approximately proportional to volume percentage (p) of content of magnetic minerals. The average empirical relation between κ and p [*Nagata*, 1961, 1965b; *Lindsley et al.*, 1966] is expressed as

$$\kappa \simeq 0.15p \qquad (16)$$

The intensity and direction of remanent magnetization J_n differ greatly in different cases.

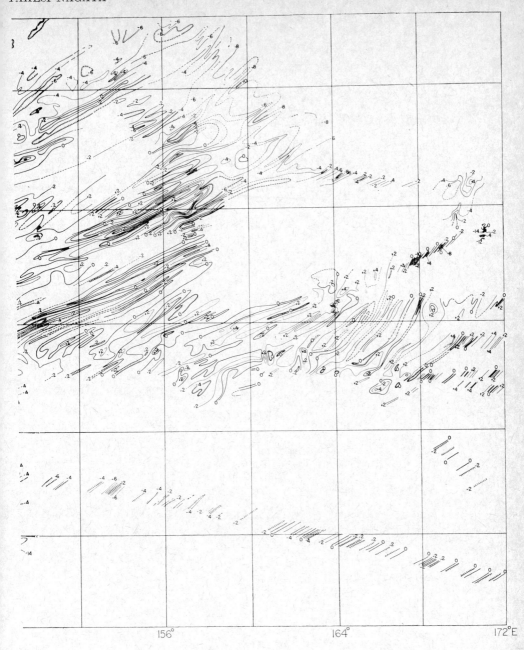

It has been concluded, however, that the direction of J_n of rocks younger than the Cretaceous is either parallel or antiparallel to that of F_0 in situ within 30° in possible deviation. The direction of J_n of older rocks is generally more deviated from that of F_0, and the sense of deviation differs in different continents (see paper by Runcorn in this Monograph, page 452). The intensity of J_n depends on the acquisition mechanism of remanent magnetization and grain size of magnetic minerals, in addition to their content. Generally speaking, $|J_n| > |\kappa F_0|$ in most igneous rocks,

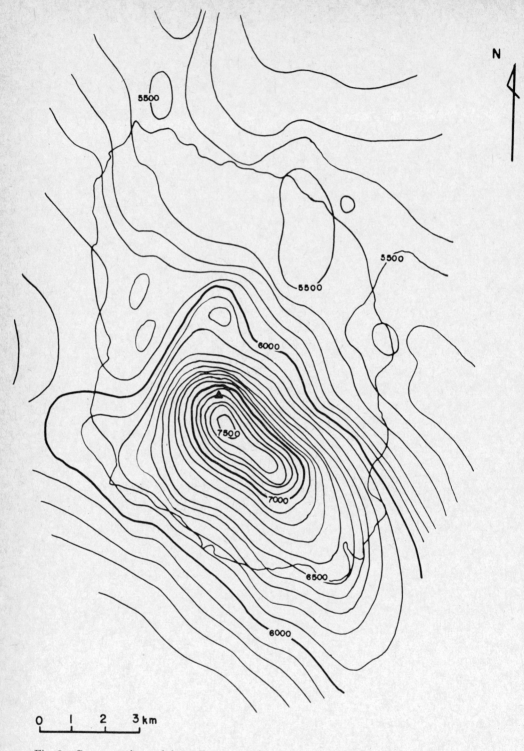

Fig. 2. Geomagnetic total force F over a volcanic island, Izu-Osima, Japan. Units, 40,000 γ.

whereas $|\mathbf{J}_n| < |\kappa \mathbf{F}_0|$ or $|\mathbf{J}_n| \ll |\kappa \mathbf{F}_0|$ in sedimentary rocks. The Q value, defined as $Q = J_n/\kappa F_0$, may be an adequate measure of $|J_n|$ for practical purposes. In igneous and thermally metamorphosed rocks over continental areas, Q amounts to about unity for well crystallized rocks, to about 10 for volcanic rocks, and to 30–50 for rapidly cooled basaltic rocks [Nagata, 1961]. The magnitude of J_n of ocean-floor basalts so far examined is generally $(3-5) \times 10^{-3}$ emu/cm^3, whereas their κ ranges from 3×10^{-4} to 5×10^{-4} emu/cm^3. Hence, their Q value is about 20 on the average [Vogt and Ostenso, 1966]. Marine geomagnetic anomalies, therefore, can be attributed mostly to the remanent magnetization, J_n, of these ocean-floor basalts.

As both the intensity and direction of Jn are affected by the geological history and petrographic structure of the rocks concerned, actual measurements of typical samples are required for reliable interpretation of observed geomagnetic anomalies.

APPROXIMATE INTERPRETATION OF GEOMAGNETIC ANOMALY

A distribution pattern of geomagnetic anomalies, $\Delta \mathbf{F}$ or Δf, cannot lead to a unique solution for the responsible geological structure. Specific cases must be examined in detail using all available geological data and magnetic properties of the rocks concerned. However, experience with observational data and their analyses may indicate the most common relationship between magnetic anomalies and the corresponding geological structure.

First, we can rewrite (15) as

$$\mathbf{J} = \kappa_A \mathbf{F}_0 \qquad \kappa_A = (1 + Q) \qquad (17)$$

where the direction of $\mathbf{J}_n$ is assumed to be either parallel or antiparallel to that of $\mathbf{F}_0$. Then, numerical values of the order of magnitude of κ_A for various observed rocks are as follows:

$\kappa_A \simeq 0.1$ emu/cm^3	Iron ore
$\simeq 10^{-2}$	Basaltic rocks
$\simeq 10^{-3}$	Andesitic rocks
$\simeq 10^{-4}$	Granitic rocks
$\simeq 10^{-5}$	Sediments

When κ_A takes a negative value for basaltic and andesitic rocks, as it occasionally does, $Q > 1$ and $\mathbf{J}_n$ is approximately antiparallel to $\mathbf{F}_0$. In other words, the possibility of the presence of reversely magnetized rocks must be taken into consideration in interpreting geomagnetic anomalies [Vine and Matthews, 1963]. If $\kappa_A > 0$, a magnetic body causes a concentration of geomagnetic flux tubes within the body, resulting in an intensification of $\mathbf{F}$ along a line of force passing through the center of the body and its neighborhood. If $\kappa_A < 0$, a magnetic body causes a reduction of $\mathbf{F}$ along the line and its neighborhood. In temperate and high latitudes, therefore, Δf takes a positive maximum value over the center and its equatorial side of a $\kappa_A > 0$ body, and, on the contrary, Δf takes a negative maximum value (i.e., a minimum one) on the same geometrical condition for a $\kappa_A < 0$ body.

The geomagnetic anomaly pattern on the land surface in igneous rock areas is generally complicated, because topographies of igneous bodies greatly affect the magnetic field in their neighborhood. In contrast, the anomaly patterns over deep sea areas are comparatively simple because the sea surface is several thousand meters above the ocean bottom. In order to see a large-scale anomaly pattern on land, one can eliminate the topographical effect by using an anomaly pattern at a high altitude obtained by airborne magnetic surveys. If anomaly patterns on a plane surface at a certain height are known, then the corresponding patterns at different heights can be computed theoretically using the assumption that $\nabla^2 W = 0$ in free space. Various practical methods for this computation have been proposed [Hagiwara, 1965]. Figure 3 illustrates an example of a theoretical reduction of the Δf pattern at 4100 meters from the observed pattern at 2100 meters over a volcanic caldera. In the figure, intense topographical effects of a small scale observed at 2100 meters in altitude almost disappear at 4100 meters, where only positive and negative anomalies on a large scale representing the subterranean distribution of igneous mass are dominant.

In this example, the large-scale anomalies correspond well to Bouguer anomalies of the gravity field, which represent the subterranean mass distribution. Such a technique as the low-pass filter is frequently useful for interpreting complicated anomaly patterns.

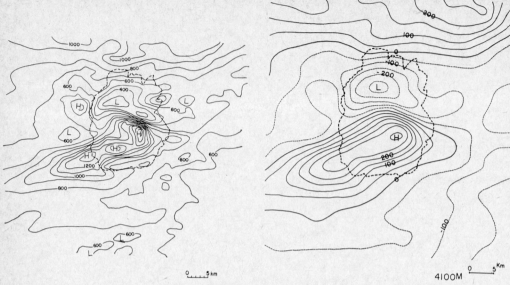

Fig. 3. Anomalies in the geomagnetic total force over the volcanic calders, Aso, Japan. *Left,* at 2100 meters altitude; *right,* at 4100 meters altitude.

REFERENCES

Affleck, J., Interrelationship between magnetic anomaly components, *Geophysics, 23,* 738–748, 1958.

Alldredge, L. R., G. D. Van Voorhis, and T. M. Davis, A magnetic profile around the world, *J. Geophys. Res., 68,* 3679–3692, 1963.

Haalck, H., Die magnetischen methoden der angewandten geophysik, *Handbook Exptl. Phys., 22*(3), 303–398, 1930.

Hagiwara, Y., Analysis of the results of the aeromagnetic surveys over volcanoes in Japan, I, *Bull. Earthquake Res. Inst. Tokyo Univ., 43,* 529–547, 1965.

Heirtzler, J. R., Marine geomagnetic anomalies, *J. Geomag. Geoelec., 17,* 227–236, 1965.

Henderson, R. G., and I. Zietz, Graphical calculation of total-intensity anomalies of three-dimensional bodies, *Geophysics, 22,* 887–964, 1957.

Jakosky, J. J., *Exploration Geophysics, 2nd edition,* Chapter 3, Trija Publishing Company, Newport Beach, California, 1950.

Lindsley, D. H., G. E. Andreasen, and J. R. Balsley, Magnetic properties of rocks and minerals, in *Handbook of Physical Constants,* revised edition, pp. 543–552, Geological Society of America, New York, 1966.

Nagata, T., *Rock Magnetism,* revised edition, Maruzen Company, Tokyo, 350 pp., 1961.

Nagata, T., Convergency of the spherical harmonic coefficients of the geomagnetic field, *J. Geomag. Geoelec., 17,* 153–155, 1965a.

Nagata, T., Q_n and Q_r of remanent magnetization of igneous rocks, *J. Geomag. Geoelec., 17,* 157–159, 1965b.

Uyeda, S., and V. Vacquier, Geothermal and geomagnetic data in and around the island arc of Japan, in *The Crust and Upper Mantle of the Pacific Area, Geophys. Monograph 12,* edited by L. Knopoff et al., pp. 349–366, American Geophysical Union Monograph, Washington, D.C., 1968.

Vine, F. J., and D. H. Matthews, Magnetic anomalies over ocean ridges, *Nature, 199,* 947–949, 1963.

Vogt, P. R., and N. A. Ostenso, Magnetic survey over the Mid-Atlantic ridge between 42°N and 47°N, *J. Geophys. Res., 71,* 4389–4411, 1966.

The Relation of Magnetic Anomalies to Topography and Geologic Features in Europe

A. Hahn and A. Zitzmann

A MAP of magnetic anomalies in Europe (Figure 1) has been prepared on the basis of available surveys (listed at the end of this section). At the boundaries of adjacent surveys, there are generally discrepancies caused by differences in the surveys, especially the components measured (ΔZ and ΔT), altitude of measurement, epoch of measurement, and standard field. No attempt at smoothing has been made except in Western Germany, where the ground and the airborne survey overlap north of 52° and west of 9°.

In some areas of outcropping basic igneous rocks, the magnetic anomalies fluctuate too much to be represented at the scale of Figure 1; these areas are marked by horizontal hatching. In the survey of France, areas of this kind do not appear, because it was measured at an altitude of 3 km, where the anomalies are much smoother than nearer to the surface.

Isoanomalies were, in general, contoured in 50 γ intervals. High narrow maximums or minimums were abandoned. In Poland (east of 15°, north of 49–51°), however, anomalies are so high and broad that it seemed reasonable to use a contour interval of 100 γ.

TECTONIC UNITS

The area under consideration can be divided into four parts corresponding to large tectonic units:

1. The Caledonian orogen includes the mountains of Norway and Scotland and governs the tectonics of England, Wales, and Ireland, including the Brabant massif. Only southwest England is part of another unit, as is indicated in Figure 2.

2. The Variscan orogen includes southwest England (Cornwall), almost all of France (Armorican and Central massif), a great part of central and southern Germany (Rhenohercynicum, Saxothuringicum), small parts of northern Switzerland and Austria, western Czechoslovakia (Moldanubicum), southern Poland, and northeastern Rumania (Dobruja).

3. The southwest border of Fennosarmatia just contacts southern Sweden and divides Poland into two halves.

4. Part of the Alpidic orogen includes the Pyrenees, the Alps, the Carpathians, and the Transsylvanian Alps.

The basement along the coast of the North Sea and the Baltic Sea between 4 and 18°E cannot yet clearly be related to one of these units.

In Figure 2 the main directions of these orogens and of their members are indicated by the position of the relevant names; some details can be seen from the direction of the big faults. A comparison of these features with the anomalies in Figure 1 shows that there is a general parallelism of the preferred directions in most parts, particularly in the areas where Paleozoic or pre-Paleozoic rocks crop out. Here, we usually find magmatic and metamorphic rocks that possess a magnetization high enough to produce the measured anomalies.

A clear correlation of a border of a tectonic unit can be seen in southwest England, where north of the exterior edge of the Variscides there are many positive anomalies, whereas to the south negative ΔT values dominate.

Greater masses of extrusive basic rocks generally produce strongly fluctuating anomalies (horizontal hatching in Figure 1).

In wide areas the folded rocks are covered by platform sediments of posthercynian age. Generally there is poor correlation between the strike of these sediments and the magnetic anomalies. One exception is to be found in the

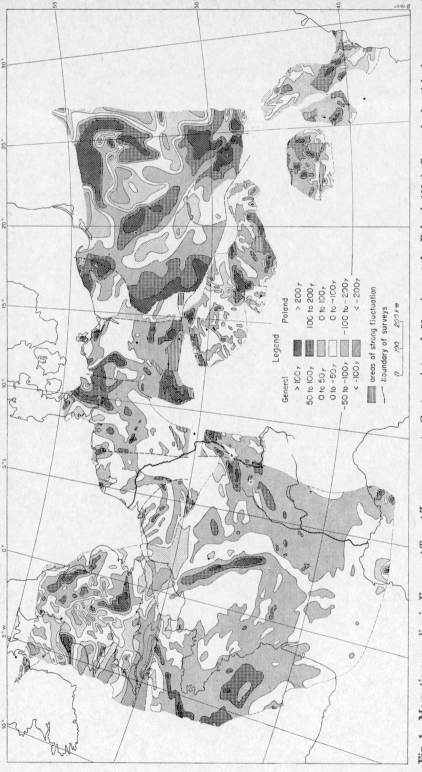

Fig. 1. Magnetic anomalies in Europe, ΔT or ΔZ measurements. Contour interval, 50 γ (except for Poland, 100γ). See end of article for sources of data.

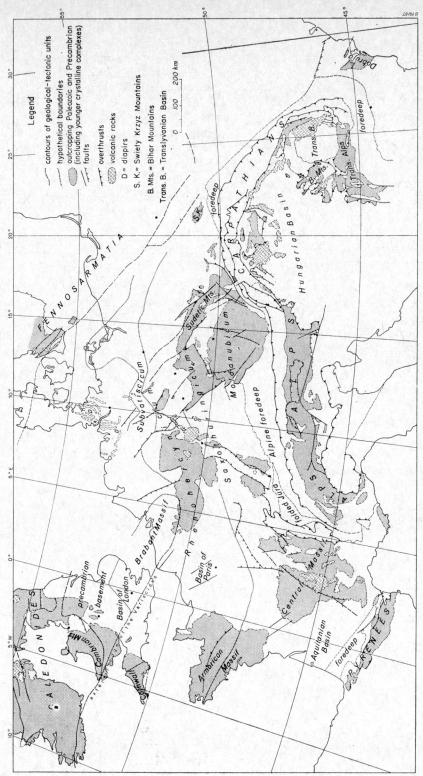

Fig. 2. General tectonic map of the area of Figure 1. Simplified after *Carte Tectonique Internationale de l'Europe*, Moscow, 1964.

Carpathians east of Vienna, in eastern Czechoslovakia, and in southern Poland. Here, the central part of the Carpathians is accompanied on both sides by bands of strong positive anomalies, where thick Tertiary flysch sediments appear. Another exception is the maximum line striking east-west around London, which coincides with the axis of the London basin.

An important feature of the magnetic field of France is a positive maximum line running NNW-SSE across the central Paris basin. In the northern prolongation of this line we find the maximum west of London running parallel to the other anomalies that are probably connected to structures of the Precambrian basement.

Depth estimates of the magnetized bodies in northern Germany (north of 52°) and in Poland led to depth values of the order of 10 km between 6°E and the assumed southwestern border of Fennosarmatia [*Mundt*, 1966]. The depth values are about 5 km northeast of this border and become reduced to the order of 1 km in northeast Poland [*Skorupa*, 1959]. The border itself is followed by wide high anomalies.

THE 'MITTELMEER-MJÖSEN ZONE'

The Mittelmeer-Mjösen zone is a graben system including parts of the valleys of the Rhone-Saône River (5°E, 45–47.5°N) and the Rhine River (8°E, 47.5–50°N) and its continuation to the north (direction 5–10°). The northern end is the Oslo graben of Permian age.

The Rhone-Saône Valley does not show any anomaly corresponding clearly to its tectonic situation. In the Upper Rhine graben we find some change in the anomalies where they pass the border fault. They indicate that long magnetic bodies striking northeast-southwest have been broken and that the different parts have been displaced in vertical and perhaps also in horizontal direction. It is interesting that around these two valleys a clear graben anomaly can hardly be encountered. This means that the dislocations of the graben do not extend to highly magnetized rocks at depth, e.g. below the Conrad discontinuity. Either the dislocations concern only zones of less than 10 km depth or the rocks in the respective zone are nonmagnetic for reason of mineral content and/or temperature. This problem is yet to be solved.

North of Frankfurt/Main is the large Vogelsberg mountain, built up by Miocene basalt producing strong fluctuations in the magnetic field. Near the coast of the North Sea and the Baltic Sea (around Hamburg) the anomalies possess a preferred north-south direction, especially in the area of the parallel salt diapirs.

THE CURIE DEPTH

The depth at which the Curie temperature of the ferrimagnetic minerals is reached is called the Curie depth. Below this interface, these minerals are paramagnetic and hence possess so small a magnetization that they can be regarded as nonmagnetic in the interpretation of magnetic anomalies. *Mundt* [1966] has attempted to estimate the Curie depth in northern Germany by means of a statistical analysis of the magnetic anomalies. He found 20–25 km with an interval of uncertainty of ±5 km.

SUMMARY

In Europe, clear relations to the surface topography and geology are found in those orogenic areas where the folded rocks crop out. Smooth Mesozoic and Cenozoic sedimentary covers above Paleozoic or pre-Paleozoic folded rocks generally have no or only a slight relation to the magnetic anomalies. Most of the graben system of the Mittelmeer-Mjösen zone has only indirect relations to the magnetic anomalies. Some parallelism is found in the area around Hamburg. The Curie depth in northern Germany is estimated to be 20–25 km ± 5 km. Considering the Curie depth, the interpretation of magnetic anomalies is related to geothermal and mineralogical studies.

SOURCES OF DATA FOR FIGURE 1

Aeromagnetic Map of Great Britain, Sheet 2, England and Wales, 1st Edition, 1965, Director General of the Ordnance Survey, Chessington, Surrey, 1965. Profile separation, 2 km or less; mean terrain clearance, 300 meters, except in southwest England, 150 meters; scale 1:625,000.

Carte Magnétique de France, Intensité du champ magnétique total F au 1.7.1964, Document diffusé par le Service de la Carte Géologique, Paris. Profile separation, 10 km; flight elevation, 3 km above sea level; scale, 1:1,000,000. From this map, a map of anomalies has been constructed by applying a quasi-linear graphically defined standard field.

Carte aéromagnetique residuelle polaire, Administration des Mines, Service Géologique de Belgique. Profile separation, 2 km; flight eleva-

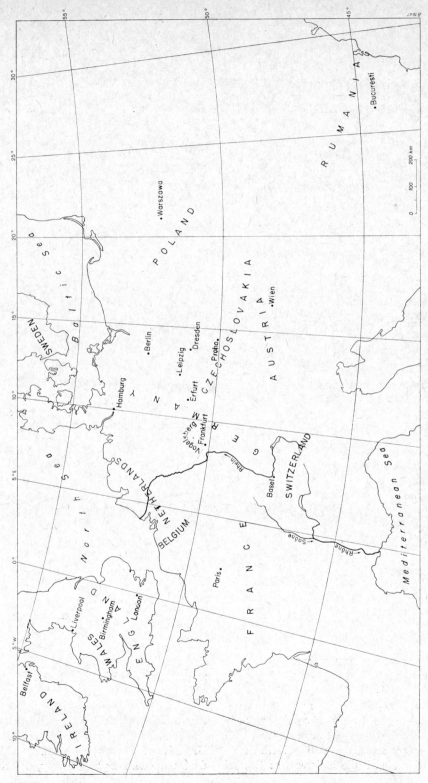

Fig. 3. Reference map for Figures 1 and 2.

tion, 600 and 900 meters above sea level; scale 1:500,000.

Map of the anomalies of the vertical intensity in the Netherlands [*Veldkamp*, 1951]. Ground survey; average station spacing, 10 km; scale, 1:500,000.

Geophysikalische Karte von Nordwest- Deutschland, 1, Magnetik.-Reichsamt für Bodenforschung, 1948. Ground survey; mean station spacing, 3 km; ΔZ measurements; scale, 1:500,000.

Parts of the airborne survey of the Federal Republic of Germany, present state, unpublished Profile separation, 2.2 km; flight elevation, 700, 1000, 1500 meters above sea level.

Geophysikalische Karte der DDR-Magnetik [*Fanselau*, 1959]. Ground survey; average station spacing, 3 km or less; ΔZ measurements; scale, 1:1,000,000.

Aeromagnetic Map of Czechoslovakia Ustřední ústav geologický, Praha, 1966. Profile separation, 2 km; mean terrain clearance, 100 meters; scale, 1:1,000,000.

The magnetic anomalies in Poland [*Pawlowski*, 1947]. Ground survey; average station spacing, 30 km; ΔZ measurements; scale, 1:1,500,000.

Five magnetic maps of parts of Rumania [*Airinei*, 1957,, 1959, 1963; *Airinei et al.*, 1963, 1964]. Ground surveys; average station spacing, 3-10 km; ΔZ measurements; scales, 1:500,000 to 1:1,000,000.

REFERENCES

Airinei, Şt., Asupra anomaliei magnetice regionale din centrul bazinului Transilvaniei, *Bul Ştiinţific, Sect. Geol. Geograph*, *2*, 209–235, 1957.

Airinei, Şt., Magnetische Gebietsforschungen in der Dobrudscha, der Südmoldau und der östlichen Rumänischen Ebene, *Rev. Geol. Geograph.*, *3*, 279–300, 1959.

Airinei, Şt., Structura Fundamentului Hercinic al Curburii Carpaţilor Orientali in Lumina Anomaliidor Cîmpurilor Gravimetric şi Geomagnetic, *Assoc. Géol. Carpato-Balkanique V-ème Congrès, 1961, Bucarest, Comm. Sci. V-éme Sect., Géophys.*, *6*, 5–33, 1963.

Airinei, Şt., B. Gavrilescu, and O. Căruţaşu, Lucrări de Recunoastere magnetică în Munţii Apuseni, *Probl. Geofiz.*, *2*, 199–224, 1963.

Airinei, St., M. Boisnard, R. Botezatu, L. Georgescu, P. Suciu, and M. Visarion, Carte de l'anomalie magnétique ΔZ de Moldavie (Roumanie), *Rev. Roum. Géol. Géophys. Géogr., Ser. Géophys.*, *8*, 125–134, 1964.

Fanselau, G., editor, Geomagnetismus und Aeronomie, Bd. 3, Berlin, 1959.

Mundt, W., Einige statistische Eigenschaften regionaler Z-und Bouguer-Anomalien und deren Beziehungen zum Krustenbau im Untersuchungsgebiet, *Jahrb. 1964 Adolf-Schmidt-Observ. Erdmagn. Niemegk Wiss. Mitt.*, pp. 121–129, 1966.

Pawlowski, St., The magnetic anomalies in Poland, *Serv. Géol. Pologne, Inst. Geol. Pologne, Bull. 44, Ser. Geophys., 1947* (2), 1–59, 1947.

Skorupa, J., Morphology of the magnetically active basement and crystalline basement in northeastern Poland, *Biul. Inst. Geol., 160* (*Ser. Geofiz., 19*), 1–51, 1959.

Veldkamp, J., Geomagnetic Anomalies in the Netherlands, *Geol. Mijnbouw, 13*, 218–223, 1951.

Aeromagnetic Investigations of the Earth's Crust in the United States

Isidore Zietz

GREAT CIRCLE SURVEY

Two special studies are described in this report. The first is a coast-to-coast survey along a great circle path. This survey covers a band 100 miles wide (flight tracks along the great circle, approximately parellel to one another, with 5-mile spacing) across the entire North American continent and the adjacent continental shelves; the center of the strip passes approximately through Washington, D. C., and San Francisco (Figure 1). This survey was specifically designed to investigate the gross features of the earth's crust by the aeromagnetic method. The survey was a joint project of the U. S. Naval Oceanographic Office, which conducted the aeromagnetic survey, and the U. S.

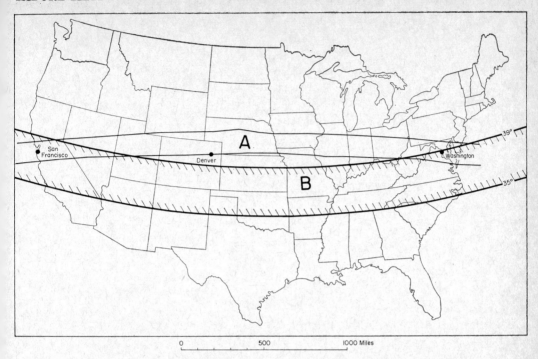

Fig. 1. Location of (A) great-circle aeromagnetic survey and (B) Transcontinental Geophysical Survey of the Upper Mantle Project. Data for the aeromagnetic survey were obtained by the U. S. Naval Oceanographic Office.

Geological Survey, which was responsible for most of the interpretation.

East of the Rocky Mountains, the flight elevation was approximately 6000 feet above sea level; in most of the area west of the Rocky Mountains, the flight elevation was approximately 16,000 feet above sea level but was reduced to 6000 feet west of the Sierra Nevada uplift. The gross structure of the North American continent is such that this survey crosses at right angles most of the major tectonic elements from the Appalachian Mountain system in the east to the Rocky Mountains, Basin and Range province, and the mountain systems of California in the west. The magnetic profiles (Figures 2 and 3) reveal anomalies of varying wavelengths. Anomalies having short wavelengths must be a reflection of contrasting lithologies in the upper part of the earth's crust, whereas the anomalies having much longer wavelengths probably reflect lithologies deep in the earth's crust.

Examination of the profiles shows a striking contrast in the magnetic character west and east of the Rocky Mountain system. To the east, there are numerous high-amplitude anomalies of short wavelength that are often superimposed on anomalies of long wavelength. The western area is characterized by fewer short-wavelength anomalies, mostly of smaller amplitude. In addition, except for the Great Valley anomaly at approximately 121°W longitude and the Sierra Nevada anomaly immediately to the east, anomalies having long wavelengths are absent in the west. This is particularly true of the Basin and Range province (approximately between longitudes 112 and 118°W). The weakly magnetic crust west of the Rocky Mountain system is relatively devoid of magnetic minerals at shallow depths and may be above the Curie temperature for magnetite (578°C) and therefore nonmagnetic at depth. These contrasting magnetic data are consistent with the data from seismic studies of the crust [Pakiser and Zietz, 1965; Pakiser and Robinson, 1966], which suggest that the conterminous United States may be divided by the Rocky Mountain system into two crustal and upper mantle superprovinces.

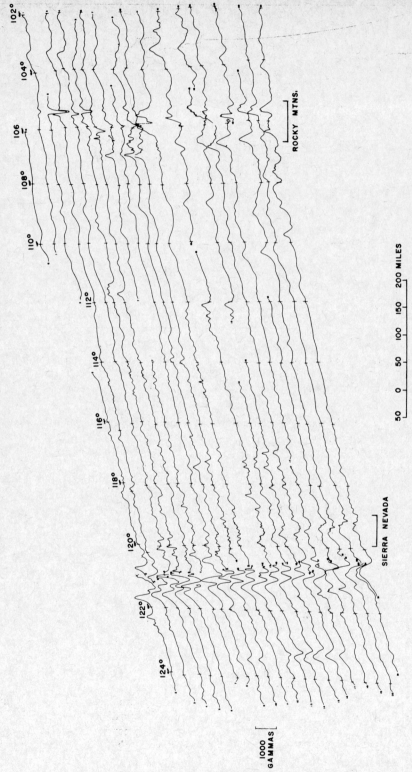

Fig. 2. Aeromagnetic profiles for the western part of the transcontinental great circle (Figure 1). Spacing of profiles, 5 miles.

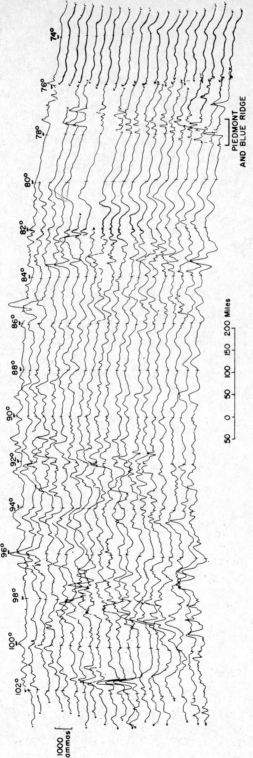

Fig. 3. Aeromagnetic profiles for the eastern part of the transcontinental great circle (Figure 1). Spacing of profiles, 5 miles.

In the eastern superprovince, the velocity of compressional waves in the upper mantle rocks is everywhere greater than 8 km/sec, the mean crustal velocity is generally greater than 6.4 km/sec, and the crust is generally thicker than 40 km. In the western superprovince, the velocity of compressional waves in the upper mantle rocks is everywhere less than 8 km/sec (except along the margin of the Pacific Ocean basin), the mean crustal velocity is generally less than 6.4 km/sec, and the crust is generally thinner than 40 km.

These observations, together with the Cenozoic geologic record of diastrophism and volcanism in the western superprovince and relative Cenozoic inactivity in the eastern superprovince, suggest a mobile upper mantle in the west and a predominantly silicic crust that is now receiving mafic and probably also silicic material from the mantle. The upper mantle in the east is relatively stable, and the now predominantly mafic crust has been extensively intruded with mafic material from the mantle. Additional mafic material has been added by extrusion of lava. The primitive continental crust that evolved from the mantle was probably silicic; it has been made slowly more mafic by addition of mafic material from the mantle and removal of silicic material from the continental surface by erosion and stream transport. A schematic geologic cross section along one of the aeromagnetic profiles is shown in Figure 4. The section was deduced from seismic determinations, and the accompanying gravity profiles were taken from the gravity map of the United States [*Am. Geophys. Union and U. S. Geol. Surv.*, 1964].

A study of the magnetic profiles clearly indicates that it is possible to distinguish magnetic patterns that are probably a reflection of crustal units in the upper part of the earth's crust. To demonstrate this more effectively, the magnetic profiles east of the Rocky Mountains were contoured, using an interval of 100 γ (Figure 5A). Except for the Blue Ridge and Piedmont provinces in the eastern part of the strip, crystalline rocks are buried everywhere under a veneer of several thousand feet of nonmagnetic sedimentary strata. Any interpretation of these data, therefore, depends to a great extent on mineralogical and petrological examination of cores obtained from drill holes that have penetrated the Precambrian basement. For the region east of Nebraska, few cores are available for examination. Fortunately, in Nebraska and central Ohio, an abundance of cores from the Precambrian makes it possible to infer gross lithologies on the basis of magnetic patterns and lineaments.

The northeast-trending magnetic features at approximately 78°W longitude reflect the mafic and ultramafic rocks of the Blue Ridge-Piedmont province. The pattern of high-amplitude anomalies in central Ohio between 82 and 84°W longitude is interpreted to represent a metamorphic complex of schists and gneisses and is probably an extension of the Grenville province of Canada. This is suggested by petrographic analysis and age determinations of the cores [*Zietz et al.*, 1966]. Fortunately, cores and cuttings reveal extensive amounts of magnetite, which probably accounts for the high-amplitude anomalies. The sharp magnetic boundary of the western edge of this unit, at approximately 84°W longitude, probably marks the western extent of the Grenville province in Ohio. West of this boundary, a lithologic study of basement cores reveals the presence of nonmetamorphosed igneous and sedimentary rocks.

A distinct magnetic pattern in eastern Iowa between 91 and 94°W longitude is clearly discernible. Unfortunately, the only cores from drill holes in this general region that penetrate the Precambrian surface are just north of the area of the strip in Figure 5. One is an olivine gabbro in northeast Iowa (W. J. Yoho, unpublished data). The other is a magnetite-rich gabbro in southeastern Minnesota [*Zietz*, 1964]. The presence of these cores suggest a gabbroic lithology at the Precambrian surface.

The northeast-trending linear anomaly between 94 and 97°W longitude is probably due to a thick section of basalts, which is part of the Keweenawan mafic belt extending from Lake Superior to Kansas. This has been supported by drilling data and by an examination of an extensive magnetic survey, which shows that the anomaly is continuous, without interruption, from northern Kansas to the Lake Superior area, where extensive amounts of basalts are exposed at the surface. This is discussed in more detail later in this paper.

There are two areas of distinct magnetic patterns in Nebraska. Although this is an area of abundant drilling, mineralogic and petro-

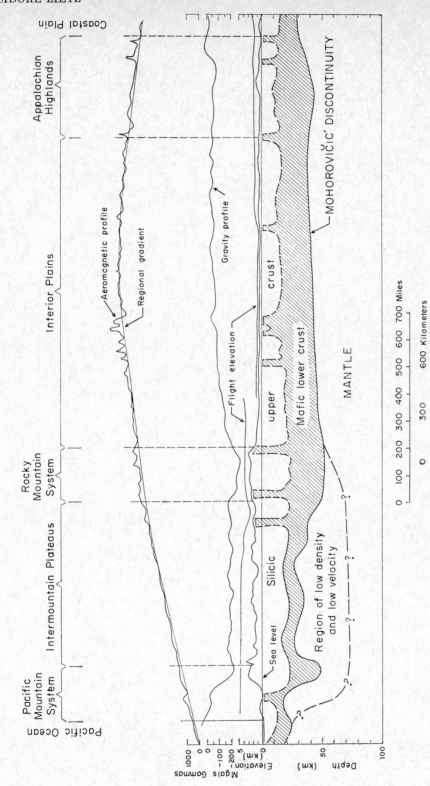

Fig. 4. Transcontinental geological and geophysical cross section of the earth's crust and upper mantle. Indicated distribution of mafic material in the silicic upper crust is schematic only. Sedimentary rocks overlying crystalline basement are not shown. Note different vertical scales above and below sea level. From Figure 5 in *Pakiser and Zietz* [1965].

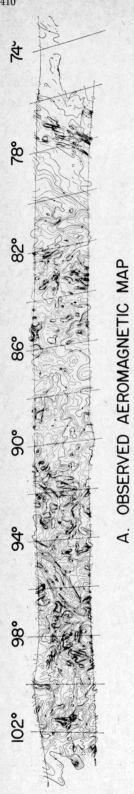  

Fig. 5. Observed and filtered aeromagnetic maps of the eastern part of the great circle survey (Figure 1).

logic analyses of the cores yield a complex mosaic in which a coherent lithologic pattern is hard to discern. The contoured magnetic data, however, make it possible to delineate two distinct areas of contrasting magnetic expression, and these form the basis for a grouping into broad lithologic units. The very anomalous area in eastern Nebraska between 97 and 99°W longitude is attributed to an area of granite and granite gneiss, whereas the magnetically flat area in central Nebraska between 99 and 101°W longitude (see also magnetic profiles in Figure 3), reflects a Precambrian surface lithology of predominantly metasedimentary rocks.

It is significant that in an area of much drilling, it is difficult to map lithologies on the Precambrian surface. In combination with magnetic data, however, our study shows that mapping of gross lithologies is possible. It should be mentioned that most of the magnetic units discussed in this report have gravity counterparts, suggesting that the more magnetic crustal units in the upper regions of the earth's crust are more dense and therefore probably more mafic than the host rocks.

A principal objective of this investigation was the determination of the magnetic character of the deeper parts of the earth's crust. The short-wavelength anomalies produced by the Precambrian rocks in the upper parts of the earth's crust were removed by taking successive equally weighted running means along each profile. The longer-wavelength anomalies that remained carry through from profile to profile, and the results for 5-point averages at 3-mile and 9-mile intervals have been contoured for the entire strip. The contoured maps east of the Rocky Mountains are shown in Figures 5B and 5C. An examination of the contoured map shown in Figure 5C shows trends and patterns quite different from those shown in the original data (Figure 5A). The maps suggest a separation of the magnetic effects of structural and lithologic units, effects due, respectively, to the shallow and deeper parts of the earth's crust. The regional aeromagnetic map for the entire survey (only the area east of the Rockies is shown in Figure 5C) shows a conspicuous east-west grain over most of the continent, except for two pairs of north-south-trending anomalies, each pair near the continental margins. The east-west grain might be attributed to the emplacement of magnetic material along an east-west system of fractures in the lower part of the earth's crust.

Specifically, a narrow east-west positive feature can be seen on the contoured magnetic map (Figure 5C) extending for approximately 500 miles and flanked by a low on the north. This feature extends from 99°W longitude in northern central Nebraska to just east of the Iowa-Illinois state line at 91°W longitude. Its narrow width indicates that the depth to its source does not exceed 40 km, which is near the base of the crust in this area. The source may, of course, be much shallower. It is on line with the inferred location of a possible rift system linking the Cape Mendocino fracture system in the Pacific Ocean off the coast of California with the wrench fault system proposed by *Drake and Woodward* [1963] in the eastern United States. The Uinta Mountains, one of the few major east-west structural units in the United States, is on line with this proposed rift zone. In the western United States, there is a broad magnetic high over the Sierra Nevada block, which is mostly granitic or granodioritic in composition. This suggests that at depth the rocks are more mafic, an observation consistent with results obtained from gravity and seismic observations [*Pakiser and Zietz*, 1965] where velocity and inferred density suggest that the deep crust is basalt or gabbro. Along the Maryland-Virginia border (Figure 5C), a broad northeast-trending anomaly probably reflects a structural unit well within the earth's crust. This feature is coincident with part of a gravity lineament that extends almost without interruption from the state of Maine to the state of Alabama [*Am. Geophys. Union and U. S. Geol. Surv.*, 1964].

There is a significant correlation of these data with measurements of heat flow in the continental United States. Preliminary values indicate that heat flow in Iowa is on the order of 1.0 $\mu cal/cm^2$ sec, in contrast to values of more than 2 $\mu cal/cm^2$ sec in the Basin and Range province (Gene Simmons, personal communication). Where heat flow is lowest, the geotherms can reasonably be assumed to be depressed, and a greater thickness of rock is below the Curie point (the temperature that establishes a lower limit for the magnetization of rocks). In the region of most intense magnetic anomalies in

Nebraska and Iowa (Figure 5C), the Curie point geotherm may be deep enough to include some upper mantle material, but in the west it may be at a much shallower depth, particularly in the Basin and Range province, so that much or all of the mafic lower crust is nonmagnetic. This is a region of great interest to crustal studies, and it is here that the landward extension of the East Pacific rise into the North American continent has been postulated.

This transcontinental aeromagnetic survey demonstrates the feasibility of using broad surveys for mapping gross lithologic units and determining major trends of basement complexes. By filtering out the short wavelength anomalies, it is possible to produce a magnetic map which must, in large part, reflect lithologic variations in the deeper parts of the crust and perhaps even in the upper mantle. Surveys of this type can contribute useful data for crustal studies such as the Transcontinental Geophysical Survey of the current Upper Mantle Project. In addition to yielding information on structure and lithology, these magnetic data can aid in interpreting other geophysical data, particularly crustal velocities, electrical conductivity, density variations, and heat flow.

MID-CONTINENT GRAVITY HIGH

The midcontinent gravity high (Figure 6), originally described by *Woollard*, [1943], is probably the most outstanding feature on the gravity map of the United States. It is an essentially continuous, high-amplitude positive gravity anomaly extending southwestward for 800 miles from Lake Superior to Kansas. The anomaly has a peak gravity value of 60 mgal and is flanked on either side by a series of lows reaching a minimum value of −100 mgal.

A detailed aeromagnetic survey covering the area was made by flying east-west lines spaced approximately 1 mile apart at a flight elevation of 1000 feet above the ground. The aeromagnetic map shown in Figure 7 was obtained by selecting contours at 100 γ intervals from the original data (E. R. King and Isidore Zietz, unpublished data).

The combined geophysical data, surface outcrops in the Lake Superior district, and drilling data suggest the existence of a thick sequence of basalts filling a trough approximately 40 miles wide and 4 miles deep, extending from the Lake Superior district (where extensive areas of basalt of Keweenawan age are known to crop out) to Kansas. For the area shown in Figure 7, basalt crops out in only a very small area in the northeast part of the map in southwestern Wisconsin. The rest of the area is covered by a sedimentary sequence which gradually thickens in a southerly direction to a maximum thickness of approximately 5000 feet in Nebraska. The presence of basalt in the Precambrian is verified by eleven drill holes in Minnesota and Iowa, all of which bottom in basalt. The flows are bounded by high angle faults on both sides and are flanked by a series of elongated basins containing red clastic sandstones 1 or 2 miles thick [*Zietz et al.*, 1966].

The geophysical data suggest that the basalt flows were subsequently faulted and folded. The trough probably represents a major rift in the older Precambrian crust and appears to be discordant with the older structures. Several large elliptically shaped magnetic anomalies (primarily lows), one in Minnesota near Minneapolis and three in Iowa southeast of Fort Dodge, Audubon, and Omaha, probably reflect the presence of later Keweenawan clastics that occupy basins at the surface of the flows.

A study of both aeromagnetic and gravity data in Lake Superior suggests that the basalt flows terminate to the north somewhere in the western part of Lake Superior. They probably do not continue into Canada, nor do they bend around to cause the large gravity high in the state of Michigan. However, in southeastern Nebraska, the sharp change in magnetic pattern suggests that the flows abruptly terminate in this area. This is supported by an abundance of drill-hole data [*Lidiak*, 1964], which reveal a restricted area of basalt in northeast Kansas surrounded by a predominantly granitic terrain.

It is possible to extract much geologic information from the combined use of the aeromagnetic and gravity data, such as a geologic map of the buried Precambrian surface and geologic cross sections at right angles to the strike of the area of the flows. Free-air gravity anomalies of the area show that the crust is locally uncompensated because of the tremendous volume of basalts in the upper surface of the crust. However, the free-air anomaly for the region containing both basalts and the flanking clastics averages to zero, suggesting

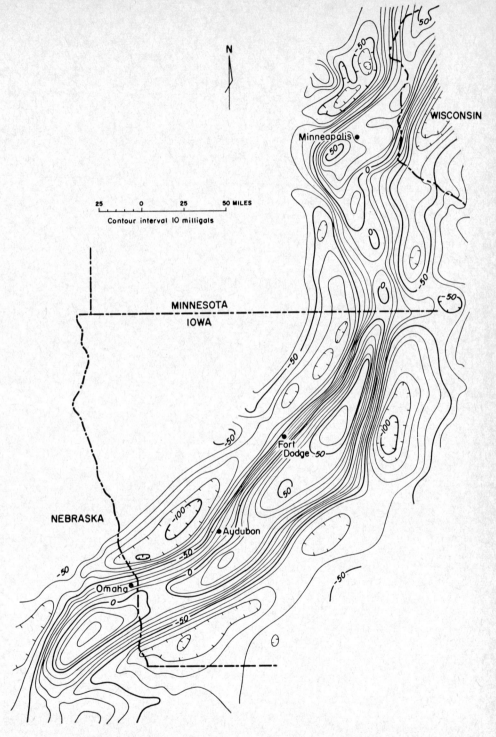

Fig. 6. Midcontinent gravity high [*Am. Geophys. Union* and *U. S. Geol. Surv.*, 1964].

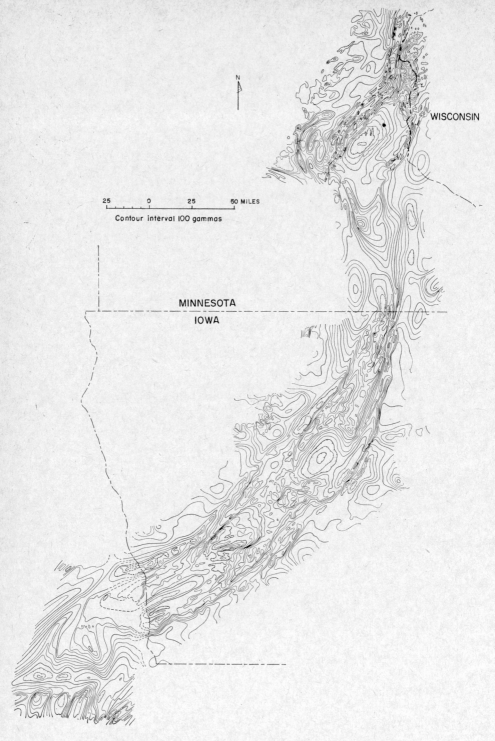

Fig. 7. Aeromagnetic map of midcontinent gravity high.

that the crust is compensated over a larger area.

REFERENCES

American Geophysical Union and U. S. Geological Survey, Bouguer gravity anomaly map of the United States (exclusive of Alaska and Hawaii), 2 sheets, scale 1:2,500,000, 1964.

Drake, C. L., and H. P. Woodward, Appalachian curvature, wrench faulting, and offshore structures, *Trans. N.Y. Acad. Sci.*, [2]*26*, 49–63, 1963.

Lidiak, E. G., Correlation of basement rocks with the midcontinent anomaly in Nebraska and Kansas (Abstract), *Soc. Exploration Geophysicists Yr. Book, 1964*, 207, 1964.

Pakiser, L. C., and Isidore Zietz, Transcontinental crustal and upper-mantle structure, *Rev. Geophys.*, *3*, 505–520, 1965.

Pakiser, L. C., and Rhoda Robinson, Composition and evolution of the continental crust as suggested by seismic observations, *Tectonophysics*, *3*(6), 547–557, 1966.

Woollard, G. P., Transcontinental gravitational and magnetic profile of North America and its relations to geologic structure, *Bull. Geol. Soc. Am. 54*, 747–790, 1943.

Zietz, Isidore, A magnetic anomaly of possible economic significance in southeastern Minnesota, *U. S. Geol. Survey Circ. 489*, 5 pp., 1964.

Zietz, Isidore, E. R. King, Wilburt Geddes, and E. G. Lidiak, Crustal study of a continental strip from the Atlantic Ocean to the Rocky Mountains, *Bull. Geol. Soc. Am.*, *77*, 1427–1446, 1966.

Relation of Magnetic Anomalies to Topography and Geology in the USSR

Tatiana Simonenko

THE magnetic field of the USSR has been mapped by an aeromagnetic survey using fluxgate magnetometers, the average distance between survey profiles not exceeding 2 km; flights were made at an altitude of about 300 meters over flat land. The anomaly $(\Delta T)a$ was calculated relative to the main geomagnetic field as a base. Mean amplitudes of anomalies in the USSR, as in other parts of the world, are several hundred or sometimes several thousand γ (not including anomalies connected with iron ore deposits); horizontal cross sections of anomalies usually do not exceed 40–50 km. The direction of the sum of induced and natural remanent magnetization of geological objects probably corresponds in general to the direction of the contemporary geomagnetic field.

The magnetic pattern clearly reflects the difference between ancient platforms and the Epihercynian platforms and folded areas of various geological age. The magnetic field of Precambrian platforms, especially the East European platform, clearly indicates a block structure. Individual basement blocks, whose horizontal dimensions reach several hundred kilometers, are distinguished by different magnetic fields. The outlines of such blocks have been sufficiently proved by independent geological and other geophysical data.

It is usual for ancient platforms to have an arching pattern of anomalies, often forming almost closed ring-shaped zones with diameters of 100–200 km. In addition, a large number of approximately isometric anomalies are observed on platforms with cross-sectional dimensions of several tens of kilometers or less, forming areas of the so-called mosaic field.

One of the peculiar features of the magnetic field of the Epihercynian (West-Siberian and Skifo-Turansk) platforms and other folded areas lies in the dominance of consistent strikes of anomalous zones over great distances, sometimes one or two thousand kilometers. (The main geological features referenced in this article are shown in Figure 1.)

The strike of the main zones of positive mag-

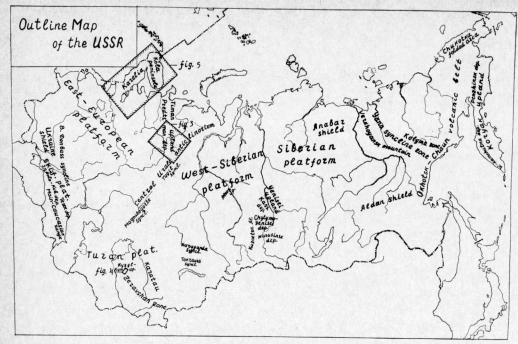

Fig. 1. Index map showing the main geological features mentioned in the text and locations of Figures 3–5.

netic anomalies in the USSR is shown in Figure 2. The disturbance lines of the structure of the principal magnetic field are expressed in detailed maps as marked variations in the strike of anomalies, breaks and horizontal displacements of anomalous zones, and local anomalies that have a discordant strike relative to the anomalies of the principal system. To a certain extent, these features of the magnetic field reflect a complex network of crustal fractures. Figure 2 does not show the degree of complexity of the anomalous field; only the boundaries of large areas with various magnetic field characteristics and the different anomalous orientations are indicated. Contours of these areas emphasize the block structure of the earth's crust.

As the sedimentary cover of the platforms is practically nonmagnetic, the anomalous magnetic field reflects the distribution of magmatic and metamorphic formations in the platform basement, as well as in folded areas of various geologic age.

This paper deals with the relationship between the anomalous magnetic field and geological structure in those areas of the USSR where folded basements are exposed or are near the earth's surface.

The most distinct relation between the anomalous field and geological structure is shown in the folded areas of the USSR. Anticlinorium zones of folded areas, irrespective of their geological age, are characterized by magnetic fields close to normal (-100, -200 γ). This is because these zones are usually represented by practically nonmagnetic folded rocks; granites related to the anticlinorium are usually just slightly magnetic. Synclinorium zones, however, with highly magnetic intrusive and effusive rocks, are usually reflected by increased magnetic fields. Zones of crustal fractures accompanied by basic and ultrabasic intrusions are characterized by intensive local anomalies.

Anticlinorium structures of folded areas are expressed by positive relief forms; consequently, over hills and mountain ranges quiet magnetic fields close to normal are observed. This is the case, for example, of the Timan and Yenisei uplands (Baikal folding), the Kuznetsk Alatau (Caledonian folds), the Urals and Karatau range (Hercynian folding), the Verkhoyansk range (Mesozoic folding), the main Caucasian range and Koryak upland (Cenozoic folding). Figure 3 shows the anomalous magnetic field for the junction of the Timan and Urals. The Timan mountain ridge and the central Urals anticlinorium are composed of ancient metamorphic micaceous-quartz schist; its magnetic susceptibility κ does not exceed $300 \cdot 10^{-6}$ cgs [*Dortman et al.*, 1964], which explains the absence of regional anomalies in this region.

Lack of marked anomalies over granite massifs in anticlinorium zones of folded areas is explained by a small percentage of magnetite (about 0.05%). Granites of the Yana-Kolyma and Chukotka folded areas in the northeastern part of the USSR are slightly magnetic, their mean magnetic susceptibility not exceeding $20 \cdot 10^{-6}$ cgs [*Pechersky*, 1964]. For granites of the eastern Ural anticlinorium zone, the values of κ do not exceed $150 \cdot 10^{-6}$ cgs; alaskitic, biotitic, and binary granites prevail here; plagiogranites and granodiorites are less common [*Dortman et al.*, 1964]. Granites of the Turkestan-Zeravshan anticlinorium zone are also slightly magnetic (κ does not exceed $100 \cdot 10^{-6}$ cgs, and natural remanent magnetization I_n is not more than $10 \cdot 10^{-6}$ cgs).

Figure 4 shows the anomalous magnetic field of one part of the Kyzyl-Kum upland, at the northwestern continuity of the Turkestan-Zeravshan zone [*Glazunova et al.*, 1961] where there are outcrops of granite and granodiorite. This part of the Kyzyl-Kum upland includes the Buckantau, Altyntau, and Kokpatas mountains. A quiet magnetic field (0 to $-50\ \gamma$) is found over the granites. Positive anomalies over granodiorites reach $500\ \gamma$; investigations of the magnetic properties of granodiorite samples showed that κ varies from 400 to $3100 \cdot 10^{-6}$ cgs, and I_n from 50 to $1000 \cdot 10^{-6}$ cgs. In view of the magnetic field characteristics over areas covered by deposits, it is possible to assume that there is a buried continuity of a granodiorite massif to the north of the Buckantau Mountains, and buried massifs of granites to the south of the Altyntau and Kokpatas Mountains; indeed, a number of boreholes have proved the presence of granites in this area.

A regular relation of nonmagnetic granites to

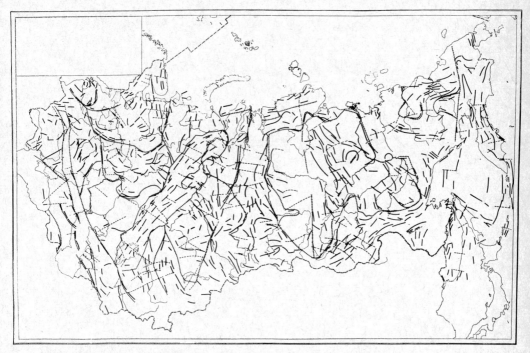

Fig. 2. Magnetic trends and patterns in the USSR from aeromagnetic surveys: solid lines, trends of the fundamental zones of positive magnetic anomalies; dashed lines, lines of disturbance of the basic structure of the field $(\Delta T)a$; crossed lines, boundaries between different types of fields $(\Delta T)a$.

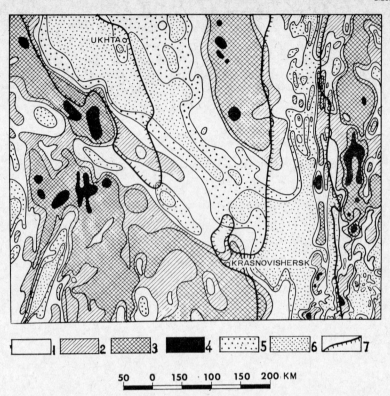

Fig. 3. Anomalies $(\Delta T)a$ in the total magnetic field for the Timan-Urals transition area: (1) -100 to $+100$ γ; (2) $+100$ to $+200$ γ; (3) $+200$ to $+500$ γ; (4) >500 γ; (5) -100 to -200 γ; (6) <-200 γ; (7) boundary of exposed folded formations.

anticlinorium zones and to positive relief forms of foldings is observed for buried folded platform basements. Hence, for example, the Alexandrov uplift, established on the basement surface in the central part of the west Siberian plain, is characterized by decreased values of $(\Delta T)a$ [Simonenko and Tolstikhina, 1965]; boreholes drilled within the uplift exposed practically nonmagnetic granites.

As was noted, synclinorium structures of folded areas are often (but not always) characterized by an increased magnetic field. The Magnitogorsk and Tagil synclinoria of the Urals are typical examples of structures with highly developed basic rocks—different effusive rocks, gabbro, and amphibole-biotite gneiss and schist —having magnetization up to 2000–5000 · 10^{-6} cgs; the $(\Delta T)a$ values over these synclinoria sometimes reach several thousand gammas (at flight altitude of about 300 meters).

Along deep fractures at the transition from synclinorium to anticlinorium zones of the Urals, rocks of mainly ultrabasic composition are widespread, their magnetization reaching 60,000 · 10^{-6} cgs if they are serpentinized [Dortman et al., 1964]. According to terrestrial survey data, anomalies of the vertical field component Z over massifs of gabbro and basic rock of the Urals reach 12,000 γ. Unmetamorphosed basic rocks do not contain magnetite and do not cause magnetic field anomalies (for example, different parts of the Krack and Kempirsai massifs in the south of the Urals).

An increased magnetic field is observed in buried synclinoria and depressions of platform areas: Krestsovsk, Predtiman, and Terek depressions in the European part of the USSR; the Karaganda and Tokrauss synclinoria in Kazakhstan; Kass, Chulymo-Yenisei, and Minusinsk depressions in western Siberia; Penzhinsk depression in northeastern USSR. There are also other synclinoria and depressions that lack magnetic anomalies, such as the Bolshoy Donbass syncline, Indolo-Kuban depression, and western Kamchatka depression.

The magnetic field character over large areas of effusive rocks within different belts should be particularly noted. Effusives are most widespread in the Okhotsk-Chaun volcanic belt. Here effusives of different composition are characterized by high magnetization (from 500 to 5000 $\cdot$ 10^{-6} cgs), causing anomalies up to

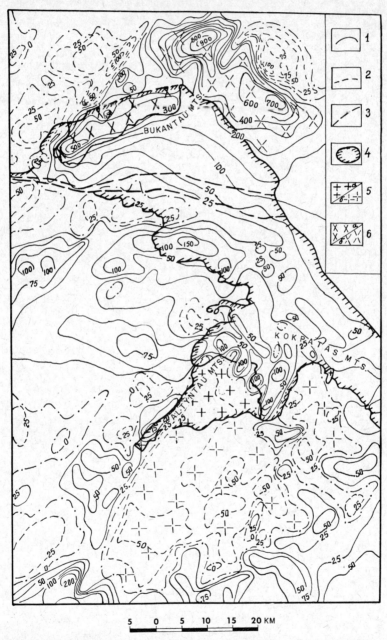

Fig. 4. Magnetic anomalies for one region in the Kyzyl-Kum upland: (1) isolines of positive $(\Delta T)a$ in gammas; (2) isolines of negative $(\Delta T)a$ in gammas; (3) fractures from geological data; (4) boundary of exposed folded formations; (5) granite massifs (a) exposed and (b) buried; (6) granodiorite massifs (a) exposed and (b) buried.

Fig. 5. Magnetic anomalies in the eastern part of the Baltic shield; symbols are the same as in Figure 3.

several thousand gammas. The anomalies of this intensity observed over basic effusive formations in Kamchatka [*Rivosh*, 1964] can be explained by the high natural remanent magnetization of Neogene-Quaternary lavas.

On Precambrian shields, as well as in folded areas, a certain regularity is observed in the zone of alternation between decreased and increased magnetic fields. However, the magnetic field characteristic for exposed ancient anticlinoria is more variable than for younger folded areas from the Baikal era to the Cenozoic era. Increased fields of $(\Delta T)a$ up to 2000–3000 γ are found in some blocks of the Kola Peninsula, the Ukraine, and the Anabar and Aldan shields. A relatively great quantity of magnetic rocks in ancient shields is probably explained by regional metamorphism and ultrametamorphism that increased the magnetite and titanomagnetite content of rocks and, consequently, magnetic susceptibility [*Dortman et al.*, 1964].

Positive anomalies of Z up to 20,000 γ are typical of charnockites of early Archean formations, especially on the Ukraine crystalline shield and on its buried southwestern slope. These formations have magnetization of about 2000–3000 · 10^{-6} cgs, depending on the degree of alteration of hypersthene gneiss.

Figure 5 shows the anomalous field $(\Delta T)a$ over the eastern part of the Baltic shield; the alternation of positive and negative magnetic fields, oriented northwest, is clearly seen. The Belomorides block, located in north Karelia and represented by late Archean gneiss formations with subordinate metasomatic granite massifs, is characterized mainly by a quiet negative field (about -100 γ), which is due to the small percentage of magnetite in its rocks. There are no marked anomalies over early Proterozoic granitoids, gneisses, and schists in Karelia and in the south of the Kola Peninsula. On the northern coast of Kandalaksha Bay there are exposed Proterozoic and Lower Archean intrusions characterized by anomalies up to 500 γ; the positive magnetic field (up to 1000 γ) is 30–40 km wide and extends across the White Sea to the northern shore of the Dvina Bay. The central part of the Kola Peninsula is characterized by intensive linear anomalies up to 2000 γ connected with volcanic depositions of the Pechengsko-Varzug synclinorium, with basic intrusions of gneisses and schists of the Keiv synclinorium (in the central part of the Kola Peninsula). Anomalies $(\Delta T)a$ up to 1200 γ are found on the northern coast of the Kola Peninsula. They are caused by the great granitoid magnetization of the Murmansk anticlinorium, which reaches 5000 · 10^{-6} cgs [*Dortman et al.*, 1964]. Increases and decreases in the magnetic field of the Baltic shield are also found to the southeast of the boundary of Precambrian exposure (Figure 5).

Rapakiwi granite massifs are characterized by a lack of magnetic anomalies; their magnetization does not exceed 50 · 10^{-6} cgs.

The establishment of certain relations between field and magnetic characteristics of exposed geological units in folded zones may, together with other geologic-geophysical data, permit prediction of the composition of buried basement rocks.

REFERENCES

Dortman, N. B., V. I. Vasilieva, A. K. Veinberg, et al., *Physical Properties of Rocks and Minerals of the USSR*, pp. 93–100, Nedra, 1964.

Glazunova, A. P., D. P. Golub, and Z. A. Makarova, Some problems of methods of aeromagnetic data interpretation in studying deep structure of western Central Asia, *Proc. All-Union Sci. Res. Geol. Inst.*, 2(46), 43–63, 1961.

Pechersky, D. M., Magnetic properties of volcanic rocks, *Proc. Northeast Sci. Res. Complex Inst.*, Magadan, 1964(9), 5–157, 1964.

Rivosh, L. A., Geomagnetic characteristics of main tectonic structures of the eastern part of the USSR, transition zone from Asia to Pacific and its abyssal bottom, *Geol. Geofiz.*, 1964(5), 38–51, 1964.

Simonenko, T. N., Anomalous magnetic field of the USSR and some tectonic problems, in *Folded Regions of Eurasia*, pp. 37–43, Nauka, Moscow, 1964.

Simonenko, T. N., and M. M. Tolstikhina, Some regularities of deep structure of the USSR territory, *Sov. Geol.*, 1965(4), 74–89, 1965.

Magnetic Intensity Field in the Pacific[1]

Victor Vacquier

MAPPING anomalies of the total magnetic intensity at sea with fluxgate and proton precession magnetometers is done to investigate the geology of the ocean floor. Anomalies in magnetic intensity occur over boundaries of bodies of rock of different magnetization. The anomalies roughly outline the areal extent of the rock bodies. To produce magnetic effects, the thickness of these bodies should be of the same order as the distance to the magnetometer, which is about 4 km over most of the ocean. If we estimate the depth at which the rocks cease to be magnetic because the Curie temperature for magnetite is reached (575°C), we find a normal depth of about 20 km. Generally, the sedimentary layer in the Pacific is only from 100 to 300 meters thick, and thus can be neglected so that one can expect magnetic anomalies to come from depths between 4 and 20 km. Over a substantial part of the Pacific Ocean the magnetic field is smooth. On the other hand, some areas of the Pacific are characterized by sets of linear anomalies 5 to 50 km wide and hundreds of km long. These lineated patterns of anomalies first discovered by *Mason* [1958] on the U. S. Coast and Geodetic Survey ship *Pioneer* and further developed by *Mason and Raff* [1961] have recently come into prominence in connection with the theory of ocean-floor spreading originally proposed by *Hess* [1962] and further elaborated by *Vine and Matthews* [1963], *Wilson* [1965], and *Vine* [1966]. The old hypothesis of continental drift is intimately connected with the concept of sea-floor spreading.

In the magnetic intensity diagram in Figure 1, the arrows mark a central magnetic anomaly on both sides of which appear narrower symmetrical side bands. The central anomaly is usually over the middle of an oceanic ridge, in this case relatively short ones called Gorda and Juan de Fuca ridges by *Wilson* [1965]. The Vine and Matthews theory of the formation of the magnetic pattern is illustrated in the bottom profile in Figure 2. Hot mantle material rises over the center of the ridge and becomes magnetized as temperature drops below the Curie temperature. It splits in two, half of it moving away from the ridge on either side of it. The geomagnetic field then reverses, and material of opposite direction of magnetization is extruded, creating the required magnetization contrast to produce the magnetic anomaly. This mechanism explains why the central anomaly is wider than the side bands; ideally the pattern should be symmetrical. The symmetry of the Juan de Fuca ridge is remarkable (Figures 2). Since the ridge is of finite length, *Wilson* [1965] takes up the motion at the ends of slippage along a horizontal fracture approximately normal to the ridge axis (Figure 3). If, as we go along this fault, there comes another ridge parallel to the first one, the same magnetic anomaly pattern is generated from it. In this manner offset segments of the ridge generate the same magnetic anomaly pattern without the pattern ever having been continuous before the formation of the horizontal faults. This combination of widening of the ridge segments and of slipping along the faults joining them was termed 'transform faulting' by *Wilson* [1965]. For the South Atlantic ridge, *Vacquier and Von Herzen* [1964] found that the zone of high heat flow coincides with the axes of the ridge segments, suggesting that the segments are growing at the same time, in accordance with the transform fault concept. However, except for the central anomaly, their data show no correlation of magnetic pattern from one profile to the next.

The assumption that the magnetic anomaly pattern originated from reversals of the geomagnetic field while the sea floor was being extruded at constant speed has been correlated with palaeomagnetic data from dated reversals of the geomagnetic field found in lavas [*Cox et al.*, 1964; *Doell and Dalrymple*, 1966] and

---

[1] Contribution of the Scripps Institution of Oceanography, new series.

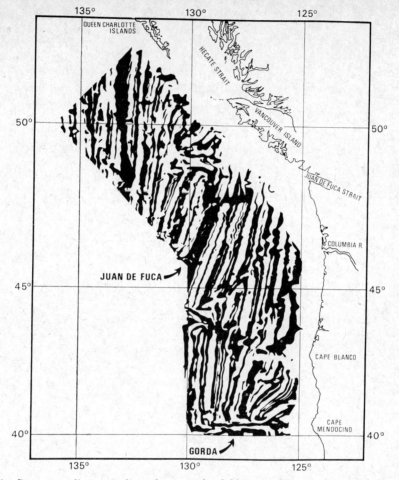

Fig. 1. Summary diagram of total magnetic field anomalies southwest of Vancouver Island [after *Raff and Mason*, 1961]. Arrows indicate Gorda and Juan de Fuca ridges from which sea floor is spreading, according to the interpretation by *Vine and Wilson* [1965].

in oceanic sediments [*Opdyke et al.*, 1966]. *Vine* [1966] and *Heirtzler* [1967] recognize the reversal pattern on the sides of presently active oceanic ridges on a worldwide scale (Figure 4). The actual correlation of the magnetic field pattern (Figure 5) with palaeomagnetic reversals as measured in lava flows and oceanic sediments extends to about 4 m.y. After that, a uniform rate of spreading is assumed for assigning an age to the anomalies.

The north-south lineated anomalies in the northeastern Pacific are intersected by the great east-west fracture zones described by *Menard* [1964]. It was shown by *Vacquier* [1959], *Vacquier et al.* [1961], and *Raff* [1962, 1966] that the magnetic pattern was displaced across the Mendocino, Pioneer, and Murray faults for a maximum distance of 1400 km (Figure 6). Before Wilson's transform fault concept, *Menard* [1964] interpreted the displacement of magnetic pattern across the faults by assuming that, before faulting, the anomaly patterns were aligned and that subsequently, due to intrusions along the crest of the East Pacific rise, the ocean floor was displaced to the west at different speeds. The east-west fracture zones would appear between the regions of different spreading rates. Quantitatively equal offsets of the bathymetric curves across the Mendocino, the Pioneer, and the Murray faults also pointed out by *Menard* [1964] strongly suggested that the magnetic anomalies, along with the bathymetric

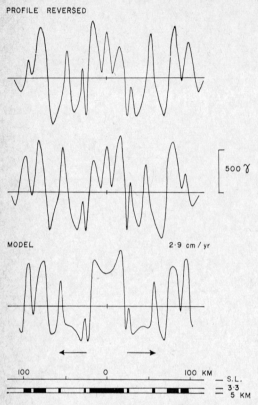

Fig. 2. Magnetic intensity profile across the Juan de Fuca ridge [after *Vine*, 1966]. Top and middle profiles are the same data plotted from southeast to northwest and northwest to southeast, respectively, to demonstrate symmetry about the axis of the ridge. Bottom profile is computed from the model drawn at the bottom of the figure. S. L. is sea level. The magnetic intensity was calculated from a two-dimensional model consisting of rectangular bodies of normal and reverse magnetizations and extending from 3.3 to 5.0 km in depth. Comparison with palaeomagnetically measured reversals of Figure 5 gives a rate of spreading of 2.9 cm/year.

contours, had at one time been continuous. However, it may also be argued that the ocean becomes deeper as one moves away from the source of the spreading.

Vine [1966] tentatively suggests that the magnetic stripes north of the Mendocino fault (see Figure 6) are spreading from the Gorda ridge, which is presently active. This implies that south of the Pioneer fault the spreading is now occurring from a ridge segment located at the longitude of the western boundary of the state of Colorado (∼110°W). The extrusion of the Pacific Ocean floor from a place so continental without breaking up western North America seems mechanically unlikely. It seems more plausible [*Vine*, 1966] to speculate that, at the end of Jurassic time, North America just started to move away from the Mid-Atlantic ridge and the main magnetic pattern of the northeast Pacific Ocean was free to form by the *Vine and Matthews* [1963] mechanism, with the combined left lateral offset of 1400 km across the Mendocino and Pioneer transform faults of *Wilson* [1965]. Thus the sequence of the magnetic stripes in the Pacific require the westward migration of North America. North of the Mendocino fault, west of 163°W, the magnetic stripes die out. This can be interpreted as reflecting the known absence of geomagnetic field reversals in the Permian. Evidently the source of spreading that produced the magnetic stripes south of the Mendocino fault was at a place that is now at least 1200 km inland. Thus, not only did the spreading oceanic crust and upper mantle slip under the continent, but the spreading ridge itself got lost. The geothermal consequences of digestion of oceanic crust under the continents have not been quantitatively explored. The geophysical anomalies in the western United States [*Schmucker*, 1964; *Pakiser*, 1965] do indicate higher temperatures than normal, which are confirmed by recent vulcanism in the Basin and Range geologic province.

In the Gulf of Alaska, *Elvers et al.* [1967] found that some magnetic lineations belonging to the old pattern need not terminate in a transform fault but that a pattern consisting of several anomalies can bend sharply and then proceed in another direction almost at right angles to its former one (Figure 7). Some of these anomalies have been traced to terminate one by one in the Aleutian trench. To account for this morphology, *Raff* [1968] proposed that the magnetic pattern was formed from three ridges issuing from a common point from which the sea floor was spreading. We see in Figure 7 only the western remains of two of them, the rest of the pattern, including the spreading center, having been absorbed by the Aleutian trench and a trench which in the past ran along the Canadian coast. The argument is reminiscent of the lost continent Appalachia proposed

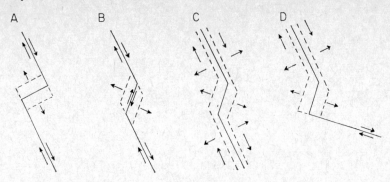

Fig. 3. Diagram illustrating variations in the transformation of mid-ocean ridges to transform faults depending upon the angles of junction and directions of motion [after *Wilson*, 1965].

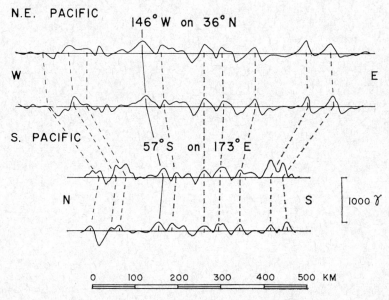

Fig. 4. The magnetic anomaly pattern correlated between profiles by *Peter* [1966] and by *Raff* [1966] in the north Pacific and *Christoffel and Ross* [1965] in the south Pacific [after *Vine*, 1966]. Vine suggests that the only difference between the two is the rate of spreading that formed them.

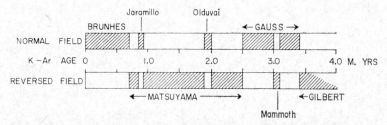

Fig. 5. Geomagnetic polarity epochs deduced from palaeomagnetic results and potassium-argon dating; from *Cox et al.* [1964] and *Doell and Dalrymple* [1966], as quoted by *Vine* [1966].

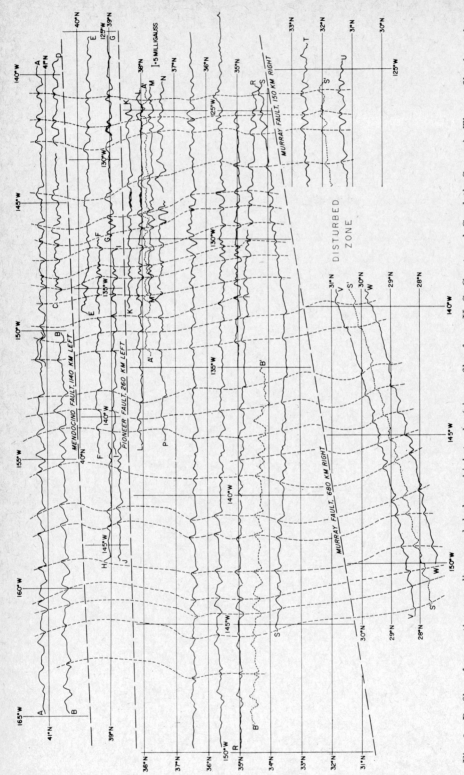

Fig. 6. Profiles of magnetic anomalies recorded along ship's tracks (six profiles from U. S. Coast and Geodetic Survey). The profiles have been shifted east-west as indicated by the longitude scales so as to align the magnetic anomalies across the major oceanic faults. Portions of profiles AA, BB, and SS are inserted on the other sides of the faults as dashed profiles $A'A'$, $B'B'$, and $S'S'$ so that the reader can make comparisons [after Raff, 1966].

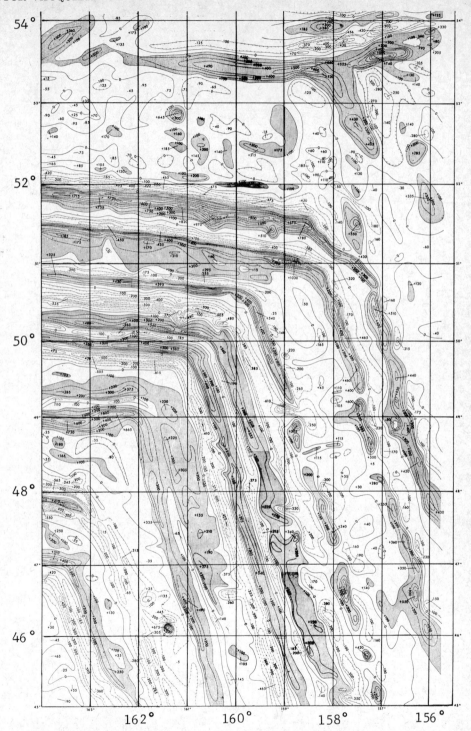

Fig. 7. The great magnetic bight of the northeast Pacific Ocean: residual magnetic anomaly map. Shaded areas: $> +100$ gamma [after *Elvers et al.*, 1967].

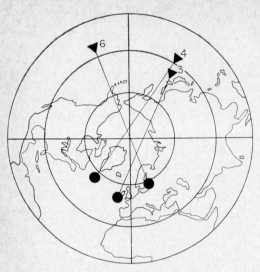

Fig. 8. Virtual geomagnetic pole positions for Cretaceous seamounts in the north Pacific Ocean [after *Vacquier and Uyeda*, 1967]. Triangles show the positions of the seamount groups. The numbers are the number of individual seamounts in each group. The circles are the positions of the virtual geomagnetic poles for each group. The distance of the VGP from the north pole is interpreted as caused by the northward migration of the seamounts.

as the source of sediments found in the Appalachian geosyncline. *Raff* [1968] gives a generalized system of spreading ridges, both visible and invisible, which could be responsible for lineated magnetic anomalies of the north Pacific. Such devices can 'explain' any geometric pattern of magnetic anomalies. A comprehensive list of references on sea-floor spreading is given by *Le Pichon* [1968].

Magnetic surveys of seamounts constitute another line of evidence for the mobility of the ocean floor in latitude. By combining magnetic and bathymetric surveys of isolated submarine uplifts under the assumption of uniformity of magnetization and of a horizontal plane surface for their bottom, it is possible to compute the direction and magnitude of their magnetization, most of which can safely be assumed to be remanent [*Vacquier*, 1962; *Richards et al.*, 1967; *Vacquier and Uyeda*, 1967]. From this direction of magnetization, a virtual geomagnetic pole position can be calculated [*Irving*, 1965]. This virtual geomagnetic pole is assumed to have been at the geographic pole at the time of formation of the seamount, and so its displacement can be interpreted as resulting from intervening drift of the seamount. The measured seamounts occur in groups, and some of them have been dated either paleontologically or by the K-Ar method from dredged specimens of rock (Figure 8) [*Vacquier and Uyeda*, 1967]. The computed virtual geomagnetic poles suggest that the seamount groups in the western Pacific have moved 37° northward since late Cretaceous and that those near Hawaii moved on the average 30° northward. Unfortunately the seamounts in the eastern Pacific have not been dated; some of them also show a northward drift, but smaller in magnitude. This suggested northward drift of the Pacific Ocean floor agrees with the generally northward displacement of the continents since Permian time obtained from palaeomagnetic measurements on land [*Irving*, 1964, p. 142]. The palaeomagnetic data can also be interpreted as demonstrating polar wandering rather than northward drift.

Possible northward drift of the floor of the northwest Pacific is compatible with recent measurements of sedimentary thickness by seismic reflection (H. W. Menard, personal communication). The equatorial sedimentary belt, which is about 10° wide and very prominent in the eastern Pacific, thins out and gradually disappears to the west. This thinning could be caused by northerly drift of the ocean floor. Abrupt thinning of sediment on crests of mid-ocean ridges also suggests that sea floor spreading as revealed by lineated magnetic patterns was not taking place before 10 m.y. ago, indicating that spreading might be episodic in addition to being world wide [*Ewing and Ewing*, 1967].

The evidence for the absorption of the oceanic crust and upper mantle is much less compelling than its generation from ridges, where the spacing of magnetic anomalies, increased heat flow, low subcrustal seismic velocities, and thin sediment attest to this dynamic process. Absorption of the crust in the Pacific presumably occurs at deep-sea trenches. However, as a rule, sediment does not pile up in trenches; this argues against the idea that the upper surface of the basaltic layer moves under the continental plate. On the other hand, where the lineated magnetic anomalies come near the continent, as they do, for example, at the west coast of North America north of the Mendocino fracture zone, the magnetic pattern abruptly disappears at the foot of the continental slope, suggesting that the rock carrying the magnetic anomalies dips under the continental mass [Raff and Mason, 1961].

An interesting possible geothermal consequence of shoving a thickness of oceanic basalt under the continental crust, and the thickness would be that above the low-velocity layer, is an increase of heat flow at some distance from the margin of the ocean. This is because the oceanic mantle, being less differentiated, contains more radioactive material than the continental mantle at the depth to which the oceanic mantle is being shoved. Given enough time, the slight excess of heat-generating material is likely to produce partial melting, which manifests itself by the familiar 'ring of fire' around the Pacific and the observed rise in terrestrial heat flow in the Bering Sea, the Sea of Okhotsk, the Sea of Japan, the Philippine Sea, the North Fiji basin, and possibly the Andes, where a rise in the geotherms is postulated from geomagnetic time variations by Schmucker et al. [1965].

REFERENCES

Christoffel, D. A., and D. I. Ross, Magnetic anomalies south of the New Zealand Plateau, J. Geophys. Res., 70(12), 2857–2861, 1965.

Cox, A., R. R. Doell, and G. B. Dalrymple, Reversals of the earth's magnetic field, Science, 144, 1537, 1964.

Doell, R. R., and G. B. Dalrymple. Geomagnetic polarity epochs: A new polarity event and the age of the Brunhes-Matuyama boundary, Science, 152, 1060, 1966.

Elvers, D. J., C. C. Mathewson, R. E. Kohler, and R. L. Moses, Systematic ocean surveys by the USCGS Pioneer 1961–1963, Operational Data Rept. C & GSDR-1, U. S. Coast & Geodetic Survey, Department of Commerce, 1967.

Ewing, M., and J. Ewing, Deep sea sediments in relation to island arcs and mid-ocean ridges, Trans. Am. Geophys. Union, 48, 216, 1967.

Heirtzler, J., Marine magnetic anomalies and a moving ocean floor, Trans. Am. Geophys. Union, 48, 216, 1967.

Hess, H. H., History of the ocean basins, in Petrologic Studies, edited by A. E. J. Engel et al., pp. 599–620, Geological Society of America, New York, 1962.

Hess, H. H., Comments on the Pacific basin, in Continental Margins and Island Arcs, Can. Geol. Surv. Paper 66-15, 311–316, 1966.

Irving, E., Paleomagnetism, John Wiley & Sons, New York, 399 pp., 1964.

Le Pichon, Xavier, Sea floor spreading and continental drift, J. Geophys. Res., 73, 3361–3397, 1968.

Mason, R. G., A magnetic survey off the west coast of the United States, Geophys. J., 1, 320–329, 1958.

Mason, R. G., and A. D. Raff, Magnetic survey off the west coast of North America, Bull. Geol. Soc. Am., 72, 1259–1265, 1961.

Menard, H. W., Marine Geology of the Pacific, McGraw-Hill Book Company, New York, 1964.

Opdyke, N. D., B. Glass, J. D. Hays, and J. Foster, Paleomagnetic studies of Antarctic deep-sea cores, Science, 154, 349, 1966.

Pakiser, L. C., Transcontinental crust and upper mantle structure, Rev. Geophys., 3, 505–520, 1965.

Peter, G., Magnetic anomalies and fracture pattern in the northeast Pacific Ocean, J. Geophys. Res., 71, 5365–5374, 1966.

Raff, A. D., Further magnetic measurements along the Murray fault, J. Geophys. Res., 67, 417–418, 1962.

Raff, A. D., Boundaries of an area of very long magnetic anomalies in the northwest Pacific, J. Geophys. Res., 71, 2631–2636, 1966.

Raff, A. D., Sea-floor spreading–Another rift, J. Geophys. Res., 73, 3699–3705, 1968.

Raff, A. D., and R. G. Mason, Magnetic survey off the west coast of North America, Bull. Geol. Soc. Am., 72, 1267–1270, 1961.

Richards, M. L., V. Vacquier, and G. D. Van Voorhis, Calculation of the magnetization of uplifts from combining topographic and magnetic surveys, Geophysics, 32, 678–707, 1967.

Schmucker, U., Anomalies in the geomagnetic variations in the southwestern United States, J. Geomag. Geoelec., 15, 193–221, 1964.

Schmucker, U., O. Hartmann, A. A. Giesecke, Jr., M. Casaverde, and S. E. Forbush, Electrical conductivity anomalies in the earth's crust in Peru, Carnegie Inst. Wash. Yr. Book 64, 309–310, 1965.

Talwani, M., X. Le Pichon, and J. R. Heirtzler, East Pacific rise: The magnetic pattern and the fracture zones, Science, 150, 482, 1965.

Vacquier, V., Measurement of horizontal displacement along faults in the ocean floor, Nature, 183, 452–453, 1959.

Vacquier, V., A machine method for computing the magnitude and the direction of magnetization of a uniformly magnetized body, Benedum Symposium on Palaeomagnetism. pp. 123–137, University of Pittsburgh, Pittsburgh, 1962.

Vacquier, V., A. D. Raff, and R. E. Warren, Horizontal displacements in the floor of the northeast Pacific Ocean, Bull. Geol. Soc. Am., 72, 1251–1258, 1961.

Vacquier, V., and R. P. Von Herzen, Evidence for connection between heat flow and the mid-Atlantic ridge magnetic anomaly, J. Geophys. Res., 69, 1093–1101, 1964.

Vacquier, V., and S. Uyeda, Paleomagnetism of some seamounts in the western Pacific and some volcanoes in Japan, 3, *Bull. Earthquake Res. Inst., Tokyo Univ., 45,* 815–848, 1967.

Vine, F. J., Interpretation of magnetic anomalies observed at sea, thesis, Cambridge University, 1965.

Vine, F. J., Spreading of the ocean floor: New evidence, *Science, 154,* 1405–1415, 1966.

Vine, F. J., and D. H. Matthews, Magnetic anomalies over mid-ocean ridges, *Nature, 199,* 947–949, 1963.

Vine, F. J., and J. Tuzo Wilson, Magnetic anomalies over a young oceanic ridge off Vancouver Island, *Science, 150,* 485–489, 1965.

Wilson, J. Tuzo, Transform faults, oceanic ridges and magnetic anomalies southwest of Vancouver Island, *Science, 150,* 482–485, 1965.

Geomagnetic Studies in the Atlantic Ocean[1]

J. R. Heirtzler

THE first marine geomagnetic measurements were made in the Atlantic Ocean, and today the Atlantic is more densely covered by magnetic profiles than any other ocean. Studies have been concentrated mainly on the Mid-Atlantic ridge system, the magnetization or nonmagnetization of seamounts, and the continental shelves.

Magnetic measurements have long been used to study the distribution of magnetic bodies within the earth's crust on land. In the late 1940's, as a result of wartime applications, magnetometers were used at sea. Although marine geomagnetic measurements were first made by aircraft, the magnetometer was soon adapted for ship-towed use [*Heezen et al.,* 1953]. *Miller and Ewing* [1956] reported the first study of a seamount, Caryn seamount, off the east coast of the United States. The large, universal magnetic anomaly of the mid-ocean ridge axis was observed over the Mid-Atlantic ridge [*Ewing et al.,* 1957], and a distinct boundary between a magnetically flat area and a region of considerable magnetic disturbance was noticed on the western edge of the Bermuda rise [*King et al.,* 1961]. The presence of great, elongated magnetic anomalies off the east coast of North America was described by *Drake et al.* [1963]. Developments in paleomagnetism have permitted an explanation of the magnetization of seamounts, anomalies associated with the mid-ocean ridges, and a tentative explanation for the rough-smooth boundary [*Heirtzler and Hayes,* 1967]. These developments in paleomagnetism can thus explain nearly all the magnetic anomalies in the deep ocean. Magnetic anomalies associated with continental margins, island arcs, and adjoining seas appear to have a different cause. The magnetic characteristics of these marine regions have been studied by the authors mentioned below.

MID-ATLANTIC RIDGE

After the discovery of the great field of linear magnetic anomalies off western North America [*Vacquier et al.,* 1961; *Mason and Raff,* 1961; *Raff and Mason,* 1961], more attention was turned to the Mid-Atlantic ridge. In 1963 the U. S. Naval Oceanographic Office, in a cooperative program with Lamont Geological Observatory, flew a detailed aeromagnetic survey over the Reykjanes ridge, a portion of the Mid-Atlantic ridge southwest of Iceland. A pattern of linear anomalies (Figure 1), symmetrically spaced about the ridge axis, was found [*Baron et al.,* 1965; *Heirtzler et al.,* 1966a]. This symmetric nature of the anomalies gave some support to the theory of formation of ridge anomalies put forward by *Vine and Matthews* [1963]. Vine and Matthews' theory was developed from the theory of ocean floor spreading, as suggested by *Hess* [1962] and *Dietz* [1961], wherein material issues from the axis of the ridge and spreads horizontally in both

---

[1] Contribution 1253 from the Lamont-Doherty Geological Observatory of Columbia University.

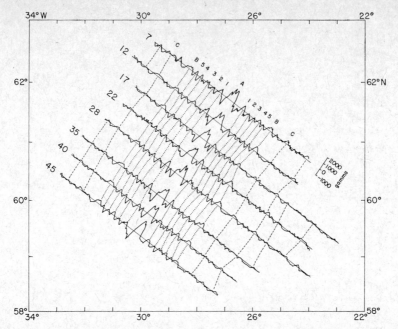

Fig. 1. Eight total intensity magnetic anomaly profiles over a section of the Reykjanes ridge southwest of Iceland [after *Talwani et al.*, 1965].

directions. Vine and Matthews proposed that the material acquired its magnetization when it was at the ridge axis, and thus it reflects the history of geomagnetic field reversals.

That the center of the ridge in the North Atlantic had magnetic characteristics similar to the rest of the Mid-Atlantic ridge was illustrated in the study of other magnetic profiles from 60°N to 42°S [*Keen*, 1963; *Heirtzler and Le Pichon*, 1965; *Matthews*, 1967]. Several of these studies indicated a longer wavelength axial magnetic anomaly upon which characteristic shorter wavelength anomalies are superimposed. The longer wavelength variation of field strength does not seem to be present on every profile across the ridge. To study these longer (few hundred kilometer) wavelength anomalies, the choice of a regional field is important and warrants careful consideration.

An interpretation of the magnetics over the Pacific antarctic ridge [*Pitman and Heirtzler*, 1966] indicated a relatively rapid motion of the ocean floor there and permitted the construction of a time scale for geomagnetic field reversals for the last 10 million years. This time scale was then applied to the magnetic pattern over the Reykjanes ridge. It was shown that the ob- servations could be accurately accounted for there with a spreading rate of 1 cm/year away from the axis. A study of a large part of the South Atlantic [*Dickson et al.*, 1968] shows a spreading rate of about 2 cm/year, and a study of the axial region between 26 and 29°N yields a speed of about 1.25 cm/year [*Phillips*, 1967]. Though many authors have found good to excellent linearity and symmetry of the anomaly pattern, that is not always the case. *Vogt and Ostenso* [1966], in working in an axial area at about 40°N, close to the median rift valley and near an area previously studied by *Hill* [1960], found an isolated body that seemed to be magnetized differently on its different sides. *Loncarevic et al.* [1966] found generally elongated magnetic bodies in a survey near 45°N and observed three pairs of seamounts situated symmetrically about the median axis. They considered the possibility of the injection of a dike swarm near the axis to account for the anomalies.

Other studies of the Mid-Atlantic ridge and its associated magnetic features include those of *King et al.* [1964] and *Avery et al.* [1968] (see Figure 2) for part of the Norwegian Sea, *Krause* [1965] for the Azores area, *Ivanov*

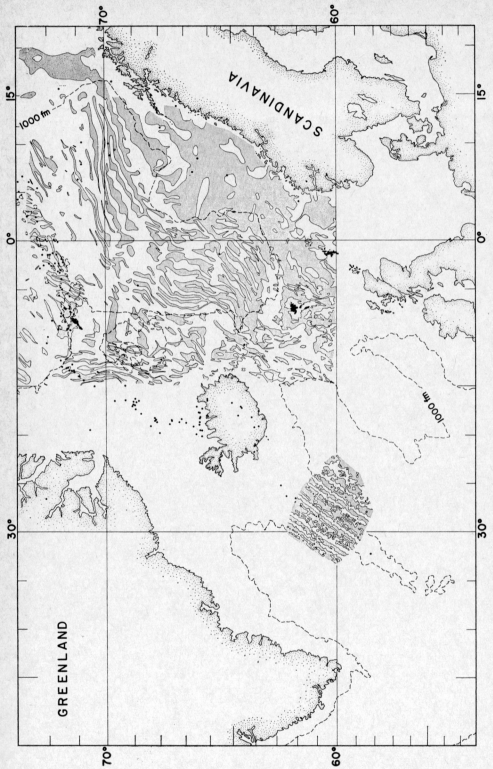

Fig. 2. Anomalies in the total field intensity over the Norwegian Sea and Reykjanes ridge [after *Avery et al.*, 1968]. Negative anomalies are shaded; dots indicate earthquake epicenters [after *Sykes*, 1965].

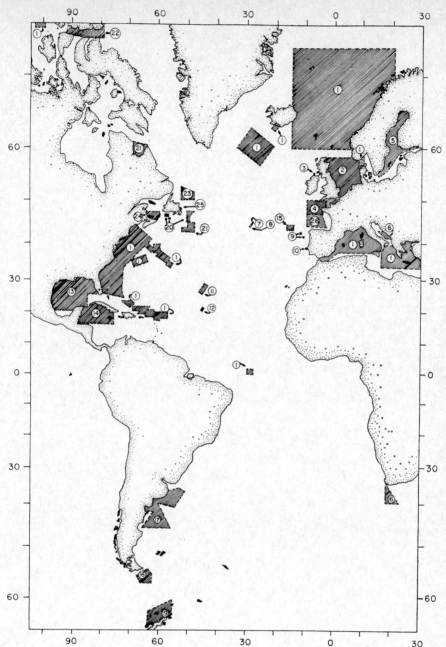

Fig. 3. Areas of the Atlantic Ocean where total magnetic field intensity has been contoured: 1, *U. S. Naval Oceanographic Office* [1967]; 2, *Oil and Gas Journal* [1963]; 3, *Riddihough* [1964]; 4, *Hill and Vine* [1965]; 5, *Ostenso* [1966]; 6, *Turajlic* [1966]; 7, *Loncarevic et al.* [1966]; 8, *Vogt and Ostenso* [1966]; 9, *Black et al.* [1964]; 10, *Allan* [1965]; 11, *Phillips* [1967]; 12, Woods Hole Oceanographic Institution; 13, *Heirtzler et al.* [1966a]; 14, Heirtzler and Gough, personal communication; 15, *Worzel et al.* [1966]; 16, *Oguti* [1964]; 17, *Capurro* [1964]; 18, *Peter* [1962]; 19, *Griffiths et al.* [1964]; 20, *Bower* [1962]; 21, Bedford Institute; 22, *Barrett* [1966]; 23, Oceanographic Institute, Dalhousie University; 24, M. J. Keen, personal communication; 25, *McCartney* [1954]; 26, Department of Geodesy and Geophysics, Cambridge University.

[1963] for a study of the depth of burial. *Matthews* [1967] has reviewed studies of magnetics and mid-ocean ridges. A series of east-west traverses of the North Atlantic is being made in a joint project between the English and the Dutch [*Vestine*, 1966]. The Marine Geophysical Survey of the U. S. Naval Oceanographic Office has surveyed four large areas in the North Atlantic. The first report from that project indicates that the area between 25 and 44°N and 51 and 77°W can possibly be divided into three magnetic provinces [*Vajk and Miller*, 1967].

CONTINENTAL MARGINS AND ISLAND ARCS

The continental margins bordering the Atlantic have also been the focus of attention. The eastern shelf of the United States has been covered in a close grid survey by the Naval Oceanographic Office, and that area will adjoin surveys of the area north of the northern Caribbean islands (see *U. S. Naval Oceanographic Office* [1967] for description of surveys). Much of the area of the Grand Banks and other waters near Canada have been surveyed by the Canadian government, Bedford Institute, and Dalhousie University. The shelves of England and France have been covered as part of continental aeromagnetic surveys of those countries, while the Imperial College and Cambridge University have completed surveys in adjoining areas [*Riddihough*, 1964; *Hill and Vine*, 1965; *Allan*, 1965]. A group of nonmagnetic seamounts were studied off the Iberian Peninsula [*Black et al.*, 1964]. In the southern hemisphere, surveys have been made off the western coast of South Africa (Van Wijk, personal communication), off the southern coast of South Africa [*Oguti*, 1964], and off the coast of Argentina [*Capurro*, 1964].

ADJOINING SEAS

All the major seas joining the Atlantic have now been surveyed. *Griffiths et al.* [1964] reported on the Scotia arc; *Peter* [1962] on the Drake Passage; *Heirtzler et al.* [1966b] and *Gough* [1967] on the Gulf of Mexico; *Hood and Godby* [1964] on the Labrador Sea, although considerably more data now exist there [*Barrett*, 1966; *Manchester*, 1967] and in Hudson Bay [*Beal*, 1967]. The North Sea was mapped in an aeromagnetic survey [*Oil and Gas Journal*, 1963], and much of the Baltic Sea and Gulf of Bothnia by the Dominion Observatory of Canada [*Serson*, 1967]. Nearly all the Norwegian and Mediterranean Seas have been mapped by the U. S. Naval Oceanographic Office.

There are nearly a hundred single magnetic profiles taken by ships across the Atlantic Ocean [*Vestine*, 1966; *Ostenso*, 1966]. To indicate the large extent of coverage, Figure 3 shows areas that have been contoured in the total field intensity and can be analyzed for crustal anomalies.

RESULTS

The identification of deep ocean magnetic anomalies with specific reversals of the earth's magnetic field has been accomplished by extrapolating from the age of known reversals and the distance of the corresponding anomaly from the mid-ocean ridge axis. Interpretation of the magnetic data in the Atlantic Ocean has yielded the following history of that ocean. The North American and European continents separated, according to proponents of continental drift, about 200–300 million years ago, and have drifted apart at a rate of 1.0 to 1.5 cm/year, to either side of the Mid-Atlantic ridge. For some time before 70 million years ago, the northern end of the Mid-Atlantic ridge went between North America and a European block. About 60–70 million years ago, the northern end of the Mid-Atlantic ridge altered its direction, becoming more easterly to pass through part of the European block. By this action a piece of this block was broken off, and it drifted to the west. That piece is Greenland. The South Atlantic data suggest that South America and Africa were probably a single land mass until 150–200 million years ago, at which time they started to drift apart at a half-rate of 1.5 to 2.2 cm/year. The magnetic data in the South Atlantic thus also support the continental drift hypothesis advocated by investigators of paleomagnetic field directions.

Acknowledgments. This work was supported in part by Office of Naval Research contract N00014–67–A–0108–0004 and National Science Foundation grant GP-5536.

REFERENCES

Allan, T. D., A magnetic survey off the coast of Portugal, *Geophysics, 30,* 411–417, 1965.

Avery, Otis E., Gordon D. Burton, and James R. Heirtzler, An aeromagnetic survey of the Norwegian Sea, *J. Geophys. Res., 73,* 4583–4600, 1968.

Barrett, D. L., Lancaster Sound shipborne magnetometer survey, *Can. J. Earth Sci., 3*, 223-235, 1966.

Baron, J. Gregory, James R. Heirtzler, and George R. Lorentzen, An airborne geomagnetic survey of the Reykjanes ridge, 1963, unpublished informal report No. H-3-65, U. S. Naval Oceanographic Office, Washington, D. C., 1965.

Beal, C. S., *Centennial Volume on Hudson Bay*, Department of Energy, Mining and Resources, Ottawa, 1967.

Black, M., M. N. Hill, A. S. Laughton, and D. H. Matthews, Three non-magnetic seamounts off the Iberian coast, *Quart. J. Geol. Soc. London, 120*, 477-517, 1964.

Bower, Margaret E., Sea magnetometer surveys of the Grand Bank of Newfoundland, Burgeo Bank and St. Pierre Bank, *Can. Geol. Surv. Paper 61-30*, 1962.

Capurro, Luis R. A., Geophysical investigations in the western South Atlantic, in *Studies on Oceanography*, University of Tokyo, Tokyo, 1964.

Dickson, G. O., W. C. Pitman, III, and J. R. Heirtzler, Magnetic anomalies in the South Atlantic and ocean floor spreading, *J. Geophys. Res., 73*, 2087-2100, 1968.

Dietz, R. S., Continent and ocean basin evolution by spreading of the sea floor, *Nature, 190*, 859, 1961.

Drake, C. L., J. Heirtzler, and J. Hirshman, Magnetic anomalies off eastern North America, *J. Geophys. Res., 68*, 5259, 1963.

Ewing, M., B. C. Heezen, and J. Hirshman, Mid-Atlantic ridge seismic belts magnetic anomalies (abstract), Comm. 10 bis Assoc. Seismol., Gen. Assembly IUGG, Toronto, 1957.

Gough, D. I., Magnetic anomalies and crustal structure in eastern Gulf of Mexico, *Bull. Am. Assoc. Pet. Geol., 51*, 200-211, 1967.

Griffiths, D. H., R. P. Riddihough, H. A. D. Cameron, and P. Kennet, Geophysical investigations of the Scotia Arc, *Brit. Antarctic Surv. Rept. 46*, 43, 1964.

Heezen, B. C., M. Ewing, and E. T. Miller, Trans-Atlantic profile of total magnetic intensity and topography, Dakar to Barbados, *Deep-Sea Res., 1*(1), 25-33, 1953.

Hess, H. H., History of ocean basins, in *Petrologic Studies: A Volume to Honor A. F. Buddington*, edited by A. E. J. Engel et al., Geological Society of America, New York, 1962.

Heirtzler, James R., Xavier Le Pichon, and J. Gregory Baron, Magnetic anomalies over the Reykjanes ridge, *Deep-Sea Res., 13*, 427-443, 1966a.

Heirtzler, J. R., Lloyd H. Burckle, and George Peter, Magnetic anomalies in the Gulf of Mexico, *J. Geophys. Res., 71*, 519-526, 1966b.

Heirtzler, James R., and Xavier Le Pichon, Crustal structure of the Mid-Ocean ridges, 3, Magnetic anomalies over the Mid-Atlantic ridge, *J. Geophys. Res., 70*, 4013-4033, 1965.

Heirtzler, James R., and Dennis E. Hayes, Magnetic boundaries in the North Atlantic Ocean, *Science, 157*, 185-187, 1967.

Hill, M. N., A median valley of the Mid-Atlantic ridge, *Deep-Sea Res., 6*, 193, 1960.

Hill, M. N., and F. J. Vine, A preliminary magnetic survey of the western approaches to the English Channel, *Quart. J. Geol. Soc. London, 121*, 484, 1965.

Hood, P. J., and E. A. Godby, Magnetic anomalies over the Mid-Labrador Sea ridge, *Nature, 4937*, 1099, 1964.

Ivanov, M. M., The relationships between the magnetic field in the Atlantic and the bottom relief of the ocean, *Geomagnetism and Aeronomy, USSR, 3*, 633-636, 1963.

Keen, M. J., Magnetic anomalies over the Atlantic ridge, *Nature, 197*, 888-890, 1963.

King, Elizabeth R., Isidore Zietz, and Leroy R. Alldredge, Genesis of the Arctic Ocean basin, *Science, 144*(3626), 1551-1557, 1964.

King, E. R., I. Zietz, and W. J. Dempsey, The significance of a group of aeromagnetic profiles off the eastern coast of North America, *U. S. Geol. Surv. Prof. Paper 424-D*, 299-303, 1961.

Krause, Dale C., East and west Azores fracture-zones in the North Atlantic, *Proceedings of the Seventeenth Symposium of the Colston Research Society*, pp. 163-172, Butterworths Publications, London, 1966.

Loncarevic, B. D., C. S. Mason, and D. H. Matthews, Mid-Atlantic ridge near 45° north, 1, The Median Valley, *Can. J. Earth Sci., 3*, 327-349, 1966.

McCartney, W., Notes on the geological interpretation of the aeromagnetic map of Conception Bay, Newfoundland, *Can. Geol. Surv., Topical Rept.*, 1954.

Manchester, K. S., Geophysical measurements between Canada and Greenland, Thesis, Dalhousie University, 1967.

Mason, R. G., and A. D. Raff, Magnetic survey off the west coast of North America, 32°N latitude to 42°N latitude, *Bull. Geol. Soc. Am., 72*, 1259, 1961.

Matthews, D. H., Mid-Ocean ridge, in *Dictionary of Geophysics*, edited by S. K. Runcorn, pp. 979-991, Pergamon Press, London, 1967.

Miller, E. T., and M. Ewing, Geomagnetic measurements in the Gulf of Mexico and in the vicinity of Caryn Peak, *Geophysics, 21*, 432, 1956.

Oguti, T., Geomagnetic anomaly around the continental shelf margin southern offshore of Africa, *J. Geomag. and Geoelec., 16*, 65-66, 1964.

Oil and Gas Journal, Biggest magnetometer survey, spans 144,000 square miles, 6 pp., Aug. 19, 1963.

Ostenso, Ned A., Geomagnetism, *Trans. Am. Geophys. Union, 47*, 303-332, 1966.

Peter, George, Magnetic total intensity measurements in the Drake Passage, Tech. Rept. 5,

CU-5-62, Nonr-Geology, Lamont Geological Observatory, 1962.
Phillips, J. D., Magnetic anomalies over the Mid-Atlantic ridge near 27°N, *Science, 157,* 920–923, 1967.
Pitman, W. C., III, and J. R. Heirtzler, Magnetic anomalies over the Pacific-Antarctic ridge, *Science, 154,* 1164–1171, 1966.
Raff, A. D., and R. G. Mason, Magnetic survey off the west coast of North America, 40°N latitude to 52°N latitude, *Bull. Geol. Soc. Am., 72,* 1267, 1961.
Riddihough, R. P., Magnetic survey off north coast of Ireland, *Nature, 203,* 747–748, 1964.
Serson, P. H., Magnetism, in *Canadian Upper Mantle Report, 1967, Can. Geol. Surv. Paper 67–41,* 98–102, 1967.
Sykes, L. R., The seismicity of the arctic, *Bull. Seismol. Soc. Am., 55,* 501, 1965.
Talwani, Manik, Xavier Le Pichon, and James R. Heirtzler, East Pacific rise: The magnetic pattern and the fracture zones, *Science, 150,* 1109, 1965.
Turajlic, R., Magnetic Institute, Grocka, Yugoslavia, 1966.
U. S. Naval Oceanographic Office, Geomagnetic Surveys, Informal Rept. IR No. 67-52, Hydrographic Surv. Dept., U. S. Naval Oceanographic Office, 1967.
Vacquier, V., A. D. Raff, and R. E. Warren, Horizontal displacements in the floor of the northeastern Pacific Ocean, *Bull. Geol. Soc. Am., 72,* 1267, 1961.
Vajk, Raoul, and E. T. Miller, Marine magnetic survey in the northwest Atlantic Ocean (abstract), *Trans. Am. Geophys. Union, 48,* 90, 1967.
Vestine, E. H., *World Magnetic Survey Notes, No. 3, Natl. Acad. Sci.,* 1966.
Vine, F. J., and D. H. Matthews, Magnetic anomalies over oceanic ridges, *Nature, 199,* 947–949, 1963.
Vogt, Peter R., and Ned A. Ostenso, Magnetic survey over the Mid-Atlantic ridge between 42°N and 46°N, *J. Geophys. Res., 71,* 4389–4411, 1966.
Worzel, J. Lamar, Manik Talwani, James R. Heirtzler, and Maurice Ewing, A marine geophysical survey using satellite navigation, A, Navigation, topography, magnetics, and gravity (abstract), *Trans. Am. Geophys. Union, 47,* 123, 1966.

Magnetic Studies over Volcanoes[1]

Alexander Malahoff

INTENSIVE magnetic studies of the geological structure of island and submarine volcanoes have been carried out largely in the Pacific region. *Minakami* [1938] and *Nagata* [1938] were among the first investigators to attempt interpretations of the magnetic fields over Japanese volcanoes. More recently, studies by *Yokoyama and Aota* [1965] and *Hagiwara* [1965] in Japan, *Malahoff and Woollard* [1966, 1968] in Hawaii, and *Steinberg and Rivosh* [1965] in Kamchatka have shown the effectiveness of airborne magnetic methods in the study of the geological structure of volcanoes. Airborne surveys can be carried out at constant elevation at relatively little cost. The most common instrument in use today is the proton precession magnetometer. Some compact instruments can be used with light aircraft (detector head suspended approximately 30 meters below the aircraft). Reading errors due to the magnetization of the aircraft are thus reduced to less than 5 γ. Seaborne surveys over submarine volcanoes, guyots, and seamounts have in the past [*Bullard and Mason*, 1963] been carried out using proton precession magnetometers with the sensing coil towed 500 feet behind the ship.

Analysis in terms of geological structure requires: (a) removal of the regional field from the observed field to form a residual total force magnetic field; (b) removal of the terrain effect from the residual field (see section by Nagata, p. 392, in this Monograph); and (c)

---
[1] Hawaii Institute of Geophysics Contribution 198.

relation of the terrain-corrected residual anomaly to source bodies in and beneath the volcanic cone or ridge.

Analytical studies by *Hagiwara* [1965], *Malahoff and Woollard* [1968], and *Steinberg and Rivosh* [1965] have conclusively shown that anomalies observed above active, dormant, and extinct volcanic centers cannot always be explained in terms of magnetization of terrain alone. Most observed anomalies can be broken down into two parts: the magnetization of the volcanic cone or ridge, and the magnetization of the volcanic plug. Over volcanic rocks, the intensity of natural remanent magnetization always predominates over the intensity of induced magnetization [*Doell and Cox*, 1965].

AREAS OF INTENSIVE STUDY

Detailed total force magnetic surveys have been made over volcanoes in Japan, Kamchatka, New Zealand, and Hawaii. Extensive shipboard studies have also been performed over seamounts and guyots of the Pacific.

Japanese Volcanoes

Hagiwara [1965] performed detailed analyses of aeromagnetic anomalies over the Hakone Volcano of Japan. The caldera of Hakone volcano is of double structure: an older caldera containing a younger crater, with a line of craters running northwest-southeast. The volcano consists predominantly of andesitic rocks. The aeromagnetic map (Figure 1) at 1700 meters above sea level indicates a complex subsurface geological structure beneath the floor of the caldera. From this map, Hagiwara computed a theoretical total intensity anomaly at 3700 meters altitude. The resultant map, with effects of near-surface bodies greatly reduced, shows a considerably simplified total force, which reflects deep-seated geologic structure. The anomaly over Hakone caldera is a complex bipole anomaly with peak-to-peak amplitudes of approximately 1000 γ.

The Towada caldera (Figure 2) is a triple volcano of the Krakatau type. After studying the 1000 γ peak-to-peak anomaly, Hagiwara concluded that the origin of the anomaly lay within a vertical body beneath the center of the caldera.

Yokoyama and Aota [1965] carried out intensive geophysical studies over the Sikotu

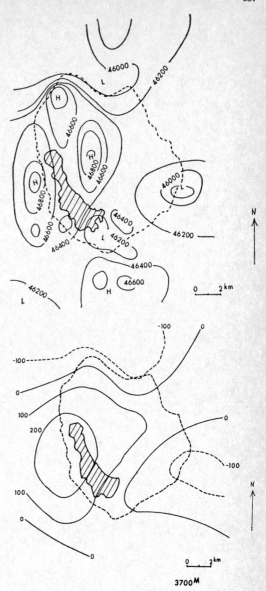

Fig. 1. Total-intensity anomaly maps over Hakone caldera, Japan [after *Hagiwara*, 1965]. Contours in gammas. *Upper:* Simplified map at 1700 meters above sea level. *Lower:* Theoretical map at 3700-meter altitude computed from map at 1700 meters.

caldera of Hokkaido (Figure 3), including a detailed survey using a boat-towed proton magnetometer over the central part of the lake. Intensities of remanent and induced magnetization as determined from samples around the caldera were used for a thorough analysis of the

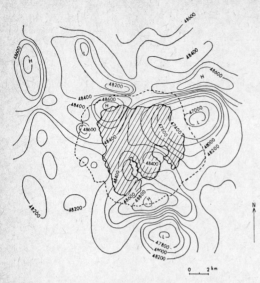

Fig. 2. Simplified total intensity anomaly map over Towada caldera, Japan, at 1400 meters above sea level. Contours in gammas [after *Hagiwara*, 1965].

anomalies. The average value of remnant magnetic intensity for rocks of the Lake Sikotu caldera was determined to be 1.0×10^{-3} emu/cm^3, and average susceptibility was also 1.0×10^{-3} emu/cm^3. The rocks of the caldera are largely andesitic and, because of the complexity of geologic structure within the caldera, produce complex bipole anomalies. The authors do not attribute the magnetic anomaly over the caldera to the lavas and other deposits of the caldera, but to intrusive rocks within the caldera; the magnetization of these rocks is considered to be much greater than the magnetization of the intruded basement rock. Quantitative analysis of the anomaly suggests that it is largely due to a vertical cylindrical body 1.5 km in diameter and buried at a depth of 2 km.

Kamchatka Volcanoes

A very thorough total-force magnetic survey of the Kamchatka volcanic region was carried out by *Steinberg and Rivosh* [1965] at flight elevations between 2000 and 3000 meters with a flight line spacing of 2 km. Since the flight elevation was kept 200–500 meters above the highest topography, terrain effects, though large, were not excessive. The total force anomalies observed above volcanic centers are bipolar in nature.

The area covered by Steinberg and Rivosh is the structural continuation of the Great Kuril Range volcanic zone and includes about 25 active volcanoes. Principal features of the magnetic field over the volcanic zone are:

1. The magnetic field is disturbed over the volcanic region.
2. There are large bipolar magnetic anomalies.
3. The direction of polarity is normal; i.e., most bipolar anomalies have maxima to the south and minima to the north or northwest.
4. Early Quaternary basaltic and andesite-basalt shield volcanoes and their lavas are largely responsible for the disturbed magnetic field. The lavas have a variable magnetic susceptibility of 1.5 to 2.0×10^{-3} cgs units and natural remanent magnetization of between 0.3×10^{-3} and 5.0×10^{-3} cgs units.
5. There is little correlation between paleomagnetic data and the direction of magnetization of the anomalies.
6. The bulk of the anomalies observed can be attributed to topographic effects (Figure 4).

The topographic effects of volcanic masses are of primary importance; disregarding them leads to erroneous conclusions about structural features within the volcanic masses. Tables 1 and

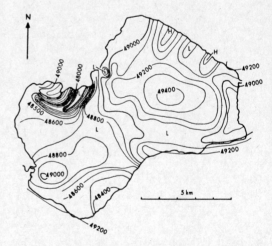

Fig. 3. Total intensity anomaly map over Lake Sikotu of the Sikotu caldera, Japan. Observed at Lake level. Contours in gammas [after *Yokoyama and Aota*, 1965].

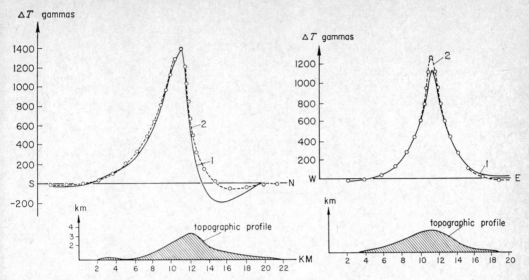

Fig. 4. Observed and calculated total force anomaly profiles over Kronotsky volcano, Kamchatka, using an effective magnetization of 4.4×10^{-3} cgs units: (1) observed anomaly; (2) computed anomaly [after Steinberg and Rivosh, 1965].

2 show the contrasts in intensities of the effective magnetization of the Kamchatka and Japanese volcanoes.

Hawaiian Volcanoes

Aeromagnetic studies over the Hawaiian volcanoes have been made by Malahoff and Woollard [1966, 1968]. As in Kamchatka, the Hawaiian volcanic centers are characterized by normally polarized, bipole anomalies (Figure 5). The magnetic surveys over Hawaiian volcanoes were carried out at an average elevation of 3100 meters above sea level, 1000 meters above the highest topography. The islands north of the island of Hawaii were thoroughly sampled for paleomagnetic determinations by Tarling [1965]. For the island of Hawaii, Doell and Cox [1965] found an average natural remanent magnetization of 5.59×10^{-3} cgs units from 148 determinations of basalt flows in a field with a dip of 33.28°. All measured polarization directions were normal; only normally polarized dipole

TABLE 1. Effective Magnetization of Kamchatka Volcanoes Computed from Aeromagnetic Anomalies*

Volcano	Altitude,† m	Height of Volcano,‡ m	ΔT, γ	Magnetization, 10^{-3} cgs units
Krasheninnikov	300	1600	1900	7.7
Ilinsky	450	1500	1400	7.2
Mutnovsky	300	1000	1350	7.2
Opala	300	2100	1700	6.2
Taunschitz	450	1600	1100	5.6
Zheltovsky	450	1900	1100	5.2
Kronotsky	350	3000	1400	4.4
Gamchen	400	1800	900	3.8
Karimsky	250	700	600	3.5
Avachinsky	450	2100	800	3.2
Kambalny	200	1800	800	2.7
Khodutka	450	1800	500	2.3

* After Steinberg and Rivosh [1965].
† Altitude of flight above crater.
‡ Height of volcanic cone above surrounding ground.

TABLE 2. Effective Magnetization of Japanese Volcanoes Determined From Magnetic Anomalies*

Volcano	Rock	Height of Volcano, m	Magnetization, 10^{-3} cgs units
Mihara	Basalt	750	30.0
Mihara	Basalt	750	50.0
Huzi	Basalt	3500	30.0
Mijake-Shima	Basalt	810	50.0
Omuro	Basalt	580	5.0
Haruna	Andesite	1270	10.0
Asama	Andesite	2530	4.0
Sakurazima	Andesite	1230	7.0
Akagi	Andesite	1830	1.0
Kamijama	Andesite	1440	20.0
Kusatu-Sirane	Andesite		8.0

* Compiled from other sources by *Steinberg and Rivosh* [1965].

anomalies were observed by Malahoff and Woollard over the volcanoes of the island. Susceptibilities of basalt samples collected by *Decker* [1963] over Kilauea volcano averaged 1.3×10^{-3} cgs units. Therefore an apparent susceptibility of 16.3×10^{-3} cgs units was assumed as an average upper limit for the magnetization of rocks of the volcanoes of the island of Hawaii (see Table 3). Two-dimensional computer techniques [*Heirtzler et al.*, 1962] were used to compute the magnetic terrain effects of the volcanoes of Hawaii, using an apparent susceptibility of 16.3×10^{-3} cgs units and 33.28° dip for the magnetic field.

Steinberg and Rivosh [1965] found that the observed magnetic anomalies in Kamchatka result from the magnetization of the physical bulk of the volcanic cones. This is not the case for the volcanoes of the Hawaiian Islands. The observed pole-to-pole magnetic anomaly over Mauna Kea volcano is 950 γ; the bipole is normally polarized. After correction for the terrain effect (an anomaly with 600 γ amplitude), the observed anomaly over the volcanic cone becomes a classical symmetrical bipole with a pole-to-pole intensity of 1600 γ. Even an unrealistically high magnetization of the bulk of Mauna Kea would not neutralize this bipole anomaly. Therefore the observed bipole ano-

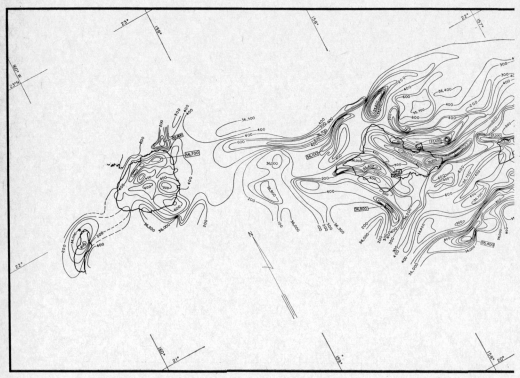

Fig. 5. Total force intensity anomalies observed over the Hawaiian Islands. Flight elevation, 3000 meters. Contours in gammas.

maly can be caused only by a body nearly vertically sided within the volcanic cone of Mauna Kea and having a greater intensity of magnetization than the surrounding basaltic lava flows. A similar situation applies to other volcanoes of the Hawaiian Islands.

New Zealand Volcanoes

The New Zealand central volcanic region was surveyed by *Gerard and Lawrie* [1955] using an airborne fluxgate magnetometer flown at an altitude of 1500 meters above sea level. *Modriniak and Studt* [1959] carried out preliminary analyses of the data using the values shown in Table 4. Small dacite and andesite masses within the central volcanic district usually produce anomalies of less than 500 γ at flight elevation of 1500 meters, showing that these masses generally have small roots or plugs. At the magnetic latitudes of the North Island of New Zealand, the observed anomalies are largely single pole positive anomalies with a minor negative pole. Most of the anomalies observed over volcanoes of the central volcanic region are normally polarized. The total force anomaly over the active volcano Ruapehu (Figure 6) at the southern extremity of the volcanic region cannot be interpreted by the magnetization of the cone itself [*Malahoff*, 1968]; one must assume either a lesser magnetized mass within the cone (such as a volcanic plug with temperatures above the Curie point), or rocks reversely magnetized and within the cone, or both conditions simultaneously.

Submarine Volcanism

A total force seaborne magnetic survey of the seaward extension of the Kilauea volcano of Hawaii (Figure 7) was made in 1965 by *Malahoff and McCoy* [1967]. Apart from the fact that the Puna ridge is the seaward extension of the Kilauea rift and has similar geomorphic features as aerial volcanism (Figure 8), the total force magnetic anomaly can best be interpreted in terms of a dike-filled volcanic plug within the volcanic ridge. Aeromagnetic and seaborne magnetic surveys over the Molokai fracture zone suggest that the oceanic floor

TABLE 3. Depth Determinations for Magnetic Anomalies over Volcanic Centers of the Hawaiian Islands*

Island	Volcano	Depth to plug,† km	Dimensions of Plug, km		ΔT, γ	Magnetization Contrast,‡ 10^{-3} cgs units	Bouguer High, mgal
			Horizontal	Vertical			
Hawaii	Mauna Loa	2.7	16.8 by 4.0	20	800	6.90	+330
	Mauna Kea	1.9	12.0	6.0	8	1500	13.8	+330
	Hualalai	1.3	8.8	4.9	15	800	6.90	+265
	Kohala	5.7	8.8	11.2	10	800	19.0	+320
Maui	East Haleakala	4.0	7.3	7.3	15	1350	16.0	+270
	West Haleakala	9.3	8.0	6.4	12	1100	18.0	+250
	West Maui	1.6	9.5	8.9	9			
Kahoolawe	Kahoolawe	2.4	10.0	12.8	2	500	7.0	+250
Lanai	Lanai	0.8	12.0	6.5	2	150	2.0	+250
Molokai	East Molokai	0.8	14.0	10.0	8	950	12.4	+270
	West Molokai	0.3	4.8	5.2	10	1000	13.9	+270
Oahu	Waianae	0.8	9.0	9.0	5	650	9.0	+310
	Koolau	1.0	20.0	10.0	16	1450	10.0	+310
Kauai	Waialeale	1.6	11.0	6.5	5	400	5.5	+310
Niihau	Niihau	0.8	8.0	8.0	6	700	8.0	+290

* Compiled by *Malahoff and Woollard* [1966].
† To top of plug, below ground level.
‡ Between the plug and the surrounding lava flows.

beneath the fracture zone is also intruded by highly magnetic dike rocks. Aerial volcanism in Hawaii appears to follow surface rifts and intersections of rifts of the Molokai-Murray fracture zone system [*Malahoff and Woollard*, 1968].

MAGNETIC PROPERTIES OF VOLCANIC ROCKS

The interpretation of magnetic anomalies observed over volcanic centers is closely associated with correct knowledge of (a) the direction of the intensity of natural remanent magnetization of the contained rocks, and (b) the susceptibility and therefore the effective induced magnetization.

With the establishment of the existence of polarity reversals within the past four million years [*Cox et al.*, 1964], it has become essential to undertake intensive sampling of the rocks contained within a volcanic mass before analyses of the observed anomalies are undertaken. Quite frequently, however, paleomagnetic evidence as to the direction of magnetization of surface rock samples may conflict with the gross direction of magnetization of the volcanic center [*Malahoff and Strange*, 1965].

The intensity of remanent magnetization is variable, decreasing with increase in the silica content and with decrease of ferromagnesian minerals in the rock. The total magnetization of rocks is the property of the rocks that governs effectively the shape and amplitude of the mag-

TABLE 4. Magnetic Properties of New Zealand Volcanic Rocks*

Rock Type	Magnetic Susceptibility, 10^{-3} cgs units			Remanent Magnetization, 10^{-3} cgM units		
	No.	Mean	Range	No.	Mean	Range
Pumice, tuff, breccia	29	0.31	0–11.3			
Ignimbrite	66	0.69	0.035–1.35	120	2.1	0.6–9.4
Rhyolite	44	0.28	0.05–1.10	10	2.0	0.1–6.06
Dacite	5	0.83	0.11–1.75	2	0.15	0.1–0.2
Andesite	2	1.32	0.95–1.70	1	1.51	
Basalt	7	2.13	1.1–3.8			

* Compiled by *Modriniak and Studt* [1959].

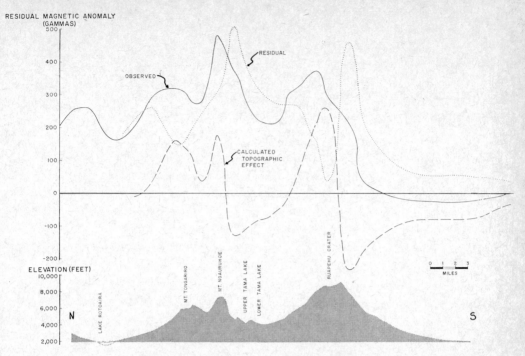

Fig. 6. Total intensity profiles over Ruapehu volcano, New Zealand [after *Malahoff*, 1968].

netic anomaly and can be expressed as

$$J = KH_0 + J_n$$

where K is the true susceptibility, H_0 the intensity of the earth's magnetic field at the place at which J is being measured, and J_n is the natural remanent magnetization.

For rocks that have the same direction of induced as remanent magnetization, the 'apparent susceptibility' K_A (see p. 397, this Monograph) can be effectively utilized in the computation of the size and shape of anomalous magnetic bodies. The apparent susceptibility can be defined as follows:

$$J \approx K_A H_0 \quad \text{where} \quad K_A \equiv K(1 + Q)$$

where Q is the Koenigsberger factor.

The observed values of magnetization for Hawaiian flow basalts as determined from rock specimens are listed in Table 5. The effective magnetization of volcanoes of Kamchatka and Japan, as determined from analyses of magnetic anomalies, are listed in Tables 1 and 2, respectively.

The magnetization of intrusive basalts appears to be of greater intensity than the magnetization of flow basalt. The difference was observed by *Malahoff and Woollard* [1966] for rocks of the Hawaiian volcanoes. Analyses of seismic [*Furumoto et al.*, 1965] and gravity studies [*Strange et al.*, 1965] over volcanoes such as the Koolau volcano of Oahu have shown that the volcanic centers are underlain by vertically dipping volcanic plugs containing high density (as high as 3.2 g/cm^3) and high velocity (7.7 km/sec) dike rocks. Analyses of the magnetic anomaly over the Koolau volcano show that the rocks of the volcanic plug also have higher average apparent susceptibility than the surrounding flow rocks. The magnetization contrast between the flow and intrusive rocks is at least of the order of 10.0×10^{-3} cgs units for Hawaiian volcanoes and can be accounted for by the greater density, the greater physical stability, and the slower rate of cooling of the intrusive rocks than of the extrusive rocks.

It appears, therefore, that aeromagnetic methods as applied to the study of the structure of volcanoes form a powerful tool that, with the aid of gravity and seismic studies, can help define the internal structure of volcanic centers.

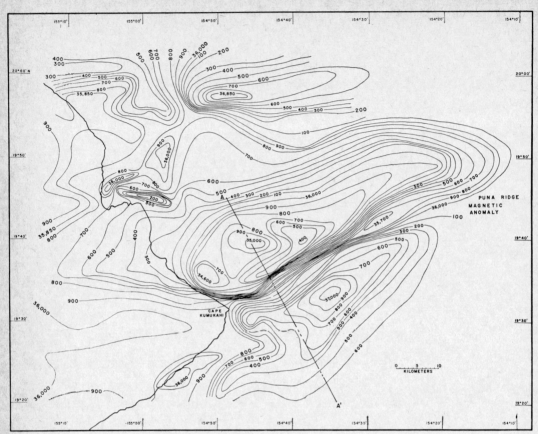

Fig. 7. Total intensity anomalies over the Puna submarine volcanic ridge, Hawaii [after *Malahoff and McCoy*, 1967].

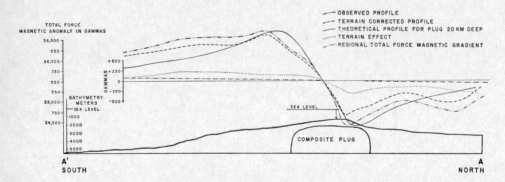

Fig. 8. Observed and computed total force magnetic profiles over the Puna submarine ridge. See Figure 7 for location of profile.

TABLE 5. Average Values of Susceptibility (K) and Natural Remanent Magnetization (NRM, in cgs $\times 10^{-3}$) for Rocks of the Hawaii Islands*

Formation	NRM	K
Hawaii, tholeiite	11.0	3.2
Hawaii, olivine-rich basalt	5.0	0.5
Hana, east Maui	17.31	4.63
Kula, east Maui	137.30	13.28
Honomanu, east Maui	0.96	2.66
Honolua, west Maui	14.34	2.74
Wailuku, west Maui	8.19	2.01
Lanai basalt	5.88	0.92
East Molokai basalt	19.43	2.13
West Molokai	13.22	1.16
Koolau, Oahu	3.09	1.83
Waianae, Oahu	2.67	2.19
Honolulu peridotite	—	0.4
Honolulu, Oahu	4.78	3.92
Koloa, Kauai	6.45	1.24
Napali, Kauai	4.21	1.01

* Compiled from *Tarling* [1965] and *Malahoff and Woollard* [1966].

ANALYTICAL METHODS AS APPLIED TO THE STUDY OF AEROMAGNETIC ANOMALIES

Nearly all magnetic anomalies observed over the Hawaiian, Japanese, New Zealand, and Kamchatka volcanoes and other volcanoes in middle or low magnetic latitudes can be divided into two groups:

1. Local bipole anomalies related to centers of volcanism marked by surface caldera, volcanic peaks, or geologic evidence defining former vent areas.
2. Elongate bipole anomalies related to dike complexes, observable and probable, rift zones in the crust that appear to be occupied by intrusive rocks at depth, and other linear geological features connecting the centers of volcanism.

In the study of these anomalies, five factors have to be evaluated:

1. Magnetic effect of the volcanic cone.
2. Approximate size and shape of the anomalous geologic body.
3. Orientation of the earth's magnetic field at the latitude of the area being studied.
4. Depth to the top of the anomalous body.
5. Susceptibility contrast and the natural remanent magnetization contrast between the surrounding rocks and the anomalous body.

An approximation to the above parameters can be obtained by utilizing various analytical procedures based on the shape of the anomaly for the magnetic latitude. The parameters can then be further defined through theoretical computations using two- or three-dimensional techniques with computer programming. After corrections are applied (e.g., for topographic effects), the resultant computed anomaly or profile is then matched with the observed.

The horizontal size of the anomalous body can be determined by an inspection of the relationship between anomaly contours over theoretical bodies, as those computed by *Vacquier et al.* [1951], or by the use of characteristic curves as described by *Grant and West* [1965]. Depth to the top of the anomalous body can be determined by using 'depth indices' [*Vacquier et al.*, 1951; *Grant and West*, 1965]. Theoretical apparent susceptibility contrast can be determined by using such formulas as:

$$K = \Delta T_m / \Delta T_c T$$

where K is the minimum susceptibility contrast [*Vacquier et al.*, 1951], ΔT_c is the total amplitude of the intensity anomaly selected from the appropriate theoretical body as computed by Vacquier et al., T is the intensity of the regional magnetic field at the point at which the anomaly is situated, and ΔT_m is the observed amplitude of the actual anomaly. A similar method of computing the theoretical apparent susceptibility contrast is given by *Grant and West* [1965].

A model can be constructed to fit the above parameters and the vertical dimension of the model can be altered until a suitable fit is achieved between the observed and theoretical anomalies.

CONCLUSIONS

From the study of aeromagnetic anomalies observed over Japanese, Kamchatka, New Zealand, and Hawaiian volcanic centers, the following principal conclusions may be drawn:

Basaltic and andesitic volcanoes appear to have differing internal structures. The basaltic volcanoes have distinct, highly magnetized, dike-filled plugs. Andesitic volcanoes have less prominent plugs.

In order to minimize local terrain effects, the

best flight altitude for aeromagnetic studies over volcanic centers is 500 meters above the highest terrain.

Terrain effects must be taken into account. These can best be calculated by using values of the intensity of natural and remanent magnetizations as determined from rock samples collected in the field.

Modern computer techniques and available anomaly-matching techniques speed up the computation of theoretical anomalies and comparison with observed anomalies.

Paleomagnetic data, if available, must also be incorporated into the computations. Often, directions of natural remanent magnetization, as determined from surface samples, do not agree with the effective magnetization direction of the volcanic cone, as determined from the inspection of the magnetic anomaly.

The magnetic intensity of recent andesitic volcanoes as computed from observed anomalies normally does not exceed 8.0–10.0×10^{-3} cgs units for basaltic volcanoes the magnetization is within the limits 20.0–50.0×10^{-3} cgs units. Residual magnetism always exceeds induced for both basaltic and andesitic volcanoes and is much greater in magnitude for basaltic than andesitic volcanoes.

REFERENCES

Bullard, E. C., and R. G. Mason, The magnetic field over the oceans, *The Sea*, vol. 3, edited by M. N. Hill, pp. 175–217, Interscience Publishers, New York, 1963.

Cox. A., R. R. Doell, and G. B. Dalrymple, Reversals of the earth's magnetic field, *Science, 144*, 1537–1543, 1964.

Decker, R. W., Magnetic studies on Kilauea Iki lava lake, Hawaii, *Bull. Volcanol., 26,* 23–25, 1963.

Doell, R. R., and A. Cox, Paleomagnetism of Hawaiian lava flows, *J. Geophys. Res., 70,* 3377–3405, 1965.

Furumoto, A. S., N. J. Thompson, and G. P. Woollard, The structure of the Koolau volcano from seismic refraction studies, *Pacific Sci., 19,* 306–314, 1965.

Gerard, V. B., and J. A. Lawrie, Aeromagnetic surveys in New Zealand, 1949–1952, *New Zealand Dept. Sci. Ind. Res., Geophys. Mem. 3,* 1955.

Grant, F. S., G. F. West, *Interpretation Theory in Applied Geophysics,* McGraw-Hill Book Company, New York, 1965.

Hagiwara, Y., Analysis of the results of the aeromagnetic surveys over volcanoes in Japan, *Bull. Earthquake Res. Inst., Tokyo Univ., 43,* 529–547, 1965.

Heirtzler, J. R., G. Peter, M. Talwani, and E. G. Zurflueh, Magnetic anomalies caused by two dimensional structure: Their computation by digital computers and their interpretation, *Columbia Univ. Tech. Rept. 6,* 6–62, 1962.

Malahoff, A., Origin of magnetic anomalies over the central volcanic region of New Zealand, in *The Crust and Upper Mantle of the Pacific Area, Geophys. Monograph 12,* edited by L. Knopoff et al., pp. 218–240, American Geophysical Union, Washington, D. C., 1968.

Malahoff, A., and F. McCoy, The geologic structure of the Puna submarine ridge, Hawaii, *J. Geophys. Res., 72,* 541–548, 1967.

Malahoff, A., and W. E. Strange, Paleomagnetic significance of aeromagnetic surveys of the Hawaiian Islands, *Pacific Sci., 19,* 390–392, 1965.

Malahoff, A., and G. P. Woollard, Magnetic surveys over the Hawaiian Islands and their geologic implications, *Pacific Sci., 20,* 265–311, 1966.

Malahoff, A., and G. P. Woollard, Magnetic and tectonic trends over the Hawaiian ridge, in *The Crust and Upper Mantle of the Pacific Area, Geophys. Monograph 12,* edited by L. Knopoff et al., pp. 241–276, American Geophysical Union, Washington, D. C. 1968.

Minakami, T., Magnetic surveys of Volcano Kusatu-Sirane, *Bull. Earthquake Res. Inst., Tokyo Univ., 16,* 1938.

Modriniak, N., and F. E. Studt, Geological structure of the Taupo-Tarawera district, *New Zealand J. Geol. Geophys., 2,* 654–684, 1959.

Nagata, T., Magnetic anomalies around volcanic craters, *Bull. Earthquakes Res. Inst., Tokyo Univ., 16,* 288–299, 1938.

Nagata, T., *Rock Magnetism,* Maruzen Company, Tokyo, 1961.

Steinberg, G. S., and L. A. Rivosh, Geophysical study of the Kamchata volcanoes, *J. Geophys. Res., 70,* 3341–3369, 1965.

Strange, W. E., G. P. Woollard, and J. C. Rose, Analysis of the gravity field over the Hawaiian Islands in terms of crustal structure, *Pacific Sci., 19,* 381–389, 1965.

Tarling, D. H., Paleomagnetic studies of the Hawaiian lavas, *Geophys. J., 10,* 93–98, 1965.

Vacquier, V., N. C. Steenland, R. G. Henderson, and I. Zietz, Interpretation of aeromagnetic maps, *Geol. Soc. Am. Mem. 47,* 1951.

Yokoyama, I., and M. Aota, Geophysical studies on Sikotu Caldera, Hokkaido, Japan, *J. Fac. Sci., Hokkaido Univ., 7,* 104–112, 1965.

The Paleomagnetic Vector Field

S. K. Runcorn

THE geomagnetic field at any historic epoch is plotted as lines of equal values of its elements or components: declination D (isogonic); inclination I (isoclinic); and total field F and its components H, X, Y, and Z (isomagnetic). The secular variation field can be plotted as lines of equal annual change in an element (isoporic charts) or conveniently on one map as lines of equal rate of change of the vertical component Z and annual vector differences in the horizontal component. The residual field (the observed minus the best fitting dipole field) can be similarly represented. All these maps show changes of pattern since the geomagnetic field was first measured: D and I since about 1500 A.D., the intensities since 1820. These fields, which are due to motions in the earth's core, can, in the remote history of the earth, be determined through the record retained in rock strata of different geological periods. Along with the less complete records of the planetary circulation of the atmosphere found in aeolian sandstones and other wind blown deposits, and the length of the day, recorded in growth increments visible on the epitheca of fossil corals, the directions of remanent magnetization of rocks form a new branch of earth science, paleogeophysics. Study of the geological column in various continents enables paleogeographic, paleoclimatic, and paleotectonic maps to be made, but these are only qualitative studies of the earth's evolution. The paleomagnetic field record is incomplete; it is known only at certain times and places at which deposition of sediments, intrusion or extrusion of igneous rocks, or metamorphism of existing rocks occurred, causing the ambient field to be recorded [*Blackett*, 1956].

There are several ways in which the iron oxide minerals in rocks can acquire a permanent magnetization:

1. Thermoremanent magnetization is acquired as the rock cools just below the Curie points of its oxide minerals, because the coercive force is then less than the ambient geomagnetic field. A typical lava flow cools to ordinary surface temperature in about 1 year, so that its remanent magnetization records a 'spot' value of the geomagnetic field; the daily variation and magnetic storms (of such interest in observatory records) are averaged out. The cooling time for sills may be thousands of years; thus the magnetic profile through the sill records the secular variation of the geomagnetic field, a phenomenon first discovered from historical observations. Deep burial followed by uplift can magnetize or partially remagnetize sediments or metamorphic rocks, resulting in a record of some averaged geomagnetic field.

2. Depositional magnetization results from the orientation of already magnetized iron oxide particles as they settle in the geomagnetic field in a reasonably quiet fluviatile or marine environment. Consolidation into a sediment preserves the roughly parallel alignment of the particle moments. It is uncertain whether the field is impressed on the sediments at the time of deposition or at some point during the process by which the water is squeezed out by an overburden. The mechanical orienting tendency of the magnetic field may not be completed, and, if a proportion of the grains is discoidal or elongated, their alignment and therefore their contribution to the bulk magnetic moment may be controlled (partly) by that of currents and gravity during settling.

3. Chemical remanent magnetization is acquired as diagenesis takes place after the sediment is deposited as a result of heat or pressure, or it may occur as a result of weathering of lavas or sediments. The formation of the red pigment of a red sandstone from the decomposition of iron hydroxides in the interstices between the quartz grains is an example of a process favorable to magnetization. The growth of the iron oxide particles in diameter is accompanied by an increase in the relaxation time, and thus when a critical size is attained the geomagnetic field direction is retained.

4. Isothermal remanent magnetization is ac-

quired by exposure to a large field and seems to be the process by which lavas in regions of high thunderstorm activity become locally magnetized in confusing directions. There seems to be no evidence that the geomagnetic field ever attained, even for short periods, magnitudes sufficient to cause such magnetization.

5. Viscous magnetization is picked up from the ambient field over long periods, especially by particles having short relaxation times or as a result of a rise in temperature. Such a process may give a secondary component of magnetization in addition to that acquired during the formation of the rock. For particles having long relaxation times, the secondary magnetization is often directed along the mean field of the Quaternary; for short relaxation times, it may correspond to the ambient field during laboratory storage.

All these processes can be demonstrated by laboratory experiments [*Nagata*, 1961; *Irving*, 1964], and the theory of magnetism of domain and multidomain particles [*Néel*, 1955] has provided a satisfactory explanation of the key phenomenon. The theory predicts times of relaxation comparable with geological time, so that the directions of the geomagnetic field for different geological periods can be expected to be preserved in many cases, although the intensity may have diminished from that which the ancient field produced. Thus the survey of the remanent magnetization of oriented samples from various geological formations gives us the ancient magnetic field directions as observed in different continents.

Field experiments also have shown that it is possible for rocks to acquire a magnetization early in their history and retain it substantially unchanged, at least in direction, to the present. As *Graham* [1949] suggested, sedimentary or other strata originally horizontal and later folded should have parallel directions of magnetization after correction for geological tilt if their magnetizations have been stable since folding, but will be parallel now, in situ, if the original magnetization has disappeared and the rocks have been remagnetized in the earth's field before folding. Similarly, if fragments or pebbles of a rock in a conglomerate are found to have randomly directed magnetization, the lithology must have been magnetically stable over the time since the formation of the conglomerate. By such 'natural experiments' the exceedingly long periods over which rocks can retain their original magnetization have been established.

Where rocks have acquired a secondary as well as an original magnetization, the resultant directions from samples are often spread in a plane (or great circle on the stereographic projection used to plot the results). Such a great circle will pass through the mean ancient field and very often the present field, or its average, which is the field most likely to have left this unwanted imprint. Secondary magnetization, acquired at modest temperature and without marked diagenesis, may often be removed by A.C. or thermal demagnetization. In the former process the secondary magnetization component is taken through a demagnetization cycle by a field of the order of the coercive force of those grains that carry it. Experimentally the sample is rotated in an alternating field that is reduced smoothly to zero, the rotation being arranged so that no direction in the rock is preferentially oriented with respect to the applied field. The demagnetizing field strength is increased by steps until the direction of the remanent magnetization no longer changes: in lavas this usually occurs with the application of a few hundred oersteds. In thermal demagnetization, the rock samples are heated above the blocking temperature of the grains carrying the secondary magnetization and then cooled in zero field. Experimentally the process of heating is taken in steps of, say, 50°–100°C until the direction of the remanent magnetization remains unchanged; finally, the intensity decreases to zero when the Curie points of the grains originally magnetized are reached.

When the original direction of magnetization of a sample of a formation has been recovered, its direction with respect to the original horizontal and the azimuth with respect to present north are found by applying a geological dip correction, and the values of D and I are plotted on a stereographic projection. Directions from a homogeneous population (many samples from one site or distributed through one formation) can be statistically examined by *Fisher's* [1953] method, which is analogous to the Gaussian treatment of a unidimensional variate. The vector sum R of the N directions of each sample, or mean of samples from one site, represented by a unit vector, provides the best estimate of the direction of the mean geomagnetic field over the epoch considered. A cone of confidence of appropriate semi-angle can be calculated such that the true direction will lie within the cone with a probability of 95%. The scatter is determined by the precision parameter K, the best estimate of which is $(N - 1)/(N - R)$.

A survey of palaeomagnetism shows that the geomagnetic field varies with three different time scales, each discussed below.

Secular Variation, 100–10,000(?) Years

Historical observations show that in this variation neither a single harmonic nor a single period predominates. A part of it represents a drift westward of the nondipole field with a variable velocity, now $1/5°$ per year; this gives rise to a clockwise rotation of H at many stations. The possibility that the westward drift may reverse in sign in the course of thousands of years has to be remembered. The irregular amplitude of the secular variation is $10°-30°$. It is probable that the scatter of directions usually obtained when samples from one geological formation are measured is due to this phenomena, but it will be increased by experimental errors in measurement or collection, or by inhomogeneities in the magnetization of the rock. The latter should be entirely random, but the average secular variation amplitude varies with latitude, because the main field intensity at the poles is twice that at the equator; the directions form an elliptical distribution about a mean due to the horizontal or vertical components of the mean field being unequal. Thus Creer [1962] has found evidence for a variation of K of rocks of one geological period with latitude, and Creer et al. [1959] have shown the departure from a circular distribution in the directions from the Whin sill (Carboniferous). Where directions are from samples contemporaneous within a few years (as from a single lava flow) or from a rock where deposition and therefore magnetization is exceedingly slow, the secular variation may not be observed and the directions found may vary by only $1°-5°$.

Where deposition is continuous, as in varved clays and ocean bottom sediments, much more can be learned about the secular variation, in particular its spectrum. It is possible that from the sense of rotation of the palaeomagnetic vector the question whether the drift of the nondipole field is always westward or has been equally often eastward can in due course be answered.

Archeomagnetic researches, extending backward in time the curves of D and I obtained from historical measurements, also bear on these questions.

Except in the cases mentioned, a paleomagnetic survey of a rock formation samples the field over times long compared to those of the secular variation. The mean direction is therefore of fundamental geophysical significance. For Quaternary and Tertiary rocks, it has been shown that in every continent this mean is close to the field of a dipole at the earth's center aligned along the present geographic axis. The comparison of palaeomagnetic results in different continents can best be carried out by calculating pole positions corresponding to each set of results by the method introduced by Creer et al. [1957]. The pole is defined as that direction in which a geocentric dipole must lie to produce a field coincident with the mean palaeomagnetic direction at the site sampled. Quaternary and Tertiary poles cluster around the present geographical pole.

Geomagnetic Reversals

The next time scale over which the field varies requires observations over a geological period: early studies in rock magnetism revealed directions roughly opposed to that of the present field. Tertiary lavas and dykes, especially, were often found to be reversely magnetized. In a study of the lava flows of Iceland, Hospers [1953, 1954] showed that the mean directions of both the reversed and normal series of lavas were exactly 180° apart, thus definitely removing the possibility that the phenomenon was purely local in origin. It was predicted by Néel [1955] and discovered by Nagata [1961] that, because antiferromagnetism was fundamentally the cause of their magnetic properties, iron oxide minerals could acquire a remanent magnetization opposite to the ambient field; this caused doubt that the reversals found in nature were due to reversals of the polarity of the main geomagnetic field. Runcorn [1956] pointed out that the Néel-Nagata explanation predicted that systematic differences would be found in the average composition and texture of the iron oxide minerals in reversed and normal rocks, whereas the geomagnetic field reversal theory predicted that contemporaneous rocks would be magnetized in the same direction all over the world. Reversals were soon found in succession in many sedimentary rocks in the Precambrian Torridonian rocks of Scotland [*Irving and Runcorn*, 1957] and in the Chugwater of the Triassic in the United States [*Collinson and Runcorn*, 1960]. This reduced the likelihood of the former explanation. Then the discovery that Permian rocks are magnetized all in a southward direction in the United States, England, and Australia was powerful evidence for geomagnetic reversals. Finally, the improvement in accuracy in the radioactive dating of lavas has proved that geomagnetic field reversals occurred in the Cenozoic [*Cox et al.*, 1963] (Figure 1). *Opdyke et al.* [1966] and *Ninkovich et al.* [1966] have shown that a similar succession of reversed and normal zones of magnetization occurs in the ocean cores of the Antarctic and Pacific and, on the assumption of uniform rates of deposition, the same dates for reversal are found (Figure 2).

The complexity of the composition and texture of the iron oxide minerals in rocks, the subtle changes that have occurred in them, and the number of theoretically possible methods of self reversal are to be borne in mind in any one case, but the reality of reversals of the earth's magnetic field is generally accepted. Three conclusions can be reached about this phenomenon: (*a*) Reversals of the field take place quickly in a time of the order of the secular variation time scales (10^3 years). These 'intermediate' zones usually appear to have scattered magnetizations, so that the field does not vanish during reversal, though it seems to decrease. Whether the process involves mainly a rotation of the dipole or temporary development of large nondipole fields has not been settled. (*b*) Reversals do not occur at regular intervals. The last three occurred 0.7 m.y., 2.4 m.y., and 3.3 m.y. ago, but they are absent in the Permian (50 m.y.), and they seem more frequent in the Tertiary than in the Mesozoic and Palaeozoic. (*c*) 'Geomagnetic events' occur in which the field reverses for a period of the order of 10^4–10^5 years and then returns to its original polarity.

A general understanding of the secular variation and of polarity reversals can be obtained by considering the magnetohydrodynamics of the electrically conducting, fluid core. The secular variations reflect the turbulent convective motions **v** in the core; the correct time scale is obtained if the velocities are about 1 cm/sec. The axial character of the mean field results from the fact that the Coriolis term $2\rho\omega \times \mathbf{v}$ is of the order of 10^{-4}, while the viscous,

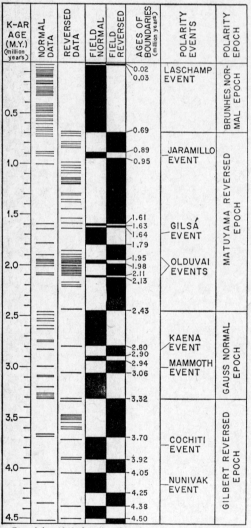

Copyright 1969 by the American Association for the Advancement of Science.

Fig. 1. Time scale for geomagnetic reversals. Each short horizontal line shows the age as determined by potassium-argon dating and the magnetic polarity (normal or reversed) of one volcanic cooling unit [after *Cox*, 1969].

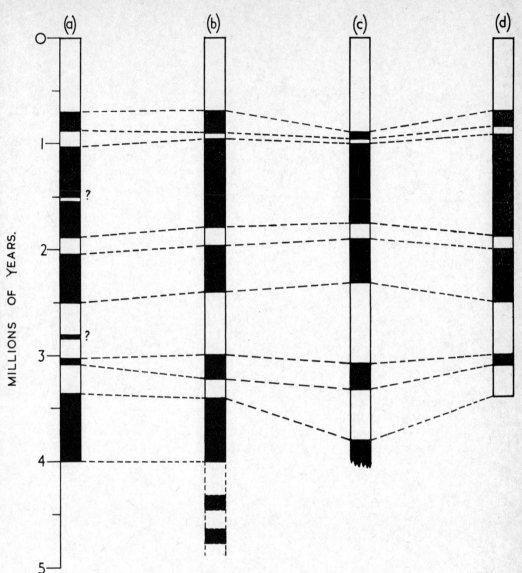

Fig. 2. Geomagnetic polarity zones (simplified) in the late Plio-Pleistocene. Black areas denote reversed polarity: (a) Radioactive correlation of lavas; (b) Pacific Ocean cores, assuming constant depositional rate; (c) Antarctic ocean cores, assuming constant depositional rate; (d) ocean floor anomalies, assuming a uniform rate of sea-floor spreading [after Vine, 1966].

acceleration, and inertia forces per unit volume must be of the order of 10^{-10}, 10^{-8}, and 10^{-8} g cm^{-2} sec^{-2}, respectively. That the mean field observed on the earth's surface is axial is a consequence of the westward drift, which shows that the core and its field rotate relative to the mantle once in 2000 years or so [Runcorn, 1959a]. Reversals seem to be the result of the fundamental absence of a preference for one polarity in the dynamo process which generates the field; the magnetic field occurs to the first power only in the equations. The question whether the oscillation requires an external trigger to set it off has not been settled. The time of the last reversal appears to be that of the Australian tektite field; the comet collision which caused the tektite field might have been such a trigger [Glass and Heezen, 1967].

Polar Wandering and Continental Drift

Paleomagnetic observations have now been carried back into the early Precambrian, to perhaps half the earth's age. The variations so observed have a time scale of hundreds of millions of years and have an origin entirely different from those discussed above.

It was found both for Europe [*Creer et al.* 1957] and for North America [*Runcorn*, 1956; *Collinson and Runcorn*, 1960] that the mean paleomagnetic directions varied between different geological periods and, when transformed into pole positions, produced a curve around the North Pacific. The poles from North American rocks were displaced about 20° west of those from Britain. *Irving and Green* [1957] showed that the polar wandering curve for Australia is greatly displaced from the others. *Creer* [1964a, b] has obtained a polar wandering curve for South America. In all cases the pole positions depend only on time and cannot very plausibly be explained by deviations of the palaeomagnetic vectors due to local causes, although some possible processes (such as magnetostriction effects arising from crustal tectonics) have been suggested and may often contribute small errors and occasionally larger ones. If continental drift has never occurred, these polar wandering curves should be coincident. Because they converge only from the late Mesozoic onward, as Figure 3 shows, displacements of the continents of thousands of kilometers in the last 100–200 m.y. have been inferred. The dipolar, though not the axial, character of the mean field before the Tertiary has been called into question. If the field were not dipolar, the colatitude (θ) of the site with respect to the ancient pole would not be given by the dipole formula $\cot \theta = \frac{1}{2} \tan I$; the contemporaneous paleodeclinations for two sites would then give the pole. Results from a third

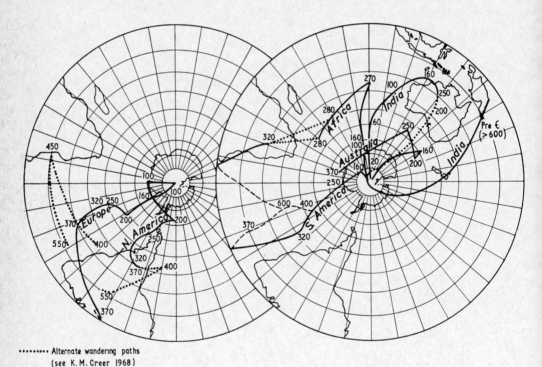

Fig. 3. Polar wandering curves for North America [*Collinson and Runcorn*, 1960]; Europe (*Creer et al.* [1957], with later small additions by *Hospers and van Andel* [1968]; Africa [*McElhinny et al.*, 1968]; India [*Athavale et al.*, 1969]; Australia [*Irving and Green*, 1957]; and South America [*Creer*, 1965].

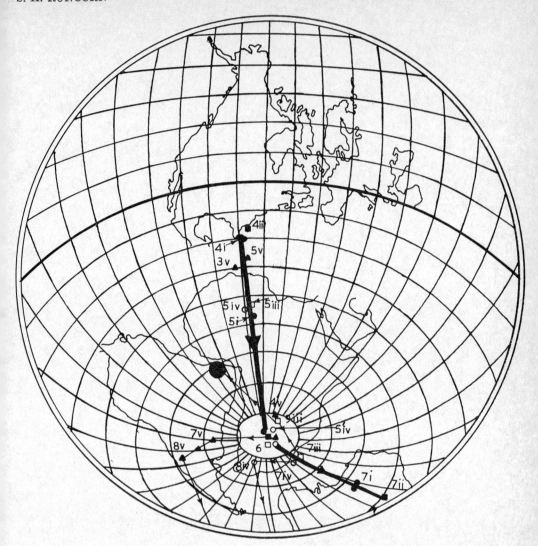

Fig. 4. Polar wandering curves for Upper Paleozoic continental distribution (Upper Paleozoic poles superimposed) [after *Creer*, 1967]. Solid circles, South America; open circles, Europe; solid diamonds, Antarctica; solid squares, Africa; solid triangles, North America; open squares, Australia; open triangles, India. *Circled numbers show position of the present pole; small roman numerals show poles of other ages, thus:* (1) South America, i; (2) Africa, ii; (3) Australia, iii; (4) Europe, iv; (5) North America, v; (6) India, vi; (7) Antarctica, vii. *Ages of poles:* (1) Precambrian; (2) Cambrian; (3) Ordovician; (4) Silurian; (5) Devonian; (6) Carboniferous; (7) Permian; (8) Triassic; (9) Jurassic; (10) Cretaceous; (11) Tertiary; (12) Quaternary.

continent test the hypothesis that an axial nondipole field has existed over periods of geological time (assuming the fixity of the continents), but it does not succeed as an explanation of the paleomagnetic data [*Runcorn*, 1959b].

On the other hand the paleolatitudes λ determined paleomagnetically for a continent by the dipole formula $\tan \lambda = \frac{1}{2} \tan I$ seem to fit the palaeoclimatic data for that continent, as *Blackett et al.* [1960] showed for the paleontological and lithologic distributions and *Opdyke and Runcorn* [1960] showed for paleowind data.

Reconstructions of the continents for different geological times are now possible: *Creer* [1968] finds that Wegener's general concept of the existence in the late Paleozoic of two protocontinents Laurasia (Europe and North America) and Gondwanaland (Australia, Africa, India, Antartica, and South America) is supported. It is not clear whether large relative continental displacements occurred before the dispersal of these protocontinents in the Mesozoic and Tertiary, or whether, as Figures 4 and 5 show, polar wandering alone explains the earlier Paleozoic and Precambrian paleomagnetic field changes [*Creer*, 1968]. The continents are not now solid blocks and would not have drifted as such. Relative movements of parts of existing continents have occurred in mountain building. Thus the comparison of pole positions based on studies in different countries are of great significance. *Van der Voo* [1969] has thus demonstrated a rotation of Spain relative to Europe. The paleomagnetic angles D and I (or the corresponding pole coordinates) fix the ancient latitude of a site and its orientation to the pole, but not its

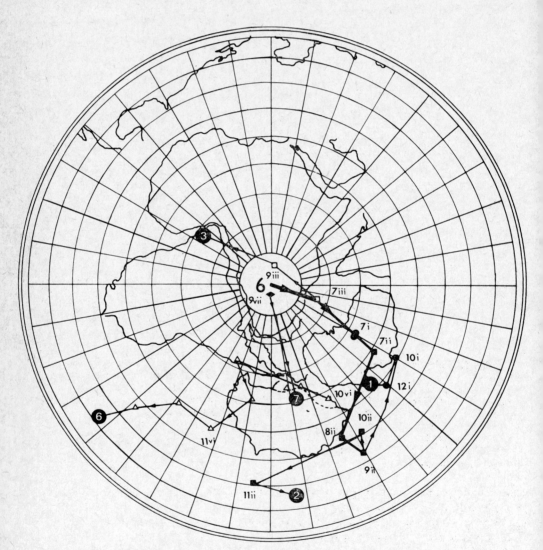

Fig. 5. Polar wandering curves for Upper Palaeozoic continental distribution [after *Creer*, 1967] with India and Antarctica placed so that their Mesozoic polar wandering paths converge to those of the other continents. For explanation of symbols, see Figure 4.

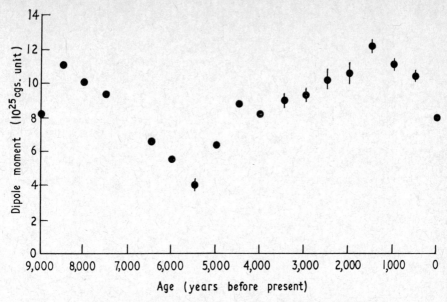

Fig. 6. Summary of archeomagnetic data [*Smith*, 1967].

longitude. Therefore there is, formally, an infinity of positions of a continent for one geological period; however, in practice, continental reconstructions are indeterminate to a limited degree only, as *Irving* [1964] emphasized.

Following a suggestion by L. Egyed, paleomagnetism has been used to test the hypothesis that the earth's radius has changed with time (rates of 0.5–1.0 mm/year being suggested in the expanding earth hypothesis). If, through

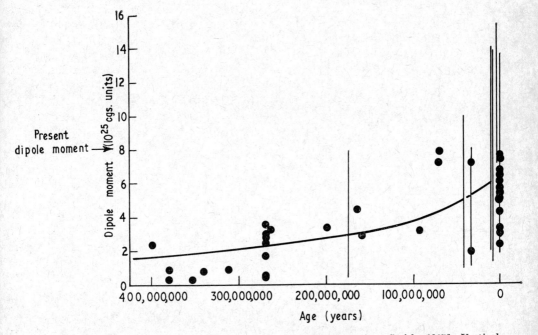

Fig. 7. Change of dipole moment through geological time [after *Smith*, 1967]. Vertical lines give range of many determinations.

absence of tectonic movements, the distance between two points on one continent has remained unchanged since a geological period for which paleomagnetic results (D_1, I_1, D_2, I_2) are available at these points, then the difference in paleolatitudes is

$$\tan^{-1}\left(\tfrac{1}{2}\tan I_1\right) - \tan^{-1}\left(\tfrac{1}{2}\tan I_2\right) = d/R$$

where d is the distance between the paleolatitudes through the points, and where R is the paleoradius of the earth. No significant change in the earth's radius has yet been discovered by the method ($\not> 6\%$ since the Permian). This and other uses of paleomagnetic data to test theories of tectonic movements and to test the nondipole hypothesis are hampered by the imprecise dating available for sedimentary rocks, on which pre-Tertiary paleomagnetism largely depends; the statistical problems which result are very difficult. It may be wise to improve techniques rather than to stretch existing data beyond a reasonable limit.

To determine the paleointensity of the earth's field is considerably more difficult than to determine its direction, because allowance must be made for the decay of the remanent moment. Progress has been made by the methods of archeomagnetism by *Thellier and Thellier* [1946, 1959] who show that 2000 years ago the field intensity was twice that at present. This discovery of another aspect of the secular variation is interesting in connection with the decrease of 7% in the dipole moment observed in the last 100 years. These studies, which involve cooling specimens in a known field in the laboratory and measuring the moment acquired, have been extended to lavas by Nagata. Figures 6 and 7 summarize data thus far obtained.

REFERENCES

Athavale, R. N., R. K. Verma, M. S. Bhalla, and A. Pullaiah, Drift of the Indian subcontinent since Precambrian times, in *Palaeogeophysics*, edited by S. K. Runcorn, Academic Press, London, 1969.

Blackett, P. M. S., *Lectures on Rock Magnetism*, Weizmann Science Press, Jerusalem, 131 pp., 1956.

Blackett, P. M. S., J. A. Clegg, and P. H. S. Stubbs, An analysis of rock magnetic data, *Proc. Roy. Soc. London, A, 256*, 291–322, 1960.

Briden, J. C., Palaeomagnetic polar wandering curve for Africa, in *Palaeogeophysics*, edited by S. K. Runcorn, Academic Press, London, 1969.

Collinson, D. W., and S. K. Runcorn, Polar wandering and continental drift: Evidence from paleomagnetic observations in the United States, *Bull. Geol. Soc. Am., 71*, 915–958, 1960.

Cox, Allan, Geomagnetic reversals, *Science, 163* (3864), 237–245, 1969.

Cox, A., and R. R. Doell, Review of paleomagnetism, *Bull. Geol. Soc. Am., 71*, 645–768, 1960.

Cox, A., R. R. Doell, and G. B. Dalrymple, Radiometric dating of geomagnetic field reversals, *Science, 140*, 1021–1023, 1963.

Creer, K. M., The dispersion of the geomagnetic field due to secular variation and its determination for remote times from paleomagnetic data, *J. Geophys. Res., 67*, 3461–3476, 1962.

Creer, K. M., Palaeomagnetism and its application to S. American rocks, *Bol. Paranaense Geograf., 10*, 93–138, 1964a.

Creer, K. M., A reconstruction of the continents for the Upper Palaeozoic from palaeomagnetic data, *Nature, 203*, 1115–1120, 1964b.

Creer, K. M., Palaeomagnetic data from the Gondwanic continents, *Phil. Trans. Roy. Soc. London, A, 258*, 27–40, 1965.

Creer, K. M., A synthesis of world-wide palaeomagnetic data, in *Mantles of the Earth and Terrestrial Planets*, edited by S. K. Runcorn, pp. 351–382, Interscience, London, 1967.

Creer, K. M., Arrangement of the continents during the Palaeozoic era, *Nature, 219*, 41–44, 1968.

Creer, K. M., D. W. Collinson, and S. K. Runcorn, editors, *Methods in Palaeomagnetism*, Elsevier Publishing Company, Amsterdam, 609 pp., 1967.

Creer, K. M., E. Irving, and A. E. M. Nairn, Paleomagnetism of the Great Whin sill, *Geophys. J., 2*, 306–323, 1959.

Creer, K. M., E. Irving, and S. K. Runcorn, Geophysical interpretation of paleomagnetic directions from Great Britain, *Phil. Trans. Roy. Soc. London, A, 250*, 144–156, 1957.

Fisher, R. A., Dispersion on a sphere, *Proc. Roy. Soc. London, A, 217*, 295–305, 1953.

Glass, B., and B. C. Heezen, Tektites and geomagnetic reversals, *Nature, 214*, 372, 1967.

Graham, J. W., The stability and significance of magnetism in sedimentary rocks, *J. Geophys. Res., 54*, 131–167, 1949.

Hospers, J., Reversals of the main geomagnetic field, I, II, *Proc., Koninkl. Ned. Akad. Wetenschap. Ser. B, 56*, 467–491, 1953.

Hospers, J., Reversals of the main geomagnetic field, III, *Proc., Koninkl. Ned. Akad. Wetenschap. Ser. B, 57*, 112–121, 1954.

Hospers, J., and S. I. van Andel, Palaeomagnetic data from Europe and North America and their bearing on the origin of the North Atlantic Ocean, *Tectonophysics, 6*, 475–490, 1968.

Irving, E., *Paleomagnetism*, John Wiley & Sons, 399 pp., New York, 1964.

Irving, E., and R. Green, Palaeomagnetic evidence from the Cretaceous and Cainozoic, *Nature, 179*, 1064–1065, 1957.

Irving, E., and S. K. Runcorn, Analysis of the paleomagnetism of the Torridonian sandstone series of northwest Scotland, 1, *Phil. Trans. Roy. Soc. London, A. 250,* 83–99, 1957.

McElhinney, M. W., J. C. Briden, D. L. Jones, and A. Brock, Geological and geophysical implications of paleomagnetic results from Africa, *Rev. Geophys., 6,* 201–238, 1968.

Nagata, T., *Rock Magnetism,* Maruzen, Tokyo, 350 pp., 1961.

Néel, L., Some theoretical aspects of rock magnetism, *Advan. Phys., 4,* 191–243, 1955.

Ninkovich, D., N. Opdyke, B. C. Heezen, and J. H. Foster, Paleomagnetic stratigraphy, rates of deposition and tephrachronology in North Pacific deep-sea sediments, *Earth Planetary Sci. Letters, 1,* 476–492, 1966.

Opdyke, N. D., B. Glass, J. D. Hays, and J. Foster, A paleomagnetic study of antarctic deep-sea sediments, *Science, 154,* 349–357, 1966.

Opdyke, N. D., and S. K. Runcorn, Wind direction in the western United States in the late Paleozoic, *Bull. Geol. Soc. Am., 71,* 959–972, 1960.

Runcorn, S. K., Rock magnetism—geophysical aspects, *Advan. Phys., 4,* 244–291, 1955.

Runcorn, S. K., Paleomagnetic survey in Arizona and Utah: Preliminary results, *Bull. Geol. Soc. Am., 67,* 301–316, 1956.

Runcorn, S. K., On the theory of the geomagnetic secular variation, *Ann. Geophys., 15,* 87–92, 1959a.

Runcorn, S. K., On the hypothesis that the mean geomagnetic field for parts of geological time has been that of a geocentric axial multipole, *J. Atmospheric Terrest. Phys., 14,* 167–174, 1959b.

Smith, P. J. Intensity of the earth's magnetic field in the geological past, *Nature, 216,* 989–990, 1967.

Thellier, E., and O. Thellier, Sur l'intensité du champ magnétique terrestre en France, a l'epoque galloromaine, *Compt. Rend., 222,* 905–907, 1946.

Thellier, E., and O. Thellier, Sur l'intensité du champ magnétique terrestre dans le passé histerique et géologique, *Ann. Geophys., 15,* 285–376, 1959.

van der Voo, R., Paleomagnetic evidence for the rotation of the Iberian Peninsula, *Tectonophysics, 7*(1), 5–56, 1969.

Vine, F. J., Spreading of the ocean floor: New evidence, *Science, 154,* 1405–1415, 1966.

Magnetic Anomalies and Crustal Structure

Ned A. Ostenso

MANY significant advances have been made in the use of magnetic observations and surveys to decipher the earth's structure. Certainly in the forefront of this advance is the use of paleomagnetism to elucidate continental drift and ocean floor spreading, to develop an added tool for geochronology, and to suggest a controlling mechanism for biological evolution. These topics are discussed in other sections of this volume. Important contributions, largely resulting from international cooperation, have been made to a better understanding of crustal structure through magnetic data.

INSTRUMENTATION AND ANALYTICAL TECHNIQUES

Many of the recent insights into crustal structure have been possible only through advances in instrumentation and data analysis. Salient progress in instrumentation includes an automated standard magnetic observatory [*Alldredge and Saldukas, 1964*] with telemetering capability [*Washkurak and Sawatzky, 1966*]; development of highly sensitive rubidium vapor [*Bender, 1960*], cesium vapor [*Giret and Malnar, 1965*], and metastable helium [*Keyser et al., 1961*] magnetometers; and construction of integrated digital recording airborne systems [*Wold, 1964; Giret, 1965; Evenden et al., 1967*] with associated computer methods for analyzing aeromagnetic data [*Wold and Wolfe, 1966*]. A recently developed and potentially powerful tool is that of measuring the vertical gradient of the magnetic field [*Hood and McClure, 1965*], which is now possible from airborne surveys by virtue of the greatly increased magnetometer sensitivities [*Hood*, 1965; *Slack et al.*, 1967; *Royer*, 1967; *Langan*, 1966; *Jensen*, 1965]. This system significantly increases resolving power, eliminates temporal corrections, and, by allowing better use of Laplace's and Euler's equations, permits discrimination between intrabasement and shallower anomaly sources [*Breiner*, 1966].

Some of the more significant advances in analytical techniques include the general use of high-speed computers [*Henderson*, 1960; *Heirtzler et al.*, 1962; *Vogel*, 1963; *Reford and Sumner*, 1964; *Talwani*, 1965; *Bhattacharyya*, 1966; and *Robinson*, 1967], the in situ determination of remanent from induced magnetization [*Goldstein and Ward*, 1966; *Ross and Lavin*, 1966; *Hood and Sangster*, 1967], surface fitting [*Wold and Wolfe*, 1966], field continuation [*Henderson*, 1966], and spectral [*Spector and Bhattacharyya*, 1966; *Gudmundsson*, 1967] and Fourier [*Gudmundsson*, 1966; *Bott*, 1967; *Bhattacharyya*, 1965] analysis.

REGIONAL MAGNETIC SURVEYS

Activity in magnetic surveying is so intense that it is nearly impossible to indicate the state of coverage. *Ostenso* [1966] shows the bulk of global effort, but does not include the recent survey of Switzerland, aeromagnetic flights over England, Scotland, Wales, and Northern Ireland, a detailed F survey of Finland, or the survey of Japan and adjacent waters. As a result of these efforts: (1) the U. S. Naval Oceanographic Office has published 1965 epoch charts of F, H, Z, D, and I for the United States, the world, and the polar regions; (2) the Dominion Observatory published 1965 epoch charts of D, H, Z, F, I, X, Y, G, U, and V for Canada; (3) the Instituto Geografico y Catastral of Spain published 1960 epoch charts of H, D, and Z for the Iberian peninsula; and (4) the Hydrographic Office and Geographical Survey Institute published 1965 epoch charts of D, H, I, F, Z, X, and Y for Japan. Several countries have produced charts of detailed magnetic surveys, Canada being most notable [*Serson*, 1967].

The USSR has compiled a magnetic chart from airborne and ground surveys that has sufficient detail to reflect features of crustal structure [*Simonenko*, 1964; *Andreyev*, 1966]. The magnetic pattern clearly shows the block structure in the basement of the Precambrian platforms. The blocks are frequently bounded by narrow zones of intense positive anomalies which are interpreted to reflect crush zones. In the epi-Hercynian platforms and folded areas, anomaly zones are consistently parallel to the trend of exposed and buried fold complexes. Some anomalies indicate previously unknown tectonic structures. In southeastern central Kazakhstan there is no obvious relationship between magnetic anomalies and surface geology [*Puchkova and Ladynin*, 1966]. Rocks of similar composition and age frequently are characterized by anomalies of different intensity, sign, and pattern. Even the regional magnetic anomalies show no relationship to deep crustal structure.

Reporting on the Transcontinental Geophysical Survey, *Pakiser and Zietz* [1965] suggest that the United States is divided by the Rocky Mountain system into two crustal and upper mantle superprovinces. To the east the crust is relatively thick and strongly magnetic, and a significant part of the upper mantle may be below the Curie temperature. The velocity and density of both the crust and upper mantle in this relatively stable region are high. In contrast, the crust in the western superprovince is relatively thin and weakly magnetic. The velocity and density of the crust and upper mantle are low, and recent diastrophism, plutonism, and volcanism have been widespread. Linear, high amplitude, long wavelength anomalies over the Sacramento Valley and the Sierra Nevada suggest that the lower part of the crust in these areas is more mafic, which is consistent with seismological and gravity data. Mathematically smoothed profiles and contour maps of the region east of the Rockies [*Zietz et al.*, 1966] outline major lithologic units deep within the crust and possibly within the upper mantle. The very broad and less numerous anomalies in the east trend more or less north-south and are significantly different from the more numerous anomalies to the west.

An aeromagnetic survey has been made of the Great Lakes at approximately 6-mile line spacing. The results of the Lake Superior survey [*Wold and Ostenso*, 1967; *Hinze et al.*, 1967] reflect the faulted asymmetrical synclinal structure of the lake.

Whitham [1956b] has reported on magnetotelluric, aeromagnetic, and other studies of the geomagnetic variation anomaly affecting the northern margin of the Queen Elizabeth Islands in the Canadian arctic archipelago. Observed data can be explained by models involving a 100-km uplift of the 1400 to 1500°C isotherm to within 20-30 km of the earth's surface. This solution is not unique and may be in error by as much as 60%. However, most of the parameters of the problem have been defined, and aeromagnetic surveys plus other data do not support lithologic changes as an explanation.

An aeromagnetic survey over an orogenic belt of early Precambrian rocks in northwestern Ontario showed a series of positive and negative bands of anomalies. *Bhattacharyya and Morley* [1965] suggested that this banding could be caused by a magma zone cooling slowly from the outside toward its center while the earth's main magnetic field underwent several reversals. Eighty-five selected anomalies were analyzed to determine the horizontal dimensions and the depths of the tops and bottoms of the causative deep crustal bodies and to determine the magnitude and direction of their polarization. The bodies ranged from 0.8 to 5.1 km in horizontal dimensions and the depth to their tops varied from 4.8 to 8 km. The mean depth to their bottoms was 20 km, which was interpreted to be the Curie isotherm. Polarization directions of the positive anomalies were mutually consistent and almost reverse to those of the negative anomalies, but they were widely different from the direction of the present earth's field. This suggests that over this large region the components of magnetization along the present field are, at depth, small in comparison with natural remanent magnetization.

Hood (personal communication) reports that granite intrusions in Nova Scotia, which themselves have a very low intensity of magnetization, are bounded by magnetic anomalies. He suggests that the natural remanent magnetization of the country rock, which already has significant susceptibility (i.e., 0.5% magnetite or more) is considerably increased at the time of intrusion by heating through the Curie temperature of the constituent magnetic materials which acquire an augmented intensity of magnetization by subsequent cooling in the earth's magnetic field. Also, contact metamorphism of the country rock may have produced some magnetite. Hood suggests that this mechanism of enhancing anomalies at the margins of granitic intrusions may be a common occurrence and that the intrusions should exhibit concomitant gravity lows due to the low density of the granite.

A total of 25,000 km of seaborne and airborne magnetic tracks have been completed over Hudson Bay. The depths to crystalline basement were found to be least around the margins of the Bay, with the greatest calculated depths in excess of 30 km occurring in the central part of the Bay [*Hood*, 1964]. However, it was recognized that part of the sediments may be Proterozoic, because substantial outcroppings of Proterozoic sediments are found in southern and eastern parts of the Bay. An area of intense magnetic anomalies, believed to reflect iron formations, is centered at about 58°37½'N, 92°40'W [*Hood et al.*, 1966].

CRUSTAL RIFTING

Extensive rifting of the ocean floor off western North America was discovered from detailed surveys conducted during the IGY. Later surveys showed that the Mid-Atlantic ridge is also offset in many places by roughly east-west oriented rifts. These features are discussed elsewhere in this monograph. From magnetic data, *Drake et al.* [1963] discovered a major transcurrent fault normal to the eastern United States coastline at about 40°N. The fault was more than 1000 km long, and geological and other geophysical evidence indicate a right lateral displacement of 160 km, which is opposite in direction to those off the west coast. This rift was projected inland to affect the Appalachian structure [*Drake and Woodward*, 1963]. *Gilliland* [1962], mainly on the basis of geologic inferences, suggested that the Mendocino fracture zone may continue across North America near 40°N and may be as old as Precambrian. Using cross correlation techniques, *Fuller* [1964] presented strong evidence for continuation of the eastern Pacific rifts deep into continental North America and possibly connecting with the east coast rift. He found no evidence for consistent amounts of offset along the fractures and considered them to be of deep origin and partly decoupled from the upper crust. Data from the transcontinental aeromagnetic survey showed an 800-km-long anomaly trending east-west, which *Zietz et al.*

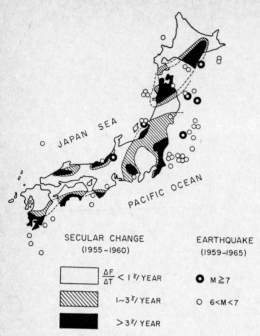

Fig. 1. Anomalous areas of secular change during the interval 1955–1960, and the distribution of recent (1959–1965) large earthquakes (after Y. Harada, personal communication).

[1966] attribute to a zone of rifting extending across Nebraska and Iowa at about 40°N. They suggest that the feature is probably less than 40 km in depth.

Extensive international cooperative effort has been stimulated by the Upper Mantle Project on the East African rift system and the Rhinegraben. These activities, including the contribution of geomagnetic studies, have been summarized in special reports [*University College*, 1965; *Rothé and Sauer*, 1967].

GEOMAGNETIC EFFECTS OF EARTHQUAKES

Magnetic surveying in Japan has become sufficiently detailed that abnormal magnetic activity has definitely been correlated with earthquake activity [*Rikitake et al.*, 1966a, b]. Surveys before and after the Niigata earthquake in June 1964 showed anomalous changes in secular variation and disturbances of the local field by several tens of γ plus large changes in declination [*Fujita*, 1965]. The same phenomenon was later observed with the Matsushiro earthquake swarms from August 1965 through 1966 [*Hayashi*, 1967]. It has been concluded that regions having secular change rates larger than 2 or 3 γ/year (Figure 1) can expect to experience a large earthquake within a few years [*Tazima and Fujita*, 1968].

In the United States, Stanford University, the California Institute of Technology, and the U. S. Geological Survey are attempting to test the piezomagnetic effect of fault zones. Five rubidium vapor magnetometers have been installed along the San Andreas fault, and three other instruments have been used at sites in the western United States [*Kovach and Breiner*, 1967; *Alldredge*, 1966].

CONTINENTAL MARGINAL ANOMALY

A strikingly persistent linear anomaly has been observed near the edge of the continental shelf off the east coast of North America [*Drake et al.*, 1963; *Hood and Godby*, 1965; *Hood*, 1966; *Geddes and Watkins*, 1966; *Zietz et al.*, 1966], in the Pacific [*Krause*, 1965; *Gaynanov*, 1964], and in the Arctic Ocean [*Shaver and Hunkins*, 1964; *Demenitskaya and Karasik*, 1966; *Hunkins*, 1966; *Roots*, 1966]. The anomaly off the eastern United States has been traced from Argentia, Newfoundland, to Jacksonville, Florida. Its width ranges from 30 to 80 km and its amplitude from 150 to 600 γ. Geophysical data suggest that the basement ridge is a buried, quiescent island arc and that the magnetic anomaly is caused by intrusive and extrusive rocks emplaced during an earlier tectonically active phase in the arc's development [*Watkins and Geddes*, 1965].

SUMMARY

Significant advances have been made in the use of magnetic data to interpret crustal structure. However, much remains to be done. The need for extensive detailed surveys, particularly utilizing the advantages of vertical gradient measurements, is apparent. The possible use of piezomagnetic effects to forecast earthquakes should be investigated carefully. Finally, international cooperation should be marshalled to provide long unbroken flight lines that can be analyzed by the Fourier methods of *Alldredge et al.* [1963] and *Hahn* [1965] and the statistical technique of *Whitham* [1965a]. The harmonic analysis of magnetic wavelengths by *Alldredge* [1965] clearly demonstrates the hiatus between anomalies produced in the crust and upper mantle and those originating from the earth's core. It would be desirable to extend such analysis to higher frequency anomalies to demonstrate quantitatively whether, as King et al. [1964] suggested, the magnetic disturbance produced by the ocean crust is diagnostically different from that produced by a continental shield. Such an analysis should be conducted for all major geologic provinces. A circumglobal flight track that is not a great circle route could be continued indefinitely without being repetitious. Hence an essentially infinitely long unbroken line of data could be obtained that would provide a statistically significant sampling of all geologic provinces and tectonic features.

REFERENCES

Alldredge, L. R., Analysis of long magnetic profiles, in *The Symposium on Magnetism of the Earth's Interior*, J. Geomag. Geoelec., 17, 173–186, 1965.

Alldredge, L. R., Magnetic surveys, in *ESSA Symposium on Earthquake Prediction*, pp. 123–127, U. S. Government Printing Office, Washington, D. C., 1966.

Alldrege, L. R., and I. Saldukas, An automatic standard magnetic observatory, J. Geophys. Res., 69, 1963–1970, 1964.

Alldredge, L. R., G. D. Van Voorhis, and T. M. Davis, A magnetic profile around the world, J. Geophys. Res., 68, 3679–3692, 1963.

Andreyev, B. A., Osnovnyye strukturnyye elementy fundamenta vostochnykh rayonov SSSR (Fundamental structural elements of the basement of the eastern regions of the USSR), *Dokl. Akad. Nauk. SSSR*, 170, 402–405, 1966.

Bender, P. L., Measurement of weak magnetic fields by optical pumping methods, *Bull. Ampere*, 9, fac. spéc., 621–628, 1960.

Bhattacharyya, B. K., Two-dimensional harmonic analysis as a tool for magnetic interpretation, Geophysics, 829–857, 1965.

Bhattacharyya, B. K., Quantitative interpretation of aeromagnetic information, Can. Geol. Surv. Paper 66-42, 22–23, 1966.

Bhattacharyya, B. K., and L. W. Morley, The delineation of deep crustal magnetic bodies from total field aeromagnetic anomalies, in *The Symposium on Magnetism of the Earth's Interior*, J. Geomag. Geoelec., 17, 237–252, 1965.

Bott, M. H. P., Solution of the linear inverse problem in magnetic interpretation with applications to oceanic magnetic anomalies, Geophys. J., 10, 313–323, 1967.

Breiner, S., Quantitative application of geomagnetic gradiometer data (abstract), Geophysics, 31, 1191, 1966.

Demenitskaya, R. M., and A. M. Karasik, Magnetic data confirm that the Nansen-Amundsen basin is of normal oceanic type, in *Continental Margins and Island Arcs*, Can. Geol. Surv. Paper 66-15, 191–196, 1966.

Drake, C. L., J. Heirtzler, and J. Hirshman, Magnetic anomalies off eastern North America, J. Geophys. Res., 68, 5259–5275, 1963.

Drake, C. L., and H. P. Woodward, Appalachian curvature, wrench faulting, and offshore structures, N. Y. Acad. Sci. Trans., [2]26, 49–63, 1963.

Evenden, G. I., F. C. Frischknecht, and J. L. Meuschke, Digital recording and processing of airborne geophysical data, in *Geological Survey Research, 1967*, U. S. Geol. Surv. Profess. Paper 575-D, 79–84, 1967.

Fujita, N., The magnetic disturbances accompanying the Niigata earthquake, J. Geodet. Soc. Japan, 11, 8–25, 1965.

Fuller, M. D., Expression of E-W fractures in magnetic surveys in parts of the U.S.A., Geophysics, 29, 602–622, 1964.

Gaynanov, A. G., O nekotorykh osobennostyakh stroyeniya zemnoy kory perckhodnykh zon Tikhogo okeana po geofizicheskim donnym (On some features of the structure of the earth's crust in the transition zones of the Pacific Ocean according to geophysical data), Mosk. Univ. Geofiz. Issled. Shornik, 1, 228–241, 1964.

Geddes, W. H., and J. S. Watkins, Atlantic shelf magnetic anomalies, in *Continental Margins and Island Arcs*, Can. Geol. Surv. Paper 66–15, 48–46, 1966.

Gilliland, W. N., Possible continuation of the Mendocino fracture zone, Science, 137, 685–686, 1962.

Giret, R. I., Some results of aeromagnetic surveying with a digital cesium-vapor magnetometer, Geophysics, 30, 883–890, 1965.

Giret, R., and L. Malnar,, Un nouveau magnétométre aérien, le magnétométre á vapeur de caesium, Geophys. Prospect., 13, 225–239, 1965.

Goldstein, N. E., and S. H. Ward, The separation of remanent from induced magnetism in situ, Geophysics, 31, 779–796, 1966.

Gudmundsson, G., Interpretation of one-dimensional magnetic anomalies by use of the Fourier-transform, Geophys. J., 10, 87–97, 1966.

Gudmundsson, G., Spectral analysis of magnetic surveys, Geophys. J. 10, 325–337, 1967.

Hahn, A., Two applications of Fourier's analysis for the interpretation of geomagnetic anomalies, in *The Symposium on Magnetism of the Earth's Interior*, J. Geomag. Geoelec., 17, 195–226, 1965.

Hayashi, T., Geodetic surveys in the area of Matsushiro earthquake swarms, Bull. Geog. Surv. Inst., 11, part 2, 20–25, 1967.

Heirtzler, J. R., G. Peter, M. Talwani, and E. G. Zurflueh, Magnetic anomalies caused by two-dimensional structure: Their computation by digital computers and their interpretation, Lamont Geol. Obs. Tech. Rept. 6, CU-6-62, 1962.

Henderson, R. G., A comprehensive system of automatic computation in magnetic and gravity interpretation, *Geophysics, 25,* 569–585, 1960.

Henderson, R. G., Field continuation and the step model in aeromagnetic interpretation, *Geophys. Prospect., 14,* 528–546, 1966.

Hinze, W. J., N. W. O'Hara, J. W. Trow, and G. B. Secor, Aeromagnetic survey of eastern Lake Superior, in *The Earth beneath the Continents, Geophys. Monograph 10,* edited by J. S. Steinhart and T. J. Smith, pp. 95–110, American Geophysical Union, Washington, D. C., 1967.

Hood, P. J., Sea magnetometer reconnaissance of Hudson Bay, *Geophysics, 29,* 916–921, 1964.

Hood, P., Gradient measurements in aeromagnetic surveying, *Geophysics, 30,* 891–902, 1965.

Hood, P., Magnetic surveys of the continental shelves of eastern Canada, in *Continental Margins and Islands Arcs, Can. Geol. Surv. Paper 66-15,* 19–32, 1966.

Hood, P. J., and E. A. Godby, Magnetic profile across the Grand Banks and Flemish Cap off Newfoundland, *Can. J. Earth Sci., 2,* 85–92, 1965.

Hood, P. J., and D. J. McClure, Gradient measurements in ground magnetic prospecting, *Geophysics, 33,* 403–410, 1965.

Hood, P. J., and D. F. Sangster, The Carey Foster in situ susceptibility meter, *Can. Geol. Surv. Paper 64-22,* 17 pp., 1967.

Hood, P. J., P. Sawatzky, and M. E. Bower, Aeromagnetic survey of a portion of central Hudson Bay, *Can. Geol. Surv. Paper 66-1,* 19–21, 1966.

Hunkins, K., The arctic continental shelf north of Alaska, in *Continental Margins and Island Arcs, Can. Geol. Surv. Paper 66-15,* 197–205, 1966.

Jensen, H., Instrument details and applications of a new airborne magnetometer, *Geophysics, 30,* 875–882, 1965.

Keyser, A. R., J. A. Rice, and L. D. Schearer, A metastable helium magnetometer for observing small geomagnetic fluctuations, *J. Geophys. Res., 66,* 4163–4169, 1961.

King, E., I. Zietz, and L. R. Alldredge, Genesis of the Arctic Ocean basin, *Science, 144,* 1551–1557, 1964.

Kovach, R. L., and S. Breiner, A search for the piezomagnetic effect along the San Andreas Fault, *U. S. Progress Report on the Upper Mantle Project,* p. 77, Natl. Acad. Sci.—Natl. Res. Council, 1967.

Krause, D. C., Tectonics, bathymetry and geomagnetism of the southern continental borderland west of Baja California, Mexico, *Geol. Soc. Am. Bull., 76,* 617–650, 1965.

Langan, L., A survey of high resolution geomagnetics, *Geophys. Prospecting, 14,* 487–503, 1966.

Ostenso, N. A., Geomagnetism, *Trans. Am. Geophys. Union, 47,* 303–332, 1966.

Pakiser, L. C., and I. Zietz, Transcontinental crustal and upper-mantle structure, *Rev. Geophys., 3,* 505–520, 1965.

Puchkova, L. I., and A. V. Ladynin, O svyazi magnitnykh anomaly so stroyeniyem zemnoy kory v yugo-vostochnoy chasti Tsentral'rogo Kazakhstana (On the relationship of magnetic anomalies to crustal structure in the southeastern part of Central Kazakhstan), *Akad. Nauk. SSSR Sibirsk. Otd., Geol. Geofiz., 12,* 126–134, 1966.

Reford, M. S., and J. Sumner, Review article—Aeromagnetics, *Geophysics, 29,* 505–516, 1964.

Rikitake, T., Y. Yamazaki, M. Sawada, Y. Sasai, T. Yoshino, S. Uzawa, and T. Shimomura, Geomagnetic and geoelectric studies of the Matsushiro earthquake swarm, *Bull. Earthquake Research Inst. Tokyo Univ., 44,* 1735–1758, 1966b.

Rikitake, T., T. Yukutake, Y. Yamazaki, M. Sawada, Y. Sasai, Y. Hagiwara, K. Kawada, T. Yoshino, and S. Takafumi, Geomagnetic and geoelectric studies of the Matsushiro earthquake swarm, *Bull. Earthquake Res. Inst. Tokyo Univ., 44,* 1335–1370, 1966a.

Robinson, E. S., Use of fan filters in computer analysis of magnetic anomaly trends, *U. S. Geol. Surv. Profess. Paper 575-D,* 113–119, 1967.

Roots, E. F., The northern margin of North America: A progress report on investigations and problems, in *Continental Margins and Island Arcs, Can. Geol. Surv. Paper 66-15,* 188–190, 1966.

Ross, H. P., and P. M. Lavin, In situ determination of the remanent magnetic vector of two-dimensional tabular bodies, *Geophysics, 31,* 949–962, 1966.

Rothé, J. P., and K. Sauer, editors, The Rhinegraben Report, *Mem. Serv. Carte Geol. d'Alsace Lorraine, 26,* 146, 1967.

Royer, G., Two years' survey with cesium vapour magnetometer, *Geophys. Prospecting, 15,* 174–193, 1967.

Serson, P. H., Magnetism, *Can. Geol. Surv. Paper 67-41,* 85–116, 1967.

Shaver, R., and K. Hunkins, Arctic Ocean geophysical studies: Chukchi cap and Chukchi abyssal plain, *Deep-Sea Res., 11,* 905–916, 1964.

Simonenko, T. N., Anomal'noy magnitnoye pole SSSR i nekotoryye voprosy tektoniki (The anomalous magnetic field of the USSR and some problems of tectonics), *Dokl. Sov. Geol., 2,* 104–110, 1964.

Slack, H., V. M. Lynch, and L. Langan, The magnetic gradiometer, *Geophysics, 32,* 877–892, 1967.

Spector, A., and B. K. Bhattacharyya, Energy density spectrum and autocorrelation function of anomalies due to simple magnetic models, *Geophys. Prospecting,* 243–254, 1966.

Talwani, M., Computation with the help of a digital computer of magnetic anomalies caused by bodies of arbitrary shape, *Geophysics, 30,* 797–817, 1965.

Tazima, M., and N. Fujita, Magnetic survey in the area accompanying the earthquake, *Bull. Geog. Surv. Inst.,* in press, 1968.

University College, *East African Rift System,* University College, Nairobi, 116 pp., 1965.

Vogel, A., The application of electronic computers to the calculation of effective magnetization, *Geophys. Prospecting,* 11, 51–58, 1963.

Washkurak, S., and P. Sawatzky, The Serson direct-reading proton free-precession magnetometer, *Can. Geol. Surv. Paper 65-31,* 33–75, 1966.

Watkins, J. S., and W. H. Geddes, Magnetic anomalies and possible orogenic significance of geologic structure of the Atlantic shelf, *J. Geophys. Res.,* 70, 1357–1361, 1965.

Whitham, K., On the depth of magnetic sources derived from long magnetic profiles, in *The Symposium on Magnetism of the Earth's Interior, J. Geomag. Geoelec.,* 17, 253–262, 1965a.

Whitham, K., Geomagnetic variation anomalies in Canada, in *The Symposium on Magnetism of the Earth's Interior, J. Geomag. Geoelec.,* 17, 481–498, 1965b.

Wold, R. J., The Elsec-Wisconsin digital recording proton precession magnetometer system, *Univ. Wis., Geophys. Polar Res. Center Rept. 64-4,* 83, 1964.

Wold, R. J., and N. A. Ostenso, Aeromagnetic, gravity, and sub-bottom profiling studies in western Lake Superior, *The Earth beneath the Continents, Geophys. Monograph 10,* edited by J. S. Steinhart and T. J. Smith, pp. 66–94, American Geophysical Union, Washington, D. C., 1967.

Wold, R. J., and T. R. Wolfe, Computer methods of analyzing aeromagnetic data, *Univ. Wis., Geophys. Polar Res. Center Rept. 66-1,* 1966.

Zietz, I., E. R. King, W. Geddes, and E. G. Lidiak, Crustal study of a continental strip from the Atlantic Ocean to the Rocky Mountains, *Bull. Geol. Soc. Am.,* 77, 1427–1448, 1966.

Conductivity Anomaly of the Upper Mantle

Tsuneji Rikitake

SYSTEMATIC studies of the geomagnetic variation anomaly began in Japan and Germany about 1950. By the middle of the 1960's, studies of anomalies of this kind, which probably reflect heterogeneous distribution of the electrical conductivity in the earth's crust and upper mantle, were being made in many countries. An international symposium on this topic was held during the IUGG Berkeley Assembly in 1963, and a detailed review of the mantle conductivity anomaly was presented by *Rikitake* [1966] on the basis of the information available to 1964.

ANOMALIES IN JAPAN

It has long been known that ΔZ (change in the vertical geomagnetic field) is surprisingly large at Kakioka, one of the standard magnetic observatories in Japan, amounting to as much as 60% of ΔH (change in the horizontal field) at times of geomagnetic bay or similar change. Since the early 1950's, much effort has been made to clarify the geomagnetic variation anomaly by setting up a number of satellite magnetic observatories in Japan. As the results of observation have been published, mostly by the present author and his collaborators in a series of papers in the Bulletin of the Earthquake Research Institute, and have been summarized elsewhere [*Rikitake,* 1959, 1966], the anomaly will be only briefly described here.

The Central Japan anomaly, the area over which we observe anomalously large ΔZ, covers the Pacific side of the central part of the Japan Islands in a roughly elliptical shape. The anomaly is best observed in geomagnetic bays and similar changes. ΔZ points downward when ΔH is positive. Marked parallelism between ΔH and ΔZ is found at observatories in the area. The $\Delta Z/\Delta H$ ratio is almost unity at the center of the area.

The above characteristics of the Central Japan anomaly are most clearly seen when a geomagnetic change takes place in a north-south direction. When the field changes in the east-west direction, as sometimes occurs for a solar flare effect (s.f.e.), the anomaly practically disappears. The anomaly cannot be observed for slower variations like solar daily variation (S_q) and storm-time variation (Dst). On the contrary, it has been suggested that these variations bring out another anomaly, which probably indicates a low conductivity at a depth of several hundred kilometers beneath Japan.

Close examination of the Central Japan anomaly makes it clear that ΔH values at observatories in southern Japan are larger than those in central Japan for bay-like events [Rikitake, 1965] and short-period fluctuations during a magnetic storm [Sasai, 1966]. Taking into account the general northward increase in ΔH for geomagnetic variations of this kind, ΔH is a few tens of per cent larger southwest of the central area of the anomaly, and becomes somewhat smaller toward the northeast margin of the anomaly.

A spectral analysis technique is also applied to the study of the anomaly [Sasai, 1966]. A marked tendency for ΔZ to increase at observatories in the center of the area and to decrease at those in the margins is demonstrated at a period of 160 min or thereabouts. Figure 1 shows the $\Delta Z/\Delta H$ changes at a number of Japanese observatories as the frequency of variation increases.

During the periods of the Upper Mantle Project and the Years of the Quiet Sun, observation of geomagnetic change was stressed in the hope of obtaining a more precise understanding of the Japanese anomaly. As a result of observations at eighteen observatories well distributed over Japan, an outstanding anomaly in geomagnetic variation was discovered in the northeastern part of Honshu Island, the mainland of Japan, although its extent is smaller than that of the Central Japan anomaly. The anomaly has been called the Northeastern Japan anomaly by Y. Kato (personal communication, 1966). Kato and his coworkers conducted intensive variographic observations over northeastern Honshu Island. For baylike events having periods shorter than 3 hours, they found that ΔZ at a station at the northern end of Honshu Island is directed upward when ΔH is positive. At an observatory on Hokkaido Island about 200 km to the northeast, however,

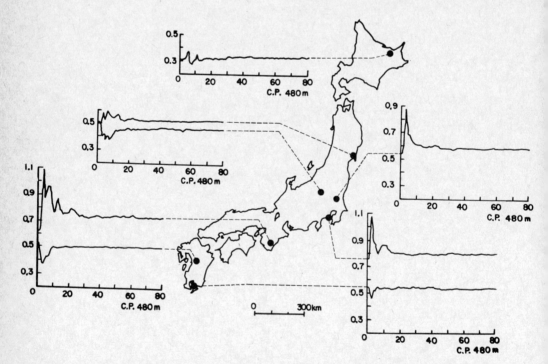

Fig. 1. $\Delta Z/\Delta H$ (ordinate) versus frequency at the observatories in Japan as obtained by the spectral analysis of a magnetic storm on September 20, 1959 [after Sasai, 1966].

ΔZ indicates a downward change. Some 200 km south of the station they observed practically no change in the vertical field. Observations at stations farther south showed a gradual increase of downward ΔZ which appears to be the continuation of the Central Japan anomaly. Figure 2 shows a schematic distribution of $\Delta Z / \Delta H$ over northeastern Japan for a bay-like variation changing in the north-south direction. It is striking that the Northeastern Japan anomaly disappears when the period of a geomagnetic variation is longer than 3 hours.

Questions have often been raised about the possibility of accounting for the Japanese anomaly by the magnetic field produced by electric currents induced in the seas surrounding Japan. Should the ocean effect be very pronounced, ΔZ should be in phase with ΔD (change in declination), and the ocean effect should be enhanced when the geomagnetic field changes in an east-west direction in northeast Japan, because the edge of the deep ocean runs from north to south there. But this is not the case. The author's estimate [*Rikitake*, 1967] suggests that the ocean has little effect on land if we take gradual increase of the sea depth into account, although a much larger effect would certainly appear over the deep sea. *Parkinson's* [1964, 1966] experiments with his terrella model similarly do not support an explanation of the Japanese anomalies by an ocean effect. It appears to the present author that the Japanese anomalies in geomagnetic variation are hardly explained by the effect of an ocean; therefore, the cause of the anomaly should be sought in an extremely anomalous distribution of the electrical conductivity beneath Japan, probably in the upper mantle. No conclusive explanation has been put forward, although a highly speculative 'branch circuit' model was proposed by *Rikitake* [1959]. Attempts to correlate the anomaly with other geophysical elements such as seismic-wave velocity, heat flow, and the like have so far been unsuccessful. Although the Japanese anomalies in geomagnetic variation are most puzzling, it seems certain that they reflect a complicated distribution of the electrical conductivity under Japan.

ANOMALIES ON THE AMERICAN CONTINENTS

A number of important contributions to the study of geomagnetic variation anomaly have

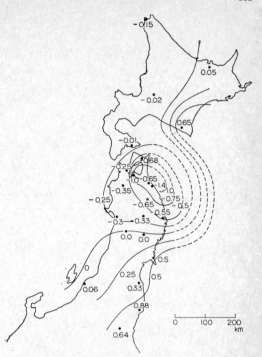

Fig. 2. Schematic distribution of $\triangle Z / \triangle H$ that indicates the Northeastern Japan anomaly (after Kato, unpublished).

been achieved on the American continents. In the Canadian arctic region, two kinds of anomaly have so far been found and intensively examined in spite of the logistic difficulties. The Mould Bay anomaly, as observed around Mould Bay on Prince Patrick Island, is characterized by an unusual attenuation of the vertical field fluctuations of short period. Spectral analyses of geomagnetic fluctuations led to the conclusion that an anomalous conductor of moderate thickness (e.g., 5 to 20 km), having the conductivity of 10^{-11} emu, can account for the anomaly, if it is assumed to exist at a depth of about 25 km [*Whitham*, 1963, 1964] (T. Yukutake and J. H. De Laurier, personal communication, 1965). If such a high value of conductivity is interpreted in terms of temperature, a high-temperature isotherm ($\sim 1400°C$) must occur at a depth of 10 $\sim$ 20 km there; heat flow measurements by *Law et al.* [1965] do not indicate the necessary high values. Variographs around Mould Bay indicate that the anomaly covers an area some 200 by 300 km with a sharp boundary where the surface of the conducting layer deepens from 40 to about 100 km within a distance of 50 km [*Whitham*, 1965]. An earthquake swarm took place near Mould Bay in early 1965.

Whitham [1964, 1965] and others have been intensively studying an anomaly at Alert, a station on the northern end of Ellesmere Island. A strong tendency for the magnetic vectors of short-period bay-like variations to be confined to a northwest-southeast direction has been found there. A possible explanation of the anomaly would be to assume an upheaval of about 100 km of the 1400°C isotherm to within 25–30 km of the earth's surface [*Rikitake and Whitham*, 1964], the length of the upheaved portion being estimated as a few hundred kilometers. Magnetotelluric observations recently made at Alert [*Whitham and Andersen*, 1965] proved, however, that the above model does not account for the phase relation between geomagnetic changes and earth currents. The present interpretation of the Alert anomaly, as well as the Mould Bay anomaly, is therefore unsatisfactory, although it is certain that they are caused by something extraordinary in the crust or upper mantle.

Variographic work has been carried out along east-west profiles in western Canada [*Hyndman*, 1963; *Lambert and Caner*, 1965; *Whitham*, 1965]. An enhancement of ΔZ for geomagnetic fluctuations at a 10- to 60-min period range was found at a station on the Pacific coast of Vancouver Island. Lambert and Caner presented an interpretation of the anomaly by taking into account a dipping of the high-conducting mantle toward the inland in addition to the true ocean effect. Less well defined anomalies were also reported further inland. Lambert and Caner attempted to correlate the anomalies with the seismic low-velocity layer.

Schmucker [1964] conducted an extensive observation of geomagnetic variation anomaly over the southwestern United States. The Californian Coast anomaly (named by Schmucker) seems to be one of the best examples of the edge effect of an ocean. It was demonstrated that ΔZ for baylike geomagnetic changes takes on a large amplitude at coastal stations, getting gradually smaller inland. ΔZ resembles ΔD rather than ΔH. It is well known that geomagnetic variation vectors tend to be confined in a plane, provided that the effect of electromagnetic induction is considerable, and a customary way of representing the direction of such a plane is to use what we call the Parkinson vector [*Parkinson*, 1959]. Its direction in the horizontal plane indicates the direction of maximum upward tilt of the preferred plane, and its length is proportional to the sine of the angle of tilt. The Parkinson vectors at many stations in California are approximately perpendicular to the boundary of the continental shelf; without exception they point to the ocean.

Larsen and Cox [1966] made telluric observations off California and preliminarily suggested that a high-conducting layer lies fairly close to the sea bottom.

Schmucker [1964] also found the Texas anomaly. ΔZ for most of the geomagnetic variations is extremely small at Tucson, Arizona. But, at a station about 400 km east of Tucson, ΔZ becomes considerably larger. Schmucker suggested a sharp drop in the top surface of the conducting mantle from a depth of 160 to 320 km there. According to V. Vacquier (personal communication, 1966), heat flow values are large immediately above the upheaved part of the conducting mantle and small above the depressed part.

Much effort has in recent years been made in extending variographic work to South America. *Schmucker et al.* [1964] reported that very little anomaly was found for S_q in Peru, but during a magnetic storm ΔZ differs distinctively from station to station, as can be seen in Figure 3. It is concluded that the subsurface conductivity structure in southern Peru must be very complicated. No pronounced coastal anomaly of the Z variations has been found, although one of the stations is very close to the deep sea. ΔZ for a bay there is opposite in sign from that due to the usual coastal anomaly. In order to account for such a reverse coast effect, we are forced to imagine that the high-conductivity mantle under the ocean off Peru is deeper than that under the land.

The present author also understood (U. Schmucker, personal communication, 1966) that a reversal in ΔZ has been observed between coastal stations and those near the eastern slope of the Andes by joint variographic work of the Instituto Geofisico del Peru and the Instituto Geofisico Boliviano with the Department of Terrestrial Magnetism, Carnegie Institution, in

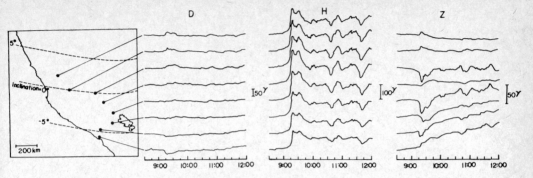

Fig. 3. The sudden storm commencement and subsequent short-period fluctuations simultaneously observed at Peruvian observatories on September 21, 1963 [after *Schmucker et al.*, 1964].

Washington, D. C. The reversal is accompanied by an intensified ΔH indicating a local concentration of internal induction currents at about 60 km depth, which flow along a zone of high conductivity under the crest of the Andes. No marked coastal anomaly has been found.

OTHER ANOMALIES

Parkinson [1966] refined his terrella model, a spherical sheet that simulates the oceans with copper plates, by representing shallow seas by nets of copper wire. According to his experiments with a realistic inducing field applied to the model, most of the former conclusions were supported although, at some places, the effect of shallow seas seems fairly large. It is impossible to account for the Central Japan anomaly by the model. Most of the Australian anomalies [*Parkinson*, 1964] seem to be explained by the oceanic conduction plus a conducting path north of Australia, which is probably not provided by the shallow water between Australia and Indonesia.

E. C. Bullard (personal communications, 1966) noticed an anomaly similar to that at Mould Bay in Scotland. E. A. Lubimova (personal communication, 1966) pointed out a high terrestrial heat flow accompanied by a geomagnetic variation anomaly near Lake Baikal. In addition to the well-known North German anomaly [*Schmucker*, 1959], a number of geomagnetic variation anomalies have been found in Germany and surrounding countries. *Wiese* [1965] has been successful in correlating the anomalies to major tectonic structures.

A number of important findings in regard to geomagnetic variation on islands have recently been brought forth [*Mason*, 1963, 1964; *Elvers and Perkins*, 1964; *Sasai*, 1967]. It has become clear that ΔZ varies from place to place on an island. Reversals in sign of ΔZ are often found for rapid geomagnetic fluctuations between northern and southern stations on an island. Such an island effect is at present thought to be caused by the electric currents induced in the sea surrounding the island. Judging from the fact that an island effect can be observed on an island that is very close to the mainland [*Sasai*, 1967], the possibility that the underground structure might play some role on the effect cannot be ruled out. We must carefully examine whether geomagnetic variations observed on an island represent a typical feature in the district concerned.

CONCLUSION

Intensive investigations of geomagnetic variation anomalies have in recent years brought to light a number of new anomalies like the Northeastern Japan anomaly, the Peruvian anomaly, and the Andes anomaly. Interpretation of these anomalies, as well as those already established, has by no means been satisfactory. In spite of extensive studies of the Central Japan anomaly over a ten-year period, no realistic explanation has been put forward. In the case of the Alert anomaly, new observations of telluric currents make the interpretation even more difficult. It is therefore premature to say anything definite about the cause of anomalies, although it is almost certain that the major characteristics of the anomalies can be accounted for by undulations of the highly conducting regions of the mantle. This suggests a corresponding undulation of isotherms within the upper mantle. Only a few inland anomalies have so far been successfully correlated within the heat flow variations.

Very few studies have attempted to correlate the mantle conductivity anomaly with other geophysical elements. Most of the mechanical properties of rocks are not sensitive to a slight change in temperature, but a relatively small temperature difference at 1000–1300°C gives rise to a large increase or decrease in the electrical conductivity of rocks. This is one explanation of a mantle conductivity anomaly being observed over a region in which no outstanding anomalies in seismic velocity and gravity have been found. Therefore, the geomagnetic method provides an independent means of detecting an anomalous condition within the earth.

The edge effect of an ocean is a puzzling problem. We are not sure what percentage of an observed coastal anomaly can be accounted for by the electric currents induced in the ocean. It has also been pointed out that the high-conducting layer in the mantle would play an important role in an ocean effect, provided that the layer is undulating. The finding of a reverse coastal effect in Peru makes the matter more complicated. Extensive application of the spectral analysis techniques would be required in order to separate the true ocean effect from the anomalies actually observed. Observations with ocean-bottom variographs would provide very useful information about the mantle conductivity beneath the ocean.

REFERENCES

Elvers, D., and D. Perkins, Geomagnetic research on spatial dependence of time variations across Puerto Rico, *Trans. Am. Geophys. Union, 45*, 46, 1964.

Hyndman, R. D., Electrical conductivity inhomogeneities in the earth's upper mantle, M.A.Sc. Thesis, Univ. British Columbia, 1963.

Lambert, A., and B. Caner, Geomagnetic depth-sounding and the coast effect in western Canada, *Can. J. Earth Sci., 2*, 485–509, 1965.

Larsen, J., and C. Cox, Lunar and solar daily variation in the magnetotelluric field beneath the ocean, *J. Geophys. Res., 71*, 4441–4445, 1966.

Law, L. K., W. S. B. Paterson, and K. Whitham, Heat flow determinations in the Canadian Arctic archipelago, *Can. J. Earth Sci., 2*, 59–71, 1965.

Mason, R. G., Spatial dependence of time variations of the geomagnetic field in the range 24 hr to 3 min on Christmas Island, Geophys. Dept. Publ. 63-3, Imp. Coll. Sci. Technol., London, pp. 1–20, 1963.

Mason, R. G., Magnetic effects at Canton Island of the 1962 high altitude nuclear tests at Johnston Island, Geophys. Dept. Publ. 64-1, Imp. Coll. Sci. Technol., London, pp. 1–12, 1964.

Parkinson, W. D., Directions of rapid geomagnetic fluctuations, *Geophys. J., 2*, 1–14, 1959.

Parkinson, W. D., Conductivity anomalies in Australia and the ocean effect, *J. Geomag. Geoelec., 15*, 222–226, 1964.

Parkinson, W. D., Terrella model experiment, preliminary results, *Australia Bur. Min. Resources, Geol. Geophys., Record 1966/70*, 1–7, 1966.

Rikitake, T., Anomaly of geomagnetic variations in Japan, *Geophys. J., 2*, 276–287, 1959.

Rikitake, T., Some characteristics of geomagnetic variation anomaly in Japan, *J. Geomag. Geoelec., 17*, 95–97, 1965.

Rikitake, T., *Electromagnetism and the Earth's Interior*, Elsevier Publishing Company, Amsterdam, 308 pp., 1966.

Rikitake, T., Electromagnetic induction within non-uniform plane and spherical sheets, *Bull. Earthquake Res. Inst., Tokyo Univ., 45*, 1229–1294, 1967.

Rikitake, T., and K. Whitham, Interpretation of the Alert anomaly in geomagnetic variations, *Can. J. Earth Sci., 1*, 35–62, 1964.

Sasai, Y., The anomalous behaviour of geomagnetic variations of short period and its relation to the subterranean structure, 11, Spectral analysis of geomagnetic disturbances, *Bull Earthquake Res. Inst., Tokyo, Univ., 44*, 167–178, 1966.

Sasai, Y., Spatial dependence of short-period geomagnetic fluctuations on Oshima Island, *Bull. Earthquake Res. Inst., Tokyo Univ., 45*, 137–157, 1967.

Schmucker, U., Erdmagnetische Tiefensondierung in Deutschland 1957–1959: Magnetogramme und erste Austwertung, *Abhandl. Akad. Wiss. Goettingen, Math. Physik. Kl., 5*, 1–51, 1959.

Schmucker, U., Anomalies of geomagnetic variations in the southwestern United States, *J. Geomag. Geoelec., 15*, 193–221, 1964.

Schmucker, U., O. Hartmann, A. A. Giesecke, Jr., M. Casaverde, and S. E. Forbush, Electrical conductivity anomalies in the earth's crust in Peru, *Carnegie Inst. Wash. Yr. Book 63*, 354–362, 1964.

Whitham, K., An anomaly in geomagnetic variations at Mould Bay in the Arctic archipelago of Canada, *Geophys. J., 8*, 26–43, 1963.

Whitham, K., Anomalies in geomagnetic variations in the Arctic archipelago of Canada, *J. Geomag. Geoelec., 15*, 227–240, 1964.

Whitham, K., Geomagnetic variation anomalies

Magnetotelluric Studies of the Electrical Conductivity Structure of the Crust and Upper Mantle

T. R. Madden and C. M. Swift, Jr.

THERE has been an increasing interest in recent years in deep electrical measurements. Although the resolution of conductivity structure from surface measurements is poor, the very strong dependence of conductivity on temperature makes the electrical measurements an important tool for the investigation of areas of unusual thermal activity. The iron content and state of oxidation of some silicate minerals also has a strong influence on the electrical properties of the minerals; thus anomalous conductivities are not necessarily due to temperature effects. At the lower temperatures and pressures of the crust, the electrical properties are usually dominated by the amount and salinity of the interstitial water [*Brace et al.*, 1965]. Large contrasts exist between the conductivities of sedimentary rocks and the conductivities of igneous and metamorphic rocks, and thus electrical measurements are also of importance in studying the structure of sedimentary basins.

To make electrical measurements of the crust and mantle, the probing currents must penetrate into the zones being studied. This means that the measurements must be made at distances of up to hundreds of kilometers away from the current sources, and the frequencies must be low enough to allow the electromagnetic fields to penetrate to the depths of interest. The distance an electromagnetic field will penetrate a conductor before being attenuated to $1/e$ of its amplitude at the surface is known as the 'skin depth' (see Table 1).

Many methods are available for studying the electrical properties of the crust and upper mantle. Conventional resistivity surveys have been conducted on a large scale [*Slichter*, 1934; *Migaux et al.*, 1960; *Cantwell et al.*, 1965; *Keller et al.*, 1966] with very large current generators. The logistic problems of such measurements are considerable; thus these methods are usually limited to investigating shallower structures.

It is possible to make useful interpretations of the electrical fields generated by ocean and tidal currents [*Longuet-Higgens*, 1949], but only special areas can be studied in this fashion.

The most widely applied techniques for deep electrical measurements have involved measurements of the electromagnetic fields produced by the dynamo effects in the ionosphere, by Alfvén waves generated in the magnetosphere or at its boundaries, and by the drift motions of ionized particles trapped in the earth's magnetic field and also perhaps driven by magnetospheric electric fields. The electrical studies using these natural electromagnetic fluctuations have concentrated on two techniques. In one, the three components of the magnetic fluctuations are simultaneously measured at several points over the area of interest. This method is known as the 'geomagnetic variation method'; in another sec-

TABLE 1. Skin Depths in Kilometers for Electromagnetic Fields

Frequency, cps	Conductivity, mhos/meter					Period, sec
	10^{-3}	10^{-2}	10^{-1}	1	10	
1	15.9	5.0	1.59	.50	.16	1
10^{-1}	50	15.9	5.0	1.59	.50	10
10^{-2}	159	50	15.9	5.0	1.59	10^2
10^{-3}	503	159	50	15.9	5.0	10^3
10^{-4}	1590	503	159	50	15.9	10^4
10^{-5}		1590	503	159	50	10^5
10^{-6}			1590	503	159	10^6
	1000	100	10	1	0.1	
	Resistivity, ohm-meters					

tion of this Monograph (p. 463), Rikitake discusses some results of such investigations. The other technique consists of measurements of the horizontal components of both the electric and the magnetic fluctuations, and is known as the 'magnetotelluric method' [*Tikhonov*, 1950; *Cagniard*, 1953]. This method is the subject of our review. A modification of the two methods called the 'telluric method' is also used. In the telluric method simultaneous measurements of the horizontal electric fields are made at several points.

An extensive bibliography of the application of all these methods has been assembled by *Fournier* [1966].

When the magnetotelluric method was first proposed, it appeared to offer a great simplification over the geomagnetic variation method, because interpretations were possible from measurements at only a single station. The variation of skin depth with frequency introduces the necessary degrees of freedom to allow interpretations in terms of a conductivity-depth profile. This is not a basic difference between the methods, however, and similar interpretations can be made with magnetic measurements if one has a priori knowledge of the source field geometry [*Eckhardt et al.*, 1963]. It must also be remembered that data from a single station cannot resolve the horizontal variations of conductivity, so that the magnetotelluric method must also use data from several stations in order to study nonhorizontally layered conductivity structures.

In the magnetotelluric method, the magnetic measurement is proportional to the total current flowing in the ground; once this is determined the electrical measurement can be directly related to a resistivity. Thus the interpretation of magnetotelluric data is more direct than that of the magnetic variation data; this is its main advantage. At times, it may be a disadvantage, as the surface conductivities can be highly variable, and the measured electric field will depend on the exact location of the ground contacts. Also, the presence of local current sources, such as the ground returns on certain electric train systems, will more easily mask the natural electrical signals than it will the natural magnetic fluctuations. These disadvantages can be reduced by using long distances between the ground contacts in making the electric field measurements. Rather than choosing between the two methods, one should take advantage of the redundancies inherent in making both types of measurements in order to reduce the ambiguities that arise because of errors in the data and because of the limited number of observing sites. One natural way to achieve this is to include vertical magnetic field measurements in a magnetotelluric survey.

MATHEMATICAL THEORY

One-Dimensional Conductivity Structures

The theory of magnetotellurics for a layered medium is well developed and can be reviewed here very briefly. For a fuller discussion we recommend *Wait*'s [1962] excellent review. The spherical layered media theory is presented by *Srivastava* [1966].

The assumption of horizontal layering makes the problem one dimensional. We can describe the electromagnetic fields in terms of plane waves of two different polarizations at fixed frequencies. If we further set the Z axis as pointing down and the propagation vector to lie in the X-Z plane, we use Maxwell's equations for the TE mode

$$E = E_Y e^{ik \cdot R - i\omega t} \qquad H = (H_X, H_Z)e^{ik \cdot R - i\omega t}$$

to obtain

$$ik_Z E_Y = -i\mu\omega H_X \qquad (1)$$

and for the TM mode

$$H = H_Y e^{ik \cdot R - i\omega t} \qquad E = (E_X, E_Z)e^{ik \cdot R - i\omega t}$$

to obtain

$$ik_Z H_Y = -\sigma' E_X \qquad (2)$$

where

$$k = \hat{i}k_X + \hat{k} k_Z$$
$$k^2 = k_X^2 + k_Z^2 = i\mu\omega\sigma + \epsilon\mu\omega^2 = i\mu\omega\sigma' \quad (3)$$
$$\sigma' = \sigma - i\epsilon\omega$$

We see from (1) and (2) that the ratio of the horizontal components of E and H is constant for any plane wave propagating through a layer. Since the horizontal components of E and H must also be continuous across a layer boundary, the propagation of these waves is completely analogous to the propagation of voltage and current along a transmission line [*Schelkunoff*, 1943]. In analogy to transmission line theory, the ratio $E_T:H_T$ is known as the impedance Z (E_T, H_T are the horizontal components of E, H). The impedance associated with a plane wave in a uniform line is known as the characteristic impedance Z_c. When both primary and reflected waves are present, the measured impedance will be different from the characteristic impedance.

When we have a homogeneous half-space, a source above the half-space can produce only a downgoing wave within the medium. In this case the measured impedance will be equal to the characteristic impedance of the medium, and if the horizontal wavelengths and mode type are known, the propagation constant of the medium can be determined from the measured value of Z. When the horizontal wavelengths become large compared to the skin depth, the determination of k becomes independent of the mode type or of the horizontal wavelength.

$$k_X \ll k \quad k_Z \cong k$$
$$Z_{TE} \cong -\frac{i\mu\omega}{ik} = -i\sqrt{\frac{i\mu\omega}{\sigma'}}$$
$$Z_{TM} \cong \frac{ik}{-\sigma'} = -i\sqrt{\frac{i\mu\omega}{\sigma'}} \quad (4)$$

The assumption that $k_x \ll k$ is valid in most applications of magnetotellurics. Thus we can define an apparent conductivity from the measured impedance

$$\sigma_{\text{app}} = -i\mu\omega/Z^2 \quad (5)$$

At the frequencies involved in magnetotellurics, the displacement current is unimportant and we consider $\sigma' = \sigma$. Thus, when the medium is homogeneous and the horizontal wavelengths are long enough, the apparent conductivity is equal to the actual conductivity. For a layered medium whose conductivity varies with depth, the apparent conductivity will vary with frequency.

Consider a single layer overlying a homogeneous half-space. From transmission line theory, the impedance of a uniform section terminated by an impedance Z_T is given as

$$Z = Z_C \frac{Z_T - iZ_C \tan(k_Z l)}{Z_C - iZ_T \tan(k_Z l)} \quad (6)$$

where l is layer thickness. At high frequencies

$$\tan k_Z l \to i \quad Z \to Z_C \quad (7)$$

At low frequencies

$$\tan k_Z l \to 0 \quad Z \to Z_T \quad (8)$$

Thus the high frequency measurement depends only on the properties of the top layer, while the low frequency measurement depends on the properties of the underlying medium. This is the principle of magnetotelluric sounding.

The two-layer analysis can easily be extended to many layers by cascading the solutions, working from the bottom up. The impedance determined at the top of the nth layer is used as the terminating impedance for the $(n-1)$ layer in repeated use of (6). A smooth conductivity profile can be approximated to any degree of accuracy by such multilayered models; the greater the number of layers, the better the representation.

A very similar analysis can be made for a spherical model. The plane waves are replaced by spherical waves, and the TE and TM modes are replaced by the toroidal E and poloidal H mode and the poloidal E and toroidal H mode [*Eckhardt et al.*, 1963]. The impedance is no longer constant as a function of depth for a homogeneous region, so that the equivalent transmission line will be a nonuniform transmission line. The horizontal wavelengths are determined by the order of the spherical harmonics that describe the waves. When the skin depth is small compared to the horizontal wavelength and the radius of the earth, the solutions resemble the plane-wave solutions. Because of the rapidly increasing conductivity with depth within the earth, we find that the plane-wave

TABLE 2. Apparent Resistivity in ohm-meters for Cantwell-McDonald Model and Toroidal E Mode

Frequency, cps	Plane Wave	Spherical Harmonic Order n				
		1	2	4	9	18	36
10^{-2}	610	610	610	610	610	590	530
10^{-3}	140	140	140	140	140	130	110
10^{-4}	59	59	59	58	55	45	21
10^{-5}	28	28	27	25	18	8.2	2.3
10^{-6}	7.1	6.8	6.4	5.3	2.7	0.88	0.23

solutions agree with the more correct spherical-wave solutions for reasonable earth models even at periods of as long as 24 hours. This is demonstrated in Table 2 for a conductivity model that will be discussed further in this review. The plane-wave solution represents the solution over a flat earth for a wave of infinite horizontal wavelength.

Two-Dimensional Conductivity Structures

Though Table 2 shows that some attention must be paid to the spherical mode number in interpreting magnetotelluric results, actual measurements show that in most areas the source fields have wavelengths long enough to be considered infinite. A much more serious difficulty exists, however, in attempting to use layered media results to interpret measurements when the actual conductivity does not have a simple layered geometry. In fact, it is just the deviations from a layered symmetry that are often the most interesting features. In such situations we find the magnetotelluric impedances are very different from those of layered media structures. The mathematical and numerical difficulties of treating three-dimensional magnetotelluric problems are very great, and until computing facilities are an order of magnitude better than at present, we shall have to be satisfied with considering only two-dimensional problems. Some cases can be treated by model studies [*Rankin et al.*, 1965]. Fortunately, many important geologic features are quite well described as two-dimensional features. The criterion for two-dimensionality is that a feature remain consistent along strike for a distance large compared to the skin depth. Thus a feature that is much longer than its depth of burial below the surface can be reasonably well described by two-dimensional models for magnetotelluric calculations, as long as we avoid the ends of the feature.

The magnetotelluric impedances over two-dimensional models have quite different behaviors depending on the orientation of the field components with respect to strike. When the horizontal wavelengths are long enough to be ignored, the impedance is a second-order tensor or matrix with principal axes parallel and perpendicular to strike.

$$\begin{bmatrix} E_X \\ E_Y \end{bmatrix} = Z \begin{bmatrix} 0 & 1 \\ -1 & 0 \end{bmatrix} \begin{bmatrix} H_X \\ H_Y \end{bmatrix} \qquad (9)$$

For E perpendicular to strike, $E_y = 0$, we can derive from Maxwell's equations

$$\frac{\partial H_Y}{\partial Z} = -\left(\sigma + \frac{ik_Y^2}{\mu\omega}\right)E_X \qquad H_X = \frac{k_Y}{\mu\omega}E_Z$$

$$\frac{\partial H_Y}{\partial X} = \left(\sigma + \frac{ik_Y^2}{\mu\omega}\right)E_Z \qquad H_Z = -\frac{k_Y}{\mu\omega}E_X$$

$$\frac{\partial E_X}{\partial Z} - \frac{\partial E_Z}{\partial X} = i\mu\omega H_Y \qquad (\text{curl } H)_y = 0$$
$$(10)$$

Let

$$-E_Z = I_X \qquad \sigma\left(1 - \frac{k_Y^2}{k^2}\right) = Z$$
$$E_X = I_Z \qquad (11)$$
$$H_Y = V \qquad -i\mu\omega = Y$$

and we obtain from (10)

$$\frac{\partial V}{\partial X} = -ZI_X$$
$$\frac{\partial V}{\partial Z} = -ZI_Z \qquad \nabla V = -ZI$$
$$\text{or} \qquad (12)$$
$$\frac{\partial I_X}{\partial X} + \frac{\partial I_Z}{\partial Z} = -YV \qquad \text{div } I = -YV$$

For H perpendicular to strike, $H_Y = 0$, we can carry out a similar analysis and again obtain (12) by setting

$$H_Z = I_X \qquad -i\mu\omega\left(1 - \frac{k_Y^2}{k^2}\right) = Z$$
$$-H_X = I_Z \qquad (13)$$
$$E_Y = V \qquad \sigma = Y$$

In the limit of long horizontal wavelengths

($k_y \to 0$), these modes also separate out in the measurements as the modes become linearly polarized, (E_x, H_y) and (E_y, H_x). The (E_y, H_x) mode has an anomaly H_z field associated with it, while the (E_x, H_y) mode does not.

For certain simple geometries, the system of equations in (12) can be separated and analytic solutions obtained. Solutions for faults, dykes, and wedges have been obtained in this fashion. [*d'Erceville and Kunetz*, 1962; *Rankin*, 1962; *Weaver*, 1963]. It is important to remember, however, when solving for the case E parallel to strike, that the air must be included in the model, since Z for the air does not represent an infinite contrast to Z for the ground, as we see in (13). For E perpendicular to strike, we can terminate the model at the air interface with the boundary condition $V = H_Y = $ constant.

Equations 12 are two-dimensional transmission surface equations, and they can always be solved numerically or by analogue models using a lumped circuit network approximation. The node spacing of the network approximation should be a fraction of the skin depth in order to maintain accurate results. For numerical solutions, one can use either relaxation methods or direct solution methods. Overrelaxation techniques will work on these equations, but at present an analysis of the optimum factors is lacking (C. Cox, personal communication). Because of the simple topology of the network interconnections, very efficient direct solution methods can also be used [*Madden and Thomp-*

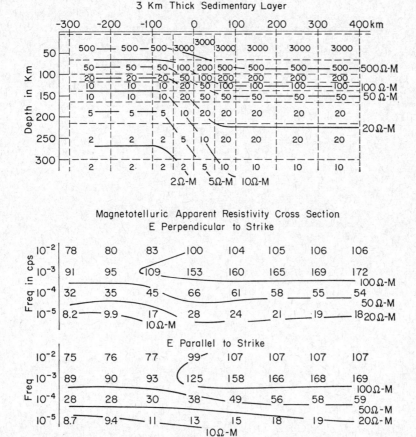

Fig. 1. Theoretical magnetotelluric profiles over a mantle conductivity structure; 3-km covering of 10 Ω-m sediments.

son, 1965]. Again it should be stressed that an air layer must be included between the source field and the ground when solving for the field polarized with E parallel to strike.

In Figures 1 and 2 we show such calculations on a two-dimensional model that represents an anomalous upper mantle feature. In Figure 1 a uniform sedimentary layer 3 km thick with a resistivity of 10 Ω-m has been included. The node spacing is also indicated on this figure, except for some very closely spaced nodes just beneath the surface.

When the electric field is perpendicular to strike, there is a tendency for the electric field to increase when approaching the transition zone from the more resistive side, and to decrease when approaching it from the conductive side. When the electric field is parallel to strike, this tendency disappears and there is a smooth transition from one side to the other. These effects result in a variable anisotropy for the computed apparent conductivities or apparent resistivities in the vicinity of the transition region.

The sedimentary layer masks the mantle feature at the higher frequencies. If thickness or conductivity variations had been introduced in the sedimentary layer, very strong anisotropies would result at the higher frequencies owing to these sedimentary features. This makes the interpretation of upper mantle conductivities in sedimentary areas much more difficult than in igneous or metamorphic areas. The vertical magnetic field anomalies associated with sedimentary basins can also be very large. The magnetotel-

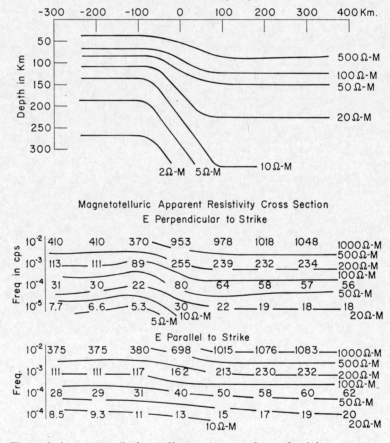

Fig. 2. Theoretical magnetotelluric profiles over a mantle conductivity structure; no sedimentary layer.

luric measurements clearly show the existence of such features, and thus can help in improving the interpretation of the magnetic variation anomalies.

At the lower frequencies the sedimentary layer influence is much less pronounced and a clearer picture of the mantle inhomogeneities is possible. Unfortunately, the technical problems of obtaining reliable magnetotelluric measurements at periods of longer than 24 hours are considerable.

DATA ANALYSIS AND OTHER PRACTICAL CONSIDERATIONS

Present electronic technology has greatly reduced the difficulty of measuring the electric and magnetic fluctuations, and we shall not concern ourselves here with instrumental problems. For mantle studies, periods of longer than a minute are desirable and a magnetic sensitivity of 0.1 γ is adequate. When it is important to know the sedimentary conductivities, higher frequencies may be necessary, and much greater sensitivities are needed. Most investigators use induction coils for recording the higher frequency fluctuations.

The measurement of the electric fluctuations is even simpler. It is important to make the electric antenna long enough to be insensitive to near surface conductivity variations that are not of interest in the study. It is not unreasonable to make the antenna length comparable to the depth of interest. Such long lines require the use of telephone cables.

A magnetotelluric survey can be made one station at a time, thus eliminating the need for simultaneous measurements. This implies that the horizontal wavelength structure of the electromagnetic waves is not important, and from Table 2 this involves some assumptions about the minimum wavelengths present. This subject has been discussed often in the literature [*Price*, 1962; *Quon*, 1963; *Wait*, 1954], but surprisingly little observational evidence is introduced to study the quantitative aspects of the problem. In Figure 3 we show the results of a correlation study of the magnetic field components at Tucson, Arizona, and Dallas, Texas (about 1300 km apart). The shortest wavelengths implied from these results are about 10,000 km, and they only occurred at the high frequency end. The data could be explained by a 2.5-min delay between Tucson and Dallas, and, since this was just the sampling time for digitization, it might well be an accounting error rather than a true delay. The diurnal and semidiurnal components do involve real delays, but the wavelengths are long enough. The amplitude variations between the stations are real, but again not great enough to undo the long wavelength assumptions.

In other areas, especially near the auroral and equatorial electrojets, it is probably not correct to ignore the finite dimensions of the horizontal variations of the source fields.

Despite the evidence that in midlatitudes the electromagnetic source fields have long wavelengths, many magnetotelluric measurements are plagued by seemingly inconsistent apparent resistivity determinations. These effects are often due to the tensor nature of the magnetotelluric impedance and the statistical nature of the source fields.

Figure 4 shows the results of a running analysis of magnetotelluric data that are quite typical of results where the impedance is anisotropic. The raw $E:H$ ratios can vary, as the balance between the various source field components changes, whenever the impedance has off-diagonal components. The effect is especially marked if the diagonal term is small compared to the off-diagonal term. In principle the impedance matrix should remain constant, but the existence of this matrix does involve certain assumptions concerning the source fields. The random nature of much of the fluctuations allows one to test for this consistency in Z.

Since Z is a matrix and E and H are vectors, one cannot determine Z from a single determination of the $E:H$ ratios. It is a common practice to choose independent estimates of the $E:H$ ratios in order to solve for Z. It is also a common practice to pick data sections in which the data look more consistent. From a statistical point of view this procedure is bad on two counts; it does not use all the data, and it introduces a bias. Because of the random nature of the fields, $E(\omega)$ and $H(\omega)$ for any finite bandwidth are not constant, but have amplitudes and phases and polarizations that vary with time. If we consider time as another dimension, equation 9, instead of being underdetermined, is overdetermined. The least-squares solution [*Lanczos*, 1961] can then be written as

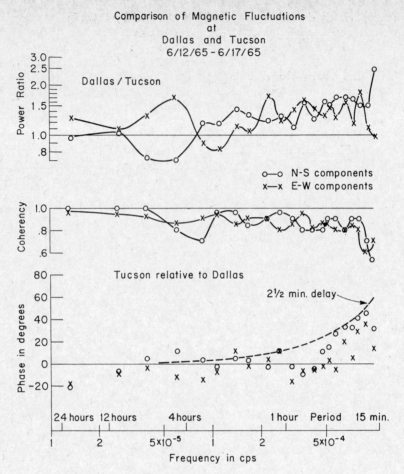

Fig. 3. Wavelength study of geomagnetic fluctuations. Comparison of two midlatitude stations 1300 km apart.

$$Z = \begin{bmatrix} \langle E_X H_X^* \rangle \langle E_X H_Y^* \rangle \\ \langle E_Y H_X^* \rangle \langle E_Y H_Y^* \rangle \end{bmatrix}$$
$$\cdot \begin{bmatrix} \langle H_X H_X^* \rangle \langle H_X H_Y^* \rangle \\ \langle H_Y H_X^* \rangle \langle H_Y H_Y^* \rangle \end{bmatrix}^{-1} \begin{bmatrix} 0 & -1 \\ 1 & 0 \end{bmatrix} \quad (14)$$

where $\langle AB^* \rangle$ is crosspower between A and B.

The assumption that (9) is consistent can now be checked by determining the coherency between the observed E and the E predicted from (9). Z becomes indeterminate if the determinant of the H crosspower matrix is zero. This arises when the coherency between H_X and H_Y approaches unity [*Madden and Nelson*, 1964].

All simple theoretical models predict Z to be Hermitian, but it may not be so in general. Thus we cannot necessarily find simple principal axes for Z.

OBSERVATIONAL RESULTS AND CONDUCTIVITY MODELS

At present there are not enough magnetotelluric results, nor a wide enough geographic coverage, to be able to systematize the data as has been done in seismology. Most results have shown strong anisotropies, which are usually indicative of lateral inhomogenicities, but lack the horizontal coverage necessary to outline the lateral variations. Such data cannot properly be interpreted. Many of the older magnetotelluric observations also lack adequate Fourier and coherency analysis, and can only report raw apparent resistivities. *Fournier et al.* [1963] have

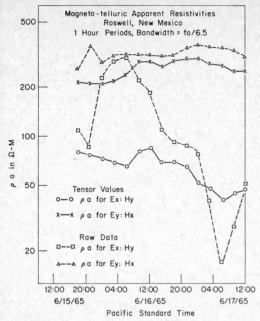

Fig. 4. Running analysis of magnetotelluric impedances; 8-hour averaging time.

assembled much of the reported magnetotelluric data, which represent perhaps several hundred soundings. Because of the difficulties, one cannot know how to use all these data. Figure 5 shows soundings from three locations that are not atypical of the other measurements. Certain general results can be seen in almost all the reported data. The major feature is a decrease of apparent resistivity for periods of longer than 10^4 sec. Data taken in sedimentary regions show a reversal of this trend at the short-period end of the spectrum. These results are indicative of the fact that the crust is more resistive than the sedimentary rocks, but the mantle shows a rapid decrease of resistivity with depth. This decrease must be significant at depths of less than 100 km.

This result is not unexpected in view of the temperature estimates that are made for the upper mantle and laboratory measurements of the effect of temperature and pressure on the conductivity of silicate minerals. This can be seen more quantitatively by comparing the predicted resistivities of an olivine mantle with the Cantwell-McDonald profile in Table 3. The Cantwell-McDonald profile is a merging of profiles that satisfy some magnetotelluric results [*Cantwell and Madden*, 1960] at high frequencies and explains the damping of the secular variations of the magnetic field [*McDonald*, 1957], as well as magnetic variation data at periods of two weeks and six months [*Eckhardt et al.*, 1963]. The magnetotelluric results of such a profile are also shown in Figure 5. The olivine model is based on *Ringwood's*

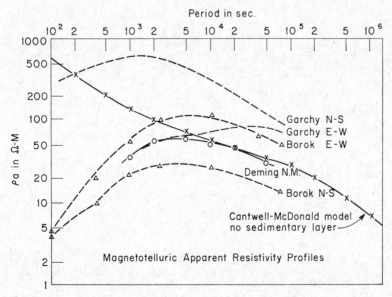

Fig. 5. Observed and theoretical magnetotelluric apparent resistivity profiles.

TABLE 3. Temperature and Electrical Resistivity Profiles (resistivities in ohm-meters)

Depth, km	Temp., °C	13% Fayalite Olivine Model		Cantwell-McDonald Model
		No Phase Transition	Phase Transition	
20				10.000
40	300			10.000
60	660	3000	3000	10,000
80	820	800	800	50
100	950	300	300	45
125	1100	100	100	40
150	1200	50	50	36
200	1400	20	20	31
300	(1600)	12	(8)	24
500	(1750)	10	(1.5)	14
750	(1850)	6	(0.2)	2
1000	(1900)	5	(0.05)	0.2
1500		(4)	(0.04)	0.033
2000		(3.6)	(0.036)	0.015
2500				0.009

[1966] continental temperature profile, and Hamilton's [1965] measurements on olivine conductivities extrapolated for a 13% fayalite olivine. The effect of a phase transition of the olivine to a spinel structure on the conductivity is also estimated in Table 3 [Akimoto and Fujisawa, 1965]. The magnetotelluric data and the laboratory conductivity measurements are not conclusive enough to allow interpretations of mantle composition or temperatures. The results shown in Table 3 are intriguing enough, however, to encourage further work along these lines. The deep mantle is not well sampled by the magnetotelluric method, and the Cantwell-McDonald profile depends on the secular variation analysis for the deep conductivity estimates. It is clear, however, that a phase transition in the olivine model is necessary if the olivine model is to correspond to the observations for the deep mantle.

The magnetotelluric measurements will perhaps be most useful in making quantitative comparisons of the upper mantle under various regions and to outline the transition zones separating such regions. To date there are few such studies reported, but with an increasing interest in the method this situation will soon change. The data from Deming, New Mexico, in Figure 5 come from the middle of one such transition zone. The southwestern United States is anomalously conductive, as was first noted by Schmucker [1964]. A still uncompleted analysis of magnetoteluric data taken by the authors over this region indicates that the upper mantle conductivities under Arizona may be 50 to 100 times higher than those of the upper mantle under eastern New Mexico. Deming appears to have isotropic impedances, but this is due to its location very close to the center of the transition zone (Figure 2). The apparent resistivities shown are intermediate to those found to the west and to the east. If the eastern end of the profile is more typical, the Cantwell-McDonald model would appear to be somewhat too conductive for the typical upper mantle. The theoretical examples shown in Figures 1 and 2 represent a somewhat similar situation without the complications of sedimentary layer variations. The anomalous electrical properties of this area are associated with high heat flows (R. Roy, personal communication) and low seismic velocities, all indicative of a considerable temperature anomaly under the crust. It is hoped that future work will help outline other such areas and lead to a better understanding of the dynamic processes involved in producing such anomalies. It is also desirable to extend these measurements to the ocean regions, as our ideas about the ocean basins are undergoing great changes at present, and temperature-sensitive measurements may play an important role in guiding our theories.

Acknowledgments. We acknowledge the contribution of fellow workers in this field who have helped to shape our ideas on the subject. We especially thank Dr. T. Cantwell and Dr. K. Vozoff of Geoscience, Inc., and Professors H. Smith and F. Bostick of the University of Texas. We also wish to thank the Office of Naval Research, the American Chemical Society, and the National Science Foundation, who have supported some of our researches in this area.

REFERENCES

Akimoto, S., and H. Fujisawa, Demonstration of the electrical conductivity jump produced by the olivine-spinel transition, *J. Geophys. Res.*, 70, 443–449, 1965.

Brace, W. F., A. S. Orange, and T. R. Madden, The effect of pressure on the electrical resistivity of water-saturated crystalline rocks, *J. Geophys. Res.*, 70(22), 5669–5678, 1965.

Cagniard, L., Basic theory of the magnetotelluric method of geophysical prospecting, *Geophysics*, 18, 605–635, 1953.

Cantwell, T., and T. Madden, Preliminary report on crustal magnetotelluric measurements, *J. Geophys. Res.*, 65, 4202–4205, 1960.

Cantwell, T., P. Nelson, J. Webb, and A. Orange, Deep resistivity measurements in the Pacific Northwest, *J. Geophys. Res.*, 70, 1931–1937, 1965.

d'Erceville, I., and G. Kunetz, The effect of a fault on the earth's natural electromagnetic field, *Geophysics*, 27, 651–665, 1962.

Eckhardt, D., K. Larner, and T. Madden, Long-period magnetic fluctuations and mantle conductivity estimates, *J. Geophys. Res., 68,* 6279–6286, 1963.

Fournier, H. G., S. H. Ward, and H. F. Morrison, Magnetotelluric evidence for the low velocity layer, Tech. Report on Contract Nonr. 222(89), No. 4, Issue 76, Space Sciences Lab., Univ. Calif., Berkeley, 1963.

Fournier, H. G., Essai d'un historique des connaissances magnetotelluriques, *Inst. Phys. Globe,* Note 17, Universite de Paris, 1966.

Hamilton, R. M., Temperature variation at constant pressures of the electrical conductivity of periclase and olivine, *J. Geophys. Res., 70,* 5679–5692, 1965.

Keller, G. V., L. A. Anderson, and J. I. Pritchard, Geological survey investigations of the electrical properties of the crust and upper mantle, *Geophysics, 31,* 1078–1087, 1966.

Lanczos, C., *Linear Differential Operators,* D. Van Nostrand Company, Princeton, 1961.

Longuet-Higgins, M. S., The electrical and magnetic effects of tidal streams, *Monthly Notices Roy. Astron. Soc., Geophys. Suppl., 5*(8), 295–307, 1949.

Madden, T., and P. Nelson, A defense of Cagniard's Magnetotelluric method, ONR Report, Project NR-371-401, Geophysics Lab., MIT, 1964.

Madden, T., and W. Thompson, Low frequency electromagnetic oscillations of the earth-ionosphere cavity, *Rev. Geophys., 3,* 211–254, 1965.

McDonald, K. L., Penetration of the geomagnetic secular field through a mantle with variable conductivity, *J. Geophys. Res., 62,* 117–141, 1957.

Migaux, L., J. L. Astier, and P. H. Revol, Un essai de determination experimentale de la resistivite electrique des couches profonder de l'escorce terrestre, *Ann. Geophys., 16,* 555–560, 1960.

Price, A. T., The theory of magnetotelluric methods when the source field is considered, *J. Geophys. Res., 67,* 1907–1918, 1962.

Quon, C., Electromagnetic fields of elevated dipoles on a two-layered earth, M. Sc. thesis, Univ. Alberta, 1963.

Rankin, D., The magnetotelluric effect in a dyke, *Geophysics, 27,* 666–676, 1962.

Rankin, D., G. D. Garland, and K. Vozoff, An analog model for the magnetotelluric effect, *J. Geophys. Res., 70,* 1939–1945, 1965.

Ringwood, A. E., Mineralogy of the mantle, in *Advances in Earth Science,* edited by P. Hurley, pp. 357–399, MIT Press, Cambridge, 1966.

Schelkunoff, S. A., *Electromagnetic Waves,* D. Van Nostrand Company, Princeton, 1943.

Schmucker, U., Anomalies of geomagnetic variations in the Southwestern United States, *J. Geomag. Geoelec., 15,* 193–221, 1964.

Slichter, L. B., An electrical problem in geophysics, *Tech. Eng. News, MIT* 8–9, 1934.

Srivastava, S. P., Theory of the magnetotelluric method for a spherical conductor, *Geophys. J., 11,* 373–387, 1966.

Tikhonov, A. V., Determination of the electrical characteristics of the deep strata of the earth's crust, *Dokl. Akad. Nauk, SSSR, 73,* 295–297, 1950.

Wait, J. R., On the relation between telluric currents and the earth's magnetic field, *Geophysics, 19,* 281–289, 1954.

Wait, J. R., Theory of magnetotelluric fields, *J. Res. NBS, 66D,* 509–541, 1962.

Weaver, J. T., The electromagnetic field within a discontinuous conductor with reference to geomagnetic micropulsations near a coastline, *Can. J. Phys. 41,* 484–495, 1963.

6. Magmatism and Metamorphism

The Ultramafic Belts

Peter J. Wyllie

ONE petrological approach to upper mantle studies is to determine which ultramafic rocks represent mantle material. Ultramafic rocks occur in a variety of field and petrographic associations: in recent reviews, *Wyllie* [1967b, 1968] outlined eleven petrographic associations, some with subdivisions. In this summary, it is convenient to consider these in four larger groups: (1) layered, stratiform, and other intrusions involving gabbro or diabase, together with accumulations or concentrations of mafic minerals; (2) the alkalic rocks, including kimberlites, mica peridotites, members of ring complexes, and ultrabasic lava flows; (3) the several serpentinite-peridotite associations often classified together as alpine-type intrusions; (4) serpentinites and peridotites of the oceanic regions.

It is generally agreed that the ultramafic rocks of group 1 were formed from mantle-derived basaltic magma, and therefore they can provide only indirect evidence of mantle chemistry and mineralogy. The other groups are associated with major tectonic features of the earth's crust, and there is evidence that these rocks include representatives of the upper mantle. Kimberlites and alkalic ultrabasic ring complexes occur in stable or fractured continental regions; their distribution appears to be controlled by deep-seated tectonics with linear trends. Alpine-type ultramafic rocks are distributed along deformed mountain chains and island arcs, usually with associated gabbros or basic volcanic rocks. The occurrence of serpentinites and peridotites along mid-oceanic ridges suggests a third type of ultramafic belt.

KIMBERLITES AND ALKALIC ULTRABASIC ROCKS IN CRATONIC REGIONS

Recent reviews of these rocks have been presented by *Upton* [1967], *von Eckermann* [1967], *Davidson* [1967a], and *Dawson* [1967a, b]. Their petrogenesis is complicated by processes involving the concentration of alkalis and volatiles in residual liquids. Most investigators agree that kimberlites originate deep in the mantle. Contamination of a mantle-derived ultrabasic magma with crustal material is often invoked to explain their unusual chemistry [*Turner and Verhoogen*, 1960, pp. 249 and 396]; an alternative proposal involves reaction of a primary carbonatite magma from the mantle with 'granitic' crustal rocks to yield the alkalic ultrabasic magmas [*Dawson*, 1967b]. Some kimberlites and mica peridotites may be emplaced as crystal aggregates transported by gases, or a carbonatite magma, at temperatures of 600 to 700°C [*Watson*, 1967; *Franz and Wyllie*, 1967].

Kimberlite diatremes are crowded with xenoliths, including peridotites, garnet peridotites, and eclogites, which are generally interpreted as primary mantle material. *Harris et al.* [1967] proposed that the nodules represent fragments from a pyroxene-peridotite layer in the mantle, and deeper garnet-peridotite. *O'Hara*'s [1967] pyroxene grid indicates that the eclogite and garnet eclogite nodules equilibrated at depths of 100 to 140 km on the thermal gradient for shield regions. *Davidson* [1967b] noted the variable mineralogical composition of the nodules and the problems that this poses for interpretations of mantle mineralogy.

ALPINE INTRUSIONS IN OROGENIC REGIONS

The petrogenesis of these rocks is complex because it involves metamorphic processes: indeed, *den Tex* [1965] discussed them in terms of their metamorphic lineages rather than their igneous origin. The argument about the possible existence of ultrabasic magmas in this association is now apparently resolved in favor of solid emplacement [*Hess*, 1966, pp 5–6]; many petrologists consider that alpine intrusions represent parts of the solid, or partially fused, mantle that have flowed into or through the crust along the unstable orogenic belt [*de Roever*, 1957; *Thayer*, 1960, 1967; *Green*, 1967]. Temperatures of intrusion, or of re-intrusion, may range from that of basaltic magmas to low temperatures well within the serpentine stability field [*Green*, 1967; *Wyllie*, 1967b]. Geochemical evidence suggests that many alpine ultramafic rocks are derived from a part of the mantle that attained a residual character during an early period of mantle differentiation [*Murthy and Stueber*, 1967].

Thayer [1960, 1967] discussed and deplored the conceptual divorce of ultramafic and mafic rocks, and he listed six criteria that characterize the intrusive peridotite-gabbro complexes of alpine type. Reconsideration of the petrogenesis of ultramafic and mafic rocks together [*Miyashiro*, 1966] has contributed to the formulation or revival of several hypotheses. The ophiolite hypothesis interprets the complexes as massive, differentiated submarine lavas ranging in composition from ultrabasic to basic, together with some intrusive rocks. *Maxwell* [1968] discussed the great composite sheets of the Mediterranean and Himalayas, concluding that these represented extrusions of mantle rocks breaking through the sea floor. They may be as thick as 8 to 10 km.

Criteria for distinguishing alpine ultramafic rocks from those of layered intrusions have been proposed by *Thayer* [1960, 1967], but *Smith* [1958] suggested that there is a continuous series of peridotite-gabbro associations between the stratiform rocks and the alpine rocks. Several alpine peridotites have recently been interpreted as gravity-stratified crystal cumulates from basic magma [*Challis*, 1965; *Miyashiro*, 1966, quoting Nagasaki], and *O'Hara* [1967] suggested that the only feature that alpine intrusions have in common is an orogenic setting that causes tectonic transport and re-intrusion. His pyroxene grid does indicate that alpine ultramafic rocks have equilibrated under a wide range of conditions.

One group of ultramafic rocks occurring in orogenic belts has features distinguishing it from alpine and from stratiform peridotite-gabbro complexes. These are the belts of cylindrical peridotite bodies characterized by a crude concentric zoning of dunite, pyroxenites, and hornblende peridotite. *Taylor* [1967] reviewed occurrences in Alaska and the Ural Mountains; similar rocks in Japan were described by *Onuki* [1966]. Taylor concluded that they were formed by the successive intrusion of liquid ultrabasic magmas of different compositions, all with high contents of FeO, CaO, and H_2O.

This summary review indicates some of the problems involved in using alpine ultramafic rocks as a basis for discussions about the upper mantle. A more detailed review of the current hypotheses for the origin of these rocks is given in *Wyllie* [1967].

MID-OCEANIC RIDGES

In 1962, Hess developed his thesis that the main crustal layer beneath the oceans may be serpentinized peridotite, like that dredged from fault scarps on the Mid-Atlantic ridge, from the Puerto Rico trench, and collected from St. Peter's and St. Paul's rocks on the Mid-Atlantic ridge; in 1964, he reviewed the significance of this association of oceanic ultramafic rocks. Genetic links between the oceanic association and the alpine ultramafic belts are indicated by *Hess'* [1960] proposal that the serpentinites of Puerto Rico represented uplifted oceanic crust, which 'may be altered mantle rocks exposed at the surface' (p. 235), and by *Dietz'* [1963] suggestion that a spreading ocean floor could cause tectonic incorporation of the serpentinite of the oceanic crust into overlying sediments of the continental rise, thus producing ultramafic rocks of alpine type when the sedimentary pile was metamorphosed.

Maxwell [1968] drew attention to similarities between the basic pillow lavas and peridotites which appear to characterize the mid-oceanic ridges, and the extrusion of mantle material in the ophiolite complexes of alpine mountain belts. Continued study of the mid-oceanic ridges may well confirm that they are the locus of ultramafic belts quite as extensive as the alpine ultramafic belts, but separated from them by an ocean basin.

DISCUSSION

For petrological information about the mantle, we must examine the rocks in the ultramafic belts, which occupy regions where hot mantle material and magma have approached or reached the surface as a result of major tectonic processes, with linear controls.

The alkalic rocks in the cratonic ultramafic belts do not provide direct representatives of the mantle because they have been strongly fractionated. For example, in kimberlites, the abundance of elements such as K, Rb, Sr, Ba, U, Th, C, and H indicates a general enrichment of up to 200-fold compared with any likely mantle parent (P. G. Harris, personal communication). However, the xenoliths carried upward in the kimberlite diatremes are promising candidates for mantle samples.

The alpine ultramafic rocks of the orogenic belts probably include representative mantle material, but the prospect that some of these intrusions are derived from cumulates in basic stratiform intrusions, or in volcanic conduits, indicates that caution is required in their interpretation. The effects of metamorphism blur the petrogenesis of the alpine peridotite-gabbro associations.

The ultramafic rocks of the mid-oceanic ridges, when they are adequately sampled, may perhaps provide the best prospects for direct correlation with upper mantle material. There are no stratiform intrusions known in this belt to confuse interpretation, and the history of the ultramafic rocks has not been obscured by orogenic metamorphism.

Acknowledgment. This review developed from research supported by National Science Foundation under Grant GA-1289.

REFERENCES

Challis, G. A., The origin of New Zealand ultramafic intrusions, *J. Petrol., 6,* 322–364, 1965.

Davidson, C. F., The kimberlites of the U.S.S.R., in *Wyllie* [1967a], pp. 251–261, 1967a.

Davidson, C. F., The so-called 'cognate xenoliths' of kimberlite, in *Wyllie* [1967a], pp. 342–346, 1967b.

Dawson, J. B., A review of the geology of kimberlite, in *Wyllie* [1967a], pp. 241–251, 1967a.

Dawson, J. B., Geochemistry and origin of kimberlite, in *Wyllie* [1967a], pp. 269–278, 1967b.

den Tex, E., Metamorphic lineages of orogenic plutonism, *Geol. Mijnbouw, 44*(4), 105–132, 1965.

de Roever, W. P., Sind die Alpinotypen peridotitmassen vieleicht tektonisch verfrachtete Bruchstücke der Peridotitschale?, *Geol. Rundschau, 46,* 137–146, 1957.

Dietz, R. S., Alpine serpentinites as oceanic rind fragments, *Bull. Geol. Soc. Am., 74,* 947–952, 1963.

Franz, G. W., and P. J. Wyllie, Experimental studies in the system $CaO-MgO-SiO_2-CO_2-H_2O$, in *Wyllie* [1967a], pp. 323–326, 1967.

Green, D. H., High temperature peridotite intrusions, in *Wyllie* [1967a], pp. 212–222, 1967.

Harris, P. G., A. Reay, and I. G. White, Chemical composition of the upper mantle, *J. Geophys. Res., 72,* 6359–6369, 1967.

Hess, H. H., Caribbean research project: Progress report, *Bull. Geol. Soc. Am., 71,* 235–240, 1960.

Hess, H. H., History of ocean basins, in *Petrologic Studies: A Volume to Honor A. F. Buddington,* edited by A. E. J. Engel et al., pp. 599–620, Geological Society of America, New York, 1962.

Hess, H. H., The oceanic crust, the upper mantle and the Mayaguez serpentinized peridotite, in *A Study of Serpentinite*, edited by C. A. Burk, Natl. Acad. Sci.—Natl. Res. Council Publ. *1188*, 169–175, 1964.

Hess, H. H., editor, *Caribbean Geological Investigations*, Geol. Soc. Am. Mem. *98*, 310 pp., 1966.

Maxwell, J. C., Continental drift and a dynamic earth, Am. Sci., *56*, 35–51, 1968.

Miyashiro, A., Some aspects of peridotite and serpentinite in orogenic belts, Jap. J. Geol. Geography, Trans., *37*, 45–61, 1966.

Murthy, V. R., and A. M. Stueber, Potassium-rubidium ratios in mantle-derived rocks, in *Wyllie* [1967a], pp. 376–380, 1967.

O'Hara, M. J., Mineral paragenesis in ultrabasic rocks, in *Wyllie* [1967a], pp. 393–403, 1967.

Onuki, H., On the iron-rich peridotites in the Sanbagawa metamorphic belt of the Kanto Mountains, J. Japan. Assoc. Mineral. Petrol. Econ. Geol., *55*, 39–47, 1966.

Smith, C. H., Bay of Islands igneous complex, western Newfoundland, Can. Geol. Surv. Mem., *290*, 132 pp., 1958.

Taylor, H. P., Jr., The zoned ultramafic complexes of southeastern Alaska, in *Wyllie* [1967a], pp. 97–121, 1967.

Thayer, T. P., Some critical differences between alpine-type and stratiform peridotite-gabbro complexes, Intern. Geol. Cong., 21st, Copenhagen, 1960, Rept. Session, Norden, *13*, 247–259, 1960.

Thayer, T. P., Chemical and structural relations of ultramafic and feldspathic rocks in alpine intrusive complexes, in *Wyllie* [1967a], pp. 222–239, 1967.

Turner, F. J., and J. Verhoogen, *Igneous and Metamorphic Petrology*, 2nd ed., McGraw-Hill Book Company, New York, 694 pp., 1960.

Upton, B. G. J., Alkaline pyroxenites, in *Wyllie* [1967a], pp. 281–288, 1967.

von Eckermann, H., A comparison of Swedish, African, and Russian kimberlites, in *Wyllie* [1967a], pp. 302–312, 1967.

Watson, K. D., Kimberlites of eastern North America, in *Wyllie* [1967a], 312–323, 1967.

Wyllie, P. J., editor, *Ultramafic and Related Rocks*, John Wiley & Sons, New York, 464 pp., 1967a.

Wyllie, P. J., Petrogenesis of ultramafic and ultrabasic rocks: Review, in *Wyllie* [1967a], pp. 407–416, 1967b.

Wyllie, P. J., The origin of ultramafic and ultrabasic rocks, Upper Mantle Symposium, Intern. Geol. Congr., Tectonophysics, *7*(5–6), 1969.

Batholiths and Their Orogenic Setting

Ahti Simonen

BATHOLITHS of granitoid rocks occur in the folded belts of the earth's crust where interaction between the upper mantle and the crust has been active. Because of the close relationship between batholith emplacement and orogenic folding, the batholiths are commonly classified into preorogenic, synorogenic, late orogenic, postorogenic, and anorogenic groups. The batholiths, whose emplacement is due to intrusion, are crystallized either from juvenile or anatectic magmas rising upward into the folding belt. Juvenile granitic magmas derive from the upper mantle, and anatectic magmas have originated by partial remelting of the granitic crust and geosynclinal sediments in the zones of tectonic activity where deviations from the thermal equilibrium state of the crust have taken place. The metasomatic granitization of older rocks has been caused by granitic material migrating upward in the earth's crust. This has played an important role in the origin and emplacement of the batholiths in the migmatite front. Granite magmas and the granitic material causing the granitization are derived from greater depths by upward migration of light granitic elements. The ultimate origin of the granite and the granitic crust is connected with gravitational differentiation which has been taking place throughout the geological history of our planet.

Silicic plutonic rocks, mainly quartz diorites, granodiorites, and granites, are very abundant in folded, orogenic areas of various ages. Silicic plutonic rocks with diverse petrographic characteristics form bodies or plutons of different size and shape in the folded and regionally metamorphosed areas. The most common are wide and irregular masses, called batholiths, with areal extent ranging from about 40 to thousands of square miles. The side walls of the batholiths are vertical or inclined steeply outward, and show no evidence of the existence of a floor. The roof of many batholiths is dome-shaped; there may be downward projections or roof pendants of country rocks. Contacts between the batholiths and the adjoining country rocks may be either concordant or discordant; furthermore, they may be sharp or transitional. The rock material of the batholiths may be fairly homogeneous or composite, consisting of many different varieties of the silicic plutonic rocks collectively called granitoids. The characteristics of batholiths, collected from various textbooks of geology, show that the term 'batholith' is commonly used in a very broad sense to include all granitoid plutons of considerable size having steep side walls.

OROGENIC SETTING

The large plutons of batholithic dimensions are especially abundant in the basement complexes representing deeply eroded sections of Precambrian folded areas. In the Baltic shield, the oldest part of the basement, which is more than 2600 million years old, consists almost exclusively of silicic plutonic rocks with a granodioritic bulk composition. In the Svecofennidic area of the Baltic shield, folded 1800–1900 million years ago, the silicic plutonic rocks occupy about 60–70% of the ancient folded area. Enormous batholithic masses of silicic plutonic rocks occur also in all younger folded belts of the earth, although they do not play such a predominant role in the younger structures as in the deeply eroded Precambrian folded areas. The above-mentioned evidence shows that the granitic crust of the continents was already present during the oldest geological times of the earth and, furthermore, that the granitoids are most abundant in the deepest sections of the folded areas. It may also be possible that during the oldest geological ages plutonic activity was more effective than during later times.

Geological studies in folded, orogenic areas of various ages have shown an intimate association between the orogenic, mountain building movements and the emplacement of batholiths. The commonly used tectonic classification of the granitoid rocks has been developed on the basis of the relationship between the emplacement of the plutons and the orogenic folding movements. The plutons that are older than a period of folding are preorogenic. The plutons, usually with batholitic dimensions, that are contemporaneous with the folding are divided into two subgroups, synorogenic and late orogenic. Both synorogenic and late orogenic batholiths were emplaced while the folding was still in progress, and their structures are in harmony with the structure of the adjoining country rocks. In addition, there are in the folded zones massive postorogenic plutons, batholiths, stocks, and dikes cutting the adjoining country rocks sharply, but belonging to the folding period. Some batholiths and stocks of massive granitoids, which sharply cut the ancient platforms and their sedimentary cover, have no relationship at all with the orogenic folding movements. These granitoid masses are called anorogenic, and they have been emplaced along deep faults cutting the earth's crust. The orogenic setting and the main characteristics of the tectonic groups of batholiths will be briefly described below. The few examples to be interpreted are taken from the deeply eroded, folded areas of the Fennoscandian or Baltic shield.

Preorogenic plutonic masses, whose rock material is gneissose and cataclastic in structure, belong to an ancient substratum of the folded belt remobilized and rejuvenated during the orogenic movements. They form either massifs brought up from the basement by thrust movements and overridden by nappes or domes and cupolas, remobilized anew and moved upward from the basement during the subsequent orogenic revolution. The operating factors in the emplacement of the preorogenic plutonic masses are orogenic thrust movements and the vertical movements of old granitoid masses. Preorogenic gneiss massifs and domes form concordant bodies in the folded belt, and they were often granitized and migmatized during the remobilization and doming.

Synorogenic batholiths were emplaced during the main period of folding. The orogenic deformation and folding continued after the emplacement, causing remarkable gneissose texture of the synorogenic plutonic rocks. Gneissose varieties are especially abundant at the margins of the plutons. The synorogenic batholiths form elongated and ovoidal masses, concordant with adjoining country rocks. They have steep side walls, and their structures (foliation and lineation) coincide very well with those of the surrounding schists, as is shown in the structural diagrams of a Precambrian, Svecofennidic folded area in Finland presented in Figure 1. Synorogenic batholiths, consisting mainly of quartz diorites and granodiorites associated with minor bodies of basic plutonic rocks, are extremely abundant in the folded belts of the Baltic shield. They occur especially in the anticlinal zones of the folded area, whereas the adjoining schists and gneisses have been depressed into synclines between the anticlinal batholiths (Figure 2). The synorogenic plutonic rocks have been emplaced as lenticular masses, in particular along the bedding and foliation planes in the anticlines of the folded belt.

Late orogenic batholiths were emplaced after the main period of folding, but while folding was still in progress. Their emplacement has caused

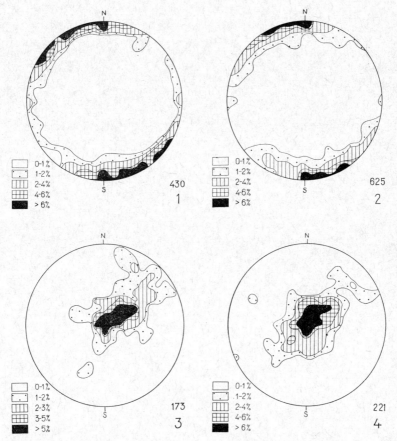

Fig. 1. Structural diagrams of plutonic rocks and schists in the Svecofennidic schist belt of the Tammela-Kalvola area, Finland. Foliation, schistosity, and lineation plotted on the Schmidt net, lower hemisphere: (1) foliation of plutonic rocks (430 samples); (2) schistosity of schists (625 samples); (3) lineation of plutonic rocks (173 samples); (4) lineation of schists (221 samples).

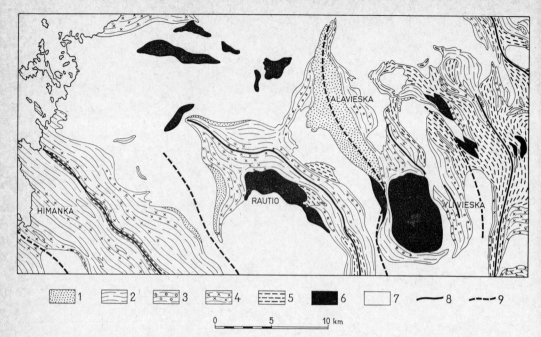

Fig. 2. Anticlinal batholiths in the Svecofennidic schist belt of the Vieska area, Pohjanmaa (west central Finland, ca. 64°N, 24°E.): (1) quartz-feldspar schists; (2) micaceous schists; (3) conglomerates; (4) porphyroblastic mica schists; (5) metavolcanics; (6) subsilicic plutonic rocks; (7) silicic plutonic rocks; (8) syncline; (9) anticline. Based on the map of *Salli* [1964].

many structural effects in adjoining country rocks; the resultant structures strongly superimpose on the older tectonics of the folded belt produced during the main period of folding. The structural elements (foliation and lineation) of the late orogenic plutons and their schist aureoles do not coincide with those of the principal fold belt. The manner of occurrence of the late orogenic batholiths is not so greatly dependent on the tectonic structures of the folded belt as is that of the concordant synorogenic plutonic masses. The late orogenic plutonic rocks are mainly granites and are more massive than the synorogenic gneissose rock types. However, sharp separation of the orogenic plutonic rocks into synorogenic and late orogenic groups is not always possible, because there may be many transitional varieties ranging from gneissose synorogenic types to more massive late orogenic plutonic rocks. In the deep sections of the Precambrian folded areas, the late orogenic plutonic rocks occur as dome-like bodies or form migmatites with the schists and synorogenic plutonic rocks. The emplacement of the late orogenic granite domes is characterized by diapiric updoming, cross folding, highly plastic migmatite structures, and metasomatic granitization. The main phases of the structural evolution of the late orogenic migmatite granite domes in the Precambrian Svecofennidic fold belt in southwestern Finland are presented in Figure 3. Granitization, which caused the origin of the migmatite granite, took place in the lowest levels of the gently folded belt (Figure 3, I). Granite diapirs rose upward in anticlines. Gneiss synclines sank between the granite domes and were compressed into steep isoclinals. The gneisses on the tops of the rising domes were stretched and flattened (Figure 3, II). Compression between the rising domes, situated one after the other in the principal direction of the fold belt, caused cross folding (Figure 3, III). The varying tectonic forms may be caused by an oblique cut through the fold belt (Figure 3, IV). A deeper level is characterized by gneiss basins in the granite area, whereas an upper level is represented by granite domes surrounded by continuous gneiss belts.

Postorogenic batholiths are clearly discordant, cutting sharply the structures of the adjoining folded belt or plutonic masses contemporaneous with folding. They belong, however, in space and time, to a period of mountain folding. Postorogenic plutonic rocks are mainly massive granodiorites and granites that show no influence of orogenic movements in their structures. The postorogenic batholiths are extremely rare in deep sections of the Precambrian folded areas, but they are abundant in younger folded belts of the earth.

Anorogenic plutons are entirely disharmonious, and their emplacement is not related at all to the period of folding but to the deep faults. Anorogenic granites are massive. They are commonly associated with subsilicic plutonites and show contact metamorphic aureoles where they have intruded into unmetamorphosed strata. In Fennoscandia, for example, large rapakivi granite massifs and granite stocks in the Oslo graben belong to this category of plutons.

As to the orogenic setting of the batholiths, plutons are most abundant in the deep sections of the Precambrian folded areas, in which a very intimate relationship between the orogenic folding and the emplacement of the batholiths occurs. The batholiths of the Precambrian folded areas are mainly synorogenic or late orogenic, whereas those of the younger folded belts are mainly late orogenic or postorogenic. This is due to the different manner of occurrence of the batholiths within different sections of the mountain chain. *Buddington* [1959] has classified the granite plutons according to their emplacement in the epizone, mesozone, or catazone of the earth's crust. He concluded that the plutons emplaced in the epizone are wholly discordant and correspond to the anorogenic or postorogenic groups. The plutons emplaced in the mesozone are complex, in part discordant, and in part concordant, and correspond to the postorogenic or late orogenic groups. The plutons in the catazone are predominantly concordant and represent the synorogenic group.

ORIGIN AND EMPLACEMENT

The orogenic setting and emplacement of batholiths are closely connected with the problem of origin of granitoid rocks. Most petrologists now agree that granitic rocks may originate in two principal ways: by magmatic and metasomatic processes.

Geologic evidence shows that the rocks of many batholiths have crystallized from silicate melt. Many experimental data on the crystallization of silicate systems have been presented, especially by *Tuttle and Bowen* [1958]. Some authors have suggested that granite batholiths are the last products of the crystallization differentiation of basaltic magmas injected from the depth of the earth into the roots of the folded belts. The plutonic rocks of the folded areas are, however, predominantly granitoid in composition and the gabbroic rocks occur only sporadically. This fact does not support the origin of plutonic rocks as a product of differentiation from a basaltic parent magma. The total bulk of granitoid rocks, in comparison with that of associated basic rocks, suggests that the parent magma of the orogenic granitoids was probably granodioritic in composition.

Because of the distribution of the temperature in the earth's crust, especially in the folding belts of great tectonic activity, conditions are favorable to the development of granodioritic or granitic magmas. That part of the sima zone

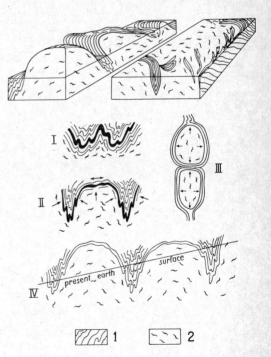

Fig. 3. Stereogram and sketches showing the evolution of migmatite-granite domes in the Svecofennidic belt of southwestern Finland: (1) gneisses; (2) migmatite-granites. According to *Edelman* [1960].

that is not entirely solidified may contain a juvenile pore solution of the granitic magma. Furthermore, the granite magma may have originated by partial remelting or anatexis of the sial crust and geosynclinal deposits in the areas of folded belts. The problem of the origin of both juvenile and anatectic granitic pore magmas has already been discussed by *Eskola* [1932], and *Winkler* [1965] has recently summarized his experimental data on the process of anatexis.

Granitic pore magma is lighter than its environment and it has a tendency to rise upward. Especially during the folding movements, the granitic pore magmas have been squeezed out and intruded upward as completely or partially liquid masses into the folding belt, and their close connection with the original birthplace becomes obscure. *Eskola* [1932] said metaphorically that 'granitic magmas might be characterized as the sweat that oozes out from the body of mother earth during the convulsions of orogeny.'

The emplacement of the magmatic batholiths took place either from juvenile or anatectic magmas, and mechanisms of emplacement may be magmatic stoping and forceful injection. Magmatic stoping is characteristic of some postorogenic and anorogenic batholiths and stocks in which the sinking blocks of country rocks in the magma chamber have given way to the upward intrusion of the magma. The best evidence of magmatic stoping is the cauldron subsidence of many anorogenic granite stocks. The forceful injection of magma is, however, the most important mechanism in the emplacement of the batholiths in the folded belts. The granite or granodiorite magma has intruded into the folding belt under the influence of gravity and orogenic forces. It has pushed the country rocks aside and upward, causing dome-like arch structures. The granite tectonic study developed by *Cloos* [1936] and *Balk* [1937] shows that the structural elements (flow lines, platy flow lines, dikes, joints, and faults) of many batholiths are systematically related to the intrusion of the magma.

Granitization is a metasomatic process whereby a rock has been changed toward a granitic composition and texture. Field evidence, especially in the migmatitic Precambrian basement complex, shows that rocks of different primary composition and origin, both metamorphic and plutonic rocks, have a tendency to change metasomatically toward more granitic rocks. Various rock types have passed into granite, if they have been entirely granitized. However, rock types close to granitic composition, such as arkosites and metagraywackes, are most easily granitized, whereas rock types, such as limestones, quartzites, and ultrabasics, deviating considerably from granitic composition, are most resistant to granitization and occur commonly as remnants in the granitized area. Many examples of partly granitized rocks occur on the scene of granitization representing transitional stages from primary rock to granite which designate the extreme end product of a complete granitization. The granite formed by granitization usually contains ghost-like, nebulitic remnants of the pre-existing rocks. The relic stratigraphy and structures of many granite batholiths indicate that in the place of the present granite there had been older solid rocks, and the batholith has been emplaced by metasomatic processes.

Simonen [1960] pointed out that granites originated by granitization usually contain more potassium than the eutectoid granites originated by the crystallization of granite magma. Metasomatic granites do not correspond to a silicic end member of a series ranging from basic to acid types, but form a unique group without magmatic ancestors.

In many cases, a marked difference has been found between the chemical composition of a pregranitization rock and a metasomatic granite. This indicates that great quantities of granitic material have moved into the environment of granitization. It is most probable that the introduced material has been a juvenile or anatectic granite melt rising upward in the earth's crust and causing the development of the migmatite front in the depths of orogenic belts. The crystallization of introduced granite has continued until the hydrothermal range, causing postmagmatic potassium metasomatism. *Eskola* [1956] has concluded that the main process in the origin of metasomatic kaligranite with noneutectoid composition is postmagmatic potassium metasomatism closely connected with magmatism.

Granitization is common in the deeply eroded migmatitic root zones of the folded areas.

Sometimes it has taken place without considerable increase of volume, but many large granitized masses of batholithic dimensions have been updomed diapirically (see Figure 3), indicating high mobility. Granitization is characteristic of the late orogenic and final stages in the evolution of the orogenic belt.

CONCLUDING REMARKS

Batholiths occur in the folded belts of the earth's crust where tectonic activity and interaction between the crust and upper mantle has been most effective. The processes in the upper mantle have given rise to the origin of juvenile granite magmas intruded either into the geosynclinal zones or along deep faults cutting platforms. Furthermore, regional metamorphism, anatexis, and granitization, which have taken place in the folded belts, mark considerable deviations from the thermal equilibrium state of the crust. The temperature conditions of these processes have been caused by interaction between the upper mantle and crust.

The ideas of different authors on the origin of granitoid rocks deviate considerably, and it is difficult to evaluate the relative importance of the different processes in the origin of granitic rocks. Many geologists admit that primary granite magmas and the material causing the granitization are derived by upward migration of granite elements from the source at greater depths. The ultimate origin of granite is connected with large scale gravitational differentiation which has been taking place throughout the whole geological history of our planet. It is probable that juvenile granites derived from the upper mantle have given material to the origin and growth of the granitic crust. The juvenile granitic rocks have been necessary to give suitable material to the processes of anatexis and granitization, which play an important role in the origin of granitoid rocks and in the geochemical differentiation of the granitic elements in the earth's crust.

REFERENCES

Balk, R., Structural behavior of igneous rocks, *Geol. Soc. Am. Mem.*, *5*, 177 pp., 1937.

Buddington, A. F., Granite emplacement with special reference to North America, *Bull Geol. Soc. Am.*, *70*, 671–747, 1959.

Cloos, H., *Einführung in die Geologie; ein Lehrbuch der inneren Tektonik*, Gebr. Borntraeger, Berlin, 503 pp., 1936.

Edelman, N., The Gullkrona region, Southwest Finland, *Bull. Comm. Géol. Finlande*, *187*, 87 pp., 1960.

Eskola, P., On the origin of granitic magmas, *Min. Petr. Mitt.*, *42*(5–6), 456–481, 1932.

Eskola, P., Postmagmatic potash metasomatism of granite, *Bull. Comm. Géol. Finlande*, *172*, 85–100, 1956.

Salli, I., The structure and stratigraphy of the Ylivieska-Himanka schist area, Finland, *Bull. Comm. Géol. Finlande*, *211*, 67 pp., 1964.

Simonen, A., Plutonic rocks of the Svecofennides in Finland, *Bull. Comm. Géol. Finlande*, *189*, 101 pp., 1960.

Tuttle, O. F., and Bowen, N. L., Origin of granite in the light of experimental studies in the system $NaAlSi_3O_8$-$KAlSi_3O_8$-SiO_2-H_2O, *Geol. Soc. Am. Mem.*, *74*, 153 pp., 1958.

Winkler, H. G. F., *Petrogenesis of Metamorphic Rocks*, Springer-Verlag, Berlin, 220 pp., 1965.

The Origin of Basalt Magmas

D. H. Green and A. E. Ringwood

THERE is convincing evidence that basalt magmas are derived by melting processes in the upper mantle. Basalt magmas are thus potentially an important source of information on the chemical and mineralogical constitution of the upper mantle.

In the following sections, recent advances in our knowledge of the melting and fractionation behavior of natural basalts at high pressure are briefly reviewed. These studies lead to an understanding of the processes of magma generation and of the nature of the parent material from which the magmas are derived, i.e. the 'primary' unmelted upper mantle. The generation of basaltic magmas and the possibilities of consequent fractionation within the mantle lead to the conclusion that the upper mantle is inhomogeneous, particularly in minor elements.

THE DIVERSITY OF NATURAL BASALTS

Natural basalts are quite variable in mineralogy and in chemical composition, but there is wide recognition of two major basaltic magma series: the tholeiitic series and the alkali basalt series [*Yoder and Tilley*, 1962; *Macdonald and Katsura*, 1964]. Members of these series can be distinguished by mineralogical criteria; e.g., the presence of Ca-poor pyroxenes or quartz in the tholeiitic series or the presence of Ca-rich titan-augite or nepheline in the alkali basalt series, or by chemical criteria, principally the relative abundances of alkalies ($Na_2O + K_2O$) and silica. Laboratory studies and studies of natural examples of fractional crystallization have shown that, at low pressure, these magma series are separated by a 'thermal divide' and fractionation processes operating at low pressure cannot in general produce members of one magma series from a parent magma of the other series [*Yoder and Tilley*, 1962]. This has led to the concept of at least two parent basaltic magma types (Table 1), both derived from the mantle but different in chemical composition and leading to divergent and distinctive fractionation series.

Recent recognition of high alumina basalt as a volumetrically significant magma type has persuaded some authors to give this magma status similar to that of the alkali olivine basalt magma and olivine tholeiite magma or even to conclude that high-alumina olivine basalt is the only primary magma directly derived from the mantle, other magmas being derived therefrom by fractionation or other shallow crustal processes [*Engel et al.*, 1965]. Other authors consider that high alumina basalts are variants and derivatives of the olivine tholeiite and alkali olivine basalt magmas. It has been shown [*Green et al.*, 1967; *Green and Ringwood*, 1967] that the threefold division of basaltic magmas is a logical consequence of the control by load pressure on the nature of magmatic liquids produced in the mantle and on the directions of fractionation of such liquids in their ascent from within the mantle to surface extrusion.

PARTIAL MELTING OF THE UPPER MANTLE

The composition of the upper mantle accepted in the following discussion is that of the 'pyrolite model' [*Ringwood*, 1966; *Green and Ringwood*, 1967]; i.e., a composition capable of providing 20–40% of typical basaltic magmas by fractional melting, leaving residual peridotite or dunite. Because of the time scale of crystal settling processes in comparison with plausible rates of energy supply to source re-

TABLE 1. Normative Compositions of Primitive Magmas*

Constituent	Tholeiitic Picrite	Olivine Tholeiite	Alkali Olivine Basalt	High-Alumina Basalt (Oceanic Tholeiite)
Or	0.5	0.5	4	1
Ab	13	15	18	22
Ne	—	—	2	—
An	26	27	26	33
Di	15	17	16	21
Hy	5	13	—	7
Ol	36	22	26	10
Ilm	3	4	5	4
Mt	1	1	2	1
Ap	0.5	0.5	1	1
$100\ Mg/Mg + Fe^{++}$	78	72	63	62

* Primitive magmas are liquids that may have been derived directly by partial melting in the mantle and that have not undergone appreciable fractionation at shallow crustal levels.

gions of basalt, the generation of magmas from such a parent rock will always be a partial melting process with extraction of a liquid fraction leaving a residue of refractory phases (olivine, pyroxenes, chromite). The relative proportions of liquid and crystals will vary, depending on the energy available for melting and the efficiency of the mechanism for extraction of a liquid fraction from the more abundant residual crystals. Although a number of processes have been advocated as the prime cause of melting in the upper mantle (e.g. relief of pressure or localization of energy during earthquakes, accumulation of radioactive heat, etc.), the authors favor a mechanism related to 'convective' or 'advective' movement in the gravitationally unstable zone of the upper mantle (i.e., <200–300 km depth). Under suitable rheological conditions, a source mass of solid pyrolite may begin to rise diapirically (in the manner of a salt dome) within the upper mantle, probably beginning from the low-velocity zone. The rising diapir is large enough in relation to its velocity that it cools adiabatically and does not interact appreciably by thermal conduction with the surrounding mantle. The adiabatic gradient, approximately 0.3°C/km, is much smaller than the gradient of the pyrolite solidus, approximately 3°C/km, so that partial melting may occur in the diapir as its T, P curve intersects the pyrolite solidus. The degree of partial melting increases as the crystal and liquid diapir rises further, although the T, P curve is deflected from the adiabatic gradient by the effects of latent heat of melting. Up to this stage, although the amounts and compositions of crystalline and liquid phases change with changing pressure, it is considered that chemical equilibrium is maintained between liquid and residual crystals of the parent pyrolite; i.e. the basaltic liquid is continually buffered by residual olivine and pyroxene. At some depth, called the depth of magma segregation, the degree of partial melting is extensive enough (?20–40%) and the tectonic environment is such that the liquid segregates from residual crystals. From this stage, the magma is no longer in equilibrium with the residual crystals with which it was originally saturated. Instead it may fractionate independently by cooling and crystal settling as it rises toward the earth's surface. It has been shown [*Green and Ringwood*, 1967] that the degree of partial melting and the depth of magma segregation are the two prime factors determining the composition and magma type of basalts derived directly by partial melting of the pyrolite of the upper mantle.

Magma Segregation at 0–15 Kilometers

At low pressure, magnesian orthopyroxene melts incongruently to yield forsterite and an oversaturated, quartz-rich liquid. The mineralogy of pyrolite at near-solidus (initial melting) temperatures and depths of 0–15 km consists of olivine, orthopyroxene, clinopyroxene, plagioclase, and chromite. Magma segregation at this depth with a sufficiently small degree of partial melting may produce oversaturated quartz tholeiitic magma, leaving residual dunite or wehrlite. It is improbable that many quartz tholeiites are derived in this way, i.e. by direct melting and magma segregation from parental pyrolite at very shallow depths. On the other hand, some mixed gabbro-peridotite complexes of alpine orogenic zones may conceivably originate by deformation and partial magma segregation from pyrolite at shallow levels. At depths of more than approximately 15 km, the orthopyroxene → olivine + liquid reaction relationship does not occur, and liquid fractions in equilibrium with residual olivine are always olivine normative.

Magma Segregation at 15–35 Kilometers

Within the range 15–35 km, the mineralogy of pyrolite at near-solidus temperatures consists of olivine, moderately aluminous pyroxenes, and minor plagioclase. Partial melting to yield 20–25% basalt leaves residual olivine, low-alumina enstatite, and possibly some subcalcic augite. The liquid produced has the composition of a high-alumina olivine tholeiite with around 10% normative olivine. It is inferred that the extensive occurrences of high-alumina olivine tholeiites along the mid-oceanic ridges are a consequence of magma segregation at the relatively shallow depths of 15–35 km. The mid-oceanic ridges are regions of high heat flow and may be loci of extremely active convective or advective movement in the upper mantle.

Magma Segregation at 35–70 Kilometers

At 35–70 km, plagioclase and garnet are absent from the near-solidus mineralogy of pyrolite. Partial melting of variable extent involves equilibrium between liquid and olivine, aluminous enstatite, and aluminous subcalcic augite. With a low degree ($\leq 20\%$) of partial melting, the liquid produced has the composition of an alkali olivine basalt magma that may segregate, leaving olivine, aluminous enstatite, and minor aluminous subcalcic augite as residual phases. With an increasing degree of partial melting, the aluminous subcalcic augite and aluminous enstatite, in particular, enter the liquid phase, continuously changing composition from that of alkali olivine basalt through olivine basalt to an olivine-rich tholeiite.

Thus, at depths of 35-70 km, a liquid segregating after about 25-35% partial melting of pyrolite is of olivine tholeiite type (with Al_2O_3 about 12%) and leaves residual olivine and enstatite (Al_2O_3-poor). The series of liquids derived from pyrolite by different degrees of partial melting at 35-70 km is lower in SiO_2 and Al_2O_3 content than the series of liquids developed at 15-35 km with similar degrees of melting of the parent rock. This is because the ratio of pyroxene to olivine, particularly orthopyroxene to olivine, in the residual crystals is always higher at 35-70 km than at 15-35 km for a given degree of partial melting. In addition, the residual pyroxene at 35-70 km contains more Al_2O_3 in solid solution, resulting in lower Al_2O_3 content of the liquid fractions.

Magma Segregation at 100 Kilometers or More

At depths of about 100 km or more, garnet appears in the near-solidus mineral assemblage of pyrolite, and the pyroxenes decrease in Al_2O_3 content and in abundance with increasing pressure. At depths of around 100 km, a relatively small degree of partial melting causes the disappearance of garnet, and liquids are in equilibrium with olivine and aluminous pyroxene. Liquids derived in this way, i.e. with olivine and aluminous pyroxene as liquidus phases, have chemical compositions approaching picrites with more than 30% normative olivine. Extremely undersaturated but Mg-rich magmas such as olivine nephelinites and olivine-melilite nephelinites may originate by low degrees of partial melting with magma segregation at depths of 100 km or more, although this hypothesis has not been experimentally verified [*O'Hara*, 1965; *Kushiro and Kuno*, 1963].

FRACTIONATION OF BASALTIC MAGMA
AT VARIOUS PRESSURES

O'Hara [1965] has argued that most basalts observed at the surface have chemical compositions and exhibit crystallization sequences that have been determined by crystal fractionation processes acting at low pressure. Excellent documentation can be found in the studies of the Hawaiian Kilauea and Mauna Loa eruptions of recent and historic time [*Murata and Richter*, 1966]. O'Hara further argues that such magmas have probably undergone appreciable fractionation at other depths during their movement to the surface. The nature of the liquidus (initial crystallization) phases of basaltic magmas, and thus the direction of fractionation, are very strongly dependent on load pressure. Thus, although the degree of partial melting and depth of magma segregation determine the initial magma composition, this may be further modified by fractionation under a range of load pressure conditions.

An olivine-rich tholeiitic magma, segregated from pyrolite at 35-70 km, can yield three distinctive derivative magma types by fractionation at different depths. The processes and products of such fractionation are summarized in Figures 1 and 2. Thus it is possible for a single 'primary' magma of olivine-rich and hypersthene normative type to produce:

1. A continuous magma series from olivine-rich tholeiite to alkali-olivine basalt to olivine basanite by fractionation at 35-70 km.
2. A continuous magma series from olivine-rich tholeiite to high-alumina olivine tholeiite by fractionation at 15-35 km.
3. A continuous magma series from olivine-rich tholeiite to quartz tholeiite by fractionation at less than 15 km.

Magmas lying along the higher pressure fractionation trends may be tapped off and further fractionate under lower pressure conditions (Figures 1 and 2). It must be emphasized that no sharp division between the magma series is anticipated, but rather a continuity in composition, since all may have the same parent liquid and there is a transition from one regime

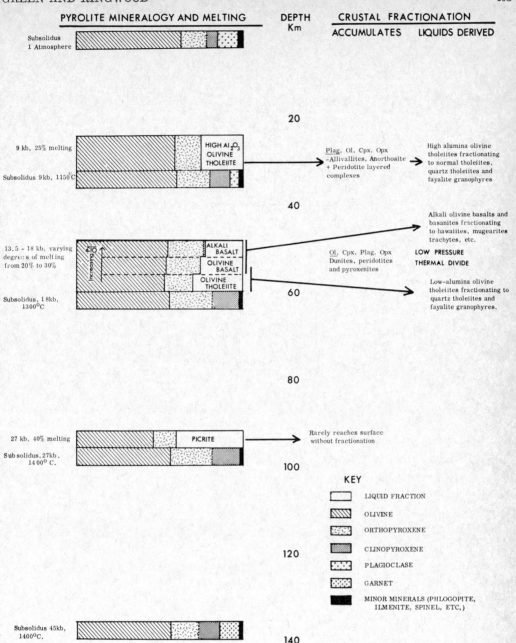

Fig. 1. The variation in the near-solidus mineralogy of pyrolite, the degree of partial melting, and nature of both the liquid and the refractory residuum at various depths within the mantle [*Green and Ringwood*, 1967].

of fractionation to another. However, the superposition of low pressure fractionation effects on magmas of the alkali-olivine basalt type and on those of the olivine tholeiitic type causes a marked divergence of derivative liquids due to the 'thermal divide' between alkali olivine basalts and tholeiitic basalts at low pressure [*Yoder and Tilley*, 1962].

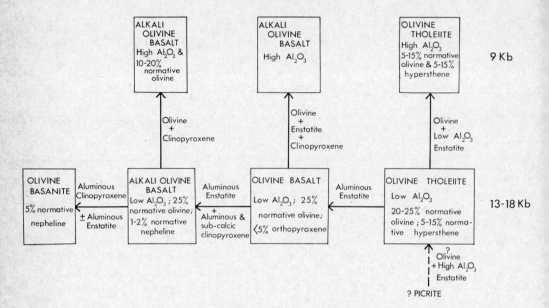

Fig. 2. The effects and directions of fractionation of basaltic liquids at moderate to high pressures [*Green and Ringwood*, 1967].

INTERACTION BETWEEN BASALTIC MAGMA AND WALL-ROCK ENVIRONMENT

The variations in major element chemical composition among basalts are determined directly by the three factors previously discussed:

1. Degree of partial melting of the pyrolite of the upper mantle.
2. Depth of magma segregation from residual crystals.
3. Conditions and extent of fractionation of magma after segregation from residual crystals.

However, there is a group of minor and trace elements (including K, Ti, P, U, Th, Ba, Rb, Sr, Cs, Zr, Hf, and the rare earth elements) that show relative abundances among basalts (particularly between alkali olivine basalts and olivine tholeiites) that are inconsistent with the crystal fractionation relationships as simply outlined above. This group of elements has been called 'incompatible elements' [*Green and Ringwood,* 1967], referring to their inability to substitute to any appreciable extent in the major minerals of the upper mantle (olivine, aluminous pyroxenes). In an environment with little temperature contrast between magma and wall rock (as with a magma near its source in the mantle), cooling of a magma with consequent fractionation also involves complementary processes of reaction with the wall rock, including selective melting and extraction of the lowest melting fraction. The incompatible elements are considered to be highly concentrated in the lowest melting fraction of pyrolite. The highly selective sampling of large volumes of wall rock during slow ascent of a magma near its depth of segregation (wall rock reaction) is considered a normal process complementary to crystal fractionation in the mantle. Magma generation in the mantle is rarely a simple, closed-system, partial melting process as far as the abundances of incompatible elements are concerned. This process of reaction with wall rock has the corollary that considerable volumes of the pyrolite of the upper mantle may undergo highly selective leaching of particular elements without actually being melted to the extent of yielding their potential 20–40% of basaltic magma.

BASALTS AS SAMPLES OF THE UPPER MANTLE

From the preceding sections it is apparent that the generation of basaltic magmas is a complex and highly variable process. Basalts are a partial sample of the upper mantle, but there may be little relevance in the simple model of magma generation in which liquid composition is determined at the source and thereafter remains essentially a closed system except for depletion by crystal fractionation. There are thus great uncertainties in the use of element abundance or isotopic abundance data from basalts to deduce element abundances and geochemical pre-histories for the mantle source region. It is considered that a self-consistent model of magma generation and relationships can be derived from the pyrolite model composition for the upper mantle. The experimental studies on magma genesis and on pyrolite compositions [*Green and Ringwood*, 1967] provide strong support for this model. Nevertheless, it must be emphasized that the processes of magma generation by variable partial melting, the local crystal accumulates formed by magma fractionation at depth, and the highly selective nature of the wall-rock reaction process all indicate that the earth's upper mantle, wherever it has acted as either source or country rock to ascending basaltic magmas, must be inhomogeneous in chemical composition and in mineralogy. Inhomogeneity will be most marked in contents of incompatible elements and this in turn, over geological time, must produce local and regional variations in radiogenic isotope abundances, e.g. in Sr and Pb isotopes.

REFERENCES

Engel, A. E. J., C. G. Engel, and R. G. Havens, Chemical characteristics of oceanic basalts and the upper mantle, *Bull. Geol. Soc. Am.*, *76*, 719–734, 1965.

Green, D. H., and A. E. Ringwood, The genesis of basaltic magmas, *Contrib. Mineral. Petrol.*, *15*, 103–90, 1967.

Green, T. H., D. H. Green, and A. E. Ringwood, The origin of high-alumina basalts and their relationship to other basaltic magma types, *Earth Planetary Sci. Letters*, *2*, 41–51, 1967.

Kushiro, I., and H. Kuno, Origin of primary basalt magmas and classification of basaltic rocks, *J. Petrol.*, *4*, 75–89, 1963.

Macdonald, G. A., and T. Katsura, Chemical compositions of Hawaiian lavas, *J. Petrol.*, *5*, 82–133, 1964.

Murata, K. J., and D. H. Richter, Chemistry of the lavas of the 1959–60 eruption of Kilauea volcano, Hawaii, *U. S. Geol. Surv. Prof. Paper*, *537-A*, 1–26, 1966.

O'Hara, M. J., Primary magmas and the origin of basalts, *Scottish J. Geol.*, *1*, 19–40, 1965.

Ringwood, A. E., The mineralogy of the upper mantle, in *Advances in Earth Science*, edited by P. M. Hurley, pp. 357–399, MIT Press, Boston, 1966.

Yoder, H. S., and C. E. Tilley, Origin of basalt magmas: An experimental study of natural and synthetic rock systems, *J. Petrol.*, *3*, 342–532, 1962.

Plateau Basalts

Hisashi Kuno

PLATEAU basalts are major units of basaltic rocks on the earth's surface. The lavas are fed from numerous extensive fissures or localized vents scattered over a wide area. Very fluid lavas are extruded without explosive eruptions. The volume of a single plateau basalt unit may range from 400 to nearly 800,000 km^3. This stands in contrast to the volume of a single stratovolcano in the orogenic belts, which is usually less than 500 km^3, ranging to about 1000 km^3. The two largest plateau basalt units, the Columbia River and Deccan, were extruded during periods of about 9 and 15 million years, respectively. In both regions, the rate of extrusion of magma is only 0.1 km^3 per 1000 years per 1000 km^2. This does not differ greatly from the rate of extrusion in the Japanese volcanoes. It therefore appears that the essential condition for the formation of plateau basalts is not the unusually large concentration of heat energy within a limited region of the upper mantle but a supply of heat energy over a wide region for a prolonged time.

The great volume of plateau basalts and their apparent uniformity in composition have led geologists to suppose that they represent undifferentiated or primary magmas. Petrologists now believe that most of the plateau basalts are formed from magmas representing fairly advanced stages of fractional crystallization of primary magmas.

Considering the degree of crystallization of primary magmas necessary to produce the plateau basalt magmas, and also the proportion of the primary magmas to be produced by partial melting of the mantle peridotite, the volume of mantle material involved in the formation of the plateau basalts can be estimated. For the Columbia River basalts, it would be equivalent to a plate about 80 km thick underlying the area covered by the basalt. This thickness is a small fraction of the mantle.

DISTRIBUTION AND AGE

Cenozoic plateau basalts exhibit a conspicuous topographic feature: they form flat plateaus extending over several tens to several hundreds of kilometers. Such plateaus are found in the stable continental regions beyond the Circum-Pacific orogenic belt (see Table 1): southern Vietnam (Tertiary to Pleistocene), eastern Manchuria (Miocene to Recent [Kuno, 1953]), northern and central British Columbia (Tertiary to Pleistocene?), the Columbia River plateau in Washington and northern Oregon (middle Miocene to older Pliocene [Waters, 1955, 1961, 1962]), the Snake River plateau in Idaho (Pleistocene? to Recent), the Patagonian plateau in southern Argentina (Miocene to Recent [Tyrrell, 1932]), and possibly the eastern plateau region of Graham Land, West Antarctica.

Other well-known plateaus are the Deccan plateau of India (late Cretaceous to early Eocene [Washington, 1922; Subramaniam and Sahasrabudhe, 1964]), a plateau on the eastern side of the Dead Sea rift (Pleistocene), and a vast region extending from western Scotland to Greenland (Eocene) where several separate plateau units may be present.

There are also many plateau basalts of older ages, which, however, do not exhibit distinct topographic features because of deep erosion or burial underneath younger sediments. Well-known examples are: the basalt lavas associated with Triassic Karroo dolerites of South Africa; those associated with the Triassic dolerite sills in New Jersey; Triassic or Jurassic lavas of the Parana basin in southern Brazil and Uruguay; the late Permian to early Triassic lavas of Tunguska basin in Siberia; and younger Precambrian lavas of northern Michigan and Minnesota, which are associated with the Duluth gabbro and other sheet-like intrusions. More examples are likely to be found among Precambrian formations.

MODE OF ERUPTION

The outstanding feature of plateau basalts is the scarcity of pyroclastic materials interbedded with the lavas. Individual lava flows are

TABLE 1. Dimensions and Ages of Cenozoic Plateau Basalts

Name and Location	Area, km^2	Thickness, km	Volume, km^3	Age
Plateau around Baekdoo-san (Hakuto-san) Volcano, Manchuria-Korea border [Kuno, 1953]	22,000	0.5 (average)	11,000	Miocene (?) to Recent
Plateau south of Tung-ning, eastern Manchuria-Soviet border [Kuno, 1953]	5,340	0.015–0.150	416	Miocene or Pliocene
Columbia River Plateau, NW United States [Waters, 1955, 1961, 1962]	220,000*	1 (average)	195,000*	21.3–12.1 m.y.
Deccan Plateau, India [Subramaniam and Sahasrabudhe, 1964]	518,000	1.0 1.5 2.0	518,000 777,000 1,036,000	Late Cretaceous to early Eocene, 15 m.y. (?)

* Personal communication, A. C. Waters, 1967.

generally from 2 to 50 meters thick and sometimes extend for distances of several tens of kilometers. These facts imply that very fluid magma is extruded without explosive eruption.

The lavas are sometimes fed from numerous extensive fissures in various parts of the plateau. If they are fed from localized vents, the plateau can be described as a group of numerous flat shield volcanoes. The Columbia River and Deccan plateaus were produced by fissure eruption, whereas the small plateaus in eastern Manchuria (Table 1), the Snake River plateau, and the Patagonian plateau [Tyrrell, 1932] were formed by eruption from numerous vents.

The distribution of fissures and vents throughout the plateau would imply that the magma production took place nearly over the whole region underneath the plateau.

Tyrrell [1932] pointed out that the fissure eruption generally produces much larger plateaus than does the eruption from localized vents, and that the rocks from fissure eruption are generally SiO_2-oversaturated basalt (tholeiite), whereas those from vents are alkalic, SiO_2-undersaturated basalt (alkali olivine basalt). However, the relation between the type of eruption and the type of magma is not so simple (see next section).

PETROLOGY

Among the plateau basalts, there is a complete gradation in composition from tholeiite* to alkali olivine basalt, Al_2O_3 being about 15% throughout this variation. This is a striking contrast to the basaltic rocks of the circum-Pacific volcanoes, where high-alumina basalt ($Al_2O_3 >$ 16.5%) is present as a group intermediate between comparatively alkali-poor tholeiite (Al_2O_3 about 15%) and alkali olivine basalt (Al_2O_3 about 15%). However, high-alumina basalt is not entirely absent among the plateau basalts.

Comparatively small plateau basalt units, such as those in eastern Manchuria, are almost exclusively made up of alkali olivine basalt, together with some hawaiite and mugearite derived therefrom [Kuno, 1953]. However, the plateau south of Tung-ning (Table 1) is composed of tholeiite and high-alumina basalt.

The plateau basalts of western Scotland and southern Vietnam include both alkali olivine basalt and tholeiite, which are completely gradational to one another.

The two largest plateau basalt units, the Columbia River and Deccan, are largely made up of tholeiite, but alkali olivine basalt, transitional to tholeiite, occurs in the lower part of the Columbia River basalt succession (the Picture Gorge basalt). In the Deccan plateau, some alkalic lavas are also found near its western and northwestern margin.

Small amounts of andesite, dacite, and rhyolite occur in association with tholeiite in Deccan, western Scotland, and Parana. Mugearite and trachyte are also associated with alkali olivine basalt in western Scotland.

If such salic rocks are disregarded, the plateau basalts in general are apparently uniform in chemical composition. This fact, taken in conjunction with their great volume and mode of extrusion, led geologists in the past to suppose that the plateau basalts represent undifferentiated magma, the parent of many igneous rocks. But if this were the case, the chemical analyses of the plateau basalts should plot within a limited area in the triangular diagram MgO, $FeO + Fe_2O_3$, $Na_2O + K_2O$. However, as can be seen in Fig. 1, there is a wide variation in the proportion of these three components, both among the group of tholeiite and high-alumina basalts and among the group of alkali basalts. Such a wide variation is seen even among the lavas of a single plateau basalt unit.

The variation trend of the group of tholeiite and high-alumina basalts, including the salic rocks such as andesite, etc., is similar to the tholeiite series of the Japanese volcanoes (the pigeonitic rock series) rather than to the Japanese calc-alkali series (the hypersthenic rock series, Figure 2). Both the tholeiitic and high-alumina plateau basalts and the pigeonitic rock series show moderate concentration of $FeO + Fe_2O_3$ in the middle stage of fractionation [Kuno, 1968].

The parental magmas of the tholeiitic and high-alumina plateau basalts may have compositions close to the Hawaiian olivine tholeiite or to the average Hawaiian oceanite plotted in

---

* Tholeiite is a type of basalt comparatively high in SiO_2 and low in alkalies ($Na_2O + K_2O$); alkali olivine basalt is comparatively low in SiO_2 and high in alkalies. High-alumina basalt is intermediate between the the two types in SiO_2 and alkalies but higher in Al_2O_3.

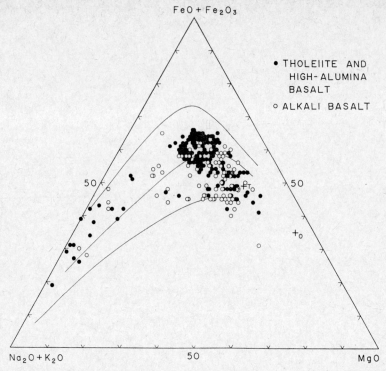

Fig. 1. MgO:FeO + Fe$_2$O$_3$:Na$_2$O + K$_2$O ratios of plateau basalts. Solid circles: tholeiite, high-alumina basalt, andesite, dacite, and rhyolite. Open circles: alkali olivine basalt, more mafic alkali basalt, hawaiite, mugearite, and trachyte. Crosses marked T and O are olivine tholeiite (Pele's hair of 1959 eruption of Kilauea Iki, Hawaii) and average oceanite of Hawaii, respectively. Curves are reproduction of those in Figure 2. This figure illustrates the effect of crystallization of basalt magmas. According to petrologists' experience, basaltic rocks that would be regarded as derived from primary magmas plot within a narrow area of the triangular diagram. If some crystals are removed from the primary magma, causing differentiation, the magma composition should shift sensitively away from the MgO corner.

Figure 1. The parental magma of the group of alkali olivine basalt may be a little more alkalic. The parental magma of the Japanese pigeonitic rock series (P.T. of Figure 2) is distinctly less alkalic.

As can be seen from the population of points in relation to the postulated compositions of the parental magmas in Figure 1, most of the plateau basalts were formed from much differentiated magmas.

The distance from the FeO + Fe$_2$O$_3$/Na$_2$O + K$_2$O edge of the triangle (Figure 1) to any point, namely the value MgO × 100/MgO + FeO + Fe$_2$O$_3$ + Na$_2$O + K$_2$O, can be used as a measure of degree of solidification by which a parental magma produced the magma represented by the point. This is called the solidification index (SI).

Parental magmas of many igneous rock series, namely the primary magmas produced in the mantle, appear to have SI about 40 (T in Figure 1 and P.T. in Figure 2), or a little higher than 40 (O in Figure 1). If we take 40 as the general value for most primary basalt magmas, rocks with SI about 20 would correspond to magmas produced by 50% solidification, namely magmas of the middle stage of fractionation.

Numbers of chemical analyses of plateau basalts whose SI values are higher than 41, between 40 and 36, between 35 and 31, and so on, are plotted in the frequency distribution diagram of Figure 3. Similar diagrams for the pigeonitic and hypersthenic rock series of the Quaternary volcanoes of Japan are shown in Figures 4 and 5, respectively.

The most striking fact seen in Figure 3 is that

the rocks of the middle fractionation stage, which are still called 'basalt' in the ordinary petrographic classification, occur most frequently among the tholeiitic and high-alumina plateau basalts.

Figures 4 and 5 show that rocks of the middle fractionation stage also occur most frequently. This is not surprising, because the predominance of andesite in the island arc volcanoes is well known. However, many rocks of the pigeonitic rock series, whose SI lies between 30 and 21, can be called 'basalt' petrographically. Thus, from the petrogenetical viewpoint, the tholeiitic and high-alumina plateau basalts are essentially the same as the pigeonitic rock series of the island arc volcanoes. The only difference is their mode of eruption.

VOLUME AND RATE OF EXTRUSION OF MAGMA

The area covered by the plateau lavas can easily be measured; estimation of average thickness is usually difficult.

According to *Waters* [1962] (personal communication, 1967) the total volume of the Columbia River basalts is about 195,000 km³. On the other hand, Kuno's estimate of the total volume, calculated from the area covered by the lavas (Table 1) and assuming the average thickness of the lava pile to be 1 km, is 220,000 km³. Thus 200,000 km³ would be a reasonable estimate.

According to *Subramaniam and Sahasrabudhe* [1964], the Upper, Middle, and Lower Deccan traps have thicknesses of 457, 1220, and 152 meters, respectively. However, the sum of these values (1927 meters) may not represent the average thickness of the whole Deccan trap. In Table 1, three alternative estimates of average thickness are given. If we take 1.5 km as the most reasonable thickness, the total volume is 777,000 km³.

In both the Columbia River and Deccan plateaus, more than several tens of successive lava flows are represented even in a single cliff

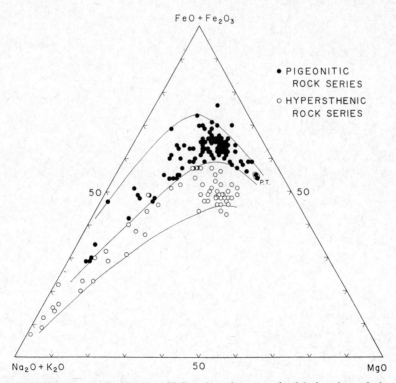

Fig. 2. MgO:FeO + Fe₂O₃:Na₂O + K₂O ratios of nonporphyritic basalt, andesite, dacite, and rhyolite of the Izu-Hakone region, central Japan [*Kuno*, 1968]. The curves show the general limits of the variation of the pigeonitic and hypersthenic rock series. Two points marked *P.T.* represent parental magma of the pigeonitic rock series.

section. They were extruded over a period of 9 million years in the Columbia River plateau (according to K/Ar dating by *Gray and Kittleman* [1967]) and of about 15 million years in the Deccan traps (late Cretaceous to early Eocene according to *Subramaniam and Sahasrabudhe* [1964]. By dividing the total volume by the time length and the total area covered by the lavas, the rate of extrusion of magma to the surface is obtained as $0.1 \text{ km}^3/10^3 \text{ years} \cdot 10^3 \text{ km}^2$ both for the Columbia River and Deccan plateau basalts.

This value, especially for the Deccan plateau, could be doubled or reduced to half by changing the estimate of the average thickness or the time of extrusion or both. However, it is almost impossible to make it more than five times the present value.

In the islands of Hokkaido, Honshu, and Kyushu, Japan, about 5.2 km^3 of basalt, andesite, and dacite were erupted during the past 1000 years from various volcanoes scattered over an area of about $200,000 \text{ km}^2$. The rate of extrusion calculated from these values is $0.026 \text{ km}^3/10^3 \text{ years} \cdot 10^3 \text{ km}^2$. This is the same as the rate of extrusion in all the volcanoes of the three islands of Japan during the whole Quaternary time ($0.025 \text{ km}^3/10^3 \text{ years} \cdot 10^3 \text{ km}^2$ as calculated from the data given by *Sugimura et al.* [1963]). During the early to middle Miocene (about 6 m.y.), $150,000 \text{ km}^3$ of basalt, andesite, dacite, and rhyolite were erupted over an area of about $131,000 \text{ km}^2$ in Japan [*Sugimura et al.*, 1963]. The rate of extrusion is therefore $0.19 \text{ km}^3/10^3 \text{ years} \cdot 10^3 \text{ km}^2$.

Thus the rate of extrusion of magma to produce the plateau basalts is not so different from

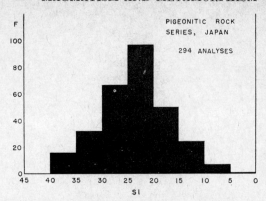

Fig. 4. Frequency (F) of rocks of the pigeonitic rock series with different solidification indices (SI).

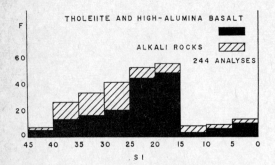

Fig. 3. Frequency (F, represented by the numbers of chemical analyses) of plateau basalts with different solidification indices (SI).

that in the island arc volcanism. It appears then that the essential condition for the formation of a great basalt plateau is not an unusually high concentration of heat energy within a limited region of the upper mantle, but a concentration over a vast region for a long time.

If we assume that the average Columbia River basalt magma was produced by 50% solidification of primary magma, and if the primary magma was produced by partial melting of 5% of the mantle peridotite, we can estimate the volume of the peridotite involved in the partial melting.

It is unlikely that all the residual magma left after 50% solidification was drained from the magma reservoir to extrude to the surface. If we assume that only half the residual magma was transferred to the surface, the primary magma that produced $200,000 \text{ km}^3$ of the Columbia River basalts would have had a volume of $800,000 \text{ km}^3$.

It is possible, if not proved, that most of the interstitial liquid or the primary magma produced by the partial melting was separated from the host peridotite and was transferred to shallower levels of the mantle or to the crust. Thus the volume of the mantle involved in the partial melting would have been $800,000 \text{ km}^3 \times 100/5 = 16,000,000 \text{ km}^3$ or a little more. The distribution of fissures implies that magma production took place over a region underlying the whole Columbia River plateau. Then, if we suppose a plate of peridotite involved in the partial melting with lateral extension approximating the area of the plateau ($220,000 \text{ km}^2$), the thickness

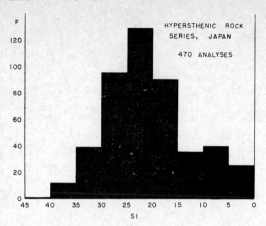

Fig. 5. Frequency (*F*) of rocks of the hypersthenic rock series with different solidification indices (*SI*).

of the plate would be about 80 km. If the area of the plate is half the area of the plateau, its thickness would be about 160 km. These values are indeed only a small fraction of the total thickness of the mantle.

It is not implied that the mass actually involved in producing the plateau basalt had the plate shape used for the calculation, nor is there any indication that all the mass involved melted simultaneously. Instead, it is most likely that only a small volume of peridotite melted at one time (owing to transportation by convection to a shallow level in the mantle), and that this process was repeated many times over a period of 9 million years.

REFERENCES

Gray, J., and L. R. Kittleman, Geochronometry of the Columbia River basalt and associated floras of eastern Washington and western Idaho, *Am. J. Sci.*, *265*, 257–291, 1967.

Kuno, H., Plateau basalt lavas of eastern Manchuria, *Proc. Pacific Sci. Congr., Pacific Sci. Assoc.*, 7th, *2*, 375–382, 1953.

Kuno, H., Differentiation of basalt magmas, in *Basalts: The Poldervaart Treatise on Rocks of Basaltic Composition*, vol. 2, edited by H. H. Hess and Arie Poldervaart, pp. 623–688, Interscience, New York, 1968.

Subramaniam, A. P., and Y. S. Sahasrabudhe, Geology of Greater Bombay and Aurangabad-Ellora-Ajanta area, *Intern. Geol. Congr., 22nd, India, 1964, Guide to Exc. Nos. A-13 and C-10*, 1–12, 1964.

Sugimura, A., T. Matsuda, K. Chinzei, and K. Nakamura, Quantitative distribution of late Cenozoic volcanic materials in Japan, *Bull. Volcanol.*, *26*, 125–140, 1963.

Tyrrell, G. W., The basalts of Patagonia, *J. Geol.*, *40*, 374–383, 1932.

Washington, H. S., Deccan traps and other plateau basalts, *Bull. Geol. Soc. Am.*, *33*, 765–804, 1922.

Waters, A. C., Volcanic rocks and the tectonic cycle, *Geol. Soc. Am. Spec. Paper 62*, 703–722, 1955.

Waters, A. C., Stratigraphic and lithologic variations in the Columbia River Basalt, *Am. J. Sci.*, *259*, 583–611, 1961.

Waters, A. C., Basalt magma types and their tectonic associations, Pacific Northwest of the United States, *The Crust of the Pacific Basin, Geophys. Monograph 6*, pp. 158–170, American Geophysical Union, Washington, D.C., 1962.

Andesitic and Rhyolitic Volcanism of Orogenic Belts

Alexander R. McBirney

BASALTS have been the subject of such intensive research and discussion in recent years that studies of other types of effusive igneous rocks have received only limited attention. With the increasing interest in volcanic rocks of orogenic systems, it has become apparent that these rocks, especially andesites and rhyolites, offer a special challenge to petrologists. New insights gained through recent research in this area have served to emphasize some of

ANDESITES

Andesite is by far the most characteristic volcanic rock of orogenic systems. In many areas it surpasses even basalt in volume. In the geologic past, andesitic volcanism has occurred during at least one stage of nearly every large orogenic sequence. The importance of andesites in older continental systems has commonly been overlooked because of a tendency to describe most basic lavas and fragmental rocks as basalts. Where closer studies have been made, as, for example, in the Archean complexes of the Canadian shield [*Wilson et al.*, 1965], the volcanic rocks have been found to resemble those of modern calc-alkaline orogenic series.

Andesitic volcanoes have usually been localized in shifting linear or arcuate belts during two or three successive tectonic episodes, each or any of which may have terminated with intrusion of plutons of similar chemical composition. In Cenozoic time, andesitic volcanism has been the dominant feature of the circum-Pacific system of youthful mountain ranges and island arcs. It has also been important, though less conspicuous, in intracontinental orogenic regions, such as the Carpathian Mountains of Central Europe and the Colorado Plateau and western Rocky Mountains of North America. Within the deep ocean basins, however, true andesites seem to be lacking. Certain oceanic lavas of intermediate silica content have been called andesite, but they differ from orogenic andesites in their high iron and low alumina contents, and more closely resemble icelandites, craignurites, and other intermediate rocks believed to be products of differentiation of tholeiitic basalt. The term 'andesite line' coined by Born is often used to designate the boundary between the main Pacific basin and the circum-Pacific orogenic belts.

Despite the volumetric importance of andesites, there is a lack of agreement about the compositional and chemical characteristics of the rocks. Most petrographers agree that andesites are effusive rocks of intermediate silica content and a color index (percentage of dark ferromagnesian minerals) lower than 40. Quartz is not an important modal mineral. According to some petrographers, the plagioclase should be more sodic than labradorite, but in many provinces very calcic plagioclase is characteristic of andesites and even dacites, and in these calcic suites, color index is the criterion most commonly used to establish a division between andesite and basalt.

Considerable confusion has resulted from attempts to classify andesites and related rocks according to idealized systems based on their chemistry. Certain petrologists, especially in Europe, limit the silica content of andesite to that corresponding to 10% normative quartz and term any rock more siliceous than this a dacite. This restricts andesites to a much narrower range than that used by most geologists and petrographers working with the rocks in the field. As a consequence, some average compositions that have been calculated on the basis of idealized systems are quite different from the true composition of the rocks to which the name andesite is commonly applied. This difference is illustrated by the contrast seen in Table 1, which gives two commonly cited average compositions of andesites. That of *Daly* [1933]

TABLE 1. Average Composition of Andesite Compiled by *Daly* [1933] and *Nockolds* [1954]

Component	Daly (87 analyses)	Nockolds (49 analyses)
SiO_2	59.59	54.20
TiO_2	0.77	1.31
Al_2O_3	17.31	17.17
Fe_2O_3	3.33	3.48
FeO	3.13	5.49
MnO	0.18	0.15
MgO	2.75	4.36
CaO	5.80	7.92
Na_2O	3.58	3.67
K_2O	2.04	1.11
H_2O	1.26	0.86
P_2O_5	0.26	0.28
Normative composition		
Qz	14.5	5.7
Or	12.2	6.7
Ab	30.4	30.9
An	25.0	27.2
C		
Wo	0.9	4.2
En	6.9	10.9
Fs	2.0	5.3
Mt	4.9	5.1
Il	1.5	2.4
Ap	0.7	0.7

TABLE 2. Average Compositions of Andesites and Similar Rocks from Different Geologic Associations

Component	Calcic Andesites[a]	Calc-Alkaline Andesites[b]	Alkali-Calcic Andesites[c]	Iron-Rich 'Andesites'[d]
SiO_2	58.68	58.65	58.05	58.31
TiO_2	0.81	0.79	1.10	1.71
Al_2O_3	17.29	17.43	17.15	13.77
Fe_2O_3	2.97	3.21	3.30	3.37
FeO	3.96	3.48	2.54	6.48
MnO	0.12	0.10	0.13	0.23
MgO	3.14	3.28	2.16	2.27
CaO	7.13	6.26	5.13	5.58
Na_2O	3.24	3.82	4.57	3.91
K_2O	1.27	1.99	3.60	1.88
Total H_2O	1.20	1.06	1.64	1.01
P_2O_5	0.17	0.18	0.43	0.46
Total	99.98	100.25	99.80	98.98
Molecular norms				
Ap	0.36	0.38	0.91	1.01
Il	1.15	1.11	1.56	2.48
Or	7.67	11.88	21.59	11.65
Ab	29.72	34.65	41.66	36.65
An	29.52	24.80	15.90	15.02
Mt	3.17	3.39	3.50	3.66
Di	4.39	4.31	5.67	8.60
Hy	9.85	9.22	3.53	8.20
Q	14.16	10.27	5.67	12.73

[a] Average of 89 calcic andesites from island arcs.
[b] Average of 29 calc-alkaline andesites from continental margins.
[c] Average of 13 alkali-calcic andesites from continental interior regions.
[d] Average of 22 iron-rich 'andesites,' icelandites, craignurites, etc., from oceanic and nonorogenic regions of continents.

includes only analyses of rocks called andesite in the source reference, while that of *Nockolds* [1954] is calculated from analyses of rocks conforming to a predetermined definition of andesite (with less than 10% normative quartz) regardless of the name used by the original worker who studied the rock. Nockolds' average, though more rigorously defined, obviously differs from what most geologists and petrographers call andesite.

Daly's average is also inadequate, however, because it lumps together andesites of very different environments and compositions. A single average of such a diverse group of rocks obscures many petrologic features that are only recognizable when related rocks of individual provinces and volcanic episodes are considered. These differences are illustrated by the first three compositions in Table 2. Each is an average of a specific compositional range of rocks within the broader limits of what most petrographers call andesite. The averages have marked differences in CaO and alkalis, and although they are almost identical in SiO_2, they differ significantly in their normative quartz content.

The three averages are of andesites from the calcic, calc-alkaline, and alkali-calcic series defined by *Peacock* [1931]. In the calcic series, the weight per cent of alkali exceeds that of CaO only in rocks with silica contents greater than 61%. Consequently, the andesites in such suites have very calcic plagioclase, commonly bytownite or even anorthite, despite the fact that the rocks may have 15 or 20% normative quartz. Andesites of this type are typical of volcanic belts rising from moderately deep seas or very near the edge of a thin continental crust in regions such as the Marianas, Japan, and the Antilles. The lavas of volcanoes in interior regions of continents vary more widely in time and place, but are commonly alkali-calcic; that is, total alkali exceeds CaO when the silica content is as low as 56% or less. Most andesitic volcanoes near continental margins that are visibly underlain by considerable thicknesses of sialic crust erupt lavas that belong to the more common calc-alkaline series; in these rocks, alkali exceeds CaO when silica percentages are greater than 56.

These differences do not seem to be related to the thickness or composition of the crust, because in some regions there are no marked differences in the products of volcanoes along the strike of belts, such as the Aleutian or Kuril Islands, where considerable differences in the thickness and nature of the underlying crust are found from one end of the arc to the other [*Coats*, 1962; *Gorshkov*, 1962]. In Central America, however, the andesitic lavas of Nicaragua, which have risen through what appears to be a thin and relatively basic crust, are less siliceous than those in the Guatemalan highlands, which are visibly underlain by a thick series of old metamorphic and granitic rocks. In both regions the volcanoes are spaced equally close to the Middle American trench and continental margin.

Transitions across the strike of volcanic belts are commonly very abrupt. In the Japanese Islands, for example, they can be detected in the space of a few tens of kilometers [*Kuno*, 1959]. The same seems to be true of the Kuril volcanoes and those of New Zealand.

Gorshkov [1962] has pointed out that the character of andesitic suites is largely independent of local tectonic conditions and structural or lithologic differences in the crust. He concludes that andesites must be derived below the base of the crust, their distinctive chemical character being inherited from subcontinental mantle conditions rather than from assimilation of crustal rocks.

GENESIS OF ANDESITE

Patterns of regional and temporal trends in andesitic suites are far from consistent. Some individual volcanic centers or groups of volcanoes produce great volumes of remarkably homogeneous andesite with only minor amounts of basalt or siliceous differentiates. Others erupt a complete range of rock types, while still others, perhaps the majority, produce two groups, a main andesitic series followed by a bimodal group of later rocks consisting of basalt and either dacite or rhyolite, with few, if any, intermediate compositions.

In the Cascade Range, for example, the principal volcanoes can be divided into two groups [*McBirney*, 1968]. One has rock associations of a 'coherent' type in which the lavas have a uniform andesitic composition or, at most, a limited range of continuous variation throughout the active history of the volcano. The other is a 'divergent' type in which uniform andesitic rocks making up most of the volume of the main cone are followed by later basic and siliceous rock, commonly the products of small flank eruptions. The contrast between the coherent and divergent types is apparent in Figure 1. Most of the volcanoes of the coherent type are in the central Cascades, where the Quaternary volcanic belt crosses a line of Tertiary stocks, while the volcanoes of the divergent type are at the northern and southern ends of the chain. There appears to be no consistent relation to the nature of the basement rocks.

In the Japanese Islands, two parallel but distinct lines of differentiation have been observed in certain well studied volcanic suites. In the Hakone district, for example, *Kuno* [1950] distinguished two series, both interpreted as products of a single parental magma. The rocks of the 'pigeonitic rock series,' which are relatively rich in iron and characterized by groundmass pigeonite, are thought to be products of crystal fractionation of tholeiitic basalt. The 'hypersthenic rock series,' marked by less iron enrichment, abundant xenocrysts, and groundmass hypersthene, is said to be the product of contamination of the same parent magma with sialic crustal rocks. Similarly, alkali basalts along the inner side of the archipelago have produced alkaline differentiates (mugearites and trachytes) thought to be produced by crystal fractionation and alkali-calcic andesite, dacite, and rhyolite by assimilation of crustal rocks [*Aoki and Oji*, 1966].

These processes, crystal fractionation and assimilation of crustal rocks, are two of the mechanisms most commonly suggested as possible explanations for the chemical and petrographic features of andesites; a third hypothesis calls

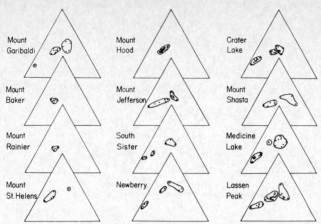

Fig. 1. Variation diagrams showing typical associations of the rocks of the High Cascade andesitic volcanoes. Points show the relative content of alkalis (lower left corner), magnesia (lower right corner), and iron (as FeO, upper corner) in rocks of the principal volcanoes. Mount Baker, Mount Rainier, and Mount Hood illustrate the coherent type of series in which the lavas are uniformly andesitic. Most of the other volcanoes are of the divergent type in which the main andesitic rocks of the principal vent are followed in time by basalts of erratic composition and rhyolite or dacite pumice eruptions or domes [*McBirney, 1968*].

upon a primary andesitic magma derived from an independent source in the lower crust or upper mantle. None of these adequately explains all andesites.

Some of the factors cited as evidence of the inadequacy of crystal fractionation, at least as the sole effective process, include the absence or scarcity of associated basalts that would represent a parent magma and the restricted compositional range of andesites in many volcanic associations. Most calculations of the composition of minerals that must be removed from basalt to produce andesite differ from observed phenocrysts both in composition and proportion.

Attempts to overcome these difficulties usually appeal to various degrees of contamination by sialic rocks, but only in rare cases has clear petrographic evidence of contamination been found. Although rocks in all stages of assimilation should be abundant, recognizable xenoliths are rare or totally absent in the great majority of andesite lavas. The volume of material that must be assimilated is often prohibitively large and of a composition unlike common basement rocks or any fraction likely to be derived from them by partial melting [*Taylor*, 1965].

Recent studies of the rubidium and strontium isotopic compositions of continental rocks indicate an additional problem. If a primary basaltic magma is contaminated with old sialic basement rocks, it should acquire an abnormally high proportion of radiogenic strontium. This will be especially evident where the crustal rocks are ancient metamorphic and granitic rocks with high rubidium concentrations. Since the isotopes of strontium have identical behavior in most magmatic processes, a high Sr^{87}/Sr^{86} ratio produced by decay of Rb^{87} in the crustal rocks should be imparted to the contaminated basalt and andesite. This has not been found. Most andesites have isotopic ratios that are identical to or only slightly more radiogenic than their associated basalts, regardless of the character of the basement series [*Pushkar*, 1968; *Hedge*, 1966].

Perhaps the strongest argument favoring a primary andesitic magma is the great volume and uniformity of andesites erupted in a given province through long episodes of volcanism. Experimental data on andesitic compositions are inadequate to demonstrate the mechanism by which such magmas could be formed, and the process involves so many poorly understood physical-chemical factors that it cannot be properly evaluated.

RHYOLITIC VOLCANISM

Although rhyolites are commonly the most siliceous members of andesitic suites, their volumetric importance in most orogenic rock series is minor. Rhyolitic and rhyodacitic lavas and tuffs erupted from mature volcanoes seem to be genetically related to more basic rocks with which they are intimately associated. Similarly, domes of obsidian in basalt-rhyolite complexes seem to be derived from a more basic magma, although not necessarily by crystal fractionation. However, the combined volume of these rhyolites is very small compared to that of widespread ignimbrite sheets, most of which are erupted from fissures that are largely independent of well-defined volcanic centers.

Rhyolitic ignimbrites are as characteristic of certain types of continental tectonic activity, especially epeirogenic uplift, as andesites are of the orogenic belts of continental margins. True rhyolites of post-Archean age seem to be largely confined to continental regions underlain by thick basement series of older sialic rocks. Rhyolites found in oceanic regions are invariably small in volume, have a comenditic or pantelleritic composition high in Na_2O and low in Al_2O_3, and are probably products of differentiation of tholeiitic basalts.

The extreme composition and great volume of rhyolitic ignimbrites places a severe constraint on the possible mechanisms by which they can be explained. An origin by crystal fractionation is ruled out by the huge volume of basic magma required to produce such a differentiate. Most recent workers have concluded that large-scale melting of the crust offers the only plausible source, but there is also evidence that at least some rhyolites, such as those of ancient shield areas, may have been derived directly from a primitive mantle.

There is considerable experimental evidence to demonstrate the possibility of producing large volumes of siliceous magma by partial melting of a wide range of common crustal rocks. Such an origin is supported by inclusions in ignimbrites of partly fused basement rocks or xenocrysts of strained quartz and other minerals demonstrably derived from older rocks. The high potassium, alumina, and silica content of most rhyolites can also be explained as the result of fusion of clay or mica-rich quartzose sediments or metamorphic rocks [*Brown*, 1963], although other explanations can also account for the major element composition, and a unique interpretation may not be possible on the basis of present experimental data [*McBirney and Weill*, 1966].

Another clue to the origin of rhyolites is found in their rubidium and strontium isotopic composition. The Rb/Sr and Sr^{87}/Sr^{86} ratios of ignimbrites are notably higher than those of intermediate and basic volcanic rocks and correspond more closely to those of sialic crustal material [*Hedge*, 1966]. Studies of the rhyolites of Central America [*Pushkar*, 1968; *McBirney and Weill*, 1966] have shown a correlation between the proportion of radiogenic strontium in rhyolitic ignimbrites and the character of the basement rocks; high Sr^{87}/Sr^{86} ratios are found only in regions underlain by Paleozoic metamorphic and plutonic rocks having high rubidium contents and similarly high Sr^{87}/Sr^{86} ratios. The oxygen isotopic composition of the same rhyolitic ignimbrites resembles that of metamorphic and plutonic rocks and is notably richer in O^{18} than basic igneous rocks or rhyolites believed to be the products of differentiation of basic magmas.

Some very ancient siliceous volcanic rocks, such as the Fennoscandian leptites, are the oldest known crustal rocks in deeply eroded Precambrian shield areas and were erupted at a time when little sialic crustal material existed. The magmas must have been produced from relatively basic rocks or directly from a source in the upper mantle. A few of these old rhyolites have been shown to be the effusive equivalents of intrusive rocks of granitic composition, and it has been suggested that they may have originated during an initial stage of formation of the continental crust [*Gastil*, 1958; *von Eckermann*, 1936].

REFERENCES

Aoki, K., and Y. Oji, Calc-alkaline volcanic rock series derived from alkali-olivine basalt magma, *J. Geophys. Res.*, *71*, 6127–6135, 1966.

Brown, G. M., Melting relations of Tertiary granitic rocks in Skye and Rhum, *Mineral. Mag.*, *33*, 533–562, 1963.

Coats, R. R., Magma type and crustal structure in the Aleutian arc, in *The Crust of the Pacific Basin, Geophys. Monograph 6*, edited by Gordon A. Macdonald and Hisashi Kuno, pp. 92–109, American Geophysical Union, Washington, D. C., 1962.

Daly, R. A., *Igneous Rocks and the Depths of the Earth*, McGraw-Hill Book Company, New York, 598 pp., 1933.

Gastil, G., Older Precambrian rocks of the Diamond Butte quadrangle, Gila County, Arizona, *Bull. Geol. Soc. Am., 69*, 1495–1514, 1958.

Gorshkov, G. S., Petrochemical features of volcanism in relation to the types of the earth's crust, in *The Crust of the Pacific Basin, Geophys. Monograph 6*, edited by Gordon A. Macdonald and Hisashi Kuno, pp. 110–115, American Geophysical Union, Washington, D. C. 1962.

Hedge, C. E., Variations in radiogenic strontium found in volcanic rocks, *J. Geophys. Res., 71*, 6119–6126, 1966.

Kuno, H., Petrology of Hakone volcano and the adjacent areas, Japan, *Bull. Geol. Soc. Am., 61*, 957–1014, 1950.

Kuno, H., Origin of Cenozoic petrographic provinces of Japan and surrounding areas, *Bull. Volcanol.*, [2] *20*, 37–76, 1959.

McBirney, A. R., Petrochemistry of the Cascade andesitic volcanoes, *Oregon Dept. Geol. Mineral Ind. Bull., 62*, 16–27, 1968.

McBirney, A. R., and D. F. Weill, Rhyolite magmas of Central America, *Bull. Volcanol., 29*, 435–448, 1966.

Nockolds, S. R., Average chemical composition of some igneous rocks, *Bull. Geol. Soc. Am., 65*, 1007–1032, 1954.

Peacock, M. A., Classification of igneous rock series, *J. Geol., 39*, 54–67, 1931.

Pushkar, P. D., Strontium isotope ratios in volcanic rocks of three island arc areas, *J. Geophys. Res., 73*, 2701–2714, 1968.

Taylor, S. R., Geochemistry of andesites and the growth of continents, *Nature, 208*, 271–273, 1965.

von Eckermann, H., The Loos-Hamra region, *Geol. Foren., 58*,(2), 129–343, 1936.

Wilson, H. D. B., P. Andrews, R. L. Moxham, and K. Ramlal, Archean volcanism in the Canadian shield, *Can. J. Earth Sci., 2*, 161–175, 1965.

Mafic and Ultramafic Inclusions in Basaltic Rocks and the Nature of The Upper Mantle

Hisashi Kuno

MAFIC and ultramafic inclusions in basaltic rocks can be regarded as cores of deep drilling carried out by nature. They are unique samples of the upper mantle and lower crust. On the basis of their elastic properties and mineralogy, these core samples can be arranged according to their postulated depth relations. Their chemistry, including the radioactive element contents, may offer a key to the source of various basalt magmas and the observed heat flows on the continents and ocean floors. This paper is largely concerned with the mineralogy and major element contents of these inclusions, their depth relations and origin, and the constitution of the upper mantle as deduced from the inclusion study.

INTRODUCTION

Nearly 200 localities of mafic and ultramafic inclusions are known throughout the oceans and the orogenic and nonorogenic regions of the world [*Forbes and Kuno*, 1965].

Peridotite,[1] the most common inclusion, is often associated with inclusions of pyroxenite, gabbro, and granulite. Eclogite inclusions are invariably associated with peridotite inclusions; they occur only in a limited number of places, such as Salt Lake crater near Honolulu, Delegate in Australia, and diamond-bearing kimberlite pipes in South Africa.

Peridotite and eclogite inclusions are confined to alkalic basaltic rocks, but gabbro inclusions are also found in tholeiite. No adequate explanation has yet been given for this mode of occurrence.

Ross et al. [1954] first pointed out a striking

---

[1] 'Peridotite' is used here for a group of ultramafic rocks including dunite, lherzolite, wehrlite, and harzburgite.

uniformity in mineral assemblage and individual mineral composition of the world peridotite inclusions. However, *White* [1966] and *Kuno* [1968] demonstrated the existence of a wide variation in the mineralogy and bulk composition of Hawaiian inclusions.

The peridotite inclusions have been interpreted either as fragments of the primordial mantle material from which basaltic magmas can be produced by partial melting, as residual solid material left after partial melting of a primordial mantle, or as crystal accumulates in basaltic magmas.

CLASSIFICATION AND MINERALOGY

Although a great variety of rock types can be distinguished in individual inclusion samples, some of them are completely gradational to one another and are closely associated in many localities. Such rock types can be regarded as genetically related and are grouped in the lherzolite, wehrlite, and eclogite series. Gabbro and granulite show erratic distribution and may or may not be related to some of these series.

The lherzolite series is uniform in mineral assemblage, being made up of olivine (about 60% of the rock), orthopyroxene, clinopyroxene, and chromium-bearing spinel, but shows a wide range in bulk chemical composition.

The wehrlite series comprises dunite (olivine, with little or no clinopyroxene), wehrlite (olivine and clinopyroxene), and clinopyroxenite. Dark spinel may or may not be present. Orthopyroxene is typically absent in this series, though it may be present in small quantity.

The eclogite series is characterized by the association of magnesium-rich garnet and aluminum-rich pyroxene. The latter may be omphacite (aluminum- and sodium-bearing clinopyroxene) alone or omphacite and orthopyroxene. Some olivine and/or dark spinel may also be present. Some aluminum-rich pyroxenite may belong to this series.

All inclusions containing plagioclase as an essential constituent are included in the groups of gabbro and granulite, the former being distinguished from the latter by having typical igneous rock texture but no sign of recrystallization. The association olivine and calcic plagioclase is characteristic of gabbro and is unstable in the P-T condition of the granulite formation [*Kushiro and Yoder*, 1966]. Granulite occurring as inclusions appears to be largely metamorphosed gabbro. In Hawaii and Japan, some gabbro inclusions appear to be genetically related to the wehrlite series, as judged from their mineralogical gradation and field association.

There are also rare types: harzburgite (olivine and orthopyroxene), orthopyroxenite, and garnet-bearing lherzolite and harzburgite. The garnet-bearing types are found in the kimberlite pipes in South Africa and also in Itinomegata crater, Japan [*Kuno*, 1967].

A small amount of plagioclase may be present in some rocks of the lherzolite and wehrlite series. Common accessory minerals in the lherzolite and eclogite series rocks are phlogopitic biotite and pargasitic hornblende [*Kuno*, 1967].

The lherzolite and eclogite series can be distinguished from the wehrlite series by the clinopyroxene composition, which is lower in CaO and higher in Na_2O + K_2O for lherzolite and eclogite than for the wehrlite series [*Kuno*, 1968] (Figure 1).

CHEMISTRY

In Figure 2, oxide percentages in the bulk compositions of the lherzolite and eclogite series inclusions are plotted against $MgO/\Sigma FeO$. This ratio is the most reliable measure for the stage of fractional crystallization of magmas and the degree of partial melting.

With the decrease of $MgO/\Sigma FeO$ in the lherzolite series, Al_2O_3, total FeO, CaO, Na_2O + K_2O, TiO_2, and MnO increase, whereas MgO decreases and Cr_2O_3 first increases and then decreases. In the eclogite series, Al_2O_3, total FeO, Na_2O + K_2O, TiO_2, and MnO increase, whereas SiO_2, MgO, CaO, and Cr_2O_3 decrease.

Distinct differences between the two series occur in the variation trends of SiO_2, total FeO, MgO, CaO, and Cr_2O_3. Thus they are genetically different from each other.

The garnet-bearing lherzolite of Japan has a composition well within the range of the lherzolite series. Garnet-bearing peridotites of South Africa plot on the higher $MgO/\Sigma FeO$ side of the lherzolite series (Figure 2). They show a trend slightly different from that of the

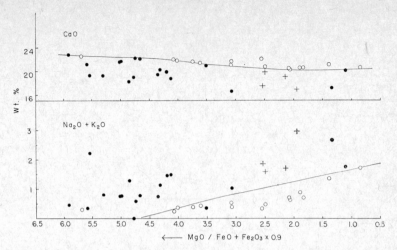

Fig. 1. CaO and $Na_2O + K_2O$ contents of clinopyroxenes: closed circles, lherzolite series inclusions from various localities; crosses, eclogite series inclusions from various localities; open circles, wehrlite series inclusions of Hawaii and Japan.

lherzolite series in MgO, $Na_2O + K_2O$, and TiO_2.

DEPTH RELATIONS OF THE SERIES AND THIER ORIGINS

The compressional wave velocity V_p of the upper mantle is generally 8.0–8.2 km/sec. This corresponds to the values for peridotite, pyroxenite, and eclogite; gabbro and granulite have lower V_p, and therefore cannot be considered as a major constituent of the mantle.

The lherzolite series inclusions are widespread and abundant throughout the world. Therefore, they are assumed to be the main constituent of the uppermost part of the mantle. In Salt Lake, lherzolite and eclogite occur in contact with each other in the same inclusions, and therefore are present at the same depth. This eclogite is interpreted as having been derived from a depth between 40 and 60 km [Green, 1966; Kuno, 1968]. The garnet-bearing peridotites of South Africa appear to have been derived from a depth between 100 and 150 km, as judged from the compositions of pyroxenes [Boyd and England, 1964; Davis and Boyd, 1966]. This rock type would extend down to the level where the olivine-spinel transition takes place (about 400 km).

The lherzolite layer probably extends to the Moho in the oceanic regions and also in most parts of the continental regions. However, where the thermal gradient in the crust is high, as it is in some orogenic belts, the uppermost part of the lherzolite layer may be subjected to phase transition to form plagioclase-bearing lherzolite, resulting in a comparatively low velocity ($V_p = 7.7$ km/sec) just below the Moho, as is observed in Japan [Kuno, 1967].

Pyroxenes in lherzolite usually contain exsolution lamellae. Some lherzolite shows lineation due to arrangement of spinel crystals in parallel strings and also to preferred crystal lattice orientation of olivine. These features imply that the lherzolite was originally formed at high temperatures and was cooled slowly to lower temperatures corresponding to those on the geotherm of the individual regions, and also that it has been subjected to deformation and recrystallization before it was captured by the magma.

Oxide percentages of the lherzolite series of the Horoman intrusion, Japan, which was most probably formed by crystal accumulation in some basaltic magma, are plotted in Figure 3 (the closed curves are reproduced from Figure 2). The points for the Horoman lherzolite series fall within the curves in Figure 2, indicating the variation trend of the inclusion lherzolite series and garnet peridotite. This fact would indicate that the latter was also formed by crystal accumulation. The compositions of the magmas which precipitated the lherzolite series

510

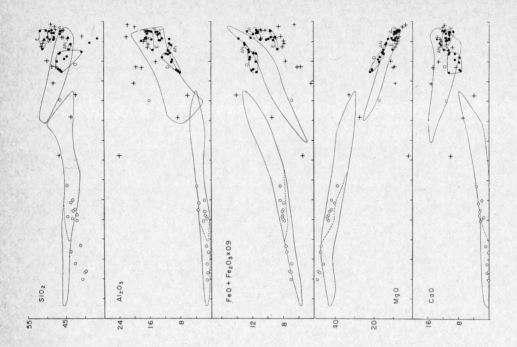

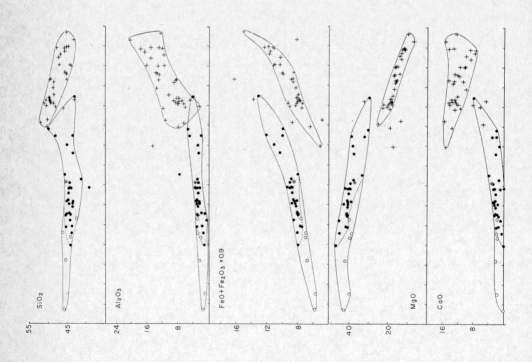

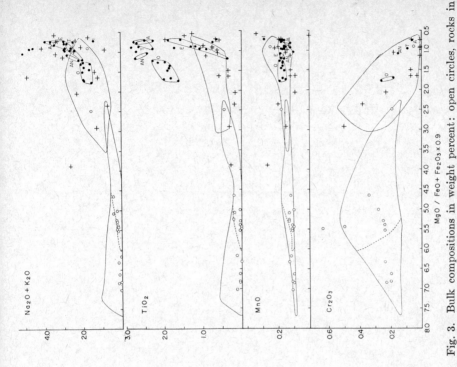

Fig. 2. Bulk compositions in weight percent: closed circles, lherzolite series inclusions from various localities; crosses, eclogite series inclusions from various localities; open circle, garnet-bearing peridotite inclusions in kimberlite pipes of South Africa; circled dot, hypothetical primordial mantle material 'pyrolite' (see text). The general ranges of variation are shown by closed curves. (Several unpublished analyses of South African eclogites by K. Aoki are included.)

Fig. 3. Bulk compositions in weight percent: open circles, rocks in different layers of the layered Horoman lherzolite intrusion, Hokkaido, Japan; crosses, gabbro and granulite inclusions from various localities; closed circles, some representative basaltic lavas; O, oceanite of Hawaii; AN, ankaramite of Hawaii; T, olivine tholeiite of Hawaii; A, average alkali olivine basalt of Hawaii; I, primary tholeiite of the Izu-Hakone region; E, average basalts of different oceanic basins [Engel and Engel, 1964]. Other unnamed points are chilled margins of tholeiite and high-alumina basalt intrusions. Closed curves are reproduced from Figure 2.

can be estimated on the assumption that they lie on the extension of the average trend lines for the oxides of the series on the lower $MgO/\Sigma FeO$ side. The estimated compositions of such magmas are quite unlike those of any of the known basaltic lavas [*Kuno*, 1968]. The compositions of some representative basaltic lavas of the world are plotted in Figure 3. It can be seen that the points for basalt do not lie near the extensions of the average trend lines for Al_2O_3, total FeO, and MgO.

This is not the only solution of the problem, however; the compositions of the magmas do not necessarily lie on the extensions of the trend lines, and the series may have been formed from some known basaltic magmas.

It is also possible that the lherzolite series variation was produced by different degrees of partial melting of a parent rock that may be represented by one of the low $MgO/\Sigma FeO$ members of the series. However, some difficulty still remains in correlating the estimated compositions of the magmas that would have been produced by partial melting with those of known lavas.

Whatever the origin of the variation may be, the low $MgO/\Sigma FeO$ members of the series may be a potential source of common basaltic magmas, provided that the amounts of the magmas produced are small.

The composition of the hypothetical primordial mantle material 'pyrolite' (see Table 1, p. 7) is plotted in Figure 2. It corresponds to the middle member of the lherzolite series, although its $Na_2O + K_2O$ and TiO_2 contents are higher. In calculating the 'pyrolite' composition, a Hawaiian olivine tholeiite, which is higher in $Na_2O + K_2O$ and TiO_2 content than most other tholeiites, was used. Accordingly, the pyrolite composition could be modified so as to fit the series completely by using the average tholeiite composition.

The close correspondence of the pyrolite composition with that of the lherzolite series is in harmony with the assumption that the lherzolite and its high-pressure modification constitute the main part of the upper mantle.

The wehrlite series rocks of Hawaii and Japan show distinct banding of olivine-rich and clinopyroxene-rich layers. This feature, taken in conjunction with their igneous rock texture, indicates that the series was formed by crystal accumulation, but not by different degrees of partial melting. The dissimilarity of the clinopyroxene compositions (compared to those of the lherzolite and eclogite series, Figure 1) suggests shallower depth of formation. The Hawaiian wehrlite series is interpreted as representing lower parts of solidified magma reservoirs in the cores of the volcanoes [*Kuno*, 1968]. Such reservoirs possibly extend from the crust to the mantle. The wehrlite series of other regions may also be relics of magma reservoirs related to past volcanic activity.

Compositions of gabbro and granulite inclusions associated with the peridotite and eclogite inclusions in Hawaii, Japan, and Australia are plotted in Figure 3. The points show much more scatter than the eclogite points; therefore, the eclogite series cannot be interpreted as having been formed by phase transition of the gabbro and granulite owing to deep burial.

The mineralogy and chemistry of the Salt Lake eclogite series have been interpreted as representing successive stages of crystallization of some basaltic magma, possibly oceanite (olivine-rich tholeiite) [*Kuno*, 1968]. The eclogite series of Australia and South Africa differ little in chemistry from the Salt Lake series, and therefore may have been formed by a similar process.

In view of the limited occurrence of eclogite inclusions, this series constitutes local pockets within the mass of lherzolite and garnet-bearing peridotite where magmas were trapped. The depths of such pockets would be between 40 and 60 km in Hawaii [*Green*, 1966; *Kuno*, 1968], and perhaps between 100 and 150 km in South Africa.

The inclusion study does not support the idea that the Moho discontinuity is a phase transition boundary between gabbro or granulite and eclogite.

The model of the upper mantle proposed by *Ringwood* [1966] is partly justified by the inclusion study. There is no indication of the presence of the hornblende-olivine rock or Ringwood's 'ampholite' at the top of the oceanic mantle. The 'dunite and peridotite' layer, which Ringwood believed to exist at the top of the continental mantle, may correspond to the high $MgO/\Sigma FeO$ members of the lherzolite series, if these were formed as residue of partial melting. If so, the low $MgO/\Sigma FeO$

members of the same series would occur preferentially in the oceanic mantle.

REFERENCES

Boyd, F. R., and J. L. England, The system enstatite-pyrope, *Carnegie Inst. Wash. Yr. Book 63*, 157–161, 1964.

Davis, B. T. C., and F. R. Boyd, The join $Mg_2Si_2O_6$–$CaMgSi_2O_6$ at 30 kilobars pressure and its application to pyroxenes from kimberlites, *J. Geophys. Res., 71*, 3567–3576, 1966.

Engel, A. E. J., and C. G. Engel, Igneous rocks of the East Pacific rise, *Science, 146* (3643), 477–485, 1964.

Forbes, R. B., and H. Kuno, Peridotite inclusions and basaltic host rocks, in *Upper Mantle Symposium, New Delhi, 1964*, edited by C. H. Smith and Theodor Sorgenfrei, pp. 161–179, Berlingske Bogtrykkeri, Copenhagen, 1965.

Green, D. H., The Origin of the 'eclogites' from Salt Lake crater, Hawaii, *Earth Planetary Sci. Letters, 1*, 414–420, 1966.

Kuno, H., Mafic and ultramafic nodules from Itinome-gata, Japan, in *Ultramafic and Related Rocks*, edited by P. J. Wyllie, pp. 337–341, John Wiley & Sons, New York, 1967.

Kuno, H., Mafic and ultramafic nodules in basaltic rocks of Hawaii, *Geol. Soc. Am., Mem. 115*, 1968.

Kushiro, I., and H. S. Yoder, Jr., Anorthite-forsterite and anorthite-enstatite reactions and their bearing on the basalt-eclogite transformation, *J. Petrol., 7*, 337–362, 1966.

Ringwood, A. E., Mineralogy of the mantle, *Advances in Earth Science*, edited by P. M. Hurley, pp. 357–399, MIT Press, Cambridge, 1966.

Ross, C. S., M. D. Foster, and A. T. Myers, Origin of dunites and olivine-rich inclusions in basaltic rocks, *Am. Mineral., 39*, 693–737, 1954.

White, R. W., Ultramafic inclusions in basaltic rocks from Hawaii, *Contrib. Mineral. Petrol., 12*, 245–314, 1966.

Petrology of the Precambrian Basement Complex

K. R. Mehnert

INVESTIGATIONS of the upper mantle are of considerable importance for the interpretation of most endogenetic processes of the upper crust. Therefore the intermediate zones between the mantle and the crust attain special significance in all attempts to establish genetic correlations.

According to present knowledge, the lowest parts of the crust are nowhere exposed at the surface in coherent geological units. Only relatively small portions that might have originated under the PT conditions of the lower crust can be locally observed, and it might be assumed that they have been dragged from below into the upper regions by geotectonic processes. They can thus be utilized as models of processes occurring regionally in the deeper parts of the crust.

Conclusions about the composition and structure of the lower crust hence have to be drawn from extrapolation of the processes observed in the deepest parts of the upper crust. Therefore the study of the Precambrian basement and its high-grade transformation processes can be regarded as one of the keys for the investigation of the relations between mantle and crust.

GENERAL PRINCIPLES

Holmes [*Rankama*, 1963] drew attention to the many misinterpretations of geological correlations based simply on petrographic methods. Only the introduction of 'absolute' geochronological methods has made it possible to correlate the rather complex age relations in a reasonable manner.

The Precambrian embraces some 3000 million years, the Cambrian and post-Cambrian about

600 million years. Since the latter constitutes the 'proper' geological history of the earth, it is not certain the extent to which geological processes of the post-Cambrian can be extrapolated backward into the relatively long period of Precambrian time.

Furthermore, petrogenetic processes of the Precambrian regions are often unrecognizably superimposed on one another (telescoped). Hence it is rather difficult to separate the individual stages in such polygenetic and often polymetamorphic rock systems. By normal petrographic methods, only the sequence of events can be reconstructed; far less can be said about the rhythm, i.e., the intervals of time between them.

In principle, the degree of metamorphism of a rock series is not strictly related to its sequence of age. In practice, however, this correlation is often still in use because of the lack of more precise arguments; it is justifiable to some extent if the older rocks have undergone more and stronger metamorphic alterations than the younger ones. But there are many instances, well documented by geochronological data, where old metamorphic rocks have indeed been remetamorphosed again, and have even been partially remobilized by younger orogenic cycles. According to the rule that continents grow from an oldest core by successive addition of younger shells, it can nevertheless reasonably be assumed that the oldest zones of the continents do indeed retain much of their original character.

It is important to know the extent to which chemical mobilization has modified the original composition of rocks, i.e., to what extent primary minerals, especially those used for geochronological determinations, are preserved. Investigations of some accessory minerals (zircon, etc.) have shown that essential parts of these minerals can survive rather intense mobilization.

PRECAMBRIAN BASEMENT

The term 'basement' implies a rock series that forms a base for overlying, younger rock systems; as a rule, an unconformity separates the two series. However, this relatively simple description frequently proves to be an oversimplification. A neat separation of the basement and the overlying sediments is usually restricted to regions of the platform type; in the orogenic belts, both rock systems are involved in folds, nappes, etc., both exhibiting a more or less metamorphic or even granitized state. Hence the term basement can here no longer be applied in its original sense, since older and younger rocks have undergone the same metamorphic transformations.

For the oldest orogenic cycles of the crust, the contrast in petrographic character, folding style, etc., between basement and cover as a rule becomes more and more doubtful. Former discontinuities are often blurred by later deformation, or they disappear between magmatic intrusions of various ages. Strongly deformed old rock systems often show subparallel alignment of their heterogeneous rock assemblages, whose primary character and origin is not easily detected (banded gneisses). Their mainly tectonic origin, i.e. by deformation of primary heterogeneities of any kind, can often be confirmed by the observation of local isoclinal folds of the rock layers involved. However, through metamorphic differentiation, or even more extensive migration of mobile components, the banded structure can be enhanced, but it can also be obliterated. Finally, owing to partial anatectic phenomena or feldspar blastesis, any correlation of primary lithological or stratigraphical units becomes practically impossible (migmatic rock complexes).

Thus attempts at a stratigraphical correlation of Precambrian rock series across large areas is generally fraught with uncertainty. Even lithological differences and correlations are uncertain: the best known example of this difficulty is the distinction of the Svecofennidic and Karelidic rock systems in Finland, which in spite of their considerable lithological differences yielded roughly the same geochronological data. Presumably the Karelidic zone contains older components, the absolute age of which was 'foreshortened' by regional mobilization.

Summaries of recent geological and comprehensive geochronological investigations of the oldest rocks series in the Precambrian shields include: *general reviews*—Holmes [1964], *Ahrens* [1955], *Rankama* [1963], *Lotze and Schmidt* [1966]; *regional reviews*—Polkanov and Gerling [1960], *Vinogradov et al.* [1960] for the USSR; *Simonen* [1960] for Finland; *Magnusson* [1965] for Sweden; *Gastil* [1960] and

Engel [1963] for North America; Nicolaysen [1962] for South Africa; Aswathanarayana [1964] for India; Wilson [1958] for Australia. A great many special papers have been submitted to the Proceedings of the International Geological Congresses (sect. 1, fasc. 1, pp. 1–154, 1952; sect. 9, part 9, pp. 1–206, 1960; sect. 10, vol. of abstracts, pp. 143–167, 1964).

SEDIMENTARY ROCK TYPES OF PRECAMBRIAN SHIELDS

It is not easy to determine exactly the amount of sediments or even the proportions of different sedimentary rock types in the Precambrian shields because of their present, partially metamorphic state. The original proportion of rock types has changed, since some rocks are more susceptible to metamorphic alteration than others. Thus, in a rock series of mixed sedimentary-metamorphic origin, some rock types may appear to be systematically missing, because they present themselves in a considerably altered metamorphic state; they may even have been selectively changed in composition by metasomatizing fluids.

Sedimentary rocks of extreme composition are mostly rather easy to recognize, i.e., quartzites or carbonates. Pure quartzites are comparatively resistant to metamorphism and anatexis. For this reason, quartzites often predominate in mixed sedimentary-metamorphic regions. 'Impure' quartzites, however, containing feldspar, micas, and other minerals, in general respond easily to metamorphism. Arkoses and graywackes obviously react even more readily.

The widely distributed biotite- or biotite-hornblende gneisses of intermediate composition probably originated from the latter rock group. Especially the fine-grained, rather homogeneous gneisses have presumably retained their mineral composition, including their high feldspar content, from such parent rocks. This can be confirmed by the rhythmic alternation of the anorthite content of plagioclase, or of the plagioclase-orthoclase ratio, especially if it varies according to differences in grain size or other petrographic details obviously corresponding to the primary stratigraphic character of the rock.

Pelitic rocks, now present in the form of phyllites, mica-schists or Mg-Fe-Al-rich gneisses, often form layers alternating with rocks of former psammitic composition. On the whole, however, meta-pelites are less common in the oldest sedimentary rock systems of the crust compared with metapsammites. Moreover, it should be pointed out that very similar or nearly identical rocks can originate by metamorphic processes, i.e. by depletion of feldspar. On the one hand, the reaction potash feldspar → muscovite can engender an enrichment of the mica content. On the other, a depletion of feldspar is possible through partial mobilization, whereby mica and other Mg-Al-rich mafic components are again enriched (for example, cordierite, garnet, sillimanite). Consequently, in rocks undergoing strong metamorphism, the original components of metapelitic origin are always liable to be overestimated.

Coarsely clastic rock types of metapsephitic origin are also relatively rare. Of special interest are conglomeratic layers, since they permit conclusions regarding the composition of the pre-conglomeratic land surface. They often yield much information about older magmatic activity, which is only preserved as relics in the pebbles of the conglomerate and otherwise not recognizable in that petrographic state. Sometimes such occurrences serve as the only indication of an older orogenic cycle. In addition, metaconglomerates generally permit the study of the influence of metamorphism on the different rock types included as pebbles in the conglomerate. This study yields a synoptic summary of the metamorphic conditions within a very restricted space.

Limestones are rather infrequent as a rule in the Precambrian basement. They occur as relatively rare intercalations and are generally rather strongly deformed with respect to the country rock by tectonic processes. Primary, undisturbed contacts are for this reason uncommon. Pronounced mineralogical changes mostly take place at the margins, i.e. a zoned formation of Ca-rich to Ca-poor silicates from the core toward the rim (grossularite, hessonite, or andradite followed by diopside and Ca- or Al-rich hornblende). Among carbonates, dolomites predominate in certain Precambrian regions; they are much rarer in post-Cambrian regions. This is a primary and very distinctive feature of the Precambrian era, for which no generally accepted explanation can as yet be given. Presumably the predominance of Mg with respect to Ca has to be taken as a peculiarity of the Precambrian sediments or of their diagenetic alteration products connected with the erosion of very old Mg-rich magmatic rocks.

Another remarkable feature of sedimentary rocks of the Precambrian basement is the widespread occurrence of iron-rich quartzites and jaspilites (banded iron ores). These constitute sequences of SiO_2-rich and Fe-rich layers without any normal clastic or heavy mineral component. This points to a predominantly chemical origin, probably due to weathering solutions. In some instances, a supply by volcanic volatiles is probable, and in others, a purely sedimentary precipitation without the action of volatiles is more likely. The restriction of these enormous deposits of iron ore to the Precambrian era is supposed by some authors to be due to the lack of land vegetation and a correspondingly different composition of the weathering solutions. Nevertheless, it is probable that the Precambrian era was generally characterized by low-temperature volcanic exhalations or migrating solutions of volcanic origin.

An especially interesting group of sediments consists of the so-called 'black shales.' They were formed from pelitic bituminous sediments belonging to a more or less pronounced sapropel facies. These sediments are often relatively rich in metals, such as gold, copper, lead, zinc, and others, which probably have likewise been introduced into the sediment by volcanic exhalations. These ore-bearing black shales sometimes occur as educts (parent rocks) of metamorphic or even anatectic rock series. Here their ore content can be mobilized and re-concentrated as pseudo-hydrothermal ore deposits or as infiltrations of the country rock. These complex 'polymetallic' ore deposits are economically very important and constitute a type of deposit essentially characteristic of the Precambrian.

METAMORPHIC ROCK TYPES OF PRECAMBRIAN SHIELDS

Another typical phenomenon of the Precambrian basement is the occurrence of metamorphic rocks, even of high-grade transformation, side by side with rocks exhibiting sedimentary relict features. This close association leads to the assumption that strong tectonic processes are responsible for their present position. Quite evidently rather steep geothermal gradients existed in relatively restricted domains. Since closely spaced intercalations of this kind occur repeatedly in the Precambrian shields, the *average* temperature gradient over large distances is rather uniform. Hence metamorphic zoning, which is so characteristic of younger orogenic zones, is generally less distinct in Precambrian shields.

All mineral facies from the greenschist to the granulite facies can thus occur rather closely side by side. The main mineral facies, however, is generally represented by the amphibolite facies, which is remarkably widespread in Precambrian areas. With regard to subfacies, there is perhaps a preponderance of temperature-dominant facies, while those dominated by pressure are generally less widespread.

High-grade mineral facies, like the granulite and charnockite facies, only occur in some regions of the Precambrian shields, where PT conditions characteristic of rather deep zones of the crust prevailed locally, or even regionally. However, these high-grade facies on the whole are not so intensely developed as one might suppose in areas that belong to the most thoroughly eroded zones of the continents. Hence it can be concluded that the oldest cores of the continents do not necessarily represent the regions of highest PT conditions.

The origin of granulites and related rocks is still a matter for debate. Evidently they are formed under high temperature and high pressure conditions and probably at a relatively low concentration of H_2O, i.e. under pressure-temperature-composition (PTX) conditions that according to present knowledge, are liable to prevail in the deepest parts of the crust. The parent rocks can be of varying origin, magmatic as well as sedimentary. Normal granulites are rocks rich in quartz and feldspar, but 'basic' members rich in orthopyroxene and garnet also exist. The latter closely correspond to rocks of the charnockite series and are regarded as the metamorphic products of gabbroidal rocks or hornfels-like sediments. 'Acid' charnockites on the whole have a magmatic character and often exhibit intrusive phenomena.

This group of rocks is characterized by a 'dehydration' process (OH-bearing → OH-free silicates). *Scheumann* [1961] therefore refers to this process of 'granulitization' as being typical for the deepest portions of the Precambrian basement. Geochemically it provides the counterpart to the 'granitization' of higher levels. In contrast to granulitization, granitization supplies H_2O, including alkalis and other substances dissolved therein; it is characterized by the reaction OH-free → OH-bearing silicates. For instance, garnet is transformed into biotite by granitization. In connection with this process, an enrichment of feldspar usually occurs (generally orthoclase or microcline). Processes of this kind are termed 'feldspathization' and belong to the high-temperature region of the amphibolite facies. It is likely that the alkali-rich fluids infiltrating the upper crust from below originated in the lower parts of the crust, where such fluids are mobilized by the process of granulitization.

In the high-temperature region of the amphibolite facies, rocks of intermediate to acidic composition exhibit the first signs of incipient melting. It has been established experimentally that melting begins at about 650°C at 4 kb H_2O pressure. This corresponds to a depth of about 12–15 km, assuming a normal geothermal gradient. However, within orogenic belts the gradient is much steeper than normal. Thus anatectic melts of leucogranitic or pegmatitic composition can be formed in the innermost zones of orogenic regions even at relatively high levels, as can be shown by supercrustal rocks deposited shortly after the main orogenic period. The H_2O content, which on the whole is relatively small in regional-metamorphic terrains, can evidently be concentrated by migration and form mobile solutions after the crystallization of a hydrous anatectic melt.

It is a widely accepted working hypothesis that most granite plutons now exposed at the surface once originated in that way (palingenetic origin of granites). Only rarely, however, has this mode of origin been in fact proven by investigation of all mobilization stages between the parent rock and its palingenetic magma.

MAGMATIC ROCK TYPES OF PRECAMBRIAN SHIELDS

The most common group of magmatic rocks of the Precambrian shields is composed of granites and their related rock family (granodiorites, biotite-quartz diorites, trondhjemites, quartz monzonites, etc.). The formation of this 'granitoid' rock group is quantitatively still unsettled. It can however, be regarded as certain that only a very limited part of the 'granitic layer' of the crust originated by direct gravitative differentiation from rocks of the upper mantle. Even the oldest parts of the continental core zones contained extensive masses of granitic rocks. Their formation must therefore be derived from processes that occurred during the pregeological epochs of the earth's evolution.

It has been noted above that the oldest rocks very frequently consist of sedimentary and metamorphic rocks of a composition that on the average is similar to granite (arkoses, graywackes, and their metamorphic products). These sediments are often intercalated with layers of volcanic origin, e.g. quartz-porphyries, tuffs, tuffites, and ignimbrites. Hence the oldest rock series are in their main part characterized by a volcanic-sedimentary mixed series of acid to intermediate rocks, including the corresponding ore deposits [*Mehnert*, 1968].

Another part of the Precambrian magmatic rock series consists of basic and ultrabasic rocks, generally in the form of dikes or sills, stocks or masses of small to medium size, or as lavas, agglomerates, and tuffs. The petrographic character distinctly corresponds to magmas derived from the upper mantle, i.e. the primary composition is mainly tholeiitic (to peridotitic). Transitions to intermediate and acid rocks do exist. Differentiations to alkali basalt, so characteristic in post-Cambrian rocks, are but rarely found in the Precambrian.

Structure and texture of these basic magmatites correspond to the various stages of their cooling history, depending on the depth of emplacement. Frequently, however, these relict features are more or less obliterated by metamorphic transformations. As a rule, it is possible to establish complete sequences of age relations of basic dikes, whereby generally the older rocks exhibit an advanced stage of metamorphism as against the younger ones.

In other cases the origin of basic metamorphic rocks is not so certain. Basic to intermediate metamorphic rocks can originate both from magmatic and sedimentary parent rocks. In this case the trace elements can on occasion afford arguments as to their origin. Pt, Cr, Co, Ni, Ti, and related elements are generally characteristic of magmatic parent rocks, whereas higher contents of Ca, Al, and Si, especially if they occur in a rhythmic layered sequence, are arguments in favor of a sedimentary origin.

Of special importance in the Precambrian are large massifs of basic magmatic rocks forming layered masses of apparently dishlike shape (lopoliths). Well-known examples are the massifs of Sudbury, the Bushveld, the Duluth massif, and the Stillwater complex. Typically they consist of a main mass of norite or gabbro exhibiting pronounced layering of basic or ultrabasic masses and often ore-bearing zones. The very considerable content of Fe, Cr, Ni, Pt, and related elements rather points to an origin from the mantle. The nature of their emplacement in the upper crust is still uncertain.

Comparison with post-Cambrian regions strongly suggests that the activity of basic magmatism was stronger in Precambrian times. The frequent occurrence of basic dikes is particularly characteristic of the Precambrian basement. This and the predominance of volcanism in general point to the fact that the crust was thinner than today and hence more easily pierced by fissures. Consequently the mobile rock material from the mantle constitutes an important part of the crustal rock series of the Precambrian.

It can safely be assumed that these phenomena become more and more pronounced with increasing depth of the crust. Possibly the greater density of the lower crust, as indicated by the greater velocity of seismic waves, can be interpreted as due to the emplacement of basic magmatic rocks. A simple in situ differentiation between the mantle and the crust is highly unlikely, since the upper mantle is certainly in the solid state at the Mohorovicic discontinuity. The intermediate layers between the crust and the mantle were therefore probably formed by successive intrusion phenomena of mobile mantle material into the primordial crust.

The oldest crust of the earth presumably had an acid to intermediate composition at a very early stage. The oldest, probably palingenetic magmatic rocks have a granitic or granodioritic to biotite-quartz dioritic composition. This leads to the assumption that they originated from an even older rock series of mixed volcanic-sedimentary character, probably formed in pregeological times. Since these rocks are more or less in a polymetamorphic and partially anatectic state, little can be said about the geochemical and petrogenetic evolution of the crust in those pregeological times.

REFERENCES

Ahrens, L. H., Oldest rocks exposed, *Geol. Soc. Am. Spec. Paper, 62,* 155–168, 1955.

Aswathanarayana, U., Age determination of rocks and geochronology of India, *Intern. Geol. Congr., India, Spec. Publ.,* 1–24, 1964.

Engel, A. E. J., Geologic evolution of North America, *Science, 140,* 143–152, 1963.

Gastil, G., The distribution of mineral dates in time and space, *Am. J. Sci., 258,* 1–35, 1960.

Holmes, A., The oldest known minerals and rocks, *Edinburgh Geol. Soc. Trans., 14,* 176–194, 1948.

Lotze, F., and K. Schmidt, *Präkambrium,* Enke, Stuttgart, 388 pp., 1966.

Magnusson, N. H., The Pre-Cambrian history of Sweden, *Quart. J. Geol. Soc. London, 121,* 1–30, 1965.

Mehnert, K. R., *Migmatites and the Origin of Granitic Rocks,* Elsevier Publishing Company, Amsterdam, 393 pp., 1968.

Nicolaysen, L. O., Stratigraphic interpretation of age measurements in Southern Africa, in *Petrologic Studies: A Volume to Honor A. F. Buddington,* edited by A. E. J. Engel et al., pp. 569–598, Geological Society of America, New York, 1962.

Polkanov, A., and E. Gerling, Use of the K-A and Sr-Rb methods in determining the age of the Precambrian rocks of the Baltic shield, *Trans. Lab. Precambrian Geol., Acad. Sci. USSR, 9,* 7–41, 1960.

Rankama, K., *The Precambrian,* Interscience Publishers, New York, 279 pp., 1963.

Scheumann, K. H., Ueber die Genesis des saechsischen Granulits, *Neues Jahrb. Mineral. Abhandl., 96,* 162–171, 1961.

Simonen, A., Pre-Quaternary rocks in Finland, *Bull. Comm. Geol. Finlande, 191,* 1–49, 1960.

Vinogradov, A. P., L. V. Komlev, S. I. Danilevich, V. G. Savonenko, A. I. Tugarinov, and M. S. Filippov, Absolute geochronology of the Ukrainian Precambrian, *Intern. Geol. Congr. Norden, 9,* 116–132, Berlingske Bogtrykkeri, Copenhagen, 1960.

Wilson, A. F., Advances in the knowledge of the structure and petrology of the Pre-Cambrian rocks of southwestern Australia, *Roy. Soc. Western Australia J., 41,* 57–83, 1958.

Age Relationships of the Precambrian Basement Rock Complex

Hisashi Kuno

THE Precambrian era is about 4 billion years long and therefore represents nearly 90% of the life of the earth. Because of the strong deformation and metamorphism to which most of the Precambrian rocks have been subjected and also because of the absence of fossils by which stratigraphic correlation can be made, the history of this long time interval is little known. However, this situation is now being greatly changed by the recent development of the technique of absolute age determination.

It is probable that, during this period, the accumulation of sediments and volcanic materials on the earth's surface was nearly continuous, though with local hiatus. However, age determinations give only the ages of orogenesis during which the rocks were metamorphosed and granite intrusion took place. Several separate ages of orogenesis have been recognized in different continents.

It is most important to construct a map for each continent showing the distribution of rocks of different ages in the Precambrian. This kind of map was recently published for the middle to the eastern part of the United States; the Precambrian rocks under the younger sediments were obtained from a great number of wells [*Goldich et al.*, 1966]. The map clearly indicates that the continent has grown from the core to the margin. If similar maps are constructed for other continents, they will provide an important clue to the origin and growth of the continents. Also, a comparison of the distribution patterns of rocks of different ages in separate continents would either support or refute the continental drift theory.

REFERENCE

Goldich, S. S., W. R. Muehlberger, E. G. Lidiak, and C. E. Hedge, Geochronology of the midcontinent region, United States, 1, Scope, methods, and principles, *J. Geophys. Res., 71,* 5375–5388, 1966.

Metamorphism and Its Relation to Depth

Akiho Miyashiro

REGIONAL METAMORPHISM AND METAMORPHIC BELTS

Metamorphism refers to the mineralogical and structural changes of solid rocks due to the conditions (especially temperature and pressure) differing from those under which the rocks originated. In current usage, changes on and near the surface of the earth, such as weathering and diagenesis, are excluded. We can classify metamorphism into a number of categories on the basis of geological causes or other factors. Geological and geophysical importance differs greatly in different categories.

Contact metamorphism is caused by the thermal effect of an intrusive igneous body on the surrounding rocks. An increase in temperature causes recrystallization in essentially solid rocks. Regional metamorphism, the most important type, is caused by the thermal effects and pressures during orogeny.

In early stages of orogeny, a thick pile of sedimentary rocks is accumulated in a geosyncline. The pile is not only folded and faulted, but also heated so that metamorphic recrystallization takes place. Individual parts of the geosynclinal pile are transformed to mineral assemblages corresponding to varying temperature and pressure within the pile. In this process, the soft pile of geosynclinal sediments is transformed into a hard crystalline mass.

In petrographic nomenclature, regional metamorphic rocks are mainly schists and gneisses derived from pelitic and psammitic sediments. When igneous rocks are intercalated in the sedimentary pile, they can also be metamorphosed. Such a complex of regional metamorphic rocks is exposed in the dissected axial zone of an old orogenic belt. The temperature and pressure of metamorphic recrystallization usually appear to increase toward the central line of the zone. Such zones of regional metamorphic rocks, frequently more than 1000 km long, are called metamorphic belts.

Many, but not all, metamorphic belts are accompanied by abundant igneous rocks, granitic and basaltic. Basaltic volcanism usually takes place in the geosynclinal stage, whereas granite emplacement usually takes place during or after the main phase of regional metamorphism. The cores of continents are usually composed of complexes of such regional metamorphic and granitic rocks. Thus, regional metamorphism can be regarded as a continent-forming process.

HIGH AND LOW P/T TYPES OF REGIONAL METAMORPHISM

The temperature T and pressure P would normally increase downward within the deformed geosynclinal pile. The value of dP/dT, however, appears to differ greatly in different cases. Hence, there appears to exist a great diversity in the P/T ratio of regional metamorphism. This has become clearly recognized only recently [*Miyashiro*, 1961].

Low P/T type of regional metamorphism produces andalusite in rocks of appropriate compositions (Figure 1). The metamorphic belts of this type are usually accompanied by very abundant granitic rocks. Examples include: the Ryoke-Abukuma metamorphic belt of Japan; the Paleozoic belts in southeastern Australia; and the Precambrian Svecofennides of Finland.

High P/T type of regional metamorphism produces glaucophane, lawsonite, and the jadeite-quartz assemblage in rocks of appropriate chemical compositions. It is sometimes called glaucophanitic metamorphism. The metamorphic belts of this type are nearly or completely devoid of granitic rocks, whereas the geosynclinal pile concerned usually contains abundant basaltic volcanics. Serpentinite is usually abundant. Typical examples occur mostly, though not exclusively, in younger orogenic belts. The Sanbagawa belt of Japan and the Franciscan metamorphic complex of California belong to this type.

Intermediate metamorphism is characterized by the occurrence of kyanite and the absence of glaucophane. Typical examples are in the Caledonides of Scotland and Norway and in the Appalachians of the northeastern United States. Petrological study of this type of regional metamorphism began toward the end of the nineteenth century, much earlier than the other two. Hence, this type was for some years incorrectly regarded as the only, or the normal kind of regional metamorphism.

RELATIVE MINERALOGICAL SCALES OF TEMPERATURE, ROCK PRESSURE, AND WATER PRESSURE

The relative values of temperature prevailing during metamorphism are revealed by the mineralogical investigation of resultant metamorphic rocks. Most metamorphic minerals contain some water (mostly as OH^-). With rising temperature, metamorphic reactions proceed to drive water away from minerals. Thus, in one metamorphic terrain, smaller water contents represent equilibria at higher temperatures, provided that all other conditions remain the same. Thus, we could draw approximate isotherms, usually called isograds, in metamorphic terrains.

The lower limit of the temperature of metamorphism is that of the beginning of recrystallization where a metamorphic terrain grades into an unrecrystallized region. In some cases, however, the metamorphic terrain grades into an area of intense diagenesis. In this case, the lower limit is only a matter of definition of

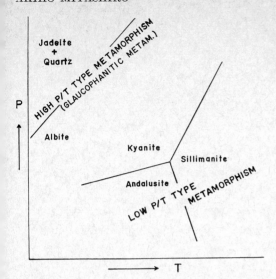

Fig. 1. Pressure and temperature in regional metamorphism and the stability relation of some metamorphic minerals.

the term metamorphism. The upper limit of the temperature of metamorphism is that of the beginning of melting, i.e., anatexis. Here, with rising temperature, metamorphism grades into the formation of anatectic rocks.

The relative values of pressure acting on the solid phase are more difficult to determine. With rising pressure the volume of the solid phases tends to decrease. Thus, rocks with higher densities tend to form at higher pressures.

It appears that water is more or less mobile in metamorphic terrains, but the magnitude of its mobility is highly debated. *Korzhinskii* [1959] and *Thompson* [1955, 1957] consider that water is often highly mobile and can be regarded as a perfectly mobile component in idealized theoretical treatment. On the other hand, *Turner and Verhoogen* [1960] and *Winkler* [1965] consider that the mobility is not so great. In the former case, the water pressure (fictive or real) is much lower than the pressure acting on the solid phases, whereas in the latter, the water pressure is similar to the pressure acting on the solid phases.

If the pressure acting on the solid phases is kept constant, and the water pressure is decreased, the temperature at which the water-bearing metamorphic minerals lose water should decrease. Hence, it is vital to know the relation between the solid pressures and water pressures. This is a difficult problem, however.

It is widely considered that CO_2 is also more or less mobile in metamorphic terrains, whereas O_2 is almost completely fixed.

NUMERICAL VALUES OF TEMPERATURE AND PRESSURES

In the past fifteen years, many laboratory studies have been made to clarify the quantitative stability relations of metamorphic minerals. For discussion of the important role of such experimental works on the study of metamorphism, readers are referred, for example, to *Winkler* [1965] or *Fyfe and Turner* [1966]. However, because the magnitude of water pressure during metamorphism is not clear, there is a limitation in the application of experimental results to natural rocks.

In this respect, the experimentally determined stability relations of anhydrous minerals are of special importance, because they depend only on the temperature and pressure acting on the solid phases, but not on the water pressure. We could use them for estimation of the temperature and pressure of metamorphism. Examples include the following equilibria:

jadeite + quartz = albite
jadeite = albite + nepheline
aragonite = calcite
andalusite = kyanite = sillimanite

Andalusite, kyanite, and sillimanite are polymorphs, having the composition Al_2SiO_5, and are widespread in a large variety of pelitic metamorphic rocks. Hence, the determination of their stability relations is a matter of cardinal importance for the whole framework of the temperature and pressure relations in metamorphic petrology. Practically all the recent writers agree that andalusite is stable at lower pressures, kyanite at higher pressures, and sillimanite at higher temperatures, and that there exists a triple point between their stability fields (Figure 1), but there is a large uncertainty about the numerical values of the triple point (Table 1).

If we provisionally accept the pressure value of *Althaus* [1967] for the triple point of Al_2SiO_5 (about 6.6 kb), it corresponds to the lithostatic pressure at a depth of about 25 km. Probably in more than half the regional

TABLE 1. Temperature and Pressure of the Triple Point of Al_2SiO_5

Reference	Temperature, °C	Pressure, kb
Bell, 1963	300	8.0
Khitarov et al., 1963	390	9.0
Newton, 1966	520	4.0
Weill, 1966	410	2.4
Holm and Kleppa, 1966	432	5.9
Althaus, 1967	595	6.6

Holm and Kleppa's value is based on calorimetric work, Weill's on solubility measurement, and all others on synthetic experiments.

metamorphic terrains, recrystallization occurred at depths greater than this. If we accept the stability relation of the jadeite-quartz assemblage as experimentally determined by Birch and LeComte [1960], the high P/T type of regional metamorphism occurred commonly at a pressure of 10 kb or more. This corresponds to the lithostatic pressure at a depth of about 37 km or more.

Thus, regional metamorphism appears to take place at various depths in the whole continental crust. The bulk of the continental crust may have been produced or modified in this way.

ENERGY AND MATERIALS FROM THE MANTLE

The regional metamorphism of a geosynclinal sedimentary pile is an endothermic process. A large amount of thermal energy is stored in the newly formed metamorphic minerals and is also used for expelling water and CO_2 from the minerals. The thermal energy must come from greater depths, mostly from the mantle.

As was mentioned above, abundant basaltic volcanic materials occur in geosynclinal piles, especially in the regional metamorphic terrains of the high P/T type. Such basaltic materials appear to have been derived from the partial melting of the mantle. Granitic rocks occupy a large fraction of the continental crust, and appear to be genetically related to regional metamorphism. Where does the granitic material come from? The direct origin of granitic magmas may lie in the melting of pre-existing rocks, for example, geosynclinal silicic sediments. Ultimately such silicic materials must have been derived from the mantle. The mantle is probably composed of some ultrabasic materials, undersaturated with SiO_2, whereas the granitic rocks of the crust are oversaturated with SiO_2. How can the oversaturated rocks originate from the undersaturated mantle?

The mantle can produce basaltic magmas by partial melting. Ordinary fractional crystallization of basaltic magmas could produce residual liquids oversaturated with SiO_2. The amount of such liquids, however, appears to be too small to account for the origin of the continental crust; Ringwood, in another section of this Monograph (p. 13), discusses a proposed hypothesis for the origin of calc-alkaline magma in the mantle.

REFERENCES

Althaus, E., Experimentelle Bestimmung des Stabilitätbereichs von Disthen (Cyanit), *Naturwiss., 54*, 42–43, 1967.

Bell, P., Aluminum silicate system: Experimental determination of the triple point, *Science, 139*, 1055–1056, 1963.

Birch, F., and P. LeComte, Temperature-pressure plane for albite composition, *Am. J. Sci., 258*, 209–217, 1960.

Fyfe, W. S., and F. J. Turner, Reappraisal of the metamorphic facies concept, *Contrib. Mineral. Petrol., 12*, 354–364, 1966.

Holm, J. L., and O. J. Kleppa, The thermodynamic properties of the aluminum silicates, *Am. Mineral., 51*, 1608–1662, 1966.

Khitarov, N. I., V. A. Pugin, C. Bin, and A. B. Slutskii, Relations between andalusite, kyanite and sillimanite in the field of moderate temperatures and pressures (in Russian), *Geokhimiya 1963*(3), 219–228, 1963.

Korzhinskii, D. S., *Physicochemical Basis of the Analysis of the Paragenesis of Minerals,* Consultants Bureau, New York, 1959.

Miyashiro, A., Evolution of metamorphic belts, *J. Petrol., 2*, 277–311, 1961.

Newton, R. C., Kyanite-andalusite equilibrium from 700° to 800°C, *Science, 153*, 170–172, 1966.

Thompson, J. B., Jr., The thermodynamic basis for the mineral facies concept, *Am. J. Sci., 253*, 65–103, 1955.

Thompson, J. B., Jr., The graphical analysis of mineral assemblages in pelitic schists, *Am. Mineral., 42*, 842–858, 1957.

Turner, F. J., and J. Verhoogen, *Igneous and Metamorphic Petrology,* 2nd edition, McGraw-Hill Book Company, New York, 1960.

Weill, D. F., Stability relations in the Al_2O_3-SiO_2 system calculated from solubilities in the Al_2O_3-SiO_2-Na_3AlF_6 system, *Geochim. Cosmochim. Acta, 30*, 223–237, 1966.

Winkler, H. G. F., *Petrogenesis of Metamorphic Rocks,* 1st edition, Springer-Verlag, Berlin, 1965.

7. Tectonics

Crustal Movements and Tectonic Structure of Continents

V. E. Khain and M. V. Muratov

MOVEMENTS resulting in deformation of the earth's crust are referred to as tectonic movements.

During geologic time, these movements have altered the distribution of land and sea, mountains and lowlands; they are also believed to have caused other changes in the inorganic life of the earth, though to a lesser degree.

Relative changes in the sea level and accurate leveling data give clear evidence that some areas of the crust are at present slowly rising, whereas others are slowly sinking. The rate of movement varies in the range from a few tenths of a millimeter to several tens of millimeters per year (see Figure 1, p. 380).

Areas subjected to a continuous secular subsidence during the past geologic time were mainly the sites of sea basins. The character of sedimentary rocks accumulated in the basins indicates that these basins were chiefly shallow. Their depth being rather constant, we can judge about the amplitude of warping only by thick sedimentary rocks accumulated in the basin; taking into consideration the time the sediments took to accumulate, we can also calculate the rate of warping.

At first accumulation of sediments in deeper basins, mainly within the mobile zones of the earth's crust, called geosynclines, lagged behind the rate of warping, but later the amount of fragmental product brought from the land and received by the basin increased. For this reason, in the long run, the thickness of the accumulated sediments (including products of volcanic activity) approximately corresponds to the degree of subsidence. If the rate of uplift of the crust exceeds the rate of surface denudation, relief develops. If elevation proceeds more slowly, relief remains of the lowland type. Deformations of surfaces that were originally flat and almost horizontal and that were generated, for example, in the coastal zone by the agencies of marine erosion, or in the mainland depths by a lasting denudation (marine and river terraces, 'denudation surfaces') allow us to estimate the amplitude and rate of elevation of the earth's crust. However, surfaces that have undergone denudation leveling before Jurassic are an exception in the present relief; our knowledge of the events that occurred during the older geologic epochs (200 m.y. ago and earlier) is restricted to the observations of only subsidences of the earth's crust (in the form of accumulated sedimentary rocks), the amplitude and rate of its uplifts being judged only by the character and volume of fragmental products removed from the uplift areas [Ronov, 1949a].

Thus, whereas present vertical movements of the earth's crust can be judged by the immediate geodetic changes, data on the comparatively recent (from the geological point of view) movements can be obtained by observing the relief of the earth's surface. The history of old vertical movements of the earth's crust occurring during previous geologic time can be reconstructed only by studying the distribution of thick sedimentary rocks of a certain period, by investigating their composition and character, which indicate the depth of the ancient basin and the distance between the given place and the shore, and by measuring their depths, which show the amplitude of tectonic warping of the earth's crust compensated by the accumulation of sediments (Figure 1).

Rates of sinking and rising calculated for past geologic time and based on the data of the thickness of sediments are invariably lower by 1 order (or even 2 or 3 orders) than those measured in modern geodetic surveying over a period of tens of years [Gzovsky, 1964]. Rates for long periods are measured in tenths or hundredths of a millimeter per year. Because vertical movements of the earth's crust often

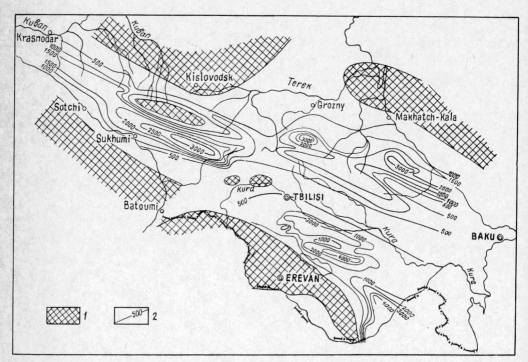

Fig. 1. Thicknesses of Dogger deposits on the Caucasus [after Kirillova et al., 1960]: (1) areas where deposits are absent; (2) isolines of thickness in meters.

change their rate and sign in time, i.e., they are of an oscillating character, with a short period; the algebraic sum of their movements for long periods is smaller than that for short periods. Nevertheless, the prevailing direction of vertical movements, judging by the Russian plain, remains the same over many tens and even hundreds of millions of years, i.e., the same regions are in general continuously subject (all the minor vibrations summed up) either to uplift or subsidence.

The fact that vertical movements maintain their direction for a long time is responsible for irreversible changes in the structure of the upper part of the earth's crust. The sedimentary layer in the areas of continuous uplift is either thinner than in other places, or else has been completely eroded, exposing the basement rocks. Examples include the so-called crystalline shields, the Baltic, the Canadian, and others, which have no sedimentary cover and expose on the surface thick masses of gneisses, crystalline schists, and granites.

We can distinguish zones or areas characterized by alternation of wide and gently sloping uplifts and depressions from zones characterized by an alternation of relatively narrow and extremely contrasting and intensive uplifts. In the former case the cross section of the uplifts and depressions, mainly of a round or irregular shape, can be as great as hundreds of kilometers, occasionally over one thousand kilometers, the amplitude of warping in the subsiding areas for several hundred million years being as little as a few kilometers. Development of this kind is typical of cratons or 'ancient' platforms, i.e., of areas which have been immobile since the middle or late Proterozoic. In the latter case the uplifts and depressions are of an elongated form. They are only tens of kilometers wide (their length being a few hundreds of kilometers), and the amplitude of vertical movements in the course of up to one hundred million years is as large as twelve to fifteen kilometers and more. This development is characteristic of mobile zones of the earth's crust referred to as geosynclines.

Gradual and wide uplifts and depressions on the platforms result in the so-called anteclises and syneclises, respectively. Narrow and contrasting uplifts and depressions in mobile zones are given different terms by various geologists. For example, if the whole mobile zone is referred to as a geosynclinal area or a geosynclinal system, individual uplifts or depressions within this area are called geanticlines or geosynclines, respectively. When the whole mobile zone is termed a geosyncline, individual uplifts and depressions are named intra-geanteclines and intra-geosynclines. Other terms are also in current use.

Where uplifts gradually change into depressions, and vice versa, standing waves develop in the earth's crust; hence the term undulatory or undulatory-oscillating movements is used to differentiate between these and other movements of the earth's crust. If, however, abrupt changes of the sign and rate of movement are recorded on the bordering zones between the uplifts and depressions, one can speak of 'block' or 'block-undulatory' movements [*Khain*, 1939, 1958; *Beloussov*, 1962].

East European and American scientists prefer the term epeirogenic for the tectonic movements of this type as a whole. The uplifts of more than average intensity resulting in the formation of a mountain relief are more often called orogenic movements [*Gilbert*, 1890; *Stille*, 1924]. This term, however, can be misleading, as it is also used in an interpretation which has become very popular (especially in German literature) to determine folding and fracturing as well (see, for example, *Sonder* [1956]). To avoid misunderstanding the term, 'orogenic movement' or 'orogenesis' should be restricted to its direct meaning for describing mountain formation. Orogenic movements are, consequently, a variety of undulatory or block (or epeirogenic) movements whose rate is much higher than that of denudation processes, and for this reason these movements always result in the formation of high relief.

Platforms are characterized by gradual change from uplifts to depressions and vice versa, whereas in mobile zones (geosynclines) some uplifts and depressions are bounded by faults. These may exist for a long geologic time and divide the earth's crust into blocks (with fixed boundaries) that move vertically relative to one another. The faults which divide sections moving at various rates and in different directions exist for a long time and are referred to as 'abyssal faults,' as they run through the whole of the earth's crust and penetrate to great depths [*Hobbs,* 1911; *Sonder,* 1938; *Peive,* 1945; *Shatsky,* 1945, 1947]. Some data suggest that the abyssal faults date back to the late Archean period of geologic history (about 2.5 b.y. ago) and that they were very significant at early stages of the crust formation. They become less important with a decreased mobility of the crust when geosynclinal conditions change to platform conditions. But at some periods, deep faults make themselves locally felt in the bordering zones dividing uplifts from depressions. This suggests that the whole of the earth's crust (and some of the upper mantle as well) is cut by deep faults and is of a block structure. This affects vertical movements of the earth's crust mainly in mobile zones, whereas in quiet platform areas deep faults remain temporarily healed and manifest themselves but little. They revive, however, owing to tectonic activization of platforms. Some geologists maintain that deep faults, or at least some of them, cannot be attributed to contrasting vertical movements of the earth's crust, but are an initial result of strains in the earth caused by some phenomena related to the planet as a whole. The resulting net of deep faults influences the arrangement of the uplift and depression areas of the earth's crust and governs their direction and shape. This point of view is supported by the evidence of some regularity in the arrangement of great tectonic zones on the earth's surface (Figure 2). For example, in the area from the Carpathians in the west to the Baikal Lake in the east, northwestern strikes of tectonic zones of various ages prevail. It was emphasized that northwestern and northeastern strikes of folded zones are more abundant among other similar zones of various age and among the boundaries between platforms and geosynclines, etc., whereas latitudinal and meridional strikes occur less frequently.

Along with undulatory and block movements of the crust wherein the prevailing movements keep to the same sign for a long time, geological records register simultaneous uplifts and depressions of structural components with both positive and negative lasting tendencies, as well as stable and mobile geosynclines with adjacent platforms. These movements are expressed by variations in the level of the ocean relative to land, manifesting themselves by general transgressions and regressions of the sea, by formation of sea and river terraces, and by sedimentary bedding. Such movements change their direction quite often and are referred to in Russian literature as 'general oscillations' or pulsations [*Khain,* 1939; *Beloussov,* 1962]. *Stille* [1924] suggested the term 'undation,' instead of undulation, which he used to distinguish movements classified by us as undulatory or block-forming and by *von Bubnoff* [1954] as dictyogenesis. The European authors, however, generally do not discriminate between undulatory or block movements and general ones.

General oscillations occur in a periodic manner, which clearly manifests itself by a rhythmic (cyclic) structure of some sedimentary formations. Flysch formation is characterized by short-period rhythms, in which sands, clays, marls, and sometimes limestone alternate, lasting for about 0.5 thousand years; these rhythms make up longer cycles of the order of 1–2 and 8 thousand years. Rhythms of salt-bearing rock masses indicate periods of oscillations of the order of 25–35 thousand years, and rhythms of coal-bearing rock masses of the order of 100–150 and 400–600 thousand years. Besides these, there is evidence of much longer periods lasting from 10–15 and 30–40 m.y.

All the cycles are estimated by periodical changes in the character of sedimentary rocks. For example, coarse fragmental deposits marking the beginning of a cycle will usually change upward into finer and still finer deposits which finally turn into marls or limestones and become coarse fragmental deposits.

The reason for these numerous alterations in deposition within one section is presumed to be a repeated change in the direction of the prevailing movements from rising to sinking and vice versa. The character of the boundary zone between two cycles suggests often a temporary interruption in the accumulation of sediments and erosion of some deposits of the previous cycle.

The longest cycles of general oscillations last for about 200 m.y. The time since Precambrian is conveniently divided into three cycles: Caledonian (early Paleozoic), Variscan (middle and late Paleozoic), and Alpine (Mesozoic and Cenozoic). At the beginning of every cycle the surface of continents is high; then the surface begins to sink, allowing the sea to spread. The character of sediments generally changes from clastic deposits to limestones. Depression maximum is followed by the second stage of the cycle expressed by a new increase of uplifts and greater significance of clastic rocks. At the end of every cycle the continent surface restores its high position *Khain and Ronov*, 1960; *Ronov*, 1961] (Figure 3).

The longest cycles can be found almost simultaneously both in geosynclines and on platforms (there is some time lag for platforms), and are responsible for the fact that some definite sedimentary rocks prove to have been predominant on continents during the same periods (for example, gypsums, salts, and red sandstones were abundant during the Permian; coal-bearing deposits are typical for the middle and upper Carboniferous period; limestones are associated with the late Cretaceous period).

During the first half of the cycle both platforms and geosynclines are subject to sea transgressions, the difference being that seas on platforms are shallow and land is flat almost without exception, whereas seas developed on geosynclines are locally quite deep, and land is expressed by elevated narrow ridges and chains of uplifted islands.

At the close of a long cycle, vast but flat lowland areas develop on platforms, whereas geosynclines become the site of mountain ranges dissected by deep depression. The epoch of mountain formation concluding every tectonic cycle in the geosyncline development is marked by an especially intensive erosion and by accumulation of thick coarse clastic and sometimes red formations of sedimentary rocks, which are also called molasse formations.

The general rule is that mountains develop in a geosyncline (frequently in the zones subjected to most intensive sagging at the beginning of the cycle, i.e. in the place of intrageosynclines), whereas peripheral zones are characterized by the formation of foredeeps. It is here that coarse deposits of molasse formations accumulate. Mountain ranges in the geosyncline first develop in the form of narrow ridges, but later they not only become higher but also expand. Foredeeps migrate further and further outwards in front of the widening mountain ranges and roll over the edge of the adjacent platform. In addition to foredeeps, the end of a tectonic cycle is marked by the development of depressions

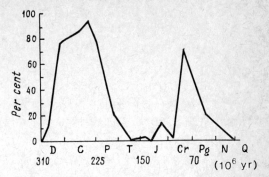

Fig. 3. Periodicity of vertical movements of the earth's crust [after *Ronov*, 1949b]. Curve indicates the percentage of limestones among sedimentary deposits on the Russian plain for different geological periods (see Figure 5 for abbreviations).

which receive rocks of the similar molasse type between mountain ranges (intermontane troughs, see Figure 4).

Undulatory movements contrast most at the beginning and the end of a tectonic cycle, the middle of the cycle being characterized by smaller contrasts [*Khain and Ronov*, 1960; *Ronov*, 1961].

Long cycles become more complicated owing to cycles of still greater order. For this reason general oscillation of the crust is a complicated process made up of superimposed vibrations of various durations and amplitudes (Figure 5). On the other hand, there is reason to presume that there are still longer cycles (megacycles) than those cited above as the longest, embracing periods from 600–800 m.y.

Tectonic movements occurring in the upper part of the crust are evidenced by deformation of sedimentary and metamorphic rocks expressed as folds and faults. Morphologically the folds and faults are highly variable: they differ in size, amplitude of displacement, and their form as viewed in cross section. For example, some folds are expressed by round or oval dome-like uplifts of strata, their diameter being in the range from hundreds of meters to tens of kilometers, and their amplitude varying from tens of meters to several kilometers. Other folds are in the form of alternating upward and downward bends elongated in one direction (linear folds). They greatly differ in dip of their limbs and in size: folds can be expressed by bending measured in centimeters or kilometers. Bends

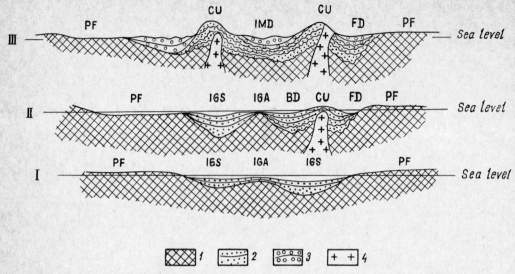

Fig. 4. Scheme of the development of a geosyncline (after V. Beloussov): I–III, stages; *PF*, platform; *IGS*, intra-geosyncline; *IGA*, intra-geanticline; *BD*, border deep; *FD*, foredeep; *CU*, central uplift; *IMD*, intermontane deep; 1, basement formed during previous tectonic cycles; 2, different sedimentary formations; 3, molasse formation; 4, granites.

in some folds are similar and parallel to each other; in others, bends vary greatly, some of them bending gently, others showing a great degree of folding (disharmonic folds).

Tectonic fractures may be both vertical and inclined. If fractures are inclined, rocks lying on either side of the fracture can move either by descent of the upper wing along the lower one (normal fault), or vice versa (overthrust). The former is associated with crustal extension, the latter with crustal shortening. Displacement amplitude can vary from a few centimeters to several kilometers (in the case of gently sloping overthrusts it may be as great as some tens of kilometers). Vertical or steeply inclined fractures can also be associated with horizontal displacement, one wing slipping horizontally relative to another one (strike-slip faults). Folds result from stresses generated in the earth's crust. Horizontal layers may be lifted by some forces operating on a certain section of the crust to obtain the form of a dome or a steeply sloping boxlike fold with a flat top. If a series of horizontal layers is subjected

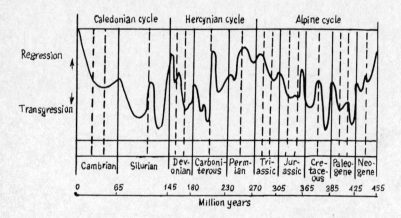

Fig. 5. Complexity of vertical movements of the crust [after *Stille, 1944*].

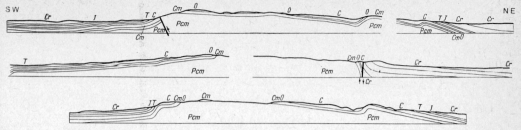

Fig. 6. Block folds from the Rocky Mountains. Cm, Cambrian; Pcm, Precambrian, [after *Darton*, 1906, 1910].

to horizontal compression, the layers curve to become a series of elongated folds.

Crustal fracturing is also the result of tectonic stresses. Many fractures occur together with folds and are regarded as their further development. For example, doming of layers by upward forces may result in the formation of vertical or steeply sloping faults dissecting the bench of layers into blocks moving vertically relative to each other. When the effective forces producing folding of layers are parallel to the latter, the folds may be accompanied by inclined overthrusts, allowing displacement along their slopes to be added to folding.

Processes of fold and fault formation operate differently because of heterogeneous media, whose properties may change in the process of deformation, because of irregularity of the effective forces, and the over-all effect of body gravitational forces. Folds and faults are being studied by detailed investigation into the folded zones, as well as by modeling, the 'equivalent' plastic materials being selected according to criteria suggested by the physical theory of similarity [*Beloussov*, 1960, 1961; *Beloussov and Gzovsky*, 1965; *Ramberg*, 1967]. Results of the investigations suggest that there are different mechanisms of fold and fault formation. The domelike or boxlike fold due to the effect of vertical forces acting on the layers is a local reflection of crustal movements of the same type as undulatory or block-undulatory movements (Figure 6).

These folds frequently develop in the sedimentary layer overlying fractured zones in the platform basement and are considered to be a reflection of the shifts along the faults continuing at greater depths.

Injection folds generated by the local squeezing of one stratum from under another, resulting in the break and deformation of the latter, are also widespread. The rock being squeezed out is always of a higher-than-average plasticity, whereas the overlying one is usually permeated by fractures. Squeezing out can be attributed to 'inversion of density,' i.e., to the fact that the underlying (plastic) rock is less dense than the overlying rock, hence tries to float up. The resulting movements resemble convection of liquids; i.e., light material rises to the surface and locally penetrates through heavy material.

Injection folds can be either superficial or deep. The former are attributed to salt and clay interbedding in sedimentary formations, whose density is much higher than that of the underlying plastic rocks. Thus salt domes or diapir folds develop alongside pointed comb-shaped anticlines, divided by wide flat synclines due to clay squeezed out from under synclines and forced into the anticline core (Figure 7).

Deep injection folds develop in deeper layers of the crust subject to regional metamorphism and granitization. These processes are responsible for the change in the rock density and in the formation of rocks lighter than the overlying ones not affected by granitization. (Granites and metamorphic rocks occurring in a solid crystalline state are, as a rule, heavier than the majority of sedimentary rocks. During the process of their formation they were lighter, however, owing to the partial melting of rocks in the process of granitization.) Light rocks float up, like salt, penetrating into other rocks and solidifying as masses in the form of vast overturned drops, 'stumps,' or 'mushrooms.' These masses, i.e. granitic and gneissic domes or ramparts, penetrate the overlying rocks and force them apart (Figure 8). Structures of this type are most commonly observed on the deeply

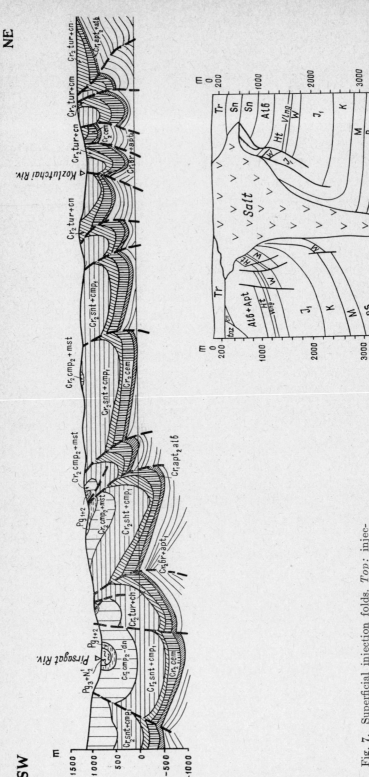

Fig. 7. Superficial injection folds. *Top*: injection folds in southeastern Caucasus (after A. Vikhert, unpublished); *Bottom*: diapiric salt dome (F.R.G.) [after Bentz, 1949].

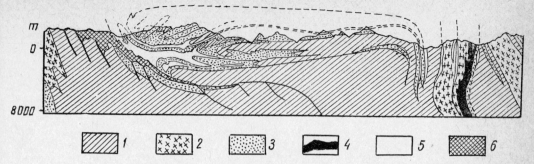

Fig. 8a. Deep injection folds in the Pennine nappes in Alps: (1) orthogneisses; (2) migmatites; (3) paragneisses; (4) late-Tectonic intrusions; (5) Mesozoic sediments on Pennine nappes; (6) Mesozoic sediments on St. Gotthard massif. Pennine nappes (recumbent folds) are due to squeezing out of orthogneisses [after *Nabholz*, 1953].

eroded crystalline shields. Linearly elongated folds of general crumpling are comprised of alternating convex and concave bends and require for their formation a horizontal compression (Figure 9). Some time ago an explanation of this problem was offered by the hypothesis of general compression of the globe (contraction hypothesis) which later proved wrong. The data available indicate that horizontal compression is of a local character, involving only a relatively narrow zone of the crust, and never the whole of it. Such zones of compression are usually found in geosynclines on the boundary between the zones of intensive rising and sinking of the crust. The ability of stratified rocks to flow slowly in response to gravity down even very gentle slopes (creep of rocks) suggests that layers are crumpled into linear folds in the places where they pile up after sliding down the slope of undulatory or block-undulatory uplifts in a geosyncline. Layers compressed between two rising abyssal diapir folds can be subject to crumpling. A crustal block rising during block-undulatory movements may, because of gravity, expand in its upper part to crumple adjacent layers with its flanks. Some scientists offer different explanations of the development of compression folds. The existence of these folds is not enough to suggest the existence of initial horizontal movements in the crust. It is not unlikely that horizontal compression is a secondary result of vertical undulatory or block movements which disturb gravitational equilibrium of sedimentary formations and cause their horizontal displacements.

General crumpling and compression are typical of geosyncline folding. However, our knowledge of the concrete mechanism of the development of geosyncline folding is insufficient. Some geosyncline folds and overthrusts have undoubtedly been generated by gravitation, i.e. they are the result of gravity creep down the uplifted folds. It is also quite obvious that other folds of this type result from penetration of intrusive masses of granites from below forcing the rocks apart and crumpling them. Folds of this kind may also be attributed to regional compression compensating for the extension of the adjacent section of the geosyncline, or may be associated with differentiated horizontal displacements of great crustal blocks along regional strike-slip faults, the existence of which is much disputed.

Development of folds on platforms and partially in foredeeps and intermontane troughs is a very slow process which takes geologic periods to accomplish and often occurs simultaneously with accumulation of sediments. The rate of folding, however, is not always the same. If considered as a whole, this irregular process is very much the same as that of slow development of undulatory crustal movements main-

Fig. 8b. Migmatite domes of irregular shape in Greenland [after *Haller*, 1955].

Fig. 9. Compression folds in a cross section through a part of the Carpathian ridge [after Băncilă, 1958]; (1 and 2) Miocene; (3) Oligocene; (4) Eocene; (5) Senonian (Upper Cretaceous); (6) black slates.

taining the same direction for a long time. Folds of general crumpling in geosynclines take comparatively short periods to form (hundreds of thousands or more often some millions of years). They are 'phases of folding.'

The most intensive folding of every long tectonic cycle is associated in time with the second half of the cycle, i.e. with the time when uplifts prevail in the geosyncline. Such observations indicate that, during every stage, the zone of intensive folding was but a narrow belt usually on the boundary between rising and sinking zones. During the second half of the geosynclinal cycle, zones of uplift become broader; the folding gradually covers all the geosyncline. Hence, early stages of folding are typical for inner parts of geosyncline, whereas later ones are found mostly on the margin of the geosyncline.

By the end of a long tectonic cycle and because of development of folding, a geosyncline changes into a zone of folding: thus Caledonian geosynclines give place to Caledonian zones of folding, Variscan geosynclines become the sites of Variscan zones of folding, and Alpine geosynclines turn into Alpine zones of folding (Figure 2).

Great horizontal displacements (along strike-slip faults) should be identified as a special type of movement. The most frequently cited examples of strike-slip faults are: Great Glen in Scotland, San Andreas in California, Thalasso-Ferganian in Central Asia, and the Alpine in New Zealand. Some geologists maintain that horizontal displacements along these faults measure in hundreds of kilometers. Other authors, however, express serious doubts about it; there is good reason for doubt, because it is at present difficult to measure the displacements. Only displacements of the order of 1–2 km that have occurred during Recent geologic time and caused obvious offsets of river valleys and other changes in the present relief are beyond any doubt. The assumptions for greater displacements are still a matter of debate (see p. 704).

Another type of displacement is planetary extension. This is associated with so-called rifts, i.e., the zones of greatest fractures in the earth's crust. These zones mostly occur on the ocean floors, where they are associated with the median ocean ridges. The system of rifts encircles the whole globe, forming a belt about 60,000 km long (Figure 2). In some places they emerge onto the continents, especially in East Africa (see p. 539).

Wherever rifts occur, they are characterized by quite obvious tension of the earth's crust. Their association in time with the zones of crustal uplift suggests that tension is not an independent phenomenon but is generated by forces acting vertically and making the crust curve upward. (The reader is referred to other chapters for more details about the world rift system.)

This résumé indicates that vertical stresses in the tectonosphere (i.e., in the crust and upper mantle) are repeatedly transformed into horizontal ones, and vice versa. Horizontal displacement of matter, its flow, and forces acting at one level can be responsible for vertical displacement, either elevation or subsidence, occurring on a higher level, which in its turn causes horizontal displacement, compression or tension, in the overlying story. For all these reasons one cannot regard either vertical or horizontal movements independently.

The development of the crust at least on continents since the Proterozoic tends to change from a mobile, geosyncline state to a relatively quiet, platform nature. Late Proterozoic (Riphean) geosynclines covered greater areas than Caledonian ones; the latter were broader than Hercynian (Variscan) geosynclines, and these exceeded Mesozoic (especially Alpine) ones, which formed very narrow sinuous zones encircling vast angular platforms that had accordingly grown in size since the Late Precambrian. At the end of every long tectonic cycle, new zones of folding were progressively generated at the edges of ancient platforms and later turned into relatively quiet regions identified as young platforms, or were involved in a new orogenesis.

The tectonic history of the earth's crust of the earlier Archean period is not well enough understood, because the metamorphism of rocks hinders estimation of their age. However, in the early stages of its development, the whole crust seems to have been more mobile. It is very likely that relatively quiet regions developed in different places, but later were involved in intensive movements. These initial stages of the crustal history are marked by an intensive magmatism.

Beginning with the middle Proterozoic, some stable platforms developed; they were at first small, but grew larger later, and the earth's crust began gradually to calm, its intensive vertical movements and deformations becoming increasingly localized and occurring mainly in geosyncline zones that separated platforms.

Ancient platforms are North American, East European, Siberian, Chinese-Korean, South Chinese (sometimes considered as a single Chinese platform), South American, African-Arabian, Indian, Australian, and Antarctic. Tarim and Kolima massifs, which can also be classified as median massifs, are sometimes considered to be independent small platforms. Among the young platforms are the Ural-Yenisei, Appalachian-Piedmont, and West European platforms. The young platform of the Ural-Yenisei has, after the Hercynian tectonic cycle, welded together two ancient platforms: the East European and the Siberian. The Appalachian-Piedmont, developed during early Mesozoic in the place of the Hercynian and still older Appalachian geosyncline area, adding to the area of the old North American platform. The West European also developed in the place of Hercynian geosyncline and still older geosyncline areas of western Europe, increasing the area of the ancient East European platform (Figure 2).

A characteristic feature of ancient platforms is their two-layer structure, which manifests itself by the existence of a pre-mid-Proterozoic basement made up of tightly folded and metamorphosed rocks (gneisses, metamorphic slates, marbles, quartzites) permeated by granites and various igneous rocks. This basement developed at the time when the place which later became the site of the ancient platform was subject to conditions near to geosynclinal, i.e., before the appearance of the platform. The surface of the basement having been eroded, it is covered by an undisturbed mantle of sedimentary rocks accumulated during the stage of platform development.

The oldest parts of the basement of the ancient platforms developed earlier than 2500–2600 m.y. ago. They are made up of gneisses and crystalline schists that have undergone granitization accompanied by intensive formation of migmatites. Within these sections there are still older ones, probably the most ancient sections of the earth's crust which are expressed in the form of individual massifs of amphibole gneiss, amphibolite, and other rocks whose age is estimated to be from 3000 to 3600 m.y. Remnants of these most ancient sections were found in the region of Lake Superior in Canada, in South Africa, in the Tsharsk-Olekminsk region of Siberia, on the Baltic shield (the Saam massifs), and in other places. A slightly metamorphosed, very ancient mantle of sedimentary rocks is preserved in some of the regions (in South Africa and in Canada the mantle is made up of green schists about 2000 m.y. old).

During the first half of the Proterozoic period, thick series of sedimentary rocks accumulated between the most ancient parts of platforms in geosynclinal troughs probably bordered by ruptures. These sediments were later folded, metamorphosed, and permeated by magmatic intrusions. The above processes seem to have occurred in two stages: 2100 to 2600 m.y. ago, and during the interval from 1750 to 2100 m.y. ago. Included in the more ancient group are, for example, the folded system of Saksaganid and Kursk on the East European platform, and the Aldan-Anabar system on the Siberian platform; the Svecofenno-Karelian on the Baltic shield and the Hudsonian of the Canadian shield are classified as a younger group. In more recent time the ancient platforms were not subject to either severe folding or metamorphism; hence their development was fully completed about 1600 to 1700 m.y. ago.

Folded systems of the upper Proterozoic

separate some mid-Proterozoic massifs from the more ancient ones only within South American and African platforms. A relatively narrow Brazil zone of folds developed as a mobile area between the eastern and western blocks of the Brazil platform until the very close of Precambrian. It became a zone of folding that connected more ancient parts of the South American platform only after the folding of 500 to 750 m.y. ago. The same happened to the folded area, extending from the shores of the Red Sea to Lake Victoria, with one branch toward Katanga and the Atlantic coast, the other toward Mozambique, which at the end of the Proterozoic was not mobile any longer.

Deposition of platform sedimentary mantle on ancient platforms came generally after the surface of the metamorphic basement was uplifted and eroded. The mantle rocks are arranged discordantly on the platform, their positions proving the very quiet nature of the tectonic regime of ancient platforms. These are classified into extremely gently sloping uplifts (anteclises) and depressions (syneclises) widely distributed in space; this results in accumulation of sediments of various thickness in some places of the platform, while in others the eroded basement becomes exposed on the surface. Areas of ancient platforms receiving sediments are called platforms 'sensu stricto,' whereas those with the exposed crystalline basement are referred to as shields.

Platforms sensu stricto are abundant in the northern ancient platforms, i.e., on the North American, East European, Siberian, Chinese-Korean, and South Chinese platforms. Considerable areas on southern platforms are free from sedimentary cover.

The arrangement and form of anteclises and syneclises (as well as of platforms and shields) on ancient platforms do not depend on the structure of the crystalline basement. Edges of syneclises and anteclises intersect structural trends in the basement. For example, the southeastern fold trends prevailing on the Baltic shield are discordantly cut by the edge of the Moscow syneclise. Meridional fold trends in the basement of the small Ukranian shield are cut both from the north and the south almost by latitudinal edges of the adjacent syneclises.

Having developed, syneclises and anteclises on ancient platforms usually remain fairly undisturbed for a considerable geologic time. In this case the stages of development are distinguished by general recurrent submergences and uplifts of the platform as a whole. The uplift of the early Cambrian (the close of the Baikal cycle) was followed by the subsidence lasting during Cambrian and Ordovician, which during Silurian or early Devonian changed into a new uplift. This is the Caledonian stage. After this came the Devonian and Carboniferous subsidence and Permo-Triassic uplift (Hercynian stage). Then a new submergence occurred during Jurassic, Cretaceous, and Paleogene to be followed by an uplift at the close of Paleogene and Neogene (the Alpine stage).

The early stages of the development of ancient platforms are characterized by the development of the so-called aulacogenes, i.e. large grabens whose lower layers are locally greatly deformed, with some lava flows and magmatic instrusions along the faults. Subsequently, strata were formed which cover these grabens, filling vast syneclises.

The size and configuration of syneclises and anteclises on ancient platforms have also undergone some changes later. For example, the large Voronezh-Ukrainian anteclise on the east European platform divided in two during Devonian time, giving place to a depression of the Dniepr-Donetz graben in the opening. The uplift in the region of the Volga (Volga-Ural anteclises) which existed previously was converted into a depression during the Carboniferous, and the vast depression, which was later divided into two syneclises, Moscow and Caspian, extends from the Pechora River in the north to the Caspian Sea in the south. All these changes, determining the thickness of sediments, contribute to the final structure of the platform and above all to the relief of the crystalline basement.

More common dislocations on the ancient platform have to a certain degree originated by the dissection of the basement. A series of faults surrounds synclines and intersects them. These faults in the sedimentary mantle are often expressed in the form of step-like bendings of layers called flexures. Other types of dislocation on ancient platforms are domes and welts. All of them are generally sloping, but sometimes welts are huge. For example, the Zhiguli welt on the east European platform is about 150 km long and some tens of kilometers wide, the uplift being about 1 km. Like many other welts, large and small, this one is asymmetric (its southern limb is gently sloping, whereas the northern one is steeply dipping to become vertical in some places and changing into a fault) which indicates that the dislocations are the result of faults in the basement and vertical displacements along them. The same is suggested by fairly frequent platform welts of a box and step-like form, which are distinguished by comparatively steep limbs consisting of a number of flexures bordering horizontal or very gently sloping wide crests. In the deepest syneclises receiving thick deposits of sedimentary rocks accompanied by salt-bearing series, the salt domes develop. Examples are deep Peri-Caspian and Dniepr-Donetz syneclises of the East European platform and the Gulf syneclise adjoining the Gulf of Mexico (the southern part of the North American platform). The Peri-Caspian syneclise down to the roof of the metamorphic basement is more than 16 km deep. The Gulf syneclise is approximately of the same depth. Dislocations of this type are the result of long and gradual processes and often take several geological periods to develop.

Evidence of magmatic activity is widely found on ancient platforms. The most abundant is plateau-basalt eruptions that cover vast areas and exhibit a marked similarity in composition (pyroxene basalts): for example, the Deccan plateau basalts in India of upper Cretaceous and Eocene age, which are about 3000 meters thick and cover the area of 500,000 km^2; plateau basalts of Parana syneclise in South America (end of the Triassic-early Cretaceous); plateau basalts of Tungus syneclise in central Siberia (of late Paleozoic, early Mesozoic age).

Kimberlite volcanic pipes known for their diamond deposits are sometimes found on ancient platforms (Africa and Siberia). Platform intrusions include also large funnel-shaped intrusions in South America and Canada that are basic in composition, greatly differentiated, and stratified, and dome-shaped alkaline intrusions, also greatly differentiated, known for their deposits of rare earths (for example, the Khibini intrusions on the Kola Peninsula).

Lying between the ancient platforms are belts of folding of geosynclinal origin (Baikalian to Alpine age), which gradually turned into young platforms of different ages.

Geologists recognize on the surface of the earth six belts of folding. The Pacific belt encircles the Pacific Ocean and continues as the Mediterranean belt, which branches off in Indonesia and extends over southern Asia and Europe, dividing the ancient platforms of India and Africa-Arabia from the platforms of the north. Adjacent to it from the north is the Ural-Mongolia belt separating the Siberian platform from the East European. The Atlantic belt includes the folded systems along the coasts of the Atlantic Ocean in Europe, North America, and Greenland. To this belt can also belong the Baikalides of Brazil and west Africa. The fifth lies to the north of the Canadian shield on the Arctic archipelago. The remnants of the sixth zone are the Cape mountains of South Africa, and perhaps the Sierras of Buenos Aires province in Argentina.

Every belt of folding is divided into several geosynclinal folded systems, the long course of geosynclinal history being reflected in the final structure of the system of folding. All the events within one tectonic cycle are divisible into two stages. During the first stage, subsidence prevails and geosyncline depressions form and develop, whereas the second (often called orogenic) stage is characterized by uplifts, on the background of which deep intramontane depressions and marginal troughs are formed. During the first stage, accumulation of a thick succession of sediments in the intra-geosyncline troughs is interrupted by lava flow, mainly of basic (basalt) composition (initial volcanism). During the same stage, intrusive basic sheets and the so-called small intrusions shaped like elongated lenslike bodies of serpentinite, gabbroic, or plagiogranite composition are formed.

Very important events occur during the period of great change in the life of a geosyncline, i.e., when instead of subsidence, uplifts begin to prevail. At this time the rocks in the geosynclinal depressions are subject to folding. The folding affects at first the uplifts in the intra-geosynclinal depressions and their margins and gradually (stage by stage, or spasmodically) involves larger and larger areas, the process being accompanied by the growth and widening of uplifts. At the same time the sedimentary rocks accumulated in deep geosynclinal troughs and undergoing uplift become subject to metamorphism that grows more intensive with depth, changing from the greenschists facies to the amphibolitic and sometimes even to the most intensive granulitic facies. All these processes occur simultaneously with the development of granitic intrusions and batholiths, the latter chiefly through the process of partial remelting of sedimentary and metamorphic rocks (anatexis).

The cycle of geosynclinal development is completed by an intensive orogeny. The uplift of mountains within the geosynclinal system is followed by the subsidence of the intramontane depressions and the foredeeps. The intramontane depressions are formed within the geosynclinal system, while the marginal depressions overlie the edge of the neighboring platform. Both formations are filled by thick sedimentary layers formed from the erosion of the neighboring mountain uplifts and called 'molasse formations' accompanied by extrusions of andesite, dacite, and basalt (subsequent volcanism). The last phases of folding are manifested in the foredeeps and intramontane depressions and result in the formation of folded structures along the fracture zones. Later the andesitic and basaltic lava erupts through fractures in growing mountain ranges and through central volcanoes (final magmatism). Blocks of crust separated from each other by faults sink or rise like piano keys, dividing the highland into blocks differing in the degree of elevation.

The internal structure of fold systems formed as a result of geosynclinal development is primarily characterized by the alternation of large elongated uplifts and troughs. We can distinguish anticlinoria as structural uplifts and synclinoria as structural depressions. Both of them are hundreds of kilometers long and tens of kilometers wide. It is quite natural that ancient deposits, some of which have originated during previous tectonic cycles, can be seen in the core of anticlinoria, whereas synclinoria are, in contrast, characterized by the youngest deposits lying on their surface.

Folding makes the structure of both anticlinoria and synclinoria more complicated, being usually more intensive in anticlinoria than in synclinoria. Large thrusts, and sometimes overthrust sheets (nappes), originate on anticlinoria.

The anticlinoria exhibit more intensive and highly varied igneous activity and regional metamorphism.

The median massifs, which represent parts of the foundation on which the geosynclinal systems developed, are located between separate geosynclinal depressions and their systems. They are often overlain by the sedimentary cover, which is not dislocated, or only slightly distorted. Median massifs can be considered as small platforms included in the folded region.

The structure and history of development of the fold belt can be fully elucidated only for the tectonic cycle that was the last in the geosynclinal development and just preceded the formation of a platform in the place of a fold system so that subsequent movements did not change the fold structures. The oldest zones of folding are of the Upper Proterozoic or Baikal age. The areas which did not experience folding or metamorphism after the Proterozoic developed into post-Proterozoic (epi-Baikalian) young platforms, whereas young epi-Caledonian platforms formed in the places where the last folding and metamorphism occurred at the close of the lower Paleozoic (Caledonian folding). The parts subject to the last folding and metamorphism during upper Paleozoic (Hercynian folding) have become the sites of young post-Hercynian platforms. Some geologists avoid the term 'young platforms' by using 'regions of completed folding' instead, thus laying emphasis on the significant difference between the ancient and young platforms and suggesting that only the former should be referred to as proper platforms (cratons).

Young platforms are much more mobile than ancient ones. Elevations and depressions of the first order, anteclises and syneclises, stand out more distinctly on them. Anteclises and syneclises are of an elongated oval form, in contrast to round and irregular forms of ancient platforms, their length and width being the same and their vertical amplitudes somewhat less than those of anticlinoria and synclinoria. The arrangement of anteclises and syneclises of young platforms largely follows the latest arrangement of anticlinoria and synclinoria in the zone of folding existing in the same place before the formation of the platform.

Evidence of the most pronounced block folding is found on young platforms. An example is extra-Alpine West Europe, characterized by the so-called 'Saxonian' tectonics.

Eruptives, mainly in the form of alkaline intrusions and flows, abound in places on some young platforms (the extra-Alpine young West European platform is also an example).

Somewhat different are the characteristic features of other parts of young platforms, such as the West Siberia lowland, the Piedmont lowland on the Atlantic coast of North America, and the Great Artesian basin in Australia.

The West Siberian lowland is a vast gentle depression that developed during the Meso-Cenozoic, burying beneath it the Hercynian and Caledonian-Baikalian system of folding. The depression is from 550 to 1500 km wide, the maximum depth of the folded Paleozoic basement underlying Meso-Cenozoic sediments reaching 10 km. This vast gentle basin (syneclise), filled with sediments, is divided into several individual depressions and elevations that are also very gentle. There are also widespread, more peculiar gentle domelike elevations of strata. General submergence of the entire West Siberian depression, which began in the Jurassic, came after the formation, during the Triassic, of separate enclosed and fairly intensively sinking basins, bordered by faults. These basins look very much like aulacogenes whose formation marks the start of submergence of the ancient East European platform. The entire structure of the extremely tranquil West Siberia lowland resembles more the structure of ancient than young platforms.

The histories of the Piedmont lowland and of the East Siberian lowland are similar up to the time when the first enclosed basins began to form. The Piedmont differs from the West Siberian in that the entire width of the Siberian depression is seen, whereas the far edge of the Piedmont (if any existed at all) is under the Atlantic Ocean.

The Australian Great Artesian basin, shaped like a large gentle sag, developed later, and is of late Lower Cretaceous age. It rests on the ancient Australian platform and partly on Baikalian, Caledonian, and even Hercynian zones of folding.

These regions should be regarded as a special tectonic category, which has not as yet been given a definite name. They are called either 'large superimposed basins' or 'interior basins.'

Tectonically activated regions, or those referred to as regions of post-platform mountain building, can also be classified as a special category. Included in this group are parts of young and ancient platforms that have in recent geologic time been subjected to a considerable tectonic revival and dissected into regions of uplift and depression whose amplitudes exceed those of both ancient and young platforms.

The most illustrious example is the Tien Shan, which was partially transformed into a young platform with calm tectonic movements after the end of the Caledonian cycle, the process being completed after the Hercynian cycle. During the Neogene, there began a progressive dissection into uplifts and depressions strikingly contrasting one another, the relative range of vertical movements having become by the present time as great as 10 km, the gradient of movement across the uplift and depression margins being about 400 meters/km. The uplifts became the highest mountain ranges, whereas the depressions received thick Neogene and Quaternary deposits.

The area of the recent tectonic activation spread from here and involved the Hindu-Kush, Kuen-lun, Nanshan, and Tibet. The recent revival involved also the Altai-Sayan and Peri-Baikal regions. It is very likely that in the same category fall also the events of the recent geologic history in the Eastern African system of grabens (see p. 539). This was the place where, during the late Mesozoic, vast uplifts developed, exceeding normal anteclises, such as Ethiopia-Arabian and East African, which were dissected by grabens during the Neogene. This process was accompanied by volcanic activity which has not yet ceased. The Baikal system of grabens has been subject to similar conditions and its history is almost analogous.

The characteristic feature of the regions of tectonic activation is a high degree of seismicity. In the USSR, the region with the highest seismicity is Tien Shan. The Baikal region and the whole of the zone of East-African grabens are also known for their considerable seismicity.

Conclusion

We have tried to give a brief summary of the main data on types of tectonic movements of the earth's crust and the most general regularities of its development. We must admit that these data are not complete. They are unequal among the continents; the southern continents especially have not been adequately studied.

In the last century, however, data have accumulated exponentially. A problem of vital importance now is to summarize the world's data on types of present tectonic structures and on the history of crustal development. This is a task to be undertaken by scientists of many countries. It is highly desirable that these summaries be done in accordance with a uniform program, so as to yield substantiated conclusions.

A collective preparation of the Tectonic Map of the World by the International Union of Geological Sciences is one of the fruitful endeavors. Another urgent step for the scientists of the world is a compilation of an atlas of paleogeographic and paleotectonic maps of the earth.

REFERENCES

Băncilă, I., *Geologia Carpatilor Orientali*, Ed. Stiint., Bucharest, 367 pp., 1958.

Beloussov, V. V., Tectonophysical investigations, *Bull. Geol. Soc. Am.*, 71(8), 1960.

Beloussov, V. V., The origin of folding in the earth's crust, *J. Geophys. Res.*, 66(7), 2241–2254, 1961.

Beloussov, V. V., *Basic Problems in Geotectonics*, McGraw-Hill Book Company, New York, 809 pp., 1962.

Beloussov, V. V., and M. V. Gzovsky, Experimental tectonics, *Phys. Chem. Earth*, 6, 1965.

Bentz, A., in *Erdol und Tektonik in Norwestdeutschland*, Hannover, 1949.

Darton, N., Bald Mountain-Dayton folio, Wyoming, and Cloud Peak Fort folio, Wyoming, *Geol. Atlas USA*, N141 and N142, U.S. Geological Survey, Washington, 1906.

Darton, N. Laramie-Sherman folio, Wyoming, *Geol. Atlas USA*, N173, U.S. Geological Survey, Washington, 1910.

Gilbert, G. K., Lake Bonneville, *U. S. Geol. Surv., Monograph 1*, 1890.

Gzovsky, M. V., Tectonic-physical comparison between the most recent tectonic movements and seismicity, gravitational anomalies, magnetism and abyssal processes in the USSR, in *Activated Zones of the Earth's Crust, Latest Tectonic Movements and Seismicity*, pp. 58–77, Nauka, Moscow, 1964.

Haller, E., Der zentrale metamorphe Komplex von NE Grönland, *Medd. om Grönland*, 73(3), 1955.

Hobbs, W. H., Repeating patterns in the relief and in the structure of the land, *Bull. Geol. Soc. Am.*, 22(2), 1911.

Khain, V. E., Oscillation rhythm of the earth's crust, *Bull. Moscow Soc. Natural., Geol. Sect.*, 17(1), 1939.

Khain, V. E., Block-undulatory (folded-block) structure of the earth's crust, *Bull. Moscow Soc. Natural. Geol. Sect.*, 33(4), 1958.

Khain, V. E., and A. B. Ronov, World paleogeography and lithological associations of the Mesozoic era, *Int. Geol. Congr. Rept., 21st Sess.*, p. 12, Copenhagen, 1960.

Kirillova, I. V., E. N. Lyustikh, V. A. Rastvorova, A. A. Sorsky, and V. E. Khain, Analysis of the geotectonic evolution and seismicity of the Causasus, Moscow, 1960.

Muratov, M. V., Structural complexes and development stages of the earth's crust, *Bull. USSR Acad. Sci., Geol. Ser.*, 1963(6), 1963.

Muratov, M. V., Main epochs of folding and megastadia of the development of the earth's crust, *Geotectonica*, 1965(1), 1965.

Nabholz, W., Das mechanische Verhalten der granitischen Kernkörper der tieferen pennischen Decken bei der Alpinen Orogenese, *Compt. Rend., Congr. Geol. Intern., Algeria*, 1952, 3(3), 9–23, 1953.

Peive, A. V., Abyssal fracturing in geosyncline regions, *Bull. USSR Acad. Sci., Geol. Ser.*, 1945(5), 1945.

Ramberg, H., *Gravity, Deformation and the Earth's Crust*, Academic Press, London, 214 pp., 1967.

Ronov, A. B., History of accumulation of sediments and oscillatory movements in the European part of the USSR, *Trans. Inst. Geophys. USSR Acad. Sci.*, 1949(3), 130, 1949a.

Ronov, A. B., Some general regularities of oscillatory movements of continents, in *Problems of Tectonics*, Gosgeoltekhizdat, Moscow, 1949b.

Ronov, A. V., Some general regularities of oscillatory movements of continents (according to the data obtained by the volume method), in *Collected Works: Problems of Tectonics*, Gosgeoltekhisdat, Moscow, 1961.

Shatsky, N. S., Essay on tectonics of Volga-Urals oil-bearing area and the adjacent part of the western slope of the South Urals, *Studies Geol. Structure USSR, New Ser.*, 2(6), 1945.

Shatsky, N. S., On structural relations of platforms and folded geosyncline areas, *Bull. USSR Acad. Sci., Geol. Ser.*, 1947(5), 1947.

Sonder, R., Die Lineamenttektonik and ihre probleme, *Ecl. Geol. Helv.*, 31(1), 1938.

Sonder, R., *Mechanik der Erde*, Stuttgart, 1956.

Stille, H., *Grundfragen der vergleichenden Tektonik*, 1924.

Stille, H., Geotektonische Gliederung der Erdgeschichte, *Abh. Preuss. Akad. Wiss., Math.-Nat. Klasse*, 3, 1944.

von Bubnoff, S., *Grundprobleme der Geologie*, Akademie Verlag, Berlin, 234 pp., 1954.

Continental Rifts

V. V. Beloussov

THE extent and structural singularity of the East African system of great grabens[1] have long attracted the attention of geologists. Presently this system is known to be merely a small part of the world rift system, the larger part of which is confined to the ocean floor, where the rifts are connected with mid-ocean ridges. The total extent of this system exceeds 60,000 km [*Heezen*, 1960].

The fact that rifts are confined to the regions of both continental and oceanic crust and the basaltic volcanism associated with rifts indicate that their formation is connected with the upper mantle rather than with the crust.

The East African rifts are considered to be the largest system of continental rifts; they extend over 6500 km from the frontiers of Turkey and Syria on the north to the lower course of the Zambezi River on the south (Figure 1). This system includes the Syrian and Lebanese grabens; the grabens of the Dead Sea, Wadi al Arabah, Gulfs of Aqaba and Suez, Red Sea, and Gulf of Aden; and the Ethiopian graben. Two belts of grabens extend southward, forming two arcs embracing the Lake Victoria district from east and west. The eastern belt (the Gregory rift) extends from Lake Rudolf southward along the depression with the chain of small lakes (Lakes Baringo, Nakuru, Naivasha, Magadi, Natron). Further south, in the vicinity of Lakes Manyara and Eyasi, the single graben is replaced by the series of unilaterally dipping structural blocks. The western branch is represented by the belt of the larger lakes (Lakes Albert, Edward, Kivu, Tanganyika). The two branches join at the northern end of Lake Nyasa. The system ends in the valley of the Shire River, the left tributary of the Zambezi [*Baker*, 1965; *Cahen*, 1954; *Cloos*, 1939; *Dixey*, 1956; *Krenkel*, 1925; *Mohr*, 1961; *Pallister*, 1965.]

---
[1] The widely used term 'rift' suggests that these structures have been formed as a result of extension. A purely morphological term is 'graben.'

The Rhine and Baikal rifts (Figures 2 and 3) are isolated structures [*von Bubnoff*, 1936; *Dorn*, 1951; *Florensov*, 1960]. Some authors assign the same type of rifts to the Gulf of California, while others attach the Basin and Range province of the Rocky Mountains to the world system of rifts.

The widths of different rifts, both continental and oceanic, are compared in Table 1. The values are surprisingly similar, being restricted mainly between 35 and 60 km (see also *Girdler* [1963]).

Structurally, a rift is always a complex graben. The total vertical displacements of grabens are, naturally, very different. In the East African rift the maximum displacement is believed to be 3000 meters, ranging generally between 1000 and 2000 meters [*Mohr*, 1961; *Pulfrey*, 1960a; *Cahen*, 1954; *Dixey*, 1956]. In the Rhine grabens the maximum vertical displacement amounts to 3000 meters [*von Bubnoff*, 1936; *Dorn*, 1951], and in the Baikal grabens it sometimes exceeds 5000 meters [*Pavlovsky*, 1937, 1948; *Salop*, 1967; *Florensov*, 1960].

The total displacement commonly consists of a great number of small displacements confined to the particular faults that complicate the sides and floors of grabens. For the most part, these are the longitudinal faults creating the step-like pattern of the sides of grabens. They also cut the floors of grabens into numerous 'slices,' which are sometimes as narrow as a few hundred meters. Directions of displacements along the particular faults are different, resulting in the formation of secondary grabens and horsts of different orders on the rift floors. Individual faults do not extend too far. They are generally characterized by echelon arrangement replacing one by another every few tens of kilometers. Faults often form a broken line, so that the borders of grabens assume a dented form. Transverse faults also have rather widespread occurrence.

Nearly everywhere the faults of continental

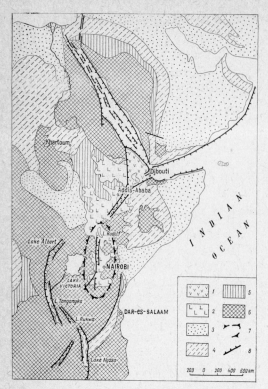

Fig. 1. East African rift system: 1, Quaternary lavas and tuffs of intermediate and basic composition; 2, basic effusive rocks of the Oligocene and Miocene; 3, Cenozoic; 4, Mesozoic; 5, Paleozoic; 6, Precambrian; 7, appropriate limits of the vault uplift reflected in the deformation of the Pre-Miocene surface of leveling; 8, basic faults.

grabens are either vertical or steeply dipping in the direction of the downthrown side (normal faults).

There is some evidence that the continental rifts are all young structures formed during the Neogene. Indications of the older movements can be found only in some places. Thus, the formation of the Dead Sea graben began in the early Miocene or late Oligocene, the movements having continued till the early Pleistocene [*Picard*, 1943]. The faults of the Gulf of Aqaba are of Pleistocene age. The graben of the Gulf of Suez began to form in the late Oligocene; during the Miocene, the subsidence spread southward, along the Red Sea, thus reaching Bab el Mandeb in the Pliocene, when the connection between the Red Sea and Indian Ocean was established for the first time [*Said*, 1962; *Whiteman*, 1965]. In Ethiopia the formation of faults began in the late Oligocene, but the main movements occurred in the Miocene [*Mohr*, 1961]. In Kenya, the Gregory rift originated in the middle Miocene, the culmination of movements that occurred in the Pliocene and Pleistocene [*Baker*, 1965; *Pulfrey*, 1960b]. The formation of the western branch of the East African rift began in the early Miocene. After the period of quiet tectonic conditions in the Pliocene, the intense crushing and subsidence were renewed in the Pleistocene [*Cahen*, 1954; *McConnell*, 1951].

There is some evidence that, in the southern end of the East African rift system, the faults began to originate as early as in late Carboniferous time and continued to grow during Jurassic and early Cretaceous time [*Dixey*, 1956]. Interrelations between these old movements and the younger ones, which gave rise to the recent grabens, are obscure.

In the Rhine rift, the first indications of subsidence without fractures are dated from the late Eocene to early Oligocene. The subsidence along fractures began in the middle Oligocene [*Dorn*, 1951].

In the Baikal rift region, the first downwarpings without fractures began in the Miocene.

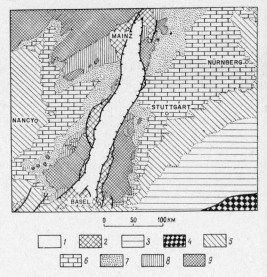

Fig. 2. Upper Rhine rift zone: 1, Graben trough; 2, Marginal blocks and basins; 3, Molasse trough; 4, Alps; 5, Jurassic; 6, Middle and Upper Triassic; 7, Lower Triassic; 8, Permian; 9, Pre-Permian basement.

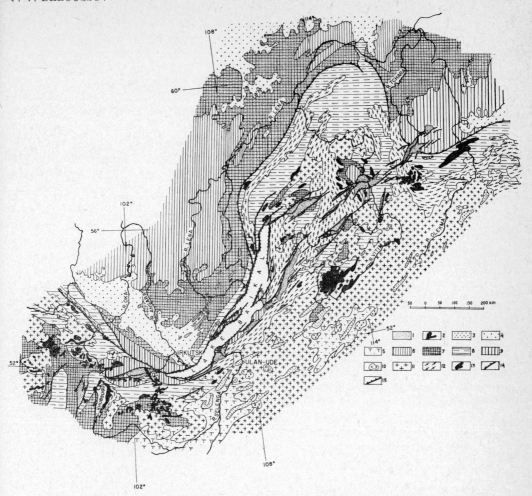

Fig. 3. Baikal rift zone: 1, Quaternary and Neogene deposits; 2, Quaternary and Neogene basalts and trachybasalts; 3, Lower Cretaceous; 4, Jurassic; 5, Triassic and Jurassic eruptive rocks in Transbaikal region; 6, Ordovician; 7, Cambrian; 8, Proterozoic; 9, Archean; 10, Mesozoic granitoids; 11, Paleozoic granitoids; 12, Proterozoic granitoids; 13, ultrabasic and basic intrusive rocks of Precambrian and Early Paleozoic; 14, faults along the borders of the Siberian platform; 15, Quaternary and Neogene faults in the rift zone.

The subsidence along the fractures occurred during the Pliocene and Pleistocene, when sedimentary series as thick as 5 km accumulated in some places [*Pavlovsky*, 1937, 1948; *Salop*, 1967; *Florensov*, 1960].

At the present time, all rifts appear to be zones of very high seismic activity, which indicates that the displacements along faults are still occurring.

The common feature of all continental rifts is their relation to arched uplifts of the earth's crust [*Cloos*, 1939]. The northern part of the East African rift system, which is sometimes considered to be an independent Eritrean system (including grabens of Syria, Palestine, the Gulfs of Aqaba and Suez, the Red Sea, and Ethiopia), was formed along the axis of the Arabian-Ethiopian structural uplift. This uplift is a vast anteclise with a crystalline Precambrian core surrounded by gently dipping Jurassic, Cretaceous, and younger sediments. The central part of the anteclise is uplifted in the form of a plateau at an altitude of 2 to 3 km. On the periphery of the anteclise, the altitude decreases to 1 km. The structural uplift of the roof of the Precambrian basement amounts to 5 km. The

TABLE 1. Widths of Continental and Oceanic Rifts

Rift	Width, km
Gulf of Aqaba	50
Dead Sea	35
Gulf of Suez	35
East African (common)	50
Rhine	40
Baikal	50
Mid-Atlantic	25–50
Mid-Indian	25
Red Sea (the inner graben)	60

anteclise being 2000 km wide, its transverse slope averages 5 meters/km. The uplift was formed in the late Eocene, that is, just before the graben began to develop [*Mohr*, 1961].

The East African rift system proper is confined to the East African arched uplift (anteclise) that separates the Congo drainage area from the coastal plains of the Indian Ocean. The anteclise has an oval configuration, being elongated from north to south. It is about 2000 km long and 1200 km wide. The central part of the anteclise subsided somewhat relative to its margins and is occupied by the basin of Lake Victoria. Altitudes of the central part range from 1200 to 1400 meters. Altitudes of two arcuate marginal belts embracing the central subsidence exceed 1500 meters and for the most part 2000 meters. The two branches of grabens are confined to these marginal elevated belts.

The pre-Miocene denudation surface rising in the direction of the rift has been mapped within the zone of the Gregory rift (Figure 4). It is elevated above the coast of the Indian Ocean up to 2600 meters, and on the side of Lake Victoria its elevation amounts to 750 meters. Within the graben, this surface subsides locally down to sea level. Near the western branch of the East African rifts, the elevation of the same pre-Miocene surface ranges from 2 to 4 km [*Bishop and Trendall*, 1967; *McConnell*, 1951; *Pulfrey*, 1960*b*].

The pre-Miocene denudation surface is accompanied by the middle Pliocene one, which is a hundred meters lower. Studies of these surfaces indicate that the general uniform uplift occurred during the Miocene and early Pliocene. At that time, the graben evidently existed only as a plastic downwarp. The intense uplift of a belt adjoining the graben as well as intense subsidence of the graben along the fractures took place later, in the late Pliocene and Pleistocene. Upward movements along the graben periphery were contemporaneous with downward movements inside the graben.

The Rhine rift system is in an area of vast arched uplift that originally united the Vosges, Black Forest, and Rhine Slate Mountains. This uplift is up to 250 km wide and 350 km long. It originated in the early Mesozoic, but its sharp elevation occurred simultaneously with the formation of grabens, in the late Miocene [*Dorn*, 1951].

The Baikal rift is along the axis of the Baikal arched uplift formed as a lowland in the Mesozoic. This lowland was intensely uplifted in the Pliocene and Pleistocene, i.e., simultaneously with the formation of grabens [*Florensov*, 1960].

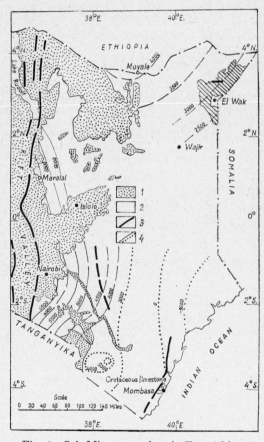

Fig. 4. Sub-Miocene surface in East Africa: 1, Tertiary-Recent lavas; 2, Cretaceous sediments; 3, faults; 4, contours of the late-Cretaceous peneplain.

The formation of continental rifts is closely connected with volcanic activity. The East African grabens are accompanied by volcanic activity, but not everywhere. This volcanic activity may have preceded the formation of grabens or could have occurred simultaneously with it. For example, enormous masses of basalt were erupted in Ethiopia after the formation of an arch and before the formation of a graben (in the Oligocene). During the subsidence of the graben, small eruptions occurred on its floor [Mohr, 1961].

The volcanic rocks from Syria to Ethiopia are typical representatives of olivine-basaltic series with accompanying trachyte. Southward, eruptions have either preceded or accompanied the formation of grabens. The oldest of them, dated early Miocene, are represented by phonolite. The nature of rocks changed in the Plio-Pleistocene. At that time, olivine basalt and trachyte similar to lavas of northern regions alkaline lavas appeared again [Bowen, 1938; were erupted. Later on, phonolite and basic Baker, 1965; Dixey, 1956; Pulfrey, 1960a].

There are no signs of Tertiary volcanic activity along the western branch of the East African rifts. However, extremely violent volcanic activity (eruptions of basalt and trachyte) evolved there in the Pleistocene [Cahen, 1954].

There are practically no signs of young volcanism in the region of the southern end of the East African rifts [Pallister, 1965].

Olivine-basaltic series and alkaline ultrabasic rocks of Oligocene to Pleistocene age are known in the zone of the Rhine rift [Dorn, 1951].

The formation of the Baikal rift was accompanied by eruptions of alkaline basalt and trachyte [Florensov, 1960].

Continental rifts strike chiefly from north to south. The rocks of the East African Precambrian basement are characterized, on the whole, by the same strike. However, there is only general conformity between the strikes of the rift and basement. Notable divergence and intersection of these strikes can be observed when they are considered in detail. The same relations between young rifts and structures of the Precambrian basement are also characteristic of the Baikal region. The Rhine grabens are superimposed on the Paleozoic and Precambrian structures discordantly; the older rocks strike from southwest to northeast, whereas the different parts of grabens trend from south-southwest to north-northeast and from southeast to northwest.

The geophysical properties of continental rifts are rather similar. Rifts are characterized by negative Bouguer anomalies (about 50 mgal) and increased heat flows [Bullard, 1964; Girdler, 1963]. Seismological observations show that recent earthquakes in the rift zones are everywhere connected with extension of the earth's crust (see section by Balakina et al., p. 166 this Monograph).

The geophysical features of rift zones will be considered in other chapters in detail. Later on, we shall also return to the problems of conditions leading to origin of these peculiar structures.

REFERENCES

Baker, B. H., The rift system in Kenya, in *East African Rift System, Upper Mantle Committee —UNESCO Seminar, Nairobi, 1965*, pp. 82–84, University College, Nairobi, 1965.

Bishop, W. W., and A. F. Trendall, Erosion surfaces, tectonics and volcanic activity in Uganda, *Quart. J. Geol. Soc. London, 122*, 385–420, 1967.

Bowen, N. L., Lavas of the African rift valleys and their tectonic setting, *Am. J. Sci.*, [5] *35, A*, 1938.

Bullard, E. C., The flow of heat through the Earth, *ICSU Rev., 6*, 78–83, 1964.

Cahen, L., *Geologie du Congo Belge*, Liege, Vaillant-Carmane, 1954.

Cloos, H., Hebung-Spaltung-Vulkanismus, *Geol. Rundschau, 30*, 4A, 1939.

Dixey, F., The East African rift system, *Overseas Geol. Res. Bull. Suppl., 1*, 1956.

Dorn, P., *Geologie von Mitteleuropa*, E. Schweizerbart, Stuttgart, 1951.

Florensov, N. A., Mesocenozoic and Cenozoic basins of the Baikal region (in Russian), *Tr. Vost.-Sibirsk. Filiala, Akad. Nauk SSSR, Ser. Geol., 19*, 1960.

Girdler, R. W., The relationship of the Red Sea to the East African rift system, *Quart. J. Geol. Soc. London, 114*, pt. 1, 1958.

Girdler, R. W., Geophysical studies of rift valleys, *Phys. Chem. Earth, 5*, 1963.

Heezen, B. C., The rift in the ocean floor, *Sci. Am.*, Oct. 1960.

Krenkel, E., *Geologie Africas*, T. I., Borntraeger, Berlin, 1925.

McConnell, R. B., Rift and shield structure in East Africa, *Intern. Geol. Congr., 18th*, part 14, 1951.

Mohr, P. A., *The Geology of Ethiopia*, Asmara, 1961.

Mohr, P. A., The Ethiopian rift system, *Bull.*

Geophys. Obs. Univ. College, AddisAbaba, 3(1), 1962.

Pallister, J. W., The rift system in Tanzania, in *East African Rift System, Upper Mantle Committee-UNESCO Seminar, Nairobi, 1965,* pp. 86–91, University College, Nairobi, 1965.

Pavlovsky, E. V., The basin of Lake Baikal (in Russian), *Izv. Akad. Nauk SSSR, Ser. Geol., 1937*(2), 1937.

Pavlovsky, E. V., Geological history and geological structure of the Baikal mountainous region (in Russian), *Tr. Inst. Geol. Nauk, Akad. Nauk SSSR, Ser. Geol. 99*(81), 1948.

Picard, L., Structure and evolution of Palestine with comparative notes on neighbouring countries, *Bull. Geol. Dept., Hebrew Univ., 4*(2, 3, 4), 1943.

Pulfrey, W., The geology and mineral resources of Kenya (revised), *Geol. Surv. Kenya Bull., 1960*(2), 1960a.

Pulfrey, W., Shape of the sub-Miocene erosion level in Kenya, *Geol. Surv. Kenya Bull., 1960*(3), 1960b.

Said, R., *The Geology of Egypt,* Elsevier Company, Amsterdam, 1962.

Salop, L. I., *Geology of the Baikal Region,* vol. 2 (in Russian), Izd. Nedra, Moscow, 1967.

von Bubnoff, S., *Geologie von Europa, Bd. II, Das ausseralpine Westeuropa, Teil. 3, Die Structur des Oberbaues und das Quartär Nordeuropas,* pp. 1135–1601, Borntraeger, Berlin, 1936.

Whiteman, A. J., A summary of present knowledge of the Rift valley and associated structures in Sudan, in *East African Rift System, Upper Mantle Committee—UNESCO Seminar, Nairobi, 1965,* pp. 34–46, University College, Nairobi, 1965.

Wrench (Transcurrent) Fault Systems

H. W. Wellman

WRENCH faults, also known as transcurrent or strike-slip faults, are those with mainly strike-slip displacement. There are two opposed kinds: dextral and sinistral. Faults with mainly dip-slip displacement are also of two opposed kinds: normal and reverse. Many wrench faults have an appreciable dip-slip component, and thus have direct importance in mountain building.

Surfaces that were once horizontal are a commonplace of geology and provide reference surfaces for determining the dip-slip component of displacement. Far less common than reference surfaces are the reference lines, essential for determining wrench faulting and for determining the true direction of displacement of any fault. The true nature of most active faults is known from the relatively abundant topographic reference lines, but, because of the general absence of geological reference lines, this is not true for most inactive faults, many of which have been assumed to be purely dip-slip faults merely because of the absence of any reference lines. On the other hand, there is conclusive geophysical evidence, in the form of displaced magnetic anomaly belts, of large wrench displacements on the sea floor [*Menard,* 1964]. The faults are straighter and longer than any of the wrench faults on land, but their present sense of displacement is unknown, and they may or may not be active.

The following account is of the major wrench faults only, special attention being given to active faults, those now active or known to have been active during the last thousand or so years. They are listed in Table 1, and are shown on a world map in Figure 1. Activity is shown by displacement of topographic features, and the degree of activity can be judged by the age of the features displaced by the fault, and less accurately by the degree of definition of the fault trace.

Except in dense forests, active faults are sharply defined on air photographs by topographic irregularities that cannot be due to differential erosion of rocks of different hardness brought together by faulting in the distant past. Trailing streams, sag ponds, shutter ridges, and other unnamed kinds of disturbed drainage and topography are the main diagnostic features of active faulting [*Allen*, 1965].

Provided that the age of the features is known, the average rate of displacement can be determined when fault-displaced features can be matched and the displacement measured. For many faults, displacement at the rate of a centimeter or so a year has been taking place for many thousands of years.

The crust of the earth is density stratified, and the density differences limit the throw of normal and reverse faults. Wrench faults are not limited in this way, and there is no theoretical reason why unlimited displacement cannot take place.

Because strike-slip displacement is less easily recognized than dip-slip displacement, wrench faults are almost certainly under-represented in the literature. The relative importance of the different kinds of faults is probably best shown by the faulting that has taken place during historical earthquakes.

According to *Richter* [1958], the nature of displacement is known for about 30 historical earthquakes, about 14 being mainly dip-slip, and about 16 mainly strike-slip. At almost all places the length along which the displacement took place was greater for the wrench faults than for the dip-slip faults. Of the long lineations on the earth's crust, many have been proved to be wrench faults, but none to be essentially dip-slip faults. Hence as length is a measure of importance, wrench faults are more important than dip-slip faults.

For any particular wrench fault, the greater the displacement to be measured, the greater is the age of the reference line needed to define the displacement. Reference lines differ in kind according to their age, and several different kinds of reference lines are needed to determine the full history of any wrench fault. Man-made features—roads, railways, walls, and rows of trees—are the most useful reference lines for determining displacements of 5 mm to 10 meters that have taken place during the last hundred or so years. Topographic features such as streams and ridges, and particularly terrace edges, are most useful for determining displacements of from 1 meter to 1 km of the last 100,000 or so years. Conglomerates with distinctive rocks derived from a restricted source area on the opposite side of the fault have proved to be most useful for determining displacements of from 2 to 50 km. Most of the examples are Pliocene to mid-Pleistocene in age. Conglomerates and source areas are facies boundaries of a particular kind, and older and larger wrench displacements on land are mostly inferred from facies boundaries of a more general kind. The best known examples are from the Alpine fault of New Zealand and the San Andreas fault of California. Postulated displacements for facies boundaries that are Jurassic or older range up to 700 km.

Reference lines may have formed during wrench faulting or before wrench faulting started, but a fundamental assumption for all reference lines is that they were straighter before wrench faulting than they are now. If they formed during faulting, then the most useful lines are those that formed quickly, relative to the rate of wrench displacement. Terrace edges are probably the most useful topographic features. They form quickly and, for several faults, give evidence of continuing movement [*Wellman*, 1955; *Sugimura and Matsuda*, 1965]. It is not unusual to find that a series of streams crossing a particular fault have been equally offset by 10 to 100 meters. Most of the reported examples are in periglacial regions. It is inferred that the streams are the same age, and probably came into being because of the sudden climatic change at the end of the last glaciation some 15,000 years ago.

The most important geological feature of wrench faults, and the most difficult to establish, is displacement of 100 km or more. It is generally inferred from one or more dog-legged facies boundaries. In extreme cases two opposite explanations are possible:

1. The fault is younger than the facies boundary, which was originally straight, the dog-leg being caused by the wrench faulting.
2. The fault is older than the facies boundary, which was never straight, the dog-leg being caused by contemporaneous dip-slip displacement controlling deposition of sediments.

For most faults the probabilities have to be balanced, simplicity of postulates and the principal of uniformitarianism often being the only tests. Belief in large wrench displacements depends largely on knowledge of the rate of active

TABLE 1. Major Wrench Faults of the World (Mostly Active)

No. on Figure 1	Name of Fault Zone	Sense of Displacement[a]	Year of Major Earthquakes[b]	Maximum Wrench Displacement during Earthquakes,[b] M	Present Displacement Rate,[c] mm/yr	Estimated Rate,[d] mm/yr	Long Term Rate,[e] mm/yr	Minimum Age of Reference Line with Total Displacement	Total Inferred Displacement, km	Fault Length (well defined), km	Fault Length (possible), km	Shape[f]	References
1	Dead Sea	S			(7)	5	1	Upper Cretaceous	100	350	700		Quennell [1959]
2	Anatolia, Turkey	D	1939–1953	4.3		20	20?	Miocene	350?	1,600		Z?	Pavoni [1961], Richter [1958]
3	Zagros, Iran	D?				1				1,500			Wellman [1965]
4	Duruneh, Iran	S				5				700		S	Wellman [1965]
5	Shahrud, Iran	S	1962	0.1		10				1,000	1,500	S	Wellman [1965]
6	Herat, Afghanistan	D	1892	1.0		10				1,000		S	Wellman [1965]
7	Chaman, Afghanistan	S				15		10?	Lower Tertiary	500?	800		Z?	Richter [1958]
8	Taofu, China	?	1923							300			Gutenberg and Richter [1954], Richter [1958]
9	Hoetanapan, Sumatra	D	1892	4						20	2,000?		Kennedy [1946], Richter [1958]
10	Philippine	S?				10				1,100	1,500	Z	Allen [1965], Richter [1958]
11	Longt. Valley, Taiwan	S	1951	2						200			Richter [1958]
12	Alpine, New Zealand	D	1848, 1888	2.5	(30)	30	5	Base Cretaceous	480	1,100	2,000	S	Allen [1965], Richter [1958], Wellman [1955]
13	Magellanes, Chile, Argentina	?								600	2,500		Katz [1964]
14	Atacama, Chile	D				2				900		Z?	Allen [1965], St. Amand and Allen [1960]
15	Oca and El Pilar	D?						5	Cretaceous	475	1,200	2,000	Z	Rod [1956]
16	Bartlett, Caribbean	S?							Cretaceous	400	1,000		S	Rod [1956]
17	San Andreas, California	D	1906, 1940	4.5	50	30	5	Base Cretaceous	500	1,000	1,700		Allen [1965], Richter [1958], Whitten [1960]
18	Denali, Alaska	D	1822?			10			200?	1,200		C	St. Amand [1957]
19	Kamchatka	?				5				500		Z	Richter [1958]
20	Median tect. line, Japan	?								600			Richter [1958]
21	Bolnai, Mongolia	S	1905			20				400	1,000	C	Florensov and Solonenko [1963], Richter [1958]

No.	Locality	Shape	Year				Age				Shape2	Reference
22	Bogdo, Mongolia	S	1957	8.8				30	300	1,000	C	Florenson and Solonenko [1963], Richter [1958]
23	Kebin, USSR	?	1911									Bogdanovitch [1914], Richter [1958]
24	Ferghana, USSR	D						20	500			Burtman [1963]
25	Great Glen, Scotland	S				Paleozoic	104	0 0.2	200		S	Kennedy [1946]

a D, dextral; S, sinistral.
b Most values from *Richter* [1958].
c Secular rate from geodetic observations or (in parentheses) rate from displaced topographic features of estimated age. Rates may not apply to full length of fault.
d Present rate estimated by author on basis of fault trace. Faults are ranked in order of disturbance of minor topographic features shown on air photographs and are given values in agreement with those faults for which present rate of displacement is known.
e Long-term average rate of displacement inferred from age of reference lines showing maximum displacement.
f C, simple arc; S, recurved 'S' shape; Z, recurved mirror image.

faulting. If the average rate of wrench displacement of a particular fault is a centimeter or so a year for the last ten thousand or more years, displacements of 10 km are not improbable for the early Pleistocene, and displacements of 100 km are not improbable for the Miocene. The last reversal in tectonic stress conditions for a region is the limit at which extrapolation becomes improbable.

Few, if any, of the major active faults have a straight surface trace. Most are recurved arcs, with either an 'S' or its mirror image form, which, for convenience, is termed 'Z' in Table 1. Some are simple arcs ('C' in Table 1), but their full extent may not be known. Table 1 shows that there is no relation between S or Z shape and dextral (D) or sinistral (S) movement. For the Z-shaped recurved Alpine fault zone, curvature is directly related to dip-slip displacement. The central part of the fault is pure dextral; to the north the western side, which contains the center of curvature, is upthrown relative to the eastern side, and to the south the reverse is true. The same relation between upthrow and center of curvature is shown by the Chaman, Anatolia, and Bolnai faults, and probably by the Philippine fault. On the other faults, vertical displacement is either small, as for the San Andreas fault, or irregular, or unknown.

The kinematic problems of wrench faulting are brought out by the construction of local or worldwide shear nets showing actual or inferred dextral and sinistral faults (or by the construction of analogous tectonic nets showing inferred compression or relative shortening). Because the earth is round, the simplest tectonic net has at least two, and most nets drawn have several, special centers which the lines of the net either encircle or radiate from. The centers are of three main kinds: (1) translation centers (any direction of translation); (2) rotational centers (clockwise or counterclockwise rotation); (3) change of area centers (expansion or contraction). Any combination of the three kinds is possible. The first kind is illustrated by the relative southeastward movement of the Borneo region between the sinistral Philippine fault and the dextral Hoetanapan fault of Sumatra. An eastward translatory movement is possible for the northern and southern Antilles. For translatory movement, it is assumed that the wrench faults pass into reverse faults in

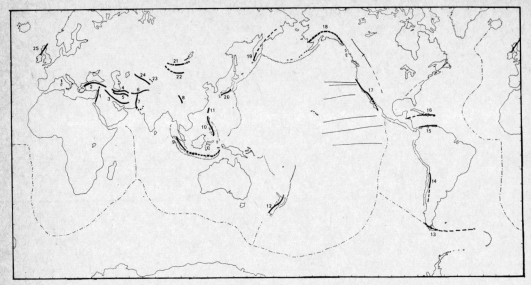

Fig. 1. World map of major wrench faults. Heavy lines are faults listed in Table 1; lighter lines are other possible major wrench faults; dot-dash lines are oceanic ridges; wavy lines are oceanic trenches. Mercator's projection with 10° graticule on margin.

the direction of translation, and into normal faults in the opposite direction. Translatory movement is of the kind that has been postulated for continental drift, but the translatory movements mentioned above are not part of any generally accepted drift pattern.

The second kind is illustrated by the somewhat hypothetical counterclockwise rotation of the Pacific Ocean on the dextral San Andreas, Atacama, Alpine, and Denali faults [Benioff, 1957]. For the simplest pattern of rotation there is no 'end problem,' the faults or their hypothetical extensions forming a complete circle.

The third kind is represented by the Dasht-I-Lut contraction center in eastern Iran [Wellman, 1965] and probably by an expansion center in Honshu, Japan. Wrench faults spiral out from such centers, dextral faults counterclockwise and sinistral faults clockwise for contraction centers and in reverse sense for expansion centers. The other ends of the faults may die out or spiral into other centers.

For thousands of years there have been repeated displacements along the major active wrench faults. Some of the displacements may have been gradual, but most have probably been sudden, and many major earthquakes are directly related to sudden displacements along major wrench faults. Indeed, several well defined wrench faults were not discovered until renewed displacement took place during an earthquake (for example Wairarapa fault, New Zealand; Bogdo and Bolnai faults, Mongolia). It is certain that the major active wrench faults will be the sites of many future earthquakes, and, if only for the purpose of earthquake zoning, it is desirable that all should be mapped. It is almost certain that many active faults remain to be discovered. Aerial photography makes their mapping easy, and all the land areas of the world should be systematically scanned for the major active faults.

REFERENCES

Allen, C. R., Transcurrent faults in continental areas, *Phil. Trans. Roy. Soc. London, A, 258*, 82–89, 1965.

Benioff, H., Circum-Pacific tectonics, *Publ. Dominion Obs. Ottawa, 20*(2), 325–402, 1957.

Bogdanovitch, Ch., et al., Earthquake of 22 December 1910 to the South of Tian Shan (in Russian), *Comité géol. [N.S.] 89*, 1914.

Burtman, V. S., *Akad. Nauk SSSR, Geol. Inst. Tr., 80*, 128–151, 1963.

Florensov, N. A., and V. P. Solonenko, *Gobi-Altai Earthquake*, Akad. Nauk. SSSR, 391 pp., 1963.

Gutenberg, B., and C. F. Richter, *Seismicity of the Earth and Associated Phenomena*, Princeton University Press, Princeton, New Jersey, 1954.

Katz, H. R., Strukturelle Verhältinisse in den Südlichen Patagonischen Anden und deren beziehung zur Antarktis, Eine Diskussion, *Geol. Rundschau, 55,* 1125–1213, 1964.

Kennedy, W. Q., The Great Glen fault, *Quart. J. Geol. Soc. London, 102,* 41-76, 1946.

Menard, H. W., *Marine Geology of the Pacific,* McGraw-Hill Book Company, New York, 1964.

Pavoni, N., Die Nordanatolische Horizontalverschiebung, *Geol. Rundschau, 51,* 1–16, 1961.

Quennell, A. M., Tectonics of the Dead Sea rift, paper presented at 20th Intern. Geol. Congr. (Mexico, 1956), *Assoc. Serv. Geol. Africains, Actas Trabajos,* pp. 385–405, 1959.

Reid, H. F., Sudden earth movements in Sumatra in 1892, *Bull. Seismol. Soc. Am., 3,* 72–79, 1913.

Richter, C. F., *Elementary Seismology,* W. H. Freeman and Company, San Francisco, 1958.

Rod, Emile, Strike-slip faults of northern Venezuela, *Bull. Am. Assoc. Petrol. Geologists, 20(2),* 1956.

St. Amand, P., Circum-Pacific orogeny, *Publ. Dominion Obs., Ottawa, 20(2),* 403-411, 1957.

St. Amand, P., and C. R. Allen, Strike-slip faulting in northern Chile (abstract), *Bull. Geol. Soc. Am., 71,* 1965, 1960.

Sugimura, A., and T. Matsuda, Atera fault and its displacement vectors, *Bull. Geol. Soc. Am., 76,* 502-522, 1965.

Wellman, H. W., New Zealand Quaternary Tectonics, *Geol. Rundschau, 43,* 248–257, 1955.

Wellman, H. W., Active wrench faults of Iran, Afghanistan, and Pakistan, *Geol. Rundschau, 55,* 716–773, 1965.

Whitten, C. A., Horizontal movement in the earth's crust, *J. Geophys. Res., 65,* 2839–2844, 1960.

Zak, I., and R. Freund, Recent strike-slip movements along the Dead Sea rift, *Israel J. Earth-Sci., 15,* 33–37, 1966.

Continental Margins[1]

Charles L. Drake

KNOWLEDGE of the structure and history of development of continental margins is critical to the understanding of basic geologic processes. Ancient geosynclines, now transformed into mountain systems and in various states of erosional reduction, tended to be formed along the margins of continents or in intracontinental troughs. Much of the present tectonic activity takes place along continental margins. We have difficulty in finding exact parallels to the intracontinental troughs in regions of present sediment accumulation, but it is possible to find continental margins that bear strong resemblance to reconstructions of earlier marginal geosynclines at different stages of development. Furthermore, current ideas about the development of both continents and ocean basins call for major activity along the margins.

It is generally accepted that there are two principal types of continental margins, Atlantic and Pacific, which differ significantly in nature. The Atlantic type is characterized by stable continental blocks on the landward side that have been deformed little, if any, since Paleozoic time. The transition from oceanic to continental crustal dimensions is in general rather abrupt and in a number of areas great thicknesses of sediment have collected. The Atlantic margins are not usually marked by a deep sea physiographic trench, although a trench-like feature is frequently found buried beneath the sediments of the continental rise. Often the Paleozoic and earlier structures are truncated by the continental margin, but there are also examples of structures paralleling the margin.

The Pacific type is marked by young tectonic belts paralleling the shores and virtually surrounding the Pacific Ocean. Shallow focus earthquakes are very common, and intermediate and deep focus earthquakes are abundant, particularly along the southern and western margins of the ocean. Volcanic activity occurs or has occurred historically along the entire margin and, with the exception of western North America, the entire Pacifice is bounded by deep sea physiographic trenches. The transition from ocean to continent is much more variable than

---

[1] Contribution 1230 from the Lamont-Doherty Geological Observatory of Columbia University.

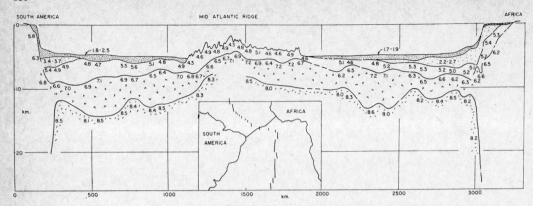

Fig. 1. Seismic refraction section across the equatorial Atlantic from Sierra Leone to Brazil [after *Leyden et al.*, 1967].

in the Atlantic, and sedimentary accumulations do not appear to be as great.

The structure of the continental margins has been studied in some detail in several areas using seismic refraction measurements supplemented by gravity and magnetic observations and samples from the ocean floor. *Drake and Nafe* [1968] in an examinaion of the average seismic properties of the crust and upper mantle, divided the marginal areas into shallow (water depth less than 1000 meters) and deep (water depth greater than 1000 meters) parts and showed that the properties of these were neither those of continents nor those of the ocean basins but were intermediate between them. *Kosminskaya and Zverev* [1968], in a parallel study, used different divisions (continental, subcontinental, suboceanic, and oceanic), but noted similar gradational changes in character.

These conclusions were based on many seismic refraction measurements across the margins. These are difficult to interpret in the marginal areas because of rapid changes in bottom topography and in the configuration of the underlying layers. Interpretation is further complicated by the fact that in the marginal areas continental materials are mingled with oceanic materials having similar ranges of seismic compressional velocity, and these cannot be differentiated. This is well illustrated by the generalized section across the equatorial Atlantic Ocean shown in Figure 1. Velocities in the range 5.2–5.4 km/sec under the African shelf and slope can be correlated with Lower Paleozoic sedimentary rocks on shore, while velocities in the same range under the ocean basins are usually assumed to be volcanic. On the mid-oceanic ridge the velocities are slightly lower, but of the same order, and here they are undoubtedly volcanic. Equivalent difficulties are encountered when one tries to establish the limits of the deeper crustal layers and their configuration as the edge of the shelf is approached.

The most extensively studied Atlantic margin is that of eastern North America. This margin is paralleled on shore by the Appalachian mountain system, which extends from the southern United States through the Maritime provinces of Canada. Total sedimentary thickness has been measured by seismic refraction methods in the water-covered parts of the margin, and the configuration of the crystalline rock surface has been mapped [*Drake and Nafe*, 1968]. In Figure 2 the seismic data have been combined with outcrop and borehole data to make a composite map. The marine contours mark the shallowest occurrence of velocities in excess of 5.6 km/sec, while those on land are based variously on Precambrian to pre-Cretaceous surfaces.

Elongated, sediment-filled troughs are found paralleling the coast, both on the continental shelf and in deep water. The sediment accumulation is considerable, reaching 10 km in places under the continental rise and 6–7 km under the shelf. The shallow structure and the character of the sediments along this margin have been favorably compared with reconstruc-

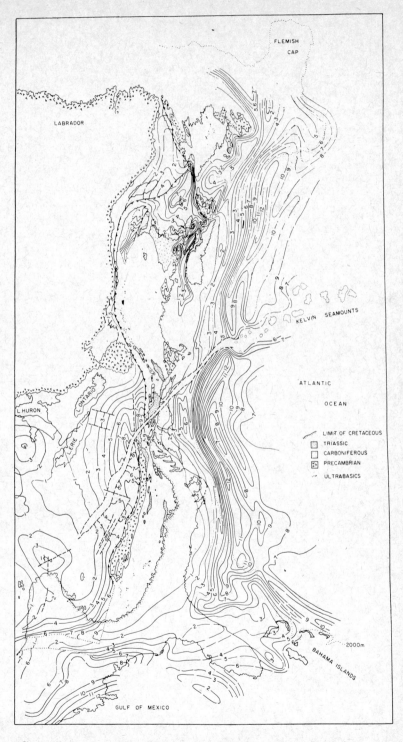

Fig. 2. Crystalline rock surface off eastern North America. Marine contours are on shallowest occurrence of velocities in excess of 5.6 km/sec. Land contours are after Basement Map of North America (1966) modified by Woodward [after *Drake et al.*, 1968].

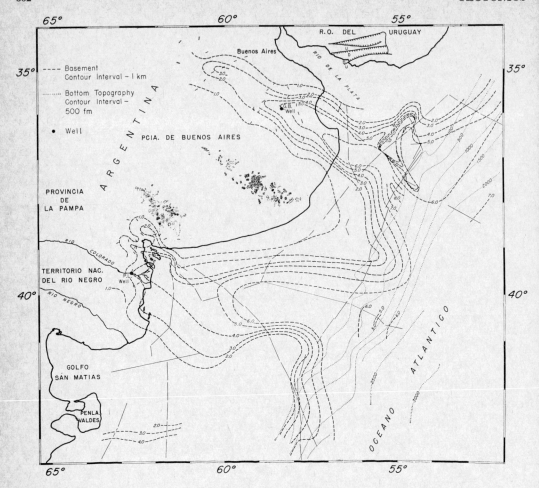

Fig. 3. Contour map of basement under Argentine continental margin [after *Ewing et al.*, 1963]

tions of the Appalachian geosyncline before its deformation [*Drake et al.*, 1959]. The deeper structure, particularly that of the top of the mantle, is more difficult to determine by seismic refraction methods. The Mohorovicic discontinuity deepens gradually toward the continental slope to 17–18 km and then must plunge rather abruptly to continental depths under the slope and the upper part of the rise. *Worzel* [1965] has used gravity measurements together with the seismic observations to determine the configuration of the mantle surface. In these calculations it was assumed that densities within each layer were constant and that the anomalies were due to the configurations of the layers. All the sections calculated show a rapid thickening of the crust to 30–35 km as the shelf is approached.

Linear magnetic anomalies are found paralleling the shore on the continental shelf [*Drake et al.*, 1963; *Taylor et al.*, 1967], but the continental rise is marked by a quiet magnetic zone quite different from the shelf or from the ocean basins. *Heirtzler and Hayes* [1967] suggested that this could be related to the ocean floor spreading hypothesis and might represent ocean crust created in late Paleozoic time. *Drake and Nafe* [1967] extended this and suggested that the zone might well represent the original crust of a small proto-Atlantic Ocean. The apparent absence of this zone from the South Atlantic Ocean [*Dickson*, 1968] could be interpreted as resulting from the absence of an original ocean basin here before postulated continental drift.

Studies in other parts of the Atlantic margin have yielded results similar in general, though differing in detail, to those from North America. With the exception of the Caribbean and Scotia arcs and the Mediterranean area, the margins have been little affected by tectonic activity since the end of the Paleozoic. Sedimentary thicknesses tend to be smaller along other parts of the margin [*Hill and Laughton*, 1954; *Ewing and Ewing*, 1959; *Leyden et al.*, 1968] and the details of the structure are often different. On the Argentine margin, for example, structures on the shelf are perpendicular rather than parallel to the edge of the shelf [*Ewing et al.*, 1963] (Figure 3).

On the Pacific margin of the United States, *Thompson and Talwani* [1964] constructed a section extending 500 km seaward and shoreward from San Francisco using seismic refraction, gravity, magnetic, and Rayleigh wave phase velocity data (Figure 4). This area has been subjected to much more recent deformation than has the Atlantic coast, and the reduced accumulation of sediments on the margin must represent a reduction in time as well as in physical dimensions. It is also complicated by the immediate proximity of the northward extension of the East Pacific rise. The crust thickens in an irregular manner over the entire section, and in the coastal area is of the order of 20 km in thicknesses as compared with 30–35 km under the Atlantic coast. In the upper mantle, velocities in the range 7.2–7.6 km/sec indicate the presence of anomalous material under the continental part of the profile [*Pakiser*, 1963]. This was taken into account in the gravity calculations, and density differences were allowed both in the lower crust and in the upper mantle. The inhomogeneities in the mantle extend to a depth of some 50 km. While the structure is necessarily simplified, it must more closely approximate the real conditions at depth than would be the case if uniform

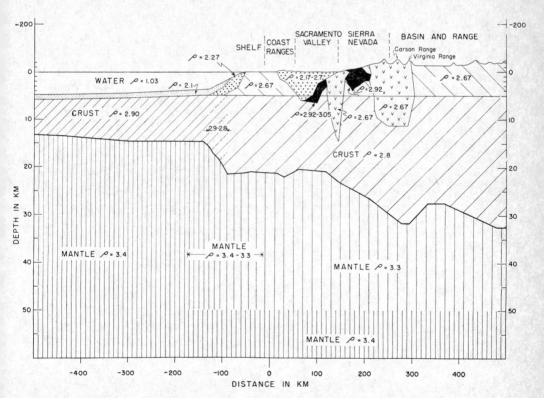

Fig. 4. Structure section from Pacific basin to central Nevada based on seismic refraction, gravity, and Rayleigh-wave phase velocity data [after *Thompson and Talwani*, 1964.]

layers were assumed. Thompson and Talwani note that the anomalous upper mantle, if it represents normal mantle altered in some way, could explain about 1 km of the uplift that took place over much of the region in Cenozoic times. Such anomalous mantle, marked by velocities of 7+ km/sec, is characteristic of other regions of present or past tectonic activity.

Kosminskaya and Riznichenko [1964] described the results of detailed seismic refraction measurements over the Pacific margin of Siberia, a region of deep focus earthquake activity (Figure 5). Magnetic and gravity observations were made simultaneously with the seismic measurements. In this region transitions are found from typically continental crust through intermediate to oceanic crust. Geological arguments indicate to Kosminskaya and Riznichenko that, where proper conditions exist, it is possible to pass from less differentiated oceanic crust to more differentiated continental crust and vice versa, mainly through changes in the thickness of the lower crust or upper mantle where velocities are of the order of 7 km/sec. This concept was supported by *Drake and Nafe* [1968], who suggested that material in the velocity range 7.2–7.7 km/sec might be both created and removed during the course of tectonic activity.

Kosminskaya and Riznechenko suggest, as did Thompson and Talwani, that density differences of the order of 0.5 g/cm³ exist in the mantle to depths of 50 km. This is supported by the lack of a clear relationship between depth to the Mohorovicic discontinuity and Bouguer gravity anomaly, a lack which cannot be wholly explained by variations in average crustal density.

Thus the margins on the two sides of the Pacific bear strong resemblances in structure although they differ considerably in physiography, local geology, and extent of earthquake activity. They are both in contrast with the Atlantic sections, where the transition zone between ocean and continent is not marked by extensive areas of intermediate crustal properties. This is in accord with recent hypotheses of movement of the continents and evolution of the ocean basins which call for the creation of new crustal material along the axes of the mid-oceanic ridges [*Hess*, 1962; *Dietz*, 1961]. Unless the earth as a whole is expanding, the creation of new crust creates a space problem, and it is necessary to find a mechanism and a location for the elimination of old crust. This is difficult to accomplish in the Atlantic Ocean, because its margins are stable and have been for a considerable period of time. Crust might be destroyed along the margins of the Pacific, which is surrounded by tectonically active zones at the present time, if a suitable mechanism could be found.

Such a mechanism is suggested by recent work of *Oliver and Isacks* [1967]. Studies of deep focus earthquakes in the Fiji-Tonga-Raratonga area revealed that there were marked differences in the attenuation of shear waves propagating along paths within and without the zone of high seismicity which dips to the west and extends to more than 700 km in depth

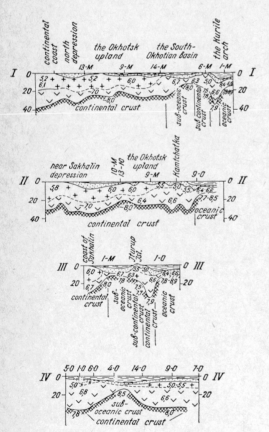

Fig. 5. Sections across the Kuril island arc and the Sea of Okhotsk based on deep seismic sounding and gravity data [after *Kosminskaya and Riznichenko*, 1964].

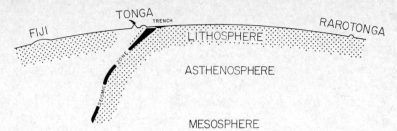

Fig. 6. Section from Fiji to Rarotonga showing the anomalous region of low attenuation (high Q) following the zone of high seismicity [after *Oliver and Isacks*, 1967].

beneath the Tonga-Kermadec arc. The attenuation is far less (or Q, the reciprocal of the specific attenuation factor, is far greater) in the seismically active zone than is the case at comparable depths on either side (Figure 6). Similar effects have been found in studies from Japan [*Katsumata*, 1960; *Honda*, 1932]. Oliver and Isacks suggest that these attenuation differences may indicate that the deep anomalous zone is continuous with the uppermost mantle east of Tonga. This interpretation would not be inconsistent with hypotheses calling for down-moving convection currents on the flanks of island arcs. If one can correlate low attenuation (or high Q) with strength, then it is not unreasonable to suggest that the lithosphere has been dragged or thrust or has settled beneath the Tonga arc. This in turn would allow for the elimination of excess crust created in the mid-oceanic ridges. If such a process is accepted for the Tonga area, then it cannot be rejected for other active areas of the Pacific, and a certain mobility for the lithosphere is clearly implied.

The Atlantic margins are stable, even though the Mid-Atlantic ridge is included with the others in ocean floor spreading hypotheses, and it is difficult to invoke the mechanism of Oliver and Isacks there. A possible solution to this problem is to invoke continental drift in addition to ocean floor spreading and to make the rate of drifting of the continents bordering the Atlantic equal to the rate of formation of new crustal material in the Mid-Atlantic ridge. This would imply that all crustal elimination would take place in the Pacific and that the Atlantic margins have remained essentially unchanged since their separation and would account for the differences in structure mentioned earlier.

Clearly much of the above is highly speculative, and it is by no means universally accepted. It is equally clear that the continental margins cannot be considered alone, but only in relationship to the development of the continents and the ocean basins. Much remains to be done in these important areas. The similarity of the crustal structure of some marginal areas to ancient geosynclines is marked, but little is known about the deeper structure or the forces necessary to cause tectonism. Conversion of oceanic crust to continental crust and of continental to suboceanic or subcontinental crust along the margins appears possible, but again the mechanism is unknown. Studies of deep earthquakes suggest a promising mechanism for the disposal of excess crustal material created during postulated evolution of the ocean basins, but this is not yet supported by other types of data pertaining to the shallow crustal layers. Last, it should be noted that the majority of the detailed data pertaining to continental margins are limited to only a few areas. A great deal of work remains to be done to ensure that generalizations drawn on a few areas apply to all.

Acknowledgment. This work was supported by Office of Naval Research contract N00014-67-A-0108-0004.

REFERENCES

Dickson, G., Magnetic anomalies and ocean floor spreading in the South Atlantic, unpublished thesis, Columbia University, 1968.

Dietz, R. S., Continent and ocean basin evolution by spreading of the sea floor, *Nature, 190*(4779), 854–857, 1961.

Drake, C. L., J. I. Ewing, and H. Stockard, The continental margin of the eastern United States, *Can. J. Earth Sci., 5*(4, part 2) 993–1010, 1968.

Drake, C. L., M. Ewing, and G. H. Sutton, Continental margins and geosynclines: The east coast of North America north of Cape Hatteras, *Phys. Chem. Earth*, 5, 110–198, 1959.

Drake, C. L., J. Heirtzler, and J. Hirshman, Magnetic anomalies off eastern North America, *J. Geophys. Res.*, 68, 5259–5275, 1963.

Drake, C. L., and J. E. Nafe, Geophysics of the North Atlantic region, paper presented at the UNESCO-IUGS Symposium on Continental Drift Emphasizing the History of the South Atlantic Area, Montevideo, Uruguay, October 1967.

Drake, C. L., and J. E. Nafe, The transition from continents to oceans from seismic refraction data, in *The Crust and Upper Mantle of the Pacific Area, Geophys. Monograph 12*, pp. 174–186, American Geophysical Union, Washington, D.C., 1968.

Ewing, J., and M. Ewing, Seismic-refraction measurements in the Atlantic Ocean basins, in the Mediterranean Sea, on the mid-Atlantic ridge, and in the Norwegian Sea, *Bull. Geol. Soc. Am.*, 70, 291–318, 1959.

Ewing, M., W. J. Ludwig, and J. I. Ewing, Geophysical investigations in the submerged Argentine Coastal Plain, 1, Buenos Aires to Peninsula Valdez, *Bull. Geol. Soc. Am.*, 74, 275–292, 1963.

Heirtzler, J., and D. Hayes, Magnetic boundaries in the North Atlantic, *Science*, 157(3785), 185–187, 1967.

Hess, H., History of ocean basins, in *Petrologic Studies: A Volume to Honor A. F. Buddington*, edited by A. E. J. Engel et al., pp. 599–620, Geological Society of America, New York, 1962.

Hill, M. N., and A. S. Laughton, Seismic investigations in the eastern Atlantic, *Proc. Roy. Soc. London, A*, 222, 348–356, 1954.

Honda, H., On the types of the seismograms and the mechanism of deep earthquakes, *Geophys. Mag.*, 5, 301–326, 1932.

Katsumata, M., The effect of seismic zones upon the transmission of seismic waves, *Kensin-Siho*, 25(3), 1960.

Kosminskaya, I. P., and Y. V. Riznichenko, Seismic studies of the earth's crust in Eurasia, in *Research in Geophysics*, edited by H. Odishaw, vol. 2, *Solid Earth and Interface Phenomena*, pp. 81–122, MIT Press, Cambridge, 1964.

Kosminskaya, I. P., and S. M. Zverev Deep seismic sounding of transition zones from continents to oceans, in *The Crust and Upper Mantle of the Pacific Area, Geophys. Monograph 12*, American Geophysical Union, Washington, D. C., 1968.

Leyden, R., R. Sheridan, and M. Ewing, A seismic refraction section across the equatorial Atlantic, paper presented at UNESCO-IUGS Symposium on Continental Drift Emphasizing the History of the South Atlantic Area, Montevideo, Uruguay, October 1967, in press, 1968.

Oliver, J. E., and B. Isacks, Deep earthquake zones, anomalous structures in the upper mantle, and the lithosphere, *J. Geophys. Res.*, 72, 4259–4275, 1967.

Pakiser, L. C., Structure of the crust and upper mantle in the western United States, *J. Geophys. Res.*, 68, 5747-5756, 1963.

Taylor, P. T., I. Zietz, and L. S. Dennis, Geologic implications of aeromagnetic data for the eastern continental margin of the United States paper presented at Session on Geomagnetic Anomalies, 14th General Assembly, IUGG, Zurich, September 1967.

Thompson, G. A., and M. Talwani, Crustal structure from Pacific basin to central Nevada, *J. Geophys. Res.*, 69, 4813–4837, 1964.

Worzel, J. L., *Pendulum Gravity Measurements at Sea, 1936–1959*, Interscience Publishers, New York, 422 pp., 1965.

The Crust and Upper Mantle beneath the Sea

Peter R. Vogt, Eric D. Schneider, and G. Leonard Johnson

IN THE abyss the upper mantle makes its closest approach to the earth's surface. The thin 5- to 9-km layer that overlies the M discontinuity beneath the ocean basins alternately is referred to as oceanic crust and mantle rind. In fact there are good reasons to believe that the upper mantle may even lie exposed to the oceanic waters in the great world-encircling mid-oceanic ridge. Thus the ocean floor offers an exceptionally detailed view of the upper mantle. We must, however, learn to distinguish processes and distributions relating to the ocean above from those resulting from processes in the upper mantle below. For this reason, the discussion begins with the sediment bodies and their modes of origin.

'Could the waters of the ocean be drawn off, the very ribs of the solid earth would be brought to light' [*Maury*, 1855].

SEDIMENTS AND PHYSIOGRAPHY

The Sedimentary Cover of the Sea Floor

The crust of the ocean floor is mantled by sediments that have modified the shape of its original igneous configuration into zones reflecting the age of the basin, nearness to continents and sediment sources, and zones of biologic productivity. The contrast in morphology between the mid-oceanic ridge and the areas adjacent to the continents (continental rises and abyssal plains) does not lie in the tectonic processes that formed them but in the sediment processes that have subsequently shaped these areas of the sea floor (Figures 1 and 2).

The scope of this paper does not permit a thorough discussion of marine sediments and processes of sedimentation, but general patterns of oceanic sedimentation must be known if we are to understand the shape of the sea floor which fathometers have probed for the past several decades.

Marine geologists from the earliest oceanographic expeditions distinguished the major sedimentary types found in the ocean. They were quick to note the difference between the biologically produced foram oozes and the land-derived terrigeneous sands and silts, but these early workers did little work on the processes of marine sedimentation and their effect on the sea floor.

There are many classifications of marine sediments; most classifications encompass sediment component, physical properties, and sedimentary process. The following is a brief classification of marine sediments.

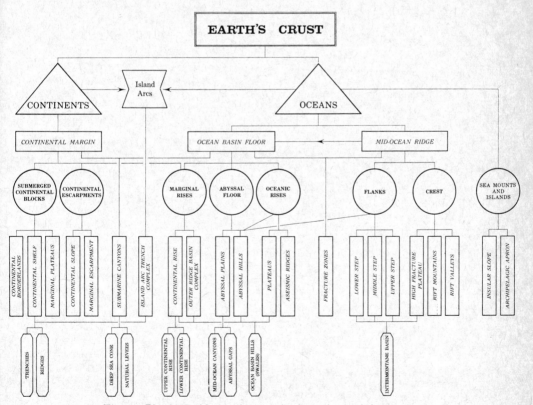

Fig. 1. Physiographic provinces and subprovinces of the sea floor.

558

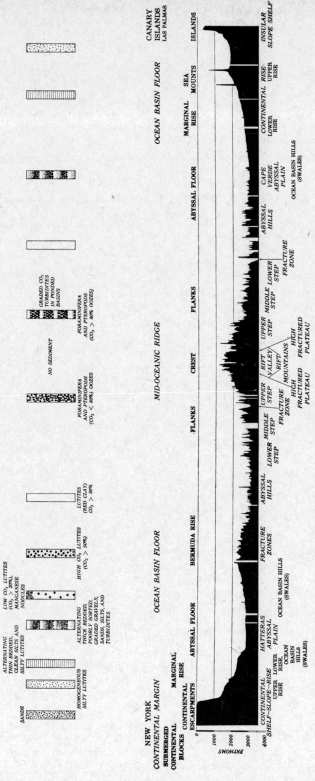

Fig. 2. Profile at 100:1 vertical exaggeration across the North Atlantic relating physiographic provinces and sediment types.

A. Mineralogical content
   1. Oceanic, terrigeneous, or extraterrestrial components
      a. Mineral grains, quartz, feldspars, pyroxenes, etc.
      b. Clay minerals
   2. Biologic components
      a. Foraminifera shells, $CaCO_3$
      b. Diatoms and radiolarians, etc., $SiO_2 \cdot H_2O$
      c. Coccoliths and discoasters, $CaCO_3$
      d. Fish bones and scales, complex phosphates
   3. Chemical residuals, precipitates, or alterations
      a. Manganese nodules
      b. Clay minerals
      c. Phosphates
      d. Carbonates
      e. Halides
      f. Sulphides
B. Physical properties
   1. Size
   2. Color
   3. Density
   4. Acoustical velocity
   5. Sorting
   6. Skewness
C. Sedimentary process or transport mechanism
   1. River transport
   2. Wind transport
   3. Ice transport
   4. Wave transport
   5. Bottom current
   6. Surface current
   7. Cataclastic explosion
   8. Mass slumping
   9. Biologic sedimentation and reworking
   10. Particulate settling

Studies of marine sediments in recent years have not only given geologists clues on rates of sedimentation and sedimentary processes in the deep sea, but have provided critical insights into paleoclimatic studies, in-fall of extraterrestrial particles and the recent history of magnetic polarity reversals [*Glass et al.*, 1967].

The sedimentary distribution in the oceans is not only dependent on the multiplicity of variants seen in the above sediment classification, but is also dependent on:

Age of the underlying crust.
Biologic productivity.
Proximity to sediment sources and sediment distribution systems.
Secondary erosion and redeposition.

The continental rise consists principally of land-derived erosion products typified by silty clays (Figure 3), transported across the continental shelf, and in places shaped and redistributed by deep, contour-following currents [*Heezen et al.*, 1966b]. The abyssal plains are also a dumping ground of terrigenous sands and silts (catastrophically emplaced by down-slope turbidity currents) interbedded within slowly accumulating pelagic deposits (Figure 9).

In the open ocean, the contribution of land-derived components declines and the type of sediment is more directly related to the organic production of the surface waters above. The sedimentary pattern on oceanic rises varies with the depth and proximity to ocean sediment distribution systems. For example, the Bermuda rise (see Figure 2) is mantled by low carbonate red clays that have been redistributed by the northerly flowing antarctic bottom water into a complex series of dunes and erosion systems or erosion surfaces [*Heezen et al.*, 1966b, c]. On the other hand, the Manihiki plateau in the south central Pacific is covered with a thick sequence of calcareous sands and oozes [*Heezen et al.*, 1966a]. At depths deeper than about 5000 meters, the carbonate fraction of the pelagic rain is dissolved by lower temperature and greater CO_2 pressure, leaving residual clay platelets (less than 2 μ in diameter) as the major deep-sea component.

Until the first seismic refraction measurements were made at sea, there was little information available on the oceanic sediment thicknesses. Estimates based on the rates of continental erosion and carbonate productivity multiplied by the geologic age of the continents indicated that the average thickness of the sediments in the oceans should be of the order of several kilometers. Seismic refraction, reflection, and surface-wave methods have revealed a sediment cover thinner than that calculated from these assumptions. The average sediment thickness over the mid-ocean ridges and in many deep basin areas of the Pacific and Indian oceans is of the order of 100–200 meters, and abyssal plain thicknesses average about 1 km. Only in the continental rise regions very close to large land masses have sediment thicknesses as great as 3–4 km been measured.

Rates of deep-sea sedimentation and considerations of the probable volumes of erosion products suggest that much greater thicknesses should be observed if it is to be maintained that the oceanic crust has an age comparable to that

Fig. 3. Sections of typical continental rise sediment cores. Cores were taken on the lower continental rise near the edge of the Hatteras abyssal plain where bottom current action is prevalent. Note the clean, well sorted winnowed sands with dark heavy mineral layers (lower right).

of the earth. In order to explain the inferred lower rates of sedimentation in the past, it has been suggested (T. Saito, personal communication) that foraminifera, which account for a significant portion of the calcareous oozes, only evolved after the Mesozoic; however, it appears likely that coccolithophidae, the other significant carbonate component, may have existed in Paleozoic time. Several authors [e.g., *Holmes*, 1965] have suggested that lower relief on the continents during past geological times limited terrestrial erosion rates and consequently limited the deposition in the oceans. In order to explain the lack of unconsolidated sediments in the ocean floor, *Hamilton* [1959] suggested that the seismic 'second layer' (V_p = 3.5 to 6 km/sec) represents lithified sediments. However, coring and dredging to date have shown the top of this layer to be largely altered volcanic basement. In recent years much effort has been given to coring and dredging samples from outcrops of old sediments on the ocean floor. At present the oldest unconsolidated sediments found on the ocean floor are the Upper Jurassic marls raised by the JOIDES drilling program north of San Salvador island in the Atlantic. Seismic profiler records from this site reveal 200 meters of sediment below the site of this core. Reasonable sedimentation rates then suggest that the base of the sediment column (in this part of the Atlantic) may be mid-Mesozoic in age.

The lack of sediment in oceans, the pattern of general thickening of sediments toward the continents, and the lack of known pre-Mesozoic sediments in the deep basins perhaps sheds light on general tectonic processes within the earth. These observations have been explained by several contrasting hypotheses:

1. The ocean basins have existed essentially in their present form throughout geological time. There are two explanations for the thinness of the present sedimentary blanket and, in particular, the absence of pre-Mesozoic sediments. The older sediments (*a*) have been covered by volcanic extrusions or (*b*) have been lithified to form layer 2 of the oceanic crust.
2. The ocean basins have *not* existed essentially in their present form throughout geologic time. Three main hypotheses have been advanced:
   a. *Continental drift* on an earth of constant radius [*Wegener*, 1924], the latest episode having begun in the Mesozoic. Pre-Mesozoic sediments were compressed on the leading edges of the drifting continents.
   b. *Expanding earth* (with consequent relative continental displacement), beginning in the Mesozoic [*Carey*, 1958]. No ocean basins (therefore no sediments) existed before the Mesozoic.
   c. *Sea-floor spreading*, with creation of new oceanic crust at the axis of the mid-oceanic ridge and destruction of crust (including sediments) along certain tectonic lines characterized by island arcs, deep-sea trenches, or orogenic belts. Rates of spreading have been calculated for about the last 5 m.y. from magnetic anomalies and geomagnetic reversals; if these rates are extrapolated, oceanic crust older than Mesozoic (together with its cargo of sediments) would have been destroyed [*Dietz*, 1961; *Hess*, 1962; *Vine and Matthews*, 1963; *Morgan*, 1968].

We do not attempt to review all aspects of these hypotheses. We note the following: the hypothesis of permanency of oceans does not satisfactorily explain the various facts offered in support of continental drift and sea-floor spreading (e.g., the linear magnetic anomalies paralleling the mid-oceanic ridge, the matching shelf edges of the opposite sides of the Atlantic, and the continuity of pre-Mesozoic geological provinces if the continental fragments are reassembled). Additional doubt is cast upon hypothesis 1*b* by dredgings in areas where layer 2 outcrops on the ocean floor: all samples of this layer have been either sedimentary rocks or volcanics, and no samples have been dated as pre-Mesozoic.

Hypothesis 2*a* has essentially been superseded by hypotheses 2*b* and 2*c*. The expanding-earth hypothesis faces at least the following difficulties: the absence of a known mechanism for the large density change necessary to account for the expansion; the requirement that the net gain of free water be such as to maintain a relatively constant sea level; the absence of any apparent change in the character of orogenic activity before and after the Mesozoic. The hypothesis does not require large-scale destruction of crust and sediments; it is consistent with the observed virtual absence of sediments on the crest of the mid-oceanic ridge, and with evidence that new oceanic crust forms only at the ridges.

The ages of the oldest sediments increase away from the axis of the mid-oceanic ridge at rates essentially consistent with rates of sea-floor spreading estimated from magnetic anomalies [*Heirtzler et al.*, 1968]. Continuous seismic profiling of the ocean floors has revealed that there is little or no deformation of the sedimentary strata [*M. Ewing et al.*, 1966a]; this is consistent with the hypothesis that crust is created in one region, destroyed in another, and that the sediments are carried along as passive cargo. In parts of the Peru-Chile trench, sediments do not appear markedly deformed [*Scholl et al.*, 1968]. Figure 15 shows deformation at the west edge of the Puerto Rico trench.

Much new evidence in support of the hypothesis of sea-floor spreading is not directly related to sedimentation; these aspects will be discussed in subsequent sections of this review.

Physiography of the Ocean Floor

The two major features on the crust of the earth are the continents and the ocean floor (Figures 1 and 2).

Contrasts in seismic velocities, gravity anomalies, and topography suggest that the transition is virtually completed beneath the marginal escarpment of the continental block (Figure 4) and certainly occupies a belt no wider than 100 km. The oceans are basically subdivided into three major morphological units: the continental margin, the ocean basin floor, and the mid-oceanic ridge.

The continental margin provinces are: continental shelf, epi-continental marginal seas (e.g., Gulf of Maine), continental marginal plateaus (e.g., Blake plateau, Yermak plateau), continental slope, marginal escarpments, and the inner slopes of trenches. These provinces mark the edge of the continental block. The transition can be further subdivided into continental rise and a marginal trench with its outer ridge. The ocean basin floor consists of the abyssal plains, oceanic rises, and isolated seamounts. The third major physiographic unit is the mid-oceanic ridge, which comprises roughly one third of the entire oceanic area. In the North and South Atlantic, between Africa and Antarctica, and between Australia and Antarctica, the ridge occupies the central third of the ocean basin.

Continental margin. Along many coasts of the Atlantic Ocean, a seaward thickening wedge of relatively undisturbed Cretaceous and younger sediments rests unconformably on a pre-Cretaceous basement. Boreholes indicate that these shelves have never been deeply submerged. The edge of the shelf, off the east coast of the United States, has been sinking at about 5 cm/1000 years [*Vogt and Ostenso*, 1967].

The rates of subsidence of the continental margin appear to be similar to rates of subsidence of a parcel of oceanic crust after it leaves the crest of the mid-oceanic ridge [*Vogt and Ostenso*, 1967].

Recent geophysical data have strongly supported the hypothesis that the ocean floor and the continents are welded together in a horizontal sense [*Morgan*, 1968], but similarity of subsidence history suggests that we may envision the continents and oceans as welded in a vertical sense also.

As the basins deepen in their voyage away from the youthful ridge crest, so do the continental margins subside, in response to thermal contraction within the originally hot, expanded lithosphere, and in response to isostatic adjustments caused by erosion on land and sediment and water loading at sea. Apparently, these vertical movements are flexures of a regional scale, without much vertical dislocation by faulting.

The original continental slope, perhaps formed at an early stage of continental rifting, has since been modified by sedimentary and erosional processes (Figure 4).

Seismic reflection records from the east coast of the United States suggest that the location and form of the continental slope is controlled by sediment supply and by ocean current erosion.

Continental margins near regions of active deformation are complex, as for example the California borderlands where tectonic activity complicates the coastal geology [*Menard*, 1964].

Ocean basin floor. Recent studies of sediment transport in the western North Atlantic and in the western South Atlantic indicate that the continental rise and associated sediment drifts are built by contour-following currents that carry and deposit sediments parallel to the regional isobaths [*Heezen et al.*, 1966b; *Schneider et al.*, 1967] (Figures 5, 6, and 7). Until recently, the role of deep sea currents in the erosion, transportation, and deposition of abyssal sediments and in the shaping of the great sediment bodies of the deep sea has not been appreciated. It was generally assumed that deep ocean currents are too weak to measure and too

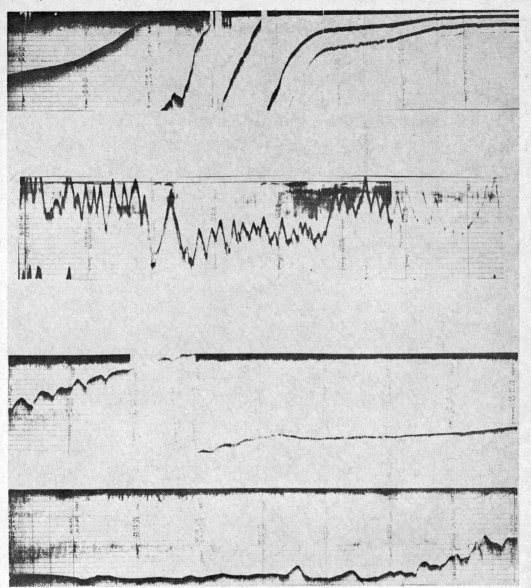

Fig. 4. Photographs of precision depth recorder (PDR) records. *Top:* Continental shelf and slope east of New York. *Second:* Multitude of minor canyons cutting the continental slope between Norfolk and Wilmington canyons. *Third and bottom:* Lower continental rise hills.

irregular in direction to have significant geological effects. Recent studies lead by Heezen and his students have now confirmed the fact that these deep ocean currents are truly of geological importance in the deep sea (Figure 6). In the western North Atlantic extensive studies have been made of sediment cores, echograms, and ocean bottom photographs [*Heezen et al.*, 1966b; *Schneider et al.*, 1967] (Figures 5 and 7). These data show that the sediments on the upper continental rise consist of lutites that are being deposited on a sea floor devoid of strong bottom currents. Below an abrupt change in regional slope (which marks the boundary between the upper and lower continental rise), a swift bottom current flows to

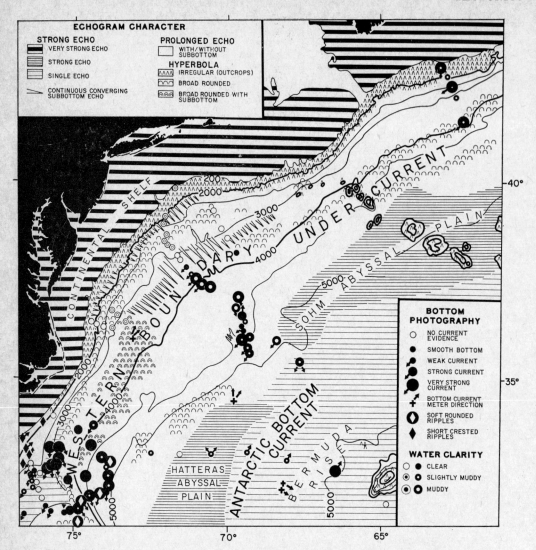

Fig. 5. Direction of sediment transport and microphysiography of the Atlantic margin of the United States. Contours in meters. The continental shelf and abyssal plains are typified by strong coherent echo returns, and on the continental rise zones of prolonged echoes and hyperbolas trend parallel to the regional contours. Above 3500 meters tranquil, current-free bottom is seen; below 3500 meters swift contour currents transport sediment, and current-controlled sedimentation appears to have shaped the continental rise and Blake-Bahama outer ridge. Photographs on the Hatteras abyssal plain show tranquil sea bottoms or variable weak northerly flowing currents, whereas swifter northerly flow of the Antarctic bottom water is observed on the western Bermuda rise [*Schneider et al.*, 1967].

the south parallel to the regional isobaths. Beneath this current the surface sediments are distinctly coarser grained; long cores show many coarse silt laminations in the sedimentary column (Figure 3). Farther down slope on the lower continental rise, the currents are weaker and variable in direction. In the area of the lower continental rise hills, a swift southwesterly current is again observed. The adjacent abyssal plains are generally devoid of strong currents.

In the western North Atlantic, cores reveal a distinct suite of sediments that can be associated with the continental rise. Cores from the

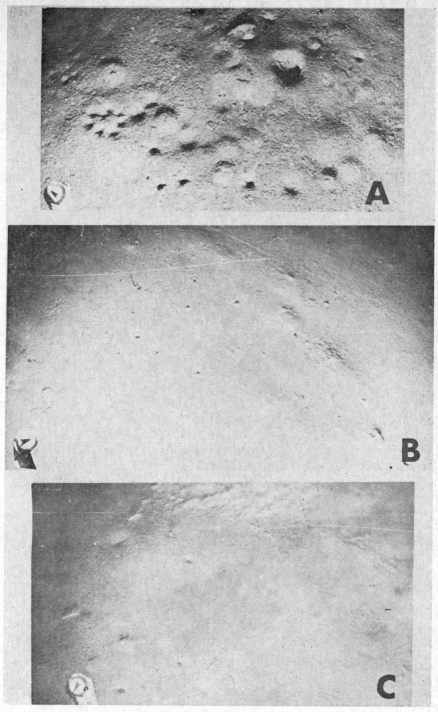

Fig. 6. (A) Trident station 28, 36°35'W, 4444 meters, showing strong current to the north-northeast with smoothed bottom and distinct sediment tails in the lee of high objects. This photograph, taken east of the Gulf Stream axis, perhaps indicates an intermittent scouring of the Gulf Stream to the bottom. (B) Current-swept bottom at the base of the lower continental rise showing smoothed sediment and sediment streamers to the southwest; little evidence of life is seen. Trident station 31, 37°05'N, 69°41'W, 4336 meters. (C) Scoured and rippled coarse-textured bottom at the base of the upper continental rise. Ripples and sediment streamers show a southwesterly bottom current; water is noticeably sediment-laden and murky. Trident station 15, 37°20'N, 70°34'W, 4190 meters [*Schneider et al.*, 1967].

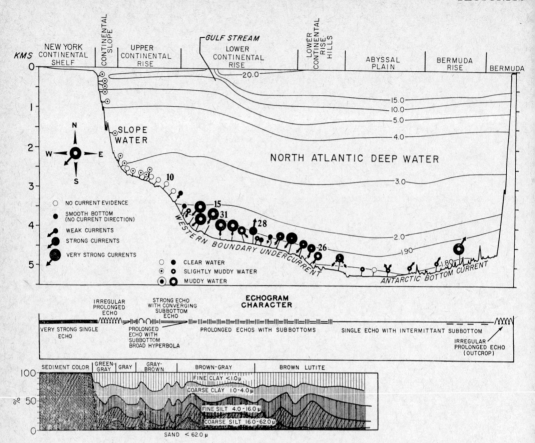

Fig. 7. Profile of bottom current direction, bottom sediment, echogram character, and isotherms of potential temperature. Cold water isotherms lap up on the continental rise (temperature 2.00°C) and on the Bermuda rise (temperature 1.80°C). Strong ocean currents occur at the break in slope between the upper and lower continental rise, in the lower continental rise hills, and above 5000 meters on the Bermuda rise. Prolonged bottom echo sequences and an increase in the surface sand and silt-sized sediments are associated with the current-swept zones on the lower continental rise [*Schneider et al.*, 1967].

upper continental rise are devoid of laminations of silt or sand, except when found in submarine canyons. Cores on the lower continental rise are marked by hundreds of fine silt laminations per 10 meters of core. Cores on the lower continental rise south of the Gulf of St. Lawrence are also marked by distinct zones of bright red clays or lutites. *Conolly et al.* [1967] and *Hollister and Heezen* [1967] have shown that these red sediments have been transported southward parallel to the contours by ocean bottom currents from the region of the Gulf of St. Lawrence as far as the Bahamas.

The effects of deep-sea currents are observed not only in the sediments but also in the macro- and micromorphology of the continental rise. The submarine canyons that are conspicuous on the continental slope (Figure 4) north of Cape Hatteras and that are often present on the continental rise become narrow and almost indistinguishable features on the lower continental rise. Such canyons (usually perpendicular to the slope) may run parallel to regional contours in areas of swift bottom current [*Rona et al.*, 1967] (Figure 5). Zones of microtopography (i.e., irregular sub-bottom echos, small regular marked hyperbolas, or zones of wedging sub-bottoms) are also parallel to the regional contours and therefore seem to result from deep ocean currents (Figure 5).

South of Cape Hatteras, the continental rise becomes detached from the continental slope and is transformed into a complex outer ridge which encloses a marginal basin. The Blake-Bahama outer ridge, which seismic reflection records showed to be a large pile of drift sediments [*Ewing and Ewing*, 1964], has been formed by the southward flowing western boundary current of North Atlantic deep water. Seismic reflection records and cores show distinct zones of preferential erosion and deposition caused by swift ocean bottom currents which have sculpted the present shape of this outer ridge. Therefore, the morphology of the continental rise and the outer ridges is the result of sedimentary processes relating to the ocean above, and the form of these features reveals nothing of the upper mantle below.

Seismic reflection records from the North Atlantic show several distinct reflection horizons. The most prominent of these reflectors is horizon A, which extends undisturbed beneath the Blake-Bahama outer ridge, the continental rise, and eastward across the Bermuda rise toward the Mid-Atlantic ridge (Figure 8) [*J. Ewing et al.*, 1966]. This interface is generally smooth with a slight westerly dip. The layer that directly underlies this interface crops out on the sea floor northwest and north of San Salvador Island; cores from this layer contain Maestrichtian coccoliths.

The continuation of these horizons beneath the outer ridges and continental rise to the north suggests that these massive sediment drifts are largely post-Mesozoic in age and that the process responsible for building these large sedimentary wedges or drifts must have become more vigorous in post-Cretaceous time. This event may be linked to the initiation of a strong thermal haline circulation in the western Atlantic basin in response to the widening of the basin.

Abyssal plains. Abyssal plains are those flat areas of the sea floor where regional gradients are less than 1:1000. Studies of these exceedingly flat regions suggest that they were formed by turbidity current deposits (turbidites) that covered the pre-existing topography. The following features of abyssal plains support an origin by turbidite deposition: (1) abyssal plains always adjoin topographic terrain capable of supplying terrigenous material by gravity-induced flows; (2) there are thick sequences of graded silt and sand layers (Figure 9), and lateral grain size decreases with distance from the source; (3) seismic reflection records reveal that an irregular basement below the abyssal plains is being buried by a sequence of horizontal layers. On the seaward side, abyssal plains abruptly abut against the abyssal hills.

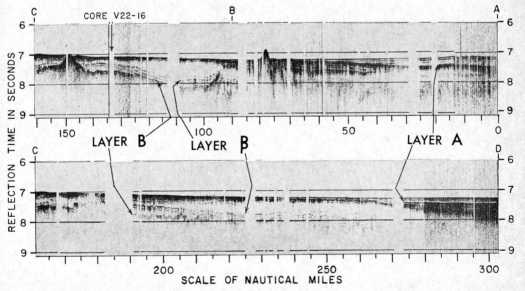

Fig. 8. Seismic reflection records from area of horizon A outcropping north of San Salvador [from *J. Ewing et al.*, 1966].

Fig. 9. Portions of core SHL 80-7 from the Caicos outer ridge. Layers are predominately graded calcareous turbidites. Primary structures include cross laminations, convolute laminations, and parallel bedding. Note contrast in grain size at base of the bed [*Schneider et al.,* 1967].

The continental rise, abyssal plain boundary is generally more gradual. Boreholes penetrating the sedimentary column at the seaward boundary of the abyssal plains would probably encounter a purely pelagic layer between the layered turbidites and the volcanic basement. The pelagic layer was deposited while that parcel of crust was still on the flanks of the ridge, out of reach of turbidity flows. Oceanic crust near the continental margin would be expected to lack such a lower pelagic sequence, because sedimentation was probably always dominated by close terrigenous sources.

Submarine canyons appear to act as conduits for sediment-laden turbidity currents. Upon reaching the abyssal plain the turbidity current apparently fans out into a sheet flow, depositing its coarser constituents first. The distal portions of abyssal plains usually contain thick, acoustically transparent clay layers.

Mid-ocean canyons. Mid-ocean canyons are enigmatic features of the ocean basin floor that have no apparent connection with adjacent land drainage systems; the northwest Atlantic mid-ocean canyon is the best known [*Heezen et al.*, 1959].

The canyon is generally box-shaped with a flat floor and steep walls (Figure 10). A maximum depth from levee to floor is approximately 100 fathoms near 52°N. The western levee is conspicuously higher than the eastern; this has been ascribed to the tendency of the Coriolis force to deflect an overflowing current to the right [*Heezen et al.*, 1959]. The canyon dies out to the south in the Sohm abyssal plain.

It seems possible that these large canyons were active during the lower sea levels of the Pleistocene and that in the past 10,000 years their incision across the continental rise has been filled in by the deposits of contour currents flowing along the continental rise.

It has been suggested by *Heezen et al.* [1959] that the mid-ocean canyons may represent fundamental faults, but we see no evidence of tectonic action and conclude again that these features relate to oceanic processes and not to the crust and mantle below.

Oceanic rises. Rising from the abyssal floor are broad, relatively smooth elevations measured in hundreds of square miles. These ocenaic rises are not obviously related either to the continental margins or to the mid-oceanic ridge. The Bermuda rise is a typical example (Figure 2). The lack of seismicity and the subdued relief suggest that they are tectonically quiescent features.

Microcontinents. The aseismic Lomonosov ridge and Jan Mayen ridge of the Arctic [*Johnson and Heezen*, 1967a, b] may be examples of microcontinents. The Walvis ridge in the South Atlantic and the Ninetyeast ridge of the Indian Ocean are probably in the same category. One of the best known is the Masacarene plateau of the Indian Ocean, which lies parallel to the mid-oceanic ridge and midway between it and Madagascar. This flat-topped massive bank is floored by a Precambrian granitic basement that outcrops in the Seychelles Islands [*Fisher et al.*, 1967]. The microcontinents have slowly subsided from sea level to their present depths beneath the sea.

Seamounts. By definition, a seamount is any submerged peak that rises 1000 meters above the adjacent sea floor. Smaller but isolated features are called knolls; when such features occur continuously over large, deep-ocean areas they are commonly called abyssal hills. Conical seamounts are assumed to be volcanic in origin. Southwest of Hawaii in the mid-Pacific, there is a large region of seamounts whose flat summits lie from 900 to 1700 meters below sea level. Shallow-water Cretaceous fossils have been dredged from tops of some of these wave-cut submerged islands (guyots), thus establishing that part of the Pacific was 4000 meters deep in the Cretaceous [*Hamilton*, 1956]. The aver-

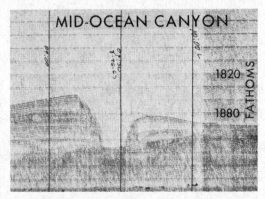

Fig. 10. Seismic profiler record taken with a 10-cubic-inch air gun across the northwest mid-ocean canyon. Length of record is approximately 22 miles. Canyon crossed at 58°05'N, 51°40'W.

age subsidence rates have been of the order of a few centimeters per 1000 years [*Vogt and Ostenso*, 1967]. Spreading and consequent descent from the mid-oceanic ridge lead to similar subsidence rates if the regional profile of the ridge represents a steady state. Virtually all seamounts lie in linear patterns, suggesting that major faults in the earth's crust may have controlled their distribution [*Menard*, 1959]. Examples of such seamount chains include the Hawaiian Islands, the New England seamounts, and the volcanoes on the Cameroon line.

Fracture zones. Oceanographic exploration during the last two decades revealed a vast network of fracture zones on the deep-sea floor. *Menard and Dietz* [1952] identified and thoroughly investigated the Mendocino fracture zone (the first discovered of many such features in the Pacific). The Pacific fracture zones are remarkably evenly spaced and are very close to being parallel.

Many fracture zones cut the Mid-Atlantic ridge [*Heezen et al.*, 1961; *Heezen et al.*, 1964a, b; *Johnson and Heezen*, 1967b; *Heezen and Tharp*, 1968] (Figs. 11, 12, and 13). The crest of the ridge is offset along these fracture zones from a few miles to more than 100 miles; adjacent fractures are parallel. The strike in the North Atlantic is northwest-southeast; in the South Atlantic northeast-southwest. Fracture zones in the Indian Ocean [*Heezen and Tharp*, 1965] are primarily oriented NNE-SSW with the exception of the east-west fractures offsetting the Carlsberg ridge; in the South Pacific, the fractures strike WNW-ESE.

Topographically, fracture zones are long thin bands of terrain conspicuously rougher than the surrounding sea floor. Deep linear troughs are

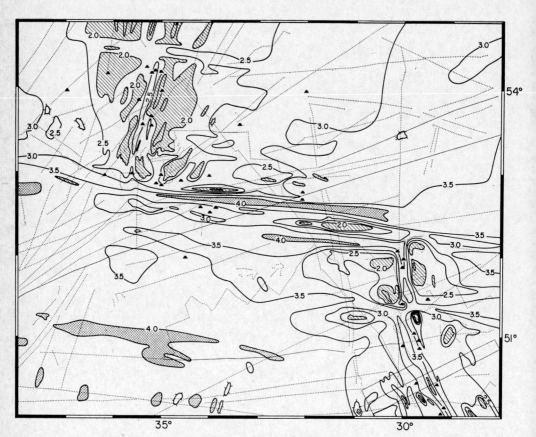

Fig. 11. Bathymetric sketch of fracture zones on the Mid-Atlantic ridge. Cross hatching indicates areas shoaler than 2000 meters. Grid pattern areas are deeper than 4000 meters Black triangles are earthquake epicenters [*Johnson*, 1967].

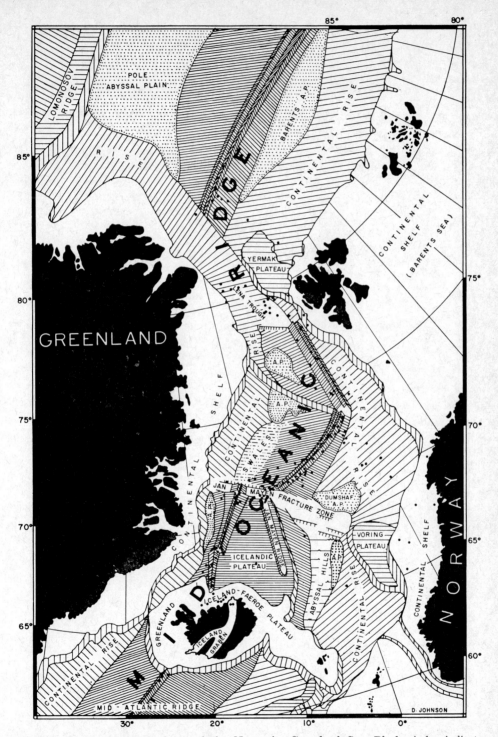

Fig. 12. Physiographic provinces of the Norwegian-Greenland Sea. Black circles indicate earthquake epicenters [*Johnson and Heezen*, 1967a].

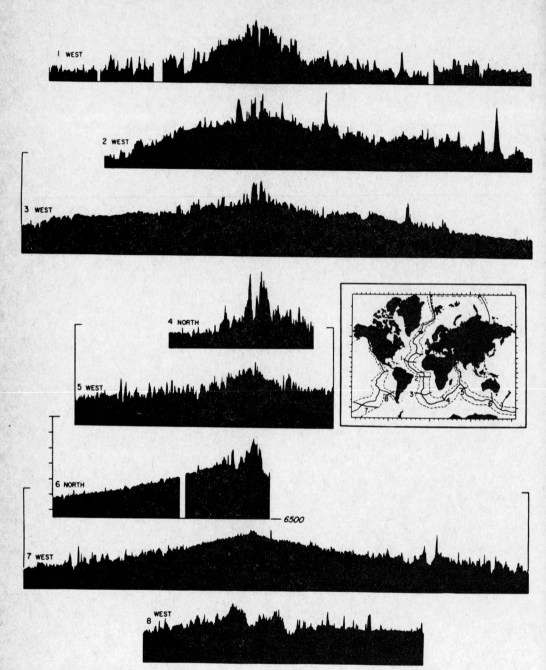

Fig. 13. Eight typical profiles of the mid-oceanic ridge in the North Atlantic, South Atlantic, Indian, and South Pacific oceans. The smooth relief across the crest of the mid-oceanic ridge in the South Pacific contrasts markedly with the rougher topography on the other ridges [after *Heezen*, 1962]. Vertical exaggeration 100:1. Locations of profiles are shown on inset.

common [*Menard*, 1964]. The Mendocino fracture zone consists of a large asymmetrical ridge with a depression at the foot of the escarpment. The average depth difference across such a fracture zone may be of the order of 1 km. The Greenland fracture zone [*Johnson and Eckhoff*, 1966] appears similiar. Another characteristic is a bordering pair of ridges enclosing a trough or zone of very irregular topography. The Murray fracture zone is of this type. Atlantic and Indian ocean fracture zones resemble this variety in general plan; however, they are considerably more rugged, with the flanking scarps rising 2000 meters or more above the sea floor and very deep troughs along the axis. The Romanche fracture zone, with depths in excess of 7500 meters, is a dramatic example.

Fracture zones exhibit some vertical throw; however, they appear to be produced by predominately horizontal movement. The North Pacific exhibits a regular pattern of north-south striking magnetic anomalies. Offsets of the magnetic anomalies indicate that Mendocino has an apparent 1170-km left lateral throw [*Vacquier et al.*, 1961]; Pioneer, 250-km left lateral [*Vacquier*, 1959]; Murray (eastern end), 150-km right lateral; and Murray (western section), 640-km right lateral [*Raff*, 1962]. Fracture zones involve the oceanic crust and upper mantle and are therefore of utmost importance in considerations of the structure and history of the upper mantle (Figure 14).

Mid-oceanic ridge. The mid-oceanic ridge is one of the principal active features of the earth (Figure 13). This globe-encircling structure extends through the Arctic, Atlantic, Indian, Antarctic, South Pacific, and the eastern extremity of the North Pacific Ocean for a total length of more than 40,000 miles [*Heezen and Ewing*, 1963]. Morphologically the ridge is a broad fractured arch rising 1–3 km above the adjacent sea floor with a total width of 1500 km in most areas. In the Atlantic and Indian oceans the ridge can be subdivided into two distinct morphological units. There is a central part or crest that is more or less distinctly elevated, rising above the gently sloping flank portions. The crestal province is commonly cut in half by a prominent median or rift valley. Earthquake activity is all shallow and only moderately high in comparison to the circum-Pacific belt. The crest is raised topographically one-half to one kilometer above the flanks.

A study of many topographic profiles across the flanks of the Mid-Atlantic ridge suggested to *Heezen et al.* [1959] that the flanks of this ridge can be subdivided into 'steps' several hundred km wide, which are separated from each other by topographic offsets or can be distinguished from one another by differences in character of the bottom. The flanks are in general characterized by moderately rough topography, with a series of intermontane basins which generally become deeper from the crest to the ocean basin floor (Figures 14, 15, and 16); there is a notable symmetry in topographic profiles with respect to bottom roughness, regional slope, and sediment thickness across the Atlantic and other branches of the mid-oceanic ridge (Figure 16). There is also a remarkable symmetry of the ridge with respect to the ocean basins and adjoining continental margins: in the Atlantic and Indian oceans, the ridge occupies approximately the central third of the ocean, with its crest near the median line of the ocean. Although the mid-oceanic ridge is usually found in the central part of the world's ocean, it enters the continental margins at several places. One of these locations is off the Lena Delta in Siberia, where the mid-oceanic epicenter belt extends across Siberia to Lake Baikal. The ridge enters the Gulf of California where it merges with the fracture zones of the western United States (such as the San Andreas), only to reappear off the coast of Oregon [*Menard*, 1960]. Another section extends northward through the Red Sea. A southerly branch strikes south from the Gulf of Aden to form the rift valleys of eastern Africa. The mid-oceanic ridge was first established as a world-encircling mountain system by combining topography and seismicity at a few well explored areas and then by tracing it as a continuous belt of shallow focus seismicity [*M. Ewing and Heezen*, 1956]. Recent precise positioning of these shallow-focus earthquakes confirms the original supposition that they originate under the axial rift or within active parts of fracture zones, i.e. between offset ends of the ridge [*Sykes*, 1967].

An extension of this world-encircling ridge into the Greenland Sea and Arctic Ocean was first hypothesized on the basis of a well defined

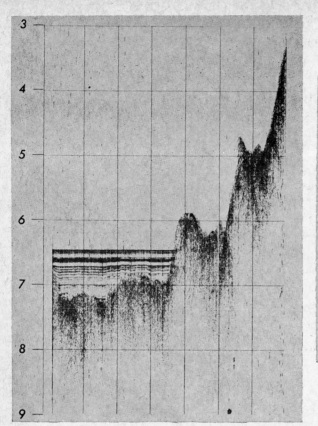
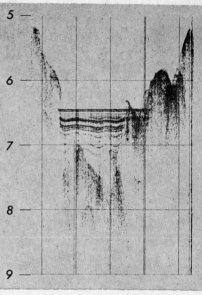
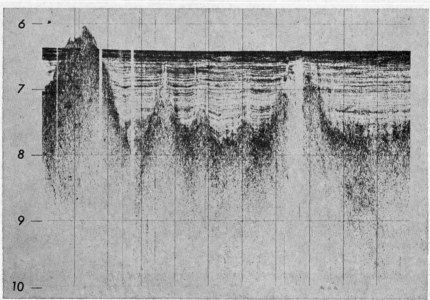

Fig. 14. *Top left:* Typical rugged Mid-Atlantic ridge topography. Note flat-lying sediment in intermontane basin approximately 800 meters thick. Record length approximately 30 miles. *Top right:* Crossing of fracture zone at 8°N in the equatorial Atlantic. Note slight deformation of sediment fill. Record length approximately 22 miles. *Bottom:* Thick abyssal plain sediment covering rough basement topography. Only one abyssal hill remains unburied. Record length approximately 48 miles. Horizontal lines are travel time in seconds.

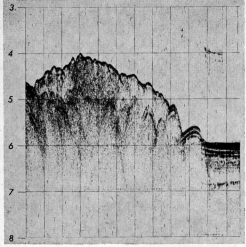
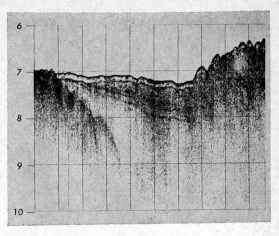
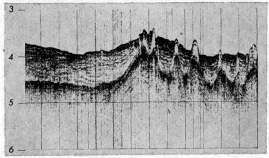

Fig. 15. *Top left:* Highly compressed and contorted sediment pile of the Barbados ridge. Note uplifted abyssal plain type sediments on lower part of ridge. Record length approximately 50 miles. *Top right:* Basement and sedimentary layers dipping beneath the Barbados ridge. As to the left, the sediments of the ridge are highly contorted. Record length approximately 42 miles. *Bottom:* Apparent diapir structures at the base of the continental slope off Spanish Sahara. Record length approximately 54 miles. Horizontal lines are travel time in seconds.

pattern of earthquake epicenters and a few spot soundings [*Elmendorf and Heezen*, 1957; *M. Ewing and Heezen*, 1956]. The topographic form of the ridge was revealed by precision depth records obtained during the pioneer polar crossings of the nuclear submarines *Nautilus* and *Skate* [*Dietz and Shumway*, 1961; *Johnson and Eckhoff*, 1966]. The Arctic mid-oceanic ridge does not bisect the entire Arctic basin but lies midway between the monolithic, aseismic Lomonosov ridge and the Eurasian continental margin [*Johnson and Heezen*, 1967a].

Within 10° of the Greenwich meridian, the ridge has a width of approximately 270 km (Figure 17, profiles 5 and 6). Flanking topographic highs border a depressed axial valley which is strikingly developed in the area north of Spitsbergen. The mid-oceanic ridge off the Laptev and Kara seas (Figure 17, profiles 2–4) is a relatively subdued feature.

The crestal zone of the mid-oceanic ridge between Iceland and Jan Mayen is only 20 miles in width at the southern portion, increasing to 50 miles west of Jan Mayen (Figures 12 and 18). That part of the mid-oceanic ridge which strikes northeast from Jan Mayen toward Bear Island is commonly called Mohns ridge. This section lies nearly equidistant between the parallel parts of the Greenland and Norwegian continental slopes. Its rugged, linear components are bilaterally symmetrical about the ridge axis. A narrow ridge and trench runs nearly due north along 8°E from the eastern extremity of Mohns ridge to the southwestern end of the Arctic mid-oceanic ridge off northern Greenland.

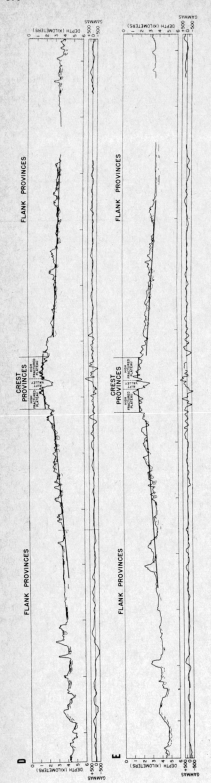
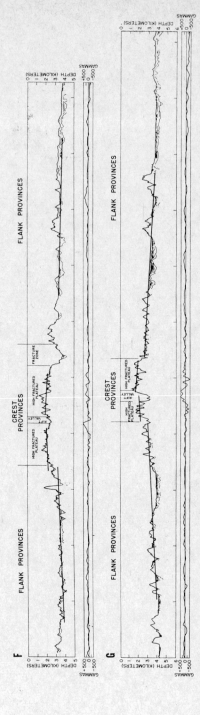

Fig. 16. Profiles D and E cross the Mid-Atlantic ridge at 36°N. Crestal province differs from adjacent flanks in basement roughness, magnetic signature, elevation, and sediment thickness. Angle between ridge and profiles is about 80°. Tic marks at base of profiles are 100 km apart. Profiles F and G cross the Mid-Atlantic ridge at 34°N. Angle between ridge and profiles is about 60°. Fracture zone was crossed on east flank of F [Schneider and Vogt, 1968].

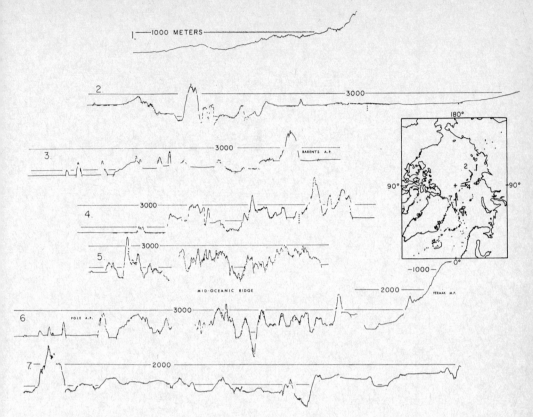

Fig. 17. Tracings of precision echograms (PDR) obtained by submarines in the Arctic. Transducer depth is not computed. Small dots on index chart are earthquakes [*Johnson and Heezen*, 1967a]. A. P., abyssal plain. Tracings are numbered in sequence; 2, 4, and 7 are numbered on inset.

The Mid-Atlantic ridge extends through Iceland, the central graben of Iceland representing an extension of the mid-oceanic rift valley. Volcanism is active in the central graben, which is cut by numerous active fissures.

The Mid-Atlantic ridge is the best known part of the mid-oceanic ridge. A typical trans-Atlantic profile is shown in Figure 2. The most rugged topography of the Mid-Atlantic ridge is found along the crest provinces. A prominent rift valley lies between flanking mountains; beyond these mountains, depths increase by 1–2 km to the high fractured plateau.

The Mid-Indian Ocean ridge continues uninterrupted from south of Africa to just south and east of Rodriguez Island. The ridge bifurcates there, with one branch entering the Gulf of Aden and the other striking southeast between Australia and Antarctica. The ridge is characterized by the same rugged topography that is found in the Atlantic. The connection of the mid-oceanic seismic belt with the seismic belt of the African rift valleys and the similarity of their morphologies suggest that they are structurally related and have a similar origin.

The topography of the mid-oceanic ridge is more subdued in the eastern Pacific. The rift valley and rift mountains are very small or absent (Figure 13). The East Pacific rise is a broad arch 2000–4000 km wide and 2–3 km high. Near Easter Island, the rise bifurcates; one branch extends to southern Chile and the other, after crossing the equator, enters the Gulf of California.

Investigations of Pacific abyssal hills suggest that they are intimately related to the mid-oceanic ridge [*Menard and Mammerickx*, 1967; *Moore and Heath*, 1967]. Menard found that individual abyssal hills to the west of North and Central America are actually long sinuous

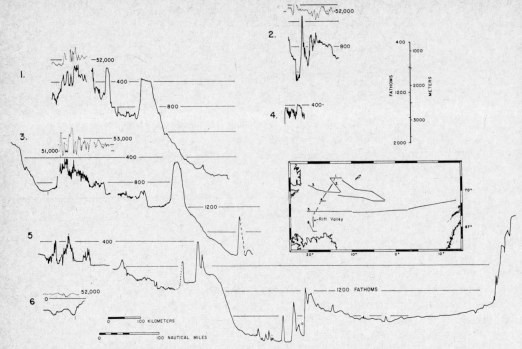

Fig. 18. Six profiles from the Icelandic plateau; the mid-oceanic ridge is clear on the left-hand margin of profiles 1, 3, and 5. Dotted lines are total magnetic intensity; solid lines are topography; vertical exaggeration, 100:1.

ridges. Both topographic and magnetic signatures are roughly symmetrical about the East Pacific rise. In this region most of the hills and valleys are elongated north-south (parallel to the mid-oceanic ridge).

Sediments of the mid-oceanic ridge. Sediment thickness across the mid-oceanic ridge varies, but sediment types are rather constant (Figures 14, 15, and 16). The depth of the surface of deposition seems to be the main factor in controlling sediment composition. Above about 4500 meters carbonate oozes consisting of foraminiferas, coccoliths, and pteropods are the major constituents. Below 4500 meters, red clays predominate (Figure 2). Seismic reflection data, cores, and bottom photographs reveal the scarcity of sediment on the crest of the mid-oceanic ridge. Early seismic reflection data showed a marked lateral discontinuity of sediment thickness on the Mid-Atlantic ridge. *Ewing and Ewing* [1967] conclude that the proposed discontinuity is of worldwide extent. Their favored explanation for this phenomenon is a lapse or halt in crustal expansion during the Tertiary, followed by a later resurgence. This lateral discontinuity in sediment thickness is evident in seismic reflection data despite marked differences in absolute thickness along the strike of the ridge (Figure 16). A crossing on the Mid-Atlantic ridge at 12°N revealed only 200 to 400 meters of sediment on the ridge flanks, whereas the thickness at 46°N ranges from 600 to 1400 meters. However, in both regions the ratio between sediment thickness on the crest and that on the immediately adjacent flanks remains betwen 1:2.5 and 1:5 [*Schneider and Vogt*, 1968]. Air gun and sparker seismic records across the ridge crest show little sediment in this province (less than 100 meters).

IGNEOUS CRUST BENEATH THE SEA FLOOR

In recent decades the study of the oceanic crust has progressed from isolated reconnaissance sampling through the stage of geophysical descriptions to the current sea-floor expansion reconstructions, which mark the beginning of a new era of genetic and historical analysis. The description of the tectonic fabric of the ocean floor and the recognition that the mid-oceanic ridge forms a single network of tensional rifts was climaxed by such regional works as *The Floors of the Oceans* [*Heezen et al.*, 1959]. Geophysical descriptions of crustal structure in which genetic hypotheses were not actively invoked are exemplified by papers such as that of *Talwani et al.* [1965]. The growth of the oceanic ridges by tensional faulting, intrusion, and volcanism in the axial zone was deduced from morphological comparisons [*Heezen et al.*, 1959; *Heezen*, 1959b, 1960]. A broad hypothesis which followed [*Dietz*, 1961; *Hess*, 1962] proposed that the entire oceanic crust is generated at the axis of the mid-oceanic ridge and is probably destroyed along fold-mountain belts and oceanic island arc trenches. The successful testing of this hypothesis by use of magnetic anomalies and field reversals [*Vine and Matthews*, 1963] has initiated a flood of genetic-historic studies and will result in a general reinterpretation of earlier data, as well as the evolution of physical-mathematical models to describe the creation of oceanic crust.

Accurate descriptions of the sea-floor by surface ship echo sounding are now being supplemented by deep-towed sounding devices, which greatly enhance the resolution of small-scale features on the deep-sea floor [*Spiess et al.*, 1968]. The past ten years have allowed us to follow the igneous basement beyond the area where it is exposed to regions, such as the abyssal plains, where it is covered by sediments, and to delineate the structure within the sedimentary layer itself.

Descriptions of the deeper igneous crust have been largely founded on seismic refraction, gravity, and magnetic data. Models for the oceanic crust have been evolved largely from the first two. Two-dimensional refraction profiles yield the thickness, P-wave velocity, and interface slopes of the main crustal layers. Surface-wave dispersion models yield average structures that should be, and are, compatible with the local refraction measurements. Densities are inferred from velocities by empirical curves [*Nafe and Drake*, 1961; *Birch*, 1960], although the applicability of these curves to the deeper crust and upper mantle, where matter may be close to the melting point of some fractions, is questionable [*Oxburgh and Turcotte*, 1968]. Models for the crust and upper mantle are found that also must satisfy the added constraint imposed by the observed anomalous gravity field [*Talwani et al.*, 1965]. Magnetic measurements have been used for many years to determine the probable magnetization of the igneous crust and the depth to its upper surface.

However, in the last five years, the Vine-Matthews interpretation of magnetic measurements in terms of sea-floor spreading has led to a dramatic re-evaluation of basic hypotheses regarding the genesis of the oceanic crust.

The heat flow through the ocean floor has been very sparsely sampled in comparison with the other geophysical quantities. Recent tabulations (see p. 93) confirm earlier conclusions regarding the general equality of oceanic and continental heat flow and the high heat flow over the mid-oceanic ridge. On the whole, heat-flow data presently represent a constraint on possible models of mantle convection and ocean-floor spreading [*Langseth et al.*, 1966; *McKenzie*, 1967]. Perhaps when many more data are available crustal conductivity models may offer an additional constraint to models of the oceanic crust as deduced from gravity and refraction seismology [*Talwani et al.*, 1965].

Mid-Oceanic Ridge and the Oceanic Crust

Marine magnetic anomalies and the reconstruction of crustal movements. Recent widespread application of the Vine-Matthews hypothesis [*Vine and Matthews*, 1963] to total magnetic field anomalies over the oceanic crust has elevated the basic concept of sea-floor expansion [*Dietz*, 1961; *Hess*, 1962] from the status of hypothesis virtually to fact. Studies of magnetic anomaly patterns in the North Atlantic [*Vine*, 1966; *Avery et al.*, 1968; *Talwani et al.*, 1968], in the South Atlantic [*Dickson et al.*, 1968], in the Indian Ocean [*Le Pichon et al.*, 1968], in the Pacific [*Pitman et al.*, 1968], and in the Arctic [*Vogt et al.*, 1969] reveal a relatively simple pattern in the movement of crustal blocks. A summary of the deduced history of sea-floor movements is given by *Heirtzler et al.* [1968]. In general the direction, magnitude, and history of the movements strongly support the large body of geological evidence accumulated in favor of Wegener's concept of continental drift [*Wegener*, 1924].

Magnetic field anomaly profiles, long the stepchild of marine geophysics, have played the cardinal role in this breakthrough. The central hypothesis [*Heezen*, 1960a; *Dietz*, 1961; *Hess*, 1962] is that the oceanic crust has been created by dike injection, with some vulcanism and normal faulting, at the axis of mid-oceanic ridges. *Vine and Matthews* [1963] noted the fairly symmetric anomalous magnetic profiles across some mid-oceanic ridges. They assumed the sea-floor axial accretion hypothesis and the likelihood that, as mantle derivatives are injected into the axial crust and cool, they must acquire a magnetization depending on the ambient strength and direction of the geomagnetic field. The chronology of geomagnetic polarity reversals (based on potassium-argon dated lavas) has recently been extended to 4.5 million years (see Figure 1, p. 450). It was also known that when averaged over periods of the order of 0.1 m.y. the geomagnetic poles tend to coincide with either the north or south rotational poles [*Irving*, 1964]. The order of magnitude of sea-floor motion also had to be several centimeters per year if continental movements were to be explained in this manner. Vine and Matthews synthesized their 'magnetic tape recorder' model of mid-ocean ridges from this earlier knowledge.

The broad, conspicuous axial anomaly long known to exist over the ridge axis [*Heezen et al.*, 1953; *Heezen et al.*, 1959] is caused by normally polarized intrusives injected since the last field reversal (about 0.7 m.y. ago). The width of this body must, therefore, be equal to twice the expansion half-rate times 0.7 m.y. The pattern of flank anomalies has thus been generated by a time sequence of polarity reversals and a spreading rate history. The linear magnetic anomalies known from large parts of the northeastern Pacific (Figure 19) could in this way be explained in terms of two variables: (1) the speed of ocean-floor expansion and (2) the time history of geomagnetic polarity reversals. In its simple form the tape-recorder model requires that each dike bisect its predecessor, thus giving rise to a sequence of alternately polarized crustal blocks symmetrical about the axis of crustal growth. (The resulting model magnetic signature is actually only approximately symmetrical about the axis, unless the ridge trends north or is located at one of the poles.) The wide applicability of this simple model is rather amazing (Figure 20 illustrates our application of the hypothesis to the mid-oceanic ridge north of Iceland). Ridge segments in which the spreading rate in one direction is significantly different from the rate in the other direction have not been observed. This suggests that the mid-oceanic ridge is a continuously rejuvenated zone of crustal and upper mantle weakness; it may migrate with respect to the deeper mantle [*Vogt et al.*, 1969].

A new dike will bisect its predecessor simply because the crust is hottest and weakest there [*Schneider and Vogt*, 1968; *Morgan*, 1968]. The geometry of the ridge with respect to adjacent continents in some areas may imply a drifting axis of growth. For example, if the southern continents have been drifting away from a fixed Antarctica while maintaining a median ridge, and if the earth has not expanded, then the ridge must also have been drifting outward from Antarctica and the apparent ocean floor growth rates are only relative to the moving axis of this ridge [*Wilson*, 1965].

Departures of the observed magnetic signatures from the simple tape-recorder model are most pronounced on slowly growing segments of the ridge such as the Mid-Atlantic ridge (1 or 2 cm/year). The axial magnetic high on that ridge had been known for some time [*Heezen et al.*, 1959; *Heirtzler and Le Pichon*, 1965], but the symmetry of the magnetic pattern remained hypothesis until the U.S. Naval Oceanographic Office conducted a detailed aeromagnetic survey over Reykjanes ridge (Figure 21) south of Iceland. The more irregular appearance of the field over slow-expansion segments of the ridge is in part related to the greater topographic relief on the magnetized basement. Such high relief, created by block faulting and vulcanism, seem to be a consequence of the mechanics of slow expansion [*Menard*, 1967b; *van Andel and Bowin*, 1968].

The topographic effect on the short-wavelength magnetic spectrum (3 to 5 km) was shown by *Vogt and Ostenso* [1966], who were able to reproduce a number of observed anomalies by assigning magnetization intensities of about 0.01 emu to model 'seamounts' in the crestal province of the Mid-Atlantic ridge between 42° and 46°N. Such high intensities of magnetization were consistent with the properties of dredged basalts in the survey area [*Vogt and Ostenso*, 1966]. More recent in-

Fig. 19. Linear magnetic anomaly patterns in northeastern Pacific, from *Raff and Mason* [1961]. Positive anomalies are black. Interruptions in patterns are associated with fracture zones, whose positions are shown by thin lettered lines. These magnetic lineations are now interpreted in terms of strips of alternately magnetized crust [*Vine and Matthews*, 1963].

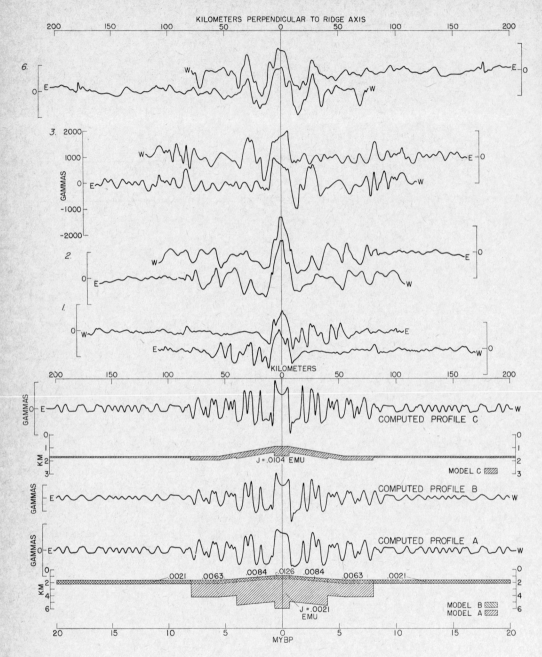

Fig. 20. Residual (anomalous) total magnetic field profiles projected perpendicular to youthful branch of mid-oceanic ridge north of Iceland (index map appears on p. 584). *Right:* Shipborne traverses, bathymetry (shaded), and magnetic model B. *Left:* Airborne traverses and magnetic models A, B, and C (vertical scale for models on left is smaller than scale for model B, right; model B completely overlaps the upper part of model A).

Magnetic scales (in gammas) are identical for all profiles. Each profile is plotted forward and backward to illustrate symmetry. Spreading half-rate is 1.0 cm/year in model according to reversal chronology of *Heirtzler et al.* [1968]. Note uncertain correlation beyond about 4 m.y.

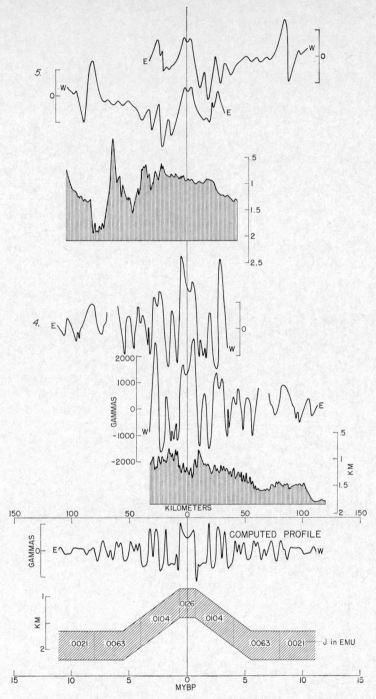

Differences between three possible models describing variation of magnetization with depth are too small to be discriminated in the observed profiles. Decay of magnetization intensity with time appears most likely. This part of the ridge is unusually shallow and therefore exhibits large amplitudes [after *Vogt et al.*, 1969].

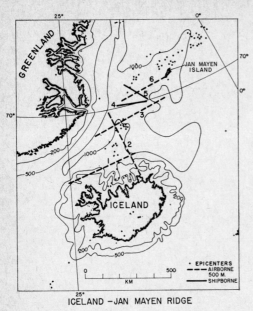

Fig. 20. (cont.) Index map for magnetic profiles on preceding pages.

vestigations confirm these results. *Talwani et al.* [1968] have shown a magnetic intensity profile along the crest of Reykjanes ridge lying completely within the boundaries of the central normally magnetized block. This profile correlates well with the topography and suggests that the layer responsible for the magnetic lineations is highly magnetized (0.03 emu) and thin (≈ 500 meters).

Some peaks near the ridge crest showed no recognizable anomaly [*Vogt and Ostenso*, 1966]. This fact finds a ready explanation if these peaks are complex volcanic cones composed largely of ash and lava breccia of low true susceptibility, although individual fragments might exhibit high thermoremanent magnetization (TRM). The work of *van Andel and Bowin* [1968] suggests that some nonmagnetic peaks may also be composed largely of greenschists or greenstones.

Some of the most imposing submarine escarpments are those associated with fracture zones (Figures 11 and 16). Recent data collected by the present authors along the fracture zone offsetting the Mid-Atlantic ridge at 24°N shows essentially no magnetic anomaly associated with scarps several kilometers high.

The thickness of the magnetized layer is still uncertain, as is illustrated by the model profiles of Figure 20. For a given height above the top of the layer, the standard dike model produces anomalies that are relatively invariant when magnetization is increased while the layer thickness is decreased. Therefore, within rather wide limits, only a combination of layer thickness and magnetization intensity can be determined from a fit of the model to the observed profile [*Vine*, 1966]. Although the top of the magnetized layer must be within or at the top of the second layer, the bottom could extend only to a few hundred meters [*Talwani et al.*, 1968], to the base of the second layer, or even to the level of the Curie isotherm under the ridge crest (of the order of 5 to 10 km below the sea floor). A shallow layer would be consistent with the magnetic properties of dredge samples and with *Hess'* [1962] proposition that the main constituent of the oceanic layer is relatively nonmagnetic serpentinite.

The tape-recorder model concerns itself with TRM acquired at the axis of the mid-oceanic ridge. The Curie isotherm probably slopes smoothly downward away from the axis (Figures 22 and 23). Any deeper TRM, acquired away from the axis, probably has little effect on the magnetic signature because there the polarity intervals are recorded as thin, very gently inclined sheets.

Ordinary induced magnetization is usually ignored for purposes of sea-floor spreading. This is justified if true susceptibility is a function of depth only, because no induced anomaly is then effected. In the case of the anomalous noise that is added to the simple Vine-Matthews pattern by magnetized sea-floor topography, the significance of induced magnetization versus TRM has not been established. Many studies have shown that the ratio Q between remanent and induced magnetizations is very high (of the order of 10 to more than 100) for the rapidly chilled subaqueous lava fragments commonly dredged from the sea floor [*Vogt and Ostenso*, 1966]. This high Q is the result of exceptionally low true susceptibility combined with a TRM that is in general typical for basalt lava and dolerite and diabase dikes in both continental and oceanic crust [*Nagata*, 1961]. Within the oceanic magnetized layer, it seems probable that induced magnetization is only very low in a (thin?) surface zone of rapidly chilled lavas and then increases to a magnitude ($Q = 1$) comparable to that of the TRM at depths.

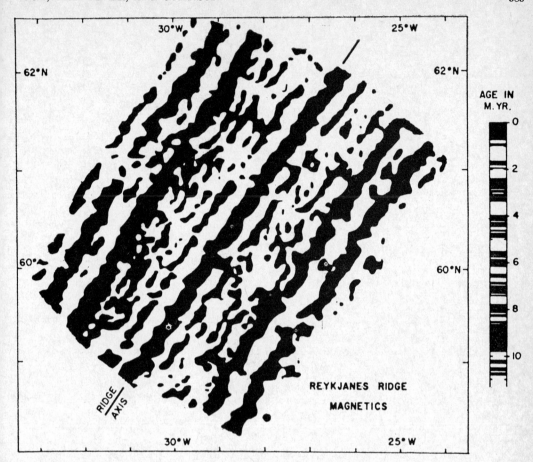

Fig. 21. Linear magnetic anomaly pattern over Reykjanes ridge (*Vine* [1966] as reproduced by *Maxwell* [1968]), based on a U. S. Navy aeromagnetic survey. Black indicates positive anomalies. Symmetry of pattern about ridge axis and correlation of profiles with those computed from independently established reversal history (shown at right in figure) was instrumental in turning tide of professional opinion in favor of the Vine-Matthews hypothesis and continental drift in general.

Detailed knowledge of the configuration of the magnetized layer awaits deep-towed magnetometers, deep drilling, and the recovery of oriented samples.

In addition to the perturbation of magnetized topography, some departures of the observed profiles from the simple model could be explained if dikes do not necessarily bisect their predecessors but are injected with a normal distribution about the ridge axis.

At about 45°N on the Mid-Atlantic ridge the observed profiles could be reproduced if the standard deviation of dike injection is about 5 km [*Matthews and Bath*, 1967].

Using another approach, *Harrison* [1968] showed that the standard deviation is less than 3 km on both the slow Reykjanes ridge segment (1 cm/year) and the fast East Pacific rise sector of the mid-oceanic ridge (4.4 cm/year). Apparently 95% of the dikes are injected within 6 km of the axis [*Harrison*, 1968]. Thus, on ridge segments where a rift valley is developed, most if not all the dikes and their associated flows, if any, are contained by the walls of the valley. This slim width of the tectonic axis of the mid-ocean ridge is consistent with the narrowness of the axial seismic belt [*Sykes*, 1967], considering the present accuracy of epicenter location (±20 km, or better).

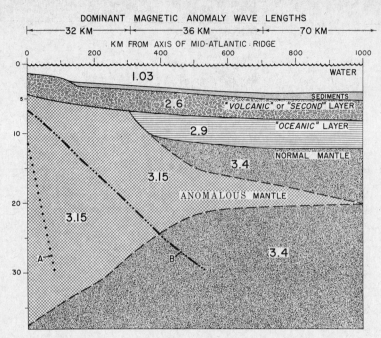

Fig. 22. Crustal half-section of Mid-Atlantic ridge, simplified [after *Talwani et al.*, 1965]. Representative model densities shown for each layer were deduced from P-wave velocities. Solid lines are interfaces with seismic control. Dashed interfaces, inferred from gravity data only, are highly uncertain. Dotted lines are two possible Curie isotherms: A is based on model for broad axial negative magnetic anomaly [*Heirtzler and Le Pichon*, 1965]; B was interpolated from thermal crustal spreading models [*Langseth et al.*, 1966; *Oxburgh and Turcotte*, 1968].

The exceptionally high amplitude of the axial magnetic anomaly has not been satisfactorily explained. An acceptable fit between the standard tape-recorder and observed profiles in most areas requires that the axial body either have about twice the magnetization of the rest of the model or else be significantly thicker [*Vine*, 1966; *Talwani et al.*, 1968]. Several explanations are proposed. *Matthews and Bath* [1967] found that if the standard deviation of dikes is as great as ±5 km from the axis, the high axial amplitude can be due to a young slab of purely normal polarity, whereas the effective magnetization of every other magnetic event recorded by the crust has been reduced by pollution during later epochs of opposite polarity.

Harrison [1968] argues that this effect is insufficient, and proposes two alternative hypotheses: (1) a real decay of magnetization with time, owing to an alternating-field demagnetization by the earth's reversing field, or (2) exceptionally high geomagnetic intensities during recent millenia. Other mechanisms are possible.

If the central body is subsequently faulted and uplifted, some of the magnetization could be removed by the action of weathering at the surface and within the water-filled fractures. The thinner the magnetized layer, the more significant such an effect can be. *Schilling et al.* [1968] dredged pillow lavas from the crest of Reykjanes ridge. Very high remanent magnetization (0.06 to 0.09 emu) was found in the small axial rift, but on the flanks altered pillows revealed more typical magnetizations of 0.025 to 0.04 emu.

A low-amplitude negative anomaly with a much longer wavelength (100 km) than those due to the stripes of alternate polarity can be observed on many profiles across the Mid-Atlantic ridge and may be due to a rise in the Curie isotherm under the axis. A possible model boundary for the Curie isotherm, calculated by *Heirtzler and Le Pichon* [1965], is indicated by curve A in Figure 22. This can be compared to curve B, the approximate Curie isotherm expected for steady-state convection on the Mid-

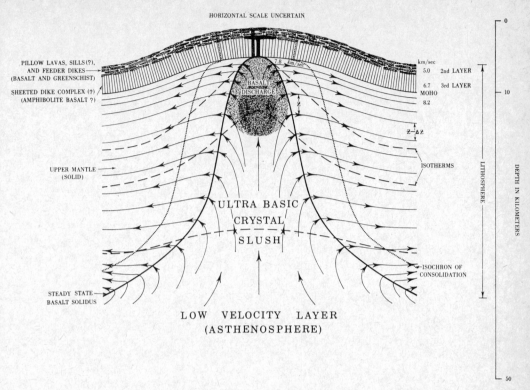

Fig. 23. Schematic cross section of crust and upper mantle under rapidly spreading segment of mid-oceanic ridge based on hypothesis that crust represents basaltic derivatives [*Bott*, 1965; *Cann*, 1968] and that basalt liquidus reaches base of crust (Moho) if spreading is fast enough. Black intrusive at axis suggests that seismic uniformity of oceanic layer may be partly due to sills injected at base of volcanic layer. Material moves as two rigid plates except in axial intrusive zone [*Morgan*, 1968]. In the upper mantle (upper right), a vertical distance Z is shown decreasing to $Z - \Delta Z$ on account of thermal contraction and solid-state phase changes. Major uncertainty is width of ultrabasic chamber [*Cann*, 1968].

Atlantic ridge [*Langseth et al.*, 1966; *Oxburgh and Turcotte*, 1968]; the difference between A and B suggests that either the axial zone of intrusion is much narrower than supposed, or spreading is not steady state, having only recently resumed. Perhaps A is not the correct model for the Curie isotherm. The broad magnetic negative could be due to a gradual decrease in magnetization intensity away from the axis, but at shallow depths.

The reconstruction of crustal movements from an application of the Vine-Matthews magnetic tape-recorder technique has now begun. Consider first the pattern of ridge growth since the late Pliocene and presumably now actively in progress. The geomagnetic polarity history for which absolute dates exist at present extends back only about 4 m.y. This history is independently known from a study of magnetization versus absolute age in deep-sea sediment cores [*Glass et al.*, 1967; *Ninkovich et al.*, 1966]. If the Vine-Matthews model is applied to this most recent sequence of anomalies, it is found that sea-floor expansion has had the following properties: Crustal growth is symmetrical about the ridge axis; growth half-rates range from about 1 cm/year in the Red Sea and North Atlantic to 6 cm/year on the equatorial East Pacific rise. The magnitude of growth rates is consistent with average values predicted from studies of continental drift. The sea-floor expansion rates vary regularly along the strike of the ridge. For each ridge segment active now, and for the past 4 m.y., sea-floor growth rates have been essentially constant with time and did not change with distance from the

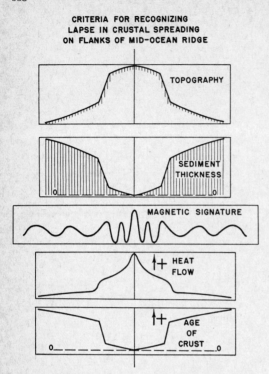

Fig. 24. Criteria for recognizing lapse in crustal spreading on flanks of mid-oceanic ridge [*Schneider and Vogt*, 1968]. Relatively abrupt changes in regional topography, sediment thickness, heat flow, and age of crust mark position of lapse. Taken by itself, the change in magnetic signature (after anomalous field has been continued upward to a constant height above the magnetized layer) could indicate (1) abrupt change in spreading rate; (2) abrupt change in geomagnetic polarity reversal rate; (3) lapse in spreading superimposed on a gradually changing reversal rate; or (4) change in spreading direction, with the longer wavelengths over the flanks only apparent; or (5) a combination of any of (1), (2), or (4) with a lapse.

axis. The variation of rates along the ridge is consistent with the assumption that the crust is divided into rigid plates which grow where they are bounded by the mid-oceanic ridge and are destroyed along their leading edge, where trenches and compressional orogens are developed [*Morgan*, 1968]. This plate model is an extension of the old idea that continents behave as rigid slabs in their travels over the earth's surface [*Wegener*, 1924] and follows directly from *Wilson* [1965]. The rigid plate assumption is based on the regularity of the oceanic crust, the lack of seismicity, and the relatively undisturbed character of the sediments in oceanic areas not close to trench or ridge structures, as has been known for some time [*Heezen*, 1962]. Recent work [*Heirtzler et al.*, 1968] reveals magnetic lineations far from the ridge axis, and their general symmetry about the present ridge axis. These facts support the proposition that, except for the effects of sedimentation and isolated vulcanism, the upper crust is not deformed or altered between the time it leaves the immediate ridge axis and the time it becomes incorporated in island-arc-trench or folded mountain structures. With deformation thus restricted to the edges of large stable plates, the pattern of crustal accretion rates as well as the orientation of fracture zones must be relatively simple. *Morgan* [1968] noted that the instantaneous movement, on a sphere, of any portion of a spherical surface with respect to another is defined by an angular velocity and a pole of rotation. The pole of rotation can be determined for the variation of crustal growth rates along the strike of the ridge bordering two plates. The rate must vary as the cosine of the plate's latitude with respect to the pole of rotation. Thus, the rate is zero at the pole and reaches a maximum at the equator of rotation. By a second method, the rotational pole can be determined from fracture zones: if the rotational pole remains fixed, the trace of movement is reflected by fracture zones lying on small circles having the pole of rotation as a common pole.

Morgan [1968] and *Heirtzler et al.* [1968] have shown that the orientation of fracture zones and the distribution of crustal accretion rates measured by the Vine-Matthews method are generally consistent with the simple model of crustal movements outlined above. With the exception of the Indian Ocean, whose opening seems to relate to a pole in Libya, the oceans of the world are opening about poles in the area of the Labrador Sea. The angular rates of opening are (in 10^{-6} degrees per year): Atlantic Ocean, 0.36; Indian Ocean, 0.40; South Pacific Ocean, 1.08 [*Heirtzler et al.*, 1968]. The proximity of many rotational poles to the earth's geographical pole invites speculation that these are related [*Heirtzler et al.*, 1968].

The history of movements before about 4 m.y. is being worked out without the benefit of an independent absolute time scale for geomagnetic polarity reversals. It will be difficult to extend such an absolute reversal history further back in geologic time than the late or middle Tertiary, owing to the short duration of polarity events compared to the relative errors in isotope dating [*Heirtzler et al.*, 1968].

Studies of magnetic anomalies indicate, however, that the same sequence of polarity history has been recorded by most active segments of the mid-oceanic ridge. Plots of distance of this anomaly sequence from the axis of one ridge segment against the distance from another are smooth curves, often nearly linear for periods of the order of ten million years at current rates of axial growth. Unless major interruptions have occurred simultaneously along the entire ridge, these curves indicate that the ratio of expansion rates for any pair of ridges has remained constant or varied slowly and regularly with time. For various reasons, *Heirtzler et al.* [1968] elected to deduce their relative expansion history on the still questionable assumption that the South Atlantic basin has been growing at a constant rate in time, while other ocean basins have grown at nonconstant rates. However, the deduced history of sea-floor expansions (and continental drift) satisfied most of the many constraints available at this time. If there have been local or global stoppages, they must be short compared to the age of the rift in order that past spreading rates not be excessive. The possible types of evidence for recognizing interruptions in the history of sea-floor growth have been discussed by *Schneider and Vogt* [1968] and are summarized schematically in Figure 24.

The JOIDES drill holes in the South Atlantic (A. E. Maxwell, personal communication) rule out the possibility of the Miocene quiescent period discussed by *J. Ewing and M. Ewing* [1967] and *Schneider and Vogt* [1968]. The South Atlantic has been widening at a fairly constant rate, at least since 75 m.y. ago. Thus the reversal chronology of *Heirtzler et al.* [1968] can be confidently applied in the dating of the ocean floor.

Seismicity. Recent work by *Sykes* [1967] and others supports older observations that most earthquake epicenters lie either within the Alpine-Himalayan and Circum-Pacific belts, which appear from geological structures to be largely compressional, or the mid-oceanic ridge system, which has been interpreted as tensional. Intermediate and deep earthquakes are almost entirely restricted to the compressional belts, the mid-oceanic ridge activity being wholly shallow-focus [*Heezen and Ewing*, 1963] and confined to narrow belts.

Where the axial rift valley is well developed, namely, for crustal expansion half-rates less than about 2 cm/year [*van Andel and Bowin*, 1968], the epicenter belt probably coincides with the median valley within the accuracy of epicenter determination (± 20 km or better). Where the ridge is offset by a fracture zone, there is usually a concentration of epicenters on the fracture zone between the offset ridge ends. Little or no seismic activity is observed on the extension of these fracture zones (Figure 2, p. 151). Where the rift crosses over onto continental crust (e.g., the East African rift, the Siberian extension of the Arctic mid-oceanic ridge, and perhaps the Great Basin of North America, the earthquakes become irregularly scattered over a belt several hundred kilometers wide [*Sykes*, 1967]. This difference between continental and oceanic segments of the rift suggests differences in the mechanics of rifting which may be related to differences in crustal thickness and properties. It appears that the thick geologically complex continental crust widens primarily by normal faulting, whereas the thin, geologically much more homogeneous oceanic crust grows largely by injection of basaltic dikes. Active continental rifts are probably always youthful if they are but incipient oceanic rifts.

Wilson [1965] postulated that many fracture zones along which the mid-oceanic ridge has been offset (Figure 1, p. 141) are not to be generally understood as strike-slip faults younger than the ridge. Instead, both ridge segments and fracture zones represent a single, continuously rejuvenated zone of weakness (Figure 25). Pure ridge segments are line sources where mantle derivatives rise to accrete to the edges of the crust and then move away at right angles to the ridge axis in a direction parallel to the fracture zones, as part of the crustal conveyor belt. In rare cases the ridge segments are oblique to the fracture zones, and the magnetic signature then yields only the component of spreading rate perpendicular to the ridge segment [*Morgan*, 1968].

The distribution and the apparent first motion

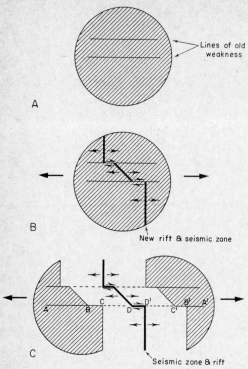

Fig. 25. Transform fault hypothesis [*Wilson*, 1965], by which the locus of ridge activity is the continuously rejuvenated line of *original* fracture. Pure transform faults are displacement lines. Oblique segments intermediate between fracture zone and ridge may be rare; magnetic anomalies would only indicate the component of spreading vector perpendicular to segment. In the North Atlantic, the original break has not been preserved in detail.

of earthquakes along the mid-oceanic ridge [*Sykes*, 1967] support the hypothesis of *Wilson* [1965] depicted in Figure 25. The evidence is discussed in more detail by Sykes in the present volume (pp. 150 ff.). In summary, the first motion solutions for earthquakes on fracture zones are compatible with the transform fault concept (see Figures 2 and 3, p. 151), and events under the ridge segments suggest a tensional regime. Earthquakes are almost entirely limited to that segment of a fracture zone between the offset ends of the ridge. This pattern supports the transform fault model (Figure 25) rather than the now discounted view that the fracture zones are strike-slip faults along which the ridge was displaced after its formation.

The tendency for fracture zones to make perpendicular intersections with ridge segments [*Sykes*, 1967; *Morgan*, 1968] requires explanation. Perhaps the initial rift consists of a set of en echelon tear fissures. These proto-rifts would prefer pre-existing lines of weakness in the crust [*Wilson*, 1965], as is illustrated in Figure 25. At a later time faults connect these en echelon fissures, with a right-angle intersection usually releasing the stress more effectively than any other. This original ridge-fracture zone pattern is then preserved by the process of sea-floor accretion (Figure 25).

Even if oblique ridge segments are somehow created, they may not represent a stable configuration [*Morgan*, 1968] for long periods. Consider an oblique ridge segment such as the one shown in Figure 25, and let L be its length, S the spreading rate parallel to the fracture zones, and α the angle between ridge and fracture zone. The spreading rate perpendicular to this oblique ridge is then only $S \sin \alpha$. If the oblique segment is replaced by a staircase pattern of ridge segments and perpendicular fracture zones, such as the pattern along the equatorial Mid-Atlantic ridge [*Sykes*, 1967], the length of the rift will be reduced from L to $L \sin \alpha$, whereas the local spreading rate will be increased from $S \sin \alpha$ to S. Because the oblique ridge is spreading more slowly perpendicular to its crest, the crust under the axis would cool and freeze to a greater depth. More work is expended to fracture this thicker crust. Further, the intrusive material, whose total volume is the same for both cases, has to be injected into a relatively longer and cooler rift in the oblique case. The thickness of a parcel of lithosphere at the ridge axis depends on the heat lost before the parcel moves away from the axis. If this loss is primarily by conduction, analogy to a downward-freezing ice sheet suggests that the thickness is proportional to $(S \sin \alpha)^{1/2}$.

This minimum energy argument requires that relatively little work be expended along the active fracture zones. If this is not the case it is hard to visualize why the oblique rift, whose total length is L, should be replaced by a staircase pattern whose total length, including fracture zones, is $L (\cos \alpha + \sin \alpha)$, a quantity always greater than L. If the staircase pattern requires less work, the observed concentration of seismic events along the active fracture zones, compared to the ridge segments [*Sykes*, 1967], then leads to the interesting consequence that very little of the work done in moving the plates apart is manifested as brittle fracturing on a seismic scale. The relative ease of slippage along the fracture zones may derive from continual longitudinal contraction of the relatively hot, low-density phase lithosphere near the axis. This thermal contraction would tend to open the fracture zones and reduce friction along their walls.

It should be noted that the rift, once formed, represents a zone of weakness along which injection of mantle material and away from which sea-floor spreading would continue, even though the distribution of the original causative forces might have changed somewhat.

The nature and origins of the oceanic crust as deduced from explosion seismology. Earlier studies of the crust under ocean basins and the mid-oceanic ridge have recently been summarized by *Oxburgh and Turcotte* [1968] in a paper dealing with mantle convection from which the following is partly adapted.

Broadly speaking, the oceanic areas can be divided into areas of 'normal' and 'abnormal' crust (Figure 26). Normal crust consists of ½ to 1 km low-velocity sediments overlying 1 or 2 km of 'second layer' characterized by a highly variable *P*-wave velocity whose average value is about 5 km/sec. Below the second layer is the oceanic, or third, layer (6.7 km/sec, 5 km thick), which is remarkably uniform in velocity and thickness beneath most of the ocean floor. The M discontinuity is generally a little more than 10 km below sea level where normal crust exists, and mantle velocities (8.0–8.2 km/sec) are normal.

The most common variety of abnormal crust is characterized by the absence of a third layer (Figure 26). Water depths are generally shallower, and the second layer somewhat thicker, than for normal crust. This thickened second layer rests directly on abnormal mantle characterized by low velocities (of the order of 7:5 km/sec) with greater scatter in individual values.

In the present discussion it is assumed that most oceanic crust has been generated by sea-floor accretion at the axis of the mid-oceanic ridge. The validity of this theory is not entirely without question, but recent evidence has given it a very high degree of credibility. The origin for the difference in the two types of crust appears to be related to sea-floor expansion rates at which these types were generated. At least under the axis of the mid-oceanic ridge the low-velocity mantle may be a transient state that becomes normal mantle upon cooling at some distance from the ridge axis. This will be discussed in greater detail below.

The normal oceanic crust rests on upper mantle characterized by compressional wave velocities of 8.0–8.2 km/sec. Data on normal crust are summarized in Table 1.

Layer 2 is also known as the 'volcanic layer'; where it has been sampled in outcrops on the crest of the mid-oceanic ridge, it is igneous. This layer is common to both kinds of crust.

Seismic data on the second layer are difficult to evaluate for a number of reasons. The layer's thinness results in a very narrow detection range in which first-arrival measurements can be obtained. First arrivals from the layer are often not detected, and a wide range of seismic velocities has been assigned to the layer. The large degree of scatter in the velocities (3.5 to 6 km/sec) is not necessarily all due to real variations in elastic constants and density. The topography on the second layer is rough; significant diffraction or scattering effects are to be expected. The paucity of first arrivals may introduce interpretational errors. If the layer is a complex assemblage of unequally weathered flows, feeder dikes, fault and explosion breccia, and intercalated sediment ponds, a wide range of velocities is hardly surprising.

There seem to be systematic variations of layer thickness with ocean basin and water depth [*Hill*, 1957]. The second layer appears thicker in shallower water, and thinner under the Pacific than under the Atlantic Ocean. The significance of those relationships, if real, to the sea-floor accretion hypothesis remains to be evaluated. There may of course be systematic hidden variables, such as regional variations in sediment thickness and basement topography. Assuming that the effect of such variation is small, and that the shallow stations are representative of the mid-oceanic ridge in the two oceans, we can then apply to the second-layer thickness data [*Hill*, 1957] *Menard*'s [1967b] observation that thickness is inversely related to accretion rate. On the whole, these rates

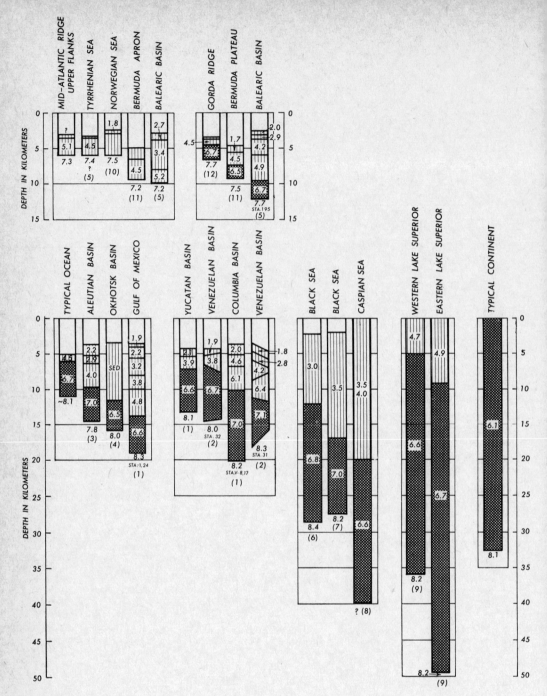

Fig. 26. Crustal sections for abyssal basin, ridges, rises, small ocean basins, and continent, determined by seismic refraction measurements. Sections are grouped according to thickness of layer with P-wave velocity of about 6.7 km/sec (oceanic layer). Within each group, the sections are arranged in order of increasing depth to Moho so as to suggest a relationship between amount of subsidence and sediment load. Figure in part arranged after *Menard* [1967a]. Sources: (1) *J. Ewing et al.* [1960]; (2) *Officer et al.* [1959]; (3) *Shor* [1964]; (4) *Aver'yanov et al.* [1961]; (5) *Hersey* [1965]; (6) *Neprochnov et al.* [1959]; (7) *Neprochnov et al.* [1964]; (8) *Gagel'gants et al.* [1958]; (9) *O'Brien* [1968]; (10) *J. Ewing and M. Ewing* [1959]; (11) *Heezen et al.* [1959]; (12) *Shor et al.* [1968].

TABLE 1. Characteristics of Normal Oceanic Crust*

Layer	V_p, km/sec	Thickness, km	Probable Composition
1	1.45 to 2.0	0.45 but variable	Unconsolidated sediments
2	5.07 ± 0.63	1.71 ± 0.75	Indurated sediment and volcanics
3	6.69 ± 0.26	4.86 ± 1.42	Basalt (mainly intrusive) and/or serpentinite; amphibolite; gabbro

* Values are taken from Hill [1957] and Raitt [1963].

were therefore higher in both Atlantic and Pacific oceans when the crusts presently flooring the deep basins of those oceans were created. A second consequence is that sea-floor growth rates have generally been higher in the Pacific than in the Atlantic Ocean; recent work on magnetic anomalies [Heirtzler et al., 1968] seems to confirm this.

Layer 3, also called the basaltic or oceanic layer, has been much more satisfactorily observed than the second layer [Raitt, 1963]. The upper limit of the random standard error in the velocity measurement may be as small as 0.11 km/sec [Raitt, 1963]. If the error is this small, there may be small but real velocity variations. Nevertheless, the thickness and velocity of the layer are remarkably uniform over most of the world ocean. Serpentinite [Hess, 1962] and basalt or gabbro have been proposed as the principal constituent. Geological objections to the serpentinite hypothesis have been given by Cann [1968], who has argued that the oceanic layer is amphibolite. Thermal arguments [Oxburgh and Turcotte, 1968] also favor a basaltic composition. Sea-floor accretion rates vary at least by a factor of three where the oceanic layer is observed, yet the layer thickness is always the same. Production of this layer by branches of the mid-oceanic ridge must keep close pace with the crustal-spreading rate, and any theory for the third layer must account for this fact.

The same two velocities (about 6.7 and 8 km/sec) tend to occur across the M discontinuity under both continents and ocean basins. As the oceanic M is apparently 'unrolled' from the mid-oceanic ridge, it would be surprising if some entirely unrelated process created the continental M discontinuity. We can therefore speculate that the entire Moho is of mid-oceanic origin. If so, perhaps the basaltic layer often observed under the continents comprises relics of ancient oceanic crust, long since loaded with sediments, folded, metamorphosed, and thus transformed to continental crust.

Abnormal oceanic crust, as previously defined, occurs mainly under certain branches of the mid-oceanic ridge, whose structure has been reviewed in great detail by Talwani et al. [1965], Heirtzler and LePichon [1965], M. Ewing et al. [1966a], and Langseth et al. [1966]. These works deal primarily with the East Pacific rise, now known to be growing from its axis at half-rates of 2 to 6 cm/year, and the Mid-Atlantic ridge, growing at 1 to 2 cm/year [Heirtzler et al., 1968]. It should be noted that the 1965 and 1966 papers were written before the sea-floor spreading hypothesis gained general acceptance.

The East Pacific rise (Figure 27) exemplifies fast-growing branches of the ridge and appears to be a generator of normal oceanic crust. The occurrence of low-velocity mantle under its crest could well be a steady-state phenomenon. That is, it may be constantly created under the axis and converted to normal mantle under the flanks.

The mantle under the ridge axis is most likely an ultrabasic crystal slush through which basaltic fluids must rise to feed the growing layers 2 and 3 [Hess, 1962; Bott, 1965; Oxburgh and Turcotte, 1968]. This slush then probably solidifies and becomes 'normal' mantle as it withdraws from the axis.

Abnormal crust underlies the Mid-Atlantic ridge within about 300 km of its axis (Figure 28). Under some ridges, such as Reykjanes [Talwani et al., 1968], velocities typical of the oceanic layer are found. In this region the crust is still too young, probably less than 50 m.y. old [Schneider and Vogt, 1968], for pelagic sediments to have masked the igneous basement. Dredging and bottom photographs of this basement window reveal that the outcropping second layer is largely volcanic in origin [Engel and Engel, 1964; van Andel and Bowin, 1968]. Layer 2 is common to both types of crust, but seems thicker under the abnormal kind (Figures 27 and 28). As was previously discussed, the thickness of this layer seems approximately inversely proportional to sea-floor accretion rates as inferred from the magnetic anomalies and

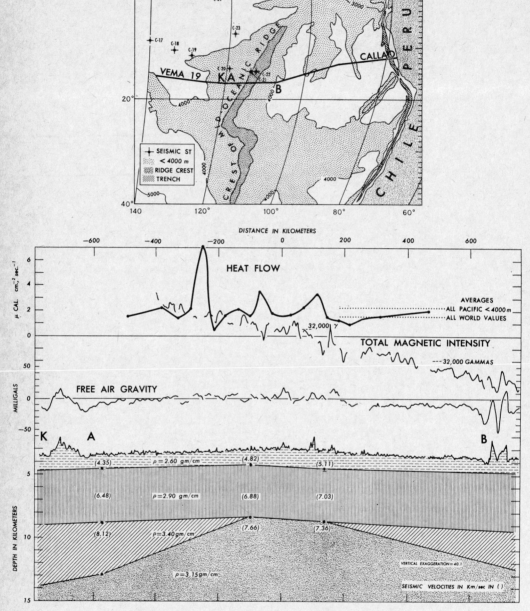

Fig. 27. Geophysical signatures across East Pacific rise, and crustal structure inferred from seismic refraction and gravity measurements [from *Talwani et al.*, 1965]. Compare with Figure 28, and note the apparent continuity of oceanic layer under ridge and narrow width of low-velocity mantle in contact with oceanic layer. Regional free-air anomaly is only slightly positive over crest, indicating isostatic compensation. Seamounts at A and B, equidistant from axis, appear uncompensated. Note that second layer is smoother and thinner than on Mid-Atlantic ridge. Symmetry about ridge axis of anomalous magnetic field is not obvious on a single profile! *Talwani et al.* [1965] used their own data and those from *Raitt* [1956] and *Menard* [1960] to make this figure. Heat-flow data from *Langseth et al.* [1966].

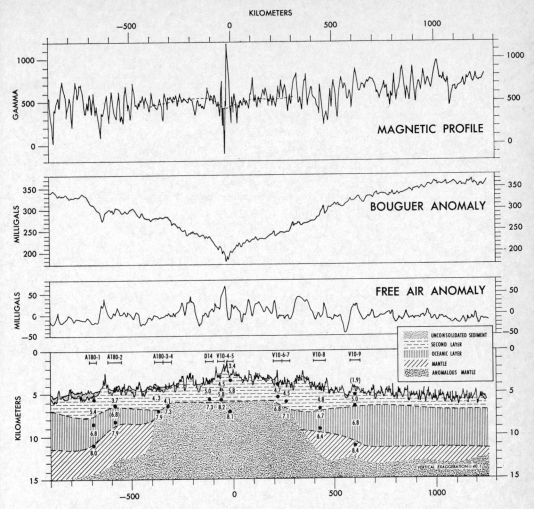

Fig. 28. Representative geophysical signatures across Mid-Atlantic ridge [from *Talwani et al.*, 1965; *Heirtzler and Le Pichon*, 1965]. Most of the data are from north temperate latitudes. Lowest profile shows crustal structure and P-wave velocities deduced from seismic refraction measurements. Note rough basement, absence of sediment in the immediate crestal zone, absence of oceanic layer within 400 km of crest, and great scatter in P-wave velocities in anomalous mantle and second layer. Free-air anomalies of short wavelength are almost absent in Bouguer profile, suggesting uncompensated surface topography with density 2.7 g/cm³. Broad Bouguer low reflects regional compensation. Magnetic profile shows change to broad, high amplitude anomalies where oceanic layer appears, suggesting faster sea-floor spreading in past. Significance of high axial anomaly amplitude and broad low (shown by dashed line) is discussed in text. Compare this figure with Figures 27 and 29.

polarity reversals. In other words, the rate of generation of this layer appears to be constant in time [*Menard*, 1967b]. This observation could be explained if the rate at which basaltic fluids drain upward out of the mantle is much faster than and independent of the accretion rate. A more detailed summary of crustal generation and destruction is given in a later section dealing with geology and petrology.

In the abnormal crustal section, the volcanic layer directly overlies low-velocity mantle (Figure 28), characterized by P-wave velocities in the range 7.2 to 7.6 km/sec, which are absent below the distal flanks [*Talwani et al.*, 1965].

The crustal section (Figures 22 and 28) across the Mid-Atlantic ridge suggests two basic interpretations:

If the appearance of the ridge does not change greatly with time, it is a steady-state system. We then would require a physical process whereby the low-velocity mantle somehow becomes segregated into oceanic layer and normal mantle as it cools and withdraws from the axis. If direct application of the Vine-Matthews hypothesis would reveal constant growth at, say, 1 cm/year for 100 m.y., such a steady-state explanation for ridge structure would have to be sought. However, considerable indirect evidence now suggests a discontinuous character to sea-floor accretion. Although the JOIDES drill holes in the South Atlantic have ruled out the possibility of global stoppages of the order of 10 m.y., the magnetic anomaly pattern in the North Atlantic reveals a number of spreading epochs characterized by distinct spreading rates and/or directions. The analysis of these data is being done by C. Anderson, O. Avery, D. Bracey, R. Higgs, and the senior author. The oceanic area east of Reykjanes ridge has been surveyed most thoroughly. Three spreading episodes are recognized between Rockall bank and Reykjanes ridge. Spreading began about 60 m.y. ago and proceeded at a rate of 1.1 to 1.2 cm/yr until about 42 m.y. ago. At that time rates fell to 0.7–0.8 cm/yr, the spreading direction changed, and transform faults were generated every 100 km. This slow spreading lasted until about 18 m.y. ago, when spreading rates increased to 1.1–1.2 cm/yr again. Although the direction of spreading did not change, the transform faults disappeared, and Reykjanes ridge became an oblique ridge, i.e., one that is not spreading perpendicular to its crest. It appears that the oblique and staircase ridge configurations represent two possible modes, the latter being preferred, in the minimum energy sense, when spreading rates are very low, say below 0.8 cm/yr. For such slow spreading the crustal thickness becomes excessive in the oblique case, and transform faults are created to decrease the length of the spreading axis and increase the spreading rate perpendicular to the axis.

This unusual magnetic anomaly pattern, which the 3-mile line spacing places beyond doubt, may yield valuable insight into the relation

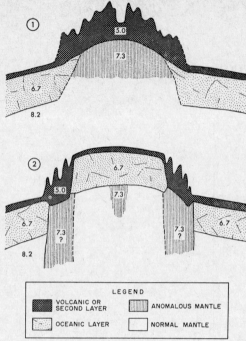

Fig. 29. Hypothetical structures for two spreading histories. In (1) half-rates were greater than about 2.5 cm/year before, and less than that value after the lapse. In (2) the rates exceeded this value except during a period of slower spreading, when crust similar to that at the axis of (1) was generated. There is some evidence that in some respects the two models correspond to the Mid-Atlantic ridge and to the East Pacific rise [*Schneider and Vogt*, 1968]. Numbers show representative P-wave velocities. The slow crust could reflect closely spaced fracture zones.

between spreading rate and crustal structure. Detailed seismic refraction work near the axis of Reykjanes ridge shows that, although spreading is relatively slow (1 cm/yr perpendicular to the crest), a normal oceanic crust is being generated [*Talwani et al.*, 1968]. The magnetic lineations are well developed, topographic relief is only moderate, and fracture zones are few. On the other hand, the spreading epoch 42–18 m.y. ago was characterized by close spaced fracture zones and an anomaly pattern that would not have been resolved even at 10-mile line spacing. Fracture zones comprise over 30% of the crustal area created during this interval. Refraction lines inadvertently sampling parts of the fracture zone crust may well yield the abnormal crustal section that tends to be associated with slowly spreading ridges. Thus the dependence of crustal structure on spreading history (Figure 29) may be related to the tendency for fracture zones to be more closely spaced on slow ridges.

The detailed survey east of Reykjanes ridge (O. Avery and R. Higgs, personal communication) shows that transform faults may be created and destroyed during the course of evolution of an oceanic rift. Therefore Wilson's hypothesis (Figure 25) that the present configuration of, say, the Mid-Atlantic ridge ought to resemble the shape of the original break is only an approximation. Similarly the large offset of the Mid-Atlantic ridge at 24°N did not exist in the lower Mesozoic, when a linear sequence of magnetic anomalies between the East Coast magnetically smooth zone and Bermuda was generated (C. Anderson, personal communication).

Heat flow. The equality of oceanic and continental heat flow has been known for some years but is still the object of intensive speculation [*Elsasser*, 1967] (see section by Von Herzen and Lee, p. 93). *Bullard* [1963] summarized the techniques and problems of measuring heat flow.

Oceanic data are far more numerous than terrestrial data owing to the thermal stability of the abyssal depths far from sediment sources. A recent compilation [*Langseth et al.*, 1966] indicates some regional variations between ocean basins. Thus, the worldwide average in heat flow units (1 HFU = 1×10^{-6} cal cm^{-2} sec^{-1}) is about 1.1 [*Lee and Uyeda*, 1965], but regional averages reach 1.5 in equatorial parts of the Atlantic Ocean. The significance of this regional variation in terms of sea-floor growth has not been evaluated.

Heat flow on the mid-oceanic ridge qualitatively supports the concept of sea-floor accretion, but models which accurately reproduce the observed pattern have not been calculated [*Langseth et al.*, 1966; *Oxburgh and Turcotte*, 1968]. High values (though with a great deal of scatter) do tend to occur on the ridge, as would be expected if hot mantle derivatives are injected there. High values occur along axially centered belts 200 km wide over the Mid-Atlantic ridge and 600 km wide over the East Pacific rise. This conforms to magnetic evidence [*Heirtzler et al.*, 1968] that growth rates are about 3 times higher on the East Pacific rise. The faster the expansion, the farther the crust can be carried away from the zone of intrusion before it is cooled.

If sea-floor accretion has been continuous, there are difficulties in adapting models to the observations. The total heat released seems to be considerably lower than expected, especially over the Mid-Atlantic ridge [*Langseth et al.*, 1966].

McKenzie [1967] has shown that if the lithosphere is assumed to be only 50 km thick (100 km in the models of *Langseth et al.* [1966]), the model anomalies are close to the observed ones. The mantle under ridge axes need not be hotter than elsewhere to account for the anomalies.

Heat-flow values tend to be low close to trenches. This conforms with the concept that trenches are the sinks where crust is destroyed by being dragged down into the upper mantle. However, models that fit the observations in detail have not been offered, and the low values may instead reflect rapid sedimentation.

Gravity. The ocean floor is basically in isostatic equilibrium [*Vening Meinesz*, 1948; *Talwani et al.*, 1965]. The only large-scale exception is the trenches over which free-air values as low as ~400 mgal occur [*Vening Meinesz*, 1948].

Where the sea floor is rough, free-air anomalies tend to mirror the topography, but topographic correlations usually do not appear on a Bouguer profile (Figure 28) if the density is assumed to be about 2.7 g/cm³. Therefore, it is concluded that topographic irregularities with a wavelength of a few tens of kilometers tend to be isostatically uncompensated. That is, they are supported by the crust. Loads of intermediate size, such as the great volcanic pile forming the Hawaiian Islands, have depressed the sea floor in their vicinity. Apparently, such features are regionally compensated [*Vening Meinesz*, 1948; *Woollard*, 1966].

The shape of the Bouguer curves over the mid-oceanic ridge indicates it is isostatically supported by a low-density root high in the upper mantle. A free-air high over Reykjanes ridge, only about 200 km wide, indicates some lack of compensation, although the Bouguer anomaly is still negative [*Talwani et al.*, 1968]. One possible crustal model for the Mid-Atlantic ridge is shown in Figure 22, simplified from *Talwani et al.* [1965]. The model fits available gravity and refraction data; dashed lines indicate density interfaces for which seismic control is lacking and which, therefore, are highly uncertain.

A similar gravity-seismic model for the East Pacific rise is shown in Figure 27.

Seismic surface waves. The use of seismic surface waves in deducing the structure of the oceanic crust and mantle has been summarized by *Oliver and Dorman* [1963] and Dorman in this Monograph, p. 257. The dispersion of these long (10- to 2000-km) waves after they have crossed hundreds or thousands of kilometers of crust depends on the average geometry and properties along their path. The significance of surface-wave studies is not restricted to the mantle. Some interesting average results have included determinations of the mean thickness of unconsolidated sediments of the ocean floor and mean thickness of the oceanic crust. The derived models are consistent with seismic refraction and reflection data, that indicate a worldwide uniformity of the average oceanic crust. It is possible to predict from the dispersion observations whether a given path is continental or oceanic. The oceanic character of the crust under the Arctic basin was demonstrated in this manner [*Båth and Vogel,* 1958]. There are some regional variations in Love-wave dispersion curves even when the paths are entirely oceanic. For example, lower velocities are recorded at Honolulu from shocks southeast of Hawaii than from other azimuths. Anomalous crustal and upper mantle structures under the East Pacific rise may be the cause [*Oliver and Dorman,* 1963].

Petrology and geology of the oceanic crust. Some areas of the mid-oceanic ridge crests have been subjected to detailed petrological studies. Hypotheses are sought to comprehensively explain how the various petrologic, topographic, and geophysical features of the crust are formed.

The general association of petrology with water depth was discovered by *Engel and Engel* [1964] and others, and is valid even for local areas such as the crest of the Mid-Atlantic ridge at 45°N [*Aumento,* 1967]. Olivine tholeiite with low to intermediate alumina content (14 to 16%) is common abyssal basalt typically recovered from the deeper parts of the ridge, including the rift valley. Higher on the mountains of the ridge occur transitional olivine tholeiites with rare high alumina equivalents. The highest seamounts and the islands are capped with alkali olivine basalts and their differentiates.

Aumento [1967] found that petrology of dredge specimens varied continuously from the olivine tholeiites flooring the valley to alkali basalts capping the crestal peaks 5 to 10 km from the valley axis. All rock types could be produced by a complex system of fractional melting and crystallization of a parent pyrolite material. A typical volcanic cycle commences with outpourings of olivine tholeiitic lavas on the floor of the median, or rift valley. Subsequent extrusions capping the tholeiites are progressively enriched in alkalis. After about 1 m.y. the former valley floor, uplifted and capped with alkali basalt, has become transformed into crest mountains now 5 or 10 km from the valley axis [*Aumento,* 1967]. Therefore the topographic crests represent more or less periodic peaks in the time rate of alkali basalt generation, superimposed on a more even outflow of tholeiitic basalt. The dominant topographic wavelength is about 30 km on the Mid-Atlantic ridge [*Heirtzler and Le Pichon,* 1965], suggesting a period of 3 m.y. in the typical volcanic cycle.

Is the rift valley a relatively steady-state feature? If so, the edges of the ever widening valley are constantly being built up by extrusion, in such a way that a valley always exists. According to this view, there can be no alkali basalts under the valley floor. This model is intuitively supported by the fact that a median valley is found almost everywhere along the crest of the slow ridges *at the present time*. If the valley is steady state, vulcanism must occur out to 5 or 10 km from the ridge axis and thus distorts the Vine-Matthews model. As was discussed above, dikes have apparently been injected with a standard deviation of 6 km or less from the tectonic axis of the Mid-Atlantic ridge at about 45°N (where a rift valley exists). Therefore, on the steady-state hypothesis the late alkali basalt extrusives 5 to 10 km from the axis are either very thin or have a low effective magnetization. The latter may be the case if these last extrusives, which form the hill crests, are largely piles of randomly oriented cobbles and ash and therefore produce no large net magnetic anomaly [*Vogt and Ostenso,* 1966].

The close topographic similarity between the rift valley and known downfaulted graben structures on land was pointed out by *Heezen and Ewing* [1963]. A graben is produced when a downward-narrowing wedge of lithosphere collapses along normal boundary faults. If the axial valley is a graben it must represent failure of this lithosphere under tension, the width of the valley (20 to 30 km) being a rough upper limit to the thickness of the lithosphere at the axis.

We then come to an alternate model by which both the occurrence of alkali basalts on crests and the dominant wavelength of 30 km are explained (Figure 30). We begin with an uplifted crestal province. The cumbersome intrusion of a thick ultrabasic mass into the upper mantle adds to the regional tension at the surface; a wedge-shaped graben begins to develop. Early derivatives from the intrusion are olivine tholeiitic lavas that flood the sinking floor of the rift valley. Continued crustal widening is accomplished as dikes intrude the valley floor. Eruptions of steadily more alkalic basalts raise the constantly widening valley floor (Figure 30). While the crust is widening in this manner, the 7.3-km/sec layer is widening with the continuing ascent of the ultrabasic intrusion. The crust and upper mantle consolidate by cooling and downward freezing. The old valley walls are by now 50 to 100 km apart and the center of the valley has been replaced by a high crestal volcanic pile. Limited volcanism and normal faulting may extend some distance from the axis during this stage. Meanwhile, a new ultrabasic intrusion is displacing the previous one at depth, and a new pair of normal faults, intersecting the volcanic pile about 20 to 30 km apart, gives birth to a new rift valley. Initial stress releases due to this normal faulting will accelerate differential melting of the ultrabasic pluton, giving initial olivine tholeiite lavas. This partial melting will in turn make the pluton more mobile.

The sequence of events outlined above could generate the dominant 30-km topographic wavelength found [*Heirtzler and Le Pichon*, 1965] on the flanks of the Mid-Atlantic ridge. The correlation between petrology and water depth is explained without resort to vulcanism more than a few kilometers from the rift axis. The model is therefore still consistent with the Vine-Matthews hypothesis [*Matthews and Bath*, 1967], even though the appearance of the axis changes with time.

In the model described above (Figure 30), the petrologic cycle is controlled by the diapiric emplacement of ultrabasic plutons. The first fluids to escape are olivine tholeiites. As the diapir rises, parent olivine tholeiite magmas have time to differentiate by gravitation and diffusion, yielding the later alkali basalts. Such a chimney effect [*Poldervaart and Green*, 1965] may occur in combination with some of the continuous differentiation reactions proposed by *Aumento* [1967].

The topographic waves are thus produced at a rate of one per three million years if the spreading rate is 1 cm/year. The dominant factor in determining the 30-km spacing may be (1) a cyclic mode of intrusion of the ultrabasic diapirs, with one about every 3 m.y. or (2) the width of a bridge of lithosphere that can be built before its central span collapses under tension. During a given 3-m.y. cycle, about 30 km of new crust is formed on each side of the axis. Of this amount, perhaps 10% is stretching of the crust by normal faulting, leading to rift valley formation, and the rest of the widening is accomplished by the injection of dikes (Figure 30).

This model predicts that tholeiite lavas may overlie earlier downfaulted alkali basalts, even underneath the margins of the rift valley. As the rift valley is so common at the present time, the ultrabasic plutonic cycle must be closely coupled along the length of the ridge. This model also implies that heat flow may vary with time, and that the present heat flow is not necessarily average.

Another petrologic feature that requires explanation is the occurrence of metamorphosed basalts near the tops of some hills at the ridge crest [*Cann and Vine*, 1966; *van Andel and Bowin*, 1968]. This metamorphism must have taken place 1 or 2 km below the sea floor, given typical values for present ridge crest heat flow and the conductivity of basalt [*Melson and van Andel*, 1966].

Van Andel and Bowin [1968] have invoked gravity sliding to remove the overlying basalts to expose the greenstones. A hypothesis that faces fewer obstacles [*Cann*, 1968] is that conjugate normal faulting has exposed the deeper crustal layers. It sems likely to us that such a period of more pronounced normal faulting would occur between C and D in Figure 30, i.e., after the ultrabasic pluton has solidified but

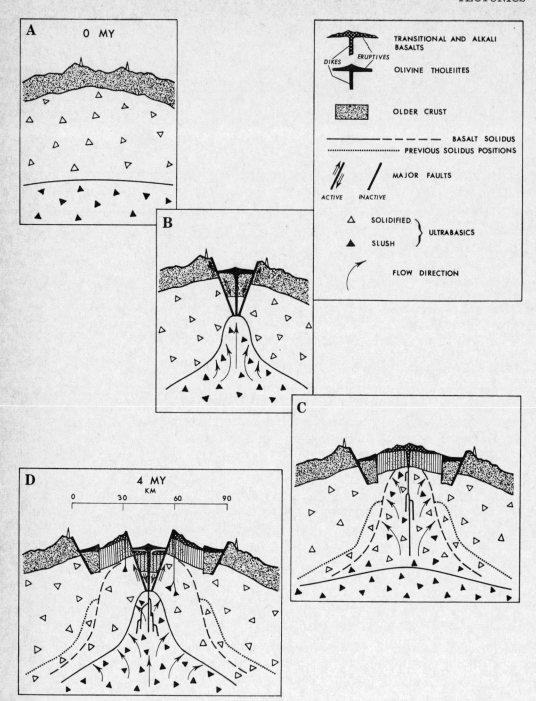

Fig. 30. Schematic cross section of slow-spreading segment of mid-oceanic ridge at several stages of 3-m.y. cycle, based on hypothesis that rift valley is not a steady-state phenomenon, but results from inability of basalt liquidus to maintain itself at base of crust. Intrusion of ultrabasic slush thus becomes an irregular process under slow-spreading ridge segments (compare with Figure 23).

before a new rift valley and pluton have developed. In the model of Figure 30, about half or two-thirds of the generated crust has spent part of its life below the floor of the rift valley; metamorphism of crustal basalt should be most favored there.

Cann [1968] has postulated that the greater fluidity of the mantle under the ridges characterized by fast spreading allows a smaller turning radius than is possible under slow ridges. This means a narrower zone of tension, with igneous activity filling the available fissures and incipient rift valleys. By contrast the broader turn of the more viscous material under slow ridges distributes tension over a zone that is wider than the zone of igneous sources. By Cann's model, igneous crust is made at the immediate axis and then passes through a 'chopping block' of normal faulting that extends out to ±20 km from the axis. The rift valley block develops in the stagnant zone between the divergent flow lines below the crust. This model is a steady-state one that does not claim to account for the generation of topographic and petrologic waves on the flanks of the ridge. It does provide a constant supply of metamorphic exposures, but the frequent occurrence of earthquakes out to 20 km from the ridge axis has not been demonstrated [*Sykes*, 1967].

In addition to the basalt types and their low-grade metamorphic equivalents, ultrabasic masses occasionally reach the sea floor. The most notable example is St. Paul's rocks on the Mid-Atlantic ridge, but ultramafic cobbles have been dredged from fracture zones [*Matthews et al.*, 1965] and within the rift valley [*Udintsev*, 1966]. Shearing along fracture zones may expose the ultrabasic basement there, and small inclusions are doubtless carried up by the rising basaltic fluids. The dunite of St. Paul's rocks is mylonitized [*Shand*, 1949], and the ultrabasic fragments analyzed by *Udintsev* [1966] reveal dynamo-metamorphism. It is therefore probable that the larger masses of such rocks were intruded as nearly consolidated crystal slushes (perhaps at stage C in Figure 30) and were then exposed by conjugate normal faulting, as was proposed for the metamorphosed basalts by *Cann* [1968]. The shearing and mylonitization could then reflect either the emplacement of this semiconsolidated slush or the conjugate faulting, or both.

Within 300 km of the axis of the Mid-Atlantic ridge, low-velocity mantle underlies the second layer (Figure 28) and low-density roots must extend still further at depths of about 20 km to explain the negative Bouguer anomaly [*Talwani et al.*, 1965]. It is unknown whether low-velocity mantle is eventually converted to normal mantle, or whether it is a stable assemblage formed, for example, when spreading rates are low. *Ringwood* [1966] and *Cann* [1968] suggest that the creation of dense, normal mantle involves a low-temperature (say, below 500°C) hydration reaction from plagioclase pyrolite to ampholite (olivine and amphibole).

The boundary between low-velocity mantle and underlying normal mantle may be a solid-state phase transformation of plagioclase pyrolite (lower-density phase) to pyroxine pyrolite (higher-density phase), as was suggested by *Ringwood* [1966] and supported by *Cann* [1968]. More specifically, the proposed reaction is olivine plus aluminous orthopyroxene plus aluminous clinopyroxene yielding olivine plus orthopyroxene plus clinopyroxene and plagioclase.

Alternatively, the upper low-velocity mantle may be a stable assemblage created during slow spreading. Ultrabasic diapirs rising at the axis cool and freeze before all basaltic fractions have been drained off. Some of the basaltic fraction corresponding to the oceanic layer under fast ridges remains in the upper mantle, giving it a permanently low *P*-wave velocity. Although this explains why low-velocity mantle and oceanic layer are nearly mutually exclusive on the Mid-Atlantic ridge (Figure 28), a phase reaction of some sort is still required to explain the narrow band of low-velocity mantle under the axis of the East Pacific rise (Figure 27).

Cann [1968] has argued that the mantle under the East Pacific rise has a higher water content, which would make the reaction from pyroxene pyrolite to ampholite start at a higher temperature for a given pressure. The upper boundary between low-velocity mantle and normal mantle is thereby displaced toward the center of the ridge, as is observed.

The composition of the oceanic, or third, layer identified by refraction seismology is still uncertain. A more detailed summary of the seismic characteristics and distribution of the layer was given in an earlier section. The layer is present under most of the oceanic crust and, where it is encountered, exhibits a remarkably constant thickness and P-wave velocity; however, the layer is thinner under the Gorda ridge [*Shor et al.*, 1968] and between 300 and 500 km from the axis of the Mid-Atlantic ridge [*Talwani et al.*, 1965]. No continuous third layer exists, for example, within about 300 km of the axis of the slowly growing Mid-Atlantic ridge [*Talwani et al.*, 1965], under the Bermuda rise [*Heezen et al.*, 1959], and in the Western Mediterranean [*Hersey*, 1965]. Exceptionally thick oceanic layers occur under some small ocean basins (Figure 26).

As *Raitt* [1963] pointed out, the observed 6.7-km/sec velocity places the material compositionally between acidic granite and ultrabasic dunite, with a rock of gabbroic composition perhaps most reasonable. *Hess* [1962] suggested that the layer is partly serpentinized peridotite formed from that part of the upper mantle whose temperature was above 500°C at the ridge axis; the base of the serpentinized layer would therefore be a frozen relict of this isotherm. *Cann* [1968] has offered three basic objections to the serpentinite hypothesis:

1. Where layer 3 is not being generated (under the Mid-Atlantic ridge), the 500°C isotherm ought to be within the crust. This would imply a higher heat flow on this ridge than on the East Pacific rise, where layer 3 is being generated. Instead, the opposite is observed [*Langseth et al.*, 1966].
2. The range of possible velocities for serpentinized peridotite (5 to 8 km/sec) is difficult to reconcile with the remarkably uniform third-layer velocity (6.7 km/sec), which would mean that exactly 5 km of mantle is everywhere exactly 70% serpentinized.
3. An inverse relationship between the thickness of layer 2 and that of layer 3 [*Le Pichon et al.*, 1965] suggests that the layers are but two states of a chemically homogeneous crust formed by basaltic liquids fractionated from the upper mantle.

Cann [1968] argues that the oceanic layer is of basaltic composition and is mineralogically an amphibolite (hornblende ± plagioclase).

In order for amphibolite to form at a depth of 2 km on the crest of the East Pacific rise, a heat flow of 8×10^{-6} cal/cm sec is required if the conductivity of second-layer basalts is 0.004 cal/cm sec °C. In other words, the top of the oceanic or third layer may be a relict metamorphic boundary, approximately the position of the 400°C isotherm at the axis of rapidly growing mid-oceanic ridges. The second layer can then be subdivided mineralogically (but not seismically) into a basaltic layer (above the position of the 250°C isotherm) and a greenschist layer (between the 250°C and 400°C isotherms). It may be significant that if the boundary between the second and oceanic layers is the axial 400°C isotherm, then the M discontinuity is the melting point for basalt. In a following section we suggest a thermomechanical model for the generation of normal oceanic crust by fast-spreading segments of the ridge. A requirement for this model is that the basalt liquidus maintain itself at such a shallow depth as the base of the crust at the axis. A byproduct of this model is that the axial 400°C isotherm may coincide with the top of the oceanic layer, where Cann's model requires it to be.

An alternate view is that the second layer is composed of flows, breccia, and ash, as well as intercalated ponds of pelagic sediment. The third layer, compositionally the same, may then be a continuous sheeted basaltic dike complex such as the unit on Cyprus that is supposed to be uplifted ancient oceanic crust [*Gass and Masson-Smith*, 1964]. If many of the dikes injected into the base of the crust do not reach the surface, the second layer must also be subject to stretching by normal faulting. The thickness of the second layer may also be a measure of the depth to which gaping fissures can open. Any dike which rises to this level can easily proceed to the surface and feed a lava flow. By this theory the seismic boundary between layer 2 and layer 3 reflects a rapid change in the percentage of high-velocity dike rock. Finally, while a dike is rising in a vertically layered complex, horizontal injection in the form of sills is discouraged. However, when it reaches the base of horizontally layered extrusives, a sill may readily develop at the interface. Once sills exist at the interface, a uniform seismic velocity is ensured by the smoothness and petrological uniformity of the sills. Sills continue to prefer planes bounding pre-existing sills rather than cutting across the vertical dikes below the interface and the irregular volcanics above it (Figure 23).

We have summarized in preceding paragraphs some possible mechanisms by which the crust might be created and what its composition may be. Most models suggest that the nature of the crust, but not its bulk chemical composition, may vary with growth rate of the ridge. Reactions and phase changes may operate on the lithosphere as it withdraws from the axis, but presumably the upper crustal layers are not appreciably altered in their gross chemical make-up. The recent findings of *McBirney and Gass* [1967] are therefore surprising, because they show a systematic variation away from the axis of the Mid-Atlantic ridge and the East Pacific rise. Basalts collected within the high heat flow belt on both ridges tend to be oversaturated with silica, whereas undersaturated alkaline rocks appear on the flanks beyond the belt [*McBirney and Gass*, 1967]. If this effect is real, we must conclude that the composition of mantle derivatives or the character of differentiation processes changed about 10 m.y. ago (extrapolating backward the late Pliocene crustal spreading rates of *Heirtzler et al.* [1968]).

The possibility of lateral and vertical variations in upper mantle composition (and, therefore, crustal composition) was recently underscored by the work of *Dickinson* [1968], who examined the petrochemistry of circum-Pacific andesitic suites. He found that (1) intra-oceanic arcs typically erupt lavas of lower potash content than marginal continental ranges do, but (2) for each arc relative potash contents rise away from trenches across island arcs in a transverse direction [*Kuno*, 1966; *Dickinson*, 1968]. This is in effect a correlation between the over-all level of potash content in a given suite and the vertical depth from the sites of eruption to the inclined seismic zone dipping beneath the volcanic arc or chain [*Dickinson*, 1968]. The latter relationship, combined with the fact that curves for potash versus silica tend to be sharply defined and straight for each suite of andesites, leads Dickinson to the conclusion that andesitic magmas are derived from the mantle without significant contamination upon passage through a sialic crust. Therefore, the systematic variations in the petrochemistry of andesite suites from island arc to island arc may reflect variable conditions within the mantle [*Dickinson*, 1968]. We can speculate that these variations are ultimately related to conditions that existed under the axis of the mid-oceanic ridge at the time of creation of the crust and upper mantle that is now being dragged under the island arcs.

A hypothesis for the generation of typical oceanic crust. The authors propose a qualitative scheme for the creation of various general features of the typical oceanic crust (Figure 23). The scheme incorporates various features of hypotheses and supporting data put forward in recent years by many workers. We proceed from the oversimplified view (1) that the mantle is an ultrabasic crystal aggregate whose interstices are filled with liquid or solid basalt, and (2) that the crust is composed of basalt dikes (oceanic layer) overlain by volcanics (second layer). *Bott* [1965] and others have discussed how the rising mantle undergoes partial melting under the mid-oceanic ridge to produce a crust of basaltic composition. Basically our hypothesis amounts to this: Under fast-spreading parts of the ridge the melting surface for basalt reaches the base of the oceanic layer under the axis of injection. This ensures that the sluggish intrusion of ultrabasic crystal slush is a *continuous* process. Tension on the lithosphere causes discontinuous fissuring of the crustal roof over the ultrabasic chamber. The tension fissures fill with basaltic fluids to build the oceanic layer. Occasional overflows feed the second layer. When spreading rates fall below a certain critical value, the melting surface of basalt cannot *continuously* maintain itself at the base of the oceanic layer. Now the ultrabasics must also be discontinuously intruded (Figure 30). The mechanics of intruding thick masses of ultrabasics give rise to cycles of volcanism and normal faulting that are superimposed on a steady crustal spreading rate. The features of the slow-spreading Mid-Atlantic ridge are thereby produced (rough topography, no oceanic layer, abnormal mantle, rift valley; see Figure 30), as was discussed in a previous section.

Consider some details of this hypothesis for conditions under fast ridges. Assume first that spreading is just fast enough that the dome-shaped ultrabasic chamber, defined by the melting surface for basalt, is tangent to the base of the oceanic layer at the top (Figure 23). The crust is a frozen roof to the ultrabasic chamber. As the two sides of the chamber withdraw, ultrabasic material is constantly accreting onto the sides of the chamber. In a steady-state system, the chamber does not change its size or shape despite the fact that crystal slush is passing through its walls to form hard mantle by the freezing of interstitial basalt. Under the ridge axis the crust forms a brittle roof over the chamber, a roof which must fracture at intervals. Basalt liquid constantly rising from the ultrabasic slush collects at the top of the chamber and is immediately injected into an available tension fissure. The oceanic layer is thereby incremented. Occasionally, the injected lava reaches the surface before consolidating and thereby feeds the volcanic layer.

The melting surface will tend to maintain itself at the base of the basaltic layer if basalt generation or temperature should increase (say, by an increase in spreading rate) above the minimum level. Any tendency for the melting to proceed into the crust would create a lens of liquid basalt which would tend to be drained by the resulting increase in lava emissions at the surface. If such a lens consolidates before it is drained, massive gabbro bodies may occasionally be formed at the base of the oceanic layer. Such a formation occurs on Cyprus between the dike complex and the ultrabasics [*Gass and Masson-Smith*, 1963]. Even temporary solidification of the uppermost mantle could be tolerated. Normal faulting and tension cracks would reduce the stresses and therefore the melting point at the base of the crust. Enough basalt might be melted from the base of the crust in this manner to fill the fissure. Both increases and slight decreases in heat supply might therefore cause an overturn of the crustal basalt, i.e. basalt at the base would be melted and deposited within and on top of the crust. The thickness of crust is not changed by such processes.

The hypothesized tendency of the basalt liquidus to maintain itself at the Moho under the axis of fast-spreading ridges is basically due to different rates at which heat can be removed from liquid basalt in dikes and the same basalt in the ultrabasic slush. The basalt dike is a thin, tabular mass in contact on both sides with the perhaps substantially cooler wall rocks in the crust. Thermal gradients are high and the dike freezes rapidly. By contrast, heat removal from the much broader ultrabasic slush is more inefficient, and downward freezing does not readily occur.

A hypothesis is still required to explain why crustal spreading rates on fast ridges vary by a factor of 2 or 3 [*Heirtzler et al.*, 1968], but crustal thickness remains remarkably constant below most of the world ocean [*Raitt*, 1963]. In other words, the rate of generation of basaltic liquids must keep pace with spreading rate. The following circumstances could account for this fact: There is little or no variation, either in time or over the earth, in the composition and gross physical state of the partly molten mantle that rises to fill the void between two withdrawing plates of lithosphere. A parcel of partly molten mantle, perhaps from the low-velocity zone, may undergo changes of state as it is carried into the chamber under the ridge. Inside the chamber a small basaltic fraction drains off, rises, and is eventually injected into the crust as part of a dike. The remainder of the parcel is frozen into the sides of the chamber and becomes part of the normal mantle.

The observation that crustal generation keeps pace with spreading rate then has a simple explanation: The rate at which basaltic liquid can rise through the crystal slush is great enough compared to the upward movement of the slush (which must be of the same order as the spreading rate unless the slopes of the chamber sides differ very greatly from 1:1) that each parcel is completely drained of its cargo of free liquid basalt. If this were not the case, a more rapidly spreading ridge would create a lower density mantle and a thinner crust than a slowly spreading one. The uniform P-wave velocities at the top of the normal oceanic mantle support the view that all available basaltic fluids have escaped before final consolidation occurs. The ratio between the mass flux through the roof of the chamber and that passing through its sides to form solidified upper mantle so remains constant for fast spreading ridges, independent of spreading rate.

Each parcel of mantle has been drained of available basaltic fluid. The mechanism by which basaltic liquids rise may also be described as a constant falling of the ultrabasic crystals within the input chamber. The rise of the slush will be accompanied by partial melting, which increases the void space filled by liquid and produces a thick suspension in which the crystals are constantly settling in their attempt to build a rigid framework.

Even a fairly continuous framework of crystals might allow convective migration of basaltic fluids within the chamber, although such movements obviously cannot change the amount of basalt in the interstices and hence do not feed the crustal layers. On the other hand, it is not certain that all the basalt originates in a zone of concentrated melting about 30 km deep [Cann, 1968]. But if significant amounts of basalt liquids migrate through the slush from, say, parts of the low-velocity layer at some distance from the input chamber, it would be hard to explain the lack of variation in crustal thickness with spreading rate on fast ridges.

The hypothesis outlined above requires that the liquidus of basalt occur at the base of the oceanic crust at the axis of the East Pacific rise. Taking the average crustal molecular conductivity as 0.006 cal/cm sec °C [Langseth et al., 1966], the conductive heat-flow maximum ought to be 9 HFU. Indeed heat-flow values this high have been reported from the East Pacific rise, although the average is lower [Langseth et al., 1966]. It is interesting that practically the same heat flow (8 HFU) is required by Cann's [1968] argument that the interface between the second and third layer is a metamorphic transition from basalt to amphibolite.

Small Ocean Basins

Shape, sediments, distribution, and tectonic environment. Crustal structures calculated from seismic refraction measurements in small ocean basins have been summarized by Menard [1967a], whose work is the basis for the following summary. The crustal sections are often significantly different from normal continents and ocean basins (Figure 26). The possibility that some of these transitional sections may represent oceanic crust in various stages of conversion of continental to oceanic crust by some process other than sea-floor accretion at the axis of rifted ridges cannot yet be discounted.

Menard [1967a] grouped small basins according to the characteristics of their boundaries: (a) the Aleutian and Okhotsk basins are separated from the main ocean basins by simple volcanic island arcs; (b) more complex arcs separate the basins of the East Indies and the Caribbean, and the Japan and South China basins; (c) the Mediterranean Sea, Black Sea, and the Gulf of Mexico are almost surrounded by continent. Yet the small basins share many details of physiography. None of them lie far from a continental edge, and in fact 35% of the continental area drains into the small basins, whose total area is only 1% of the oceanic area. It is therefore hardly surprising that about one sixth of all identifiable ocean basin sediment may be in the small basins [Menard, 1967a]. Most of the basins have similar size, topography, and depth. All contain at least some areas of abyssal plain, and many have large areas of almost flat floor. The topography of the basins is strikingly similar. Steep sides in most places rise directly from almost level floors except when turbidity currents have deposited transitional fans as in the Gulf of Mexico and Balearic basin. In a number of cases the steep sides are in fact simple or complex fault scarps (the Gulf of Mexico, Aleutian, and South China basins, and the Celebes and related basins).

The authors believe that the small ocean basins share a common tectonic environment; Menard [1967a] disagrees. These basins, without exception, lie either entirely within or along the margins of the Alpine-Himalayan or Circum-Pacific deformational belts. The northwest Pacific basins generally occur on the continental side of island arcs, whereas the Mediterranean-Black Sea basins are bounded by Alpine folded belts. The East Indian and Caribbean groups of basins appear intermediate, occurring where island arc structures interweave with Alpine-Cordilleran ones.

In Figure 26, small basins have been assigned to several groups according to the thickness of the oceanic layer ($V_p = 6.7$ km/sec) as deduced from refraction seismology. Several types of oceanic crustal section have been arranged among the small-basin groups on the same criterion. In one group the basins apparently consist of ordinary oceanic igneous crusts loaded with one to ten kilometers of sediments. The velocity and thickness of the oceanic layer generally are well within the limits for this layer under the great ocean basins, 6.69 ± 0.26 km/sec and 4.86 ± 1.42 km [Raitt, 1963]. Examples are the Aleutian, Okhotsk, and Gulf of Mexico basins. A similar group (Figure 26) has a somewhat thicker oceanic layer.

Under small basins (shown at the lower right

in Figure 26), the basaltic layer is very thick (10 to 20 km) and the sediment column is also thicker (5 to 20 km) than in the first group. Mantle P-wave velocities are normal under these first two types of basins. Members of this group are the Black and Caspian basins, in which the basaltic layer is several times thicker than the typical oceanic layer. Some of this thickening may be due to consolidated sedimentary rock, such as limestone.

In the upper left of Figure 26 is shown the Tyrrhenian section, where sediments and second layer lie directly on low-velocity mantle [*Hersey*, 1965]. Upon removal of sediment, this type of section is nearly identical to the Bermuda rise or the Mid-Atlantic ridge within 300 km of its axis. A similar group at the upper right differs only in that a thin oceanic layer is present. The two groups at the top of Figure 26 suggest that, wherever the oceanic layer is thin or absent, mantle velocity is abnormally low.

The Japan Sea is without counterpart, being underlain by 2 km of sediments, a 7-km-thick sialic layer ($V_p = 6.2$ km/sec), and a normal mantle 12 km below sea level [*Andreyeva and Udintsev*, 1958].

The subsidence problem. Consider first the basins that appear to be nothing more than segments of oceanic crust depressed to great depths by thick sediment burdens. Water depths also decrease with increasing sediment burden, supporting the view that the subsidence is in some way a response to the sediment load. The free-air gravity anomalies measured over several ocean basins are small, usually within ±25 mgal [*Menard*, 1967a]. Thus, whatever the process of small basin subsidence may be, isostatic balance is maintained. There is not any certain example of a basin floor that has sunk without accumulation of sediment.

Nevertheless, it is easily shown that the total subsidence of these basins is far in excess of that expected from simple isostatic considerations [*Menard and Smith*, 1966]. This is still the case after reasonable allowance has been made for increases in the density of the sediments and the igneous crust. Apparently mantle density must increase with increasing sediment load.

Recent numerical investigations [*van de Lindt*, 1967; *Joyner*, 1967] show that the extraordinary subsidence and later uplift of geosynclinal basins can be due to the combined effects of isostasy and vertical migrations of a phase boundary. Both works consider the subsidence of a *continental* crust whose M discontinuity is the basalt-eclogite transition. The results are not directly applicable to the problem of small basin subsidence, because the sequence of normal oceanic layer basins (Figure 26) clearly indicates that any phase interface must lie below the Moho. A fruitful model to which theoretical approaches might be directed could assume that the phase interface occurs at the top of the low-velocity layer and involves melting and refreezing of the basaltic fraction in an ultrabasic crystal slush.

In view of the interest of phase changes to the problem of small basin subsidence and, possibly, the problem of conversion of oceanic to continental crust, we will summarize the results of *van de Lindt* [1967] and *Joyner* [1967]. The methods assume a crust bounded below by a phase transition and overlain by a depositional basin. Thermal and mechanical equilibrium are initial conditions. As sediments are deposited, the basin floor subsides in accordance with the requirements of isostasy. The sediment load increases the pressure at the phase boundary, causing the boundary to migrate upward. This rise in the phase discontinuity is manifested as a further subsidence of the sea floor, in addition to that required by isostasy. This process of positive feedback is arrested only after enough sediments have accumulated to act as a thermal blanket over the crust, which now begins to warm. Eventually the warming effect predominates and the basin is uplifted above sea level. Cyclic alternations of uplift and deposition are possible.

In van de Lindt's work, a trough initially 2.7 km deep accumulated more than 10 km of sediments after a few tens of millions of years. A maximum uplift of 3 km was reached after about 100 m.y. Joyner obtained similar results. Seven to ten kilometers of clastic sediments were deposited in a basin initially only 1.5 km deep. Deposition could last as long as 40 m.y. for clastic sediments and 200 m.y. or longer for limestone. The most important parameter is the initial water depth, the models being relatively insensitive to other parameters, including the slope of the phase transition curve (temperature versus pressure).

Reflection profiles indicated diapirs in the Balearic and Gulf of Mexico basins [*Menard, 1967a*]; JOIDES drilling in the latter has supported this interpretation. Our discovery of diapirs in crust with Mesozoic sea floor spreading anomalies in the eastern Atlantic (Figure 15) means that small basins are not peculiar in this respect. In both cases the diapirs occur near terrestrial salts of Triassic to Jurassic age. Low-latitude and restricted circulation probably characterized the lower Mesozoic Atlantic and attached basins. The salt may be a 'hot hole' (Red Sea rift) or other deep precipitate. Alternatively, shallow water salts may have moved onto deep ocean crust by sliding from seaward-dipping shelf deposits or by migrating up from adjacent deeply buried deposits.

Small basins with very thick basaltic layers. Consider next those basins underlain by abnormally great thicknesses (10 to 20 km) of a layer whose velocity lies on the high side for the oceanic layer and which overlies normal mantle (V_p = 8.0 to 8.2 km/sec). On top of this quasi-oceanic layer lies a similar thickness of sediments. Representatives of this group include the Columbia basin, the central Black Sea basin, and the Black Sea basin south of the Crimea (Figure 26). The total crustal thickness in this group approaches that of the typical continent, and these basins may indeed represent the final stages of conversion from oceanic to continental crust. If so, the great thickness of this quasi-oceanic layer must somehow be explained in terms of later basaltic intrusions augmenting the original layer thickness. The thick basaltic layer under central Asia may include remnants of the oceanic floor once between India and Siberia and now thickened by compression [*Holmes, 1965*]. The great subsidence and loading by sediments would be expected to cause metamorphism and granitization at the base of the sedimentary column, leading to typically continental seismic velocities of 6.1 km/sec. If crustal sections involving buried oceanic layers represent stages in the conversion of oceanic to continental crust, we may ask whether there are, in the geological record, examples of such crusts that were arrested at some stage before complete conversion. An example may be the Lake Superior syncline.

Geological and geophysical data [*O'Brien, 1968*] suggest that the crust under Lake Superior consists of about 0.5 km of water overlying 4 to 10 km of lithified late Precambrian to early Paleozoic sediments characterized by P-wave velocities of 4.5 to 5.0 km/sec (Figure 26). Below the sediments rests a thick layer, seismically identical to the oceanic layer, which is probably composed of late Precambrian (Keweenawan) flood basalts and related intrusives. The thickness of this layer varies from 20 to 30 km in the western end of the lake to more than 40 km below the eastern end.

By comparison, two crustal sections of the Black Sea crust (Figure 26) revealed a 17-km-thick basaltic layer [*Neprochnov et al., 1959*] under the central part of the basin and 11 km south of the Crimea [*Neprochnov et al., 1964*]. A 20-km-thick basaltic layer was found under the Caspian Sea [*Gagel'gants et al., 1958*]. Normal mantle velocities occur under both the Black Sea and Caspian Sea and the Lake Superior crusts. The Superior syncline is smaller than most small ocean basins, but corresponds in size to the Andaman and Sulu basins in the East Indies group. The Superior syncline may have once been more extensive, the flanks having been uplifted and removed by erosion.

The crustal sections under the Black and Caspian basins could be transformed into a Superior-type section merely by uplift and erosion of 5 or 10 km of sediments. Alternatively, it would have to follow that a thick basaltic layer could be evolved without large-scale subsidence and sediment loading. The 1000-m.y. age of the Superior syncline demonstrates that a crustal section similar to that under the Black and Caspian Sea can exist for a long time without being converted to typical continental crust.

Comparison of crustal sections from the Western Mediterranean, Mid-Atlantic ridge, and Bermuda rise. Small ocean basins whose crustal sections resemble those of active mid-oceanic ridges form a third category (Figure 26). In the Tyrrhenian basin [*Hersey, 1965*], 3 km of water overlie a similar thickness of sediment and/or extrusives (V_p = 4.4 km/sec). This sequence directly overlies low-velocity mantle (V_p = 7.4 km/sec).

The oceanic layer is entirely missing, and the sedimentary layer is thin. Another basin whose crust seems most similar to the Mediterranean is the Aleutian basin, which could be developed from the former type by (1) the addition of 2 km sediment accompanied by a similar amount of subsidence, and (2) partial melting of the Tyrrhenian abnormal mantle (7.4 km/sec) and segregation into a 7.0-km/sec 'oceanic' layer and 7.8-km/sec mantle.

Alternatively, we observe that the Tyrrhenian section, including the water column [Hersey, 1965], is practically identical to sections obtained in the Norwegian Sea [J. Ewing and M. Ewing, 1959]. More generally, the Tyrrhenian crust resembles that observed within 300 km on either side of the Mid-Atlantic ridge [Talwani et al., 1965].

The Balearic basin of the western Mediterranean Sea similarly exhibits an abnormal crust [Hersey, 1965]. One long refraction profile revealed 7 km of sediments and possible extrusives, overlying a 7.2-km/sec mantle. Another profile showed a similar crust with an additional 3 km of oceanic layer overlying a 7.7-km/sec mantle. The Balearic sections could have been generated from the Tyrrhenian type by subsidence and sediment filling (Figure 26).

The similarity of the Balearic (upon removal of sediments) and Tyrrhenian sections to the type observed under the crest of the mid-oceanic ridge system (Figure 26) suggests that the western Mediterranean may be a complex rift of the mid-oceanic type [Menard, 1967a].

Another type of feature exhibiting abnormal crust of the western Mediterranean type (viz., 3 to 5 km of sediments and volcanics, the oceanic layer thin or absent, and a low-velocity mantle) is exemplified by the Bermuda rise (Figure 26). To make this rise compatible with the view that the Atlantic basin has been created by crustal accretion from the Mid-Atlantic rift, we must suppose that either (1) the subsedimentary Bermuda rise crust was created at the axis of the Mid-Atlantic ridge, or (2) it represents typically oceanic crust somehow transformed at a later time. The Cretaceous reflector A continues under the Bermuda rise; still older sediments lie below horizon A [J. Ewing et al., 1966]. If the igneous crust and upper mantle predates the sediment cover, the age of the crust must be of the order of 100 m.y. Applications of the sea-floor accretion hypothesis to the Mid-Atlantic ridge support such an age. Current estimates of the accretion history in the North Atlantic [Phillips, 1967; Schneider and Vogt, 1968; Talwani et al., 1968] also suggest that the crust under the Bermuda rise may have been created of the order of 100 m.y. ago. The alternative view is that a pre-existing oceanic layer was somehow absorbed into the mantle to the accompaniment of regional uplift. If the disturbance of layered horizons as old as A and older was concomitant with this uplift, the Bermuda rise is a post-Cretaceous feature in which older basin crust was altered and uplifted.

Recent structural interpretations of phase velocity data reveal that an unusual upper mantle underlies at least one crustal section of this sort. Berry and Knopoff [1967] found a shear velocity minimum of 4.10 ± 0.05 km/sec under the central part of the northeast-trending Balearic basin. The roof of the central low-velocity channel is 50 ± 6 km below sea level and drops to 100 ± 20 km around the margins. The 7.7-km/sec low-velocity mantle that occurs 11 km below sea level under the central parts of the basin [Hersey, 1965] is shown from surface wave data to lie 31 ± 4 km under the margins. Higher-velocity material ($V_s = 4.80 ±$ 0.15 km/sec) lies between the 7.7-km/sec layer and the low-velocity channel, but extends to such great depths that a transitional rather than truly oceanic crust is indicated by Berry and Knopoff [1967]. These authors state that the extent of the low-velocity channel, although poorly known, may be continuous with a similar channel previously found to underlie the Alps. If so, a shallow mantle structure has been found which seemingly cuts across widely discordant topographic and geological features.

This result indirectly contrasts with the distribution of seismicity under the Tonga-Kermadec island arc [Sykes, 1966]. The westward turn of the Tonga trench at its northern end is shared by the entire seismic zone, down to its maximum focal depth of 650 km, and by the volcanic belt. Therefore at least in some parts of the world the tectonic processes associated with crustal structures can extend deep into the upper mantle.

Although the Alpine and Balearic basin crusts differ, it may be significant that immediately to the east of the Alps lies a large mass of anomalous mantle, the Ivrea body [Fuchs et al., 1963], that resembles the mantle under the western Mediterranean [Hersey, 1965] both in its P-wave velocity (7.5 km/sec) and its minimum depth (10 km below the surface).

Therefore, both the low-velocity channel

under the Alps and the Ivrea body may reflect the fact that a small ocean basin such as the Tyrrhenian or Balearic basins recently existed near the present location of the western Alps.

The similarity (Figure 26) between the crust under the western Balearic basin and Mid-Atlantic ridge crusts seems to extend into the upper mantle and thus supports the possibility [*Menard*, 1967a] that the western Mediterranean basin is a complex rift in the crust of the earth. The 50-km depth to the top of the low-velocity layer under the central Balearic basin found by Berry and Knopoff is about the same as under the oceans [*Oliver and Dorman*, 1963].

The difference between the thickness of the layer under the ocean floor [*Oliver and Dorman*, 1963] and that found by Berry and Knopoff (130 to 150 km) is so small that it may not be significant. The minimum shear velocity within the Mediterranean channel (4.10 km/sec) compares with an average for this layer under the oceans. However, no surface wave paths along the mid-oceanic ridge have been analyzed; still lower shear velocities seem probable because of comparatively higher temperatures and partial melting in the vicinity of the axis.

The boundary between the $V_p = 7.7$-km/sec layer and the 8.17-km/sec normal mantle lies of the order of 20 or 30 km below sea level in the surface-wave models of *Berry and Knopoff* [1967]. A similar boundary occurs at similar depths under the Mid-Atlantic ridge, according to the gravity models of *Talwani et al.* [1965] shown in Figures 22 and 28.

Heat flow. Few heat-flow measurements have been made in small ocean basins; both high and low values (compared to the oceanic average) occur. Interpretation is encumbered by two basic uncertainties. First, the thermal stability characteristic of deep ocean basins can hardly be expected in an environment of catastrophic sedimentation or even overturn of surface water [*Menard*, 1967a]. Second, sedimentation at the high rate of 10–30 cm/1000 years for several 100 m.y. would tend to depress the observed heat flow to roughly half the equilibrium value after sedimentation ceased [*Von Herzen and Uyeda*, 1963]. The equilibrium heat flow of a basin in which 25 km continental sediments overlie an oceanic crust must be considerably higher than the oceanic average.

The heat flow from the floors of some small basins landward of trenches in the northwestern Pacific is about twice the oceanic average [*Vacquier et al.*, 1966; *McKenzie and Sclater*, 1968]. These data are discussed in the next section.

Trenches and Island Arcs

Crustal structure. The crustal structure and physiography of trenches were reviewed by *Fisher and Hess* [1963]. Some significant recent advances include precise machine recomputations of earthquake foci and epicenters under island arcs [*Sykes*, 1966], and the discovery that the seismic zone lies near the top of a perhaps 100-km-thick layer of material characterized by lower attenuation (i.e., high Q) than is typical for mantle at corresponding depths [*Oliver and Isacks*, 1967].

The main geophysical features of trenches, summarized by *Fisher and Hess* [1963], have been known for several decades. All of the surveys confirm the existence of a strip of large negative isostatic anomalies centered slightly shoreward of the trench axis, and commonly smaller positive anomalies closer to the island group or mainland and also seaward of the trench. According to *Vening Meinesz* [1955] the anomalies reflect plastic deformation of crustal material in the mobile belts, with thickening and down-buckling under compression. In any case a force must be sought whose vertical component is sufficient to maintain a pressure differential of 1 kb at the level of the trench floor [*Fisher and Hess*, 1963]. Apparently the trench-island arc structure and the Cordilleran and Alpine-Himalayan fold belts reflect crust compressed or dragged into the upper mantle. These mobile belts then represent boundaries along which adjacent stable crustal blocks are in collision. The rate (integrated over the globe) of crustal destruction along these boundaries must equal the rate of crustal generation at the axis of the mid-oceanic ridge if the size of the earth remains substantially unchanged. The lack of deformation in unconsolidated sediments that have partially filled the Peru-Chile trench is an unresolved mystery if the crust is being actively swallowed there [*Scholl et al.*, 1968]. Perhaps the destruction of crust has been confined to the Andes Mountains or the landward walls of the trench during the time the sediments were deposited in the trench.

Most geological and geophysical evidence is at least in qualitative accord with the prevailing concept that the island arcs and fold mountains are the crustal sinks corresponding to the mid-oceanic sources (Figure 15).

Heat-flow data are among the most elusive in their interpretation. Although some low values have been measured in the immediate vicinity of trenches, a broad low seems to be either missing or obscured by other effects. The seas on the continental side of the northwest Pacific trenches (the Okhotsk, Japan, and the Philippine basins and the Fiji plateau) are characterized by abnormally high heat flow, the opposite of what would be expected above a cold descending convective limb. This region of high heat flow extends from about 100 to 800 km landward from the trench axis, with the median value about 2 HFU, or roughly twice the oceanic average [*Vacquier et al.*, 1966; *McKenzie and Sclater*, 1968].

Dissipative shear heating could cause such a positive heat-flow anomaly only (1) if the shearing is restricted to a shallow depth (i.e., not along the Benioff seismic zones dipping under the arcs), or (2) if the effective mantle conductivity is very high compared to 0.01 cal/cm sec °C [*McKenzie and Sclater*, 1968].

Such a high conductivity could be the result of partial melting, i.e., mobile basaltic fluids. A second possible cause for the positive heat-flow anomaly behind the Northwest Pacific island arcs is the injection of hot magmas into and onto the crust. However, this requires such a high rate of intrusive activity that, for example, the entire crust under the Japan Sea could have been formed in this manner in a mere 10 m.y. [*McKenzie and Sclater*, 1968].

Oxburgh and Turcotte [1968] have examined possible causes for the high heat flow and volcanism landward of the trenches. They conclude that heat is generated by viscous friction within the Benioff seismic zone and is lost by conduction. By allowing for a variable viscosity, those authors were able to obtain values for heat production and local temperature anomalies high enough for local magma generation. The magmas feed surface volcanism as well as lubricate the viscous slip zone which is thought to coincide with the Benioff plane.

There is the suggestion of a negative heat-flow anomaly in the immediate vicinity of circum-Pacific trenches, but its origin remains obscure. Heat-flow values for the Northwest Pacific island arc structures have been presented by *Vacquier et al.* [1966] and others. When all these data are plotted together, the 50% line drops to about 0.6 HFU over the trench axis. It is not clear whether this is significant, and if it is, whether descending convection currents or merely rapid sedimentation is responsible [*McKenzie and Sclater*, 1968].

Heat flow through the trench floors of the Middle America trench [*Bullard et al.*, 1956] and the Peru-Chile trench [*Von Herzen*, 1959] is low (0.17 to 0.47 HFU), as would be expected if cold crustal layers enter the mantle there. The possibility that the low heat flow reflects rapid sedimentation has not been discounted.

Trenches appear to lack diagnostic magnetic signatures. The field is often quite undisturbed, and the anomalies observed are due to susceptibility contrast rather than basement structure (published results summarized by *Fisher and Hess* [1963]). Recent work has revealed a 200-km-wide undisturbed belt between the eastern Mediterranean island arc and Africa [*Vogt and Higgs*, 1968]. A similar belt, 100 km wide, straddles the Aleutian trench [*Hayes and Heirtzler*, 1968]. Several geologic features are common to most trenches (published results summarized by *Fisher and Hess* [1963]). In both island-arc and continental-margin trench systems, a row or double row of volcanoes, active at present or in the fairly recent geologic past, lie subparallel to the trench trend. The volcanoes lie 150 to 350 km landward from the trench axis, the most common separation being about 200 km. Such volcanoes, for example, comprise the islands of western Tonga, much of the Marianas, Bonin, Kuril, and Aleutian groups, the peaks on the east shores of the southern Philippines and the Kamchatka Peninsula, and the mountains bordering the west coast of Central America. A volcanic arc can even be recognized in the Aegean Sea north of the Cretan arc and its trenches, illustrating the probably common origin for the Alpine-Himalayan and circum-Pacific tectonic belts. The island volcanoes thus do not rise from the deep-sea floor but from the crests or back slopes of geanticlinal ridges bordering the trenches. Several hypotheses have been advanced to account for the location of volcanic chains. *Gunn* [1947] suggested that the spacings and structures could be due to an elastic lithosphere failing by shear, with the continental slab overriding the oceanic one to form a trench.

As the stresses at the base of the landward lithosphere are less than at corresponding depths under the oceanic slab, compressed magmas can be injected into the continental side more readily. The hypothesis fails in areas where oceanic crust seems to exist on both sides of a trench. *Fisher and Hess* [1963] propose that magmas rise on the concave side of an arc, because here there will be a tendency for open fractures to form at right angles to the direction of compression, i.e., at right angles to the trench axis. Deformation and plastic flow in the region where the mantle and crust bend down into the earth prevents direct vertical rise of the fluids. This hypothesis does not explain the regularity of distance from volcanic arc to trench axis. There ought to be, at least in some cases, volcanic lineaments along the fractures, that is, perpendicular to the trench. Finally, the volcanic lines behind the linear Tonga-Kermadec and Peru-Chile trenches are not explained by this hypothesis.

It may be significant that the volcanic belt is above the central part of the sloping Benioff plane, a part characterized by less seismicity than is found along the plane near the surface or at depths of 400–700 km. Partial melting may lubricate the fault planes and enhance deformation by other mechanism than brittle fracture.

The petrology of the volcanic belts shows significant variations from arc to arc and within each arc [*Kuno*, 1966; *Dickinson*, 1968].

The crustal structure of island arcs has been generally less thoroughly studied than that of the mid-oceanic ridge. The Puerto Rico island arc, which may not be typical, is the most thoroughly investigated. In general there is an abrupt thickening of the crust near the inside edge of the trench. The second layer, probably largely volcanic, becomes about 5 km thick under the island. The oceanic layer seems to merge into a layer, of the order of 5–15 km thick, characterized by seismic velocities ranging from that of the oceanic layer (6.7 km/sec) to 7.7 km/sec.

Typical mantle velocities have been found by refraction seismic studies of island arcs in only a few instances. *Fisher and Hess* [1963] believe that abnormal crust under arcs is formed as follows: The oceanic layer descending into the mantle below the trench is thought to be composed of partially (70%) serpentinized peridotite, i.e., a rind on the peridotite mantle. As the layer is heated above 500°C in the mantle, deserpentization occurs and the water thus released rises under the island arc and there reacts with the normal peridotite mantle to form the thick low-velocity mantle layer (40% serpentine, $V_p = 5.5$ km/sec) observed under the islands.

The narrow range of velocities and thicknesses of the oceanic layer [*Raitt*, 1963], in spite of variations in the observed sea-floor growth rates [*Heirtzler et al.*, 1968], make it difficult to believe that the layer is exactly 70% serpentinized peridotite. If the layer is basalt [*Oxburgh and Turcotte*, 1968], the source of water would be removed. The thick second layer could then simply be volcanics (erupted early in the trench formation) overlain by consolidated sediments, while the peridotite mantle has been polluted by many andesitic and basaltic intrusions, including feeder dikes. Both serpentization and pollution by andesite and basalt would effect uplift of the islands by reducing the upper mantle density. In any case, compression is not the only explanation for the observed crustal thickening and uplift of the arcs.

The petrology of many islands bordering the trenches supports the argument that serpentinized peridotite forms the foundations and roots of those islands. Exposures of altered ultrabasics are common in island groups bordering the East Indian, western Pacific, and Antillean trenches [*Fisher and Hess*, 1963]. A recent dredge haul at a depth of 9.5 km on the near-shore flank of the Tonga trench yielded 27 kg dunite and 10 kg basalt. Another 10 kg showed various alternations of serpentine, and there were several large fragments of dunite breccia [*Fisher and Engel*, 1968].

Similar rocks occur along the Alpine-Himalayan belt. The Troodos complex on Cyprus [*Gass and Masson-Smith*, 1963] is an uplifted mass of variously serpentinized peridotite overlain by what seems to be a thin oceanic crust.

This crust is composed of a lower sheeted intrusive complex composed largely of north-striking basaltic dikes, and an upper pillow lava sequence. The ultrabasic mass seems to have been intruded as a semi-consolidated crystal slush.

Ultrabasic masses, the so-called ophiolite complexes, occur all along the Alpine-Himalayan belt [*Maxwell*, 1968]. These ophiolites and the thick basaltic layer under Central Asia may constitute part of the Tethyan ocean crust that was destroyed by the northward movement of Africa and India [*Holmes*, 1965].

Ultrabasic masses are another feature the Alpine-Himalayan belt shares with the island arcs. Whatever the exact mechanism by which crust (continental or oceanic) is compressed and dragged into the mantle, pockets of ultrabasics (perhaps mobilized by the melting of interstitial basaltic fractions) are caught up in the compression of the crustal layers and are spared the fate of returning to the mantle.

Seismicity. The accuracy of epicenter and focal depth locations has been increased to ± 20 km or better (see section by Sykes, p. 148) with the use of more recording stations and least-squares solutions by digital computers. The general pattern deduced earlier [*Benioff*, 1954] has been confirmed. Seismic activity appears to be confined to Benioff planes which begin at the trench floor and dip underneath the arc, usually in a landward direction, at angles of 45° and 60°.

The more accurate work by *Sykes* [1966] involved 1500 earthquakes in the Tonga-Fiji, Kermadec, Kuril-Kamchatka, and Caribbean regions. Structural features as small as 20 km could be resolved. Near its northern end the Tonga trench curves abruptly to the west. The entire focal surface, including earthquakes as deep as 650 km, shares this abrupt curve, as does the volcanic arc to a less pronounced degree. This remarkable feature indicates that the tectonic processes responsible for the earthquakes, volcanoes, and the trench are intimately related for depths from 0 to 650 km. Earthquake foci in the Tonga-Fiji region are confined to a zone about 50 to 100 km thick that dips under the arc. The zones may actually be thinner than 20 km for the deeper shocks [*Oliver and Isacks*, 1967]. The focal surface is at least in some regions a continuous zone extending from the surface to depths as great as 650 km [*Sykes*, 1966]. Earlier work indicated a steepening of the seismic surface with increasing depth, but under the arcs studied by Sykes the dip was independent of depth. The intensity of activity along any vertical section perpendicular to the arc does depend on depth. Sykes found a downward exponential decrease on which is superimposed a relatively sharp secondary maximum of deep-focus shocks. Activity is sharply terminated below this lower maximum. The depth of the deep-focus maximum may vary markedly along the strike of some arcs. For example, near the southern end of the Tonga trench the depth of maximum deep-focus activity decreases from about 625 km to 475 km within a horizontal distance of 500 km. Such variations make it unlikely that the deep-focus maximum is related to a phase change or to a particular mantle layer unless the physical or chemical properties of the mantle have large lateral variations [*Sykes*, 1966].

The sharp turn at the north end of the Tonga trench is associated with a change from west-dipping to east-dipping focal surfaces. The trench system begins again about 1000 km southwest of the turn, and continues northwestward from there in the form of the New Hebrides trenches. These trenches are remarkable exceptions to the rule that seismic surfaces dip away from the Pacific basin.

The case for the old concept of oceanic crust underthrusting island arc structures has recently been strengthened by *Oliver and Isacks* [1967]. Regional seismic data for deep and shallow earthquakes associated with the Tonga-Kermadec arc show that there is in the mantle an anomalous zone whose thickness is of the order of 100 km and whose upper surface is approximately defined by the seismic zone (the 'Benioff planes') that dips to the west beneath the island arc and extends to depths of about 700 km. The authors reach this conclusion primarily from the low apparent S-wave attenuation within the zone, compared to the mantle at similar depths elsewhere. Seismicity occurs near the top of the zone, i.e., in the brittle crustal layer which has been dragged deep into the upper mantle. If the oceanic crust, derived from mantle derivatives at the axis of the mid-oceanic ridge, is returned to the mantle in this fashion, the average mantle composition perhaps does not change greatly with time, although if the sediment blanket accompanies the oceanic crust on its 'homeward voyage' into the mantle, a slight increase in acidity with time would be expected. Indeed, *Armstrong* [1968] finds from isotope abundances in rocks that the crust must be continually recycled through the mantle. Present regional variations in the character of the upper mantle may simply reflect the remains of such sheets of oceanic lithosphere which have been swallowed by the upper mantle at other times and places.

As *Oliver and Isacks* [1967] recognized, the present configuration of a seismic high-Q zone is not necessarily steady state. A strong argument for steady-state conditions could be made if it could be shown that (1) sea-floor spreading rates have been relatively constant for 100 m.y., and (2) the crustal 'sink' corresponding to the mid-oceanic ridge 'source' was indeed the trench-island arc feature. There is, however, evidence for a worldwide slow-down or stoppage in sea-floor spreading which ended about 10 m.y. ago [*J. Ewing and M. Ewing*, 1967; *Schneider and Vogt*, 1968]. The amount of crust generated during this interval, measured along any great circle, can be estimated from the sea-floor isochron maps of *Heirtzler et al.* [1968]. The crust generated in the last 10 m.y. is seen to be of the same order (500 to 1000 km) as the 'destroyed' crust represented by the seismic high-Q zones. This fact would have to be coincidence if there had not been a major rejuvenation of sea-floor spreading about 10 m.y. ago.

There is no direct way, as yet, to measure the rates of destruction of the lithosphere along the sinks, i.e., the island arc–compressional orogen lines. However, these rates can be computed from spreading rates measured at the dilatational boundaries of the plates, i.e., along the mid-oceanic ridge [*Isacks et al.*, 1968]. Such a computation neglects possible changes in the radius of the earth and the possibility of some dislocation within the plates. However, the good correlation between various trench parameters and the computed destruction rates [*Isacks et al.*, 1968] suggest that these effects are small.

Acknowledgments. The authors wish to thank Dr. B. C. Heezen and Dr. W. J. Morgan for reading the manuscript and offering suggestions for its improvement. Many of the figures were generously made available by several authors.

REFERENCES

Andreyeva, I. B., and G. B. Udintsev, Bottom structure of the Sea of Japan, from the *Vityaz* expedition data, *Izv. Akad. Nauk SSSR, Ser. Geol., 1958*, 1–15, 1958.

Armstrong, R. L., A model for the evolution of strontium and lead isotopes in a dynamic earth, *Rev. Geophys., 6*, 175–199, 1968.

Aumento, F., The Mid-Atlantic ridge near 45°N, 2, Basalts from the area of Confederation Peak, *Can. J. Earth Sci., 5*, 1–21, 1967.

Aver'yanov, A. G., P. S. Veizman, E. I. Galperin, S. M. Zverev, M. A. Zaionchkovski, I. P. Kosminskaya, R. M. Krakshina, G. G. Mikhota, and Yu. V. Tulina, Deep seismic sounding in the zone of transition from the Asian continent to the Pacific Ocean during the IGY, *Izv. Akad. Nauk SSSR, Ser. Geofiz., 1961*, 169–184, 1961.

Avery, O. E., G. D. Burton, and J. R. Heirtzler, An aeromagnetic survey of the Norwegian Sea, *J. Geophys. Res., 73*, 4583–4600, 1968.

Båth, M., and A. Vogel, Surface waves from earthquakes in northern Atlantic-Arctic Ocean, *Geofis. Pura Appl., 39*, 35–54, 1958.

Benioff, H., Orogenesis and deep crustal structure —Additional evidence from seismology, *Bull. Geol. Soc. Am., 65*, 385–400, 1954.

Berry, M. J., and L. Knopoff, Structure of the upper mantle under the Western Mediterranean basin, *J. Geophys. Res., 72*, 3613–3626, 1967.

Birch, F., The velocity of compressional waves in rocks to 10 kilobars, 1, *J. Geophys. Res., 65*, 1083–1102, 1960.

Bott, M. H. P., Formation of oceanic ridges, *Nature, 207*, 840–843, 1965.

Bullard, E. C., The flow of heat through the floor of the ocean, in *The Sea*, vol. 3, pp. 218–232, Interscience Publishers, New York, 1963.

Bullard, E. C., A. E. Maxwell, and R. Revelle, Heat flow through the deep sea floor, *Advan. Geophys., 3*, 153–181, 1956.

Cann, J. R., Geological processes at mid-ocean ridge crests, *Geophys. J., 5*, 331–341, 1968.

Cann, J. R., and F. J. Vine, An area on the crest of the Carlsberg ridge: Petrology and magnetic survey, *Phil. Trans. Roy. Soc. London, A, 259*, 193–217, 1966.

Carey, S. W., The tectonic approach to continental drift, in *Continental Drift—A Symposium*, pp. 177–355, Geology Department, University of Tasmania, Hobart, 1958.

Conolly, J. R., H. D. Needham, and B. C. Heezen, Late Pleistocene and Holocene sedimentation in the Laurentian channel, *J. Geol., 75*, 131–147, 1967.

Dickinson, W. R., Circum-Pacific andesite types, *J. Geophys. Res., 73*, 2261–2269, 1968.

Dickson, G. O., W. Pitman, and J. R. Heirtzler, Magnetic anomalies in the South Atlantic and sea-floor spreading, *J. Geophys. Res., 73*, 2087–2100, 1968.

Dietz, R. S., Continent and ocean basin evolution

by spreading of the sea floor, *Nature, 190,* 854–857, 1961.

Dietz, R. S., and G. Shumway, Arctic basin geomorphology, *Bull. Geol. Soc. Am., 72,* 1319, 1961.

Elmendorf, C. H., and B. C. Heezen, Oceanographic information for engineering submarine cable systems, *Bell System Tech. J., 36,* 1047–1093, 1957.

Elsasser, W. M., Interpretation of heat flow equality, *J. Geophys. Res., 72,* 4768–4770, 1967.

Engel, A. E. J., and C. G. Engel, Composition of basalt from the Mid-Atlantic ridge, *Science, 144,* 1330–1333, 1964.

Ewing, J., J. Antoine, and M. Ewing, Geophysical measurements in the western Caribbean Sea and in the Gulf of Mexico, *J. Geophys. Res., 65,* 4087–4126, 1960.

Ewing, J., and M. Ewing, Seismic-refraction measurements in the Atlantic Ocean basins, in the Mediterranean Sea, on the Mid-Atlantic ridge, and in the Norwegian Sea, *Bull. Geol. Soc. Am., 70,* 291–318, 1959.

Ewing, J., and M. Ewing, Sediment distribution on the mid-ocean ridges with respect to spreading of the sea floor, *Science, 156,* 1590–1592, 1967.

Ewing, J., J. L. Worzel, M. Ewing, and C. Windisch, Ages of horizon A and the oldest Atlantic sediments, *Science, 154,* 1125–1132, 1966.

Ewing, M., and J. Ewing, Distribution of oceanic sediments, in *Studies on Oceanography,* edited by Kozo Yoshida, p. 568, University of Washington Press, Seattle, 1964.

Ewing, M., and B. C. Heezen, Mid-Atlantic ridge seismic belt, *Trans. Am. Geophys. Union, 37,* 343, 1956.

Ewing, M., X. Le Pichon, and J. Ewing, Crustal structure of the mid-ocean ridges, 4, Sediment distribution in the South Atlantic Ocean and Cenozoic history of the Mid-Atlantic ridge, *J. Geophys. Res., 71,* 1611–1636, 1966a.

Ewing, M., T. Saito, J. I. Ewing, and L. Burckle, Lower Cretaceous sediment from the Northwest Pacific, *Science, 152,* 751–755, 1966b.

Fisher, R. L., and C. G. Engel, Dunite dredged from the nearshore flank of Tonga trench on Expedition Nova 1967 (abstract), *Trans. Am. Geophys. Union, 49,* 217–218, 1968.

Fisher, R. L., and H. H. Hess, Trenches, in *The Sea,* vol. 3, edited by N. N. Hill, pp. 411–436, Interscience Publishers, New York, 1963.

Fisher, R. L., G. L. Johnson, and B. C. Heezen, Mascarene plateau, Western Indian Ocean, *Bull. Geol. Soc. Am., 78,* 1247–1266, 1967.

Fuchs, K., St. Mueller, E. Peterschmitt, J. P. Rothe, A. Stein, and K. Strobach, Krustenstruktur der Westalpen nach refraktions—seismischen Messungen, *Beitr. Geophys., 72,* 149–169, 1963.

Gagel'gants, A. A., E. I. Gal'perin, I. P. Kosminskaya, and R. M. Krakshina, Structure of the earth's crust in the central part of the Caspian Sea, determined by deep seismic data, *Dokl. Akad. Nauk SSSR, 123,* 520–523, 1958.

Gass, I. G., and D. Masson-Smith, The geology and gravity anomalies of the Troodos massif, Cyprus, *Phil. Trans. Roy. Soc. London, A, 255,* 417–467, 1963.

Glass, B. P., Glassy objects (microtektites?) from deep-sea sediments near the Ivory Coast, *Science, 161*(3844), 891–893, 1968.

Glass, B., D. B. Ericson, B. C. Heezen, N. D. Opdyke, and J. A. Glass, Tektites and geomagnetic reversals, *Nature, 216,* 437–442, 1967.

Glass, B. P., and B. C. Heezen, Tektites and geomagnetic reversals, *Sci. Am., 217,* 32–38, 1967.

Gunn, R., Quantitative aspects of juxtaposed ocean deeps, mountain chains, and volcanic ranges, *Geophysics, 12,* 238–255, 1947.

Hamilton, E. L., Sunken islands of the Mid-Pacific mountains, *Geol. Soc. Am. Mem. 64,* 96 pp., 1956.

Hamilton, E. L., Thickness and consolidation of deep-sea sediments, *Bull. Geol. Soc. Am., 70,* 1399–1424, 1959.

Harrison, C. G. A., Formation of magnetic anomaly patterns by dyke injection, *J. Geophys. Res., 73,* 2137–2142, 1968.

Hayes, D. E., and J. R. Heirtzler, Magnetic anomalies and their relation to the Aleutian island arc, *Trans. Am. Geophys. Union, 49,* 207–208, 1968.

Hays, J. D., and N. D. Opdyke, Antarctic radiolaria, magnetic reversals, and climatic change, *Science, 158,* 1001–1011, 1967.

Heezen, B. C., Dynamic processes of abyssal sedimentation, *Geophys. J., 2,* 142–163, 1959a.

Heezen, B. C., Paleomagnetics, continental displacements and the origin of submarine topography, *Intern. Ocean. Congress, 1,* 26–28, 1959b.

Heezen, B. C., The rift in the ocean floor, *Sci. Am., 203,* 98–110, 1960a.

Heezen, B. C., Ocean basin formation, in *McGraw-Hill Encyclopedia of Science and Technology,* vol. 13, p. 223, McGraw-Hill Book Company, New York, 1960b.

Heezen, B. C., The deep-sea floor, in *Continental Drift,* edited by S. K. Runcorn, pp. 235–288, Academic Press, New York, 1962.

Heezen, B. C., and M. Ewing, The mid-oceanic ridge, in *The Sea,* edited by M. N. Hill, vol. 3, pp. 388–410, Interscience Publishers, New York, 1963.

Heezen, B. C., and A. S. Laughton, Abyssal plains, in *The Sea,* edited by M. N. Hill, vol. 3, pp. 312–364, Interscience Publishers, New York, 1963.

Heezen, B. C., and M. Tharp, The Atlantic floor, in *North Atlantic Biota,* edited by A. Love, p. 430, Pergamon Press, London, 1963.

Heezen, B. C., and Marie Tharp, Tectonic fabric of the Atlantic and Indian oceans and continental drift, *Phil. Trans. Roy. Soc. London, A, 258,* 90–106, 1965.

Heezen, B. C., and R. E. Sheridan, Lower Cretaceous rocks (Neocomian-Albian) dredged from Blake escarpment, *Science, 154,* 1614–1617, 1966.

Heezen, B. C., and M. Tharp, Physiographic chart

of the North Atlantic, Lamont Geological Observatory, Columbia University, 1968.

Heezen, B. C., M. Ewing, and E. T. Miller, Transatlantic profile of total magnetic intensity and topography, Dakar to Barbados, *Deep-Sea Res.*, *1*, 25–33, 1953.

Heezen, B. C., M. Tharp, and M. Ewing, The floors of the oceans, 1, The North Atlantic, *Geol. Soc. Am., Spec. Paper 65*, 1–122, 1959.

Heezen, B. C., M. Tharp, and R. D. Gerard, Equatorial Atlantic fracture zones, *Geol. Soc. Am. Spec. Paper 68*, 195–196, 1961.

Heezen, B. C., E. T. Bunce, J. B. Hersey, and M. Tharp, Chain and Romanche fracture zones, *Deep-Sea Res.*, *11*, 11033, 1964a.

Heezen, B. C., R. D. Gerard, and M. Tharp, The Vema fracture zone in the equatorial Atlantic, *J. Geophys. Res.*, *69*, 733–739, 1964b.

Heezen, B. C., B. Glass, and H. W. Menard, The Manihiki plateau, *Deep-Sea Res.*, *13*, 445–458, 1966a.

Heezen, B. C., and C. D. Hollister, and W. F. Ruddiman, Shaping of the continental rise by deep geostrophic contour currents, *Science*, *152*, 502–508, 1966b.

Heezen, B. C., E. D. Schneider, and O. H. Pilkey, Sediment transport by the Antarctic bottom current on the Bermuda rise, *Nature*, *211*, 611–612, 1966c.

Heirtzler, J. R., G. O. Dickson, E. M. Herron, W. Pitman, and X. Le Pichon, Marine magnetic anomalies, geomagnetic field reversals, and motions of the ocean floor and continents, *J. Geophys. Res.*, *73*, 2119–2136, 1968.

Heirtzler, J. R., and X. Le Pichon, Crustal structure of the mid-ocean ridges, 3, Magnetic anomalies over the Mid-Atlantic ridge, *J. Geophys. Res.*, *70*, 4013–4033, 1965.

Hersey, J. B., Continuous reflection profiling, in *The Sea*, vol. 3, pp. 47–72, Interscience Publishers, New York, 1963.

Hersey, J. B., Sedimentary basins of the Mediterranean Sea, in *Submarine Geology and Geophysics*, pp. 75–92, Butterworth and Company, London, 1965.

Hess. H. H., The oceanic crust, *J. Marine Res.*, *14*, 423–439, 1955.

Hess, H. H., History of the ocean basins, in *Petrologic Studies*, pp. 599–620, Geological Society of America, New York, 1962.

Hill, M. N., Recent geophysical exploration of the ocean floor, *Phys. Chem. Earth*, *2*, 129–163, 1957.

Hollister, C. D., and B. C. Heezen, Contour current evidence from abyssal sediments, *Trans. Am. Geophys. Union*, *48*, 142, 1967.

Holmes, A., *Principles of Physical Geology*, Ronald Press Company, New York, 1288 pp., 1965.

Horn, D. R., R. M. Horn, and M. N. Delach, Correlation between acoustical and other physical properties of deep-sea cores, *J. Geophys. Res.*, *73*, 1939–1957, 1968.

Irving, E., *Paleomagnetism and Its Application to Geological and Geophysical Problems*, John Wiley & Sons, New York, 399 pp., 1964.

Isacks, B., J. Oliver, and L. R. Sykes, Seismology and the new global tectonics, *J. Geophys. Res.*, *73*, 5855–5900, 1968.

Johnson, G. L., North Atlantic fracture zones near 53°, *Earth Planetary Sci. Letters*, *2*, 445–448, 1967.

Johnson, G. L., and O. B. Eckhoff, Bathymetry of the North Greenland Sea, *Deep-Sea Res.*, *13*, 1161–1173, 1966.

Johnson, G. L., and B. C. Heezen, The Arctic mid-oceanic ridge, *Nature*, *215*, 724–725, 1967a.

Johnson, G. L., and B. C. Heezen, The morphology and evolution of the Norwegian-Greenland Sea, *Deep-Sea Res.*, *14*, 755–771, 1967b.

Joyner, W. B., Basalt-eclogite transition as a cause for subsidence and uplift, *J. Geophys. Res.*, *72*, 4977–4998, 1967.

Kuno, H., Lateral variation of basalt magma across continental margins and island arcs, in *Continental Margins and Island Arcs*, *Can. Geol. Surv. Paper 66–15*, edited by W. H. Pool, 317–336, 1966.

Langseth, M. G., X. Le Pichon, and M. Ewing, Crustal structure of the mid-ocean ridges, 5, Heat flow through the Atlantic ocean floor and convection currents, *J. Geophys. Res.*, *71*, 5321–5355, 1966.

Lee, W. H. K., and S. Uyeda, Review of heat flow data, in *Terrestrial Heat Flow*, *Geophys. Monograph 8*, edited by W. H. K. Lee, pp. 87–190, American Geophysical Union, Washington, D.C., 1965.

Le Pichon, X., J. Ewing, and R. Houtz, Deep sea sediment velocity determination made while reflection profiling, *J. Geophys. Res.*, *73*, 2597–2614, 1968.

Le Pichon, X., and J. R. Heirtzler, Magnetic anomalies in the Indian Ocean and sea-floor spreading, *J. Geophys. Res.*, *73*, 2101–2117, 1968.

Le Pichon, X., R. E. Houtz, C. L. Drake, and J. E. Nafe, Crustal structure of the mid-ocean ridges, 1, Seismic refraction measurements, *J. Geophys. Res.*, *70*, 319–339, 1965.

MacDonald, G. J. F., Dependence of the surface heat flow on the radioactivity of the earth, *J. Geophys. Res.*, *69*, 2933–2946, 1964.

Mason, R. G., A magnetic survey off the west coast of the United States, *Geophys. J.*, *1*, 320–329, 1958.

Matthews, D. H., and Jennifer Bath, Formation of magnetic anomaly pattern of Mid-Atlantic ridge, *Geophys. J.*, *13*, 349–357, 1967.

Matthews, D. H., F. J. Vine, and J. R. Cann, Geology of an area of the Carlsberg ridge, Indian Ocean, *Bull. Geol. Soc. Am.*, *73*, 675–682, 1965.

Maury, M. F., *Physical Geography of the Sea*, Harpers and Row Publishers, New York, 389 pp., 1855.

Maxwell, J. C., Continental drift and a dynamic earth, *Am. Sci.*, *56*, 35–51, 1968.

McBirney, A. R., and I. G. Gass, Relations of

oceanic volcanic rocks to mid-ocean rises and heat flow, *Earth Planetary Sci. Letters*, *2*, 265–276, 1967.

McKenzie, D. P., Heat flow and gravity anomalies, *J. Geophys. Res.*, *72*, 6261–6273, 1967.

McKenzie, D. P., and J. G. Sclater, Heat flow inside the island arcs of the Northwestern Pacific, *J. Geophys. Res.*, *73*, 3173–3179, 1968.

Melson, W. G., and T. H. van Andel, Metamorphism in the Mid-Atlantic ridge, 22°N latitude, *Marine Geol.*, *4*, 165–186, 1966.

Menard, H. W., Deformation of the northeastern Pacific basin and the west coast of North America, *Bull. Geol. Soc. Am.*, *66*, 1149–1198, 1955.

Menard, H. W., Minor lineations in the Pacific basin, *Bull. Geol. Soc. Am.*, *70*, 1491–1496, 1959.

Menard, H. W., The East Pacific rise, *Science*, *132*, 1737–1746, 1960.

Menard, H. W., *Marine Geology of the Pacific*, McGraw-Hill Book Company, New York, 271 pp., 1964.

Menard, H. W., Transitional types of crust under small ocean basins, *J. Geophys. Res.*, *72*, 3061–3073, 1967a.

Menard, H. W., Sea-floor spreading, topography, and the second layer, *Science*, *157*, 923–924, 1967b.

Menard, H. W., and R. S. Dietz, Mendocino submarine escarpment, *J. Geol.*, *60*, 266–278, 1952.

Menard, H. W., and J. Mammerickx, Abyssal hills, magnetic anomalies and the East Pacific rise, *Earth Planetary Sci. Letters*, *2*, 465–472, 1967.

Menard, H. W., and S. M. Smith, Hypsometry of ocean basin provinces, *J. Geophys. Res.*, *71*, 4305–4325, 1966.

Moore, T. C., and G. R. Heath, Abyssal hills in the central equatorial Pacific: Detailed structure of the sea floor and sub-bottom reflectors, *Marine Geol.*, *5*, 161–179, 1967.

Morgan, J., Rises, trenches, great faults, and crustal blocks, *J. Geophys. Res.*, *73*, 1959–1982, 1968.

Nafe, J. E., and C. L. Drake, Physical properties of marine sediments, Tech. Rept. 2, CU-3-61 NObsr 85077 Geology, Lamont Geological Observatory, Palisades, New York, 1961.

Nagata, T., *Rock Magnetism*, 2nd edition, Maruzen Company, Tokyo, 350 pp., 1961.

Neprochnov, Iu. P., V. P. Goncharov, and A. F. Neprochnova, Seismic data on the structure of the earth's crust in the central section of the Black Sea, *Dokl. Akad. Nauk SSSR*, *129*, 408–411, 1959.

Neprochnov, Iu. P., A. F. Neprochnova, S. M. Zverev, V. I. Mironova, R. A. Bokur, and A. V. Chekunov, New data on structure of the earth's crust of the Black Sea basin south of the Crimea, *Dokl. Akad. Nauk SSSR*, *156*(3), 561–564, 1964.

Ninkovich, D., Pleistocene volcanic eruptions in New Zealand recorded in deep sea sediments, *Earth Planetary Sci. Letters*, *14*, 89–102, 1968.

Ninkovich, D., and B. C. Heezen, Physical and chemical properties of volcanic glass sands from Pozzuolana ash, Thera Island and from upper and lower ash layers in the eastern Mediterranean deep sea sediments, *Nature*, *213*, 582–584, 1967.

Ninkovich, D., B. C. Heezen, J. R. Conolly, and L. H. Burkle, South Sandwich Tephra in deep-sea sediments, *Deep-Sea Res.*, *11*, 605–619, 1964.

Ninkovich, D., N. Opdyke, B. C. Heezen, and J. H. Foster, Paleomagnetic stratigraphy, rates of deposition and tephrachronology in North Pacific deep-sea sediments, *Earth Planetary Sci. Letters*, *1*, 476–492, 1966.

O'Brien, P. N. S., Lake Superior crustal structure: A reinterpretation of the 1963 seismic experiment, *J. Geophys. Res.*, *73*, 2669–2689, 1968.

Officer, C. B., J. I. Ewing, J. F. Hennion, D. G. Harkrider, and D. E. Miller, Geophysical investigations in the eastern Caribbean: summary of 1955 and 1956 cruises, *Phys. Chem. Earth*, *3*, 17–109, 1959.

Oliver, J., and J. Dorman, Exploration of suboceanic structure by the use of seismic surface waves, in *The Sea*, vol. 3, pp. 110–133, Interscience Publishers, New York, 1963.

Oliver, J., and B. L. Isacks, Some evidence on the structure of the mantle near an island arc (abstract), *Trans. Am. Geophys. Union*, *48*, 219, 1967.

Oxburgh, E. R., and D. L. Turcotte, Mid-ocean ridges and geothermal distribution during mantle convection, *J. Geophys. Res.*, *73*, 2643–2661, 1968.

Phillips, J. D., Magnetic anomalies over the Mid-Atlantic ridge near 27°N, *Science*, *157*, 920–923, 1967.

Pitman, W. C., III, and J. R. Heirtzler, Magnetic anomalies over the Pacific-Antarctic ridge, *Science*, *154*, 1164–1171, 1966.

Pitman, W., E. M. Herron, and J. R. Heirtzler, Magnetic anomalies in the Pacific and sea floor spreading, *J. Geophys. Res.*, *73*, 2069–2085, 1968.

Poldervaart, A., and J. Green, Chemical analyses of submarine basalts, *Am. Mineralogist*, *50*, 1723–1727, 1965.

Raff, A. D., Further magnetic measurements along the Murray fault, *J. Geophys. Res.*, *67*, 417–418, 1962.

Raff, A. D., and R. G. Mason, Magnetic survey off the west coast of North America, 40°N to 52½°N, *Bull. Geol. Soc. Am.*, *72*, 1259–1265, 1961.

Raitt, R. W., Seismic-refraction studies of the Pacific Ocean basin, 1, Crustal thickness of the central equatorial Pacific, *Bull. Geol. Soc. Am.*, *67*, 1623–1640, 1956.

Raitt, R. W., The crustal rocks, in *The Sea*, vol. 3, pp. 85–102, John Wiley & Sons, New York, 1963.

Ringwood, A. E., *Advance in Earth Science*, pp. 357–399, MIT Press, Cambridge, 1966.

Rona, P. A., E. D. Schneider, and B. C. Heezen, Bathymetry of the continental rise off Cape Hatteras, *Deep-Sea Res.*, *14*, 625–633, 1967.

Schilling, T. G., D. C. Krause, and J. G. Moore, Geological, geochemical, and magnetic studies of the Reykjanes ridge near 60°N (abstract), Trans. Am. Geophys. Union, 49, 201, 1968.

Schneider, E. D., P. J. Fox, C. D. Hollister, H. D. Needham, and B. C. Heezen, Further evidence of contour currents in the western North Atlantic, Earth Planetary Sci. Letters, 2, 351–359, 1967.

Schneider, E. D., and B. C. Heezen, Sediments of the Caicos outer ridge, The Bahamas, Bull. Geol. Soc. Am., 77, 1381–1398, 1966.

Schneider, E. D., and P. R. Vogt, Discontinuities in the history of sea-floor spreading, Nature, 217, 1212–1222, 1968.

Scholl, D. W., R. Von Huene, and J. R. Redlon, Undeformed sediment in Peru-Chile trench, Science, 159, 869–871, 1968.

Shand, S. J., Rocks of the Mid-Atlantic ridge, J. Geol., 57, 89–92, 1949.

Shor, G. G., Jr., Structure of the Bering Sea and the Aleutian ridge, Marine Geol., 1, 213–219, 1964.

Shor, G. G., Jr., P. Dehlinger, H. K. Kirk, and W. S. French, Seismic refraction studies off Oregon and northern California, J. Geophys. Res., 73, 2175–2194, 1968.

Spiess, F. N., B. P. Luyendyk, and J. D. Mudie, Geological observations of abyssal hills using a deeply towed instrument package (abstract), Trans. Am. Geophys. Union, 49, 213, 1968.

Sykes, L. R., The seismicity and deep structure of island arcs, J. Geophys. Res., 71, 2931–3006, 1966.

Sykes, L. R., Mechanism of earthquakes and nature of faulting on the mid-oceanic ridges, J. Geophys. Res., 72, 2131–2153, 1967.

Talwani, M., B. C. Heezen, and J. L. Worzel, Gravity anomalies, physiography, and crustal structure of the Mid-Atlantic ridge, Publ. Bur. Central Seismol. Intern., A, Trav. Sci., 22, 81–111, 1961.

Talwani, M., X. Le Pichon, and J. R. Heirtzler, East Pacific rise: The magnetic pattern and the fracture zones, Science, 150, 1109–1115, 1965.

Talwani, M., C. Windisch, M. Langseth, and J. R. Heirtzler, Recent geophysical studies on the Reykjanes ridge, Trans. Am. Geophys. Union, 49, 201, 1968.

Uchupi, E., and K. O. Emery, Structure of continental margin off Atlantic Coast of the United States, Bull. Am. Assoc. Petrol. Geologists, 51, 223–224, 1967.

Udintsev, G. B., Results of Upper Mantle Project studies in the Indian Ocean by the Research Vessel Vityaz, in The World Rift System, Geol. Surv. Can. Paper 66-14, 148–172, 1966.

Vacquier, V., Measurement of horizontal displacement along faults in the ocean floor, Nature, 183, 452–453, 1959.

Vacquier, V., A. D. Raff, and R. E. Warren, Horizontal displacements in the floor of the northeast Pacific Ocean, Bull. Geol. Soc. Am., 72, 1251–1258, 1961.

Vacquier, V., S. Uyeda, M. Yasui, J. Sclater, C. Corry, and T. Watanabe, Studies of the thermal state of the earth: Heat flow measurements in the northwestern Pacific, Bull. Earthquake Res. Inst., Tokyo Univ., 44, 1519–1535, 1966.

van Andel, T. H., and C. O. Bowin, Mid-Atlantic ridge between 22° and 23°N latitude and the tectonics of mid-ocean rises, J. Geophys. Res., 73, 1279–1298, 1968.

van de Lindt, W. J., Movement of the Mohorovicic discontinuity under isostatic conditions, J. Geophys. Res., 72, 1289–1297, 1967.

Vening Meinesz, F. A., Gravity expeditions at sea, Publ. Neth. Geod. Comm., Waltman, Delft, 4, 1923–1938, 1948.

Vening Meinesz, F. A., Plastic buckling of the earth's crust: The origin of geosynclines, Geol. Soc. Am. Spec. Paper 62, 319–330, 1955.

Vine, F. J., Spreading of the ocean floor: New evidence, Science, 154, 1405–1415, 1966.

Vine, F. J., and D. H. Matthews, Magnetic anomalies over oceanic ridges, Nature, 199, 947–949, 1963.

Vogt, P. R., and R. H. Higgs, Aeromagnetic survey of the eastern Mediterranean (abstract), Trans. Am. Geophys. Union, 49, 156–157, 1968.

Vogt, P. R., and N. A. Ostenso, Magnetic survey over the Mid-Atlantic ridge between 42°N and 47°N, J. Geophys. Res., 71, 4389–4411, 1966.

Vogt, P. R., and N. A. Ostenso, Steady-state crustal spreading, Nature, 215, 810–817, 1967.

Vogt, P. R., N. A. Ostenso, and G. L. Johnson, Magnetic evidence for sea-floor spreading north of Iceland, in preparation, 1969.

Von Herzen, R. P., Heat flow values from the southeastern Pacific, Nature, 183, 882–883, 1959.

Von Herzen, R., and A. E. Maxwell, The measurement of thermal conductivity of deep-sea sediments by a needle-probe method, J. Geophys. Res., 74, 1557–1564, 1959.

Von Herzen, R., and S. Uyeda, Heat flow through the eastern Pacific Ocean floor, J. Geophys. Res., 68, 4219–4252, 1963.

Wageman, J. M., and G. L. Johnson, A study of the eastern flank of the Mid-Atlantic ridge, J. Geophys. Res., 72, 1175–1182, 1967.

Wegener, A., The Origin of Continents and Oceans, E. P. Dutton and Company, New York, 1924.

Wilson, J. T., Hypothesis of earth's behavior, Nature, 198, 925–929, 1963.

Wilson, J. T., A new class of faults and their bearing on continental drift, Nature, 207, 343–347, 1965.

Woollard, G. P., Crust and mantle relations in the Hawaiian area, in Continental Margins and Island Arcs, Geol. Surv. Can. Paper 66-15, edited by W. H. Pool, 294–310, 1966.

Worzel, J. L., Extensive deep-sea sub-bottom reflections identified as white ash, Proc. Natl. Acad. Sci., U.S., 45, 349–355, 1959.

8. *Experimental and Theoretical Geophysics*

Electrical Conductivity of the Mantle

D. C. Tozer

THIS article is divided into two parts: the first is concerned with geophysical observations and their interpretation in terms of an electrical conductivity of the mantle and the second is a very brief survey of laboratory studies that are thought to be relevant to the further interpretation of the conductivity data in terms of the physical and chemical state of the mantle.

INTERPRETATION OF GEOPHYSICAL OBSERVATIONS

A number of techniques are used to determine the electrical conductivity of the earth's interior, but it is the original purely magnetic method that has given us the most information about the distribution of mantle conductivity. *Berdichevesky* [1960] gives a general discussion and bibliography of the mixed electrical and magnetic (magnetotelluric) method of conductivity determination that is much used in geophysical crustal surveys.

The principles of the method of conductivity determination in the earth that uses only magnetic observations [*Lahiri and Price*, 1939; *Eckhardt et al.*, 1963; *Takeuchi and Saito*, 1963] will be described here only in enough detail to emphasize the deficiencies of the analysis that are relevant to the further interpretation of these data. Magnetic field observations are made simultaneously at as many widely spaced stations on the earth's surface as possible and over an extended length of time. With the assumption that the relevant underlying equations of the electromagnetic field are linear, it becomes physically meaningful to concentrate on the varying part of the field and make appropriate spatial and temporal analyses of this field in terms of complete sets of orthogonal functions. In particular, it is possible to explore the possibility of representing this field as the gradient of a scalar potential and of dividing the surface field into parts of internal and external origin. Since several such analyses have been made using rather old data from poorly placed observatories, it cannot be said that the adequacy of such a representation has yet been demonstrated with any great precision. Nevertheless, it has been possible to demonstrate for those field variations with a high degree of temporal coherence (most significantly the diurnal variations), that a scalar potential representation is a fair description of the observations. To the same accuracy it can be shown that the external and internal parts of these fields can be thought of as stationary fields moving about the earth's axis in a fixed amplitude and phase relationship. Although significant departures from this simple picture occur locally, it is sufficient to indicate that the phenomenon could be understood as one of electromagnetic induction in an external inducing field, and that the earth was at least approximately axisymmetric in its electromagnetic properties. It is perhaps an essential task for the immediate future, with the more comprehensive modern data now available, to check the hypothesis that the sources of the varying surface geomagnetic field are localized in regions that do not intersect this surface.

To proceed further from a statement of the approximate electromagnetic symmetry of the earth to a knowledge of the conductivity distribution is a difficult problem with no unique solution. It is worth pointing out that the internal part of the surface magnetic field is a volume effect of eddy currents, and it is not surprising that the magnetic induction method is a blunt tool for revealing structure that is either much smaller or larger than a length that corresponds to the skin depth in homogeneous conductors. If one had measurable inducing fields with a broad continuum of frequencies and length scales to probe the earth, this characteristic would work in one's favor, and it might be possible to achieve a near unique solution to the conductivity distribution. However, it is practical only to observe the natural geomagnetic variations; since these are usable at just a few discrete frequencies in a narrow frequency range, the conductivity solution is subject to large systematic errors. Any statement of the electromagnetic properties of the earth derived from induction arguments should include, in addition to a discussion of observational and data processing errors, an attempt to evaluate systematic errors and, most important, a description of the extent and location of the induced current system for any model considered to be a satisfactory explanation of the data. Only in this way is it possible to see which parts of the model are fixed and which are relatively arbitrary. For example, it is possible to infer immediately from *Lahiri and Price*'s [1939] analysis of the daily variation that the spherical symmetry of their conductivity model only refers to the 'average' properties at a depth of 500–800 km.

From the point of view of further interpretation of the conductivity distribution, the effect of oceans and sedimentary layers is particularly unfortunate. Quite apart from the difficulties of adequately representing the oceans in the induction problem, the presence of a lower conductivity at greater depths is particularly difficult to elucidate. Lahiri and Price's original analysis gave a thickness of about 600 km to the layer, with conductivity less than the average surface layer conductivity. It was impossible to distinguish, with diurnal and longer period variations, a conductivity of this region anywhere in the range $0 < \sigma < 10^{-4}$ ohm^{-1} cm^{-1}. It is unfortunate that the upper limit of this range cannot be somewhat reduced, since the role of conductivity in reducing the ambiguity in estimates of the temperature begins to be exercised at about a conductivity of 10^{-4} ohm^{-1} cm^{-1}. Since the data take the form of an upper limit on the conductivity, which is an increasing function of temperature for mantle-type materials, the temperature is least restricted for those models of the mantle that assume the most insulating phases. Laboratory studies show that a likely candidate for the most insulating material that is geochemically plausible for the upper mantle is single-crystal olivine consisting of ~90% forsterite and ~10% fayalite. With such a material, the temperature is limited to less than 1700°K within the low conductivity layer, although this estimate is rather uncertain below 300 km because of possible phase change phenomena. Such an estimate comes at the lower limit of plausible temperature estimates from conduction theory and certainly requires appreciable amounts of the heat flow to be carried by radiation. However, if the conductivity of mantle material is appreciably greater than that of the single crystal 90% forsterite phase, then the electrical conductivity indicates temperatures less than any conduction equation solution. Without postulating different phases, an increase of $\sim 10\%$ in fayalite content or a crystal size of much less than a millimeter would reduce the upper limit of temperature by at least 100°C. The development of polycrystallinity, a higher iron content, or the presence of other opaque phases enhances the electrical conductivity but decreases the thermal conductivity by increasing the absorption and scattering of radiation. At the moment, the experimental situation is not well enough understood to say categorically that an inconsistency exists in explaining the conductivity data with a conduction theory of heat transfer, but we are very near it and any appreciable reduction of the upper limit of the conductivity of this layer would soon decide the matter. For some time I have held the view that all the current conduction estimates of mean temperature at various depths in the mantle are too high on account of their hydrodynamic instability and will be considerably reduced when we have better knowledge of the rheology of the mantle [*Tozer*, 1967].

From the point of view of further interpretation, the evidence from shorter period variations that the conductivity may be higher than 10^{-2} ohm^{-1} cm^{-1} at depths of less than 100 km in very localized parts of the upper mantle is extremely interesting. Although the problem of determining conductivity structures with such low symmetry is even more difficult than the problem outlined above, any evidence of conductivities of this magnitude at such depths is indicative of a major thermal or compositional difference from the surroundings. The interpretation of abnormal magnetic behavior as a conductivity anomaly under parts of Peru and Japan lends support to the idea that the temperature in these active regions may be some hundreds of degrees higher than the average at these depths and that one must use volcanic lava temperatures only as indicators of abnormally hot conditions in the mantle. However, there are other anomalies, e.g., the German anomaly and Mould Bay in the Canadian Arctic that do not fit into this scheme, although the evidence in these cases for the anomaly being associated with the mantle is meager. These are more probably compositional in origin.

CONDUCTION IN GEOPHYSICALLY INTERESTING MATERIALS

A number of variables are needed to specify the electrical conductivity of oxide systems, which of course form the major part of any acceptable compositional model of the interior. Temperature, state of oxidation, polycrystallinity, and pressure are all known to be involved, but from the point of geophysical interpretation in the mantle, I have attempted to place them in decreasing order of importance. However, in the crust the effect of pressure is relatively much greater on account of its effect on the porosity. It has been known for a long time [*Wyllie*, 1963] that the conductivity of porous rocks would be determined by electrolytic conduction in the pores. The experiments of *Brace et al.* [1965] and *Parkhomenko and Bondarenko* [1963] confirm the belief that direct measurements of the conductivity of rocks at atmospheric pressure and varying temperatures are a waste of time as far as interpretations of the conductivity of the deep crust and mantle are concerned. Applications of pressure corresponding to only a few tens of kilometers burial reduce the conductivity by factors up to a thousand, especially if the rocks are saturated with water.

Probably the most significant information relevant to mantle studies has come from natural and synthetic single-phase studies of mantle materials and of the very simplest oxide components of these materials. There is now fairly clear evidence that, at the extreme forsterite end of the olivine series, the conductivity is a surface phenomenon, at least at temperatures less than 10^3 °C, and is probably due to ionic motion. Increasing the fayalite content to 10% increases the conductivity, but the exact nature of this change is not yet understood, because the material is insufficiently controlled in its state of oxidation and purity. A factor of the order of a hundred would seem to be involved. The increase of conductivity with polycrystallinity has been clearly demonstrated only for forsterite; a comparable change for the iron-bearing olivines will depend on whether volume electronic conduction dominates. In general, the effect of pressure is to reduce the effects of polycrystallinity [Hamilton, 1965]. The conductivity of pure fayalite is $\sim 10^4$ times greater than the geophysically interesting olivines at temperatures less than 900°C, and it is thought that this may conduct by migration of valence states of iron through the crystal [Bradley et al., 1962]. It has also been demonstrated [Akimoto and Fujisawa, 1965] that the olivine-spinel transformation in fayalite is associated with an increase of conductivity of about 10^2, although the lack of reversability may again indicate a contribution to the conductivity of grain boundaries formed at the transition. If a comparable change occurs during phase change for the olivines containing 10% fayalite, the arguments for revising the temperature in the mantle downward becomes quite strong. There now seems little doubt that, if the success of *Shankland* [1967] in producing good single crystals of forsterite can be followed by preparation of single crystals of controlled stoichiometry throughout the olivine and pyroxene series, we shall be in a strong position to begin understanding the conductivity of basic rocks; further progress will depend on the development of electromagnetic induction theory in the earth.

REFERENCES

Akimoto, Syun-Iti, and Hideyuki Fujisawa, Demonstration of the electrical conductivity jump produced by the olivine-spinel transition, *J. Geophys. Res.*, 70, 443–451, 1965.

Berdichevsky, M. N., *Electricheskaya razvedka metodem tellutricheskikh tokov*, USSR, Gostoptekhizdat, Moscow, 1960; English translation in *Quart. Color. School Mines*, 60(1), 1965.

Brace, W. F., A. S. Orange, and T. R. Madden, The effect of pressure on the electrical resistivity of water-saturated crystalline rocks, *J. Geophys. Res.*, 70, 5669–5678, 1965.

Bradley, R. S., A. K. Jamil, and D. C. Munro, Electrical conductivity of fayalite and spinel, *Nature*, 193, 965–966, 1962.

Eckhardt, D., K. Larner, and T. Madden, Long-period magnetic fluctuations and mantle electrical conductivity estimates, *J. Geophys. Res.*, 68, 6279–6287, 1963.

Hamilton, Robert M., Temperature variation at constant pressures of the electrical conductivity of periclase and olivine, *J. Geophys. Res.*, 70, 5679–5692, 1965.

Lahiri, B. N., and A. T. Price, Electromagnetic induction in nonuniform conductors and the determination of the conductivity of the earth from terrestrial magnetic variations, *Phil. Trans. Roy. Soc. London, A*, 237, 509–540, 1939.

Parkhomenko, E. I., and A. T. Bondarenko, An investigation of the electric resistivity of rocks at pressures up to 40,000 kg/cm² and temperatures up to 400°C, *Izv. Akad. Nauk SSSR, Ser. Geofiz.*, 1963(12), 1106–1111, 1963.

Shankland, T. J., *Bull. Am. Ceram. Soc.*, 46, 1160, 1967.

Takeuchi, H., and M. Saito, Electromagnetic induction within the earth, *J. Geophys. Res.*, 68, 6287–6293, 1963.

Tozer, D. C. Towards a theory of Mantle Convection, in *The Earth's Mantle*, edited by T. F. Gaskell, Academic Press, London, pp. 327–353, 1967.

Wyllie, M. R. J., *The Fundamentals of Well Log Interpretation*, 3rd edition, Academic Press, New York, 238 pp., 1963.

Heat Conductivity in the Mantle

Sydney P. Clark, Jr.

THE determination of temperatures within the earth remains an outstanding geophysical problem. Only two quantities that are directly relevant to its solution can be observed at the surface: the mean temperature and the heat flow. The worldwide average of heat flow is 1.5×10^{-6} cal/cm² sec, and it is clear from modern data on the radioactivity of rocks that only a small fraction of the observed thermal flux can originate from radioactive decay in the outermost 100–200 km of the earth. This conclusion is clearest in oceanic regions, where heat production in the thin crust is of minor importance, but less direct arguments indicate that the conclusion is of more widespread validity.

Thus we must find a means of getting the heat to the surface from greater depths. This cannot be done by conduction if the thermal properties are the same as those representative of surface rocks, for a thermal gradient in excess of 10°C/km would be required, the increase in temperature in the upper mantle would be some thousands of degrees, and the melting interval of the mantle would be exceeded. The mantle is known to be dominantly solid, and we must therefore find a more efficient means of heat transfer than is exhibited by rocks under near-surface conditions, in order to avoid the necessity of steep thermal gradients at depth in the earth.

In subsequent sections we will examine a number of possible mechanisms of heat conductivity in the mantle. The ones of greatest geophysical interest are those that show promise of becoming very efficient at depths in the range 100–200 km, since it is in this part of the earth that the geothermal curve is predicted to intersect the melting curve if the near-surface value of the thermal conductivity persists throughout the upper mantle. The ability of the various mechanisms to meet this criterion is discussed in the final section of this paper.

'ORDINARY' HEAT CONDUCTIVITY

At ordinary temperatures heat is transferred in dielectric crystals principally by the thermal vibrations of atoms. These can be resolved into a system of traveling waves, or phonons, which transfer energy down a thermal gradient. The principal factors that lead to thermal resistance are interactions of phonons with crystalline imperfections and impurities, or with each other. The first types of collisions lead to a thermal resistivity (the reciprocal of the conductivity) that is independent of temperature, but the temperature dependence of the resistivity arising from phonon-phonon interactions is still the subject of controversy. Conservation of energy and momentum forbid two-phonon interactions, and *Pomeranchuk* [1941] showed that three-phonon processes could lead to no thermal resistance in perfect, infinite, elastically isotropic solids. He argued that the difficulty could be removed if account is taken of four-phonon processes and arrived at a resistivity proportional to $T^{5/4}$ at high temperatures. *Herring* [1954] later showed that, because of the elastic anisotropy of real crystals, three-phonon processes would often lead to thermal resistance. His result implies a resistivity proportional to T at high temperatures, which is in good agreement with experiments [*Klemens*, 1958; *Kingery et al.*, 1954]. Thus, although the theoretical problem is not satisfactorily resolved, experiment supports a resistivity equal to $a + bT$, where a and b are constants. A notable exception to this law are some plagioclase feldspars, in which the resitivity decreases with temperature [*Birch and Clark*, 1940]. This is also true of glass, and it is tempting to attribute the behavior of the feldspars to the very large unit cell, which may appear glasslike to the short-wavelength phonons that predominate at high temperatures. Values of a and b for a number of interesting materials for which the resistivity is a linear function of temperature are given in Table 1.

TABLE 1. Thermal Resistivity as a Function of Temperature (Resistivity in cm sec °C/cal is expressed as $a + bT$, with T in °K.)

Material	Temp. Range, °C	a, cm sec °C/cal	b, cm sec/cal	Reference
NaCl	0–400	−22	0.330	Birch and Clark, 1940
MgO	100–800	−9	0.053	Kingery et al., 1954
Al_2O_3	100–800	−12	0.065	Kingery et al., 1954
SiO_2*	100–400	26	0.162	Birch and Clark, 1940
$MgAl_2O_4$	100–1000	8	0.051	Kingery et al., 1954
$ZrSiO_4$	100–800	55	0.039	Kingery et al., 1954
Mg_2SiO_4	100–600	36	0.118	Kingery et al., 1954
$(Mg, Fe)SiO_3$	100–300	84	0.093	Birch and Clark, 1940

* Single crystal with heat flowing perpendicular to optic axis.

The effect of pressure on ordinary thermal conductivity has been little studied either theoretically or experimentally. For geophysical purposes it is usual to connect the conductivity with density and the velocity of sound, and then to estimate its change with depth from the variations of these parameters. *Klemens* [1958] gives the relationship $R \propto \gamma^2 T \rho^{1/3}/\theta^3$, where ρ is density, γ the Grüneisen constant, and θ the Debye temperature, for the resistivity arising from phonon-phonon interactions. *Birch* [1952] estimated the variation of γ with compression and found that it decreases, but less rapidly than ρ increases. Nevertheless, $\gamma^2 \rho^{1/3}$ decreases by about 30% for the maximum strain encountered in the mantle. On the other hand, θ is proportional to $\rho^{1/3} V_s$ [*Birch*, 1952], and both of these quantities increase by more than 60% through the mantle. Hence θ^3 rises by a factor of about 8, and the quantity b of Table 1 decreases by roughly an order of magnitude. The change in the upper mantle is much smaller, however, since the velocities and densities rise by 15% or less. The decrease in b would then be less than a factor of 2.

Information on the temperature-independent term in the resistivity is much less definite. As a rough generality, it can be taken to be proportional to the velocity, and therefore to increase by 15% in the upper mantle and about 60% in the entire mantle. It may not be worth while to apply such a small correction to the values of a in Table 1, since they were obtained by lengthy extrapolation of high-temperature data to 0°K. Their accuracy is not high, although they give a good representation of the measurements in the stated range of validity, and the negative values are devoid of physical reality. The contribution of this term to the resistivity is small in any case.

RADIATIVE TRANSFER

Electromagnetic radiation may transfer heat through nonopaque solids, especially at high temperatures. The contribution to the heat conductivity is given by $16n^2\sigma T^3/3\bar{\epsilon}$, where n is the refractive index, σ the Stefan-Boltzmann constant, and $\bar{\epsilon}$ the mean extinction coefficient averaged over all wavelengths of light [*Clark*, 1957a]. The extinction coefficient contains contributions from both absorption and scattering; its definition is such that the intensity along a pencil of radiation varies with distance according to $e^{-\bar{\epsilon} x}$. If we insert a typical value of 1.7 for n, the expression for the radiative conductivity reduces to $22 \times 10^{-12} T^3/\bar{\epsilon}$. Evidently there is a strong tendency for this term to increase with temperature, but its importance relative to other conduction mechanisms will be controlled by the magnitude of $\bar{\epsilon}$.

The absorption coefficients of solids vary greatly with wavelength. At long wavelengths, usually greater than 5 μ, intense absorption results from photon-phonon interactions. These have little effect on radiative transfer, since a large fraction of the electromagnetic energy is located at these wavelengths only when the temperature is so low that radiative transfer is unimportant anyway. At the opposite end of the spectrum, strong absorption arises from optical excitation of electrons across the fundamental energy gap of nonmetallic crystals. In simple oxides such as MgO, Al_2O_3, or SiO_2 the energy required is several electron volts, and the absorption takes place at wavelengths considerably less than 0.3 μ. Only at very high temperatures does this process limit radiative transfer. Between these regions the simple oxides hardly absorb at all, which accounts for their lack of color and high transparency when well crystallized.

The presence of transition-metal ions, of which the most important terrestrially is Fe^{2+}, changes the situation radically. These ions have partially filled d-electron orbitals, which are degenerate in the free ion. A crystalline field of octahedral symmetry splits the d orbitals into two energy levels, each of which is also degenerate. Optical excitation of electrons from one d level to the other produces weak absorption (the intensity of the band is 5–50 cm^{-1} in common minerals) at a wavelength of about 1 μ. Much more intense absorption at wavelengths shorter than about 0.3 μ results from the excitation of electrons from oxygen levels to the unfilled d levels. This latter absorption was formerly attributed to excitation of electrons across the fundamental energy gap of the crystal, but Clark [1959] showed that the absorptions at 1 μ and at less than 0.3 μ, which are prominent in iron-bearing natural olivines, are totally absent in pure synthetic Mg_2SiO_4.

Neither absorption maximum connected with d-electron orbitals is a sharp spectral feature. The peak at 1 μ has broad wings, and the 0.3 μ peak has a pronounced long-wavelength tail. Superposition of the two features leads to an absorption minimum in the visible part of the spectrum, which accounts for the green color of Fe^{2+}-bearing minerals. The wings and tail presumably originate partly from thermal motions of the atoms and partly from distortion of the octahedral field, which removes the degeneracies of the d orbitals to a slight extent. But whatever the source, the lack of sharpness of these features is critical in determining the importance of radiative transfer. The electromagnetic spectrum can be thought of as a bank of resistors in parallel, with low resistance associated with regions of high transparency. The total resistance of such an array is mainly determined by the values of the lowest individual resistances and is insensitive to the higher values.

The few relevant experimental data on absorption are confined to room temperature and atmospheric pressure [Clark, 1957b]. Absorption coefficients of olivine and pyroxene are 1 cm^{-1} or less between 2 and 5 μ and are between 1 and 5 cm^{-1} in the visible. The radiative conductivity is predicted from these data to rise to 0.1 cal/cm sec °C at temperatures between 800 and 1500°C, but this can only be a first approximation to the situation in the earth, since the effects of temperature and pressure on $\bar{\epsilon}$ are not included. The pertinent experiments have not been done; theory can only provide a qualitative guide to the effects to be expected. Increased temperature should further broaden the peaks and result in increased absorption at intermediate wavelengths. The chief effect of high pressure should be to shift the positions of the peaks. The peak at 1 μ is expected to shift to shorter wavelengths because compression increases the crystal field strength and hence the splitting of the d orbitals. The 0.3 μ peak should shift to longer wavelengths owing to the reduction in the width of regions of forbidden energy at the expense of allowed regions. Hence pressure should increase the absorption in the 'window' at visible wavelengths, but it might reduce the absorption in the near infrared. Little can be said about the magnitudes of these effects until quantitative experimental results become available.

A fundamentally different source of absorption may become important at high temperatures and pressures, although it is unlikely to be appreciable in the upper mantle. Electrically conducting materials may absorb over a wide range of wavelengths; for mantle materials the absorption per centimeter is roughly 200 times the conductivity per ohm per centimeter. It is to be emphasized that this relation is true only if the conductivity and absorption are both evaluated at the same wavelength, or frequency. In the deep mantle the conductivity may exceed 1 Ω^{-1} cm^{-1} at low frequencies [McDonald, 1957], so that considerable absorption could result, provided that the conductivity does not change markedly between low and optical frequencies. The frequency effect cannot be estimated quantitatively, since we lack detailed knowledge of the conduction mechanism. In the upper mantle the conductivity is orders of magnitude less, and no great absorption is anticipated from this source.

The contribution of scattering to the extinction coefficient is probably minor throughout the mantle. The principal source is grain boundaries, but, if the crystal size is of the order of millimeters or more, the scattering coefficient will be less than 1 cm^{-1}.

OTHER CONDUCTION MECHANISMS

Two other possible means of heat conduction in the earth are by mobile charge carriers, principally electrons and holes, and by excitons. The Wiedemann-Franz law relating thermal conductivity to electrical conductivity and temperature must be modified in the semiconducting mantle to take account of the fact that excitation energy as well as kinetic energy is carried by the electrons and holes. Their contribution to thermal conductivity may not be negligible in the lower mantle if σ exceeds 1 Ω^{-1}cm^{-1}, but in the upper mantle this process is unimportant.

Excitons are electron-hole pairs bound together by their coulombic attraction. They are electrically neutral, and hence carry no current, but they are capable of acquiring their excitation energy in a hot region, migrating down a thermal gradient, and then giving up the excitation energy. Thus they are potentially capable of transferring large amounts of heat, because their content of energy may be comparatively large and their mean free path comparatively long.

Lawson and Jamieson [1958] quote the relationship $K = 6.64 \times 10^{-3} l_0 T^2 [(E/kT + 2)^2 + 2]e^{-E/kT}$ for the conductivity contributed by excitons, where l_0 is the mean free path of the excitons, E their excitation energy, and k is Boltzmann's constant (a misprint in Lawson and Jamieson's equation has been corrected). For $l_0 = 10^{-5}$ cm, which is very uncertain, and $E = 1.19$ ev, they found $K = 2 \times 10^{-5}$ cal/cm sec °C at 1000°K (a numerical slip has been corrected), and $K = 0.02$ cal/cm sec°C at 2000°K. If $E = 3$ ev, which in our present view of the electronic structure of iron-bearing silicates might be a more reasonable estimate in the upper mantle, K is reduced to 6×10^{-14} and 3×10^{-6} cal/cm sec °C at the same temperatures. The theory of heat transfer by excitons is none too well founded (see the comments by Price following Joffe's [1956] paper), and even these estimates may be far too high. Despite the strong dependence of conductivity on temperature provided by this mechanism, it can hardly be given serious consideration until more compelling evidence for its terrestial importance has been provided.

HEAT TRANSFER BY MOVING MATERIAL

Terrestrially important motions of material may occur during hydrothermal activity, 'degassing,' volcanism and magmatism, mountain building and deformation, and what is often loosely termed convection. When account of relative motion is taken, the conduction equation becomes $\nabla(K\nabla T) - \rho c \mathbf{v} \cdot \nabla T + A = \rho c \partial T/\partial t$, where A is the heat production term, and $\mathbf{v}$ is the velocity of motion. If $\mathbf{v}$ is zero, the other terms in the equation are found to be of the order of 10^{-13} to 10^{-14} cal/cm^3 sec in terrestrial problems. The radial component of ∇T is 10^{-4} to 10^{-5} °C/cm, ρc is of the order of unity, and therefore the second term in the equation is comparable with the others if the radial velocity is of the order of 10^{-9} cm/sec or 0.03 cm/year. In situations where liquids are percolating through solids, the porosity times the radial velocity must be this large.

Velocities of this order of magnitude are readily imagined in geological processes. If the continents grew by vertical segregation over geologic time, the mean rate of motion is of the order of 10^{-8} cm/year. The growth of the island of Hawaii in Cenozoic times has taken place at a much greater mean rate. Since neither the growth of the continents nor of Hawaii took place uniformly, considerably higher velocities must have existed at certain times and places. Circulatory motion, such as convection, implies velocities considerably greater than 0.03 cm/year. Motion of material must have transported significant amounts of heat, at least locally and temporarily, during the earth's history.

CONCLUSIONS

Of the processes of heat transfer considered, only radiative transfer and mass transport appear particularly promising as mechanisms to increase the effective thermal conductivity substantially between the surface and depths of 100–200 km. Electrons and holes are unimportant in this connection, at least in the upper mantle. Heat conduction by phonons is unlikely to increase appreciably at shallow depths, because the positive pressure coefficient is small and the pressure effect is counterbalanced by the large increase in temperature in this part of the earth. In the deep mantle this mechanism may assume increased importance, particularly if other processes prevent the temperature from becoming too high. Excitons also appear unlikely to cause a major increase in conductivity at shallow depths. They too may be more important in the deep mantle, since the excitation energy may be reduced at very high pressure. Unlike phonon conduction, this mechanism is favored by high temperatures, but its importance cannot be assessed reliably until the process is better understood.

The effectiveness of radiative transfer depends on the extinction coefficient, its change with temperature, and its change with pressure. The temperature-dependence is likely to be the most critical factor, and it cannot be estimated quantitatively from present information. If $\bar{\epsilon}$ rises to 10 cm^{-1} in the earth, the radiative contribution to the thermal conductivity is 0.007 cal/cm sec °C at 1500°K and 0.027 cal/cm sec °C at 2000°K. These values are already substantial, and even larger contributions would result if $\bar{\epsilon}$ is lower.

Mass transport of material will transfer important amounts of heat if the mean velocity of the motion is high enough. The critical value is about 10^{-9} cm/sec, and much higher velocities, particularly in the radial direction, would make this process dominant in the thermal regime of the earth. Mass transport in fact represents the earth's 'ultimate weapon' against overheating, since if other mechanisms, including radiative transfer, fail to provide sufficiently rapid escape of heat, the mantle either weakens to the point where flow can readily occur or it partially melts. Motions then automatically provide for the efficient removal of heat to the surface.

Acknowledgment. This article was written while I was visiting the University of Hawaii; I wish to thank the staff of the Hawaii Institute of Geophysics for their hospitality.

REFERENCES

Birch, F., Elasticity and constitution of the earth's interior, *J. Geophys. Res.*, *57*, 227–286, 1952.

Birch, F., and H. Clark, The thermal conductivity of rocks and its dependence upon temperature and composition, *Am. J. Sci.*, *238*, 529–588 and 613–635, 1940.

Clark, S. P., Radiative transfer in the earth's mantle, *Trans. Am. Geophys. Union*, *38*, 931–938, 1957a.

Clark, S. P., Absorption spectra of some silicates in the visible and near infrared, *Am. Mineral.*, *42*, 732–742, 1957b.

Clark, S. P., Optical and electrical properties of silicates, *Carnegie Inst. Wash. Year Book 58*, 187–189, 1959.

Herring, C., Role of low-energy phonons in thermal conduction, *Phys. Rev.*, *95*, 954–965, 1954.

Joffe, A. F., Heat transfer in semiconductors, *Can. J. Phys.*, *34*, 1342–1355, 1956.

Kingery, W. D., J. Francl, R. L. Coble, and T. Vasilos, Thermal conductivity, 10, Data for several pure oxide materials corrected to zero porosity, *J. Am. Ceram. Soc.*, *37*, 107–110, 1954.

Klemens P. G., Thermal conductivity and lattice vibrational modes, *Solid State Phys.*, *7*, 1–98, 1958.

Lawson, A. W., and J. C. Jamieson, Energy transfer in the earth's mantle, *J. Geol.*, *66*, 540–551, 1958.

McDonald, K. L., Penetration of the geomagnetic secular field through a mantle with variable conductivity, *J. Geophys. Res.*, *62*, 117–141, 1957.

Pomeranchuk, I., Thermal conductivity of dielectrics at temperatures higher than the Debye temperature, *J. Phys. USSR*, *4*, 259–268, 1941.

Magnetic Properties of Rocks

John Verhoogen

DEFINITIONS; DIAMAGNETISM; PARAMAGNETISM

The magnetization J of a rock can be written as the sum of two terms,

$$J = J_i + J_r$$

representing, respectively, the induced and remanent (residual) magnetization. J_r is zero in rocks containing only diamagnetic and paramagnetic minerals; in such rocks, J_i is proportional to the external magnetic field H that induces the magnetization. The ratio $\kappa = J_i/H$ is the susceptibility per unit volume. The specific susceptibility X is κ/δ, where δ is the density of the substance. The magnetization J (magnetic moment per unit volume) is measured in the same units as the magnetic field, oersteds or gammas; $1\ \gamma = 10^{-5}$ oe. The susceptibility κ is dimensionless.

All substances are diamagnetic, i.e., they have a negative susceptibility of the order of $10^{-6}/g$ or less, which arises from the distortion of the electron orbits in a magnetic field. In the earth's field (~ 0.5 oe), a body of rock with density $\delta = 3$ g/cm³ will acquire a diamagnetic magnetization $J_{di} \sim -1.5 \times 10^{-6}$; the strongest field that such a body could produce at the earth surface, $2\pi J_{di}$, would be about $1\ \gamma$ and is therefore negligible.

Substances that contain atoms with incomplete shells (iron transition group, rare earths) are paramagnetic, at least above their Curie or Néel temperature (see below). The paramagnetic susceptibility of a substance containing per unit volume N atoms (or ions) with moment μ is

$$\kappa_p = N\mu^2/3KT \qquad (1)$$

where K is Boltzmann's constant, and T is the absolute temperature. The atomic moment μ is of the order of a few Bohr magnetons (1 Bohr magneton = 0.927×10^{-20} ergs/oe). The paramagnetic susceptibility of common silicates minerals is proportional to their iron content [*Syono*, 1960]. Typically, the specific susceptibility of an olivine $(Fe_{0.1}Mg_{0.9})_2SiO_4$ is 1.8×10^{-5} [*Nagata et al.*, 1957]. Again, paramagnetism of rocks contributes only a few gammas to the magnetic field at the earth's surface, and is of little geophysical importance.

FERROMAGNETISM; ANTIFERROMAGNETISM; FERRIMAGNETISM

In paramagnetic substances, the magnetic moments of individual atoms are uncoupled; each atom behaves independently of the others. When the crystal structure is such that atoms can interact, e.g., by overlap of orbitals of neighboring atoms, coupling may occur and atomic moments behave collectively. Neighboring atoms usually couple with parallel or antiparallel moments, corresponding, respectively, to ferromagnetism or antiferromagnetism; more complicated arrangements (e.g., helicoidal) of the moments also occur. In both ferromagnetic and antiferromagnetic substances, there is a temperature (Curie temperature T_c, Néel temperature T_n) above which the coupling energy becomes less than the thermal energy KT and ceases to be effective; the substance then becomes paramagnetic. A ferromagnetic substance has, at any temperature below T_c, a 'spontaneous' or 'saturation' magnetization J_s that varies with T and that equals, at $T = 0$, the sum of all atomic moments. The spontaneous magnetization decreases with increasing temperature, first rather slowly, then more rapidly, and vanishes at $T = T_c$. The spontaneous magnetization of an antiferromagnetic substance is always zero.

'Parasitic' antiferromagnetism may occur in antiferromagnetic substances with low crystallographic symmetry, when atomic moments are not exactly antiparallel. The spontaneous magnetization is very small; for instance, in $\alpha\ Fe_2O_3$ (hematite) it is about 0.5/g, as against 350/g if all moments are parallel. This magnetization is perpendicular to the general alignment of the moments; it vanishes at the Néel temperature.

A ferrimagnetic substance is one in which magnetic atoms or ions occur on two distinct ferromagnetic sublattices coupled antiferromagnetically; the spontaneous magnetization is the difference between the spontaneous magnetizations of the two sublattices. Its variation with temperature can be quite complicated. In some substances, e.g., magnetite (Fe_3O_4), it is much like that of a ferromagnetic; in other substances, it may either increase or decrease in certain temperature intervals, and the sign of the spontaneous magnetization itself may reverse in certain temperature intervals, as it does below room temperature in some naturally occurring titanomagnetites (solid solutions of Fe_3O_4 and Fe_2TiO_4) [Schult, 1965].

The spontaneous magnetization must usually be measured in a strong field, because in weak or zero fields the substance demagnetizes itself in order to minimize its magnetostatic energy by forming 'domains' that are spontaneously magnetized regions that differ as to the orientation of their magnetization. In zero external field, the domain pattern would be such as to produce a zero net magnetization. In general, magnetocrystalline anisotropy restricts the possible direction of spontaneous magnetization in a domain to a few preferred crystallographic directions, e.g., [111] in magnetite.

MAGNETIC MINERALS

The main carriers of magnetism in rocks are iron oxides or sulfides. The common ones fall in the following groups:

1. Members of the α series range in composition from hematite (α-Fe_2O_3) to ilmenite ($FeTiO_3$); they have a corundum structure. Hematite has parasitic antiferromagnetism between the Morin transition (about 250°K) and the Curie, or Néel, point (948°K). Ilmenite is antiferromagnetic below 55°K. They form solid solutions with an upper critical mixing temperature of about 1250°K. Unmixing below that temperature is very slow. The magnetic properties of the solid solution are complicated and sensitive to heat treatment [Ishikawa and Akimoto, 1958; Ishikawa and Syono, 1963]. Phases that are ordered with respect to the distribution of Fe and Ti are ferrimagnetic; disordered phases are antiferromagnetic; it is not clear how the Fe^{2+}–Fe^{3+} ordering affects the magnetic properties. Some of the solid solutions are 'self-reversing' (see below). $MgTiO_3$ is often present (up to 15 or 20 mole %) in ilmenite-rich α phases; V_2O_3 is also common.

2. Members of the β series range in composition from magnetite (Fe_3O_4) to ulvospinel (Fe_2TiO_4) also called titanomagnetite; they have a spinel structure. Magnetite, the formula of which can be written $Fe^{3+}[Fe^{2+}Fe^{3+}]O_4$ to emphasize the difference between the two sublattices, is ferrimagnetic, with $J_s = 98/g$ at $T = 0$ and $T_c = 858°K$ (580°C). Ulvospinel $Fe^{2+}[Fe^{2+}Ti^{4+}]O_4$ is antiferromagnetic with $T_n = 120°K$. Solid solutions, with an upper critical temperature of about 900°K, are ferrimagnetic, with J_s and T_c varying almost linearly with composition; phases with more than 80 mole % of Fe_2TiO_4 are paramagnetic at room temperature. A large amount of substitution of other bivalent and trivalent cations can occur in the spinel structure; Mg, Al, Cr, Mn are common in natural β phases. The Curie temperature and spontaneous magnetization of substituted magnetites are usually less than those of magnetite itself.

3. Members of the γ series, metastable iron-titanium oxides with spinel (defect) structure, range in composition between those of the α and β series. The end-member γ-Fe_2O_3 (maghemite), with composition $Fe^{3+}[Fe_{5/3}^{3+}\square_{1/3}]O_4$, where $\square$ stands for a vacancy, is ferrimagnetic, with J_s corresponding to the uncompensated moment of $Fe^{3+}_{2/3}$. Its Curie point is not exactly known because, when heated, it reverts to the stable α form (hematite). γ phases with low Fe^{3+}/Fe^{2+} ratios break down to assemblages consisting of intergrowths of an α phase rich in Ti and a β phase poor in Ti, the exact compositions of which can be read from the chart prepared by *Buddington and Lindsley* [1964].

4. Pyrrhotite FeS_{1+x} has an NiAs structure. Again, magnetic properties are complicated and sensitive to composition and heat treatment. For x small (<0.1), the mineral is antiferromagnetic; ferrimagnetism appears in ordered specimens with $0.1 < x < 0.15$, approximately. The Curie temperature ($\sim 600°K$) depends on composition.

SUSCEPTIBILITY; INDUCED MAGNETIZATION

The induced magnetization of a ferro- or ferrimagnetic substance is not proportional to the applied field except at very low fields ('initial' susceptibility); it also depends on the past magnetic history of the specimen. The initial susceptibility of an unmagnetized specimen is controlled by the domain structure and the ease with which domain walls can move; this depends on the presence of impurities, vacant lattice sites, internal stresses, and dislocations. The susceptibility is also very sensitive to grain size. In grains with a diameter of 10^{-6} cm or less, thermal fluctuations may suffice, in a short time and in a small field, to produce an almost perfect alignment of the moments of each grain, so that the susceptibility is large ('superparamagnetism') and given by equation 1, in which μ is now the moment VJ_s of the grain with volume V. The relaxation time, however, increases exponentially with grain size, so that grains still small enough to consist of a single domain (diameter less than about 10^{-5} cm in Fe_3O_4) may have a lower susceptibility than smaller superparamagnetic grains. In larger, multidomain grains, the specific susceptibility increases with diameter; representative values for the specific susceptibility of magnetite might range from about 1×10^{-2} for single-domain grains to 0.1 for grains with diameters of 5×10^{-2} cm, and to about 1 for well-formed large crystals. The specific susceptibility of hematite is about 2×10^{-5} at room temperature. Susceptibility of large grains usually increases with increasing temperature (by a factor of 2 or so) before decreasing to its paramagnetic value at the Curie point.

The susceptibility κ_r of a rock containing a fraction p (by volume) of a magnetic substance is roughly proportional to p and to the susceptibility κ of the substance itself. The induced magnetization of the magnetic grains in the rock depends, however, on the effective field H_e acting inside the grains, which may differ from the applied field H_a because of demagnetizing fields. Neglecting all factors other than the shape of the grains, the demagnetizing field of a grain is $-NJ$, where N is the demagnetizing factor corresponding to the shape of the grain and the direction of magnetization (N is always less than 4π; it is nearly 4π for a flat disk magnetized along its shortest dimension and 0 for a long needle magnetized along its axis). Since $J_i = \kappa H_{eff}$, and $H_{eff} = H_a - NJ_i$, the apparent susceptibility κ_a of a grain is $\kappa_a = J_i/H_a = \kappa/(1 + N\kappa)$ and

$$\kappa_r = \frac{p\kappa}{1 + N\kappa} \quad (2)$$

It is found that the susceptibility of many igneous rocks can be represented by (2), with $\kappa \sim 0.4$ and $N \sim 3.5$ to 4.0 [*Nagata*, 1961]; however, deviations by a factor of 10 from the predicted value are common, reflecting mostly the effect of grain size on κ.

If the grains are not equidimensional and have a preferred orientation, the rock itself becomes magnetically anisotropic, and the susceptibility depends on the direction of the applied field. In igneous rocks and many sediments, the variation of κ_r with direction is usually much less than 10%; it may be greater in some metamorphic rocks.

Strictly speaking, the magnetization induced in a body of rock with susceptibility κ_r depends also on the shape of the body itself. This effect is rarely important, except in rocks in which κ_r is exceptionally large.

It is found theoretically and experimentally [*Grabovskii and Parkhomenko*, 1953; *Kalashnikov and Kapitsa*, 1952; *Kapitsa*, 1955; *Kern*, 1961; *Nagata*, 1966] that the susceptibility varies under uniaxial compression σ as

$$\kappa = \frac{\kappa_0}{1 + S\sigma}$$

where κ_0 is the susceptibility at zero stress and is of the order of 1×10^{-10} cm^2/dyne. S is proportional to the magnetostriction coefficient and inversely proportional to the square of the spontaneous magnetization. More complicated expressions for S, involving N and the magnetocrystalline anisotropy, have also been given [*Nagata*, 1966; *Stacey*, 1962].

The susceptibility of basalts usually falls in the range 1 to 8×10^{-3}; that of submarine flows tends to be smaller ($<1 \times 10^{-3}$) [*Ade-Hall*, 1964], possibly as a result of more rapid cooling and finer grain size. Granites and other acidic intrusives of low iron content have a susceptibility one-tenth as large (10^{-4} or less), although some coarse-grained granitic rocks may have a susceptibility of the order of 10^{-3}. The susceptibility of sediments is more commonly in the range 10^{-5} to 10^{-6}. Rocks with exceptional content of magnetite (ores) may have susceptibilities as high as 10^{-1} to 1.

REMANENT MAGNETIZATION

Many rocks, particularly igneous ones, carry a permanent or remanent magnetization J_r that is many times greater than that ($\kappa_r F$) induced by the earth's field F. The ratio $Q = J_r/\kappa_r F$ is known as Koenigsberger's ratio and is commonly between 1 and 10, and sometimes as large as 100 or more. It follows that magnetic anomalies are more likely to be caused by J_r than by J_i. The latter is parallel to the present earth's field; the former, except in very recent rocks, may have a direction differing by as much as 180° from that of the present earth's field. The subject of paleomagnetism is based on the study of J_r.

Two properties of remanence are of particular importance in rock magnetism and paleomagnetism. The first is its intensity; the second is its stability or coercivity, that is, its resistance to demagnetization by (1) time, (2) thermal treatment, and (3) treatment in an alternating field, the intensity of which is gradually reduced to zero (AC demagnetization). The separation of magnetizations acquired by a rock at different times, and the determination of the original direction of the earth's field when the rock formed, are possible only if later magnetizations are less stable than the original one. Stability of remanence generally bears an inverse relationship to susceptibility. Very small grains in the superparamagnetic state have high susceptibility but no remanence. Somewhat larger, but still single-domain grains have low susceptibility but high stability of remanence; large multi-domain grains have a larger susceptibility, but their remanence is less stable. Remanence can be acquired in a number of ways. The geophysically most important ones are as follows:

Isothermal Remanent Magnetization (IRM)

IRM is produced by application of an external field for a short time at constant temperature. IRM increases (but not linearly) with increasing applied field until saturation is reached; the saturated value of IRM, produced by fields of a few thousand oersteds, is typically of the order of one-half the spontaneous magnetization. Fields of 1 oe or less generally produce only a negligible amount of IRM (10^{-5} or less). IRM is therefore important only in rocks that have been exposed to local, intense fields, such as may be produced by lightning discharges. IRM in weak fields has relatively low stability and can generally be erased by an AC field of peak amplitude equal to the intensity of the field that produced it.

Viscous Remanent Magnetization (VRM)

The remanence acquired at constant temperature in weak fields is found to increase roughly proportionally to the logarithm of the time during which the field is applied. More exactly, it is proportional to the square of the logarithm of time. The time-dependent part of IRM is called VRM. It increases with increasing intensity of the applied field (in weak fields), and with increasing temperature of the sample during application of the field. Stability with respect to AC demagnetization increases in the same way, so that a VRM acquired over a long period of time and at elevated temperatures may be very stable [Briden, 1965; Sholpo and Pechnikov, 1965]. VRM decreases with increasing time since the field is removed at about the rate at which it is acquired.

Thermal Remanent Magnetization (TRM)

TRM is generated by cooling from the Curie point to room temperature in an applied field H_a. Partial TRM (PTRM) is the remanence acquired when the field is applied only during part of the cooling.

TRM is proportional to $\tanh aH_a$, where a is a constant for a given rock, and is therefore very nearly proportional to H_a if H_a is small, but departures from linearity can be seen in some lavas at fields as low as 1 or 2 oe. Saturation can be reached in fields of 50 oe or less. In weak fields, TRM is stronger than IRM acquired in the same field, but in strong fields, the two become very similar [Roquet, 1954]. In single crystals of α-Fe_2O_3 and other substances with parasitic antiferromagnetism and strong magnetocrystalline anisotropy, a field of only a few oersteds may induce a TRM of the same order as the spontaneous magnetization itself [*Syono et al.*, 1962]; in magnetite the same field would produce a TRM of the order of $10^{-3} J_s$.

Most TRM is usually acquired in a temperature interval of 100 or 150° below the Curie point. PTRM has the interesting property that the total TRM produced by cooling from T_c to room temperature T_0 turns out to be the sum of all PTRM's acquired in the intervals $T_c - T_1, T_1 - T_2, \cdots, T_n - T_0$. Conversely, reheating to temperature $T_i < T_c$ and subsequent cooling in zero field destroys only that part of the original TRM, or PTRM, that was acquired below T_i. This remarkable property of TRM [*Thellier*, 1951], which is found to hold in many (but not all) lavas, can sometimes be used to determine the intensity of the earth's field that produced their natural magnetization [*Thellier and Thellier*, 1959].

TRM in weak fields is much more stable with respect to thermal demagnetization than IRM; it is also more stable with respect to AC demagnetization. An AC field of 200 oe or more may be necessary to destroy 50% of the remanence acquired by some magnetite-bearing rocks in the earth's field, whereas fields as high as 1000 oe may hardly affect the TRM of hematite-bearing rocks. Stability decreases, however, with increasing grain size. It is this remarkable stability of TRM that permits separation in igneous rocks of the original TRM acquired at the time they cooled from any magnetization (e.g., VRM) acquired later [*Thellier and Rimbert*, 1954, 1955].

Experimentally produced TRM in rocks is found to depend on a number of factors (grain size, etc.) and may vary by a factor of 10 or more from one specimen to the next. It is roughly proportional to the coercivity, or inversely proportional to the susceptibility; for instance, large and well annealed grains with high susceptibility acquire a smaller TRM than smaller grains containing impurities or dislocations. The mechanism of acquisition of TRM is well understood for single-domain grains [*Néel*, 1949], but not for multi-domain grains. Thermal fluctuations may play a role, as also does the temperature dependence of the magnetocrystalline energy and magnetostriction coefficients [*Everitt*, 1961, 1962; *Stacey*, 1963; *Verhoogen*, 1959].

It may happen that a rock will acquire TRM directed at 180° from the field that produced it ('self-reversal'). As noted above, self-reversal is known to occur in some α phases [*Ishikawa and Akimoto*, 1958; *Ishikawa and Syono*, 1963; *Uyeda*, 1962]; it may perhaps also be produced in β and γ phases by ordering or partial oxidation [*Verhoogen*, 1962], or by magnetostatic interaction between two constituents with different Curie temperatures [*Uyeda*, 1962].

Chemical Remanent Magnetization (CRM)

A stable and strong remanence may be induced by a weak field applied at the time of nucleation and growth or recrystallization of a magnetic phase. The mechanism appears to be very similar to that which produces TRM in single-domain particles [*Kobayashi*, 1959]. Some of the remanence of red sediments may be CRM acquired by dehydration and recrystallization of geothite to hematite, or of lepidocrocite to maghemite. Similarly, some of the remanence of igneous rocks may be CRM produced by low-temperature oxidation.

Detrital Remanent Magnetization

Grains of magnetic minerals that carry a remanence (e.g., TRM) acquired earlier may become oriented by the earth's field as they settle in quiet water and become imbedded in sediments. Much of the remanence of sediments may be acquired in this way, although it has been shown [*Collinson*, 1966] that in some red beds much of the remanence resides in the red pigment and is presumably CRM.

Piezo Remanent Magnetization (PRM)

A pre-existing remanence is decreased by uniaxial compression in exactly the same way as the susceptibility [*Kern*, 1961]. It has also been reported [*Nagata and Kiroshita*, 1965] that an additional component of isothermal remanence can be acquired under stress if the stress is released while the field is still applied. Theoretically, a nonhydrostatic stress of sufficient magnitude should, by magnetostriction, alter the direction of magnetization, but there is no clear evidence in rocks of an irreversible effect by which the remanence would not be parallel to the field that produced it. The effect would be difficult to separate from that of the anisotropy (preferred orientation of grains) that is likely to occur in deformed rocks.

GENERAL COMMENT

Nagata [1961] gives an extensive treatment of the magnetic properties of rocks; a good summary appears in *Irving* [1964].

The interpretation of magnetic anomalies, either on a small scale (geophysical exploration) or on a large scale (structure of the crust and upper mantle) is complicated by several factors:

1. The magnetization of a rock usually has two components, induced and remanent, the latter being commonly several times greater than the former, particularly in strongly magnetized rocks.

2. The magnitude of both induced and remanent magnetization depends on the mineralogical composition of the rocks and is very sensitive to grain size. As a general guide line, one can expect a magnetization of the order of 10^{-5} to 10^{-6} or less in sediments, 10^{-5} to 10^{-4} in granitic rocks and gneisses, and 5×10^{-4} to 5×10^{-2} in basalts; in any case, it is generally impossible to predict its value within a factor of 10.

3. The direction of the induced component is, of course, that of the present earth's field; that of the remanent field may, and commonly does, depart from that direction by a large angle, up to 180°. The direction of remanence will depend on the rock's fabric (anisotropy), and on the time at which it acquired various components of its remance (TRM, CRM, VRM, etc.). Because of these factors, and also because of past displacements of continents, and of intracontinental deformation, it is still impossible to predict with any accuracy what the direction of magnetization of a given rock may be.

Anomalies of the Continental Crust

Magnetic anomalies in continental areas, including those with wavelengths of several hundred kilometers [*Alldredge and Van Voorhis*, 1961; *Alldredge et al.*, 1963; *Schwahn*, 1966], can almost certainly be accounted for by the magnetic properties of the continental crust down to a depth of 30 km or so, perhaps slightly more in shield areas with low heat flow. As the Curie temperature of most rocks is less than 600°C, it is probable that the upper mantle in continental regions is purely paramagnetic.

More information on magnetic properties of metamorphic rocks, particularly amphibolites, is desirable.

Oceanic Crust

Very little is known of the composition of the oceanic crust, and correspondingly little can be said of its magnetic properties. The remarkable linear magnetic anomalies [*Vine*, 1966] seem to require a remanent magnetization of the order of 5×10^{-3} [*Pitman and Heirtzler*, 1966], which is close to the median for submarine basalts [*Ade-Hall*, 1964]. Not much is known of the magnetic properties of serpentinites which may form part of the oceanic crust. Serpentinites formed by hydration of dunites and peridotites usually contain some fine-grained magnetite formed by breakdown of iron-bearing olivine; their remanence, however, is small and unstable. Oceanic sediments are usually weakly magnetized (10^{-6}) [*Opdyke et al.*, 1966].

Upper Mantle in Oceanic Areas

Again, lack of knowledge of the petrology of the upper mantle precludes an extended discussion of its magnetic properties. Eclogites are usually essentially paramagnetic, as most of their iron resides in pyroxenes and garnets. The magnetic constituent of peridotites is commonly picotite $(FeMg)(AlFeCo)_2O_4$, with dominant Fe and Al, which has a Curie temperature slightly lower than that of magnetite; it is thus unlikely that the upper mantle would be magnetic much below a depth of the order of 30 km, in regions of low or average heat flow, and a shallower depth in regions of high-heat flow.

REFERENCES

Ade-Hall, J. M., The magnetic properties of some submarine oceanic lavas, *Geophys. J.*, 9, 85–91, 1964.

Alldredge, L. R., and G. D. Van Voorhis, Depth to sources of magnetic anomalies, *J. Geophys. Res.*, 66, 3793–3800, 1961.

Alldredge, L. R., G. D. Van Voorhis, and T. M. Davis, A magnetic profile around the world, *J. Geophys. Res.*, 68, 3679–3692, 1963.

Briden, J. C., Ancient secondary magnetizations in rocks, *J. Geophys. Res.*, 70, 5205–5221, 1965.

Buddington, A. F., and D. H. Lindsley, Iron-titanium oxide minerals and their synthetic equivalents, *J. Petrol.*, 5, 310–357, 1964.

Collinson, D. W., Carrier of remanent magnetization in certain red sandstones, *Nature*, 210, 516–517, 1966.

Everitt, C. W. F., Thermoremanent magnetization, 1, Experiments on single domain grains, *Phil Mag., 6,* 713-706, 1961.

Everitt, C. W. F., Thermoremanent magnetization, 2, Experiments on multidomain grains, and 3, Theory of multidomain grains, *Phil. Mag., 7,* 583-597 and 599-616, 1962.

Grabovskii, M. A., and E. I. Parkhomenko, On changes in the magnetic properties of magnetite under high compressive pressures, *Izv. Akad. Nauk SSSR, Ser. Geofiz., 5,* 405-417, 1953.

Irving, E., *Paleomagnetism,* John Wiley & Sons, New York, 399 pp., 1964.

Ishikawa, Y., and S. Akimoto, Magnetic property and crystal chemistry of ilmenite ($FeTiO_3$) and hematite (αFe_2O_3) system, 2, Magnetic property, *J. Phys. Soc. Japan, 13,* 1298-1310, 1958.

Ishikawa, Y., and Y. Syono, Order-disorder transformation and reverse thermo-remanent magnetism in the $FeTiO_3$-Fe_2O_3 system, *J. Phys. Chem. Solids, 24,* 517-528, 1963.

Kalashnikov, A. G., and S. P. Kapitsa, Magnetic susceptibility of rocks under elastic stresses, *Dokl. Akad. Nauk SSSR, 86,* 521-523, 1952.

Kapitsa, S. P., Magnetic properties of eruptive rocks exposed to mechanical stresses, *Izv. Akad. Nauk SSSR, Ser. Geofiz., 6,* 489-504, 1955.

Kern, J. W., The effect of stress on the susceptibility and magnetization of a partially magnetized multidomain system, *J. Geophys. Res., 66,* 3807-3816, 1961.

Kobayashi, K., Chemical remanent magnetization of ferromagnetic minerals and its application to rock magnetism, *J. Geomag. Geoelec., 10,* 99-117, 1959.

Nagata, T., *Rock Magnetism,* revised edition, Maruzen, Tokyo, 339 pp., 1961,

Nagata, T., Magnetic susceptibility of compressed rocks, *J. Geomag. Geoelec., 18,* 73-80, 1966.

Nagata, T., and H. Kiroshita, Studies on piezomagnetization, 1, Magnetization of titaniferous magnetite under uniaxial compression, *J. Geomag. Geoelec., 17,* 121-135, 1965.

Nagata, T., T. Yukutake, and S. Uyeda, On magnetic susceptibility of olivines, *J. Geomag. Geoelec., 9,* 51-56, 1957.

Néel, L., Théorie du trainage magnétique des ferromagnétiques en grains fins avec applications aux terres cuites, *Ann. Geophys., 5,* 99-136, 1949.

Opdyke, N. D., B. Glass, J. D. Hays, and J. Foster, Paleomagnetic study of Antarctic deep-sea cores, *Science, 154,* 349-357, 1966.

Pitman, W. C., and J. R. Heirtzler, Magnetic anomalies over the Pacific-Antarctic ridge, *Science, 154,* 1164-1171, 1966.

Roquet, J., Sur les rémanences magnétiques des oxydes de fer et leur intérêt en géomagnétisme: 1 and 2, *Ann. Geophys., 10,* 226-247 and 282-325, 1954.

Schult, A., Über die Umkehr der remanenten Magnetisierung von Titanomagnetiten in Basalten, *Beitr. Z. Mineral. Petrog., 11,* 196-216, 1965.

Schwahn, W., Spatial spectrum of the main part of the geomagnetic field in the northern part of Eurasia, *Geomag. Aeron.* (English edition), *6,* 83-89, 1966.

Sholpo, L. Ye., and V. S. Pechnikov, *Nastoyashcheye i Proshloye Maynitnogo Polya Zemli,* pp. 162-167, Akad. Nauk SSSR, Inst. Fiz. Zemli, Moscow, 1965.

Stacey, F. D., Theory of the magnetic susceptibility of stressed rock, *Phil. Mag., 7,* 551, 1962.

Stacey, F. D., The physical theory of rock magnetism, *Advan. Phys., 12,* 45-133, 1963.

Syono, Y., Magnetic susceptibility of some rock forming silicate minerals such as amphiboles, biotites, cordierites and garnets, *J. Geomag. Geoelec., 11,* 85-93, 1960.

Syono, Y., S. Akimoto, and T. Nagata, Remanent magnetization of ferromagnetic single crystals, *J. Geomag. Geoelec., 14,* 113-124, 1962.

Thellier, E., Propriétés magnétiques des terres cuites et des roches, *J. Phys. Radium, 12,* 205-218, 1951.

Thellier, E., and F. Rimbert, Sur l'analyse d'aimantations fossiles par action de champs magnétiques alternatifs, *Compt. Rend., 239,* 1399, 1954.

Thellier, E., and F. Rimbert, Sur l'analyse d'aimantations fossiles par action de champs magnétiques alternatifs, *Comp Rend., 240,* 1404, 1955.

Thellier, E. and O. Thellier, Sur l'intensité du champ magnétique terrestre dans le passé historique et géologique, *Ann. Geophys., 15,* 285-376, 1959.

Uyeda, S., Thermoremanent magnetism and reverse thermoremanent magnetism, in *Proceedings Benedum Symposium on Earth Magnetism* edited by T. Nagata, University of Pittsburgh Press, Pittsburgh, 1962.

Verhoogen, J., The origin of thermoremanent magnetization, *J. Geophys. Res., 64,* 2441-2449, 1959.

Verhoogen, J., Oxidation of iron-titanium oxides in igneous rocks, *J. Geol., 70,* 168-181, 1962.

Vine, F. J., Spreading of the ocean floor: New evidence, *Science, 154,* 1405-1415, 1966.

Phase Equilibrium Studies Relevant to Upper Mantle Petrology

M. J. O'Hara

STUDIES have been made in three related fields:

1. Polymorphism and melting of phases likely to form a major part of the upper mantle.
2. Changes of subsolidus phase assemblage in complex compositions, mineral facies in basic and ultrabasic rocks.
3. Crystal-liquid equilibria in complex systems, composition and possible evolution of basic magmas produced by partial melting of the mantle.

Accuracy of pressure and temperature measurement in the piston-cylinder apparatus that has been used for most of the studies referred to below will not be discussed in detail here. There are doubts as to whether mean stress is equivalent to hydrostatic stress, the friction correction to be applied, pressure effects on thermocouple emf, thermocouple contamination, and the influence of time on the temperatures recorded for particular equilibria. Very little weight should be attached to the absolute values of pressure and temperature recorded in the literature. Where temperatures have been estimated from power consumption, the temperature values are subject to a large uncertainty. In the following account references are given to the most recent work in the field only.

POLYMORPHISM AND MELTING RELATIONSHIPS IN SIMPLE PHASES

Orthosilicate

The melting equilibrium for forsterite has been determined to 50 kb [*Davis and England*, 1964]. Forsterite melts congruently. An inversion to spinel structure has been demonstrated at 150–170 kb [*Akimoto and Ida*, 1966; *Ringwood and Major*, 1966a], while the inversion of members of the forsterite-fayalite series at lower pressures has been demonstrated by *Akimoto and Fujisawa* [1966] and *Ringwood and Major* [1966a]. *Lindsley* [1966a] has determined the melting of fayalite to 40 kb and discussed consequences of the olivine-spinel inversion for MgO/FeO fractionation in the mantle.

Ca-Free Metasilicate

Melting and polymorphism in $MgSiO_3$ to 50 kb have been studied [*Boyd and England*, 1965]. Enstatite melts congruently at pressures above a few kilobars. *Lindsley* [1965] has studied relations in $FeSiO_3$ and intermediate Mg-$FeSiO_3$ solid solutions to 50 kb. *Ringwood and Major* [1966b] report breakdown of Of_{100}-Of_{25}-En_{75} pyroxenes to spinel + stishovite at pressures near 200 kb, and suggest a pyroxene-ilmenite structure inversion in $MgSiO_3$ at still higher pressures. The stabilities of Al_2O_3-bearing, enstatite-rich pyroxenes have been studied by *Boyd and England* [1964] and *MacGregor and Ringwood* [1964].

Ca-Bearing Metasilicate

Melting of diopside to 50 kb has been determined [*Boyd and England*, 1963] and is congruent. No high pressure inversions have yet been reported. The diopside-enstatite system has been studied at 30 kb [*Davis and Boyd*, 1966]. Jadeite stability has been reinvestigated by *Newton and Smith* [1967], its melting by *Bell and Roseboom* [1965], and jadeite-diopside solid solutions by *Kushiro* [1965a] and *Bell and Davis* [1966]. Acmite stability and melting has been studied by *Gilbert* [1966] to 50 kb, and data for the jadeite-acmite system are given by *Newton and Smith* [1967]. Stability of $CaAl_2SiO_6$ [*Hays*, 1966] and of diopside-$CaAl_2SiO_6$ solid solutions [*Kushiro*, 1965a] have been investigated to 50 kb, and extensive solid solution along the diopside-pyrope, diopside-Al_2O_3, diopside-$MgAl_2SiO_6$, and diopside-kyanite joins are reported by *O'Hara and Yoder*

[1967], while more than 5% Mg_2SiO_4 solubility in diopside at 20 kb has been observed [*Kushiro*, 1964].

Aluminous diopsides and enstatites are stable at 10–40 kb but crystallize as Al_2O_3-deficient garnet structures at about 100 kb [*Ringwood*, 1967].

Plagioclase Feldspars

Neither albite nor anorthite is stable at high pressures [*Newton and Smith*, 1967; *Hays*, 1966], and the melting relations of plagioclases becomes incongruent and complex above 10 kb [*Lindsley*, 1966b].

Garnets

Stabilities of pyrope, almandine, grossular and of the pyrope-almandine and pyrope-grossular garnets have been studied by *Boyd and England* [1962], *Yoder and Chinner* [1960], *Chinner et al.* [1960], and *Hays* [1966]. *Ringwood* [1967] reports Al_2O_3-deficient garnet structures, and *Clark et al.* [1962] have suggested a possible inversion of garnet to corundum structure at very high pressure. The stability of intermediate grossular-pyrope garnets relative to the alternative diopside-corundum or pyroxene solid solution assemblages has been discussed by *O'Hara and Mercy* [1966].

Spinels

Little work has been done on the stability of the (MgFe) (Al, Cr, Fe)$_2O_4$ spinels at high pressures. No structural inversions are to be expected at upper mantle pressures, but progressive increase in the solubility of silicate spinels Si(MgFe)$_2O_4$, culminating in very extensive ranges of solid solution above 200 kb, is to be expected.

CHANGES OF SUBSOLIDUS PHASE ASSEMBLAGES

Changes of subsolidus phase assemblages (i.e., changes of mineral facies) in ultrabasic and basic rocks are primarily connected with the instability at high pressures of the plagioclase-pyroxene and plagioclase-olivine assemblages characteristic of low pressures, and their replacement by garnet-olivine and diopside-kyanite assemblages.

Experimental data on assemblages in peridotites as a function of pressure and temperature have been reviewed by *MacGregor* [1967] and *O'Hara* [1967a, b]. Stabilization of merwinite-diopside-forsterite assemblages at moderate pressures in silica-poor ultrabasics, replacing monticellite- and melilite-bearing assemblages, has been reported by *Kushiro and Yoder* [1964].

Kushiro and Yoder [1966] have studied the breakdown of anorthite-enstatite assemblages to pyroxene + garnet + quartz, and *Kushiro* [1965a] the breakdown of anorthite-diopside assemblages to clinopyroxene-garnet-quartz, two of the essential transformations in the transition from pyroxene hornfels through granulite to eclogite mineral facies. *Green and Ringwood* [1967a] have studied the same transition in natural basic rock compositions.

LIQUIDUS STUDIES

Yoder and Tilley [1962], *Davis and Schairer* [1965], *Kushiro* [1965b], *O'Hara* [1965], *Green and Ringwood* [1967b], and *O'Hara and Yoder* [1967] have studied the effect of high pressure on the composition and evolution of the partial melting products of peridotites in natural and synthetic systems.

RESULTS OF EXPERIMENTAL STUDIES

Peridotites composed of magnesian olivines, clino- and orthopyroxenes, and an alumina-rich phase will be the stable ultrabasic rocks within the upper mantle. The alumina-rich phase will be plagioclase at low pressures, spinel at intermediate pressures, and garnet at high pressures, but at high temperatures and intermediate-high pressures greatly increased alumina solubility in pyroxene will reduce the amount of the Al_2O_3-rich phase in any given bulk composition and may even eliminate it entirely.

Eclogite is the high pressure equivalent of basaltic or gabbroic compositions, but there is a broad transition zone of pressure within which granulite facies assemblages are developed. The pressure at which eclogite facies mineralogy is developed depends critically upon the degree of silica saturation.

Both eclogite and garnet-peridotite may be stable in the deeper parts of the continental crust, provided that temperatures are low.

Increasing pressure progressively reduces the primary phase volume of olivine crystallization from liquids in basalt-peridotite systems, to 40 kb at least. Partial melting products of peridotite become progressively richer in normative olivine (i.e. picritic) and, therefore, less like extruded magmas with increasing pressure. Common extrusive basalt magmas are not primary, but have fractionated and reacted with their wall rock en route to the surface.

The primary phase volume of orthopyroxene, which is restricted to quartz-normative compositions at low pressure, extends first into the olivine-normative volume, and then through the plane of critical undersaturation in silica into nepheline normative compositions at intermediate pressures. At high pressures this primary phase volume contracts and no longer extends into nepheline normative volumes. Consequently, the partial melting products of olivine-orthopyroxene lherzolites must be quartz-normative at low pressure, olivine tholeiitic at moderate pressure, nepheline-normative picritic at intermediate to high pressure, and hypersthene-normative picritic at high pressure. Olivine-orthopyroxene nodules in alkali basalts are, therefore, probably derived from an intermediate pressure zone, but no consensus exists as to whether they are cognate or accidental xenoliths. The eclogite nodules in kimberlite could be crystal accumulates from a picritic magma formed by partial melting of a garnet-peridotite mantle at high pressure. The residual liquid of such a process would have geochemical affinities to kimberlite and alnöite.

REFERENCES

Akimoto, S., and H. Fujisawa, Olivine-spinel transition in the system Mg_2SiO_4-Fe_2SiO_4 at 800°C, *Earth Planetary Sci. Letters, 1,* 237–240, 1966.

Akimoto, S., and Y. Ida, High-pressure synthesis of Mg_2SiO_4 spinel, *Earth Planetary Sci. Letters, 1,* 358–359, 1966.

Bell, P. M., and B. T. C. Davis, Investigation of a solvus in the system jadeite-diopside, *Carnegie Inst. Wash. Yr. Book, 65,* 239–241, 1966.

Bell, P. M., and E. H. Roseboom, Phase diagram for the system nepheline-quartz, *Carnegie Inst. Wash. Yr. Book, 64,* 139–141, 1965.

Boyd, F. R., and J. L. England, Mantle Minerals, *Carnegie Inst. Wash. Yr. Book, 61,* 107–112, 1962.

Boyd, F. R., and J. L. England, Effect of pressure on the melting of diopside, $CaMgSi_2O_6$, and albite, $NaAlSi_3O_8$, in the range up to 50 kb, *J. Geophys. Res., 68,* 311–323, 1963.

Boyd, F. R., and J. L. England, The system enstatite pyrope, *Carnegie Inst. Wash. Yr. Book, 63,* 157–161, 1964.

Boyd, F. R., and J. L. England, The rhombic enstatite- clinoenstatite inversion, *Carnegie Inst. Wash. Yr. Book, 64,* 117–120, 1965.

Chinner, G. A., F. R. Boyd, and J. L. England, Physical properties of garnet solid solutions, *Carnegie Inst. Wash. Yr. Book, 59,* 76–78, 1960.

Clark, S. P., J. F. Schairer, and J. De Neufville, Phase relations in the system $CaMgSi_2O_6$-$CaAl_2Si_2O_6$-SiO_2 at low and high pressure, *Carnegie Inst. Wash. Yr. Book, 61,* 59–68, 1962.

Davis, B. T. C., and F. R. Boyd, The join $Mg_2Si_2O_6$-$CaMg$ Si_2O_6 at 30 kilobars pressure and its application to pryroxenes from kimberlites, *J. Geophys. Res., 71,* 3567–3576, 1966.

Davis, B. T. C., and J. L. England, The melting of forsterite up to 50 kilobars, *J. Geophys. Res., 69,* 1113–1116, 1964.

Davis, B. T. C., and J. F. Schairer, Melting relations in the join diopside-forsterite-pyrope at 40 kilobars and at one atmosphere, *Carnegie Inst. Wash. Yr. Book, 64,* 123–126, 1965.

Gilbert, M. C., Acmite, *Carnegie Inst. Wash. Yr. Book, 65,* 241–244, 1966.

Green, D. H., and A. E. Ringwood, An experimental investigation of the gabbro to eclogite transformation and its petrological applications, *Geochim. Cosmochim. Acta, 31,* 767–834, 1967a.

Green, D. H., and A. E. Ringwood, The genesis of basaltic magmas, *Contrib. Mineral. Petrol., 15,* 103–190, 1967b.

Hays, J. F., Lime-alumina-silica, *Carnegie Inst. Wash. Yr. Book, 65,* 234–241, 1966.

Kushiro, I., The system diopside-forsterite-enstatite at 20 kilobars, *Carnegie Inst. Wash. Yr. Book, 63,* 101–708, 1964.

Kushiro, I., Clinopyroxene solid solutions at high pressures, *Carnegie Inst. Wash. Yr. Book, 64,* 112–117, 1965a.

Kushiro, I., The liquidus relations in the systems forsterite-$CaAl_2SiO_6$-silica and forsterite-nepheline-silica at high pressures, *Carnegie Inst. Wash. Yr. Book, 64,* 103–109, 1965b.

Kushiro, I., and H. S. Yoder, Breakdown of monticellite and akermanite at high pressures, *Carnegie Inst. Wash. Yr. Book, 63,* 81–83, 1964.

Kushiro, I., and H. S. Yoder, Anorthite-forsterite and anorthite-enstatite reactions and their bearing on the basalt-eclogite transformation, *J. Petrol., 7,* 337–362, 1966.

Lindsley, D. H., Ferrosilite, *Carnegie Inst. Wash. Yr. Book, 64,* 148–150, 1965.

Lindsley, D. H., Pressure-temperature relations in the system FeO-SiO_2, *Carnegie Inst. Wash. Yr. Book, 65,* 226–230, 1966a.

Lindsley, D. H., Melting relations of plagioclase at high pressures, *Carnegie Inst. Wash. Yr. Book, 65,* 204, 1966b.

MacGregor, I. D., Mineralogy of model mantle compositions, in *Ultramafic and Related Rocks,* edited by P. J. Wyllie, pp. 382–393, J. Wiley & Sons, New York, 1967.

MacGregor, I. D., and A. E. Ringwood, The natural system enstatite-pyrope, *Carnegie Inst. Wash. Yr. Book, 63,* 161–163, 1964.

Newton, R. C., and J. V. Smith, Investigations concerning the breakdown of albite at depth in the earth, *J. Geol., 75,* 268–286, 1967.

O'Hara, M. J., Primary magmas and the origin of basalts, *Scott. J. Geol.*, *1*, 19–40, 1965.

O'Hara, M. J., Mineral facies in ultrabasic rocks, in *Ultramafic and Related Rocks*, edited by P. J. Wyllie, pp. 7–18, J. Wiley & Sons, New York, 1967a.

O'Hara, M. J., Mineral parageneses in ultrabasic rocks, in *Ultramafic and Related Rocks*, edited by P. J. Wyllie, pp. 393–403, J. Wiley & Sons, New York, 1967b.

O'Hara, M. J., and E. L. P. Mercy, Exceptionally calcic pyralspite from South African kyanite eclogite, *Nature*, *212*, 68–69, 1966.

O'Hara, M. J., and H. S. Yoder, Formation and fractionation of basic magmas at high pressures, *Scott J. Geol.*, *3*, 67–117, 1967.

Ringwood, A. E., The pyroxene-garnet transformation in the earth's mantle, *Earth Planetary Sci. Letters*, *2*, 255–263, 1967.

Ringwood, A. E., and A. Major, Synthesis of Mg_2SiO_4–Fe_2SiO_4 spinel solid solutions, *Earth Planetary Sci. Letters*, *1*, 241–245, 1966a.

Ringwood, A. E., and A. Major, High-pressure transformations in pyroxenes, *Earth Planetary Sci. Letters*, *1*, 351–357, 1966b.

Yoder, H. S., and G. A. Chinner, Almandite-pyrope-water system at 10,000 bars, *Carnegie Inst. Wash. Yr. Book*, *59*, 81–84, 1960.

Yoder, H. S., and C. E. Tilley, Origin of basalt: An experimental study of natural and synthetic rock systems, *J. Petrol.* *3*, 342–532, 1962.

Phase Transitions

A. E. Ringwood and D. H. Green

DURING the past few years, the development of experimental static high-pressure, high-temperature techniques and their application to the earth sciences have resulted in some major advances in an understanding of the constitution of the earth's interior. It is now possible to duplicate in the laboratory the range of pressure-temperature conditions existing in the earth down to a depth of about 700 km. Thus the stability fields of mineral assemblages and of individual minerals throughout this region can be determined and the properties of new mineral phases can be measured, leading to an explanation of such primary geophysical properties as the seismic velocity and density distributions. One of the most important results of this work has been an understanding of the key role of pressure in determining the stabilities and structures of minerals.

In the present review, we will interpret the term 'phase transitions' in a broad sense and consider not only transitions of individual minerals, but also more complex transformations exhibited by assemblages of minerals. We propose to discuss the following topics: (*a*) the gabbro-eclogite transformation; (*b*) the stability fields of mineral assemblages in the outer mantle (i.e., to a depth of about 400 km); and (*c*) phase transitions in the transition zone (400–900 km).

THE GABBRO-ECLOGITE TRANSFORMATION

It is well known that rocks of basaltic chemical composition may exhibit several distinct mineral assemblages. Two such assemblages are those of gabbro (plagioclase + pyroxene) and eclogite (garnet + omphacitic pyroxene). One of the first workers to appreciate the relationships between these rock types was *Fermor* [1913, 1914], who pointed out that eclogite ($\rho \sim 3.5$ g/cm³) must be a high-pressure form of gabbro or basalt ($\rho \sim 3.0$ g/cm³). This led him to propose that the earth's outer mantle was composed of eclogite, and the transformation between gabbroic crustal rocks and the eclogitic mantle (infracrust) would have some important tectonic implications. *Holmes* [1926a, b; 1927] showed that the seismic velocity of eclogite was similar to that observed in the upper mantle, and argued that the crust-mantle boundary, now defined as the Mohorovicic (M) discontinuity, was caused by a transformation from gabbro to eclogite. Holmes also emphasized the tectonic consequences of this model: changes in temperature at the crust-mantle boundary would cause transformation of gabbro to eclogite or vice versa, resulting in crustal uplift or depression. This hypothesis has been widely discussed by many authors during the last 15 years. It has provided a considerable incentive for the experimental determination of gabbro-eclogic stability relationships.

Ringwood and Green [1964, 1966] and *Green and Ringwood* [1967a] have recently carried out such an investigation and have discussed the geophysical implications of the results. These authors chose several representative basalts (Table 1), subjected them to a wide range of closely spaced P, T conditions, and investigated the nature of the mineral phases produced and their stability fields. This work has led to a detailed understanding of many important aspects of the gabbro-eclogite transformation, particularly relating to the transitional mineral assemblages that intervene between the gabbro and eclogite stability fields. Some of the results are shown in Figure 1, in which we compare in simplified form the phase assemblages found in the typical basalts studied at 1100°C as a function of pressure. Although the pressures required for appearance and disappearance of phases vary between individual basalts, there is an important qualitative resemblance between the sequence of phase assemblages displayed by all rocks with increasing pressure. For each basalt there are clearly three principal mineral stability fields corresponding

TABLE 1. Chemical Compositions (%) of Basaltic Glasses Used in Experimental Work

Component	High Alumina Basalt	Quartz Tholeiite	Alkali-Poor Quartz Tholeiite	Alkali Olivine Basalt	Oxidized Alkali Olivine Basalt	Alkali-Poor Olivine Tholeiite
SiO_2	49.90	52.16	49.88	45.39	45.39	46.23
TiO_2	1.29	1.86	2.14	2.52	2.52	0.07
Al_2O_3	16.97	14.60	13.89	14.69	14.69	14.52
Fe_2O_3	1.52	2.46	2.84	1.87	9.82	0.54
FeO	7.60	8.39	9.65	12.42	4.18	11.80
MnO	0.16	0.14	0.16	0.18	0.18	0.30
MgO	8.21	7.36	8.48	10.37	10.37	12.45
CaO	11.41	9.44	10.82	9.14	9.14	13.00
Na_2O	2.78	2.68	1.84	2.62	2.62	0.81
K_2O	0.16	0.73	0.08	0.78	0.78	0.01
P_2O_5		0.18	0.22	0.02	0.02	0.03
Cr_2O_3						0.24
	100.00	100.00	100.00	100.00	99.71	100.00
100 Mg/Mg + Fe, mole	68	61.0	61.0	60	82	66
Norms						
Qz		2.5	2.8			
Or	1.0	4.8	0.5	4.5	4.5	
Ab	23.5	22.1	15.4	18.0	22.0	6.8
Ne				2.2		
An	33.4	25.5	29.3	26.2	26.2	35.9
Di	18.9	17.1	19.6	15.7	15.0	23.5
Hy	9.4	20.6	23.7		6.8	9.6
Ol	9.3			25.8	8.4	23.0
Ilm	2.5	3.6	4.2	4.8	4.8	0.2
Mt	2.2	3.6	4.2	2.9	6.7	0.7
Ap		0.4	0.5			
Haem					5.2	
Chrom						0.3
Normative Olivine	Fo_{62}			Fo_{66}	Fo_{100}	Fo_{65}
Normative	An_{59}	An_{52}	An_{64}	An_{58}	An_{54}	An_{83}

Analyst: A. J. Easton, Dept. of Geophysics and Geochemistry, Australian National University, Canberra

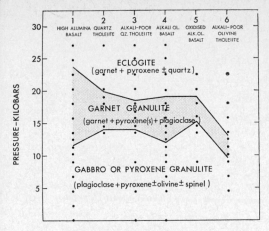

Fig. 1. Principal mineral assemblages of several basalts (Table 1) as a function of pressure at 1100°C. Solid circles denote experimental runs.

closely with naturally observed mineral assemblages. The low-pressure assemblage is that of gabbro or pyroxene granulite and is characterized by the presence of pyroxene + plagioclase ± olivine ± quartz ± spinel according to the particular bulk chemistry. In each basaltic composition, as pressure increases, a point is reached at which garnet appears in the mineral assemblage. With further increase in pressure, the proportion of garnet steadily increases, while the proportion of plagioclase correspondingly decreases. Thus we have here a field of coexisting garnet, aluminous pyroxene, and plagioclase. At the highest pressures, sodic plagioclase breaks down to form jadeite, which enters into solid solution in the pyroxenes, while the aluminous components (Tschermak's molecules) of the pyroxene are exsolved to form more garnet. These transformations mark the beginning of the eclogite mineral assemblage, characterized by the coexistence of pyrope-rich garnet, omphacite ± quartz.

Thus we see that in all the basalt compositions studied, the transformation from gabbro or pyroxene granulite to eclogite proceeds through an intermediate mineral assemblage characterized by coexisting garnet, pyroxene, and plagioclase. This possesses an extensive stability field varying from 3.5 to 12 kb in width, and is identical with the natural garnet-clinopyroxene-granulite subfacies recognizable in some metamorphic terranes.

The results plotted in Figure 1 show that rather modest changes in chemical composition cause large changes in the pressures and width of the gabbro-eclogite transformation. Thus, the pressures at which garnet first appears vary between 9.6 and 15.2 kb, while the pressure required to cause the final disappearance of plagioclase varies between 13.0 and 23.3 kb. The reasons for the variation are readily explicable in terms of the chemical and mineralogical equilibria involved in the transformation and have been discussed in detail by *Green and Ringwood* [1967a]. The pressures required for the incoming of garnet are smaller in undersaturated rocks than in oversaturated rocks (Figure 1, Table 1), while the pressure required for the final disappearance of plagioclase is decreased in basaltic compositions that are poorer than usual in soda. Changes in FeO/MgO ratio have an important influence over the pressure required for the incoming of garnet, these pressures being inversely proportional to the FeO/MgO content of the rock.

Within the garnet granulite transition zone, as the pressure increases, garnet regularly increases in abundance while plagioclase regularly decreases. To a first approximation the changes in mineralogy across the garnet granulite zone occur at a uniform rate and, accordingly, the corresponding changes in seismic velocity and density will also change regularly throughout the transition zone. This observation, coupled with the large width of the transitional intervals, renders it almost impossible to explain the Mohorovicic discontinuity in terms of a gabbro-eclogite transformation. (The effective transition widths in the earth are considerably expanded by the tendency of the earth's temperature-depth curves (geotherms) to intersect the garnet granulite field boundaries at low angles.)

Since most of the experimental results have been obtained at 1100°C, it is desirable to determine the temperature gradient of the transition (dP/dT) if maximum use is to be made of the information at lower temperatures. The results of experiments on a quartz tholeiite over a range of temperatures are given in Figure 2. These results establish a temperature gradient for the transformation that permits extrapolation to lower temperatures. The slopes of the boundaries for the incoming of garnet and the disappearance of plagioclase tend to converge

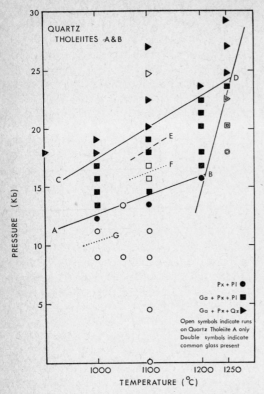

Fig. 2. Mineral assemblages observed in quartz tholeiites (Table 1) over a range of temperature and pressure conditions. The eclogite field lies above CD and the garnet granulite field lies below AB. Area between AB and CD represents the garnet granulite stability field. Boundary BD is the solidus. Px, pyroxene; Pl, plagioclase; Ga, garnet; Qz, quartz.

at lower temperatures. This appears to be a real phenomenon, although it may not be as marked as it appears in Figure 2. The mean transition gradient given by Figure 2 is 21 bars/°C.

An alternative method of establishing the gradient of the transition zone is by comparison with known gradients in simple systems closely related to the gabbro-eclogite transformation. It is reasonable to assume that the gradients in these simple systems will be generally similar to that of the gabbro-eclogite transformation. As was discussed previously, and more fully by Green and Ringwood [1967a], the principal equilibria occurring near the garnet granulite/eclogite boundary are the breakdown of sodic plagioclase to form jadeite + quartz and the breakdown of aluminous pyroxenes to form garnet and low-alumina pyroxenes. The gradients of several of these equilibria (Table 2) have been established experimentally, and some are plotted in Figure 3. The average slope of the nine equilibria in Table 2 is 20 bars/°C, which is essentially identical with the average gradient of the quartz tholeiite (21 bars/°C, Figure 2). This constitutes strong support for the applicability of this gradient.

In Figure 3, the high-temperature experimental results on stability field boundaries in the quartz tholeiite composition have been extrapolated to lower temperatures using the mean gradient of 21 bars/°C as discussed above. The temperature on the garnet granulite-eclogite boundary at a pressure corresponding to the base of the normal continental crust (10 kb) is 670°C. If the temperature at the base of the crust is lower than 670°C, then eclogite would be the stable form of a basalt of this composition throughout the crust.

The temperature distribution in the crust is determined principally by the surface heat flux and the depth distribution of radioactivity. Many models have been investigated by Birch [1955], Clark [1961, 1962], Clark and Ringwood [1964], and others. Their results show that in stable continental regions of normal crustal thickness, and characterized by heat flows between 0.8 and 1.5 μcal/cm² sec, the temperature at the base of the crust is usually less than 670°C for most reasonable assumptions as to radioactivity distribution. This conclusion is practically certain for Precambrian shields characterized by mean heat flows of 1.0 μcal/cm² sec. The temperature at the base of the crust of these regions is probably less than 450°C. Such temperature distributions, taken in conjunction with the experimental results discussed above, indeed imply that eclogite is the stable modification of the quartz tholeiite composition throughout very large regions of normal continental crust. Most of the basalts that we have studied require *smaller* pressures to reach the eclogite stability field than the quartz tholeiite (Fig. 1); accordingly the conclusion that eclogite is thermodynamically stable throughout the normal crust is generalized and reinforced.

This conclusion is somewhat surprising, and doubtless it will receive careful scrutiny. Nevertheless, we believe that it is justified by the

TABLE 2. Gradients of Simple Equilibria Related to the Garnet Granulite-Eclogite Transition

Reference	Equilibrium	dP/dT, bars/°C
Birch and LeComte [1960]	Albite = jadeite + quartz	20
Kushiro [1965]	Albite + diopside = omphacite (40% Jd) + quartz	19
Boyd and England [1964]	Aluminous enstatite (6% Al_2O_3) = enstatite (6% Al_2O_3) + pyrope	34
Robertson et al. [1957]	Albite + nepheline = 2 jadeite	18
Kushiro and Yoder [1965]	Clinopyroxene + orthopyroxene + anorthite + spinel = garnet + clinopyroxene (overall composition: 1 forsterite + 1 anorthite)	16
Kushiro and Yoder [1965]	Clinopyroxene + orthopyroxene + forsterite + spinel = garnet + forsterite + clinopyroxene (overall composition: 1 forsterite + 2 anorthite)	25
Kushiro and Yoder [1964]	Clinopyroxene + orthopyroxene + quartz = garnet + quartz + pyroxene (overall composition: 2 enstatite + 1 anorthite)	10 ? *
MacGregor [1964]	4 Enstatite + spinel = forsterite + pyrope	17
Hays [1966]	3 Anorthite = grossularite + 2 kyanite + quartz	24

* This gradient is based upon very limited experimental data and has a large uncertainty.

experimental data. If correct, it would have profound tectonic consequences, since its corollaries are that basalts and gabbros in the earth's crust are metastable, and under appropriate kinetic conditions, should transform to eclogite. Consider the possible future of present geologic structures characterized by great thicknesses of basaltic lavas resting upon the crust, e.g. Hawaii. These vast structures will be stable as long as their internal temperature distributions are high enough to maintain them in the gabbro stability field, or in the medium grades of the garnet granulite stability field. However, we can look forward to the time when volcanic activity dies down and a supply of convective heat from the mantle is no longer maintained. When this stage is reached, there will be a general cooling of the entire structure, and a state of thermal conduction equilibrium with a 'normal' temperature-depth distribution will ultimately be reached. Under favorable kinetic conditions, transformation into eclogite will probably occur in the deeper levels. Because the density of eclogite (3.5 g/cm³) is higher than that of the ultramafic rocks believed to constitute the upper mantle (3.3 g/cm³), the entire structure may eventually subside into the mantle, causing complex crustal deformation. The subsiding eclogites may undergo fractional melting during sinking, resulting in production of large volumes of intermediate and acidic magmas [*T. Green and Ringwood*, 1966]. The magmas will rise to the surface, and it is possible that they will form the nucleus of a new continental block. It is interesting to speculate that 'continental' islands such as Fiji may have originated from basalt piles similar to Hawaii.

Before leaving this subject, we should mention the significance of the low pressure needed to produce an eclogitic assemblage in the low-alkali olivine tholeiite composition (Figure 1, Table 1). It is practically certain that this rock would occur as eclogite under the P, T conditions throughout the stable continental crust. The low pressure mineralogy of this rock is basic plagioclase (An_{83}) 42.7%, olivine (Fo_{85}) 23%, and pyroxene 33.1%. These minerals characteristically crystallize from common basaltic magmas under low-pressure, crustal conditions. It is now apparent that most of the very large flows and intrusions of basalts, dolerites, and gabbros in the crust have undergone extensive high-level fractionation and that often the visible basic rocks represent only about half the original primary magma, the other half having crystallized out at deeper levels as olivine, basic plagioclase, and pyroxene [*Hess*, 1960; *O'Hara*, 1965; *Green and Ringwood*, 1967b]. Thus, beneath most high-level occurrences of basic rocks in the crust, there should exist an equally extensive accumulation of precipitated early

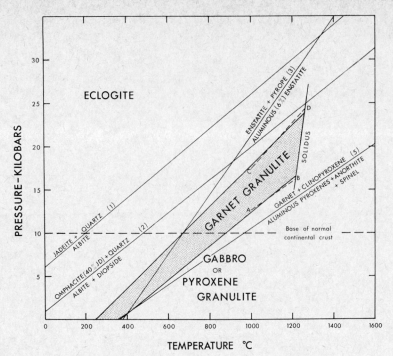

Fig. 3. Extrapolated stability fields of eclogite, garnet granulite, and pyroxene granulite-gabbro for the quartz tholeiite (2) composition. The equilibrium curves of some simple transitions related to the gabbro-eclogite transition are plotted for comparison.

crystals with a bulk composition resembling that of the alkali-poor olivine tholeiite. Cumulates of this type will be especially likely to transform to eclogite after adequate cooling, and, if large enough, may sink back into the mantle. Basic cumulates may thus play an important role in the initiation of tectonic phenomena. Furthermore, the model offers a possible answer to one of the objections sometimes levelled at hypotheses that derive other common crustal rock types of extensive crystallization fractionation of basic magmas. This objection concerns the inordinately large volume of precipitate crystals that must accumulate below the fractionating magmas. If, however, these cumulates can be transformed to eclogite that is removed by subsidence deep into the mantle, the objection loses much of its force.

MINERALOGICAL ZONING AND STABILITY FIELDS IN THE OUTER MANTLE

The chemical composition of the upper mantle is discussed in another section of this Monograph (p. 1), and a brief summary of the preferred model will suffice here. The upper mantle is believed to be of ultramafic composition and chemically zoned. Beneath continents there is a layer of dunite-peridotite of variable thickness but perhaps averaging between 100 and 300 km. This passes downward into a more primitive rock-type 'pyrolite' or pyroxene-olivine rock. Pyrolite is defined as a material that, when fractionally melted, yields a basalt magma and an unmelted refractory residue equivalent to a normal dunite or alpine peridotite. A basalt:peridotite ratio in the vicinity of 1:3 is indicated by various lines of evidence. The composition of pyrolite according to this model is given on page 7. Beneath oceans, the dunite-peridotite layer is thin or it may be absent, so that the primitive pyrolite extends to the suboceanic Moho. To a first approximation, the mean chemical composition of the upper mantle and crust over any extensive region of the earth and down to depths of a few hundred kilometers is identical, whether the surface is continental or oceanic. Continents are considered to have segregated more or less vertically owing to differentiation by fractional melting of the primitive pyrolite. The suggested chemical model is shown on page 6.

Having set up a model for the chemical composition of the upper mantle, we must next investigate the effects of the range of P, T conditions occurring in the outer mantle on the mineralogy of dunite, peridotite, and pyrolite. From evidence to be discussed later, it appears that the dominant components of the dunite-peridotite zone, olivine and orthopyroxene, are stable throughout the P, T range of the outer 400 km of the mantle. On the other hand, the mineralogical assemblages displayed by pyrolite are sensitively dependent on temperature and pressure.

Rocks approaching the pyrolite composition rarely occur at the surface of the earth. They are found most commonly as inclusions in diamond pipes, as inclusions in basalts, and as high-temperature peridotites. *Green and Ringwood* [1963] observed that naturally occurring rocks approaching the pyrolite composition displayed four distinct mineral assemblages clearly indicative of different P, T conditions of crystallization and equilibration. If the upper mantle is indeed composed of rocks approaching the pyrolite composition, it appears that large-scale mineralogical zoning controlled by the P, T stability fields of these assemblages may be present.

The next step is to locate the P, T fields of these assemblages by direct experiment. *Ringwood et al.* [1964] (see also *Ringwood* [1966b]) attempted to locate the boundary between pyroxene, pyrolite, and garnet pyrolite on the basis of phase relationships determined in simplified garnet-pyroxene systems [*Boyd and England*, 1964; *MacGregor and Ringwood*, 1966]. It is clear from recent work by *Green and Ringwood* [1967c] on the ideal pyrolite composition, and by *Ito and Kennedy* [1967] on a garnet peridotite, that the early estimates placed the boundary between pyroxene pyrolite and garnet pyrolite at pressures that were too high by about 10 kb on the average. About 3 kb of this is due to a -10% pressure correction that has been applied to later results because of nonuniform pressure distribution within the pressure vessel [*Green et al.*, 1966]. The reason for the rest of the discrepancy is not understood, but it is presumably connected with the compositional complexity of the pyrolite system, which is causing behavior different from that expected on the basis of simplified pyroxene-garnet systems.

An outline of pyrolite stability fields based on our most recent results is given in Figure 4. The plagioclase pyrolite field has been drawn on the basis of work by Green (manuscript in preparation) on simplified systems and may require revision in the future. The boundary between garnet pyrolite and pyroxene pyrolite was determined directly on the pyrolite composition and has been reversed. It is important to note that it is the *first appearance* of garnet that is marked by this boundary. The transition between pyroxene pyrolite and garnet pyrolite (containing Al-poor pyroxenes) is at least 10 kb wide above 1000°C. As pressure increases (at a given temperature), an increasing proportion of garnet is formed by the breakdown of aluminous pyroxenes on the right-hand side of the boundary until all the potential garnet originally in solid solution as the pyroxenes has been crystallized according to the equilibrium

$$3MgSiO_3 \cdot xAl_2O_3 = xMg_3Al_2Si_3O_{12} + 3(1-x) MgSiO_3$$

Aluminous enstatite = Pyrope garnet + Enstatite

An analogous reaction can be written for aluminous diopsidic pyroxene.

The stability field of the olivine + amphibole assemblage is mainly controlled by the partial pressure of water vapor and temperature. Under conditions where P_{H_2O} is load pressure, the upper limit to the thermal stability of this assemblage may exceed 1000°C. However, in an environment where water pressure is much smaller than load pressure, the stability field will be displaced to lower temperatures. Since quantitative data on the stability of amphiboles under conditions where $P_{H_2O} < P_{load}$ are not available, the stability field of ampholite can be indicated only in very general terms in Figure 4. The occurrence of ampholite is probably limited to the uppermost layer of the suboceanic mantle. This would require a water content of 0.5 to 0.8% in this region, which does not seem unreasonable in view of the abundance of water above the crust.

From Figure 4 it appears improbable that

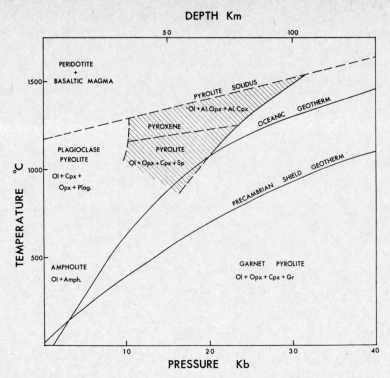

Fig. 4. Preliminary experimental determination of the stability fields of garnet pyrolite, pyroxene pyrolite, plagioclase pyrolite, and the pyrolite solidus, Ol, olivine; Cpx, clinopyroxene (Ca-rich); AlCpx, aluminous clinopryoxene (Ca-rich); Opx, orthopyroxene; AlOpx, aluminous orthopyroxene; Plag, plagioclase; Amph, amphibole; Sp, spinel ($MgAl_2O_4$-rich); Gr, garnet (pyrope-rich).

plagioclase pyrolite occurs in normal oceanic regions. However, in local areas of high heat flow, where the crust is thin and the partial pressure of water low, plagioclase pyrolite may be the stable assemblage. This might be the case in parts of Japan and the western United States where the crust is thin, the heat flow high, and the seismic velocity at the Moho is smaller than 8 km/sec. Plagioclase pyrolite may also occur beneath some island arc areas characterized by low mantle seismic velocities and beneath mid-oceanic ridges.

Because of the revised boundary between garnet pyrolite and pyroxene pyrolite stability fields, the occurrence of the latter assemblage in the mantle is much more restricted than was suggested in the model of *Ringwood et al.* [1964]. The new results (Figure 4) indicate that, in the suboceanic mantle, pyroxene pyrolite would transform to garnet pyrolite at about 70 km, depending on temperature distribution. Thus, if ampholite occupies the upper 30 km,

pyroxene pyrolite would occur only between 30 and 70 km. *MacGregor* [1965] and *Ito and Kennedy* [1967] have reached similar conclusions. It is unlikely that pyroxene pyrolite would be stable beneath stable continental areas. The occurrence of this assemblage in the form of alpine peridotite in the cores of folded mountain chains indicates formation temperatures much higher than those expected under normal, stable continental crusts. Garnet pyrolite would be the stable assemblage in such regions.

THE TRANSITION ZONE

The transition zone, which extends between about 400 and 900 km, is characterized by an abnormally high rate of increase of seismic velocities with depth. *Bullen* [1936] demonstrated that the density in this region increased much more rapidly with depth than would be expected in a uniform self-compressed layer. This implied that the region was inhomogeneous. *Birch* [1952] compared the observed variation of seismic velocities with depth with the variation that would be expected on thermodynamic grounds in a homogeneous medium. From this comparison, Birch concluded that the rate of increase of velocities with depth between 400 and 900 km was much higher than would occur in a homogeneous self-compressed layer, thus confirming the view that this layer was inhomogeneous. However, the rate of increase of velocities with depth between 900 and 2900 km agreed closely with expectations for a homogeneous, self-compressed layer. When the elastic ratio of the material of this region was extrapolated to surface conditions using an equation of state, Birch found that it was far higher than those of common silicates such as olivine and pyroxene, but that it agreed well with the elastic ratios of closely packed oxide phases such as spinel, corundum, rutile, and periclase. Accordingly, he suggested that the region below 900 km consisted of dense closely packed polymorphs of normal ferromagnesian silicates possessing 'oxide' structures. The inhomogeneous region between 400 and 900 km was then interpreted as a region over which the gradual transition from normal to closely packed silicates took place.

Clearly then, an experimental verification of Birch's hypothesis would require the demonstration that the characteristic outer-mantle minerals olivine, pyroxene, and garnet are indeed unstable at very high pressures and transform to new closely packed phases possessing the required elastic properties within a specific P, T range corresponding to conditions between 400 and 900 km.

Since the pressures required are very high (between 130 and 350 kb) and beyond the range of static high pressure-temperature apparatus until recently, the evidence was first gathered by indirect methods, making use of the techniques of thermodynamics, crystal chemistry, and the extrapolation of phase boundaries determined at lower pressures. Investigations of germanate compounds that frequently serve as high pressure models for the corresponding silicates have been particularly rewarding. Since a germanate usually requires a much smaller pressure to undergo a given transition than its corresponding isostructural silicate, the study of germanates and germanate-silicate solid solutions offers the possibility of obtaining useful information about phase transformations that may occur in silicates at pressures beyond the range of available experimental techniques.

A detailed review of the application of these indirect techniques was given by *Ringwood* [1966b]. This covered experimental work carried out during the period 1956–1963. Generally speaking, these results provided strong evidence in favor of Birch's hypothesis, and enabled the construction of widely self-consistent models of the earth's mantle. Nevertheless, there was still a strong incentive to develop apparatus that would reproduce the P, T conditions of the region 400–900 km so that the inferred new silicate polymorphs could be synthesized directly and their properties measured. This latter stage has now been reached with the development of apparatus capable of yielding pressures greater than 250 kb simultaneously with temperatures of about 1000°C. Many new transitions in silicates have been discovered, and a direct understanding of the mineralogical constitution of the transition zone is beginning to emerge. Brief reviews of results to date in some important transitions are given below.

The Olivine-Spinel Transition

Olivine-spinel polymorphism was discovered by *Goldschmidt* [1931] in the compound Mg_2GeO_4. *Bernal* [1936] suggested, by analogy with Mg_2GeO_4, that common olivine might transform in the mantle under sufficiently high pressure to a spinel phase possessing a substantially increased density. This suggestion was adopted by *Jeffreys* [1937] as the basis for an explanation of the 20° discontinuity.

Bernal's hypothesis was strongly supported by the results of *Ringwood* [1958b, 1962, 1963], who discovered that the olivines Fe_2SiO_4, Ni_2SiO_4, and Co_2SiO_4 transformed to spinel structures at pressures between 15 and 70 kb in the temperature interval 600 to 700°C. The transformations were accompanied by density increases of about 10%. A quantitative estimate of the pressure required for the Mg_2SiO_4 olivine-spinel transition was obtained by *Ringwood* [1956, 1958a] from a thermodynamic investigation of solid solution equilibria at atmospheric pressure in the system $Ni_2GeO_4 - Mg_2SiO_4$. He found that the transition should occur within the pressure interval 175 ± 55 kb at

1500°C. This pressure corresponded to a depth of approximately 500 ± 140 km, which is within the transition zone. A more elaborate investigation of equilibria in the same system over the pressure range 0–90 kb at 600°C was reported by *Ringwood and Seabrook* [1962]. Extrapolation of phase boundaries in this system yielded a pressure of 130 ± 20 kb for the olivine-spinel at 600°C.

Final confirmation of an olivine-spinel transition in the mantle was obtained very recently when *Ringwood and Major* [1966a] succeeded in synthesizing a continuous series of spinel solid solutions ranging in composition from pure Fe_2SiO_4 to $85Mg_2SiO_4$, $15Fe_2SiO_4$ (Figure 5). The latter composition is close to the composition of olivine in the earth's mantle. The depth of the transition in the earth's mantle is estimated to lie between 400 and 500 km. This estimate will be refined when problems of pressure calibration are solved. The olivine-spinel transition involves a density change of 10.3%. A transformation was also obtained by the above authors in pure Mg_2SiO_4 to a phase that was about 8% denser than the olivine. The X-ray diffraction pattern of this phase showed a marked resemblance to spinel; however, line-splitting and extra lines were present. It is possible that this phase is a distorted spinel and that distortion may perhaps have occurred during release of pressure. On the other hand, *Akimoto and Ida* [1966] claim to have synthesized a true Mg_2SiO_4 spinel at pressures over 150 kb, at 800°C. However, in view of the fact that the maximum degree of transformation they obtained was about 50%, and that only 6 lines were indexed, 4 of which (including the two strongest) agreed with the spacings of the phase recognized by Ringwood and Major, the question of identification of Akimoto and Ida's phase should not be regarded as finally settled.

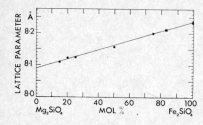

Fig. 5. Lattice parameters of Mg_2SiO_4-Fe_2SiO_4 spinel solid solutions.

Quartz-Coesite-Stishovite Transitions

The first high-pressure polymorph of quartz-coesite was discovered by *Coes* [1953] in the course of his pioneering investigations in high-pressure mineralogy. The density of coesite is about 2.9, and its structure shows that the silicon ions remain in fourfold coordination [*Zoltai and Buerger*, 1959].

Stishov and Popova [1961] showed that at a pressure of approximately 130 kb at 1600°C, coesite transformed to a new phase possessing the rutile structure. The density of the new phase, stishovite, was close to 4.3 g/cm³, the large increase in density being caused by the change in silicon coordination from 4 to 6. The importance of this discovery is twofold. First, it is possible that stishovite may occur as a distinct phase in the mantle. The pressure required for its formation is reached in the upper regions of the transition zone. Second, the demonstration that the coordination of silicon can change from 4 to 6 under high pressure greatly increases the range of transformations possible in the deeper regions of the mantle.

Transitions in Simple Pyroxenes

An investigation by *Ringwood and Seabrook* [1963] of the stability of several germanate pyroxenes at high pressure has provided considerable insight into the probable behavior of their silicate analogs at much higher pressures. In every case, the germanate pyroxenes were found to be unstable at high pressures and transformed to one or more denser phases. Two modes of transformation were observed. $MgGeO_3$ and $MnGeO_3$ pyroxenes transformed to polymorphs possessing ilmenite structures, which were 15–18% denser than the original pyroxenes. On the other hand, $CoGeO_3$, $FeGeO_3$, and $(Mg_{0.75} Ni_{.25}) GeO_3$ transformed into mixtures of spinels + GeO_2 (rutile), which were about 10% denser than the original pyroxenes:

$$2MGeO_3 \text{ (pyroxene)} \rightarrow M_2GeO_4 \text{ (spinel)} + GeO_2 \text{ (rutile)}$$

Recently, *Ringwood and Major* [1966b] have shown that $FeSiO_3$ and $CoSiO_3$ transform similarly to their germanate analogs, i.e. into mixtures of Fe_2SiO_4, Co_2SiO_4 (spinels) + SiO_2 (stishovite) at pressures around 100–120 kb.

These authors [*Ringwood and Major, 1966c*] have also studied the system $FeSiO_3$-$MgSiO_3$. These pyroxene solid solutions transform completely to spinel-stishovite mixtures over the composition range FS_{100} to $FS_{50}MS_{50}$ and partly from $FS_{50}MS_{50}$ to $FS_{15}MS_{85}$. In the more magnesian compositions, large amounts of untransformed $MgSiO_3$ remain even at 200 kb. This compound is much more stable than Mg_2SiO_4 at high pressures.

The transformation mode of pure $MgSiO_3$ is not yet established. It may break down to a spinel-stishovite mixture, or it may transform directly to an ilmenite structure. From an investigation of the solid solubility of $MgSiO_3$ in the ilmenite polymorph of $MgGeO_3$ at high pressures, *Ringwood and Major* [1966c] estimated that a pressure of about 250 kb would be required for the pyroxene-ilmenite transition in $MgSiO_3$.

Transformations in Pyroxenes in the Presence of Alumina

Ringwood and Major [1966c] investigated the behavior of $MgSiO_3$ compositions containing 5 and 10% alumina at high pressures. It was anticipated that the presence of Al_2O_3 in solid solution might lower the pressure needed to stabilize the ilmenite structure. The results were unexpected. At pressures of 150 kb and above, the $MgSiO_3$, 10% Al_2O_3 glass crystallized completely to a garnet structure, apparently consisting of a solid solution between $Mg_3Al_2Si_3O_{12}$ (pyrope) and an end-member $Mg_3(MgSi)Si_3O_{12}$. One quarter of the silicon atoms in the latter end-member are octahedrally coordinated. The solid solution of enstatite in the garnet becomes significant at pressures between about 90 and 100 kb. Similar results were obtained on glasses of compositions $CaMgSi_2O_6$, 10% Al_2O_3 and $CaSiO_3$, 10% Al_2O_3. At first it was suspected that these garnets might represent products of metastable crystallization of glass; however, subsequent investigations (unpublished) have yielded strong evidence that the garnets were thermodynamically stable.

These results are important to an understanding of transformations in the mantle, where the ratio Al_2O_3 to total pyroxene may well be in the vicinity of 10%. It appears that most of the pyroxene minerals will transform to the new garnet structure characterized by octahedrally coordinated silicon at depths between about 350 and 500 km (making an allowance for the effect of temperature on the transition). The transformation involves a density increase of about 10%.

The Garnet-Ilmenite Transition

In the preceding section we saw that in the complex chemical system of the mantle, the direct $MgSiO_3$ pyroxene-ilmenite transition may not occur because of the intervention of the pyroxene-garnet transition at much lower pressures. This raises the question of the stability of garnets at high pressure. Ringwood and Major (unpublished) are currently studying the system $(G)Mg_3Al_2Ge_3O_{12}$, $(S)Mg_3Al_2Si_3O_{12}$ at high pressure. The former compound does not form a garnet structure but instead, at modest pressures, it displays the ilmenite structure, essentially a solid solution between $MgGeO_3$ (ilmenite) and Al_2O_3. However, at compositions between $G_{80}S_{20}$ and S_{100}, garnets are stable at relatively low pressures. Between $G_{80}S_{20}$ and $G_{50}S_{50}$ compositions, the garnets have been completely or partially transformed to the ilmenite structure at high pressures. It is believed that in the mantle silicate garnets will ultimately transform in an analogous fashion to ilmenite structures, essentially a solid solution between $MgSiO_3$ (ilmenite) and Al_2O_3.

Perovskite-Type Silicates

Calcium metasilicate ($CaSiO_3$) is a significant component of the earth's mantle, and a study of its behavior at high pressure is accordingly of interest. *Ringwood and Major* [1967] have found that the germanate analog $CaGeO_3$ transforms from a wollastonite structure, first to a garnet $Ca_3(CaGe)Ge_3O_{12}$ structure, and finally to a perovskite structure at high pressures. A study of the solid solubility of $CaSiO_3$ in $CaGeO_3$ (perovskite) is currently in progress. At 175 kb, about 35 mole % of $CaSiO_3$ enters into solid solution in $CaGeO_3$ (perovskite). Extrapolation of densities of the germanate-silicate solid solutions leads to a density of 4.10 g/cm³ for $CaSiO_3$ (perovskite). This is substantially higher than the mean density of the isochemical oxide mixture of $CaO + SiO_2$ (stishovite), 3.77 g/cm³. It is estimated that $CaSiO_3$ perovskite should become stable between about 200 and 250 kb. A run at 250 kb on $CaSiO_3$, however, produced a new phase with a complex structure and a density of about 3.5 g/cm³. This phase possessed some curious optical properties, and it is possible that it may represent a retrogressive transformation product from a denser phase (perhaps perovskite?) that was produced in the experiment, but that did not remain kinetically stable when pressure was released.

SUMMARY

Many of the results quoted above are very new and some of the investigations are still in progress. It is too soon to attempt a comprehensive and detailed survey of the bearing of these transformations on our understanding of the constitution of the earth's mantle. However, the following conclusions appear justified at the present stage.

There is no doubt that the principal minerals of the outer mantle, olivine and pyroxene, are unstable at high pressures and transform to denser phases at pressures equivalent to those between depths of about 300 and 600 km. Thus the essential part of Birch's hypothesis is undoubtedly correct; namely, that the abnormal variations of seismic velocities and density in the transition zone are caused by phase transformations.

The first major transformation to occur with increased depth in the mantle is probably the pyroxene-garnet transition. This is followed and perhaps overlapped by the olivine-spinel transformation, so that at 600 km the mineral assemblage consists of garnet + spinel and is about 10% denser than the original olivine + pyroxene.

Below 600 km, garnet is expected to transform to an ilmenite structure, $CaSiO_3$ to a perovskite structure, and Mg_2SiO_4 (spinel) either to a mixture of $MgSiO_3$ (ilmenite) + MgO (periclase) or to a new Mg_2SiO_4 structure characterized by sixfold or eightfold coordination of Mg and sixfold coordination of Si, e.g. the high pressure structures of Mn_2GeO_4 and Ca_2GeO_4. There is also a possibility that $MgSiO_3$ might finally transform from an ilmenite to a perovskite structure.

Present estimates of the density and elastic properties of the above assemblage of closely packed phases match well the inferred properties of the lower mantle [*Clark and Ringwood,* 1964].

REFERENCES

Akimoto, S., and Y. Ida, High pressure synthesis of Mg_2SiO_4 spinel, *Earth Planetary Sci. Letters, 1,* 358–359, 1966.

Bernal, J. D., Discussion, *Observatory, 59,* 268, 1936.

Birch, F., Elasticity and constitution of the earth's interior, *J. Geophys. Res., 57,* 227–286, 1952.

Birch, F., Physics of the crust, *Geol. Soc. Am. Spec. Paper, 62,* 101–118, 1955.

Birch, F., and P. LeComte, *Am. J. Sci., 258,* 209, 1960.

Boyd, F., and J. L. England, The system enstatite-pyrope, *Carnegie Inst. Wash. Yearbook, 63,* 157–161, 1964a.

Boyd, F., and J. L. England, *Carnegie Inst. Wash. Yearbook, 63,* 157, 1964b.

Bullen, K. E., The variation of density and the ellipticities of strata of equal density within the earth, *Monthly Notices Roy. Astron. Soc., Geophys. Suppl., 3,* 395–401, 1936.

Clark, S. P., Geothermal studies, *Carnegie Inst. Wash. Yearbook, 60,* 185–190, 1961.

Clark, S. P., Temperatures in the continental crust, in *Temperature, Its Measurement and Control in Science and Industry,* vol. 3, edited by C. M. Herzfeld, Reinhold Publishing Corporation, New York, 779–790, 1962.

Clark, S. P., and A. E. Ringwood, Density distribution and constitution of the mantle, *Rev. Geophys., 2,* 35–88, 1964.

Coes, L., A new dense crystalline silica, *Science, 118,* 131–133, 1953.

Fermor, L. L., Preliminary note on garnet as a geological barometer and on an infra-plutonic zone in the earth's crust, *Records Geol. Surv. India, 43,* 1913.

Fermor, L. L., The relationship of isostasy, earthquakes and vulcanicity to the earth's infra-plutonic shell, *Geol. Mag., 51,* 65–67, 1914.

Goldschmidt, V. M., Zur Kristallchemie des Germaniums, *Nachr. Ges. Wiss. Goettingen, Math-Physik, Kl. I, 2,* 184–190, 1931.

Green, D. H., and A. E. Ringwood, Mineral assemblages in a model mantle composition, *J. Geophys. Res., 68,* 937–945, 1963.

Green, D. H., and A. E. Ringwood, An experimental investigation of the gabbro to eclogite transformation and its petrological applications, *Geochim. Cosmochim. Acta,* in press, 1967a.

Green, D. H., and A. E. Ringwood, Genesis of basalt magmas, *Beit. Mineral. Petrog.* in press, 1967b.

Green, D. H., and A. E. Ringwood, Stability fields of aluminous pyroxene peridotite and garnet peridotite, *Earth Planet. Sci. Letters, 3,* 151–60, 1967c.

Green, T. H., and A. E. Ringwood, Origin of the calc-alkaline igneous suite, *Earth Planetary Sci. Letters, 1,* 307–316, 1966.

Green, T. H., A. E. Ringwood, and A. Major, Calibration of a piston-cylinder high pressure apparatus, *J. Geophys. Res., 14,* 3589–3594, 1966.

Hays, J. F., Abstracts, 79th Annual Meeting, Geological Society of America, p. 89, 1966.

Hess, H. H., Stillwater igneous complex, *Geol. Soc. Am. Mem., 80,* 117–185, 1960.

Holmes, A., Contributions to the theory of magmatic cycles, *Geol. Mag., 63,* 306–329, 1926a.

Holmes, A., Structure of the continents, *Nature, 118,* 586–587, 1926b.

Holmes, A., Some problems of physical geology and the earth's thermal history, *Geol. Mag., 64,* 263–278, 1927.

Ito, K., and G. C. Kennedy, Melting and phase relations in a natural peridotite to 40 kbars, *Am. J. Sci., 265,* 510–538, 1967.

Jeffreys, H., On the materials and density of the earth's crust. *Monthly Notices Roy. Astron. Soc., Geophys. Suppl., 4,* 50–61, 1937.

Kushiro, I., *Carnegie Inst. Wash. Yearbook, 64, 112,* 1965.

Kushiro, I., and H. S. Yoder, *Carnegie Inst. Wash. Yearbook, 63,* 108, 1964.

Kushiro, I., and H. S. Yoder, *Carnegie Inst. Wash. Yearbook, 64,* 89, 1965.

MacGregor, I. D., *Carnegie Inst. Wash. Yearbook, 63,* 156, 1964.

MacGregor, I. D., Stability fields of spinel and garnet peridotites in the synthetic system $MgO\text{-}CaO\text{-}Al_2O_3\text{-}SiO_2$, *Carnegie Inst. Wash. Yearbook, 64,* 126–134, 1965.

MacGregor, I. D., and A. E. Ringwood, The natural system enstatite-pyrope, *Carnegie Inst. Wash. Yearbook, 63,* 161–163, 1964.

O'Hara, M. J., Primary magmas and the origin of basalts, *Scottish J. Geol., 1,* 19–40, 1965.

Oxburgh, E. R., Upper mantle inhomogeneity and the low-velocity zone, *Geophys. J., 8,* 456–462, 1964.

Ringwood, A. E., The olivine-spinel transition in the earth's mantle, *Nature, 178,* 1303–1304, 1956.

Ringwood, A. E., The constitution of the mantle, 1, Thermodynamics of the olivine-spinel transition, *Geochim. Cosmochim. Acta, 13,* 303–321, 1958a.

Ringwood, A. E., The constitution of the mantle, 2, Further data on the olivine-spinel transition, *Geochim. Cosmochim. Acta, 15,* 18–29, 1958b.

Ringwood, A. E., Prediction and confirmation of olivine-spinel transition in Ni_2SiO_4, *Geochim. Cosmochim. Acta, 26,* 457–469, 1962.

Ringwood, A. E., Olivine-spinel transformation in cobalt orthosilicate, *Nature, 198,* 79–80, 1963.

Ringwood, A. E., The chemical composition and origin of the earth, in *Advances in Earth Science,* edited by P. M. Hurley, pp. 290–356, MIT Press, Cambridge, 1966a.

Ringwood, A. E., Mineralogy of the mantle, in *Advances in Earth Science,* edited by P. M. Hurley, pp. 357–398, MIT Press, Cambridge, 1966b.

Ringwood, A. E., and D. H. Green, Experimental investigations bearing on the nature of the Mohorovicic discontinuity, *Nature, 201,* 566–567, 1964.

Ringwood, A. E., and D. H. Green, An experimental investigation of the gabbro-eclogite transformation and some geophysical implications, *Tectonophysics, 3,* 383–427, 1966.

Ringwood, A. E., and A. Major, Synthesis of $Mg_2SiO_4\text{-}Fe_2SiO_4$ solid solutions, *Earth Planetary Sci. Letters, 1,* 241–245, 1966a.

Ringwood, A. E., and A. Major, High pressure transformation in $CoSiO_3$ pyroxene and some geophysical implications, *Earth Planetary Sci. Letters, 1,* 209–211, 1966b.

Ringwood, A. E., and A. Major, High pressure transformations in pyroxenes, *Earth Planetary Sci. Letters, 1,* 351–357, 1966c.

Ringwood, A. E., and A. Major, Some high pressure transformations of geophysical interest, *Earth Planetary Sci. Letters,* in press, 1967.

Ringwood, A. E., I. D. MacGregor, and F. R. Boyd, Petrological constitution of the upper mantle, *Carnegie Inst. Wash. Yearbook, 63,* 147–152, 1964.

Ringwood, A. E., and M. Seabrook, Olivine-spinel equilibria at high pressure in the system $Ni_2GeO_4\text{-}Mg_2SiO_4$, *J. Geophys. Res., 67,* 1975–1985, 1962.

Ringwood, A. E., and M. Seabrook, High pressure transformations in germanate pyroxenes and related compounds, *J. Geophys. Res., 68,* 4601–4609, 1963.

Robertson, E. C., F. Birch, and G. J. F. MacDonald, *Am. J. Sci., 255,* 115, 1957.

Stishov, S. M., and S. V. Popova, New dense polymorphic modification of silica, *Geokhimiya, 1961*(10), 837–839, 1961.

Zoltai, T., and M. J. Buerger, The structure of coesite, the dense high-pressure form of silica, *Z. Krist., 111,* 129–141, 1959.

Equation of State at High Pressure

V. N. Zharkov and V. A. Kalinin

AN UNDERSTANDING of the earth's interior depends on information about the behavior and properties of matter under high pressures and temperatures. An equation of state that connects pressure (P), volume (V), and temperature (T) is thus very important.

The properties of materials under pressures of several dozen kilobars were first studied by *Bridgman* [1949]; he obtained the isotherms for many elements and compounds. Compressibility of solids under static conditions is usually measured by a cylinder-piston device. Exhaustive experimental information on this subject is given by *Birch* [1966]. Other installations are also used in combination with X-ray devices, but such investigations are still few.

Advances in high pressure physics provide dynamical methods of compressing solids by shock waves. This method consists of experimentally determining kinematic parameters such as velocity of a shock wave (U_s) and mass velocity behind the shock front (U_p). A linear equation was empirically established between these values:

$$U_s = C + SU_p \qquad (1)$$

where C and S are experimental constants. Using conservation laws, we proceed from kinematic parameters to $P(V)$, i.e.

$$P_H = \frac{\rho_0 C^2 (1 - x)}{[1 - S(1 - x)]^2} \qquad (2)$$

where C and S are as in (1), ρ_0 is the density of the matter ahead of the wave front, $x = \rho_0/\rho$ is dimensionless volume of compressed solid, and P_H is the Hugoniot pressure.

The patterns of temperature and pressure dependencies on the Hugoniot curve are shown in Figures 1 and 2. When $x \geq 0.85$, temperature increases negligibly and the Hugoniot curve is close to the isotherm ($T \approx 300°K$). When compression is increased, temperature rapidly increases and the shock-wave adiabat lies above the isotherm. Hence, for $x \geq 0.85$, equation 2 expresses $P(V)$ for $T \cong$ constant.

Dynamic methods of compressing solids are described in detail by *Altshuler* [1965] and *Rice et al.* [1958]; their experimental data are given by *Birch* [1966].

The range of pressure and temperature within the earth's interior is $P \leq 3.5$ mbar, $T \leq 6 \times 10^3$ °K. Unfortunately, it is not possible to derive a theoretical equation of state for this domain. Hence, various semiempirical methods, especially the potential method, are widely used. The potential method consists of a prior determination of some qualitative dependency of crystal potential energy on the lattice constant (or on the volume of the solid); then the numerical parameters are determined from experimental data. The potential method was first applied by *Mie* [1903], *Grüneisen* [1908], and *Born and Göppert-Mayer* [1933] to obtain the equations of state for ionic crystals. *Davydov* [1956] extended it to study different solids, and used the experimental data more completely. A number of potential methods have previously been reported, including those in which potential is formally represented as a finite power series of volume.

A nonpotential method, used to derive the zero isotherm ($T = 0°K$), was described by *Rice et al.* [1958] and *Altshuler* [1965]. In this method, the equation containing an experimental Hugoniot curve is numerically integrated. However, the results have poor accuracy at small compressions, and are not convenient for extrapolation.

EQUATION OF STATE AT ABSOLUTE ZERO

With respect to interatomic forces, crystalline solids can be roughly classified into four groups: ionic, molecular, and valence crystals, and metals. We shall consider only ionic crystals and metals, since molecular crystals are of no interest in geophysics, and, though valence crystals

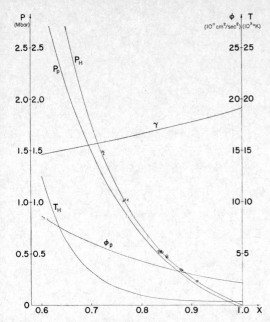

Fig. 1. Shock-wave data for nickel: circles, experimental points (P_H) used to determine the equation of state [Rice et al., 1958]; crosses, additional experimental points (P_H) [McQueen and Marsh, 1960].

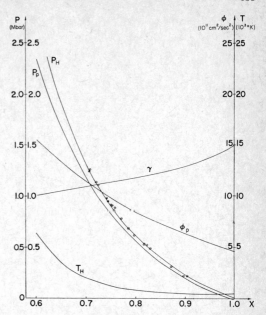

Fig. 2. Shock-wave data for periclase: circles, experimental points (P_H) used to determine the equation of state [McQueen and Marsh, 1966].

play some part in the earth's constitution, it is difficult to determine their potential.

The principal forces in ionic crystals are the Coulomb attraction and exponential repulsion forces. Equation 3 is an expression of potential energy of ionic crystals, which takes these two into account:

$$E_p(x) = \frac{3A}{b\rho_0} e^{b(1-x^{1/3})} - \frac{3K}{\rho_0} x^{-1/3} \quad (3)$$

where ρ_0 is the density at zero pressure, $x = \rho_0/\rho = V/V_0$ is dimensionless volume ($V = 1/\rho \propto r^3$) and A, b, K are constants to be determined. Van der Waals' energy and other less important interaction energies are effectively included in these two terms. Detailed information on this question is given by Born and Huang [1954].

Using the thermodynamic relation

$$P = -\left(\frac{\partial F}{\partial V}\right)_T = -\rho_0 \left(\frac{\partial F}{\partial x}\right)_T \quad (4)$$

we obtain for the pressure at $T = 0$:

$$P_p(x) = Ax^{-2/3} e^{b(1-x^{1/3})} - Kx^{-4/3} \quad (5)$$

since, at $T = 0$, the free energy and the internal energy are the same.

Attraction forces in metals are governed by Coulomb interactions between free electrons and positive lattice ions, as well as by the exchange energy of free electrons. These two terms can be combined into one: c/r. Coulomb repulsion of lattice ions is proportional to $1/r$ and decreases the value of c in attraction forces. Repulsion in metals is due to overlapping of electron shells of lattice ions and can be expressed by $ae^{-r/\rho}$, which is similar to respective interaction in ionic crystals. Fermi's kinetic energy of free electrons also yields the repulsion and depends on the volume ($dV^{-2/3}$).

Thus, the specific potential energy of metal is

$$E_p(x) = \frac{3A}{b\rho_0} e^{b(1-x^{1/3})}$$

$$+ \frac{3D}{2\rho_0} x^{-2/3} - \frac{3K}{\rho_0} x^{-1/3} \quad (6)$$

where A, b, D, and K are constants to be determined.

In metals with small overlapping of electron shells of lattice ions (for example, in alkali metals), repulsion forces are determined by

Fermi's kinetic energy of conductivity electrons. In this case the first term in (6) is small and can be neglected; its effect is observable only at high pressures. When overlap forces are significant, the second term in (6) should be omitted; the potential energy for metal then has a form similar to that for ionic crystals (3). Thus, the potential pressure in metals is expressed by (5).

By adding in (5) the pressure caused by the crystal vibrations at $T = 0$, we obtain a zero isotherm. This pressure is several kilobars under normal conditions, and increases proportionally to Debye temperature as the volume decreases.

Although many types of potential exist, let us consider only one equation of state often used in geophysical applications:

$$P = \tfrac{3}{2}K_0(x^{-7/3} - x^{-5/3})$$
$$\cdot [1 - \xi(x^{-2/3} - 1) + \cdots] \quad (7)$$

where K_0 is the compression modulus at $P = 0$ (i.e., $x = 1$), and ξ is a correction factor. This equation was derived by Birch [1952] using Murnaghan's theory of finite strain. In practice, only the first few terms of the expansion are used; this increases the discrepancy between theoretical and experimental isotherms as x decreases. Evidently an increase in pressure requires that more and more terms be included, which is practically impossible to do.

EQUATION OF STATE FOR PHONON DOMAIN; GRÜNEISEN'S PARAMETER

If the temperature of a solid increases, its energy also increases owing to excitation of atomic or ionic thermal vibrations. Below the melting point, the mean amplitudes of thermal vibrations are at least one order smaller than the lattice constant. Thus, it is possible to expand the potential energy of a crystal as a power series of displacement up to the quadratic term. Expressing this energy in normal coordinates, we find that a crystal of N atoms is mechanically equivalent to a system of $3N$ independent harmonic oscillators, and the free energy of a crystal is

$$F = E_p + \tfrac{1}{2}\sum_\alpha \hbar\omega_\alpha$$
$$+ kT\sum_\alpha \ln(1 - e^{-\hbar\omega_\alpha/kT}) \quad (8)$$

where h is Planck's constant, k is Boltzmann's constant, and ω_α is the frequency of oscillator α. Summation is performed over all $3N$ frequencies. The first term in (8) is equal to the potential part of the internal energy of a solid, the second describes the zero vibrations of the crystal, and the last corresponds to atomic thermal vibrations. All the thermodynamic parameters can, in principle, be obtained from (8). However, it is not generally possible to calculate the terms in (8), since their values essentially depend on a particular frequency distribution in the spectrum of a solid. Only in extreme cases of low or high temperatures can such a summation be carried out.

In a number of cases, the Debye theory could be used as a good approximation. According to this theory, the frequencies of a solid are continuously distributed from zero to some maximal value, ω_{max}, with the spectral density proportional to ω^2; the free energy (8) is given by

$$F = E_p(x) + NkT\left[\frac{9}{8}\frac{\theta}{T}\right.$$
$$\left. + 3\ln(1 - e^{-\theta/T}) - D\left(\frac{\theta}{T}\right)\right] \quad (9)$$

where θ is the characteristic temperature related to ω_{max} by

$$k\theta = \hbar\omega_{max}$$

and

$$D\left(\frac{\theta}{T}\right) = \frac{3T^3}{\theta^3}\int_0^{\theta/T}\frac{z^3\,dz}{e^z - 1}$$

is the Debye function. The equation of state for free energy (9) is

$$P = P_p(x) + \frac{\gamma\rho_0 Nk}{x}\left[\frac{9}{8}\theta + 3TD\left(\frac{\theta}{T}\right)\right] \quad (10)$$

where $P_p(x)$ is potential pressure and γ is the Grüneisen parameter defined by

$$\gamma = -\frac{\partial \ln \theta}{\partial \ln x} \quad (11)$$

From (9) for the internal energy of the crystal, we obtain

$$E = E_p(x) + Nk\left[\frac{9}{8}\theta + 3TD\left(\frac{\theta}{T}\right)\right] \quad (12)$$

The second term on the right in (12) corresponds to lattice vibrational energy. Denoting this energy by E_T, (10) could be rewritten as

$$P = P_p(x) + \frac{\gamma E_T}{V} \quad (13)$$

which is usually called Mie-Grüneisen's equation of state.

From (13), using thermodynamic relations, we obtain

$$\gamma = \frac{\kappa x K_T}{\rho_0 C_v} \quad (14)$$

where $\kappa = (1/x)(\partial x/\partial T)_p$ is the coefficient of thermal expansion, $K_T = -x(\partial P/\partial x)_T$ is the isothermal modulus of compressibility, and C_v is the specific heat capacity. The $\gamma(x)$ dependency on the basis of (14) was not confirmed experimentally; hence, the γ in (13) has to be considered an unkown function.

Thus, to determine (10), two functions, $P_p(x)$ and $\gamma(x)$, need to be evaluated. This could in principle be done by using two independent experimental curves, of which one should pertain to the low-temperature domain to make the second term of the right-hand side of (10) small enough compared with $P_p(x)$, while the second curve should cover the high-temperature domain to make the second term at least comparable with $P_p(x)$ (temperature should not be very high, otherwise high-temperature corrections will play a significant part in the equation of state, which will be discussed later). However, at present the corresponding experimental data are not available, and thus some functional dependency between $P_p(x)$ and $\gamma(x)$ is assumed.

Slater [1939] derived a formula for $\gamma(x)$ strictly on the Debye theory, assuming that the Poisson ratio does not depend on volume. This formula is derived from (15) for $m = 0$ and $\delta = 0$. However, in this case, poor agreement of Debye theory with experiment is not excluded. There is no doubt that the spectral density of a solid should not always be proportional to ω^2. The changes in vibrational spectrum when volume changes are revealed by γ, and it is hardly possible to describe these changes with an incorrect spectrum.

Besides Slater's formula, several other expressions for $\gamma(x)$ were advanced, of which the most common is

$$\gamma = -\frac{x}{2}\frac{\partial^2(P_p x^{2m/3})/\partial x^2}{\partial(P_p x^{2m/3})/\partial x} + \tfrac{1}{3}(m-2) + \delta \quad (15)$$

If $m = 1$ and $\delta = 0$, we obtain the *Dugdale and McDonald* [1953] formula from (15). And when $m = 2$ and $\delta = 0$, we have the $\gamma(x)$ dependency derived by *Vaschenko and Zubarev* [1963] from the free volume theory.

The normalization constant δ introduced by *Zharkov and Kalinin* [1962a] can be determined when normal conditions are applied and the γ values in (14) and (15) coincide. δ gives a correct value to γ at least at one point ($P = 0$, $x = 1$), and since γ is weakly dependent on x, it should introduce no great error into other values of x. Integrating (11) and using γ from (15) we get

$$\theta = \theta_0 \exp \int_x^1 \frac{\gamma}{x} dx \quad (16)$$

where θ_0 is the characteristic temperature at $P = 0$ and $x = 1$.

Thus, the right-hand side of (10) now contains only one unknown function, $P_p(x)$. To evaluate $P_p(x)$ by the potential method, it is necessary to compute A, B, and K. One of these, K for example, can be determined from the condition that $P = 0$ when $T = T_0$ and $x = 1$, which gives

$$K = A + \rho_0\gamma_0 Nk\left[\frac{9}{8}\theta_0 + 3T_0 D\left(\frac{\theta_0}{T}\right)\right] \quad (17)$$

where the values with zero indices relate to normal conditions. A and b must be determined from experimental data. For this purpose we use the least squares method and determine them from the minimum condition of the function

$$M = \sum_{i=1}^{n} (P_{ei} - P_i)^2 \frac{1}{\sigma_i^2} \quad (18)$$

where P_{ei} is the experimental pressure at (x_i, T_i), P_i is the theoretical pressure at the same point, $1/\sigma_i^2$ is the weight of the given point, σ_i is the mean root square error, and n is the number of experimental points.

If the isotherm is chosen as the initial experimental data, then P_i in (18) will be replaced by P from (10), using γ from (15) and θ from (16). Parameters calculated by this method (for $m = 1$ in equation 15) are shown in Table 1.

TABLE 1. Potential Energy Parameters Calculated from Isotherm (Static) Data (Equation 15, $m=1$)*

Substance	ρ_0, g/cm^3	γ_0	θ_0, °K	A, 10^5 bars	b	K, 10^5 bars	δ
Grossularite†	3.53	1.0	900	2.63835	17.63317	2.84280	−2.278
Diopside†	3.29	0.9	900	2.22356	17.09137	2.36518	−2.282
Andradite‡	3.86	1.1	800	4.62820	11.94568	4.83050	−1.206
Periclase‡	3.56	1.4	800	10.73735	6.72871	10.97963	0.007
Pyrite‡	5.02	1.5	600	4.17271	12.73215	4.33773	−0.938
Olivine†	3.32	1.2	900	25.53967	3.44127	25.72847	0.522
Magnetite‡	5.20	1.4	600	26.68296	3.93331	26.87287	0.581
Orthoclase†	2.56	0.5	1000	3.45932	6.78571	3.52548	−0.902
Nickel‡	8.86	1.91	375	4.76458	13.94740	4.99224	−0.737

* A, b, K, and δ are based on experimental data of Bridgman as cited in first column.
† Bridgman, 1948.
‡ Bridgman, 1949b.

Along with the empirical equation of shock-wave adiabat (2), we can obtain the corresponding theoretical equation, namely:

$$E_H = E_0 + \frac{P_H}{2\rho_0}(1 - x) \quad (19)$$

where E_0 is the specific energy of a solid before the shock wave, and E_H is the energy behind the wave. Assuming that a solid compressed by a shock wave is in thermodynamic equilibrium, we obtain from (10), (12), and (19):

$$P_H = \frac{\{P_p(x) + (\rho_0\gamma/x)[E_0 - E_p(x)]\}}{1 - [\gamma(1 - x)/2x]} \quad (20)$$

Now, if we replace P_{ei} in (18) by experimental pressures on the shock-wave adiabat, P_i will express P_H in (20). Corresponding parametric values (for $m=1$ in equation 15) are shown in Table 2.

Theoretical isotherms (10), computed from Table 2, agree well with experiment. In static compressions the change in volume is not large; therefore, there is no reason to expect a noticeable discrepancy between theoretical and experimental curves in this region.

Theoretical shock-wave adiabats (20), calculated from Table 1, are in most cases in good agreement with the experimental Hugoniot curves. However, a significant discrepancy observed for some solids is probably due to insufficient accuracy of corresponding static values. It is known that Bridgman's isotherms sometimes do not agree with each other even for the same solid. To obtain an equation of state, the following condition should hold: the smaller the interval of x variation, the higher the accuracy of experimental data.

To obtain the optimum parametric values in the equation of state, both the static and dynamic data on compressibility are used simultaneously. These parameters (for $m=1$ in equation 15) are shown in Table 3.

For subsequent use, we require the adiabatic compression modulus K_s, which can be readily determined from (10) as

$$K_s = -x\left(\frac{\partial P}{\partial x}\right)_s = K_p(x)$$
$$+ \frac{\gamma\rho_0}{x}\left(\gamma + 1 - \frac{\partial \ln \gamma}{\partial \ln x}\right)$$
$$\cdot Nk\left[\frac{9}{8}\theta + 3TD\left(\frac{\theta}{T}\right)\right] \quad (21)$$

TABLE 2. Potential Energy Parameters Calculated from Shock-Wave (Dynamic) Data (Equation 15, $m=1$)

Substance	A, 10^5 bars	b	K, 10^5 bars	δ
Corundum*†	41.44586	4.26289	41.72016	0.701
Periclase†	7.59873	8.40077	7.84101	−0.293
Olivine†	11.82059	4.99197	12.00938	0.140
Nickel‡	14.49045	6.25763	14.71811	0.606

* $\rho_0 = 3.99$, $\gamma_0 = 1.6$, $\theta_0 = 600$.
† McQueen and Marsh, 1966.
‡ Rice et al., 1958.

TABLE 3. Potential Energy Parameters Calculated from Static and Dynamic Data
(Equation 15, $m = 1$)

Substance	A, 10^5 bars	b	K, 10^5 bars	δ
Periclase*†	9.06280	7.58615	9.30506	−0.148
Olivine†‡	18.77562	3.95477	18.96441	0.375
Nickel*§	8.17398	8.99049	8.40164	0.116

* Bridgman, 1949b.
† Bridgman, 1948.
‡ McQueen and Marsh, 1966.
§ Rice et al., 1958.

where

$$K_p(x) = \frac{Ax^{-2/3}}{3}(bx^{1/3} + 2)e^{b(1-x^{1/3})} - 4/3 Kx^{-4/3} \quad (21')$$

HIGH-TEMPERATURE CORRECTIONS

With an increase in temperature, the two-term Mie-Grüneisen equation of state becomes inadequate, because at high temperatures such effects as anharmonicity of atomic vibrations in a lattice and thermal excitation of conduction electrons in metals cannot be ignored. High-temperature corrections to the equation of state have been considered by the present authors [Zharkov and Kalinin, 1968]. Here we confine ourselves to only two of these corrections.

At high temperatures, the free energy of the crystal, when these effects are taken into account, will be expressed by

$$F = F_D(x, T) - \frac{A_a(x)}{\rho_0} T^2 - \frac{A_e(x)}{\rho_0} T^2 \quad (22)$$

Here $F_D(x, T)$ should be taken from (9). From (22) we have

$$P = P_D(x, T) + (A_a \gamma_a + A_e \gamma_e) T^2 / x \quad (23)$$

where $\gamma_a = \partial \ln A_a / \partial \ln x$ and $\gamma_e = \partial \ln A_e / \partial \ln x$ are introduced by analogy with Grüneisen's parameters of (11). If $P_D(x, T)$ is known, then, assuming $P = 0$ in (23), we can determine

$$(A_a \gamma_a + A_e \gamma_e) = -\frac{xP_D(x, T)}{T^2} \quad (24)$$

Other values could be determined analogously, but we shall not consider them here.

EQUATION OF STATE AFTER PHASE TRANSITION

The sharp discontinuities in the earth's interior are ordinarily attributed either to different chemical composition on both sides of the discontinuity or to some phase transition occurring within a chemically uniform solid. The former hypothesis cannot be verified practically, but the latter may be qualitatively analyzed. Therefore the following three types of phase transitions are important in geophysics: (1) polymorphic transitions accompanied by a change in crystal symmetry as well as by a jump in the volume; (2) transitions associated with metallization; and (3) melting.

Polymorphic Transitions

We are at present interested only in equations of state, in particular, in the equation of state after phase transition. For this discussion, we take the polymorphic transition as an example. Theoretical investigation of phase transitions is a rather complex task, and so we shall not discuss the problem in detail but shall start from the experimental fact that a high pressure state does exist.

Assume that the transition occurs under P_t pressure and T_t temperature and that the latter is not very high, so that the high temperature corrections in (22) can be neglected. Denote the values relating to a low pressure phase by the index 1, and those of a high pressure phase by 2. At the P_t, T_t point, the volume of the solid reduces from x_1 to x_2.

Now let us consider the phase transition when a solid is compressed by a shock wave. The shock-wave adiabat for iron is shown in Figure 3. AB represents the shock-wave adiabat for the low pressure phase. Under p, T conditions, observed at B, phase transition occurs. The BmC arc corresponds to a mixture of phases, and two

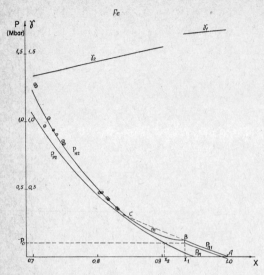

Fig. 3. Shock-wave data for iron: circles, experimental points (P_H) used to determine the equation of state [Rice et al., 1958; McQueen and Marsh, 1960].

shock waves are observed within the substance. Point C lies on the intersection of ABC with the shock-wave adiabat. Above this point we have the shock-wave adiabat for a high pressure phase. The equation for this state is analogous to (20):

$$P_{H_2} = \frac{P_{p2}(x) + \frac{\rho_0 \gamma_2}{x}[E_0 - E_{p2}(x)]}{1 - \gamma_2(1-x)/2x} \quad (25)$$

where E_0 and ρ_0 are the parameters of low pressure phase at $x = 1$. E_{p2} should be determined from the same energy level as E_0. Thus, we return to Figure 3, from which we obtain

$$E_{p2}(x_2) = E_{p1}(x_1) + P_0(x_1 - x_2) \quad (26)$$

Assuming that the equation of state for the low pressure phase is known, the pressure $P_0 = P_{p1}(x_1)$ can easily be found. Moreover, from Figure 3 we have

$$P_{p_2}(x_2) = P_0 \quad (27)$$

Equations 26 and 27 can be used to determine x_2 and K_2 from (3). A_2 and b_2 can be estimated from (18) using experimental data on P_{H2}. As for the δ_2 constant in (15), we proceed as follows: assuming $\delta_2 = 0$, or assuming $\theta_2(x_2) = \theta_1(x_1)$, we obtain the normalizing condition

$$\frac{\gamma_2(x_2)}{x_2} = \frac{\gamma_1(x_1)}{x_1} \quad (28)$$

When $\delta_2 = 0$, the following parameters were obtained for gabbro:

$A_2 = 2.0252 \cdot 10^6$ bars
$b_2 = 4.3216$
$K_2 = 2.5036 \cdot 10^6$ bars
$x_2 = 0.7766$

When (28) holds, the parameters for iron are:

$A_2 = 8.0222 \cdot 10^5$ bars
$b_2 = 7.9960$
$K_2 = 8.8681 \cdot 10^5$ bars
$x_2 = 0.9018$
$\delta_2 = 0.0201$

Metallization

Metallization of substances under pressure has great significance in explaining the nature of the earth's core. Two questions arise: (1) can the silicates of the earth's interior transform into a metal state under a pressure of 1.4 Mbar, and (2) how can the equation of state be derived for a metallized phase occurring in this transition?

The second problem is comparatively easily solved in view of what we have said above. If the temperature observed during metallic transition is not too high (so that the high-temperature corrections can be neglected), the equation of state for the metallized phase is derived as for the polymorphic transition given above.

At present, no definite answer can be given to the first question. Recently, *Trubitsin and Ulinich* [1964] determined the transition pressure for lithium hydride (~24 Mbar) and for helium (~90 Mbar). *Trubitsin* [1966] also estimated that the molecular hydrogen became a metal under a pressure of several megabars. *Drickamer* [1963] discovered that selenium goes into the metallic state under a pressure of 128 kbar, and iodine, at 235 kbar. However, we cannot evaluate at present the transition pressures for such complex compounds as silicates.

Altshuler [1965] did not observe any phase transition with noticeable volume changes above 1.4 Mbar when most important silicates were compressed by shock waves up to several megabars. Of course, this does not mean that these experiments completely repudiate Lodochnikov-Ramsay's hypothesis on the phase nature of the interface between the mantle and the earth's core, but because of these experiments, we begin to doubt its validity.

Melting

Melting is a phase transition of the first order where a crystalline solid transforms into a liquid. In melting, as in any other phase transition, phases could be in equilibrium only under some specific conditions:

$$T_1 = T_2 \quad P_1 = P_2 \quad \mu_1 = \mu_2$$

where μ is the chemical potential. It is these conditions that govern the phase equilibrium curve [in the given case, it is the melting curve $T_m(P)$]. However, the present theories on the liquid state do not allow us to express the chemical potential of a liquid with sufficient accuracy. Therefore various semi-empirical methods (based only on some properties of the crystalline phase) are widely used to determine melting curves. We shall discuss Lindemann's method, the method of critical concentration of thermal defects, and Simon's empirical equation.

Lindemann assumed that, in the melting curve, the ratio of mean root squares for thermal displacement of an atom from the equilibrium point $\bar{U}^2$ to the square of lattice constant a^2 remains constant. It can be shown that at high temperatures

$$\bar{U}^2 = \frac{kT}{3mN} \int_0^\infty \frac{Z(\omega)}{\omega^2} d\omega \quad (29)$$

where m is the atomic mass, k is Boltzmann's constant, N is the number of atoms in the crystal, and Z is the spectral density. Choosing Debye spectral density as $Z(\omega)$ and expressing Debye temperature through adiabatic modulus of compression K_s, Poisson ratio σ, and crystal density ρ, we get

$$\bar{U}^2/a^2 = G \frac{T_m \rho}{K_s \zeta(\sigma)} = \text{const} \quad (30)$$

where G is a constant for the given substance. From the assumption that $\bar{U}^2/a^2$ is constant on the melting curve, it follows that (30) contains the melting point T_m instead of some arbitrarily chosen temperature. If we denote the parameters under normal pressure by zero indices, then we get from (30) a convenient formula for relative variations of T_m

$$T_m = T_{m_0} \left(\frac{\rho_0}{\rho}\right)\left(\frac{K_s}{K_{s_0}}\right)\left[\frac{\zeta(\sigma)}{\zeta(\sigma_0)}\right] \quad (31)$$

It is generally accepted that σ does not vary ($\sigma = \sigma_0$). Thus, if we know the equation of state, $T_m(p)$ can be derived from (21) and (10).

The method of critical concentrations of the defects was proposed by *Zharkov* [1959] on the assumption that the concentration of thermal defects, C_d, on the melting curve, remains constant.

$$C_d(T_{m0}, P = 0) = C_d(T_m, P_m) \quad (32)$$

where zero index denotes the melting point under normal pressure ($P \approx 0$). The $C_d(T, P)$ dependency is expressed by

$$C_d(T, P) = \exp \frac{s_2 T - h}{kT} \quad (33)$$

where s_2 is the thermal entropy for the formation of one defect, and h is the enthalpy for generating the thermal defect.

From (32) and (33), by stipulating some computation simplifications, the following expression can be obtained:

$$T_m = T_{m_0} \exp \left\{ \int_{P_p(1)}^{P_p(x)} \frac{L \, dP}{K_p(x)} \right\} \quad (34)$$

where $K_p(x)$ is the potential part of compressibility modulus (21). L is estimated either from the initial slope of experimental curve, $T_m(P)$, or from the Clausius-Clapeyron equation

$$\frac{dT_m}{dP} = \frac{T_m \Delta V}{q} = T_m L \frac{1}{K_T(1)}$$

using the known values of volume change (ΔV) in melting, heat of melting q, and compressibility modulus $K_T(1)$. The magnitude of L is usually 2 or 3.

Simon's empirical equation

$$P_m = a\left[\left(\frac{T_m}{T_{m0}}\right)^c - 1\right] \quad (35)$$

where a and c are experimental constants, has received wide application. In the review by *Babb* [1963] a vast amount of experimental data on the melting curves was analyzed on the basis of (35). The success of Simon's equation in describing the experimental results is based on the fact that a and c make it possible to explain satisfactorily the comparatively small part within the melting curve. However, in extrapolating the equation to the higher pressures that sometimes occur in geophysics, one might face some unexpected errors. Detailed information on this subject has been given by Zharkov and Kalinin [1968].

GEOPHYSICAL APPLICATIONS

In addition to a number of indirect applications in the study of the earth's interior, equations of state can be used to interpret geophysical data. Birch [1952] and Magnitsky [1952] started such investigations. Below we shall consider a quantitative method of comparing the experimental data on compressibility of solids under high pressures obtained in the laboratory with available geophysical information. If equations of state for a number of rocks and minerals are known, then this method can be used to determine the chemical compositions of deeper layers of the earth.

As initial geophysical data, we take the well-known values of distributions $\phi(h) = K_s/\rho = V_p^2 - 4/3 V_s^2$, pressure $P(h)$, and density $\rho(h)$ at a depth h. The function $\phi(h)$ is satisfactorily determined from seismological data, and $P(h)$ practically does not vary with $\rho(h)$ within reasonable limits. The density distribution $\rho(h)$ is derived from the Adams-Williamson equation and is less accurate than $\phi(h)$ and $P(h)$. This implies that the physical properties of

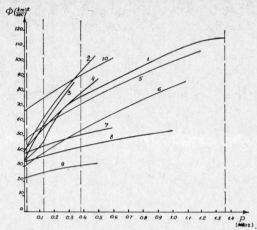

Fig. 5. ϕ—P diagram; notations are as in Figure 4.

the material should be compared with geophysical data on the ϕ-P plane. However, in geophysics the function $P(\rho)$ is preferred because it is not possible to calculate ϕ without knowing the equation of state.

Figure 4 shows both the $P(\rho)$ distribution for the upper mantle in Bullen's model A' and the theoretical $P(\rho)$ curves for a number of solids calculated from (10) when the parameters were chosen from Tables 1–3. Similarly Figure 5 shows geophysical distribution $\phi(P)$ and curves $\phi(P)$, obtained from (10) and (21) for the same solids (calculations made for $T = 300°K$).

If the temperature $T(h)$ inside the earth were known with some accuracy, then this method would be considered as conclusive. Thus we could produce an artificial mixture having the same mechanical properties as the matter of earth, by mixing different minerals in specific proportions. Unfortunately, $T(h)$ is not known with certainty, and, as can be seen in Figures 1 and 2, the thermal part of pressure at temperatures of several thousand degrees constitutes a major part of the total pressure.

Temperature can be neglected in the following case. Suppose we must verify whether a particular earth's layer consists of a given substance or mixture. Let the equation of state be known for this substance. Then dividing (21) by ρ and eliminating temperature with the help of (10), we obtain

$$\phi = \frac{K_p}{\rho} + \frac{x}{\rho_0}[P - P_p(x)]\left(\gamma + 1 - \frac{\partial \ln \gamma}{\partial \ln x}\right)$$

(36)

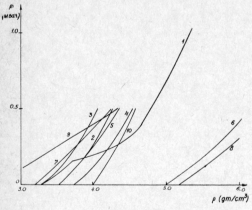

Fig. 4. P—ρ diagram: (1) Bullen's model A'; (2) grossularite; (3) diopside; (4) andradite; (5) periclase; (6) pyrite; (7) olivine; (8) magnetite; (9) orthoclase; (10) corundum.

If the layer really consists of the assumed substance, then by substituting the pressure distribution for this layer in (36) instead of p, we must obtain a ϕ agreeing with geophysical distribution. If these distributions do not coincide, then the layer cannot consist of the given substance.

Zharkov and Kalinin [1962b] used this method to compare the equations of state for gabbro and dunite with the geophysical properties of the D layer of the earth and obtained a negative result, i.e. the D layer cannot consist of either gabbro or dunite.

REFERENCES

Altshuler, L. V., Application of shock waves in high-pressure physics, *Usp. Fiz. Nauk*, 85(2), 197–258, 1965.

Babb, S. E., Parameters in the Simon equation relating pressure and melting temperature, *Rev. Mod. Phys.*, 35(2), 400–413, 1963.

Birch, F., Elasticity and constitution of the earth's interior, *J. Geophys. Res.*, 57(2), 227–286, 1952.

Birch, Francis, Compressibility; elastic constants, in *Handbook of Physical Constants*, edited by S. Clark, *Geol. Soc. Am. Mem. 97*, 97–173, 1966.

Born, M., and M. Göppert-Mayer, Dynamische Gittertheorie der Kristalle, *Handbuch Phys.*, 24(2), 623–794, 1933.

Born, M., and Kun Huang, *Dynamical Theory of Crystal Lattices*, Clarendon Press, Oxford, 425 pp., 1954.

Bridgman, P. W., The compression of 39 substances to 100,000 kg/cm², *Proc. Am. Acad. Arts Sci.*, 76, 55–70, 1948.

Bridgman, P. W., *Physics of High Pressure*, G. Bell, London, 445 pp., 1949a.

Bridgman, P. W., Linear compression to 30,000 kg/cm² including relatively incompressible substances, *Proc. Am. Acad. Arts Sci.*, 77, 187–234, 1949b.

Davydov, B. I., On the equation of state of solids, *Izv. Akad. Nauk SSSR, Ser. Geofis.*, 1956(12), 1411–1418, 1956.

Drickamer, H. G., The electronic structure of solids under pressure, in *Solids under Pressure*, edited by W. Paul and D. Warschauer, pp. 357–384, McGraw-Hill Book Company, New York, 1963.

Dugdale, J. S., and D. K. C. McDonald, The thermal expansion of solids, *Phys. Rev.*, 89(4), 832–834, 1953.

Grüneisen, E., Zusammenhang zwischen Kompressibilität, thermischer Ausdehnung, Atomvolumen und Atomwärme der Metalle, *Ann. Phys.*, 26, 393–402, 1908.

Magnitsky, V. A., On the density and compressibility of the earth's shell, *Vopr. Kosmogonii Akad. Nauk SSSR*, 1, 15–33, 1952.

McQueen, R. G., and S. P. Marsh, Equation of state for nineteen metallic elements from shock-wave measurements to two megabars, *J. Appl. Phys.*, 31, 1253, 1960.

McQueen, R. G., and S. P. Marsh, see pp. 153–159 in *Birch*, 1966.

Mie, G., Zur kinetischen Theorie der einatomigen Körper, *Ann. Phys.*, 11, 657–697, 1903.

Rice, M. H., R. G. McQueen, and J. M. Walsh, Compression of solids by strong shock waves, *Solid State Phys.*, 6, 1–63, 1958.

Slater, J. C., *Introduction to Chemical Physics*, McGraw-Hill Book Company, New York, 521 pp., 1939.

Trubitsyn, V. P., and F. P. Ulinich, Metallization theory of crystals, with small atomic numbers, under high pressures, in *Solids under Conditions of Pressure and Temperature of the Earth's Interior*, pp. 80–193, Nauka, Moscow, 1964.

Trubitsyn, V. P., Phase transition in a hydrogen crystal, *Fiz. Tverd. Tela.*, 8(3), 862–865, 1966.

Vaschenko, V. Ya., and V. N. Zubarev, Grüneisen coefficient, *Fiz. Tverd. Tela*, 5(3), 886–890, 1963.

Zharkov, V. N., Melting temperatures of the earth's shell and iron under high pressures, *Izv. Akad. Nauk SSSR, Ser. Geofiz.*, 1959(3), 465–470, 1959.

Zharkov, V. N., and V. A. Kalinin, Grüneisen parameter for high pressures, *Dokl. Acad. Nauk SSSR*, 145(3), 551–554, 1962a.

Zharkov, V. N., and V. A. Kalinin, Equation of state for gabbro and dunite under high pressures, *Izv. Acad. Nauk SSSR, Ser. Geofiz.*, 1962(3), 288–306, 1962b.

Zharkov, V. N., and V. A. Kalinin, Equation of state for solids under high pressures and temperatures, Nauka, Moscow, 312 pp., 1968.

The Mohorovicic Discontinuity[1]

D. P. McKenzie

MOHOROVIČIĆ [1910] noticed that the travel times from near earthquakes could be understood only if the compressional wave velocity increases discontinuously with depth. He suggested a depth of about 50 km in central Europe for the discontinuity which now bears his name. The Mohorovicic discontinuity, or Moho, has since been recognized all over the earth, though in tectonically unstable regions it often becomes vague and ceases to give clear refractions. The material above the Moho is the earth's crust, whose thickness varies widely from 5 km under the deep oceans to perhaps 70 km under the Andes.

Two explanations have been suggested for the sudden increase in velocity at the Moho. The first is that there is a change in composition from perhaps gabbro in the lower crust to peridotite in the upper mantle. The other suggestion is that the Moho is the phase boundary between rocks having the same chemical composition, with either eclogite below and gabbro above, or peridotite below and serpentinite above. The second suggestion has many attractions, but is not supported either by high pressure experiments on rocks or by seismic studies of the structure of the crust and upper mantle. However, the second phase change has not been excluded as the cause of the oceanic Moho. In general, as information about the crust and upper mantle has accumulated, the importance of the Moho has decreased.

The principal source of information about the depth to the Moho and the contrast in compressional wave velocity V_p across it is seismic refraction experiments. Such studies can rarely be used to investigate shear wave velocities V_s. Since the phase and group velocity of surface waves are principally determined by the shear wave structure, short period Love and Rayleigh waves may also be used to determine the depth of the Moho where refraction studies have not been made. *Drake and Nafe* [1968] have collected and summarized the available refraction results (see *McConnell et al.* [1966] and *Kosminskaya and Riznichenko* [1964] for references), and their conclusions about the regional variation in V_p in the crust and upper mantle agree in general with the less extensive results from surface waves (see section by Brune in this Monograph, p. 230). Beneath Precambrian shields of low elevation and ocean basins distant from active ridges, V_p is between 6.5 and 7.2 km/sec in the lower crust, and it increases to 7.8–8.5 km/sec in the upper mantle. The depth at which this change takes place varies from 30 to 60 km beneath shields and from 5 to 8 km beneath the ocean floor. The models for V_s that fit the surface wave observations in these regions have a velocity of 3.7 to 3.8 km/sec above the Moho and 4.8 km/sec below. The depths to the mantle are within the range determined from refraction studies.

Ocean basins and shields are the stablest regions of the earth's surface. Less stable regions, such as continental margins and tertiary orogenic belts, tend to have greater scatter in velocities in both the crust and upper mantle. The base of the crust is generally taken as the upper surface of the layer with a V_p of 7.8 km/sec and is usually fairly definite, though the occurrence of velocities between 7.2 and 7.8 km/sec may confuse the interpretation. Crustal thicknesses determined in this way vary from 15 km under continental margins to at least 60 km under high mountain ranges. The surface wave model also requires an intermediate velocity of 4.0 km/sec between the two in stable areas.

In very active and unstable regions, such as the midocean ridges and the Basin and Range province in the western United States, refracted arrivals are often difficult to observe and may come from a layer with a velocity of

---

[1] Division of Geological Sciences, California Institute of Technology, contribution 1463.

7.7 km/sec or less [*Herrin and Taggart*, 1962; *Le Pichon et al.*, 1965]. The surface wave models for V_s in such regions have no high velocity layer between the crust and the low velocity layer in the mantle. Thus, though the velocities are low, they are probably those of upper mantle and not crustal rocks. Defined in this way, the crust is thinner beneath active regions than beneath more stable continental and oceanic regions. The elevation of the Basin and Range province and the oceanic ridges is compensated by a low density upper mantle and not by a thick crust. The topography of the Moho is often determined from the measurement of the acceleration due to gravity [*Worzel and Harrison*, 1963]. The calculation assumes the gravity field to be due to the density contrast across the Moho. Any knowledge of crustal structure obtained from seismology can be used to improve the model, but the basic assumption is that the density of the upper mantle is constant. This is now known not to be true, though the upper mantle density variations are believed to be regional and not local. Thus gravity measurements may be used to interpolate between seismic refraction stations, but not between different regions.

The regional variations in crustal thickness may well be related to the stress field. Beneath the Basin and Range province the Moho is shallower than beneath other more stable continental regions to the east [*Pakiser*, 1963; *James and Steinhart*, 1966]. The same is true of crustal thicknesses under mid-ocean ridges as compared with ocean basins [*Le Pichon et al.*, 1965]. Both regions are probably in tension, since active normal faults are common. In contrast, the crust is thickened beneath regions believed to be in compression. In continents this effect is well known and fold mountains are described as having low density roots. In island arcs, detailed seismic refraction studies [*Talwani et al.*, 1959; *Kosminskaya et al,*. 1963] show that the crust is thickened beneath the islands. The negative gravity anomalies occur over the trenches and are not produced by a change in crustal thickness.

Regions with low upper mantle velocities are often areas of high heat flow. On mid-ocean ridges this correlation is well established, and recent measurements of heat flux in the Basin and Range province [*Birch*, 1966] give around 2 $\mu cal/cm^2$ sec, or considerably above the world average. Seismic attenuation also appears to be correlated with low upper mantle velocities. A study of high frequency surface waves by *Sutton et al.* [1967] shows that both P_g and L_g have an effective Q of at least 600 in the eastern United States, compared with perhaps 250 in the Basin and Range province. Though this difference may be due to scattering from inhomogeneities in the crust, it could also be explained by regional variations of Q in the upper mantle.

The sharpness of the transition from the crust to the upper mantle is poorly determined, despite its importance in understanding the nature of the Moho. Spectral analysis of refracted arrivals suggests that the transition layer is not more than ½ km thick in stable areas [*Nakamura and Howell*, 1964]. A thin transition layer is also supported by the existence of reflected arrivals from the continental [*Johnson*, 1965; *Dix*, 1965] and oceanic Moho [*Helmburger*, 1968]. In active regions the transition is probably less sharp.

An attractive and simple explanation of the Mohorovicic discontinuity is that it is a phase boundary between phases of different density but the same composition (see *Bullard and Griggs* [1961] for earlier references). The phase change from gabbro above to eclogite below was the first possibility to be suggested. If the phases are in equilibrium at the Moho, regional uplift is easily produced by increasing the temperature gradient. Since the eclogite to gabbro transition is endothermic, the Moho then migrates downward and the compensation of the thicker crust produces uplift without horizontal transport of light materials. Unfortunately, both the seismic results and experiments on rocks of basaltic composition at high pressure demonstrate that the gabbro-eclogite phase change cannot account for the sharpness and regional variation in crustal thickness. *Hess'* [1955] suggestion that the oceanic Moho is the boundary between serpentinite in the crust and peridotite in the upper mantle cannot be dismissed by the same arguments.

Bullard and Griggs [1961] gave two arguments against a phase change at the Moho. Their first objection depended on the approximate equality of oceanic and continental heat flows. Since the heat production in the continental crust is considerably greater than that in the oceanic, the temperature gradient beneath continents must be less than that beneath oceans. If the Moho is a phase boundary between gabbro and eclogite, its depth must increase with the temperature gradient, and it should therefore be deeper beneath oceans. Any attempt to escape from this difficulty by forcing the phase transition through both oceanic and continental Moho requires the dense phase to form the crust in oceanic regions. Such a structure would be gravitationally unstable and is not in agreement with the seismic velocities. This difficulty can be avoided only if the composition of the lower crust and mantle is not the same beneath continents and oceans.

Their other argument is more general and depends on the phase rule. The exact composition of the mantle is not known, but all rocks suggested so far contain at least two component minerals. In such solids the phase rule requires that phase changes take place over a temperature range at a given pressure. Though the thickness of the Moho has not been accurately determined, it is unlikely that a diffuse boundary can explain the seismic observations.

The second argument now has strong experimental support. *Ringwood and Green* [1966] have synthesized eclogite from basalt of various compositions at high pressure and temperature. At 1100°C the transition takes place over a pressure range of between 3 and 12 kb, or a Moho thickness of 10 to 40 km. The transition temperature and pressure suggest that eclogite may be stable throughout the continental crust. Crustal basalts and gabbros are then required to be in metastable equilibrium. Green and Ringwood conclude that neither the continental nor the oceanic Moho can be the phase boundary between gabbro and eclogite. They believe that the Moho is the boundary between rocks of basaltic composition above and peridotites below.

Hess [1955, 1962] avoided the objections of Bullard and Griggs to a phase change at the Moho when he suggested that the oceanic lower crust was serpentinite, the hydrated phase of peridotite. The two phases can only be in equilibrium at the Moho if the temperature gradient is about three times greater than the oceanic mean. Hess proposed that all the non-sedimentary oceanic crust originates along the mid-ocean ridges, where this condition may well be satisfied. The oceanic crust and upper mantle are then produced by freezing this equilibrium as the sea floor moves away from the ridge. Unlike the gabbro-eclogite phase change, the hydration of olivine involves only one mineral and therefore the phase boundary is sharp. Since the two phases are in equilibrium only at the ridges, the thickness of the crust will not reflect changes in the temperature gradient. Hess' hypothesis is supported by the probable absence of the lowest crustal layer under the mid-ocean ridges [*Le Pichon et al.*, 1965], but the range of velocities observed at the base of the crust is less than would be expected if it were to be composed of serpentinite (Figure 1). Thus Hess' suggestion cannot be dismissed, though it must be restricted to the oceanic crust and does not have the same tectonic possibilities as the gabbro-eclogite transition. Indeed, except where it is produced and destroyed, Hess' oceanic crust behaves as a rock with a composition different from that of the upper mantle.

The Mohorovicic discontinuity is seismically the most important boundary in the crust and upper mantle. However, it is no longer believed to have a similar importance in gravity and tectonics because density variations in the upper mantle are compensated. Therefore, the lower boundary of the rigid layer or lithosphere must lie within the mantle, not at the Moho. The change in mechanical properties is now believed to occur at about the top of the low velocity layer [*Gutenberg*, 1959; *Anderson*, 1962]. Such a transition is probably not sharp, and for this reason alone is unlikely to occur at the Moho. This mechanical model applied to the sea floor suggests that the oceanic upper mantle to a depth of perhaps 50 km moves with the crust as spreading takes place from the ridges. These arguments and the experimental evidence against the gabbro-eclogite transition reduce the tectonic importance of the Moho. However, regional uplift may well be produced by phase changes within the upper mantle rather than at the Moho. If the resulting density changes are above the depth at which isostatic compensation takes place, they must result in changes of surface elevation. Various phase changes in peridotites are pos-

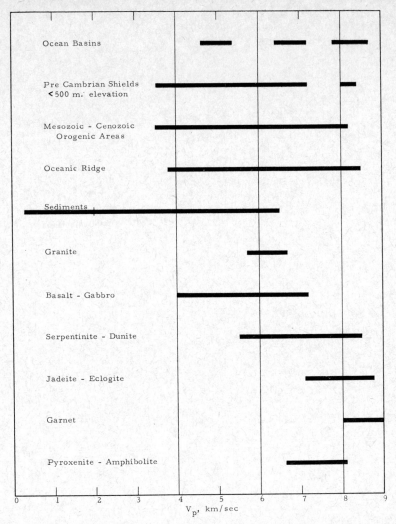

Fig. 1. The range of velocities in ocean basins and shields agrees with a basaltic lower crust underlain by an upper mantle of peridotite. Elsewhere the range of velocities covers all possible compositions.

sible, but, if uplift is to take place in regions of high heat flow, the reaction must be endothermic when the density decreases. Such a phase change would also account for the regional variation in both mantle velocities beneath the Moho and in the thickness of the low velocity layer.

Perhaps the most important conclusion from recent seismic experiments is that the Moho is only one of several seismic boundaries in the crust and upper mantle. It is probable that some of the other diffuse boundaries are more important in tectonics than is the Moho.

REFERENCES

Anderson, D. L., The plastic layer of the earth's mantle, *Sci. Am.*, July 1962.

Birch, F., Heat flow measurements in the last decade, in *Advances in Earth Sciences*, edited by P. M. Hurley, MIT Press, Cambridge, 1966.

Bullard, E. C., and D. T. Griggs, The nature of the Mohorovicic discontinuity, *Geophys. J.*, 6, 118–123, 1961.

Dix, C. H., Reflection seismic crustal studies, *Geophysics*, 30, 1068–1084, 1965.

Drake, C. L., and J. E. Nafe, The transition from continents to oceans from seismic refraction data, in *The Crust and Upper Mantle of the Pacific Area, Geophys. Monograph 12*, American Geophysical Union, Washington, D. C., 1968.

Gutenberg, B., *The Physics of the Earth's Interior*, Academic Press, New York, 1959.

Helmburger, D. V., The crust-mantle transition in the Bering Sea, *Bull. Seismol. Soc. Am., 58*, 179–214, 1968.

Herrin, E., and J. Taggart, Regional variations of P_n velocity and their effect on the location of epicenters, *Bull. Seismol. Soc. Am., 52*, 1037–1046, 1962.

Hess, H. H., The oceanic crust, *J. Marine Res., 14*, 423–439, 1955.

Hess, H. H., History of the ocean basins, in *Petrologic Studies: A Volume to Honor A. F. Buddington*, edited by A. E. J. Engel et al., Geological Society of America, New York, 1962.

James, D. E., and J. S. Steinhart, Structure beneath continents: A critical review of explosion studies 1960–1965, in *The Earth beneath the Continents, Geophys. Monograph 10*, edited by J. S. Steinhart and T. Smith, American Geophysical Union, Washington, D. C., 1966.

Johnson, L. R., Crustal structure between Lake Mead, Nevada, and Mono Lake, California, *J. Geophys. Res., 70*, 2863–2872, 1965.

Kosminskaya, I. P., and Y. V. Riznichenko, Seismic studies of the earth's crust in Eurasia, in *Research in Geophysics*, vol. 2, pp. 81–122, MIT Press, Cambridge, 1964.

Kosminskaya, I. P., S. M. Zverev, P. S. Veitsman, Yu. V. Tulina, and R. M. Krakshina, Basic features of the crustal structure of the Sea of Okhotsk and the Kuril-Kamchatka zone of the Pacific Ocean from deep seismic sounding data, *Bull. Acad. Sci. USSR, Geophys. Ser.*, English Transl., no. 1, 11–27, 1963.

Le Pichon, X., R. E. Houtz, C. L. Drake, and J. E. Nafe, Crustal structure of the mid-ocean ridges, 1, Seismic refraction measurements, *J. Geophys. Res., 70*, 319–339, 1965.

McConnell, R. K., Jr., R. N. Gupta, and J. T. Wilson, Compilation of deep crustal structure seismic refraction profiles, *Rev. Geophys., 4*, 41–100, 1966.

Mohorovičić, A., Das Beben vom 8. X. 1909, *Jahrb. Meteorol., Observatorium Zagreb für das Jahr 1909, 9*, Pt. IV, Sec. 1, 63 pp., 1910.

Nakamura, Y., and B. F. Howell, Jr., Maine seismic experiment: Frequency spectra of refracted arrivals and the nature of the Mohorovicic discontinuity, *Bull. Seismol. Soc. Am., 54*, 9–18, 1964.

Pakiser, L. C., Structure of the crust and upper mantle in the western United States, *J. Geophys. Res., 68*, 5747–5756, 1963.

Ringwood, A. E., and D. H. Green, An experimental investigation of the gabbro-eclogite transformation and some geophysical implications, *Tectonophysics, 3*, 383, 1966.

Sutton, G. H., W. Mitronovas, and P. W. Pomeroy, Short period energy radiation patterns from underground nuclear explosions and small magnitude earthquakes, *Bull. Seismol. Soc. Am., 57*, 249–267, 1967.

Talwani, M., G. H. Sutton, and J. L. Worzel, A crustal section across the Puerto Rico trench, *J. Geophys. Res., 64*, 1545–1555, 1959.

Worzel, J. L., and J. C. Harrison, Gravity at sea, in *The Sea*, vol. 3, edited by M. N. Hill, pp. 134–174, Interscience Publishers, New York, 1963.

Low-Velocity Layers in the Upper Mantle

V. A. Magnitsky and V. N. Zharkov

DEFINITIONS AND REGIONAL FEATURES

It was established by seismic investigations as early as the first quarter of the twentieth century that the mantle of the earth as a whole is characterized by a rise in the velocities of the longitudinal (V_p) and transverse (V_s) seismic waves from values of about 8 km/sec and 4.5 km/sec, respectively, under the Moho to 13.6 km/sec and 7.3 km/sec at the boundary of the core (depth about 2900 km). However, *Gutenberg* [1926] suggested that the monotonic increase of seismic velocities with depth is interrupted in the uppermost parts of the mantle at depths of approximately 100 km. The results of investigations dealing with the propagation of longitudinal and transverse waves were summarized by *Gutenberg* [1959]. He assumed that the decrease in the velocity of both wave types starts on the average at depths of about 50 km, with the velocities reaching minimum value at a depth of some 150 km and reattaining the initial value (before the beginning of the decrease) at a depth of 250 to 300 km. The maximum decrease in the velocities reaches 3% for longitudinal waves, and 5% for transverse waves. The region is called the 'low-velocity layer.'

A low-velocity layer would be a waveguide (channel) [*Redwood*, 1960]. Attempts were made to find channelized waves; their existence was first reported by *Caloi* [1953, 1954] and *Press and Ewing* [1954, 1955], and later in the data of the USSR seismic network [*Magnitsky and Khorosheva*, 1961].

Investigations of the earth's free oscillations, resulting from large earthquakes, as well as long-period surface waves, have left little doubt as to the existence of a low-velocity layer in the upper mantle. A summary of the principal results has been given by *Anderson* [1967]. However, many questions pertaining to this problem still remain. First, it is not clear whether the low-velocity layer is uninterrupted. It is possible that there are areas in the globe where no such layer exists. In a number of areas, such as Iceland and the North Atlantic Ocean [*Tryggvason*, 1964], Japan and the Kuril arc [*Fedotov et al.*, 1964], and some regions in the western United States [*Pakiser*, 1963], unusually low velocities of seismic waves have been observed immediately below the Moho (for example, P-wave velocities from 7.4 to 7.8 km/sec). In these areas there appears to be no low-velocity layer in the strict meaning of the term; instead, a monotonic increase in velocity with depth would be expected, although the increase begins with abnormally low velocities.

Second, there are regional variations in the position and the properties of the layer where it does exist. It is well established that the low-velocity layer differs in depth and in magnitude of the velocity decrease for oceanic and continental areas, and for areas of tectonic activity [*Anderson*, 1967]. In addition, there are indications of differences between the Atlantic and Indian Oceans on the one hand, and the Pacific Ocean on the other [*Aki and Press*, 1961], as well as between the southeastern and southwestern parts of the Pacific Ocean and its central part [*Kovach and Press*, 1961; *Santo*, 1961].

Another question to be resolved is the nature of the boundaries of the low-velocity layer: Are they characterized by a gradual change or a jump in velocities? The uncertainty is related to the fact that the drop in velocities is very small, which makes the detection of boundaries rather difficult.

There is no doubt about the existence of a low-velocity layer for transverse waves, but the question regarding the longitudinal waves is far from being clear [*Nuttli*, 1963]. It appears that in some large regions, for example, the Canadian shield, there is no layer of low velocities for longitudinal waves. Instead, there is, at most, an interval of depths where the velocities of longitudinal waves practically do not increase with the depth [*Brune and Dorman*, 1963].

The question of the intensity of the decrease in velocities in the layer is far from being resolved. The decrease in velocities varies for different areas; moreover, in one area different methods yield different, often incompatible results. One of the most interesting solutions, based on the joint use of travel times of transverse waves and the dispersive curves of Love waves, shows in the central part of the United States a layer of low velocities in transverse waves only 60 km wide with a maximum drop in velocities of only 1% [*Ivanova et al.*, 1964].

However, a part of the upper mantle within a range of 50 to 300 km exhibits not only a drop in the velocities of seismic waves, but also a number of other peculiarities that will be mentioned below.

THERMODYNAMIC PARAMETERS

For evaluation of the thermodynamic parameters of the subcrustal layer, the following geophysical characteristics should be used:

$$V_s \sim 4.4\text{--}4.7 \text{ km/sec} \quad V_p \sim 8\text{--}8.2 \text{ km/sec}$$
$$\Phi = K_s/\rho = V_p^2 - \tfrac{4}{3} V_s^2$$
$$T \sim (1.5\text{--}2) \times 10^3 \text{ °K} \quad \rho \sim 3.3\text{--}3.5 \text{ g/cm}^3 \quad (1)$$

where V_s and V_p are the velocities of transverse and longitudinal seismic waves, ρ is density, K_s is the adiabatic bulk modulus and T is the absolute temperature. The problem of the thermodynamic characteristics of the earth's mantle has been considered by *Zharkov* [1959a]. For evaluation of the parameters, it is possible either to proceed from the experimental data obtained under laboratory conditions, or to use a

Debye or an Einstein model of a solid body and to calculate all the quantities by means of the appropriate formulas. In either case it should be borne in mind that silicates belong to substances with polyatomic elementary cells, and that apart from acoustic frequencies which can be defined by an acoustic Debye temperature θ, there also exist optical frequencies, as well as their corresponding characteristic temperatures θ_{oi} ($i = 1, 2 \cdots, \nu - 1$; ν is the number of atoms in an elementary cell), which can be far higher than θ. For example, for olivine, $\nu = 28$. An estimate of θ_{oi} can be obtained from experimental data on infrared absorption. For many silicates, including olivine, the principal infrared absorption band corresponds to the length of a wave of $\lambda \approx 9 \mu$, which corresponds to $\theta_{oi} \sim 1520°$K, while $\theta \sim 500°$K.

In extrapolating the values of thermodynamic parameters to the conditions occurring in the subcrustal zone, it is always assumed that one deals with a classical case, where temperature T is higher than the Debye temperature θ. However, as the optical temperatures of silicates are high, whereas the experimental data were obtained at appreciably lower temperatures, a considerable error may result from extrapolation for the coefficient of thermal expansion α, heat capacities C_p and C_v, and Grüneisen's parameter γ (the extrapolated values will be smaller than the real ones). In addition, the temperature dependence of the moduli of elasticity may be somewhat reduced for the same reason. Zharkov's [1959a] thermodynamic magnitudes were evaluated by means of formulas for an Einstein model of a solid body with due regard to the greater values of θ_{oi}. For a depth $l \sim 100$ km, the thermodynamic parameters are

$$C_p/C_v^* = \beta_T/\beta_s = K_S/K_T = 1.19$$

$$C_p \sim 1.32 \times 10^7 \text{ ergs/g deg}$$

$$C_v/C_{v,0} \sim 0.95$$

$$C_{v,0} = 1.17 \times 10^7 \text{ ergs/g deg} \quad (2)$$

$$\beta_T \sim 0.86 \times 10^{-6} \text{ bar}^{-1}$$

$$\alpha \sim (5-7) \times 10^{-5} \text{ deg}^{-1}$$

$$\gamma \sim 1.6-1.9$$

$$(\partial p/\partial T)_v \sim 70 \text{ bar/deg}$$

where $C_{v,0}$ is the heat capacity after Dulong and Petit.

DISTRIBUTION OF PRESSURE AND TEMPERATURE

The assumption regarding hydrostatic equilibrium is a satisfactory approximation. Then

$$\frac{dp}{dl} = g\rho \quad p = p_0 + \int_0^l g\rho \, dl \quad (3)$$

$$p_0 \sim 1 \text{ bar}$$

If the distribution of density in the crust (the upper mantle) is defined, we determine at once the distribution of pressure by means of (3).

The question of the distribution of temperature T and of its gradient is more complicated and uncertain. To determine the temperature in the outer layers of the earth down to a depth of 100 to 150 km, it is possible to apply the stationary equation of heat conductivity for models of a flat earth. Then

$$\frac{d}{dl}\left(\chi \frac{dT}{dl}\right) = -P(l) \quad (4)$$

$$\left(\chi \frac{dT}{dl}\right)_l = \left(\chi \frac{dT}{dl}\right)_{l=0} - \int_0^l P(\xi) \, d\xi \quad (5)$$

$$T(l) = T_0 + \left(\chi \frac{dT}{dl}\right)_{l=0} \int_0^l \frac{d\xi}{\chi(\xi)}$$

$$- \int_0^l \frac{d\eta}{\chi(\eta)} \int_0^\eta P(\xi) \, d\xi \quad (6)$$

where χ is the coefficient of heat conductivity, and P is the generation of heat in a unit of volume, which in the problem under consideration can be regarded as being independent of time. In outer layers, χ can be expressed by the formula

$$\chi = \frac{\chi_0 T_0}{T} + \frac{16}{3} \frac{\sigma \nu^2 T^3}{\bar{\alpha}} \quad (7)$$

The first term in (7) is the lattice part of the coefficient of heat conductivity, and the second is the radiative part of the latter. χ_0 is the constant determined from experimental data, ν is the refractive index which depends slightly on the depth (for rock, $\nu^2 \sim 3$), σ is the Stefan-Boltzmann constant, and $\bar{\alpha}$ is the absorption coefficient, a rather uncertain quantity which may be $\sim$10–100.

The essential peculiarity of (7) is that it

leads to a minimum of the thermal conductivity coefficient at depths $l \sim 100$ km. This means that a heat-insulating layer in the earth is situated in the zone of reduced seismic velocities. The variable P, which along with χ determines the distribution of temperatures in the outer layers, is by itself a very important parameter: it determines the energetics of the earth's outer layers.

The stratified model of the earth $[P(l), \chi(l)]$ being given, we are in a position to determine the corresponding distribution of temperatures by means of (4)–(7). This procedure is ambiguous, since it proceeds from parameters that are known to be quite uncertain.

In this connection it is more reliable for practical purposes to apply the method of datum-mark points. Let us represent the dependence of the temperature gradient on depth (dT/dl) by the interpolative formula

$$\bar{\tau} = \bar{\tau}_0 + bz + cz^2$$

$$T = T_0 + \bar{\tau}_0 z + \tfrac{1}{2} bz^2 + \tfrac{1}{3} cz^3 \quad (8)$$

$$\bar{\tau} = \frac{dT}{dz}$$

where $z = l/\bar{l}$ is dimensionless depth ($\bar{l} = 100$ km), while the coefficients b and c are to be determined by some datum-mark points (conditions) that can be obtained from experimental data. For example, with $z = 0$, T is equal to the average day temperature of the earth's surface T_0, and $\bar{\tau}$ is equal to the average thermal gradient at the surface $\bar{\tau}_0$. The foci of volcanoes are at depths $l \sim \bar{l} = 100$ km, while the melting points of lava are known to be $\sim 1200\,°\mathrm{C}$. Consequently,

$$T_1 = 1200\,°\mathrm{C} \quad \text{for} \quad z = 1$$

Further, let us suppose it is known that the minimum on the velocity distribution curve is at depth z_2; at that depth $\bar{\tau}$ must have some critical value of temperature gradient $\bar{\tau}_2$ (The question of the magnitude of $\bar{\tau}_2$ will be taken up in greater detail below). As a result we can express b and c in (8) by parameters T_0, T_1, and $\bar{\tau}_0$, $\bar{\tau}_2$. Then

$$b = \frac{3z_2{}^2[\bar{\tau}_0 - (T_1 - T_0)] - (\bar{\tau}_0 - \bar{\tau}_2)}{z_2(1 - \tfrac{3}{2} z_2)} \quad (9)$$

$$c = \frac{\tfrac{3}{2}(\bar{\tau}_0 - \bar{\tau}_2) - 3z_2[\bar{\tau}_0 - (T_1 - T_0)]}{z_2(1 - \tfrac{3}{2} z_2)} \quad (10)$$

Figure 1 gives the temperatures obtained from (8) and (10).

In determining the distribution of temperatures, there arises the important question of the lower and upper limit of temperatures. Adiabatic temperatures are assumed to be the lower limit

$$T_l = T_{100} \exp \int_{l_{100}}^{l} \frac{g\alpha}{C_p} \, dl$$

$$\approx T_{100}\left\{1 + \left(\frac{g\alpha}{C_p}\right)_{100}(l - l_{100})\right\}$$

$$T_{100} \sim 1500\,°\mathrm{K} \quad (11)$$

and the melting-point curve to be the upper one. The melting-point curves of a number of silicates are presented in Figure 2. They can be described by the simple relationship

$$T_m = T_{m0} + ap \quad (12)$$

A relationship of this type also ensues from the theory of melting-point curves [*Zharkov*, 1959b]:

$$T_m = T_{m0}(1 + L\beta_{T0} p) \quad (13)$$

where L is a constant of the order 2–3. Equation 13 gives τ_m for the gradient of the melting-point curve:

$$\tau_m = \frac{dT_m}{dl} = T_{m0} L \beta_T g \rho \quad (14)$$

$$\tau_m \sim 3 \text{ deg/km} \quad (15)$$

is obtained upon substituting the numerical values; the temperature gradients at depths of 100 km or more cannot appreciably exceed this value, since the temperatures in this region are close to the melting-point. Otherwise the temperature curve would intersect the melting-point curve, which is impossible.

CONDITION FOR THE ABSENCE OF CONVECTION IN A CHEMICALLY INHOMOGENEOUS MEDIUM

In the case of a chemically homogeneous medium, the requirement for the absence of convection is that the temperature gradient be smaller than adiabatic (τ_a). In case the substance becomes heavier with depth owing to a change in the chemical composition, the corresponding critical temperature gradient will rise and contain a term proportional to the

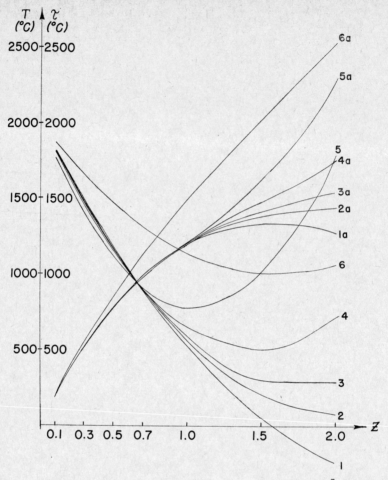

Fig. 1. Temperature $T(z)$ and temperature gradient $\tau(z)$ versus depth z ($=l/\bar{l}=$ depth/100 km). See equations 8 and 10.

Curve		Conditions, °C		
$T(z)$	$\tau(z)$	$\tau(0)$	$T(1)$	$\tau(\frac{3}{2})/\tau(0)$
1a	1	2000	1200	1/40
2a	2			1/10
3a	3			3/20
4a	4			1/4
5a	5			1/2
6a	6		1500	1/2

concentration gradient dc/dr [Zharkov, 1964]:

$$\tau_{ac} = \tau_a + \frac{\Delta\rho}{\rho\alpha}\frac{dc}{dr} \quad \frac{dc}{dr} = -\frac{dc}{dl} \quad \tau_a = \frac{Tg\alpha}{C_p}$$
(16)

$$\frac{1}{\rho} = V = cV_1 + (1-c)V_2 \quad \Delta\rho = \rho_2 - \rho_1$$
(17)

Let us determine what value of dc/dr is required to double the adiabatic gradient $\tau_{ac} = 2\tau_a$:

$$\left(\frac{dc}{dr}\right)_0 = \frac{\rho \cdot \alpha^2 Tg}{\Delta\rho \cdot C_p} \sim 1.9 \times 10^{-9} \text{ cm}^{-1}$$

$$= 1.9 \times 10^{-2}/10^2 \text{ km} \quad (18)$$

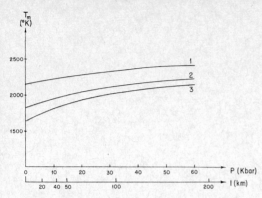

Fig. 2. Melting-point curves for (1) forsterite, (2) enstatite, and (3) diopside.

We assumed that

$$\Delta \rho / \rho \sim 0.2 \quad \alpha \sim 5 \times 10^{-5} \quad T \sim 2 \times 10^3$$

Hence, a change in the concentration by 2% at the depth interval of 100 km is sufficient to double the magnitude of the adiabatic gradient in the zone of reduced velocities. To raise it to $\sim$7.5 deg/km, it is necessary to change the concentration by $\sim$10% in an interval of 100 km. The dependence of the adiabatic gradient on phase transitions was analyzed by Verhoogen [1965].

CHANGES IN THE VELOCITIES OF SEISMIC WAVES WITH DEPTH

Choosing p, T, and c as thermodynamic variables, we have

$$\frac{dv}{dl} = \left(\frac{\partial v}{\partial p}\right)_{T,c} \rho g + \left(\frac{\partial v}{\partial T}\right)_{p,c} \frac{dT}{dl}$$

$$+ \left(\frac{\partial v}{\partial c}\right)_{p,T} \frac{dc}{dl} \quad (19)$$

Assuming the concentration to be constant, we obtain the following condition when a layer of reduced velocities originates in a chemically homogeneous medium [Magnitsky, 1965]:

$$\tau_b = \left(\frac{dT}{dl}\right)_b > -\rho g \frac{(dv/dp)_{T,c}}{(\partial v/\partial T)_{p,c}} \quad (20)$$

By means of (19) and (17), it is easy to generalize (20) for variable chemical composition. Let us represent the last term in (19) as

$$\left(\frac{\partial v}{\partial c}\right)_{p,T} \frac{dc}{dl} = -\left(\frac{\partial v}{\partial \rho}\right)_{p,T} \Delta \rho \frac{dc}{dl} \quad (21)$$

Then the condition $(dv/dl) \leq 0$ will give for the temperature gradient

$$\tau_{bc} = \tau_b + \tau_c \quad \tau_c = \frac{(\partial v/\partial \rho)_{p,T} \Delta \rho}{(\partial v/\partial T)_{p,c}} \frac{dc}{dl} \quad (22)$$

The influence of the change in concentration in (19) can be estimated in the following way. The rise in v with a change in depth of 100 km is of the order of 2%. Further, according to Birch [1952], the value of (K_T/ρ) under normal conditions for forsterite (Mg_2SiO_4) is 36 km^2/sec^2, and for fayalite, 26 km^2/sec^2. Assuming Poisson's ratio to be constant and modeling the subcrustal layer with a dunite composition, we can see that a change in the concentration from zero to unity leads to a change in velocity by 15%. Consequently, a drop in velocity of $\sim$2%/100 km could be caused by a change in the composition of $\sim$13% for 100 km. The latter value is too great, and so the corresponding change in composition can hardly lead to the origin of the low-velocity layer.

HYPOTHESIS OF RELAXATION

The explanations of the low-velocity layer as a purely temperature effect and as a result of chemical composition alteration encounter some difficulties. Another explanation, relaxation, was suggested by Zharkov [1964]. Essentially, it consists of the following: with sufficiently high temperatures, which depend on the frequency of vibrations, the boundaries of the grains in polycrystals are capable of viscous gliding with relation to one another. As a result, a polycrystalline body can be defined by the coefficient of viscosity η_r, which, like the coefficient of viscosity of an amorphous body, depends exponentially on temperature: $\eta_r \sim \exp(H/kT)$, where H is the energy of activation for atomic-diffusive gliding along the boundaries of the grains (of the order of 1 ev), and k is Boltzmann's constant. The high-temperature creep of polycrystals differs from that of amorphous bodies only in that it proceeds without obstacle to deformations of $\sim 10^{-5}$, owing to an accumulation of stresses in the angular regions of the grains. Accordingly, relaxation of the shear modulus μ_R/μ_v (μ_R and μ_v are the relaxing and nonrelaxing shear moduli) in polycrystals is finite; i.e., the modulus cannot become zero as in amorphous bodies and liquids. The process of relaxation is accompanied by a marked rise in inner friction of Q^{-1} (Q is the dissipative function).

The variables Q^{-1} and μ (dynamic shear modulus) are functions of the product $l\nu$ exp (H/kT):

$$Q = Q[l\nu \exp (H/kT)] \qquad (23)$$
$$\mu = \mu_v \phi[l\nu \exp (H/kT)]$$

where l is the average size of the grain and ν is frequency. Experimentally, μ_v is determined as the shear modulus of a monocrystal. If, when values P_1 and T_1 are achieved, there is relaxation in the medium, and modulus μ attains the value μ_R, which can be considered the cause of appearance of low velocities, then, for the low velocities to be stable for some range of depths, a change in P and T with depth is necessary such that the product in (23) will remain constant:

$$l\nu \exp (H/kT) = \text{constant} \qquad (24)$$

(the change in μ_v can be ignored).

The dependence of H on P is known:

$$H(p) = H_1[1 + L\beta_1(p - p_1)] \qquad (25)$$

where β is compressibility, $L = |\partial \ln H_1/\partial \ln V|$, and V is volume.

By substituting $H(p)$ from (25) into (24), we obtain the following condition of constancy of μ_E:

$$(T - T_1)/T_1 = L\beta_1(p - p_1) \qquad (26)$$

From (26) we obtain an expression for the critical temperature gradient:

$$\tau_r = \frac{dT}{dl} = TL\beta_T \rho g \qquad (27)$$

Assuming that $\rho \sim 3.5$, $T \sim 2 \times 10^3$, $g \sim 10^3$, and

$$\beta_T \sim 0.85 \times 10^{-12} \qquad L \sim 3 \pm 1$$

we obtain

$$\tau_r = (0.6 - 1.2) \text{ deg/km} \qquad (28)$$

It is evident that (27), obtained for a special relaxation mechanism, is nevertheless of general significance and is valid for any diffusive-relaxation mechanism. The value in (28) seems to be quite reasonable as compared with the melting-point gradient (15).

HYPOTHESIS OF AMORPHIZATION

The region of the upper mantle in the depth range 50 to 300 km is not only a zone of relatively reduced velocities of seismic waves and of their most pronounced absorption. It is also characterized by: the minimum of the annual number of earthquakes; the primary sources that feed volcanoes; the processes of migration of matter, which lead to a reestablishment of isostatic equilibrium; the minimum of viscosity of the mantle's substance; and the maximum of electric conductivity. The hypothesis of partial amorphization of the mantle's substance at these depths may account in a most plausible way for all these peculiarities of the low-velocity layer. This is also in satisfactory agreement with the fact that precisely at these depths, the temperature in the mantle comes closest to the melting-point temperature.

In the case of partial amorphization, the reduction in velocities can be explained simply by the very property of the amorphous state. Amorphization is accompanied by a decrease in density and in the moduli of elasticity. As a consequence, there is a reduction in the velocities of seismic waves.

Because there is an increase of volume during amorphization, the simplest way to appraise the effect of amorphization is to assume that the effect is approximately equal to the change in velocity during a similar change in volume. Assuming that the change in density with amorphization is 3%, we find that such an effect corresponds to a change in pressure of 30 kbar, and a change in the velocities of longitudinal and transverse waves of ~6% [Clark, 1966] (this reasoning is precise for longitudinal waves, but it is rather approximate for transverse waves). The fact that the actual drop in the velocities is smaller means that only a part of the mantle's substance is amorphized, which should have been anticipated considering the melting points of minerals; however, it is rather difficult to appraise the percentage of amorphized material. Assume that the amorphized part influences the velocities to an extent equal to its percentage of the zone in which amorphization occurs and that the decrease in density amounts to 4–7% for rocks such as diabase, gabbro, and olivine;

then amorphization does not exceed 20% if the actual drop in the velocities amounts to 3%, and not more than 7% if the drop in the velocities is 1%. The latter value is close to the percentage of partial melting in initial meteoritic material [*Vinogradov*, 1961]. *Ringwood* [1966] estimates 20% melting. The hypothesis of amorphization provides numerical values that are in rather good agreement with geochemical estimates of partial melting.

If 10–20% of the mantle's substance is amorphized, the layer of low velocities should react to long-acting forces as a Maxwell solid, while the overlying crust and the subjacent parts of the mantle with a viscosity of the order of 10^{26} poises will behave like a non-viscous, elastic body (possibly acquiring the properties of a Maxwell solid for processes with periods of tens of millions of years). But for a layer of low velocities with viscosity of 10^{20} poises, the fluid properties begin to manifest themselves for periods of tens and hundreds of years.

For an increase in deformation with a constant velocity, we have the following expressions for stresses:

$$\sigma = \dot{\epsilon}_0 \mu \tau (1 - e^{-t/\tau}) \qquad (29)$$

for a Maxwell's body, and

$$\sigma = \dot{\epsilon}_0 \mu_R t + \dot{\epsilon}_0 \mu_R \Delta \tau (1 - e^{-t/\tau})$$

for a standard linear body, where τ denotes the period of relaxation, t is the elapsed time since the beginning of the process, μ_R is the static modulus, μ is the instantaneous shear modulus, and $\Delta \tau \approx 0.3\tau$.

It is evident from the above formulas that the reactions of these media to deformation differ greatly. Whereas in a standard linear body the stress grows in time until it reaches a critical value and the body is destroyed, the stresses in a Maxwell solid cannot exceed the critical value for a given rate of deformation:

$$\sigma = \dot{\epsilon}_0 \mu \tau = \dot{\epsilon}_0 \eta$$

This may provide an adequate explanation for the fact that earthquakes are much less frequent in the low-velocity layer than in deeper layers. It is evident that to attain the necessary stresses in the asthenosphere, quite large rates of deformation are required. These apparently do not occur under platforms, but only in tectonically active areas.

Assuming that the gradient of velocity of vertical movements for platforms is 0.002 "/ year and that it is of the same order as the rate of deformation [*Magnitsky*, 1965], we find, with the parameters established for the medium, that the critical stress that can then appear in the asthenosphere does not exceed 10^5–10^6 dynes/ cm^2, while earthquakes originate with stresses of 10^7–10^9 dynes/cm^2. As the velocities in geosynclinal regions are higher by an order or two [*Magnitsky*, 1965], the required stresses appear there.

Nabarro-Herring's diffusive mechanism, combined with ordinary viscous flow in the amorphous parts of the medium, can be proposed as a mechanism of viscous flow that yields a viscosity of 10^{20} poises (the value has been obtained by the rise of the Baltic shield).

Another possible mechanism of viscous flow in the layer of low velocities may consist of the migration of grain boundaries by exchange with the amorphous material of the same composition [*Andrade and Aboav*, 1964]. Then coefficient K_μ in the formula for the process of creep

$$\epsilon = \mu(1 - e^{-mt}) + K_\mu t \qquad (30)$$

is

$$K_\mu = \text{const } n \cdot \sigma^2 \left(1 - \frac{T}{T_c}\right)^{-3/2} \qquad (31)$$

where T_c is some critical temperature at which the surface tension at the crystal-amorphous medium boundary becomes zero, and n is the number of grains per unit of length. As the temperatures in the asthenosphere are very close to the melting points of the minerals that compose it, this mechanism should operate actively. In this case, the medium cannot be considered as a Maxwell solid. For diffusive processes, the coefficient K_μ is linked with stress by the relationship:

$$K_\mu = \frac{\sigma}{\eta}$$

where viscosity η does not depend on stresses. This is in agreement with the definition of a Maxwell solid. Unlike the Maxwell solid, viscosity in Andrade's mechanism depends on stress as:

$$\eta = \text{const } \frac{1}{n\sigma} \left(1 - \frac{T}{T_c}\right)^{3/2}$$

As the relaxation period in the asthenosphere is measured for Maxwell's flow in tens and hundreds of years, this mechanism cannot be responsible for the absorption of seismic waves and for the small value of Q.

DISTRIBUTION OF DISSIPATIVE PARAMETERS

The idea that the distribution of the dissipative function Q in the earth's mantle can be established by the attenuation of the earth's free oscillations was advanced by Zharkov [1962b] and has been analyzed in detail [Anderson and Archambeau, 1964; Zharkov et al., 1967]. The analysis was made on the assumption that Q does not depend on frequency, which is far from being evident (see p. 273). Q appears to be minimum (~ 100) in the depth range $38 \leq l \leq 300$ km.

PHYSICAL INTERPRETATION

Let us consider first the properties of a rheological model of a linear standard solid body which well-defines viscous gliding along the boundaries of grains.

Assuming that we can ignore the relaxation of bulk modulus K, as compared with the relaxation of shear modulus μ, we have

$$Q_s^{-1} = tg\delta = \frac{\mu_v - \mu_R}{\bar{\mu}} \frac{\omega\bar{\tau}}{1 + (\omega\bar{\tau})^2} \quad (32)$$

$$\mu = \mu_v - \frac{\mu_v - \mu_R}{1 + (\omega\bar{\tau})^2} \quad \bar{\tau} = \tau_0 e^{H/kT}$$

where

$$\bar{\mu}^2 = (\mu_v \mu_R) \quad \frac{\mu_v}{\mu_R} = \frac{\tau_\sigma}{\tau_\epsilon} \quad \bar{\tau} = (\tau_\sigma \tau_\epsilon)^{1/2}$$

τ_σ is the time of relaxation of deformation at a constant stress; τ_ϵ is the time of relaxation of stress at constant deformation.

The results obtained by means of (32) differ from those observed in experiments only in that the experimental peak of absorption Q^{-1} and the region of change μ/μ_v are somewhat wider than follows from (32). This divergence can be explained, assuming that in reality there is a spectrum of times of relaxations with approximately similar H, but differing in τ_0.

Of interest are the cases of (32) having extreme high and low frequencies:

$$\mu_v - \mu_\omega = \frac{\mu_v - \mu_R}{(\omega\bar{\tau})^2} \quad (33a)$$

$$Q_s^{-1} = tg\delta = \frac{\mu_v - \mu_R}{\bar{\mu}\omega\bar{\tau}} \quad \omega\bar{\tau} \gg 1$$

$$\mu_\omega - \mu_R = (\mu_v - \mu_R)(\omega\bar{\tau})^2 \quad (33b)$$

$$Q_s^{-1} = tg\delta = \frac{\mu_v - \mu_R}{\bar{\mu}} \omega\bar{\tau} \quad \omega\bar{\tau} \ll 1$$

Similar formulas for longitudinal waves are:

$$Q_p^{-1} = L_v \frac{(\mu_v - \mu_R)}{\bar{\mu}\omega\bar{\tau}} \quad (34a)$$

$$L_v = \frac{4}{3} \frac{\mu_v}{K + \frac{4}{3}\mu_v} = \left(\frac{v_{sv}}{v_{pv}}\right)^2 \cdot \frac{4}{3} \quad \omega\bar{\tau} \gg 1$$

$$Q_p^{-1} = L_R \frac{(\mu_v - \mu_R)}{\bar{\mu}} \omega\bar{\tau} \quad (34b)$$

$$L_R = \frac{4}{3} \frac{\mu_R}{K + \frac{4}{3}\mu_R} = \frac{4}{3} \left(\frac{v_{sR}}{v_{pR}}\right)^2 \quad \omega\bar{\tau} \ll 1$$

In the case of a layer of reduced velocities

$$\frac{\mu_v - \mu_R}{\mu_v} \gtrsim 0, 1$$

$$\frac{Q_s^{-1}}{Q_p^{-1}} \sim \frac{3}{4} L_{(v,R)}^{-1} \sim 2, 3 \quad (35)$$

It can be shown that if dissipation is due to relaxation of the shear modulus, relationship 35 is of a general nature. If one considers the experimental data on the attenuation of seismic body and surface waves, as well as the free oscillations of the globe, one gains the impression that the attenuation of longitudinal oscillations is smaller in relation to that of transverse oscillations, and that evaluation 35 is acceptable.

According to (32), (33), and (34), Q^{-1} is proportional to

$$\Delta = \frac{(\mu_v - \mu_R)}{\mu_v}$$

If we refer to experimental data, we can see that for polycrystal aluminium, with $\Delta \sim 0.3$, $(Q^{-1})_{max} \sim 0.09$.

For a layer of reduced velocities, $\Delta \lesssim 0.1$, and, consequently,

$$(Q_s^{-1})_{max} \lesssim 0.03 \quad (Q_p^{-1})_{max} \lesssim 0.01 \quad (36)$$

It is thus evident that the relaxation hypothesis leads to reasonable results in relation to dissipative parameters as well. The question regarding the nature of absorption is still quite uncertain, and for this reason it is of interest to look for other possible explanations. One of them may be provided by the mechanism of dissipation suggested by *Zener* [1948] and *Isakovich* [1948] that adiabatic temperature differences will arise in a body consisting of crystal grains due to boundary conditions during the passage of elastic waves. The same temperature differences will exist between the adjacent grains, causing large temperature gradients in boundary layers and leading to dissipation of energy. The formula for Q [*Landau and Lifshitz*, 1954] is:

$$Q^{-1} = \frac{2T\alpha^2 \rho v^2 \chi^{1/2}}{l \cdot C \cdot \omega^{1/2}}$$

The formula is valid for the case:

$$\chi/l^2 \ll \omega \ll v/l$$

Using the appropriate parameters for the mantle, we obtain:

$$Q^{-1} \approx l^{-1} \omega^{-1/2} \times 10^{-2}$$

where l is the grain size, v is the velocity of seismic waves, χ is thermal diffusivity, and C is heat capacity.

Hence, the mechanism not only explains in a satisfactory way magnitude Q with the dimensions of the grains being of the order of centimeters; both Q and the absorption coefficient γ depend on frequency as $\omega^{1/2}$, in agreement with the conclusions of *Gutenberg* [1958] and *Knopoff* [1964] that, within the range of seismic frequencies, both Q and γ are practically independent of frequency.

DIFFUSIVE VISCOSITY

At high temperatures, solid polycrystalline bodies are capable of flowing like a viscous liquid. The transfer of matter is then effected by self-diffusion, and the viscosity itself is called diffusive. The theory of diffusive viscosity was developed by *Nabarro* [1948] and *Herring* [1950]. *Zharkov* [1960b] drew a generalization of diffusive viscosity for the case of high pressures and applied it to the physics of the earth's mantle. The diffusive viscosity coefficient η is:

$$\eta = A \frac{kT}{D \cdot a} \left(\frac{h}{a}\right)^2 \quad (37)$$

where h is the average size of crystal grains, a is the parameter of the lattice, k is Boltzmann's constant, T is the absolute temperature, A is a constant ($\sim 1/30$), and D is the coefficient of self-diffusion. The dependence of η on temperature and pressure is

$$D = D_0 \exp[-H(P)/kT] \quad (38)$$

In complex substances, such as silicates, η (37) will be determined by the self-diffusion of the slowest diffusing particles, apparently represented by oxygen ions O^{--}. It can be assumed qualitatively [*Zharkov*, 1959c] that the magnitude of activation energy $H(P)$ is proportional to the melting point $T_m(P)$:

$$H(P) \sim T_m(P) \quad (39)$$

This allows us to write (38) and (37) in a convenient way for analysis:

$$D(p, T) = D_0 \exp -\xi\left(\frac{T_m(p)}{T}\right) \quad (40)$$

$$\xi \sim 20\text{--}30 \qquad \vartheta = \frac{T}{T_m}$$

that is, the coefficient of self-diffusion and diffusive viscosity are approximately constant at the melting-point curve.

From the post-glacial uplift of Fennoscandia, the approximate viscosity of the subcrustal layer (see p. 386) is

$$\eta_{100} \sim 10^{20} \text{ to } 10^{21} \text{ poises} \quad (41)$$

It can be reasonably supposed that it is the viscosity of the subcrustal layer that is diffusive. Then, by normalizing η at the depth $l \sim 100$ km, it is easy by means of (40) to deduce its variation in the subcrustal layer. In the region where the substance of the subcrustal layer is nearest to the melting point (ϑ at its maximum), viscosity is minimal. Then ϑ begins to decrease and viscosity starts increasing correspondingly, and, according to *MacDonald* [1963] and *McKenzie* [1966] η amounts to $\sim 10^{26}$ poises. Thus we come to the idea that the layer of reduced viscosity is in the low-velocity layer. This is of fundamental significance both for the problem of the mechanism of earthquakes at intermediate depths and for the dynamics of the earth's crust and of the subcrustal layer.

ELECTRICAL CONDUCTIVITY

Investigation into electric conductivity is a promising field in modern geophysics from which one can hope to obtain a number of fundamental results, especially in relation to the problem of the distribution of temperatures in the upper mantle.

According to the experimental data, at moderate temperatures ($T < 1000\text{--}1200°C$), the electrical conductivity of rocks is due to impurities: electrons, holes, and ions. At the temperature range of $T \sim 1000\text{--}1200°C$, natural ionic electrical conductivity begins to prevail, and at higher temperatures it predominates. A. T. Bondarenko (see *Volorovich* [1966]) gives the following evaluations for the electrical conductivity coefficient (σ, ohm^{-1} cm^{-1}) at $T \sim 1000\text{--}1200°C$:

Gabbro, basalt, peridotite, olivinite	2×10^{-4} to 10^{-3}
Granite, dunite	7×10^{-6} to 10^{-4}*
Eclogite	2×10^{-3} to 7×10^{-3}

The dependence of the coefficient σ on pressure in the region of natural ionic conductivity was investigated by Zharkov [1958]. It appeared that, like the coefficient of self-diffusion (40), coefficient σ may be expressed as

$$\sigma(p, T) = \sigma_0 \exp -\xi_1[T_m(p)/T]$$
$$= \sigma_0 \exp(-\xi_1/\vartheta) \quad (42)$$
$$\vartheta = \frac{T}{T_m(p)} \quad \xi_1 \sim 15\text{--}20$$

Formula 42 makes it possible to analyze the distribution of electrical conductivity in a layer of reduced velocities. As depth increases, electric conductivity rises sharply. At a certain depth there exists the conducting layer, the region of natural ionic conductivity with the value

$$\sigma \sim 2 \times 10^{-4} - 5 \times 10^{-3} \text{ ohm}^{-1} \text{ cm}^{-1} \quad (43)$$

The depth of the conducting layer can be regarded as a reference temperature point ($T \sim 1100\text{--}1200°C$). Electrical conductivity continues to rise with an increase in ϑ, and at the point $\vartheta = \vartheta_{max} < 1$ it passes through the maximum.

---

* In Bondarenko's high temperature experiments, some samples underwent chemical transformations, which may have introduced an appreciable error into the value given.

Subsequently, as temperature T 'leaves' the melting point temperature $T_m(p)$, the dimensionless temperature $\vartheta = T/T_m(p)$ drops, leading to a reduction in electrical conductivity with depth. The reduction in electrical conductivity continues to the C layer at depths of 300–400 km. Electric conductivity again rises sharply in the C layer and probably changes into the mechanism of natural electron conductivity (see p. 463).

REFERENCES

Aki K., and F. Press, Upper mantle structure under oceans and continents from Rayleigh waves, *Geophys. J.*, 5(4), 292–305, 1961.

Anderson, D. L., Latest information from seismic observations, in *The Earth's Mantle*, edited by T. F. Gaskell, pp. 355–420, Academic Press, London, N.Y., 1967.

Anderson, D. L., and C. B. Archambeau, The anelasticity of earth, *J. Geophys. Res.*, 69, 2071–2084, 1964.

Andrade, E. N., and D. A. Aboav, The flow of polycrystalline cadmium under simple shear, *Proc. Roy. Soc., London, A, 280*(1382), 353–382, 1964.

Birch, F., Elasticity and constitution of the earth's interior, *J. Geophys. Res.*, 57(2), 227–286, 1952.

Birch, F., The velocity of compressional waves in rocks to 10 kilobars, 1, *J. Geophys. Res.*, 65, 1083–1102, 1960.

Birch, F., The velocity of compressional waves in rocks to 10 kilobars, 2, *J. Geophys. Res.*, 66, 2199–2224, 1961a.

Birch, F., Composition of the earth's mantle, *Geophys. J., 4,* 295–311, 1961b.

Bolt, B. A., Theoretical phase velocities for a lunar seismic experiment, *J. Geophys. Res.*, 66(10), 3513–3518, 1961.

Brune, J., and J. Dorman, Seismic waves and earth's structure in the Canadian shield, *Bull. Seismol. Soc. Am.*, 53(1), 167–209, 1963.

Caloi, P., Onde logitudinale e transversali gnidate dell, asteno-sfera, *Rend. Acad. Naz. Lincei*, [8] 15(6), 1953.

Caloi, P. L., L'astenosfera come canaleguida dell' energia sismica, *Ann. Geofis.*, 7(4), 491–501, 1954.

Clark, S. P., editor, *Handbook of Physical Constants*, Geol. Soc. Am. Mem. 97, 1966.

Fedotov, S. A., N. N. Matveyeva, R. Z. Tarakanov, and T. B. Yanovskaya, Velocities of longitudinal waves in the upper mantle in the area of the Japanese and Kuril Islands, *Izv. Akad. Nauk SSSR, Ser. Geofiz.*, 1964(8), 1185–1191, 1964.

Gutenberg, B., Untersuchungen zur Frage bi zu welcher Tiefe die Erde kristallin ist, *Z. Geophys.*, 2(1), 24–29, 1926.

Gutenberg, B., Attenuation of seismic waves in the earth's mantle, *Bull. Seismol. Soc. Am.*, 48(3), 269–282, 1958.

Gutenberg, B., *Physics of the Earth's Interior*, Academic Press, New York, 240 pp., 1959.
Herring, C., Diffusional viscosity of polycrystalline solid, *J. Appl. Phys.*, *21*(5), 437, 1950.
Isakovich, M. A., Theory of sound absorption in polycrystals, *J. Exp. Theoret. Phys.*, *18*(4), 386–391, 1948.
Ivanova, Z. S., V. I. Keilis-Borok, A. L. Levshin, and M. G. Neigaus, Love waves and the structure of the upper mantle, *Geophys. J.*, *9*(1), 1–7, 1964.
Kê T'ing-Sui, Experimental evidence of the viscous behavior of grain boundaries in metals, *Phys. Rev.*, *71*, 533–546, 1947a.
Kê T'ing-Sui, Stress relation across grain boundaries in metals, *Phys. Rev.*, *72*, 41–46, 1947b.
Knopoff, L., Q, *Rev. Geophys.*, *2*, 625–660, 1964.
Kovach, R. L., and F. Press, Rayleigh wave dispersion and crustal structure in the eastern Pacific and Indian Oceans, *Geophys. J.*, *4*, 202–216, 1961.
Landau, L. D., and E. M. Lifshitz, *Mechanics of Continuous Media*, Leningrad, 1954.
MacDonald, G. J. F., The deep structure of oceans and continents, *Rev. Geophys.*, *1*, 587–665, 1963.
Magnitsky, V. A., Waveguides in the earth's crust and in the subcrustal layer, *Bull. Moscow Soc. Nature, Geol. Sect.*, *33*(4), 15–23, 1958.
Magnitsky, V. A., *Inner Structure and Physics of the Earth*, Izd. Nedra, Moscow, 379 pp., 1965.
Magnitsky, V. A., and V. V. Khorosheva, The waveguide in the mantle of the earth and its probable physical nature, *Ann. Geofis.*, *14*(1), 87–94, 1961.
McKenzie, Dan P., The viscosity of the lower mantle, *J. Geophys. Res.*, *71*, 3995–4010, 1966.
Nabarro, F. R. N., Deformation of crystals by the motion of single ions, Report on a conference on the strength of solids, Phys. Soc., London, 75, 1948.
Nuttli, O., Seismological evidence pertaining to the structure of the earth's upper mantle, *Rev. Geophys.*, *1*(3), 351–400, 1963.
Pakiser, L. C., Structure of the crust and upper mantle in the western United States, *J. Geophys. Res.*, *68*(20), 5747–5756, 1963.
Press, F., and M. Ewing, P and S velocities at great distances (abstract), *Bull. Geol. Soc. Am.*, *65*, 1348, 1954.
Press, F., and M. Ewing, Waves with P_n and S_n velocity at great distances, *Proc. Natl. Acad. Sci. U.S.*, *41*(1), 24–27, 1955.
Redwood, M. R., *Mechanical Waveguides*, Oxford University Press, London, 1960.
Ringwood, A. E., Mineralogy of the mantle, in *Advances in Earth Science*, edited by P. M. Hurley, pp. 357–399, MIT Press, Cambridge, 1966.
Santo, T. A., Division of the southwestern Pacific area into several regions in each of which Rayleigh waves have the same dispersion character, *Bull. Earthquake Res. Inst., Tokyo Univ.*, *39*(4), 603–630, 1961.
Tryggvason, E., Arrival times of P waves and upper mantle structure, *Bull. Seismol. Soc. Am.*, *54*(2), 727–736, 1964.
Verhoogen, J., Phase changes and convection in the earth's mantle, in *Symposium on Continental Drift*, pp. 276–283, Royal Society, London, 1965.
Vinogradov, A. P., Origin of the matter of the earth's crust, *Geokhimiya*, *1961*(1), 1–32, 1961.
Volarovich, M. P., editor, *Electric and Mechanical Properties of Rock at High Pressures*, Nauka, Moscow, 1966.
Zener, C., *Elasticity and Anelasticity of Metals*, Chicago, 1948.
Zharkov, V. N., Electric conductivity and the temperature of the earth's mantle, *Izv. Akad. Nauk SSSR, Ser. Geofiz.*, *1958*(4), 458–470, 1958.
Zharkov, V. N., Thermodynamics of the earth's mantle, *Izv. Akad. Nauk SSSR, Ser. Geofiz.*, *1959*(9), 1414–1419, 1959a.
Zharkov, V. N., The melting point of the earth's mantle and of iron at high pressures, *Izv. Akad. Nauk SSSR, Ser. Geofiz.*, *1959*(3), 465–470, 1959b.
Zharkov, V. N., The physical nature of waveguides (layers with reduced velocity) in the upper layers of the mantle at depths of 50–200 km, *Dokl. Akad. Nauk SSSR*, *125*(4), 771–774 1959c.
Zharkov, V. N., The influence of pressure on the diffusion coefficient in solid bodies, *Tr. Inst. Fiz. Zemli Akad. Nauk SSSR*, *11*(178), 1960a.
Zharkov, V. N., Viscosity of the earth's interior: Diffusive processes and diffusive viscosity of the earth's mantle, *Tr. Inst. Fiz. Zemli Akad. Nauk SSSR*, *11*(176), 36–60, 1960b.
Zharkov, V. N., Physics of the earth's core, *Tr. Inst. Fiz. Zemli Akad. Nauk SSSR*, *20*(187), 1962a.
Zharkov, V. N., Natural oscillations of the earth: Attenuation, *Izv. Akad. Nauk SSSR, Ser. Geofiz.*, 159–170, *1962*(2), 1962b.
Zharkov, V. N. The layer of reduced velocities and the adiabatic gradient of temperatures in the earth's mantle, *Izv. Akad. Nauk SSSR, Ser. Geofiz.*, *1964*(9), 1281–1291, 1964.
Zharkov, V. N., Electric conductivity of the lower mantle, *Fiz. Zemli*, *1966*(9), 3–11, 1966.
Zharkov, V. N., V. M. Lubimov, A. A. Movchan, and A. I. Movchan, Torsional oscillations of the anelastic earth, *Geophys. J.*, *14*(1–4), 179–188, 1967.

The Transitional Layer in the Mantle

V. A. Magnitsky

THE C layer (so-named by Bullen) is a transitional zone separating the upper mantle (B layer) from the lower (D layer). Velocities (V_p) of longitudinal waves within the B layer are in the range 7.8–9.0 km/sec; those of transverse waves (V_s) are 4.4–4.9 km/sec. In the lower mantle, V_p ranges from 11.3 to 13.6 km/sec, V_s from 6.3 to 7.3 km/sec. Thus vertical gradients of velocities in the C layer are 0.46 km/sec per 100 km for V_p and 0.28 km/sec per 100 km for V_s. For the lower mantle, corresponding gradients are 0.11 and 0.05 [Clark, 1966]. If the gradient of velocities in the lower mantle is caused by an increase in the density of a chemically homogeneous medium under pressure with an adiabatic increase of temperature, then the large gradients in the C layer indicate changes of chemical composition or phase transitions under increasing pressure.

In interpreting seismic data, it is convenient to use Φ (Figure 1):

$$\Phi = \frac{K_s}{\rho} = V_p^2 - \tfrac{4}{3} V_s^2 \qquad (1)$$

Two density sections of the mantle are given in Figure 2. These two models, in combination with the velocity section of Gutenberg, are in good accord with P and S propagation data and with observations of free oscillations of the earth [Clark, 1966; Anderson, 1967; Pekeris, 1966].

Figures 1 and 2 indicate that there are essential physico-chemical changes within the C layer and that this layer, unlike the D layer, is far from being homogeneous. This is also testified by the increase of electrical conductivity of the C layer (by a factor of 10^3–10^4).

The gradient of density increase for the C layer is 0.17 to 0.18 g/cm³ per 100 km of depth, in the D layer only 0.05 g/cm³ per 100 km. Taking the latter value as normal for a homogeneous layer, the C layer density has been found to increase by 0.60–0.65 g/cm³, or by 17%, owing to physico-chemical changes.

The gradient of Φ for the C layer is 4.8 km²/sec² per 100 km, for the D layer 1 km²/sec². Thus the Φ value of the C layer increases by 19 km²/sec², or by 40%, owing to physico-chemical changes [Bullen, 1963; Birch, 1964].

The mineral composition of the B layer is determined by using the data on the composition of meteorites, xenoliths coming from the mantle, and igneous rocks, and by comparing the density and velocity data on seismic waves within the B layer with corresponding values obtained in laboratories for various rocks and minerals. This leads to the conclusion that the B layer consists of minerals characteristic of ultrabasic and partly basic rocks: olivine, pyroxene, jadeite, omphacite, and garnet (see p. 1).

From Figure 1, Φ_0 for the B layer is 32–35 km²/sec². Temperature changes have not been taken into account while reducing to a zero pressure. The correction for temperature difference is:

$$\Delta \Phi = \alpha \frac{K}{\rho} \left(I - \frac{dK}{dp} \right) \Delta T$$

[Magnitsky, 1965]. Assuming the coefficient of thermal expansion $\alpha = 6 \cdot 10^{-5}$ and the temperature of the B layer $t = 1000$–$1500°C$, Φ_{00} values for layer B (reduced to zero pressure and room temperature) are in the range of 38–43 km²/sec². Experimental values of Φ and ρ are given in Table 1. Comparison of Table 1 with Φ_{00} for layer B shows agreement with the mineral composition indicated above. For layer D, however, Φ_{00} values reduced to zero pressure and room temperature are in the range 51–53 km²/sec², i.e., higher than almost all the values shown in Table 1. Similarly, the density of the D layer, reduced to the same initial conditions, is in the range 4.0–4.5 g/cm³, that is, higher than density values of common ultrabasic rocks and their mineral constituents.

NATURE OF THE TRANSITION

The following principal explanations are suggested for the transition from the B layer to

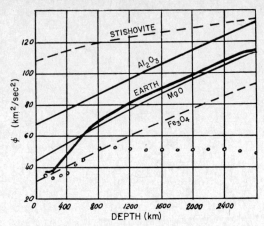

Fig. 1. $\phi = K_s/\rho$ for the earth's mantle (after Gutenberg [*Clark*, 1966]) and for some minerals. Values of ϕ_0 for the mantle are reduced isothermically to zero pressure by the method of finite deformation [*Magnitsky*, 1965].

the D layer:

1. The D layer differs from the B layer in chemical composition.
2. The transition to the D layer is caused by decomposition of silicates and alumosilicates of the mantle into oxides.
3. The transition to the D layer is related to phase transitions of the principal minerals of the mantle to densely packed modifications which are stable under high pressures but unstable in conditions of the earth's crust and layer B.

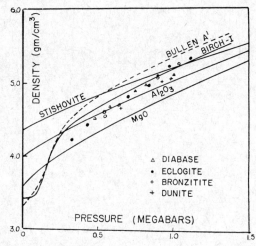

Fig. 2. Density of minerals and of the earth's mantle versus pressure [after *Bullen*, 1963; *Birch*, 1964].

TABLE 1. Experimental Values of Φ and ρ (zero pressure, room temperature)*

Mineral	Φ, km²/sec²	ρ, g/cm³	
Olivine	37	3.36	(3.2–4.1)
Diopside	38	3.26	(3.2–3.4)
Hypersthene	35	3.42	(3.3–3.5)
Pyrope	50	3.67	(3.5–3.8)
Grossularite	44	3.48	(3.4–3.6)
Jadeite	40	3.33	(3.3–3.5)
Spinel	56	3.70	(3.5–4.1)
Periclase	46	3.58	
Corundum	67	4.02	
Rutile	50	4.25	
Stishovite	107	4.35	
Magnetite	33	5.08	

* After *Clark* [1966].

4. Transition to the D layer is caused by a change of the type of chemical bonds.

The last three explanations are interrelated.

Change of Chemical Composition

Change of chemical composition should involve an increase in the density of 0.6 g/cm³, with simultaneous increase of seismic wave velocities and Φ values.

For a very large number of minerals and rocks, there has been established an approximate linear dependence of V_p upon density for constant mean atomic weight [*Birch*, 1961]. Changes of atomic weight are primarily due to variations in the iron content. The following data are obtained for the dependence of ρ and V_p on mean atomic weight m:

$$\frac{\partial \rho}{\partial m} = 0.13 \qquad \frac{\partial v}{\partial m} = -0.79$$

Thus, when the mean atomic weight changes, the seismic wave density and velocity vary in opposite directions. Therefore it is impossible to attribute the transition from the B layer to the D layer exclusively to changes of chemical composition.

Decomposition of Silicates and Alumosilicates into Oxides

If a mixture of oxides equivalent to a given silicate has a smaller volume, in accordance with Le Chatelier's law, that mixture could become a steady-state system with increasing pressure.

Table 1 shows that a mixture of the required density can be easily made of oxides. The mixture of stishovite (40–45%), periclase (40%), corundum (5–10%), and magnetite (10%) will have a density of 4.1–4.2 g/cm^3; this value corresponds to the density of the D layer, reduced to the initial conditions. Behavior of components of the mixture with an increase of pressure up to values prevailing within the D layer has been investigated by using shock waves. Figure 2 gives the curves of density ρ versus pressure P for periclase, corundum, and stishovite (zero isotherms) [Altshuler et al., 1965; Clark, 1966]. By introducing the corrections for some temperature increase in the interior of the earth, the curve slopes become slightly reduced; however, the correction values do not exceed 1–2%, or 0.1 g/cm^3, at depths of more than 1000 km; that is, they are within the accuracy of density data for the earth's interior. Comparison of these curves with the models of Bullen and Birch shows good agreement at initial densities, but subsequent divergence, as the density curve for stishovite is essentially less sloped than that of the indicated models of the earth. Since the stishovite accounts for 40–45% of the oxide mixture, a curve representing a mixture will be somewhat less sloped. But this divergence cannot be a decisive argument against the hypothesis of an oxide mixture for the composition of layer D. Density variation with depth for adopted models of the earth is obtained by using the Adams–Williamson equation, assuming the composition of each respective layer to be constant and the temperature adiabatic. An assumption that the temperature of the D layer increases faster than for the adiabatic curve or that the composition of the oxide mixture varies slightly with depth is sufficient to obtain the variation of density with depth in agreement with the hypothesis.

The hypothesis can be checked more strictly against seismic velocities (or Φ values) that are obtained by observation. Figure 1 gives the curves of Φ versus pressure (depth) for periclase and corundum, in accordance with their equation of state and for stishovite and magnetite, with somewhat smaller accuracy. It follows from Figure 1 or Table 1 that the mixture of stishovite (40–45%), periclase (40%), corundum (5–10%), and magnetite (10%) has an absolutely unacceptable initial Φ_0 value of 67–69 km^2/sec^2.

The Φ curve for stishovite is less sloped than a similar curve for the earth's interior within the D layer. Hence it seems that the hypothesis of oxide composition of the D layer does not agree with the data on seismic wave velocities. In order to obtain required initial Φ_0 values by varying the composition, it is necessary to admit the following composition of upper parts of the D layer: stishovite, 20%; periclase, 45%; corundum, 5%; magnetite, 30%. But besides being geochemically improbable, this composition brings initial density to an unbelievable value of 5.5 g/cm^3. Moreover, a decomposition of silicates and alumosilicates into oxides failed to be discovered experimentally up to pressures of 200–250 kbar [Ringwood and Major, 1966b].

Transitions to Closely Packed Lattices

Experimental results confirm transitions to closely packed lattices under sufficiently high pressures [Ringwood and Major, 1966a, b; Akimoto and Ida, 1966] (see section by Ringwood and Green in this Monograph, p. 645). The transition from olivine structure to spinel structure occurs for forsterite (Mg_2SiO_4) at pressures of 150–170 kbar and temperatures of 800–900°C, and for fayalite (Fe_2SiO_4) at 38 kbars, 600°C. In both of these cases the density is increased by 10%.

Enstatite passes into ilmenite structure under a pressure of about 250 kbar. Enstatite in the presence of aluminum oxide forms solid solutions of garnet structure that later pass into ilmenite. Olivines of spinel structure seem also to become of ilmenite structure with a further pressure increase. It is possible to have a further transition to perovskite structure.

Thus the transition to the density of layer D can be provided by a number of phase transformations to more closely packed lattices with partial transition of Si from fourfold to sixfold coordination.

Experiments with shock waves for dunite, eclogite, pyroxenite rocks, and peridotite all showed a jump in the density under pressures of 200–300 kbar, but the resultant densities are slightly lower than those for the models of the earth. Some results of these experiments are shown by dots in Figure 2. An insufficient increase of density is probably connected with an incomplete transition under the effect of the shock wave on such complicated materials as rocks.

Unfortunately, the evaluation of experimental data has been inadequate as yet to check Φ.

The principal difficulty concerning the application of this hypothesis relates to the thickness of the transition layer C: a depth range of 500–600 km and corresponding pressure range of 250 kbar. Phase transitions should occur, even in solid solutions, within very limited ranges, practically as a jump. Figure 3 is a diagram of transition of solid solutions of two components: A (for example, Mg_2SiO_4) and B (for example, Fe_2SiO_4) from modification I with a lower density (olivine) to modification II with a higher density (spinel). Concentration of A component is indicated by X. Transition from phase I to phase II with decreasing temperature and under zero pressure in the TOX plane is started on curve AaB and finished on curve AbB. Both the phases coexist within a 'cigar,' formed by curves AaB and AbB. In X, P, T space, phase I is separated from phase II by cylinder $ABB'A'$, with the section in the shape of a cigar. At constant temperature T_1 (point D) the transition is carried out through the 'cigar' CF, formed by cutting the cylinder with plane $CDMEF$, parallel to plane XOP. A corresponding diagram is given in Figure 4. It presents the transition from olivine to spinel in $(Mg,Fe)_2SiO_4$ solid solution calculated according to data of Ringwood and Major [1966a] at 900°C.

In Figure 3, a section of the cylinder cut by a plane of constant concentration X' (MKL plane) gives two curves aa' and bb', which only for the two extreme concentrations (0 and 1) join into one line AA' and BB', respectively. If curve 1 in Figure 5 represents the temperature of the earth's interior as a function of pressure (depth), the transition from phase I to phase II will begin at pressure P_1 and end at pressure P_2. If point Q in Figure 5 is the intersection of curve 1 with plane $CDMEF$, then, from Figure 4, at this point phase I will exist in concentration X_1 and phase II in concentration X_2, thus:

$$\frac{I}{II} = \frac{X_2 - X'}{X' - X_1}$$

If pressures P_A and P_B of the transition of pure components A and B at temperature T_1 from modification I to modification II are known, and the change ΔV of volume per mole with such transition is also known, in this case curves X and $\tilde{X}$, separating phase I and phase II from the 'cigar' in Figure 4, can be calculated by the formula of chemical thermodynamics:

$$\ln \frac{\tilde{X}}{X} = (P - P_A) \frac{\Delta V_A}{2RT_1}$$

$$\ln \frac{1 - \tilde{X}}{1 - X} = (P - P_B) \frac{\Delta V_B}{2RT_1}$$

If temperatures T_A and T_B of the transition under constant pressure of pure A and B components and heats H_A and H_B of the transition are known, curves AaB and AbB in Figure 3 can be calculated on the basis of the following formula:

$$\ln \frac{\tilde{X}}{X} = \left(\frac{1}{T} - \frac{1}{T_A}\right) \frac{\Delta H_A}{2R}$$

$$\ln \frac{1 - \tilde{X}}{1 - X} = \left(\frac{1}{T} - \frac{1}{T_B}\right) \frac{\Delta H_B}{2R}$$

where X and $\tilde{X}$ are molar fractions of component A in phases I and II, respectively.

From the formulas above and data obtained

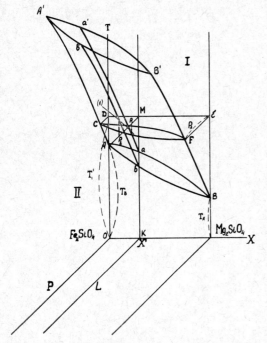

Fig. 3. Diagram of phase transitions in solid solutions.

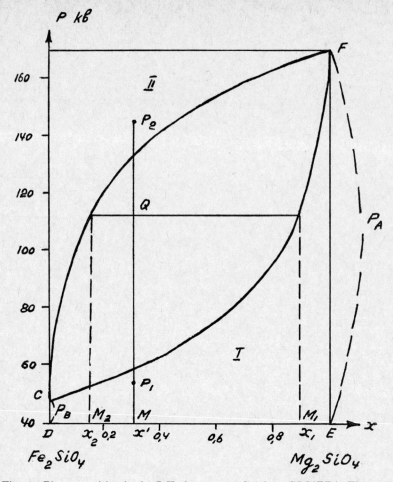

Fig. 4. Phase transition in the P,X plane at 900°C (plane $CDMEF$ in Figure 3).

from experiments on solutions of $(Mg, Fe)_2SiO_4$, curves (corresponding to Figure 4) giving percentage of spinel phase depending on initial content of Fe and pressure (depth) are shown in Figure 6. From the above curves it follows that almost the whole transition occurs within the small depth range of 30–40 km.

If we postulate a transition in some binary solution of components A and B to be spread over depths from 200 to 900 km and adopt the density variations required by the models of the earth, we would find that with a content of one of the components up to 33% the whole transition actually occurred in two jumps: a small jump at depths of 350–400 km and a sharp transition at 850–900 km, with almost no transition between these two depths. This contradicts the models of the earth in which the main part of the transition takes place at intermediate depths [Meijering and Rooymans, 1958].

It is hard to suppose that sharp transitions with the change of the density by some 10% within the depth range of 20–50 km could not have been discovered by seismological observations. Some attempts have been made by a number of authors to reveal a fine structure of the C layer. Rather uncertain indications of jumps in velocities for different depths have been found, depending on the method and material used and area of investigation [Magnitsky, 1965]. Thus taking into account the complexity of mantle composition, several jumps in velocity and density quite probably occur within the transition layer.

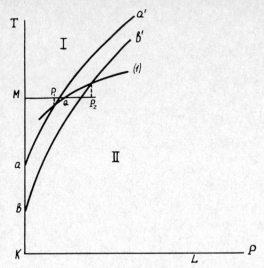

Fig. 5. Phase transition in the P,T plane (plane MKL of Figure 3).

Change of the Type of Chemical Bond

Members of the sequence—ionic crystals: valent crystals:metals—differ in the type of chemical bond. In ionic compounds the bonds depend on Coulomb forces of attraction of oppositely charged ions. In valent compounds the bonds depend on the exchange of covalent forces. Metals have their own bonding forces, but there are no impassable gaps between these types. In fact, in each compound there are forces of all the types, and if one of them prevails the compound is attributed to one or the other group of the sequence. Actually there is always a whole chain of compounds forming a transition from one group of the sequence to the other. Thus silver halides form the whole range of transitions from almost idealy ionic AgF to almost purely valent AgI. Under normal conditions, SiO_2 has almost equal valent and ionic bonds. Graphite, indium, and gray tin are typical intermediate substances between valent crystals and metals. Variations of temperature and pressure change the relationship between different types of forces and may transfer the substance from one group to the other. Some transitions from valent crystals to metals have been well studied. Phosphorus, iodine, selenium, and carbon turn into metals under pressure [Paul and Warschauer, 1963], but the transition of rocks and minerals of the mantle into the metal phase takes place under too high pressures to be related to the C layer.

The transition from the ionic to the covalent bond is important for the explanation of the C layer. An increasing overlapping of electronic clouds of outer electrons takes place during compression, owing to shortening of the distances between atoms. This leads to ever increasing delocalization of outer electrons and to strengthening of the role of exchanging forces of valent bonds instead of ionic [Paul and Warschauer, 1963; Pauling, 1960].

Using expressions for the equation of state for ionic and valent bonds [Davydov, 1956], the curves of energy were obtained for the B and D layers of the earth's mantle on the basis of experimental Φ values and density of Bullen's model A' [Magnitsky and Kalinin, 1959]. The transition from purely ionic bonds in layer B to covalent ones in D leads to a density increase of 18% and Φ increase of 19 km^2/sec^2, these values being in close agreement with the data for the earth. Such a change of the type of

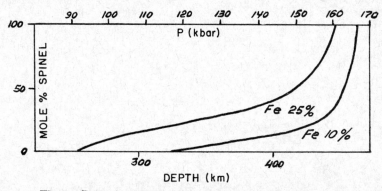

Fig. 6. Rate of transition from olivine to spinel in the mantle.

bonds is not necessarily accompanied by change of the type of lattice, but it is quite possible. Hence, it is most probable that both a quick but continuous increase in ρ and Φ takes place under increasing pressure, caused by a change of ionic bonds into covalent, and the superposed jump with ρ and Φ increase, the type of lattice being changed accordingly. This consideration may be confirmed by the fact that, while ionic bonds prevail in minerals of a type of olivine and pyroxene, a covalent type of bond prevails in minerals with a very close packing such as spinel [Goodenough and Loeb, 1955].

Thus both qualitative considerations and quantitative estimation lead to the conclusion that a most probable explanation of the physical nature of the transitional C layer in the earth's mantle is an assumption that within this layer there take place both the reconstruction of lattices of minerals of the mantle to more closely packed forms of a type of spinel, ilmenite, rutile, and perovskite, and the simultaneous transition from a prevailing ionic type of bond to a covalent one. That the last effect would be responsible for a great increase of electrical conductivity of the C layer is confirmed by experiments of Vereshchagin et al. [1962] on the replacement of ionic conductivy by semi-conductivity.

REFERENCES

Akimoto, S. and Y. Ida, High-pressure synthesis of Mg_2SiO_4 spinel, Earth Planetary Sci. Letters, 1(5), 358–359, 1966.

Altshuler, L. V., R. F. Trunin, and G. V. Simakov, Shock compression of periclase and quartz and chemical composition of the lower mantle of the earth, Phys. Earth, 1965(10), 1–6, 1965.

Anderson, D. L., Latest information from seismic observations, in The Earth's Mantle, edited by T. F. Gaskell, pp. 355–420, Academic Press, London, 1967.

Birch, F., Composition of the earth's mantle, Geophys. J., 4, 295–311, 1961.

Birch, F., Density and composition of mantle and core, J. Geophys. Res., 69(20), 4377–4388, 1964.

Bullen, K. E., Introduction to the Theory of Seismology, Cambridge University Press, London, 381 pp., 1963.

Clark, S. P., Handbook of Physical Constants, Geol. Soc. Am. Mem. 97, 587 pp., 1966.

Davydov, B. I., On the equation of state of solids, Izv. Akad. Nauk SSSR, Ser. Geofiz., 1956(12), 1411–1418, 1956.

Goodenough, J. C., and A. L. Loeb, Theory of ionic ordering crystal distortion and magnetic exchange due to covalent forces in spinels, Phys. Rev., 98(2), 391–408, 1955.

Magnitsky, V. A., Internal Constitution and Physics of the Earth, Nedra, Moscow, 1965 (NASA Tech. Transl. F395, 1967).

Magnitsky, V. A., and V. A. Kalinin, Properties of the earth's mantle and physical nature of the transition layer, Izv. Acad. Nauk SSSR, Ser. Geofiz., 1959(1), 87–95, 1959.

Meijering, J. L., and C. J. M. Rooymans, On the olivine-spinel transition in the earth's mantle, Koninkl. Ned. Akad. Wetenschap. Proc., B, 61(5), 333–344, 1958.

Paul, W., and D. M. Warschauer, Solids under Pressure, McGraw-Hill Book Company, New York, 478 pp., 1963.

Pauling, L., The Nature of the Chemical Bond and the Structure of Molecules and Crystals, Cornell University Press, Ithaca, 644 pp., 1960.

Pekeris, C. L., The internal constitution of the earth, Geophys. J., 11(1–2), 85–132, 1966.

Ringwood, A. E., Mineralogy of the mantle, in Advances in Earth Science, edited by P. M. Hurley, pp. 357–399, MIT Press, Cambridge, 1966.

Ringwood, A. E., and A. Major, Synthesis of Mg_2SiO_4–Fe_2SiO_4 spinel solid solutions, Earth Planetary Sci. Letters, 1(4), 241–245, 1966a.

Ringwood, A. E., and A. Major, High pressure transformations in pyroxenes, Earth Planetary Sci. Letters, 1(5), 351–357, 1966b.

Vereshchagin, L. F., A. A. Semerchan, S. V. Popova, and N. N. Kuzin, Variations of electrical resistance of some semiconductors at pressures to 300,000 kg/cm^2, Dokl. Akad. Nauk SSSR, 145(4), 757–760, 1962.

9. Special Problems

Continental Drift and Convection[1]

Leon Knopoff

BY NOW a considerable amount of information has been collected bearing on present ideas concerning the spreading of the sea floor and continental drift. Whether the ideas of sea-floor spreading and continental drift can be used to explain geologic processes in regions remote from oceanic ridges, trenches, and rift zones, and whether a process of convection must be invoked to explain the spreading and drifting observations, is as yet undetermined. The pertinent results will first be summarized briefly. By itself, none of these data is convincing; taken together, they raise the probabilities considerably that large-scale horizontal motions take place and have taken place in the past.

Continental Drift

The notion of continental drift was inspired in large part by the remarkable apparent fit of the present profiles of the facing coasts of South America and Africa [*Wegener*, 1915]. The two continents were presumed to have been originally one protocontinent until about 150–200 million years ago; the date of the rupture was approximated by the ages of older common fossils on both shores. With the discovery of the Mid-Atlantic ridge, further geographic comparisons were possible; to a striking degree of accuracy, the central ridge is half-way between the shelves of the continents along the margins. *Bullard et al.* [1965] have tested the fit of the continents around the Atlantic at the 500-fathom contour and find the fit to be so good as to be outside the realm of chance. Support of this result is to be found in the observation that the contact between rocks of ages 2×10^9 years and 0.6×10^9, found on both the African and South American coasts, is in excellent accord with the geographical fit [*Hurley et al.*, 1967].

Vine and Matthews [1963] have shown that elongated magnetic anomalies in the Atlantic have a remarkable symmetry about the median ridge. Further, the distances to the anomalies

---

[1] Publication 736, Institute of Geophysics, University of California, Los Angeles.

from all the oceanic ridge axes are in excellent correlation [*Vine*, 1966] with a chronology for reversals of the earth's magnetic field [*Cox et al.*, 1964], with an apparent uniform 'velocity' for the distance-time relationship. Unfortunately, the chronology is available for only 4×10^6 years. Velocities range from about 1 cm/year for the Atlantic to more than 4 cm/year for the other oceans. Thus the correlation with the time scale is available for the central 100–350 km of the zone of magnetic anomalies. Alternations in the sign of magnetic stripes are found well beyond the distance corresponding to the chronologies; these can be used only as markers at present.

If the floor of the sea is spreading at oceanic ridges, the mechanism of earthquakes occurring on fracture zones should be one of transform faulting instead of transcurrent faulting [*Wilson*, 1965]. *Sykes* [1967], using Byerly's fault-plane method [*Stauder*, 1962], has shown that the sense of motion, if parallel to the principal geomorphic trend, is that for transform faulting and opposite to that for transcurrent faulting.

Oliver and Isacks [1967] showed that the deep ocean trenches are the loci where the surface area created at ridges is destroyed. The trenches are the places where plates of surface matter, perhaps 50–100 km thick, are being thrust into the mantle. Deep-focus earthquakes, usually associated with deep sea trenches, are thus possibly the consequence of thermal stresses set up by the plunging of cold, near-surface matter into a warm mantle. Below a depth of 600 to 700 km for the most rapidly moving blocks, the downward-thrusting block probably approaches a state of thermal equilibrium with its environment [*Griggs and Baker*, 1968]. The depth of deep-focus shocks is well-correlated with the velocity of the downward moving block [*Isacks et al.*, 1968] for the different geographic regions.

Paleomagnetic measurements of the remanent magnetic field in sediments serve to give the direction of the magnetic field of old rocks. Comparison of rocks of the same age from different continents give divergent magnetic poles. These can be brought into coincidence by invoking both continental drift and polar migration. The method involves assumptions regarding (1) the undisturbed nature of the samples since deposition, (2) the dipolar nature of the magnetic field, (3) the coupling of the magnetic and rotational poles, and (4) lack of scatter in the data. Under these assumptions, the magnetic data, as far back as 1.5×10^9 years ago, indicate consistency of continental drift with the geomorphic fit [*Creer*, 1965].

Finally, the work of *McKenzie and Parker* [1967], *Le Pichon* [1968], and *Morgan* [1968] has shown that the surface regions between ridges and trenches move as more-or-less rigid plates. The surface plates are bounded by sources of area at ridges, by sinks of area at trenches, and by transform faults where area is neither created nor destroyed. The edges of the plates are delineated by a map of earthquake epicenters [*Barazangi and Dorman*, 1969]; earthquakes occur where adjacent plates move relative to one another. At sources of area, the earthquakes are tensile in character; at sinks of area they are, in part, compressional; and at transform faults, they are strike-slip.

Note added in proof. A. E. Maxwell [*Trans. Am. Geophys. Union, 50*, 113, April 1969] reported results from the Deep Sea Sampling Program showing that the deepest sediments on the sea floor have ages proportional to their distance from the axis of the mid-ocean ridge in the South Atlantic. Results include sediments to ages about 90 million years, thus extending the previous chronology twentyfold and indicating that spreading has continued steadily at about 2 cm/yr for at least 90 m.y., and probably has been steady since the onset of spreading when the primeval protocontinent, Gondwanaland, ruptured.

Flow in the Mantle

A model has been described above of large-scale motions at the surface. Matter emerges from the interior at rifts and disappears into the interior at trenches. But as yet, this picture describes the motion of surface or near-surface matter only. No attempt has yet been made to describe the complete kinematics in terms of the return path of the flow, required by the continuity conditions for conservation of matter.

What is unknown is the rheological state of the mantle. It was formerly suspected that the deep mantle, below about 1000 km, had a viscosity of the order of 10^{26} cgs, a value sufficiently high to inhibit large-scale motions in the deep interior [*Munk and MacDonald*, 1960; *McKenzie*, 1968]. Recently it has been shown that this conclusion can be explained in an alternate way by renormalization of the second-order harmonic in the gravitational field as determined from the perturbation of satellite orbits [*Goldreich and Toomre*, 1969]. Thus we cannot say, at this time, whether there is significant motion in the deeper mantle.

The inference that the lower mantle had a very high viscosity was a consequence of the size of the second-order harmonic in the gravitational field; this was not appropriate to account for the observed polar flattening of the earth computed on the basis of the dynamics of a rotating fluid. A significant second-order term has always presented difficulties from the point of view of convection theory, since this must be produced by some process symmetric with respect to the axis of rotation; the other harmonics need not be so, and can, in principle, be generated by instabilities in the motion, as will be outlined below. *McKenzie* [1968] considered the possibility that the nonequilibrium flattening might be produced by mantle-wide poloidal currents caused by the failure of the isotherms to coincide with the gravitational equipotentials. If these do not coincide in a rotating fluid, flow must take place according to a theorem of von Zeipel. McKenzie concluded that this type of flow is not sufficient to account for the flattening and prefers the hypothesis that the flattening is associated with a deep-seated viscosity.

Inspection of the distribution of the perturbation gravitational fields as a function of spherical harmonic order number casts further light on the contribution of the second-order harmonic. The result shows that these coefficients, including the second-order term, vary as $1/n^2$ [*Anderle and Smith*, 1968]. This suggests, but not conclusively, that the process which produced the higher-order perturbations in the gravitational field also produced the second-order one.

The low-velocity channel appears to be a worldwide phenomenon occurring at different depths beneath the surface. Under more or less inactive areas, including shields, it is at a depth of about 115 km [*Brune and Dorman*, 1963]. Under more active regions, including mountains and oceans, it is at shallower depths [*Berry and Knopoff*, 1967; *Knopoff et al.*, 1966]. In addition, there appears to be a significant lateral variation of velocity in the channel. The reduction of velocity at these depths in some parts of the world may be due to partial melting. The increased fluidity would imply a somewhat reduced viscosity in this region. *Anderson* [1966] has shown that there is a correlation between viscosity and the specific attenuation factor $1/Q$ for seismic waves in about the same region as the low-velocity channel. The classical measurements of viscosity from elastic rebound of land masses due to the removal of a surficial load may in fact reflect the viscosity of the low-viscosity zones. If the low-velocity channel is a zone of reduced viscosity, it seems likely that the surficial plates described above move about using the low-velocity/low-viscosity zone as a lubricant. Thus the low-velocity channel might act as a zone of decoupling of large motions near the surface from those in the deeper interior where the motions are probably significantly reduced.

It has been shown [*Knopoff*, 1964, 1967a] that significant inhomogeneity in the form of phase transitions in the mantle might inhibit vertical motion across these boundaries if the transition is sharp enough. In addition to the boundaries of the low-velocity channel, significant discontinuities in the upper mantle are found at depths of about 360 and 600 km [*Johnson*, 1967]. These discontinuities, which seem to be well correlated with phase changes to high density polymorphs of magnesium silicates [*Anderson*, 1967], are imperfectly explored as yet; the degree of sharpness is not yet known, although the preliminary cross sections that are drawn through these regions do not appear to be so sharp as to inhibit vertical motions.

Anderson [1967] suggests that the mantle below 600 km has a higher iron-to-magnesium ratio compared with the region above this depth. This result depends significantly upon an equation of state derived from velocity-density relations for materials at laboratory pressures plus data derived from shock-wave studies. The laboratory data cannot be extrapolated accurately to give velocity-density relations at pressures corresponding to 600 km or more of depth; there are such significant uncertainties in the shock wave reductions [*Knopoff and Shapiro*, 1969] as to make the shock wave data unusable at this time. If the suggestion that the lower mantle is richer in iron than the upper mantle should prove to be correct, then vertical mass transport across the interface of the lower and upper mantle is forbidden, since convection is basically a homogenizing process; stirring would destroy the chemical stratification.

Scale of Mantle Motions

At this time, the evidence is extremely weak either for or against the idea that motions take place in a mantle-wide process. This point is raised because in several theories of convective processes the horizontal scale of convection is strongly dependent upon the vertical dimensions of the region involved in the calculations. If one imagines the surface plates to be the upper limbs of some large-scale convective process, with hot matter rising at the ridges and cold matter descending at the trenches, the distances between the trenches and ridges must be related to the vertical scale of the region of circulation. The horizontal distances between ridges and trenches seem more likely to be of the order of the depth to the core than, say, to the deepest major transition in the upper mantle, namely 600 km. Thus, if this point of view is taken, it is crucial that no barriers to the circulation be present anywhere in the mantle, whether they be of the type involving chemical stratification or of the type involving polymorphic phase changes over short ranges of depth. *Verhoogen* [1965] has shown that it is highly likely that most polymorphic transitions of the silicates probably do not present impenetrable barriers to vertical convection.

Driving Mechanism

It is of interest to see whether a model of a driving process that will generate these large horizontal motions at the surface can be constructed consistent with the observations mentioned thus far and consistent with the potential constraints on the motion. Three published models have been claimed, at various times, to be pertinent to the problem of continental drift. These are the models of (1) a layer of fluid heated from below (considered originally by Lord Rayleigh), (2) a fluid heated from within, and (3) a fluid with horizontal temperature gradients applied to it. The first model, that of a layer heated from below (summarized by *Knopoff* [1964]), involves marginal instability; circulation does not start until the vertical temperature gradient exceeds a certain critical value. At the condition of marginal instability, the form of the convective flow is prismatic with a plan form of regular polygons. The horizontal scale factor is of the order of two to three times the vertical scale factor. If the superadiabatic vertical temperature gradient is significantly supercritical, the motion becomes irregular but, because of a high ratio of viscosity to thermal diffusivity, it probably retains approximately the ratio of dimension of plan form to vertical scale in some statistical sense.

The model of the fluid heated from within [*Tozer*, 1965] has certain similarities to the first model. It too involves marginal instability, but one governed by a different dimensional number than the first model. The scale factors appear to be of the same order as those for the first model.

In the presence of horizontal temperature gradients [*Allan et al.*, 1967], the fluid is always unstable, in contrast to the two foregoing models. The horizontal scale is once again of the order of the thickness of the layer for a delta function-like temperature distribution at the surface (Knopoff and Smith, unpublished results). Unfortunately, the nonlinear problem of heat transport has not yet been included in the calculations for this model, nor has it been taken into account adequately in other models.

Excluding the problems of the possible irregularity of the convection patterns in the mantle, which may itself be defeating to the argument of applying the above models, it would seem that the distance from the East Pacific rise to the Japan trench is significantly larger than the scale of the motion allowable by the above theories, even assuming that mantle-wide convection is possible. In addition, no mechanism is available to allow for the likelihood that the sources and sinks of area, the ridges and trenches, migrate. Migration is required to allow several sources of area to be adjacent to one another without the occurrence of intervening sinks. An example of this problem is to be found in the adjacent ridges of the southern Mid-Atlantic, the East African rift, and the Mid-Indian Ocean (Figure 1). No intervening trench is to be found between these three sources of area. These three rifts can all be sources of spreading if the distance between the rifts is continually increasing. Indeed, the four adjacent plates abutting on the three rifts mentioned above must all be increasing in net

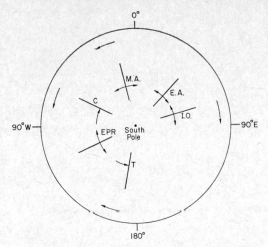

Fig. 1. Schematic diagram of the southern hemisphere showing major rift zones and trench structures. The inner arrows show motions relative to these features; the outer arrows show the motions of the features relative to one another. The East African rift has been taken as stationary in an inertial frame of reference, but this is arbitrary. *EA*, East African rift; *MA* Mid-Atlantic ridge; *IO*, Indian Ocean; *EPR*, East Pacific rise; *C*, Chilean trench; *T*, Tonga trench. The actual motions are not azimuthal as shown.

area, since the trenches at their outer edges, the Chile and the Tonga trenches, are oriented in the sense of disappearance of Pacific area. Hence these two trenches must be moving toward the East Pacific rise despite the spreading from the latter, since the net area of the Pacific will be reduced, to compensate for the increase in area of the four other plates.

A final consideration in the problem of mantle-wide convection is that the time constant to set up steady convection of this type is extremely long. Using rates of sea-floor spreading of the order of centimeters per year, the time for this process to become stable is of the order of 1 billion years [*Knopoff*, 1967b] or longer if the viscosity at depths is great; this value is approximately the time it takes a particle to complete one circuit of a regular convection cell. This time scale is long enough compared with the time for postulated rupture of the protocontinent of the southern hemisphere that it would seem unlikely that this model is appropriate.

It is therefore quite unlikely that mantle-wide convection is a pertinent model for the earth. The effects of possible chemical stratification, the scale of distance, and especially the scale of time make this quite improbable. To restrict the circulation to the upper mantle only increases the difficulty with regard to the horizontal scale.

Episodic Movements

Additional geophysical evidence lends support to the point of view that convection is not a steady-state process. This is the observation that the spreading process may be episodic. Two independent sets of observations bear on this point. *Ewing and Ewing* [1967] have shown that there exists a rather abrupt change in the thickness of sediments of all the major oceans at a distance of the order of a few hundred kilometers from the axes of these ridges. If there has been continuous steady spreading, then an abrupt change in the rate of deposition of sediments in the oceans occurred several million years ago. On the other hand, it may be that the parts of the sea floor more remote from the axes of the ridges have been relatively stationary with respect to the location of the ridges for some time. Ewing and Ewing surmise that the latter explanation is the more likely and that the present episode of sea-floor spreading began about ten million years ago. This was preceded by a quiet interval in which no relative motion took place. An earlier episode of spreading may have terminated about forty million years ago. It is possible that the Atlantic Ocean opened up and spread to virtually its present shape in episodes ending about a million years ago and has only recently begun to widen again. The recent onset of spreading seems to have started more or less simultaneously in all oceans.

Grommé et al. [1967] find that earlier epochs of continental drift probably took place over quite short periods of time. Paleomagnetic pole positions for rocks from different continents seem to show consistent ages over long periods of time. Grommé et al. concluded that several spasmodic episodes of sea-floor spreading may have taken place in the lifespan of the earth and that the time during which the earth's surface is in a state of large-scale motion is probably small compared to the time during which it is in a stationary state.

If we assume that the motion is episodic, the transient process requires a somewhat different picture for the circulation. It is very likely that static stresses accumulate during the quiet period until they exceed certain critical stresses. When the critical state is reached, a major rupture takes place and the two adjacent plates are set in motion in opposite directions. Their momentum probably carries them a certain distance; ultimately the plates slow down because of viscous drag forces. A further quiet episode sets in once again. If this model is accepted, then it would seem likely that the source of the material to fill up the space between the diverging plates must be local. One possibility is that the material comes from the low-velocity channel itself.

The source of the instability is to be found in the static episode. The cause of the stresses that initiate the onset of spreading is a source of speculation at present, but it is undoubtedly thermal in nature. Once the surface parts of the earth begin to move, they readjust their positions to some new configuration, which is presumably in a lower energy state. The system then freezes and the static stresses begin to build up once again, much as some giant relaxation oscillator. A model somewhat along these lines was suggested by *Griggs* [1939], but in connection with a mantle-wide convection process. It seems useful to revive this picture in the context of the new information about sea-floor spreading.

REFERENCES

Allan, D. W., W. B. Thompson, and N. O. Weiss, Convection in the earth's mantle, in *Mantles of the Earth and Terrestrial Planets*, edited by S. K. Runcorn, pp. 507–512, Interscience Publishers, New York, 1967.

Anderle, R. J., and S. J. Smith, Observation of twenty-seventh and twenty-eighth order gravity coefficients based on Doppler observations, *J. Astronaut. Sci.*, 15, 1–4, 1968.

Anderson, D. L., Earth's viscosity, *Science*, 151, 321–322, 1966.

Anderson, D. L., Phase changes in the upper mantle, *Science*, 157, 1165–1173, 1967.

Barazangi, M., and James Dorman, World seismicity map of ESSA Coast and Geodetic Survey epicenter data for 1961–1967, *Bull. Seismol. Soc. Am.*, 59, 369–380, 1969.

Berry, M. J., and L. Knopoff, Structure of the upper mantle under the Western Mediterranean basin, *J. Geophys. Res.*, 72, 3613–3626, 1967.

Brune, J. N., and J. Dorman, Seismic waves and earth structure in the Canadian shield, *Bull. Seismol. Soc. Am.*, 53, 167–210, 1963.

Bullard, E. C., J. E. Everett, and A. G. Smith, The fit of the continents around the Atlantic, *Phil. Trans. Roy. Soc. London, A*, 258, 41–51, 1965.

Cox, A., R. R. Doell, and G. B. Dalrymple, Reversals of the earth's magnetic field, *Science*, 144, 1537–1543, 1964.

Creer, K. M., Palaeomagnetic data from the Gondwanic continents, *Phil. Trans. Roy. Soc. London, A*, 258, 27–40, 1965.

Ewing, J., and M. Ewing, Sediment distribution on the mid-ocean ridges with respect to spreading of the sea floor, *Science*, 156, 1590–1592, 1967.

Goldreich, P., and A. Toomre, Some remarks on polar wandering, *J. Geophys. Res.*, 74(10), 1969.

Griggs, D., A theory of mountain building, *Am. J. Sci.*, 237, 611–650, 1939.

Griggs, D. T., and D. W. Baker, Origin of deep focus earthquakes, in *Properties of Matter*, John Wiley & Sons, New York, in press, 1968.

Grommé, C. S., R. T. Merrill, and J. Verhoogen, Paleomagnetism of Jurassic and Cretaceous Plutonic rocks in the Sierra Nevada, California, and its significance for polar wandering and continental drift, *J. Geophys. Res.*, 72, 5661–5684, 1967.

Hurley, P. M., F. F. M. deAlmeida, G. C. Melcher, U. G. Cordani, J. R. Rand, K. Kawashita, P. Vandoros, W. H. Pinson, Jr., and H. W. Fairbairn, Test of continental drift by comparison of radiometric ages, *Science*, 157, 495–500, 1967.

Isacks, B., J. Oliver, and L. R. Sykes, Seismology and the new global tectonics, *J. Geophys. Res.*, 73, 5855–5899, 1968.

Johnson, L. R., Array measurements of P velocities in the upper mantle, *J. Geophys. Res.*, 72, 6309–6325, 1967.

Knopoff, L., The convection current hypothesis, *Rev. Geophys.*, 2, 89–122, 1964.

Knopoff, L., Thermal convection in the earth's mantle, in *The Earth's Mantle*, edited by T. F. Gaskell, pp. 171–196, Academic Press, New York, 1967a.

Knopoff, L., On convection in the upper mantle, *Geophys. J.*, 14, 341–346, 1967b.

Knopoff, L., and J. N. Shapiro, Comments on the interrelationships between Grüneisen's parameter and shock and isothermal equations of state, *J. Geophys. Res.*, 74, 1439–1450, 1969.

Knopoff, L., S. Mueller, and W. L. Pilant, Structure of the crust and upper mantle in the Alps from the phase velocity of Rayleigh waves, *Bull. Seismol. Soc. Am.*, 56, 1009–1044, 1966.

Le Pichon, X., Sea-floor spreading and continental drift, *J. Geophys. Res.*, 73, 3661–3697, 1968.

McKenzie, D. P., The influence of the boundary conditions and rotation on convection in the earth's mantle, *Geophys. J.*, 15, 457–500, 1968.

McKenzie, D. P., and R. L. Parker, The North Pacific: An example of tectonics on a sphere, *Nature*, 216, 1276–1280, 1967.

Morgan, W. J., Rises, trenches, great faults, and

crustal blocks, *J. Geophys. Res., 73,* 1959–1982, 1968.
Munk, W. H., and G. J. F. MacDonald, *The Rotation of the Earth,* Cambridge University Press, Cambridge, 1960.
Oliver, J., and B. Isacks, Deep earthquake zones, anomalous structures in the upper mantle, and the lithosphere, *J. Geophys. Res., 72,* 4259–4275, 1967.
Stauder, W., The focal mechanism of earthquakes, *Advan. Geophys., 9,* 1–76, 1962.
Sykes, L. R., Mechanism of earthquakes and nature of faulting on the mid-oceanic ridges, *J. Geophys. Res., 72,* 2131–2153, 1967.
Tozer, D. C., Heat transfer and convection currents, *Phil. Trans. Roy. Soc. London, A,* 258, 252–271, 1965.
Verhoogen, J., Phase changes and convection in the earth's mantle, *Phil. Trans. Roy. Soc. London, A, 258,* 276–283, 1965.
Vine, F. J., Spreading of the ocean floor: New evidence, *Science, 154,* 1405–1415, 1966.
Vine, F. J., and D. H. Matthews, Magnetic anomalies over oceanic ridges, *Nature, 199,* 947–949, 1963.
Wegener, A., *Die Entstehung der Kontinente und Ozeane,* Vieweg u. Sohn, Braunschweig, 1915.
Wilson, J. T., A new class of faults and their bearing on continental drift, *Nature, 207,* 343–347, 1965.

Problem of Convection in the Earth's Mantle

E. N. Lyustikh

WHETHER convection exists in the earth's mantle is still uncertain. Much theoretical attention is paid to investigating whether convection is possible in the mantle, whether it can extend over the entire mantle or only a part of it, whether it is continuous or intermittent, what the most probable number of cells is, whether these cells are arranged in one or several layers, and what causes changes in the convection pattern.

If the mantle is composed of a Newtonian fluid with homogeneous composition and uniform viscosity, then a steady-state convection extending over the entire mantle is possible. *Chandrasekhar* [1953] computed this convection by supposing uniform boundary conditions for the mantle surface. This supposition simplifies calculation, but it is far from reality, since the temperature of the uppermost part of the mantle under continents is higher than under oceans; also the mechanical stresses at the mantle surface differ. From Chandrasekhar's calculation, a quantity n is taken commonly to denote the order of the spherical harmonic controlling convection at a given radius R of the earth's core. However, n depends not only on R, but also on other parameters and boundary conditions which do not correspond to those existing on the earth; therefore this n cannot be used to represent the real earth. Moreover, neither n nor the number of cells is what we need to determine, but the size of a cell, as was done by *Rayleigh* [1916] for plane convection. If the area of the part of each cell at the earth's surface is equal to 1, 1/2, 1/4, 1/6, 1/8, 1/12, or 1/20 the earth's area, then a steady-state convection forming one, two, or more cells in the form of a regular polyhedron is possible. In other cases, the convection pattern, in all probability, will change slowly without ever becoming stable. This will especially be true if the earth's area is not an integral multiple of the area of the cell.

The situation is complex, because convection sharply increases the heat transfer out of the earth, and cooling of the mantle causes a strong increase in viscosity which in turn retards convection. Viscosity in general increases with depth under the effect of increased pressure; this is one of the reasons why Chandrasekhar's calculations are not applicable to the earth. *Lyustikh* [1965a, b, c, 1967] expounded arguments against such calculations and against recent methods of connecting the convection hypothesis with the continental drift hypothesis. We shall consider only a few of these objections. The use of spherical functions to analyze the connection between global tectonics and convection is misleading. This is seen from the work of *Coode* [1966], who expanded the linear elements of global tectonics into series of spherical harmonics and plotted the harmonic content curves, obtaining a maximum for seismic oceanic ridges on the fourth harmonic; when East African rifts were added, the maximum was on the fifth harmonic. This shows the instability of the position of the maximum on the harmonic content curve and casts doubts on the results of Coode, who concluded that a shift from $n = 4$ to $n = 5$ indicates a radical rearrangement of the convection system. To explain this shift, it is sufficient to suppose that East African rifts originated later than the other seismic ridges: it seems evident that a rearrangement of the convection system over the entire globe is not necessary for the formation of these rifts.

The question about convection is closely connected with the question about composition of the earth's core. Convection penetrating the entire mantle is not compatible with an iron core. An iron core having practically no radioactive elements becomes heated more slowly than the mantle; therefore an inverse temperature gradient is inevitable at the bottom of the lower mantle, and convection cannot extend over this part of the mantle. On the other hand, if convection is present, the vertical temperature gradient in the mantle differs very little from the adiabatic gradient. This means that the temperature at the core boundary does not exceed 3000°C; an iron core cannot exist in a liquid phase at such a temperature. Therefore, the hypothesis connecting the convection system with growth of the iron core is untenable.

If the core is composed of silicates, the transition curve on the phase diagram will have a negative gradient, i.e. heating will facilitate the transition into a more dense (metallic) state [*Magnitsky*, 1965]. Therefore the current of mantle material cannot pass through the core, and one-cell convection is not possible. The growth of such a core could have led to the same consequence as the growth of a metallic core, i.e. to a rearrangement of convection pattern, but, for the core to grow, its temperature must increase, and this is excluded in the presence of convection. At the same time, convection does not preclude the melted state of a silicate core, since the melting point of such a core seems to be low enough [*Zharkov*, 1966], lower than the temperature corresponding to the adiabatic gradient in the mantle.

What we have said above applies only to a viscous mantle having no yield point. However, it is highly probable that the mantle is composed of material having fundamental strength. Horizontal nonuniformity in temperature is necessary to initiate convection and determines the number of convection cells. In this case convection would be interrupted by such long intervals that we have to seek a different mechanism to account for tectonic phenomena. For different cells, the intervals may be different, and the movement will not take place simultaneously; consequently the question of size and number of cells becomes all the more complicated.

Bott [1964] pointed out that ascending convection currents should be situated under continents (because the temperature is higher under continents), descending currents under oceans. If the continents are capable of shifting as a whole, they should shift toward descending currents until these currents are under continents. Then the descending currents would become heated, ascending currents cooled; convection would be stopped and reversed. Thus, this mechanism facilitates a continuous wandering of continents without any growth of core.

A comparatively small increase in density with depth due to changes in chemical composition of the mantle may prevent convection. For example, if the density increases by 2% from the top to the bottom of the mantle for this reason, then a difference of 2000°C above the adiabatic gradient would not be sufficient to start convection [*Lyustikh*, 1965b]. However, if convection exists, it makes the mantle homogeneous (with the exception of the parts of the mantle that rotate as solid bodies, which is possible in a plastic mantle). Thus, nonuniformity in the mantle and heat convection are not compatible. Though it is difficult to comprehend how vertical heterogeneity in the mantle composition could originate, yet such heterogeneity in the upper parts of the mantle is clearly substantiated. The vertical gradient is high and a layer of low viscosity (asthenosphere) begins at a depth of 80–100 km. If the mantle had been uniform, the asthenosphere would have been less dense than the colder layer over it; equilibrium would have been unstable, and under slight disturbances the whole asthenosphere would have been thrust upward to the base of the crust or even higher. It is therefore unavoidable to suppose that the density increases with depth because of chemical composition, at least in this part of the mantle.

Continental drift is pointed out commonly as evidence of the existence of convection in the mantle, yet continental drift itself is not established (see p. 708). The question about the thickness of the drifting layer is especially difficult for the drift hypothesis. If only the continental crust is not drawn into convection circulation [*Dietz*, 1962], then it is not clear why it is not broken into pieces. Indeed, the underlying substratum is stronger than the crust, and convection currents have different velocities and directions in different places. If the entire layer above the asthenosphere (i.e., about 100 km in thickness) drifts, it is difficult to understand where this layer disappears in front of the drifting continent and how it emerges behind the continent.

Even if continental drift were established, it would not solve the question about convection, since drift may be due to the sliding of continents over the inclined surface of the mantle [*van Bemmelen*, 1966].

To judge the role of convection in structural geology, it is useful to estimate the amount of energy that can be obtained from the convection mechanism. The efficiency of an ideal heat engine is $\eta = \Delta T/T_M$, where ΔT is temperature difference (including adiabatic), and T_M is the maximum temperature. Since the conditions in the mantle are very close to ideal ones, we can apply this formula in steady-state convection theory. Taking $\Delta T = 1500°K$ and $T_M = 2500°K$, we get $\eta = 0.6$. However, it can be shown that no more than one tenth of the total mechanical energy developed in a convection cell can be transmitted to the crust [*Lyustikh*, 1960]. Thus, the final efficiency is not more than 6%. Of course, the energy spent on convection is less than the total heat energy escaping from the earth. If the latter is approximately equal to 10^{28} ergs/year, then no more then 10^{27}–10^{26} ergs/year are transferred to tectonic processes by convection. The total energy of all seismic waves is about 10^{25} ergs/year. The values are not contradictory if we assume that an appropriate fraction of tectonic energy is transferred to seismic wave energy. If the mantle is plastic and convection is intermittent, then the efficiency should be less because much energy escapes from the earth by conduction during the pauses of convection. Some other mechanism must be sought to maintain tectonic movements during the pauses. Some authors assume the development of minute convection cells during these intervals, but these small cells do not have the power to cause such movements; moreover, it is rather doubtful that such cells can exist in the presence of fundamental strength [*Lyustikh*, 1957].

'Chemical convection' can give much more energy, if the differentiation of the earth's material causes hydrostatic disequilibrium. Thus, for example, when the crust is differentiated from the upper mantle, a denser layer remains which may sink to the boundary of the core. In this process considerable energy is released, a part of which can be consumed in tectogenesis [*Lyustikh*, 1961*a*]. If thermal convection takes place only in the lower mantle [*Bernal*, 1961], then it would cause only insignificant shifts of individual upper mantle blocks but not continental drift; it would hardly provide sufficient energy for tectonic processes. Therefore it may be necessary to suppose differentiation of lower mantle material and lifting of the less dense matter into the crust. Chemical convection can also be assumed in the earth's core; if crust material comes from the core, it would provide sufficient energy for tectonics [*Lyustikh*, 1961*b*].

Theory and calculations of convection are still in an early stage of development, primarily related to simple models. They must be extended to more complex models that take fuller account of the known structure and inhomogeneities of the earth's interior.

REFERENCES

Bernal, J. D., Continental and oceanic differentiation, *Nature, 192*(4798), 123–125, 1961.

Bott, M. H. P., Convection in the earth's mantle and the mechanism of continental drift, *Nature, 202*(4932), 583–584, 1964.

Chandrasekhar, S., The onset of convection by thermal instability in spherical shells, *Phil. Mag., 44*(350), 233–241, 1953; Correction, *44* (357), 1129–1130, 1953.

Coode, A. M., An analysis of major tectonic features, *Geophys. J.*, *12*(1), 55-66, 1966.
Dietz, R. S., Ocean-basin evolution by sea floor spreading, in *Continental Drift*, edited by S. K. Runcorn, pp. 289-293, Academic Press, New York, 1962.
Lyustikh, E. N., On convection in the mantle according to Pekeris' calculations, *Izv. Akad. Nauk SSSR, Ser. Geofiz.*, *1957*(5), 604-615, 1957.
Lyustikh, E. N., Convection in the earth's mantle, *Izv. Akad. Nauk SSSR, Ser. Geofiz.*, *1960*(1), 3-6, 1960.
Lyustikh, E. N. The energy of gravitational differentiation of the earth's mantle, *Ann. Geofis.*, *14*(2), 169-172, 1961a.
Lyustikh, E. N., Hypothesis of differentiation of the earth's mantle and tectonic generalizations, *Sov. Geol.*, *4*(6), 28, 1961b.
Lyustikh, E. N., Neomobilism and convection in the earth's mantle, 1, Pros and cons of mobilism and convection, *Byul. Mosk. Obshchestva Ispytatelei Prirody, Otd. Geol.*, *40*(1), 5, 1965b.
Lyustikh, E. N., Neomobolism and convection in the earth's mantle, 2, Agreement between the convection and continental drift hypothesis, *Byul. Mosk. Obshchestva Ispytatelei Prirody, Otd. Geol.* *40*(2), 5, 1965c.
Lyustikh, E. N., Criticism of hypotheses of convection and continental drift, *Geophys. J. Roy. Astron. Soc.*, *14*(1-4), 347-352, 1967.
Magnitsky, V. A., *Vnutrenneye Stroyeniye i Fizika Zemli*, Izd. Nedra, Moscow, 1965.
Rayleigh, Lord, On convection currents in a horizontal layer of fluid when the higher temperature is on the under side, *Phil. Mag.*, *32*(192), 529-546, 1916.
van Bemmelen, R. W., On mega-undulations: A new model for the earth's evolution, *Tectonophysics*, *3*(2), 83-127, 1966.
Zharkov, V. N., Viscosity of the earth's interior, *Tr. Inst. Fiz. Zemli, Akad. Nauk SSSR*, *11*(178), 36, 1966.

Convection in the Mantle

S. K. Runcorn

THE theory of convection (from the Latin *to carry with*) in the earth's mantle is an attempt to explain how the internal energy, thermal or chemical, can be the cause of dynamic processes of geology and geophysics. The postulated flow is usually supposed to have a cellular pattern of large scale, that is, the mean flow vectors change only over distances of the order of the earth's radius and over times comparable with the geological periods (hundreds of millions of years). There are two ways by which evidence for such a phenomenon has been obtained. First, there is geological and paleogeophysical evidence for such flow over the last 1-100 m.y. For example, the movements and stresses in crustal layers associated with mountain building appear to be an order of magnitude greater than those that can result from the cooling of the earth. Convection in the mantle seems the most reasonable physical process in the earth's interior to explain this.

Second, certain geophysical measurements reveal that the mantle has properties that are not radially symmetrical. If the earth was once fluid, or if it was formed cold as a randomly accreted sphere, it would be expected to be uniform. Can the convection hypothesis explain the acquisition of these nonsymmetrical properties?

If convection in the mantle is found to be probable, after such evidence is weighed, then its origin must be sought. By thermal convection we mean instability which results from heating within the mantle or from the core, or which is caused by the outward escape of internal heat by conduction through the crust. An alternative cause is chemical convection in which the downward segregation of heavier material in the mantle releases gravitational energy. Order of magnitude calculations show that density variations of 1 part in 10^5 will produce flow at a rate of 1 cm/year. This is equivalent to a temperature difference of 1°C [*Runcorn, 1962a*].

If the convection hypothesis is accepted as a working hypothesis, other questions can be considered. Is there a turbulent flow superposed on the steady pattern? Is the flow Newtonian to a first approximation? How deep into the mantle does the flow extend? Has it been continuous or discontinuous through the earth's life? Can changes in the pattern of convection be traced through geological time?

In attempting to answer these questions, we try to find diagnostic evidence. Much evidence can be interpreted otherwise than by convection. Historically, convection was first suggested by A. Holmes as a mechanism for mountain building, the continental crust being compressed by the descending currents, and later by F. A. Vening Meinesz as a mechanism for ocean trenches, the oceanic crust responding differently and being dragged downward by the same type of mantle flow. Lately, the uprising limbs of the convection pattern have been considered the cause of oceanic ridges, their central rifts being interpreted as demonstrating tension in the ocean floor. Thus the hypothesis of convection replaces that of earth contraction as an explanation of the compressional features, and the hypothesis of an expanding earth, which was advanced to explain the more recently discovered tensional features, is no longer required.

A method of mathematical analysis has recently been developed which strongly supports the idea that those compressional and tensional tectonic features that are presently active have a single physical cause. *Coode* [1966] represented both sets of tectonic features as linear distributions of Dirac delta functions, the scalar distribution function representing the rest of the earth's surface being put equal to zero. He then carried out a spherical harmonic analysis of this scalar surface function; he calculated the harmonic content for each degree n, that is, the root mean square of the coefficients of terms of the nth harmonic for all orders m. Figure 1 compares the harmonic content diagrams for (b) the oceanic ridge system and the continental rift valleys (tensional features) with (c) the Tertiary mountain belts and oceanic trenches (compressional features). Each curve peaks at $n = 5$, a circumstance which would be strange if the tensional and compressional features had a different physical origin.

It is, in any case, generally agreed that the contraction theory of mountain building, supposing the earth to be cooling by conduction, fails by an order of magnitude, although Lyttleton has suggested that, if the core is a silicate

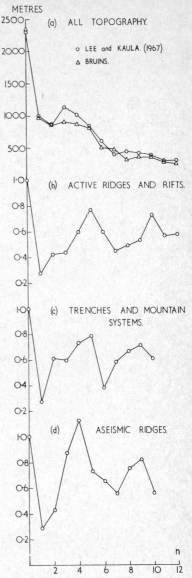

Fig. 1. Harmonic content of topographic and tectonic features of the earth, where n (abscissa) is the degree of spherical harmonic: (a) RMS variability of all topography; ordinate in meters [after *Lee and Kaula*, 1967]; (b, c, and d) Spherical harmonic content of tectonic features (ordinate is normalized for $n = 0$) [after *Coode*, 1966]; (b) seismic ridges and rifts (without the East African rift); (c) trenches and recent mountain systems; (d) aseismic ridges.

which has undergone a phase change to a metallic state as a result of high pressures and temperatures, sufficient contraction might be obtained.

The tectonic arguments for convection cannot be regarded as decisive, for they only demand movement of a hundred kilometers at most. If the evidence for continental drift is accepted, then the convection hypothesis becomes essential to explain it. Wegener's theory supposes that over the past 100–200 m.y. the continents have been displaced by the order of 1000 km. Quantitative evidence for drift is given by the differences between the polar wandering curves calculated from the remanent magnetization vectors for rocks of different continents. Possible reconstructions of the continents are those that bring the polar wandering curves into agreement. Paleomagnetic data agree with paleoclimatic indications that, before the Permian, Europe and North America were a single continent (Laurasia), which was in low latitude during the upper Paleozoic, and that Australia, South America, Africa, India, and possibly Antarctica were grouped around the pole in the upper Paleozoic as Gondwanaland. The continental blocks were then grouped in a simpler pattern than today. *Vening Meinesz* [1951] carried out a spherical harmonic analysis of the topography of the present globe and showed that the harmonics $n = 1, 3, 4$, and 5 were markedly larger than the others, as curve a of Figure 1 shows. He argued that, as the fact of isostasy proved that the continental blocks were floating in the mantle, the continents lay above descending convection currents, and he therefore concluded that the convection pattern is described by harmonics of similar low degrees.

This principle, which may appropriately be called after Vening Meinesz, that the continents move to positions over the descending convection currents seems clearly to be supported by the fact that most tensional features appear in the ocean basins (the exceptions are the East African rift valleys and possibly the Lake Baikal line, which has been conjectured to be a similar rift in the Asian continent). This association of ocean basins and tension would be the natural result of the continents being moved away from the rising convection current. Similarly, in the non-Pacific hemisphere, the compressional features of the Tertiary and Quaternary periods exist only in the continental blocks. However, compressional features in the Pacific floor must occur if the convection patterns have a wavelength of the order of a few thousand kilometers, that is, if there are more than about 2 cells.

Runcorn [1962b] drew the conclusions, first, that the occurrence of continental drift is conclusive evidence for the existence of convection currents in the mantle and, second, that the geologically recent occurrence of drift shows that the pattern of currents must have changed drastically about 100–200 m.y. ago. He argued that, by Vening Meinesz' principle, the positions of Laurasia and Gondwanaland near the equator and pole, respectively, in the upper Paleozoic point to the dominant harmonic being of fourth degree at that time. Thus the change in the convection pattern must have involved the fourth degree harmonic decaying and being replaced by a strong fifth degree harmonic, for which there is evidence in Coode's analysis. If the convection is mantle-wide and if the convection is marginally stable, Runcorn showed that the change $n = 4$ to $n = 5$ might be due to a growth of the earth's core radius by about 1% since the Paleozoic. This theory attempts to remove from continental drift its enigmatic character as a unique and drastic change in the earth's crust, occurring so recently as 5% of the earth's age before the present, not associated with any other event in the earth's history. Supposing a relation between the development of the core and changes in the convection pattern, Wegener's continental drift is to be understood as the latest of four lengthy episodes of crustal upheaval. If the convection pattern during the earliest history of the earth was $n = 1$, the sialic material would have been concentrated in one place; the absence of continents in the Pacific hemisphere may be a relic of this early distribution. Runcorn drew attention to the concentration of the ages of the igneous and metamorphic rocks of the shield areas of North America, Europe, Africa, and Australia about 1000 m.y., 1800 m.y., and 2600 m.y. These must be worldwide periods of enhanced tectonic activity and might be explained as the result of earlier changes of the convection pattern from $n = 1$ to $n = 2$ to $n = 3$ to $n = 4$, successively. It is concluded that the core began to segregate only 3000 m.y. ago for, accepting the accretion origin of the earth, the initial temperatures throughout the earth would be too low to allow solid state creep; thus the iron and silicate objects from which the earth accreted were uniformly distributed throughout the earth and only began to separate when the temperature rose sufficiently for convection to begin.

During the period from the origin of the earth, 4600 m.y. ago, to the beginning of core separation, the disseminated radioactivity, if in concentrations similar to those found in chondritic meteorites, would heat the earth to about present temperatures. Thereafter convection would maintain the temperature distribution throughout the convecting part of the mantle near the adiabatic gradient. The convection hypothesis thus radically alters the evolution of the earth and especially its thermal history. The idea that the iron core forms gradually was first suggested by *Urey* [1952] for geochemical reasons *Runcorn* [1965] shows that the gravitational energy released by the separation of the core is probably twice as great as that produced over the earth's life by radioactivity. Thus convection has partly a thermal origin and partly a chemical one.

Progress in understanding convection has been inhibited by the lack of evidence from present geophysical observations of a nonelastic behavior of the mantle. The equations of classical elasticity have sufficed to explain the whole spectrum of seismological waves with periods up to hours. Although Maxwell, Kelvin, and others pointed out the possibility of elastoviscous behavior of solids, *Jeffreys* [1962] has shown that any simple model cannot be widely applied to the planets without contradictions. The decay of the Chandlerian nutation after excitation has a time constant estimated to be some years, and the buildup of stress over tens or hundreds of years, which produces deep focus earthquakes down to 700 km depth, shows that, over longer periods than are encountered in seismology, the mechanical behavior of the mantle is not wholly elastic. Nor of course can its short-term behavior be described in terms of a Newtonian viscosity. Solid-state physics predicts that over these intermediate time scales, the stress and strain relations may be exceedingly complex. Over millions of years, however, it is not impossible that the behavior of the mantle might become simply that described by a Newtonian viscosity μ. Indeed the only experiments nature has performed for us, the uplifts of Fennoscandia after the glaciers melted and of Lake Bonneville after it drained, are fully consistent with rigid continental blocks floating in a viscous mantle. The value of μ thus obtained for the upper mantle is 10^{21} poise (see p. 386).

Elsasser [1963] and *Orowan* [1965] have argued that laboratory experiments on creep all show strong nonlinear relationships for stress versus rate of strain and conclude that convection in the mantle is likely to have very different properties from the classical theories, which can be tested experimentally, of thermal convection in fluid layers at marginal stability. Laboratory experiments on solids are all concerned with rates of strain (10^{-9} sec^{-1}) rapid compared to those of convection in the earth's mantle. The latter, calculated by dividing the flow velocity (1 cm/year) by the length scale of the convection pattern, gives creep rates of 10^{-15} sec^{-1}.

We assume a viscous mantle as a first approximation: over long periods it seems inevitable [*Runcorn*, 1962b] that the strain in a solid will relax through the fundamental tendency of a solid to be restored to minimum free energy by thermal agitation. The factor exp $(-E/kT)$, where T is the temperature, k is Boltzmann's constant, and E is the excitation energy, suggests that the rapid rate of increase of temperature in the earth's crust provides an explanation for the fact, long known by geologists, that the crust, even over millions of years, acts like an elastic solid, fracturing when the stresses exceed the breaking stress but not flowing under the influence of lesser stresses, while the region below about 50 km depth (that of isostatic equilibrium, termed the asthenosphere) behaves more nearly like a viscous fluid.

Because of the very large number of mechanisms responsible for creep processes in the laboratory, it seems premature to attempt to predict which is responsible for convection in the earth. A possible rapid rise of μ with depth has been suggested which would confine the convection currents to the upper mantle, possibly the seismologists' low-velocity layer. It seems unlikely that a layer discovered by seismic waves of short period should necessarily be the same as one of low viscosity, defined in the sense used here for periods of millions of years. The only evidence for higher viscosity in the lower mantle is based on a theory of *Munk and MacDonald* [1960] to explain why the observed value of J_2, determined from the precession of earth satellites, is different from that on the Darwin Radau theory of the ellipticity of a spheroid in hydrostatic equilibrium, rotating with the present angular velocity of the earth.

Thus as a first approximation we use the Navier-Stokes equation to describe the convective velocity field $\mathbf{v}$ in the mantle:

$$\rho_0 \frac{d\mathbf{v}}{dt} + \rho_0(\mathbf{v}\cdot\nabla)\mathbf{v} + 2\rho_0\mathbf{v}\times\boldsymbol{\omega} - \mu\nabla^2\mathbf{v}$$
$$= -\nabla p + \mathbf{g}\rho$$

where $\mathbf{g}$ is gravitational acceleration, $\boldsymbol{\omega}$ is the earth's angular velocity, ρ_0 is the unperturbed density, ρ is the density field, the variations over a spherical surface giving the buoyancy forces driving the convection. *Runcorn* [1964] shows that the first three terms are many orders of magnitude too small to be taken into account, the acceleration, inertia, and Coriolis terms being, respectively, 10^{-19}, 10^{-18}, and 10^{-8} of the viscous term for mantle-wide convection. Assuming velocities of 10–100 cm/year, these omitted terms would only be of the same order as the viscous one if the flow has a length scale of 10^{13} km or if the viscosity is 10^3 poise! For mantle-wide convection, the viscous force per unit volume is 10^{-2} g cm^{-2} sec^{-2}; thus the non-radial variation of ρ will be 1 part in 10^5.

We now ask whether there are any worldwide, present geophysical measurements that detect convection in the mantle and could be used to determine the present flow pattern.

The density variations in the mantle, which cause the buoyancy forces, will make the external gravitational potential U depart from that expected for hydrostatic equilibrium, represented by the international gravity formula. The analysis by Jeffreys and later Uotila of surface observations and the observations of satellite orbits have shown that this is so, and the variations are about 1 part in 10^5, as Figure 2 shows. These are, in order of magnitude, explained by the convection theory. *Runcorn* [1966, 1967a] shows that, for constant μ and g/r, the equations

$$\nabla^2 U = -4\pi G \rho$$

$$\rho = -\frac{\mu r}{g}\nabla^4 W$$

where G is the gravitational constant and W is the scalar function related to the poloidal velocity field $\mathbf{v} = \nabla\times\nabla\times\mathbf{r}\,W$, can be solved to provide a relationship between the coefficients of the external gravitational potential and the values of d^2W/dr^2 over the boundaries. If the internal boundary can be ignored, one can calculate from the geoid the stresses that the convection currents are exerting on the rigid crustal shell. These stresses can be shown to be of the order of those required to overcome the strength of the crust: thus, in order of magnitude, an adequate theory of tectonics has been found. The pattern shows compression around the Andes and Rockies and in Europe and around the Tonga, Java, and Japan trenches. It shows tension in the East Pacific and Indian oceans along the East Pacific, the Carlsberg rise, and the Indian-Antarctic rise and in the Atlantic Ocean paralleling but markedly displaced westward of the Mid-Atlantic ridge. The pattern shows tension in the East African rift system and very strongly across Asia through the Lake Baikal region. The stresses also indicate a northward movement of India against Asia and of Australia, such as has been shown from the paleomagnetism of these continents since the Mesozoic. The hypothesis of a rigid mantle, in which density variations were 'frozen in' during the early history of the earth either as a result of the upward separation of sial to form the continents or because the moon separated from the earth or through unknown causes is, of course, an alternative, but it explains neither tectonics nor continental drift.

The other physical measurement that might detect convection is the present worldwide distribution of heat flow; however, because the distribution is still poorly known, its spherical harmonic coefficients are of dubious accuracy. *Girdler* [1967] has shown that the regions of higher than average heat flow correlate with low geoid height and vice versa, a result to be expected if the heat sources just underneath the crust were the uprising convection currents. However, the correlation is only just significant and does not apply to the Indian Ocean area. It perhaps should be doubted whether the correlation would be a good one unless the pattern of convection had been constant far longer than the time constant of thermal diffusion through the rigid crust, through which heat is conducted

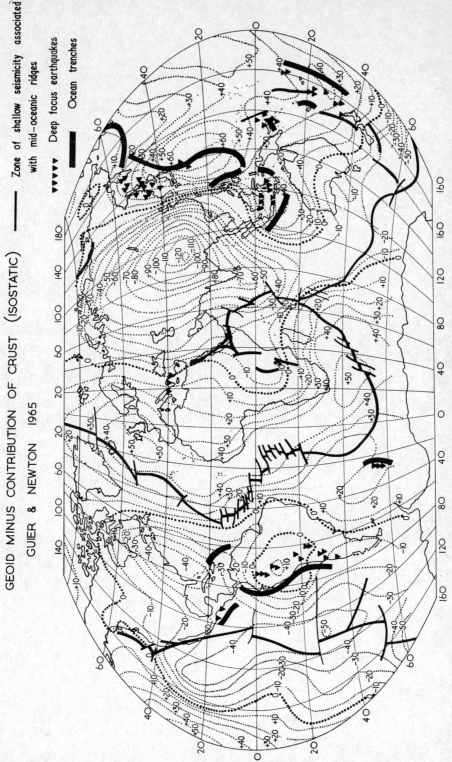

Fig. 2. Geoid of *Guier and Newton* [1965] and earth's main features; gravitational effect of the top 100 km, arising from the vertical distributions of mass due to isostatic compensation, has been removed. Contour interval, 20 meters [after *Runcorn*, 1967b].

to the surface. If the crustal thickness is 50 km, this time constant is of the order of 100 m.y. and the facts of continental drift imply that the convection pattern changed about this time. Seismic travel-time anomalies are 2 orders of magnitude greater than those arising directly from the density variations. However, *Horai and Simmons* [1968] examine their global distribution but fail to find a clear correlation with the geoid, though there is some evidence that they relate to the thermal field.

Finally, one should ask whether present observations of changes in latitude and longitude detect polar wandering and continental drift. The International Latitude Service observations since 1900 show that the pole is moving at a rate of 0.003 inch per year (or about 1°/m.y.) but show no significant relative displacements between the continents [*Markowitz and Guiton*, 1968]. Polar wandering of this rate could possibly be related to convection, as the orders of magnitude of the velocities are comparable.

REFERENCES

Coode, A. M., Analysis of major tectonic features, *Geophys. J.*, *12*, 55–66, 1966.

Elsasser, W. M., Early history of the earth, in *Earth Science and Meteoritics*, edited by J. Geiss and E. D. Goldberg, pp. 1–30, North Holland, Amsterdam, 1963.

Girdler, R. W., A review of terrestrial heat flow, in *Mantles of the Earth and Terrestrial Planets*, edited by S. K. Runcorn, pp. 549–566, John Wiley & Sons, New York, 1967.

Guier, W. H., and R. R. Newton, The earth's gravity field as deduced from the Doppler tracking of five satellites, *J. Geophys. Res.*, *70*, 4613–4626, 1965.

Horai, K., and G. Simmons, Seismic travel-time anomaly due to anomalous heat flow and density, *J. Geophys. Res.*, *73*, 7577–7588, 1968.

Jeffreys, H., *The Earth*, Cambridge University Press, Cambridge, 420 pp., 1962.

Lee, W. H. K., and W. M. Kaula, A spherical harmonic analysis of the earth's topography, *J. Geophys. Res.*, *72*, 753–758, 1967.

Markowitz, Wm., and B. Guinot (Eds.), *Continental Drift, Secular Motion of the Pole, and Rotation of the Earth, IAU Symp. 32*, D. Reidel, Dordrecht, Holland, 1968.

Munk, W. H., and G. J. F. MacDonald, *The Rotation of the Earth*, Cambridge University Press, Cambridge, 323 pp., 1960.

Orowan, E., Convection in a non-Newtonian mantle, continental drift and mountain building, *Phil. Trans. Roy. Soc. London, A, 258*, 284–313, 1965.

Runcorn, S. K., Towards a theory of continental drift, *Nature*, *193*, 311–314, 1962a.

Runcorn, S. K., Convection currents in the earth's mantle, *Nature*, *195*, 1248–1249, 1962b.

Runcorn, S. K., Satellite gravity measurements and a laminar viscous flow model of the earth's mantle, *J. Geophys. Res.*, *69*, 4389–4394, 1964.

Runcorn, S. K., Changes in the convection pattern in the earth's mantle and continental drift: Evidence for a cold origin of the earth, *Phil. Trans. Roy. Soc. London, A, 258*, 228–251, 1965.

Runcorn, S. K., Satellite gravity observations and convection in the mantle, in *World Rift System*, edited by T. N. Irvine, *Can. Geol. Surv. Paper 66-14*, 364–370, 1966.

Runcorn, S. K., Flow in the mantle inferred from the low degree harmonics of the geopotential, *Geophys. J.*, *14*, 375–384, 1967a.

Runcorn, S. K., Wandering continents, in *The Earth's Mantle*, edited by T. F. Gaskell, pp. 475–492, Academic Press, London, 1967b.

Urey, H., *The Origin of the Earth and Planets*, Oxford University Press, London, 1952.

Vening Meinesz, F. A., A remarkable feature of the earth's topography, origin of continents, and oceans, *Proc. Koninkl. Ned. Akad. Wetenschap., Ser. B*, *54*, 212–228, 1951.

Interrelations between the Earth's Crust and Upper Mantle

V. V. Beloussov

IN THIS monograph, the basic results of study of the earth's crust and upper mantle—their constitution, structure and processes of development—are set forth. These results, obtained by various methods, illuminate the processes of the earth's interior.

Now we must draw some conclusions; this is no small task. In fact, it cannot, at present, be fully done: there are too many gaps in our knowledge about the deep interior, too many uncertainties about the interrelation among various phenomena.

We can try to present at least partial conclusions, especially for the most important basic data and those data most likely to aid in the solution of an important problem set forth in this monograph: to discover the explanation and mechanism of endogenous geologic processes, tectonic, magmatic, and metamorphic.

Our task is complicated not only by the incompleteness of our knowledge, but by disagreement of interpretation among the specialists. The same basic data can be interpreted in various ways; they can be assigned different significance, depending on the concepts of the investigator. The reader can easily note the diversity of viewpoints of the specialists who have contributed to this monograph.

Of course, we shall try to be objective and to reflect equally the divergent viewpoints. This goal may not be fully attainable; the author of this section also has his own point of view from which he cannot totally divorce himself. The best course is not to dwell upon results, but to indicate the unsolved problems and the means of their solution. As there are many more unsolved problems than solved, we shall have much to consider.

Composition of the Continental Crust

The first problem that concerns us is the composition of the earth's crust and upper mantle; unfortunately, our knowledge is very scanty. Paradoxically, we study many physical properties and processes of these regions without knowing of what materials they consist. Our definite knowledge ends in the uppermost part of the 'granitic' layer on continents; this material crops out at the surface in crystalline shields and has been sampled by many drill holes through the covering sedimentary rocks in the east European and North American platforms. About half the upper part of the 'granitic' layer consists of granite; the other half of metamorphic and igneous rocks, mainly gneiss. The chemical composition is approximately that of granite but, petrologically, it would probably be more correct to call it the 'granite-gneiss' layer.

We must use indirect means to investigate the deeper regions under continents, as well as all consolidated layers under the oceans. The methods become more indirect with increasing depth.

Comparison of velocities of seismic waves in the lower part of the continental crust and in the third (main) layer of the oceanic crust with elastic properties of rocks led seismologists to designate the so-called 'basaltic' layer. In most crustal models, that layer comprises the entire lower part of the continental crust (separated from the granitic layer by the Conrad discontinuity) and extends uninterrupted under the oceans where, because there is no granitic layer, it constitutes the entire consolidated crust.

First, in this usage, the word 'basalt' is surely inexact. 'Basalt' refers to an igneous rock of basic composition formed at or near the earth's surface. *Gabbro* rock is of the same composition but found at depth (and consequently is coarse-grained). It would therefore be preferable to speak of the 'gabbro' layer, if only to dispel the thought that the granitic layer is a younger layer formed on a previously erupted layer of basalt.

Second, the basalt solution is not the only possible one. Modern seismic, volcanological, and laboratory studies make it clear that basalts originate in the mantle, at depths not less than 50 km and perhaps as great as 200 km. Only temporary chambers of basalt magma, on its way to the surface, could be found in the crust.

At the same time, the seismic velocities observed for the basaltic layer (6.5–7.0 km/sec) can be correlated not only with basic igneous rocks, but also with high-grade metamorphic rocks (granulite facies): plagioclase gneiss with garnet, cordierite, and other dense minerals.

The last alternative is particularly interesting. The migration of water is closely associated with the process of regional metamorphism: the content of water in the initial deposits can vary widely, by tens of percent. But the higher the metamorphic grade, the less is water retained. In the granitic layer, rocks of the amphibolite phase of metamorphism predominate, with minerals still having a significant percent of water. Petrological and laboratory data show that the presence of water is essential for the process of granitization in rocks of the amphibolite facies. Rocks of the granulite facies contain practically no water. They outcrop in very few places; it is completely reasonable to suppose that they are found primarily at greater depth, beneath the granite layer, i.e., at the depth of the basalt layer. Such a distribution of metamorphic facies seems completely logical on the basis of the concept that regional metamorphism is set in motion by the flow of heat from below, from the mantle to the crust. Then it is reasonable to expect a higher degree of metamorphism in the deeper layers of the crust, and upward migration of water in the crust.

The 'metamorphic version' of the model of the crust deserves attention for other reasons. Seismic studies show variations in different localities of crustal structure, total thickness, and the thicknesses of the granite and basalt layers. The character of the Conrad discontinuity also varies: in some places it is expressed by a sharp jump in seismic velocities; in others, a gradual increase. And there are variations in second-order seismic discontinuities that have been identified in the granitic and basaltic layers.

Although this question has not been fully resolved, we can say with confidence that variations in crustal structure are observed from one tectonic zone to another. The thickness and layering of the crust are different under folded zones of various ages, under platforms and shields, under zones of tectonic activity, etc. Different tectonic zones of the continents reflect different stages of development of the earth's crust; they exist side-by-side indicating that evolution of the crust takes place at an uneven rate.

Observations of the differences of structure of the crust show that not only total thickness but also thickness of the individual layers of the crust and the character of the layering change in space and also with time. One should therefore speak first of all about the changes of density and elastic properties of the materials of the crust and about changes in position in the crust of materials with different densities and elastic properties. It is easy to imagine such changes for metamorphic rocks because metamorphic processes, both progressive and regressive, lead to changes in both density and elastic properties of rocks. Progressive metamorphism increases density and elastic wave velocities; regressive metamorphism decreases both. Because the metamorphic processes that change the density of rocks cause a corresponding change of volume, these processes can explain some features of deformation observed in the upper layers of the crust, for example the 'platform folds' that are expressed in gentle uplifts of circular or oval cross section.

However, both basic and ultrabasic magmatic rocks (gabbro and peridotites) can change density, volume, and elastic properties as a result of the processes of amphibolitization and serpentinization connected with the influx of water.

Moreover, increases and decreases in the thickness of the earth's crust as a whole clearly indicate the existence of exchange of material between the upper mantle and the crust. Such an exchange presupposes either injection into the crust or additions to the crust from below of material from the mantle—either basic or ultrabasic magmatic material. Indeed without recourse to the transport of heated material from the mantle to the crust, it is difficult to explain the increased heat flow in the crust which significantly exceeds the normal value of heat flow. At present, high values of heat flow, sometimes exceeding the normal value by several times, are observed in volcanic regions. But from determinations by geological methods of the most probable depths of formation of various metamorphic minerals it is possible to conclude that in geosynclines of various tectonic cycles in the epochs of development of regional metamorphism and granitization, heat flow exceeded the normal value by 3–5 times.

From these considerations, it can be concluded that the basaltic layer of the continental crust consists of a complex mixture of metamorphic rocks, mostly of high-grade metamorphism (perhaps in some places having undergone regressive metamorphism), and of basic and ultrabasic igneous rocks from the mantle, with amphibolitization and serpentinization in places. In addition, judging by the results of experimental investigations of the conditions of the formation of eclogite carried out by A. E. Ringwood, it is reasonable to presuppose eclogite inclusions in the lower layers of the crust. It is possible that the layer of transitional seismic velocities (7.2–7.4 km/sec), which is usually interpreted as a layer of a mixture of material of the crust and mantle, is in reality a layer in which eclogite plays a significant role.

Such a conclusion about the composition of the lower part of the continental crust cannot of course be considered finally determined, but at the present time it can hardly be made more precise. Probably the question about the composition of deep layers of the continental crust will not be solved without direct penetration of these layers by drilling or by some other method that permits us to obtain specimens of these rocks from depths or, in any case, to obtain primary information about the mineralogical composition of these rocks. Deep drilling on continents, even if only to reach the upper part of the basaltic layer and to determine its composition, represents an urgent scientific and technological task.

Oceanic Crust and the Mantle of the Earth

Composition of the consolidated part of the oceanic crust is also an unsolved problem whose final solution will require deep drilling. It is within the limits of present technology to drill a hole completely through the crust and to reach the upper part of the mantle.

The second oceanic layer (having seismic velocities of approximately 5.5 km/sec) is apparently formed entirely or chiefly of plateau basalts which can be traced on the floor of the oceans and samples of which have been excavated from oceanic basins in various regions. However, it is unknown whether densified and metamorphosed sedimentary rocks are present in this layer. Solution of this question is of exceptional interest for illuminating the time of formation and the history of development of oceanic depressions. New seismic reflection techniques permit the reflecting horizons of a sedimentary layer to be traced from the deep oceanic basins all the way to the continental shelves, to the point where paleontological methods of age determination are applicable. These results show that the oldest oceanic sediments (lying on the second layer) almost everywhere belong to the Cretaceous system. Only in a few regions of the Pacific Ocean is it possible by extrapolation to assume the presence of upper or middle Jurassic sediments. If there were older sedimentary deposits in the oceans, they should be included in the second layer.

Drilling through the sedimentary layer and the second layer under the ocean floor is exceptionally important.

The third oceanic layer is called basaltic, but the oceanic basaltic layer, like the continental, is not necessarily basalt. At present there are two points of view about its composition, namely, that it consists either of basic magmatic rocks (gabbro) or of serpentinized ultrabasic rocks of the upper mantle.

The recovery of pieces of serpentinite from the floor of the ocean confirms that serpentinized rocks play a role in the structure of this basal layer of the oceanic crust, but it remains in doubt that this layer is entirely composed of serpentinite. The upper temperature boundary of the process of serpentinization is $+500°C$. In addition, it is completely clear that the base of the oceanic crust (i.e., the Moho) lies significantly above the $500°$ isotherm. Why then does the process of serpentinization not propagate all the way down to that isotherm, i.e., to a depth of about 20 km instead of terminating at a depth under the ocean bottom of 5–6 km? In that case, what is the nature of the Moho? And how can it be so sharp when serpentinization should lead to a gradual transition from the crust to the mantle? The attempt to answer these questions had led some investigators in the direction of supplementary hypotheses about the origin of the oceanic crust; these hypotheses will be discussed later. In any event, the question about the composition of the third layer of the oceanic crust remains open.

The answer to the question 'Does the upper mantle consist of eclogite or peridotites?' seems to be in favor of the second solution. An eclogite upper mantle caused excessively great difficulties in explaining the production of ultrabasic igneous rocks and led to a significant contradiction between the heat flow actually observed at present from the mantle and that which would be expected if the mantle contained the radioactivity of eclogite. Numerous inclusions of peridotite in extruded basalt and the almost complete absence of inclusions of eclogite also speak in favor of a peridotitic composition. We can even put the question more precisely: Of what varieties of peridotite is the upper mantle composed at various depths?

The production of basaltic magma from a peridotitic upper mantle is now related to partial melting. Results of experimental investigations and certain geological deductions indicate that the composition of the molten basalt should depend on the depth at which the complete separation of the basaltic liquid from the crystalline substrata took place. Absolute values cannot be assigned, but it is clear that supersaturated basalts, which apparently serve as the initial rock for geosynclinal spilites, are separated at smaller depths and that undersaturated olivine alkaline basalts and picrites characteristic of the continental platform are separated at greater depths.

This conclusion has great geological significance. As a result questions can be formulated about the relation of individual structural units of the earth's crust with depth. For example, the relation of the more acid basalts to geosynclines shows that the final separation of basaltic melt from ultrabasic crystalline material occurs under geosynclines at relatively small depths, from 20–30 km beneath the base of the crust, whereas alkaline basaltic lavas characteristic of the continental rift zones separate at greater depths (60 or more km).

It is particularly interesting to know to what final results the evaluation of the question of composition of oceanic basalts leads. Until recently it was accepted that typical alkaline olivine basalts were found in the oceans. Newer investigations, especially of C. and A. Engel, have indicated that alkaline basalts occupy only the summit of volcanic mountains in the oceans, whereas the deeper slopes of the mountains and the entire floor consists of tholeiitic basalt. It is essential to obtain more information on this matter.

For region C of the upper mantle it is necessary to postulate a transition of the chief mineral peridotite (olivine) to a denser structural form (spinel, corundum) and a similar modification of silica to denser forms (coesite, stishovite[1]).

The peridotitic composition of the upper mantle apparently requires that the Moho be a chemical, not a phase discontinuity. However, in the light of new data it seems quite probable that there is eclogite in the upper mantle—if not as a continuous layer, at least as inclusions resulting either from the cooling of basaltic magma under pressure or the transformation into metamorphic rocks of the granulite facies. Therefore, the possibility that the Moho is a phase discontinuity *in places* cannot be ruled out.

From the equality of heat flow on continents and in the oceans, it follows that the mantle has a larger radioactive content under oceans than under continents. If at least two thirds of the continental heat flow originates in the crust where radioactive sources are concentrated (especially in the 'granitic' layer), then, because of the thinness of the oceanic crust and presumed lower radioactive content, almost all the oceanic heat flow must come from the mantle. This conclusion about the difference in composition between continents and oceans of the crust and mantle to depths of several hundreds of kilometers is of basic importance for understanding the structure and development of the outer parts of the earth.

Processes in the Crust and Upper Mantle

We have noted that comparison of deep structure of various tectonic zones leads to the conclusion that the thickness and density of the crust have changed during geologic time. Recent conclusions based on seismic and gravimetric data establish the fact that changes in density affect not only the crust, but also the upper mantle. The earlier idea of isostasy requiring that every major surface elevation have a corresponding thickening of the crust (mountain roots), appears to answer only part of the circumstances. In many regions surface elevations do not have a corresponding root at the base of the crust; in these regions, isostasy requires a corresponding decrease of density in the upper mantle. Similarly, there are regions of depression of the earth's surface that do not have 'antiroots' (i.e., upward flexures of the Moho): they require an increase in density in the upper mantle.

We do not know why the Caucasus and the Alps have roots, or why the Fergana depression does not have antiroots. A more careful comparison of the tectonic nature and deep structure of various regions is required.

However, this does not mean that, if we learn more precisely the relation of deep structure with vertical movements of the crust, we shall

---

[1] *Stipoverite* in the Russian literature, after S. M. Stishov, S. V. Popova, and L. F. Vereshchagin.

arrive at a simple scheme in which vertical movements represent a direct reaction to changes of density and, consequently, in volume within the crust or in the upper mantle. Recently M. E. Artemiev showed that almost everywhere vertical movements of the earth's crust are anti-isostatic. That means that, where isostatic anomalies indicated that the crust should sink, it is rising, and vice versa. This kind of contradiction between isostatic forces and current tectonic movements is particularly manifested in young geosynclines. The deep mechanism of the vertical movements of the earth's crust proves to be more complicated than the influence of changes in density and volume directly under the observed region; we do not know the details of this complexity.

The exceptions to the 'rule of anti-isostasy' include regions recently freed of continental ice sheets, such as Scandinavia and Canada. Uplift in these regions is in accordance with isostatic forces.

Low-Velocity Channel

The low-velocity zone in the upper mantle is of great significance in relation to deep processes. The existence of the low-velocity channel is not in doubt, but we have very limited information about its depth, thickness, and structure in different geological regions. The data are extremely schematic: they indicate that it ranges in depth from 100 to 250 km under continents, from 50 to 400 km under oceans. There are indications that the low-velocity channel is weakly expressed under crystalline shields, and is perhaps completely absent in some places. There is basis for supposing that the low-velocity zone becomes thicker under active tectonic zones, with upward projection in its surface. Under the Kuril volcanic island arc, a projection of the low-velocity channel reaches the base of the crust; lava flows through this projection to the volcanoes. The low-velocity zone has greater effect on velocities of transverse waves than on longitudinal. There are records in which decrease in velocity was noted only for transverse waves.

In the Pamirs and the Kurils, it is concluded that there are several low-velocity channels in the upper mantle, one above the other.

The low-velocity channel can be explained as a chemical or phase change. A. E. Ringwood proposes a chemical change, and ascribes a pyroxenitic-peridotitic composition to the low-velocity channel as a counterbalance to the overlying dunites and underlying garnet peridotite. Ringwood proposes that the pyroxenitic peridotite of the low-velocity channel is initial mantle material, whereas the overlying dunite represents the residual after removal of basalt by partial melting. From this point of view, melting of basalt began in the uppermost part of the mantle and gradually proceeded downward. The low-velocity layer is thinnest where this process is most advanced, and thickest where differentiation of the mantle is in its initial stage. The latter conditions may be supposed for the oceanic mantle where the channel therefore is thickest.

Inasmuch as decrease in seismic velocity can be related to increase in mean atomic weight, an increase in the iron content could explain the appearance of the low-velocity channel.

More often, the low-velocity channel is interpreted as a result of the predominance at a certain depth of the effect of temperature over pressure. This effect is especially great if there occurs either a transition from the crystalline to amorphous state, or partial melting of the upper mantle, in connection with which droplets or films of liquid melt appear among the crystals. This latter point of view is interesting in connection with the possibility of relating melting of basalt from ultrabasic mantle material to the low-velocity channel. This point of view is not contradicted by what is now known or supposed about distribution of temperature with depth in the upper mantle, and the partial melting of ultrabasic rock at various pressures. However, Birch (pp. 27 ff.) indicates obstacles to this hypothesis. The question about the low-velocity channel requires further investigation.

From the point of view of interaction between the upper mantle and the crust, it is highly important to know whether the mean density of the material of the low-velocity channel is lower than the density of the overlying material (between the channel and the base of the crust). Some hypotheses presume such a lower density of the material of the low-velocity channel; consequently, they suppose also a mechanical instability in the upper part of the channel due to this 'density inversion.'

With this condition, disturbances can be expected at the top of the channel because of the tendency of the material in the channel to rise, and the overlying material to sink. The character of the motion is reminiscent of convection (or better, 'advection'; the conditions are not right for a full circuit, since the light material, having risen, would remain above, the heavy material would remain below). Formation of domes by salt diapirs is fully analogous to this process. Something similar to diapirs must enter the overlying layers from the low-velocity channel. If melting of basalt occurs in the channel, the basalt melt might become isolated, collecting in independent liquid masses (astenoliths) that make their way upward either by floating or fusing. The ascent of basalt asthenoliths could be a main instrument of compositional thermal and mechanical interaction between the mantle and the crust. As a result of asthenoliths, the crust is filled with basaltic magma which is differentiated in the process of crystallization and assimilation of crustal rocks, and this gives rise to the whole gamut of igneous rocks. With heat brought up from depths of 100–200 km, asthenoliths intensely warm the crust, initiating regional metamorphism and granitization. Finally, immense masses of basalt, having been formed under the base of the crust, must tend to lift the crust, thus causing 'undulatory oscillating movement.'

On the basis of this hypothetical mechanism, a well-rounded scheme of deep processes can be formulated that rather successfully explains the basic peculiarities of development of vertical movements of the earth's crust, magmatism of geosynclines, and their transformation into platforms, as well as the general features of the tectonic development of platforms. However, this is evidently quite hypothetical; not less evident is the need for more precise knowledge about the low-velocity channel in order to improve our interpretations.

Horizontal Compression

Is the deformation in geosynclines due to a 'general' horizontally directed compression superimposed on the geosyncline from outside? Or is such a general compression not necessary to explain the folding? This is a decisive question in the study of deep processes.

For a long time, geologists were brought up on the contraction hypothesis, postulated on cooling of the earth, shrinking of its surface, and consequent compression of geosynclines between the hard masses of platforms. They saw the results of this compression in the folded formations.

Although the contraction hypothesis is not accepted by the majority of specialists now, many remain convinced that the explanation of folding in geosynclines requires external compressive forces, which result either from the relative horizontal motions of continents, or from the drag of the descending branch of a convection current in the mantle. This idea assumes significant horizontal displacements of upper mantle material, which lead in turn to the proposal of enormous convections cells on a continental scale.

There is another viewpoint about folding. Some kinds of folding (block folding and injection folding) cannot be related to external horizontal compression. Only one kind of folding, called geosynclinal or holomorphic, requires horizontal compression of the layers; the limited distribution of this type of folding (usually surrounded by other types of folding in the geosyncline), as well as many other structural features, suggests that the horizontal compression is local, being related either to the pressure of masses of rocks rising from depth through the upper layers of the crust, or to layers sliding down slopes of tectonic uplifts under the influence of gravity. This point of view does not require major horizontal motions in the mantle; it is satisfied by vertical circulation of material and a limited horizontal motion, just sufficient to permit passage of two oppositely directed vertical currents.

Resolution of this problem requires continued detailed structural investigations in folded zones, clarification of the character of changes in folding with depth, study of the history of development of folding against a background of vertical movements in the earth's crust, and solution of a series of tectonic problems.

Major Horizontal Displacements

There is the related question: Are there major strike-slip faults in the earth's crust? The highly visible horizontal dislocations along the San Andreas fault in California (expressed by dislocations in river beds and other features) are quite small, not exceeding one kilometer.

The evident displacement along the Alpine fault (South Island of New Zealand) is still smaller. However, on the basis of more general geological data, many investigators conclude that the horizontal motions along these two faults during geological time amount to hundreds of kilometers. Even greater horizontal displacements are indicated by magnetic anomalies along large latitudinal faults in the eastern Pacific Ocean. The proposed displacement in some places exceeds 1000 km. According to the interpretation, a large number of shears cut across the mid-oceanic ridges, causing lateral displacements in sections. The Great Glen fault in Scotland is considered to have a displacement of more than 100 km.

The idea of large shear displacements has many contradictions. Some geological peculiarities near these faults seem to confirm the large displacements; others decisively contradict it. For example, geological features in many places along the San Andreas fault are so similar on opposite sides of the fault, that it is difficult to believe in their accidental co-location. Two granite massifs situated on opposite sides of the Great Glen fault (supposedly representing two separated parts of one massif) are petrographically quite different; this naturally leads to doubt about the simple shear interpretation.

The principle difficulty for large horizontal displacements arises in explaining what happens to the large volume of material that should accumulate at both ends of the fault. For example, where the latitudinal faults of the eastern Pacific emerge onto the continent, there are tectonic faults, but without significant horizontal displacements. Where, for example, did the material disappear that was carried 1200 km horizontally toward the shore along the Mendocino fault?

Difficulties of this type led to the concept of the 'transform fault' in which horizontal displacements are compensated by sinking of crustal material into the mantle on one side, and creation of new crust (from mantle material) on the other. This 'exchange' of material between mantle and crust might indeed compensate horizontal motion of any amplitude. However, it is essential that observations confirm the disappearance and creation of crust at opposite ends of the fault; at present, such observations are unknown.

The problem of large horizontal motions, like the problem of folding, needs to be carefully studied with maximum objectivity.

Formation of Oceans and Continents

We spoke above about the age of sediments on the ocean floor. Remarkably, most oceanic sediments are Cretaceous or younger; in a few places, they may be older, but not older than Jurassic. Under the sediments lies the second layer, apparently formed by outpourings of basalt, thus indicating the existence in some epoch of universal basaltic volcanism throughout the entire modern ocean basins. The traces of that violent epoch (whose duration is unknown) contrast sharply with the peaceful deposition of sediments on the surface of the basalt.

If older sediments (lower Mesozoic or older) ever existed in the oceans, they now lie within the basalt of the second layer. It is not impossible that older material of the oceanic crust is concentrated in the third layer. However, if we consider that most investigators believe that there are no sediments in the present oceans older than Cretaceous (or at most Jurassic), we must recognize that the present oceanic crust is quite young, having appeared at the end of the Paleozoic or beginning of the Mesozoic. Any older oceanic crust was somehow destroyed by this 'renovation.' The proposed mechanism of this destruction and renovation are discussed below.

Mid-Oceanic Ridges

The most important elements of oceanic structure are the mid-oceanic ridges. The more typical ones are in the Atlantic and Indian oceans. Almost unbroken chains of 'rifts'—depressions with steep sides and uneven floors—extend along the crest of these ridges.

On the slopes and especially on the crest of a ridge, the layer of loose sediments, which is quite thin in the open ocean, thins out and disappears altogether. Moreover, near the ridge, only younger (Neogene and Quaternary) sediments are found. Basalts on the crest of the median ridges are, by absolute age determination, Neogene or younger. Thus, the mid-Atlantic and mid-Indian ridges represent zones of development of volcanism right to the present day, whereas volcanism in the neighboring depressions was terminated in the Cretaceous.

The Pacific median ridge (southern and eastern Pacific) is eccentrically located and differs in appearance from the Atlantic and Indian ridges, although it is an uninterrupted continuation of these ridges. There was a proposed hypothesis that there was earlier a typical median ridge in the Pacific that broke up in the Cretaceous and almost entirely sank. Its remains can be seen in numerous ranges of mountains of volcanic origin. Very young basaltic eruptions, up to the present, are observed in these ranges, whereas (judging by the age of the sediments) volcanism in the surrounding depressions terminated at the beginning of the Cretaceous or earlier.

The median ridges have a characteristic deep structure. Combined gravity and seismic observations indicate the existence of lenses of material (up to 20 km thick) of seismic velocity 7.2–7.6 km/sec beneath the ridge; these lenses may be a mixture of mantle and crustal material. The basal (third) oceanic crustal layer thins out over these lenses with 'transitional' velocity. Another interpretation proposes that these lenses are composed of partially melted material of the mantle and crust. This interpretation fits very well with the high heat flow associated with the median ridge.

There are ridges of another type in the ocean: 'aseismic' or 'fault block.' They are mainly rectilinear crests with steep slopes (or one steep slope) and flat summits. Study of their deep structure indicates a thicker than normal oceanic crust with a massive series of basaltic lavas of the type of upper crustal material. These ridges may have formed as a result of intensive outpouring of basalts through immense fissures penetrating the oceanic crust.

Guyots

Guyots are interesting phenomena, found mostly in the central Pacific. No one doubts that their flat summits were formed by erosion at sea level; this conclusion is strengthened by shallow-water sediments found on many guyots. The sediments are of different ages, but none older than the second half of the early Cretaceous. Consequently, guyots, having sediments of such age, were formed as volcanic peaks before the end of the lower Cretaceous, were truncated by erosion, and, from the end of the lower Cretaceous, began to sink.

The summits of some guyots have sunk two kilometers below present sea level. The question arises: Did each guyot sink independently or was there a general sinking of a vast region? Independent sinking is indicated by the great differences in the depths of individual guyots; some quite near one another differ in depth by many hundreds of meters. The second possibility seems equally probable when we note that the circular trenches observed around some guyots never achieve the dimensions necessary to compensate for the sinking of the guyots. The different depths of the guyots can be explained by different times of formation that are compatible with an observed direct relation between the depth of the guyot and the age of its accumulated sediments.

If we adopt the second point of view, the depth of the central Pacific was not greater than 3 km at the beginning of the Cretaceous, and has increased by 2–2½ km in the last 100 million years.

Oceanic Margins

Observations of oceanic margins yield important information about the history of the oceans. There are two types of oceanic margin: Atlantic and Pacific. The Atlantic type occurs almost everywhere in the Atlantic, Arctic, and Indian oceans. Exceptions are the Antilles, Southern Antilles, and Great Sunda island arcs, which belong to the Pacific margin type. The Atlantic margins everywhere coincide with Meso- and Cenozoic platforms of the adjacent continents. All geosynclines that previously existed on the continental margins completely ended their development by the beginning of the Mesozoic, and gave way to the platform regimes. The second characteristic feature of the Atlantic margin is the discordant structural boundary between the continental pre-Mesozoic and the oceans: Paleozoic and older structures of the continents are cut off by this boundary; the ocean seems superimposed on older continental structures.

The Atlantic-type margin is accompanied by a well-defined shelf formed of shallow-water Mesozoic and Cenozoic sediments, deposited unconformably on the eroded surface of Paleozoic (or older) rocks. Its structure shows that the principal feature of its development was a gradual tilting of the basement surface toward the ocean. The edge of the shelf is now depressed 2–3 km from its original position. In some places, the depression began in the Jurassic; in the lower Cretaceous it occurred everywhere. There is basis for interpreting 'marginal plateaus' (such as the Blake and Iberian) as peripheral parts of the shelf that separated from it and experienced greater sinking. Only in a few places has volcanism accompanied the Mesozoic and Cenozoic depression of the Atlantic margins.

The Pacific-type margin encircles the Pacific Ocean and occurs in a few other places (mentioned above as exceptions to the Atlantic type). Margins of this type coincide with Mesozoic and Cenozoic geosynclines on the adjacent continental margins. Although, as a whole, the Mesozoic and Cenozoic orogenic structures of the surrounding continents are parallel to the ocean boundary, there is sharp truncation of these structures in important details, including their youngest (Pleistocene) elements. A characteristic feature of margins of this type is the frequent occurrence of faults along which blocks of the earth's crust have subsided. This process is accompanied by intense volcanism. The Pacific margin is everywhere accompanied by deep-water trenches and inclined fracture zones that (as indicated by deep-focus earthquakes) descend deep into the mantle.

Where the continent directly borders the deep ocean, the transition from continental crust to oceanic is abrupt (usually within 200 km). There are transitional types of crust. The transitional suboceanic crust is formed in enclosed seas such as Bering, Okhotsk, Japan, and the inland seas (Mediterranean, Black, Caribbean, Gulf of Mexico, and southern part of the Caspian). They have practically oceanic crusts, but with a much thicker sedimentary layer. The transitional subcontinental type of crust is characteristic in narrow zones in the transition from continent to ocean and in island arcs. This type is similar to continental crust, but is thinner.

Data now available are much too scanty to resolve the question of the origin of continents and oceans. The existing theories on this subject cannot be reconciled.

Continental Growth

One hypothesis (that dates back to the time when the difference in continental and oceanic structure was first recognized) proposes that the oceanic crust represents a lesser degree of differentiation from the mantle than the continental; thus the continents represent a more advanced stage of evolution. Geosynclines represent transitional elements in which the process of differentiation is occurring very energetically. It is supposed that continents gradually grow at the expense of the oceans, passing through the geosynclinal stage. The youngest geosynclines occupy the continental margins, whereas the continental 'core' lies at its center.

This simple hypothesis is in complete agreement with such phenomena as the equality of continental and oceanic heat flow. The North American continent is constructed very nearly as required by the hypothesis: the rocks of the oldest 'consolidation' are found in the center of the platform (the Canadian shield) and the margins of the continent are occupied by Paleozoic and younger geosynclines. Thus, the transition from older to younger geosynclines of the Pacific margin proceeds outward from the center of the continent toward the ocean.

However, this hypothesis is contradicted by many facts. It cannot be reconciled with the truncation of continental structure on the Atlantic-type margins, which is widely observed on the coasts of South America, Africa, western Europe, India, western Australia, and around the Arctic Ocean. All indications are that on these margins, the continent did not grow, but on the contrary, collapsed; the oceans captured segments that had previously been continents. This direction of development is confirmed by many paleogeographic data indicating the previous existence of dry land in various places that are now covered by oceans or deep seas. In the enclosed and inland seas, it is particularly difficult to dispute the previous existence of dry land—and, geologically speaking, quite recently. The Sea of Japan was formed on previously dry land in the Miocene; the Sea of Okhotsk in the Pliocene; dry land existed in the area of the western Mediterranean until the end of the Oligocene or beginning of the Miocene. In these seas, the crust now has a transitional character; if we accept that major areas of dry land should be underlain by continental crust, then it appears that in these regions the process of development is not proceeding from oceanic through transitional to continental, but the reverse.

If we accept the absence of oceanic sediments older than Mesozoic as an established fact, this also ties in with the hypothesis that the oceans represent the oldest formations from the undifferentiated crust.

Continental Drift

The hypothesis of large horizontal displacements endeavored to respond to some obstacles to the differentiation hypothesis. This hypothesis was proposed by A. Wegener in the early part of the present century; it virtually disappeared from the scene, only to be revived in the last two decades, mainly because of paleomagnetic investigations. However, the hypothesis has been greatly modified from its original form.

The hypothesis of continental drift made a sharp distinction between the Atlantic and Indian oceans, and the Pacific. The first were considered secondary, resulting from the drift of the continents; the Pacific was primary. The enclosed and inland seas were also secondary. This hypothesis explained the truncation of continental structure on margins of the Atlantic type by the splitting and drifting apart of the continents. Paleogeographic data indicating the existence of dry land where now there are seas and oceans were tied in with the idea that the dry land did not sink, but was moved aside.

However, this hypothesis was confronted by a series of fatal questions that could not be answered. For example, there is no indication that the Pacific Ocean is older; all oceans seem to be of the same age. The division of oceans into primary and secondary was refuted. The equality of continental and oceanic heat flow contradicts the hypothesis. The long-continuing, persistent locations—unchanged for hundreds of millions of years—of regions of slow uplift and subsidence (anteclises and syneclises) on the ancient continental platforms (for example, the Moscow and Congo syneclises) contradict the idea of shifting continental crust on the mantle because it is clear that the causes of this uplift and subsidence lie at a depth of many hundreds of kilometers in the mantle. The persistence of tectonic zones can be reconciled with continental drift only if it is supposed that not only the crust moves, but the crust together with a section of the mantle hundreds of kilometers thick. This is also required by the differences in the continental and oceanic mantle revealed by heat flow data. But to assume motion of blocks of crust and mantle several hundred kilometers thick is to encounter insurmountable mechanical and physical-chemical difficulties.

Moreover, this hypothesis, in attaching primary importance to the present-day shape of continents and oceans, simply brushes aside such questions as the causes and regularities of development of oscillatory movements in the earth's crust, folding of various types, magmatism and metamorphism, i.e., all that constitutes the basis of activity in the continental crust. Some of these phenomena are totally ignored; others are distorted by schematic representation.

Sea-Floor Spreading

'Sea-floor spreading' is a modification of the hypothesis of continental drift. On the basis of the latest data on the young age of oceanic sediments, this hypothesis concludes that the oceanic crust is periodically renovated. The most recent such renovation occurred at the beginning of the Mesozoic. In this sense, all oceans are equal: all experienced renewal of their floors at the same time. Renewal occurs by emergence of mantle material through fissures in the rift valleys of the median ridges. This material, upon reaching the surface, flows outward on both sides, thus creating new oceanic crust and a new ocean floor. The old crust is carried under the neighboring continents and is dragged back down into the mantle by the branch of a convection cell descending under geosynclines or deep-water trenches, and is dissolved in the mantle. The currents from the median ridge move the continents until they reach a balance point between the influences of opposing currents. An unavoidable attribute of this hypothesis is the acceptance of serpentinitic composition of the basal layer of the oceanic crust: it cannot be gabbro or basalt if it is formed as a result of the elevation to the surface of ultrabasic mantle material.

A set of serious questions can be posed for

this hypothesis. We shall mention some of them.

If the oceanic crust is serpentinitic, yet the upper temperature limit of serpentinization is 500°C, why is the oceanic Moho so persistently at a depth at which we know this temperature is not attained, and why is the Moho so sharp? How is a serpentinized crust formed in seas where there are no rift valleys (e.g., Okhotsk, Japan, Caribbean) and where the suboceanic type of crust is found? How is the equality of mean heat flow on continents and oceans explained? Why is there no trace of deformation on the ocean floor near the continents where the floor 'flows under' the continent? How can the long persistence of location of tectonic zones on continental platforms be explained? If geosynclines are located over the descending, i.e., the cooler branches of convection currents, how can the indications of high heat flow in geosynclines, i.e., the regional metamorphism and granitization typical of geosynclines, be explained?

In addition, though it may be easy to imagine a system of convection currents for the Atlantic Ocean alone, no one has succeeded in constructing a successful scheme of convection currents for the earth as a whole. For example, Africa represents an insurmountable problem: in Africa there is no meridional folded zone under which currents from the mid-Atlantic and mid-Indian oceans might meet and combine into one descending current. The East African rift, a direct structural continuation of the mid-Indian Ocean rift, introduces inexplicable complications. Why does no mantle material emerge through it? It is well known that ordinary continental crust lies at the bottom of the East African rift! A construction of subcrustal currents that is hydrodynamically absolutely impossible would be necessary to explain the supposed motion of Hindustan from north to south, the formation of folded zones in the Mediterranean area, the relation of the median ridge in the Pacific to the surrounding folded zones, and many others.

If such difficulties arise when we attempt to explain the distribution of young folded zones as dependent upon the distribution of rifts, then a much greater number of difficulties arise when we remember that geosynclines were distributed quite differently in earlier tectonic cycles, and that rifts of those cycles are completely unknown to us. How were structures formed in earlier tectonic cycles? And how did geosynclines of different cycles continue in their development according to one main scheme of dependence and interrelation with tectonic and magmatic processes if they were continually wandering from place to place, carried about by convection currents? The proponents of this hypothesis try to convince us that volcanoes are carried along on the spreading sea floor, continuing their activity all the while. How is it that they do not become disconnected from their sources under the crust?

In spite of all this, there are phenomena that support the hypothesis of continental drift, and its latest modifications. Aside from the parallelism of the Atlantic coasts, which it is very tempting to explain as a division of one continent, paleomagnetic results are recognized by all as the main evidence in support of the hypothesis. The divergence of data about the location of the pole in different geological periods in Europe and North America can be explained by a gradual separation of the continents. There are other arguments, paleogeographic, structural, and paleoclimatic, basically pertaining to the explanation of the upper-Paleozoic glaciation in the southern hemisphere. The structural arguments draw attention to the similarity of geological structure on opposite sides of the Atlantic Ocean and to the continuation of certain structural elements that began on the opposite side.

As many authors have noted, the geological and geographical arguments can be interpreted in various ways; the paleomagnetic data are the main support of the hypothesis. In addition, the alternating normal and reverse magnetization parallel to the rift valley is interpreted by many investigators as confirmation of sea-floor spreading; the rate of spreading is calculated by comparing the distance between segments of opposite polarity with the chronology of magnetic inversions.

However, in spite of the positive evidence of magnetic observations, the hypothesis of continental drift encounters such a decisive conflict with other geophysical and geological data that the questions arise: are the paleomagnetic data being interpreted correctly? Is it a correct basis for interpretation to assume that the earth's magnetic field always had the same structure as at present? In the study of remanent magnetism, have the influences experienced by the rocks after their formation been adequately taken into account? Finally, attention is drawn to the vast scatter in individual paleomagnetic determinations among which a statistical mean must be found. Upon detailed areal investigation, the polar disposition of anomalies on the ocean floor turns into a rather disorderly combination of points.

Expansion Hypothesis

There is also the expansion hypothesis. According to it, the original earth was small enough that the entire surface was covered by the present continents. As the earth expanded, this one continent was fractured, the pieces were forced apart, and mantle material emerged between them, forming the ocean floor.

The objection to continental drift, based on equality of oceanic and continental heat flow, applies in this case. Since the separations among continents differ, yet the earth retained its spherical form, the present distribution of continents could not have been achieved without relative horizontal displacements. Therefore, we come back to the hypothesis of continental drift with all its contradictions. We do not even raise the questions about the large and rapid growth of the earth (from the end of the Paleozoic to the present). The ideas expressed in the literature on this subject have not emerged beyond the limits of highly undefined assumptions.

Oceanization

The hypotheses of continental drift and sea-floor spreading are directly opposed by the hypothesis of 'oceanization,' which allows the formation of oceans as a result of destruction of continental crust by splitting the crust into blocks which sink and are dissolved in the mantle. It is supposed that as a result of long, moderate radioactive heating, the material of the upper mantle was partially melted, leading to its differentiation and to formation everywhere of continental type crust. At the end of the Paleozoic or beginning of the Mesozoic, increased radioactive heating of the earth's interior led to complete melting of ultrabasic material in some regions of the mantle. This was accompanied by violent volcanic processes, during which basic and ultrabasic intrusions penetrated the earth's crust, broke it into fragments and, with its heat, initiated metamorphism with expulsion of water and increase in density. The solidification of the intrusions also led to an increase in density. As a result of this increased density, blocks of the crust sank into the upper mantle, gradually melting and dissolving. At the surface, these blocks were replaced by water and basaltic extrusions.

The latest experimental results of Ringwood and his colleagues indicated significantly broader conditions than had previously been supposed for the formation of eclogites, as a result of metamorphism in the continental crust. Since eclogites are denser than the upper layers of the mantle, they may sink. Ringwood himself suggests that fragments of eclogite, formed upon cooling of basalts and other rocks at the base of the crust, sink into the mantle and that their melting leads to formation of andesite magma. In this way the enormous volume of andesite magma around the Pacific Ocean can be explained.

The hypothesis of oceanization overcomes practically all objections introduced against the hypothesis of continental drift and sea-floor spreading. It explains the equality of heat flow on continents and oceans. It escapes the conflict between continental and oceanic geology, because the continents remain in one place and develop in relation to processes occurring in the mantle directly beneath them; similarly, the oceans are related to processes beneath them. The tectonosphere is divided into vertical blocks which develop more or less independently. In the oceans of the Atlantic type, the median ridges represent zones in which the process of oceanization is still continuing, whereas beyond the ridges, it has terminated. In the regions of development of the borders of the Pacific type, the borders themselves represent an arena of increased development of oceanization, which here enters into conflict with the opposing process of differentiation and accumulation of continental crust taking place in the geosynclines surrounding the Pacific. The island arcs and ridges surrounding the Pacific Ocean represent regions of this collision of opposing deep processes that thus cause violent seismic and volcanic activity. The steep, deep trenches surrounding the Pacific separate regions of development of processes of continental differentiation and oceanic homogenization.

The intermediate types of crust are interpreted by this theory as reflecting the transitional stage on the way from continental to oceanic crust. The gradual basification of the crust explains the sinking of the oceanic margins and of the entire Pacific Ocean with its guyots.

The objections to this hypothesis are related mainly to the mechanisms of oceanization. To postulate sinking of fragments of the continental crust deeply enough into the mantle would require the elevation of a large volume of ultrabasic material into the crust, an amount probably exceeding the volume of the continental crust. Moreover, all this had to occur rapidly (in geological terms), in less than 100 million years. This appears to be a major cataclysm; it is strange that there is no evidence of it on dry land beyond the ocean boundaries. There is no explanation for the localization of the process of destruction of continental crust in exactly those regions where oceans are now found. Until now, no place has been found in which one can observe such intermediate stages of oceanization that would clearly demonstrate the mechanism of the process. The intermediate situation would be particularly instructive for understanding how the granitic layer is destroyed while the basaltic layer is preserved in the development of suboceanic crust. It has been pointed out that the argon content of the atmosphere contradicts the process proposed by this hypothesis. To many it seems doubtful that there were no oceans of the present form before the end of the Paleozoic; rather there were shallow seas on a continental crust. A sudden increase in the volume of water on the earth's surface at the end of the Paleozoic or beginning of the Mesozoic is difficult to understand.

Future Investigations

The existence of so many hypotheses about the development of the earth, each with its contradictions, more than demonstrates the inadequacy of our knowledge. We do not have enough facts to bring internal order to our observations and to develop a successfully based scheme of deep processes. The defects in our knowledge are so great that to enumerate them and to indicate the necessary further investigations would require a repetition of all the sections of this monograph.

We shall mention only the most urgent. Deep drilling is necessary for study of the composition of the crust. Further investigations of silicates under high pressure and temperature are necessary to clarify the conditions of their melting and crystallization, differentiation and homogenization, and phase transitions. We need to investigate the natural conditions and nature of the processes of metamorphism.

For the continents, we need a new systematization of data on the relations in space and time among tectonic, metamorphic, and magmatic processes, and also the comparison of deep structure of different tectonic zones.

For the oceans, we urgently need information about the composition and age of the second and third crustal layers. Special investigations should illuminate the now completely unknown details about the transition from continental crust through intermediate types to oceanic. It is of special interest to study the relation between structures of the oceanic and continental crusts. The Maldive-Laccadive underwater range apparently abuts directly against the Deccan plateau with its basaltic flows. Chains of underwater volcanoes in the Gulf of Guinea as well as the Walvis ridge apparently continue in the form of a chain of magmatic manifestations on the African continent. The island of Madagascar having continental crust is extended to the south in the form of an underwater range characterized by oceanic crust. The same phenomenon is observed south of New Zealand.

The region of the Seychelles Islands is exceptionally interesting. Continental crust was found under them, but to the south, along the Mascarene Ridge, we find a rapid transition through crust of the volcanic-ridge type to oceanic crust. Might not definitive evidence in support of oceanization be found here?

Toward a Consolidated Theory

In evaluating the general state of sciences of the solid earth, we can state that there is a definite asymmetry: in the latest conclusions, enormous attention is devoted to the new results of oceanic studies, while the results of earlier extensive investigations of geologists and geophysicists on continents are often forgotten.

Psychologically, this is understandable, because there have been great successes in oceanic studies in recent years. They have opened up a new world of phenomena. The new and unusual results are hypnotic; they overshadow much that is older and more ordinary.

In the two centuries of its existence, geology accumulated an immense quantity of material on the structure and regularities of development of continents. A significant amount of geophysical and geochemical information was also accumulated. Some of these results are briefly summarized in other sections of this Monograph. They are reinforced by a vast arsenal of facts; it is impossible to doubt their reality. The continents cannot be considered only as lifeless frames around the oceans: they live a complex but ordered tectonic and magmatic life.

Inadequate attention to continental geology leads to a loss in the historical method of the sciences of the earth. The material of historical geology is concentrated almost entirely on the continents; because of understandable circumstances, the historical approach is only beginning to penetrate into the study of the structure of oceanic depressions. From the point of view of the historical approach, it is unjustifiable to convert all the geological life of the continents into formation of young, alpine, folded zones. The tectonic history of continents is much more complicated. This approach would not have permitted the interpretation of the formation of Meso-Cenozoic geosynclines and rift zones as two facets of a single process, because the geologic data indicate that the rifts began to be formed not earlier than the Paleogene, whereas the Meso-Cenozoic geosynclines were already developing at the beginning of the Mesozoic.

We must attempt to build generalized conceptions on a joint and balanced utilization of oceanic as well as continental geological, geophysical, and geochemical data. If these data contradict each other in some way, then the contradiction cannot be artificially avoided by arbitrary presentations; it must be carefully studied, and its reasons and nature must be thoroughly explained. There should be a full and equal respect both for new data about the oceans and for the achievements of classical science about the continents. We cannot lose sight of the fact that the data about the structure and development of the oceans is much more schematic than analogous data about the continents. It does not follow from this, that for generalizations embracing oceans and continents the information about the continents should undergo the kind of schematization that would bring them to the level of data about the oceans. In fact, excessive schematization often represents the basic trouble that leads generalizations into a blind alley. It would be very regrettable if the regularities and mechanisms of geological processes worked out by continental geology were lost in order to accommodate our presentations to the more schematic data about the oceans.

Only by uniting data of continental and oceanic geology and geophysics can we hope in a reasonable period of time to form the basis of a genuine theory of deep processes, a theory that will have not only scientific but also practical significance because it will open new doors to the wealth of minerals and energy of the earth's interior.

Author Index

References are indexed for the pages on which an author's work is cited. References at the end of articles are not indexed if the author was cited in the text. Small **boldface** numbers indicate inclusive pages of articles in this monograph. *Italics* indicate 'et al.' citation.

Aboav, D. A., 671
Abramovici, F., 266, 267, 268
Adams, J. A. S., 4
Adams, L. H., 20, 33, 34, 63, 68
Ade-Hall, J. M., 629, 632
Affleck, J., 393
Agger, H. E., 184
Agranovsky, L. E., 197
Ahrens, L. H., 514
Airinei, St., 404
Akamatsu, H., *96*
Aki, K., 8, 156, 162, 173, 239, 665
Akimoto, S., 478, 621, 628, 631, *631*, 634, 646, 678
Aksenovich, G. I., 197, 202
Aldrich, L. T., *212, 213, 214*, 253
Alekseev, A. S., 202, 246
Alexander, Shelton S., 241, 245
Algermissen, S. T., 244
Allan, D. W., 63, 68, 74, 75, 686
Allan, T. D., 433, 434
Alldredge, L. R., 392, *431*, 457, 460, *461*, 632
Allen, C. R., 128, 131, 545, 546
Almeida, F. F. M. de; see de Almeida, F. F. M.
Andel, S. I. van; see van Andel, S. I.
Andel, T. H. van; see van Andel, T. H.
Alterman, Z., **265–272**, 266, 267, 268, 270, 271
Althaus, E., 521, 522
Altshuler, L. V., 69, 650, 656, 678
Andel; see van Andel
Anderle, R. J., 685
Andersen, F., 466
Anderson, Don L., *4*, 7, *8*, 8, 18, 25, 26, 30, *30*, 31, 32, 114, 257, 260, 263, 266, 267, 269, 271, 274, 662, 665, 672, 676, 685
Anderson, L. A., 469
Anderson, Orson L., 19, 19, 20, 21, 22, 23, 31
Andrade, E. N., 671
Andreasen, G. E., *394*
Andrews, P., *502*
Andreyev, B. A., 458
Andreyeva, I. B., 606
Andrić, B., 191
Antoine, J., 592
Aoki, K., 504
Aota, M., 436, 437, 438
Archambeau, C., 173, 672
Aric, K., 184
Arkhangel'skaya, V. M., 241

Armstrong, R. L., 14, 58, 60, 613
Aronov, L. E., *197, 202*
Aronson, J. R., 27
Arrhenius, G. O. S., 48
Artemyev, M. E., 384, 387
Artyushkov, E. V., **379–390**, 383, 385, 387, 388
Asano, S., 103
Asanuma, T., *103*
Asbel, I. J., 247
Astier, J. L., *469*
Aswathanarayana, U., 515
Athavale, R. N., 452
Aumento, F., 598, 599
Avery, Otis E., 431, 432, 579
Aver'yanov, A. G., 592
Avilova, N. S., *44*

Babb, S. E., 657
Backus, G., 251, 256, 275
Baker, B. H., 539, 540, 543
Baker, D. W., 684
Baker, M., *365*
Balakina, L. M., **166–171**, 166, 167, 168
Balk, R., 488
Balsley, J. R., 394
Băncilă, I., 532
Bancroft, A. M., 253
Bancroft, Dennison, 28
Banwell, C. J., 96
Barazangi, M., 684
Baron, J. Gregory, 430, *430, 433*
Barrett, D. L., 433, 434
Barth, T. F. W., 54
Bath, Jennifer, 585, 586, 599
Båth, M., 117, 122, 139, 180, 185, 241, 266, 270, 598
Beal, C. S., 434
Beck, A. E., 70
Behnke, Cl., 178, *178*, 179, *179*, 182
Bell, P., 522, 634
Bellamo, S., *184*
Beloussov, V. G., 199, 201, 202
Beloussov, V. V., **539–544, 698–712**, 28, 38, 39, 46, 51, 53, 64, 73, 207, 282, 384, 387, 525, 526, 529
Belyaevsky, N. A., **195–208, 312–319**, 38, 196, 316
Bemmelen, R. W. van; see van Bemmelen, R. W.
Bender, P. L., 457
Benfield, A. E., 88
Benioff, H., 128, 173, 548, 612

Ben-Menahem, A., 158
Benseman, R. F., 96
Benson, W. H., 3
Bentz, A., 530
Berckhemer, H., 173, 276
Berdichevsky, M. N., 618
Berg, J. W., 242
Bernal, J. D., 645, 691
Berry, M. J., 8, 253, 254, 608, 609, 685
Berzon, I. S., *201*
Betlej, K., 194
Bezrukov, P. L., 48
Bhalla, M. S., *452*
Bhattacharyya, B. K., 458, 459
Bin, C., 522
Binns, R. A., 5
Birch, Francis, **18–36**, 2, 7, 19, 20, 21, 22, 23, 24, 25, 28, 30, 31, 32, 33, 34, 35, 63, 64, 66, 69, 78, 93, 104, 186, 522, 579, 622, 623, 640, 641, *641*, 644, 650, 652, 658, 661, 669, 676, 677
Birch, F. S., 96
Bishop, W. W., 542
Black, M., 433, 434
Blackett, P. M. S., 447, 453
Blayney, J. L., 173
Bleie, J., 184, 185
Bodvarsson, G., 100, 103, 104, 107, 150
Bogdanovitch, Ch., 547
Boisnard, M., *404*
Bokur, R. A., *592, 607*
Boldizsar, T., 95, 96
Bollinger, G. A., 156, 157, 158, 173
Bolt, Bruce A., 244, 257, 266, 267, 674
Bonchkovsky, V. F., 173
Bondarenko, A. T., 620
Bonnet, G., 108
Borisov, A. A., *38*, 196, *316*
Borok, V. I. Keilis-; see Keilis-Borok, V. I.
Born, M., 650, 651
Bostrom, K., 99
Botezatu, R., *404*
Bott, M. H. P., 458, 587, 593, 603, 690
Bowen, N. L., 4, 5, 19, 487, 543
Bower, Margaret E., 433, 459
Bowin, C. O., 580, 584, 589, 593, 599
Boyd, Francis R., 67, 509, 634, 635, *635*, 641, 643, *643, 644*
Brace, W. F., 173, 176, 469, 620
Bradley, R. S., 621
Breiner, S., 458, 460
Bremaecker, J. C. De, see De Bremaecker, J. C.
Briden, J. C., *452*, 630
Bridgman, P. W., 20, 650, 654, 655
Brisbin, W. C., 367, 369
Brock, A., *452*
Brooks, B. G., *242, 244*
Browne, W. R., 147
Brune, James N., 230–242, 4, *145*, 158, 238, 239, 263, *263*, 271, 665, 685
Bryan, G. M., *342*
Bubnoff, S. von; see von Bubnoff, S.
Buddington, A. F., 487, 628
Buerger, M. J., 646

Bullard, E. C., 5, 6, 11, 88, 93, 296, 297, 375, 436, 543, 597, 610, 661, 683
Bullen, K. E., 24, 644, 676, 677
Bultitude, R. J., 9, 10
Bunce, E. T., *570*
Bune, V. I., *282*
Burckle, Lloyd H., *434*, 614
Burke-Gaffney, T. N., 145
Burtman, V. S., 547
Burton, Gordon D., *431, 432, 579*
Byerly, P., 173

Cagniard, L., 470
Cahen, L., 539, 540, 543
Caloi, P. L., 18, 114, 665
Cameron, H. A. D., *433, 434*
Campbell, F. E., *46*
Caner, B., 466
Cann, J. R., 587, 592, 593, 599, 601, *601*, 602
Cantwell, T., 469, 477
Capurro, Luis R. A., 433, 434
Carder, D. S., 145, 215, 242, 244
Carey, S. W., 561
Carpenter, E. W., 184
Cărutasu, O., *404*
Casaverde, M., *466, 467*
Červený, V., 249
Chadwick, P., 387
Challis, G. A., 481
Chander, R., 158
Chandrasekhar, S., 689
Chao, E. T. C., 44
Chekunov, A. V., *189*, 191, *191, 193, 202, 592*, 607
Chernyshova, V. I., 51
Chinner, G. A., 635
Chinnery, M. A., *4, 8, 30*, 173, 245
Chinzei, K., 107, 500
Chirvinskaya, M. V., *193*
Choroshева, V. V.; see Khorosheva, V. V.
Choudhury, M., 178
Chow, T. J., 60
Christensen, N. I., 19, 20, 22, 30
Christoffel, D. A., 425
Chudakova, L. N., *202*
Cisternas, Armando, 241
Clairaut, A. C., 293, 294
Clark, D. M., 244
Clark, H., 622, 623
Clark, Sydney P., Jr., 622–626, 4, 5, 11, 18, 30, 64, 66, 67, 623, 624, 635, 640, 648, 670, 676, 677, 678
Clarke, F. W., 37, 50, 53, 54
Clarkson, N., *342*
Cleary, J., 4, 145, 146, 147, 244
Clegg, J. A., *453*
Cloos, H., 488, 539, 541
Closs, H., **178–188**, 179, 180, 182, 186, 187
Coats, R. R., 13, 504
Coble, R. L., *622, 623*
Coes, L., 646
Colette, B. J., 184
Collinson, D. W., 450, 452, 631
Conolly, J. R., 566
Coode, A. M., 690, 693

AUTHOR INDEX

Cook, K. L., *242*, 302
Cordani, U. G., *58, 683*
Coron, S., **304–312**, 304, 309
Corry, C. E., 89, *89, 98, 609, 610*
Couch, R. W., *352, 354, 355, 359, 360, 361*
Cowan, G. H. McTaggart-; see McTaggart-Cowan, G. H.
Cox, Allan V., 422, 425, 437, 439, 442, 450, 684
Cox, C. S., 93, 466
Cram, I. H., Jr., 290
Crampin, S., 265, 266, 267, 270, 271
Creer, K. M., 449, 452, 453, 454, 684
Crittenden, M., 386

Dachev, Ch. I., 193
Dadieva, V. D., *191*
Dalrymple, G. B., 58, 422, *422*, 425, *425*, 442, *450, 684*
Daly, R. A., 38, 44, 502
Dampney, C., *147*
Danes, Z. F., 373
Danilevich, S. I., *514*
Darton, N., 529
Darwin, G., 296
Davidson, C. F., 480, 481
Davis, B. T. C., 5, 26, 27, 509, 634, 635
Davis, T. M., *392, 460, 632*
Davydov, B. I., 650, 681
Dawson, G. B., 96
Dawson, J. B., 3, 480, 481
de Almeida, F. F. M., *58, 683*
De Bremaecker, J. C., 247
Decker, E. R., 79
Decker, R. W., 440
Dehlinger, P., **352–363**, 290, 352, 354, 355, 359, 360, 361, *361, 592, 602*
Delach, M. N., 615
Delitsin, I. S., *65*
Delsemme, A., 108
Demenitskaya, R. M., **312–319**, 37, 39, 52, 313, 314, 316, 318, 460
Demidenko, Ju. B., 194
Dempsey, W. J., *430*
Den, N., *103*
De Neufville, J., *635*
Dennis, L. S., 552
DeNoyer, J., 173
den Tex, E., 481
d'Erceville, I., 473
de Roever, W. P., 481
de Sitter, W., 294, 296, 297
de Visintini, G., *178, 184*
Dickinson, W. R., 603, 611
Dickson, G. O., 431, 552, *562, 579, 582, 588, 589, 593, 597, 603, 604, 611, 613*
Dietz, R. S., 152, 430, 482, 554, 561, 570, 575, 579, 580, 691
Dix, C. H., 661
Dixey, F., 539, 540, 543
Dmitriyev, L. V., 51
Doe, B. R., 58, 60
Doell, R. R., 422, *422*, 425, *425*, 437, 439, *442, 450, 684*
Dohr, G., 180

Dolan, W. M., 242
Dorman, James, **257–265**, 2, 4, 238, 239, 257, 263, 265, 266, *342*, 598, 609, 665, 684, 685
Dorn, P., 539, 540, 542, 543
Dortman, N. B., 417, 418, 421
Doyle, H. A., 29, *146*, 147
Dragašević, T., 191
Drake, Charles L., **549–556**, 2, 30, *221, 315*, 349, 411, 430, 459, 460, 550, 551, 552, 554, 579, *602, 660, 661, 662*
Drickamer, H. G., 656
Druzhinin, V. S., *202*
Dubois, C., *89*
Duda, S. J., 115, 116, 117, 118, 120
Dürbaum, H., 179, 180
Dugdale, J. S., 653

Eaton, J. P., 171
Eckermann, Harry von; see von Eckermann, Harry
Eckhardt, D., 470, 471, 477, 618
Eckhoff, O. B., 573, 575
Eckroad, S. W., *27*
Edelman, N., 487
Elder, J. W., 104
Elmendorf, C. H., 575
Elsasser, W. M., 69, 597, 695
Elvers, D. J., 424, 427, 467
El Wakeel, S. K., 42, 48
Emery, K. O., 617
Emslie, A. G., *27*
Engel, A. E. J., 4, 11, 44, 48, 51, 54, 490, 511, 515, 593, 598
Engel, C. G., *11, 44*, 48, *48*, 51, *51*, 54, *490*, 511, 593, 598, 611
England, J. L., 26, 27, 67, 509, 634, 635, *635*, 641, 643
Epinatieva, A. M., *201*
Erceville, I. d'; see d'Erceville, I.
Ericson, D. B., *290, 559, 587*
Eskola, P., 488
Evenden, G. I., 457
Everett, J. E., 4, *683*
Everingham, I. B., 147
Everitt, C. W. F., 631
Ewing, John I., **220–225**, 221, 222, 223, 224, 225, 389, 428, *551, 552*, 553, *553, 562*, 567, 578, *579*, 589, 592, *592, 593*, 608, 613, 614, 687
Ewing, W. Maurice, *2*, 2, *92, 93, 94*, 148, 150, 220, 221, 222, *222*, 223, 224, *224, 239*, 241, 264, 265, 266, *266*, 269, 270, 290, *349, 369, 370, 371*, 389, 428, 430, *430*, 433, 550, 552, *552*, 553, *553*, 562, 567, *567*, 569, 573, *573*, 575, 578, 579, *579, 580, 586, 587*, 589, 592, *592, 593, 593, 594*, 597, 598, *602, 605*, 608, *608*, 613, 665, 687

Facca, G., 96
Fairbairn, H. W., *4, 58, 59, 683*
Fanselau, G., 404
Faure, G., *4*, 46, *59*
Fedotov, S. A., 246, 247, 665
Fedynsky, V. V., 319
Fermor, L. L., 637
Fersman, A. E., 50

Filipov, Yu. V., 380
Filippov, M. S., *514*
Finetti, I., 184
Fisher, R. A., 448
Fisher, R. G., 96
Fisher, R. L., 569, 609, 610, 611
Fleischer, M., 44
Fleming, R. H., *90*
Flinn, E. A., 145
Flint, R. F., 383
Florensov, N. A., 282, 539, 541, 542, 543, 546, 547
Förtsch, O., 184
Forbes, R. B., 507
Forbush, S. E., *466, 475*
Foster, J. H., *423, 450, 587, 632*
Foster, M. D., *507*
Fournier, H. G., 470, 476
Fowler, W. A., *64, 65, 66*
Fox, P. J., *562, 563, 564, 565, 566, 568*
Francis, T. J. G., *252*
Francl, J., *622, 623*
Franz, G. W., 481
Fraser, G. D., 173
French, W. S., *361, 592, 602*
Freund, R., 549
Fricker, P. E., 64, *64, 69,* 71
Frischknecht, F. C., 457
Fritsch, J., *179, 180*
Fritz, J. N., *19*
Frolov, A. I., *388*
Frolova, N. V., 46
Fuchs, K., 179, 608
Fujii, N., *27,* 64, 71
Fujisawa, H., 478, 621, 634
Fujita, N., 460
Fukutomi, T., 101
Fuller, M. D., 459
Furumoto, A. S., 304, 365, 443
Fyfe, W. S., 521

Gabriel, V. G., 241
Gadomska, B., 194
Gaffney, T. N. Burke-; see Burke-Gaffney, T. N.
Gadomska, B. Vojtczak-; see Vojtczak-Gadomska, B.
Gagelganz (Gagelgants), A. A., *197, 202,* 592, 607
Galabov, I., 380
Galanopoulos, G. A., 140
Galdin, N. E., *19,*
Galfi, G., 194
Galperin, E. I., *197,* 202, *202, 592, 607*
Gamburtsev (Gamburzev), G. A., 201, 319
Gaposhkin, E. M., 300
Gardner, L. W., 252
Garetsky, R. G., 282
Garkalenko, I. A., *189, 191, 202*
Garland, G. D., 321, *472*
Gaskell, T. F., 18, 222
Gass, I. G., 602, 603, 604, 611
Gast, P. W., 4, 58, 60, 64
Gastil, G., 46, 506, 514
Gavrilescu, B., *404*
Gaynanov, A. G., 460
Geddes, Wilburt, *80, 408, 412, 458, 459,* 460, *460*

Gedney, L. D., 131
Gemperle, M., *352, 354, 355, 359, 360, 361*
Geneslay, R., 180
Georgescu, L., *404*
Gerard, R. D., *570*
Gerard, V. B., 441
Gerasimov, I. P., 380
Gerling, E., 514
Giese, P., 178, *178,* 179, 182, 186
Giesecke, A. A., Jr., *429, 466, 467*
Gilbert, F., 154, 275
Gilbert, G. K., 525
Gilbert, M. C., 634
Gilliland, W. N., 459
Gilman, R., 173
Girdler, R. W., 73, 103, 150, 539, 543, 696
Giret, R. I., 457
Girin, Yu. P., *37, 40, 51*
Gladun, V. A., 388
Glass, B. P., *403, 450,* 451, 559, *559,* 587, 614, *632*
Glass, J. A., *559, 587*
Glazunova, A. P., 417
Godby, E. A., 434, 460
Godin, U. N., 197, 202
Göppert-Mayer, M., 650
Goldberg, E. D., 48
Goldich, S. S., 519
Goldreich, P., 685
Goldschmidt, V. M., 37, 50, 53, 54, 645
Goldstein, N. E., 458
Golub, D. P., *417*
Golubeva, N. V., 134–139, 139
Goncharov, V. P., *592, 607*
Gonshakova, V. I., *19*
Goodacre, A. K., 213
Goodenough, J. C., 682
Gorczyński, L., 194
Gordon, D. W., *215, 242,* 244
Gorshkov, G. S., 504
Goryatchev, A. B., 282
Gough, D. I., 434
Grabovskii, M. A., 629
Graham, J. W., 448
Grant, F. S., 445
Gray, J., 500
Green, D. H., 489–495, 637–649, 2, 3, 5, 6, 7, 8, 9, 10, 11, 13, 481, 490, *490,* 491, 493, 494, 495, 509, 512, 635, 638, 639, 640, 641, 643, 662
Green, J., 599
Green, R., 147, 452
Green, T. H., 10, 13, 490, 641, 643
Griffin, J. J., 99
Griffiths, D. H., 433, 434
Griggs, D. T., 661, 684, 688
Grim, P., *93*
Grommé, C. S., 687
Grossvald, M. G., 383
Grout, F. F., 46
Grow, J. A., 374, 377
Grüneisen, E., 650
Gubin, I. E., *134,* 282
Gudmundsson, G., 458
Guier, W. H., 697
Guinot, B., 698

AUTHOR INDEX

Gulatee, B. L., 339
Gunn, R., 611
Gupta, M. L., *74*
Gupta, R. N., *208, 220, 660*
Gurvich, I. I., 207
Gutenberg, B., 8, 24, 25, 26, 28, 30, 115, 117, 118, 120, 124, 134, 135, 142, 150, 247, 249, 277, 383, 546, 662, 664, 673
Guterch, A., *191*, 194
Gzovsky, M. V., 282, 384, 524, 529

Haalck, H., 393
Hadjebi, B., *180*
Hadley, J. B., 173
Hänel, R., 188
Hagiwara, K., *103*
Hagiwara, T., **714–176**, *145, 173, 176*
Hagiwara, Yukio, 397, 436, 437, 438, *460*
Hahn, A., **399–404**, 460
Hales, Anton L., 4, 29, 244
Hall, J. M. Ade-; see Ade-Hall, J. M.
Haller, E., 531
Halunen, A. J., Jr., 96
Hamilton, E. L., 222, 561, 569
Hamilton, George, *342*
Hamilton, Gordon R., *220*
Hamilton, Robert M., 478, 621
Hamilton, W., 173
Hamza, V. M., *74*
Harding, S. T., *244*
Harkrider, D. G., 266, 271, *592*
Harris, P. G., 6, 481
Harrison, C. G. A., 585, 586
Harrison, J. C., 367, 369, 661
Hart, S. R., **58–62**, 58
Hartmann, O., *429, 466, 467*
Hasegawa, Y., 386
Hashin, Z., 28, 33
Haskell, N. A., 158
Havens, R. G., *11, 44, 48, 51, 490*
Hayashi, T., *382*, 460
Hayes, Dennis E., 430, 552, 610
Hays, J. D., *423, 450*, 614, *632*
Hays, J. F., 634, 635, 641
Hayford, J. F., 296
Healy, James, 104
Healy, John H., **208–220**, *213*
Heath, G. R., 577
Heck, N. H., 130
Hédervári, P., 107
Hedge, C. E., 4, 60, *60*, 505, 506, *519*
Heezen, B. C., 148, 149, 150, 151, 223, *290*, 342, 390, 430, *450*, 451, 539, 559, *559*, 562, *562*, 563, *563, 564*, 565, 566, *566*, 568, 569, *569*, 570, 571, 572, 573, 575, 577, 579, 580, *587*, 588, 592, 598, 602
Hehn, Kl., *180*
Heier, K. S., 4, *4*, 64, 66
Heirtzler, J. R., **430–436**, 393, 423, 430, *430*, 431, *431, 432*, 433, *433*, 434, 440, 458, *459, 460*, 552, *552*, 562, 579, *579*, 580, 582, 584, *587, 586*, 588, 589, 593, *593, 594*, 595, *595, 596*, 597, *597*, 598, 599, *601, 602*, 603, 604, *608, 609*, 610, 611, 613, 632

Heiskanen, W., 304, 383, 387
Helmburger, D. V., 661
Helmert, F. R., 303
Helterbran, Wayne, *244*
Henderson, R. G., 393, *445*, 458
Hennison, J. F., *592*
Henriksen, S. W., 296, 297
Henschel, H., 194
Herrin, Eugene T., **242–246**, 79, 131, 242, 244, 661
Herring, C., 622, 673
Herron, E. M., *562, 579, 582, 588, 589, 593, 597, 603, 604, 611, 613*
Hersey, J. B., *220, 570*, 592, 602, 606, 607, 608
Herzen, R. P. Von; see Von Herzen, R. P.
Hess, H. H., 3, 5, 11, 12, 152, 222, 250, 251, 422, 430, 481, 482, 554, 561, 579, 580, 584, 592, 593, 602, 609, 610, 611, 641, 661, 662
Higgins, M. S. Longuet-; see Longuet-Higgins, M. S.
Higgs, R. H., 610
Hill, D. P., 212, 213, 214
Hill, M. N., 220, *222*, 431, 433, *433*, 434, *434*, 553, 591
Hill, R., 19
Hinz, K., 180, *186*
Hinze, W. J., 458
Hirschleber, H., 185, 186
Hirshman, J., *430, 459, 460, 552*
Hjelme, J., *185, 186*
Hobbs, W. H., 526
Hodgson, J. H., 154, 155
Hollister, C. D., *223, 559, 562, 563, 564, 565*, 566, *566, 568*
Holm, J. L., 522
Holmes, A., 63, 68, 514, 561, 607, 612, 637
Holopainen, P. E., 305, 311
Holub, K., 278
Honda, H., 148, 154, 155, 160, 555
Hood, P. J., 434, 457, 458, 459, 460
Hopson, C. A., *60*
Horai, K., **95–109**, 96, *96*, 698
Horn, D. R., 615
Horn, R. M., 615
Hospers, J., 449, 452
Hotta, H., *103*
Houtz, R. E., *221, 315, 579, 602, 661, 662*
Howard, L. E., 104
Howell, B. F., Jr., 661
Hoyle, F., *64, 65, 66*
Hrdlička, A., 191
Hristov, K., 380
Huang, Kun, 651
Huene, R. von; see von Huene, R.
Hughes, H., *4, 59*
Hunkins, K., 460
Hurley, P. M., 4, 46, 58, 59, 683
Hyndman, R. D., 4, 466

Ichikawa, K., *103*
Ichikawa, M., **160–165**, 160, 162
Ida, Y., 634, 646, 678
Iida, C., *5*
Ilyukhin, M. N., *37, 40, 51*
Innes, M. J. S., 128

Iosif, T., 247
Iriyama, Jun, 69, 70
Irving, E., 428, 448, *449*, 450, 452, *452*, 455, 580, 632
Isacks, B. L., 145, 554, 555, 609, 612, 613, 684
Isakovich, M. A., 673
Ishikawa, Y., 628, 631
Ito, Keisuke, 5, 7, 643, 644
Ivakin, B. N., 276
Ivanov, M. M., 431
Ivanova, Z. S., 665
Ivantishin, M. N., *191*
Izsak, Imre G., 287

Jackson, W. H., 211, 212, 213, *213*, 214, 215
Jacobs, J. A., 63, 68, 74, 75
Jaeger, J. C., **145–147**, *4*, 82, 147
Jakosky, J. J., 393
James, David E., 208, 243, 661
Jamieson, J. C., 625
Jamil, A. K., *621*
Jardetsky, W., 241
Jarosch, H., *266*
Jeffreys, H., 63, 68, 249, 294, 296, 297, 298, 300, 301, 645, 695
Jensen, H., 188, 458
Joffe, A. F., 625
Johnson, B. D., *147*
Johnson, G. L., **556–617**, 569, *569*, 570, 571, 573, 575, 577, *580*, *583*
Johnson, L. R., 661, 685
Johnson, M. W., *90*
Jones, B. R., 290
Jones, D. L., *452*
Jones, H. S., 294, 296
Jordan, J. N., *215*, *242*, 244, *244*
Joyner, W. B., 606
Judd, W. R., 173

Kääriäinen, E., 380, 386
Kalashnikov, A. G., 629
Kalinin, V. A., **650–659**, 653, 655, 658, 659, 681
Kalyuzhnaya, L. T., *189*, *191*, *202*
Kaminuma, Katsutada, 239
Kanamori, H., 19, 20, 27, 30, 31
Kanasewich, E. R., 58
Kane, M. F., *374*
Kapcan, A. D., 277
Kapitsa, S. P., 629
Kappelmeyer, O., 82, 179
Karasik, A. M., 460
Kárník, V., **139–144**, 139, 142, 278
Karus, E. V., *201*
Kasahara, K., 164, 173
Katsumata, M., 555
Katsura, T., 490
Katz, H. R., 546
Kaula, W. M., 301, 302, 303, 341, 349, 350, 693
Kawada, K., *460*
Kawashita, K., *58*, *683*
Kazakov, G. A., *37*, *40*, *51*
Keen, M. J., 431
Keilis-Borok, V. I., 248, 665
Keller, G. V., 469
Kelvin, L., 63, 68

Kennedy, G. C., 5, 7, 643, 644
Kennedy, W. Q., 546, 547
Kennet, P., *433*, *434*
Kern, J. W., 629, 631
Kê T'ing-Sui, 675
Keyser, A. R., 457
Khain, V. E., **523–538**, 37, 39, 282, 524, 525, 526, 527
Khalevin, N. I., 202
Khan, M. A., **293–304**, 294, 295, 296, 297, 298, 301
Kharechko, G. E., 194
Kharin, D. A., *134*
Khilinskiy, L. A., *189*, *191*, *202*
Khitarov, N. I., 522
Khlebnikova, Z. V., 37
Khorosheva, V. V., 277, 665
King, Elizabeth R., *80*, *408*, *412*, 430, 431, *458*, *459*, *460*, 461
Kingery, W. D., 622, 623
Kinoshita, W. T., 365
Kirillova, I. V., 282, 524
Kirk, H. K., *361*, *592*, *602*
Kiroshita, H., 631
Kishii, T., 101, 103
Kislovskaja, V. V., 277
Kittleman, L. R., 500
Kivioja, L. A., 345, 346
Klemens, P. G., 622, 623
Kleppa, O. J., 522
Klíma, K., 276
Knopoff, Leon, **273–276**, **683–689**, 8, 31, 64, 154, 173, 238, 239, *263*, 273, 274, 275, 608, 609, 673, 685, 686, 687
Knorring, O. von; see von Knorring, O.
Knothe, Ch., 194
Kobayashi, K., 631
Kohler, R. E., *424*, *427*
Kojima, G., 97, *102*
Kolotukhina, S. E., 39
Komlev, L. V., *514*
Kormer, S. B., 69
Kornfeld, P., 270, 271
Korzhinskii, D. S., 521
Kosminskaya, I. P., **177**, **195–208**, 38, 52, 103, *197*, 200, 202, 316, 550, 554, *592*, 607, 660, 661
Kovach, Robert L., 257, 266, 267, 269, 274, 460, 665
Kovylin, V. M., *46*, *48*, *49*
Kozai, Y., 295, 298, 300
Kozlovskaya, S. V., *69*
Krakshina, R. M., *103*, *197*, *202*, *592*, *607*, *661*
Krause, Dale C., 431, 460, *586*
Krenkel, E., 539
Krestnikov, V. N., *384*
Krilov, S. V., *198*, *202*
Krivoy, H. L., 365, *365*
Kroenke, L. W., 304, 365
Kuenen, P. H., 49, 54, 107, 384
Kumarapeli, P. S., 130
Kunetz, G., 473
Kuno, H., **495–501**, **507–513**, 519, 5, 101, 492, 496, 497, 504, 507, 508, 509, 512, 603, 611
Kuo, J., 241, 263
Kushiro, I., 5, 492, 508, 634, 635, 641
Kuzin, I. P., 247

AUTHOR INDEX

Kuzin, N. N., *682*

Labrouste, Y., 179, 180, *180*, 186
Lachenbruch, A. H., 93
LaCoste, L., 342
Ladynin, A. V., 458
Lagaay, R. A., 184
Lahiri, B. N., 618, 619
Lambert, A., 466
Lambert, I. B., 4, *4*
Lanczos, C., 475
Landau, L. D., 673
Landergren, S., 42, 48
Landisman, M., 178, 179, 184, *265*, *270*, 377
Lane, M. P., 365
Langan, L., 458, *458*
Lange, B., 194
Langseth, M. G., 88, 91, 92, 93, 579, *579*, *584*, 586, *586*, 587, 593, *593*, 594, *596*, 597, *597*, 602, 605, 608
Larner, K., *470*, *471*, *477*, *618*
Larsen, J., 466
Lasukov, G. I., 389
Laughton, A. S., *433*, *434*, 553, 614
Lavin, P. M., 458
Law, L. K., 465
Lawrie, J. A., 441
Lawson, A. W., 625
Lazareva, A. P., 168
Lebedev, T. S., 194
LeComte, Paul, 522, 641
Ledersteger, K., 297
Lee, W. H. K., 88–95, 16, 64, 65, 67, 71, 72, 73, 74, 75, 78, 82, 87, 89, 92, 93, 94, 96, 97, 102, 107, 220, 597, 693
Leibenson, L. S., 63, 68
Le Pichon, Xavier, 341–351, *92*, *93*, *94*, 221, *222*, *239*, 315, *369*, *370*, *371*, 389, 428, *430*, 431, *431*, *433*, 562, 579, *579*, 580, *582*, 586, *586*, *587*, *588*, *589*, 593, *593*, *594*, 595, *595*, *597*, 598, 599, *601*, 602, *602*, *603*, *604*, *605*, *608*, *609*, *611*, *613*, 661, 662, 684
Levin, B. Yu., 64, 67, 69
Levshin, A. L., *665*
Leyden, R., *223*, 550, 553
Lidiak, Edward G., *80*, *408*, 412, *412*, 458, *459*, *460*, *519*
Liebscher, H. J., 180, 184
Lindsley, D. H., 394, 628, 634, 635
Lindt, W. J. van de; *see* van de Lindt, W. J.
Lifshitz, E. M., 673
Lisitsyn, A. P., *48*
Litvinenko, I. V., 202
Livanova, L. P., *193*
Loeb, A. L., 682
Loncarevic, B. D., 431, 433
Longuet-Higgins, M. S., 469
Lorentzen, George R., *430*
Lotze, F., 514
Lovering, J. F., 16, 66
Lubimov, V. M., *672*
Lubimova, E. A., 63–77, 82–88, 27, 63, 64, 65, 68, 69, 70, 71, 74, 75, 86, 93, 103
Ludwig, W. J., *552*, *553*

Lukk, A. A., 246, 247
Luyendyk, B. P., *579*
Lynch, V. M., *458*
Lysenko, V. A., 194
Lyustikh, E. N., 689–692, 69, 282, 313, 390, *524*, 689, 690, 691

Mabey, D. R., *365*
McBirney, A. R., 501–507, 504, 505, 506, 603
McCartney, W., 433
McClure, D. J., 457
McConnell, R. B., 540, 542
McConnell, R. K., Jr., *27*, 64, 69, 200, 201, 208, 220, 660
McCoy, F., 441, 444
McDonald, D. K. C., 653
Macdonald, G. A., 490
MacDonald, G. J. F., 4, 11, 25, 26, 64, *64*, *65*, *66*, 70, 71, 72, 75, 93, 274, 345, 615, *641*, 673, 684, 698
McDonald, K. L., 477, 624
MacDonald, R. R., 365
McElhinney, M. W., 452
McEvilly, T. V., 238
MacGregor, I. D., 634, 635, 641, 643, *643*, 644, *644*
Machesky, L. F., 316, 365
McKenzie, D. P., 660–664, 386, 579, 597, 609, 610, 673, 684, 685
MacKenzie, J. K., 28
McManus, Dean A., 353
McNitt, J. R., 104
McQueen, R. G., 19, *650*, 651, *651*, 654, *654*, 655, *655*, 656, *656*
McTaggart-Cowan, G. H., 200, 201
Madden, T. R., 469–479, *469*, *470*, *471*, 473, 476, 477, *477*, *618*, *620*
Magnitsky, V. A., 664–675, 676–682, 39, 64, 69, 86, 318, 382, 387, 388, 658, 665, 669, 671, 676, 677, 680, 681, 690
Magnusson, N. H., 514
Major, A., 634, *643*, 646, 647, 678, 679
Major, Maurice, *263*
Makarenko, F. A., 74
Makarova, Z. A., 417
Malahoff, A., 364–379, 436–446, 304, 366, 374, 376, *376*, *436*, 437, 439, 441, 442, 443, 444, 445
Malnar, L., 457
Mammerickx, Jacqueline, 577
Manchester, K. S., 434
Manghnani, M. H., 365
Mansfield, R. H., *242*, *244*
Manyuta, M. G., 194
Margotieva, M. V., *197*, 202
Markov, K. K., 389
Markowitz, Wm., 698
Marsh, S. P., *19*, 651, 654, 655, 656
Marshall, B. V., 93
Markov, V. K., 65
Maruyama, T., 173
Marvin, R. F., 58
Mason, Brian, 50
Mason, C. S., *431*, *433*
Mason, R. G., 422, 423, 429, 430, 436, 467, 581
Masson-Smith, D., 602, 604, 611

Masuda, A., 66
Mathewson, C. C., *424, 427*
Matsuda, T., *107, 500,* 545
Matthews, D. H., 11, 397, 422, 424, 430, 431, *431,* 433, *433, 434,* 579, 580, 581, 585, 586, 599, 601, 683
Matveeva, N. N., 246, *246, 247, 665*
Maury, M. F., 557
Maxwell, A. E., 11, 88, *93, 610,* 684
Maxwell, J. C., 481, 482, 585, 612
Mayer, M. Göppert-; see Göppert-Mayer, M.
Mayeva, S. V., 64, 67, 71, 72, 75
Mead, W. J., 54
Medi, E., 373, 374
Medvedev, S. V., 282
Mehnert, K. R., **513–518,** 517
Meijering, J. L., 680
Meinesz, F. A. Vening; see Vening Meinesz, F. A.
Meissner, R., 178, 180, 184, 186
Melcher, G. C., *58, 683*
Melson, W. G., 599
Menahem, A. Ben-; see Ben-Menahem, A.
Menard, H. W., 128, 148, 149, 150, 152, 352, 423, 544, *559,* 562, 570, 573, 577, 580, 591, 592, 594, 595, 605, 606, 607, 608, 609
Mercy, E. L. P., 635
Merrill, R. T., *687*
Mescherikov, Yu. A., **379–390,** 383
Metzger, A., *185*
Meuschke, J. L., *457*
Meyer, Robert P., 30, 242, 251
Meyers, W. B., 173
Michot, P., 46
Mickey, W. V., *244*
Mie, G., 650
Migaux, L., 469
Migdisov, A. A., 37, 40
Mikhailovskaya, M. S., *37*
Mikhota, G. G., *197, 202, 592*
Mikolajczak, A., 194
Miller, E. T., 430, *430, 434,* 580
Miller, D. E., *592*
Minakami, T., 436
Mironova, V. I., *592, 607*
Misharina, L. A., **166–171,** *166,* 167, 168
Mitronovas, W., *661*
Mituch, E., 192, *193*
Miyabe, N., 381
Miyamura, S., **115–124,** 121, 124, 143, 163, 381
Miyashiro, A., **519–522,** 481, 520
Mizoue, M., 119, 120, 121, 123, 124, 381
Mizutani, H., 19, 20, *27,* 30, 31
Moberly, R., 374, 376
Modriniak, N., 374, 375, 441, 442
Moe, E. E., *365*
Mogi, K., 143
Mohorovičić, A., 660
Mohr, P. A., 539, 540, 542, 543
Mooney, H. M., 266, 267
Moore, J. G., 58, *97, 102,* 365, *586*
Moore, T. C., 577
Morelli, C., 188, 373, 374
Morgan, J. W., 66
Morgan, W. J., 561, 562, 580, 587, 588, 589, 590, 684

Moritz, H., 284
Morkovkina, V. F., 45
Morley, L. W., 459
Morris, G. B., *252*
Morrison, H. F., 476
Moses, R. L., *404, 427*
Movchan, A. A., 672
Movchan, A. I., *672*
Moxham, R. L., *502*
Moye, D. G., 146
Mudie, J. D., *579*
Muehlberger, W. R., *519*
Mueller, St., 178, 179, 180, 184, *238, 239, 263, 265, 608, 685*
Mundt, W., 403
Munk, W. H., 345, 684, 698
Munro, D. C., *621*
Murata, K. J., 171, 492
Muratov, M. V., **523–538,** *37,* 538
Murauchi, S., 103
Murthy, V. R., 481
Muysson, J. R., *46*
Myers, A. T., *507*

Nabarro, F. R. N., 673
Nabholz, W., 531
Nafe, J. E., 30, 31, *221, 315,* 550, 552, 554, 579, *602,* 660, *661, 662*
Nagashima, K., *5*
Nagata, T., **391–398,** 392, 394, 397, 436, 448, 449, 584, 627, 629, 631, *631,* 632
Nairn, A. E. M., *449*
Nakamura, K., 106, *107, 500*
Nakamura, Yosio, 661
Narens, H. D., *242*
Needham, H. D., *562, 563, 564, 565, 566, 568*
Néel, L., 448, 449, 631
Neigaus, M. G., *665*
Nelson, P., *469,* 476
Neprochnov, J. P., (Iv. P.), 249, 592, 607
Neprochnova, A. F., *592, 607*
Nersesov, I. L., 246, 247, *384*
Ness, N. F., 25, 26
Nesterenko, G. V., 44
Nettleton, L. L., 364
Neufville, J. De; see De Neufville, J.
Newton, R. C., 522, 634, 635
Newton, R. R., 697
Nezolenova, E. A., *202*
Niazi, M., 131
Nickel, H., *179, 180*
Nicolaysen, L. O., 515
Niewiarowski, E., 380
Nikolaev, V. A., *389*
Nikolayev, N. I., 282
Ninkovich, D., 450, 587
Nishimura, E., 173
Niskanen, E., 383
Nixon, P. H., 3
Nockolds, S. R., 39, 44, 502, 503
Nordquist, J. M., *128, 131*
Noyer, J. De; see De Noyer, J.
Nurmia, M., *185*
Nuttli, Otto, 115, 244, 257, 665

AUTHOR INDEX

Oborn, L. E., 368
O'Brien, P. N. S., 188, 592, 607
Officer, C. B., 592
Ogawa, J., *382*
Oguti, T., 433, 434
O'Hara, M. J., **634–637,** 5, 481, 492, 634, 635, 641
O'Hara, N. W., *458*
Oji, Y., 504
Okamura, R. T., 365
O'Keefe, J. A., 296, 297
Oldroyd, J. G., 28
Oliver, Jack E., 145, 176, 241, 257, 261, 265, 266, *266,* 269, 270, 554, 555, 598, 609, 612, 613, *613,* 684, *684*
Onuki, H., 481
Opdyke, N. D., 423, 450, *450,* 453, *559, 587,* 632
Orango, A. S., *469, 620*
Orowan, E., 695
Ostenso, N. A., **457–463,** 397, 431, 433, 434, 458, 562, 570, 580, *580, 583,* 584, 598
Otsuka, M., 244
Oxburgh, E. R., 8, 13, 579, 586, 587, 591, 593, 597, 610, 611, 649

Pakiser, L. C., 39, 53, 80, 208, 211, *211,* 212, 213, *213,* 214, 245, 374, 405, 409, 411, 424, 458, 553, 661, 665
Pallister, J. W., 539, 543
Papazachos, B. C., 140
Parker, R. L., 56, 684
Parkhomenko, E. I., 620
Parkinson, W. D., *147,* 465, 466, 467
Paterson, W. S. B., *465*
Pattenden, G. E., *46*
Patterson, C. C., 59, 60
Paul, W., 681
Pauling, L., 681
Pavlenkova, N. I., *189, 191*
Pavlovsky, E. V., 39, 539, 541
Pavoni, N., 546
Pawlowski, St., 404
Peacock, M. A., 503
Pechersky, D. M., 417
Pechnikov, V. S., 630
Peck, D. L., 97, 102
Peive, A. V., 526
Pekeris, C. L., *266,* 676
Penttilä, E., 185
Perkins, D., 467
Perrier, G., 184, 187
Petelin, V. P., *48*
Peter, George, 425, 433, 434, *434, 440, 458*
Peterschmitt, E., 142 *608*
Peterson, M. N. A., 99
Petkov, I. N., 193
Petrashen, G. I., 277
Petrov, V. P., *65*
Petrushevsky, B. A., **279–282,** 282
Phillips, J. D., 431, 433, 608
Phinney, R. A., 245
Picard, L., 540
Pichon, Xavier Le; *see* Le Pichon, Xavier
Pilant, W. L., *238, 239, 685*
Pilkey, O. H., *559*

Pinson, W. H., *4, 58, 59, 683*
Pitman, W. C., III, 431, *431, 562, 579, 582, 588, 589, 593, 597, 603, 604, 611, 613,* 616, 632
Plafker, G., 173
Plaumann, S., 180, *180,* 182, *186*
Poincare, H., 304
Poldervaart, A., 37, 44, 49, 52, 53, 599
Polkanov, A., 514
Pollard, D. D., 361, 362
Polyak, B. G., **82–88,** 73, *74,* 75, 87, 96, 97, 107, 108
Pomeranchuk, I., 622
Pomerantseva, I. V., *197, 202*
Pomeroy, Paul, 241, *661*
Popova, S. V., 646, *682*
Porkka, M. T., 185, *185*
Powers, H. A., 5
Pozhgay, K., 192, *193*
Press, Frank, **171–173,** 8, 173, 176, *220,* 241, 266, *374,* 665
Preuso, S. B., 303
Price, A. T., 475, 618, 619
Pritchard, J. I., *469*
Prodehl, C., *178, 179,* 182
Pros, Z., *276*
Prosen, D., 194
Pshennikov, K. V., 167
Puchkova, L. I., 458
Pugin, V. A., *522*
Pulfrey, W., 539, 540, 542, 543
Pullaiah, A., *452*
Pushkar, P. D., 505, 506
Puzyrev, N. N., 198, 201, 202

Quennell, A. M., 546
Quon, C., 475

Radau, R., 294
Radu, C., 142, 248
Radzhabov, M. M., 197
Raff, A. D., 422, 423, *423,* 425, 426, 428, 429, 430, *430,* 573, *573,* 581
Raitt, R. W., **225–230, 250–256,** 222, 225, 230, 252, 359, 360, 593, 594, 602, 604, 605, 611
Ramberg, H., 529
Ramlal, K., *502*
Rand, J. R., *58, 683*
Randall, M. J., 154
Rankama, K., 513, 514
Rankin, D., 472, 473
Rao, G. V., *74*
Rao, R. U. M., *74*
Rastvorova, V. A., 282, *524*
Rayleigh, Lord, 689
Read, Lesley, **145–147**
Reay, A., *481*
Redford, M. S., 458
Redlon, J. R., *562, 609*
Redwood, M. R., 665
Reed, G. W., 17, 66
Reich, H., 180
Reid, H. F., 549
Reilly, G. A., *46*
Reilly, W. I., 368, 373
Reisner, G. I., *384*

Reitzel, John, 79
Revelle, R., 11, 88, *93, 610*
Revol, P. H., *469*
Reynolds, R. T., 64, 69, 71
Rezanov, I. A., 46, 282
Rice, J. A., *457*
Rice, M. H., 650, 651, 654, 655, 656
Richards, M. L., 428
Richter, C. F., 115, 117, 124, *128, 131*, 134, 135, 140, 142, 150, 277, 492, 545, 546, 547
Riddihough, R. P., 433, *433*, 434, *434*
Rikitake, T., 463–469, 107, 176, 460, 463, 464, 465, 466
Riley, J. P., 42, 48
Rimbert, F., 631
Ringwood, A. E., 1–17, 637–649, 2, 3, 4, *4*, 5, 6, 7, 8, 9, 10, 11, 13, 15, 18, 26, 27, 28, 29, 30, 34, 66, 477, 490, *490*, 491, 493, 494, 495, 512, 601, 634, 635, 638, 639, 640, 641, 643, *643*, 644, 645, 646, 647, 648, 662, 671, 678, 679
Ritsema, A. R., 110–115, 157, *184*
Rivosh, L. A., 421, 436, 437, 438, 439, 440
Riznichenko, Yu. V. (J. V.), 143, 201, 248, 276, 277, 319, 554, 660
Robertson, E. C., 641
Robertson, E. I., 96, 366, 367
Robinson, E. S., 458
Robinson, G., *256*
Robinson, Rhoda, 39, 53, 405
Rod, Emile, 546
Roever, W. P. de; *see* de Roever, W. P.
Rogers, J. J. W., 16, 64
Roller, J. C., 214, 215
Romney, Carl, 242, 244, 247
Rona, P. A., 566
Ronov, Alexander B., 37–57, 37, 38, 39, 40, 51, 54, 524, 527
Rooke, J. M., *3*
Roots, E. F., 460
Rooymans, C. J. M., 680
Roquet, J., 630
Rose, J. C., 290, 304, *355, 362*, 376, *443*
Roseboom, E. H., 634
Ross, C. S., 507
Ross, D. I., 425
Ross, H. P., 458
Rothé, J. P., *180*, 460, *608*
Roy, R. F., 78–81, 79
Royer, G., 458
Rubey, W. W., 3, 5, 6, 11, 14, 50, 53
Ruddiman, W. F., *223, 559, 562, 563*
Runcorn, S. K., 447–457, 692–698, 387, 449, *449*, 450, 451, 452, *452*, 453, 692, 695, 696, 697
Ruprechtová (Ruprekhtova), L., 166, 247
Ruskol, E. L., 70
Rustanovich, D., 170
Ryabinin, Yu. N., *65*
Ryaboi, V. Z., *199*, 202
Ryall, Alan, 131, *145*, 242
Rybalka, V. M., *202*

Safronov, V. S., 69
Sahama, Th. G., 33, 34
Sahasrabudhe, Y. S., 496, 499, 500

Said, R., 540
St. Amand, P., *128, 131*, 546
St. John, V. P., *147*
Saito, M., 239, 618
Saito, T., 614
Saldukas, I., 457
Salli, I., 486
Salop, L. I., 539, 541
Sangster, D. F., 458
Santo, T. A., 237, 238, 665
Sasai, Y., *460*, 464, 467
Sass, J. H., 104
Sato, T., *103*
Satô, Y., *265, 270*
Sauer, K., 460
Saull, V. A., 130
Sauramo, M., 383, 385, 386
Savage, J. C., 158, 212, 214
Savarensky, E. F., 134–139, 134
Savonenko, V. G., 514
Sawada, M., *460*
Sawatzky, P., 457, *459*
Sazhina, N. B., 319
Schairer, J. F., 635, *635*
Schearer, L. D., *457*
Scheidegger, A. E., 253
Scheinmann, Yu. M., 316
Schelkunoff, S. A., 471
Schenk, V., 178, 180, 278
Scheumann, K. H., 516
Schilling, T. G., 586
Schmidt, K., 514
Schmucker, Ulrich, 79, 424, 429, 466, 467, 478
Schneider, E. D., 556–617, *559*, 562, 563, 564, 565, 566, *566*, 568, 576, 578, 580, 588, 589, 593, 596, 608, 613
Schneider, G., 173
Scholl, D. W., 562, 609
Schreiber, Edward, 19, *19*, 20, *21*, 22
Schukin, Yu. K., 196, 319
Schult, A., 628
Schwahn, W., 632
Schwarzchild, M., 15
Sclater, J. G., *89, 98*, 609, *609*, 610, *610*
Seabrook, M., 646
Secor, G. B., *458*
Sederholm, J. J., 46
Seidl, D., 263
Sellevoll, M., 185, *185, 186*
Semenenko, N. P., 191
Semerchan, A. A., *682*
Semikhatov, M. A., 39
Serson, P. H., 434, 458
Shamina, O. G., 248, 276, 277
Shand, S. J., 601
Shankland, T. J., 621
Shapiro, J. N., 685
Shatsky, N. S., 526
Shaver, R., 460
Shaw, D. M., 46
Shebalin, N. V., 249
Shelagin, V. A., 86
Sherburne, R. W., *244*
Sheridan, R., *550, 553*, 614

AUTHOR INDEX

Shimazu, Yasuo, 69, 70
Shimomura, T., *460*
Shimozuru, D., 28
Shirokova, E. I., **166–171**, 166, *166*, 167, 170
Shmidt, O. Yu., 68
Sholpo, L. Ye., 630
Shor, G. G., Jr., **225–230**, 252, 361, 362, 592, 602
Shumway, G., 575
Shurbet, G. L., 2, 314, 349
Simakov, G. V., *19*, *678*
Simmons, Gene, **78–81**, 19, 20, 22, 30, 33, 34, 95, 104, 698
Simonen, A., **483–489**, 488, 514
Simonenko, T. N., **415–421**, 418, 458
Sitter, W. de; see de Sitter, W.
Skinner, B. J., 23, 34
Skornyakova, N. S., *48*
Skorupa, J., 403
Slack, H., 458
Slater, J. C., 653
Slavin, V. I., *189*
Slemmons, D. B., *131*, *145*
Slichter, L. B., 63, 68, 469
Slutskii, A. B., *522*
Small, J. B., 381
Smirnov, Ya. B., 73, *74*, 75, 87
Smirnova, N. P., *44*
Smith, A. G., *683*
Smith, C. H., 481
Smith, D. Masson-; see Masson-Smith, D.
Smith, J. V., 634, 635
Smith, P. J., 455
Smith, S. J., 685
Smith, S. M., 606
Smith, Stewart W., 238, 239
Smith, T. Jefferson, 212, 213, 214, 253
Soga, N., 19, 20, 21,22, 23
Sollogub, V. B., **189–195**, 189, *189*, 191, *191*, 193, 202
Solodkova, I. I., *37*
Solonenko, V. P., 282, 546, 547
Soloviev, S. L., 139
Soloviev, S. P., 38, 39, 44, 46
Sonder, R., 525, 526
Sørnes, A., 184, 185
Sorsky, A. A., 282, 524
Spector, A., 458
Spence, W. J., 244
Spiess, F. N., 579
Spikhina, L. M., 194
Springer, D. L., 244
Srivastava, S. P., 470
Stacey, F. D., 629, 631
Starkova, A. G., *69*
Stauder, W. S. J., 154, 156, 157, 158, 173, 684
Steenland, N. C., *445*
Stefánsson, R., 156
Stegena, L., 194
Steiger, R. H., 60
Stein, A., *180*, *186*, *608*
Steinberg, G. S., 436, 437, 438, 439, 440
Steinhart, John S., 30, 208, *212*, *213*, 214, *214*, 242, 243, 245, 251, *253*, 661
Steketee, J. A., 173
Stelzner, J., 247, 248

Stepp, J. C., 244
Stevens, A. E., **153–160**, 154, 155, 156, 157, 158
Stewart, F. H., 40
Stewart, S. W., *211*, *213*
Stille, H., 525, 526, 528
Stishov, S. M., 646
Stockard, H., *551*
Stoneley, R., 266
Strang van Hees, G. L., 342
Strange, W. E., 286, 288, 313, 314, 316, 349, 355, 362, 365, 366, 442, 443
Strelkov, S. A., 389
Strobach, K., *608*
Stroev, P. A., *388*
Strutt, R. J., 63
Struve, E. A., 44, 45
Stubbs, P. H. S., *458*
Studt, F. E., 374, 375, 441, 442
Stueber, A. M., 481
Subbotin, S. I., 189, *191*, 319
Subramaniam, A. P., 496, 499, 500
Suciu, P., *404*
Sudo, K., *101*, *103*
Suggate, R. P., 368
Sugimura, A., 103, 107, 164, 500, 545
Summers, A. L., 64, *64*, 69, *71*
Sumner, J., 458
Sutton, G. H., *2*, *30*, *265*, 304, *349*, 374, 376, 377, *552*, 661, *661*
Sverdrup, H. U., 90
Swallow, J. C., *222*
Swift, C. M., Jr., **469–479**
Sykes, Lynn R., **148–153**, 128, 148, 150, 151, 154, 155, 242, 266, 269, 432, 573, 585, 589, 590, 591, 601, 608, 609, 612, *613*, 684, *684*
Syono, Y., 627, 628, 631

Taggart, J., 131, 242, 244, 661
Tajima, H., 369, 372
Takafumi, S., *460*
Takeuchi, H., 239, 266, 386, 618
Talwani, Manik, **341–351**, 2, 30, 222, 239, 342, 349, 369, 370, 371, 376, 377, 429, 431, *433*, *440*, 458, *458*, 553, 579, 584, 586, 593, 594, 595, 596, 597, 601, 602, 608, 609, 661
Tanaoka, I., *102*
Tanni, L., 305, 311
Tarakanov, R. Z., 246, *246*, *247*, *665*
Tarling, D. H., 439, 445
Tatel, Howard E., 242
Tatsumoto, M., 59, 60
Taylor, H. P., Jr., 481
Taylor, P. T., 220, 552
Taylor, S. R., 46, 53, 505
Tazima, M., 460
Teng, T., 158
Tex, E. den; see den Tex, E.
Tharp, M., 148, 149, 150, 151, 342, *569*, 570, *570*, *573*, 579, *580*, *592*, *602*
Thayer, T. P., 3, 6, 481
The, G., *147*
Thellier, E., 456, 631
Thellier, O., 456, 631
Thompson, G. A., 553

Thompson, J. B., Jr., 521
Thompson, N. J., *443*
Thompson, William Baldwin, 473
Thompson, William Bell, *686*
Thorndike, E. M., 224
Thüna, P. C. von; see von Thüna, P. C.
Tikhonov, A. V., 470
Tilley, C. E., 5, 490, 495, 635
Tilton, G. R., 17, 58, 60, *60*, 66
Tisserand, F., 304
Tobin, D. G., 128, 150
Tobyáš, V., *278*
Tocher, Don, 250
Toksöz, M. Nafi, 4, 8, 30, 158, 260, 263, 266
Tolstikhina, M. M., 418
Toomre, A., 685
Tozer, D. C., **618–621**, 72, 620, 686
Trendall, A. F., 542
Trifonov, P., *189, 191, 202*
Trow, J. W., 458
Trubitsyn, V. P., 656
Trunin, R. F., 19, *678*
Tryggvason, E., 180, 185, 221, 239, 665
Tsuboi, C., 173, 314
Tsubokawa, J., 382
Tugarinov, A. I., 39, 46, *514*
Tugolesov, D. A., 37
Tulina, Yu. V., *103, 592, 661*
Turajlic, R., 433
Turchanenko, N. Z., *193*
Turcotte, D. L., 8, 13, 579, 586, 587,, 591, 593, 597, 610, 611
Turner, F. J., 3, 481, 521
Tuttle, O. F., 487
Tuve, Merle A., 242
Tykhonov, A. N., 63
Tyrrell, G. W., 496, 497

Uchman, I., *191*, 194
Uchupi, E., 617
Udias, A., 157
Udinstev, G. B., *37, 46, 48, 49,* 51, *93*, 601, 606
Uffen, R. J., 67
Ulinich, F. P., 656
Upton, B. G. J., 480
Urey, H. C., 53, 68, 69, 70, 695
Urry, W. D., 63, 74, 75
Usami, T., 270
Ushakov, S. A., *388*
Uyeda, Seiya, **95–109**, 16, 64, 71, 73, 74, 75, 82, 87, 89, *89*, 92, 93, 94, 96, *96* 97, 98, 99, 101, *101*, 102, 103, *103*, 107, 369, 394, 428, 597, 609, *609, 610,* 627, 631
Uzawa, S., *460*

Vacquier, V., **422–430**, 89, *89,* 93, 94, 98, 101, 103, 394, 422, 423, 428, *428,* 430, 445, 573, 609, 610
Vajk, Raoul, 434
Valek, R., 312
Valle, P. E., 25
van Andel, S. I., 452
van Andel, T. H., 580, 584, 589, 593, 599
van Bemmelen, R. W., 691

van de Lindt, W. J., 606
van der Voo, R., 454
Vandoros, P., *58, 683*
Vaněk, J., **246–250, 276–279,** 247, 248, 249, 276, 278
Van Voorhis, G. D., *392, 428, 460,* 632, *632*
Vaschenko, V. Ya., 650, 653
Vasilieva, V. I., *417, 418, 421*
Vasilos, T., *622, 623*
Vees, R., *180*
Veinberg, A. K., *417, 418, 421*
Veitsman (Veizman, Weizman), P. S., *103, 592, 661*
Veldkamp, J., 404
Vening Meinesz, F. A., 98, 343, 355, 361, 387, 597, 609, 694
Vereshchagin, L. F., 682
Verhoogen, J., **627–633,** 3, 5, 64, 68, 107, 481, 521, 631, 669, 686, *687*
Verma, R. K., 19, 20, 22, 74, 250, *452*
Vesanen, E., 185
Vestine, E. H., 434
Vine, F. J., 11, 148, 152, 369, 397, 422, 423, 424, 425, 430, 433, 434, 451, 561, 579, 580, 581, 584, 585, 586, 599, *601,* 632, 683
Vinogradov, A. P., 37, 39, 44, 46, 50, 51, 52, 53, 514, 671
Vinogradov, S. D., 143
Visarion, M., *404*
Visintini, G. de; see de Visintini, G.
Vogel, A., 458, 598
Vogt, Peter R., **556–617,** 397, 431, 433, 562, 570, 576, 578, 580, 583, 584, 588, 589, 593, 596, 598, 608, 610, 613
Voitkevich, G. V., 46
Vojtczak-Gadomska, B., 191
Volarovich, M. P., 674
Volvovsky, B. S., *199, 201, 202*
Volvovsky, I. S., **195–208,** *38,* 196, *199, 201, 202, 316*
von Bubnoff, S., 526, 539
von Eckermann, Harry, 480, 506
Von Herzen, R. P., **88–95,** 91, 92, 93, *93,* 94, 97, 99, 103, 369, 422, 609, 610
von Huene, R., *562, 609*
von Knorring, O., *3*
von Thüna, P. C., *27*
Voo, R. van der; see van der Voo, R.
Voorhis, G. D. Van; see Van Voorhis, G. D.
Vozoff, K., *472*
Vvedenskaya, A. V., **166–171,** 154, 166, *166,* 167, 173
Vyalov, S. S., 388

Wadati, K., 173
Wageman, J. M., 617
Wager, L. R., 5
Wagner, P. A., 3
Wait, J. R., 470, 475
Wakeel, S. K. El; see El Wakeel, S. K.
Walker, F., 44
Walker, G. P. L., 150
Walsh, J. M., *650, 651, 654, 655, 656*
Walthall, F. G., 4, 60
Waniek, L., 178, 180, 276, *276,* 278
Ward, S. H., 458, *476*
Warren, D. H., **208–220,** 213
Warren, R. E., *423, 430, 573*

AUTHOR INDEX

Warschauer, D. M., 681
Washington, H. S., 496
Washkurak, S., 457
Wasserburg, G. J., 58, 64, 65, 66
Watanabe, T., *89, 101, 103, 609, 610*
Waters, A. C., 496, 499
Watkins, J. S., 460
Watson, K. D., 481
Watt, P. A., *147*
Weaver, J. T., 473
Webb, J., *469*
Weber, J. R., 213
Wedepohl, K. H., 50
Wegener, A., 561, 579, 588, 683
Weill, D. F., 506, 522
Weiss, N. O., *686*
Weizman, P. S.; see Veitsman, P. S.
Wellman, H. W., **544–549**, 545, 546, 548
Wellman, P., 365
West, G. F., 253, 254, 445
Weston, A. A., 128
Wetherill, G. W., 58
White, D. E., 104, 106
White, I. G., *481*
White, R. W., 508
White, W. R. H., 212, 214
Whiteman, A. J., 540
Whitham, K., 458, 460, 465, *465*, 466
Whittaker, E. T., 256
Whitten, C. A., 383, 546
Wickens, A. J., 154
Wiese, H., 467
Williams, A. F., 3
Willmore, P. L., 253
Wilshire, H. G., 5
Wilson, A. F., 515
Wilson, H. D. B., 502
Wilson, J. Tuzo, 4, 46, 150, 151, 152, 155, *208, 220,* 352, 422, 423, 424, 425, 580, 588, 589, 590, *660,* 684
Windisch, C., *224, 567, 579, 584, 586, 592, 593, 596, 597, 608*
Winkler, H. G. F., 488, 521
Wiseman, J. D. H., 6
Witkind, J. J., 173
Wold, R. J., 457, 458
Wolfe, T. R., 457, 458
Woodward, H. P., 411, 459
Woollard, G. P., **125–133, 283–293, 320–341,** 30, 128, 133, 245, 285, 289, 290, 301, 304, *304,* 313, 314, 316, 320, 324, 349, 355, *355,* 361, *362,* 365, *365,* 366, *366,* 367, *376,* 412, 436, 437, 439, 442, 443, *443,* 445, 597
Worzel, J. Lamar, 2, *2, 30, 220, 224, 290,* 302, 314, 342, 343, 349, *349,* 355, 362, 369, 376, 377, 433, 552, *567, 592, 608,* 661, *661*
Wu, F. T., 158
Wyllie, M. R. J., 620
Wyllie, P. J., **480–483,** 480, 481, 482
Wyrzykowski, T., 380

Yamasaki, K., 5
Yamazaki, Y., 460
Yanovskaya, T. B., 246, *246,* 247, *247, 665*
Yanshin, A. L., *37,* 83
Yaroshevsky, Alexei A., **37–57,** 38, 53, 54
Yasui, M., 89, 96, 101, 103, *103, 609, 610*
Yegorkin, A. V., 207
Yoder, H. S., Jr., 5, 33, 34, 490, 495, 508, 634, 635, 641
Yokoyama, I., 107, 369, 372, 436, 437, 438
Yoshino, T., *460*
Yukutake, T., *102, 460, 627*
Yurov, Yu. G., 202

Zaionchkovsky, M. A., *197, 202, 592*
Zak, I., 549
Zener, C., 673
Zharkov, V. N., **650–659, 664–675,** 67, 68, 386, 653, 655, 657, 658, 659, *665,* 666, 667, 668, 669, 672, 673, 674, 690
Zietz, Isidore, **404–415,** 80, 208, *393,* 405, 408, 409, 411, 412, *430, 431,* 445, 458, 459, 460, *461, 552*
Zitzmann, A., **399–404**
Zoltai, T., 646
Zotov, P. P., 70
Zubarev, V. N., 650, 653
Zurflueh, E. G., *440, 458*
Zverev, S. M., 46, 48, 49, *103,* 197, 200, 202, 550, *592,* 607, *661*

Subject Index

Boldface entries and small **boldface** numbers indicate pages on which extended discussion begins.

Achondrite model, 60
Adams-Williamson equation, 33, 658, 678
Age
  of the crust, 70, 73
  of the earth, 70
  relationships Precambrian basement, 519
Airy concept of isostasy, 285, 320
Airy-Heiskanen hypothesis, 242, 345, 368
Alaska earthquakes, 1964-65, 156
Alert anomaly, 466, 467
Alpine crustal type, 233
Alpine fault, 112, 545, 705
Alpine intrusions in orogenic regions, 481
Alpine orogene, 180, 399, 532
Alpine structure, SPLAN-50, 238
ALPS model, 263
Amorphization, hypothesis of, 670
Amplitude-depth curve, 277
Amplitude-distance curves, 247, 277
Amplitudes, variations of body-wave, 244
Anatolian fault system, 112, 140, 144
Anderson velocities, 25, 30
Andes anomaly, 467
Andesite
  average composition of, 502, 503
  genesis of, 504
  line, 364, 502
  volcanoes, High Cascades, variation diagrams, 505
Andesitic and rhyolitic volcanism, 501
Andrade's mechanism, 671
Anistropy, 240
  crust, 235
  mantle, 250, 267
Antiferromagnetism, 627
Archeomagnetic data, summary of, 455
Argentine continental margin, 552
Atlantic Ocean, seismic model, 220
Atlantic type margin, 549, 706
Attenuation, 247, 273, 386, 555, 609, 661, 672, 685
Aulacogenes, defined, 196, 534
Australian anomalies, 467
Australian tektite field, 451

Bad Harzburg, alternate base value for Potsdam, 292
Baikal
  anomaly, 467
  rift, 103, 152, 281, 537, 573, 696
  map, 540–541
Banda Sea earthquake, 1964, 156
Barbados Ridge, 575
Basalt
  abyssal, 58
  alkali series, 490
  distribution and age of plateau, 496

Basalt (*continued*)
  diversity of, 490
  mafic and ultramafic inclusions, 507
  magmas, origin of, 11, 489
  mineral assemblages as a function of pressure, 639
  mode of eruption of plateau, 496
  petrology of plateau, 497
  plateau, 495
  plateau, dimensions and ages, 496
  tholeiite series, 490
  samples of the upper mantle, 495
  volume and rate of extrusion, 499
Basaltic glasses, chemical composition, 638
Basaltic layer, 38, 48, 191, 194, 428, 607, 699, 700, 711
  hypotheses, 39
Basaltic magma
  fractionation of, 492
  interaction with wall-rock, 494
Basaltic shell, volume and composition, 51
Basement map of North America, 551
Basin-range, crustal type, 232
Batholiths
  anticlinal, 486
  origin and emplacement, 487
  orogenic setting, 483
Benioff plane, 610, 611, 612
Bernal's hypothesis, 645
Birch's hypothesis, 645, 648
B-layer, 676
Block folds, 529, 704
Block movements, 525
Block structure, 86, 125, 194, 200, 204, 380, 387, 415, 458, 525, 710
Bouguer anomaly, crustal types, 231
Bowen's hypothesis, 5
Brune—Dorman model, 239
Bullard's hypothesis, 5, 11
Bullen's model A', 270, 658, 681

Caledonian orogen, 399
Californian Coast anomaly, 466
Caloi's hypothesis, 277
Canadian shield
  CANSD structure, 238
  model, 263
Cann's model, 601, 602
CANSD model, 238, 239, 263
Carpatho-Balkan Geological Association, Geophysical Commission of the, 189
Cantwell-McDonald model, 472, 477, 478
Chandlerian nutation, 695
Charleston, South Carolina, earthquake of 1886, 128, 130
Chemical remanent magnetization (CRM), 631
CIT11A model, 25

Chondrite model, 7, 45, 60, 64, 67, 108
Clairaut's method, 294
Clark melting curve, 71
Clausius-Clapeyron equation, 657
C-layer; see transition zone
Compression folds, 532
Computers, use of, 111
Conduction
　in geophysically interesting materials, 620
　radiative, 27
Conductivity
　electrical, in the crust and upper mantle, 78, 469, 618, 674
　of ocean sediments, 89
Conductivity anomaly, 79, 463, 478, 620
Conductivity models, 476
Conductivity structures
　one-dimensional, 470
　two-dimensional, 472
Conrad discontinuity, 179, 182, 184, 185, 186, 190, 191, 193, 194, 699, 700
　vertical reflections, 180
Continental accretion, 58
Continental crust, average chemical composition, 48
Continental drift, 58, 125, 152, 171, 173, 387, 422, 434, 452, 553, 555, 561, 579, 683, 690, 698, 708
Continental formation, 59
Continental growth, 58, 707
Continental margins, 549, 562
　and island arcs, 434
Contraction hypothesis, 531, 704
Convection, 64, 94, 171, 236, 348, 387, 555, 667, 683, 689, 692, 704
Core, iron versus silicate, 68
Coulomb interactions, 651
Crust
　abundance of minerals, 50
　average chemical composition, 53
　composition, 37, 55, 699
　　initial data and calculation methods, 37
　continental, 40
　continental, subcontinental, and oceanic, 37
　formulas for thickness, 314
　horizontal compression, 704
　intermediate layer, 186
　layers
　　ratios of rock-forming elements, 52
　　volume and chemical composition, 49
　　and mantle, dynamics of, 112
　mean density of the normal continental crust, 2
　oceanic, 40, 556
　subcontinental, 40
　and upper mantle
　　evolution, 11
　　interrelations, 698
　　processes, 702
　volume and composition, 51
Crustal columns, average, 214
Crustal cross-sections, 179, 182, 184, 186, 192-3, 200, 203-4, 209, 221, 278, 359-360, 362, 370-371, 376-377, 409, 550, 553-554, 592, 594-595

Crustal evolution, 239
Crustal models, 200, 202, 204, 239
Crustal movements, 523
　correlation with geophysical fields, 384
　horizontal, 383, 704
　recent, 379
　vertical, 379, 702
Crustal structure, 215
　anomalous, 212
　magnetic anomalies, 457
　oceanic, generalized, 236
　from surface waves, 230
Crustal thickness
　versus Bouguer anomalies, 316
　map, 37, 206, 318
Crustal types, 239
　comparison of geophysical and geological characteristics, 231
　four, 554
　from surface waves, 231
　transitional and composite, 237
　two, 240
Crust-mantle models, 265
Crust-mantle processes, isotope geochemistry, 58
Curie depth, 403
Curie isotherm, 459, 584, 586
Curie temperature, 391, 403, 411, 422, 441, 447-48, 458-59, 627

Darwin-Radau theory, 696
Debye model, 666
Debye temperature, 623, 652, 657, 666
Debye theory, 652
Deep ocean basin, crustal type, 235
Deep seismic sounding (DSS), 195, 202
　defined, 177
Density
　value for the crust, revision of the, 285
　variations in the upper mantle, 219
　versus velocity, 30
Detrital remanent magnetization, 631
Diamagnetism, 627
Diapric structure, 530, 607
Differentiation
　continuous and discontinuous upward migration, 64
　depths affected, 53
Diffusive viscosity, 673
Discontinuity, 20°, 249; see also Conrad, Förtsch, Moho
Dispersion curves, 240
D-layer 659, 676
Drilling, 507, 701, 711
　ocean, 58, 289, 290, 561, 607, 684
Duda's catalog, 115

Early Rise, Project, 214, 218, 219
Earth
　figure
　　dynamic methods, 294
　　geometric methods, 293
　　mass, anomalies from satellite perturbations, 293; see also Satellite
　pear shape, 295

Earth (*continued*)
   thermal history, 63, 71
Eclogite hypothesis, 1
Earthquake
   catalog, 115
      Asia, 138
      Europe, 139
      P nodal solutions, 153
   Charleston, 128, 130
   deep, 112, 115, 121, 134, 142, 150, 161, 246, 266, 274, 278, 589, 612, 684, 707
      focus in Spain, 122, 142
   depth
      of foci, 132
      groups, 116
   distribution, 113
   energy-depth relation, 118
   energy index, 116
   epicenters, 151
      Asia, 135
      Australia, 146
      and geological structural patterns in the United States, 129
      Mediterranean region, 141
      North America, 127
      preliminary determination; *see* preliminary determination of epicenters
   fault plane solutions, 114
   focal mechanism, 153, 156
      from surface waves, 156
      from S waves, 156
   focal stresses, 154
   foci
      depth and geological association, 132
      migration, 113
      pressure areas, 114
      stresses in the, 384
      tension areas, 114
      type of movements, 113
   focus, mathematical models, 154
   frequency-depth, 113, 118, 150
   Gansun, 279
   geomagnetic effects, 176, 460
   Hebgen Lake, 242
   Helena, 130
   intermediate, 161
      and deep-focus, 148
      and shallow, 150
   isoseismal contours, 130
   Kern County, 130
   Kosoma, 130
   large, 116
   magnitude, 115
      M vs m, 120, 142
   magnitude-frequency relation, 118, 120, 143
   Matsushiro, 460
   mechanism, 113, 153, 160
   medium, 122
      and small, 118
   micro, 113
      and ultra-micro, 145
   minima in depth-frequency relation, 116
   Mongolian, 281

Earthquake (*continued*)
   Nankaido, 165
   nests, 113
   New Madrid, 130
   Niigata, 171, 175, 382, 460
   nodal planes, 166
   Oklahoma, 130
   prediction, 113, **174**
      geomagnetic changes, 176
   premonitory strain, 171, 174
   radiation patterns, 153, 162
   shallow and intermediate depth, 142
   shallow-focus, 152, 162
   stress directions, 150, 160, 166
      Arctic-Atlantic belt, 168
      Mediterranean-Asiatic belt, 166
      Mongolian-Baikal zone, 167
      Pacific belt, 168
   swarm, 164, 165, 175, 460, 466
   tectonic activity in North America, 125
   and tectonics, 279
   Texas-Mexico border, 130
   Tottori, 175
   Verensk, 281
East African rifts, 103, 121, 130, 281, 364, 374, 378, 460, 537, 539, 589, 593, 686-87, 690, 694, 696, 709
   map, 540
East Pacific Rise, 91, 93, 102, 150, 152, 155, 228, 369, 371, 412, 423, 577, 579, 601, 687, 696
Einstein model, 666
Electrical conductivity; *see* **conductivity**
Ellipsoid
   international reference, 296
   1967 satellite, 284, 296
Endogenous geologic processes, 699
Episodic movements, 687
Equation of state
   at absolute zero, 650
   at high pressure, 650
   after phase transition, 655
   for phonon domain, 652
European Seismological Commission, 139, 178, 189
Excitons, 625
Expanding earth hypothesis, 455, 561, 710
Explosion seismology, 177
   Atlantic Ocean, 220
   crustal model, 180
   Europe
      southeastern, 189
      western, 178
   Indian Ocean, 225
   instruments, 197
   interpretation, 177, 198
   North America, 208
   observational method, 197
   observation lines
      Alpine region, 181
      Germany, 183
      Project Early Rise, 218
      Scandinavia, 185
      USSR, 197
   Pacific Ocean, 225

SUBJECT INDEX

Explosion seismology (*continued*)
  piece-continuous profiling, 197
  propagation of energy, 180
  reflection methods, 179
  refraction methods, 179
  types of waves, 201
  USSR, 195
Extinction coefficient, 623, 626
Fairbairn's packing index, 31
Fault model, earthquake, 154
Faults
  displacement
    in earthquakes, 545
      rate, 545
  major transcurrent, 544
  strike-slip, 532, 704
  transcurrent, 112, 150, 155, 459, 544, 684
    versus transform, 150
  transform, 150, 155, 422, 425, 590, 684, 705
  wrench, 544
Fermi's kinetic energy, 651
Ferrimagnetism, 627
Ferromagnetism, 627
Fit of the continents, Atlantic, 58, 683, 709
Flattening
  international reference ellipsoid, 295
  satellite ellipsoid 1967, 284, 295, 296
Flora experiment, 251
Flow in the mantle, 684
Förtsch discontinuity, 184
Folding, six belts of, 535
Free oscillations, 18, 171, 264, 266, 273, 665

Gabbro-eclogite transformation, 637
Gansun earthquake in 1920, 279
Gardner-Willmore method, 253
Garnet granulite-eclogite transition, gradients of equilibria, 641
Garnet-illmenite transition, 647
Gaussian theorem, 372, 376
Geochronological methods, 58, 513
Geoid; see also ellipsoid, gravity, satellite
  Guier and Newton, 697
Geoidal anomalies defined by satellite, 286
Geoidal undulations, 300
Geologic features of crustal types, 231
Geologic structural pattern, North America, 125
Geomagnetic data, reduction, 391
Geomagnetic effects of earthquakes, 176, 460
Geomagnetic polarity
  epochs, 425
  zones, 451
Geomagnetic reversals, 422, 434, 442, 449, 559, 580, 684
  time scale, 450
Geomagnetic studies in the Atlantic Ocean, 430
Geomagnetic variation method, 469
Geosynclinal sediments, composition similar to metamorphic rocks, 39, 54
Geosyncline, scheme of development of, 528
German anomaly, 467, 620
Gnome explosion, 131, 242
Gobi-Altai earthquake, 134
Gondwanaland, 54, 454, 694

Granite magma, origin, 522
Granitic layer, 38, 39, 48, 49, 51, 178, 184, 186, 191, 194, 699, 702, 711
Granitization, 54, 483, 517, 529, 533, 700
Granitoid masses, 484
Gravitational effect of undulations of the geoid, 285
Gravitational energy
  density, 70
  liberation, 69
Gravitational field from satellites, 341
Gravity
  anomalies, 320, 352, 383
    Atlantic, 314
    Bouguer map of U.S., 408, 411, 413
    versus depth of Moho, 317
    Europe, 304
    Pacific, 352
    relation to elevation, 326
      Alaska, 329
      Africa, 337
      Canada, 329
      Central America, 333
      Europe, 304
      India, 334
      Mexico, 333
      South America, 334
      USSR, 312
      USA, 326
    USSR, 312
    volcanic regions, 364
  field
    power spectra of the, 303
    regional, 345
  formula
    for crustal thickness, 314
    international, 283, 350
    1967, 284
  global pendulum observations, 290
  high, mid-continent, 412
  international, standardization program, 291
  measurements
    standardization, 283
    surface-ship, 342
  normal crust, maps, 313
  oceanic, 341, 352, 597
  regional variations, 320
  over seamounts, 367
  standards, revision in, 290
  tares (defined), 290
Great island arc crustal type, 234
Griggs model, 688
Grüneisen parameter, 68, 623, 652, 655, 666
Gutenberg-Richter catalog, 115
Gutenberg velocities, 25–28, 30, 267, 270
Guyots, 569, 706

Hashin relation, 28
Hashin's theory, 33
Hawaiian volcanoes, magnetic studies, 439
Hayward fault, slippage on, 173
Heat; see also Temperature, Thermal conduction

Heat (*continued*)
    electrons and holes, 625
    excitons, 625
    phonons, 622
  conductivity in the mantle, 622
  flow
    accuracy versus representativeness, 91
    and age of tectogenesis, 87
    catalogue, 89
    Caucasus and Crimea, 85
    Central Europe, 84
    conduction, 95
    continental and oceanic, 11, 73, 88, 93, 347, 579, 597, 708
    correlation with other geophenomena, 73, 93
    crustal types, 231
    data, $5° \times 5°$ grid, 90
    Eurasia, 82
    field and anomalies, 83
    geothermal areas, 95
    hot spring areas, 105
    Iceland, 103
    Japan, 101
    mass transportation, 95
    mean (arithmetic and geometric), 73
    mid-oceanic ridges; *see* Mid-oceanic ridges
    North America, 78
    oceanic, 88, 597
      measurement techniques, 88
      statistics, 92
    Pacific, 98
    province, 78, 209
    regions of high, 99
    relationship to age of tectonic structures, 74
    shields, 73
    statistics for geological features, 97
    unit (HFU) defined, 73
    values for major geological features, 87
    volcanic areas, 95
  generation, 64
  transfer
    mass transport, 625
    nonconductive, 103
    radiative, 623
Hebgen Lake, Montana, earthquake in 1959, 242
Herrin-Taggart Pn map, 131
Hess' hypothesis, 662
HFU (heat flow unit) defined, 73
Higher modes, 265
Hugoniot pressure, 650

Ice cap; *see* Pleistocene ice cap
Incipient melting, 7
Inclusions in basalt, 507
Indian Ocean
  contrast to Pacific, 228
  seismic studies, 225
Initial heating, 68
Injection folds, 529, 704
International Association of Geodesy, 284, 290
International Association of Seismology and Physics of the Earth's Interior, 139
International Astronomical Union, 296

International Central Seismological Bureau (BCIS), 139
International Geophysical Year, 126, 459
International Latitude Service, 698
International Seismological Summary, 139
International Union of Geodesy and Geophysics, 139, 295, 296, 299, 463
International Union of Geological Sciences, 578
International Year of the Quiet Sun, 464
Inversion of data, 231, 246; *see also* seismic data
Iron core, 68
Island arc, 609
  Japan, and orogeny, 163
Isochrons, fit in southern hemisphere, 58, 683
Isoseismal contours in the United States, 130
Isostasy; *see* Airy, Pratt
Isostatic adjustment of the crust, 385
Isostatic anomaly, 387
  map
    India, 339
    United States, 288
Isostatic equilibrium, 217, 229, 289, 312, 320, 322, 345, 362, 379, 702
Isostatic movements, glacial, 383; *see also* Pleistocene ice sheet
Isothermal remanent magnetization (IRM), 630
Isotope
  geochemistry of crust-mantle processes, 58
  tracers, 58
Japan
  anomaly, 463, 467, 620
  structure 6EJ, 239
Japanese volcanoes, magnetic studies, 437
Jeffreys-Bullen A model, 267
JOIDES; *see* Drilling, ocean

Kamchatka volcanoes, magnetic studies, 438
Kimberlite, 480
  pipes, 2, 535
Kimberly eclogites, 364
Knopoff model, 239, 275
Knopoff's scaling law, 31
Koenigsberger's ratio, 443, 630
Lake Superior
  experiment, 214, 253
  survey, 458

Lateral movements, asthenosphere, 388
Lateral variations, 78, 112, 242, 257, 275, 342, 347, 476, 685
  lower mantle, 245
Laurasia, 54, 454, 694
Lead isotope aberration, 59, 60
Le Chatelier's law, 677
Lindemann's method, 657
Low velocity channel
  crustal, 178, 240
  in granitic layer, 186
  second, 246
Low velocity zone, 4, 8, 10, 11, 18, 25, 35, 112, 178, 217, 234, 236, 240, 246, 249, 257, 263, 265, 276, 608, 609, 664, 685, 688, 695, 703
  partial fusion, 27, 490

SUBJECT INDEX

Low velocity zone (continued)
   pressure and temperature distribution, 666
   thermodynamic parameters, 665
   transverse waves versus longitudinal waves, 665
   variation of iron content, 33
Love waves, defined, 230
Lubimova's C concentration, 65

Mackenzie-Hashin relation, 28
Mafic and ultramafic inclusions in basalt, 507
Magma
   normative composition of primitive, 490
   segregation, 491
Magmatic rock
   types of Precambrian shields, 517
   volume, mass, composition, 44
Magnetic anomalies, 405, 422, 430, 436, 562, 632
   continental margin, 460
   and crustal structure, 457
   Europe, 399
   interpretation, 391
   patterns, 392
      linear, 393, 422, 423, 427, 430, 432, 552, 581, 585, 684
   Puna ridge, 444
   and tectonic units, 399
   USSR, 415
   USA, 404
Magnetic 'apparent susceptibility', 443
Magnetic bight, northeast Pacific Ocean, 427
Magnetic charts, types, 447
Magnetic field, remanent, 684
Magnetic instrumentation and analytical techniques, 457
Magnetic intensity field in the Pacific, 422
Magnetic investigations of the United States, 404
Magnetic maps, U. S., 410
Magnetic minerals, 628
Magnetic profiles
   Juan de Fuca ridge, 424
   Reykjanes ridge, 431
   U. S., 406
Magnetic patterns in the USSR, map, 417
Magnetic properties of rocks, 627
Magnetic secular variation, 449
Magnetic studies
   submarine volcanism, 441
   over volcanoes, 436
Magnetic survey, 430
   great circle (U. S.), 404
   regional, 458
   over Reykjanes ridge, 580
   transcontinental, 459
Magnetic susceptibility, 393, 417, 443, 627, 629
Magnetism, volcanic regions, 396
Magnetization
   of earth's crust and upper mantle, 391
   induced, 393, 458, 627, 629
   remanent, 393, 458, 627, 630
      chemical (CRM), 447, 631
      detrital, 631
      depositional, 447
      isothermal (IRM), 447, 630
      natural, 439, 443, 459

Magnetization (continued)
      piezo (PRM), 631
      thermal (TRM), 447, 584, 630
      viscous (VRM), 448, 630
   of rocks, 393
Magnetotellurics
   data analysis, 475
   mathematic theory, 470
   studies, crust and upper mantle, 469
Mantle
   material, primary, 481
   model, 239, 263
      chondritic abundances, 7
   motion, driving mechanism and scale, 686
   oceanic, 556
   types, two, 240
Matsushiro earthquake swarm, 164, 165, 460
melting-point curves, 67
metallization, 387, 656
Metamorphic belts, 519
Metamorphic grade, 699
Metamorphic isograds, 520
Metamorphic rocks
   composition similar to geosynclinal sediments, 39, 54
   geosynclinal origin, 39
   volume, mass, composition, 44
   types of Precambrian shields, 516
Metamorphism
   contact, 519
   geosynclinal rocks, 54
   mineralogical scales, 520
   regional, 700
   relation to depth, 519
Microcontinents, 569
Microearthquakes, 145, 174, 175
Mid-continent, crustal type, 232
Mid-Atlantic ridge, 142, 150, 155, 169, 222, 224, 237, 286, 302, 315, 345, 370, 378, 424, 430, 459, 579, 601, 683, 686-687, 696
   fracture zones, 570
   heat flow, 93, 97, 103
   gravity, 342, 369
   profiles, 576
   topography, 574
Mid-ocean; see also Ocean
   canyons, 569
Mid-oceanic ridge, 11, 93, 95, 112, 155, 319, 364, 482, 526, 554-556, 561, 573, 579, 660, 686-687, 705; see also East Pacific Rise, Mid-Atlantic
   Atlantic profiles, 572
   earthquakes, 154
   high heat flow, 73, 93, 97, 102, 597
   map, 149
   sediments, 578
   seismicity, 148
mid-oceanic seismic zones, 121
Mie-Grüneisen equation, 653, 655
Migmatite domes, 531
Migmatite-granite domes, evolution, 487
Minerals, potential energy parameters, 654, 655
Mineralogical zoning and stability fields in the outer mantle, 642

Mittelmeer-Mjösen Zone, 403
Models
  crust and mantle; see crust and mantle
  plane layered, 266
Moho
  average oceanic depth, 221
  correlation with ocean depth, 229
  definition, 230
  formulas for depth, 314
  map, 206, 318
  relation to surface elevation, 205
  vertical reflections, 180
Mohole, 228
Mohorovicic discontinuity (Moho, M.), 1, 6, 14, 26, 37, 66, 125, 133, 177-87, 190, 215, 228, 230, 236, 242, 249, 257, 264, 273, 285, 289, 313, 316, 320, 326, 349, 376, 409, 509, 512, 518, 552, 556, 587, 591-593, 602-606, 637, 639, 642, 644, 660, 664-65, 701-02, 709
Mongolian earthquakes of 1905, 281
moon core, 70
Morin transition, 628
Mould Bay anomaly, 465, 620
Muiskoje earthquake, 134
Munk-MacDonald theory, 695

Nabarro-Herring's diffusive mechanism, 671, 673
Nafe-Drake density-velocity relationship, 30, 349, 376, 579
Nankaido earthquake, 165
Navier-Stokes equation, 696
Néel-Nagata theory, 448, 450
Néel temperature, 627
New Madrid, Missouri, earthquake of 1811, 130
New Zealand volcanoes, magnetic studies, 441
Niigata earthquake, 171, 175, 382, 460
Nodal plane solutions, 154, 160

Ocean; see also Mid-ocean
  basins, small, 605
  floor, physiography of, 562
  sediment cores, 560
  trenches, seismicity, 612
Oceanic abyssal plains, 567
Oceanic areas, anomalous, 228
Oceanic cores, 568
Oceanic crust, 221, 556, 578, 632, 701
  characteristics, 593
  nature and origins, 591
  petrology and geology, 598
  hypothesis for the generation of typical, 603
Oceanic crustal structure, generalized, 236
Oceanic fracture zones, 570
Oceanic gravity, 597
Oceanic heat flow, 597; see also Heat flow
Oceanic layers. 42, 46-48, 54, 222-226, 561, 587, 591, 593, 601-602, 699, 701, 706, 711
  horizon A, 223-224, 567, 608
  median values, 230
  two intermediate, 228
Oceanic magnetic anomalies, 579
Oceanic magnetic models, 582
Oceanic margins, 706

Oceanic precision depth recorder (PDR) records, 563
Oceanic profile, Atlantic, 558
Oceanic provinces, 557
Oceanic rises, 569
Oceanic sediments, 221
  and physiography, 557
Oceanic seismic surface waves, 598
Oceanic seismicity, 589
Oceanic subsidence, 606
Oceanic upper mantle, 221, 556, 632, 701
Oceanization, 710
Oceans
  and continents, formation of, 705
  hypothesis of permanency, 561
  three major morphological units, 562
Olivine
  elastic properties, 19
  layer, temperatures, 25
  physical properties versus iron content, 34
  spinel transition, 645
Orgueil meteorite, 66
  model, 65, 67, 75
Orogenic belts, volcanism, 501

Pacific; see East Pacific Rise
  margin type, 549, 706
  mobile ring, 281
  model 8099, 263
  Ocean
    contrast to Indian, 228
    seismic studies, 225
  ring of fire, 429
Paleomagnetic vector field, 447
Paramagnetism, 627
Parkinson's model, 465, 467
Parkinson vectors, 466
Partial melting of the upper mantle, 27, 490
Partition of energy among surface-wave modes, 271
Patterson-Tatsumoto model, 59
PDE; see Preliminary determination of epicenters
Peridotite
  alpine, 3
  densities and velocities, 7
  hypothesis, 1
Perovskite-type silicates, 647
Peru anomaly, 466, 467, 620
Petrology, phase equilibrium studies, 634
Phase equilibrium studies, 634
Phase transitions, 637
  equation of state after, 655
  melting, 657
Phonons, 622
  conduction, 27
Piezo remanent magnetization (PRM), 631
Planck's constant, 652
Plate tectonics, 13, 588, 684
Plateau basalts, 495
Pleistocene ice cap, 128, 331, 380, 383, 385, 388, 695, 703
Poisson's relation, 393
Pn velocity

Pn velocity (*continued*)
  crustal types, 231
  map, United States, 243
Polar wandering, 452, 694, 698
Polymorphic transitions, 655
Polymorphism and melting relationships, 634
  Ca-bearing metasilicate, 634
  Ca-free metasilicate, 634
  garnets, 635
  orthosilicate, 634
  plagioclase feldspars, 635
  spinels, 635
Potsdam
  absolute value of gravity, 283
  gravity system, revision of the, 284
Pratt concept of isostasy, 320
Pratt-Hayford compensation, 242, 245
Precambrian basement, age relations, 519
Precambrian basement complex, petrology, 513
Precambrian shields
  magmatic rock types, 517
  metamorphic rock types, 516
  sedimentary rock types, 515
Preliminary Determination of Epicenters (PDE), 115
Press-Ewing hypothesis, 277
Protocontinents, 454, 687
P-wave velocity, regional variations of, 242
Pyrolite
  mineralogy, density, and velocity, 7, 9
  model, 6–10, 26, 66, 490, 493–495, 512, 643
  partial melting, 10
  stability fields, 8, 644
Pyroxene
  transformations in the presence of alumina, 647
  transitions, 646

Q; *see* Attenuation
Quartet experiment, 251
Quartz-coesite-stishovite transitions, 646
Quebec earthquake of 1935, 131

Radioactive age, 58
Radioactive decay schemes, 59
Radioactive heat production, 66
Rayleigh wave
  defined, 230
  velocity, continental versus oceanic, 261
Recent movements, mechanisms, 387
Regional-metamorphism, 517, 519
Relaxation, hypothesis of, 669
Resistivity surveys, 469
Reykjanes ridge, 430, 432
Rhine rift map, 540
Rhyolitic volcanism, 501, 505
rifts; *see also* Baikal, East African, East Pacific, Mid-Atlantic, Mid-ocean
  continental, 539, 589
  Rhine (map), 540
  widths, 542
  world system, 229, 532, 539
  map, 149
Rikitake model, 465
Rikitake-Yokoyama hypothesis, 108

Ringwood-Green model, 11
Roberts Victor pipe, 3
Rocks, magnetic properties, 627
Rosh's law, 69
Rubey's hypothesis, 5
Rubey's theories, 11

Saito and Takeuchi model, 239
San Andreas fault, 704, 112, 126, 128, 132, 383, 460, 545, 573
Santo's curves, 237
Satellite
  data, geoid and gravity, 284, 286, 293, 336, 349
  ellipsoid 1967, 284, 295, 296, 299
  orbital perturbations, 293, 685
Scotland anomaly, 467
Sea-floor spreading, 11, 12, 14, 58, 125, 148, 150, 152, 173, 222, 235, 237, 240, 369, 374, 422, 428, 430, 451, 482, 552, 555, 561, 579, 584, 587, 591, 613, 683, 684, 687, 708
Seamounts, 569
  first study of, 430
  gravity measurements, 367, 369
  magnetic, 436
  nonmagnetic, 434
  northward migration of, 428
Sedimentary rocks
  composition contrast with granitic layer, 49
  high calcium content, 50
  types of Precambrian shields, 515
  volume, mass, composition, 40
Sedimentary shell, chemical constitution, 49
Seismic array, 111
  large aperture, 111
Seismic body waves, amplitude curves, 247
Seismic channel waves, 240
Seismic data
  acquisition, 111
  processing, 111
Seismic delay times, 243, 255
  map, 244
Seismic energy transmissibility, 131
Seismic experiments; *see* Early Rise, Flora, Gnome, Lake Superior, Quartet
Seismic instrument development, 110
Seismic layers, tectonic zones of USSR (table), 196
Seismic maps, 139
Seismic models, 248
  Atlantic Ocean, 220
  of the upper mantle, 276
Seismic PL waves, 241
Seismic profile, 223
Seismic regions of Europe (map), 140
Seismic surface waves, 230
  oceanic, 598
  tectonic associations, 257
Seismic velocities, change with depth, 669
Seismic waves
  attenuation of; *see* Attenuation
  higher modes, 240, 265
  leaking modes, 241
Seismic zones
  mid-oceanic, 120, 148

Seismic zones (*continued*)
  stress state, 169
Seismicity, 115
  continental Asia, 134
  European area, 139
  geological criteria of, 280
  mid-oceanic ridge system, 148
  North America, 126
  oceanic, 589
  regional, 121
  Sea of Okhotsk, 134
  southeast Australia, 145
  worldwide, 118
Seismograph network, 110
  Asia, 134
  Canada, 111, 155, 159
  USSR, 111, 155, 159
  Worldwide Standard, 110, 115, 148, 155, 159
Seismographs, ocean bottom, 111
Seismology
  inverse problems, 112
  model, 178
  upper mantle investigations, 110
  zero frequency, 171
Seismomagnetic effects, 176
Seismotectonic map of Europe, 282
Shallow water mid-ocean crustal type, 237
Shield
  and basement, average composition and rock types, 46
  crustal type, 232
  heat flow, 73
Shockwave data, 19, 69, 651, 654, 656, 685
Silicate core, 68
Simon's empirical equation, 657
Six-(6EJ) model, 239
Skin depth, defined, 469
Smith model, 239
Solid-Earth Geophysics: Survey and Outlook, 115
Solidification index, 498
South America-Africa age relation; *see also* Isochron, 58
SPLAN-50 model, alpine structure, 238, 239
Stipoverite (shishovite), 702
Strain release, 142, 143
Stratisphere, defined, 37
Strength of the earth, 300
Stress distribution, earth, 166, 387
Sun, luminosity increase, 15
Superprovinces, **79**, 209, 405, 458
Surface waves, 230
  data on the upper mantle, 257
  dispersion data, 258
  higher-mode, 265
Symposium
  Earthquake Prediction, 114, 176
  Geophysical theory and computers, 115
  Non-elastic processes in the mantle, 115
  Seismic models, 115, 278

Tape recorder model, 580, 584, 587
Tectonic activity, North America, 125
Tectonic characteristics of crustal types, 231
Tectonic features, harmonic content, 302, 690, 693

Tectonic Map of the World, 538
  foldout, 528
Tectonics, earthquakes and, 279
Tectonic, seismo-, map of Europe, 282
Tectonic structure of continents, 523
Tectonic zones, areas, volumes, thicknesses, and masses, 38
Telluric method, 470
Temperature
  –depth distribution, 7, 27, 622
    chondrite model, 72
    silicate core, 71
  gradient, perturbations and corrections, 82
  initial, 69
  oceanic, 28
  and pressure distribution, 666
Texas anomaly, 466, 478
Thermal; *see also* Heat
Thermal balance models, 63
Thermal conduction, equation, 63
Thermal evolution, uniform model, 71
Thermal history of the earth, 63, 70, 695
Thermal remanent magnetization (TRM), 630
Thermal resistivity as a function of temperature, 623
Thickness of crustal types, 231
Tholeiite, defined, 497
Tholeiites, quartz, mineral assemblages, 640
Tottori earthquake of 1943, 175
Transcontinental Geophysical Survey, 208, 405, 412, 458
Transition zones
  change of chemical composition, 677
  change of the type of chemical bond, 681
  decomposition of silicates and alumosilicates into oxides, 677
  geologic provinces, 214
  transitions to closely packed lattices, 678
  upper mantle (region C), 19, 219, 644, 674, 676, 702
Travel-time curves, 178, 189, 195, 213
  theoretical, 178
Trenches, 609
Tryggvason model, 239
Tschermak's molecules, 639

Ultramafic belts, 480
Ultrabasic rocks, 480
Unesco, 174
Uplift, post-glacial; *see* Pleistocene ice cap
Upper mantle
  anisotropy, 250
  approximation as an olivine layer, 18
  chemical model, 6
  chemical zoning, 7
  composition
    and evolution, 1
    geochemical and petrological evidence, 3
  density and composition, 18
  evolution models, 60
  inhomogeneity, 387
  isotope abundance, 67
  mineralogical zoning, 7, 642

SUBJECT INDEX

Upper Mantle (*continued*)
  model, 275, 276
  P velocity (Herrin-Taggart map), 131
  structure, 112, 247
    Eurasia, 246
  surface wave data, 257
  velocity distribution, 246
  viscosity, 385
  zoning, 6
Upper Mantle Project, 78, 110, 112, 187, 208, 257, 405, 412, 460, 464

Vancouver Island anomaly, 466
Van der Waals' energy 651
Variscan orogen, 139, 399, 532
Vela Uniform Program, 208
Velocity versus density, 30, 213
Velocity-depth distribution, 179; *see also* **Crustal Models, Mantle**
Vening Meinesz' principle, 694
Verensk earthquake of 1911, 281
Vertical movements, 524
Vine-Matthews hypothesis, 11, 422, 430, 579, 585, 596, 598–99
Viscous remanent magnetization (VRM), 630
Vogt-Schneider-Johnson hypothesis, 603
Volatiles, excess, 50
Volcanic areas, heat flow, 95
Volcanic regions
  gravity anomalies, 364

Volcanic regions (*continued*)
  magnetization, 396
Volcanic rocks
  magnetic properties of, 442
  volume, mass, composition, 40
Volcanism
  andesitic and rhyolitic, of orogenic belts, 501
  energetics, 107
  magnetics of submarine, 441
Volcanoes
  distribution, 98
  effective magnetization, 439, 440
  volcanoes, magnetic studies, 436
von Zeipel theorem, 685

Wasserburg model, 66, 75
Wasserburg mixture, 65, 67
Water, total in crust and hydrosphere, 52
Wegener's theory, 694, *see also* Continental drift and Sea-floor spreading
Wickens-Hodgson catalog, 154
Wiechert-Herglotz method, 178
Wiechert model, 68
Wiedemann-Franz law, 625
Worldwide seismograph network; *see* seismograph network

Zero frequency seismology, 171
  signal threshold, 172

Geophysical Monograph Series

WALDO E. SMITH, *managing editor*

1 Antarctica in the International Geophysical Year
A. P. CRARY, L. M. GOULD, E. O. HULBURT, HUGH ODISHAW, AND WALDO E. SMITH, *editors*

2 Geophysics and the IGY
HUGH ODISHAW AND STANLEY RUTTENBERG, *editors*

3 Atmospheric Chemistry of Chlorine and Sulfur Compounds
JAMES P. LODGE, JR., *editor*

4 Contemporary Geodesy
CHARLES A. WHITTEN AND KENNETH H. DRUMMOND, *editors*

5 Physics of Precipitation
HELMUT WIECKMANN, *editor*

6 The Crust of the Pacific Basin
GORDON A. MACDONALD AND HISASHI KUNO, *editors*

7. Antarctic Research: The Matthew Fontaine Maury Memorial Symposium
H. WEXLER, M. J. RUBIN, AND J. E. CASKEY, JR., *editors*

8 Terrestrial Heat Flow
WILLIAM H. K. LEE, *editor*

9 Gravity Anomalies: Unsurveyed Areas
HYMAN ORLIN, *editor*

10 The Earth beneath the Continents: A Volume in Honor of Merle A. Tuve
JOHN S. STEINHART AND T. JEFFERSON SMITH, *editors*

11 Isotope Techniques in the Hydrologic Cycle
GLENN E. STOUT, *editor*

12 The Crust and Upper Mantle of the Pacific Area
LEON KNOPOFF, CHARLES L. DRAKE, AND PEMBROKE J. HART, *editors*

13 The Earth's Crust and Upper Mantle
PEMBROKE J. HART, *editor*